LIQUID- AND SURFACE-BORNE PARTICLE MEASUREMENT HANDBOOK

LIQUID- AND SURFACE-BORNE PARTICLE MEASUREMENT HANDBOOK

EDITED BY

JULIUS Z. KNAPP

Research & Development Associates, Inc.
Somerset, New Jersey

THOMAS A. BARBER

Baxter Healthcare Corporation
Round Lake, Illinois

ALVIN LIEBERMAN

Particle Measuring Systems, Inc.
Fremont, California

Marcel Dekker, Inc. New York • Basel • Hong Kong

Library of Congress Cataloging-in-Publication Data

Liquid- and surface-borne particle measurement handbook / edited by Julius Z. Knapp, Thomas A. Barber, Alvin Lieberman.
p. cm.
Includes index.
ISBN 0-8247-9386-2 (alk. paper)
1. Particles—Measurement—Handbooks, manuals, etc. 2. Drugs—Quality control—Handbooks, manuals, etc. 3. Fouling—Measurement—Handbooks, manuals, etc. I. Knapp, Julius Z. II. Barber, Thomas A. III. Lieberman, Alvin.
RS190.P37L57 1996
615'.19—dc20 96-13063
CIP

The publisher offers discounts on this book when ordered in bulk quantities. For more information, write to Special Sales/Professional Marketing at the address below.

This book is printed on acid-free paper.

Marcel Dekker, Inc.
270 Madison Avenue, New York, New York 10016

Current printing (last digit)
10 9 8 7 6 5 4 3 2 1

PRINTED IN THE UNITED STATES OF AMERICA

Preface

Accurate particle measurements are essential to the control of quality in many products. For pharmaceuticals, the determination that particle contamination lies within specified limits is a legal requirement prior to sale and clinical use. For high-performance hydraulic systems, particle contamination measurement and control is an operational requirement. These hydraulic systems are used to control high-powered, complex mechanisms including those that operate advanced avionic systems. In the manufacture of products ranging from pigments to photographic film base, the determination of particle size and concentration is essential to successful production.

The information and techniques needed to measure and understand particle contamination have been developed by specialists in many disciplines. As a result, the reports of their work are distributed in the literature of their respective disciplines. This handbook provides a convenient, intelligible, single reference source for the particle measurement practitioner. It has been assembled by a selected team of specialists with the capability to make many facets of particle measurement understandable and usable.

This handbook provides an accurate description of current particle measurement information and technology. It is our hope that it will not only make available the knowledge that is presently needed, but also provide a sound foundation upon which new particle measurement developments can be successfully added.

Julius Z. Knapp
Thomas A. Barber
Alvin Lieberman

Contents

Preface *iii*
Contributors *vii*

1. Particle Characterization in Liquids: Background Information 1
   Alvin Lieberman
2. Microscopy: The Setup and Operation of the Polarized-Light Microscopy Lab for Particle Identification 29
   John Gustav Delly
3. Image Analysis (Automated Microscopy) for Particle Analysis 61
   Thomas A. Barber
4. The Electrical Sensing Zone Method (The Coulter Principle) 113
   Roy W. Lines
5. Particle Sizing Using Ensemble Averaging Techniques 155
   Bruce B. Weiner
6. In Situ Particle Measurement 173
   Alvin Lieberman
7. Holographic Methods 197
   Brian J. Thompson
8. Instrumental Analysis of Particulate Matter 235
   Thomas A. Barber
9. Evaluation and Validation of Nondestructive Particle Inspection Methods and Systems 295
   Julius Z. Knapp and Lee R. Abramson
10. Validation of Counting Accuracy in Single-Particle Counters: Application of a New Coincidence Model 451
    Julius Z. Knapp and Lee R. Abramson

11. Particles in Semiconductor Manufacturing Process Liquids 569
Donald C. Grant and Alvin Lieberman

12. A Systematic Approach for Particle Counting Petroleum Oils 585
Barry M. Verdegan, Brian W. Schwandt, and Christopher E. Holm

13. Sources of Particulate Matter in Pharmaceutical and Medical Products 637
Thomas A. Barber

14. Physiological Consequences of Injected Particulates 661
Lawrence John Pesko

15. Diagnostic Applications of Microspheres 687
Leigh B. Bangs

16. Reference Materials in Particle Measurements 709
Renliang Xu

17. Selective Particle Capture with Transparent Sieves in the Evaluation of Contamination in Parenteral Solutions 721
Julius Z. Knapp

18. Particle Size Distribution Analysis Using Light Scattering 745
Renliang Xu

19. Industry's Role in the Development of Standards and Controls for Particulate Matter in Small Volume Injectables 779
John A. Syverson

20. Pharmaceutical Clean Laboratory Operations 791
Thomas A. Barber

21. Particles in Fluids and on Surfaces: Specialized Bibliography 825
Alvin Lieberman

Index 871

Contributors

Lee R. Abramson Statistical Consultant, Washington, D.C.

Leigh B. Bangs Bangs Laboratories, Inc., Carmel, Indiana

Thomas A. Barber Corporate Research and Technical Services, Baxter Healthcare Corporation, Round Lake, Illinois

John Gustav Delly Émile Chamot Laboratory, McCrone Research Institute, Chicago, Illinois

Donald C. Grant CT Associates, Inc., Excelsior, Minnesota

Christopher E. Holm Corporate Research, Nelson Industries, Inc., Stoughton, Wisconsin

Julius Z. Knapp Research & Development Associates, Inc., Somerset, New Jersey

Alvin Lieberman Particle Measuring Systems, Inc., Fremont, California

Roy W. Lines Coulter Electronics Ltd., Luton, Bedfordshire, England

Lawrence John Pesko Department of Pharmaceutical Care, St. Vincent Medical Center, Toledo, Ohio

Brian W. Schwandt Corporate Research, Nelson Industries, Inc., Stoughton, Wisconsin

John A. Syverson* Pfizer, Inc., New York, New York

Brian J. Thompson University of Rochester, Rochester, New York

---

*Retired.

Barry M. Verdegan Corporate Research, Nelson Industries, Inc., Stoughton, Wisconsin

Bruce B. Weiner Brookhaven Instruments Corporation, Holtsville, New York

Renliang Xu Coulter Corporation, Miami, Florida

1

Particle Characterization in Liquids: Background Information

Alvin Lieberman

Particle Measuring Systems, Inc., Fremont, California

I. APPLICATION AREAS FOR PARTICLE MEASUREMENT

The instrument widely used to characterize particulate contamination in pharmaceutical products is the optical single particle counter (OPC) operating on light extinction effects. Even with its wide use, problems occur in operating individual OPCs and in correlating data from several OPCs, especially those with different designs. These problems arise from a combination of OPC design differences and operational difficulties in handling the liquids to be characterized and with the OPC in use. Some of these difficulties may be reduced if the operator is aware of the genesis of the instrument itself and of the requirements for careful operation of the OPC and for correct handling of the materials being measured. A summary of the background areas from which the operational requirements have been developed should aid in reducing overall problems and should help to understand reasons for some of the points and procedures to be discussed later.

Much of the application and sample handling methodology used in measuring particles in pharmaceutical liquids is derived from previous work in particle measurement in other areas. As part of the discussion on particle counter use in characterizing contaminants in pharmaceutical liquids, a brief summary of the basis for particle characterization and of the instruments used in the past in other areas is useful in understanding how and why the present instruments have evolved. It will be seen that special problems in particle measurement have often resulted in modifications in the instrument design and operation. It will also become apparent that the evolutionary process is a continuing phenomenon and that the development of new instruments and improvement in operation of existing instruments continues even now. Some areas of liquid production and use where particle size and quantity control has been and still is important will be mentioned.

Even before adaptation of the OPC to pharmaceutical product contaminant measurement, an extensive background experience existed in terms of manual and instrumental particle characterization in many industrial areas. The instruments in use operate on many different principles and their operations and applications are widely varied. In all cases, the instrument data quality is strongly affected by the sample acquisition, handling and storage procedures. These vary with the nature of the material to be characterized, as well as with the method of measurement used by the selected sizing system. The instruments may respond to one or another physical property of the particles in addition to size. Selection of a particular measurement method or instrument should be based on its capability to respond to the particle property of greatest importance in the application area of the process to be defined. In some situations, however, the selection has been based primarily upon past history, availability, cost or personal preference of the laboratory operators. In all situations, the problems of sample handling are still paramount in optimizing the quality of data generated by any instrument. Some of the measurement systems in use, the areas where they are applied, and the development history of the optical particle counter will be discussed in more detail. Some industrial liquid handling and production areas where particle sizing is of concern are mentioned, along with some mention of measurement techniques used there.

A. Product Powders and Liquids

The manufacture and processing of foods, drugs, paints and pigments, ceramics, plastics, metals and many other materials are affected by the physical and chemical properties of the finely divided materials used in these processes. Some examples may help in appreciating these material property effects. In many instances, reaction rates in solution are controlled by the available concentration of one of the components. If one component dissolves too slowly or too rapidly, the rate of solution can be modified by controlling the available surface. This can be done by providing the soluble component in the form of larger or smaller particles. Reaction rate control can be affected by the particle size of the reactants. Obviously, it is necessary to know the particle size of the critical component(s). A wide variety of particle measurement methods have been used in the several industries mentioned here. They range from visual examination of the particles, measurement of sieve size or use of sedimentation rate systems and application of a variety of optical devices. The measurement system choice is usually based on observation of the particle physical property of most concern in the production process.

Where powders are to be handled in air, oxidizable materials can burn very easily. Finely divided organic materials, such as starch, flour, sugar, etc., can burn extremely fast or even explosively. Similarly, many metals, after sufficient size reduction, can oxidize rapidly. Product particle size information is necessary in order to select optimal combustion controls. These problems become especially serious for particles smaller than 5–10 μm in diameter. Wet sieving and sedimentation in liquid measurement methods are often used to characterize powders of these types in order to keep them immersed below the surface of an inert liquid while measuring particle size.

When moving bulk powder in a wet or dry suspension, power required to maintain flow through tubes or ducts can be reduced when the powder is present in a range of sizes rather than a single size. The bulk density of a polydisperse suspension is usually greater than that of a monodisperse suspension since small particles tend to fill in open spaces between large ones. Thus, fluid pressure can be applied to more solids since there is less free space for fluid flow with polydisperse materials. This situation can be helped by mixing in a small amount of over or undersized particles; powder flow rate can also be improved with little increase in required power. Rough sizing by sieving or porosimeter methods is adequate to characterize powders for this purpose. Information on overall particle volume and on particle shape is important here. Detailed size distribution data at the extremes of the distribution are not necessary.

The appearance and texture, as well as the color tone of many materials is strongly affected by particle size of the powder or colloidal suspension used in their preparation. Cosmetic compounds are applied so that their texture is acceptable both for color and appearance, as well as for the "feel" of the materials. Reflection and absorbance of light is strongly affected by particle size; thus the smoothness of appearance, as well as color is controlled by particle size of the material. The preparation process frequently involves grinding and blending batches to exactly the correct particle size distribution for this purpose. These materials are sized by microscopic measurement of deposited sample powders or by air elutriation methods.

Metals and ceramic materials are often produced by chemical reaction and/or sintering of finely divided powders. Abrasive and polishing compounds are frequently produced by similar processes. The physical properties of these products are strongly dependent on the size distribution of the powders used for compounding specific alloys and materials. This is particularly true for controlling crystalline structure of intermetallic compounds, for ensuring strength and quality of many modern ceramic materials, and for producing abrasive or polishing powders that will not contain oversize fragments or excessive fine powders. These product powders are often defined by sieving or by wet sedimentation methods. Particle shape is quite important for these products.

B. Process Effluents

Metal refining and processing, paper production, petroleum processing, glass production, etc., frequently involve high-temperature reactions which produce fine particle emissions to the environment. The transport and deposition of these materials and inhalation hazards are affected strongly by particle size. These process effluents, exhausted through a stack or in any gas stream, are controlled by ordinances that specify allowable effluent opacity or particulate mass loading with some particle size range limits imposed. Particle size of the material in the effluent will materially affect the visibility and/or travel of the exhaust plume as well as deposition rates and effects on the atmosphere. This is due to the particle size sensitivity of the optical extinction coefficient when the particle size is close to the wavelength of visible light. For any mass concentration, the plume opacity will be affected mainly by particle size, composition effects being trivial. Combustion product particles are frequently defined by microscopic observation of collected particles or by direct observation of plume opacity.

C. Information Storage

Information storage and exchange by optical or magnetic devices are strongly affected by particle size. Sensitivity and resolution of photographic images depend on the size of the film emulsion materials. Similarly, the quality of photographic images and laser printer output depends on size control of the toner powder used to form the images. A very common information storage system used in most computer operation involves magnetic recording, where the information storage density and accuracy depend upon the size of the magnetic particles on the tape or disk. These materials are frequently defined by using electron microscopic observation of deposited samples collected during manufacture or by quasi-elastic light scattering techniques.

D. Agricultural Applications

Airborne control of insect infestation by powders and aircraft spray is a typical agricultural application where particle size control is necessary. The powder or spray droplet size must be closely controlled. The droplets must be large enough so that they will settle to the ground and crops to be treated without drifting away from the target area; they must be small enough so that they will deposit upon the insect surface for maximum effect. The same considerations apply to application of germination control materials or of any liquid crop treatment material. Individual spray nozzle operation is frequently defined by measuring Fraunhofer scattering from droplets in the spray cone. Performance of delivery systems is frequently defined by examining emitted droplet cloud sizes with optical means or by collecting the droplets and examining their images with microscopic means. In one application, the collected droplets were frozen, and the solid spherical ice particles were sieved at low temperature for size definition.

E. Energy Production

Energy production by combustion of fossil fuels always results in emission of ash particles. This applies to combustion of fuel oil droplets or of powdered coal. The efficiency of the combustion process, as well as the nature of the fuel residue emission material, depends on the particle size of the fuel. Cyclone burners for coal require that the coal size be reduced to 5–10 μm for maximum combustion efficiency, while oil burning systems operate best with droplets no larger than 50 μm. The widespread use of fuel injection systems for automotive purposes and the requirements of aircraft turbine combustor fuel injectors are additional examples of the need for particle size control in energy production areas. Once the fuels are burned, environmental considerations in controlling combustion product emissions require that the particle size of the fly ash must be known and controlled so that the emission cleaning and processing elements will operate correctly. When nuclear energy power plants are operating, all emissions from the facility are sampled to ensure that no particulate radioactive emissions have occurred. The same measurement methods as used for manufacturing process emissions are used for observation of power plant emissions.

F. Food and Beverage Products

Beverage production, from potable water through almost all beverages, requires control of particle content. This may involve flocculation of fine silt in a potable water supply or controlling sedimentation of solids during aging of wines. Appearance, processing time and effort, as well as taste, of these liquids is strongly affected by particle content in their processing steps. Particle content in water both before and after treatment is often defined by optical determination of turbidity. The original procedure was based on the length of the light path through a container of water which just caused the image of the flame of a "standard" candle to disappear when the flame was viewed through the suspension. Turbidity was defined in terms of "Jackson turbidity units" (JTU), named after an original investigator [1]. Turbidity is presently defined with photoelectronic instruments much more sensitive and reproducible than the Jackson turbidimeter.

The taste, as well as the texture, of many food products is affected by the size of the particles which make up that food. The obvious examples of particle size effectiveness in terms of food character are in the preparation of processed vegetables, such as mashed potatoes or vegetable mixtures in a food processor. Chocolate is another prime example of this situation. The texture and mouth feel of different chocolate types is directly affected by the particle size of the cocoa and sugar powders in that material. A variety of measurement methods are used to characterize the particles in food products. These materials are easily affected by temperature and moisture changes and the measurement method must be selected so that these effects do not cause problems in operation of the method and particle properties are not altered by the method.

II. PARTICLE ARRAY CHARACTERIZATION DEVICES

Particle measurement can be carried out by observing a cloud or an array of particles and determining a mean size and the breadth of the particle size distribution. Alternately, individual particles in a suspension can be observed and the suspension size data can be derived by summing data from each of those particles. Particle array measurement procedures normally characterize particles in terms of a fraction of a total population with the specific property being measured by the procedure. For example, percentage by weight of a powder sample with longest dimensions within a specified range can be defined by precision sieves; the percentages by volume or mass of a sample above or below specific values are characterized by gravitational or centrifugal sedimentation devices; the total fraction with projected area greater or less than that of a sphere of specific diameter is defined by photometric devices. This methodology differs from the measurement procedures of single particle observation devices. In that type of system, each individual particle is observed and the size function associated with the measurement method is defined. The number of particles in each of several size fractions is measured and population fractions are developed by mathematically processing the data, rather than by observing the sampled population directly. In the case of the array systems, a measurement procedure can be selected to produce data which show the influence of a particular size fraction directly. In the case of the single particle measurement systems, individual over- or undersize particles which might cause problems can be detected easily. The choice of a particular measurement method is mainly dependent on the objective of measurement, although the material properties and the availability of instrumentation and time will affect that choice.

A. Screening, Wet and Dry

A method of separating solids from liquids has long been the use of vibrating screens. Performance of these commercial screens is frequently determined by use of test sieves to define the particle size distribution of the solids either retained on the screens or which have passed through those screens. The test sieves have opening dimensions and configurations that can be defined by relatively uniform apertures in woven wire or cloth sheets. The aperture shapes are typically square, rectangular or diamond shaped, depending on the screen weave. Figure 1 illustrates a few standard rectangular screen weaves. The sieves are usually characterized in terms of a "mesh" number, related to the number of parallel wires per inch in the weave of the sieve. Sieves can be obtained with a wide range of mesh numbers. In the United States, one series of mesh numbers is defined so that the openings of successive standard sieves progress from a base of a millimeter by a factor of the fourth root of 2:1. In selecting sieves from this series, customary practice is to take each sieve in a given range, every second sieve, or every fourth sieve. For example, a 10 mesh sieve has 2.00-mm openings, a 14 mesh sieve has 1.40-mm openings and a 18 mesh sieve has 1.00-mm openings. The smallest aperture in this series is 25 μm, for a 635 mesh sieve. Table 1 illustrates standard mesh number and aperture size relationships for NIST standard sieves.

For determination of particle size distribution data, a stack of screens is used, each screen with a selected mesh size. The procedure is relatively easy to use and a number of standard methods have been developed for powder characterization. As in any particle size measurement procedure, careful sample preparation, identification of the test procedure and use of good judgment in establishing data tolerances are all required [2]. Cascaded screen assemblies can be used to characterize size distributions. The screen with the largest apertures is at the top of the stacked cascade and the aperture size decreases with each lower screen. The powder, either dry or in liquid suspension, is placed onto the topmost screen. Under the influence of gravity or a pressure differential along with transverse and/or vibratory motion of the screen, the liquid and particles smaller than the screen openings passes through those openings. Mechanical or ultrasonic vibration has been used to help move powders through the screens. The quantity of powder passed through the last screen is observed and the procedure is halted when the passing powder quantity rate of increase is statistically insignificant. The weight or volume fraction of particles equal to and larger than the screen openings is determined by weighing the material retained on each screen. Weight fraction data are usually used for dry powder measurements;

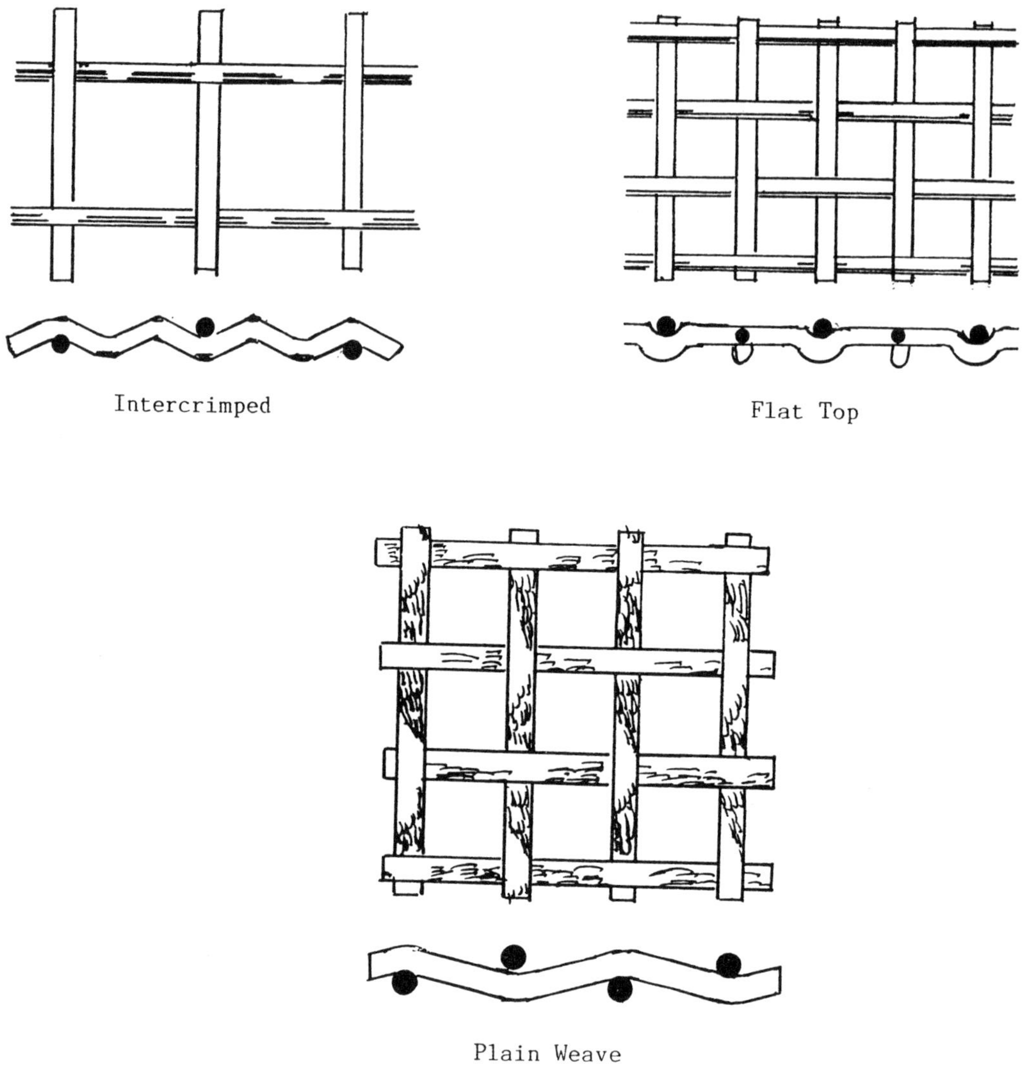

Figure 1 Sieve aperture configurations.

volume fraction data are usually used for liquid systems. Most sieve measurements are carried out using single batches of particles. Following each measurement, the sieve(s) are cleaned and a fresh batch is characterized. Where it is necessary to carry out many measurements with sieves, automated sieving machines are available. These permit continuous measurement of successive material batches with automatic sample feeding, agitation, continuous sieve weighing, followed by sieve cleaning. Data are reduced to the desired format.

Woven wire screens are used to characterize particles in the size range from approximately 30 μm to several millimeters in diameter. Precision etched metal screens with square or circular apertures have been used for measurements of particles as small as 2–5 μm [3]. For these small apertures, the particles are usually suspended in liquid in order to minimize agglomeration effects and to aid in the screening operation. Recently, thin glass screens with very well-defined cylindrical apertures and larger open area than present in the etched metal screens have been

Table 1 Sieve Opening Dimensions for Standard Mesh Numbers

Mesh no.	Opening (mm)	Mesh no.	Opening (mm)
5 in.	125	No. 10	2.00
4.24 in.	106	No. 12	1.70
4 in.	100	No. 14	1.40
3½ in.	90	No. 16	1.18
3 in.	75	No. 18	1.00
2½ in.	63	No. 20	0.850
2.12 in.	58	No. 25	0.710
2 in.	50	No. 30	0.600
1.75 in.	45	No. 35	0.500
1½ in.	37.5	No. 40	0.425
1.06 in.	26.5	No. 45	0.355
1 in.	25.0	No. 50	0.300
⅞ in.	22.4	No. 60	0.250
¾ in.	19.0	No. 70	0.212
⅝ in.	16.0	No. 80	0.180
0.530 in.	13.2	No. 100	0.150
½ in.	12.5	No. 120	0.125
7/16 in.	11.2	No. 140	0.106
⅜ in.	9.5	No. 170	0.090
5/16 in.	8.0	No. 200	0.075
0.265 in.	6.7	No. 230	0.063
¼ in.	6.3	No. 270	0.053
No. 3½	5.6	No. 325	0.045
No. 4	4.75	No. 400	0.038
No. 5	4.00	No. 450	0.032
No. 6	3.35	No. 500	0.025
No. 7	2.80	No. 635	0.020
No. 8	2.36		

produced for better size fractionation of particles. With these small aperture screens, the particle transfer operation is aided by sonic pulsation of the liquid to move material through the apertures.

Sieving can be used to measure particles as small as 1–2 μm in diameter, but precision micromesh sieves, wet sieving and skilled personnel are required for accurate measurement of particles much smaller than 10–20 μm in diameter. Automated systems are available and can be used for particles in any size ranges with computer control for the entire operation. Sieving is especially adaptable for measurements of irregular powders and can be used with concentrated slurries containing up to several percent by weight of solids.

B. Flow Restriction Observation

When liquid flows through small apertures, force is required to cause passage through the apertures. As the apertures decrease in dimension or as the liquid viscosity increases, the required force also increases or the flow rate will decrease. If the liquid is contaminated with particulate material of approximately the same dimension as that of the apertures, then particles near the aperture size will deposit in those apertures, resulting in reduced passageway dimensions. As a result, if constant pressure is maintained on the liquid, then the flow will be reduced. The flow reduction will depend upon both the quantity and size distribution of the particulate material in the liquid. An ASTM test procedure has been developed on the basis of this phenomenon to characterize gross cleanliness of reagent-grade waters in terms of a filterability index [4]. Figure 2 shows the system described in the ASTM method for determining flow reduction due to the presence of particulate contamination. The procedure will indicate gross differences in particle loading in the liquid being examined, but differences in contaminant particle size distribution or in the filter pore size distribution will modify the rate at which the filter in the system becomes plugged. The method is primarily useful for production of comparative data. Care is required to ensure that filter-to-filter variability is minimized.

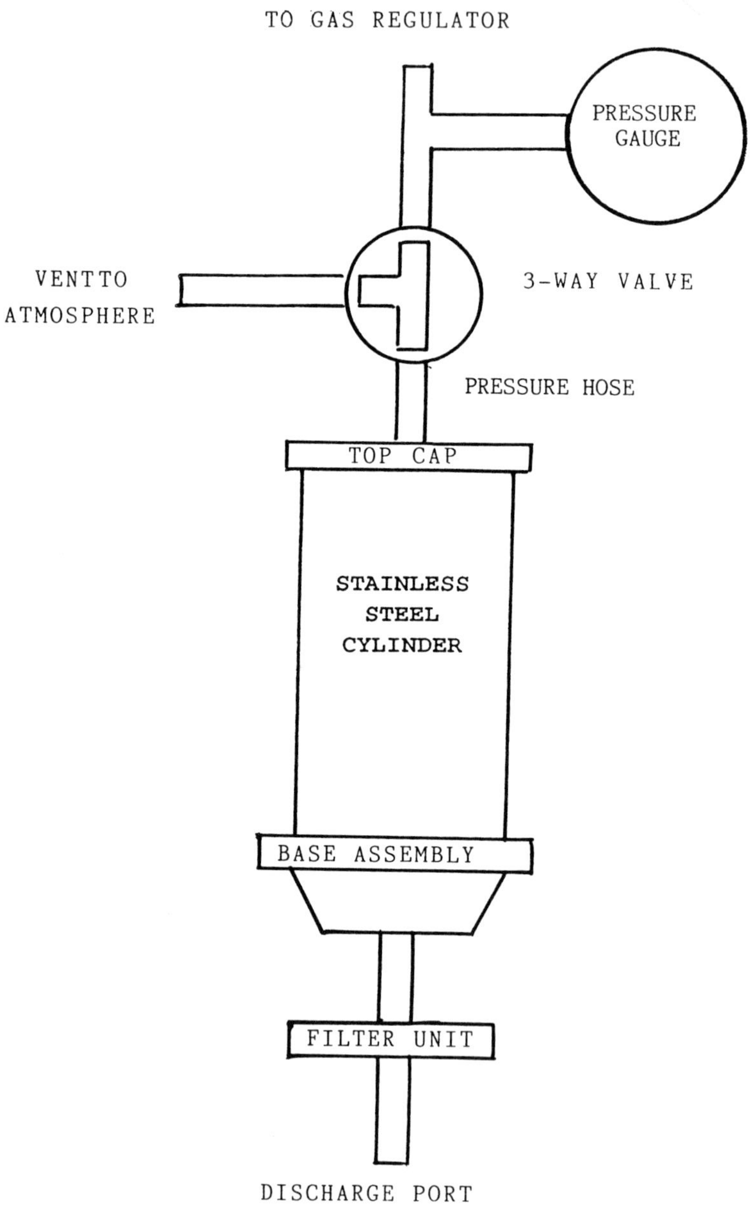

Figure 2 Flow restrictor system. (From Ref. 4.)

Even though the ASTM method indicates applicability to reagent-grade and high-purity water, the time required for characterizing clean liquids with some of the submicrometer pore size present-day membrane filters makes it impractical for many purposes. The method characterizes the particle content of the liquid being tested only in terms of its effect on a specific filter flow rate. No quantitative indication of particle mass, volume, or particle size is produced, but comparative data can be produced for contaminated liquids. This requires that the filter element used for all measurements are identical and that the liquid flow rate, temperature and pressure are carefully controlled. If the nature of the system being evaluated changes, then the quantity of material which will affect flow rate will also change. Empirical calibrations can be obtained for situations in which only the concentration of contaminants changes, but the size distributions remain constant. If the size distribution of the collected material on the filter changes, then the pressure drop will also vary, even if the mass of material does not change. The procedure has also been used to characterize some hydraulic fluids for "silt content" levels. Even though the relationship between particle size distribution and concentration is not specifically defined, the system appears to be useful for defining particulate material content in these liquids at levels of at least 10 mg/l.

C. Sedimentation Systems

When particles are dispersed in a liquid, they are exposed to gravitational forces and to viscous drag and buoyancy forces which tend to establish an equilibrium settling velocity for the particles suspended in the liquid. Both forces vary with the particle size; the terminal settling velocity, which is proportional to the square root of particle diameter, can then be transformed to a Stokes-size parameter. Particles in a suspension can be characterized by placing a sample of that suspension at the top of a column of particle-free liquid and observing the quantity of particles which have settled to a particular level in that column or to the bottom of the column as a function of time. Alternatively, a uniform suspension of particles can be placed in a vessel and concentration or suspension density changes at a series of depths within the originally homogeneous suspension are measured at a series of specified times. As the particles settle to the bottom of the cuvette, the decreasing particle concentration is measured at one or more levels by the increasing light transmission through the suspension. Figure 3 shows operation of a sedimento-

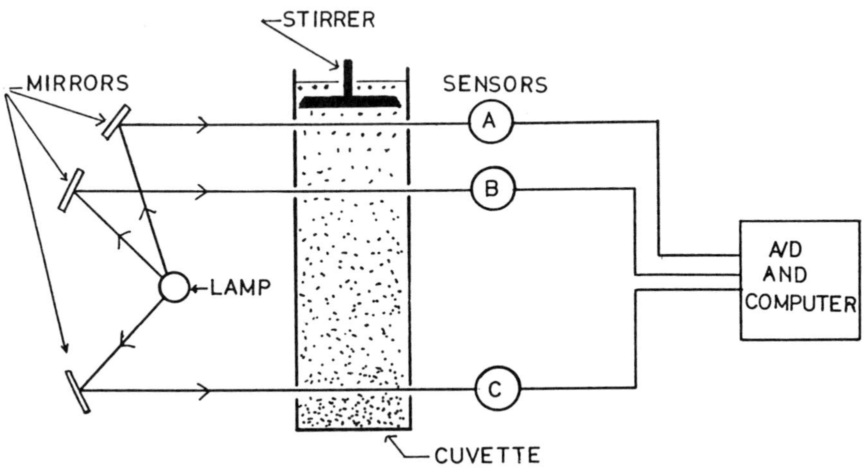

Figure 3 Gravity sedimentometer. (Courtesy N. O. Hagglund, Paar USA, Warminster, PA.)

meter where particle settling occurs under the influence of gravity. In the one shown, simultaneous measurements of particle concentration are made at three levels in the liquid to allow faster and more accurate data collection.

When measuring particle size by gravitational sedimentation rates, the sedimentation rate of small particles becomes quite low and thermal nonuniformities in the liquid cause variable flows that interfere with uniform settling of small particles; as a result, the time for collection of meaningful data restricts the operation of this system to particles no smaller than 2–5 μm. For this reason, sedimentation under the influence of centrifugal force has been used to extend the capability of measurement down to 5–10 nm in diameter. Rotational speeds up to 10,000 rpm have been used for this high sensitivity. The first centrifugal sedimentometers removed a liquid sample for determination of particle content by either weight or microscopic observation. Some centrifugal sedimentometers observe particle concentration in the liquid by X-ray transmission measurement.

The particle concentration can be determined by removing a small, fixed volume sample from a specific point within the column of liquid at specific times and weighing the particles in that liquid sample [5]. It is also possible to characterize particle content in the liquid by defining changes in transmission of radiation through the liquid. Both optical [6] and X-ray [7] transmission measurements have been used in sedimentometers. Using these techniques, data on particle suspensions in the size range from 0.01 to 300 μm can be obtained in an hour or so. These data are used to define a Stokes-diameter-size distribution of the particles present in the original suspension. Figure 4 shows a centrifugal sedimentometer where particle content at various locations along an axis of the spinning disk is determined by observing light transmission through windows at specific distances from the center of rotation. The illustration also shows a

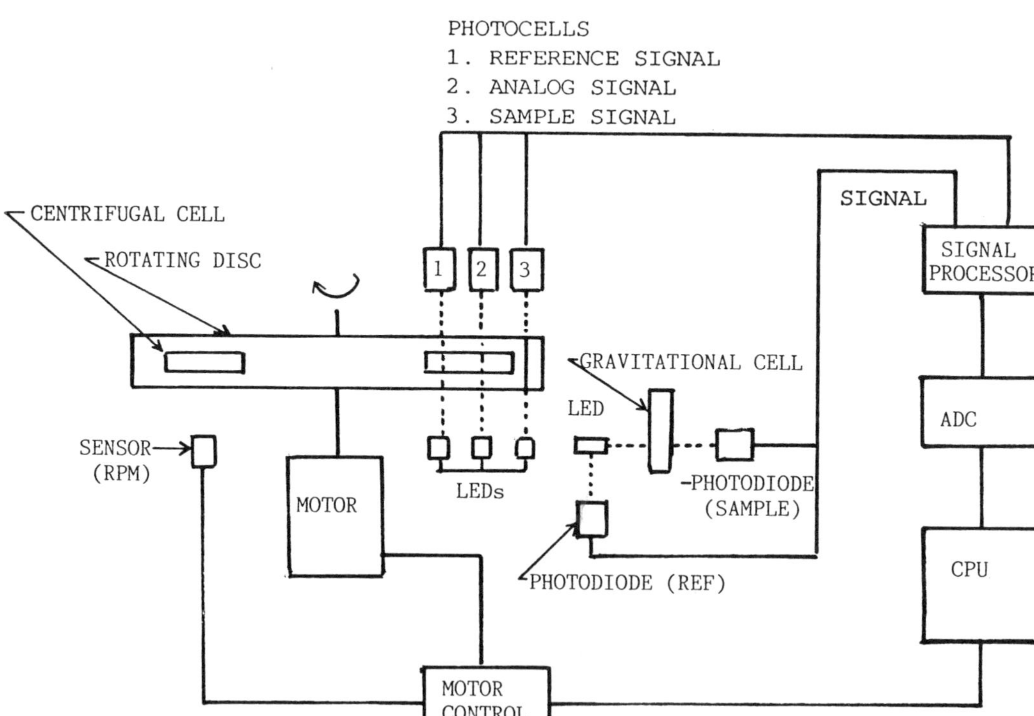

Figure 4 Centrifugal sedimentometer. (Courtesy B. Heninger, Horiba Instru., Ltd., Sunnyvale, CA.)

gravitational cell that is used to measure particles so large that they would move too rapidly for accurate measurement in the centrifugal cell. The signal processing system allows definition of a broad particle size distribution by combining data from both cells.

D. Optical Methods

There are several optical methods used for detecting particles and characterizing particle size and size distribution in particle arrays. The operating principle of these devices is relatively simple: a light beam is passed through the liquid in which the particles are present and the quantity of light absorbed by or scattered out of the direct beam by particles in a defined volume within that liquid is measured and related to the particulate content of the liquid. A wide variety of optical designs have been used for these instruments. They include both light extinction and light scattering systems.

1. Photometers and Turbidimetric Devices

These instruments define the quantity of light removed from a direct beam by a suspension of particles in liquid. They are used primarily on-line to indicate the overall particle concentration levels in the liquid. The photometers can record the reduction in light level caused by the combination of scattering out of the direct beam and by absorption of light by the particles in the liquid. The removal of light by a particle is referred to as light extinction. When reduction in light levels are measured, the instrument is referred to as an absorption analyzer or a turbidimeter. These instruments are widely used to characterize clarity of potable water during treatment. For measurement of turbidity in relatively clean water, the ratio of light scattering at a specific scattering angle to light extinction is used. Figure 5 shows the optical system for a commercial ratio turbidimeter. Instruments of this type can be used to observe turbidity in either flowing liquid streams or in batch samples. This type of device is sensitive to turbidity levels as low as 0.0001 nephelometric turbidity unit (NTU). Multirange system can be set to indicate levels as high as 2000 NTU.

The light flux reduction is a function of the total particulate projected area and the particle extinction coefficient. The particle extinction coefficient is defined as the ratio of total light flux scattered and absorbed by a particle to the light flux incident upon that particle. In a polydisperse particle suspension, the relationship between particle area and concentration will vary with the particle size distribution function. The coefficient varies with particle size and refractive index.

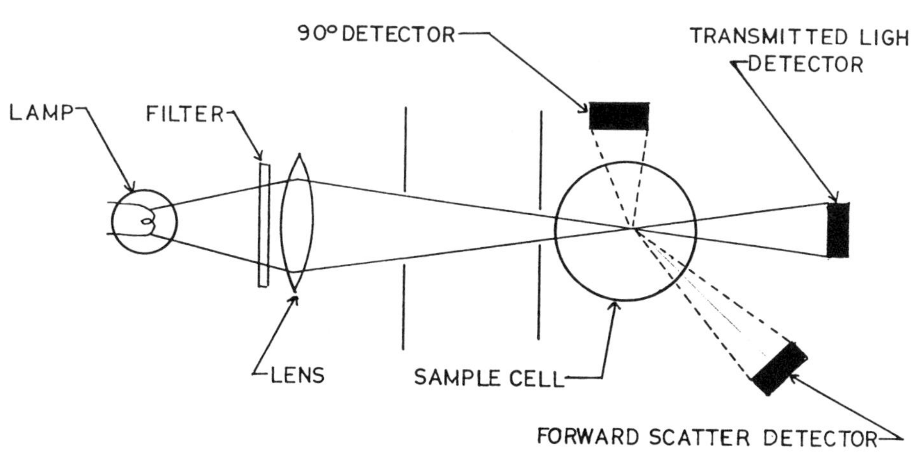

Figure 5 Turbidimeter.

Absorption analyzers (turbidimeters) are usually used to define particle concentrations up to approximately 1000 ppm. Nephelometers are available which record only the light flux which is scattered out of the beam at a specific solid angle. Depending upon the angle which is observed, the scattering analyzers can be used to define concentration levels up to 10–15% solids in liquids. Particle size information is not defined with either instrument type. They are used mainly to characterize particle content of waste liquids.

An early quantitative definition of turbidity was based on the 1900 work of Whipple and Jackson [1], who developed the Jackson candle turbidimeter for measurement of potable water quality. Empirical calibration to a turbidity scale is normally used to quantify the particle concentration. The first calibration material to be used was a part per million suspension of diatomaceous earth, but this material and the measurement method which used it relied heavily both on subjective observation and preparation methods. A flat-bottom, open-top glass tube is held vertically above a special candle designed to produce a fairly constant white light flux. Measurement was made by slowly pouring a sample of liquid to be evaluated into the tube until the visual image of the candle flame diffused to a uniform glow. The relationship between the calibration standard material concentration and the optical path length (the depth of water in the column) where diffusion occurred was used to define the water clarity in terms of JTU. Accurate measurement of turbidity below approximately 25 JTU is difficult.

Other methods of defining turbidity of treated potable water were required, since the turbidity of such water generally falls within the 0–1 JTU range. The diatomaceous earth suspension has been replaced by carefully prepared suspensions of formazin whose turbidity level is constant and permits acceptable reproducibility of measurement. This material is frequently used to calibrate optical right-angle light scattering turbidity instruments. When turbidity of water is measured with such an instrument, the results are reported as NTUs rather than JTUs. Waters with turbidity less than 1 NTU can be measured in this way [8]. The same instruments are used after calibration with controlled concentration suspensions of latex spheres for analyzing potable water clarity.

2. Fraunhofer Diffraction Analyzers

A stream of gas or liquid containing the particles to be measured is passed through a sensing zone illuminated by a collimated laser beam. Particles in the size range from approximately 0.7 to 700 μm in diameter scatter light in the forward direction at angles inversely proportional to size at intensities directly proportional to size. The scattered light pattern from each particle in the sensing zone can be thought of as producing a number of Airy patterns, each consisting of concentric bright and dark rings. The scattered light is collected by a Fourier optical system and focused upon a radial photodiode array detector. In this way the pattern of scattered light intensity as a function of scattering angle can be defined. Figure 6 illustrates the basic optical design of a typical Fraunhofer diffraction analyzer system. By assuming one of several possible particle size distribution functions, an algorithm can be used in a high-speed data processing system to deconvolute the data and convert the light intensity and scattering angle data to particle size distribution and concentration data [9]. Experience has shown that only a few particle size distribution functions are adequate to characterize almost all of the particle size distributions appearing in nature. Difficulty can arise if detailed information on particles in the tails of the distribution is needed. The instrument readout may not record these particles at all. Concentrations up to a few percent by volume can be measured over the size range of the system.

Operating procedures are quite simple. Modern instruments incorporate automated and well-controlled sample dilution and mixing systems. Either liquid suspensions or dry powder can be handled in the flow and mixing system. The scattered light measurement system is programmed to operate with almost no manipulation and the data reduction program is almost

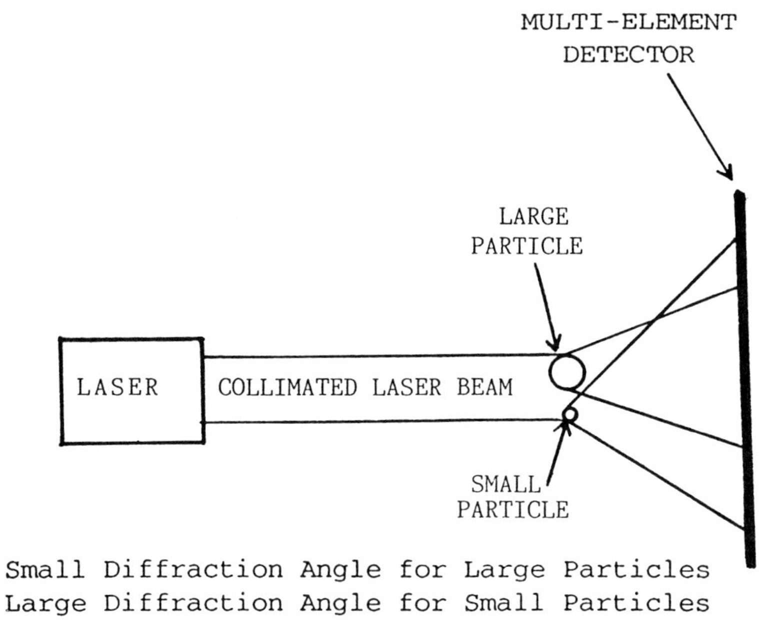

Figure 6 Fraunhofer diffraction system operation.

completely automatic. Several common size distribution function formats are programmed into the system; the system seeks a best fit to one of the formats. The advantages of the system are the capability of handling large samples in an extremely wide concentration range, the repeatability demonstrated in operation and the wide particle size range that can be handled. The system is capable of particle measurement in either gas or liquid suspensions. The major disadvantage is the assumption that the particles being measured will fit one of the assumed size distribution functions. Even this is not a problem unless detailed information on specific particles is required; most real particle size distributions fit one of the standard two- or three-constant distribution functions such as a lognormal or a log-probability function.

3. Dynamic Light Scattering Analyzers

Dynamic or quasi-elastic light scattering instruments observe the variations in intensity of a coherent light beam transmitted through a suspension of colloidal particles. The variations arise from Brownian motion of particles in the liquid. The particle motion extent can be correlated with its diffusion constant, a direct function of particle size. Measurement of the light level variation due to Brownian motion can then be correlated to the diffusion constant of the particles in the liquid. The mean size and polydispersity of the size distribution can also be derived directly from the data. If an appropriate algorithm is used, the system software can also characterize multimodal particle size distributions, even when the modes are very close in size [10].

Particles in the size range from about 0.005 to 3 μm can be sized. Particle concentrations in the range from approximately 0.01% to 1% by volume can be measured. At lower concentrations, the signal level is not adequate for accurate measurements in reasonable time. In operation, a sample volume of a few milliliters is prepared in a cuvette and placed in a laser beam. Photodetector response is transmitted directly to a microcomputer control and data processing system and results are displayed, including mean particle size, polydispersity and diffusion constant values. The advantages of the system include capability for very small particle characterization, small sample size requirements and reasonably rapid response. The disadvantages are the relatively high concentrations required for operation, the need for reliance on the algorithm used in the computer program, the effects of temperature and liquid viscosity on the diffusion constant measurement and the need for skilled operators in terms of data interpretation and system operation. Even so, this is still the only near real-time measurement method for colloidal particles as small as this device can define.

III. SINGLE PARTICLE CHARACTERIZATION DEVICES

A. Filter Surface Observation Methods

One of the oldest methods of particle characterization in liquids is that of collection and direct observation. The particles can be collected upon almost any surface where they can be observed. For particles in liquid, collection upon a membrane filter is most convenient. A known quantity of liquid is passed through a membrane filter with pore size smaller than the smallest particle of concern. Essentially all of the particles in the liquid which are larger than the filter pore size will be deposited upon the filter surface. The filter is then examined microscopically and the particle number and size distribution are recorded. Particles larger than approximately 0.5 μm are usually observed using light microscopy. The techniques are well established and documented [11,12]. Smaller particles are measured with an electron microscope. For any single measurement, the size range that can be observed conveniently is limited by depth of focus and microscope resolution to about 30 to 1. In order to minimize the possible effects of one particle masking another on the filter surface, coverage of no more than 10% of the observed substrate is recommended. Particle counts up to a few thousand per sample can be made easily, especially if an image analyzer is used. Detailed information on particle size and particle morphology can be obtained; skilled operators can also identify many particle compositions from the appearance of the deposit and a few simple procedures for determining optical properties. Deposits on a single filter surface can be subdivided for examination into several hundred fields, depending upon the filter size and particle concentration on the filter surface. Most measurements are made with filters which are either 25 mm or 47 mm in diameter.

A brief summary of the operating sequence for measurement, after microscope calibration, is

1. Set up the filter in the filter assembly.
2. Pass an known volume aliquot of the sample through the filter either by pressure or vacuum.
3. Place all or a section of the filter on the microscope stage.
4. Clear the filter with a suitable solvent.
5. Scan the filter surface visually to make sure that the deposit is well distributed over the surface.
6. Count and size the particles per unit area of filter.
7. Relate count data to total filter surface and sample volume.

An image analyzer or another automated filter surface observation system can be used to reduce personnel subjective measurement errors [13]. The systems can be adjusted to scan the entire filter surface or selected sample areas. Particle contrast limits and shape definitions can be established to define countable items. Particle size can be defined in terms of any accepted dimensional definition; the system will also define the particle size distribution for the entire sample, if desired.

The major advantages of this method are that the particles on the membrane are available for repeated observation; morphological or chemical characterization can be determined. A segment of the membrane can be removed and either destructive or nondestructive analysis can be carried out to define the composition of particles on the filter [14]. Biological culturing techniques or epifluorescence microscopy [15] can be used to determine viable microorganism concentration. The time needed for measurement can range from 15 min or so to many hours per sample. Equipment cost is relatively low unless an electron microscope or an image analyzer is used for measurement. Capital costs for other equipment are in the $5000 range. However, costs for laboratory space and personnel time may bring operating cost per sample to a relatively high level.

The problems associated with this method arise from the requirements for skilled personnel activity in both sample handling and measurement. A major problem with the method is the potential for variability in reported particle count and size data that can occur for inter- and intraoperator activities. This is important both in sample handling prior to measurement as well as if data from manual observation of the particle deposit is used for analysis.

B. Electrical Resistance Pulse Measurement

Electrical resistance counters were originally developed for blood cell counting applications and are extensively used in that area as well as to define powders in other industrial operations [16]. In operation, an electrically conductive liquid flows through a small aperture in a nonconductive barrier. Electrical current is passed through the liquid passageway and the current across the aperture is monitored. If a particle is present in the sensing zone (the aperture volume plus a small surrounding equipotential zone), then the current is decreased to an extent relatable to the particle volume. Figure 7 shows the operation of the electrical resistance particle counter. The system can be used with both conductive and nonconductive particles [17]. Conductive metal particles will usually have a layer of oxide with conductivity different from that of the liquid. Pulse sorting and counting then allows definition of the particle size distribution and number in the liquid which has passed through the aperture.

Apertures are available with diameters ranging upward from 20 μm. For the smallest aperture, 0.5 μm particles are defined with good resolution. The particle sizing range for any single aperture is usually considered to be 2% to 40% of the aperture diameter. In some circumstances, the linear range can be extended to 80% of the aperture diameter [18]. Counting speeds to 5000/s can be attained for any aperture, liquid flow varying with aperture size. In normal operation, particle concentrations are measured to the point where coincidence (more

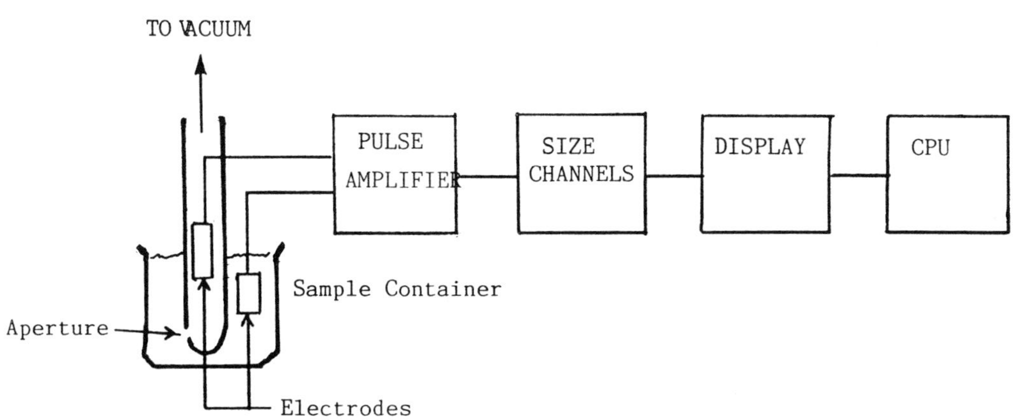

Figure 7 Electrical resistance particle counter. (Courtesy Coulter Electronics Corp., Hialeah, FL.)

than one particle in the sensing zone) probability approaches 10%. For small apertures, this may be more than 100,000 particles/ml.

In operation, a sample is placed in a sample container with a stirrer to keep particles in reasonable suspension. A tube containing the aperture is placed in the sample container and vacuum is applied to a liquid column to draw sample into the tube through the aperture. Continuous stirring and avoidance of a flat-bottom container is recommended strongly, since sample withdrawal rate may be so low that turbulence due to horizontal stirring motion and aspiration of the liquid alone may not be adequate to keep the particles well-dispersed. After the standard sample volume for the aperture in use has been counted, the data are recorded in terms of the number of pulses or specific amplitude generated during passage of a known volume of liquid. These data are directly related to particle concentration and size.

The advantages of the electrical resistance instrument are that the particle sizing resolution is excellent since the response is directly related to particle volume (resulting in output signal varying with the third power of particle diameter), the counting rate is very high, materials of almost any type can be counted with little effect of particle composition variability (even conductive metals have sufficient oxide layer so that the conductivity is much less than that of the conductive liquid), that the instruments are very repeatable and have a long history of successful operation, and that the manufacturers have provided sophisticated and usable software for almost any data transformation required. Very highly skilled operators are not necessary; the instrument vendors normally provide good training programs.

The disadvantages of the electrical resistance method are that the liquid must be conductive or that a miscible conductive solvent be available which will mix with the liquid to be tested and that will not modify the particles present in that liquid, that "large" particles can block flow of other particles through the aperture, that the long electrical path through the liquid can pick up any stray electrical noise from external sources. In very noisy areas, the instrument must be operated within a Faraday cage.

C. Acoustic Response System

Characterization of particle content in liquids usually involves removal of a sample from the main body of the liquid. This operation often modifies the particles in some way. Therefore, a nonintrusive, nondestructive analytical procedure has long been sought. Shortly after World War II, the successful application of acoustic detection techniques for submarine location led to studies of acoustic response for particle measurement in the liquids used in the aerospace industry. The Sperry Corporation developed an acoustic echo-sounding system for detection of particles in opaque liquids [19]. Liquid samples could be placed in a container fitted with a piezoelectric crystal at its base or the crystal could be located in the wall of a liquid flow line. The crystal functioned as both an emitter and receiver for the sonic energy. A pulse was emitted in a focused converging beam. After the pulse was emitted into the liquid, the crystal function was changed so that it acted as a receiver. When the energy beam reached a particle in the liquid, some energy was reflected back to the crystal. Only those pulses which were returned to the crystal after a time required for transmission to and from that band of liquid whose location represented the time between the original pulse emission and the time that the crystal was operated as a receiver would be detected. This time gate allowed definition of a particular band of liquid from which particle signals would be accepted. The reflected energy is proportional to the size of the particle which produced the echo. Operation of the acoustic response counter is shown in Fig. 8. In use, careful calibration was required because of the nonuniform energy distribution in the conical beam. Some problems were also noted because of false signals from turbulence and bubbles in the liquid. The smallest particle that could be measured by this device

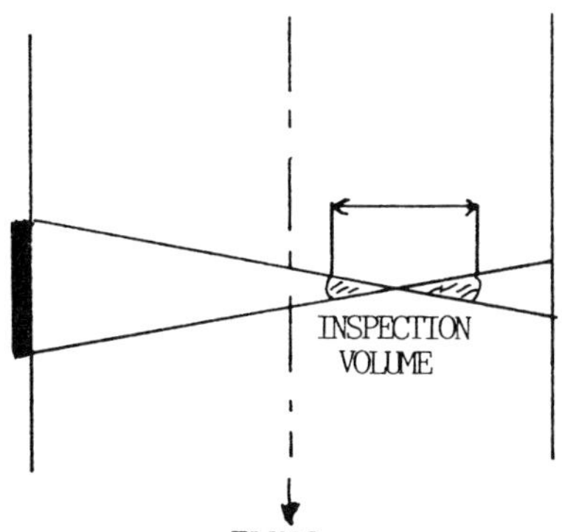

Figure 8 Acoustic response counter operation.

was some 25 μm in diameter. A system using higher-frequency (15 mHz) crystal emitting microsecond pulses was used for measurement of in vivo blood flow [20]. This device was applied to measurement of contaminant particles in other liquids, but could not detect individual particles smaller than 5 μm. In addition, sizing resolution was poor; the large angle of convergence used to provide high energy at the acoustic beam focal zone and resulting low-level signals from large particles at the outer portion of the sensitive zone made it impossible to define the size of individual particles producing signal information or the liquid volume from which the data were produced.

The major advantage of the method is that it is indeed nonintrusive and that it can detect changes in particulate contamination level. However, the disadvantages of poor size sensitivity and the difficulty of defining particle concentration in the widely divergent acoustic beam make it very difficult to define the sensitive volume consistently. Poor size resolution and the requirement that the speed of sound be known for the liquid in use in order to define particle size and sample volume makes the device strictly a gross contamination indicator rather than a true particle measurement system.

D. Optical Single Particle Counting Systems

Systems which use either light extinction or light scattering from individual particles can be used for particle counting and sizing. Although there are differences in operation of the system types, there are also similarities in operation which must be considered for optimum results from either system. As with most particle measuring operations, a sample is observed from a larger population; the sample must be truly representative of that population. In addition, these devices observe a sample during transit through a sensing zone; in some instances, the required sample flow rate may not be adequate to retain large, dense particles within the sample in acceptable spatial and temporal distribution as that in the original population. Some care in sample handling is required which may be peculiar to a particular measurement method.

The instruments do *not* image the particle being measured, but record the change in light intensity caused by the presence of the particle in the sensitive zone. The change may be caused by a single particle, an array of particles, a gas or vapor bubble, or from any other phenomenon which will modify the detected light level. If more than one particle is present in the illuminated volume, then the instrument will respond as if a single particle with the same projected area as the particle array were to be present. For this reason, the allowable sample concentration is limited by the probability that more than one particle may be present at any time within the sensing volume in use. For many optical particle counters, the sensing volumes are in the size range of 10^{-5} cm^3 in volume. A flow passage some 500 μm by 1000 μm in cross section is illuminated by a light beam 20 μm or 50 μm high across the 500-μm face. At that size, the probability of more than one particle being present at any time is acceptably low at concentrations of approximately 10,000 particles per milliliter, as long as the particles are small. If a significant portion of the particles in the liquid are in the same size range as one of the sensing volume dimensions (20 or 50 μm), then that concentration may decrease by at least a factor of 2. For very low concentrations, single particles can be distinguished as long as the signal for the smallest particle being measured is approximately four to five times larger than the random noise pulses in the system.

Optical particle counting systems are calibrated with latex spheres of refractive index 1.6 suspended in water with refractive index of 1.33 or in oil with refractive index of 1.47. Any particle passing through the sensitive volume is defined as having the same size as the latex sphere that would produce the same change in light level as was produced by the particle. Equal-sized particles in liquid of refractive index different from that of the liquid used for calibration will produce data indicating a different size for each liquid. Since the system responds only to changes in light flux level in the sensing zone, then any gas bubble will change the emitted light in the same way as a particle would. If the liquid flow velocity changes so that the duration and rise time of the pulse caused by particle presence in the sensing zone changes, then the electronic system response may also change to indicate a larger or smaller particle than actually present.

The advantages of the optical particle counting systems include the ability to count and size extremely small particles in real time in flowing liquids. Data are produced rapidly, allowing acquisition of statistically adequate information even from relatively clean liquids. The systems can be used for either laboratory or field measurements and can handle any liquid over a large range of pressures and temperatures. The systems are capable of resolving very small differences in particle size and are very reproducible in operation, as long as calibration and maintenance are kept up. Required operator skill is minimal for instrument operation, although sample acquisition and handling requires knowledge of particle physics and some understanding of fluid dynamics [21]. The instruments are capable of remote operation and detailed data processing under computer control. There are several disadvantages to the optical particle counting systems. The instruments respond to more particle properties than size alone. The particle refractive index and shape affect response and relatively small differences in instrument construction can modify performance even within the same instrument model [22]. This problem is aggravated when data are reported from aliquots of a single sample measured by instruments with different optical designs or illumination sources. Correlating different instruments can be a problem. Even with the problem areas, these instruments are widely used in many industrial areas as the major control element in defining product and/or process acceptability.

1. Light Extinction Systems

Light extinction systems are probably the most widely used single particle counting instruments for measurement of contaminant particles larger than 1 or 2 μm in diameter in clean liquids. In operation, a liquid is passed through a small windowed passageway which is illuminated with a transverse light beam. Total light flux passing through the windows is measured by a photodetector after passage of the beam through the liquid [23]. The optical system of a light extinction single particle counter is shown in Fig. 9. Although this illustration shows a gas laser light source, the original extinction counting instruments used incandescent filament illumination and most present-day instruments use light from solid-state lasers. A constant-level light beam passes

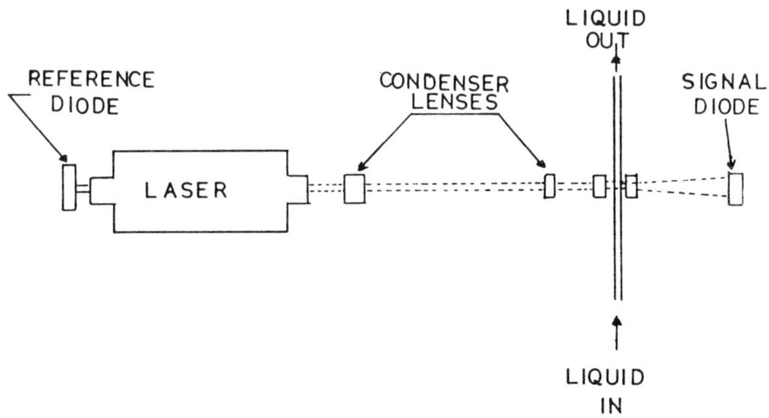

Figure 9 Extinction OPC optical system. (Courtesy Particle Measuring Systems, Inc., Boulder, CO.)

through the transparent small-dimension sample tube. If there are no particles present in the liquid stream, then the light level will remain constant. If a particle passes through the illuminated volume, then some of the light will be removed by a combination of light absorption and scattering out of the beam by the particle during its passage. This device is *not* an imaging system; only total light flux changes are measured. The amount of light removed is a function of the particle projected area as viewed by the light beam and of its differential refractive index, including both the real and imaginary components. The square law relationship of light reduction to particle area was observed early in the development of these instruments. The initial assumption was that the illumination level reduction was due only to a shadowing effect. For this reason, the instruments were described as "light blockage" devices, even though much of the observed light reduction occurs because of scattering out of the direct beam. The reduced light level pulses are counted and sized during the liquid sample transit. The amplitude of each pulse can be related to particle size by empirical calibration with reference standard particles of known size and optical nature.

These devices are used to count and size particles in sizes ranging from approximately 1 μm to several millimeters in diameter; the dynamic range for a single sensor is approximately 100 to 1. The signal amplitude varies almost directly with the square of particle size. The extinction coefficient varies noticeably for particles smaller than approximately 5 μm in diameter but settles to a value of 2 for larger particles; thus dynamic range of the electronic system used to count and size pulses with specified resolution may be the limiting factor on dynamic range. At this time, extinction sensors are in use with incandescent filament lamps, gas lasers and solid-state lasers as illumination sources. Each one emits light at a somewhat different wavelength than the others. For this reason, differences in response occur if the calibration material has optical properties different from the contaminant particles being measured when different instruments are used.

2. Light Scattering Systems

Volumetric light scattering systems are similar in many ways to the light extinction systems, which are normally always operated as volumetric measurement devices. The light extinction system observes the change in light flux within the illumination beam caused by a particle presence. The light scattering system observes the amount of light scattered out of the illumination beam by a particle. A volumetric light scattering optical system is shown in Fig. 10. The liquid sample flows through a transparent small-dimension tube that can be rectangular or

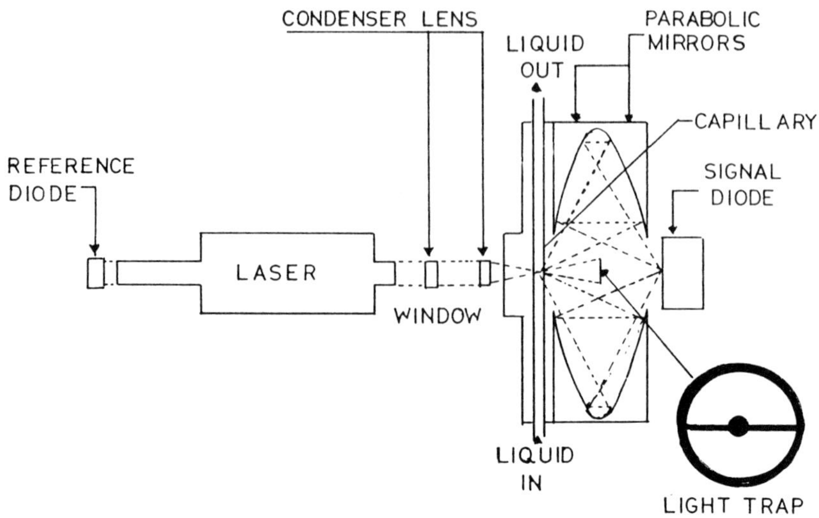

Figure 10 Volumetric light scattering OPC operation. (Courtesy Particle Measuring Systems, Inc., Boulder, CO.)

circular in cross section. Scattered light from the volume observed by the wide-angle collection system mirrors is transmitted to the photodiode. As is the case with any volumetric particle counting system, light scattered from windows and any other liquid containment surfaces in the optical system field of view is collected, adding to the optical noise level of the system.

Light scattering systems can also be operated as in situ systems, as shown in Fig. 11. This in situ system is an older design, but it can be seen that the system observes, nonintrusively, the material flowing within a line. Observation is made through windows in the line with the window surface out of the field of view of the collecting optics. The beam-splitter system indicated in this illustration is used to aid in defining the in situ volume in which particles will be counted. Operation of this system will be explained in more detail in Chapter 7.

The terms "volumetric" and "in situ" refer to the sample observation conditions used for operation of the particular system. The volumetric counter observes particles in a sample procured from an external source and then passed through the counter. Discrete samples can be obtained from a batch container or a continuously flowing sample from a sidestream sample line can be observed. The total sample flow passing through the counter is observed for particle content. Samples are passed through the sensing zone either by aspiration or pressure. The entire sample stream in the sensing zone is observed. Size sensitivity for a volumetric light scattering sensor is usually 0.1 μm with a viewed flow rate of 15 ml/min or 0.3 μm with a sample flow rate of 80 ml/min.

An in situ system is one where the sensing zone, optically defined within the stream, is used to observe a portion of a liquid stream without disturbance, as by observation through a set of windows. This type of system may observe the central 0.0001% to 5% or so of a liquid stream flowing through a process liquid line. The sensitive zone is optically defined in the in situ system. Since the sensitive zone is remote from the windows and any physical boundaries, optical interferences from such surfaces does not produce optical interference with the signal, allowing greater sensitivity to small particles. Particles as small as 0.05 μm in diameter have been counted and sized in a sample stream of some 0.25 ml/min optically sampled from the central portion of a contained larger flow of some 100 ml/min [24].

A recent development in optical particle counting has been described where interferometric

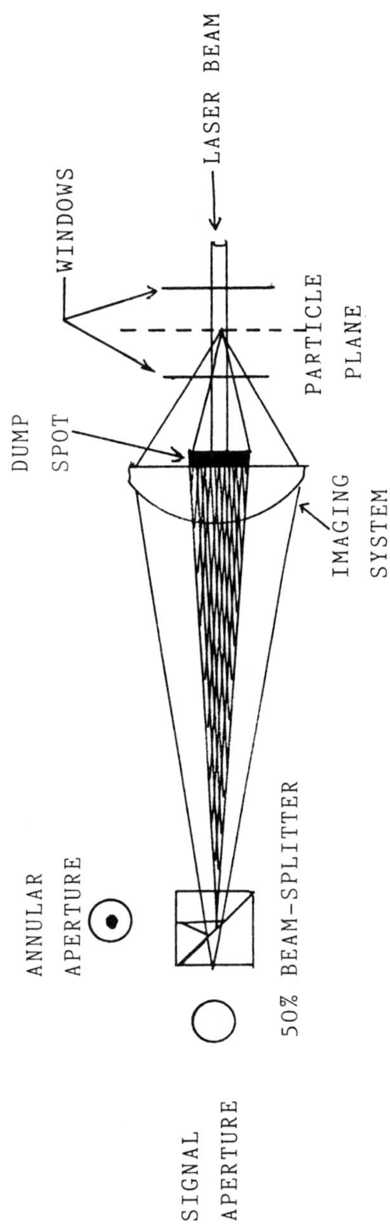

Figure 11 In situ light scattering OPC. (Courtesy Particle Measuring Systems, Inc., Boulder, CO.)

measurement is used to count and size particles in liquids [25]. This instrument does not measure the intensity of scattered light from a particle, but rather the phase shift caused by that particle. In a manner similar to operation of a Mach Zehnder interferometer, an incoming laser beam is divided into two equally intense beams by a beam splitter. One of the beams undergoes a 90° phase change. A second beam splitter combines the beams to form beams which are 180° out of phase. The intensities of the out-of-phase beams are measured. When no particles are present, these are equal. Particle presence causes a phase shift related to particle size. When the refractive index of the particle is greatly different from that of the substrate liquid, the phase shift is close to a function of the ratio of the indexes; when the two indexes are close, the phase shift is a function of the difference between the two indices. It is suggested that this effect will allow discrimination between gas bubbles and particles, since the bubbles have lower refractive index than does the liquid. Since particle indexes are normally larger than that of the liquid, the particle signals should have polarity opposite to those from bubbles. The instrument is too new to provide detailed comments as to its advantages or problems. However, the developers point out advantages in potentially distinguishing between particles and bubbles, in detecting particles with refractive index near that of the liquid. One problem appears in that the device requires very intense light, requiring the laser beam focused to a very small sensing volume, so that observed volume will also be very small. Questions have also been raised regarding the instrument's effective dynamic range when particles with significant absorption are present.

A somewhat modified light scattering approach has been used to measure extremely low concentrations of nonvolatile residues in ultrapure liquids [26]. The technique determines the concentration of residual particles after evaporating atomized droplets of the liquid. An atomizer generates a spray of nearly monodisperse liquid droplets which are evaporated. Each droplet leaves a nonvolatile residue particle whose size depends upon the quantity of nonvolatile residue originally present in the liquid. A condensation nucleus counter (CNC) measures the residue particle concentration, which is related to the original nonvolatile residue content in the liquid. The technique is sensitive to parts per billion nonvolatile residue levels; at lower levels, the nonvolatile residue content does not produce a particle large enough to be detected even by the CNC. Note that the system does not differentiate between particles and any other nonvolatile dissolved material in the liquid.

3. Development Background Information for Extinction Counters

Because of the widespread use of the optical single particle counters in industries concerned with particulate contamination, it may be of interest to review some of the historical background information on development of these devices. In the mid-1950 period, the aerospace industry became very much concerned with cleanliness of hydraulic fluid control systems in aircraft and missiles because of the difficulties in maintaining operational control with "sticking" or leaking valve assemblies. Cleanliness was monitored by procuring a liquid sample from the system, filtering it through a membrane "patch" and counting the particles collected on that patch, using manual microscopic methods. The primary standard procedure was developed under the auspices of the Society of Automotive Engineers and involved observation under oblique illumination and manual counting and sizing of the particles on the patch. The method was tedious, prone to errors and depended strongly on subjective evaluation of particle size by operators with various skill and experience levels. Round-robin testing revealed variability of several hundred percent between data from observations of the same patches by different operators.

Faced with these problems, personnel in the hydraulics laboratory at the General Dynamics missile plant in Pomona, California, investigated the possibilities of automated methods of measurement. The one which showed the most promise involved passing the liquid through a transparent tube illuminated by a transverse light beam producing an illuminated volume so small that particles in the tube would pass through the beam singly. After passing through the liquid, the light beam was collected on a photomultiplier vacuum tube. The size and intensity of the light beam were controlled sufficiently accurately so that the size of each particle could be defined in relation to the amount of light momentarily lost as it passed through the beam. Individual particles of some 5 μm in diameter could be detected in this way.

A proposal to General Dynamics for developmental funding was rejected, but the technology was released for private development. The High Accuracy Products Corporation was formed by Mr. Leon Carver, and development of the product began in August 1958. The first commercial "light blockage" counter was placed on the market in late 1959 and a few instruments were produced and sold. Even though the aerospace industry technologists were greatly interested in use of the product, their need to comply with military specifications which were written around the microscopic measurement method prevented widespread adoption of the instrument. This situation changed when the Fluid Power Research Program at Oklahoma State University adopted the instrument for measurement in many research and development applications supported by military and industrial organizations [27]. The instrument that was used there was designed to count and size particles in as many as 16 size ranges and could handle liquid samples at flow rates up to several hundred milliliters per minute. Widespread acceptance of the instrument followed in the aerospace industry and its use was expanded into monitoring the cleanliness of many other liquids and into measuring the size of many powders which could be dispersed into a liquid for analysis.

The extinction-type counter is capable of measuring a suspension of particles with varying optical properties in a single suspension since the extinction coefficient approaches an asymptote value of 2.0 for essentially all particles larger than approximately 5 μm in diameter, as shown in Fig. 12. Both the theoretical and experimental data show that the extinction coefficient oscillates about that asymptote value for transparent particles, but settles to that value very quickly for absorbing particles. For particles somewhat smaller than 5 μm, the effects of different extinction coefficient for different materials results in differences in size indication for different materials. At this time, extinction counters are used for measurement of particles 2 μm to several hundred micrometers in diameter with a dynamic range of 50 for any single sensor in concentrations up to several tens of thousand per milliliter of liquid. For these reasons, it has become a preferred method for measurement of particles of this size range in many industries. The U.S. Pharmacopeia specifies that it be used for compendial measurements of particulate contaminants in injectable products; work is in progress in the contamination technology committee of the National Fluid Power Association which will lead to its specification as the required instrument for measurement in that industry as well.

4. Development Background Information for Light Scattering Counters

Shortly after the initial commercial acceptance of the extinction counters, work on light scattering counters for measurement of particles in liquids was carried out by personnel at Royco Instrument Corporation. This work, in the 1967–1969 period, was an outgrowth of industrial application of light scattering technology to airborne particle measurement. The initial work showed that monodisperse latex particles as small as 2 μm could be measured in either water or oil. However, it became apparent quickly that the variations in scattering coefficients of different materials made it difficult to apply light scattering instruments for measurement of unknown contaminants in those size ranges.

In the 1970–1975 time period, the electronics industry began production of semiconductors by photolithography and metallic component deposition processes onto silicon surfaces. Component features were in the size range of 5–20 μm line and spacing dimensions. As a result,

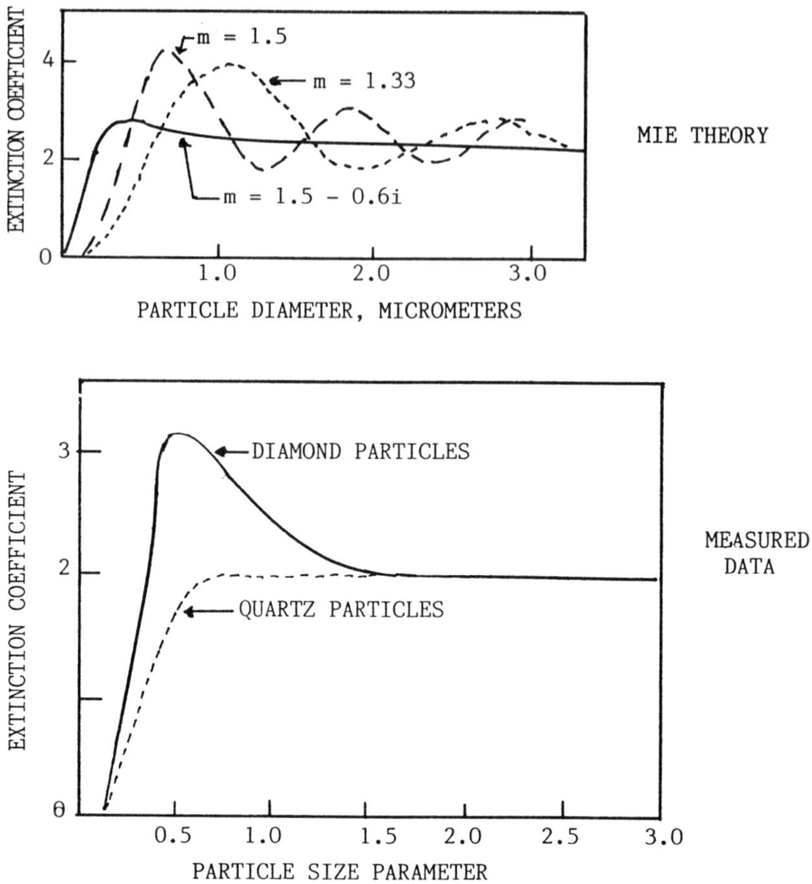

Figure 12 Theoretical and experimental extinction coefficient variation with particle size and refractive index. (From Hodkinson, J. R., 1965, The optical measurement of aerosols, *Aerosol Science* (C. N. Davies, ed.), Academic Press, London.)

submicrometer size contamination particles in the process and flushing liquids were found to be part of the problems resulting in yields of 50% or less. There was a need to measure submicrometer particulate content of such liquids. Examination of liquid samples by neither light nor electron microscopy was practical for production control operations. In 1978, personnel at Royco Instruments carried out a program to develop a liquid-borne particle counter capable of counting and sizing particles as small as 0.5 μm in diameter suspended in water or isopropanol. This development program was completed in 1978 with production of a forward scattering optical system using helium-neon laser illumination of a liquid stream contained in a stainless steel/sapphire window flow system [28]. Particle size sensitivity was achieved for 0.5-μm particles at a liquid flow rate of 100 ml/min. The optical layout of this system is shown in Fig. 13. A portion of the direct laser beam was chopped and reflected to the photodiode to maintain constant signal output as a calibration means. Further development of other light scattering optical systems using helium-neon laser illumination at larger scattering angles resulted in development of systems capable of sizing particles as small as 0.1 μm in liquid streams flowing at approximately 10 ml/min. Figure 10 illustrated an optical system with these capabilities. Larger flow rates result in decreasing particle size sensitivity for these systems.

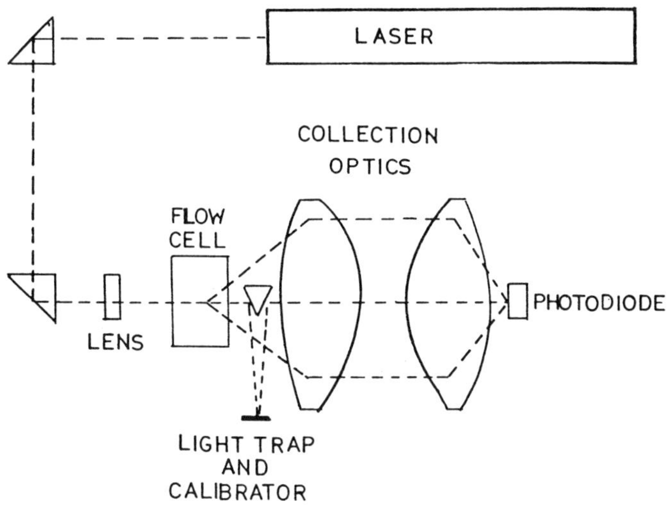

Figure 13 First light scattering OPC capable of 0.5-μm sensitivity. (From Ref. 28.)

Further work on increasing sensitivity has been carried out, using solid-state lasers to increase light flux in the sensitive volume by orders of magnitude over the level in the 0.5-μm counter just described. In 1992, size sensitivity of 0.05 μm was achieved by focusing the 780-nm radiation from a diode laser to a small volume where the liquid flow through that optically defined volume is approximately 0.3 ml/min at the center of a flow stream moving at the rate of some 300 ml/min [24]. Sample observation within the liquid flow stream and away from the window and flow containment surfaces allows reduction of background stray light and noise due to surface scattering. A solid-state laser illumination system delivers approximately 50 mW of light at the measurement point from a source with dimensions and power requirements suitable for small footprint table-top instruments. The optical system for this system is shown in Fig. 14.

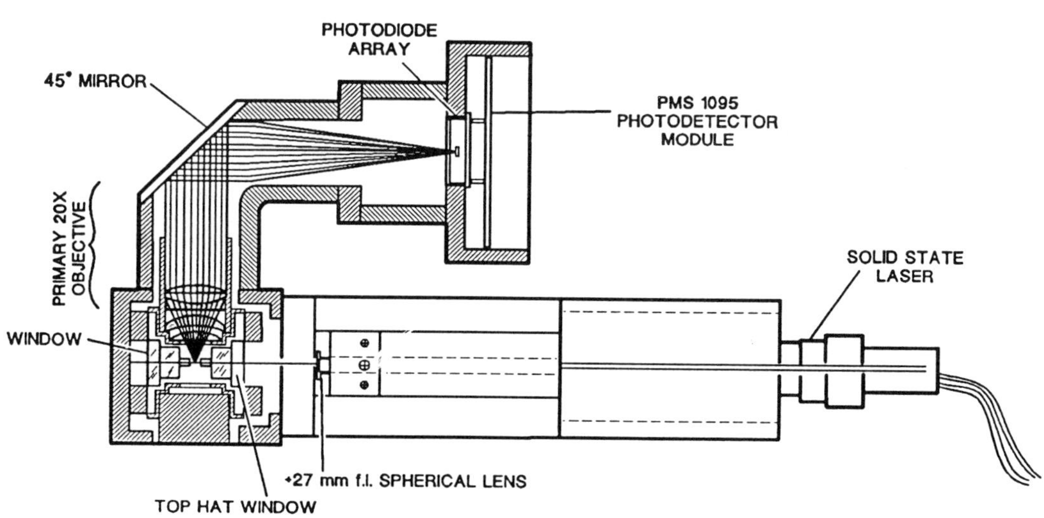

Figure 14 Right-angle optical system for 0.05-μm particle counter. (From Ref. 24.)

The right angle collection optics also allows collection of scattered light over a large solid angle to minimize oscillations in the system response curve.

IV. CORRELATION AND CALIBRATION FOR PARTICLE COUNTERS

Even with the increased sensitivity, problems of interinstrument correlation still exist. The effect of differences in illumination source wavelength and collection optics design causes differences in response from one instrument to another when the material being observed is different than the calibration material. This is particularly important when the refractive indices of the two materials are different. Correction factors can be used if the refractive index of the material being observed is known, but this situation seldom exists. In any industry, the contaminants are diverse in physical properties. Even in pharmaceutical areas, the contaminants can include glass fragments, plastic seal materials, metal fragments from flow control device operations, fibers, skin flakes, and many other materials.

At this time, calibration of optical single particle counters for use in pharmaceutical liquid observation is based on an ASTM standard method [29] (F658-87) for determining the response of the instrument to isotropic spherical particles of known size and refractive index. The method is primarily a calibration procedure to verify accurate determination of particle size and instrument sizing resolution. Unfortunately, the method is not practical for calibrating the OPC for particle concentration determination. The differences in optical nature of the spherical calibration particles and the typical contaminant particles result in differences in size and concentration correlation when OPCs of different optical designs are used. A calibration material is very similar to the particles which must be characterized in practical use is needed, but no satisfactory calibration materials have been identified. Work is in progress to develop a suitable concentration calibration method and material by the National Fluid Power Association [30]. This procedure uses spherical calibration particles in hydraulic oil to define sizing resolution of an OPC and a suspension of fine test dust in oil to define size and concentration performance of the OPC. The dust is quite similar to the contaminant material usually seen when working with hydraulic fluid power systems. However, better control has been attained in batch uniformity and better definition of the particle size distribution has been obtained. The procedure includes a requirement that the OPC being calibrated must meet specific requirements for sizing accuracy, size resolution and signal-to-noise ratio for the smallest particles measured. A sequential dilution procedure is used to define maximum concentration capability for the counter. Test data using these procedures with a variety of OPCs and liquids containing a wide range of particle concentrations indicate acceptable correlation between different hydraulic fluid power laboratories. Similar procedures will probably be required for pharmaceutical applications, although the range of material types is much greater than that for hydraulic fluid power contaminants.

REFERENCES

1. Whipple, G. C., and D. D. Jackson, A comparative study of the methods used for the measurement of turbidity of water, *Mass. Inst. Technol. Quart. 13*:274 (1900).
2. Nushart, S., Sieve analysis: making your numbers come out right, *Powder Bulk Eng. 1*(8):14–19 (1987).
3. Finch, J. A., and M. Leroux, Fine sizing by cyclosizer and micro-sieve, *CIM Bull. 75*(839):235–240 (1982).
4. Anon, ASTM, *ASTM F 569-78, Standard Test Method for Filterability Index of Reagent-Grade Water*

and High Purity Waters for Microelectronic Device Processing, Annual Book of ASTM Standards, Vol. 11.02, American Society for Testing and Materials, Philadelphia, PA, 1988, pp. 765–768.

5. Dunn, T. F., and R. Hogg, Estimation of particle size distribution using pipet-withdrawal centrifuges, *Particulate Characterization 3*(3):122–128 (1986).
6. Staudinger, G. et al., Quick optical measurement of particle distribution in a sedimentation apparatus, *Particulate Characterization 3*(4):158–162 (1986).
7. Somasundaran, P., Y. B. Huang, and C. C. Gryte, CAT scan characterization of sedimentation and flocs, *Powder Technol. 53*(1):73–77 (1987).
8. Hach, C., Measurement of color and insoluble matter in reagent chemicals, *Amer. Lab. 15*(6):42–48 (1983).
9. Daniels, C. A., and A. A. Etter, Latex particle size analysis by the Coulter nano-sizer: response to synthetic mixture of sizes, outsize particles and comparison with other methods, *Powder Technol. 34*(1):113–119 (1983).
10. Elings, V. B., and D. F. Nicoli, A recent advance in submicron particle sizing by dynamic light scattering, *Amer. Lab. 16*(6):34–39 (1984).
11. B.S.I., Methods for the determination of particle size of powders, No. 3406, part 4. *British Standards Institute,* London, England, 1963.
12. Lieberman, A., Fine particle characterization methods in liquid suspensions, *Particle Characterization in Technology* (J. K. Beddow, ed.), CRC Press, Boca Raton, FL, 1984, pp. 187–232.
13. Levy, J. D., C. J. McCarthy, and R. E. Stevens, Application of image analysis for the rapid determination of particles in parenteral solutions, *Bull. Parenteral Drug Assoc. 34*:183 (1980).
14. Balazs, M. K., Measuring and identifying particles in ultrapure water, *Microcontamination 6*(5):35–40 (1988).
15. Mittelman, M. W., Rapid enumeration of bacteria in purified waters by direct-count epifluorescence microscopy, *Proceedings 9th International Committee of Contamination Control Societies Conference,* Los Angeles, CA, 1988, pp. 531–536.
16. Kinsman, S., Instrumentation for filtration tests, *Chem. Eng. Prog. 7*:48–50 (1974).
17. Van der Plaats, G., and H. Herps. A study on the sizing process of an instrument based on the electrical sensing zone principle, *Powder Technol. 36*(1):131–136 (1983).
18. Harfield, J. G. et al., Response of the Coulter counter model ZM to spheres, *Particulate Characterization 1*(1):32–36 (1984).
19. Alberson, C., Detection of particles in liquids by the use of ultrasonics, *Proceedings of the First National Symposium on Nondestructive Testing of Aircraft and Missile Components,* San Antonio, Texas, 1960.
20. Abts, L. R. et al., Computerized discrimination of microemboli in extracorporeal circuits, *Amer. J. Surgery 135*:535 (1978).
21. Dillenbeck, K., Advances in particle-counting techniques for semiconductor process chemicals, *Microcontamination 5*(2):31–39 (1987).
22. Knollenberg, R. G., The importance of media refractive index in evaluating liquid and surface microcontamination measurements, *J. IES 30*(2):50–58 (1987).
23. Sommer, H. T. Physics and performance of light extinction particle counters, *Proceedings of 10th International Committee of Contamination Control Societies,* Zurich, Switzerland, 1990, pp. 23–27.
24. Knollenberg, R. G., and D. L. Veal, Optical particle monitors, counters and spectrometers: performance characterization, comparison and use, *J. IES, 35*(2):64–81 (1992).
25. Blackford, D. B., and D. C. Grant, Evaluating a novel liquid-borne particle counter based on interferometry, *Microcontamination 11*(2):27–32, 66 (1993).
26. Blackford, D. B., G. J. Sem, and T. A. Kerrick, Residue analysis in high purity water using a condensation nucleus counter, *Proceedings International Conference on Particle Detection, Metrology and Control,* Arlington, VA, 1990, pp. 546–560.
27. Carver, L. D., Nature of hydraulic fluid contaminants. *Proceedings of the National Conference on Fluid Power,* Chicago, IL, 1972, pp. 190–193.

28. Lieberman, A., and A. D. Bates, Optical sub-micron particle counter for liquid streams, *Proceedings of PARTEC Symposium,* Nuremberg, Germany, 1979.
29. Anon, ASTM, *Defining Size Calibration, Resolution and Counting Accuracy of Liquid-Borne Particle Counters Using Near-Monodisperse Spherical Particulate Material,* ASTM F658-87, American Society for Testing and Materials, Philadelphia, PA, 1987.
30. ANSI (NFPA)T2.9.6R1-1990, *Hydraulic Fluid Power—Calibration Method for Liquid Automatic Particle Counters Using Latex Spheres,* National Fluid Power Association, Milwaukee, WI, 1990.

2

Microscopy: The Setup and Operation of the Polarized-Light Microscopy Lab for Particle Identification

John Gustav Delly

McCrone Research Institute, Chicago, Illinois

I. INTRODUCTION

Particles are minute fragments of, usually, solid or semisolid matter. They may be of a nuisance variety (e.g., household dust, air pollution particulate matter, incinerator-stack fallout), a health hazard (e.g., asbestos fibers, hay fever pollens, silica dust), an industrial contaminant (e.g., dirt in photographic film, automobile paint, integrated circuit, magnetic tape, paper product, etc.), a pharmaceutical contaminant (e.g., glass flakes, silicone rubber, drug degradation product, thermally degraded paper fiber), or any desirable product, which just happens to be at or just below the threshold of unaided human vision. In every case, the outstanding characteristic of all types of particles is their small size, and the use of magnifiers or microscopes is the obvious approach to studying and identifying tiny particles. This approach is hundreds of years old—as old as microscopes themselves. Particles have been characterized by light microscopy long before modern instrumental methods; before X-ray diffraction. Serious particle characterization dates to the confluence of crystallography and the introduction of Nicol prisms to the microscope in the early part of the last century.

The modern particle characterization laboratory still has, as its principal investigative instrument, the polarized light microscope, along with its companion low-power stereomicroscope. The addition of X-ray diffraction, scanning and transmission electron microscopes, especially in their more recent X-ray analytical configurations, electron and ion microprobes, micro-Raman, micro-Fourier transform infrared spectrometer, and the many new kinds of microscopes currently under development have made the microscopist's job of identifying particles easier, faster, and more comprehensive. The fact remains that in a laboratory where all of these analytical instruments are available the first instrument used on any particle problem is the low-power stereomicroscope, followed by the polarized-light microscope. More than 95% of all particles can be characterized by the experienced microscopist with these two instruments alone, and until about 30 years ago that is about all there was available that could be used. In the past, if a tiny opaque particle was recovered, a portion of it was put into solution either through direct dissolution, or fusion followed by dissolution, and then individual cations, anions, and functional groups were tested for using the methods of microchemistry. Today, with the aid of an ion microprobe not only can the elements present be determined, but because of the instrument's incorporated mass spectrometer, even the isotopes of the elements present can be ascertained. Of course instruments like this cost one-half million to one million dollars. Most particle-characterization laboratories cannot afford expensive analytical instruments of this kind, so that they start with low-power stereomicroscopes and polarized-light microscopes, and then add perhaps a scanning electron microscope with X-ray capability and a Fourier transform infrared spectrometer. With these, the vast majority of particles can be characterized, and the remaining can be characterized by laboratories that specialize in this area.

It is proposed here to describe the kinds of low-power stereomicroscope and polarized-light microscope that are needed to characterize and identify particles, together with their setup, alignment, illumination, and use. Ancillary equipment, accessories, chemical reagents, and techniques also need to be discussed. To understand the equipment needed, it is first necessary to understand the characteristics of particles that are to be observed.

II. IDENTIFYING CHARACTERISTICS OF PARTICLES

There are quite a large number of identifying characteristics of particles that can be seen or ascertained with the aid of a polarized-light microscope, the most important of which include

1. *Morphology:* The shape of a particle; is it equant (about the same dimension in all directions), or acicular (needlelike), or does it resemble a rod, plate, flake, cylinder or sphere? For synthetic fibers, morphology refers to the cross-sectional shape.
2. *Size:* What are the particle's dimensions? How thick is it?
3. *Surface texture:* Is it smooth? porous? rough? scaly? Are there surface spines, ridges, furrows, sculpturing of any kind?
4. *Hardness:* Does it "squoosh" when depressed and stay that way, or does it recover its shape (elastomericity)? Is it brittle? Does it break into a few or "zillions" of pieces?
5. *Reflectivity:* Is it very dull, semidull, or highly reflective?
6. *Transparency:* Is it transparent, translucent, opaque, or some combination?
7. *Color:* Is it colorless or colored by transparent light? What color(s)? What color is it by reflected light?
8. *Magnetism:* Is the particle magnetic?
9. *Refractive index/indices:* Perhaps the single most important identifying characteristic of transparent particles. The refractive index is the ratio of the velocity of light through a particle relative to the velocity of light in a vacuum. The internal, atomic structure of the particle may result in one, two, or three principal refractive indices.
10. *Pleochroism:* A change in color or hue in a colored, transparent particle that has more than one refractive index.
11. *Dispersion staining color(s):* An optical method of imparting color to colorless, transparent particles; related to the refractive index relationship between particle and mounting medium, so that more than one color may be observed.
12. *Birefringence:* The numerical difference between the principal refractive indices. This may be determined through a measurement of the refractive indices, or through observation or interpretation of the dispersion staining colors, but is more quickly and efficiently done through observation of the retardation (interference color) in conjunction with the particle's thickness.
13. *Extinction angle:* The angle between the particle's optical and crystallographic axes.
14. *Sign of elongation:* In particles which are elongated, the location of the high and low refractive indices relative to the elongated direction.
15. *Interference figure:* Optical phenomenon observed at the objective back focal plane. Its formation is the result of the particle's orientation and effect on polarized light passing through it.

16. *Melting point:* Does the particle melt when heated? at what temperature? Does the particle sublime, decompose, or explode? Is there evidence of a polymorph, or eutectic, or addition compound, or solid solution?
17. *Chemical composition:* What are the elements, ions, functional groups that make up the particle?
18. *Fluorescence:* color and intensity of fluorescence under ultraviolet (365 nm) and blue-violet (400 nm) excitation.

There are many other characteristics of particles which may be determined, but those enumerated are usually all that are needed, and, in most cases, fewer characteristics are required to identify a substance. Now, it turns out that the first eight characteristics listed can be determined with any kind of microscope. Refractive indices, pleochroism and dispersion staining all require that the sample be illuminated with light vibrating in one direction only—the "plane polarized-light" condition. Birefringence, sign of elongation, extinction angle, and interference figures all require that the sample be viewed between two polarizers whose vibration directions are crossed relative to one another—the "crossed polars" or, formerly, "crossed Nicols" condition. Additionally, observations during heating and cooling of the particle, and certain microchemical tests require crossed polars.

The only kind of microscope that will enable us to make all of the observations necessary for identification is a true polarizing microscope—not a phase contrast microscope, nor a reflected-light (metallurgical) microscope, and certainly not a biomedical, clinical-type microscope fitted with polarizing filters, as is commonly provided. The microscope must be a true polarizing microscope, as will be described in detail shortly.

III. THE LOW-POWER STEREOMICROSCOPE

Two microscopes, ideally, are on the microscopist's bench: a low-power stereomicroscope and a polarizing microscope. The low-power stereomicroscope is almost always the first to be used because it aids in sample preparation. The particle can be seen in relation to its matrix, if any, and can be manipulated and removed from the tablet, filter paper, membrane filter, or other substrate, and mounted on a microscope slide, probe plate, stub, glass fiber, spindle-stage needle, etc. Occasionally a particle problem can be solved at this stage of the examination.

The low-power stereomicroscope is so called because its total magnification is low (5×–80×) relative to the higher-power polarizing microscope, whose lowest magnification can be 10× but is more often 40×–50×. The "stereo" in the designation comes from the fact that there are, in this microscope, what amounts to two microscopes—one for each eye, each with a slightly different viewpoint (included angle between the two systems equals about 15°) resulting in the familiar three-dimensional image. The depth of field is great; the resolving power, however, limits the useful magnification to about 50×.

Figure 1 illustrates a typical, good-quality stereomicroscope suitable for particle identification work. There are three kinds of lighting that should be available for any stereomicroscope intended for particle identification, namely transmitted, top light (epi-darkfield), and coaxial. The microscope in Fig. 1, together with a fiber-optic illuminator for toplight, fulfills these requirements.

Transmitted light is used for looking through pharmaceutical vials, ampules, and bottles, Petri dishes, and all transparent matrices. The stage insert shown in Fig. 1 is replaced with a different kind for transmitted light observations; depending on the kind of lighting desired, the insert chosen should be completely transparent, frosted, or opalized.

Top light, also known as reflected light, epi-illumination, or epi-darkfield, is used for lighting opaque specimens, or for lighting any kind of specimen from above and to the side. Top

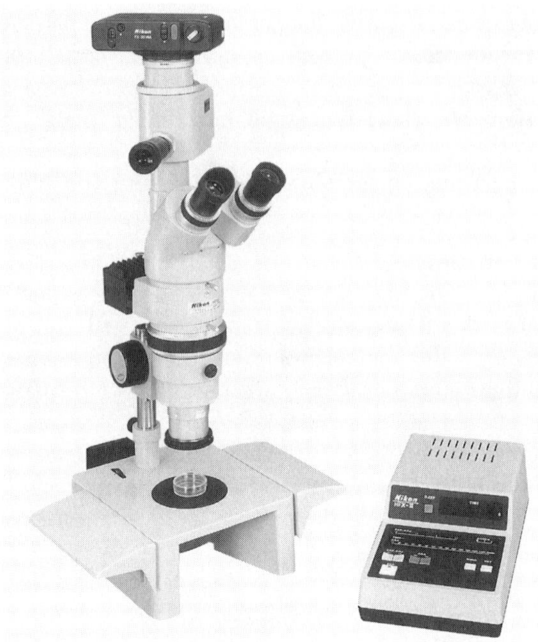

Figure 1 Low-power stereomicroscope suitable for particle identification. This microscope has both transmitted light and coaxial illumination as illustrated. The addition of a fiber-optics illuminator would provide epi-darkfield illumination as well. (Courtesy Nikon, Inc., Melville, New York.)

light is often achieved through a built-in, fixed lamp directed downward, rotatable, adjustable illuminators affixed to a ring clamp above the objective, use of a ring light around the objective, or through use of two or more fiber optic bundles attached to a separate illuminator.

Coaxial illumination is rarely used, but is actually one of the most useful modes of illumination. In this kind of illumination the light comes down directly through the objectives themselves providing true incident illumination; the light is coaxial with the microscope's optical axes. The microscope in Fig. 1 is equipped for coaxial illumination. This form of illumination is absolutely indispensable in detecting extremely fine glass flakes which would otherwise be invisible in transmitted or nonincident top light. It is also the only illumination method that will provide the microscopist with the flexible-collodion film thickness needed for affixing a particle to, say, a beryllium block for X-ray analysis. The thickness of the film being made is indicated by the interference color seen by coaxial lighting.

Often two or more methods of illumination are combined. Rarely are stereomicroscopes sold configured for all three methods, but this will be found the most effective. Personally, I have my lights controlled by momentary-contact foot switches, so that I can change and vary my lighting methods without having to remove my hands from the sample or look away. Additionally, my own preference is to have the stereomicroscope on a boom stand, preferably one affixed to the left front corner of the bench rather than on a heavy base. In this way, the microscope can be swung out from over the bench to examine large objects, such as truck tires, oil paintings, automobile hoods, or anything else that would not ordinarily fit in the space restricted by the pillar. For normal use, the microscope is swung back over the bench and the nonattached, transmitted-light base is positioned under the magnifier body.

Notice in Fig. 1 that a camera is attached. I favor the use of a photographic camera, video camera, or electronic image recording device for documenting what is seen in the stereomicroscope. Polaroid is a commonly used medium for this, and for many applications the electronic image capturing systems may be adequate. So the stereomicroscope should be equipped for some sort of image capture equipment.

There are two forms of objective design in the stereomicroscope. One is the Greenough type, in which there are two separate objectives; the other is the common main objective type which uses a single lens system for the front element. All other things being equal, the Greenough type provides superior image quality, because the rays are all axial and paraxial, whereas in the common main objective, the rays for each microscope are centered at a point which is off-axis.

There are also several modes of magnification change, including stepped (through rotating or sliding different lens systems into the light path) and zoom (mechanical cam action to move lenses internally). The zoom systems are the most versatile, but the stepped systems yield higher image quality. Because of the trade-off between image quality and convenience, most microscope manufacturers offer both Greenough and common main objective, and zoom and stepped magnification.

Accessories that should be ordered with the stereomicroscope include amplifiers (2× supplementary lenses which are attached to the front of the objectives, resulting in a much-reduced free-working distance, but a doubling of the magnification), and measuring eyepieces (eyepieces with a focusing eye lens, capable of accepting different graticules for measuring and/or counting particles). Some manufacturers also offer polarizing attachments, darkfield stages and mechanical stages, all of which are useful for special applications.

General use and conventional application of the stereomicroscope is available in audiovisual format [1]. Unconventional techniques include the lighting of opaque specimens by reflecting transmitted light off of the front lens of a 2× amplifier. This technique and others applicable to particle identification have been extensively described elsewhere [2].

IV. THE POLARIZING MICROSCOPE

A polarizing microscope suitable for particle identification work is illustrated in Fig. 2. A polarizing microscope differs from a biomedical, clinical type microscope in many ways. Both may have separate light sources or built-in illumination, as illustrated in Fig. 2, but that's where the similarity ends. Light sources have in common three main parts: a filament (bulb), a lamp condenser, and a field diaphragm. The field diaphragm is near the light exit port. In the better illuminators, either the lamp condenser or the filament is adjustable so that the filament may be centered on the lamp condenser's optical axis, and focused in the plane of the aperture diaphragm. Less critical systems have a diffuser of some kind built in. There is also provision for inserting filters of various kinds in the light path.

Immediately above the light exit port at the front of the base is a rotatable polarizer graduated every 90°. The illustrated model has a ball-bearing click stop at the "zero" setting. At this setting, according to DIN standard, the vibration direction of the polarizer in east-west (left-right). A small nameplate attached to the base behind the light exit port has an arrow running left-right labeled "P" as a reminder; this is done because prior to the DIN standard some microscopes had their polarizer vibration direction north-south (up-down). More elaborate polarizers have finer graduations and a compensator slot just above the polarizer.

Above the polarizer is a centerable strain-free substage condenser. These condensers are typically of the two- or three-lens Abbe-type, with a flip-in/flip-out top lens to change the numerical aperture (angle of the illuminating cone). The numerical aperture is often 0.90 with the top lens in. Incorporated in the condenser is an adjustable iris diaphragm, sometimes called the substage diaphragm, but better called the aperture diaphragm. This diaphragm regulates the

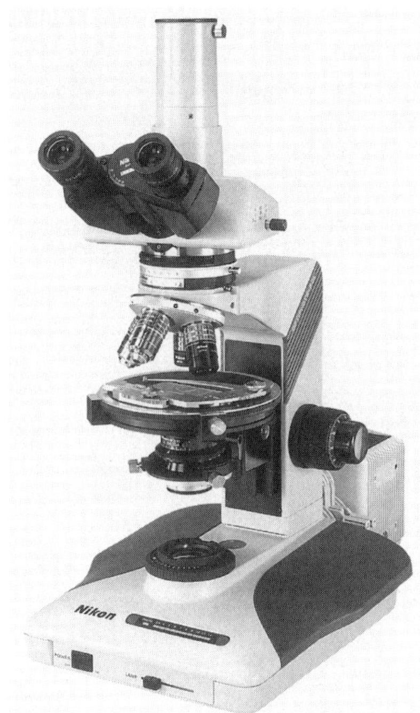

Figure 2 High-power polarized-light microscope needed for particle identification. (Courtesy Nikon, Inc., Melville, New York.)

numerical aperture downward from its maximum of 0.9 or 0.25. For oil-immersion work, the condenser needs to have a numerical aperture value greater than 1.0 (usually 1.2, 1.25, 1.3 or 1.4), and have its top lens in oil contact with the bottom of the slide. Biological microscope condensers, with the exception of those for Nomarski differential interference contrast (DIC), are not strain-free.

Above the substage condenser is the stage. The stage of a polarizing microscope is rotatable and graduated. The rotation may be effected via a grease stage, in which metal rotates in metal on a grease film, or via ball bearings. The grease stage is a cost-saving measure, and should be avoided, as it will take both hands to turn as the grease polymerizes, and it does not run so true. Ball-bearing stages may have one to three rows of ball bearings, and require only light finger pressure to rotate. These stages cost more money, but are worth the difference over time in efficiency.

Rotating stages are graduated into 360°, with one or two verniers supplied to further divide each degree into tenths of a degree. The graduations are used in conjunction with a crosshair graticule in the eyepiece for measuring internal and external crystal angles, and for the measurement of extinction angles.

There are a number of holes in the stage, many of which are threaded. These holes are for the attachment of stageclips, mechanical stages, crystal rotation apparatus (spindle stages) and universal multigimbal stages. There is also a stage-rotation lock to prevent stage rotation when a mechanical stage is being used for a systematic count. This lock, in the form of a knurled thumbscrew, is seen in Fig. 2 running through one of the two verniers in the plane of the stage. The attachable mechanical stage is also seen in the figure. Normally nothing is on the stage; the microscope slide with particle is moved manually; it is much faster, and one rather quickly gets used to moving the specimen in any direction by hand.

The stage as described is the minimum required. More elaborate, expensive stages may have further refinements such as an engagable clutch to provide slow-motion rotation of the stage—useful when working with tiny particles at very high magnification, and when determining very small extinction angles. Another refinement is the 45° engagement lever. When engaged, the lever will click stop every 45°. Birefringence measurements are made at a particle's maximum brightness position, which is 45° from extinction. For particles with oblique extinction, this lever obviates the need to read the stage at extinction, add or subtract 45°, and reset the stage to the new reading; one merely turns the particle to extinction, engages the lever, and turns the stage until it click stops. It is a valuable time-saving feature over the life of the microscope. Unfortunately, this feature quickly becomes defective—clicking continuously—when those not familiar with it disengage the lever at some point in the rotation other than at one of the clickstops.

Above the stage is the nosepiece with the objectives. The nosepiece may be made to hold anywhere from one to six or seven objectives. The objectives must all be on-axis with the stage rotation, and this is achieved differently with different manufacturers. In the Nikon microscope illustrated in Fig. 2, notice the openings in the knurled rim of the nosepiece; these are centering screws. The objectives are screwed into floating threaded rings which are mounted against a spring. Each objective ring is provided with two screws which cause the ring to bear on the spring. Once set to be on the axis with the stage rotation, the nosepiece should not be turned by grabbing the objectives, but rather by turning the knurled rim of the nosepiece.

In the Olympus polarizing microscope, all of the nosepiece objective positions except one are centerable, but the stage itself is centerable. It is best to put the highest magnification objective in the noncenterable nosepiece position, and center the stage to this objective; then center all of the other objectives in the nosepiece to the stage.

In the Zeiss polarizing microscope, such as the Standard, GFL, or Universal type, neither the nosepiece nor the stage are centerable. In this case the microscope is supplied with objectives that have centering rings built in. This makes it less convenient to use non-pol objectives.

Single-objective carriers are normally made centerable.

Much can be written about objectives themselves. The most important features are the magnification, numerical aperture, correction, mechanical tube length, cover-glass thickness required and freedom from strain. Objectives for polarized-light work need to be strainfree, just like the substage condenser. If the objectives are not strain free, the background will not be black when using crossed polars, and errors will be made in the detection of weak birefringence, and all quantitative conoscopy will be suspect. Ordinary biomedical-type objectives are not suitable.

Objectives range in magnification from 1× to 190×, and more. Typically though, the lowest magnification objective is a 4×, 5×, 6.3× (there is a variable magnification 2.5–5× objective which is very valuable for photomicrography). There are two series of magnifications, one based on tradition (4×, 10×, 40×, 100×), and another based on steps in the square root of two series, so that there will be approximately equal steps between magnifications. To complicate matters, some manufacturers make objectives in both series. 4×, 10×, 20×, 40× and 100× is a commonly supplied series and is perfectly adequate. The magnification engraved on objectives are nominal; for highest accuracy, the true magnification should be determined through use of the projected image of a stage micrometer.

The numerical aperture of an objective is an expression of its light gathering ability in angular terms. The radius of curvature of the objective front lens determines the focal length, magnification, and working distance. From this, the aperture or diameter of the front lens may be determined and expressed as an angle, known as the angular aperture. The numerical aperture is the sine of one-half of the angular aperture, times the refractive index of the medium between the cover glass and the objective (normally air). Air has a refractive index, n, of 1.00; however, the objective may be designed for use with water between the specimen and objective, in which case $n = 1.33$; or with glycerin, $n = 1.47$; or with cedarwood oil or one of its synthetic substitutes, $n = 1.515$ or $n = 1.52$.

Objectives of any given magnification are available with different numerical apertures. For example, 40× objectives are available with numerical apertures (NA) of 0.57, 0.60, 0.65, 0.70, 0.80, 0.85, 0.90 and 1.00 oil. All yield the same magnification, but as the numerical aperture goes up, so does the resolving power, because the resolving power (r), or the smallest space between any two structures that just enables one to discern that there are two structures, is determined by the wavelength (λ) of light being used, and the numerical aperture (NA)

$$r = \frac{\lambda}{2\text{NA}}$$

Thus, the higher the numerical aperture, the smaller the space discerned between structures.

The correction of the objective refers to the degree of chromatic and spherical aberration present. The vast majority of objectives in the world are "achromatic," which is the least corrected of the objectives, but they are serviceable and perfectly adequate for routine work. Higher correction of chromatic aberration is achieved in the "fluorite" objectives. They are called so because of the use of the mineral fluorite (calcium fluoride) or a modern synthetic along with glass in its makeup. Objectives will be marked "fluorite," "FL," or have the designation somewhere in its name (Fluotar, Neofluar, etc.). The most highly corrected objectives are termed "apochromats," and the name "APO" will be engraved somewhere on the objective; these yield exquisite images, but, unfortunately, are not strain free. For polarized-light work there are strainfree fluorite objectives available.

The mechanical tube length is the physical distance between the nosepiece threads where the objective shoulder seats, and the top of the eye tube, where the eyepiece fits. This distance is most commonly 160 mm, although Leitz used to be 170 mm; and metallurgical objectives require tube lengths of 185 mm, 215 mm, etc., because of the need for illuminators above the objective. The importance of observing the correct mechanical tube length is that if this length is too short, or too long, spherical aberration will be introduced and the image will be degraded.

Finally, the cover-glass thickness is engraved on the objective. This is usually 0.17 mm for transmitted light objectives (in the past, some manufacturers required 0.18 mm) and—i.e., no cover glass (NCG) for metallurgical objectives intended for viewing polished metal specimens without a coverglass. On objectives of German manufacture, the designation may be "o.D." (ohne Decke)—without cover.

Additionally, objectives may be engraved with their intended immersion medium, if not dry, e.g., water (W, or Wasser), glycerin (Glyc.), or oil (Öel), or, in some special objectives, more than one medium.

There may also be phase plates, irises, prisms, annular or central stops, or correction collars incorporated into the objective for specialized applications. A survey of all objective types available together with their features is available. [3].

For general particle identification work, the most used objectives will be strain-free achromats or fluorites of 10×, 20× and 40×, with less use of the 4× and 100×. The 40× objectives supplied with polarizing microscopes today usually have a relatively low numerical aperture, 0.65. If this is the case, then it would be a good idea to add, say a 50×/0.9, or a 60×/0.85. Higher numerical apertures are needed in polarized-light microscopy not so much for resolution increase, as for measuring large optic axial angles in interference figures.

Immediately above the nosepiece with its objectives is the analyzer unit. This module is so called because it contains a second polarizer that vectorially resolves, or analyzes, the light rays after they have passed through crystalline particles. The "analyzer" can be removed or placed in the light path by pushing or pulling a handle connected to it. This analyzer is also rotatable, and a scale with vernier is provided for reading the exact degree of rotation. This feature is used for "slightly uncrossed polars" (7°–12° from fully crossed) or "parallel polars," or in conjunction with monochromatic light and a Senarmont compensator for determining the exact retardation observed within one order. In Fig. 2 the handle of the "in" analyzer can be seen sticking out a bit on the right of the analyzer module.

Just below the analyzer (and to the right of the vernier scale) there is a slot in the analyzer module. This is known as a compensator slot. Compensators are crystalline plates (quartz, mica, gypsum, calcite, etc.) of fixed or variable (through tilting) thickness, in which the retardation and location of high and low refractive indices are known. They serve as reference crystals when determining the retardation and location of high and low indices in unknown crystals.

The analyzer module also usually includes a supplementary lens at its upper end which can be introduced into the light path by rotating a ring, or flipping a lever or sliding a carrier. This is known as the Bertrand lens, or Amici-Bertrand lens. When in position, the lens is focused on the objective back focal plane and forms a real image in the intermediate image plane, where it is magnified by the eyepiece. This lens is incorporated into polarizing microscopes so as to study the interference figures that form in the objective back focal plane when an anisotropic crystal (one having more than one refractive index) is viewed between crossed polarizers. Ideally the Bertrand lens is focusable, centerable, and incorporates a tube diaphragm, but only a few microscopes incorporate all of these features. The Nikon Optiphot-Pol illustrated in Fig. 2 is unique in having its Bertrand lens incorporated in the head; it can be seen as a black semicircular knurled disc at the front base of the trinocular head. The disadvantage of this location is that interference figures cannot be photodocumented in the trinocular phototube. The advantage is that the pinhole type Bertrand here is favored by petroleum geologists for its depth of field. In the latest version of this microscope, the Nikon Optiphot II Pol, this location is retained, but, in addition, a second Bertrand lens of the traditional type is now located in the analyzer module; the Bertrand lens is also centerable in the new version.

Above the analyzer module is the monocular, binocular, or trinocular head. Zeiss tends toward the conservative monocular tube for its full-featured treatment; most other manufacturers opt for the binocular or trinocular head. The trinocular is more versatile, because cameras can then be attached for documenting observations. A lever seen sticking out to the right of the head in Fig. 2 directs all of the light to the binocular observing tubes, or all of the light to the photo unit, or divides the light for both observation and photorecording simultaneously. The head itself incorporates a depolarizer at its base to compensate for the tilting head.

One of the two eyepieces contains a crosshair graticule, or a measuring scale, or a combination of the two. The crosshair eyepiece is fixed in a slotted eye tube via a screwhead in the eyepiece, such that the crosshairs are nonrotatable and fixed east-west/north-south. The crosshairs represent the vibration directions of the polarizer and analyzer, and are used as references for angular measurements, and observations in plane polarized light. The eye lenses are adjustable so that users with different vision problems may focus on the cross or scale.

Interpupillary distance is effected through pulling the eye tubes apart or by pushing them together, or, as in the case illustrated in Fig. 2, by folding the light path.

Differences between the eyes are compensated for by a diopter adjustment ring at the base of one of the eye tubes, or, as in the case of the microscope in Fig. 2, by providing both eyepieces with adjustable eyelenses.

Eyepieces range in magnification from 1.8×–3× for projection to 5×–25× for visual use. The 10×, 12.5×, and 16× are the most useful with the 10× or 12.5× used most often.

V. MECHANICAL ALIGNMENT OF POLARIZING MICROSCOPE

Now that the various parts have been located, and before setting up correct illumination, it is necessary to mechanically align the parts. Specifically, this refers to the centering of the objectives relative to the stage rotation. The exact procedure will depend on the microscope design: (1) all objectives centerable in the nosepiece; (2) all but one objective centerable in the nosepiece; centerable stage; (3) neither nosepiece nor stage centerable.

1. When all of the objective positions in the nosepiece are centerable, start by focusing on any specimen that consists of many fine particles; such as power-plant fly ash. When a suitable field is found with a low-power objective, rotate the nosepiece to the highest magnification objective and focus with the fine adjustment. Move the slide by hand until a small particle, preferably a sphere, is directly under the crosshair. Now rotate the stage. Ideally the tiny sphere should remain under the cross, and rotate about its own axis. If the particle describes a circle rather than rotating about its own axis, the objective optical axis is not coincident with the rotational axis of the stage. As the stage is noncenterable, the objective must be centered. Place the two centering wrenches provided into the nosepiece centering screw openings for the objective being centered. Adjust these centering screws, moving the slide by hand after each adjustment until the same or another particle lies under the cross. Again rotate the stage to check if the particle is rotating about its own axis. The circle the particle makes on rotating the stage will become smaller and smaller with each adjustment, until the particle is perfectly centered. Rotate the objective to the next lower magnification, and proceed in the same manner to center this objective. Repeat for all other objectives.

2. All but one objective centerable in the nosepiece; centerable stage. In this case, place the highest magnification objective in the single, noncenterable position in the nosepiece, focus on a fine dust particle, and move the slide until the particle lies directly under the crosshairs. Rotate the stage and observe the particle. It should remain under the cross, rotating about its own axis. If it does not, it will describe a circle and the objective optical axis is not coincident with the stage rotation. Place the centering wrenches provided into the centering screw openings *located in the stage*. Center the stage with respect to the fixed objective. Continue to recenter a particle by moving the slide by hand, and making adjustments in the stage, until the rotating particle stays under the cross and rotates about its own axis. When the objective and stage are on the same axis of rotation, change to the next lower objective and move the slide by hand until another particle lies under the cross. Rotate the stage. If the particle is not centered, *do not center the stage*. Move the centering wrenches to the *nosepiece* centering screws for the objective being used, and center the objective. Proceed the same way for all of the remaining objectives, centering in the *nosepiece*.

3. Neither nosepiece nor stage centerable. Focus on a fine dust particle with any objective, and move the slide by hand until the particle lies under the crosshairs. Rotate the stage. If the particle does not remain under the cross rotating about its own axis, the stage axis of rotation and objective optical axis are not coincident. If the particle describes a circle, place both hands on the two knurled rings surrounding the objective, and adjust them to achieve centration. Check by moving the slide by hand until a particle lies under the crosshairs, and rotate the stage. Continue to adjust the two rings until the particle remains on center during rotation. Proceed in the same way with all remaining objectives.

VI. SETUP OF KÖHLER ILLUMINATION

Once the microscope objectives and stage are mechanically aligned, correct illumination must be achieved. There are many kinds of microscope illumination, but the method described by August Köhler over 100 years ago is still the best for all forms of microscopy. The Köhler illumination method provides the maximum amount of even illumination from inhomogeneous light sources, i.e., coiled filaments. Reference should be made to Fig. 3 for the location of all of the microscope parts described earlier. In Köhler's method of illumination there are only two steps to perform after first focusing on a specimen: (1) focus the field diaphragm in the plane of the specimen, using the substage condenser (2) focus the filament in the plane of the aperture diaphragm, using the lamp condenser. The first step, and its consequences are illustrated in the ray path to the left in Fig. 3; the second step, and its consequences are illustrated in the ray path to the right in Fig. 3. In principle, Köhler illumination is simple; in practice some difficulty may be encountered.

Let us start with the first step. Focus on a specimen using the highest magnification objective, say a 40×/0.65 or 50/0.90. Reach down to the base of the microscope, and close the field diaphragm; this will be a knurled ring around the light exit port, or just behind it protruding out of the base. Looking at the specimen, you should now see a sharply focused and centered image of the diaphragm leaves of the field diaphragm. Chances are that you will not see this. Therefore, find the substage condenser focus knob that makes the substage condenser move up and down, and turn it until the leaves of the closed field diaphragm are sharp. Next, open the field diaphragm until the leaves are near the edge of the field of view, and center the image, using the two centering screws located in the substage condenser mount. When the field diaphragm is sharply focused and centered, open the field diaphragm until the leaves are just outside the field of view, no more. The function of this diaphragm is to illuminate only that part of the specimen that is in the field of view for the objective in use; no more. If it is opened more than the objective field of view, light will be reflected from specimen structures outside of the field of view into the objective, resulting in glare, and reduced contrast in the image. You have now completed the first step in setting up Köhler illumination.

The next step may lead to some difficulty. The filament should be focused in the plane of the substage aperture diaphragm, using the lamp condenser. But here are the problems: (1) The polarizer will prevent you from observing the aperture diaphragm directly, so that you must go to the second conjugate focus of the aperture diaphragm, which is the objective back focal plane. This plane can be seen by removing the eyepiece and looking down the tube, or by using a pinhole eyepiece, or by using a phase-contrast telescope, or, best, by inserting the Bertrand lens. (2) When the Bertrand lens is in place you may not see the filament because of a built-in diffuser or ground glass. Manufacturers do this as a cost-cutting measure and for convenience when dealing with the majority of microscope users who merely want to be able to flip on the light switch and focus on the specimen. Check to see if the diffuser is removable; if it is, remove it and leave it out, it is used only when observing interference figures. If the diffuser is removable you may now see the filament, but if it is not centered or not in focus you may not be able to do anything about it because (3) many manufacturers do not provide centering adjustment to the filament, or focusable lamp condenser. Let us assume, first of all, that we have a well-designed system, in which case, you start by opening the aperture diaphragm fully. Next, focus the filament by adjusting the focusable lamp condenser. If the lamp condenser is not focusable move the filament closer to or away from the lamp condenser, until the filament is in focus. Next, center the filament using the filament centering screws. The objective back focal plane should be fully and evenly illuminated, i.e., the filament image should completely fill the objective back focal plane

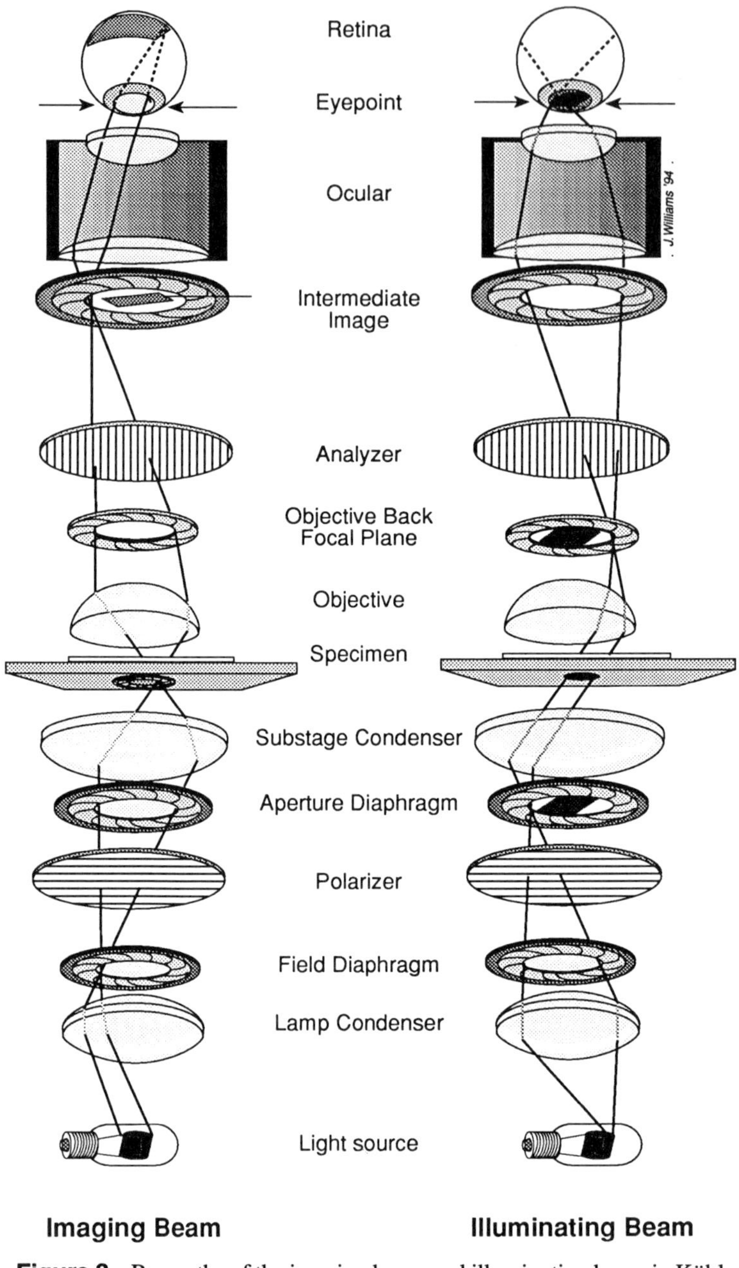

Figure 3 Ray paths of the imaging beam and illuminating beam in Köhler illumination. In actual practice, after setting up correct illumination, these two beams occur simultaneously. This method of illumination provides the maximum amount of even illumination from inhomogeneous light sources.

with as many filament coils as possible. Finally, close the aperture diaphragm, so that its image just comes into the objective back focal field of view; it will be closed about 15% or so.

If there is a built-in diffuser, notice how evenly the diffuser is illuminated. If there is a hot spot or noncentered condition, see if the bulb can somehow be moved to evenly illuminate the diffuser. Then close the aperture diaphragm about 15%.

There are many other kinds of illumination systems, including separate light sources used with horseshoe-base microscopes, and those which incorporate a concave mirror to reflect the image of the filament back to effectively double its size. There is not space here to cover all of the possibilities. There is step-by-step, fully-illustrated procedure for most of the possibilities published elsewhere [4].

Let us look at Fig. 3 for a moment to be sure we understand what is going on. There are two image-forming paths occurring simultaneously, one is the image-forming ray path (left) which traces the image of the field diaphragm and specimen throughout the system, and the other is the illuminating ray path (right) which traces the images of the filament and aperture diaphragm throughout the system. Starting with the image-forming ray path (left), note that the substage condenser focuses a real image of the field diaphragm in the plane of the specimen. This is the first step you did when you adjusted the focus of the substage condenser until you saw the field diaphragm leaves come in sharp focus while you were looking at the specimen. But the objective is focused on both the specimen and the field diaphragm, and so it forms a real, magnified image of both the specimen and the field diaphragm in the intermediate image plane (this plane, by DIN standard, lies 1 cm from the top of the eye tube where the eyepiece goes in). Next, the eyepiece further magnifies the image of the specimen and field diaphragm, and this image is projected on the retina by the eye lens. There are, then, three conjugate foci of the field diaphragm, namely, the plane of the specimen, the intermediate image, and the retina (or film, or screen). And it is the substage condenser that must be adjusted to get the field diaphragm in focus.

Next, consider the illuminating ray path (right), starting with the filament. The filament is focused in the plane of the aperture diaphragm, using the adjustable lamp condenser (or by moving the filament with respect to the lamp condenser). The substage condenser and objective then act together to relay the image of the filament and the aperture diaphragm into the objective back focal plane. Next, the eyepiece relays the image of the filament and aperture diaphragm into the eyepoint (exit pupil, Ramsden disk). There are, then, three conjugate foci of the filament, namely, the aperture diaphragm, the objective back focal plane, and the eyepoint (the point where the lens of your eye goes when using the microscope visually). And it is the lamp condenser that controls the focus of the filament.

What one is trying to avoid is getting the specimen and filament in focus in the same plane. Optically, the plane of the aperture diaphragm and the objective back focal plane are as far away as it is possible to get from the specimen plane. Köhler illumination ensures that the filament will be nowhere near the specimen. The real key to this system of illumination may be seen in Fig. 3 (right). Notice that the light is *parallel* in the plane of the specimen; this results in an even illumination on the retina—see the retina. The specimen image (left, at retina) is seen against this evenly illuminated background (right, at retina).

For those microscopes which do not have fully centerable and focusing capability, there are modifications published that require access to a machine shop [5]. For simple plug-in, low-wattage lamps with built-in diffuser, not much can be done; these are simply not of optimum quality.

Darkfield, Rheinberg illumination, and phase contrast can all be done on a polarizing microscope, provided the polarizer is removable. The Olympus polarizer is not removable, but can be easily modified so that it is [6].

VII. SUMMARY OF ADJUSTMENTS

Once the objectives are centered with respect to the stage rotation, they never have to be centered again, except for minor adjustments due to accidental bumping. The filament focus and centration only have to be done again when the bulb burns out, and a replacement bulb is installed since the filament may be in a different location in the new bulb.

Other adjustments, however, are made more often, some continuously. The field diaphragm is focused in the plane of the specimen using the substage condenser. When the slide is changed, the new specimen will probably lie in a different plane, due to differences in slide thickness (in one box, the measured slides ranged in thickness from 0.9 to 1.06 mm). This means that technically, the field diaphragm should be refocused, using the substage condenser, each time the slide is changed. In routine work this is seldom done; in critical photomicrography it is always done.

In addition to the focus, there is the matter of field diaphragm setting; it should be just outside the field of view, but the diameter of the field of view changes with each objective, being smaller for the higher magnifications. So that each time the objective magnification is changed, the field diaphragm opening must be changed, opening for lower magnification and closing for higher magnification; each time adjusting so that the field diaphragm image lies just outside the field of view.

Perhaps the most critical setting in the microscope, and the most neglected one, is the setting of the aperture diaphragm. This diaphragm controls the numerical aperture of the condenser, which controls the angle of the illuminating cone, which, in turn, controls the numerical aperture of the objective, which, ultimately, controls the resolving power, the depth of field, and the contrast. Closing the aperture diaphragm increases the depth of field and contrast, but decreases resolution. Opening the aperture diaphragm decreases depth of field and contrast but increases resolution; but if it is opened all the way, one is using the margin of the lens, where the aberrations are severest, and this leads to glare and low image quality, which is why in setting up the illumination the aperture diaphragm is closed about 15% in the objective back focal. Theoretically, the aperture diaphragm should be wide open for maximum resolving power. The more highly corrected objectives such as the fluorites and apochromats, allow the aperture diaphragm to be opened more. In actual fact, in a mixed preparation of many kinds of particles, such as in an air-pollution sample, there are transparent, translucent, and opaque particles all with different refractive indices, some below, some above, some matching or almost matching the refractive index of the medium. Thus, each particle requires a different setting of the aperture diaphragm for optimizing its image. It is on this kind of sample that one hand should be on the aperture diaphragm control lever, constantly changing it slightly as one looks from one particle to the next; always optimizing the image quality.

VIII. CALIBRATION OF THE EYEPIECE MICROMETER

Once the microscope is mechanically aligned, and Köhler illumination is set up, only one more formality needs to be taken care of before we start identifying particles, and that is the calibration of the eyepiece micrometer (ocular micrometer; eyepiece/ocular scale). The measuring of microscope specimens is known as micrometry. The microscope object is compared to a measuring scale in the eyepiece. This eyepiece measuring scale may take many forms, there being almost 100 different kinds of measuring graticules, including single scales divided into 50 or 100 divisions which may or may not be numbered, crossed scales, multiple scales, concentric circles, a series of outlined circles and opaque circles of various diameters, squares within squares, squares divided into smaller squares, and countless specialized shapes, each designed for determining lengths and widths, areas, and other parameters involved in particle counts and measurements, often to conform to some standard, such as grain size; cells; algae, mold or suspended matter per unit volume; fibers of asbestos per cubic liter of air, etc. Of these, the simple scale is of universal application. The scale is of some arbitrary length (usually 1 cm) divided into 100 parts. If the scale is diamond ruled, it will not be numbered. If it is a microphotograph, it will usually be numbered every 10 divisions. By itself, the scale divisions are meaningless; they need to be compared to some reference standard of known length. This reference standard consists of a divided scale of known length mounted on a microscope slide, known as a stage micrometer. The three most common types include (1) a 1-mm-long scale divided into 100 parts, or 0.01 mm (10 μm) per division (2) a 2-mm-long scale divided into 200 parts, or 0.01 mm (10 μm per division) and (3) a 5-mm-long scale, with reference marks at each millimeter, and one of the 5 mm divided into 100 parts, or, again 0.01 mm (10 μm) per division.

The eyepiece micrometer is compared to the stage micrometer in one of two ways: (1) The stage micrometer is moved in such a way that the eyepiece micrometer lies superimposed directly over it. Then one looks for any two places, as far apart as possible, where the two scales coincide. Record the number of stage micrometer divisions that correspond to the matching number of eyepiece micrometer divisions. (2) The stage micrometer is moved in such a way that the eyepiece micrometer is exactly superimposed on it, and the "zeros" of both scales are aligned. Next, go to the end of the eyepiece micrometer scale, and record the number of divisions on the stage micrometer that correspond to the 100th eyepiece division, estimating to tenths, if necessary.

In either case, multiply the number of stage micrometer divisions by 10 (remember that there are 10 μm per division on the stage micrometer), and divide by the corresponding number of eyepiece divisions (100 in the second case). This will yield the number of micrometers (μm) per eyepiece scale division, or, simply, μm/div. Each objective will have to be calibrated.

In the past, the values per division for each objective more or less corresponded to the focal length (mm) of the objective, so that the 10× objectives came out around 16 μm/div, the 20× was 8 μm/div, the 40× was about 4 μm/div, and the 100× was about 2 μm/div. These values varied with individuals because differences in interpupillary distance resulted in the mechanical tube length differences, and, therefore different magnification and calibration factors. In modern microscopes, the mechanical tube length is maintained regardless of interpupillary distance either because the light path is folded (Zeiss, Nikon), or there is as internal compensation for tube length (Olympus) whereby the eye tubes become longer as the interpupillary distance decreases, and shorter as the interpupillary distance increases. Because the tube length is constant for different users some manufacturers have made the eyepiece scales of such a length, that even values for each division result, for example, the calibration factor for 10× objectives is 10 μm/div; for 20× objectives it is 5 μm/div, and for 40× objective it is 2.5 μm/div. The calibration factors for each objective could be placed on a label at the base of the microscope until they are memorized.

IX. SAMPLE PREPARATION

Particles may be on or in a product, on a membrane filter, in a solution, embedded in a paint film, on a suspect's clothing or murder victim, in the adhesive of sticky tape, in the crevasses of a work of art, in the ink of counterfeit currency, etc.—the list is endless. The particle of interest needs to be recovered, either through filtration onto a small membrane filter, or ultrasonication followed by filtration, or physically removed from a substrate using a needle. There are scores of ways to recover particles, most of which are described in *The Particle Atlas* [2].

Eventually, however, the recovered particles need to be mounted for examination with the polarizing microscope. Recovering particles is usually done with the aid of a stereomicroscope and a fine needle. The fine needle consists of a piece of tungsten wire (~0.030 in.) chucked into a dissecting needle holder. The tungsten wire is thermally eroded down to a fine point, which can be to atomic dimensions. This is done in one of several ways: Sodium nitrite, $NaNO_2$, is cast into sticks by melting the granulated compound in a test tube, allowing it to solidify, and then breaking the glass to recover the molded sticks. The sticks are broken into convenient size (1 to 1½ in.). They are deliquescent, so need to be stored in small plastic bags, or, as I prefer, epoxy one end to a cork—which also serves as a handle—and insert the cork with stick attached into a shell vial or bottle that contains some indicating silica gel [7]. Now, hold the stick by the cork near an alcohol or bunsen flame, heat the tungsten needle red hot, then stroke the end of the stick with the hot wire. It will "fizz" and form a sharp point. Give the needle a twist while stroking so the needle will not be eroded lopsided. Make several needles of different tip diameter to suit particles of various sizes. Alternatively, one may use a Meeker burner to make the needles. In this case, place a small amount of sodium nitrite on the grid at the edge of the burner, where it will become molten. Stroke the needle through the molten compound after first making the needle red hot. The needles can also be made electrolytically [2].

Once the needles are made, the only other thing needed is a set of refractive index liquids. These range from 1.3 to 1.7 in steps of 0.002 and 1.7 to 2.0+ in larger steps. There are almost 100 liquids, the most useful as a starter set being the series 1.4–1.7; the others can be added later. The liquids are expensive, but if used properly, with coverglass fragments, a set will last a professional lifetime. Just as biomedical microscopists have a preference for mounting medium—Canada balsam ($n \sim 1.53$)—so do particle analysts, namely $n = 1.66$. The reason for this is that particle analysts have long made permanent slide preparations of particles by mounting them in Aroclor 5442 (The Aroclors are a series of polychlorinated biphenyls, formerly made by Monsanto, that range from mobile liquids to brittle solids). The refractive index of Aroclor 5442 is 1.66. Materials have different appearances in mounting media of different index because of relief, so it is a good idea at the outset to decide on a standard refractive index. As the most extensive particle reference to date [2] is based on 1.66, there is no good reason to change it. The choice of refractive index liquid 1.66 is to duplicate the particle's appearance in the permanent reference preparation (1.66).

Now we have the recovered particle on, say, a small membrane filter; we have our needles made, and we have a set of refractive index liquids. Begin by placing a *tiny* drop of liquid 1.66 on a cleaned microscope slide (wipe the dropper on the edge of the bottle several times, and then quickly and lightly touch the dropper to the precleaned slide). Next break an 18-mm #1½ cover glass into four or more parts (place the precleaned cover glass on a soft surface, like a piece of cardboard, and press down in the middle of the cover glass with a carbide scriber). Select one of the precleaned cover glass fragments, and have it ready to place on the drop. Using the stereomicroscope, focus on the membrane filter with the recovered particles, and go in with the fine tungsten needle and gently touch the particle. It will stick to the needle. Hold it, and transfer the particle to the tiny liquid drop. The surface tension will help wash the particle off as the needle is withdrawn from the liquid. Go back and pick up another particle, and another, until a representative amount has been transferred. Cover the droplet with the cover-glass fragment, and ensure that there is enough liquid to completely fill beneath the cover glass. If there is excess liquid, absorb it into a membrane filter fragment, or lintless tissue. Place the slide on the stage, and focus on the particles.

X. PARTICLE IDENTIFICATION: OBSERVATIONS USING ORDINARY LIGHT

Once experience has been gained, the first view of a particle will be made using slightly uncrossed polars, but to begin with, it is a good idea to start systematically using ordinary light. If a polarizer is in place, it may be left in the light path, as its presence does not affect the observations to be made. Another good idea to start with is to make a chart with particle types listed down the left side, and several columns headed with the titles: morphology, size, absorption color, pleochroism, refractive indices, dispersion staining, birefringence, sign of elongation, extinction angle, interference figure, and remarks. This will form a sort of checklist to ensure that all of the essential characteristics for identification are observed and recorded.

A. Morphology

To begin with, one of the first features that will be apparent on looking at a particle is its shape. Spheres, needles, flakes, plates, all will be immediately apparent. Describe in brief terms what is seen; include surface markings, textures, inclusions, degree of roughness or smoothness, sharp or rounded edges, type of fracture (conchoidal, rhombohedral, etc.) irregular or other outline in projected aerial view, cross-sectional shape of a fiber, ribbon, etc. Use as many descriptive terms as necessary to build up a word picture. Many organized biological structures (diatoms, pollen grains, plant hairs, insect parts) will soon become recognizable on sight; minerals and industrial dusts will take a few more observations.

B. Size

Next, measure the particles using the calibrated eyepiece micrometer. Many biological structures are identified through various keys in which their diameter, distance between structures, etc., is required. If there is a size range, measure the smallest and largest particle of the same type, and record the mode as well. If the particles are irregular, determine Martin's diameter (the dimension, parallel to the eyepiece scale, that divides a randomly oriented particle into two projected areas), for example. If a particle size distribution is needed, tabulate the particles into predetermined size classes, and count enough particles so that commonly used statistical methods may be applied to the data. Specific examples of each of these may be found in texts on quantitative polarized-light microscopy [8], in texts devoted exclusively to particle size determination, and to other chapters in this book.

C. Absorption Color

Record the color, or range of colors seen, by both transmitted and reflected light. The "color by reflected light" brings up a brief diversion. I find it necessary to make a small modification to any microscope dedicated to particle identification, and that is the addition of two small lights, such as the American Optical Company Universal Illuminator. A typical microscope, such as that one shown in Fig. 2, has a hollow stand, covered in back by a metal plate. Remove the metal plate, and drill a hole through both sides of the upright stand. Place a threaded bar between the two holes, and mount the illuminators to the sides of the microscope. Arrange the lamps so that they point down toward the specimen. Replace the back plate. To conserve bench space, I mount the transformers on a small shelf beneath the bench and to the rear so they are out of the way. These lights are controlled by a momentary-contact foot switch; my ordinary microscope light is also controlled by a foot switch of the on/off kind. The voltage is always at the maximum (12 V), an 80-A filter is in the light path, so that the color temperature of the light is always at 5500 K (daylight). Intensity is controlled by neutral density filters. Now, when colors need to be observed, either transmitted or toplights or both can be turned on without letting go of the slide, focus, etc., and all colors will be recorded under standard daylight conditions.

D. Magnetism

This is also a good time to take a small, strong magnet (of the kind mounted at the top end of some carbide scibers) and wave it back and forth around the lower end of the objective; move the slide around with the other hand; occasionally turn the toplight on with the foot switch. Anything magnetic will wave back at you while you are viewing it, and you can note the relative percentage of the sample that is magnetic, and whether the magnetic particles are scalelike, as wear particles, or if they show machining, drilling, filing, or shear marks; whether there is corrosion present, or evidence of rust, or mill scale; degree of reflectivity; or are they round, as magnetite spheres from fly ash, or grinding operations. All of these questions are asked and answered while the magnet is continuously moved back and forth, and different parts of the sample are viewed.

E. Brittleness/Elastomericity

This is also a good time to hold the slide with one hand, controlling the field of view, and with the other hand hold a dissecting needle and direct its point under the objective (start with a 10× objective). You will see the needle's shadow as you move it back and forth under the objective. Stop when it is in the middle of the field of view, and withdraw it until the point is centered in the field. Now move the particle in question to the center of the field, and depress the cover glass with the needle point directly above it. Observe how the particle responds to the lowered cover glass. Some particles, like diesel soot, carbon black, and iron oxide agglomerates, will immediately disintegrate into a "zillion" tiny (< 1 μm) particles. Others, like petroleum that has had the volatiles driven out, but is not yet fused, will be glassy, and will shatter when the cover glass is depressed. Still others, such as rubber from an enclosure, or from automobile tires, or from a conveyer belt, will flatten out easily, but will "bounce back" when the pressure is released. Perhaps there will be no change. All of these observations must be recorded and added to those already observed, as possibilities are already forming in your mind as you integrate all of your observations.

XI. PARTICLE IDENTIFICATION: OBSERVATIONS MADE USING PLANE POLARIZED LIGHT

When you have learned all you can by observing the particle using ordinary light, it is time to check for pleochroism, refractive index, and dispersion staining.

A. Pleochroism

Pleochroism ("many colors") is a change in color or hue relative to the vibration direction of polarized light. Colored crystals which are anisotropic (have more than one refractive index) often exhibit different degrees of light absorption in different directions within the crystal, resulting in the crystal showing a different color, or different intensity of color, when rotated in plane polarized light. Ordinary light will not do because the light vibrates in all directions, and therefore will show a composite of all effects, which is analytically useless. Light vibrating in only one direction is required. The procedure is simple. If the transparent particle being viewed is colored, rotate the stage 90°, (alternatively, the polarizer may be rotated 90°, but it is usual to rotate the stage) and watch for any change in color or intensity. Record any changes. If the particle is elongated, the color changes seen are referred to the shape. For example, the common asbestos fiber crocidolite (fibrous riebeckite) is characteristically blue to blue-green in absorption color. If the fiber is oriented east/west, i.e., parallel to the polarizer vibration direction, it will be showing a blue color, and this color is that characteristic of the fiber length ($n_{\parallel}$), i.e., parallel to the polarizer vibration direction, and is recorded as, $n_{\parallel}$ blue. The stage is rotated 90° so that the fiber is now oriented north/south. The color of the fiber is now gray, and this color is characteristic of the fiber's width ($n_{\perp}$), i.e., parallel to the polarizer vibration direction, and is recorded as $n_{\perp}$ gray. In the column under pleochroism, this fiber would be recorded as $n_{\parallel}$ blue; $n_{\perp}$ gray.

Kevlar fibers show $n_{\parallel}$ pale yellow-green; $n_{\perp}$ colorless. Colored anisotropic substances, do not necessarily show detectable pleochroism, however, which is what makes its observation an identifying characteristic.

B. Refractive Index

Refractive index is the ratio of the velocity of light through a substance relative to the velocity in a vacuum. For example,

$$\frac{\text{Velocity of light in vacuum } 3\times10^{10}\text{ cm/s}}{\text{Velocity of light in glass } \quad 2\times10^{10}\text{ cm/s}}$$

Notice that there are no units. Refractive index is expressed as n for cubic crystals and amorphous substances; thus, the refractive index of glass is expressed as $n = 1.5$. If the refractive indices of all solids and liquids were plotted against the number of materials with those indices, there would be formed an approximate bell-shaped curve, in which the lowest index would be around 1.3 (gases have lower indices; the refractive index of air is 1.0). Sodium fluoride is 1.32; water is 1.33; ethanol is 1.34; my favorite cognac is 1.3568 (one has to know the important things!). The peak of the curve would fall around 1.52. Diamonds are 2+, and the oxides of the heavy metals are 3+, 4+ and 5+. The exact value depends on the crystal's atomic makeup, and is therefore, different for each different material. Our aim then is to determine the values of these refractive indices for each material.

Liquids have one refractive index, designated n. Cubic crystals also have one refractive index, designated n. Tetragonal and hexagonal (including trigonal) crystals have two refractive indices, designated ε and ω, either one of which may be the higher value, i.e., $\varepsilon > \omega$ or $\varepsilon < \omega$. This provides for a further subdivision of all tetragonal and hexagonal crystals; namely, those in which $\varepsilon > \omega$ which are, by convention, termed optically positive (positive sign of double refraction), and those in which $\varepsilon < \omega$, which are termed optically negative (negative sign of double refraction). Orthorhombic, monoclinic and triclinic crystals have three refractive indices, designated α, β, γ. By definition, α is the lowest refractive index, γ is the highest refractive index, and β is intermediate in refractive index. Note that β, being intermediate, can be closer to α or closer to γ. This allows for the further subdivision of all orthorhombic, monoclinic, and triclinic crystals into two subclasses: If $\gamma - \beta > \beta - \alpha$, the crystal is said to be optically positive; if $\gamma - \beta < \beta - \alpha$ the crystal is said to be optically negative.

In principle, the determination of the refractive index is simple. Consider the case of a cubic crystal, such as ordinary sodium chloride ($n = 1.5443$). If this crystal were placed in a liquid whose refractive index were also 1.5443, the solid would be invisible. Thus, given an unknown cubic crystal, one places it in one of the refractive index liquids in the 1.3–1.7 series and compares it to the medium through the relief that is produced. The liquids are changed, until one is found in which the crystal disappears. Then, knowing the refractive index of the liquid, the refractive index of the crystal is known, and one then goes to a table of refractive indices of cubic crystals, where the value 1.5443 will be found next to sodium chloride.

In practice, there are a couple of things that must be taken into consideration. One of these is that the refractive index varies with wavelength, being, generally, higher for blue than for red. In practice, a single wavelength has to be agreed upon to record refractive index values. The wavelength selected for the standard is 589 nm, Fraunhofer's D line due to sodium. This wavelength is commonly obtained through the use of an interference filter of 589 nm; the filter is placed over the light exit port of the microscope. As several materials may have the same refractive index at 589 nm, but different indices at other wavelengths, it is common to have both blue (486 nm, Fraunhofer's F line) and red (656 nm, Fraunhofer's C line) filters for those occasions when the dispersion needs to be determined. The dispersion (ν), or Abbe number, is

$$\nu = \frac{n_D - 1}{n_F - n_C}$$

Another consideration to keep in mind is that temperature affects the refractive index. As a substance is heated, it expands; there is more space between the structural elements, light travels faster, and the refractive index goes down. As a substance is cooled, it contracts, there is more matter in the path of the light and it slows down; the refractive index goes up. Thus, some standard temperature must be agreed upon for the expression of refractive index. In the United States, the standard is 25°C; in Europe the standard is 20°C. The correct designation for sodium chloride, for example, is $n_D^{25°C} = 1.5443$. A thermometer is provided with all complete sets of refractive index liquids so that the temperature may be determined. Normally, the thermometer is placed on the stage where the liquids are being used.

In the event that the temperature is not 25°C, or if the liquids are used in a hotstage or coldstage, a correction must be applied. There is a temperature coefficient indicated on each bottle of refractive index liquid. A typical one reads $-dn/dt = 4 \times 10^{-4}$/°C. That is, there is a change in refractive index (*dn*) that amounts to 0.0004 refractive index units for each 1°C change in temperature (*dt*). Suppose one is using liquid 1.680 and the thermometer on the stage reads 21°C. That is a 4°C difference from the standard 25°C. The liquid changes index by 0.0004 unit for each degree off of standard or $4 \times 0.0004 = 0.0016$, and since it is colder than standard, the liquid will contract, and the index will increase. Therefore the 0.0016 is added to the bottle's value of 1.680, or $n_D^{21°C} = 1.6816$.

For laboratories with constant temperature doing routine work, the correction is seldom necessary, but it is essential for any patent or legal work.

If on placing an unknown crystal in a liquid the crystal and liquid do not match, it would be nice to know which is higher or lower so that one knows in which direction to go to find a matching liquid. This is done through the use of the Becke line. The Becke line, named after a German mineralogist, is a diffraction halo that forms around a particle as the result of closing the aperture diaphragm. Normally one would not want to close this diaphragm because of the loss of resolving power. However, the resolving power is not relevant here; the diffraction halo is. The Becke line diffraction halo is best produced with high numerical aperture objectives, usually the 40×. The 589-nm filter is in place, and the particle is brought into good focus. Now, the fine adjustment is suddenly turned in such a way that one is focused *above* best focus (on Nikon and Olympus microscopes this means turning the fine focus *toward* the operator, which *lowers* the stage, which is the same as focusing *above* best focus). During this operation, the Becke line will be seen to move either into the particle or into the medium. On focusing above best focus like this, the Becke line *always* moves toward whichever has the higher index ("focus high, goes high"). If one focuses *below* best focus, the opposite effect occurs. In practice, one focuses both ways to confirm the "up" observation. Start at best focus, and turn toward you, and away; toward you, and away; up, down; up, down, up. Stop at "up" and decide which way the halo of light is moving. If the Becke line moves into the particle, the particle has the higher index, and a higher liquid must be chosen to match the particle. If the Becke line moves into the liquid, the liquid is higher in index, but we are interested in the particle: if the liquid is higher, the particle is lower in index. The amount of relief, or contrast will be a visual measure of the refractive index difference. If the contrast is great, chose a new liquid many bottles away.

If there is very little contrast, and the particle is difficult to see, you are close; choose a liquid, say, two bottles away.

For particles with more than one refractive index, the particle must be oriented correctly. Clues to the correct orientation are indicated by crystal morphology and interference figures. Orientation of the crystal requires some knowledge of crystallography, which cannot be included here, but is available in one of the applied polarized-light books [2,8], or in one of the several excellent texts dealing with optical crystallography [9–14].

C. Dispersion Staining

Dispersion staining is an optical method of imparting color to colorless, transparent materials. A special objective is required for this technique—one which has a central light-occluding stop permanently mounted in the objective back focal plane, or one which has interchangeable central and annular stops of one or more diameter that are introduced into the objective back focal plane via a rotating selector wheel, or a slider. These objectives are commercially available, or they can be made [15,16] if one has access to a machine shop.

The dispersion staining objective is a 10× objective, and the central stop is by far the most useful, as it results in a black background, against which the dispersion staining colors stand out brilliantly. To use the objective, focus on a sample in the normal manner, then swing out the top lens of the condenser to lower the numerical aperture. Put in the Bertrand lens, at which time you will see the central stop in the objective's back focal plane. Close the aperture diaphragm until its image almost touches the central stop. If the central stop is not exactly centered within the closed aperture diaphragm, center the stop with the centering wrenches, or the centering screws, depending on the model of dispersion stainer used. When the stop is centered, take the Bertrand lens out, and close the aperture diaphragm slowly until the background just becomes black. If the dispersion curve of the solid crosses the dispersion curve of the liquid anywhere in the visible portion of the spectrum, the particles will be seen in brilliant colors against a black background. If the dispersion curves of the solid and liquid cross outside of the visible region, the particles will appear white against a black background.

The colors seen in dispersion staining depend on the refractive index relationship between the solid and liquid, so that the colors seen can be used to tell what the refractive indices of the solid are, without having to do the Becke line test. It is a very rapid way of obtaining refractive index data on many particles at once.

In central stop dispersion staining, the colors seen are the complements to the solar spectral colors. For example, the solar spectral colors from 400 to 700 nm are blue, green, yellow and red, and intermediates between these. The complementary colors, in the same order, are yellow, magenta, blue and cyan. The complements are plotted as their solar spectral counterparts, i.e., yellow is around 400 nm, magenta is around 540 nm, blue is around 590, and cyan is around 650 nm. On a dispersion staining graph, refractive index is plotted linearly on the vertical axis, and wavelength is plotted as $1/\lambda^2$ on the horizontal axis. If a yellow particle is seen in central stop dispersion staining, the wavelength of the color is estimated, say 420 nm, and a point is plotted on the horizontal line that represents the refractive index of the liquid that the particle is mounted in. Next, a line of positive slope is drawn through the point. Notice where this line crosses the 589 nm line (for refractive index). Come straight left and read the refractive index; it will be much higher than the liquid's refractive index. Yellow colors mean the particle's refractive index is considerably above the liquid's. If the particle's color is orange, it's refractive index is somewhat less than the yellow particle's, but still higher than the liquid. If the particle shows a magenta color, its refractive index is slightly above that of the liquid; if the particle shows a blue color, its refractive index is the same as that of the liquid. If the particle shows a cyan color, its refractive index is less than that of the liquid. Thus, a quick glance at an entire field of particles will reveal refractive index information about all, without having to perform the Becke line test.

The refractive index noted through the dispersion staining color will be that refractive index in the particle that is parallel to the polarizer. If the particle is isotropic (one refractive index) its color will not change on rotation of the stage. If the particle is anisotropic (more than one refractive index), its color will change as the stage is rotated, and information about both indices seen may be obtained. For more extended treatment, and explanation of the theory of dispersion staining, see Refs. 2 and 8.

D. Birefringence

Birefringence is the numerical difference between the principal refractive indices. For amorphous materials and crystals in the cubic system, this will be zero, because all of these materials have one refractive index only. For tetragonal and hexagonal crystals, the birefringence is $\varepsilon - \omega$ (the absolute value since either may be larger). For orthorhombic, monoclinic, and triclinic crystals, the birefringence is $\gamma - \alpha$. Birefringence, therefore, may be either $\varepsilon - \omega$ or $\gamma - \alpha$, and the general form of the equation is $B = n_2 - n_1$ where n_2 and n_1 represent $\varepsilon - \omega$ or $\gamma - \alpha$, depending on the crystal being examined. The birefringence may be determined through use of the Becke line method or dispersion staining, but there is a much faster way that is based on the particle's thickness and the interference color it shows between fully crossed polars due to its retardation.

Refer to Fig. 4, which illustrates an anisotropic crystal oriented 45° to the vibration plane of plane polarized light; the slow (high refractive index) and fast (low refractive index) directions are indicated.

When the plane polarized beam, enters the crystal from below it is immediately divided into two components, one traveling in the fast direction and one traveling at reduced velocity in the slow direction. The two components therefore exit the crystal at different times; the fast component leaves first, and the slower component follows behind. The slow component is said to be retarded relative to the fast component, and the physical distance between the two, measured in nanometers, is known as the retardation. The exact value of the retardation depends on the difference between the fast and slow rays (birefringence), and on the crystal's thickness. Specifically, $r = 1000\,Bt$, where B is the birefringence, t is the thickness (measured in micrometers), and the 1000 is a conversion factor necessary to convert the thickness in micrometers into nanometers, the unit of retardation. By transposition, we also obtain

$$B = \frac{r}{1000t}$$

$$t = \frac{r}{1000B}$$

Normally, we do not obtain thickness through calculation, because it is estimated from other particle measurements made with the calibrated eyepiece micrometer. The retardation is obtained from the color that the particle shows between fully crossed polarizers. The colors are those in the Newtonian series, and from a chart of these colors, known as the Michel-Lévy chart (Fig. 5), the number of nanometers corresponding to a particular color may be read directly. It is the birefringence that we are normally after, as this characteristic of materials. Quartz, for example, has a birefringence of 0.009, this is the numerical difference between its two refractive indices, and constitutes one of its identifying characteristics. The birefringence remains constant. The retardation color that quartz shows between crossed polars will vary with the thickness, and so

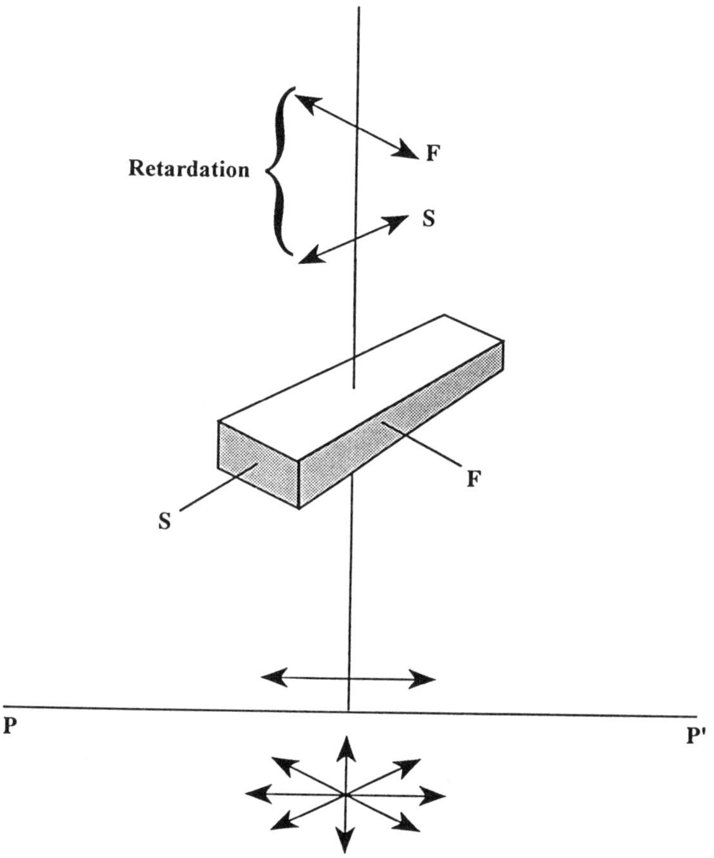

Figure 4 Retardation (measured in nanometers) results from plane polarized light passing through a crystalline material that has more than one refractive index. "S" in the slow direction; "F" in the fast direction. The slow ray here is retarded relative to the fast ray on exiting the crystal.

a field full of quartz grains is typically quite colorful because of the thickness variations in a single grain.

To obtain the birefringence of an unknown particle, we need the thickness and retardation. The thickness is obtained through direct measurement or estimation from other measurements using the calibrated eyepiece micrometer. The retardation is obtained by matching the color at the point where the thickness was determined with the color on the Michel-Lévy chart, and reading the value in nanometers. The Michel-Lévy chart is actually a graphical solution to the birefringence equation up to the limits of 50 μm for thickness, and about 1800 nm for retardation.

$$\text{Birefringence} = \frac{\text{retardation (nm)}}{\text{thickness } (\mu\text{m}) \times 1000}$$

In actual practice what is done is this: The particle of interest is viewed between crossed polarizers and oriented into its maximum brightness and birefringence position. Then its thickness is determined with the eyepiece micrometer, and its retardation color noted. Going to the Michel-Lévy chart, one enters at the thickness, moves along this thickness line to the right until one gets to the retardation color the particle is showing. At the intersection of the thickness

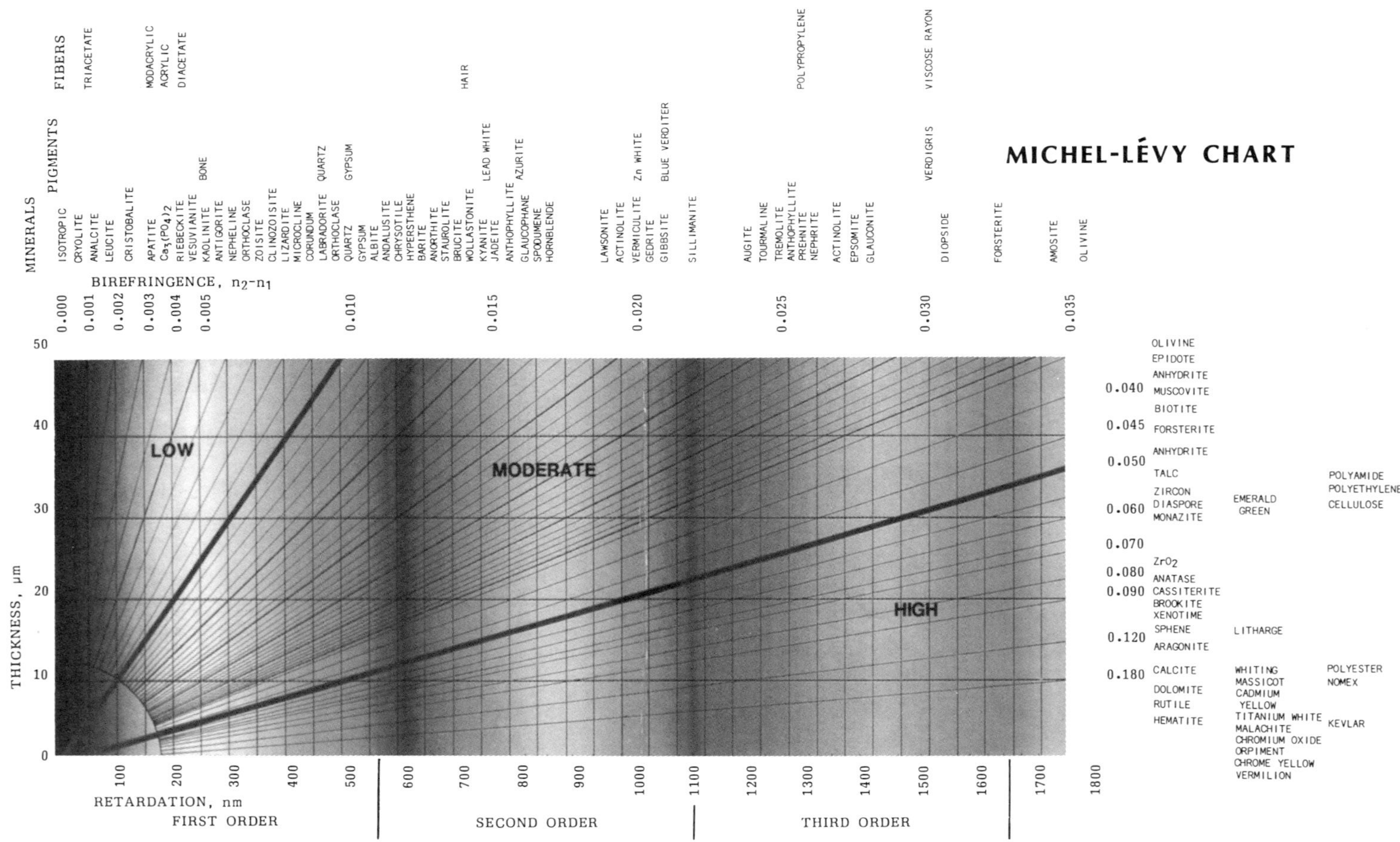

Figure 5 Michel-Lévy chart. This chart graphically relates thickness, retardation, and birefringence. The full-color version would show colors in the Newtonian series. The chart is usually used to determine birefringence of an unknown material after measuring the thickness of the material and noting its retardation color.

and retardation is a diagonal line. Follow the diagonal line up, and read the birefringence. Of course, the equations may be used to calculate the birefringence, but using the chart is much faster. For thickness and retardation beyond those indicated on the chart, one simply extrapolates. Further details on the use of this chart, including a history, can be found in Ref. 17, as well as Refs. 2, 8, and 9–13.

E. Sign of Elongation

Sign of elongation refers to the location of the high and low refractive indices in an elongated material. By convention, if an elongated material (fiber, needle, elongated cleavage fragment) has the high index parallel to its length it is said to have a positive sign of elongation, and is described as length slow. Elongated materials which have the low refractive index parallel to its length are said to have a negative sign of elongation, and are described as being length fast. Which of these one has can be determined through the Becke line method or dispersion staining, but a much faster way is through the use of compensators.

Let us start by considering the phenomenon of compensation itself. Refer to Fig. 6. In the figure (left) an anisotropic crystal is showing a retardation of 100 nm; it appears a kind of gray color, with perhaps, a touch of lavender, between crossed polars. Now, suppose that another crystal exactly like the first one is stacked on top of the first crystal, effectively doubling its thickness. Note that fast is on fast, and slow is on slow. The fast ray leaves the lower crystal first and enters the upper crystal in the fast direction; the slow ray leaves the lower crystal last and enters the upper crystal in the slow direction where it slows down even more. The result is that the slow ray gets twice as far behind the fast ray, and the net retardation is 200 nm—a clear gray color between crossed polars. If four such crystals were stacked, the total retardation would be 400 nm—a yellow color between crossed polars. Notice that we are going up-scale on the Michel-Lévy chart; the retardations are additive when slow is on slow and fast is on fast.

Now look at Fig. 6 (right). Here are the same two crystals, but this time, the upper crystal has been rotated 90°. Now the fast ray from the lower crystal enters the upper crystal in the slow direction, and gets slowed down; the slow ray enters the upper crystal in the fast direction and is speeded up. If both crystals are exactly alike, the net retardation will be zero, and if we were looking down the microscope at these crossed crystals, the color where they cross would be black, and the arms of the cross would be gray (100 nm). The retardations are subtractive when fast is on slow, and slow is on fast. The colors go down scale on the Michel-Lévy chart. The upper crystal here is said to have compensated the low-crystal completely. Compensation may be partial, as when the upper crystal is one-half the thickness of the lower crystal, for example.

From the above we derive the following conclusion: if we had in our microscope a crystal such as the lower one, and we knew its retardation and the location of its slow and fast directions, we could superimpose on it an unknown crystal, and by noting whether the resulting retardation color has gone up scale, or down scale, we would know where in the upper crystal the fast and slow directions were. And that is where our compensators come in. The first-order red compensator, is a mineral cut of such thickness that it shows a first order red color, 550 nm (530 in some instruments), between crossed polars when inserted in the light path—that is our reference crystal retardation. Further, the compensator is marked with the fast and slow directions. The slow direction (high index) will be marked Z' across the width of the compensator, and X' (fast direction; low index) parallel to the long direction of the compensator. The compensator is inserted in the microscope at a 45° angle, so that in the field of view, the slow direction (high index) will be 2–8 o'clock. That completes the reference crystal; we know its retardation and location of the slow direction.

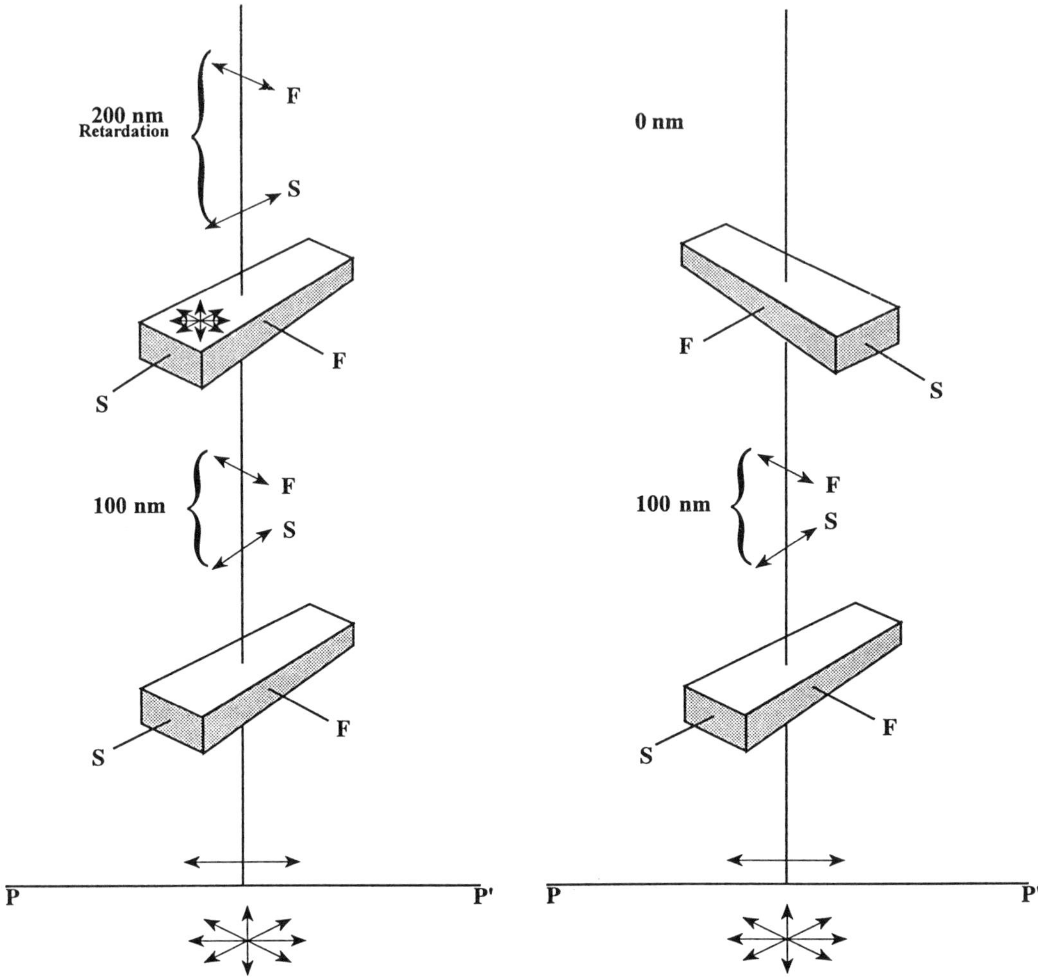

Figure 6 Principles of compensation: retardations can be added or subtracted, resulting in retardation color going up scale or down scale on the Michel-Lévy chart. If the retardation and slow direction of the lower crystal (compensator) are known, the slow and fast directions in an unknown crystal can be determined by noting the resulting retardation on superimposing the two. (Left) retardations added; (right) retardations subtracted.

To determine the sign of elongation proceed as follows: between crossed polars, orient an unknown, elongated material, say a synthetic fiber, 2–8 o'clock in the field of view, and note its retardation color, say, gray 100 nm. Now insert the first-order red compensator, and note what happens to the retardation color of the fiber. The compensator's retardation is 550 nm; the fiber's retardation, 100 nm, can only add to or substrate from the 550 nm, resulting in 650 nm (blue) or 450 nm (orange). Suppose the fiber turns blue. Blue is 100 nm up scale from 550. The only way to go upscale is to add fast on fast, slow on slow. The compensator's slow direction is 2–8 o'clock, therefore slow in the fiber must also be 2–8 o'clock, i.e., parallel to its length. The slow direction (high index) is parallel to the fiber's length, and, therefore, it has a positive sign of elongation. If the fiber's color had gone down scale from 550 nm, the fiber would have the low index parallel to its length, and would have a negative sign of elongation (length fast).

Orthorhombic, monoclinic, and triclinic crystals which are elongated and have γ running along their length will show a positive sign of elongation; those with α running along their length will show a negative sign of elongation; and those which have β running along their length will show both positive and negative sign of elongation, depending on how they are oriented about their long axis.

F. Extinction Angle

The extinction angle is the angle between the optical and crystallographic axes; the angle between the vibration directions inside the crystal and some prominent crystal face, or fiber edge. The optical axes α, β, γ are mutually perpendicular. Monoclinic and triclinic crystals have either one crystallographic axis inclined to the other two (monoclinic) or have all three crystallographic axes mutually oblique (triclinic). Therefore, in these crystals there will be views where the optical and crystallographic axes will not be parallel. The angle between the optical and crystallographic axes is the extinction angle. The extinction angle is characteristic and must, therefore, be measured. The presence of an extinction angle also serves to distinguish monoclinic and triclinic crystals from orthorhombic crystals, which always have parallel extinction (extinction angle equals 0°).

To measure an extinction angle, proceed as follows: Suppose we have an acicular (needle-like) crystal that we observe between crossed polars to be anisotropic. Move the slide until the crystal lies under the crosshairs, and line the crystal up with either the north-south or the east-west crosshair. This aligns the crystallographic axis to the crosshair. Now note whether the crystal can be seen or not. If it cannot, than it is at extinction and the internal vibration directions are parallel to the polarizer and analyzer vibration directions, and the extinction angle is 0°. This will always be the case with orthorhombic crystals and synthetic fibers.

If when the acicular crystal is aligned with a crosshair it can be seen showing some degree of brightness, the crystal's internal vibration directions are not aligned with the polarizer and analyzer in this orientation, and the crystal's extinction angle must be measured. To do this, leave the crystal aligned with the crosshair, and read the graduation on the stage, using the vernier, and record the reading. Next, rotate the stage until the crystal does go to extinction. This may mean turning clockwise or counterclockwise; extinction will occur either way, but the smaller of the two angles should be chosen. At extinction, again read the stage graduation using the vernier. Record this reading. Subtract the two readings to obtain the extinction angle. The real objective though is to find the maximum angle. Depending on orientation about the long axis the angle will range from 0° to some maximum. Therefore, in randomly oriented samples, a hundred or more particles need to be observed to be sure the maximum angle is being approached. If only a single crystal is available, it needs to be oriented about its long axis by using a spindle stage, although the crystal can usually be coaxed into a favorable orientation by continuous tapping on the cover glass with a dissecting needle.

G. Interference Figures

Interference figures are optical phenomena that appear at the objective back focal plane when a tetragonal, hexagonal, orthorhombic, monoclinic, or triclinic crystal is viewed between crossed polars. The interference figures are seen with the Bertrand lens in. Proceed in the following way: use a high numerical aperture objective (40×/0.65 minimum); put a diffuser in the light path; open the aperture diaphragm fully; put the crystal in question under the crosshairs, and make sure the polarizers are crossed; finally, insert the Bertrand lens and rotate the stage.

What is seen after proceeding as above depends on the crystal system, and orientation of the crystal. Tetragonal and hexagonal crystals will result in a centered Maltese cross figure together with concentric colored circles, when the crystals are oriented so that one is looking down the crystallographic c axis (i.e., down ε). In this particular view, the crystal will appear to be isotropic, or nearly so when the Bertrand lens is removed. This isotropic view is the one and only seen when viewing a tetragonal or hexagonal crystal down this axis; no other, and therefore, tetragonal and hexagonal crystals are also known as uniaxial crystals.

By inserting a first-order red compensator, the optic sign of the crystal may be determined: ε' lies 2–8 o'clock in the interference figure, and inserting the compensator superimposes the compensator's slow direction over ε'. It's value can thereby be determined relative to ω, in a manner similar to the addition and subtraction of retardations used with sign of elongation.

Orthorhombic, monoclinic, and triclinic crystals (and pseudocrystals, such as some polymer films) when looking down α or γ yield interference figures that appear to have two black arcs and a series of colored rings in a more or less figure eight shape; the center of the black arcs mark the exit points of the optic axes—these being two unique points down which isotropic views are seen. Crystals producing figures like this are said to be biaxial.

An imaginary line drawn between the exit points of the optic axes mark the optic axial plane. This plane also includes α and γ; β is perpendicular to this plane. In a centered view of such a biaxial crystal, we are either looking directly down γ at an α-β plane (γ being the bisector of the acute optic axial angle), or we are looking directly down α at a β-γ plane (α being the bisector of the acute optic axial angle). Which one we have can be determined with a compensator. If γ is the acute bisectrix, the crystal is optically positive; if α is the acute bisectrix, the crystal is optically negative.

Additionally, after measuring the diameter of the objective back focal plane (related to numerical aperture), and the distance between the exit points of the optic axes, the optic axial angle (an identifying characteristic) can be calculated through equations or use of a nomograph.

Next, β is determined by orienting the crystal so that β is parallel to the polarizer vibration direction, this being done by reference to the interference figure (β is perpendicular to the optic axial plane), and then using the Becke line method with refractive index liquids.

γ (or α) is next calculated through the birefringence of the α-β plane (γ-β plane for α) and then, through equations or use of other nomographs α (or γ) may be determined.

One ends up with all three refractive indices, the optic sign, and the optic axial angle, which pretty much nails the crystal down.

Of course, noncentered interference figures, crossed axial plane dispersion, and a few other complications cannot be dealt with in this brief overview, but these are the subjects of the many excellent books in optical crystallography [2,8–14].

H. Microchemistry

The chemical nature of a particle is, of course, of vital interest, and although today, with modern instrumental methods, much may be learned through nondestructive means. Simple tests for cations, anions, and functional groups, are however, still of major importance, and utilized daily in particle identification, particularly if one does not have access to the expensive analytical instrumentation, or if there is sufficient sample. Microchemistry constitutes an entirely separate branch of study, and since there are adequate currently available references [18–23] in this field, it will not be further discussed, except to urge the reader to obtain at least set 2 of Chamot and Mason chemical reagents supplied by Cargille Labs (Cedar Grove, NJ) for use with Volume 2 of Chamot and Mason's *Handbook of Chemical Microscopy* and become acquainted with the tests.

I. Fusion Methods

Fusion methods refers to hotstage, and, by extension, coldstage work, i.e., nonambient temperatures. Microscopically observing what happens to samples when they are heated is a procedure that goes back more than a hundred years.

A simple application of this technique is a melting point determination. The microscopically determined melting point is somewhat different from the ordinary capillary method, because with the microscope one can see when melting first begins, when the majority of the sample melts, and when the last crystal melts, and the temperatures can be recorded. The range of melting gives some indication as to the purity of the sample. For this kind of work a variety of heating stages have been developed over time, that incorporate some means of measuring temperature, regulating temperature, adjusting the rate of temperature increase or decrease, holding at any given temperature, and recording events. Thermometers have given way to thermocouples, and manually operated rheostats have given way to computer control of temperature regulation.

In addition to simple melting point determination, many other observations may be made. Instead of melting, for example, a sample may sublime, or decompose (slowly or violently). Polymorphism may be studied. Most materials have more than one form; some may be stable only at elevated temperatures. When allowed to cool spontaneously, such substances may be seen to go through a whole series of crystallographic changes as it transforms from one polymorph to another, on its way to achieving its stable form at room temperature. One can also determine the degree of supercooling, or do a meltback, where a portion of the sample is remelted and the form of the crystal front determined. Crystal velocities, crystal angles, refractive indices relative to the melt and interference figures may all be studied. Moreover, two or more substances may be mixed or melted side by side, and then melted together to study thymol mixed fusions, eutectics, addition compounds, and solid solutions, and from these observations construct composition diagrams. This line of investigation has proven to be especially valuable in pharmaceutical research and in explosives technology. For detailed treatment of hotstage methods, McCrone's work [24] should be consulted.

J. Contrast Methods

Substances whose refractive indices are very close to the medium they are in are extremely difficult to see. Changing their environment's refractive index seems like a simple solution, but often, as in the case of live materials, this may be impossible. Staining is sometimes helpful, but most stains must be dissolved in solvents that adversely affects living things. It is common to close the aperture diaphragm down when viewing such specimens; the resulting diffraction halos completely destroy resolving power, but at least the specimen can be seen. Darkfield techniques applied to specimens of this kind are hundreds of years old. In this technique, the central, axial rays that come from the condenser in normal transmitted brightfield are deliberately blocked out by an opaque stop located in the condenser, and the aperture diaphragm is opened all the way. If the stop diameter is chosen so as to block only those rays that directly enter the objective, all of the peripheral rays will be available to illuminate the sample from the sides, and the sample will be seen lighted against a dark background. The stop to do darkfield can be homemade very simply, or commercial darkfield condensers, may be purchased.

Rheinberg illumination, also over 100 years old, is a modification of darkfield whereby a colored filter is used instead of an opaque stop, and another, contrasting color filter in the shape of a ring, is placed around the first colored filter. The sample can then be made to appear in any color against a background of any other color. Directions for making darkfield and Rheinberg stops are available in a very inexpensive publication [4].

Phase contrast, developed in the 1930s by the Dutch physicist Zernike, is a contrast-enhancing technique that converts phase differences into amplitude differences, without having to sacrifice resolving power [25]. Nomarski differential interference contrast (DIC) is another contrast-enhancing technique that complements the phase contrast method. Very often the means to do both are provided in the same specialized substage condenser. Both techniques require special objectives.

XII. SUMMARY

After acquiring the proper kind of microscope, mechanically adjusting it, and setting up Köhler illumination, there remains only the task of spending time studying knowns, and characterizing unknowns. It is a good idea to obtain several of the particle reference sets that are commercially available, and spend some time characterizing them by the methods discussed above. Again, a chart should be made listing all particles available, and across the top, morphology, size, absorption color, pleochroism, refractive index, dispersion staining color in some standard liquid, birefringence, sign of elongation, extinction angle, interference figure, optic axial angle, etc. Then, record the property for each substance in the reference sets. Results can be checked with *The Particle Atlas* [2] which is, as the name suggests, a photomicrographic atlas picturing thousands of particles, together with all of the identifying characteristics listed. When proficiency has been attained at this, the same thing is done with the raw materials and final products in the microscopist's industrial area. References of known materials should be constantly collected, some retained in vials, and a portion made into a permanent mount on a microscope slide for characterization. Little by little further course work in specialized areas can be added—microchemistry, fusion methods, photomicrography, crystallography, etc. Fluorescence, Jamin-Lebedeff transmission interference, infrared luminescence, and other techniques of a more specialized nature can be added as needed.

The field of particle identification is experience intensive, and never ending, but extremely useful analytically, and satisfying for the analyst. It is hoped that this brief survey will give some idea of the methods of the particle analyst, and some guidance in the setup and operation of a microscopy lab for particle identification.

REFERENCES

1. Delly, J. G. *How to Use and Maintain a Stereoscopic Microscope.* Filmstrip: 48 frames; tape running time 20:55. Module 8131, Communication Skills Corp.; Prentice-Hall Media, 1976.
2. McCrone, W. C., and Delly, J. G. *The Particle Atlas,* 2nd ed., 4 vols. Ann Arbor Science, Ann Arbor, MI, 1973; Vol. 5, with Palenik, S. J. 1979; Vol. 6, McCrone, with Brown, J. A. and Stewart, I. M. 1980. *The Particle Atlas Electronic Edition* is available on CD-ROM from MicroDataware, Hayward, CA (1992).
3. Delly, J. G., and Wills, W. F. How to buy a compound microscope: an update, *Amer. Lab.* (April, 1985).
4. Delly, J. G. *Photography through the Microscope,* 9th ed., Eastman Kodak Co., Publication P-2, Rochester, NY, 1988.
5. Sirovatka, J. C., and Delly, J. G. Fully adjustable Olympus BHSP illuminator, *The Microscope 36*:327–337 (1988).
6. Delly, J. G. Conversion of Olympus polarizing condenser to non-pol, *The Microscope 39*:127–130 (1991).
7. Delly, J. G. Sodium nitrite stick storage, *The Microscope 38*:40 (1990).
8. McCrone, W. C., McCrone L. B., and Delly, J. G. *Polarized Light Microscopy,* Ann Arbor Science, Ann Arbor, MI, 1978.
9. Bloss, F. D. *An Introduction to the Methods of Optical Crystallography,* Holt, Rinehart and Winston, New York, 1961.

10. Nesse, W. D. *Introduction to Optical Mineralogy,* 2nd ed., Oxford University Press, New York, 1991.
11. Wahlstrom, E. E. *Optical Crystallography,* 3rd ed., Wiley, New York, 1960.
12. Hartshorne, N. H., and Stuart, A. *Crystals and the Polarizing Microscope.* 4th ed., Arnold, London, 1970.
13. Stoiber, R. E., and Morse, S. A. *Microscopic Identification of Crystals,* Chapman & Hall, New York and London, 1994.
14. Marshall, C. E. *Introduction to Crystal Optics,* Cooke, Troughton & Simms, York, England, 1953.
15. Delly, J. G., and Sirovatka, J. C. A dedicated central-stop dispersion staining objective, *The Microscope 36*:205–211 (1988).
16. Sirovatka, J. C. A dedicated central-stop dispersion staining objective (Nikon), *The Microscope 37*:43–47 (1989).
17. Delly, J. G. The Michel-Lévy interference color chart—microscopy's color key, *The Microscope 37*:89–102 (1989).
18. Chamot, E. M., and Mason, C. W. *Handbook of Chemical Microscopy,* Vol. 1, 4th ed., Wiley, New York, 1983; Vol. 2, 2nd ed., Wiley, New York, 1940.
19. Keune, H. *Bilderatlas zur qualitativen anorganischem Mikroanalyse,* VEB Deutscher Verlag für Grundstoff-industrie, Leipzig, 1967.
20. Geilmann, W. *Bilder zur qualitativen Mikroanalyse anorganischer Stoffe,* 3rd ed., Verlag Chemie, Weinheim/Bergstr., 1960.
21. Delly, J. G. Microchemical tests for selected cations, *The Microscope 37*:139–166 (1989).
22. Delly, J. G. The microchemical bench. 1:Chemicals, *The Microscope 40*:269–274 (1992).
23. Delly, J. G. The microchemical bench. 2:Laboratory solutions and other liquids, *The Microscope 42*:35–46 (1994).
24. McCrone, W. C. *Fusion Methods in Chemical Microscopy,* Interscience, New York, 1957.
25. Bennett, A. H., Jupnik, H., Osterberg, H., and Richards, O. W. *Phase Microscopy,* Wiley, New York, 1951.

3

Image Analysis (Automated Microscopy) for Particle Analysis

Thomas A. Barber

Baxter Healthcare Corporation, Round Lake, Illinois

I. INTRODUCTION

Historically, methods of particle analysis were developed as a means of identifying particles or for determining how particle populations (such as bulk powders) would behave in specific processes. The widely used technique of sieving a powder sample and then observing portions of each sieve fraction under a microscope has been applied for more than 100 years. Techniques of particle isolation, identification, and observation have generated concepts relating to particle size, shape, number, population distribution, and chemical composition that relate to different geometric, physical, and chemical properties of particles. This chapter briefly summarizes the principles of microscopic image analysis (IA) and discusses its application as an improved method for performing morphologic studies and population analyses on particulate matter samples. I gratefully acknowledge the assistance of Dr. John Keith Beddow, who provided invaluable insights into shape analysis, and Mr. Julius Knapp for his assistance in the preparation of this manuscript. I am particularly grateful to Dr. Beddow for supplying the mathematical descriptors for particle shape factors.

Image analysis is a generic term referring to any use of a computer-based system to process and analyze digitized images. The primary goal of the process is to obtain quantitative and/or qualitative information from an image. Excellent overviews of this endeavor are provided by Inoue [1], Baxes [2], Green [3], and Kak and Rosenfeld [4]. In this chapter, all the author's references to IA or IA systems will pertain to instruments consisting of image analyzers coupled to light microscopes; image analysis is also very useful with electron microscopes. In the following discussion of the structure and function of microscopic image analysis systems, terminology unique to the field is frequently applied; the reader is referred to Section X for explanation of terms that are not directly defined in the text.

An image analyzer interfaced to a light microscope (Fig. 1) can perform enumeration and sizing of particulate matter collected on a membrane filter or microscope slide to provide a result that is generally consistent with and often superior to that obtained by manual microscopic analysis.

To perform such an automated analysis, a microscopic field containing particulate matter is scanned by a television camera and digitized or broken down into electronically defined individual picture elements (pixels). The term "digital" in this context simply means that the

---

I must emphasize that any and all subjective judgments, opinions, and viewpoints expressed in this chapter are my own and not necessarily those of Baxter Healthcare Corporation or any organization representing manufacturers of medical devices or injectable products.

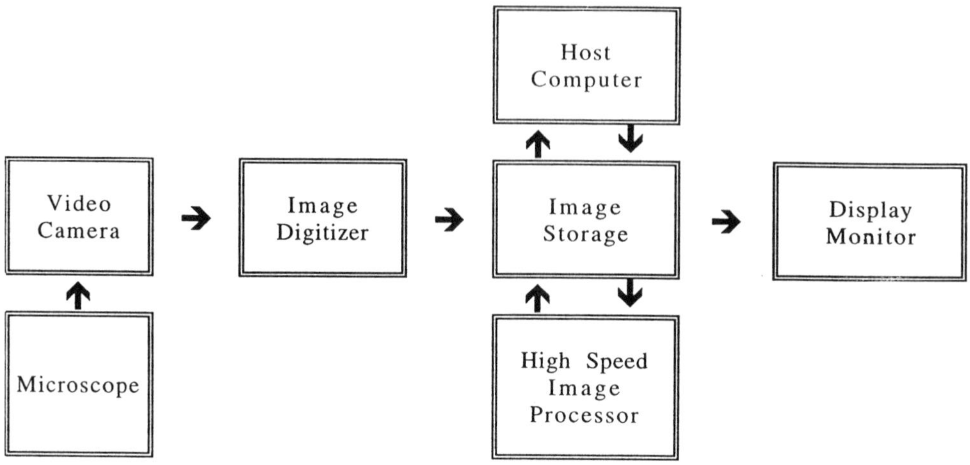

Figure 1 Microscopic image analysis system block diagram.

image contains discrete tones; every sample taken from the original analog image becomes a pixel of the discrete image. The digitized image constitutes a numerical representation of the original image that can be processed like any other data stored in computer memory; thus, digital images allow for flexible image processing using a computer. In a gray-scale video system, each pixel gives an indication of the local image contrast or the gray level. In color systems, a color tone (hue) is assigned instead of a gray level. For a typical analysis of particulate matter, the image formed is then segmented into features of interest (particles) and background (filter or microslide) based on user-established gray-level thresholds. Finally, the particulate matter of interest is sized and counted, and results are printed out in the desired format, such as a particle number versus particle size histogram.

Image analysis may also be performed based on morphologic features of particles. Morphologic descriptions obtained enable the observer to measure specific features of the particle images. Detailed morphological analysis of particle profiles can be carried out using specifically designed instrument systems [5,6]. Useful measures readily available in morphological analysis include such invariant morphological features as size, shape, roughness, microroughness, and rotational symmetries.

Image analysis systems have been evaluated a number of times in the past with regard to their applicability to enumeration of particulate matter [7–11]. In these studies, the correlation of IA data with that obtained by other counting methods has been generally poor [12]. This result is attributable to the morphologic heterogeneity of the particle populations analyzed [13], and the fact that any particle count assay, microscopic or instrumental, is an operational definition incorporating positive and/or negative biases with regard to particles of different morphologies. Further, optimally configured IA systems were not available when some of the referenced studies were conducted. More recent evaluations of presently available instrument systems [7] suggest that IA will allow light microscopic particle count assays to be automated, with significant increases in the accuracy and precision of the data obtained.

In this chapter, operational factors in image analysis are considered first, followed by discussions of automated particle count assays and morphologic analyses that emphasize factors impacting accuracy and precision. Summary data obtained with an IA system are presented to provide a comparison between manual and automated particle counting. In the interest of simplifying the discussion, black-and-white or gray-level systems are emphasized; color imaging increases versatility regarding the range of analyses and the types of special applications that can be undertaken, but it also adds complexity that cannot be dealt with in the space allotted. Similarly, no effort is made to discuss the wide range of image analysis applications that do not relate to particulate matter, such as video and photographic image analysis and processing.

II. OPERATIONAL FACTORS IN IMAGE ANALYSIS

A. System Operation

Analysis of particulate matter by automated microscopy is based on a well-defined series of sequential operations performed by the instrument system. These are

1. Video image acquisition
2. Digitization of the video image
3. Image preprocessing
4. Image segmentation
5. Particle measurement
6. Particle sizing
7. Data presentation

With an automated microscopic system, image acquisition is carried out by a video-type TV camera that accepts a portion of the magnified image field produced by the microscope. The analog output signal produced by the camera may be processed so that the resultant image can be viewed on a video monitor. In reality, the camera signal constitutes a digitized form of the image, but technically the image is considered to be digitized only after it has been converted to a gray-scale image by an image digitizer (analog-to-digital converter) with a fixed gray-scale resolution (e.g., 8 bit or 256 gray levels). Preprocessing routines are used to provide an image more suitable for analysis. Shading corrections and contrast enhancement using digital filtering are two commonly applied preprocessing steps. For purposes of measurement or counting, images of particles on a substrate are most frequently segmented into intermediate binary images in which all particles (regardless of light or dark color) are segmented into either black or white, and the background is assigned the opposite gray level (binary or monochrome segmentation). Finally, particles are measured, and the data for a given analysis is printed or sent to disk storage.

B. Video Signal Generation

The basic element in video microscopy is the formation of an electronic image from the visual (analog) image produced by illumination of the subject of interest. A video camera (Fig. 2) provides the mechanism by which the light optical image is converted to a video image that may be subjected to image analysis. The reader should be aware that a significant number of different camera systems are available; the one shown in this figure is intended only to describe basic principles of image generation.

In the camera tube in Fig. 2, the light image falls on a photoconductive layer on a target electrode. This photoconductive layer is a transducer, the resistance of which is lowered as a function of the intensity of illumination falling on its surface. The electron beam sweeps the target electrode, charging up its back surface with electrons. In the dark, the photoconductive layer is essentially an insulator, so the electric charge remains on the surface of the target. When light strikes the photoconductive layer, the resistance of the layer drops. Electron-hole pairs are formed and stored in the regions of the photoconductive layer that are illuminated. Within the dynamic range of the camera tube, the drop in resistance of the photoconductive layer is approximately proportional to the intensity of illumination.

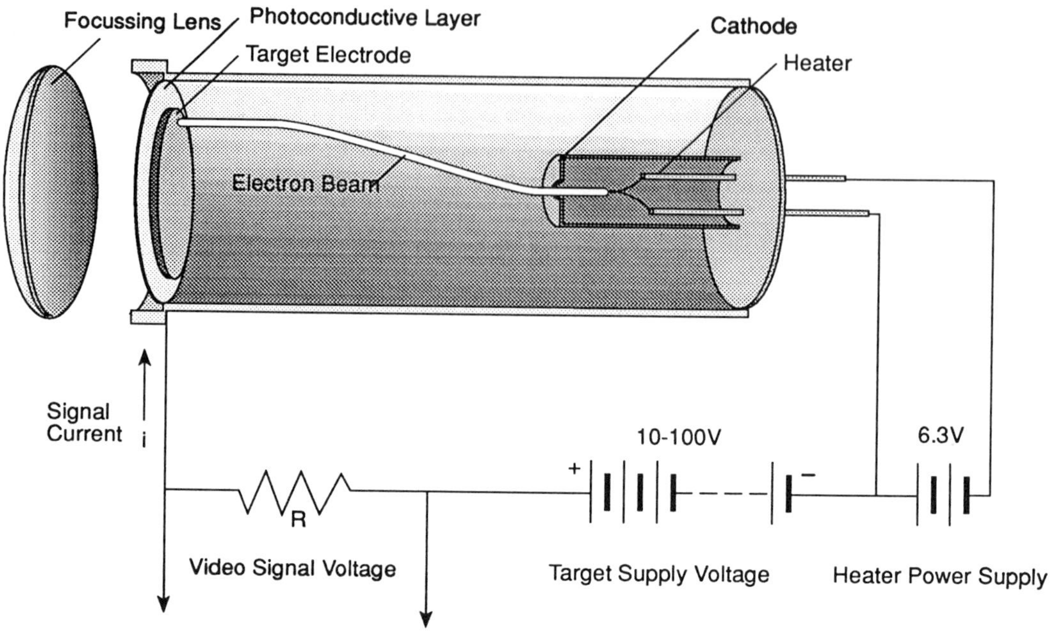

Figure 2 Video camera tube.

The target electrode of the tube in the figure is set at a potential 10 to 100 V positive relative to the electron gun or the cathode. When the electron beam lands on a particular illuminated target area, the lowered resistance of the photoconductive layer instantaneously allows a current of from a few to a few hundred nanoamperes to flow. As simplistically shown in the figure, the current (i) flows from an external positive voltage supply through a resistor (R), the target electrode, the photoconductive layer, the electron beam, the cathode, and then back to the negative terminal of the target voltage supply. The iR drop across the resistor gives rise to the video signal voltage, or the output signal. In a simplified system, the camera signal is used to modulate the signal via a beam current control grid on a video cathode ray tube (CRT or monitor) as shown schematically in Fig. 3.

The monitor is essentially a large, evacuated glass flask enclosing an electron gun, electrodes, and phosphor screen. The heated cathode in the electron gun, located in the neck of the tube, emits a beam of electrons. The high-voltage anode (10 to 25 kV positive potential) accelerates the electrons. The focusing coil and electrodes focus the electron beam onto the phosphor screen located on the faceplate or the front end of the picture tube. The accelerated electrons landing on the screen excite the phosphor. The excited phosphor in turn emits light from each spot illuminated by the beam, the brightness of each spot varying with the accelerating voltage and the current of the electron beam. The beam current is modulated by the amplitude of the camera video signal imposed on the grid as the beam traces the raster. Thus, each spot on the CRT phosphor is excited according to the amplitude of the video camera signal and the corresponding intensity of the beam, which represents image picture points (pixels).

The video image produced is based on the intensity modulation of the CRT electron beam from the signal produced by the camera. Each specific pixel location illuminated on the monitor face bears a precise spatial relationship to the corresponding image point in the camera field. The image produced by the camera and displayed on the monitor is composed of a checkerboard array of picture elements or pixels, the number of which is a critical determinant of the resolution

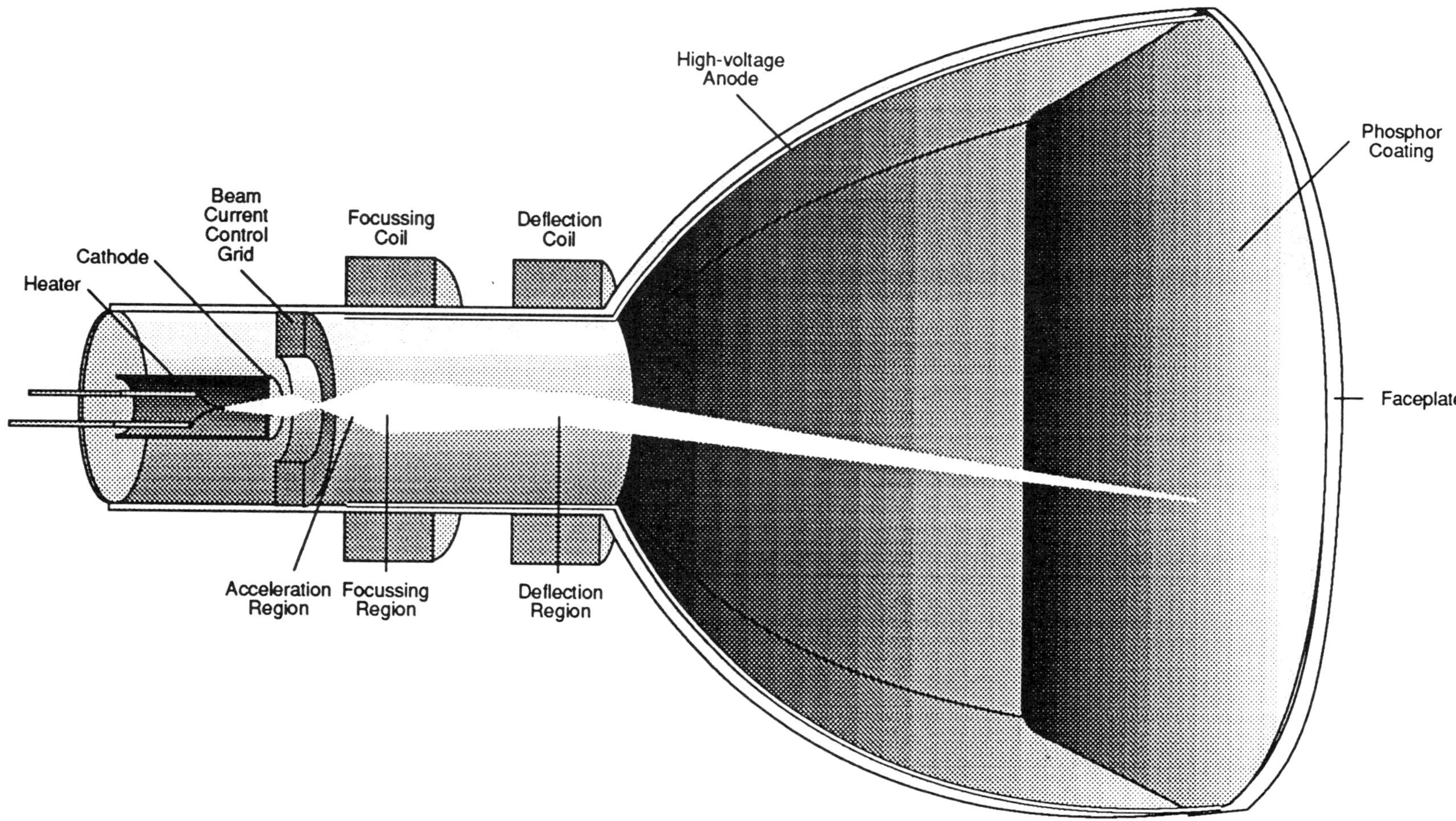

Figure 3 Operation of video monitor.

of the video system. The higher the number of pixels, the higher the level of detail attainable in the image. Again, this brief description represents a significant oversimplification of the process, but serves as a basis for discussion. The reader is referred to Inoue [1] and Baxes [2] for more detailed considerations.

C. Magnification and Resolution in Image Analysis

Particle counts and particle measurement are inextricably interrelated, since the count obtained in any given size range will depend on how accurately particles are sized. Technically, particle sizing occurs in two steps: the dimensions of a particle are measured by some means, and then the particle is assigned a size based on the measurement obtained (e.g., maximum linear dimension, equivalent circular diameter, or other). Whether our system consists of a human analyst and a microscope, or a microscope, video camera, and image processor, the measurement of a particle and the size assigned are critically affected by resolution and precision of this instrument.

Resolution describes the ability of a microscope system to portray fineness of detail in the magnified image it produces. It is commonly quantified in terms of the smallest dimension by which two object points or lines can be separated and still be observed as two objects (point resolution). If a microscope or IA system cannot resolve at a < 1-μm level, it is theoretically impossible to detect a difference between a 10- and 11-μm size particle. The theoretical resolution parameter of an IA system is complex and may be calculated based on a consideration of optical (microscopic), video, and digital resolution. The resolution of the system will be impacted by (1) the magnification and numerical aperture (NA) of the objective lens; (2) the magnification of the relay (projection) lens, which produces a focused image on the video camera target; (3) the vertical and horizontal linear resolution of the video camera tube; (4) the resolution of the digitizing process; and (5) tonal or gray-scale resolution. Ensuring that a system under consideration will be theoretically capable of a desired level of resolution involves four steps:

1. Choose an objective lens with adequate resolution (i.e., < 1 μm).
2. Select a combination of video camera and relay lenses that allows the objective lens image to be electronically recorded without degradation of resolution.
3. Magnify the image to the extent that two pixels of minimum size to be resolved are present for each feature in the video image and in the digitized image (Nyquist criterion).
4. Test the system capability with standard particles. This step is of utmost importance, since it provides a realistic assessment of system resolution based both on the calculated theoretical resolution of the system and the precision with which particles are measured and sized.

The concept of point resolution, commonly used in microscopy texts, fails to adequately explain the critical effects of IA system resolution on particle sizing and counting. Higher-resolution values not only signify that points or lines more closely spaced can be resolved, but also indicate that objects in the specimen can be more clearly visualized and more precisely measured. The degree of resolution necessary to visualize a particle is generally far below that necessary for its accurate measurement. The Airy disks produced in a microscopic image by diffraction effect constitute an area of uncertainty that is apparent as a blurring or halo at a particle's periphery. Since, by definition, higher resolution imaging results in narrower Airy disks, particles visualized with higher resolution systems will have more sharply defined margins. Thus, resolution initially affects particle sizing and detection primarily through its effect on edge sharpness and contrast. The indistinct edges of particles visualized at low resolution will affect the precision and accuracy of size measurements made by both manual and automated means. With manual measurement, the indistinct edges of poorly resolved particles cannot be precisely compared to the lines on an ocular micrometer. With an IA system, the edge uncertainty results in a lower contrast slope at particle edges. This makes any form of measurement less reproducible, and causes slight differences in illumination and contrast between particles and background to have a pronounced effect on data.

The speed of the automated IA systems in particle counting allows the use of increased magnification (from 100× to 200×) as one means of dealing with the resolution issue. Obtaining improved resolution in the manual assay by this means results in a significant increase in the amount of time required for the assay due to the increase in the number of fields to be counted per sample at the higher magnification. As is the case with the manual assay, IA systems may also make significant errors in particle sizing due to the means of measurement used, and measurement error often becomes a more important consideration than resolution. It is critical to discriminate between the effects of resolution and precision of measurement on accuracy of particle count data collected either by an IA system or by manual analysis.

The finest gradation on the linear ocular scales most often used for manual particle sizing corresponds to approximately 10 μm at 100×. This means that to designate a particle as > 10 μm or < 10 μm the analyst must mentally divide the ungraduated 10-μm increments. Mental division of the ungraduated space into less than five parts is extremely difficult to carry out. Resolution of better than 1 to 2 μm is, therefore, unlikely to be attained. Thus, on a practical basis, a particle must be < 8 μm to be reproducibly sized as < 10 μm, and > 12 μm to be dependably sized as > 10 μm. This defect, coupled with the fact that the analyst must mentally transpose particles within a field to place them against the vertically oriented scale, results in poor precision in sizing. The reproducibility of particle sizing based on this means of measurement is generally considered to be ±20% at best for 10 μm particles at 100× [14]. Errors in particle sizing using this subjective method of linear measurement can potentially result in a significant variability in test results.

D. Objective Lens Resolution

The most commonly used lens arrangement with video microscope systems involves a 20 to 40 power objective and a 1 to 1.5× relay lens, which transfers the image to the camera. Optical characteristics for typical achromatic and apochromatic objective lenses are presented in Table 1. From

Table 1 Optical Characteristics of Objective Lenses

Nominal magnification	Numerical aperture	Diameter of field (mm)	Resolving power, white light (μm)
Achromat			
5	0.10	5	3.00
10	0.25	2	1.22
20	0.50	1	0.61
40	0.85	0.6	0.36
Apochromat			
4	0.16	5	1.91
10	0.32	2	0.95
20	0.65	1	0.47
45	0.95	0.5	0.32

Values included in this table do not represent maximum performance figures for lenses; 10× apochromats with a NA of 0.4 and 45× apochromats with a NA of 1.0 are available.

this table it can be seen that the lowest power objective lens affording the desired 1-μm optical resolution is a 10× apochromatic objective. An objective lens with the lowest possible power is often desirable, since working distance, depth of field, and field width all decrease with increasing magnification; smaller image fields significantly increase the time required to scan a membrane.

Resolution calculations used to provide the data in the table are based on point source optics and transmitted light according to criteria of Abbe and Rayleigh. A pertinent consideration is that the actual resolution of microscopic or image analysis of particles on a filter performed using reflected light is less than that calculated for transmitted light.

The microscopic resolution of the IA system critically impacts the measurement of particles in IA through its effect on segmentation. Segmentation as a general term describes image processing operations that separate (segment) dissimilar image features such as particles and the background. With regard to automated particle measurement and counting, the segmentation boundary of interest is the perimeter of a particle being sized. Consistent segmentation is the key to precise and accurate particle sizing. Microscopic resolution effects on particle boundary definition and edge contrast are shown in Fig. 4. This figure illustrates edge effects noted with high-

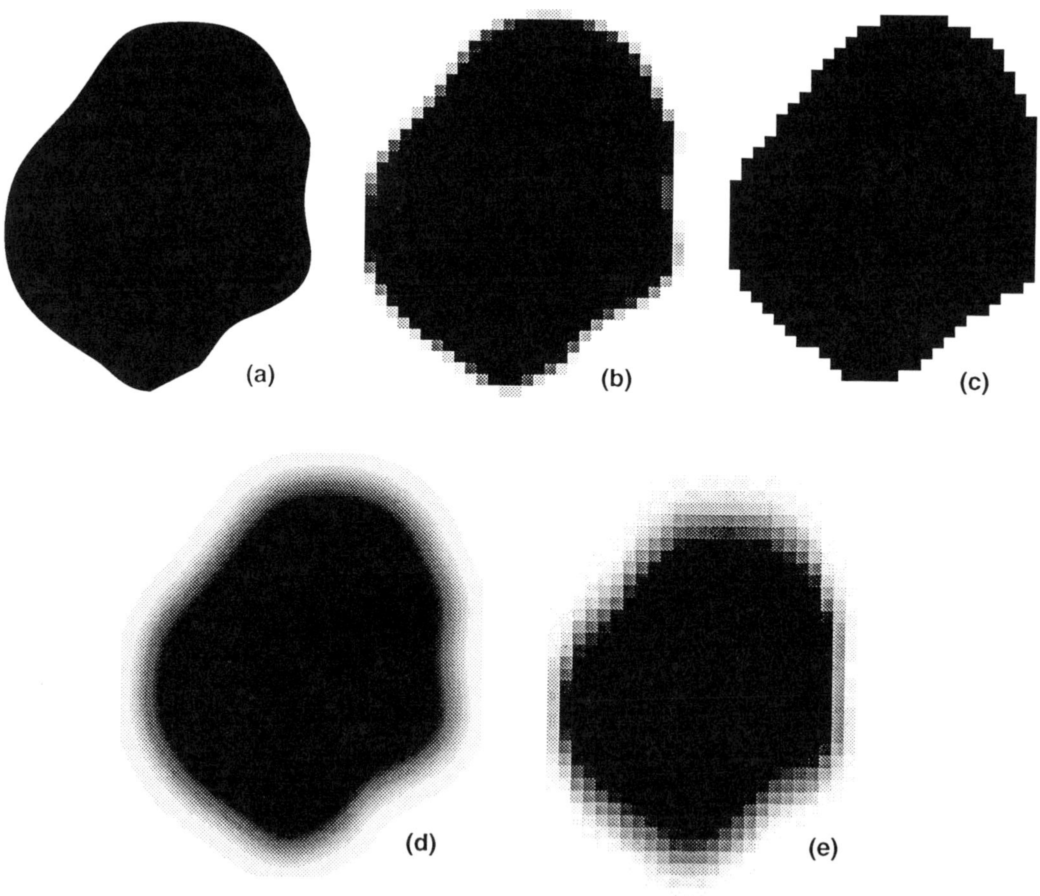

Figure 4 Effects of resolution of digitization of particle image: (a) high-resolution analog particle image; (b) high-resolution digitized image; (c) binary image of *b*; (d) low-resolution analog image; and (e) low-resolution digitized image.

and low-resolution images of the same particle (4a and 4d). The high-resolution digitized image (4b) of this particle incorporates an ideal edge pixel boundary and sharp edge contrast. Figure 4c illustrates the binary image of 4b. Figure 4e, a low-resolution digitized image, incorporates a greater edge uncertainty that will consequently make for a greater uncertainty in segmentation and measurement. Although both the high- and low-resolution images can be recognized by the system as a particle, the image of 4b will yield more reproducible measurements.

E. Camera Resolution

This selection of a camera with suitable resolution may be made by first converting both the microscope resolution and the resolution of the camera to frequency (f) response in cycles (2 lines) per millimeter, with the finest detail in the image considered to be the highest frequency component. Thus, for a 10×, 0.38 NA apochromat with 0.88-μm resolution, the frequency resolution of the objective lens is

$$
\begin{aligned}
f_I &= \frac{2\text{NA}}{\lambda} \\
&= \frac{0.76}{5.46 \times 10^{-4}\ \text{mm (green light)}} \\
&= 1400\ \text{cycles/mm}
\end{aligned}
$$

For this objective lens (in combination with an IA relay lens and a high-resolution 1024 × 1024 video camera with a 15-mm diagonal active area measurement) the resolution in terms of frequency response is

$$
\begin{aligned}
f_C &= \frac{M_{\text{objective}} M_{\text{relay}} N}{1.2D} \\
&= (10)(1)(56.8) \\
&= 568\ \text{cycles/mm}
\end{aligned}
$$

where M = magnification
N = number of lines of resolution on vertical axis
D = tube diagonal measurement

Therefore, for this system there is a significant decrease in the resolution of the objective lens due to the video imaging process. A very simple solution is to increase the relay lens magnification to 2.5×. This will effectively increase the vertical spatial frequency capability of the system to 1420 cycles/mm, which effectively matches that of the objective. This calculation is independent of the magnification required at the camera face to provide 2 pixels per the smallest dimension to be resolved (Nyquist criterion), and is merely intended to ensure that the resolution response is equivalent for lens and camera. The necessary magnification to ensure that two line pairs are present in the digitized image for each feature of interest (1 μm) may be calculated as follows:

Active area of camera face: 11 mm × 11 mm
Number of pixels in vertical direction: 1024
Desired sampling rate: 0.5 μm/pixel
Minimum necessary magnification:

$$
\frac{\text{Lines/image}}{\text{Sampling rate}} = \frac{1024/11{,}000}{0.5\ \mu\text{m/pixel}} = 21.5\times
$$

Thus, the 10× objective with 2.5× relay lens system is adequate both to provide camera resolution equal to lens resolution and adequate resolution on the camera face.

F. Image Digitization and Pixel Resolution

Another element of the IA system that critically impacts resolution is the digitization process. This resolution is typically specified in terms of numbers of horizontal and vertical lines of picture elements. The best camera and digitization resolution currently available is 1024 × 1024 lines. The rule of thumb applied is that the smallest image element to be resolved must be represented by at least two pixels in the digitized image. So, if 1 μm is to be reproducibly resolved in digitization using a 1024 × 1024 camera pixel array, no more than a 500-μm field width can be represented on the total digitized image. If a 512 × 480 matrix is used, maximum field width is approximately 250 μm. Lens field, camera field, and digitized image for the 1024 × 1024 line system are related (Fig. 5).

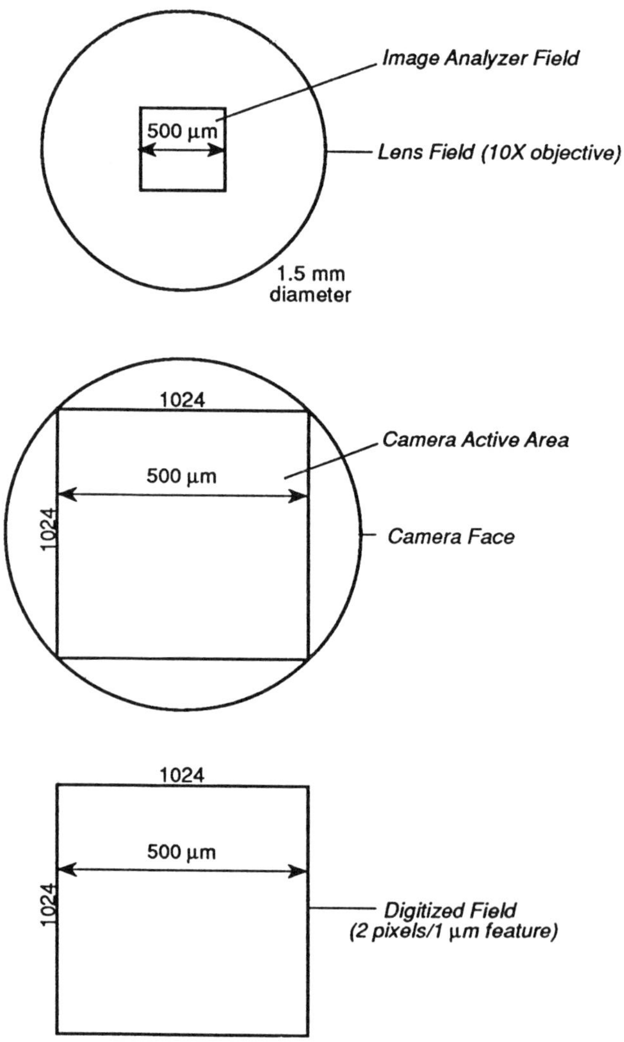

Figure 5 Image transfer and digitization.

For the electron portion of the IA system, the pixel number in the processed image determines spatial resolution. Since a digital image processor must work with digitized (discrete) data, this image is not a continuous-tone image, but rather consists of discrete gray levels at individual pixels. Each pixel is assigned a numeric value based on its brightness, ranging from black through the gray scale to white. Additionally, each pixel is assigned coordinates describing its location within the image.

Thus, an image field is digitized into a square grid of pixels, each of which is defined with a pair of coordinates referring to a vertical (column) and horizontal (row) position. Numbers range from (0-0) at the top left side of the image, to (*N*), where *N* is the number of vertical columns or horizontal rows in the image. For example, a pixel with coordinates (150100) is located at the intersection of column 150 and row 100. When dealing with the digitization of an image, there is always the question of how accurate the representation is compared to the original. The accuracy or fineness of the digitization process is defined in the spatial resolution term for the digitizing process, which is a measure of the many pixels the digital image is divided into. The greater this resolution, the more closely the digitized image will represent the original image.

G. Image Processing and Gray-Scale Resolution

In addition to spatial resolution discussed earlier, contrast or gray-scale resolution also impacts particle detection and measurement. The concept of gray-scale resolution involves the number of gray levels into which the contrast in the specimen is broken down. This measurement describes how well the digitized image portrays differences in the tonal values in the original image. Typically, high-performance light-microscopic IA systems have a brightness resolution of 8 bits (2^8) or 256 gray levels, provided that the electronic noise level is below the lower gray levels. IA systems (depending on gray-scale range) can detect up to $\frac{1}{256}$ of the range of a contrast between black and white (0.39%), whereas a 5% difference is the minimum generally detectable by the human eye. This finer discrimination is not necessarily useful in particle analyses since sharp edge contrast, which enhances both detection and sizing, is desirable. Contrast of an image is simply a measure of difference in gray levels between features of interest. The greater the gray-level number, the more accurately the digital image will represent the tonal range of the actual specimen. The differences in gray levels are used to separate image features (particles) from background (filter), and therefore play an important part in the analysis.

With regard to automated microscopic analysis of particulate matter, the fundamental problem to be solved regards separation of particles from the background. The situation is somewhat different from that in general image analysis where a wide range of tonal values may be necessary to produce a realistic image of the subject of interest, such as the landscape of the planet Mars. To enumerate particulate matter with microscopic image analysis, the issue becomes one of carefully selecting the gray level or threshold at which an image feature will be recognized by the system as a particle. For nonmorphologic analysis where one is not concerned with particles identified and measured, the ideal situation is one in which the particle and background show a maximal difference in contrast. While this situation is easily achieved when the sample is a glass slide bearing a dispersion of opaque particles, accurate enumeration and sizing of particles on the surface of a nontransparent medium (such as a filter membrane) frequently calls for manipulation of the gray scale through contrast enhancement.

Image processing is closely tied into gray-scale resolution. Technically speaking, an image has been processed when it has been changed in any fashion. Changes to the image are typically performed after it has been digitized and stored in image memory; these processing steps occur at less than video rate, and so become a limiting step with regard to speed of analysis. Preprocessing is frequently performed to correct problems of noise or shading apparent in the image. Random noise that takes on the appearance of snow can be reduced by frame averaging. This involves acquiring a number of frames and summing or averaging them together. This process reduces the effect of noise because the sum of the random noise signal increases with the square root of the number of frames averaged, while the signal amplitude is additive. Shadows or uneven illumination on the specimen negatively affect the accuracy of image segmentation. Shading can be corrected by an analog correction applied at the video camera, and/or by digitally subtracting the shaded area from the image to be analyzed. The analyzed image should show minimum shading and maximal contrast.

With regard to automated microscopy of particles, image processing is primarily useful in separating the contrast levels of particles from that of the substrate, thus enhancing particle detection. Image processing for contrast enhancement falls into two categories: point processing where the output pixel value is directly dependent upon the input pixel, and the filtering process by which the output values are determined independent of the input pixel. Point processing is carried out by manipulation of the gray-scale histogram. Every digitized image has a unique gray-scale histogram relating gray level to pixel number. In the image of Fig. 6, the dark particle image is represented by gray-scale values centered around a value of 70. The background is represented by gray-scale values centered at 120. The light particle has a gray scale of about 175. This histogram may be manipulated in a number of ways to increase or decrease contrast, or to expand the gray scale. Histogram sliding (adding the same value to each pixel) or stretching (multiplying each pixel value by a constant) are simple processing steps that increase image brightness or expand the gray-scale contrast. Enhancement may be carried out through point processing using a nonlinear intensity transform function (ITF) as shown in Fig. 7. In this case, the ITF is used to compress gray-scale values of the particles and provide a sharper response for the background gray levels. Since, in our application, the desired enhancement involves increasing contrast of dark and light particles relative to the membrane, a two-level ITF may be used that compresses the gray scale for both light and dark particles.

Filtering to increase particle edge contrast may be carried out using Laplacian or sharpening filters that fall into the category of spatial convolutions. A convolution operation on a digital image involves a point-for-point multiplication of selected pixel values of the input image times the corresponding pixels in a second image (known as a convolution mask or a kernel). Convolution masks typically contain an odd number of rows and columns and are usually square,

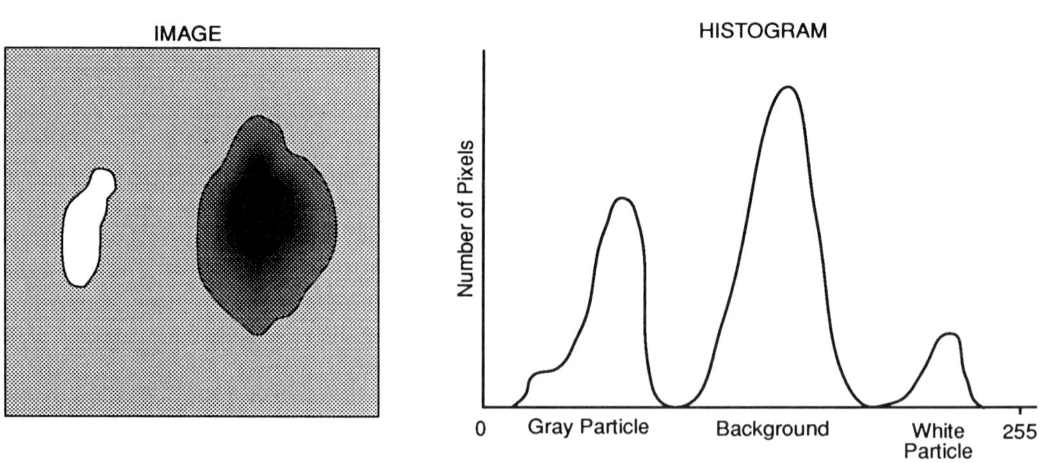

Figure 6 Gray level histogram.

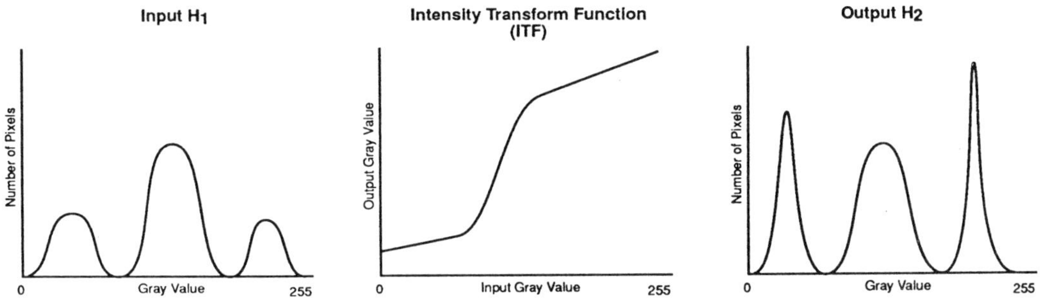

Figure 7 Intensity transform functions.

with a 3×3 mask being a common form. The convolution operation is performed individually on each pixel of the original input image and involves three sequential operations as shown in Fig. 8. In the first step, the convolution mask is overlaid on the original image so that the center pixel of the mask is matched with the input image target pixel. Then each pixel value in the original image is multiplied by the corresponding value in the mask. The gray value of the target pixel is then replaced by the sum of all the products determined in the second step. To perform

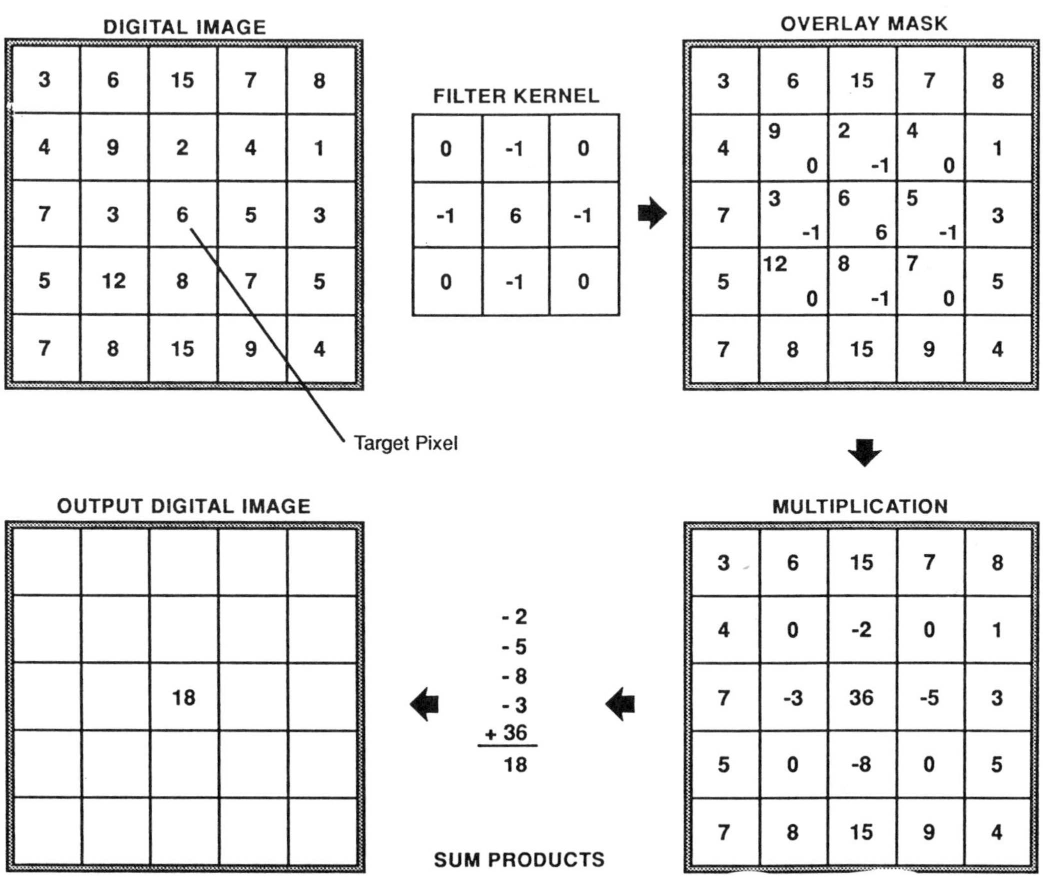

Figure 8 Point processing.

a convolution on an entire image, this operation is time consuming, since it must be repeated for each pixel in the original image.

Sharpening and Laplacian filters enhance higher spatial frequencies in an image while suppressing lower frequencies. They enhance object boundaries and also remove low-frequency variation of background shading in an image, and so can be used to correct for shading distortion in an image. These filters may also have the undesirable effect of enhancing noise.

Sharpening and Laplacian filters can be adjusted in size in order to have maximum effect on different spatial frequencies. The 3×3 masks have their greatest effect on image features that vary greatly over the spacing of a single pixel. In order to enhance spatial frequencies that vary more slowly in an image, these filters can be scaled to match the desired frequency. To double the size of a 3×3 mask, each coefficient is replaced by a 2×2 matrix having the same value (creating a 6×6 mask).

The most powerful techniques of digital image processing involve Fourier analysis. Fourier techniques are somewhat complicated, but offer a useful perspective for approaching and understanding the fundamentals of image formation. The basis for Fourier techniques is the transform, which is an operation based on a theorem that states that any harmonic function can be represented by a series of sine and cosine functions that differ only in frequency, amplitude, and phase [15]. An image is generally viewed in the spatial domain. Frequency transforms, such as Fourier, allow the frequency content (domain) of an image to be viewed. They break up an image into frequencies, or sum up frequencies to form an image in space. An image can be converted from the spatial domain to the frequency domain, and vice versa, using the Fourier transform. A fast Fourier transform (FFT) is a fast computer algorithm for implementing the Fourier transform; even with today's computers it is a time-intensive routine.

As discussed earlier, particle margins and their clear definition are of critical importance regarding particle segmentation and measurement. One specific type of image processing is extremely useful in dealing with disjointed or incomplete particle boundaries (edge segments) that may occur with agglomerated or complex particles and fibers. This method is based on an expansion-contraction (erosion-dilation) routine that defines complete particle margins where margins are indistinct or incomplete. In this routine, discontinuous edge segments are made to expand or dilate by artificially surrounding each white pixel in a binary edge image with eight white pixels. This boundary-growing process will eventually cause disconnected edge segments to contact each other. Once a connected edge is established, the dilated area is eroded to form a boundary line a single pixel in width by removing all white pixels that can be deleted without causing a dark pixel to appear between adjacent white pixels. This is one of the earliest routines developed for image processing and is available with almost all instruments.

Even simpler point processing routines require thousands of calculations: 512 pixels $\times$ 480 pixels $\times$ 8 bits of brightness = 0.25 million calculations for a simple subtraction. Many more calculations are required for a spatial filtering. Most IA systems accomplish this number crunching rapidly with dedicated pipeline or array image processors, which are computers designed to handle large arrays of numbers at high speed, typically 1/25 s for simpler operations. The final processing step performed before counting and sizing the particles is image segmentation.

H. Empirical Testing of Resolution

Table 2 shows the result of an empirical test of resolution. Ten of these values were collected in two different field areas with an irregular particle in a single rotational position; the remaining 10 values were collected after rotating the particle by approximately 20°. The mean of measurements of maximum chord for the light microscopic system was 9.93 μm; the maximum chord of the particle measured by scanning electron microscopy was 9.87 μm.

Table 2 Empirical Test of Resolution

Measurement number	Minimum projection (μm)	Maximum projection (μm)
1	7.82	9.76
2	8.12	10.92
3	8.17	10.81
4	8.45	10.71
5	8.17	10.67
6	7.91	9.47
7	8.21	10.26
8	8.04	9.76
9	8.21	10.07
10	7.91	9.39
11	8.68	10.26
12	7.46	9.19
13	8.47	10.26
14	7.36	9.35
15	7.35	9.59
16	7.50	9.99
17	7.51	9.77
18	7.20	9.35
19	7.24	9.59
20	7.20	9.34
	X = 7.84 μm	X = 9.93 μm
	SD = 0.46 μm	SD = 0.54 μm
	CV = 5.9%	CV = 5.4%

Particles in two rotational positions (SEM max. chord = 9.87 μm; SEM min. chord = 7.93 μm).

Although the theoretical resolution for the system used was approximately 1.6 μm, the actual resolution determined in the empirical test was better than 1 μm for both maximum and minimum chords. The measurement routine used was based on a calipering algorithm, with calipered measurements made following calculation of a convex perimeter for the particle. Although the Tracor (Noran) 8502 system used was not capable of sizing particles in increments of less than one pixel, data presented in Table 2 are carried to two decimal places. This is an interesting function of the measurement algorithim used. The convex perimeter calculation, in effect, rounds off the sawtooth pixel effect occurring at the edges of a digitized particle image on which measurement is being performed. The result is maximum or minimum chordal measurements in which fractions of a micron may be expressed in a continuous range of decimal values rather than in single fixed increments. Since floating-point mathematics are used in the algorithm applied to generate these data, a calipering measurement can, in theory, be carried to four to five decimal places.

I. Illumination

Both reflected and transmitted illumination should be available on a microscope used to perform particle analyses. Reflected illumination (Fig. 9a,b) is critical to analysis of particles collected in membrane filters. In episcopic illumination, a particle is illuminated by light passing through

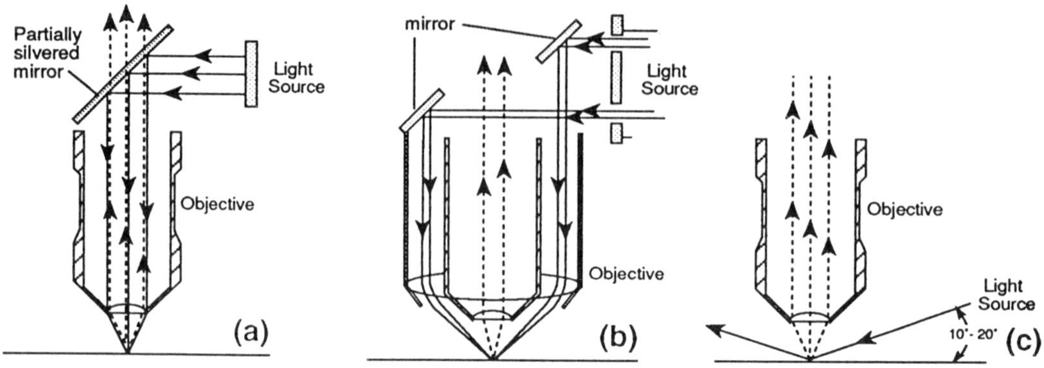

Figure 9 Reflected and episcopic illumination: (a) bright-field episcopic; (b) darkfield episcopic; and (c) low angle reflected.

the objective lens itself (Fig. 9a) or circumferentially down the objective lens assembly (Fig. 9b). The use of lateral reflected illumination from a single source (Fig. 9c) is generally undesirable, as previously discussed, since it results in shadows that must be subtracted by the system to derive the correct particle size. This manipulation adds significantly to analysis time. Darkfield episcopic illumination generally results in an image of low contrast and minimizes the apparent roughness of the filter membrane, but the image produced is difficult to use without contrast enhancement. Bright-field episcopic lighting provides excellent visualization of thin, transparent particles due to the strong specular reflection effect, but decreases contrast of irregular particles below that obtained with reflected illumination.

The best method of illumination is often found to be a combination of bright-field episcopic illumination and lateral reflected illumination from two sources spaced on either side of the sample. The use of episcopic illumination eliminates, to some extent, the orientation-dependence of the particle image produced. This system results in maximum contrast for both irregular (three-dimensional) particles and thin transparent particles. The analyst using this system can adjust the components of the illumination to suit a specific analysis. Fiber-optic illuminators are useful in obtaining optimal reflected illumination and allow a wide range of adjustments for light intensity and angular orientation.

III. PARTICLE DIMENSIONAL MEASUREMENTS

The reader must be aware that the division of particle-related image analysis endeavors pursued in this chapter is an artificial one. Measurements of particle size are related to particle shape, and particle number data derived from any form of analysis depend upon particle shape and size. As automated methods of particle shape analyses have become available over the past decade, a critical dependence of the behavior of particle populations on individual particle shapes has emerged. The application of image analysis in morphometric endeavors (i.e., the automated measurement of size and shape) is a highly specialized application of this instrumentation, and typically involves reduction of size and shape information to mathematical descriptors that are handled conveniently only by high-speed computers. Key references for additional reading include Beddow [16,17], Beddow and Meloy [18], Luerkens [19], Kaye and Clark [20], and Kaye [21]. Of specific usefulness in understanding morphometric concepts are Beddow [22] and the text by Kaye [23] on characterization of fine particles. I am indebted to Dr. John Keith Beddow of The Shakespeare Corporation for his assistance in providing the mathematical descriptors of particle shape and some of the figures used in this section.

A. Area Measurements

Initially, the concept of particle size was concerned with the dimensions of a particle. One early technique of measurement involved photographing the particle image in a microscope, developing a negative of the image, projecting the image onto a piece of graph paper, and then counting the number of squares contained within the particle boundary. Alternatively, the particle profile was traced on graph paper and the profile cut out and weighted on an analytical balance. The weighing method accounted for portions of squares not contained entirely within the boundary. A planimeter that measures area while tracing the perimeter of the profile has also been applied for this purpose. These methods form the basis of modern image analysis. A modern version of the graph paper technique (available in several commercial image analyzers) uses a video camera to obtain the magnified particle image as a pixel representation. A computer simply counts the number of pixels in the particle image to obtain the area of the particle profile.

Several different definitions of size have been proposed to measure the cross-sectional dimension of particles. The projected area diameter (equivalent circular diameter) denoted as D_p (Fig. 10) is defined as the diameter of a circle having the same area as the particle. Most classical feature measurements are not invariant to orientation. A key feature of this definition of size is that this measure is unique and does not depend on the orientation of the particle as it is presented in the field of view. It constitutes the means of measurement applied both in light extinction particle counters, and in widely applied microscopic methods of particle counting [14,24].

Analytically the projected area diameter is described as

$$\iint_A dx\,dy = \pi\left[\frac{D_p}{2}\right]^2$$

or

$$D_p = \sqrt{\frac{4}{\pi}\iint_A dx\,dy}$$

This definition of D_p requires that the area of the particle be computed or measured. There are a number of good arguments for the use of the equivalent circle diameter. It corresponds generally to the measurement made in light extinction and electrical zone sensing particle counters where particles are sized based on equivalent circular diameter (HIAC, Climet) or equivalent spherical diameters (Coulter). The relationship of this diameter to others is well covered by Schroeder and DeLuca [14].

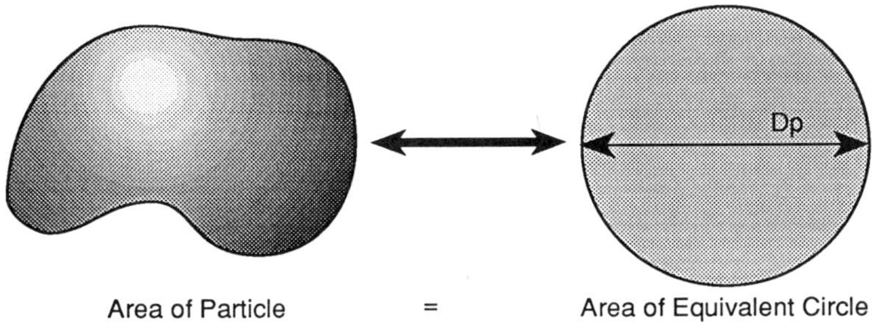

Figure 10 Projected area or equivalent circular diameter.

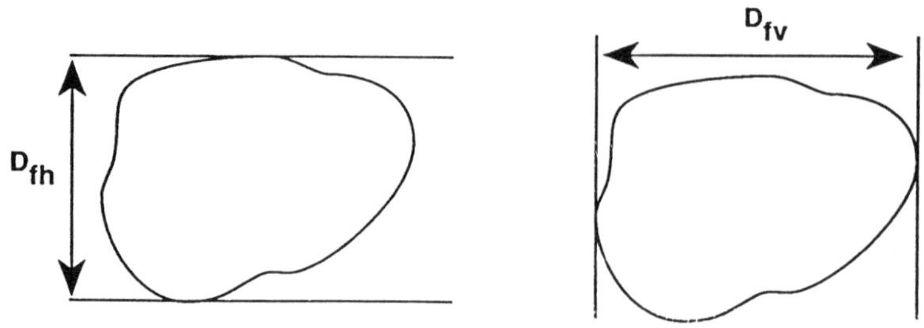

Figure 11 Horizontal and vertical Feret's diameters.

B. Chordal Measurements

Other definitions of size seen in classical analyses are Feret diameters (*Dfh* – horizontal, and *Dfv* – vertical). Feret diameters are defined as the distance between parallel tangents (Fig. 11). Feret diameters are easily measured and relate intuitively to a characteristic dimension (i.e., length) that may be important in processes such as sieving. A key feature of Feret diameters is that they are orientation dependent—when they are measured, no account is taken of the particle orientation as it occurs in the field of view. A single particle may possess an indefinite number of Feret's diameters differing or not, differing depending upon orientation when it is digitized and analyzed. If sufficient numbers of particles are measured, their random orientation will result in "averaging out," a phenomenon related to the Feret's diameter in population measurements. Two cases are illustrated in Fig. 12.

Another definition of size in classical analysis is Martin's diameter denoted by *Dm*, and is defined as the length of a chord through the particle centroid that divides the particle into two equal areas. Figure 13 illustrates the concept of Martin's diameter.

Using an image analyzer, Martin's diameter is determined by drawing successive diameters through the centroid of the profile and computing the bisected areas. With equal bisected areas, Martin's diameter is the length of the chord bisecting the two equal areas. A particle may or may not have a unique Martin's diameter. Figure 14 shows that a circle has infinite numbers of

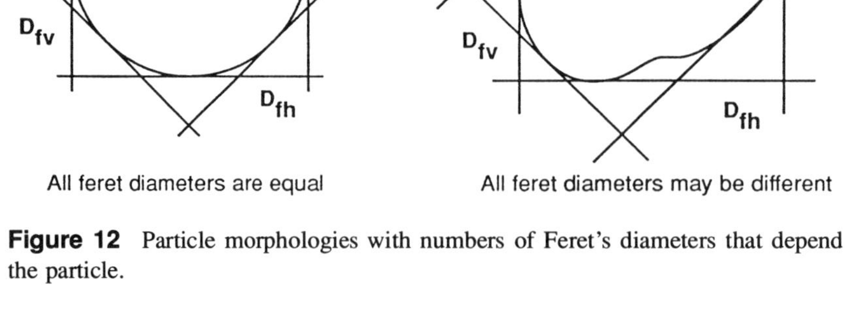

Figure 12 Particle morphologies with numbers of Feret's diameters that depend on the orientation of the particle.

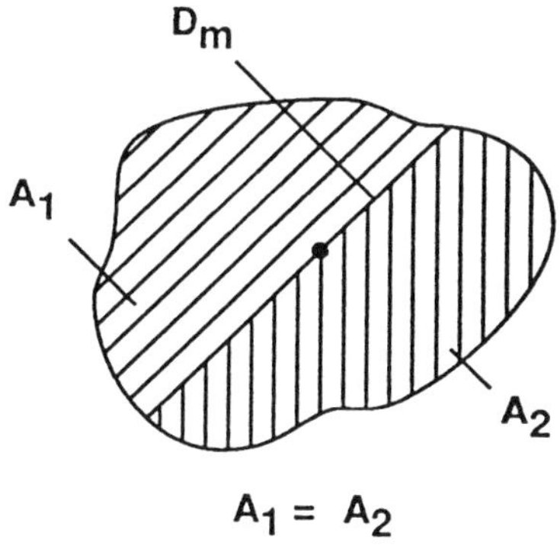

Figure 13 Martin's diameter.

Martin's diameters with the same value. Squares have a profile that may have an infinite number of Martin's diameters with or without the same value.

Other classical measures of size are limiting dimensions of a particle in decreasing order of magnitude. These are length (*L*), breadth (*B*), and thickness (*T*) (Fig. 15). Thickness is defined as the minimum distance between two parallel planes tangent to opposite surfaces. The breadth is defined as the minimum distance between parallel planes perpendicular to the planes defining the thickness. The length is the distance between parallel planes perpendicular to both sets of planes defining the thickness and breadth. The concept of thickness is an important consideration in morphometric analysis. Any light-microscopic measurement is essentially two dimensional, and we must base our assessment of shape on the two measurements that are available. Two general corrections are available: (1) assumption of a three-dimensional morphology that is predicted by the two-dimensional image; or (2) the assumption of some thickness less than that of a cuboid or equant configuration. In practice, the thickness consideration has proven to be amenable to both empirical and theoretical solutions, and shape analysis provides a powerful means of predicting or explaining the behavior of particle populations based on the two-dimensional morphology of individual particles.

Figure 16 illustrates that the measures of thickness, length, and breadth are not unique since all of the objects in the figure can have identical measurements of these parameters while having different measures of bulk.

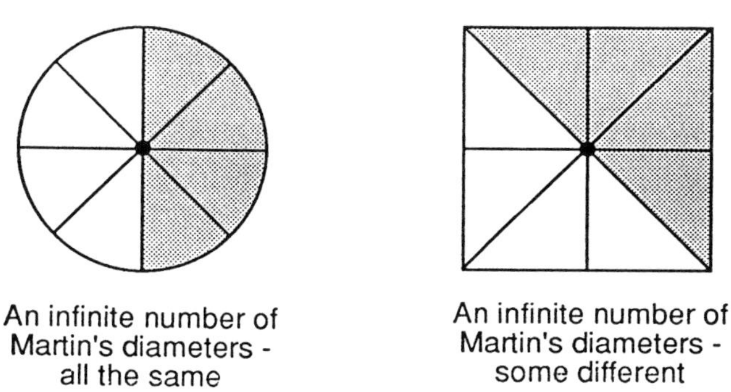

Figure 14 Martin's diameters.

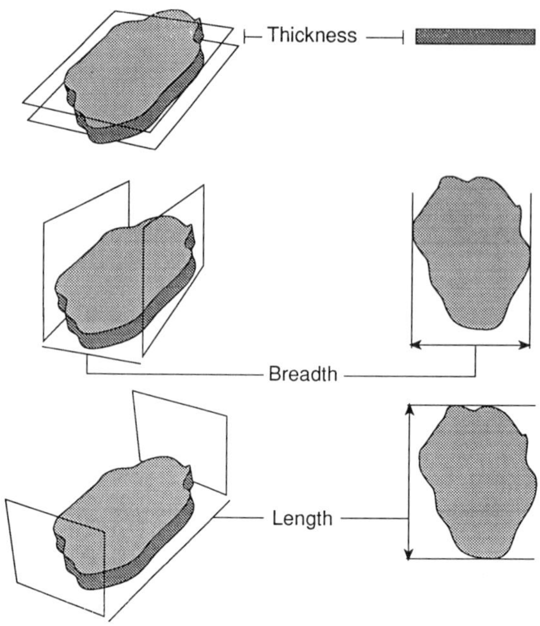

Figure 15 Length, breadth, and thickness as measures of classical size.

C. Perimeter Measurement

Another definition of size in classical analysis is the perimeter of a particle profile denoted *P*. Perimeter is defined as the distance around a particle profile (Fig. 17).

Analytically, the perimeter of a particle profile is defined as

$$\int_c ds = \int_c \sqrt{dx^2 + dy^2}$$

where *s* is total arc length/perimeter and *dx* and *dy* are coordinate differentials. Curve *c* represents the particle profile boundary.

Early techniques measured perimeter by use of a divider that stepped off a constant distance or step size along the boundary. The number of steps was recorded, and the perimeter was calculated by multiplying step size by the total number of steps. The accuracy of the

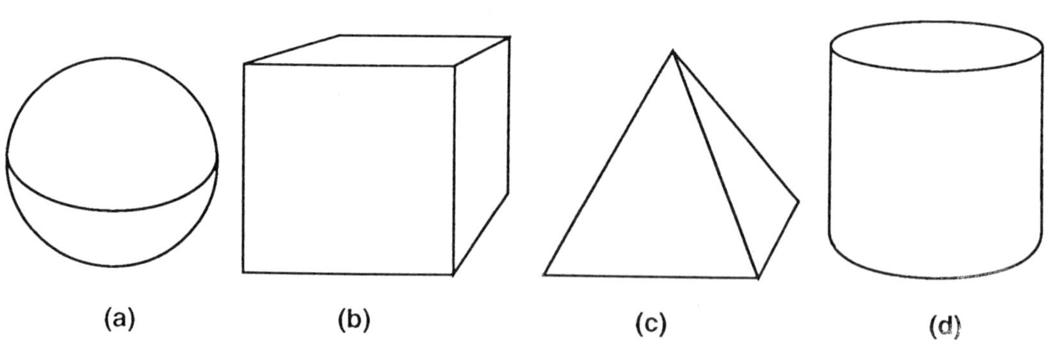

Figure 16 Identical measures of thickness, breadth, and length, but different measures of bulk.

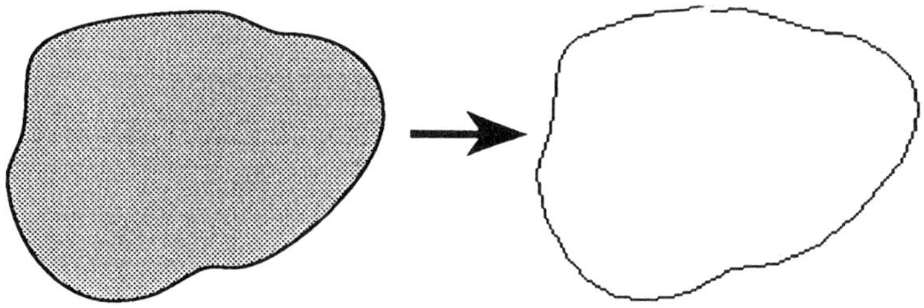

Figure 17 Particle perimeter measurement.

method depends on the step size used. Modern image analysis emulates this method by counting the number of pixels lying on the particle boundary. Perimeter is calculated by multiplying the total number of pixels on the boundary by the length of the pixel diagonal. Modern digital imaging systems approximate the perimeter from data (x,y) on the boundary using an algorithm of the form

$$p = \sum_{j=1}^{N} \sqrt{\Delta x_j^2 + \Delta y_j^2}$$

Perimeter does not depend on particle profile orientation as presented in the field of view. In differential geometry, perimeter or arc length is a natural parameter. Each size measure satisfies the concept of size as stated classically, illustrating the need to precisely state each concept for a clear operational definition.

IV. AUTOMATED PARTICLE SIZING

Generally, measurement of particle dimensions in IA is made based on one of three types of sizing algorithms that relate to the linear or area measurements discussed in the previous section:

- Intersecting chords
- Rotating coordinates (calipering)
- Grid (total area) measurement

These are shown in Figs. 18–20. Assignment of particle size is made either on the basis of maximum linear dimension (maximum Feret's diameter) as in manual assays, or on the basis of conversion of the particle area measurement to an equivalent circular diameter (diameter of a circle with equivalent area). A linear measurement made on a 10-μm particle has the potential to be in error at either end by 1 pixel. Hence for 1-μm resolution, pixels cannot be greater than 0.5 μm in width. A positive error occurs when an image feature of < 1 pixel dimension results in a whole pixel being turned on, or in a negative fashion when a dimension fractional pixel size fails to be represented in the image. Importantly, our earlier discussions of resolution assumed that the maximum dimension of a particle was accurately measured.

Measurement by intersecting chords is made as the image processor draws an array of radially dispersed intersecting chords superimposed on the particle image (Fig. 18). The angular dispersion between chords is inversely related to the number of chords (2 = 90°, 4 = 45°, 8 = 22.5°, etc.). A measurement based on intersecting chords is generally sufficient for irregularly shaped particles. As the aspect ratio of a particle increases, the probability that the maximal dimension of the particle will lie between the chords is also increased. For a fibrous particle, there are a number of orientations for which chords drawn with an arbitrary x-y orientation and

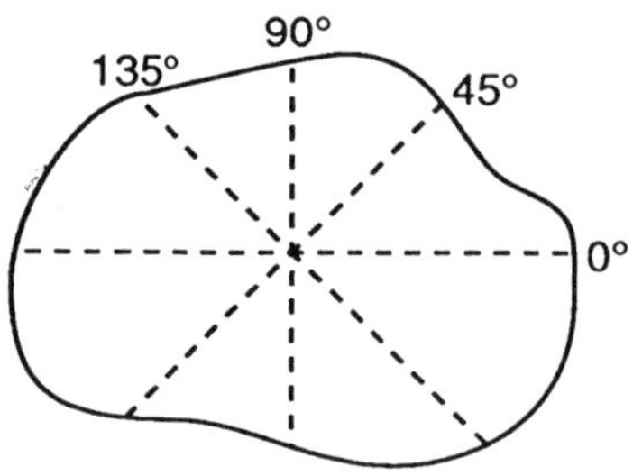

Figure 18 Measurement by intersecting chords.

angular dispersion will conclude without intersecting the end of the fiber and without providing an accurate measurement of length (Fig. 19). Thus, with fibers, the likelihood that the maximum particle dimension will be between chords is great, and increases as the total length of the fiber and its aspect ratio (length:width) increases. For an accurate measurement of maximum linear dimension, one chord of the array must contact the fiber image at both ends.

An example is a straight fiber that is 50 μm in length and 5 μm in width. If only four chords are used, a radial dispersion of 45° results. The minimal length of a chord necessary to reach one end of the fiber from the center point is 25 μm. At this distance from the origin, the linear distance between two 45° chords is approximately 19 μm, or almost four times the fiber width. In this situation, a number of orientations of the fiber would result in an erroneously low measurement. If only four chords are used, a particle 5 μm in width can be no greater than 12.5 μm in length if it is to be measured with some probability of obtaining an accurate measurement of total length. For our 50-μm × 5-μm fiber, the angular dispersion of chords can be no greater than 11.5° (18 chords) if the maximum length is to be measured. The impact of the use of a fixed chordal array for measurement of resolution and precision of particle sizing is obvious. Regardless of the ability of a system to resolve a 1-μm difference in the size of two particles, the use of a chordal measurement will result in a significant spread of measurement values when a single irregularly shaped or elongated particle is measured with different

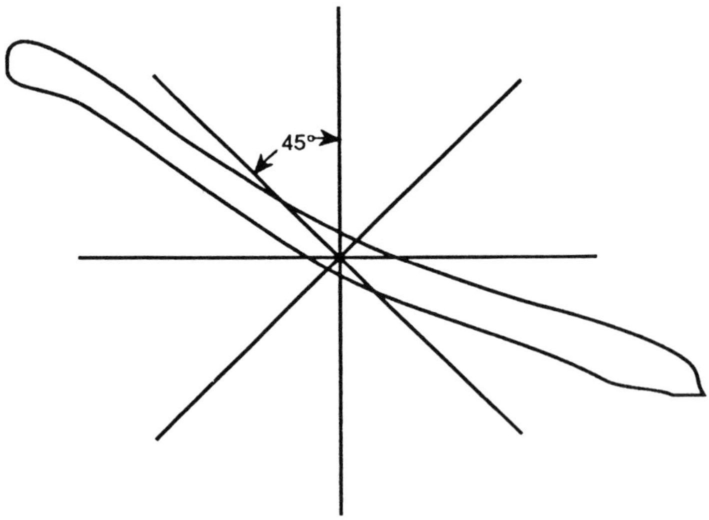

Figure 19 Measurement of fiber by intersecting chords.

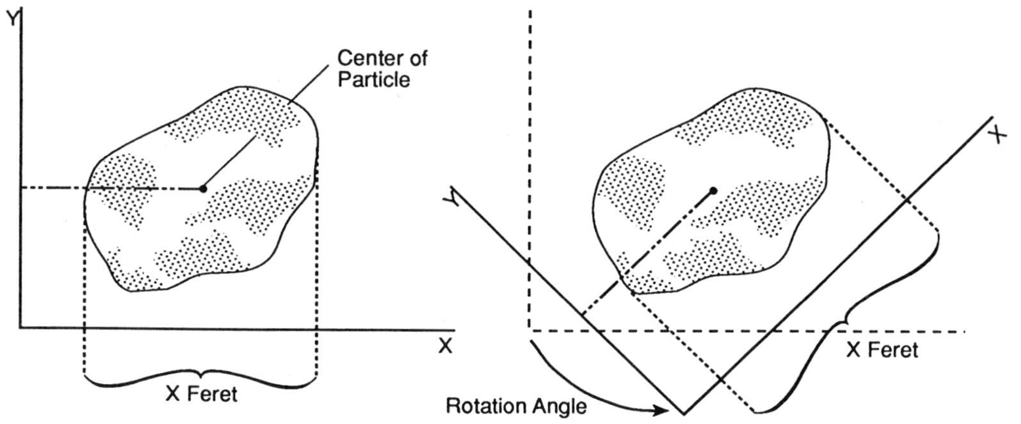

Figure 20 Measurement of particle by rotating coordinates (calipering).

orientations relative to the chord array. This measurement artifact is a critical issue, whether a static array of chords is used, or if a single chord is rotated through 360° with measurements made in fixed angular steps.

Sizing by rotating coordinates or calipering (Figs. 20, 21) imposes the same constraints on measurement and sizing precision as does the use of rotating chords. This mode of measurement is available with Noran feature analysis software. With this system, the number of diameters measured may be specified up to 180. This corresponds to 180 intersecting chords or a 1° angular dispersion. While having only a slight advantage for measurement of irregular particles, this mode of measurement has significant advantages if fibers are being measured, since a near maximum dimension will invariably be recorded. With either chords or rotating coordinates, any increase in the number of axes measured must be paid for in increased time of analysis.

Intersecting chords or rotating coordinates are generally appropriate if maximum linear measurements are of interest. Both of these measurements result in a perimeter map of the particle that will have a precision dependent on the number of chords or diameters measured. An interesting and simple alternative method of particle sizing is based on the fact that an area measurement may be made based on the sum of pixels composing the digitized image of the particle without performing any dimensional measurement (Fig. 22). Though the pixel counting

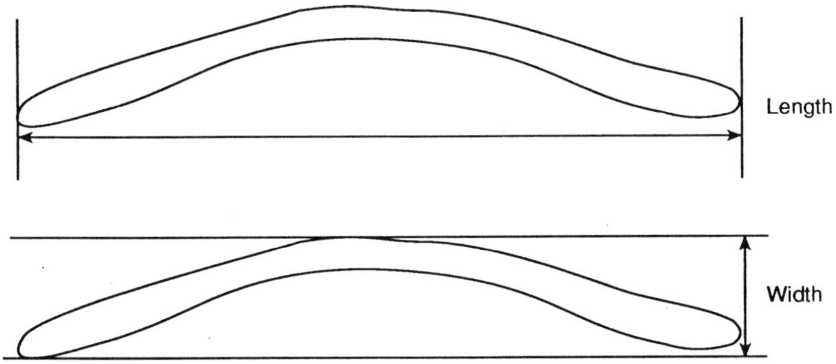

Figure 21 Measurement of fiber by rotating coordinates (calipering).

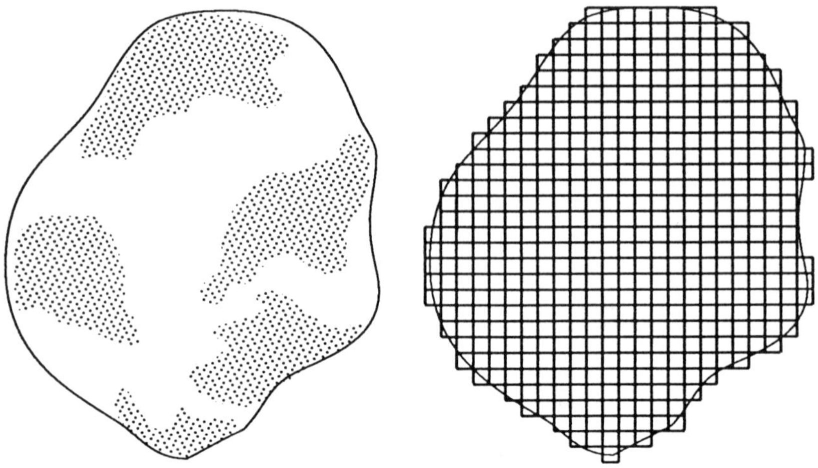

Figure 22 Pixel count measurement.

technique is the most rapid sizing technique, it is limited in the same way as the early projection methods in that fractional pixels on the boundary may either be counted as whole pixels or omitted from the count. This pixel count area measurement may be shape-normalized to the area of a circle. The diameter of the resulting circle constitutes the equivalent circular diameter of the particle.

The area measurement can be expected to show higher precision than a chordal measurement, particularly with an instrument that uses a simple sizing algorithm based on a small number of chords. This is true for two reasons. First, the precision of the chordal measurement in determining longest linear dimension decreases when irregularly shaped particles are sized and when simple sizing algorithms that use a minimum number of chords are applied. Area measurement suffers neither of these effects. Second, the statistics involved in making the area measurement are superior to the chordal measurement. A single linear chord measurement can easily be in error by two pixels (i.e., one pixel error at each end of the chord). This error will have a significant effect on how a particle is sized. In contrast, the area measurement is based on a much larger number of pixels so that a pixel error has a relatively insignificant effect on the final measurement. Furthermore, the large number of pixels used in the area measurement increases the likelihood that the loss and gain of peripheral pixels will balance out, since each occurs with equal probability. This error-canceling effect also increases the precision of the area measurement. In effect, the area measurement encompasses all possible chordal measurements that can be made on a particle. It should be observed that an area measurement may not be appropriate for elongate particles or fibers, since the equivalent circular diameter will be misleadingly low.

V. AUTOMATED PARTICLE COUNTING

The following discussion is based on the author's familiarity with a highly modified Omnicon image analysis system which was marketed for years by Bausch and Lomb and is now available from the Artek corporation. Both the software developed to drive this system and hardware components were specifically directed at the task of counting particle or membrane filters. Technical details of the use of this instrument have been described by Murphy [25].

Key issues which must be addressed in counting of particles by IA include:

- The area of membrane that cannot be counted due to the use of a square field matrix on a round filter
- Particles that overlay fields or extend from one field to another
- Agglomerated or overlapped particles

Generally, the scan grid can be arranged to miss very little area, so that the count error contribution from missed particles becomes inconsequential (Fig. 23). Agglomerated particles occur only infrequently due to the disruption of all but persistent aggregates by shear forces during agitation, pouring, and filtration of the solution tested. Thus, agglomerated particles that appear on the assay membrane should reasonably be counted as larger particles (aggregates). Overlapped particles are readily limited in number by dilution of the filtered material.

If particles on the entire surface of the analytical membrane are to be counted, it is essential that minimal magnification be used consistent with resolution requirements; the number of fields required for total coverage increases fourfold with each doubling of magnification. Whereas a field 500 μm wide and 400 μm high will fit into the filtration area of a 25-μm filter 1800 times, only 450 fields of 1000 × 800 μm will be required to cover the same area. The counting time for an analysis involving the smaller field will be approximately four times that required for the larger field, so considerable time may be saved if the lower magnification provides adequate resolution.

A significant issue is posed by particles that overlap the boundary of the counted field and extend beyond the area counted. This is particularly problematical with fibers. Generally, this may be dealt with in two ways. The simplest method involves the use of a guard region (Fig. 24).

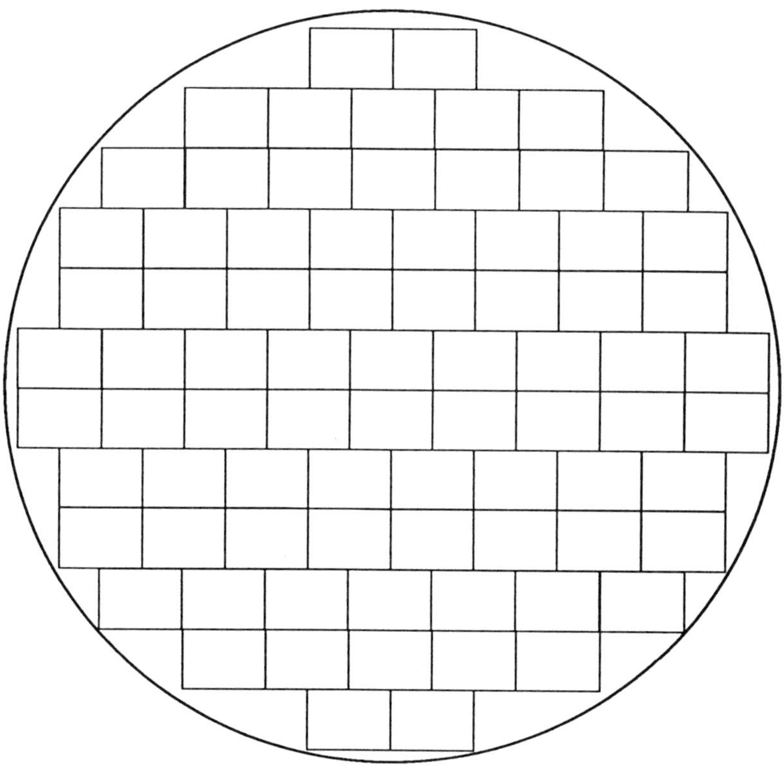

Figure 23 Adjacent fields scan matrix for filter membrane.

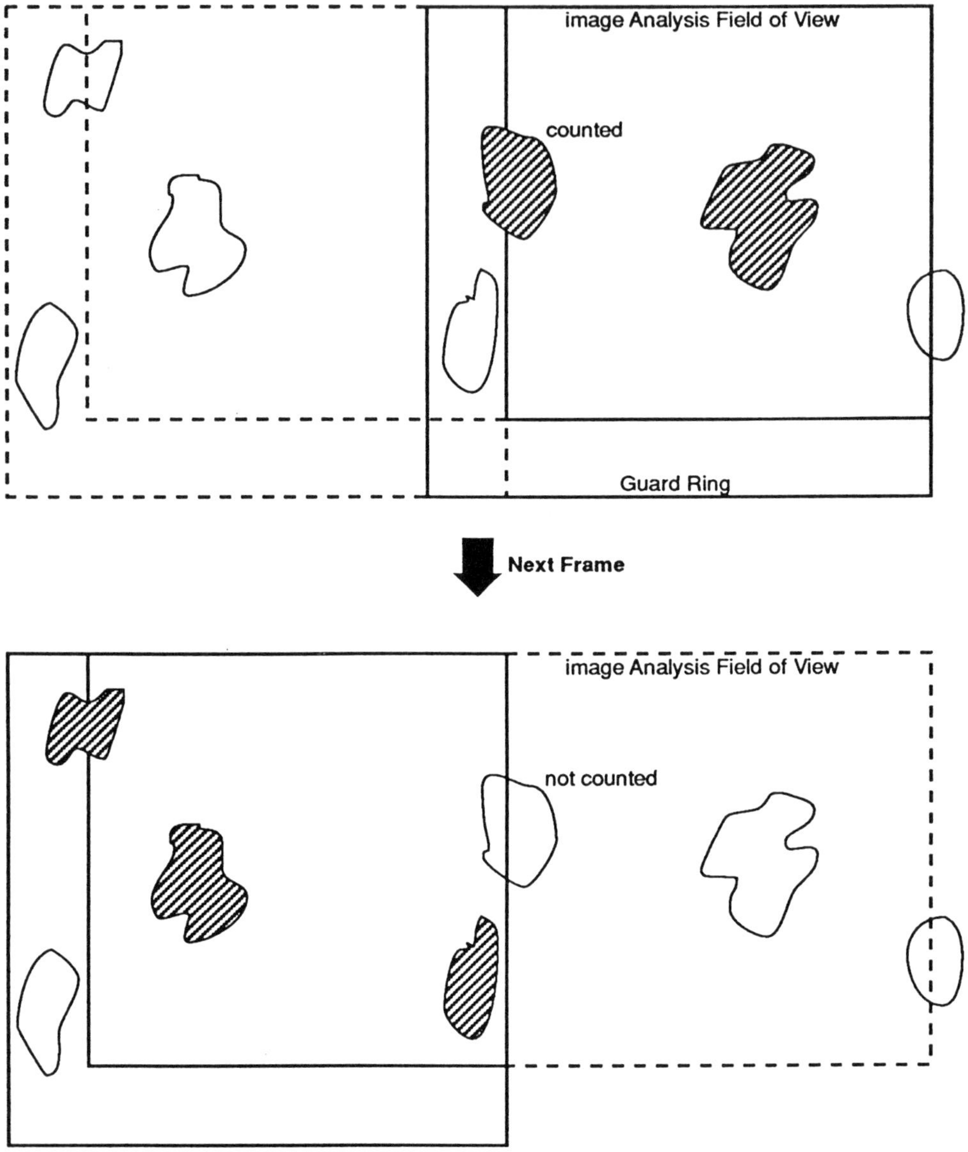

Figure 24 Guard region.

The field being analyzed is surrounded on two sides by a region into which overlapping portions of particles fall. Particles extending into this region are sized and counted, but particles extending into the analysis field from the previous field or one above are ignored. A more complex and somewhat less practical approach (Fig. 25) involves a shift of the analysis field to measure the portions of particles that extend into the analyzed field. This value is added to the portion of the particles counted in the original field to find the total size. This approach requires significant programming and computing time.

The automated stage movement for microscopic IA systems is carried out by stages that are moved in the *x-y* directions by stepper motors. The impulse of the motors is transferred to the

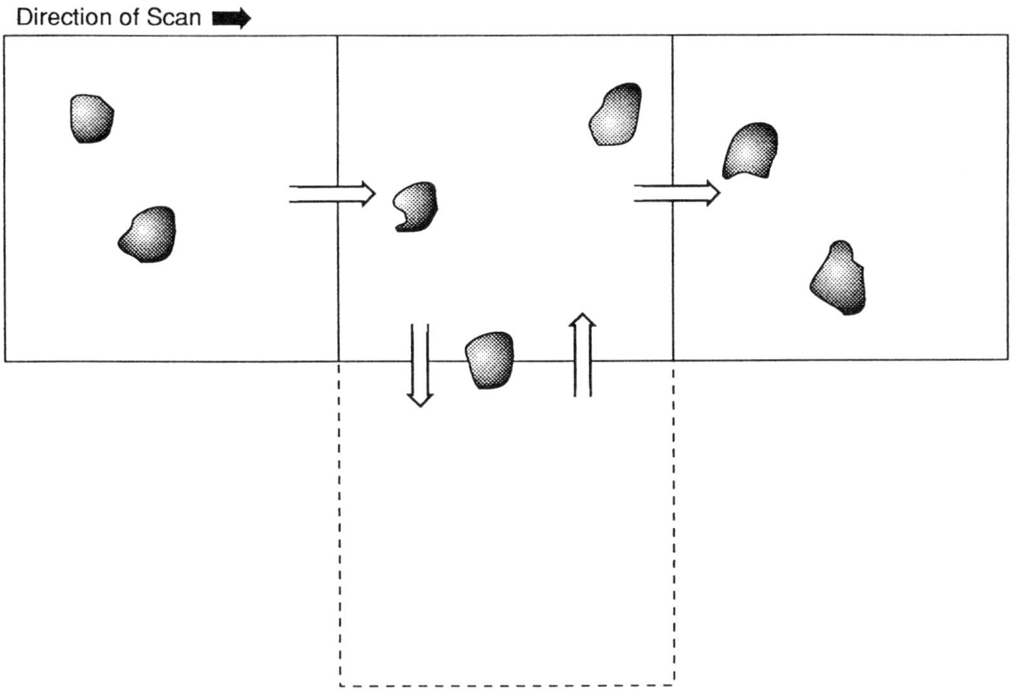

Figure 25 Field shift.

stage either by toothed belts or by a precision worm-gear drive. Stage drive precision (reproducibility of movement) will impact the extent to which the numerical analysis of particles on a filtration membrane is reproducible.

The precision of the better worm-gear drives is approximately five times that of belt drives (i.e., ±5 μm). With the latter system it is necessary to leave an area of significant width at the periphery of the filtration area unscanned to allow for variability in stage positioning. Worthy of note is that automated stage drive is of little value without automated focusing. The image in-focus position will change with wrinkles in the sample membrane, uneven thickness of sample support, and so forth. All of the autofocus mechanisms operate by maximizing image contrast. Focusing may be programmed to occur with a frequency appropriate to the sample being counted.

Particles analyzed using the Omnicon system are collected on green membrane filters that provide optimum color contrast across a wide variety of particle types. The worst technical problem encountered is that of filter surface defects (i.e., pits, scratches, or other surface irregularities) that give enough contrast to be mistaken for particles. We have used four different vendors for filters, but all supply filters of variable quality. Despite the use of green color contrast membranes, there are, invariably, small numbers of particles present that are so close to the membrane in image gray level that they are difficult to detect [25].

Regarding filter quality, it seems quite likely that filter vendors will be amenable to supplying filters of suitable quality if a commitment is made by the user to purchase them on a batch basis. A batch of cellulose acetate filters of the 25-mm size typically used will contain 5000 to 10,000 membranes and will sell for $4600 to $7000. This provides the vendor with sufficient economic incentive to exercise careful control over film casting and cutting to ensure the defect-free surfaces required. With this procedure, the buyer may demand a 1% to 2% defect rate (AQL). The defect rate can be expected to be from 15% to 30% in membranes from a routine production run. High-quality filters are a necessity to avoid prescreening of the membranes prior to use.

Since most large particles contain areas that differ in contrast level, some may register erroneously as multiple small particles. This is particularly true of fibers that often weave in and out of the microscopic focal plane. Particles in contact or closely adjacent to each other are assumed to be a single particle that is reflecting light unevenly (the most likely case) and are agglomerated for counting and sizing. Since reflected light is used with the Omnicon system, particles will cast shadows, and the shadows must be separated from the particles. Shadows that are cast due to low-angle reflected illumination are subtracted by the program, which screens the image for dark objects with the correct angular orientation to a particle. Some particles will overlap field boundaries. A particle recombination routine can compensate for this error source, but some error component in automated counts will still be attributable to this source. In any analysis, due to a summation of error effects, the Omnicon system will miss some particles and count some nonparticles. Thus, the instrument provides more precise data when relatively large numbers of particles are counted. In this situation, count errors have a minimal effect on the true count, and the number of missed particles will be compensated for by the number of nonparticles that are counted.

The many potential applications of automated microscopy in particle population analysis are obvious, based on (1) the speed of the analysis compared to manual microscopy, (2) the large number of parameters that may be measured, and (3) the range of computer data processing available. Our experience during eight year's use of the Omnicon system indicates that IA is a practical alternative to manual microscopy for population analysis. The automated analysis represents a significant improvement over the manual assay in that it is faster and gives more reproducible data. The data presented in Tables 3 through 5 serve to demonstrate the application of automated microscopy to the analysis of different types of particulate contaminants occurring in parenteral solutions. For this testing, particles were collected on 0.8 μm porosity green membrane filters. Prior to analysis of each sample, blank (background) counts were obtained on the filtration apparatus using 50 ml of 0.45-μm filtered, distilled water which was used to rinse membranes and filtration apparatus prior to and following sample filtration. Filtration was conducted at a rate of approximately 50 ml/min with the filtration funnel kept filled to half or more of its full volume during sample filtration; this procedure aids in obtaining a uniform distribution on the filter surface. To keep particles dispersed, the solutions being filtered were stirred during filtration with a magnetic stirrer.

The membrane samples obtained were examined manually at 100× magnification with 20° lateral illumination under conditions of the USP assay. A Nikon Optiphot microscope with a 10× objective lens was used (NA = 0.25; white light resolution = 1.2 μm). This lens theoretically

Table 3 Comparative Analysis of Particulate from LVP Solution Units (total count for 10 unit pool of 500-ml LVP units)

	Total particulate counts (μm)				
Analysis	> 10	> 25	> 50	> 100	Fibers
Manual microscopic	645	123	24	3	7
CV (%)	31	46	41	108	64
Omnicon system	692	144	31	4	8
CV (%)	5	11	17	23	14

CV = coefficient of variation.

Table 4 Reproducibility of Counts Using Omnicon 3500 Image Analysis System (10 runs per sample)

Analysis	Total counts (μm)			
	> 10	> 25	> 50	> 100
ACFTD	964	228	76	23
CV (%)	8	11	10	16
Glass spheres (NBS1003)	486	86	12	1
CV (%)	9	14	27	34
Nickel spheres	231	96	38	19
CV (T)	4	8	9	16
Fiber suspension	311	280	144	56
CV (%)	26	28	38	57
Mica flakes	593	372	204	131
CV (%)	24	31	34	37
Mixed latex suspension	842	586	129	44
CV (%)	6	11	14	16

satisfies an ~10% resolution requirement only for particles ≥ 10 μm in size. Analysts were allowed unlimited time to count membranes. Time of analysis for manual counting for the filters counted ranged from 7.8 to 56 min.

The Omnicon system was used with a 10× objective of the same type used by the analysts and a 2× relay lens (calculated resolution for system = 1.9 μm). Times required for analysis by the Omnicon system for the filter samples analyzed ranged from 4.5 to 14.6 min. The time required by the automated system is reflective of the time savings that results from counting membrane samples with larger numbers of particles.

The first test (Table 3) involved collection of total numbers of particles from a 5 L pool of 10 intravenous solution units of 5% dextrose injection in glass and plastic containers. The resulting analysis membrane was counted by seven different analysts, and seven times on the automated system with the instrument operating parameters being reset before each run. The data obtained from automated counts of particles collected on the filtration membranes approximated

Table 5 Sample Size Effect on IA Count Reproducibility with Mixed Particulate Suspensions (5 runs per sample)

Sample volume	Mean total count (μm)				
	> 10	> 25	> 50	> 100	Fibers
25 ml	36	18	6	2	4
CV (%)	23	28	36	87	84
100 ml	101	47	20	9	3
CV (%)	21	22	26	41	71
500 ml	384	126	51	18	8
CV (%)	12	16	18	21	43
10 unit* pool	2049	569	282	74	38
CV (%)	4	7	14	19	24

*5 L of test suspension filled into ten 500-ml units.

those obtained by manual methods and showed significantly better reproducibility. The lower variability of counts obtained with the automated system was also apparent with 11 test suspensions containing particle types that differed widely in optical and physical characteristics (Table 4). For the test, suspensions were made using six different particle standard types that covered the range of shapes, transparencies, and refractive indices of particles isolated from parenteral solutions. For each of the particle types counted, reproducibility of counts for 10 runs of each sample with the IA system was far better than the 50% CV generally accepted as the best attainable precision for manual analysis using a linear dimension measurement. Mica flakes (which are thin and transparent) and fibers (which are sized subjectively) provide highly variable data when enumerated and sized by manual microscopy; higher variability was also observed with the automated system when counting these two particle types.

As mentioned earlier, the higher counting rate of an IA system allows larger numbers of particles to be enumerated and sized in a given period of time than with manual methods. The implications of this speed advantage are significant in that particles from larger sample solution volumes can be enumerated without any appreciable increase in time over the manual microscopic method. Applied in testing a batch of product, this procedure will result in a better estimate of the particles present than assays performed on sample aliquots from the units tested. The data in Table 5 were obtained from a 10 L mixed test suspension prepared from the particle types identified in Table 4. The precision of the count data collected increased significantly as the volume of solution assay increased and higher numbers of particles were counted. This suggests a means to significantly improve accuracy in the current microscopic test methods [26]. The potential for improvement seems significant indeed, considering the fact that the Omnicon system has a slower microprocessor than current systems and was not configured for high-resolution imaging.

VI. PARTICLE SHAPE ANALYSIS

Classically, the concept of shape has been defined through the introduction of shape factors. Beddow [22], Kaye [21], and Kaye *et al.* [27] are key references for the reader wishing to pursue this somewhat complex subject. The fundamentals of shape analysis have been discussed by Beddow [22]. The following brief description of shape analysis is intended to serve only as an introduction to a complex and useful application of the computer power of an IA system. Simple consideration of the effects of particle roughness or boundary morphology have led to explanations of the behavior of populations, such as powders, unobtainable by other means.

A. Fundamentals

A general type of classical shape factor relates the fourth power of a characteristic linear dimension, d, to a geometrical particle property using

$$Q = C_k d^k \quad \text{or} \quad C_k = \frac{Q}{d^k}$$

where Q is a geometrical property such as perimeter, area, or volume, C_k is the shape factor, d is a characteristic linear dimension of the particle, and k is an exponent. The characteristic dimension, d, may be one of the classical size features, such as Martin's diameter, Feret's diameter, etc. Using this definition, perimeter, area, and volume shape factors may be defined as

Figure 26 Four different geometrical figures made from nine identical squares.

$$C_1 = \frac{\text{Perimeter}}{d^1}$$

$$C_2 = \frac{\text{Area}}{d^2}$$

$$C_3 = \frac{\text{Volume}}{d^3}$$

A key issue with the classical shape factors C_1, C_2 and C_3 is that they are not adequately sensitive to changes in shape. For example, the four different shapes in Fig. 26 have the same projected area and the same perimeter. Thus, if the projected area diameter is used to define the characteristic linear dimension, the C_1 and C_2 shape factors are identical for all four shapes, even though these four shapes are very different from each other.

These definitions are based on a comparison of shape factors with standard shape factors for regular figures (i.e., circle or sphere) in which the perimeter, area, and volume are well known, and the characteristic linear dimensions are unique and well defined (i.e., the radius). Other shape factors that have been defined by ratios of actual and idealized geometric properties include

$$\text{Sphericity} = \frac{\text{Surface area of equivalent sphere}}{\text{Actual surface area of particle}}$$

$$\text{Circularity} = \frac{\text{Circumference of circle of area equal to particle projected area}}{\text{Actual perimeter of particle}}$$

$$\text{Rugosity} = \frac{\text{Perimeter of particle profile with minor irregularities and corrugations}}{\text{Perimeter of smooth curve circumscribing particle profile}}$$

$$\frac{\text{Circularity}}{\text{Rugosity}} = \text{Circularity including effects of rugosity}$$

These shape factors do not provide a unique description of the real shape of an object. This is of interest in the case of sphericity, for example. Sphericity is often used in calculations as a shape factor by geologists and engineers, yet it has been shown empirically that sphericity is not a good shape factor. For example, in Fig. 26, all of the shapes have the same sphericity, and yet they are quite different in shape from each other. Furthermore, it has been demonstrated experimentally (in the case of a water-fluidized bed, and in the case of slurry flow up a vertical pipe) that particles with the same sphericity but different shapes behave differently.

B. Radial Measurement of Particle Shape [22]

Classical image analysis is based on the simple counting of pixels. For example, if the length of an intercept chord is required, the number of pixels, *N,* along the intercept is counted. Then, knowing the length of one pixel, *f,* the length of the line is *Nf.* Similarly, to measure the area of a figure, the number of pixels, *N,* within the boundary is counted and knowing the area, *c,* of one pixel, the area of the figure is *Nc*. This means that classical image analysis uses simple arithmetic only.

Image analysis by morphological analysis as described by Beddow [22] deals directly with the irregular boundary function that represents the profile of the object. This is achieved by directly measuring the *X,Y,* coordinates of points on the image boundary. The *X,Y* coordinates of the centroid of the profile are then determined (using the trapezoid rule, for example). Knowing the *X,Y* coordinates of the profile points and the *X,Y* coordinates of the centroid, one then calculates the radial vectors of the profile. This is shown in Fig. 27A,B. Once the radius vector *R* is known as a function of the angle Θ, the *R*- plot leads directly to the Fourier equation of the curve in its Cartesian or polar form, as shown in Fig. 27C.

Once the image has been transformed into a Fourier equation, one can use many of the techniques of modern mathematics including calculus, matrices, infinite series, moment functions, harmonic analysis, statistics, and symmetry operations to analyze the properties of the profile under examination. The power of image analysis by morphological analysis is so much more than that of classical image analysis as to amount to a significantly different approach to image analysis.

Once the analysis of Fig. 27C is completed, the Fourier equation for the curve may be determined as follows.

The Cartesian form is

$$R(\Theta) = a_0 + \sum_{n-1}^{\infty} (a_n \cos n\Theta + b_n \sin \Theta)$$

where a_0 = mean or average value of the radius
a_n, b_n = Fourier coefficients
n ranges 1 to ∞.

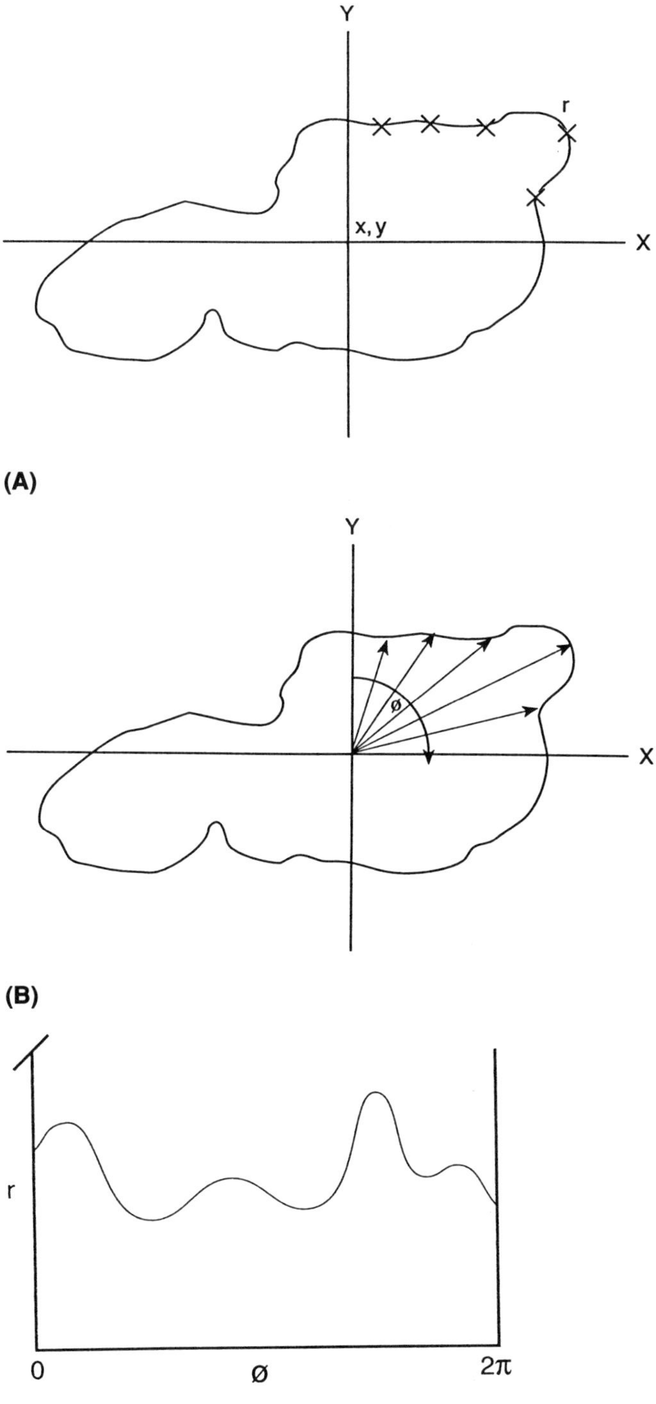

Figure 27 Measurement of particle shape by boundary function. (A) Measurement of *x, y* value for perimeter/centroid placement. (B) Determination of radius. (C) Plat of *R* versus Θ.

Figure 28 Equipaced method of fractal analysis.

The polar form is

$$R(\Theta) = A_0 + \sum_{n-1}^{\infty} A_n \cos(n\,\Theta - \propto n)$$

where A_0 = mean radius
A_n = amplitude
$\propto n$ = phase angle

The result of this procedure is the assignment of a unique and invariant shape factor to each particle analyzed.

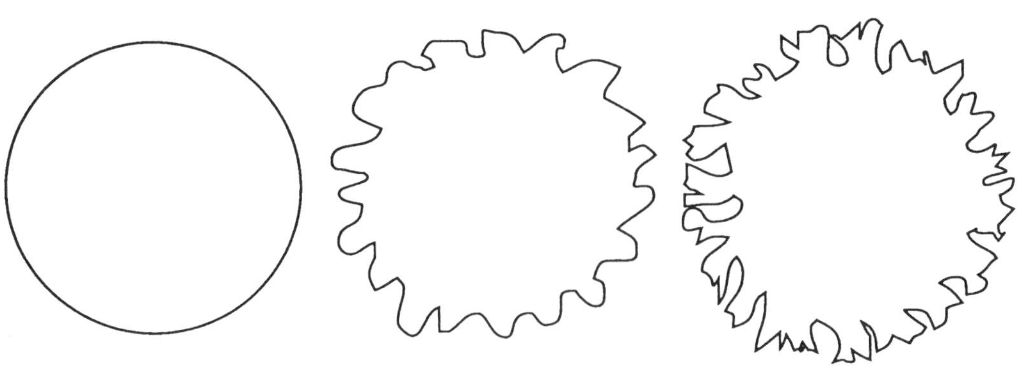

Figure 29 Euclidean (circle) versus fractile profiles.

Figure 30 Fractal intersection frequencies.

C. Fractal Analysis

Fractal analysis provides yet another means of shape analysis. The publication of Mandelbrot's book [28] on fractal geometry in 1977 provided a basis for fractal analysis as practiced today. Fractal analysis is a powerful computer-based type of shape factor analysis. In classical fractal analysis, the particle edge is approximated by a polygon with all sides but the closing side of the same length. Particle ruggedness is inferred from an increase in polygon perimeter as the length of polygon sides is reduced. A simple (equipaced) example is shown in Fig. 28. The concept of fractal analysis is not new. In theory, it relates to the methods used by Archimedes in determining the value of π. Archimedes determined the value of π by measuring the perimeters of inscribed and exscribed polygons of increasing numbers of sides. This random walk procedure allowed him to identify the value for the factor as being between $3\frac{1}{7}$ and $3\frac{10}{71}$. This method of determining particle perimeter is referred to as the equipaced method based on the equal length of the side of the exploratory polygon. With this procedure, the estimate of the perimeter decreases as the sides of the polygon used to explore the perimeter decrease in length.

Another method of characterizing profiles is based on intersection frequency of particle edges with a uniform line matrix (Fig. 29). The equivalent Euclidean shapes of the particle profiles are drawn along with the rugged variants of the basic profiles. By Kaye's method, [27] intersection frequency measurements are then carried out with these profiles by tracing them onto a graphics tablet, and then measuring the line intersection frequency on each profile using a scaling algorithm within the computer memory. The numerical result of this analysis is expressed as the ratio of the intersection frequency for the particle (I_r) to that for the Euclidean form (I_s): As the term I_r/I_s increases, the perimeter estimate increases. The author suggests that for the ongoing characterization of profiles that are not changing drastically (for example in the quality control of a product in which the fine particles have a fractal structure), it is possible to develop a method for characterizing the rugged profiles by means of the intersection frequency as suggested in Fig. 30.

VII. EVOLUTION AND DEVELOPMENT OF IA SYSTEMS

Some readers may be interested in evaluating IA systems for purchase. The historical development of commercial systems will provide the reader with an insight into the evolution of instrument capabilities and the development of features that make the instruments suited for specific tasks. The reader should keep in mind that all systems available are critically dependent on software (i.e., programming) to perform both routine and nonroutine tasks. Thus the purchaser should be assured of adequate software support by the vendor, or be able to program the instrument in-house, so that the inevitable modifications necessary to optimize performance or to configure an instrument for a specific task may be made. Some of the useful aspects of instrument performance are briefly discussed in the following sections with reference to specific systems that were or are offered commercially. A listing of performance criteria are provided in Secs. XI and XII.

A. Omnicon

The system we initially purchased for IA particle counting applications in our laboratory eight years ago was an Omnicon 3500 system. Though outdated, it is still one of the most common IA instrument systems encountered in the field. We incorporated a number of modifications to the system and used extensive programming to make it suitable for our application. Today we use this instrument routinely for particulate matter counting, such as enumerating particles flushed from medical devices and for population analysis of drug powders. This unit has also proven applicable to the analysis of contaminant particles from parenteral products.

Our Omnicon system (Fig. 31) consists of (1) a microscope with an automated stage drive and automated focus, (2) a high resolution television camera, (3) a data general eclipse S/120 computer, (4) a text terminal, (5) an image analysis video display terminal, and (6) a printer. As purchased, the system included hardware for analog-to-digital conversion of binary image and gray-level thresholding, and software for measuring quantitative characteristics of the particles. The system was also delivered with software intended to perform microscopic particle counting on filter membranes, including automatically scanning a membrane filter surface, detecting shadows, and agglomerating particles. This software proved unsuitable for our application, and new software designed to operate under the conditions of the USP assay was developed in-house. Our present software performs all functions necessary for enumerating and sizing particles on

Figure 31 Omnicon 3500 system.

membrane filters. This includes tracking particles that overlap field boundaries, and testing for defects in the surface of the membrane filters on which particles are collected.

The Omnicon design differs from that of some newer systems in a number of important respects. The system has no feature for image storage or processing, and digitization consists simply of converting each field image to a binary image composed of particles and background. Size measurement is chordal, based on eight chords (22.5 approximate dispersion). The system operates quickly (approximately five times the speed of a human analyst).

Typically, we perform analyses at magnifications of 10× to 35×. This results in less than 2 pixel elements per micron and a theoretical resolution of > 1 μm. The large number of image fields analyzed are largely due to the relatively low magnifications used (approximately 1200 μm × 2500 μm in width). One favorable feature of the instrument is its fairly high electronic resolution, which is a result of its incorporation of a high resolution camera and a 1066 × 485 digitization matrix. With the Omnicon or with other instruments, the resolution of the video portion of IA systems is limited by the low-resolution axis of the camera or digitization matrix.

Our experience with the Omnicon system has led us to conclude that it is an adequate but minimal system for automated numerical analysis of particulate matter. Although it performs no image processing, it gives more reproducible results than those of human analysts. Our success with the Omnicon, despite its very basic level of function and limited capabilities, strongly suggests that instruments with current technology may be capable of more impressive results with greatly decreased requirements for programming and operator interaction.

The modifications we made to our Omnicon system and the problems encountered in our application of this system have helped us to develop criteria by which we currently evaluate new systems. Many improvements have been incorporated in instruments commercially available today, and further improvements can be expected in the future. Those that are most significant include lower cost, greater speed of analysis, greater sizing and counting accuracy, and an increase in image processing power. Current microcomputers are much less expensive than the minicomputer used for the Omnicon system. They are also faster and have more memory. The computer on the Omnicon system is a 1.6-bit minicomputer that runs with a clock frequency of 2 MHz and has 128 kbytes of memory. Thirty-two-bit microcomputers (based on state-of-the-art microprocessors that run in at a rate in excess of 20 MHz with 20 Mbytes of memory) are currently available at a 10-fold reduction in cost. The use of microcomputers with dedicated IA chips on current generation systems has the potential for decreasing both the cost of image analysis systems and the time required for analysis.

B. NORAN 8502

The effect of incorporating evolving technology into system design is readily apparent in a comparison of the Omnicon system with the newer NORAN 8502 model (Fig. 32). This instrument represented a quantum leap over the Omnicon system in terms of decreased size and increased speed and capability. The entire system (minicomputer plus the microscope) requires only the space of an office desk. Unlike the Omnicon, the 8502 was based on a high-level microcomputer. The space required could be further reduced by mounting the microscope and attached television camera in an adjacent hood area. The 8502 is capable of performing most types of particle measurement discussed up to this point (e.g., chordal measurements, pixel counting, calipering) and offers a range of specialized modes of analysis as well (see Sec. XII). This instrument was available with automated stage drive and autofocus so that totally automated particle analyses could be performed using off-the-shelf software.

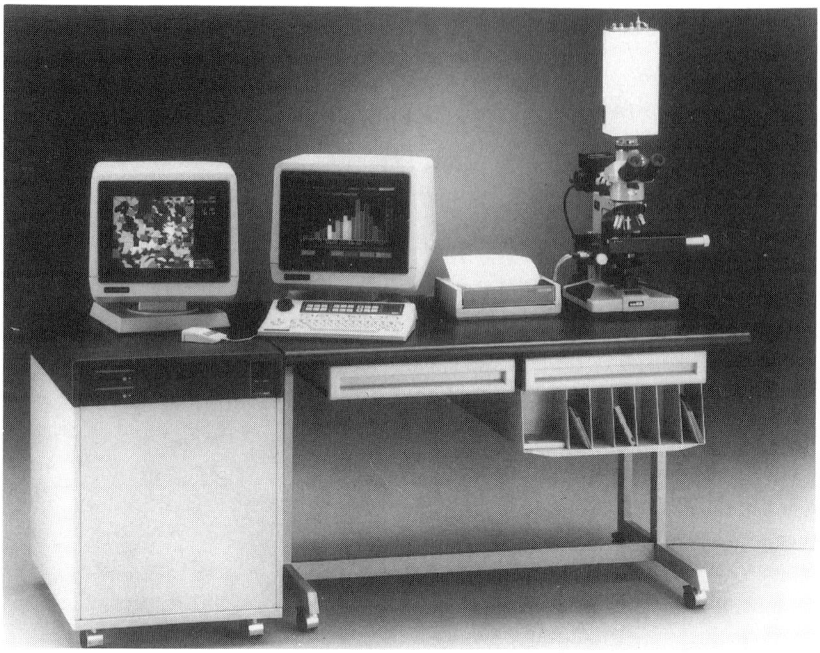

Figure 32 NORAN 8502 System.

C. Juliet III (Shakespeare Corp.)

A third system, representative of another specialized and sophisticated approach to automated particle analysis, is the Juliet III system which was marketed by the Shakespeare Corporation of Iowa City, Iowa. This instrument represented a system dedicated to morphologic analysis of particles. The system was based on a 486-level microcomputer and included proprietary software for radial array measurement of particle shape, and a LCD camera and relay lens adaptable to almost any research microscope.

Juliet III measured both size and shape directly. As opposed to some image analyzers that merely count pixels and, therefore, can only add and subtract, Juliet III obtained the boundary function which enabled it, in effect, to perform calculus. The higher mathematics allowed precise and versatile calculations of boundary functions and shape factors. Juliet III measured 50 features, many of which were derived from the theory of morphology of Luerkens [19]. These measures included roughness, microroughness, partial symmetries, moments of the radial distribution, and the grand radial distribution. This capability made it possible to analyze the geometrical characteristics of powder samples as thoroughly as one can measure their chemical composition.

Despite its capabilities for complex morphological analysis, the Juliet III system was simple in operation. The field to be analyzed was presented to the instrument at a predetermined magnification, particle images were digitized, the data stored in a named file, and data analysis performed by menu selection using the mouse control. The results could be stored and printed as desired. Measurement mode on each particle image was as follows: locate the center, perform principle axis transformation, determine *x,y* coordinates of the profile, convert to R-Θ space, determine Fourier equation of the R-Θ curve, extract coefficients, and convert to features. This morphological analysis permitted sophisticated statistical analysis and matrix algebra of processing data to be used, including the Fourier transfer of the RΘ curve as discussed earlier.

D. Leica Quantimet 600

The Leica Quantimet 600 is representative of a state-of-the-art, currently available system that includes powerful image processing and morphologic analysis capabilities (Fig. 33). The processor provides color image processing that allows the display of a wide array of hues. Users can develop their own programs in the language of choice, using a Windows user-interface and harnessing the power of a high-performance PC to the high-speed image processor. Automatic microscopes are fully integrated with this system. Software can be used to move motorized stages from field to field and to employ the analysis system as part of a feedback loop that automatically controls lighting and focus.

With the Quantimet system, image analysis based on mathematical morphology simplifies the image for segmentation and measurement. The stored image is manipulated by a gray-tone image processor based on programmable logic array technology and optimized to handle a morphologic transformer.

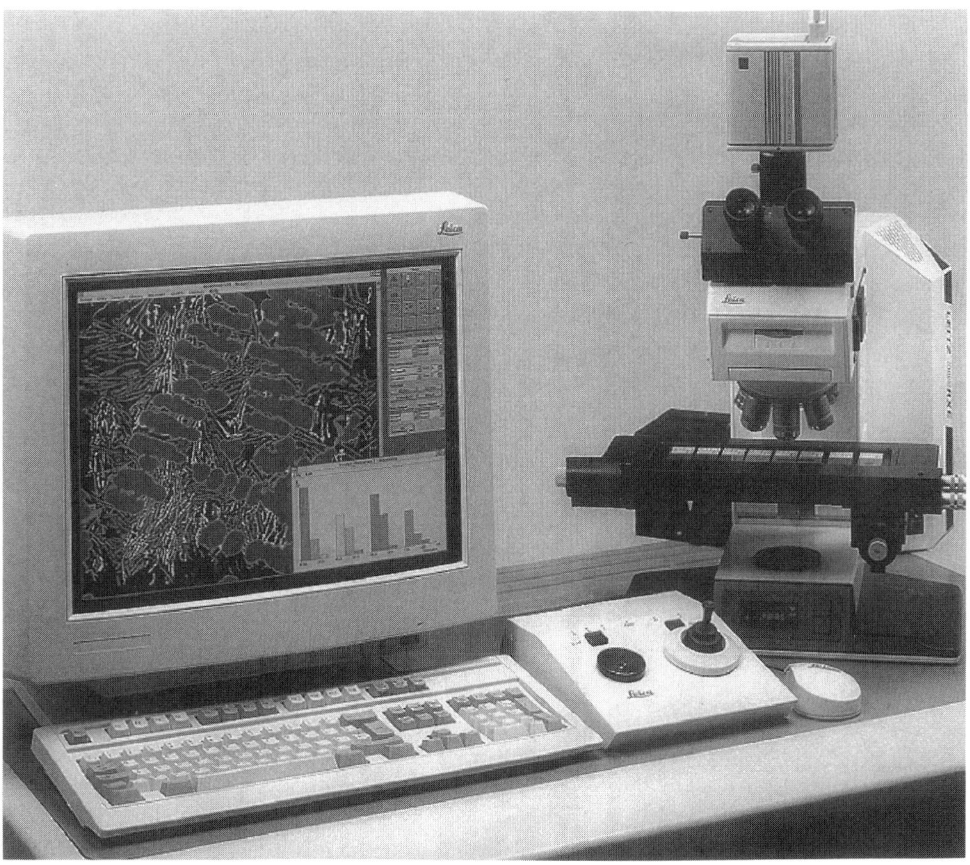

Figure 33 Leica Quantimet 600 system.

E. NORAN VOYAGER

Another currently available state-of-the-art system is the NORAN VOYAGER. The block diagram of this system in Fig. 34 both describes the system architecture and indicates the design basis of the instrument's wide range of capability. It is probably the most powerful microanalysis/digital imaging system currently available off-the-shelf. A dedicated Motorola 68030 microprocessor drives all data acquisition operations. The VMEbus architecture includes an 8192-channel × 32-bit multichannel analyzer. The analyzer section incorporates a 100-MHz, Wilkinson-type

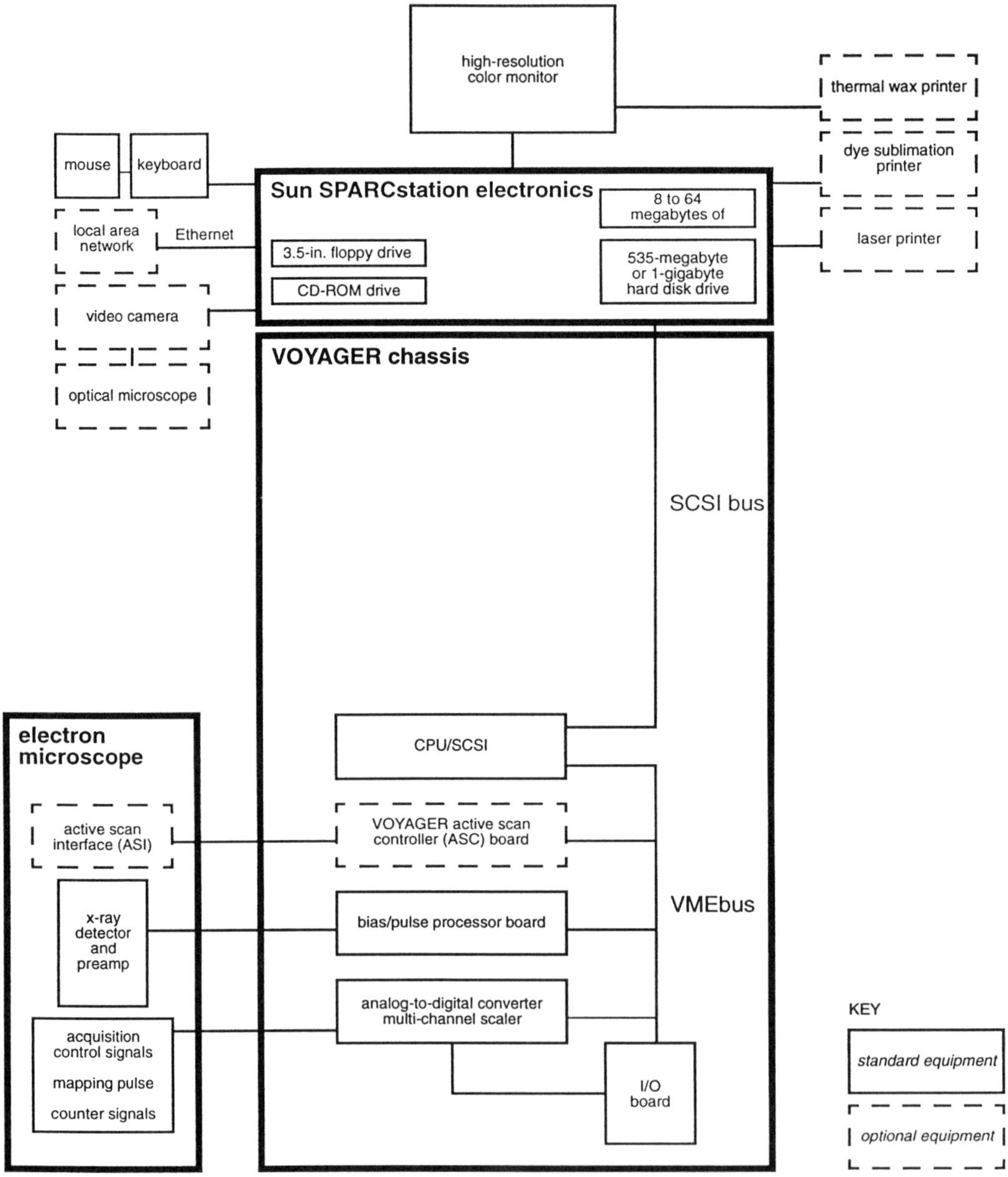

Figure 34 NORAN VOYAGER system block diagram.

analog-to-digital converter; a bias supply with automatic protection circuitry; automatic energy calibration and zero stabilization; and a microprocessor-controlled, time-variant pulse processor with four selectable time constants. This instrument is optimized for networking applications, can collect data from two image analysis systems simultaneously, and is well suited for use with scanning and transmission electron microscopes (SEM and TEM) and light microscopes.

Importantly, the VOYAGER is based on a Sun Microsystems SPARC workstation. High-speed, multitasking microanalysis and imaging operations can be performed at significantly faster rates than with PC-based, DEC-based, or "home-built" systems. With speeds up to 40 MHz and 28.5 million instructions per second (MIPs), VOYAGER's central processor is an order of magnitude faster than most current systems. Popular programs for spreadsheets, databases, word processing, report generation, and desktop publishing are available. Data can be output to laser printers, color and black-and-white hard-copy printers, and color thermal wax printers.

F. Selection of Systems for Purchase

The two currently available systems described above (Quantimet, VOYAGER) have capabilities far beyond those of instruments available a decade ago, but both have specific strengths and relative weaknesses that make them more or less useful for specific applications. The author strongly recommends that prior to consideration of an IA system for purchase, prospective purchasers become familiar with the principles of image processing and analysis. The suitability of a system for use may well depend on the availability of specialized software systems. The monograph on video microscopy by Inoue [1] is highly recommended to the prospective purchaser. This book contains all of the information necessary for a basic understanding of video technology and its application in light-microscopic IA systems.

VIII. ANALYSIS OF COLOR IMAGES

I have omitted any in-depth consideration of analysis of true color (as opposed to pseudocolor) images from this chapter. The use of color images has many advantages for microscopic work with bulk specimens, and some of these overlap with particle analysis. The most obvious advantage is that true color allows the hue to be extracted, in contrast to monochrome images where it can prove impossible to separate colors, since different hues often have the same gray-level intensity. For example, pale red and dark red can be detected while dark blue is ignored even if the intensities of colors correspond. Using monochrome images in particle analysis, particles of different colors may be presented as closely similar gray levels, and their discrimination is not possible; with color imaging, particles can be separated on the basis of color. The Quantimet 600 system discussed above has powerful capabilities for dealing with color images. The system is shown diagrammatically in Fig. 35.

The highest-quality color image definition is achieved by using a miniature-head, color camera. The camera has high resolution comparable to that achieved by many monochrome cameras. The system automatically splits a color image into red, green, and blue spectral bands, and stores each segment as a separate gray image. These components are then converted into hue, luminance, and saturation components, subjected to transformation either separately or together, and are displayed as a color image on the monitor. Example software routines are provided that perform system calibration and color detection by RGB or hue, luminance, and saturation. Along with real color software, the full range of pseudo color modes are also available, in which color is used to highlight the significance of gray shades, or to identify different classes of objects by assigning them different colors.

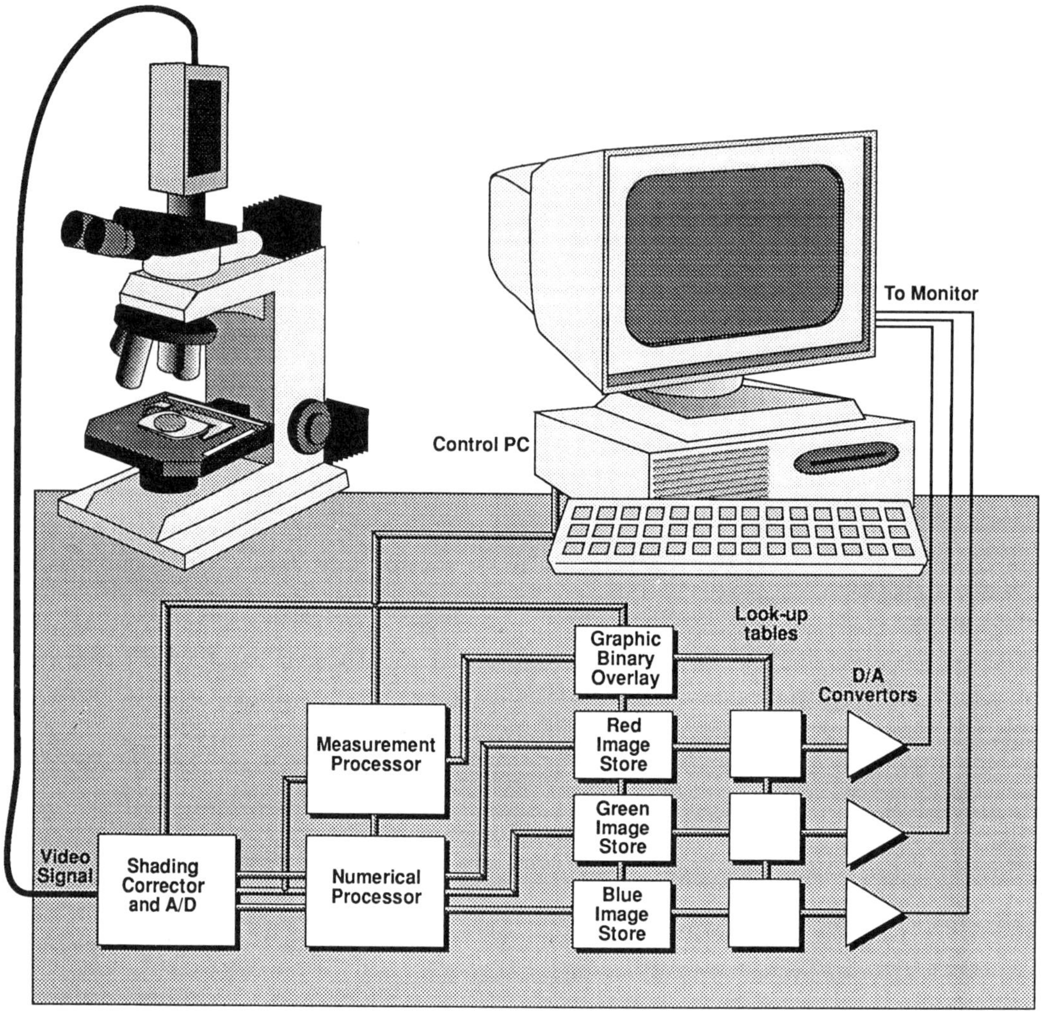

Figure 35 Color image processing (Leica Quantimet 600).

IX. SUMMARY AND CONCLUSIONS

Based on the author's experience, IA can provide an accurate and precise automated analysis of particulate matter collected onto a filter membrane. It appears that all marketed IA systems will require some modification in order to count and size particles collected on a membrane filter using reflected illumination. While each system has inherent advantages, no one system constitutes the optimum combination of features for enumerating particles. Each system available presents specific advantages, and the user must decide what qualities are most important for their intended application. Further, the user is reminded that their specific application may well represent a "customized" application for the vendor. Thus the user must assess the depth of support available for their use (e.g., software generation capability, competence of the vendor applications staff).

Image analysis-based methodology appears capable of resolving difficulties with manual microscopic counting, including both time requirements and data precision. Laboratory technicians familiar with manual microscope operation can readily be trained to operate IA systems. Factors that must be carefully weighed prior to purchase include the relative complexity of the instrumentation, cost, and the required level of technical expertise for optimizing and maintaining the purchased system.

The user acquiring a new instrument is advised to purchase a system that incorporates features deemed most important (e.g., ease of use, high-resolution imaging, speed of operation, image analysis routines available, and the ability to modify the system to minimize the effects of any shortcomings). Generally, the more expensive systems offer more resolution, computing power, and flexibility, and may more readily be customized. Features available on instruments currently available that represent distinct advances over models available several years ago include more powerful image processing capability, higher-resolution imaging, menu-driven operation, and range of software. Most current systems have been designed by their manufacturers with a wide range of applications to ensure success in the marketplace.

Evaluation and experience with our present Omnicon system suggests that two divergent paths may provide the user with usable IA systems for automated microscopic analysis of particulate matter. The versatile high-performance systems (such as the Noran VOYAGER) have adequate power to perform any of the image processing functions necessary for high-precision counting of particles, and make corrections necessary to deal with uneven lighting, background irregularities, and so forth, but provide their versatile capabilities at a significant cost. A useful alternative, given that the requisite capabilities can be developed on a PC-based system, is a dedicated system at lower cost. Such an instrument would incorporate automated stage drive, autofocus, and high-resolution imaging, with minimal image processing and storage. For the operation of this type of system, the performance of minimal image processing would result in significant time savings.

This chapter has presented an overview of some of the current capabilities of IA systems and theory of operation. Those believing that IA may have a useful application in their laboratory are urged to learn and explore the subject and carefully analyze costs prior to purchase. Capabilities of present instruments indicate that we are at the beginning of a new era in particle technology, whereby significant features of analyzed particles may be related to fundamental parameters of processing as a matter of routine. Improvement in instrument capabilities will be supported by applications developments including more powerful statistical analysis, automatic sampling, and improved lighting designs. Given that user laboratories can support the current generation systems, the next decade should bring a significant increase in the knowledge of particulate material as a result of application of image analysis to particle characterization.

X. DEFINITIONS

Brightness resolution. Brightness resolution, also known as quantization, is the depth of each individual pixel (i.e., number of bits used to represent a pixel). For example, if 8 bits/pixel are used, 256 gray levels are represented by the pixel. Brightness resolution is important in the segmentation process, whereby features (particles) are separated from background. The greater the number of bits/pixel, the wider the dynamic range that pixel represents.

Chord measurement. A measurement of a particle's longest linear dimension typically made by rotating or overlapping a number of Feret diameters about the particle's center of gravity. The accuracy of this measurement depends upon pixel size, particle shape, and number of Feret's diameters as discussed in Sec. III. Clearly, the number of chords used to make a measurement is important. A spherical particle requires only one Feret diameter or chord for accurate measurement, while rectangular or irregular particles with different aspect ratios require more diameters. For example, a 1:2 aspect requires 7 chords; 1:3 requires 10 chords; 1:5 requires 16 chords; and 1:10 (fibers) requires 32 chords: the more fiberlike the particle, the more chords needed to accurately size it.

Chord rotation angle. The angle at which the chord's diameters are rotated about at the center of the feature. Chord rotation angle is determined by the number of chords used (360°/number of chords).

Clock speed. A relative measure of the computer speed of operation. Clock speed defines the frequency (MHz) of the timing signal that determines the fundamental working rate of a computer. The larger the clock speed value, the faster the host computer will execute program instructions and the more rapidly the image processor will execute designated operations on an image.

Frame averaging. Acquiring a number of image frames to improve the signal-to-noise ratio (S/N). Summing N frames results in a S/N improvement of ~N, while averaging N frames results in a S/N improvement of $\sim 2N - 1$.

Frame rate. In images captured from video sources and displayed on TV monitors, a third form of resolution must also be considered in representing the digital image via the frame rate at which the image is updated on the screen. Thirty frames per second is standard for video images: the higher the frame rate, the less the screen flickers, and, hence, the better the resolution.

Frequency domain. Frequency domain also refers to pixels. However, those referred to are not the original digitized pixels in the image. The original pixels are transformed using the Fourier transform operator, giving a set of complex pixels representing the frequency content in the image. The Fourier transform is a powerful mathematical operator that also allows viewing the pixels as frequencies. These frequencies (like the different frequencies of a spectrum) can be manipulated in any desired manner, and then the inverse Fourier transform renders the image back to the spatial domain.

Hard disk. Computer disk storage system utilizing a rigid magnetic (hard) disk rotated at high speed. A hard disk is used for long-term storage of digital images.

Host computer. A microcomputer or minicomputer that provides the operating base for the system. The host computer serves as the user's interface to the image processor and may also perform IA functions.

Image compression. These techniques are used primarily when reduced storage space is necessary, as in the case of image transmission. The images are coded using a variety of compression algorithms. These coding techniques can be reversible or irreversible.

Image enhancement. Image enhancement is the most common application where the desire is simply to effect a blind improvement of the picture, irrespective of what the original image was. These techniques tend to attempt to improve the visual appearance of an image for a human observer without recourse to knowledge of the degradation phenomenon. The image's information is selectively emphasized or suppressed to make the output image subjectively more pleasing. Image enhancement is the most active area in the commercial image processing industry.

Image memory. RAM memory used to store digitized and processed images. Image memory is immediately addressable by the host computer and image processor for rapid manipulation of the digital image. Resultant processed images are stored in free areas of the image memory. One 512×480 digital image with 8 bits of gray scale requires 0.25 Mbyte of memory ($512 \times 480 \times 8$). Thus, 1 Mbyte of memory will accommodate up to four 8-bit images in 512×480 pixel format. Higher-resolution images require more storage area.

Image processor. A computer used to manipulate digital images. Generally, more expensive systems utilize a dedicated high-speed processor capable of handling large image arrays at or near real time (a frame in 1.30 s). Systems using the host computer or normal microcomputer coprocessor for this function will perform image processing at slower rates. Systems with an array processor or the faster pipeline processor will manipulate images at higher speeds. Note: The image processing capabilities of the system selected will powerfully impact its usefulness in performing complex, nonroutine analyses. This consideration is especially important for enumeration of particles on membrane filters with reflected light. This type of illumination rarely gives completely even illumination; this defect must be compensated for by a shading correction. Random image noise, an artifact of the system electronics and filter background, is overcome by frame averaging. The characteristics of the spatial filters and the gray-level segmentation determine the fineness with which the system can discriminate between particulate matter and background.

Image processor hardware. The computer hardware involved in processing a digital image. Unlike the host computer, the image processor hardware is not user programmable. It simply performs dedicated functions at high speed. The use of pipeline or array processors in this role greatly increases the speed of this function.

Image restoration. Image restoration attempts to restore an image to its original quality by some mathematical manipulation. The emphasis is on estimation of true image of the observed picture (i.e., the degradation in the picture is modeled and inverted to best yield the original image). Thus, some a priori knowledge of the degradation model is needed for image restoration. The simplest example of image restoration involves removing additive noise from an image.

Image segmentation. Separating the digital image into features of interest and background on the basis of gray-level differences.

Max kernel size. The largest kernel matrix that can be used in filtering. The larger the kernel the more flexible the spatial filters.

Number of chords. The number of chords rotated about the feature. Some IA systems allow the user to select the number.

Number of threshold bins. A range of gray levels, determined by an upper and a lower gray level or threshold, forms one threshold bin. Features falling within a threshold bin will be converted to a binary image and counted and sized. Two threshold bins are required to simultaneously analyze light and dark particles on a filter membrane.

Operating system, menus, and macrofiles. Software programs that tell the host computer what operations to perform. The operating system controls all basic computer functions and directs the execution of all other software. Menus are lists of options generated by a program; they make learning and running a program easier. Macrofiles are user-defined programs used to automate repetitive functions. The operating system, menus, and macrofiles determine the versatility, user friendliness, and programmability of the host computer.

Random access memory (RAM). Data storage for programs and numerical data on computer memory chips that is immediately addressable by the host computer's CPU. With more RAM, more programming instructions/data can be held in fast access RAM memory, which results in time savings in program execution. In some image analysis systems the host computer RAM also acts as image memory.

Shading correction. An image processing step performed to correct nonuniform illumination (shading). Shading is a serious defect caused primarily by nonuniform microscope illumination and nonlinearity in the video camera, creating nonuniform segmentation. Shading corrections may be applied to both analog and digital images; manipulating the latter offering the greatest range of correction.

Spatial domain. Spatial domain refers to the pixels in the digital image (i.e., spatial domain processing techniques are those that directly operate on the pixels). The most frequently used techniques in the spatial domain are those based on convolution kernels. The actual pixels in an image are manipulated and replaced by new pixels obtained as a result of some manipulation of the original pixel and its neighboring pixels. The flexibility and ease of implementation of these techniques make them very popular.

Spatial filters. Changing the gray value at each pixel by averaging its value with respect to its neighboring pixels in a matrix or kernel. Depending on the size of the kernel and the numerical pattern of the convolution mask, spatial filtering results in image sharpening, smoothing, or edge enhancement.

Spatial resolution/sampling. Spatial resolution is the number of pixels into which an image is sampled. The Nyquist theorem requires that the image be sampled at least twice in the highest spatial frequency content in the image. Images are typically sampled to a resolution of 51×512 pixels to be consistent with the television RS-170 format.

Threshold comparison. The method of comparing the segmented binary image to the gray-scale image. Essential for the accurate sizing of features is the proper determination of the threshold bins such that the features are not undersized or oversized. Three methods of comparison are used.

- **Overlay** A white or black binary image of the detected feature is laid over the top of the gray-scale image.
- **Side by side** A half-sized binary image is displayed on one half of the monitor screen, while a half-sized gray-scale image is displayed on the other side.
- **Toggle** The gray-scale image and the binary image are alternately viewed in rapid succession. Usually a single key stroke will flip-flop the two images.

XI. EVALUATION OF IA SYSTEMS

A. Performance Criteria

Currently available IA systems represent a third generation of the technology. Prior to the beginning evaluations, the user should develop a list of performance categories by which systems may be judged. General performance categories, in what the author believes to be their approximate order of importance, are listed below:

1. Measurement

a. Accuracy
b. Precision
c. System resolution
d. Speed of operation

2. Operational Factors

a. Level of unattended operation
b. Ease of use

3. Image Processing Capability

a. Shading correction
b. Image processing/enhancement
c. Background suppression
d. Contrast enhancement

4. Data Handling

a. Data reduction
b. Archiving and database management
c. Report quality
d. Statistical analysis

5. Other

a. Cost
b. Service and support
c. Upgradability

B. Additional Specific Criteria

In addition to evaluating systems with regard to performance in these general categories, we have also collected more specific information that we believe relates critically to our use of an IA system in analysis of pharmaceutical particulate matter. The latter includes

1. Counting and Sizing Accuracy

a. Measurement of calibration standards
b. Measurement of particles on membrane filters
c. Data coefficient of variation—IA versus manual analysis

2. Operation (Ease of Use)

a. Estimated operator learning curve in hours
b. Utility of screen display and on-screen help messages
c. Conciseness and clarity of documentation
d. Programming skill required

3. Image Processing: Overcoming Artifacts Related to

a. Uneven illumination
b. Dust on optics
c. Filter surface irregularities
d. Shadows created with oblique illumination

4. Data Handling

a. Customized report format
b. Statistical analysis capability
c. Stored image analysis
d. Provision for rapid storage and retrieval data

XII. SPECIALIZED MODES OF ANALYSIS

A. Frame Averaging

Frame-averaging techniques make use of the fact that signal strength can be increased by averaging image contrast levels, so that signal strength is increased while random noise is "averaged" out. The process is begun by applying matrix shading correction, and distinguishing the image into 256 gray levels. Each frame is cleaned up by successively averaging a series of images at video rate (Kalman averaging) to filter out the "noise." The resulting image is then placed in image storage or directly analyzed.

B. Deagglomeration of Spheres

The deagglomeration of spheres program is designed to calculate and plot the important dimensions of spherical objects in an image, even though a significant portion of each object of interest is hidden from view. This program significantly expands the statistical database for such applications as latex, particle, and biological cell sizing and counting because such parameters (such as diameter, surface area, and volume) can be calculated for all spherical objects in the image, even though the perimeter of the features is partially obstructed.

C. Fiber Analysis

Fiber analysis is designed for measuring various physical dimensions of long thin objects that can be slightly curved and crossed, but are not kinked or spiraled. The resulting data can be plotted as a histogram or put in tabular form. An image of the fibrous sample is either read or acquired into memory, and a filled binary of that image is created. The operator then inputs various maximum and minimum values into the program including limits for lengths, widths, slope angles, etc. These values are used by the program to define how the fiber is to be identified. When the program is run, the binary image is automatically skeletonized and a "distance map" made for each fiber. Using these data and the defined values, artifacts are corrected (such as split ends, false ends, and broken fibers), dimensions are determined, and the required fiber data is generated. Other features include

- The ability to determine when partial fibers are in reality the same fiber, even though the two fibers are offset and at different slope angles
- Provision for rejecting detected fibers based on length requirements
- Prevention of low-aspect-ratio objects from being considered as fibers
- Isolation of crossed and curving fibers is done automatically

D. Grain Boundary Reconstruction

The grain boundary reconstruction program is designed to semiautomate the process of tracing individual grain boundaries in an image so that grain morphology can be determined. Once an image of the granular sample to be analyzed is acquired, the various grains are then defined using the gray-level thresholds of the image. The binary image produced by the segmentation process is then automatically filtered to refine boundary lines and eliminate noise resulting from nonboundary lines having the same gray levels as the boundaries. Missing boundary lines can then be automatically reconstructed or manually drawn.

E. Two-Dimensional Fast Fourier Transform

Two-dimensional FFT is an image processing tool that is useful for a broad range of image enhancement and image analysis functions including (1) the elimination of background noise, (2) the enhancement of spatial details, and (3) the measurement of periodic features. Specifically, the software provides for converting an image from the normal spatial domain into the frequency domain, where various filtering and measurement functions can be performed. After image enhancement functions have been performed, the program provides for converting the image back into the spatial domain for viewing.

With this feature, an image to be analyzed or processed is first read into the image memory. Following a 16-bit complex Fourier transform, there exist real (cosine) and imaginary (sine) images. These images can be multiplied together to produce a power spectrum. First, a binary mask is created that identifies those features of the power spectrum to be eliminated. Applying this mask to the real and imaginary images eliminates those undesirable features from the FFT image. Performing a reverse FFT on this modified frequency image produces an image that is free of the deleted frequency information. Other specialized features are listed below.

F. Locked and Liberated Analysis

For some mineralogical and metallurgical studies, it is important not only to characterize samples, but to determine the acceptability of the sample morphology as well. This is the purpose for which the "locked" and "liberated" program is designed. Software computes the required morphological data from the standard particle sizing output and produces boundary data relating to each phase in the sample. The percentage of a particular phase that is liberated (either totally free or on the surface of the sample) can be determined.

G. Stereology

The stereology program calculates three-dimensional data such as volume density, mean volume, and numerical density using measurements made on a one-dimensional image of a sample. Statistically valid data is produced if there are a large number of related structures. Two types of analyses can be performed: lineal and areal. Lineal analysis is generally used when features are aggregated into large related groups. Areal analysis is more often used when particles are dispersed and appear as discrete areas.

H. Statistical Analysis

Some imaging applications programs (including complex feature analysis, critical dimension measurement, stereology, fiber analysis, and deagglomeration of spheres) produce extensive data sets. The purpose of this software package is to provide for further processing of the data produced by these application programs using advanced statistical techniques, which include inferential and descriptive statistical methodologies. The file(s) to be statistically analyzed are called into memory. The operator then selects the type of statistics package to be used. The possible types include a basic statistical package, a linear correlation test, a linear regression test, and a set of inferential statistical tests including hypothesis and confidence interval estimation-type tests. Once the type of test is selected and appropriate variables and limits are entered, the desired statistical results are calculated and ready for plotting and printing in a tabular format. Some of those included in the programs are

- Basic statistical calculations: mean, standard deviation, maximum, minimum, variance, standard error, geometric mean, harmonic mean, median, skewness, kurtosis, and autocorrelation.
- Plotting routines: histogram, histogram normal, scatter, scatter regression, cumulative histogram, cumulative histogram normal, ln(X) versus Y, X versus ln(Y), and ln(X) versus ln(Y).
- Inferential statistical techniques: one-sample t-test, two-sample t-test, one-sample chi-square test, two-sample F-test, one-sample Kolmogorov-Smirnov test, and two-sample Kolmogorov-Smirnov test.

I. Three-Dimensional Stereo Graphics Measurement

The 3-D stereo graphics measurement program is designed to display and measure topographic details of a stereo pair of images. The resulting images can be viewed with the aid of red/blue stereo glasses, and desired height profiles generated. Images for this program can be generated in several ways: (1) by tilting the sample in an SEM between image acquisitions, (2) by translating the image, such as is done in aerial photography, or (3) by stereopair generation using a confocal light microscope. After the stereo pair is generated and entered into the program, a red/green anaglyph image is created by the program. The image can then be viewed using stereo glasses, or a height profile can be generated using image shift or parallax data from the two images.

J. Complex Feature Analysis

The complex feature analysis program provides for the measurement of over 30 physical parameters of each particle in an acquired image. These measurements can be either field specific (such as field areas, number of fields, percentage areas of features, etc.) or feature specific (such as areas, Feret dimensions, centers of mass, minimum, mean, and maximum projection dimensions, and much more). The complex feature analysis also includes phase correlation and image histogramic analysis capabilities. Phase correlation is an analysis technique wherein two or more binary phases constructed from the same gray-level image are analyzed for phase attachment. The histogramic mode involves the statistical analysis of pixel intensities for a selected area or user-defined line of the image. In the area mode, the number of pixels at each intensity level are plotted in histogram form. In the line mode, the gray level at each point along a line is plotted.

K. Three-Dimensional Reconstruction

The 3-D reconstruction program is designed for viewing and analyzing the structural nature of samples. Typical applications include investigating the morphology of biological cells, the porosity of geological samples, or the granular features in solids.

The program provides for highlighting the perimeters or internal features of serial sections of objects, and then combining these serial sections into a single image on the display screen. The resulting image can be rotated, tilted, or a cross section taken so that the user can develop a good understanding of the topography and structural characteristics of the viewed sample.

TRADEMARKS

The following trademarks used in this chapter are the property of the companies indicated. A brief product description is included where appropriate; additional product information may be obtained from the vendor.

Climet®. Climet Instruments, 1320 W. Cotton Ave., Redlands, CA 92374.

HIAC/Royco®. Division of Pacific Scientific Company, Silver Spring, MD 20904.

Coulter®. Coulter Electronics, Inc., Hialeah, FL 33010.

Artek®. Artek Systems Corporation, Farmingdale, NY 14404.

Nikon Optiphot®. Nikon, Inc., Instrument Group, Garden City, NY 11530.

Omnicon®. Image Analysis System, ArtekImaging Systems, 4425 Bookfield Corp. Drive, Suite 200, Chantilly, VA 22021.

Quantimet®. Image processing and analysis system, Leica Cambridge Ltd., Clifton Rd., Cambridge, CBI 3QH, England.

Shakespeare®. The Shakespeare Corp., 901 Park Place, Iowa City, IA 53346.

Noran®. Noran Instruments, 2551 W. Beltline Hwy., Middleton, WI 53511.

Leica®. Leica Cambridge Ltd., Clifton Rd., Cambridge, CBI 3QH, England.

NORAN®. Noran Instruments, 2551 W. Beltline Hwy., Middleton, WI 53511.

Bausch and Lomb®. Bausch and Lomb Optics, 9200 Cody St., Overland Parks, KS 66214.

Microsoft®. Microsoft Corp., One Microsoft Way, Redmond, WA 98052.

VOYAGER®. Microanalysis System, Noran Instruments, 2551 Beltline Hwy., Middleton, WI 53511.

Motorola®. Motorola Electronics, 600 N. Highway 45, Libertyville, IL 60048.

Sun®. Sun Microsystems Inc., 2550 Garcia Ave., Mt View, CA 94043-1100.

Juliet III™. Image Analysis System, The Shakespeare Corp., 901 Park Place, Iowa City, IA 53346.

Windows™. Software information and data management system, Microsoft Corp., One Microsoft Way, Redmond, WA 98052.

REFERENCES

1. S. Inoue, *Video Microscopy,* Plenum Press, New York, 1986, pp. 531:13.
2. G. A. Baxes, *Digital Image Processing: A Practical Primer,* Prentice Hall, Englewood Cliffs, NJ, 1984, p. 186.
3. W. Green, *Digital Image Processing—A Systems Approach,* Van Nostrand, Princeton, NJ, 1982.
4. A. Kak, and A. Rosenfeld, *Digital Picture Processing,* Vols 1, 2, Academic Press, New York, 1982.
5. J. K. Beddow, *Particulate Sci. Technol. 11*:101–123 (1983).
6. H. Schwas, and H. E. Exner, *Powder Technol. 27*:207–213 (1980).
7. T. A. Barber, M. D. Lannis, and J. G. Williams, *J. Parent. Sci. Technol. 43*:27 (1989).
8. C. J. Digrado, *Bull. Parenter, Drug Assoc. 24*:62 (1970).
9. R. C. Hohmann, I. Conea, and K. P. Munnely, *Bull. Parenter Drug Assoc. 27*:96 (1973).
10. E. E. Sims, and R. Washington, *Inr. J. Pharm. 17*:167 (1983).
11. J. D. Levy, C. J. McCarthy, and R. E. Stevens, *Bull. Parenter. Drug Assoc. 31*:299 (1977).
12. J. R. Tricome, M. L. Lannis, and R. J. Sax, *J. Parenter. Sci. Technol. 40*:58 (1986).
13. S. Borchert, J. Abe, D. S. Aldrich, L. E. Fox, J. E. Freeman, and R. D. White, *J. Parenter. Sci. Technol. 40*:21 (1986).
14. H. G. Schroeder, and P. P. DeLuca, *J. Parenter. Drug Assoc. 34*:183 (1980).
15. E. O. Brigham, *The Fast Fourier Transform and Its Applications,* Prentice Hall, Englewood Cliffs NJ, 1988.
16. J. K. Beddow, *Particulate Science and Technology,* Chemical Publishing, New York, 1980.

17. J. K. Beddow, *Particulate Systems: Technology and Fundamentals,* Hemisphere, New York, 1983, p. 87.
18. J. K. Beddow, and T. Meloy, *Testing and Characterization of Powders and Fine Particles,* Heydon, New York, 1980, p. 216.
19. D. W. Luerkens, J. K. Beddon, L. A. Vetter, and F. London, *Theory of Morphological Analysis,* CRC Press, Boca Raton, FL, 1991, pp. 3–15.
20. B. H. Kaye, and G. G. Clark, *Part. Part. Syst. Charact. 6*:1–12 (1989).
21. B. H. Kaye, *A Random Walk through Fractal Dimensions,* VCH, New York, 1989, p. 149.
22. J. K. Beddow, *Classical Image Analysis—The Long March to Nowhere,* Tech. Rep., The Shakespeare Corp., Iowa City, IA, 1991.
23. B. H. Kaye, *Direct Characterization of Fine Particles,* Wiley, New York, 1981, p. 397.
24. *British Standard 3406,* Determination of particle size of powders. 4: Optical microscope method, British Standards Institution, London, 1963, p. 23.
25. C. Murphy, Application of image analyses in sizing and enumeration of pharmaceutical particulate matter, *Proceedings of International Conference on Liquid Borne Particle Inspection and Metrology,* Parenteral Drug Assoc., 1987, p. 183.
26. H. Kushner, I. Abrahamson, T. Barber, and J. Knapp, Error contribution in particulate determination from the viewpoint of sampling theory, *Proceedings of International Conference on Liquid Borne Particle Inspection and Metrology,* Parenteral Drug Assoc., 1987, p. 87.
27. B. H. Kaye, R. A. Trottier, and G. G. Clark, *Part. Part. Syst. Charact. 9*:209 (1992).
28. B. Mandelbrot, *Fractals, Form, Chance and Dimension,* W. H. Freeman, San Francisco, CA, 1977, p. 289.

4

The Electrical Sensing Zone Method (The Coulter Principle)

Roy W. Lines

Coulter Electronics Ltd., Luton, Bedfordshire, England

I. BACKGROUND AND SUMMARY

The first types of electronic particle and cell counters to reach the market employed optical sensing. This probably arose from the habits of viewing and counting such particles visually through a microscope. Indeed, one of the first two types was an automated microscope; the other was based on the use of a narrow beam of light focused through a narrow stream of particle-bearing fluid. The light beam in such a system is modulated by the passage of individual cells through the beam, and such effects are picked up by a photocell across the stream to produce electrical pulses for counting, and for size analysis if required.

These first optically based counters were quickly displaced for most applications by the simpler and more effective nonoptical electrical sensing zone method which remains the dominant laboratory tool for such work.

Although the first optically based counter designs were not commercially successful, one development of that time has been employed in virtually all optical counters since then. A fundamental requirement is the necessity of ensuring that the particle stream always intercepts the transverse light beam in precisely the same path for stable control of pulse amplitude and duration. This is accomplished by providing a small-diameter sample-carrying stream into the center of a larger sheath stream. This invention was due to Crosland-Taylor [1]. It is more commonly known today as "hydrodynamically focusing" the sample flow.

In those days, to obtain better particle sizing required a modified light path of small dimensions, an internal "aperture," but the laser did not then exist, and the requirements for an optical design exceeded the skill and patience available. Wallace H. Coulter had taken to exploring alternative ways of producing pulses, culminating in an electrical path of small dimensions [2]. This sensed particle volume, its sensitivity being dependent upon the ratio of the volume of the particle to the volume of the sensing zone. For example, 13 V applied across a 100 μm aperture of 13,000 Ω resistance generated a 1 mA current and a particle volume ratio of 1:13,000 for a 5 μm particle. This apparatus was the first to measure particle volume.

It has since been intensively developed for the particle counting, and the particle size distribution analysis, of almost all forms of finely divided particulate materials. The review which follows describes the principle and the methods used, as well as their application to biological and nonbiological particles.

Basically, the principle is that of the modulation of the impedance of an electrical current path of small dimensions, and this means not only by the application of a direct current, but also by an alternating field. Since around 1957 Coulter has known how to gather information from the interior of the particles electrically, however only recently has this been exploited, for instance for biological cells.

Biological cells can be generally considered as little bags of electrolyte solution contained within very-thin-walled membranes. At high applied electrical frequency, such as about 20–25 MHz, the membrane is mostly transparent, so if the applied aperture current is of the same frequency it is obvious that some of that current flows through the cells and not entirely around them. The resulting "high-frequency" pulse is slightly complex; it is a function both of particle volume and internal conductivity in relation to the conductivity of the surrounding fluid. In practice, a simultaneous direct current (dc) pulse produces a response primarily to particle volume. By dividing the high frequency (rf) signal by the dc signal, this quotient is quite a good measure of the cells' internal conduction or resistivity. One interesting application is to separate white cells from whole blood, analyze with an aperture, and plot the pulses produced on a scatter diagram to give three distinct clusters representative of three distinct populations, i.e., inherently different salt concentrations. By challenging the cells with red cell lysing chemicals one can selectively alter the conductivity of at least one of the populations. This then presents four clusters, the fourth being from eosinophils. Further refinements give a fifth cluster, due to basophils.

That technology is the basis of one type of white cell blood counter. The use of dc and rf current gives measurements therefore of two parameters which did not exist before the aperture impedance method was invented.

The electrical sensing zone (e.s.z.), or aperture impedance, method can be combined with other sensing means so that additional parameters are available for more precisely classifying the individual particles or cells. Optical sensing can be combined in several fashions, such as light absorption, light scatter, light polarization, and fluorescence, which are all feasible and practical now that reliable and more economical lasers are available.

II. INTRODUCTION

The principle of counting and sizing particles was developed by Wallace H. Coulter in the late 1940s under contract to the United States Navy as a method of counting blood cells rapidly [2,3], and was described in detail by Mattern, Brackett and Olson in 1957 [4]. Since that time, COULTER* particle counters and other devices utilizing the same principle have become established not only for counting and sizing blood cells but also for measuring the particle size distribution of any finely divided particulate matter which can be suspended in an electrolyte solution. Over 98% of automated blood cell counters in use today employ the aperture sensing system. Many hundreds, if not thousands, of different industrial particulate materials are characterized using the method. It is the only method which measures a particle volumetrically, i.e., in three dimensions. It has a lower limit of detection of below 0.3 μm. The method is the subject of American and British standards (see Sec. IX.C).

The principle, apparatus, and operation for the determination of particle size distribution has been reviewed in many publications, for example Ullrich [5], Allen [6–8], Rabinovitch [9–10], and Allen and Marshall [11]. Coulter Electronics Ltd. publishes two bibliography lists (Industrial, including pharmaceutical [12], and Medical, including hematological [13]), giving over 7000 references to the uses of the various COULTER COUNTER models. Richardson-Jones [14] has provided a comprehensive review of the application of the principle to blood cell analysis.

---

*COULTER, COULTER COUNTER, CHANNELYZER, and ISOTON are registered trademarks of Coulter Corporation.

III. THE COULTER PRINCIPLE

Figure 1 shows a schematic diagram of a simple form of apparatus. Particles suspended homogeneously at a low concentration in electrolyte solution are made to flow through a small aperture (or orifice) in the wall of an electrical insulator, which is commonly called the aperture (or orifice) tube; the aperture creates the sensing zone. In addition, a current path is established between two immersed electrodes across this aperture, setting a certain base impedance to the electrical detection circuitry. A direct current is generally used. As each particle enters the aperture, it has effectively displaced a volume of electrolyte solution equal to its own immersed volume, and the base impedance is therefore modulated by an amount proportional to the displaced volume of the particle. This results in an electrical pulse of short duration being created by each particle, the height of the pulse being essentially proportional to the volume of the particle. The pulse may be measured for instance as the change in resistance, current, or voltage across the electrodes.

The passage of a number of particles produces a train of pulses which can be observed on an oscilloscope and analyzed by counter and pulse height analyzer circuits to produce a number against particle volume, or equivalent spherical diameter, distribution. A volume or mass ("weight") against size distribution can also be measured, calculated, or computed; the "weight" percentage being possible if all of the particles have uniform density or a known density distribution across their size range.

Simple models have only one counter and size-level circuit (and so are called single-channel models); more complex instruments can obtain number and/or mass (weight) distributions automatically in up to 256 size channels within a few seconds. Counting and sizing rates of up to some 10,000 particles per second are possible, with each pulse height being measured to within 1% or 2%.

Since most particulate materials are irregularly shaped, the volumetric response is invaluable, as volume is the only single measurement which can be made of an irregular particle in order to characterize its size. In biological applications the size response is usually left calibrated in volume units (femtolitres or "cubic microns"), but industrially it is conventional to report the equivalent spherical diameter calculated from it. This volumetric method makes no assumptions about particle shape, and indeed is not greatly affected by particle shape except in extreme cases like flaky materials, such as some clays. To count the number of particles in a known volume of suspension, such as for particulate contamination studies or for a blood cell count, the sample volume must be accurately metered. The original method was by means of a simple mercury siphon and metering system, often called a manometer. A mercury-free piston displacement metering system is also available.

Modern instrument designs have a range of specific features and various data reduction handling and presentation methods, and the manufacturers should be contacted for details. Some are specific to just one application, e.g., human blood cell characterization. The volumetric sizing resolution, speed of data collection, statistical accuracy of counting, freedom from any optical response effects, and the reliability and simplicity of calibration make these devices unsurpassed for providing particle counting and size distribution analyses. The only basic limitation is that particles must be suspended in an electrically conductive fluid.

IV. PARTICLE SIZING RESPONSE

The volumetric response is, in theory, absolute. For instance, Kachel [15] stated that the electrical signal from an aperture of length-to-diameter ratio in the range of 1:1 to 2:1 depends on the particle and aperture parameters:

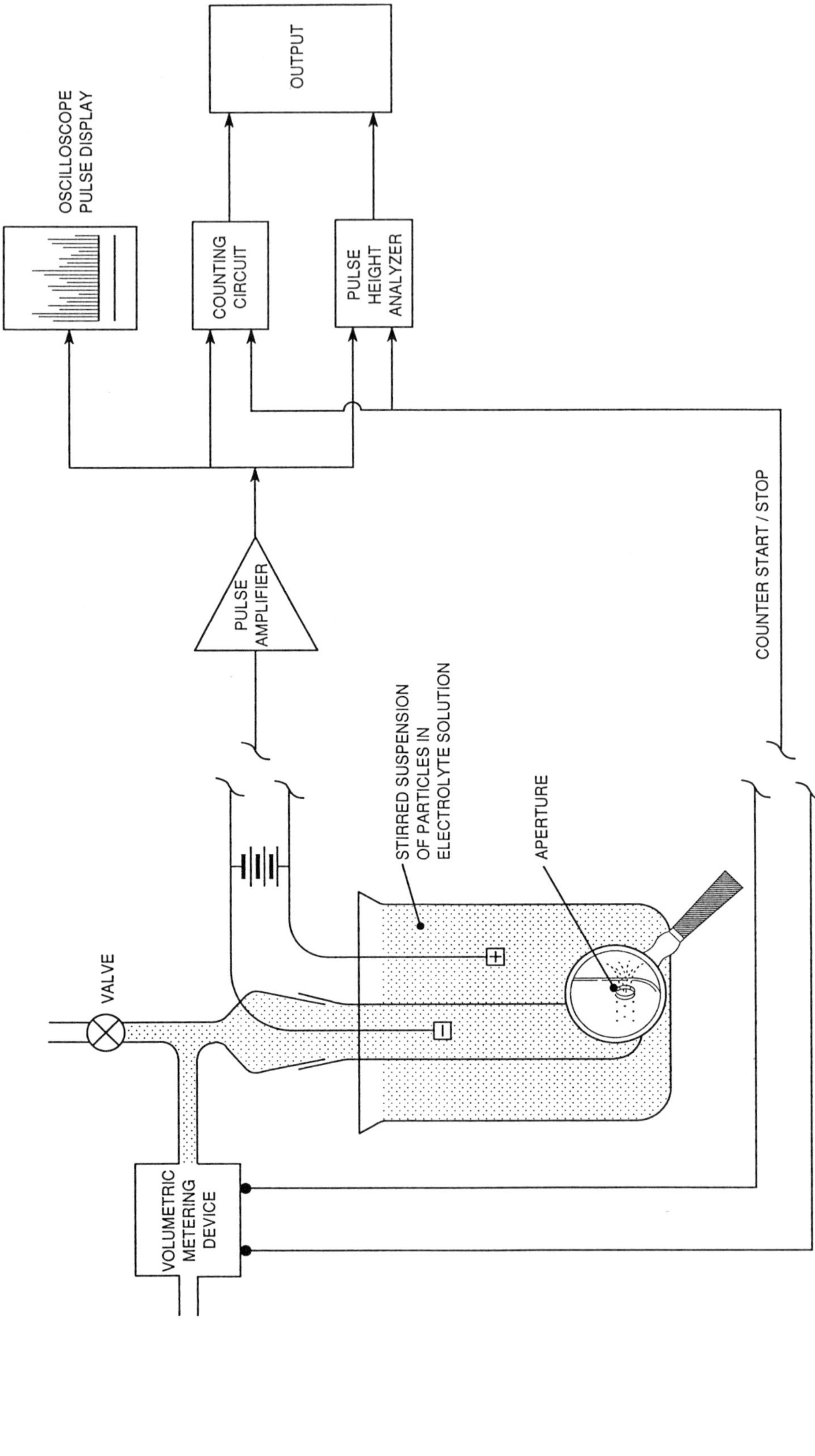

Figure 1 Schematic drawing of a simple form of analyzer.

$$U = \frac{V\rho_0 if}{\pi^2 R^4} \tag{1}$$

where

U = amplitude of voltage pulse
V = particle volume
ρ_0 = electrolyte resistivity
i = aperture current
f = particle "shape" factor
R = aperture radius

Particle resistivity has very little effect on voltage response, unless it is close to the resistivity of the fluid. If the particle resistivity changes from one millionth to one hundredth of that of the electrolyte solution there is less than 1% change in the response [16]. Colloidal charges are too small to have any effect [16]. The first such instruments from all sources were quite sensitive to electrolyte resistivity changes, an effect which was eliminated from many later designs as a result of the principle and circuitry disclosed in 1966 [17]. Table 1 shows the effect on instrument calibration response before and after the introduction of that "constant-current" invention.

Other simple, early, response theories (e.g., Coulter Electronics, Inc. [16], Batch [18], Allen [6], and Grover et al. [19–21]) predicted a dependence not only upon particle shape but also that the response would increasingly deviate from linearity as the size of the particle approached the diameter of the aperture. It was thus established in the early literature that an upper limit of sizing accuracy lay at some 35–40% of the aperture diameter. This idea held for some years, even after Saranummi [22], Kachel [15], and Scarlett [23] independently developed more comprehensive theories which showed otherwise. Saranummi's approach predicted only a small dependency on the particle to aperture ratio, while Kachel's showed no such dependency, and Scarlett concluded that the response seemed to be due only to the volume of the particle, if second order effects were small. Those effects included the field potential distribution through and around the sensing zone, and are discussed later (see Sec. VII.A).

The first experimental proof of virtual linearity of response, up to some 80% of the aperture diameter, came from Harfield and others [24,25]. They used spherical particles of polymer latex and the COULTER COUNTER model ZM. The experiment involved the use of different sizes of "monosized" latex particles each measured by a range of different apertures, and therefore required no assumed or measured "real" sizes for the particles. The experiment has not been repeated for other particle shapes as no other series of suitable model particles appears to exist, so the linearity of response for other particle shapes has not yet been verified experimentally. It is reasonable to assume, however, from all existing theory, that no significant extra alinearity will exist for nonspherical shapes.

That particle shape has a basic effect upon instrument calibration was proposed in 1957 [16] and has continued in later theories. It is, however, a constant effect and, as indicated by Harfield and others [24–26], is not dependent on the particle-to-aperture-diameter ratio. That is, there may be a different calibration factor for different shapes of particles, but that one factor will hold over the full measuring range of aperture. The effect can be studied, for instance, by calibrating the instrument with a fraction of the material itself (see Sec. V). Batch [18] showed that the calibration factor varied little with widely different shapes and resistivities. Eckhoff [27,28] also used this "mass integration" calibration method to determine the calibration factors for spherical particles (glass beads) and angular particles (cement, and the raw mix used in its preparation), and found no significant variation between them. A summary of their results is given in Table 2.

Personal experience of rigid particles confirms that moderately irregular particles, such as

Table 1 Effect of "Constant-Current" Feature on Consistency of Calibration

(i) Effect of electrolyte strength (aperture resistance) on calibration

Sodium chloride in water (g/l)	Calibration factor COULTER COUNTER Model ZB	Model A
100	1.510	6.38
50	1.510	5.30
25	1.515	4.41
10	1.515	3.50
5	1.515	3.21
2.5	1.515	2.59

(ii) Effect of electrolyte solution temperature on calibration

Electrolyte solution: ISOTON II buffered saline solution (Coulter Electronics Ltd.)

Temperature (°C)	Calibration factor COULTER COUNTER Model ZB	Model A
2	1.515	3.21
20	1.515	3.46
40	1.515	3.81
60	1.510	4.02
80	1.510	4.10

milled silica or alumina, have no measurable shape effect and that only grossly irregular shapes do, such as the flaky particles of some clays and mica. These have been found to give an apparent volume response double or even treble the physical volume; i.e., the diameter calibration factor could differ by up to 66% from the calibration factor derived from spherical latex particles.

Rodlike particles appear to give no shape effect. O'Connell and Martsch [29] microtomed wool fibers to a constant length, and measured them by means of a COULTER COUNTER instrument, so that the "volume" response would reduce to "cross-sectional area" response, from which fiber diameter could be calculated and compared with microscopy. The results were virtually identical (Table 3(i)). Personal investigation showed similar results (Table 3(ii)).

Interestingly, just as fiber diameter can be measured from the volume response of particles cut to a constant length, fiber or particle length can be measured to some degree of accuracy by observing for instance the length of the particle-generated pulses under known conditions [30,31].

The more voluminous reports which theorize a "shape" factor have considered the response of red blood cells; however, these cells are readily deformable, subject to osmotic change, and

Table 2 Effect on Calibration Using Particles of Different Materials

(i) Calibration results

Material	Particle shape	Electrical resistivity (Ω-cm) (or above)	Calibration factor (*Kd*) 400 μm aperture	200 μm aperture	100 μm aperture
Perspex	Spherical	10^{15}	17.04	7.03	3.4
Aluminum oxide	Very angular	10^{10}	(B)	6.86	3.57
Pulverized coal	Angular	10^{10}	17.0	6.75	3.4
Ground coke	Rounded	1	16.55	7.0	3.85
Fly ash	Rounded	(A)	16.9	7.03	3.46

(ii) Calibration results

Material	Calibration factor (*Kd*) mean of four to six determinations	Standard deviation of *Kd* determinations (*n*-1)
Glass beads	8.08	0.08
Cement	8.02	0.15
Raw mix	8.08	0.18

(A) Fly ash contains particles of various resistivities ranging from silica (circa 10^{18} Ω-cm) to coke (circa 1 Ω-cm).
(B) Aloxite particles appropriate to this aperture were too heavy to maintain in suspension.
Source: Part (i) from Ref. 18. Part (ii) from Refs. 27 and 28.

can be conductive, so they are not as easy to study as the rigid "industrial" particulate materials. Gregg and Steidley [32], Waterman et al. [33], Grover et al. [19–21], Kachel [34], and others have examined the response of the electrical sensing zone method to blood cells and have proposed a "form factor" of 1 for an elongated object oriented along the axis of the aperture, 1.5 for a sphere, and 2.8 for a disk perpendicular to the flow. The analogy is that somehow the electrical path flows around a streamlined envelope, or outline, of the particle, thereby making the particle appear to be larger than it is.

Certainly, evidence presented from red blood cells does support this view, but it is complicated by the above-mentioned effects, as well as often by a shape change to spherical or crenated morphometry when the cells are placed in an electrolyte diluent necessary for their analysis. All of this makes the expression "form factor" for red cells more meaningful than just "shape factor."

Marshall and Mehta [35], Eckhoff [36], Lloyd et al. [37], and Lloyd [38] were forced to investigate the effect of shape in detail experimentally by using large-scale mock-ups of the aperture, several centimeters across. If there is no error in such scaling up, their work indicates that shape does have a constant effect. Marshall and Mehta concluded that

1. Pulse height was directly proportional to particle volume for spheres.
2. Pulse height was not directly proportional to particle volume for rod and disk shaped particles, but was modified by a shape factor, dependent on their orientation within the aperture.

Table 3 Number Percentages of Fibers in Various Diameter (denier) Ranges

(i)

	Sample 1		Sample 2		Sample 3	
Fiber diameter range (μm)	Microscopy	COULTER COUNTER analyzer	Microscopy	COULTER COUNTER analyzer	Microscopy	COULTER COUNTER analyzer
10–20	52.1	54.2	30.0	32.1	14.8	17.4
20–30	44.8	42.4	57.5	55.6	62.0	56.2
30–40	3.0	3.1	11.7	11.6	20.2	23.4

(ii)

Fiber diameter range (μm)	USDA microscope[a]	COULTER COUNTER analyzer[b]
10–20	42.05	42.05
20–30	52.31	52.00
30–40	5.28	5.15
40–50	0.33	0.22
50–60	0.03	0.00

[a]Average of 8 laboratories
[b]Coulter laboratory
Source: Part (i) from Ref. 29. Part (ii) from Lines, unpublished (1967).

3. Particle volume as seen by the sensing zone was greater than its physical volume for spheres, disks and short rods. (This observation supports the electrical flow path theory noted above.)

Eckhoff's model showed that response is proportional to particle volume, including any reentrant regions. For porous particles, response is proportional to displaced volume when the pores are parallel with the field.

With these effects in mind, it is recommended that for any particular, known, material the apparatus is self-calibrated when the instrument design allows [16,39,40]. This is a "primary" calibration method, as follows.

V. CALIBRATION

The electrical sensing zone method provides two basic measurements, particle number and particle size (or volume). In the "semiautomated" instrument designs (e.g., as in Fig. 1), which feature volumetric metering of the sample suspension through the aperture, the particle number result requires no calibration. Subject to coincidence and other corrections (see Secs. VI.G and IX.A, for example) the resulting particle count can be regarded as accurate; it is a reference method.

Calibration of the particle size response is, however, always required.

As already stated, the ideal method of calibration, when it can be performed, is the "primary" method of mass integration (e.g., [40]), sometimes also called mass balance. That is, the instrument is calibrated directly with the particles of material under test, using the standard gravimetric and volumetric methods of an accurate balance, pipets and flasks. A size fraction of the particles under test, preferably not exceeding some 10:1 by particle diameter to be sure that they are all measured, is diluted to a known mass concentration in electrolyte solution. The total volume, in instrument units, of the particles measured in a known volume of suspension is related to the known mass concentration and the particles' immersed specific gravity (density), allowing the calibration factor to be calculated. In this way, the calibration procedure approaches an absolute reference method, as it eliminates any possible errors which may be caused by particle shape, porosity, conductance, or any other effect which may be present but cannot easily be isolated.

The diameter calibration factor (*Kd*) is calculated from

$$Kd = \left[\frac{6WV_m 10^{12}}{\pi V_T \rho (\Delta n V)} \right]^{1/3} \tag{2}$$

where

W = mass of sample in beaker (g)
V_T = volume of electrolyte solution in which W is diluted (cm^3)
V_m = manometer volume (cm^3)
ρ = immersed density of the particles (g/cm^3)
Δn = number of particles in a size interval
V = arithmetic mean volume for that particular size interval, in instrument units (e.g., product of threshold value, aperture current, and attenuation)

It is of interest to note, then, that an electrical sensing zone instrument can be used to determine the immersed density of a powder, if the calibration factor is known. The approach has been used to obtain the wet densities of various clays [41,42].

Of course not all particulate matter is homogeneous in terms of its density (for example, river sediments), so the method is not universal. In these cases, and also for general convenience, it is more common instead to perform a "secondary" calibration method using a narrow-size-ranged polymer latex sample, whose mean, mode or some other "size" has been measured by another technique. Suitable latex particles are obtained from several sources including, in the United States the National Institute of Science and Technology (formerly the National Bureau of Standards) [43] and, in Europe, the Community Bureau of Reference (B.C.R.) [44] (see Sec. IX.D). Whichever the source, full traceability to those reference materials is recommended.

A common method of calibration is to choose latex particles having a mean diameter of between some 5% and 20% of the aperture diameter, and to determine the instrument settings (e.g., of current, attenuation, and threshold) which halve the total particle count of the population of singlet particles. These settings are equated to the mean diameter of the latex particles. Doublets, caused by agglomeration and/or coincident particle passage through the sensing zone, are thereby prevented from causing any error. This "half-count" method is simple, rapid, and very precise.

Models fitted with a multichannel pulse height analyzer can be more quickly calibrated by relating the mode of the generated size distribution to the assayed modal value of that latex, as provided by the supplier.

Hematologically, for red blood cell sizing, calibration is performed against a commercial, stabilized, blood product of known assayed values, or against a series of specimens of normal fresh blood cells, taken into a known anticoagulant, whose mean sizes have been determined by the reference procedures of total red cell volume (centrifuged hematocrit) and total red cell count. (Total red cell count may be obtained, for example, from an electrical sensing zone analyzer fitted with an accurate manometer and used with accurate dilution methods; see Sec. IX.A). The

Table 4 Stability of Calibration Over Time

	Calibration value (Kd)		
	November 1982	March 1987	% change
Instrument 1 and 280 μm aperture	45.8	44.3	3.3
Instrument 2 and 50 μm aperture	6.45	6.45	0

calibration procedure may or may not be allowed to account for any plasma still trapped within the packed bed of red cells.

Whichever method of calibration has been chosen, it should be specified when results are reported. The one calibration will be valid for the whole range of size measurement from any one aperture or sensor. Calibration drift is uncommon (Table 4), but it is recommended that routine control tests are made. Note that, with some more basic models, the size response is dependent also upon the resistivity and temperature of the electrolyte solution, in which case these two effects must be eliminated or minimized by standardizing the electrolyte solution composition and temperature or by frequent recalibration.

VI. PRACTICAL CONSIDERATIONS

A. Particle Size Range

For routine practical purposes with most industrial particulate materials, the overall size range commonly considered for the electrical sensing zone method is about 0.5–200 μm equivalent spherical diameter, although some commercial instrument designs will provide results from about 0.4 to 1200 μm, particularly for low density particles, and experimental designs have provided results down to 0.1 μm or below and up to 1500 μm or more. More than one aperture (or sensor) size is needed in order to cover such a large size range.

The lower size limit of a sensor is set by the electrical noise generated by the applied current upon the electrically conducting fluid within the sensing zone itself, as the aperture is, in effect, a somewhat noisy resistor. The electrical noise level is often as low as 1.5% of the aperture diameter, e.g., 1.5 μm with a 100 μm aperture. Coupled with the large sizes measurable with each aperture (up to 80%), this means a dynamic measurement range of some 50:1 by diameter from each sensor. Of course not all instrument designs are such that this dynamic range can be utilized or are linear enough in their circuitry to allow such a range to be measured with accuracy.

Commercially available apertures exist down to 15 or even 10 μm in diameter, so that particle size distributions down to 0.3 μm or below can be obtained.

An interesting experimental design using the Coulter principle was created by DeBlois and Bean [45]. They created a sensing zone by selecting an individual submicrometer pore etched in an irradiated polycarbonate plastic sheet; the material now known as Nuclepore membrane. A typical aperture was some 0.5 μm in diameter and 3 μm thick. Because it was very fragile, particles were made to pass through it more by means of their electrophoretic mobility than by a differential pressure. The aperture membrane only lasted for three weeks, but during that time they established that they could accumulate a size distribution of a total of 134 polystyrene latex particles of nominal diameter 0.091 μm, obtaining a coefficient of variation from the mean diameter of 5.9%. The device was later used to count and size viruses and bacteriophage particles in the range of about 0.1–0.2 μm diameter, and also to study simultaneously their electrophoretic and electroosmotic velocities [46,47].

The upper size limit of the method is usually set by the design's ability to suspend particles uniformly in the sample beaker. Various stirrer and container designs have been proposed and are used, and the electrolyte solution's density and/or viscosity may be increased in order to assist homogeneous suspension. McCave and Jarvis [48] found that, as the rate of sampling through a large (2000 μm diameter) sensing aperture exceeded the rate of fall of sediment particles initially suspended in a large beaker, they could simply turn off the stirrer and obtain a particle size distribution directly when using a COULTER COUNTER model T multichannel analyzer.

Harfield et al. [49] approached the same problem by mounting the aperture wafer horizontally and allowing (and encouraging) the particles to fall through the sensor. In this way they were able to measure 1.5 mm steel balls. This device did not become commercially available.

B. Choice of Sensing Aperture

If the majority of the particles lie within a 50:1 diameter size range, the most suitable aperture will be chosen. For instance, a 30 μm aperture will measure, on some instrument models, from about 0.5 to 20 μm or more; a 140 μm aperture from about 2 to 85 μm or more.

If the particles cover a wider size range, two or more aperture sizes will be used, and the results overlapped to provide the full-size distribution analysis (e.g., [40]).

The smallest apertures may be more sensitive to external electrical noise and are more prone to temporary blockage, such as from inadvertent contamination or agglomerated particles. Blockages are normally not common, but will occur from time to time depending on the particulate specimen and/or the care taken with the experimental conditions. The cleanliness of the aperture is monitored by optical means, as well as by other indicators such as electronic blockage detection, an oscilloscope display of pulses, and an audible counting rate. Blockages are usually easily cleared by applying momentary back pressure to the aperture by means of a small piston, or by a rotation of the operating taps, and perhaps at the same time applying high current to the aperture to "boil" out the blockages. In case of difficulty, the aperture tube can be easily dismounted and cleared by the use of greater back pressure, brushing, soaking in acid, or, for the larger apertures, by carefully applying low power ultrasonic energy for a fraction of a second.

C. Choice of Electrolyte Solution

The electrolyte solution should have a conductivity similar to 0.2–20% w/v sodium chloride in water, sufficient to give an aperture resistance between about 1 and 100 kΩ, ideally around 5–40 kΩ. It must be compatible chemically with the sample material under test and allow proper dispersion (usually with the help of a surfactant and ultrasonic agitation) and suspension of the particles (with the help of a built-in stirrer and thickening agents if necessary).

A great many materials are water-insoluble, or virtually so. In these cases, an electrolyte solution of about 10 g/l sodium chloride in water (e.g., normal or isotonic saline) or 50 g/l trisodium orthophosphate in water is commonly used. Sometimes chloride ions inhibit dispersion and an alkaline phosphate solution has advantages, for instance in dispersing silicas, clays, and minerals. Many other solutions of inorganic salts in water have been used for particular purposes; for instance, drug solubility studies can be carried out in dilute hydrochloric acid to simulate the contents of the stomach, or the common ion effect used to depress the solubility of some materials so that they can be more easily measured. Typical concentrations of particles for measurement are of the order of 10–100 mg/100 ml, so even low solubilities can be important, though it is the solubility rate which is the determining factor as to whether the analysis can be made at all. Multichannel counters are invaluable for soluble materials as the total time of analysis is short (10–20 s). It has been found advantageous sometimes to remove the aperture illumination temporarily to prevent any small temperature rise in the electrolyte solution increasing the solubility of the particles.

Alternatively, organic electrolyte solutions may be used, such as 50 g/l lithium chloride in methanol and 50 g/l ammonium thiocyanate in isopropanol, and many others are available. The dielectric constant of the "solvent" base must be greater than about 10 to allow the dissolved salt (or acid) to dissociate and give an acceptable aperture resistance. Thus, liquids like petroleum spirit and mineral oils cannot be used without a coupling solvent.

To measure very small particle sizes (e.g., 1 μm and below) with small apertures, it is often beneficial to raise the conductivity of the electrolyte solution, where this is not detrimental to the sample. An example is to use 100 g/l sodium chloride in water rather than 10 g/l. As well as improving the signal-to-noise ratio, this also assists in preventing any "electrical translucency" effects due to the surface breakdown potential of conductive materials such as metals (see Sec. E).

The analyzer itself may be used to monitor the efficacy of dispersion. By taking particle counts near to the coarse end of the size distribution and also near to the fine end, the change in particle counts over a few minutes may be used to determine if the suspension is stably dispersed (see Table 5).

When selected, the electrolyte solution should be filtered before use so that it is essentially free of particles in the desired size range of measurement. Commonly, 0.45 μm or 0.22 μm membrane filters are used for this purpose.

D. Suspension of Large and/or Dense Particles

Large particles may not be adequately suspended, even with the stirrer at the maximum rate which does not entrain air bubbles, which would then be counted as particles. For instance, 300 μm particles of density 2.5 g/ml may be difficult to suspend. In these cases, it is customary to add a thickening agent (e.g., up to 700 ml/l glycerol or 1000 g/l common sugar) to raise the viscosity and/or density of the electrolyte solution.

Even with efficient stirring and an increase of the viscosity and/or density of the electrolyte solution, large dense particles may not stay in suspension uniformly enough to enable a representative sample to be withdrawn repeatedly through the aperture without distortion of the size distribution. Kinsman [50] suggested a round-bottom beaker and the addition of baffles. Yarde [51] described a round-bottom beaker used with a double-bladed stirrer paddle to improve the conventional squat-form beaker, and round-bottom beakers have now virtually replaced conventional beakers even for small particle analyses. Eckhoff [52] described the use of a

Table 5 Monitoring the Stability of a Suspension

Change in particle count at		
Largest particle sizes (d_{max})	Smallest particle sizes (d_{min})	Suggests
Increase	Increase	Crystallization, precipitation
Decrease	Decrease	Dissolution
Decrease	Increase	Size reduction, deflocculation
Increase	Decrease	Flocculation, "aggregation"
Decrease	No change	Settling of large particles
Increase	No change	"Creaming" of large particles
No change	No change	Stable dispersion

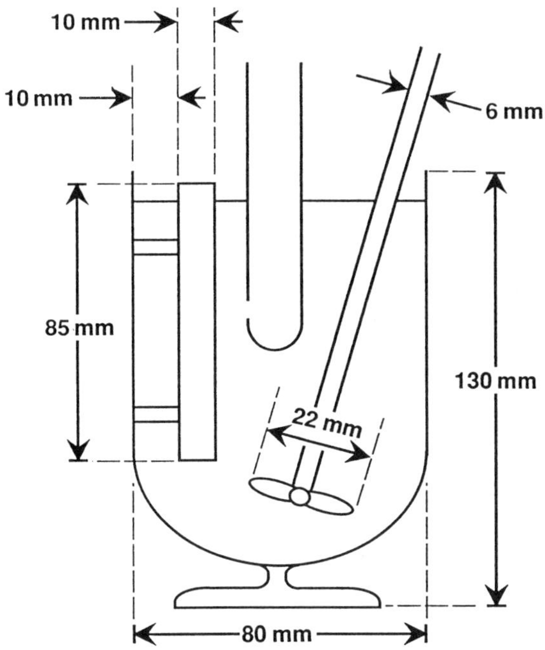

Figure 2 Design of baffled beaker (Harfield et al. [49]).

separate baffle that could be introduced into the beaker and orientated to break up the rotational component of stirring.

Harfield et al. [49] described an improved new beaker design, which combined a large-capacity (400 ml) round-bottom beaker with a single glass baffle spaced sufficiently away from the wall of the beaker to allow a free path of liquid between the baffle and the wall. Figure 2 shows the design of the beaker with approximate dimensions determined empirically to give the best mixing conditions.

Using this new baffled beaker, glass beads (specific gravity approx. 2.5 g cm^{-3}) of nominal size 100 μm could be suspended in a low viscosity electrolyte (ISOTON II diluent, Coulter Electronics Ltd.), and glass beads of nominal 370 μm size in a solution of 70% v/v glycerol/ ISOTON II diluent.

Nienow and Ang [53] and de Neve and Wuyts [54] have also published novel approaches to measuring large and/or dense particles.

E. Conducting Particles and Porous Particles

It might be expected that particles of conductive materials could not be accurately measured by the electrical sensing zone method. For this to be so, the conductivities of the particle and of the electrolyte solution would have to be very similar and there would have to be no electrical boundary between them. It is a fact that particles of highly conductive materials such as copper (e.g., Ullrich [55]), silicon, and carbon can be sized correctly providing that the voltage applied across the aperture does not allow the barrier potential at the particles' surface to be exceeded. This is about 5–20 V across the aperture, depending on material, according to Harfield [56]. The effects of the Helmholtz double electrical layer and/or an oxide layer or an adsorbed gas film on the surface of the particles help to render metal particles less conducting.

Particles which are porous and contain electrolyte solution have been found to be sized as their solid volume; i.e., the current effectively enters the particles to give a sizing response not related to their "envelope" volume. The effect of porosity and the technique for filling porous particles so as to allow the measurement of "envelope" volume are detailed by Horák et al. [57] and Van der Plaats and Herps [58].

In the case of cells enclosed by membranes, such as red blood cells, the linear response to cell volume changes at some "breakdown" value of applied voltage (e.g., Groves [59]). Blood cell analyzers are, of course, designed to operate below the breakdown potential.

F. Precision of Measurements

In any type of particle counter, the number of particles in a metered volume of suspension will vary from the number in any other identically metered volume, as expected from statistical theory. A mean count of $\overline{N}$ particles per unit volume could be expected to vary according to the standard deviation of measurement, calculated from Poisson theory as $(\overline{N})^{1/2}$. Thus a mean count of 100,000 particles should be obtained with a standard deviation of 316.2 particles, i.e., a precision (expressed as coefficient of variation, c.v.) of 0.316%.

In practice, the Poisson model is imperfect, as two particles cannot exist in the same spatial position at the same time, so precision of particle counting is expected to be better than the "theoretical" $(\overline{N})^{1/2}$. Table 6 shows that, for a clean manometric sampling system, this is so.

Clearly, it is important that any particle counter/sizer should measure a large number of particles in order to achieve the greatest statistical confidence in the results. In electrical sensing zone models, the concentration should be sufficiently great, but not so great that coincidence limitations are exceeded. Consequently, measured precision depends on both $\overline{N}$ and the value of $\overline{N}$/ml; i.e. larger measurement volumes may be preferred, at lower concentrations. Of course, in contamination studies where there are a low number of particles per unit volume, ideal counting statistics may never be achieved but coincidence corrections may be insignificant and unnecessary.

The resulting precision of particle size distributions obtained from different total numbers counted has been evaluated by Lines [60] who showed that the average standard deviation of any point on a typical weight (or, mass) percentage (wt%) size distribution was

2.12 wt% for a total distribution of 2700 particles counted
0.74 wt% for a total distribution of 10,000 particles counted
0.33 wt% for a total distribution of 100,000 particles counted.

Thus, a size distribution obtained from 100,000 particles gave a precision of significantly better than 1 wt% over its whole range.

Table 6 Precision of Particle Counts at Various Count Levels

Average count ($\overline{N}$) ($n = 31$)	Standard deviation S.D. (n-1)	Poisson S.D. $(\overline{N})^{1/2}$
106	9.46	10.3
1,076	30.8	32.8
9,755	85.8	98.8
10,005	85.0	100.3
10,500	99.4	102.5
49,444	221	222
98,559	234	314

COULTER COUNTER model ZM, nominal 10 μm latex, 31 replicate counts, clean manometric sampling system.

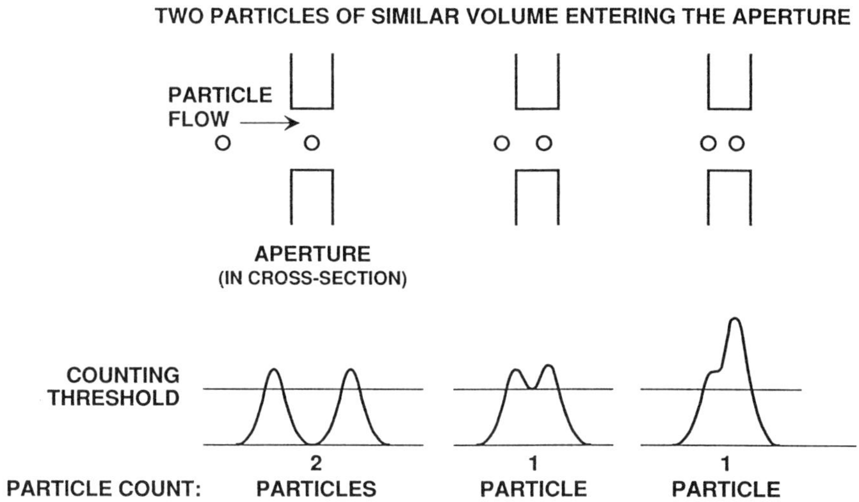

Figure 3 Effect of coincidence. 1. Primary coincidence, or horizontal interaction: two particles close together in the sensing zone count only as one particle.

G. Coincident Particle Passage

All types of sensing zone particle counters must be operated at concentrations low enough that essentially only one particle is present in the zone at any one time. Coincidence limits therefore vary according to the volume of the zone (i.e., the size of the aperture) and according to the response of the electronic circuits. If two particles are in the zone at the same time, they may only be counted as one (Fig. 3) and, in addition, if they are very close together they may be sized as one particle of volume equal to their sum (Fig. 4), exactly as if they were physically joined. By keeping concentration levels relatively low, e.g., to 10^5 particles/ml for a 100-μm electrical

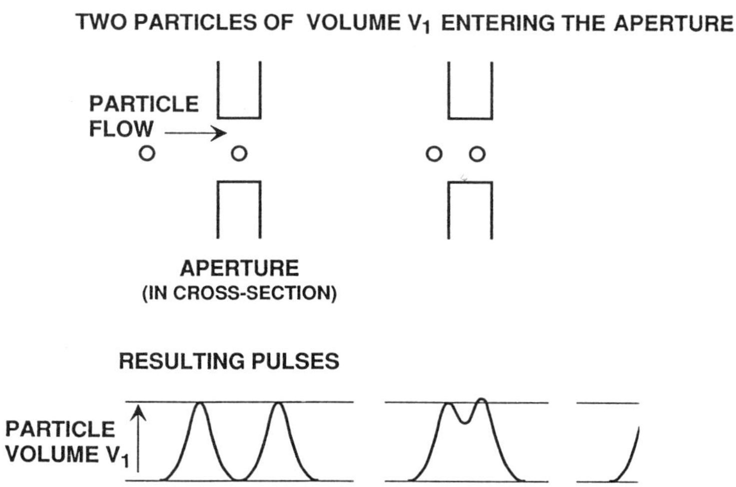

Figure 4 Effect of coincidence. 2. Secondary coincidence or vertical interaction: two particles close together in the sensing zone are sized as one particle of volume (V_2), approaching the sum of their individual volumes ($2V_1$).

sensing zone aperture, this count "loss" is small and can be easily corrected mathematically. Under these conditions, any particle sizing error caused by coincidence is usually not detectable but, if necessary, the effects can be virtually eliminated by analyzing at a lower concentration for a longer time, or by using a smaller aperture.

Several coincidence theories exist, varying in complexity (for instance, Refs. 16,61–65) but, providing that the concentration is kept fairly low, say to below the "10% coincidence limit" (where one particle in 10 is not being counted) they all give very similar corrections. A common expression [16] for the lost count due to coincidence (n'') is

$$n'' = p\left[\frac{n}{1000}\right]^2 \tag{3}$$

where

$$p = 2.5\left[\frac{d}{100}\right]^3\left[\frac{500}{V}\right] \tag{4}$$

p is the coincidence factor and n is the number of particles counted in the metered volume V (μL) using an aperture of diameter d (μm). The factor of 2.5 was derived experimentally for an aperture 75% as long as its diameter, a standard Coulter configuration for apertures above 50 μm. It can be thought of as the effective volume of the sensing zone being two and a half times its physical volume; the zone extending beyond the physical edges of the aperture itself (Fig. 5).

Harfield, Lloyd, and Cowan [66] cast doubt on the common use of the Poisson distribution model for predicting coincident occurrences and related the correction to the particle transit times, sensing volume and circuit response.

In practice, most modern analyzers make the corrections automatically, or the manufacturer provides the coincidence factor for each aperture size.

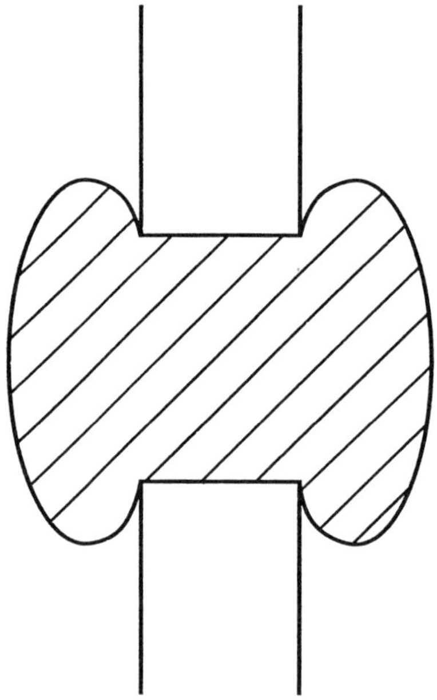

Figure 5 Cross section through the effective volume of the sensing zone (shaded).

VII. IMPROVEMENTS IN, AND DEVELOPMENTS OF, THE METHOD

A. Improving Sizing Accuracy

With standard apertures and sampling configuration, the electrical sensing zone method gives a much superior sizing accuracy compared with most other particle sizing methods. Even so, for very narrowly size ranged particles, such as polymer latexes, much greater sizing resolution can be achieved by making a few improvements.

In the standard design configuration (e.g. Fig. 1) particles are drawn through the aperture from all positions around it. Those particles passing centrally into, through, and out of the sensing zone generate pulses which are essentially proportional to their displaced volume (as in Fig. 6A). However, those particles which do not pass centrally through the electrical zone produce a slightly larger "artifact" pulse than expected for their displaced volume (e.g., Thom et al. [67], Thom [68], Grover et al. [19], Kachel [34], Atkinson and Wilson [69], and Wilson [70]) the error being most for those particles passing closest to the edge of the aperture, (as in Fig. 6B,C). This causes the size distribution to be skewed towards to the coarser sizes, the effect being most significant and hence most noticeable on very narrow size distribution materials (e.g., [26,69]). In extremely narrow size distributions, the error may even show as a second "particle" population (e.g., Fig. 7). On wider size range samples, such as most powders, suspensions and emulsions (wider than some 3:1 by diameter) the effect is not significant or observable.

For increased sizing accuracy, when needed, many of these "artifact" pulses can be screened out electronically by examining each pulse and then "editing" the unwanted ones out of the analysis, or the particles can be physically prevented from flowing nonaxially by "hydrodynam-

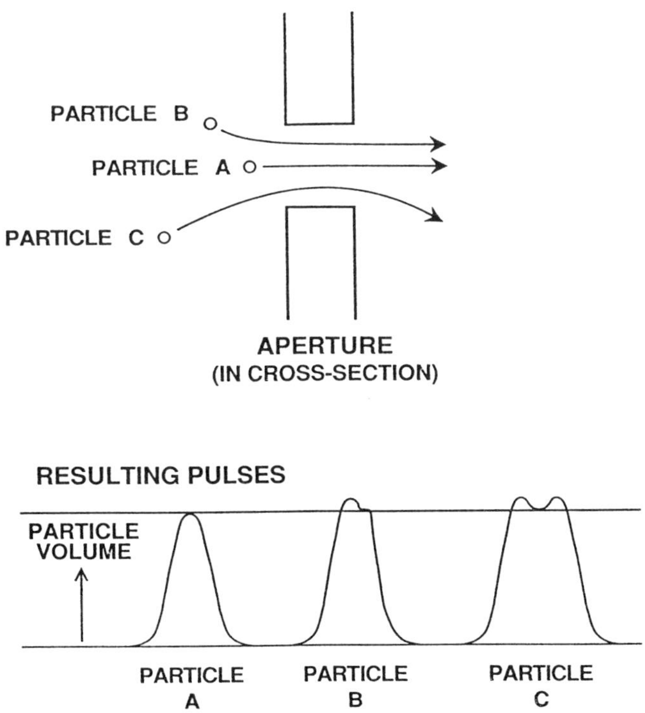

Figure 6 Pulse shape depends on the particle flow path through the aperture.

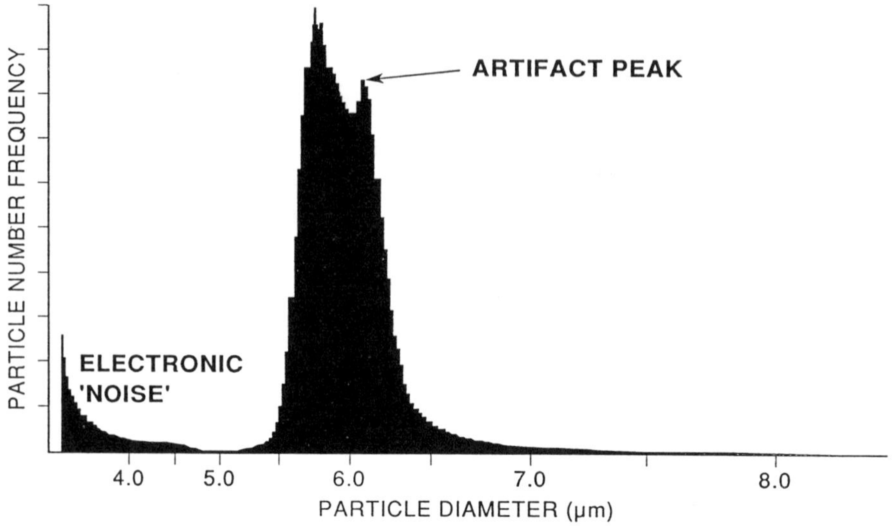

Figure 7 Narrow-size-range latex distribution showing second population artifact due to some particles' nonaxial flow through a standard aperture.

ically focusing" them into the sensing zone (Fig. 8). In both of these ways, extreme size distribution resolution and accuracy can be obtained if it is required. Figure 9 shows the same latex sample as in Fig. 7, but analyzed from edited pulses. The editing approach (e.g., Hogg [71]) is also useful in removing some of the electronic noise, allowing the smallest particles to be resolved more clearly.

Thom [67,68] introduced hydrodynamic focusing as a means of improving the sizing response of the electrical sensing zone method. Harfield and Cowan [72,73], Göransson [74], and Merkus et al. [75] have reported results from an improved experimental hydrodynamic focusing device. Göransson [74] also compared his results from narrow size-ranged materials

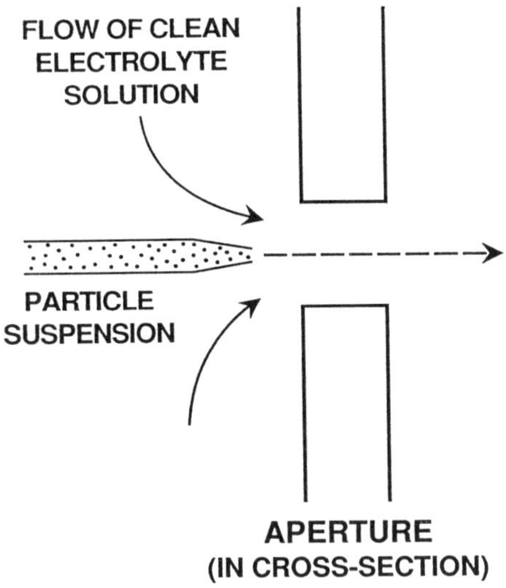

Figure 8 Hydrodynamically focused particle flow through aperture.

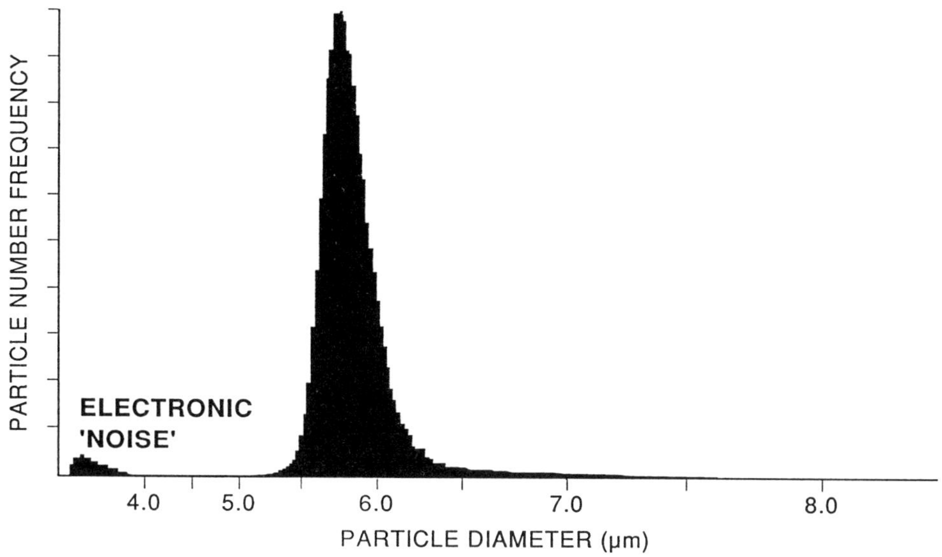

Figure 9 Narrow-size-range latex distribution recorded after electronic editing. (Compare with Fig. 7.)

with the editing method. The use of electronic editing instead of the excellent method of Thom avoids complicating the sample flow path with its susceptibility to collecting debris.

B. Aperture Design

An investigation into the e.s.z. method of particle size analysis, initiated at IIT Research Institute in 1967, resulted in the development of a novel aperture configuration and assembly [76,77]. The aperture was made funnel shaped and was mounted downstream of a sieve mesh within a flow-directional collar. This had the advantage of removing any oversize particles likely to block the sensing aperture as well as tending to focus the particle stream into the center of the aperture. The development was described in three papers [78–80], but did not become widely available.

C. Single Particle Measurement

A novel approach has recently been used at the University of Oslo by Berge and others. They constructed an electrical sensing zone device with a pressure reversal mechanism, which enabled one particle to be measured by repeated transit through the aperture. In this way they have studied single particle dissolution and particle flow dynamics [81,82], transit time and electrokinetic effects [83], the dissolution of air bubbles [84]and the pulse height response to particles transiting close to the walls of the aperture [85]. In the latter work, they found a maximum error in pulse height with respect to particle volume of 10%. This is a 3.3% diameter oversizing for those particles passing closest to the aperture walls (cf. Fig. 6).

D. Measuring Contaminants in Molten Metals

The electrical sensing zone method requires a conducting liquid within the sensor. Simensen [86] and Guthrie and Doutre [87] have dipped an aperture tube made of boron nitride into molten metals, such as aluminum, zinc and lead, applied a high current across the resulting sensing zone by way of iron electrodes, and measured the insoluble particles, or inclusions, within the molten metal. Currents of 20–60 A were used, along with apertures of up to 2 mm diameter.

E. Continuous Monitoring of Cell Cultures

It has been shown [88] that cells can be efficiently grown, in continuous culture in a bio-fermenter, on the surface of polymer beads. Only the electrical sensing zone method allows the growth to be monitored to the required degree of sensitivity, as the "bead" size increases due to the attached, growing, cells. At the end of the process, the cells are removed from the beads' surface and harvested. The signals, as processed by standard COULTER COUNTER and CHANNELYZER instrument circuitry, have been shown to give a higher resolution than a direct method of measurement using low-frequency dielectric spectroscopy [89]. The crucial point about this "cells on beads" technology is that the aperture is sensitive enough to allow the detection, during replication, of the point of confluence (i.e., the point when all of the cells run together). Then, the cells just form a monolayer covering the entire bead surface. Following on from this condition, second layer of cells would starve the first layer of its nutrients, and then a second layer would not grow in the same way as the first, being contact inhibited from proliferating.

Because the data from the aperture can be converted into a size spectrum, the most frequently occurring bead size can be monitored. Methods giving the average signal (even from an aperture) are too heavily influenced by the extremes of the culture. Harvesting at confluence not only gives the most efficient yield (i.e., the maximum monolayer per kilogram per hour of beads) but the majority of the cells will all be in exactly the same state as a result of their just contacting their near-neighbor cells.

Coulter Electronics Ltd. has cooperated in producing a unique autoclavable, self-contained, sampling, diluting, and counting probe to be fitted permanently to a biofermenter [90].

It is possible that this new technology may allow patients' own skin cells to be rapidly cultured, to be reapplied to large areas of burns.

F. On-line, In situ, and On-Stream Applications

As well as the molten metals and bio-fermenter applications described above, other specially designed devices have been described to make direct measurements.

Barnett et al. [91] and Lines et al. [92] introduced an on-line monitor system which comprised a sampling device, a control unit and a printer, to be connected to a conventional COULTER COUNTER instrument. It was used for continually monitoring the particulate contamination levels within an electrolyte solution, such as normal saline. The sampling device included three apertures, any one of which was used at any one time while the others were held in reserve in case of blockage. At preselected time intervals, samples of solution were passed through the active sensor and particle counts determined at various preset size levels.

Submersible and self-contained sensor devices to determine the numbers and sizes of individual particles directly in seawater at various depths have been reported [93–95]. These towed devices contained the necessary pumps and sensing apertures, and passed their signals to onboard COULTER COUNTER instruments for real-time processing.

G. Internal Measurements of Particles

So far, the Coulter principle has been illustrated when using an applied direct current across the aperture; this enables particle volume to be measured with very high precision and accuracy. Recently, Coulter has begun exploiting the use of alternating current (or radio frequency). This has the ability to generate a response dependent upon the composition of particles, and so can help to identify them (e.g., Coulter and Hogg [96]). One such application is to use rf to obtain information within blood cells and, in combination with a laser for further identification

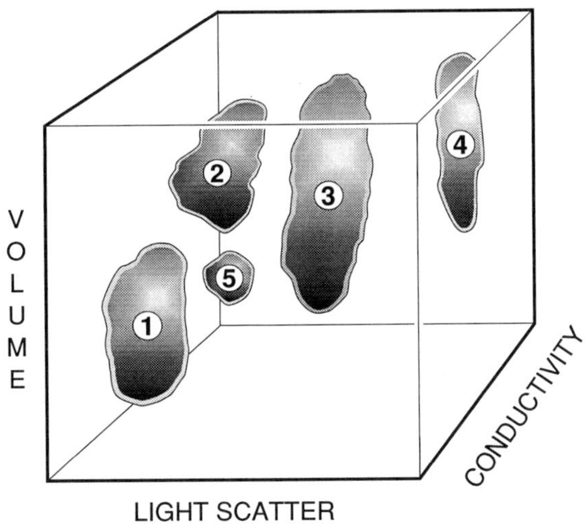

Figure 10 Three-dimensional scatterplot of the five types of white blood cells. Cell population 1 is due to lymphocytes. Cell population 2 is due to monocytes. Cell population 3 is due to neutrophils. Cell population 4 is due to eosinophils. Cell population 5 is due to basophils.

purposes, a unique triple transducer has been developed. The three detection systems enable a three-dimensional scatterplot of the white blood cells to be produced (see Fig. 10), so that different types of leukocytes may be routinely quantified. This triple transducer is commercially available, has been found valuable [97,98], and is being further developed. It is also expected that the combined use of rf and dc aperture technology will have further benefits elsewhere in particle size analysis applications.

VIII. APPLICATIONS AND ACCURACY

A. Blood Cells

The very first application of the e.s.z. method was to provide rapid and precise red blood cell counts. Blood contains three particulate types; red blood cells (erythrocytes), white blood cells (leukocytes), and platelets (thrombocytes). Since the e.s.z. method in its simplest form cannot distinguish between cell types, per se, but distinguishes between different particle volumes, simple physical and/or chemical steps are added to determine the counting process.

For red blood cell measurement, whole blood, collected into an anticoagulant, is diluted typically 1:50,000 into a suitable electrolyte solution, and the total number of cells above some 25–30 fl in, e.g. 0.5 ml, of suspension is counted. This typically may produce a raw instrument count of some 45,000 which, after coincidence correction and allowance for the dilution step, would indicate a red blood cell count (RBC) of the order of 5×10^{12}/l.

In practice, this RBC also includes any white blood cells but, as their concentration is usually 1000 times less than that of the red blood cells, their presence is insignificant. If required, however, the white blood cells, being larger, can be eliminated from the count by using an upper threshold level which electronically excludes them.

White blood cells themselves are counted by making another dilution of the blood, this time typically 1:500, from which the red cells are eliminated by chemical lysis. A solution of quaternary ammonium salts is often used for that purpose. The white cells remain, and are subsequently counted; a typical raw instrument count of 5000 per 0.5 ml suspension volume corresponds to a white blood cell count (WBC) of the order of 5×10^9/l.

Platelets are much smaller than red and white cells, and can be distinguished by size discrimination alone. One simple method is to allow the red and white cells to sediment out of the blood, and the resulting supernatant containing the platelets is diluted typically 1:3000 before counting the numbers of cells between 3 and 36 fl (relative to the red cell volume calibration). Those cells are the platelets. A raw instrument count of 20,000 per 0.1 ml of suspension indicates a platelet count (PLT) of the order of 250×10^9/l on a fairly normal blood specimen. Other methods of platelet counting, usually on routine clinical analyzers, allow platelets to be counted directly in the original whole blood dilution also used for the RBC.

As already noted (see Sec. V), the e.s.z. method provides two basic measurements, particle number and particle size (or volume). To obtain a total particle or cell count, accuracy in particle size is not important. To obtain an accurate particle size distribution, both are important.

For the particle count applications of RBC, WBC, and PLT, it was established very early on that particle count was accurate (see Secs. VIII.A.1–3). There was not then, and there is still not today, a certified reference material for particle count, comprising a known number of fine particles; although blood cell calibrator and control suspensions are now available commercially, assayed using reference instrumentation and methods. Accuracy of count therefore had to be validated against the "reference" method, which was the microscope chamber (or hemocytometer) method. In this method, a known volume of blood is placed into a thin chamber, marked out into known areas, placed on a microscope stage, and the number of cells per unit area (which is equal to unit volume) counted by eye. The method is subject to high imprecision because the numbers of cells counted is necessarily quite low. Counting even 500 cells is time consuming, and leads to a coefficient of variation (cv) no better than some 4.5%. The electronic measurement of 40,000 cells leads to a cv of the order of 0.5%. Comparing the results of one method against the other requires allowance for this statistical variation [14]. This may be achieved by presenting the 95% confidence limits predicted for the expected precision of the measurement by the two methods as a boundary envelope to the experimental data (Fig. 11).

Since then, a great many species other than man have been studied using e.s.z. blood cell analyzers (e.g., Hawkey [99]), and the method is in routine use worldwide for human hematology (e.g., Wintrobe [100]).

According to the College of American Pathologists' (CAP) regular survey data, the results from these interlaboratory hematology trials clearly show the improved precision which the automated method gives over the manual method [101]; see Table 7.

After finding that the method provided accurate blood cell counts, interest then followed in obtaining the cells' mean volume and volume distributions. Red cell mean volume (MCV) is an important diagnostic tool for thalassemia [102], while the red cell distribution width (RDW) assists the measurement of anisocytosis [103–105]. In addition, other red cell calculations can be made, which have diagnostic or quality control value; for example mean cell hemoglobin (MCH) which is calculated from hemoglobin/RBC, and mean cell hemoglobin concentration (MCHC) which is calculated from hemoglobin/(RBC $\times$ MCV). Obtaining the size distribution of white blood cells allows discrimination of two or three groups of white cell subpopulations. The technic is used in many routine clinical analyzers to quantify the numbers of lymphocytes, mononuclear cells and granulocytes [106]. An extension of the basic direct current/aperture design, by adding alternating current and simultaneous laser illumination and scattered light detection, allows five (and more) subpopulations of white blood cells to be quantified (see Sec. VII.G).

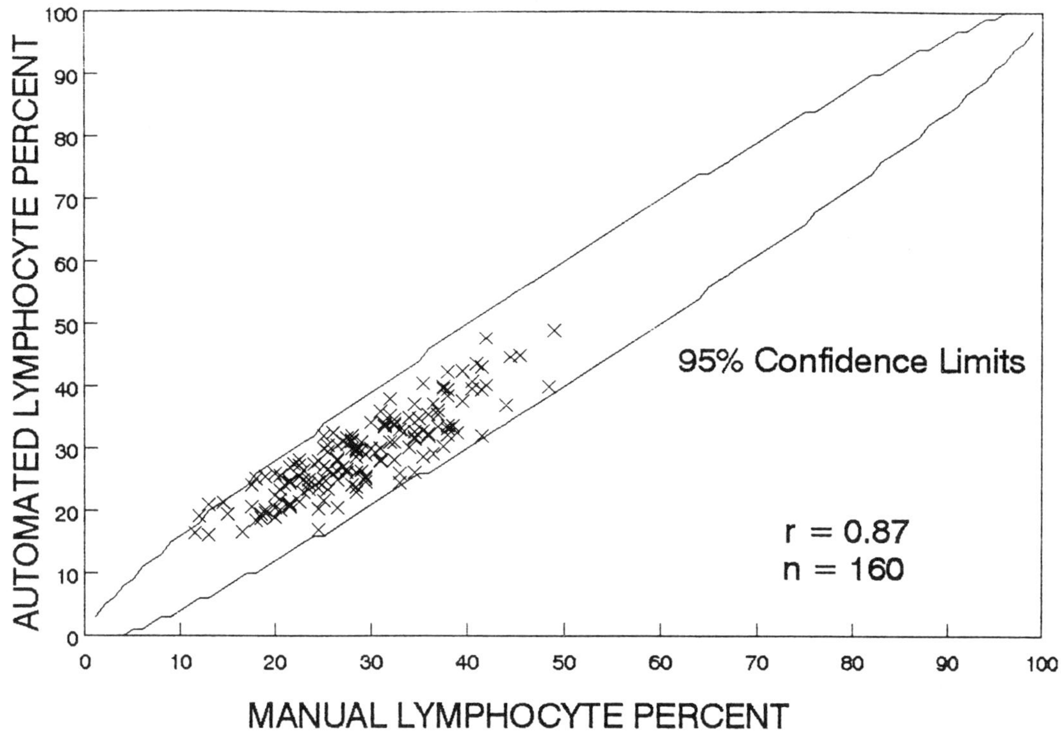

Figure 11 The results of comparing automated lymphocyte counts with 100-cell conventional count (hemocytometer) measurement in 160 patients ($r = 0.87$).

1. Red Blood Cells

Following the introduction by Coulter in 1956 of the e.s.z. method [3] and its application to red cell counting and size distribution measurement, Brecher et al. [107] evaluated the counting performance for human bloods. They compared erythrocyte results with the then standard method, the hemocytometer. The correlation of the e.s.z. method (C) on the hemocytometer method (H) was given by the equation $C = 0.98H + 0.16$, with the correlation coefficient $(r) = 0.978$.

Mattern et al. [108] also evaluated the new method, both for human and for mouse erythrocytes. They noted that obtaining red cell size distribution was a valuable aid to charac-

Table 7 Precision of Common Hematology Tests

Parameter	Manual method[a]	Automated method[a]
Red cell count	6.4%	1.0–2.7%
White cell count	17%	1.7–4.4%
Platelet count	23–27%	4.0–10.0%

[a]All data are expressed as coefficient of variation for the method. Ranges for the automated methods are representative of the various instruments' performance in CAP survey programs.

Source: From Ref. 101 by kind permission of LabMedica International.

terizing disease, such as thalassemia minor. They also documented the use of the method for obtaining white blood cell (leukocyte) counts, following elimination of the red cells by lysis using saponin.

Almost immediately, the e.s.z. method was picked up worldwide to become the standard method for red and white cell counting and sizing.

2. White Blood Cells

In their evaluation of the method in 1959, Ackeroyd et al. [109] eliminated the red cells by lysis with Triton X-100. They obtained leukocyte counts with a standard error of 2.8%, compared with the stated standard error from the hemocytometer method of 10.7%. The correlation between the two methods was found to be given by $r = 0.9998$.

Richar and Breakell in the same year [110] used saponin lysis, and obtained a correlation coefficient of 0.9647 for 56 human bloods, between the e.s.z. and the hemocytometer methods. They noted that the e.s.z. standard errors were usually in the 2–3% range.

White blood cells in milk are routinely counted to detect subclinical mastitis (e.g., Zeidler et al. [111]).

3. Platelets

In 1962, it was found possible to obtain accurate platelet counts from the platelet-rich plasma which had been separated from whole blood [112]. There was good agreement again with the hemocytometer method, and the technic was found to be simple enough for junior technicians to perform after only a short period of instruction [113]. This, too, soon became the standard method in the hematology laboratory, partly due to the work of Bull, Schneiderman, and Brecher [114]. Nowadays, platelet counting is fully automated, uses whole blood, and is a routine method.

B. Other Cells and Organisms

The e.s.z. method has been applied to counting a great many types of cells and species of organisms.

1. Bacteria

It had been found in the late 1950s that the sensitivity of the early e.s.z. analyzers could be improved in order to measure particles or cells down to 1 μm diameter or below. The use of a smaller aperture than those used for red cells, white cells and platelets was necessary for bacteria, so 30 μm, and smaller, apertures were brought into use, and background filtration of electrolyte solutions was improved.

Curby et al. [115] showed the correlation of e.s.z. results with the standard "plate" method, using *Staphylococcus aureus* and variants of *Escherichia coli*. Many others (e.g., [116]) also noted the increased precision of the e.s.z. method over plate counts, and also over the turbidity method using a spectrophotometer or nephelometer as a comparator. The e.s.z. method is now commonly used to study bacterial growth, with over 300 published references to date [13].

2. Yeast Cells

Yeast cells are relatively easier to count and size than bacteria, being nearer to 5–7 μm diameter than to 1 μm or below. The earlier methods of measurement of cell numbers and mass had been by turbidimetry and, again, by the visual counting of cells within a microscope chamber. Macrae (1964) [117] showed the advantages of the new method for examining yeast in beer, and the alcoholic fermentation of wine musts has been studied [118]. The method has been used for a great many yeast species ever since [12,13].

Koch et al. [119] evaluated a yeast test as an alternative to the standard LD_{50} test for acute toxicity of drug substances and other harmful chemicals.

3. Other Cells

Once it was established that blood cells could be readily counted and sized, it was clear that other animal and tissue cell types could also be studied, providing that single cells were available. Harris (1959) [120] showed that the COULTER COUNTER instrument count results compared well with the microscope (chamber) counts for chick skeletal muscle fibroblasts. Other workers obtained successful results with other cell lines, e.g., sarcoma ascites tumor cells [121], and with bone marrow cells, epithelial cells, endothelial cells, amniotic fluid cells, and many others [13]. The new method was firmly established by the time of a review by Daly et al. [122] in 1966, and is in widespread use today (e.g., [123]).

Another early application was in the enumeration of spermatozoa; again, the standard method at the time was the microscope and hemocytometer chamber. Glover and Phipps, 1962 [124] found a high correlation ($r = 0.96$) between the methods for specimens of bull semen. Segal and Laurence [125], studying human semen, found a correlation coefficient of 0.96, while Laurence and Carpuk [126] obtained $r = 0.82$ for guinea pig specimens. Iversen [127] obtained $r = 0.94$ for bull semen; and with human spermatozoa Gordon et al. [128], using a COULTER COUNTER model B, obtained a correlation coefficient of 0.999, slope of 0.998, over the range of concentrations of 2 to 700×10^9/l.

Adipose cells, i.e., fat cells within body tissue, can be extracted and rapidly counted and sized electronically. Hirsch and Gallian [129] showed good agreement with results from microscopy. The method was used by Brook and others [130–132] to study obesity in children.

4. Organisms

The e.s.z. method has been used to count and size a great many organisms, the measurement of which had previously also required the tedious and imprecise microscope (hemocytometer) method. Borchardt and O'Melia, 1961 [133], had studied the growth of the alga *Chlorococcum macrostigmatum*. Hastings, Sweeney, and Mullin [134] compared the instrument performance with the visual method for a range of unicellular marine organisms, e.g., *Gonyaulax polyedra* (Table 8).

A study by El-Sayed and Lee [135] noted that not also was the electronic count precision better than by hemocytometer; e.g., the precision of 30 replicate electronic counts of *Chlorella pyrenoidosa* was found to have a cv of 0.44% on a count of some 43,800 organisms, but also the precision was at least that predicted from Poisson theory (0.47% in this case).

Table 8 Comparison of Instrument Counts with Visual Counts for *Gonyaulax polyedra*

COULTER COUNTER Model A		Visual microscope	
Mean count	cv	Mean count	cv
5800	± 2%	5250	± 10%
436	± 6.3%	365	± 5.5%
218	± 2.7%	175	± 8.3%

Source: From Ref. 134.

Many organisms have been studied, for total count (or concentration), growth rates and even feeding behavior, where one species is predatory of another.

Excellent reviews of the many other applications of the e.s.z. method in marine science were published by Sheldon and Parsons [136] and Milligan and Kranck [137]. The methods are also applicable to limnology, the study of lake organisms, sediments, etc.

C. Parenteral Fluids

Since the e.s.z. method has the basic benefit that any and every particle passing through the sensing zone is counted if its volumetric size exceeds a given size level and the concentration is low, it is ideal for measuring contaminating particles in electrolyte solutions. Counting all particles in conducting parenteral fluids is a simple application, therefore. Nonconducting fluids need to be made conducting before measurement, such as by the addition of a trace of well-filtered concentrated saline solution.

The e.s.z. method has been used widely for this purpose since the 1960s (e.g., [138]) resulting in its inclusion as a test in the British Pharmacopoeia (1973) [139]. The method is discussed in recent books [140–142]. Compared with the alternative test method, light blockage, it does not ignore optically transparent particles, nor does it ignore particles not passing through the optical detection zone. The limitations of light blockage have been well reviewed by Barber [143].

The British Pharmacopoeia later [144] recognized that, if the two different methods were to be used, then two different acceptance criteria would be needed. They specified that, to pass the test for large volume (≥ 100 ml) solutions, the e.s.z. limit would be not more than 1000 particles per milliliter greater than 2 μm, and not more than 100 particles per milliliter greater than 5 μm, while the light blockage limits would be 500 and 80 particles per milliliter, respectively. It was never explained why the "lost" particles should be accepted at 50% and 20% respectively for the two size levels.

Differences between the two methods have been documented, (e.g., [145–147]). Haines-Nutt and Munton [145] found that although there was a reasonable correlation between e.s.z. (type COULTER) and light blockage (type HIAC) results for salt solutions, if sugars are present in reasonable concentration there is no relationship. They concluded that refractive index effects were the major causes of the discrepancies; some particles were invisible to the optical detection method.

Oxborrow [146] compared the counts obtained from particle counters from five manufacturers; one instrument was of the e.s.z. type and four were of the single particle optical (both light obscuration and light scatter) types. His test fluids ranged from barely conductive to very conductive, and very low particulate count to moderate particulate count. The sample types varied from contamination in otherwise clean parenterals to suspensions of mica, fiberglass, silicone and test dust in water. His count results were presented as logarithms rather than directly as particle count per unit volume, and are not readily comprehensible. In addition, his presenting summary results as averages (and averages of logarithms) also makes interpretation difficult. I have re-presented some of his results, in Table 9, as particle counts per 10 ml of sample suspension. The five instruments were from the manufacturers HIAC, CLIMET, and KRATEL (light blockage), MET ONE (light scatter), and COULTER (e.s.z.). The KCl parenteral was claimed to be at 20 mEq/ml (which at a calculated 1.5 kg/l salt concentration seems clearly to me to be an error; probably 20 mEq/l (about 1.5 g/l) was intended); the $NaHCO_3$ parenteral was at 8.4% (84 g/l; more than adequately conducting); the $CaCl_3$ (sic), said to be at 1.4 mEq/ml, was probably intended to read $CaCl_2$ at 1.4 mEq/l (about 0.08 g/l); the fiberglass, mica, silicone and A.C. Fine test dust were all prepared in water.

Table 9 Particle Counts from Five Different Counters

(i) Number of 5 μm counts per 10 ml

	HIAC	MET ONE	COULTER	CLIMET	KRATEL
KCl	1,000	1,445	162,818	1,950	3,162
$NaHCO_3$	2,951	2,630	8,128	631	5,754
$CaCl_3$ (sic)	589	724	44,668	129	1,862
Heparin Sodium			(Coulter abandoned here)		
Fiberglass					
Mica					
Silicone					
A.C. Fine					
Average		N/A			

(ii) Number of 10 μm counts per 10 ml

	HIAC	MET ONE	COULTER	CLIMET	KRATEL	MICR.
KCl	89	245	240	1,259	372	—
$NaHCO_3$	646	741	912	372	525	—
$CaCl_3$ (sic)	78	115	195	65	234	—
Heparin Sodium	43,652	33,113	57,544	19,953	47,863	—
Fiberglass	4,571	4,266	6,918	1,950	2,818	2,291
Mica	1,820	1,413	550	794	2,089	1,122
Silicone	148	2,630	468	178	589	—
A.C. Fine	1,259	3,467	214	1,820	3,890	3,162
Average	6,533	5,749	8,380	3,299	7,298	N/A

Table 9 continues on next page

Since the e.s.z. method requires conductivity, it is probable that the high counts for KCl and $CaCl_3$ (sic) recorded at 5 μm (using a 140 μm aperture) in Table 9(i) were due to the electrical noise introduced because the conductivity was too low, and it is almost certain that the low or zero conductance for the other particle suspensions was the cause of the failure to report any e.s.z. results.

However, at 10, 20, and 30 μm (see Tables 9(ii)–(iv)), the results were usually of the same order from all instruments. In some cases, Oxborrow included microscope counts for further comparison. Allowing that the sample volumes counted varied from 2 to 10 ml so that the statistical variation becomes significant on the low (say < 1000) counts, that the fibrous fiberglass and platelike mica have a different particle "size" in each dimension (which affects the count obtained at that size), and that the suspensions were not stirred, even for the 30 μm counts, the results are not out of line with others' attempts at obtaining similar results from different instruments. For the record, though the individual results in Table 9(ii)–(iv) carry the more useful information, Table 9(v) shows the overall averages of these data where comparisons can be made. Once the conductivity effect is eliminated, the variation in counts obtained between e.s.z. and light blockage seems little different from the variation obtained between light blockage and light scatter, or indeed between the three instruments all based on light blockage. The tendency was for the light blockage results to be lower than e.s.z. for the $NaHCO_3$ parenteral, heparin sodium, and silicone preparations.

Table 9 Particle Counts from Five Different Counters (Continued)

(iii) Number of 20 μm counts per 10 ml

	HIAC	MET ONE	COULTER	CLIMET	KRATEL	MICR.
KCl	5	16	15	33	10	—
$NaHCO_3$	30	27	42	32	30	—
$CaCl_3$ (sic)	6	4	2	4	11	—
Heparin Sodium	1,230	1,905	2,512	692	1,778	—
Fiberglass	1,412	724	871	2,042	2,042	1,738
Mica	245	76	95	135	195	562
Silicone	11	16	12	6	10	—
A.C. Fine	209	132	10	145	174	603
Average	394	363	445	386	531	N/A

(iv) Number of 30 μm counts per 10 ml

	HIAC	MET ONE	COULTER	CLIMET	KRATEL	MICR.
KCl	0	4	5	0	2	—
$NaHCO_3$	3	0	0	4	4	—
$CaCl_3$ (sic)	0	0	0	0	5	—
Heparin Sodium	50	51	219	85	186	—
Fiberglass	—	16	10	676	490	1,202
Mica	—	7	2	19	24	363
Silicone	—	0	5	0	2	—
A.C. Fine	—	4	0	15	4	107
Average	N/A	10	30	100	90	N/A

(v) Average Number of 10–30 μm counts per 10 ml

	HIAC	MET ONE	COULTER	CLIMET	KRATEL	MICR.
Average	N/A	2,041	2,952	1,262	2,640	N/A

Source: From Ref. 146.

Cham et al. [147] presented suspensions of latex beads in various testing solutions, and obtained the particle counts using two e.s.z. (type COULTER) and two light blockage (type HIAC/ROYCO) analyzers. The total particle counts above 2 μm, for the nominal 6 μm latex particles used, showed reasonable consistency from one COULTER COUNTER analyzer to another, but the two HIAC/ROYCO instruments were not as consistent. In addition, the HIAC/ROYCO counts above 2 μm were invariably less than those obtained from the e.s.z. analyzers. They concluded that the HIAC/ROYCO type analyzer is "not good for the determination of absolute count in large volume parenteral solutions." Some of their results are presented in Table 10.

Seville et al. [148] found that single particle light scattering counters oversized fly ash, primarily because of the particle shapes involved, but generally undersized strongly absorbing char particles, compared with the e.s.z. method. That is, both shape and optical properties are more important in those optical analyzers.

As well as large-volume parenterals, small volumes can be measured too. Lines [149,150] introduced a very-narrow-aperture tube to be used with COULTER COUNTER instruments, which could enter and sample from vials and ampules as small as 1 or 2 mL capacity.

An alternative to using such an "insertable" aperture tube is to bulk-up the small volumes

Table 10 Typical Results from Various Parenteral Solutions; Cumulative Particle Counts per ml ≥ 2 μm

	COULTER COUNTER Model		HIAC/ROYCO Model	
	TA II	MULTISIZER	(1)	(2)
Solution	Mean (S.D.)	Mean (S.D.)	Mean (S.D.)	Mean (S.D.)
Compd. sodium lactate	2328(40)	2338(133)	1652(16)	1789(39)
0.9% NaCl	2515(96)	2622(82)	2014(19)	1395(42)
Ringer solution	2581(108)	2562(101)	1531(30)	1484(20)
5% Glucose	2462(58)	2386(52)	1597(23)	1595(35)
Xylitol	2505(42)	2439(53)	1848(15)	1997(11)
Maltose	2474(55)	2544(50)	1905(25)	2063(30)

Source: Extracted from Cham et al., Drug Development and Industrial Pharmacy, Volume 15, pages 2441–2454, 1989 [147], by courtesy of Marcel Dekker, Inc.

by transferring the contents of several ampuls into a larger container for counting. That transfer could be by means of a disposable 20 ml syringe fitted with a collar to prevent slough particles being created in the wetted region or by a vacuum transfer method [151].

Contaminant particle counts can be obtained satisfactorily even if working without a special clean area. Any significant extraneous contamination is more likely to derive from inadequately cleaned surfaces such as beakers and transfer equipment, and it is in those aspects that especial care is required. Barnett [152] showed that particle fallout into the sampling beakers in an open laboratory was insignificant (Table 11).

Particles in total parenteral nutrition products have also been quantified [153–155], as has the release of particles from hemodialyzers [156].

D. Other Contamination Studies

As well as the evaluation of particulate contamination in parenteral solutions (see above), particulate contamination in other solutions and liquids can readily be quantified; such as in seawater, electroplating solutions, water in oil (roller) emulsions, and even in liquid metals, (see Sec. VII.D). The particle content of typhoid [157] and BCG [158] vaccines has been studied.

Particles accumulating in wool-scour liquors, and the recovery of clean liquors for reuse and of wool grease as a by-product, have been studied by Worth and McLaughlin [159]. Contami-

Table 11 Particle Counts After Leaving in an Open Laboratory for up to 3 h

Beaker no.	Average count of solution at time zero, above 2 μm	Average count of solution at 3 h, above 2 μm
1	32/ml	37/ml
2	24/ml	22/ml
3	32/ml	24/ml

Source: From Ref. 152.

nating particles in hydraulic or lubricating oils or other nonconducting fluids can be quantified by rendering the fluid conductive by means of dilution into a solvent electrolyte solution, or by filtration and resuspension of the particles into electrolyte solution (e.g., [160]).

Airborne contaminants also require capture, such as by filtration or by liquid impinger, before resuspension into electrolyte solution prior to counting and sizing (e.g., [161–164]). Ghadhiri et al. [165] noted the advantage that the equivalent spherical diameter from the e.s.z. method is a close approximation to the sedimentation diameter, making it a most useful size measurement parameter for airborne dusts. Ghosal et al. [166] considered the COULTER MULTISIZER to be the most appropriate device for sizing collected fly ash samples for particle diameters ≥ 1 µm. Contaminants from volcanoes and within glacial sediments have also been studied (e.g., [167,168]).

E. Water Treatment

In recent years, interest has grown in quantifying and improving water treatment processes, for instance in the study of treatment plant performance [169] and of flotation technics [170,171]. Not only for water treatment purposes, but also for research into the cardiovascular system, particularly the heart, have lipid-coated microbubbles been developed and characterized [172].

F. Powders and Other Particulate Solids Materials

The electrical sensing zone method has been very widely applied to almost every form of finely divided particulate matter. The applications are nearly endless, and include drugs, pigments and fillers, reprographic toners, foodstuffs, abrasives, explosives, clays and minerals, construction materials (e.g., cement), coating materials (e.g., epoxy powders and chromium carbide), metals, and also filter testing, crystal growth and dissolution studies [12].

For example, in the food industry, applications include sugar [173,174], chocolate and cocoa (e.g., [175]), tomato ketchup and tomato juice [176], ice cream mix, and flour (e.g., [177–179]). In chocolate, the manufacturer needs to balance the particle size not to be too coarse to give a gritty texture nor to be too fine to require an excess of the expensive cocoa "butter" required to give the low viscosity for production processes and not to give a cloyey texture to the product. The color of tomato ketchup (and most finely divided material) is influenced by the size of the cellular particles. The protein content and baking properties of flours can be estimated from their size distribution.

The pharmaceutical industry is much concerned with particles and the properties they give to a product. Pharmaceutical applications have been well reviewed, for example by Polderman [180], Thornton [181], Boylan [182], Matthews [183], and Orr and Spence [184], and include the measurement of the particle size distribution of drugs, as well as studies of their grinding (e.g., [185]), dissolution rates (e.g., [186]), crystal growth [187,188], energy savings in crystallization [174], the disintegration of tablets (e.g., [54,189,190]), deaggregation rates [191], and the study of parenteral nutrition emulsions (e.g., [192,193]).

The size of drug particles is important because of their effect on the appearance, texture or taste of the drug preparation as well as the rate of adsorption. For some drugs, for example griseofulvin, both the degree of alimentary absorption and the consequent levels in the blood are inversely related to particle size. Insulin zinc suspension (crystalline) B.P., which contains particles of some 10–40 µm, has a longer duration of action than insulin zinc suspension (amorphous) B.P. where the particles rarely exceed 2 µm. Orr and Spence [184] have used the COULTER COUNTER model TA to assay the percentage of amorphous insulin in insulin zinc suspension B.P.

After injury and the injection of a painkilling drug, the lessening of pain intensity as well as its duration may be a function of the particle size of the active ingredient. The timing of drug release is important; for instance antiworm compounds must be effective in the gut, so the particle size must be small enough for dissolution, but large enough not to have been dissolved before the gut is reached; 10–30 μm is often suitable, depending on the drug. The relevance of particle size also relates to the segregation of powders during handling or storage, capsule filling, the rheology of suspensions and tableting. It is also important for inhalation therapy, such as bronchodilator powder aerosols. The usual size range is 1–6 μm, but particles of even less than 0.5 μm deposit readily due to diffusion. Usually only a few percent of the drug actually reaches the bronchioles, the larger particles being impinged in the throat. Aerosols have been studied for example by Haile et al. [194] and Karig et al. [195].

The concentration of solids in a suspension for injection is important not only for dosage rates, but also because the thixotropic nature of the product facilitates its passage through the needle during administration. Close attention to particle size and shape will also prevent needle blockage. A suspension may be the preferred dosage form to a solution when it remains chemically more stable, or when sustained action is required (for example with zinc insulin and procaine penicillin G). The size and shape of suspended drug particles should remain unchanged during storage; but crystal growth may occur through Ostwald ripening (the slow dissolution of small particles giving rise to recrystallization onto larger ones, producing a coarsening of the particle size distribution upon storage).

Eye ointments must be kept free of large particles which could cause irritation, and the correct size distribution of particles of active ingredient in medical skin care creams is important to ensure an even covering and effectiveness.

The size distribution of most oil-in-water (o/w) emulsions can be easily measured by the e.s.z. method, (e.g. [196–199]), and emulsion stability has been frequently studied (e.g., [200]). Water-in-oil emulsions cannot be measured, as the water droplets dissolve in any nonaqueous electrolyte suitable for the oil. Multiple emulsions (w/o/w) have been successfully analyzed (e.g., [201]).

Emulsions should remain stable upon storage; a common problem is that of "creaming," when the oil droplets in an o/w emulsion slowly rise and coalesce to give an oily layer which cannot be easily redispersed. Although a large number of droplets in the disperse phase of an emulsion are often well below the detectable limit of the e.s.z. method, it is the proportion above about 0.8 μm which is of interest in stability studies. The e.s.z. method can measure the total volume of these particles, which can be expressed as a percentage of the total volume of disperse phase (which is known for instance by chemical assay).

The Coulter method has been used to detect variations in a batch crystallization process and to control closely the grinding of drugs [184,202].

There has been increasing interest in the use of particulate systems for drug transport, such as starch microparticles [203], albumin microspheres [204], oil-in-water emulsions [205–207], alginate [208], and liposomes [209].

The pharmaceutical industry also concerns itself with biological effects, such as the mode of action and optimum conditions of antiseptics and preservatives on microorganisms. The e.s.z. method has been used to study the effects of antibiotics upon bacteria, the changes in the cell division cycle and cell wall synthesis of a large range of microorganisms, and in tissue culture research.

Many pharmaceutical applications of particle counting and sizing come close to the purely medical uses; an unusual new one is in the monitoring of the breakdown of flexible tubing used in renal dialysis machines. It is not enough to show that there are particles present, for there will be some even in brand new tubing, but in order to get a reasonable life out of the tubing while not allowing oversize particles to pass into the bloodstream, the size range of particles being shedded must be closely monitored. Other applications include the deposition of polymer particles onto T-lymphocytes [210], the measurement of antibody-coated latexes [211], and calcium oxalate stone formation in urine (e.g., [212,213]).

IX. STANDARDIZATION

Standardization is the alignment of similar equipment with standard materials and with each other. The e.s.z. method measures two separate and independent parameters which can, and should, be standardized: count and size.

A. Counting

1. The Counting Response of Different COULTER COUNTER Models and Instruments to Single Stimulae (From Groves [214], with Permission)

Instrument response to a single stimulus can be provided internally by means of a square-wave generator as an electronic source of uniform pulses, or, externally, by counting the number of particles in a suitable suspension.

Pulse Generator. The counting circuits of a number of instruments were compared exactly by using one set of glassware and one 100 μm orifice tube with a square-wave generator (Coulter Electronics Ltd.). Counts were obtained within a few minutes so that electrolyte temperature as a source of manometer flow deviation could be ignored. The model Fn and F counters are not fitted to take external glassware, and the results on these instruments reflect the change of mercury head more than the counting capacity. Results are shown in Table 6.2 (sic).

Particle Suspension. The response to a suspension of particles by different instruments fitted with various combinations of glassware and orifice tubes is more relevant to actual usage conditions.

Table 6.2 The Response of Different COULTER COUNTER Models and Instruments to a Square-Wave Generator Producing a Pulse of Standard Frequency and Height

Instrument no.	Model	Duplicate counts	
1	A	20,733	20,723
2	A	20,722	20,736
3	A	20,677	20,661
4	A	20,677	20,642
5	A	20,756	20,903
6	A	20,827	20,800
7	B	21,166	21,163
8	B	20,718	20,731
9	B	20,803	20,765
10	Fn[a]	19,985	19,971
11	F[a]	20,308	20,292
Mean count = 20,671		Standard error = 63.6 (= 0.131%)	

The glassware is No. (iv) of Table 6.3 fitted with a 100 μm orifice tube.
The electrolyte is ISOTON diluent (Coulter Electronics Ltd.).
(Results were obtained by R. W. Lines, Coulter Electronics Ltd.).
[a]Integral glassware, 100 μm orifice tubes.
Source: From Ref. 214, with permission.

A dilution of 8.25 μm polystyrene-DVB copolymer latex was prepared in 0.9% sodium chloride solution in a 1-liter rubber-capped intravenous solution bottle. Using the sampling tube previously described by Groves (1969) [sic; *Analyst, 94:* 192 (1969)] the suspension was passed to the orifice and the instrument response adjusted until the mean pulse height was about halfway up the instrument oscilloscope.

Counts were then made of increasing instrument thresholds, i.e., the arbitrary markings on the controlling potentiometer. Initially, at the lowest thresholds the counts fell slowly as the threshold increased, but at thresholds corresponding to the mean size of the particles the count dropped rapidly.

This change in count is shown for results obtained on one of the instruments in Table 6.2. Extrapolation of the two linear portions of the relationship between count and threshold enables the number of particles in the suspension to be determined at a point in the size distribution curve which can be easily defined. Results are shown in Table 6.3 (sic). In both cases the reproducibility can be seen to be very high and consistent (from Ref. 214, with permission). This would be a promising start for any collaborative trial since the accuracy of the method has been demonstrated by others, e.g., Allen (1968) [7].

2. Count Reference Methods

No reference method for counting exists save the human eye. In most circumstances none is required. Excluding coincidence passage, a one pulse equals one particle relationship is acceptable throughout industry, where the essentially digital nature of modern counting circuits is well understood.

In hematology, the accepted reference count is from a semiautomatic COULTER COUNTER model, for example the ZM or ZBI [215], with manual large scale volumetric dilutions counted above or between specified particle size levels. In their appraisal of reference blood cell counting [215], the International Committee for Standardization in Haematology (ICSH) recommended a list of attributes of the reference counter method,

Table 6.3 The Response of Different COULTER COUNTER Models and Instruments to a Suspension of Polystyrene-DVB Latex Particles of 8.25 μm Mean Diameter

Instrument no. (Table 6.2)	Model	Glassware	Orifice size (μm)	Counts/0.5 ml estimated at the intercept (Groves's Fig. 6.8)
2	A	(i)	100	15,400
3	A	(ii)	100	16,200
4	A	(i)	100	15,500
5	A	(ii)	100	15,600
6	A	(ii)	100	16,500
8	B	(ii)	100	15,100
10	Fn	integral	100	17,000
11	F	integral	100	17,000
12	A	(iii)	100	15,000
13	A	(i)	100	15,000
14	B	(i)	100	13,200
15	B + M	(iv)	70	15,000
	Mean count = 15,542		Standard error = 304 (= 1.96%)	

Some instruments were those in Table 6.2 with the appropriate glassware.
Source: From Ref. 214 with permission.

(i) The count is to be performed on an accurately known volume of blood.
(ii) The count is to be made on a sufficient and representative sample.
(iii) Cells should be counted one at a time.
(iv) Each cell should be counted once and once only.
(v) The cells to be counted are to be discriminated from all other particles.
(vi) There should be no spurious (i.e., nonparticle) counts.
(vii) The method should be usable on all types of samples (except possibly for those known to cause interference with the count).

A proposal for a standard method for the determination of erythrocyte concentration in blood was also made by the Dutch authorities [216] who identified some of the variables to be considered, and quantified or eliminated. Others have provided similar considerations [217–221]. The ICSH has recently published its reference method for the enumeration of erythrocytes and leukocytes [222]. It requires the use of a semiautomated e.s.z. analyzer and notes that optical sensing counters do not meet the required criteria of a reference counter.

As a means of national assessment of the quality control of hematology laboratories, blood samples are sent monthly by the United Kingdom Department of Health (formerly DHSS) to most U.K. laboratories in order to monitor performance, and results are returned to the laboratory together with the national mean. For example, Table 12 shows results from various groups of automatic blood cell counters and the manual (hemocytometer) method on a sample of fixed human platelets suspended in an artificial plasma. Similar regular trials are organized in the United States by the College of American Pathologists.

In the dairy industry, the Breed smear (Prescott and Breed [223]) is the reference method to obtain a quantitative nucleated cell count. Cell size is of little interest, but cell concentration in milk is widely measured. This is a good example of the precision of a "reference" method being inferior to that of a routine analytical technique; in this case, the COULTER COUNTER instrument. In addition, because so few laboratories are equipped for microscopically counting large numbers of milks with a good accuracy, it is often left to control/reference laboratories to issue assays for cross-reference purposes. Unfortunately, such assayed samples are rarely of a quality suitable for a reference; i.e., they are usually routinely obtained samples with a short life, perhaps only a matter of one or two days. That same situation often applies to blood samples used in national hematology quality assurance schemes.

Table 12 DHSS United Kingdom National External Quality Assurance Scheme for Hematology. Trial No. 832352 (Material: Fixed Human Platelets)

Instrument	Mean	Coefficient of variation (%)	Number of measurements
Coulter S Plus Series	108	9.6	114
Manual (hemocytometer)	107	20.7	122
Technicon	105	15.5	39
Ortho	124	15.5	13
Coulter Semi-Auto (ZBI, etc.)	110	20.2	15
Coulter ThromboCounter	110	16.8	199
Ultraflow 100	118	24.0	16
Mean	$\overline{N} = 111.7$		
Mean weighted for observations	$\overline{N}_w = 109.1$		

Reproduced with permission.

In order to improve this situation, some commercial companies produce blood and milk control materials intended to monitor the day-to-day performance and stability of the analyzers. The products are stabilized to have an effective sealed vial life of over four weeks. They are also supplied with reference control values for monitoring the accuracy of calibration of the instruments, as most automatic blood analyzers do not have accurate volumetric metering of the specimen suspension, and so particle count will vary if the flow rate through the aperture varies.

B. Sizing

Although calibration materials can also be used for the regular quality control of the sizing accuracy performance, it is better to use a different material to control than was used to calibrate. A user might retain a well-sampled batch of his own material, if it is stable over a long time, to control the precision of the method. Various international test materials (see Sec. D) are, however, available, and these allow greater standardization between laboratories.

In addition, interlaboratory standardization is often monitored by means of "round-robin" tests using aliquots of the same test material. If very carefully controlled, they can be invaluable [224].

C. Published Standard Procedures

The use of standardized methods for sampling, dispersing, analysis and reporting is just as important for electrical sensing zone analyzers as it is for any other method.

Standards involving the e.s.z. method include

Recommendations for the use of the method which are not specific to the application, e.g., British Standard BS 3406: Part 5 [40] and AFNOR (French) Standards [225,226].

Those for specific applications, e.g., British Pharmacopoeia 1973, 1980, 1988, and 1993 [139,144,227,228], for large-volume parenteral solutions [obsoleted in 1994 by the 1994 European Pharmacopoeia], and American Standards for the measurement of airborne particulates [163,229], alumina and quartz [230], aluminum oxide [231], beryllium oxide [232], filter evaluation and testing [233–238], toners [239], and polymeric powders [240].

The International Standards Organisation (ISO) is currently preparing a Standard on the method via its subcommittee ISO/TC24/SC4.

D. Suitable Standard Reference Materials

Standard or certified reference materials (SRMs or CRMs) which have been validated for use with e.s.z. instrumentation include

From the National Institute of Standards and Technology (NIST) [43]
- SRM 1003b; glass spheres, 8–58 μm
- SRM 1690; polystyrene latex suspension, 0.895 μm
- SRM 1692; polystyrene latex suspension, 2.982 μm
- SRM 1960; polystyrene latex suspension, 9.89 μm

From the European Community Bureau of Reference (BCR) [44]
- CRM 067; quartz powder, 2.4–32 μm [241]
- CRM 069; quartz powder, 14–90 μm [241]
- CRM 165; polymer latex suspension, nominal 2 μm
- CRM 166; polymer latex suspension, nominal 4.8 μm
- CRM 167; polymer latex suspension, nominal 9.6 μm

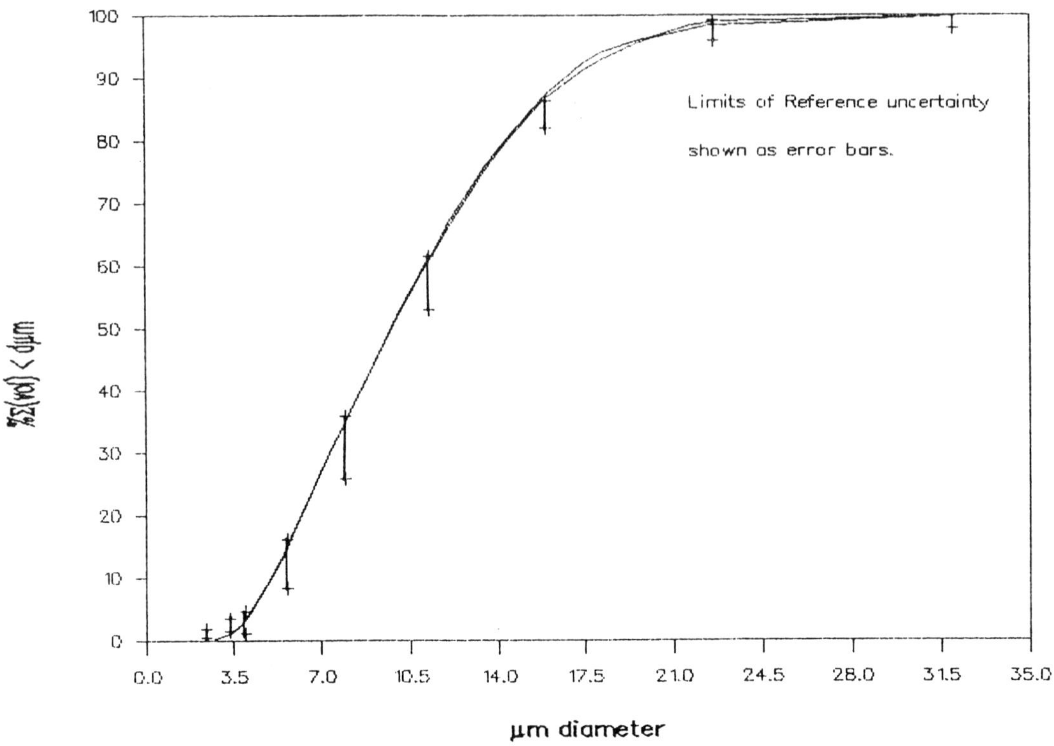

Figure 12 Analysis of BCR 67 reference material using a COULTER MULTISIZER II with edit capability.

Figure 12 shows an example of the measurement of a certified reference material using an e.s.z. analyzer. The stated limits of reference uncertainty were derived from gravitational sedimentation measurements. The upper plot is at a concentration of 0.035 g l^{-1} and the lower plot is at a concentration of 0.021 g l^{-1}.

The European Community Bureau of Reference is currently actively engaged in certifying a new series of powdered reference materials, some of the certification of which will be by the electrical sensing zone method [242,243].

REFERENCES

1. P. J. Crosland-Taylor, *Nature 171*:37 (1953).
2. W. H. Coulter, U.S. Patent 2,656,508 (1953).
3. W. H. Coulter, *Proc. Natl. Electr. Conf. 12*:1034 (1956).
4. C. F. T. Mattern, F. S. Brackett and B. J. Olson, *J. Appl. Physiol. 10*:56 (1957).
5. A. O. Ulrich, Instr. Soc. America Conf., New York City, 1960.
6. T. Allen, Proc. Conf. Particle Size Analysis, Loughborough, Society for Analytical Chemistry, London, 1966, p. 110.
7. T. Allen, *Particle Size Measurement,* 1st ed., Chapman and Hall, London, 1968.
8. T. Allen, *Particle Size Measurement,* 4th ed., Chapman and Hall, London, 1990, p. 445.
9. F. M. Rabinovitch, *Conductimetric Method of Dispersion Analysis,* Chemical Publishing House 'Khimiya', Leningrad, 1970. (in Russian)
10. F. M. Rabinovitch, *The Application of Conductimetric Particle Counters to Medicine,* Meditzina, Moscow, 1972. (in Russian)

11. T. Allen and K. Marshall, *The Electrical Sensing Zone Method of Particle Size Measurement,* University of Bradford, 1972.
12. *Industrial and Scientific Bibliography,* Coulter Electronics Limited, Luton, 1994.
13. *Medical and Biological Bibliography,* Coulter Electronics Limited, Luton, 1994, to appear.
14. A. Richardson-Jones, *Advances in Hematological Methods:The Blood Count* (O. W. van Assendelft and J. M. England, eds.), CRC Press, Boca Raton, FL, 1982, Chap. 5, p. 49.
15. V. Kachel, *Flow Cytometry and Sorting* (M. R. Melamed et al., eds.), Wiley, New York, 1979, p. 64.
16. *Theory of the Coulter Counter-Bulletin T1,* Coulter Electronics, Inc., 1957.
17. W. H. Coulter, W. R. Hogg, J. P. Moran, and W. Claps, U.S. Patent 3,259,842 (1966).
18. B. A. Batch, *J. Inst. Fuel 37*:455 (1964).
19. N. B. Grover, J. Naaman, S. Ben-Sasson, and F. Doljanski, *Biophys. J. 9*:1398 (1969).
20. N. B. Grover, J. Naaman, S. Ben-Sasson, and F. Doljanski, *Biophys. J. 9*:1415 (1969).
21. N. B. Grover, J. Naaman, S. Ben-Sasson, and F. Doljanski, *Biophys. J. 12*:1099 (1969).
22. N. Saranummi, *Techn. Res. Centre of Finland Report No. 10* (1975).
23. B. Scarlett, Proc. Conf. Partec (Nuremberg), 1979, p. 682.
24. J. G. Harfield and P. Knight, Proc. Conf. Particle Size Analysis 1981, Loughborough, Wiley-Heyden, Chichester, 1982, p. 151.
25. J. G. Harfield, R. A. Wharton, and R. W. Lines, *Part. Charact. 1*:32 (1984).
26. J. G. Harfield and R. A. Wharton, Proc. Conf. Partec (Nuremberg), 1984, p. 661.
27. R. K. Eckhoff, *J. Sci. Instr. 44*:648 (1967).
28. R. K. Eckhoff and P. Soelberg, *Betongtekniske Publikasjoner 7*:1 (1967).
29. R. A. O'Connell and R. J. Martsch, *Textile Res. J. 32*:581 (1962).
30. R. B. Valley and T. H. Morse, U.S. Patent 3,441,848 (1969).
31. R. H. Berg and R. F. Karuhn, U.S. Patent 4,290,011 (1981).
32. E. C. Gregg and K. D. Steidley, *Biophys. J. 5*:393 (1965).
33. C. S. Waterman, E. A. Atkinson, B. Wilkins Jr., C. L. Fisher, and S. L. Kimzey, *Clin. Chem. 21*:1201 (1975).
34. V. Kachel, *J. Histochem. Cytochem. 24*:211 (1976).
35. K. Marshall and R. G. Mehta, M. Sc. Thesis, University of Bradford, 1969.
36. R. K. Eckhoff, *J. Sci. Instr. (J. Phys. E.). 2*:973 (1969).
37. P. J. Lloyd, B. Scarlett and I. Sinclair, Proc. Conf. Particle Size Analysis 1970, Bradford, Society for Analytical Chemistry, London, 1970, p. 276.
38. P. J. Lloyd, Proc. Conf. Partec (Nuremberg), 1979, p. 694.
39. C. M. L. Atkinson and R. Wilson, Proc. Conf. Particle Size Analysis 1981, Loughborough, Wiley-Heyden, Chichester, 1982, p. 185.
40. *British Standard 3406: Part 5:1983;* British Standards Institution, London, 1983; Amendments No. 1, AMD 4535 (1984); No. 2, AMD 7976 (1994).
41. J. W. Gartner, Ph.D. thesis, University of South Florida, 1978.
42. J. W. Gartner and K. L. Garder, *J. Sed. Petrol. 49*:631 (1979).
43. National Institute of Standards and Technology, Standard Reference Materials Program, Room 204, Building 202, Gaithersburg MD 20899; U.S.A.; suppliers of particulate reference materials, glass beads and latex suspensions.
44. Community Bureau of Reference-BCR, Rue de la Loi 200, B-1049 Brussels, Belgium; suppliers of particulate reference materials, quartz powders and latex suspensions.
45. R. W. DeBlois and C. P. Bean, *Rev. Sci. Instr. 41*:909 (1970).
46. R. W. DeBlois and R. K. A. Wesley, *J. Virol. 23*:227 (1977).
47. R. W. DeBlois, C. P. Bean, and R. K. Wesley, *J. Coll. Interf. Sci. 61*:323 (1977).
48. I. N. McCave and J. Jarvis, *Sedimentology 20*:305 (1973).
49. J. G. Harfield, B. Miller, R. W. Lines, and T. Godin, Proc. Conf. Particle Size Analysis 1977, Salford, Heyden, London, 1978, p. 378.
50. S. Kinsman, Proc. Coulter Counter Users' Conf., Chicago, Ill., Coulter Electronics, Inc., 1961, p. 30.
51. H. R. Yarde, *J. Sci. Instr. 42*:711 (1965).
52. R. K. Eckhoff, *J. Sci. Instr. 44*;648 (1967).

53. A. W. Nienow and H. M. Ang, *Lab. Practice 21*:495 (1972).
54. R. de Neve and L. Wuyts, *Pharm. Weekblad 111*:220 (1976).
55. W. J. Ullrich, *Modern Developments in Powder Metallurgy,* Vol. 1, *Fundamentals and Methods,* Plenum Press, 1966, p. 125.
56. J. G. Harfield, Proc. Conf. Particle Size Analysis 1981, Loughborough, Wiley-Heyden, Chichester, 1982, p. 165.
57. D. Horák, L. Peška, F. Švec, and J. Štamberg, *Powder Technol. 31*:263 (1982).
58. G. van der Plaats and H. Herps, *Powder Technol. 38*:73 (1984).
59. M. R. Groves, *I.E.E.E. Trans. Biomed. Engrg. BME-27,* 364 (1980).
60. R. W. Lines, *Powder Technol. 7*:129 (1973).
61. M. Wales and J. N. Wilson, *Rev. Sci. Instr. 32*:1132 (1961).
62. L. H. Princen and W. F. Kwolek, *Rev. Sci. Instr. 36*:646 (1965).
63. L. H. Princen, *Rev. Sci. Instr. 37*:1416 (1966).
64. H. Bader, H. R. Gordon, and O. B. Brown, *Rev. Sci. Instr. 43*:1407 (1972).
65. P. J. Lloyd, Proc. First European Symp. Particle Size Measurement, Nuremberg, 1975, p. 325.
66. J. G. Harfield, P. J. Lloyd, and M. P. Cowan, Proc. Conf. Particle Size Analysis 1988, Guildford, Wiley, Chichester, 1988, p. 121.
67. R. Thom, A. Hampe, and G. Sauerbrey, *Z. ges. exp. Med. 151*:331 (1969).
68. R. Thom, *Vergleichende Untersuchungen zur Elektronischen Zellvolumenanalyse,* A.E.G. Telefunken, Ulm, W. Germany, 1972.
69. C. M. L. Atkinson and R. Wilson, *Powder Technol. 34*:275 (1983).
70. R. Wilson, *Part. Charact 1*:37 (1984).
71. W. R. Hogg, *U.S. Patent 3,668,531* (1972).
72. J. G. Harfield and M. P. Cowan, Proc. Conf. Particle Size Analysis 1988, Guildford, Wiley, Chichester, 1988, p. 151.
73. M. P. Cowan and J. G. Harfield, *Part. Part. Syst. Charact. 7*:1 (1990).
74. B. Göransson, *Part. Part. Syst. Charact. 7*:6 (1990).
75. H. G. Merkus, H. Liu, and B. Scarlett, *Part. Part. Syst. Charact. 7*:11 (1990).
76. R. F. Karuhn, R. Davies, and M. J. Clinch, U.S. Patent 3,739,258 (1973).
77. R. F. Karuhn, R. Davies, and B. H. Kaye, U.S. Patent 3,739,268 (1973).
78. R. F. Karuhn, R. Davies, and M. J. Clinch, *Powder Technol. 11*:157 (1975).
79. R. Davies, R. F. Karuhn, and J. Graf, *Powder Technol. 12*:157 (1975).
80. R. Davies, R. F. Karuhn, J. Graf, and J. Stockham, *Powder Technol. 13*:193 (1976).
81. L. I. Berge, J. Feder, and T. Jøssang, *Rev. Sci. Instr. 60*:2756 (1989).
82. L. I. Berge, J. Feder, and T. Jøssang, *J. Coll. Interf. Sci. 138*:480 (1990).
83. L. I. Berge, *J. Coll. Interf. Sci. 135*:283 (1990).
84. L. I. Berge, *J. Coll. Interf. Sci. 134*:548 (1990).
85. L. I. Berge, T. Jøssang, and J. Feder, *Meas. Sci. Technol. 1*:471 (1990).
86. Chr. Simensen, personal communication, 1985.
87. R. I. L. Guthrie and D. A. Doutre, Proc. Int. Seminar on Refining and Alloying of Liquid Aluminium and Ferro-Alloys, Trondheim, Norway, 1985, p. 147.
88. O. Hanotte, C. Dejaiffe, D. Dubois, P. Vanderpoorten, F. Menozzi, and A. O. A. Miller, *Biol. Cell 59*:175 (1987).
89. V. Degouys, I. Cerckel, A. Garcia, J. Harfield, D. Dubois, L. Fabry, and A. O. A. Miller, *Cytotechnology* 13:195 (1993).
90. D. Thebline, J. Harfield, O. Hanotte, D. Dubois, and A. O. A. Miller, Conf. 8th ESACT-32nd OHOLO, Tiberias, Israel, 1987.
91. M. I. Barnett, E. Sims, and R. W. Lines, *Powder Technol. 14*:125 (1976).
92. R. W. Lines, J. G. Harfield, W. M. Wood, and B. V. Miller, Intl. Symp. In Stream Measurements of Particulate Solid Properties, Bergen, 1978.
93. W. S. Maddux and J. W. Kanwisher, *Limnol. Oceanogr. 10*:R162 (1965).
94. R. W. Lines, Conf. Oceanology International '69, Brighton, 1969.
95. C. M. Boyd, *Neth. J. Sea Res. 7*:103 (1973).

96. W. H. Coulter and W. R. Hogg, U.S. Patent 3,503,973 (1970).
97. A. Richardson-Jones, Evaluation of the COULTER® VCS Automated Differential Counter, Coulter Electronics, Inc., 1988.
98. D. F. Barnard, S. A. Barnard, A. B. Carter, K. G. Patterson, A. Yardumian, and S. J. Machin, *Clin. lab. Haemat. 11*:255 (1989).
99. C. M. Hawkey, *Comparative Mammalian Haematology,* Heinemann, London, 1975.
100. W. M. Wintrobe, *Clinical Hematology,* 8th ed., Lea and Febiger, Philadelphia, 1981, p. 14.
101. J. A. Koepke, *LabMedica Int. XI*:16 (1994).
102. Thalassaemia Working Party of the BCSH General Haematology Task Force, *J. Clin. Pathol. 47*:289 (1994).
103. D. L. Simel, E. R. DeLong, J. R. Feussner, J. B. Weinberg, and J. Crawford, *Arch. Int. Med. 148*:822 (1988).
104. P. C. M. Bartels, Red cell size distribution histograms in haemocytometry, Ph.D. thesis, Utrecht University, 1989, p. 15.
105. J. D. Bessman and R. K. Johnson, *Blood 46*:369 (1975).
106. R. M. Rowan, *Blood Cell Volume Analysis,* Albert Clark, London, 1983.
107. G. Brecher, M. Scheiderman, and G. Z. Williams, *Am. J. Clin. Pathol. 26*:1439 (1956).
108. C. F. T. Mattern, F. S. Brackett, and B. J. Olson, *J. Appl. Physiol 10*:56 (1957).
109. J. H. Ackeroyd, M. B. Gibbs, S. Vivano, and R. W. Robinette, *Am. J. Clin. Pathol. 31*:188 (1959).
110. W. J. Richar and E. S. Breakell, *Am. J. Clin. Pathol. 31*:384 (1959).
111. H. Zeidler, A. Tolle, and W. Heeschen, *Milchwissenschaft 23*:564 (1968).
112. C. R. Sipe and E. P. Cronkite, *Ann. N.Y. Acad. Sci. 99*:262 (1962).
113. J. H. Girling, *J. Med. Lab. Technol.* 1 (1962).
114. B. S. Bull, M. A. Schneiderman, and G. Brecher, *Am. J. Clin. Pathol. 44*:678 (1965).
115. W. A. Curby, E. M. Swanton, and H. E. Lind, *J. Gen. Microbiol 32*:33 (1963).
116. G. J. Mountney and J. O'Malley, *J. Appl. Microbiol. 14*:845 (1966).
117. R. M. Macrae, Proc. European Brewery Convention, Brussels, Belgium, Elsevier, Amsterdam, 1964, p. 510.
118. J. M. Sablayrolles and P. Barre, *Biotechnol. Techn. 6*:15 (1992).
119. H. P. Koch, M. Hofeneder, and B. Bohne, *Meth. Find Exp. Clin. Pharmacol. 15*:141 (1993).
120. M. Harris, *Cancer Res. 19*:1020 (1959).
121. A. A. Hirata, *J. Immunol. 91*:625 (1963).
122. W. F. Daly, L. P. Piascik, and J. L. Bittle, *J. Pharm. Sci. 55*:426 (1966).
123. R. I. Freshney, *Culture of Animal Cells—A Manual of Basic Technique,* 3rd. ed., Wiley-Liss, New York, 1994, p. 270.
124. F. A. Glover and L. W. Phipps, *J. Reprod. Fertil. 4*:189 (1962).
125. S. J. Segal and K. A. Laurence, *Ann. N.Y. Acad. Sci. 99*:271 (1962).
126. K. A. Laurence and O. Carpuk, *Fertil and Steril. 14*:451 (1963).
127. S. Iversen, *J. Agric. Sci. 62*:219 (1964).
128. D. L. Gordon, D. J. Moor, T. Thorslund, and C. A. Paulsen, *J. Lab. Clin. Med.65*:506 (1965).
129. J. Hirsch and E. Gallian, *J. Lipid Res. 9*:110 (1968).
130. C. G. D. Brook, J. K. Lloyd, and O. H. Wolf, *Brit. Med. J.* April 1: 25 (1972).
131. C. G. D. Brook and J. K. Lloyd, *Arch. Dis. Childhood 48*:301 (1973).
132. C. G. D. Brook, *Arch. Dis. Childhood 48*:725 (1973).
133. J. A. Borchardt and C. R. O'Melia, Proc. Ann. Conf. Amer. Waterworks Assoc., 1961.
134. J. Hastings, B. M. Sweeney, and M. M. Mullin, *Ann. N.Y. Acad. Sci. 99*:280 (1962).
135. S. Z. El-Sayed and B. D. Lee, *J. Marine Res. 21*:59 (1963).
136. R. W. Sheldon and T. R. Parsons, *A Practical Manual on the use of the Coulter Counter in Marine Science,* Coulter Electronics Sales Co. of Canada, 1967.
137. T. G. Milligan and K. Kranck, *Principles, Methods and Application of Particle Size Analysis* (J. P. M. Syvitski, ed.), Cambridge University Press, 1991, p. 109.
138. M. J. Groves and J. F. G. Major, *Pharm. J. 193*:227 (1964).
139. *British Pharmacopoeia: Limit Test for Particulate Matter,* A 242 (1973).

140. M. J. Akers, *Parenteral Quality Control,* Marcel Dekker, New York, 1985, p. 178.
141. T. Barber, *Pharmaceutical Particulate Matter: Analysis and Control,* Interpharm, Buffalo Grove, IL, 1993, p. 51.
142. M. J. Groves, *Particulate Matter: Sources and Resources for Healthcare Manufacturers,* Interpharm, Buffalo Grove, IL, 1993, p. 182.
143. T. A. Barber, Proc. Int. Conf. Liquid-Borne Particle Inspection and Metrology, Arlington, VA, 1987, Parenteral Drug Assoc., Inc., Philadelphia, 1987, p. 317.
144. *British Pharmacopoeia: Injectable Preparations—Preparations for Parenteral Use,* 578 and A120, 1980.
145. R. F. Haines-Nutt and T. J. Munton, *J. Pharm. Pharmacol. 36*:534 (1984).
146. G. S. Oxborrow, Proc. Int. Conf. Liquid-Borne Particle Inspection and Metrology, Arlington, VA, 1987, Parenteral Drug Assoc., Inc., Philadelphia, 1987, p. 482.
147. T.-M. Cham, H.-M. Yu, and L.-C. Tung, *Drug Devt. Ind. Pharm. 15*:2441 (1989).
148. J. P. K. Seville, J. R. Coury, M. Ghadhiri, and R. Clift, *Part. Charact. 1*:45 (1984).
149. R. W. Lines, *J. Pharm. Pharmacol 19*:701 (1967).
150. R. W. Lines, *Bull. Parent. Drug Assoc. 21*:113 (1967).
151. R. W. Lines, *Anal. Proc. 18*:514 (1981).
152. M. I. Barnett, Proc. Int. Conf. Liquid-Borne Particle Inspection and Metrology, Arlington, VA, 1987, Parenteral Drug Assoc., Inc., Philadelphia, 1987, p. 222.
153. P. Revell, *Anal. Proc. 29*:149 (1992).
154. C. Washington and T. Sizer, *Int. J. Pharm. 83*:227 (1992).
155. C. Washington, J. A. Ferguson, and S. E. Irwin, *J. Pharm. Sci. 82*:808 (1993).
156. N. A. Hoenick, J. Thomson, J. McCabe, and D. R. Appleton, *Intnl. J. Artif. Organs 13*:803 (1990).
157. A. S. Benenson, *Bull. W.H.O. 30*:635 (1964).
158. M. J. Groves, *J. Pharm. Sci. 82*:555 (1993).
159. G. H. Worth and J. R. McLaughlin, *J. Coll. Interf. Sci. 160*:347 (1993).
160. R. Wickham, *J. Roy. Naval Sci. Serv. 19*:153 (1964).
161. F. G. Anderson, T. F. Tomb, and M. Jacobson, *U.S. Dept. Interior Bureau of Mines Report 7105* (1968).
162. J. H. Flux, D. J. Smithson, and R. N. Smithson, *J. Iron Steel Inst.* 1188 (1968).
163. *ASTM D 3365-74T* (1974).
164. M. Ahlberg, L. Berghem, G. Nordberg, S.-Å. Persson, L. Rudling, and B. Steen, *Env. Health Perspectives 47*:85 (1983).
165. M. Ghadhiri, J. P. K. Seville, J. A. Raper, and R. Clift, Proc. Partec, Nuremberg, 1986, p. 203.
166. S. Ghosal, J. L. Ebert, and S. A. Self, *Part. Part. Syst. Charact. 10*:11 (1993).
167. T. C. Huang, N. D. Watkins, D. M. Shaw, and J. P. Kennett, *Earth Planet. Sci. Lett. 20*:119 (1973).
168. L. D. Taylor and J. Gliozzi, *Antarctic Snow and Ice Studies, Antarctic Res. Ser. 2*:267 (1964).
169. *Evaluation of Particle Counting as a Measure of Treatment Plant Performance* (E. E. Hergesheimer, C. M. Lewis, and C. M. Yentsch, eds.), Amer. Water Works Assoc., Denver, 1992, p. 21.
170. J. D. Hazlett and M. A. Bergougnou, *Powder Technol. 70*:99 (1992).
171. V. Kirjavainen, *Acta Polytechn. Scand.* No. 206: 51 pp (1992).
172. J. S. D'Arrigo and T. Imae, *J. Coll. Interf. Sci. 149*:592 (1992).
173. E. A. Neidiek, *Zeit. Zuckerind. 19*:495 (1969).
174. P. W. Van der Poel, H. P. M. Heere, C. C. Bleyenberg, M. A. M. Schutter, N. H. M. der Visser, and P. M. T. van Heeschvelde, Proc. 17th General Assy., C.I.T.S., Copenhagen, 1983.
175. D. J. McDonald, Proc. Annual Mtg. Pennsylvania Manuf. Confectioners Assoc., Lancaster, PA, 1965.
176. J. G. Surak, R. F. Matthews, V. Wang, H. A. Padna, and R. M. Hamilton, *Proc. Fla. State Hort. Soc. 92*:159 (1979).
177. A. C. Stringfellow and A. J. Peplinski, *The Northwestern Miller 271*:10–12 (1964).
178. V. K. Chaudhary, S. T. Yamazaki, and W. A. Gould, *Cereal Chem. 58*:314 (1981).
179. A. D. Evers, *Anal. Proc. 19*:393 (1982).
180. J. Polderman, *Boll. Chim. Farm. 101*:105 (1962).
181. M. J. Thornton, *J. Pharm. Pharmacol. 15*:742 (1963).

182. J. C. Boylan, Proc. Ind. Pharm. Technol. Section of Academy Pharm. Sci., Miami Beach, FL, 1968.
183. B. A. Matthews, *Canad. J. Pharm. Sci. 6*:29 (1971).
184. N. A. Orr and J. Spence, *Analyst 102*:466 (1977).
185. E. E. Sims, PhD. thesis, University of Wales, 1979.
186. C. Nyström, J. Mazur, M. I. Barnett, and M. Glazer, *J. Pharm. Pharmacol. 37*:217 (1985).
187. J. E. Carless and A. A. Foster, *J. Pharm. Pharmacol. 18*:697 (1966).
188. J. E. Carless, M. A. Moustafa, and J. D. C. Rapson, *J. Pharm. Pharmacol. 20*:639 (1968).
189. J. I. Wells and M. H. Rubinstein, *Pharm. J.* 629 (1976).
190. C. Caramella, F. Ferrari, and M. C. Bonferoni, *Part. Part. Syst. Charact. 7*:131 (1990).
191. A. J. Aguigar, J. E. Zelmer, and A. W. Kimbel, *J. Pharm. Sci. 56*:1243 (1967).
192. A. L. Cripps, *Brit. J. Pharm. Practice 6*:187 (1984).
193. T. L. Whateley, G. Steele, J. Urwin, and G. A. Smail, *J. Clin. Hosp. Pharm. 9*:113 (1984).
194. D. G. Haile, G. A. Mount, and N. N. Pierce, *Mosquito News 38*:1 (1978).
195. A. W. Karig, G. E. Peck, and G. J. Sperandio, *J. Pharm. Sci. 62*:811 (1973).
196. M. Schrenzel, *Pharm. Acta Helv. 41*:218 (1966).
197. L. H. Princen, J. A. Stolp, and W. F. Kwolek, *J. Paint Technol. 39*:182 (1967).
198. L. H. Princen, J. A. Stolp, and W. F. Kwolek, *J. Paint Technol. 39*:188 (1967).
199. H. A. Hamza and N. E. Anderson, *Fine Particle Processing* (P. Somasundaran, ed.), 1988, Vol. 1, p. 388.
200. M. J. Akers and J. L. Lach, *J. Pharm. Sci. 65*:216 (1976).
201. S. S. Davis, T. S. Purewal, and A. S. Burbage, *J. Pharm. Pharmacol. 28* (Suppl.): 60p (1976).
202. M. I. Barnett and E. Sims, *Int. J. Pharm. Tech. Prod. Mfr. 4*:59 (1983).
203. P. Artursson, P. Edman, T. Laakso, and I. Sjöholm, *J. Pharm. Sci. 73*:1507 (1984).
204. I. Vural, H. S. Kas, A. A. Hincal, and G. Cave, *J. Encapsulation 7*:511 (1990).
205. V. Sovilj, L. Djaković, and P. Dokić, *J. Coll. Interf. Sci. 158*:483 (1993).
206. K. Iwamoto, T. Kato, M. Kawahara, N. Koyama, S. Wanatabe, Y. Mikaye, and J. Sunamoto, *J. Pharm. Sci. 80*:219 (1991).
207. K. J. Steffens, A. Lich, and H. Perschbacher, *Manuf. Chem. 63* (No.1):23 (1992).
208. M. Rajaonarivony, C. Vauthier, G. Courraze, F. Puisieux, and P. Couvreur, *J. Pharm. Sci. 82*:912 (1993).
209. H. Talsma and D. J. A. Crommelin, *Pharm. Technol. Int. 5*:37 (1993).
210. H. Tamai, N. Kando, T. Suzawa, and H. Yasuda, *J. Coll. Interf. Sci. 154*:210 (1992).
211. S. Stoll, V. Lanet, and E. Pefferkorn, *J. Coll. Interf. Sci. 157*:302 (1993).
212. J. F. Desmars and R. Tawashi, *Biochem. Biophys. Acta 313*:256 (1973).
213. L. C. Cao, E. R. Boevé, and F. H. Schröder, *J. Lithotripsy Stone Disease 3*:324 (1991).
214. M. J. Groves, *Parenteral Products,* Heineman, London, 1973.
215. J. A. Koepke, *Laboratory Hematology,* Vol. 2, Churchill-Livingstone, New York, 1984, p. 915.
216. CCKL Committee on Standardisation of Hematological Laboratory Methods, *Tijdschr. NVKC (Netherlands),* 12:160 (1987).
217. R. Thom, *New Approaches to Laboratory Medicine* (S. B. Rosalki, ed.), Ernst Giebeler Verlag, Darmstat, 1981, p. 3.
218. J. Jakschik and K. Schulze, Uncertainties of the Measurements of Particle Concentrations with Conductance Blood Cell-Counters due to Equipment Characteristics, Abstract, Amal Conference, Germany, 1986.
219. J. M. England et al., ICSH Expert Panel on Cytometry, *Clin. lab. Haem. 10,* 203 (1988).
220. J. Jakschik and K. Schultze, Proc. Conf. Partec 89, Nuremberg, 1989, p. 845.
221. J. G. Harfield, presented at Reference Methods in Blood Cell Counting Seminar, Düsseldorf, November 1993.
222. J. M. England et al., ICSH Expert Panel on Cytometry, *Clin. lab. Haem. 16*:131 (1994).
223. S. C. Prescott and R. S. Breed, *J. Inf. Dis. 7*:632 (1910).
224. J. G. Harfield, *Anal. Proc. 21*:162 (1984).
225. *AFNOR Provisional Standard XII-671,* AFNOR, France, 1981.
226. *AFNOR Standard NF XII-670,* AFNOR, France, 1989.

227. *British Pharmacopoeia,* 758 (1988).
228. *British Pharmacopoeia,* 748 (1993).
229. *ANSI/ASTM D 3563-76T* (1976).
230. *ANSI/ASTM C 690-75* (1975).
231. *ASTM F7-70* (1970).
232. *ANSI/ASTM C 699-79* (1979).
233. *ASTM F 662-86* (1986).
234. *ASTM F 795-88* (1988).
235. *ASTM F 796-88* (1988).
236. *ASTM F 797-88* (1988).
237. *ASTM F 1067-87* (1987).
238. *ASTM F 1170-88* (1988).
239. *ANSI/ASTM F 577-78* (1978).
240. *ANSI/ASTM D 3451-76* (1976).
241. J. E. Daborn and R. Wilson, *Report EUR 10867 EN,* Commission of the European Communities, Brussels, 1986.
242. D. R. Booker and J. M. Mitchell, *J. Aerosol Sci. 23*:S 293 (1992).
243. J. M. Mitchell, *Anal. Proc. 29*:508 (1992).

5
Particle Sizing Using Ensemble Averaging Techniques

Bruce B. Weiner

Brookhaven Instruments Corporation, Holtsville, New York

I. INTRODUCTION

Novices have many misconceptions about particle sizing. Perhaps the most common is the difference between single particle counting (SPC) and ensemble averaging (EA) techniques. Single particle counting, as the name implies, involves counting of individual particles. This is often associated with either imaging of single particles (microscopy, for example) or counting single particles as they pass through an orifice (electrical zone counters, for example). The information obtained is primarily the concentration, expressed as the particle number per unit volume of fluid. A red blood cell count is a good example. Particle size may be obtained in addition, with varying degrees of accuracy and resolution, depending on the particular SPC technique employed. From the measured sizes, a particle size distribution (PSD) can be constructed.

Single particle counters are required whenever a concentration is required. Examples include cleanroom applications, contamination in liquids, and hematology studies to name a few. To characterize the tails of a PSD, SPC is also useful, sometimes mandatory, because the few particles in the tail may be hard to determine accurately any other way. Some SPC techniques are capable, usually over limited size ranges, of very high resolution, such as the differentiation of multiple peaks.

There are, however, drawbacks to SPC techniques. These include clogging of orifices, coincidence counting errors, and, often, poor statistics when not enough particles are counted. In addition, calibration is usually necessary, and this means that subsequent measurements are based on the properties of the particular particles used for calibration. An example includes the calibration of optical clean-room counters with airborne latex particles. The results obtained on the real sample are now dependent upon the optical properties of latex, and these may differ significantly from those of the real particles.

EA techniques involve the measurement of many particles simultaneously. As a result, resolution varies from high to low, depending on the particular EA technique. Here, resolution is defined as the ability to distinguish fine details in the PSD. For example, most light diffraction and light scattering techniques yield medium-to-low resolution PSD's. With other EA techniques the distribution is partially fractionated prior to measurement, and this results in higher resolution. Instruments based on sedimentation are examples.

EA techniques do not produce absolute concentrations; sieving is the major exception.

Rather it is the relative PSD that is produced and desired, not a concentration. Yet, unlike SPC, EA techniques rarely suffer from poor statistics; often 10^5 or more particles contribute to the signal. There is no orifice to clog, and coincidence counting is not an issue, although most EA techniques do require dilute suspensions if accuracy is important. With many EA techniques calibration is not required, and the accuracy of the resulting PSD may be quite high.

The purpose of this article is to present three common and often used EA techniques: sedimentation (SD), dynamic light scattering (DLS), and Fraunhofer diffraction (FD).

II. SEDIMENTATION

In some ways the theoretical concepts behind particle sizing are difficult: the principles of diffraction, scattering, and diffusion are relatively abstract. Yet, one of the oldest methods, SD, is simple to understand. And, as a bonus, the mathematics are relatively easy to develop [1].

If a particle falls in a liquid, then the equation of motion under laminar flow conditions is

$$m_p\left(\frac{d^2x}{dt^2}\right) = m_p g - m_f g - f\frac{dx}{dt}$$

where m_p is the particle mass, m_f is the mass of an equal volume of fluid it displaces, g is the local acceleration due to free fall (typically, 981 cm/s^2), and f is the frictional coefficient. The equation of motion says that the mass of the particle m_p times its acceleration d^2x/dt^2 is equal to the sum of the forces on it: the force of gravity, a compensating force due to buoyancy, and a drag force proportional to particle velocity to account for the friction experienced by the particle as it travels through the fluid.

This second-order, homogeneous, differential equation with constant coefficients can be solved exactly. The transient part of the solution dies away very quickly, typically in less than a millisecond. The long-time part of the solution is the important part. It can also be obtained by assuming that the particle acceleration approaches zero very rapidly. With this assumption, the terminal velocity, $[dx/dt]_0$, is

$$\left[\frac{dx}{dt}\right]_0 = \frac{g\,\Delta m}{f}$$

It is useful to express the difference in mass, Δm, as the volume of the particle, V, times the difference in density between the particle and the fluid density, $\Delta\rho = \rho_p - \rho_f$. For a sphere $V = \pi d^3/6$. The frictional coefficient f equals $3\pi\eta d$ for a sphere, where η is the viscosity of the liquid through which the particle sediments. This result for the frictional coefficient was first determined by Stokes in 1891. Integrating yields the time t it takes a spherical particle of diameter d to reach a distance h below its starting position. The result is Stokes' law:

$$t = \frac{18\eta h}{g\,\Delta\rho\, d^2}$$

In a centrifuge where the rotational velocity is ω rad/s, Stokes' law becomes

$$t = \frac{18\eta \ln(R_D/R_I)}{\omega^2\,\Delta\rho\, d^2}$$

where R_D is the radial distance from the center of the disk to the position of the detector, and R_I is the radial distance from the center of the disk to the initial position of the particle. These radial positions are calculated from the volume of spin fluid initially injected into the spinning disk.

This is the liquid through which the particles sediment radially outward when $\Delta\rho$ is positive, or radially inward when $\Delta\rho$ is negative.

There are two starting techniques in SD [2]. In the line start (LIST) technique, all the particles start moving from the same position. The larger particles travel faster than the smaller ones. This gives rise to the greatest separation and, hence, the greatest resolution. At the start of the homogeneous (HOST) technique, particles should be uniformly distributed throughout the liquid. The signal at any time is the sum over all the particles that reach the detector at the same time. These include the small ones that were close, the large ones that were far away, and all those in between with a size such that they reach the detector at the same time t. Resolution is not as great as it is with the LIST technique; yet, the resolution with the HOST technique is much greater than with full ensemble averaging techniques such as DLS or FD. In either the LIST or HOST techniques the enhanced resolution is a direct result of the partial fractionation that takes place prior to detection.

In modern SD instruments there are two detection techniques: X-ray absorption and turbidity using light. In both cases a beam of radiation passes through the sample at some known distance from the surface of the liquid. The time it takes particles to reach the detector yields particle size through Stokes' law. But determining particle size yields only a range, not a distribution. Distribution information is obtained from measuring the extinction of the beam due to the presence of the particles.

In the case of X-ray detection, it is the absorbance that dominates the extinction. X-ray scattering is insignificant in particle sizing. When particles absorb X-rays, they do so in direct proportion to mass. The important thing is this: there are no light scattering corrections, no refractive index-dependent corrections that apply in the case of X-ray absorption. The mass distribution is obtained directly. Such a detection scheme is good for high-density materials such as metal powders, clays, most ceramics, and most inorganic pigments, where the signal from X-ray absorption is sufficiently strong to yield values well above the weak absorption due to the liquid.

But X-ray absorption it is too weak for most low-density organic materials such as latexes, liposomes and other pharmaceuticals agents, carbon blacks, and most organic pigments. The X-ray mass absorption coefficients for these materials are close to that of the liquids in which they are suspended. Therefore, the X-ray signals are too weak for accurate particle sizing. Since all materials scatter or absorb light, the X-ray source and detector may be replaced by a light source and, typically, a photodiode detector to gain the sensitivity necessary for good detection.

When a beam of light is partially extinguished by the presence of particles, it may be the result of scattering or absorption or some combination of both. For spheres, the extinction efficiency, Q_{ext}, can be calculated exactly from Mie scattering theory given the wavelength of light, the refractive indices of the particle and liquid, and the particle size [3]. Extinction is embodied in the turbidity, τ, a quantity calculated from the measured transmittance, T, defined as

$$T = \frac{I}{I_0} = \exp(-\tau L)$$

where I and I_0 are the measured intensities of transmitted light with and without particles, and L is the path length through the cell. Given the finite area of any real detector, the turbidity is related to the differential volume distribution by the equation

$$\frac{dV}{dd} = \frac{C\tau}{Q_{ext}}$$

where C is a constant for a fixed detector position. When the particles are large, at least many times larger than the wavelength of the light source, and the particle refractive index is substantially different from that of the liquid, then $Q_{ext} = 2$, the Fraunhofer limit. In this case the differential volume distribution is proportional to the turbidity. For particles smaller than several microns, or for particles with refractive indices close to that of the liquid, Q_{ext} is a complicated function of particle size and refractive index. (See Ref. 3 for a discussion of calculating Q_{ext}.) This is the case for liposomes and many other interesting pharmaceutical dispersions.

Figure 1 shows the cumulative undersize distribution of $BaSO_4$ obtained with the Brookhaven BI-XDC using the HOST technique and the gravity mode.

An *estimate* of the largest size measurable with sedimentation can be calculated using Reynolds' criterion for the onset of turbulent flow. Stokes' law is based on the assumption of laminar flow, and, for a sphere, turbulent flow begins when the Reynolds' number, R_e, is greater than, roughly, 0.2. The Reynolds' number is defined as follows,

$$R_e = \frac{\rho_f d[dx/dt]_0}{\eta}$$

where $[dx/dt]_0$, the terminal velocity of the particle, is given by Stokes' law. Substituting, and collecting terms yields the theoretical maximum value d_{max},

$$d_{max} \leq \left[\frac{18\eta^2 R_e}{g\rho_f \Delta\rho}\right]^{1/3}$$

For $BaSO_4$ where $\rho_p = 4.5$ g/cm^3, run in 20% sucrose, at 25°C, and $R_e = 0.2$, the maximum theoretical particle size is 73 microns, just within the upper range of the measurement shown in Fig. 1.

Figure 2 shows the differential distribution by mass for TiO_2. Again, a Brookhaven BI-XDC

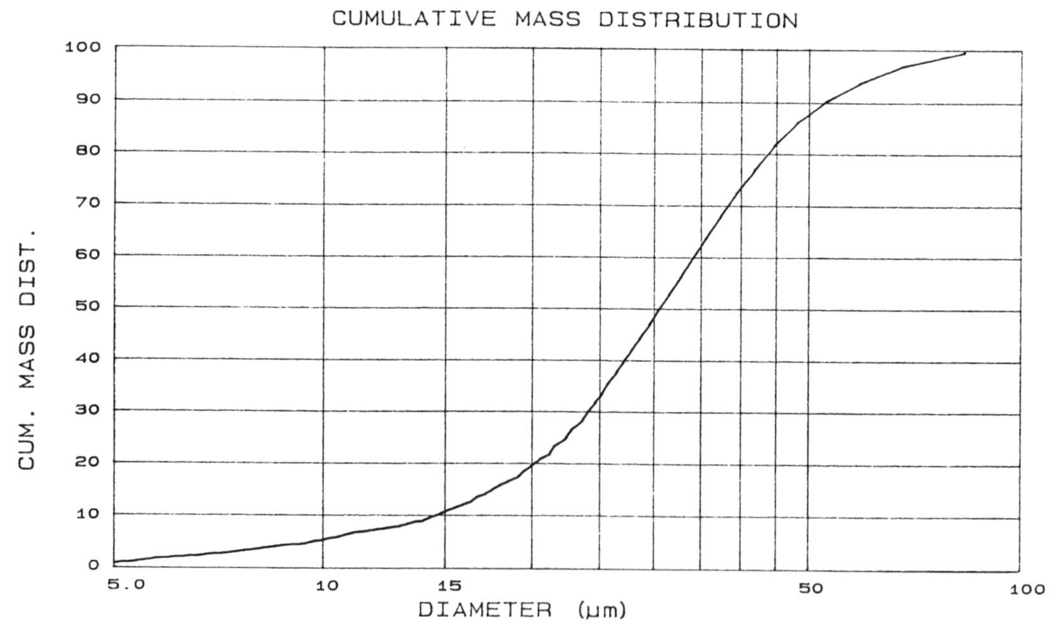

Figure 1 Cumulative undersize distribution by mass of $BaSO_4$ obtained using the HOST method on a BI-XDC, scanning X-ray sedimentometer, in the gravity mode.

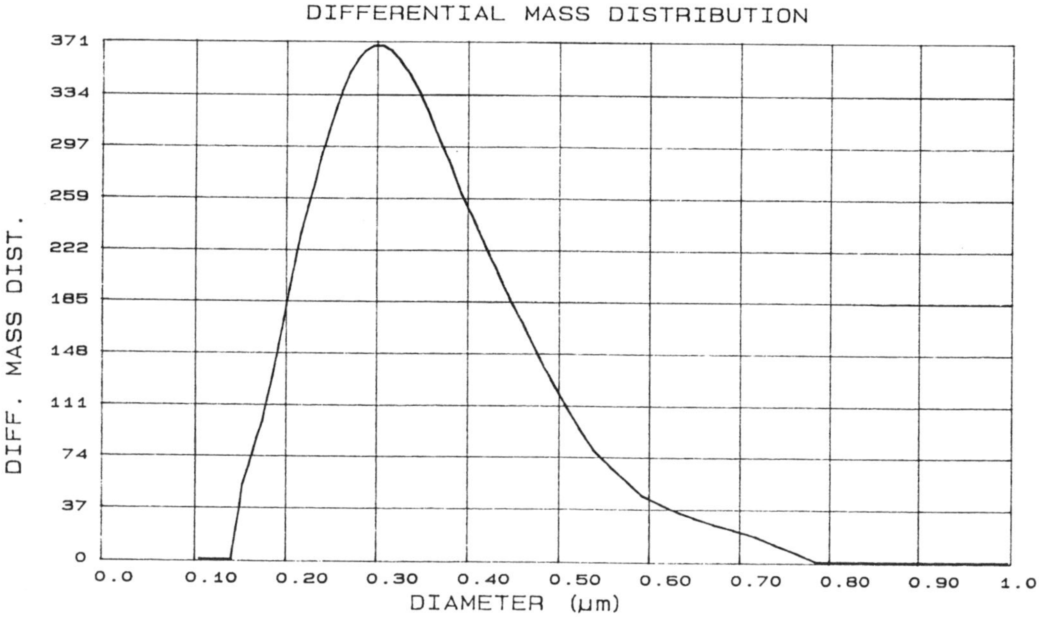

Figure 2 Differential distribution by mass of TiO_2 obtained using the HOST method on a BI-XDC in the centrifuge mode.

was used with the HOST technique; however, given the submicron character of the sample, the centrifuge was run at 1500 rpm. Such a measurement with the scanning detector head requires only a few minutes.

For broad distributions, say a size range covering more than a decade, a measurement using sedimentation becomes somewhat slow. This is a direct result of the inverse d^2 dependence in either of Stokes' laws. There are three methods for decreasing the measurement time: scanning the detector, multiple detectors, or increasing the rotational speed during the measurement in the case of the centrifuge.

Scanning detectors are routinely and successfully employed. Although some resolution is lost, scanning does provide a three- to five-fold decrease in the measurement time compared to the measurement without scanning. Multiple optical detectors, with corresponding photodiodes, have been successful in the case of gravity sedimentation, but the technique has not yet been tried with a centrifuge. With multiple detectors the difference in sensitivity between source/detector pairs must be taken into account. Multiple X-ray source/detector pairs are, at present, too expensive and require too much space.

A centrifuge with an initial rotation speed of ω_0 and a constant acceleration α, yields the following modification to Stokes' Law:

$$t = \left[\left(\frac{\omega_0}{\alpha}\right)^3 + \left(\frac{\omega_0}{\alpha}\right)^2 3G\right]^{1/3} - \frac{\omega_0}{\alpha}$$

G is equal to the time t given by Stokes' law in a centrifuge with constant rotational speed ω_0. In the limit as α goes to zero, this expression reduces to G as expected.

Such a scheme provides, in principle, a three- to four-fold decrease in total measurement time, without the need to account for differences in source/detector pairs. However, in practice, PSD results on the same sample are a function of the acceleration. This indicates that such an acceleration is impossible to provide, at the present time, without disturbing the laminar flow pattern, a tacit assumption if the Stokes results are valid.

Figure 3 shows the extremely high resolution results obtained with the Brookhaven BI-DCP, a disk centrifuge photosedimentometer, using the LIST technique. The sample was a mixture of five polystyrene latexes, and Q_{ext} was calculated over a range from 0 to 10 microns. Over this same range Q_{ext} rises sharply at first (as d^4), reaches a maximum, oscillates with ever-decreasing amplitude, and finally reaches the Fraunhofer constant limit (to within a few percent) around 10 microns. The PSD obtained without applying this correction, or applying a correction only based on d^n, with n fixed at 1 or 2, is inaccurate, perhaps highly inaccurate, even though the sizes are correct.

Figure 4 shows results down to 25 nm on carbon black using the LIST technique with a BI-DCP. The strong absorption in the visible gives rise to an extinction efficiency that is linear up to 200 nm, at which point it bends over, becoming relatively constant above a few hundred nanometers. The lower limit of this technique is approximately 10 nm, below which diffusion contributes significantly to the position of the particles. Figure 5 shows the differential mass distribution for two samples of albuterol sulfate, a bronchial dialator used in the treatment of asthma. The data were obtained using the HOST technique and a Brookhaven BI-DCP. Both samples exhibit a similar PSD. In one case the ratio of the two peaks is less than 2-to-1 apart, demonstrating, again, the high resolution of the SD technique. Neither DLS nor FD is capable, routinely, of this resolution. Whether the bimodality plays a part in the efficacy of this drug is a question that remains unanswered.

Before moving on to the next topic, it is worth pointing out two complications involved with SD. The first concerns centrifugation and the HOST technique. Since particles reaching the detector either sediment radially outward or inward, and since the signal is the sum over particles originating at different radial positions, account must be taken of radial dilution, or in the case of $\Delta\rho < 0$, radial concentration. The proper way to do this is to apply the Kamack equation [4], an integral equation. In the past, correcting for radial dilution or concentration was an arduous task; it was often neglected in the name of "fingerprinting" for purposes of quality control. Now

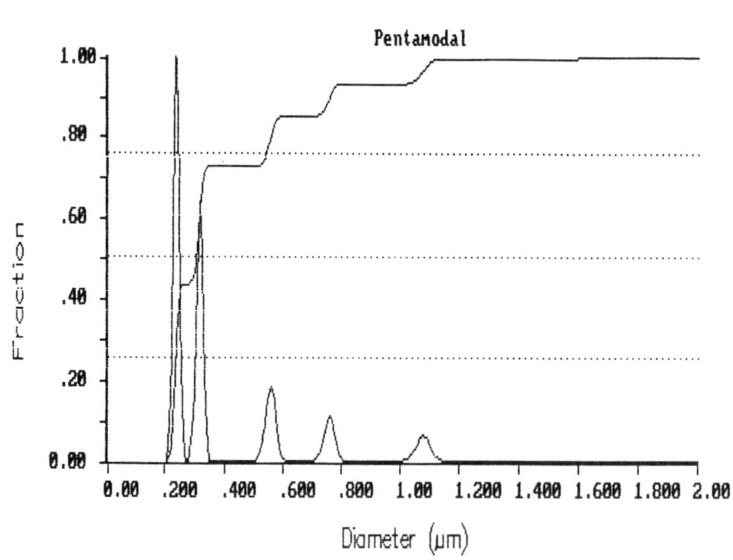

Figure 3 Differential and cumulative undersize distribution by weight (volume) of mixture of five standard polystyrene latexes obtained using the LIST method on a BI-DCP, disk centrifuge photosedimentometer.

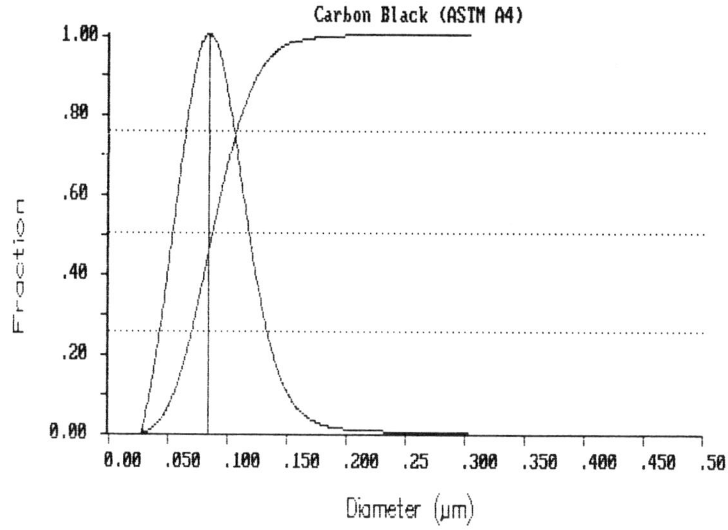

Figure 4 Differential and cumulative undersize distribution by weight (volume) of carbon black obtained using the LIST method on a BI-DCP.

that such corrections can be made almost instantaneously with a computer, there is no excuse for not making them.

The second problem concerns the determination of the true beginning and end of the experiment. In the LIST technique, where no particles appear for a few tens of seconds, it is easy to determine the start time; and the end time is easily determined when the signal returns to the original baseline. But in the HOST technique, or with a broad distribution and the LIST technique, a compromise is often made between finding the true starting time and the total

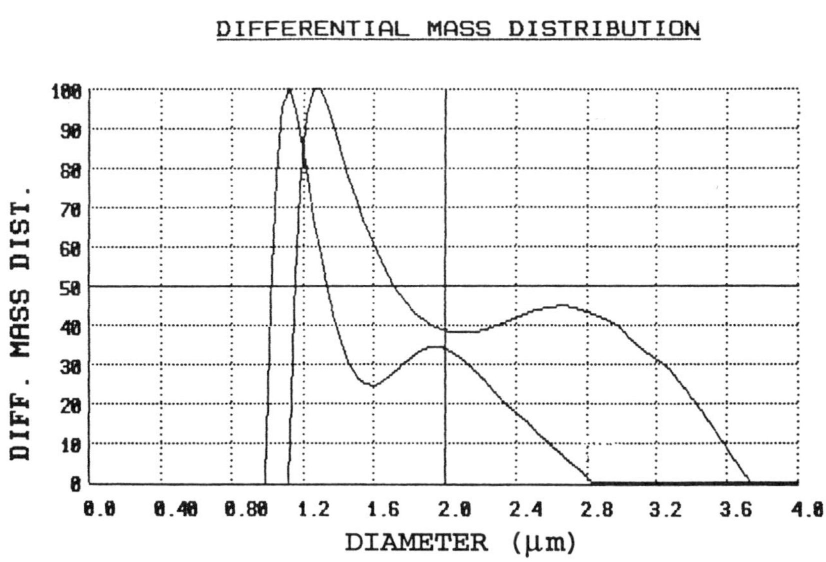

Figure 5 Differential distribution by mass (volume) of two samples of albuterol sulfate, a bronchial dialator used in the treatment of asthma, obtained using the HOST method on a BI-DCP.

experiment duration. Where accuracy and high resolution are not required, such broad distributions are best handled with other techniques.

III. DYNAMIC LIGHT SCATTERING

To the novice, particle sizing by light scattering is all inclusive. In reality, light scattering is a diverse field, and many aspects of it have been exploited for use in particle sizing. For example, with a DCP, the time it takes a particle to travel from its initial position to the position defined by the light beam passing through the sample is used to determine particle size from Stokes' law; and the extinction of the beam determines the relative concentration of particles at that size. In fact, many techniques use the principles of light scattering to determine particle size, or concentration, or both. In most of these instances it is the intensity as a function of angle that is the primary source of the information. Interestingly, in DLS the intensity plays a secondary role.

DLS applies only to particles small enough to remain suspended in a liquid. This includes dissolved macromolecules, liquid-liquid microemulsions, liposomes, self-assembly particles such as micelles, and all the solid colloids such as latexes, carbon blacks, pigments, clays, and ceramics, to name just a few.

All these systems have one thing in common: gravitational sedimentation is much less important than self-diffusion. (There is more than one type of diffusion. The type measured depends on the length scale used to observe it. In what follows it will be assumed that the particles are far enough apart that only the infinitely dilute, self-diffusion is observed. And the prefix "self" will be dropped for convenience.) Diffusion is the result of a large number of random collisions between the particle of size d_p and the surrounding solvent molecules of size d_s. The number of solvent molecules surrounding the surface of a particle is proportional to $(d_p/d_s)^2$. The water molecule, for example, is about 0.1 nm in size. For a 100-nm particle, the number of water molecules surrounding it is on the order of 1 million. Even for a 1-nm particle, the number of surrounding water molecules is on the order of 100. Such large numbers give rise to the idea of a random collision process for which statistical ideas apply quite well. And these ideas lead one to the notion of diffusion.

Scattered light intensity is proportional to the square of the scattered electric field. The scattered field is a function of the distance from the particle to the detector. This distance is fluctuating randomly as a result of the diffusion of the particles. It is, therefore, not surprising that by analyzing the fluctuations, not the magnitude, of the scattered light intensity one can determine the diffusion coefficient. Particle size, as shown below, is related to the diffusion coefficient [5].

Figure 6 is a diagram of a typical DLS experimental setup [6]. The focused beam from a vertically polarized laser impinges upon particles that are undergoing random, Brownian motion inside a temperature controlled sample cell. As a result of the random motion, the scattered light

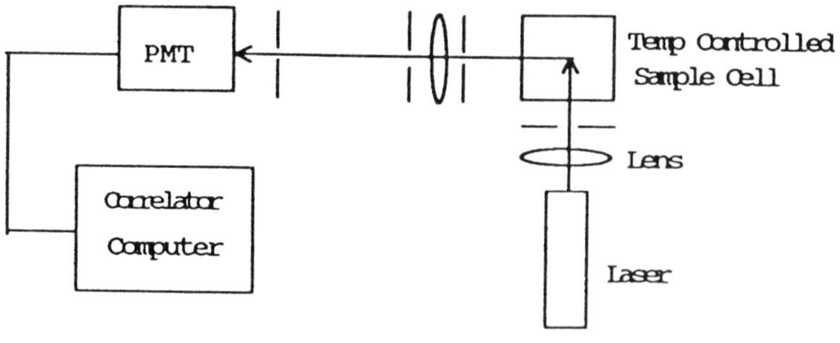

Figure 6 Typical configuration for a DLS instrument for submicron particle sizing.

fluctuates in time. Large particles give rise to low-frequency fluctuations; small particles give rise to higher-frequency fluctuations. The time dependence of the fluctuations are best analyzed by autocorrelating the scattered photons. (For this reason the technique is sometimes called photon correlation spectroscopy (PCS). Additionally, because the incident light undergoes a slight frequency broadening, the technique is also called quasi-elastic light scattering (QELS).)

The second-order autocorrelation function (ACF) of the scattered photons is given by

$$C(\tau) = \langle n(t)n(t - \tau)\rangle$$

where $n(t)$ is the number of photons counted over a sampling interval $\Delta\tau$ centered at time t, and $n(t - \tau)$ is the number of photons counted over $\Delta\tau$ but delayed in time by τ. For dilute, rigid, globular particles, measured in the self-beating mode (often and erroneously called the homodyne mode), the ACF is given by

$$C(\tau) = B\left\{1 + f\left|g^1(\tau)\right|^2\right\}$$

For a monodisperse sample $|g^1(\tau)| = \exp[-\Gamma\tau]$ and $\Gamma = D_T q^2$, where Γ, the linewidth, is the half-width at half-height of the frequency-broadened distribution of the scattered light. Since the early 1970s the ACF, which is the Fourier transform of the frequency distribution, is measured; yet, the name "linewidth" is still used.

The amplitude of the scattering wave vector, q, is equal to

$$q = \frac{4\pi n_0}{\lambda_0} \sin\left(\frac{\theta}{2}\right)$$

where n_0 is the refractive index of the liquid, λ_0 is the wavelength of the laser, and θ is the scattering angle in the liquid. In 1905 Einstein showed that the translational diffusion coefficient $D_T = k_B T/f$, where k_B is Boltzmann's constant (1.38×10^{-16} erg/deg), T is the absolute temperature in kelvins, and f is a frictional coefficient. This is the same frictional coefficient in Stokes' law for sedimentation. Thus, the final equation is called the Stokes-Einstein equation:

$$D_T = \frac{k_B T}{3\pi\eta d_H}$$

where η is the viscosity of the liquid and d_H is the hydrodynamic diameter of the particle.

There are some similarities between particle sizing using SD and particle sizing using DLS. In SD the sedimentation coefficient is related to particle size through f; in DLS the diffusion coefficient is related to particle size through f. In both SD and DLS one relies on Stokes' result for the frictional coefficient as a function of viscosity and particle size. Even if the particle is not a sphere, a measured sedimentation or diffusion coefficient can be used to calculate an equivalent spherical size. Finally, in both sedimentation and diffusion, the resulting particle size is, in principle, equal to the dry-volume size plus any attached species that move with the particle. Thus, the size calculated is really the size of this kinetic unit. In the vast majority of cases the difference between the kinetic unit size and the dry volume size is insignificant. A prominent exception is a colloidal particle stabilized by a high molecular weight polymer.

For a polydisperse sample the situation is more complicated. The first-order, electric field, autocorrelation function $|g^1(\tau)|$ is a weighted sum or integral over the exponential decays from each of the contributing particles. The weighting comes from the scattered light intensity

$$\left|g^1(\tau)\right| = \int_0^\infty G(\Gamma)\exp(-\Gamma\tau)\, d\Gamma$$

where $G(\Gamma)$ is the relative intensity corresponding to a particle with linewidth Γ. The intensity weighting has many consequences. For particles much smaller than the wavelength of light the intensity varies as d^6. Thus, it is almost impossible to register a significant signal from small particles in a broad distribution. Also, relatively few large particles, generically called "dust," may dominate the scattered light. In order to calculate the volume or mass distribution, one has to divide the intensity-weighted distribution by $d^3P(\theta)$. The particle scattering function $P(\theta)$, calculated from Mie scattering theory, is a complicated function of size, wavelength, scattering angle θ, and the refractive indices of the liquid and particle. Both the particle size and the amount have to be determined from the intensity-weighted distribution function $G(\Gamma)$, and this function is only obtained after a deconvolution of the ill-conditioned Laplace transform shown above. A sum of exponentials looks remarkably like a single exponential. For all these reasons, DLS does not yield high-resolution size distribution information.

There are two approaches to calculating distribution information from the measured ACF. The first, and most robust, is the method of cumulants. Similar to finding the moments of the distribution, the cumulant technique involves expanding the exponential in the above integral and then integrating term by term. The result is

$$\ln\left|g^1(\tau)\right| = -\bar{\Gamma}\tau + \frac{\mu_2\tau^2}{2!} - \frac{\mu_3\tau^3}{3!} + \cdots$$

From $\bar{\Gamma}$ an average diffusion coefficient can be obtained, from which an average particle size is calculated. Note that this average size is neither a weight (read volume or mass) average nor an intensity-weighted average. It is an inverse or harmonic average, weighted by the intensity of scattered light. It is often called the effective or apparent diameter. The second moment (moments and cumulants are identical up to third order) μ_2 is a measure of polydispersity. It is common to report the relative second moment $\mu_2/\bar{\Gamma}^2$, often called the "poly." It approaches zero as the size distribution approaches monodispersity.

Some novices are confused by the process of describing size distributions with moments; yet, it is quite common for the data from even a high-resolution size distribution determination to be reduced to one or two pieces of information. For example, when the median or mean or modal value is used to describe an entire distribution, or when the size at the 90th or 10th percentile is given, this represents the entire distribution by just a single number. In addition, if the spread of the distribution is given as one number—perhaps as the half or full width at half maximum, or the standard deviation, or a ratio of the 90th-to-10th percentile diameters, or by quoting the average diameter of the volume and number distributions—then the entire distribution has been reduced to just two numbers: a measure of the average and a measure of the spread. In DLS this is equivalent to quoting the effective diameter and poly.

The second approach to calculating distribution information is to deconvolute numerically the ill-conditioned Laplace transform. Several methods are available including CONTIN, nonnegatively constrained least squares, exponential sampling, maximum entropy method, singular value decomposition [7]. None are universal. The ill-conditioned nature of the problem does not allow a unique, robust solution in all cases, especially given the sensitivity of the result to small drifts in baseline (dust and/or electronic drifts) and small differences in the experimental noise from one ACF to the next. This is yet another manifestation of the relatively low resolution inherent in full ensemble averaging techniques.

Table 1 shows the typical and excellent repeatability obtained on a polystyrene latex labeled 96 ± 3.9 nm used as a reference material. These results were obtained with a Brookhaven 90Plus submicron particle sizer employing a 5-mW HeNe laser. The distribution is narrow, as shown by

Table 1

Run	Eff. diam. (nm)	Half-width (nm)	Polydispersity	Sample quality
1	98.9	5.9	0.004	9.9
2	97.5	7.3	0.006	10.0
3	96.8	17.5	0.033	10.0
4	97.9	9.3	0.009	10.0
5	97.6	6.4	0.004	10.0
Mean	97.7	9.3	0.011	10.0
Std. error	0.3	2.1	0.006	0.0
Combined	97.8	9.0	0.008	10.0

the small polydispersity index. The accuracy is also exceptionally good. Each measurement required only 60 s. Such results are quite useful for quality and process control applications.

Figure 7 shows results on a fabric softener obtained using the Brookhaven BI-90 particle sizer. The bimodal character is due to a mixture of liquid crystal-like structures and cationic-stabilized emulsion droplets. In general, however, multimodal results are always suspect due to mathematical artifacts arising from the ill-conditioned mathematics.

Figure 8 shows the PSD results for a liposomal formulation. Liposomes (vesicles) have been described as targeted drug delivery systems. This particular liposome is designed for injection into solid cancer tissue.

DLS measurements have been made on particles as small as 1 or 2 nm [8]. Of course, such results were not obtained with a 5-mW HeNe laser. They were obtained using 400 mW from a water-cooled, argon-ion laser. Special precautions were used to preclude the dust.

DLS measurements above 1 μ become increasingly more inaccurate. Larger sizes sediment.

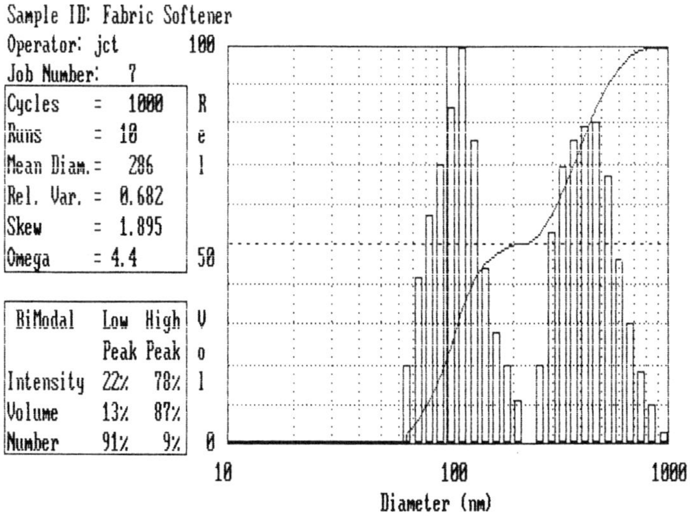

Figure 7 Differential and cumulative undersize distribution by volume of a mixture of liquid crystal-like structures and cationic stabilized emulsions (commercial fabric softener) obtained using the BI-90, a particle sizer based on the technique of DLS.

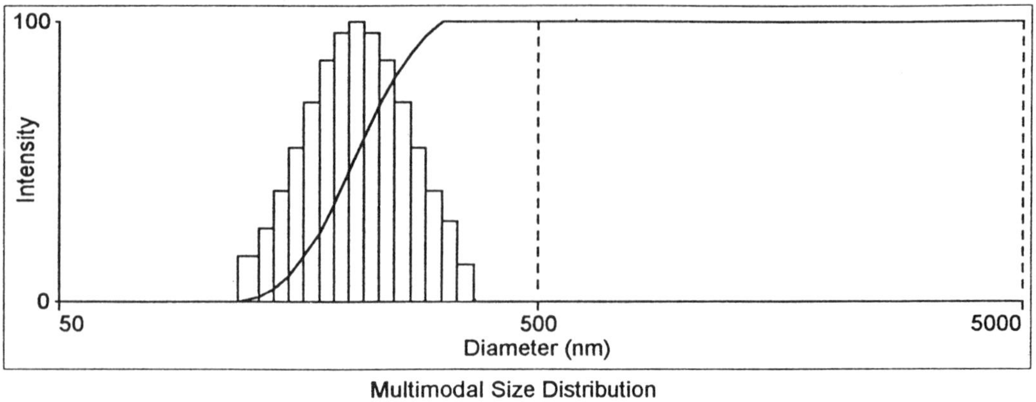

Figure 8 Differential and cumulative undersize distribution by intensity of a liposomal formulation used to treat solid cancers obtained using the 90Plus, a particle sizer based on the technique of DLS.

The scattered light is mostly in the forward direction which is affected by flare light and dust. The measurement time increases in proportion to size; a 2- or 3-min measurement at 300 nm would require 20 to 30 min at 3 μ. More subtle is the effect on the number of particles in the scattering volume. The relationship between the first- and second-order autocorrelation functions is valid only for a large number of particles in the scattering volume, but the use of the Stokes-Einstein relationship requires very dilute suspensions. These conflicting requirements are easy to arrange for particles below 0.5 μ; increasingly difficult to arrange for particles between 0.5 and 1 μ; and nearly impossible to arrange for particles above 1 μ [9].

IV. FRAUNHOFER DIFFRACTION

Since its inception as a commercial instrument in the mid 1970s, FD has become the reigning king of ensemble averaging, automatic particle sizers in the range above a few microns. To what should one attribute this phenomenal success? in a word: speed. Like DLS, measurements using FD often take a minute or less. Like DLS, there are no orifices or columns to clog; there is nothing to be calibrated. A wide range of particle sizes, typically more than two orders of magnitude, can be accommodated in one measurement. Measurements can be made both in liquids and in gases [10].

Figure 9 shows, schematically, the experimental configuration used in a typical Fraunhofer instrument.

Fraunhofer theory is really very old, and it is described in most textbooks on optics. The majority of the scattered light appears in a relatively small cone centered around the incident light for particles of a few microns or greater. For a sphere or aperture or solid disk (oriented such that the incident light is perpendicular to the face of the disk) the intensity I is related to the incident intensity I_0 by

$$I = I_0 \left[\frac{2J_1(x)}{x} \right]^2$$

where J_1 is the first-order Bessel function, and $x = 2\pi rs/\lambda_0 F$. Here r is the particle radius, s is the radial distance in the detection plane from the optical axis, λ_0 is the wavelength of the light,

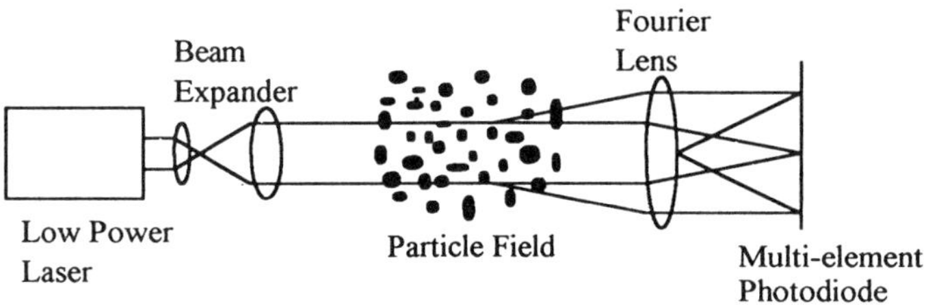

Figure 9 Typical configuration for a FD instrument for particle sizing above a few microns.

and F is the focal length of the Fourier lens. As long as the detector is located at the focal plane, then the diffraction pattern is stationary, even if the particles are moving. This feature of the optical configuration enables particle ensembles to be measured in a variety of ways: flowing, suspended, or stationary. The squared term is called the Airy function in honor of the British astronomer who first derived it for use in explaining diffraction effects seen in telescopes.

The Airy function is a series of concentric rings of light separated by dark rings. The zeros of the Bessel function give the values at which $x = 0$: the first four are 3.83, 7.02, 10.17, and 13.32. Given the distance s at which a dark ring appears, the focal length of the lens, and the wavelength of light, it is an easy matter to calculate the radius and diameter of the particle. Large particles diffract more light into the forward direction; and small particles diffract more light at higher angles. A particularly nice feature of the Fraunhofer instrument is this: whenever the theory applies, the results do not depend on particle refractive index or particle density. This is an advantage over SD instruments where the density is required, and over any instrument where the refractive index is required to account for both the particle size and amount.

Now imagine the difficulty of superimposing a large number of such patterns arising from different size particles. Except for very narrow distributions, the oscillations that made it easy to match the particle size to the zeros in the Bessel function disappear. To be sure, an angular variation still exists; however, it becomes increasingly difficult to resolve different sizes as the Airy patterns merge. In principle it is possible to capture and store the pattern for individual particles; analyze each pattern for a single particle size; then combine the results from thousands of such determinations to obtain a complete particle size distribution. In practice, such an instrument would constitute an SPC, not an EA.

At the present time no such instrument is practical. The demands on memory and speed would be too costly. Instead, the energy falling between two detector rings, one with radius $s1$ and one with radius $s2$, is analyzed. [11] For a single particle this energy is given by the expression

$$L_{s1,s2} = C\pi r^2 \left\{ \left[J_0^2(x) + J_1^2(x) \right]_{s1} - \left[J_0^2(x) + J_1^2(x) \right]_{s2} \right\}$$

where J_0 is the zero-order Bessel function and C embodies various optical constants associated with a particular light source and detector.

For a distribution of particles one sums up all the contributions for various sizes under the assumption that multiple scattering is insignificant (implying dilute systems). The result is

$$L_{s1,s2} = K \sum_{i=1}^{M} \frac{V_i}{r_i} \left\{ \left[J_0^2(x_i) + J_1^2(x_i) \right]_{s1} - \left[J_0^2(x_i) + J_1^2(x_i) \right]_{s2} \right\}$$

where V_i is the volume fraction at particle radius r_i, K is a constant, and M is the number of detector elements. Each detector element defines another $s1$-$s2$ pair. Therefore, the size distribution is calculated by analyzing a set of M equations each like the one above but with M sets of $s1$-$s2$ pairs. This is a highly nonlinear problem that reduces significantly the resolution of an FD instrument.

Attempts at matrix inversion, unless constrained with a-priori knowledge, are difficult. For this reason assumptions (often not stated explicitly) are made about the distribution. Often the resulting peaks are then fit to a simple function such as a lognormal. The final results are not well resolved, perhaps narrow peaks no closer than 3-to-1 apart can be separated. This is true despite the fact that common detector arrangements include 32, 64, and 128 separate detector elements plus, as explained later, a few detectors around 90° for characterizing submicron particles. Although it is claimed that increasing the number of detectors increases FD resolution, in comparison to a good SPC or even a fractionation technique such as sedimentation, FD is *still* a low-resolution technique.

Novices are often fooled into believing a distribution has been determined with high resolution because the number of size classes in the printed result is quite high. This is, more often than not, a result of fitting the data and interpolation than it is of the fundamentals associated with measurement. Real resolution is a function of the signal-to-noise ratio inherent in the measurement. Real resolution is limited when the fundamental equations describing the transformation between measured quantities and particle size are highly nonlinear. Think of it this way: doubling the number of size classes by interpolation cannot increase real resolution.

Figure 10 shows results obtained on a metered dose inhaler using the Sympatec HELOS

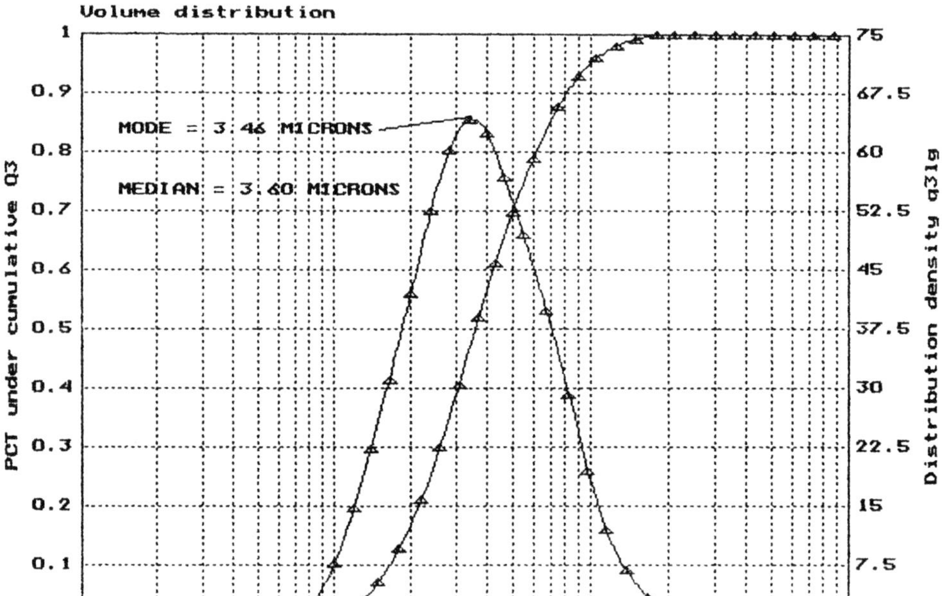

Figure 10 Differential and cumulative undersize distribution by volume from a metered dose inhaler obtained using the Sympatec HELOS instrument based on FD. (Courtesy of P. Faraday, Sympatec Inc., Princeton, New Jersey.)

instrument. The instrument provides a repeatable force on the inhaler's actuator button, thus ensuring reproducible results. The results were obtained over 100 ms, again emphasizing the speed with which such measurements can be made. Sizes above, approximately, 7 μ are trapped in the throat; sizes below, approximately, 1 μ are exhaled. Thus, the results on this particular inhaler show that roughly 85% of the volume of droplets are useful.

Figure 11 shows results obtained with the Sympatec HELOS on a double-milled, dry, pharmaceutical powder. The double milling has resulted in a bimodal with one peak at roughly 13 μ and another at 2.4 μ. The measurement time in this case was 20 s.

Notice the 10% by volume below 1 μ. In this range the Fraunhofer theory does not normally apply. Fraunhofer theory applies when the size is much larger than the wavelength, and/or the refractive index difference between the particle and suspending medium is large. By measuring the powder in air instead of liquid, the refractive index difference is larger, and less error is incurred using Fraunhofer theory. Of course, it is usually more difficult to disperse a dry powder, and the results considerably overestimate the true size distribution if this is not done properly.

In general, however, the accuracy in any analyzer using just the intensity pattern suffers more and more as the particles approach a few microns or less in size. As a result, many instruments include high-angle detectors to register the light scattered at high angles. In addition, Mie scattering theory must be applied rather than Fraunhofer theory. And this requires knowledge of the particle refractive index as well as the liquids. Even with high-angle detectors, and Mie theory, the increasing loss of variation with angle of the scattering pattern

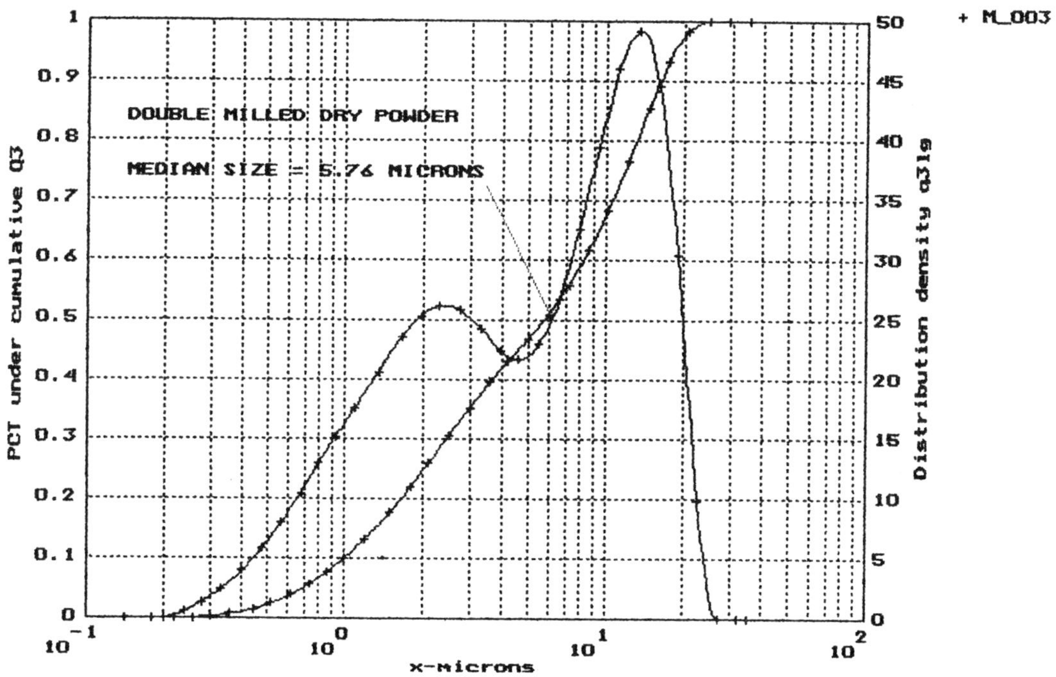

Figure 11 Differential and cumulative undersize distribution by volume of a double-milled, dry, pharmaceutical powder obtained using the Sympatec HELOS instrument. (Courtesy of P. Faraday, Sympatec Inc., Princeton, New Jersey.)

makes it very difficult to obtain accurate sizes let alone any real resolution. This is not to say that a-priori knowledge cannot be used to obtain good results; however, at some point such a-priori knowledge begs the question: Why make a measurement at all?

It is rather surprising to find that many combination "Fraunhofer and high-angle" instruments claim to make measurements down to 0.1 and even 0.04 μ. Under special circumstances, and with known samples, typically polystyrene latex in water, any algorithm can be adjusted to deliver credible answers. But what about the unknown samples of real interest?

As particles approach the colloidal size range below a few microns, the surface area per unit mass increases dramatically, varying as $1/d$. Modest errors in the volume-weighted size distribution are transformed into large errors in surface area- and number-weighted distributions.

The real value of a Fraunhofer instrument lies in its speed, versatility, repeatability, and simplicity of operation. Sizes up to several hundred microns can be measured, provided care is taken that sampling errors do not occur. It is not a high-resolution instrument such as a zone counter or disk centrifuge; nor is it particularly accurate below a few microns with unknown samples. Yet its virtues for many quality control applications, especially above a few microns, outweigh its drawbacks, and this has led to its wide acceptance by industrial customers.

V. SUMMARY

Given the ubiquitous nature of particle sizing applications in industry and research, it is not surprising that such a large number of different particle sizing techniques have been developed. EA techniques are not appropriate for applications requiring detailed knowledge of the tails of a distribution. Nor are they appropriate when absolute concentrations are required. Yet, EA techniques are very useful for quality control and many research applications, provided caution is taken in choosing the technique best able to produce reliable results. Care should be exercised when choosing an instrument that yields very repeatable results that are neither accurate enough nor provide enough resolution to solve a particular problem. Too often, in the name of quality control, very repeatable results are repeatably wrong.

REFERENCES

1. T. Allen, *Particle Size Measurement,* 4th ed., Chapman and Hall, London, 1990, Chaps. 7, 12.
2. T. Allen, A review of sedimentation methods of particle size analysis, *Particle Size Analysis* (N. G. Stanley-Wood and R. W. Lines, eds.), The Royal Society of Chemistry, Cambridge, UK, 1992, pp. 454–476.
3. B. Weiner, D. Fairhurst, and W. Tscharnuter, Particle size analysis with a disc centrifuge: importance of the extinction efficiency, *Particle Size Distribution II: Assessment and Characterization* (T. Provder, ed.), American Chemical Society, Washington DC, 1991, Chap. 12.
4. H. J. Kamack, *Br. J. Appl. Phys.* 5:1962 (1972).
5. B. Chu, *Laser Light Scattering: Basic Principles and Practice,* 2nd ed., Academic Press, New York, 1991, Chaps. II–IV.
6. B. Weiner, Particle sizing using photon correlation spectroscopy, *Modern Methods of Particle Size Analysis* (H. G. Barth, ed.), Wiley-Interscience, New York, 1984, Chap. 3.
7. Many of the techniques developed in the 1970s for size distribution analysis can be found in different chapters of *Measurement of Suspended Particles by Quasi-Elastic Light Scattering* (B. E. Dahneke, ed.), Wiley-Interscience, 1983. Methods developed in the 1980s are discussed in Chapter VII of Ref. 4.
8. R. Neuman and Z-J. Yu, *Colloids Surf. 46*:45 (1990).

9. For a more applications-oriented discussion of DLS in particle sizing see B. Weiner, Twenty seven years of QELS: a review of the advantages and disadvantages of particle sizing with QELS, *Particle Size Analysis,* (N. G. Stanley-Wood and R. W. Lines, eds.), The Royal Society of Chemistry, Cambridge, UK, 1992, pp. 173–185.
10. B. J. Azzopardi, Instrumentation for particle size analysis by far field diffraction: accuracy, limitations and future, *Particle Size Analysis* (N. G. Stanley-Wood and R. W. Lines, eds.), The Royal Society of Chemistry, Cambridge, UK, 1992, pp. 108–132.
11. B. Weiner, Particle and droplet sizing using Fraunhofer diffraction, *Modern Methods of Particle Size Analysis* (H. G. Barth, ed.), Wiley-Interscience, New York, 1984, Chap. 5.

6
In Situ Particle Measurement

Alvin Lieberman

Particle Measuring Systems, Inc., Fremont, California

I. INTRODUCTION

A. Measurement Procedure Definition

When observing particles in any liquid, measurement is usually carried out in a sample taken from the liquid, unless the total quantity of liquid is so small that the entire quantity can be observed and recovered for further use. In many cases, a sample of liquid is removed from the main body, the particulate content defined, and the sample is discarded; samples are seldom returned to the main body of liquid, even with reprocessing. The main body of liquid may be present in a reservoir, from which material is withdrawn for use as needed. Alternately, the liquid may be flowing in a processing system or in a transport line. In either of these cases, a sample of the liquid is removed for observation. The particulate content of the entire volume of liquid in the sample is measured. This technique is frequently referred to as "volumetric sample measurement." Alternately, an "in situ sample measurement" technique can be used. Here, a sample region is optically or acoustically defined and observed near the center of a flowing fluid line and no liquid is removed from the flowing line.

The detailed requirements for sample acquisition with minimum generation of artifacts or changes in the particle content of the liquid are discussed elsewhere. However, it has long been known that problems exist when samples are removed from a larger body of liquid and placed in containers or otherwise manipulated for measurement of particle content. One problem arises because of the difficulty in maintaining sample container cleanliness and freedom from particulate contamination, particularly for particles smaller than 5 μm. For almost all container materials, the fabrication, handling, and filling processes normally produce particulate contamination which is difficult to remove from the interior surfaces of the container and its closure. These particles will be suspended into any liquid sample added to the container. Samples which are not immediately analyzed may be stored under conditions which may permit change in the particulate content. The change may arise from particulate material solubility due to temperature effects during storage. Particles may also be generated if chemical reaction occurs or if viable biological material is present in the liquid or in the container.

Other problems can arise in withdrawing liquid from a moving stream or from a container. The sample withdrawal may remove liquid at a flow rate or conditions different from those when liquid is removed for normal operations or processing. In this situation, the liquid circulation in a container will be quite different during normal operation and during sample withdrawal; particle movement and sedimentation will differ in the two operations. This effect can also be expected when removing material from a moving stream unless isokinetic or at least isoaxial sample collection is accomplished. Liquid flowing in a line will usually be under some pressure. It may be necessary to contain the liquid at pressure or temperature which will not allow exposure to ambient conditions for sample removal. Otherwise, the liquid may adsorb undesirable gas or otherwise react with materials in the environment.

For these reasons, in situ observation systems have been developed for measurement of particles in liquids without sample extraction. These systems can be applied for measurement of particle content in liquid flowing under pressure or at high rates within process or sidestream sampling lines. The measurement methodology is based on transmission of either optical or acoustic energy into and out of a flow system to determine particle size and number in the liquid flowing therein. An acoustic system with a convergent-divergent cone-shaped sensing volume had been developed and was evaluated in 1981 [1]. The particle position and the varying acoustic energy level in that volume prevented any differentiation between particles with sizes varying by a factor of less than 2. The authors who evaluated the system stated ". . . the effective sensing volume for 19.8 μm spheres is 22 times that for 9.8 μm spheres." This observation indicates that it would indeed be difficult to define the concentration in which any sample is observed if there is a range of particle sizes in that sample.

In situ systems are used to observe the particle content at a location close to the center of a fluid flow without interfering with liquid flow or particle passage. The measurement should be based on observation of the effects of particle presence on a *uniform* optical or acoustic energy flux defined in a small volume within the overall liquid stream. A small liquid volume is observed in order to maximize signal-to-noise so as to permit measurement of the smallest possible particles and to permit measurement of particles in high concentrations. The liquid volume observed is typically at the central portion of the liquid flow line to allow observation of particles moving with the main liquid flow and to minimize effects of liquid-container boundary effects on the signal. Although the observed sample flow volume may be only a few milliliters per minute, the liquid in which these sample volumes are observed may be flowing at many liters per minute. Sensitivity to particles smaller than 0.1 μm in diameter can be achieved with optical systems and particles in concentrations up to thousands of particles per milliliter can be counted and sized in the small sensitive volumes. Acoustic sensors can be used to count and size particles ≥ approximately 10 μm in diameter in concentrations to a few thousand per milliliter. As mentioned previously, the sonic energy distribution in the focused beam varies with distance from the source. In order to measure small particles with acceptable sizing resolution, a more uniform energy source is required. Optical techniques have been found acceptable for this purpose. In using in situ measurement techniques, a viewing module is added to a liquid flow line and the liquid is scanned through windows in that module. The scanned portion of the liquid is defined optically rather than by a physical flow boundary. The intersection of the illuminating laser beam and the observation path is maintained within the liquid flow, as shown in Fig. 1. A laser illumination system is used to maintain maximum illumination intensity within the sampled volume of the liquid to permit particle measurement with maximum signal level. Since the beam from a laser has Gaussian intensity variation, only the central portion of the beam is used to minimize illumination intensity variation in the viewed volume. Means of defining the dimensions of the sample volume cross section will be discussed later.

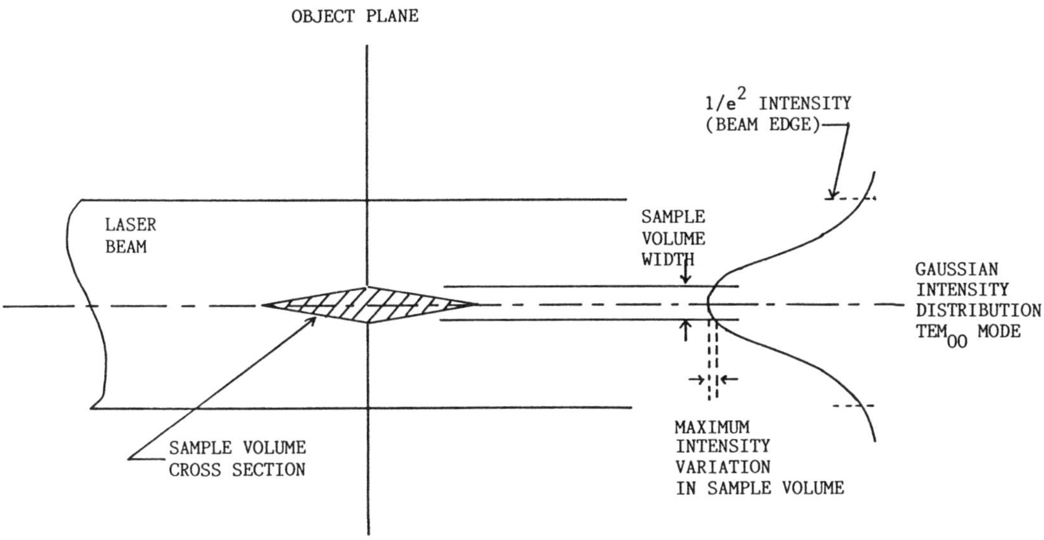

Figure 1 In situ measurement volume definition. (From Ref. 3.)

B. Advantages and Problem Areas

1. Advantages of in situ Sampling

There are several advantages to an in situ sampling system. First, the possibility of adjusting the defined volume dimensions allows measurement of particle concentrations over a very wide range. This adjustment can be accomplished remotely by adjusting an optical or acoustic element orientation externally, without directly interfering with the liquid. Even though the defined volume may be only a fraction of a percent of the total volume being sampled, the system can be used in lines with flow rates ranging from fractions of a liter per minute to several hundred liters per minute. This capability also allows measurement over a large range of particle concentrations. Skilled personnel are required to design correct modifications of an optical system, but the changes are relatively simple and allow reducing or expanding the sampled volume as anticipated particle concentration increases or decreases.

When high particle concentrations or large particles are present in the liquid being measured, the typical small-flow-rate volumetric measurement device flow passageway (some with passageways smaller than 1-mm cross section) can become clogged. The in situ sensor will not clog, since its flow cross section is that of the process line.

The particle size sensitivity of the in situ system is potentially always better than that of a volumetric sensor, since interference effects from the flow boundary walls are decreased for the in situ system. For an optical system, scattered light interference will always be present when the light passes through an interface between two materials of different refractive index. For an acoustic system, the energy will be reflected and scattered when passing through the interface. In both systems, the background noise increases, reducing system sensitivity. The extent of these effects will vary with changes in the liquid, since significant differences can occur in the refractive index or the speed of sound between different liquids. The sensing zone of the in situ system is located away from the interference zones. The stability of both signal and noise levels will be better for the in situ system than for the volumetric system since the window-liquid interface is out of the view volume of the sensor imaging system and particle-containing liquids will *always* deposit debris on container surfaces.

In situ measurement does not require sample handling and manipulation of sample containers. Sample withdrawal and liquid handling problems are thus eliminated with in situ measurement. The problems of ensuring cleanliness of sample containers and of eliminating particle losses or generation during storage are removed. These are very serious problem areas when working with particles significantly smaller than one or 2 μm in diameter. Heroic measures are required to clean sample containers of particles smaller than a few micrometers in diameter.

Isokinetic sampling errors will not occur when in situ sampling techniques are used. No inertial separation and loss of larger particles will occur, since the liquid is not moved out of the normal flow conditions during the sampling operation. This also removes the possibility of leakage and work area contamination during sample withdrawal. Time and effort to clean the work area are reduced.

2. Problem Areas of in situ Sampling Systems

There are some problem areas when working with in situ sampling systems. Perhaps the most serious problem relates to defining the sampled volume exactly. The volume is defined by establishing an edge or a line where there is a sharp change in the energy content from one side to the other. The line does not have zero thickness; it may be a fraction of a micrometer thick, but even this may be significant when defining particles that are only a few micrometers in dimension. For example, any particles within the edge of the optically defined sample volume can generate anomalous data. Questions arise as to how much of the particle should be within the sampled volume in order to be counted and how to define the dimensions of the sampled volume when all edges of that volume are not orthogonal to the liquid flow streamlines. This is an especially serious problem for particles with large aspect ratio such as flakes or fibers. Some procedures have been developed and will be discussed later which provide reasonable solutions to these problems, but they are not absolute solutions.

Unlike volumetric sample feeder devices where the sample flow rate can be well controlled, in situ measurement relies on the flow rate of the liquid in the line being measured to move liquid through the measurement zone. If the flow rate does not vary excessively, then this is not a serious problem and can be handled by using electronic pulse processing circuitry with wideband width range and good signal-to-noise ratio. With this type of system, any changes in pulse rise time and duration due to changes in flow rate will not cause any significant changes in indicated pulse amplitude when a particle is measured. However, the possibility of such flow rate changes means that the electronic system will probably be more costly than one designed to measure pulses with constant duration and rise time. Further, unless continuous measurement of flow rate during a measurement can be provided, a simple time measurement assuming constant flow rate cannot define the sampled liquid volume needed to define particle concentration if flow rate varies during a measurement. A very sophisticated electronic system would be needed to measure pulse duration and to calculate particle (and liquid) velocity from that measurement. Another flow rate problem exists when liquid flow occurs in early horizontal liquid feed lines which are not always full of liquid. This may be due to intermittent flow or to low flow rates; it may also be due to the presence of foam in a line. If an in situ sensor is exposed to a mixed gas and liquid flow, then completely invalid data will be produced each time a gas-liquid boundary passes through the sensing zone. There will be no indication of this problem except for the generation of anomalous data. This problem can be reduced by placing the in situ sensor so it observes liquid flowing in a section of line oriented vertically (if possible). In this way, the line is essentially full of liquid at the point of measurement unless dissolved gases or vapors are producing bubbles in the liquid.

Primary calibration of in situ systems with standard particles is difficult. Passage of a suspension of standard particles through such a system is awkward since access to the in situ sensor location may be difficult. Two methods have been used for primary calibration of in situ systems. In one, the sensor component is removed from the line and the windows through which it is observing process liquid; the sensor element is placed on a calibration loop where standard suspensions can be provided. In the other, a means for controlled injection and removal of calibration suspension is provided to the process line in such a way so that the process is not contaminated with calibration particles. Fortunately, the in situ sensor components are relatively stable, so that calibration is not required at frequent intervals. When systems containing "sticky" particles are measured, it is necessary to control window cleanliness.

C. Application Areas

In situ particle measurement is usually carried out when characterizing flowing liquids in process or product lines. This is particularly true when the liquid is hazardous because of composition or operating pressure requirements. Safety considerations indicate that it is best to observe flammable or toxic liquids contained within a line, rather than to open that line for sample withdrawal. This is particularly true if the liquid of concern is one with high vapor pressure. Particle measurement within a pressurized line avoids potential problems with bubble formation when determining particle content of liquids with high vapor pressure. If the liquid system to be characterized is in a contaminated environment, in situ measurement avoids the possibility of contaminant ingression to either the liquid being measured or to the sample container during sample withdrawal. In situ measurement also avoids the possibility of problems due to contamination of a sample by contamination left in inadequately cleaned sample containers.

Some examples of industrial processes where in situ measurement is carried out include characterization of process and product liquids in pharmaceutical product manufacturing, in hydraulic fluid power systems, in electronic production areas, and in power plant liquid streams.

In pharmaceutical areas, clean water is used for cleaning product containers or for diluting concentrated product or process materials. Levels of extraneous particulate contamination in this water must be strictly controlled. In situ measurement of particle content provides near real-time information on the cleanliness of these water streams as they are being used, rather than after a production run has been completed. Similarly, in situ measurement of particle content in such areas as parenteral product filling lines provides information on product cleanliness soon enough so that any problems can be seen and remedied before large quantities of potentially reject product have been prepared.

Hydraulic fluid power systems are usually operated at pressures in the range from 30 to 300 bars (1000 to 10,000 psi). The hydraulic fluid can be exposed to external contaminants during operation of valves, pistons, and other moving parts exposed to the environment. Clearances in spool valves, for example, are in the range of 2–200 μm. Ingression of atmospheric dust during operation can result in accumulation of sufficient abrasive material to degrade or destroy components quite rapidly. Sensitive components are protected by liquid filters installed at critical locations. Continued ingression of contamination results in the filters becoming plugged with debris. In this situation, either the fluid flow rate is decreased significantly or a filter bypass valve opens to allow necessary liquid flow. In the latter situation, there is some possibility of component damage from the unfiltered and contaminated fluid. In situ measurement of particulate loading in the hydraulic fluid during flow from the fluid reservoir, pump, and filter system to the operating components can be used to control filter replacement scheduling with minimum component damage and operating cost savings.

Electronic components produced since approximately 1965 are based on low-level energy transfer with semiconductor materials. This is often controlled by the presence of trace quantities of additives in very pure materials. Many computer memory systems are based on moving magnetic components with read/write heads spaced a fraction of a micrometer from a moving magnetic element. For both of these systems, contaminant particles in the submicrometer size range, as well as part per billion quantities of organic films, can degrade component operation. Optical memory disks are not quite so sensitive to particulate contaminants. These disks can be produced and operate acceptably with some particulate contamination in the micrometer size range present. The manufacturing processes for all these devices includes exposure to process chemicals and flushing away of waste materials after processing. In both cases, the process chemicals and the flushing liquid *must* be extremely clean so that no contaminants are left on the surface after cleaning. As of 1992, successful production of state-of-the-art semiconductor memory devices requires that essentially no contaminant particles larger than 0.2 μm in size be deposited on the silicon wafer during the manufacturing operation and that the quantity of contaminating 0.05-μm particles be limited to no more than 100 per liter of process liquid. Electronic products use very large quantities of very clean deionized water for flushing during the manufacturing process. The cleanliness of this material must be verified before use. An in situ particle measurement system is normally used for maximum sensitivity since it avoids background scattering from the interface between the liquid and the transparent viewing volume walls. The sensitive volume, as will be discussed later, is defined optically within the flowing stream of liquid.

For both fossil fuel and nuclear energy power plant operation, large amounts of clean liquid are required. In the fossil fuel systems, high-temperature, high-pressure steam is generated for operating turbines. The steam must be free of particulate contamination to reduce turbine blade erosion and to maximize heat transfer in the boiler by keeping heat exchanger surfaces clean. In nuclear energy systems, all water used for any purposes should not contain any particulate material which might become radioactive by exposure to the energy flux in the reactor. In situ measurement of the cooling water for particle content allows definition of its cleanliness without requiring handling of liquid which may contain radioactive materials.

Once the power plant has produced electrical power for transmission to the point of use, the voltage level is usually increased to very high levels for transmission to remote locations and is subsequently reduced for residential or industrial application. The transformers used for these purposes contain recirculating, insulating hydrocarbon oil to keep the transformer temperature at the point of maximum efficiency. This oil must be extremely clean and free of particulate material so that the possibility of any conductive debris depositing within the transformer is minimized. Otherwise destructive arc generation can occur. Since the oil may be heated during transformer operation, oxidation can occur if the hot oil is exposed to ambient air. The oils are examined for particulate content by observing cleanliness during flow through a line where temperature is controlled. In situ measurement through a sealed window maintains electrical integrity of the oil.

II. SAMPLE POINT LOCATION CONSIDERATIONS

The validity of any liquid sample data depends strongly upon the quality of the procured sample. This is especially important for in situ sampling since the nature of the liquid may vary from point to point, both axially and radially within the flow line, depending upon the liquid physical properties, the flow rate, and the line configuration. Both the liquid velocity profile and particle distribution in the stream depend upon these factors. For this reason, the sample point selection must take these factors into consideration. Even though fluid flow is most important in selecting a sample point, the potential problems of physical accessibility must be considered. Both of these factors must be considered in selecting a sample collection point to characterize liquids.

A. Fluid Flow Considerations

The ideal sample collection or observation location would be one where the liquid flow is turbulent rather than laminar, so that larger particles do not settle excessively under the influence of gravity and the smaller particles tend to remain reasonably well dispersed due to the turbulent flow. Turbulence should not be so great that particles are impacted on liquid containment walls. The extent of turbulence is expressed as a Reynolds number. This dimensionless number represents the ratio of inertia force to viscous force for fluid flow. It is defined as the product of the flow line wetted diameter, liquid velocity, and liquid density, all divided by liquid viscosity. The transition from laminar to turbulent flow occurs at a Reynolds number of approximately 2100. It is recommended that in situ sample measurement be made in a line with a Reynolds number in the range from 3000 to 20,000. At Reynolds numbers under 3000, some sedimentation can occur if the line flow decreases slightly or if a slightly expanded line section is used. At Reynolds numbers much over 20,000, line velocity may be so high as to cause excessive pressure drop and impaction of particles if changes in the direction of flow occur.

If liquid is flowing in a system with changes in line dimension, as may occur when a stream is split or combined with another stream, then both the line dimension and the liquid velocity will change. If the liquid flows through a line with fittings that change the flow direction, then eddies may form immediately after the fitting. When these situations occur, the turbulence index will also change. However, the change in direction of the liquid streamlines does not occur instantaneously. The liquid and the particle inertia result in flow in the original direction for some period of time after the line dimension or configuration has changed. For this reason, the ideal sample point would be located at a distance 12 line diameters or more after any change in flow direction or rate, so that the particles have had time to accommodate to any flow changes and are moving with the liquid streamlines when they are sampled.

Sample measurement should be as close as possible to the point of use for the liquid, so that the sampled liquid conditions are similar to those present in the process or product whose nature is of concern. As pointed out previously, the liquid flow rate may change during some processing operations. When such a change occurs, there may be a section of line which is only partially full of liquid. Depending upon the type of liquid, the flow may consist of either intermittent gas and liquid slugs or of liquid with a layer of foam on its surface. It is also possible that wavelike motion occurs with varying flow rates in the liquid. The presence of any of these conditions of two-phase flow can cause erroneous data when an in situ particle measuring system is used. The ideal sample point would be located within a line where liquid flow is vertically upward; in this way, the sampled line would be full of liquid, with minimum gas bubbles and discontinuities in the liquid. When partial flow exists in lines where in situ measurement occurs, foam and/or bubbles occur and gas-liquid interfaces are interpreted as particles.

B. Physical Access Considerations

In addition to the physical requirements for good sampling, there may be potential problems due to the physical nature of the system being sampled. Many production and processing facilities are installed in areas where space and accessibility to components may be difficult. The recommended sampling conditions may not be attainable. Ideal installation conditions, good instrument environment and maintenance of sample measurement instrumentation may be very difficult. If the only available sample point is at a location where incorrect sampling is inevitable, the operator should indicate the sample point element configuration and the probable poor quality of data from that sample point should be understood. Although accurate data will not be obtained, at least trend information can be acquired.

When a process line is installed, designers normally select production component materials to be long-lived, stable and compatible with the process and product liquids. Inert, nonreactive materials are selected that will not degrade in use or react with product. If process conditions or materials change, previously compatible materials may no longer be suitable for the new conditions. Changes in process liquids may result in the possibility of chemical reactions with components that could produce particulate debris and may even result in hazardous operations. In addition, it is also possible that processing condition changes in temperature or pressure could also result in component stresses that may release particles from component surfaces. This will probably not occur when working with selected inert metal components, but may occur if a plastic material is used in a temperature or pressure regime for which it has not been designed.

Another minor potential physical access problem can occur when sampling in a restricted area. Some particle measurement instruments have capability for local data display, as well as transmission to remote display devices; operators may wish to observe this on-the-spot information, possibly to verify performance or to make immediate adjustments in process conditions. If local visibility problems occur, then the instrument display may not be easily accessible. In this situation, a special data display terminal may have to be installed for the convenience of the operators. If this requirement exists, then plans for such installation should be made when the particle measurement device is originally procured so that interference with electronic components will not occur.

III. SAMPLING CONSIDERATIONS

A. Statistical Requirements for Sample Quantity

Defining the contamination level in a liquid requires that a representative sample of that liquid be obtained for measurement. The sample should be large enough so that statistically valid data are obtained, but not so large as to change the nature of the process material which is being sampled. Determination of sampling schedules should take into consideration the questions as to whether the process being observed is continuous or intermittent and how such changes may affect both liquid flow rates and/or particle generation processes. Whether the measurements are carried out in situ or on samples removed from the process line, the sampling operation should produce both particle size distribution and concentration data which can be related to those in the process stream.

The statistical requirements for sample quantity are based on collection of sufficient data to define particle concentration *for those particle sizes of concern* with acceptable error bounds. It is generally accepted that collection of sufficient data to define a variable with 95% confidence limits can be accomplished if 20 or more items of data are collected. When observing particle concentrations in any liquid, a power-law particle size distribution will frequently be seen. This distribution shows that the particle concentration varies inversely with a power of the particle size. In most situations, the exponent will be in the range of 2 to 3. If a measurement device capable of sizing particles over one order of magnitude in size is used, the number of the smallest particles counted will usually be at least 100 times the number of the largest particles that can be counted and may be several orders of magnitude greater. If only the total number of particles within the measurement range of the device is of interest, then the operator must realize that essentially all of the particles counted will be in the smallest size ranges measured. If only

20 particles are measured, then the number of large particles will be so few as to be statistically insignificant. If information on these materials is important, then the operator must be prepared to collect large quantities of data on small particle concentration. Therefore, a major consideration in defining sample quantity is a definition of the particle sizes of concern. Ideally, the in situ observation sample should be large enough so that a total count much greater than the minimum value of 20 should be reached in a reasonable time; a sample time of an hour or so may be considered acceptable.

B. Sample Size and Stream Flow Relationships

Most single particle measuring instruments are designed to count and size individual small particles in reasonably high concentration for maximum application versatility. This design typically results in a small instrument measurement volume, with liquid flow cross section dimensions normally smaller than a millimeter in any direction. If the in situ sensor is used for a flow line operating at several hundred liters per minute, the volumetric sample may be a small fraction of a percent of the total flow. For this reason, it is necessary to define the sensitive volume exactly and to know the total liquid flow rate accurately if accurate concentration data are to be developed. The total liquid flow rate can be measured with an external flowmeter for real-time data, needed particularly if flow rate may vary during measurement. Alternately, the sensor pulse duration can be measured; comparing this value with sensitive volume dimension in the flow direction provides a value for liquid flow velocity. This technique should be reserved for flow systems known to be operating in the turbulent flow regime. Otherwise, the parabolic velocity profile which exists for laminar flow will cause large differences between the average flow velocity and the velocity existing at the point of in situ measurement. Normally, the maximum velocity for laminar flow at the center of the stream is twice the average velocity for the entire stream flow. Figure 2 shows fluid velocity profiles for laminar and turbulent flows. The velocity profile for a laminar flow is parabolic, while the velocity profile for turbulent flow is nearly uniform over much of the stream dimension. Note that sample observation at the edges of the flow line will still indicate line velocity lower than the average flow rate, even for turbulent flow lines.

Another important reason for exact definition of the sampled volume boundary is avoidance of particle sizing errors for those particles in the main stream which may flow only partially within the sampled volume. Those particles will be undersized since only a portion of the particle will be observed by the sensor. This phenomenon occurs when the sensitive volume definition is not positive. With optical systems, masking techniques may be used to aid in improving definition of the sampled volume [2]. Figure 3 illustrates the optics of one such system. A high-resolution optical system images the scattered light from a particle which may be within the defined sensing volume. A beam splitter produces two image planes for two detectors. The reflected image prism face is masked so as to block central image transmission, and the other

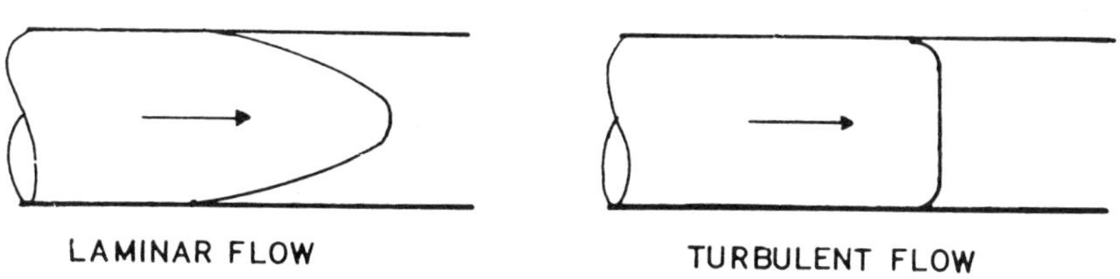

Figure 2 Velocity profiles for laminar and turbulent flow.

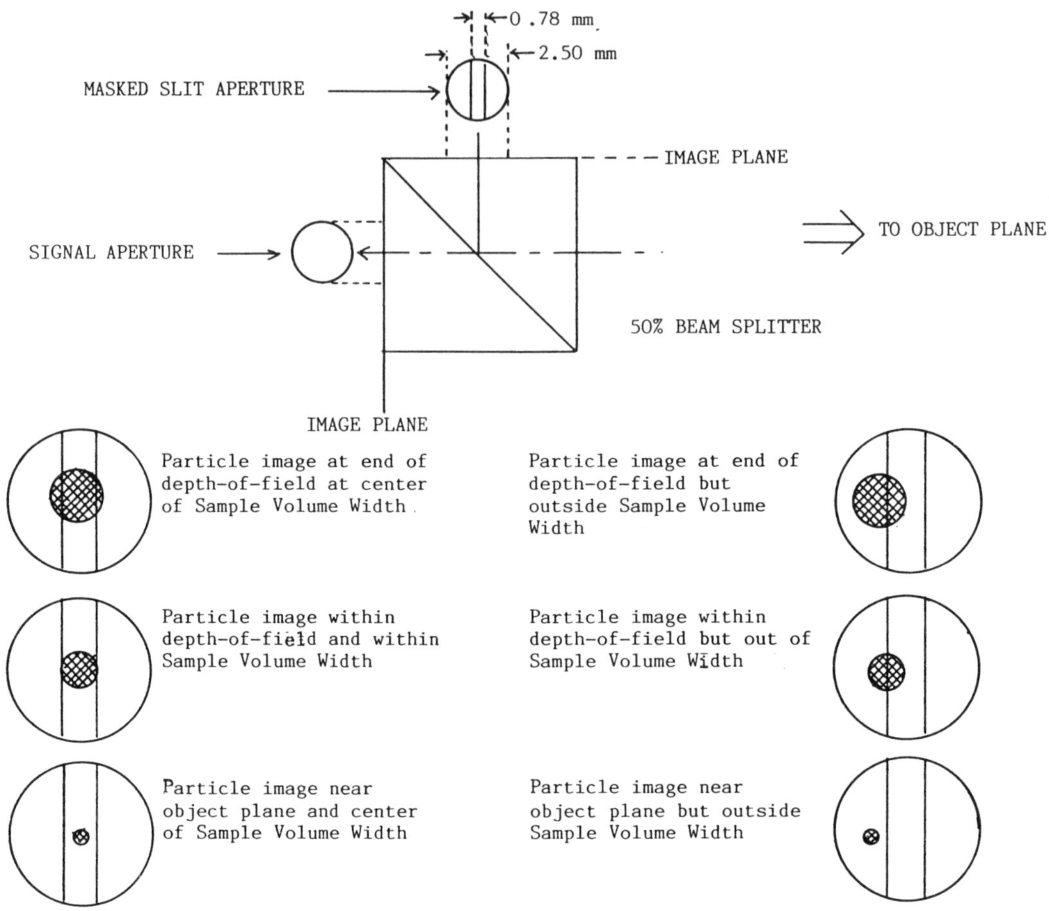

Figure 3 Optical masking volume definition. (From Ref. 3.)

detector prism face is not masked. The masked beam splitter derives two signals which are used to determine if the particle is within the sensing volume. The three particles shown at the left side of the figure will be counted and sized correctly. The large and midsized particles at the right side of the figure will produce undersized pulses. The small particle at the right side of the figure will not be counted at all.

Another more recent technique [3], as shown in Fig. 4, involves comparison of S and P polarized beams of light from a single polarized laser source to define the sampled volume. These beam components are shaped to produce beams of circular (S) and elliptical (P) cross section in the sensitive volume. The smaller circular beam is aligned so that when a particle traverses both beams, it must have crossed the center of the elliptical beam, as shown in Fig. 5. This defines the sample volume in one dimension. The other dimension, based upon system depth-of-field, truncates naturally because the circular S beam converges and diverges more rapidly than the elliptical P beam near the focal plane. When the scattered light signals from the two polarized beams are compared, only those particles within the defined sampled volume are accepted and counted if the S scattering signal exceeds the P scattering signal. Particle sizing is accomplished by analysis of the P scattering signal.

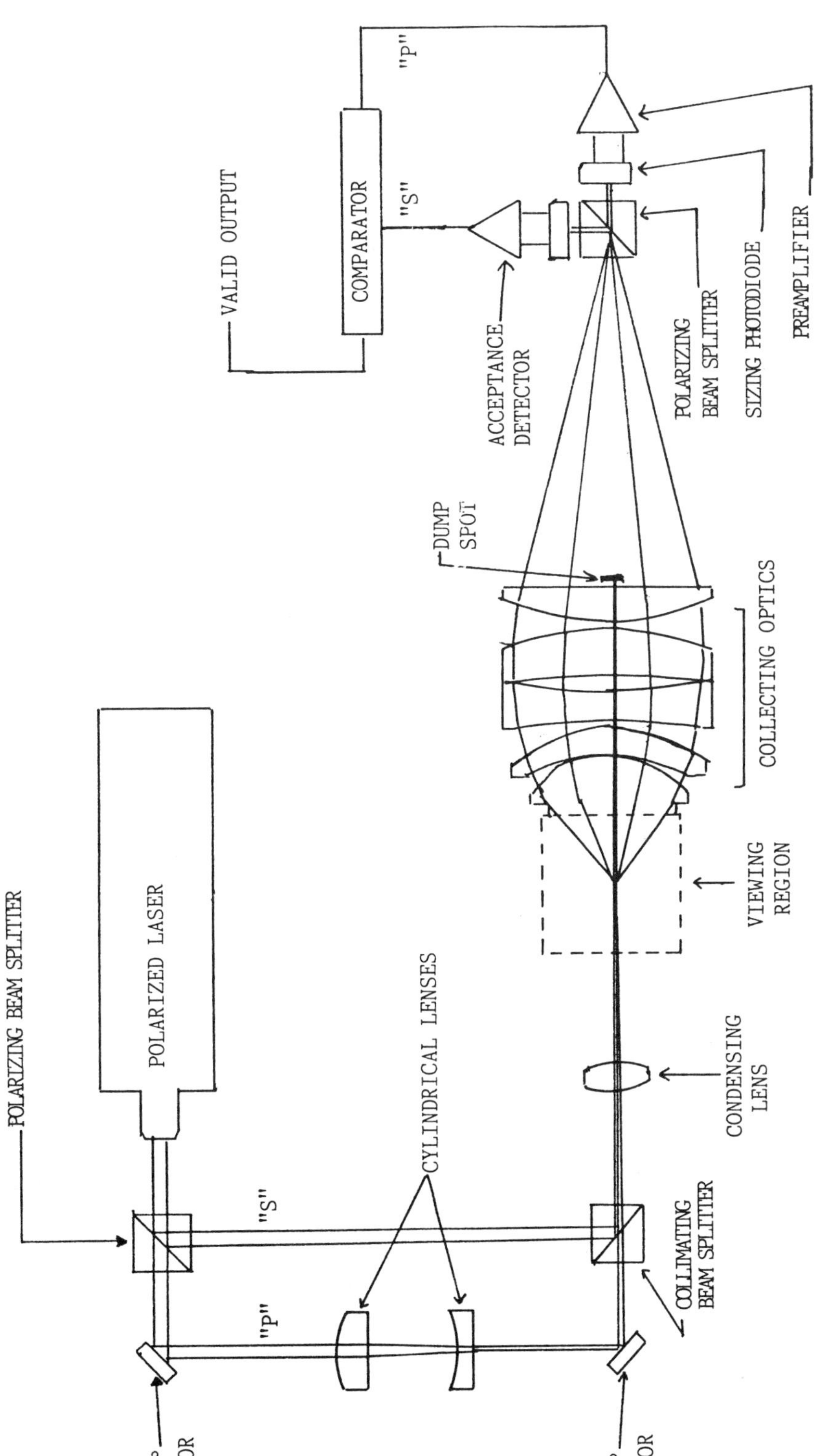

Figure 4 Polarized light in situ optical system. (From Ref. 3.)

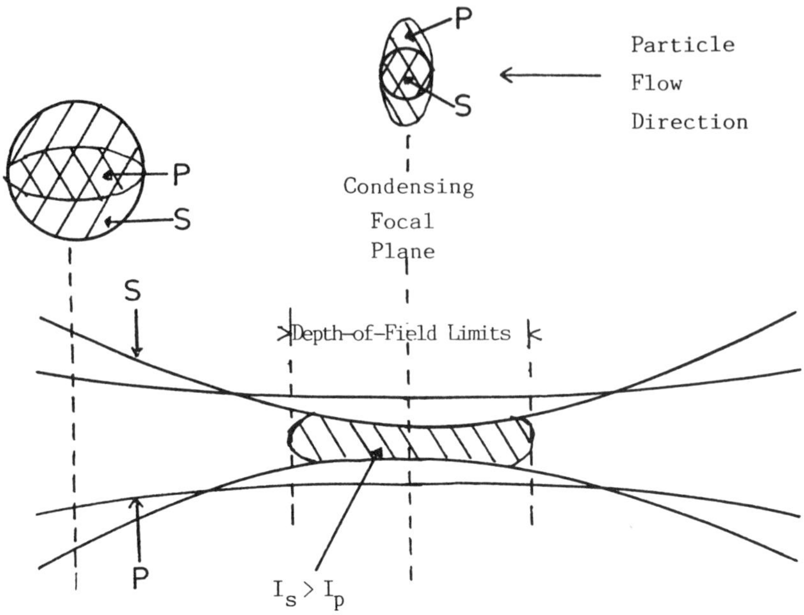

Figure 5 Sample definition for polarized light in situ optical system. (From Ref. 3.)

C. Sample Acquisition Scheduling

In situ sampling systems have a particular advantage when sample acquisition scheduling must be planned. The components are frequently permanently installed on an operating flow line. When samples are required, liquid sample port actuation, flushing and/or sample collection container preparation and handling are not required. The observing and detection portions of the device need only to be energized, some possible electronic warm-up time allowed and data collection can begin. For this reason, sample acquisition scheduling can be based solely on process needs and operator convenience. The operator can select a sample data acquisition program varying from continuous data collection or programmed routine intermittent data collection at any selected sample location or group of locations. The program can be set to operate only when a process of interest occurs. The program can also include capability for interruption and/or manual sampling processes if anomalous conditions are noted or process variations occur anywhere in the processing system.

D. Sidestream Sampling

Whenever particle measurement is to be made in a flowing fluid from which a sample is to be removed, the question of isokinetic sampling usually arises. This question arises since the need to remove a sample from the main body of fluid without changing either the main body of fluid or the particle content in the sample is paramount. When true in situ measurement occurs, no sample removal is involved; thus isokinetic sampling is not a concern. However, in some measurements, a sidestream is removed from the main body of liquid flow, and the particle content of that entire sidestream liquid is measured. In this situation, isokinetic sampling is not a concern during the measurement process, but the sidestream removal from the main body must be carried out isokinetically to ensure that the measurement data represent the conditions in the main body of liquid as well as they do that of the sidestream material. For most liquids, particle divergence from streamlines due to changes in typical process liquid flow rate or direction is important only for particles larger than approximately 5 μm. As liquid density and viscosity increase, the relaxation time (the time for a particle to accommodate to velocity changes in the liquid) for particles decreases and particles tend to remain within the liquid streamlines. For liquids with viscosity similar to that of water, particles larger than 10 μm or so diverge from the liquid streamlines rapidly.

In selecting sidestream sample probes, a fundamental decision must be made on selection of either permanently installed or removable probes which can be inserted into the liquid system through suitable valves and packing glands. A removable probe can be easily inspected and cleaned between sample collections, but the fittings and seals must remain at the sample point. These, as well as the sample probe, must be cleaned before the sample probe is installed. For liquid systems designed to operate in very clean conditions, permanently installed probes should be selected. Even though they may collect some particles from the liquid, they provide less opportunity for ingression of environmental contaminants and debris into the liquid stream.

Connections and fittings to the sampled liquid lines should be as smooth and uniform as possible. If sidestream sample lines longer than a few meters are to be used and intermittent sample flow is involved, no data should be recorded until sample flow in that line has stabilized and flow in the line from the main liquid line to the measurement point has been sufficient to flush any residual gas or liquid from that line. Most sample lines that have been in use for any period of time will have some particles deposited on their interior wall surfaces. If the line has any flexibility, then movement or vibration can cause reentrainment of those particles or particles generated from the line surfaces into the liquid flowing in the line. Sidestream sample lines should be laid out to avoid possible contact with sources of vibration or shock. In addition to particle entrainment, vibration transmitted to a particle measurement device may cause generation of artifact data.

The sidestream sample line inlet should face into the sampled liquid flow and should collect liquid isokinetically or at least isoaxially if flow rate in the main liquid line may vary. If the entire sidestream line flow is passed through the particle measurement device, a flow rate from one to 100 ml/min will probably be used. Sample probe lines ranging from 2 to 6 mm in diameter will produce adequate liquid flow velocity to minimize deposition and settling of most particles within horizontal portions of the sidestream flow line. Even so, the line should be as short and direct as possible to minimize *any* deposition. Figure 6 illustrates some sidestream probe configurations. Since the probe dimensions (particularly radius of curvature for *any* bends in the probe) are smaller than those of the line being sampled, particle deposition by centrifugal force to the probe walls must be avoided by minimizing probe line curvature. That is why straight probe lines facing into the flow with minimum curvature to the point of measurement are favored over the common pitot tube sample probe configuration, even though the pitot tube types require smaller ports in the flow line. These design recommendations are also used to optimize compressed gas sample acquisition.

Once the sidestream sample has been passed through the measurement device, disposal of the measured liquid must be accomplished. If inert, innocuous liquids are being measured, then dumping to waste is no problem. If hazardous materials are being measured, then the measured material must be disposed of safely with no serious cost problems. If the operator is sure that the measurement process has not affected the liquid, then the sidestream sampled material can be returned to the main process flow line. If the measurement process has affected the liquid, then discussions with local environmental protection and/or occupational safety agencies may be required in order to ensure selection of a safe and acceptable disposal method for sampled materials.

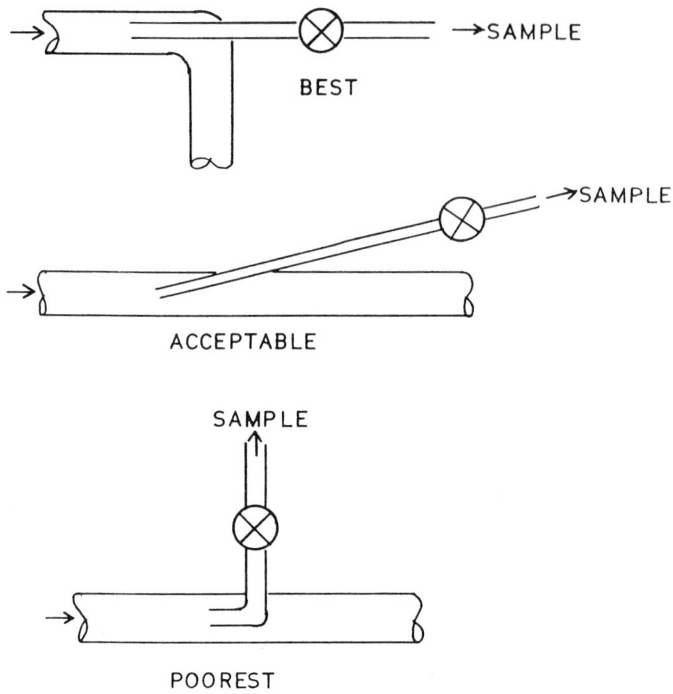

Figure 6 Sidestream probe line configurations.

IV. OPTICAL SINGLE PARTICLE COUNTING INSTRUMENT SYSTEMS

A. General System Requirements

Although several instrumental methods have been used for in situ measurement of liquid-borne particles, most data are obtained with optical single particle counting instruments at this time. Some measurements are made with electrical resistance detection methods, but the measured sample volumes are too small for use with very clean modern-day liquids and in situ sample measurement is extremely difficult with electrical resistance counters. These devices are primarily useful for liquids with concentrations sufficient so that a sample volume of no more than 2 ml will contain enough particles to provide statistically valid data. Acoustic systems can be used to indicate trends in particle content, but accurate volumetric data are almost impossible to obtain because of the difficulty in defining the volume of liquid in which particles of specific size are measured. Photometric systems are used for in situ measurement of many liquids, but the poor sensitivity of these devices restricts their use to applications where the particle content is as high as the level acceptable for potable water. For these reasons, discussion of the systems used for in situ measurement over a wide range of particle sizes and concentrations in liquids will be restricted to single particle counting optical systems. Either light extinction or light scattering systems can be used for in situ measurement. Operation of these devices are discussed elsewhere and will not be repeated here in any detail.

The major requirements for these optical systems are relatively simple. By definition, the in situ detection system is a nonintrusive and nondestructive device which will not interfere with the liquid being observed. As discussed previously, this system is applicable when the liquid flow cannot be disturbed nor can a sample be removed for measurement. The system must be capable of producing accurate particle size and concentration data for the liquid being measured. If the liquid or particle type is changed, the system must compensate for the change automatically or contain capability for correction for different particle-liquid systems. Means of size and concentration calibration for the system must be available. These can be components which allow introduction and complete removal of calibration standard suspensions to the device or a surrogate standardization signal input capability must be provided. The device may be used for either production of accurate size and concentration data at a particular time or for producing a record of long-term trend data. For this reason, it must be a stable system with long-term repeatability; it should also correlate acceptably well with similar devices used for measurement in the same application area.

B. Extinction Optical Systems (Sidestream Flow Devices)

For all practical purposes, extinction optical systems are always used with the entire liquid stream passing through the sensitive zone of the instrument. Since extinction counters are used in many routine pharmaceutical process and product liquid system measurements, in situ measurement can aid in verifying quality of that liquid. If the process liquid flow rate is similar to the extinction counter flow rate, then the counter can be plumbed directly into the flow line. Otherwise a sidestream can be diverted and passed through the counter. This in situ definition considers that the (continuous or intermittent flow) sidestream is part of the measured flow system. It is removed from the main liquid stream, passed through the extinction counter and either returned to the main line or discarded. Operating requirements for satisfactory sidestream removal and either return or disposal have been discussed and need not be repeated here. Some counters may have flow passages with dimensions less than 1 mm. If there is a possibility that large particles may be in the sidestream then large particle control measures may be required. These can include incorporation of a protective screen at the sidestream line entrance or addition of a backflushing capability. The entire sidestream which is removed from the main liquid stream must flow through the counter; for that reason, the counter must be capable of handling liquid flows of at least 25 ml/min. At lower flow rates, a sidestream line should have no horizontal segments longer than 2–3 m. The liquid residence time in line segments of greater length will be long enough to allow significant loss by sedimentation for particles larger than 10–15 μm. Reducing residence time by reducing sidestream line size may cause excessive pressure drop.

Some caution is required when attempting to correlate data from extinction counters with differing illumination sources. As of 1993, extinction counters with a variety of illumination sources are in use. There are many extinction counters still operating with incandescent filament light sources (broadband wavelength ca. 450–800 nm) and it is anticipated that these will still be used through at least 1995. Counters with helium-neon laser light sources (wavelength 633 nm) will probably still be used through at least 1999 and solid-state laser sources (wavelength 780 or 840 nm) will probably be used for 10 or more years after that time. When an extinction counter is calibrated with standard monosized latex spheres and then used to observe material with refractive index different from that of the spherical latex particles ($1.6 - 0i$), its response to those particles will vary with the illumination source wavelength. Data have been obtained [4] which show reported size differences up to a factor of 3 or more when counters with differing illumination sources and optics were used to examine aliquot samples of hydraulic oils containing A.C. Fine test dust after calibration with latex spheres. For this reason, correlation between different extinction counters should not be expected unless the counters have been calibrated with materials similar to that which they are expected to count or unless the counters are all of the same type.

C. Light Scattering Optical Systems

Light scattering optical systems for in situ measurements are often used for counting and sizing submicrometer particles. The in situ optical system observes a portion of the liquid stream which is isolated from the illumination beam entrance and exit windows. This isolation is made possible by either focusing a collecting optical system at a point away from the scattered light exit window or orienting the collecting optical system at 90° to the illumination beam axis. In this way, collection of light scattered by defects or debris on the windows is greatly reduced and instrument sensitivity is nearly that imposed by the background scattering from the liquid alone. The near-forward scattered light collection system, shown in Fig. 4, uses helium-neon laser illumination. It is capable of detecting 0.1-μm particles at observed sample liquid flow rates of a few milliliters per minute. The right-angle collection system [5] shown in Fig. 7 has a sensitivity of 0.05 μm at observed sample flow rates of a fraction of a milliliter per minute. The higher sensitivity is obtained by observing a smaller sensitive volume and using a solid-state laser source which is more powerful than the gas laser. The system of Fig. 4 observes a sample of several milliliters per minute within a stream flowing at several hundred liters per minute, using a conventional 5-mW HeNe gas laser for illumination. The system of Fig. 7 observes a sample of 0.3 ml/min flowing within a stream flowing at 300 ml/min, using a 50-mW solid-state laser for illumination.

Aside from the sensitivity and convenience in physical layout, the performance of the system will depend upon the choice of right-angle or near-forward collection optical systems. Either a gas laser or a solid-state laser can be used with both systems. Conveniently sized helium-neon gas laser power levels are in the 5–10 mW range while solid-state laser power levels are nearly an order of magnitude higher. The gas laser emits 633-nm radiation, while the diode

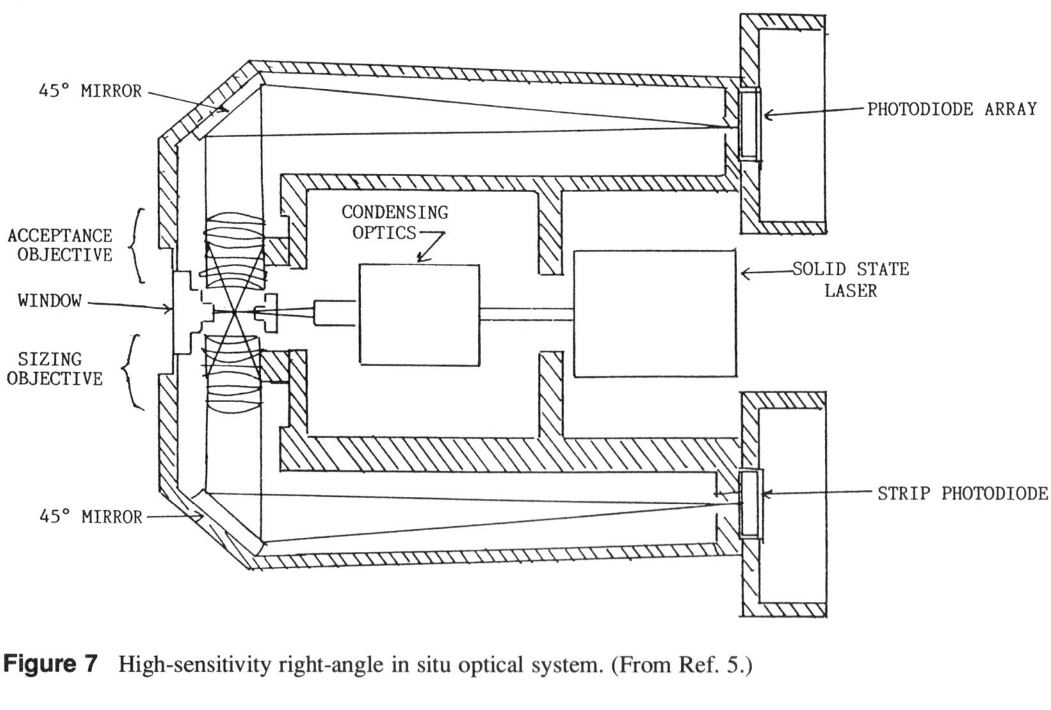

Figure 7 High-sensitivity right-angle in situ optical system. (From Ref. 5.)

laser emits 780- or 840-nm radiation. Wavelength changes over these levels will not change signal levels to a great degree; however, the scattered light signal versus particle size response function oscillates at any wavelength, particularly for particles larger than a few tenths of a micrometer; the location of the oscillations shifts somewhat with wavelength. The near-forward collection optical system is limited to a smaller solid collection angle than is the right-angle collection system. Figure 8 shows how the intensity of the oscillations is reduced to the point where instrument response to particle size is essentially monotonic for optical systems with larger solid collecting angles. Particle size resolution is significantly better for instruments with larger solid collecting angles. Changes in slope of the response curve are of minor concern, as long as the curve does not reverse.

In order to maximize signal level, the beam may be focused to a smaller cross section in order to increase intensity. The uniphase TEM_{00} laser mode emits a beam with a Gaussian energy distribution; even a multimode laser beam, normally considered to have a relatively uniform intensity beam, has a zone where intensity decreases at its edges. For this reason, the amplitude of the scattered light signal from a particle will always depend upon its location when it passes through the beam. An in situ measurement sensor does not control the location of particles which are present in the liquid. Thus a large particle may pass through the edge of the illumination beam where very low intensity exists. Only part of that large particle may be illuminated and it will still produce a signal as large as that from a smaller particle passing through the center of the beam. When a particle monitor is used, it responds to signals with amplitudes equal to and greater than a preset threshold level; these signals may be produced by particles passing through any part of the illuminated volume. For this reason, the effective sample volume in which a particle will be detected can be considered to increase to some extent with particle size. (The location of a particle is usually defined as the position of its center.) This effect is greatest for a particle monitor and is reduced for a particle spectrometer [5]. Either special optical and

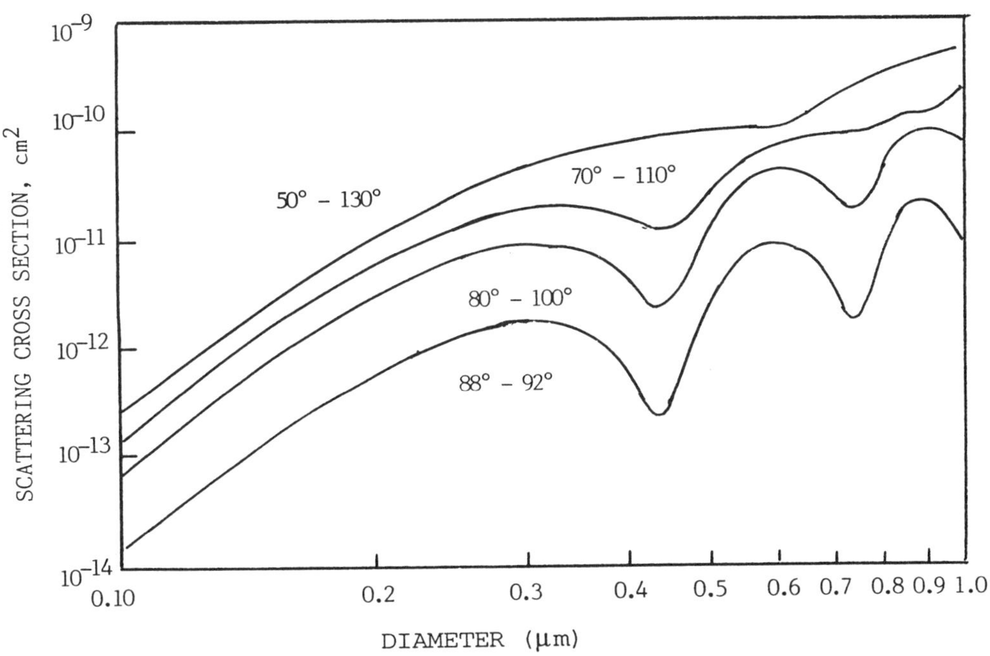

Figure 8 Scattering response for latex particles at various collection angles. (From Ref. 5.)

electronic components are used to reduce the detection volume change with particle size or the counter detection volume is reduced by discarding data from the beam edges. Figure 9 shows how effective sample volume varies with true particle size for a particle monitor as opposed to the lesser variation for a particle spectrometer. More of the laser beam is used for a monitor than for a spectrometer (see Fig. 1, which shows the Gaussian intensity distribution from a TEM_{00} laser beam as part of the means of defining the illuminated volume). When a segment of a particle much larger than the threshold limit size sensitivity for the particle monitor passes through the edge of the illumination beam for a monitor, the segment may be as large and scatter as much light as a threshold limit size particle. This can occur even though its center may not be outside of the defined sample volume for the monitor simply because the particle is sufficiently larger than the specified threshold size in a size range where response varies with the sixth power of particle size.

D. Particle Property Effects on Counter Response

The response of any optical single particle counting instrument will vary with the refractive index difference between the particle and the fluid suspension. This subject has been extensively discussed elsewhere [6] and will not be repeated here in any detail. Figure 10 illustrates how changes in the index contrast can change the scattering cross section and the resulting scattering signal for particles of varying sizes at a typical optical single particle counter (OPC) collection angle. As the index contrast approaches unity, the scattering signal approaches zero. Ideally, the in situ instrument should be calibrated with material that will provide a refractive index contrast identical to that of the contaminant material in the samples to be measured. Unfortunately, the nature of that material is seldom known in advance. For that reason, the instruments are

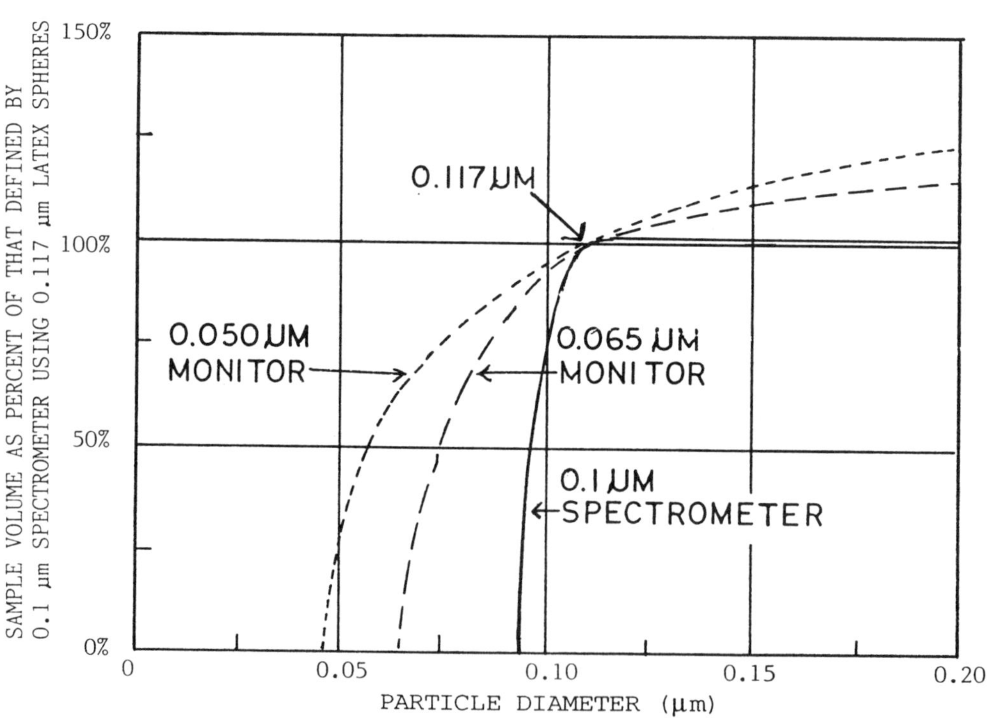

Figure 9 Effective sample volumes for counters and spectrometer. (From Ref. 5.)

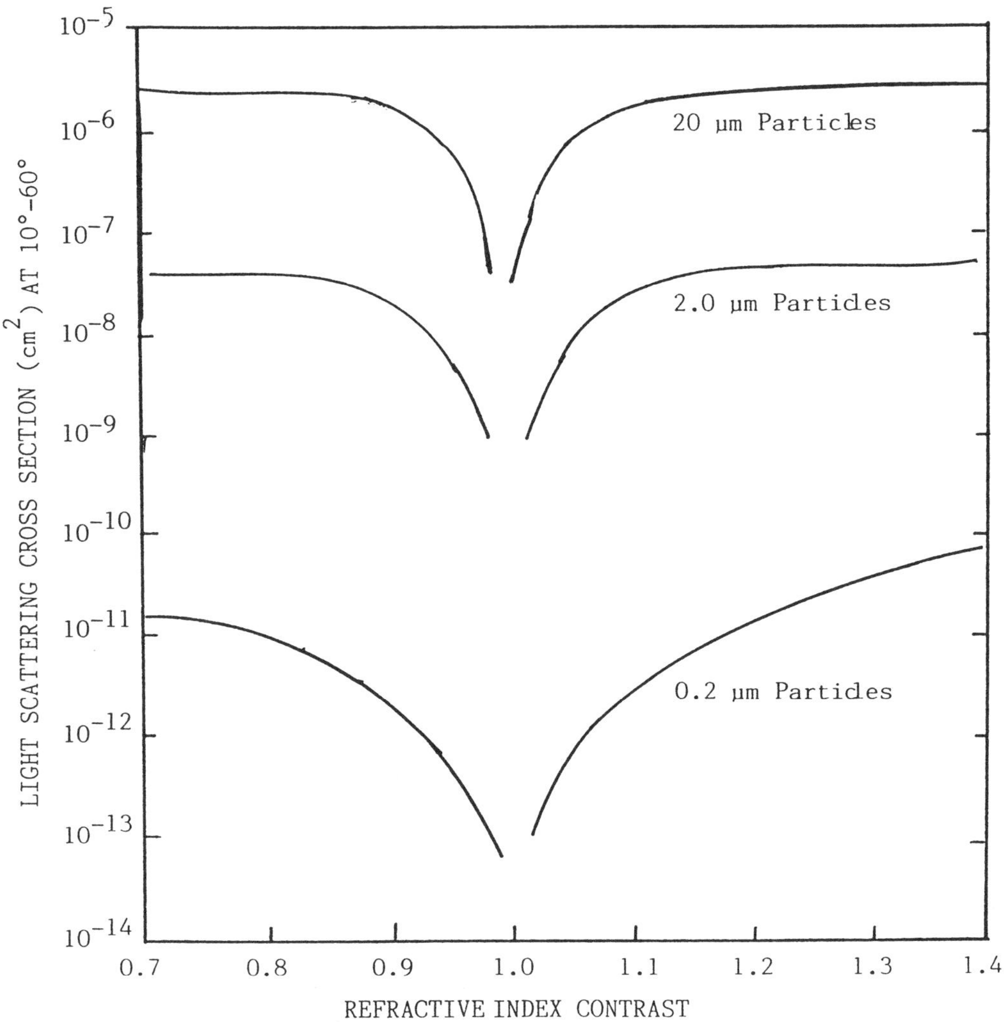

Figure 10 Scattering cross section as function of refractive index contrast. (From Ref. 3.)

calibrated with standard materials that are not necessarily similar to those that will be measured; however, the calibration material should be well characterized in terms of particle size and concentration in the calibration suspension as well as in terms of its morphology and chemical or physical properties.

Primary size calibration for liquid-borne particle counters is normally carried out using a suspension with a known concentration of monodisperse latex particles in water. For volumetric instruments, an aliquot sample of the suspension can be characterized for particle size and concentration by a referee method and another part passed through the instrument. An American Society for Testing and Materials (ASTM) procedure is available for this purpose [7]. The volume of suspension passing through a volumetric instrument can be defined exactly. For an in situ instrument, the injection point for a calibration suspension must be located upstream of the measurement volume and the injection will almost always be dispersed to some uncontrollable extent in the main liquid flow. For that reason, concentration or counting accuracy calibration for in situ instruments is difficult to carry out, although particle size calibration can be carried out for in situ instruments as easily as for volumetric instruments. An ASTM size calibration procedure is available that can be used for this purpose [8]. However, concentration calibration can be carried out only for those in situ instruments where the main liquid stream can be filled with calibration suspension or where the flow regime from the injection point to the observation volume is known sufficiently well to allow valid modelling of the dispersion of the calibration suspension.

E. Suspension Effects on Counter Response

Similarly to volumetric optical particle counters, the fundamental consideration in particle size and concentration measurement for in situ counters is the assumption that each data bit occurs as a result of one and only one particle being present in the sensing volume at any time. When a counter is used to measure a concentrated suspension of particles, then there is a finite probability that more than one particle may be present in the counter sensitive volume. Since particles are usually distributed randomly within a fluid, the instantaneous concentration at any point can vary widely about an average value. If the average concentration in the liquid is high, then the probability of particle coincidence within the sensitive volume will also be high. The counter will then report multiple particles as being individual particles with size equivalent to a function of the summed size of the particles in the sensitive volume. In that situation, the indicated particle size(s) will be greater than the true size [9].

The true particle concentration can be calculated by an equation derived from statistical probability theory which states that the ratio of indicated to true particle concentration is equal to one minus half of the product of the true concentration and the sensitive volume. This product reflects the long-term average population in the sensitive volume. The normal assumption in this determination is that the particles have dimensions small enough to be insignificant in comparison to any dimension of that volume. The maximum recommended concentration stated by counter vendors is usually based upon a coincidence error of either 5% or 10%. That recommended value will be overestimated when particles are present with dimensions similar to those of the sensitive volume.

For very high particle concentrations, the optical signal level may never decrease to the point where the electronic counting circuits can reset. The effect of this situation is shown in Fig. 11. Data are shown for two sensors counting samples from a polydisperse suspension being sequentially diluted. For very high concentrations, both sensors show only fractions of the true concentration. As seen for the 3000/ml-rated sensor, the reported concentration is approximately 10% of the true concentration when the actual concentration is approximately 10 times the recommended maximum for that sensor. Eventually the counter shows only a few particle counts, but these are produced by random variations in the overall concentration of particles in the sensitive volume instead of by individual particle passage. A similar situation occurs when a few large particles are present, but randomly varying high concentrations of subcountable particles pass through the sensitive volume causing signals similar to those from individual larger particles [10].

When an in situ or a volumetric counter is used to observe particle content of clean liquids, it is desirable that the counter be capable of measuring the smallest particles to produce maximum data in minimum time. The relative standard deviation of the data decreases, increasing precision and reliability of the data. An increase in recorded data is obtained by increasing sample size or (preferably) counter sensitivity. An increase in sensitivity can be obtained by decreasing the separation between the OPC background noise level and the signal level for the smallest detectable particle. This procedure usually results in an increase in

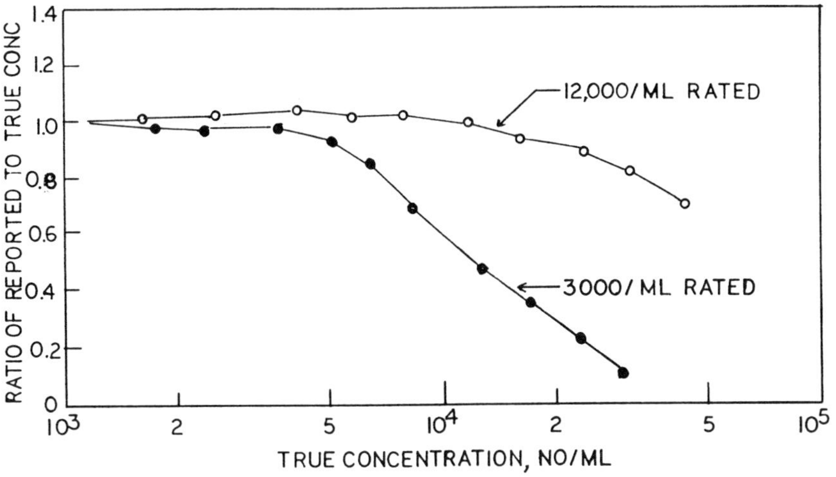

Figure 11 Coincidence effect on indicated particle concentration.

background noise counts as well as in increased sensitivity. A reduction in background noise level is a preferred method of increasing sensitivity.

Most optical particle counters are operated at maximum size sensitivity, where the signal-to-noise ratio is at a minimum. In this situation, some electronic noise pulses large enough to be counted will be present. If particle data are very low, electronic noise pulses may produce an excessive fraction of the total count data. Even though the in situ optical systems normally produce less stray light from optical surfaces than do volumetric systems, there are always some optical effects from the liquid media in which the particles are suspended and some electronic noise will be present in the data processing system. It is then necessary to ensure that the counter is not operated at a point where random noise events can exceed the minimum measurement threshold level in significant quantity.

Some background noise pulses will always be present. These arise from either electronic system noise or from light scattered from contaminant deposits in the optical system. It is therefore necessary to select a signal-to-noise level where the *detected* background noise pulses occur so seldom as to be insignificant in relation to the number of true data signals. The noise level of concern is the largest noise pulse produced during a time period typical for measurement of a "standard" sample, *not* the root-mean-square noise level. A suitable smallest detectable particle signal-to-noise ratio of at least 1.5:1 is frequently chosen to ensure that the reported data represent the system being measured rather than the measurement system performance.

F. Electronic System Operation Considerations

The OPC electronic system for a present-day instrument has two major operations. The first is to process the data signal pulses so that particle size and concentration data can be produced. The second is to monitor operation of the OPC to ensure that all components are functioning correctly. Since this discussion is not concerned with electronic component operation, those features will be discussed only briefly. The emphasis will be on results, rather than on the mechanics of operation.

Present-day OPCs process the signal pulses in multichannel pulse height analyzer systems. Most OPC pulses are generated with durations from 5 to 100 μs, depending upon specific instrument, liquid flow rate, sensitive volume dimension, and particle size. Average particle pulse frequencies can range up to several thousand per second with short-term pulse frequencies up to 10 times that level. For an OPC dynamic size range of 20:1, the pulse amplitude range can be 400:1 for larger particles (measured with an extinction counter). For very small particles the range can be up to 300,000:1 (measured with a scattering counter). The larger range may require multirange electronic systems to process data without distortion. Analog-to-digital conversion of the photodetector output signal levels is used to define pulse amplitude in up to 2048 levels. This high level of electronic pulse level resolution may be excessive in view of the limited optical resolution of many OPCs. These electronic components are also capable of pulse processing rates of up to 100 kHz over the entire dynamic range of the OPC. Some older OPCs still in use sort the signal pulses with a pulse comparator circuit into 8 or 16 amplitude bins. This type of system typically has a counting speed of 10 kHz or so. Some pulse comparator systems use fixed or adjustable resistors to select the pulse amplitude boundaries. These have been reported as subject to both long-term drift and short-term instability. Data from the pulse amplitude measuring systems are stored in pulse counting circuits with capability up to 10^6 per channel. These data can subsequently be transferred to tabular or graphical display formats or can be transferred via RS-232 interface to external storage or further data processing elements.

In addition to data conversion and processing, the OPC electronic system can be used to monitor and control of OPC. It can monitor OPC operation to verify optical and electronic component performance; with suitable elements, liquid flow data can also be recorded. It can be used to label individual measurement data with specific operation identification in terms of sample point location and sample collection time. It can collect and store data from a series of measurements based on preset sampling time or volume duration at selected sample collection intervals. It can be set to indicate warning or alarm levels for specific particle sizes or concentrations. It can be used to carry out simple statistical calculations. Data can be recorded, displayed, printed, or transmitted in a variety of formats. An example of data from two OPCs simultaneously observing a series of samples from a single polydisperse contaminant suspension is shown in Fig. 12. For these data, the suspension was circulated through a system which permitted both addition of concentrate and passage through a filter with pore size smaller than

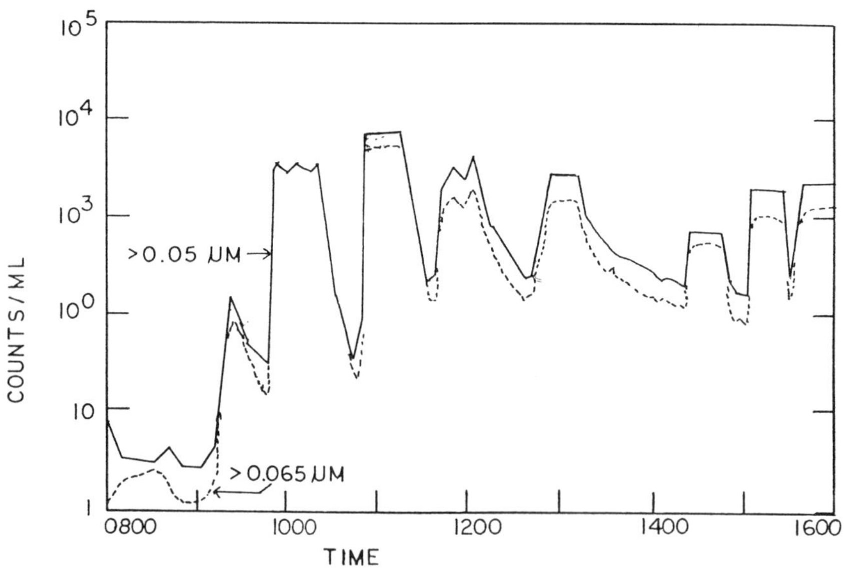

Figure 12 Comparative data for two OPC outputs.

the threshold sensitivity size of either OPC. OPC sensitivities were 0.05 μm (data shown as solid line) and 0.065 μm (data shown as broken line). Intermittent additions of concentrate were added, as shown by the increases in particle concentration for both OPCs. When the suspension flow was switched to the filtered line, both OPCs showed decreasing concentrations. As expected, the 0.05 μm OPC showed somewhat higher total particle concentration than did the 0.065 μm OPC. The higher concentration of small particles expected in most contaminant suspensions is also indicated by the data from the two OPCs.

REFERENCES

1. L. A. Gatlin and S. L. Nail, Initial evaluation of focused ultrasound for in-situ measurement of small particles in flowing water streams, *J. Parenteral Sci. Technol. 35*(5):227–230 (1981).
2. C. N. Montgomery, In-situ particle sizing in liquids, *Proceedings of the 33rd Annual Technical Meeting, Institute of Environmental Sciences,* San Jose CA, 1987, pp. 353–365.
3. R. G. Knollenberg, Monitoring systems for semiconductor fluids, *Handbook of Contamination Control in Microelectronics* (D. L. Tolliver, ed.), Noyes, Park Ridge, NJ, 1988, pp. 257–300.
4. B. N. Verdegan et al., Advances in oil filter test methods, *Proceedings of the International Fluid Power Applications Conference,* Chicago, IL, 1992, pp. 333–342.
5. R. G. Knollenberg and D. L. Veal, Optical particle monitors, counters, and spectrometers: performance characterization, comparison, and use, *J. IES 35*(2):64–81 (1992).
6. R. G. Knollenberg, The importance of media refractive index in evaluating liquid and surface microcontamination measurements, *J. Environ. Sci. 30*(2):50–58 (1987).
7. American Society for Testing and Materials, *Defining Size Calibration, Resolution and Counting Accuracy of a Liquidborne Particle Counter Using Near-Monodisperse Spherical Particulate Material, ASTM F658-87,* Philadelphia, PA, 1987.
8. American Society for Testing and Materials, *Test Method for Calibrating Liquid-Borne Particle Counters for Submicrometer Particle Sizing, F1226-89,* Philadelphia, 1989.
9. J. Raasch and H. Umhauer, Errors in the determination of particle size distribution caused by coincidences in optical particle counters, *Particle Character. 1*(1):53–58 (1984).
10. K. T. Whitby and R. A. Vomela, Response of single particle optical counters to nonideal particles, *Environ. Sci. Technol. 1*:801–814 (1967).

7
Holographic Methods

Brian J. Thompson

University of Rochester, Rochester, New York

I. INTRODUCTION

It is now almost 30 years since holography was first proposed and evaluated as a method of particulate measurement. Since that time the technique has been used in a variety of ways to solve measurement problems in many disparate fields from agriculture to cloud physics. Specific examples include; fog studies; cloud particulate measurements; explosively generated aerosols, snowflake imaging; drug contamination detection, dust erosion studies, agricultural spray measurement, cavitation studies, nozzle design and droplet distributions, fibrous particle evaluations, two-phase flow, impact and particle or droplet breakup, ocean particulates and organisms, and combustion studies. This list, while not exhaustive, is one prepared for an invited presentation in 1991 [1].

Progress in this field of optics and its applications have been the subject of a variety of review articles from time to time over the thirty year history of this field. A recent book by Vikram [2] is an excellent starting point for the study of holographic particle diagnostics, and a companion volume is the Milestone volume of reprints entitled, *Selected Papers on Holographic Particle Diagnostics,* edited by Vikram [3]. This later volume also contains a complete bibliography of papers in the field from the world's literature up to the date of going to press. Earlier review articles are also reliable reading on the development of the subject (Thompson [4], Thompson and Dunn [5], Crane et al. [6], Trolinger [7], Lee and Kim [8], Thompson [9]).

Holographic particle sizing was claimed to be the first direct application of holography [10]; a statement that came about because for many years after the invention of holography by Gabor in 1947, it remained a stimulating scientific curiosity (a situation that has certainly changed today!). Holography as applied to the particle measurement field provides a two-step imaging technique that allows a dynamic volume of particulates to be "photographed." In the first step, a hologram is made of the particle field so that each individual particle is recorded as a hologram. In the second step the holographic record is reilluminated and a stationary image of all the particles in the volume is created; an image that can be studied and restudied. Since the images of the individual particles are created at their original location in the three-dimensional space, a viewing system can systematically scan through the volume to focus on the individual particles and make the appropriate measurement of size, cross-sectional shape and position in the volume. The result is a measurement of the state of all the particles in a volume at one instant in time.

Thus, a particle size distribution function can be obtained for that particular moment in time. Taking a second, third, etc., hologram after an interval of time can give a time history of the particles in that volume. This contrasts with indirect measurements such as light scattering in which particle size distributions have to be built up over time since one particle is measured at a time. Thus, the size distribution that results is the distribution over a period of time at a specific location. Holography also has the advantage of providing a very satisfying method since a direct image is formed and there is a permanent record that can be studied and restudied.

Holography should be considered as a viable technique when direct imaging is not possible. This is usually the case for dynamic particulates existing in a volume. Consider the problem of trying to focus a camera on a small particle, of say, 10-micron size that exists in a volume whose depth dimension is 10 cm. For resolution of a 10-micron particle to an accuracy of one micron requires an imaging system that would have a depth of field of a few microns, assuming that the particle was in focus to start with! Nevertheless, holography is not a panacea, and as this author and others have pointed out, do not use holography unless other methods of imaging do not work!

So far, the discussion has been focused on particle sizing. This method can provide more information than just a size distribution. The particle cross-sectional shape is determined since it is an imaging method. The relative position of all the particles in the volume is determined also. By using multiple-exposure holography, the time history of a particle field can be determined, including the time history of each individual particle in the field—its actual movement and its rotation. Finally, velocity distribution information can also be obtained by multiple-exposure holography.

II. BACKGROUND

The concept of using holographic recording for the measurement of particle size in dynamic samples came about in a very interesting way. A group at the Air Force Cambridge Research Laboratories in Bedford, Massachusetts, led by Dr. Bernard Silverman had the idea of using pulsed lasers to obtain laser shadowgraphs of particles in a volume so that the particle size could be measured from the shadowgraphs. The laser illumination was necessary for this purpose since the exposure times to capture the shadowgraphs were extremely small. For example, it would be important for accuracy of measurement that a particle not move more than 10% of its diameter during the exposure—even this would, of course, only give 10% accuracy in the measurement of the particle size. The exposure time is then determined by the velocity of the particle; so if the particle is moving at, say 1 m s^{-1}, and the particle to be measured has a nominal size of 10 microns, then the exposure time required for 10% accuracy is 10^{-6} s. Higher velocities and higher resolution would require shorter exposure times. However, it should be noted that in this example an accuracy of measurement of 1 micron is assumed, which is going to be close to the limit of optical measurement in any event.

Figure 1 shows some of the results obtained by Silverman [11] and his colleagues and to their surprise these did not look much like the shadow pictures that they expected. That is when they brought these results to a group of us at Technical Operations in Burlington, Massachusetts. Since this group was well versed in dealing with diffraction problems, we immediately recognized these results for what they were—Fresnel diffraction patterns of the field containing the particles. There are, of course, conditions under which an approximation to a shadowgraph can be obtained, i.e., when the distance between the individual object and the recording plane is small; this will occur for larger objects that have a large angular subtense at the recording plane. For smaller objects that have a small angular subtense at the receiving plane, the shadowgraph scheme does not work at all. The results shown in Fig. 1 were taken in 1962 but were not published until relatively recently (Thompson [9]). Figure 1a shows the result for a large bead

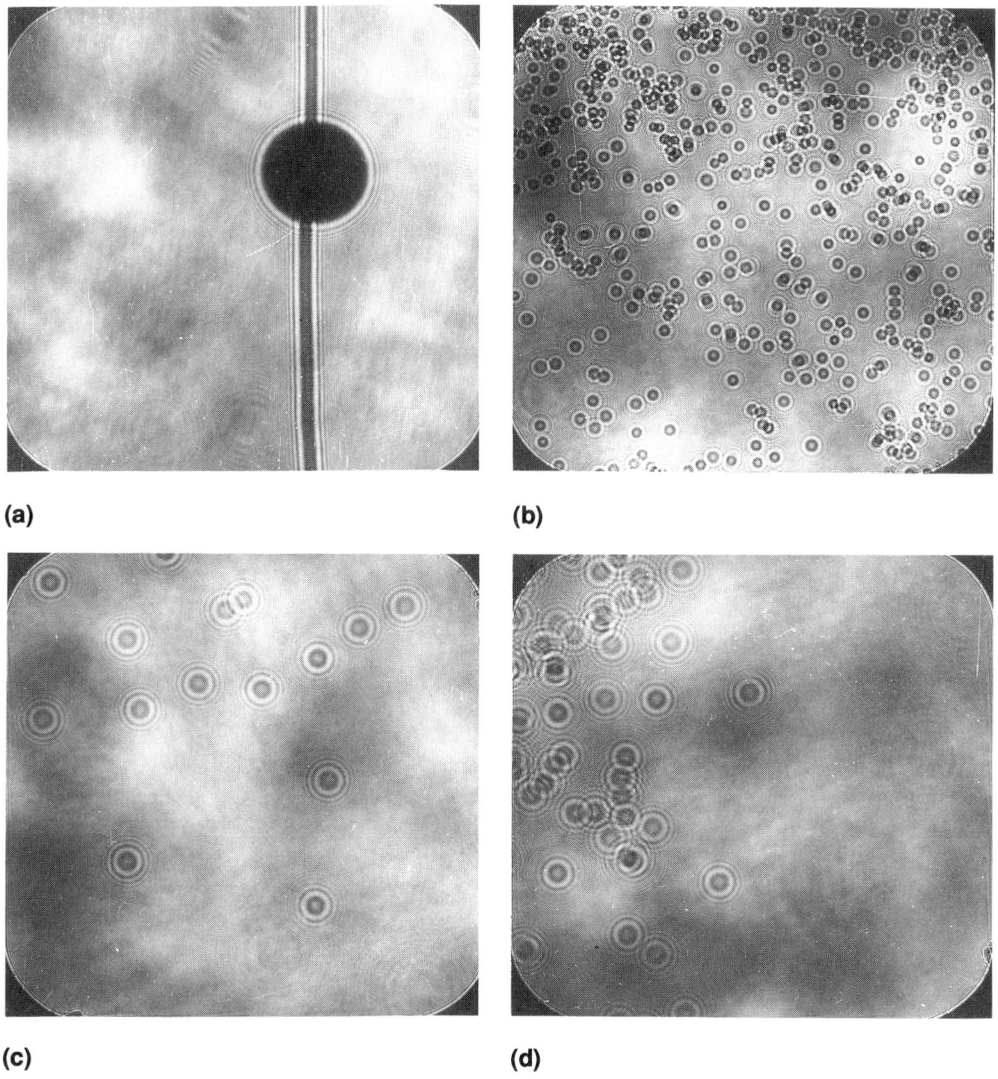

Figure 1 Original attempt to produce "shadowgraphs" of objects using a collimated beam of laser light from a pulsed ruby laser. (From Ref. 11; see Thompson [9].)

and a wire illuminated by a collimated beam of coherent light from a pulsed ruby laser. Even for this relatively large object, the very near-field Fresnel diffraction effects can be seen in the fringe patterns at the edges of the object. The object size is measurable from the record but more understanding of the phenomenon is needed than the concept of a mere shadowgraph. Thompson and Malyak [12] have shown how this type of record can be interpreted as an out-of-focus coherent image of the object. Quite accurate measurements can be made with this correct interpretation of the "image" or "shadowgraph."

Figure 1b shows the record for a fairly larger number of small glass beads with the same distance between the beads and the recording plane as in Figure 1a; again the beads are illuminated with an expanded collimated beam from a pulsed ruby laser. If the distance is increased between the beads and the recording plane then the results shown in Fig. 1c and 1d are obtained. The bead size cannot be accurately determined *directly* from these records using the techniques described in Ref. 12.

It is interesting to note that in the shadowgraph concept the better the collimation of the incident beam, the sharper the shadowgraph. However, the better the collimation, the greater the coherence of the beam and, hence, the dominance of the diffraction effect. Initially, these results shown in Fig. 1 were thought of purely in terms of Fresnel or near-field diffraction, but in fact, they are much more important than merely Fresnel diffraction patterns; they are, in fact, holograms of a rather special type. It was the analysis of these results that led to the concept and definition of in-line farfield holography, that has had such a profound influence on particle diagnostics. Thus, the story goes from shadowgraphs to near-field or Fresnel diffraction to in-line far-field holography. It is time then, in this narrative, to describe these "so-intended shadowgraphs" in the language of diffraction patterns with coherent background.

III. DIFFRACTION PATTERNS OF OPAQUE AND TRANSPARENT PARTICLES WITH COHERENT BACKGROUND

In examining the results shown in Fig. 1, my colleague George Parrent and I were determined that these recordings were of significant value in particle size measurement and that Dr. Silverman and his colleagues had inadvertently perhaps stumbled onto a very useful technique. It is important to recognize Parrent's important contributions to the initial analysis and understanding of these results. (See, e.g., Parrent and Thompson [13]) The current author had presented these concepts in an earlier paper in [14] and was destined to spend a good deal of time on these methods in the future—even unto this writing!

The first step in understanding the results shown in Fig. 1 was to move away from a purely classical Fresnel diffraction approach which, of necessity, considers the whole field. The idea was to consider the actual diffraction from each individual particle and how that diffracted field interfered with the coherent background that passed through the sample volume uninterrupted. It was quickly recognized that the diffracted field associated with the particle was itself basically a far-field (Fraunhofer) pattern providing that the distance z from the particle to the recording plane was greater than the square of the maximum dimension, d, of the particle divided by the wavelength of the illumination, λ. Thus, the far-field condition is formally defined by the inequality

$$z >> \frac{d^2}{\lambda} \tag{1}$$

This does not mean that if this inequality is not met that the resultant record is not of value (see a later discussion). On the contrary, it is of value but the evaluation of this type of pattern is best understood by following the original far-field logic. The original publications [13,14] addressed the Fraunhofer (far-field) diffraction patterns of opaque and transparent objects with coherent background. Let us examine that analysis in summary form here.

Figure 2 sets up the basic parameters where a collimated beam of light illuminates a particle described in general as the complex distribution $\psi(\xi,\eta)$. That is the field, after passing through or by the particle, will be altered (multiplied) by this complex function. In other words, in general the particle may change the amplitude and phase of the incident light. The recording plane is the x,y plane and the field recorded is the complex field $\psi(x,y)$. The receiving plane is at a distance z from the particle such that the far-field condition is obeyed, i.e., $z>>d^2/\lambda$.

Let us examine this system first of all with an opaque single particle in the ξ,η plane whose cross-sectional shape can be written as $A(\xi,\eta)$ (note $A(\xi,\eta) = |\psi(\xi,\eta)|$). The optical field in the receiving plane $\psi(\xi,\eta)$ is then given by

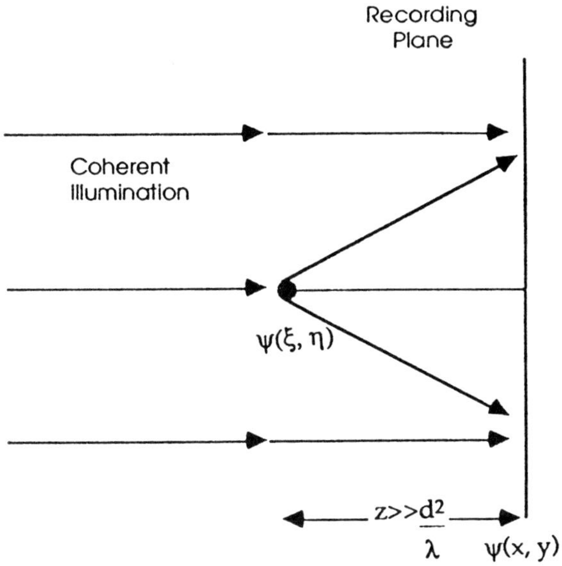

Figure 2 Diffraction by a particle illuminated with a collimated beam of coherent light.

$$\psi(x,y) = \frac{-i}{\lambda z} \exp(ikz) \int_{-\infty}^{\infty}\!\!\int \left[1 - A(\xi,\eta)\right] \exp\left\{\frac{ik}{\lambda z}\left[(x-\xi)^2 + (y-\eta)^2\right]\right\} d\xi\, d\eta \tag{2}$$

The normalized amplitude associated with this record is

$$\psi(x,y) = \exp(ikz)\left\{1 + \frac{i}{\lambda z} \exp\left[\frac{i\pi}{\lambda z}\left(x^2 + y^2\right)\right] \tilde{A}\left(\frac{x}{\lambda z}, \frac{y}{\lambda z}\right)\right\} \tag{3}$$

The record is, of course, of the normalized intensity associated with this field. Thus, the intensity, $I(x,y)$, is given by

$$I(x,y) = 1 + 2\left|\frac{\tilde{A}(x/\lambda z, y/\lambda z)}{\lambda z}\right| \sin\left[\frac{\pi}{\lambda z}\left(x^2 + y^2\right) + \phi\left(\frac{x}{\lambda z}, \frac{y}{\lambda z}\right)\right] + \frac{\tilde{A}^2(x/\lambda z, y/\lambda z)}{\lambda^2 z^2} \tag{4}$$

$|\tilde{A}|$ is the magnitude of the amplitude distribution and ϕ the phase distribution associated with $\tilde{A}$. The first term shown here as unity in this normalized result is actually the level of the background illumination. The second term contains both the amplitude and phase of the diffracted field associated with the particle cross section. The phase $\phi(\,)$ is contained within the sine term which is very important because it records the interaction between the plane background wave and the diverging spherical wave that arrives at the recording plane and has a center of curvature at the particle location. The third term in the equation is the intensity distribution associated with the Fraunhofer diffraction pattern of the particle cross section.

Figure 3 shows the result associated with Eq. (4) for a long thin particle of width $2a$. The essential elements of the result are clearly seen in the plot. The rapid variation is the sine term of Eq. (4); since the sine varies as the square of the spatial coordinate, the fringes get closer and closer together as the spatial parameter increases. The envelope on this pattern is the $\tilde{A}(x)$ term which in this particular case is

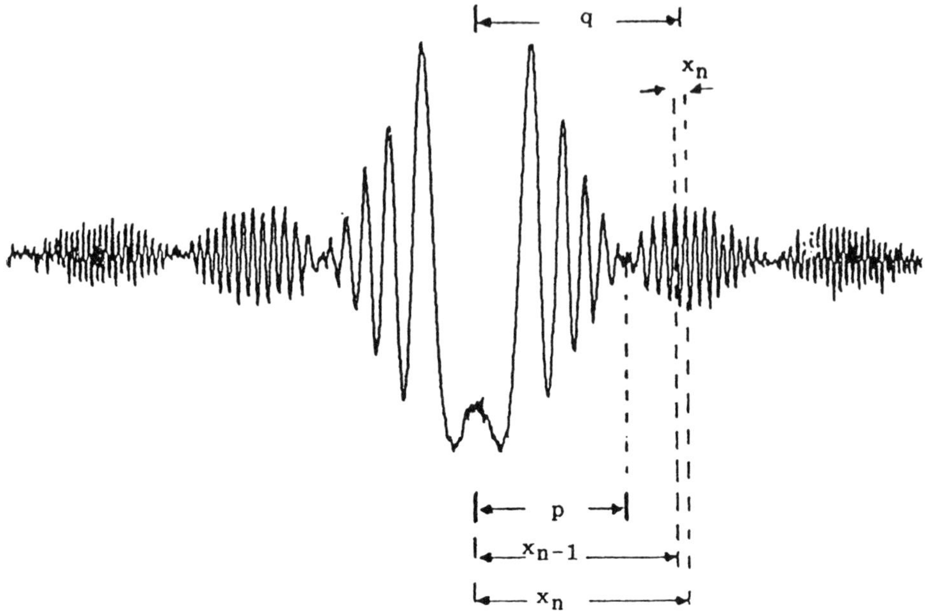

Figure 3 Diffraction by an opaque long thin particle in an otherwise transparent field; the intensity profile as a function of position (Eq. (4)). (From Ref. 9.)

$$\tilde{A}(x) = \operatorname{sinc}\left(\frac{kax}{z}\right) \tag{5}$$

where $k = 2\pi/\lambda$, a is the half-width of the particle, and sinc $\alpha = \sin(\alpha)/\alpha$. Since the sinc function has well defined zero values these are easily located in the pattern. It is of interest to note that various parameters can be determined directly from measurements made on this record, i.e., the location of the particle, z, and its cross-sectional dimension. If a measurement is made at a distance q from the center of the pattern, then since the fringes vary as sin $(kx^2/2z)$,

$$z = \frac{(x_n - x_{n-1})^2}{2\lambda} \tag{6}$$

where x_n and x_{n-1} are the actual distances from the center of the record for adjacent fringe maxima. A measurement made from the center of the pattern to the first zero of the envelope function gives a distance p. Then, the width of the object, $2a$, is

$$2a = \frac{z\lambda}{p} = \frac{(x_n - x_{n-1})^2}{2p} \tag{7}$$

Thus, the distance z and the width of the object, $2a$, are both determined.

While these observation are important in the understanding of this type of diffraction record, they are not necessarily or usually used as the diagnostic method. Rather, the record of Eq. (4) is recognized as a particular type of hologram with its attendant virtues. It will be recognized that the analysis described above is for a particle in an otherwise transparent field, or for a series of particles existing in a volume but in which the particles density is such as to allow for a significant background beam. Each particle will produce its own individual pattern as is apparent by reference to Fig. 1b and 1c.

IV. IN-LINE FAR-FIELD HOLOGRAPHY

The recordings discussed in the previous sections are best described as holograms, but holograms of a particular kind and with some interesting properties. A hologram, it will be recalled, is the record of an interference pattern formed by the interaction of a light diffracted (or scattered) by an object with a coherent background or reference wave. Thus, the records discussed here so far are the interaction between the far-field (Fraunhofer) diffraction pattern of a particle with a coherent in-line background illumination. In the circumstance when the far-field condition is not met then the interaction is between the near-field (Fresnel) diffraction pattern of the particle and the in-line coherent background.

Equation (4) shows an intensity that, when recorded, is a hologram. The property of a hologram is that when the record is reilluminated, an image of the original object, in this case a particle, can be formed. This property can be seen conceptually by considering the object as an idealized point so that $\tilde{A}$ is a constant, C, and $\phi = 0$. The intensity record is then of the form

$$I(x,y) = 1 + 2C \sin\left(\frac{\pi(x^2 + y^2)}{\lambda z}\right) + C^2 \tag{8}$$

Equation (8) can be rewritten as

$$I(x,y) = A + \frac{2C}{i}\left[\exp\frac{\pi(x^2 + y^2)}{\lambda z} - \exp - \frac{\pi(x^2 + y^2)}{\lambda z}\right] \tag{9}$$

where $A = 1 + C^2$ and is a constant; the first exponential term is a spherical wave diverging from the original point object as if it was still there and the second exponential term is another spherical wave that converges to a point a distance z in front of the record. These are the terms that produce the two images characteristic of holographic image formation. The image formed at the original location is a virtual image and the other one is real.

With this concept in mind an expression for the far-field distribution in the image plane can be developed from Eq. (4). Let α, β be the coordinates in the image plane as formed when the hologram is illuminated. The actual field in the image plane, $\psi(\alpha,\beta)$ is

$$\begin{aligned}\psi(\alpha,\beta) = \exp(ikz)\Bigg\{1 - A(\alpha,\beta) - \frac{1}{2\lambda z}\exp\left[\frac{i\pi}{2\lambda z}(\alpha^2 + \beta^2) - \frac{\pi}{2}\right]\tilde{A}\left(\frac{\alpha}{2\lambda z}, \frac{\beta}{2\lambda z}\right) \\ + \frac{1}{\lambda^2 z^2}\tilde{A}\left(\frac{\alpha}{2\lambda z}, \frac{\beta}{2\lambda z}\right)\tilde{A}^*\left(\frac{\alpha}{2\lambda z}, \frac{\alpha}{2\lambda z}\right)\Bigg\}\end{aligned} \tag{10}$$

where the asterisk denotes a complex conjugate. The required image is $A(\alpha,\beta)$ and the remaining terms are, again, a hologram of the original object, but formed at the distance $2z$ and comes about because of the other image.

V. IMPLEMENTATION OF THE METHOD

The hologram of a dynamic field of particles or droplets is formed by illuminating that sample with a collimated beam of coherent light usually in the form of a pulse, short enough to freeze the motion of the particles during the recording process but energetic enough to create a satisfactory recording. It will be noted that the hologram is a record of a variation of intensity about the background level and hence, it is this background level that determines the

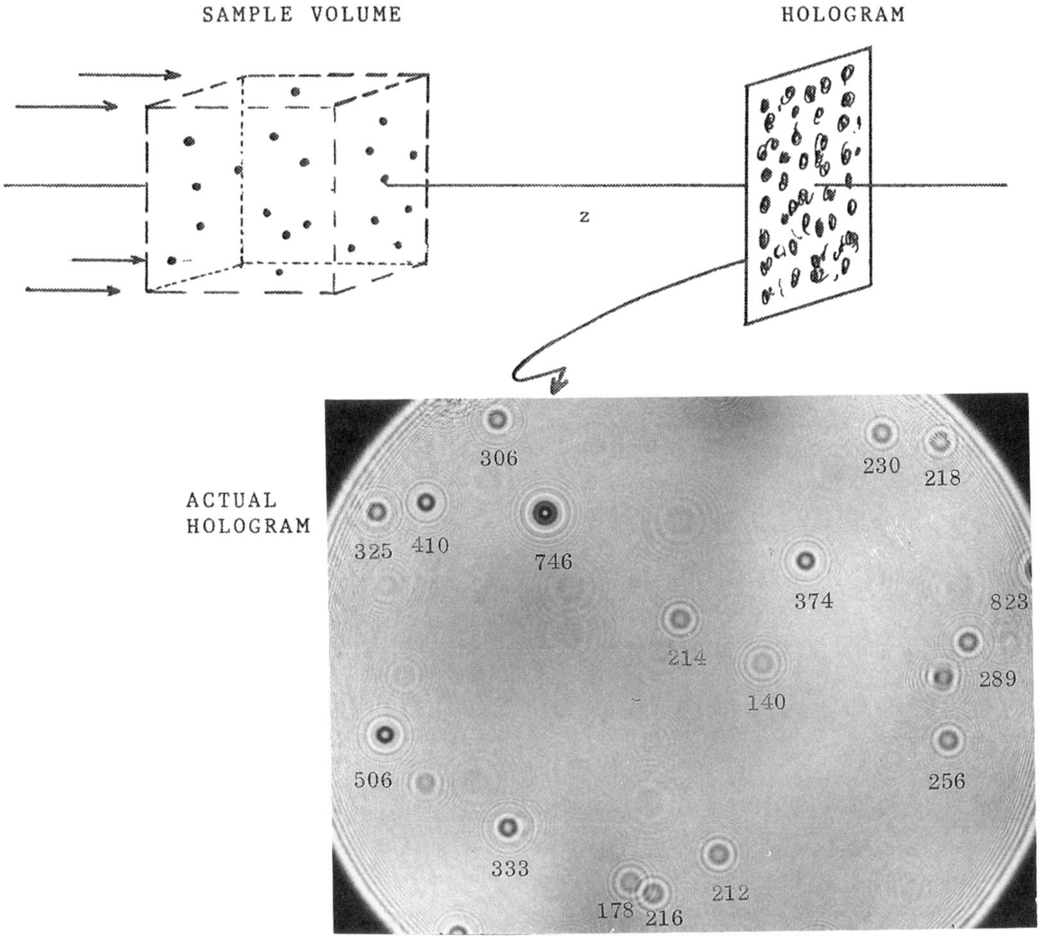

Figure 4 In-line method of recording a hologram of a volume of particles. The lower illustration is the first recorded hologram of a set of particles. (Parrent and Thompson [13])

exposure time. Figure 4, top, illustrates the method of hologram recording, with the lower part of the figure showing the first hologram of particles made deliberately. The size of each particle in microns is shown.

After the hologram has been recorded, it can be reilluminated with a collimated beam of coherent light; a continuous wave laser is usually used in this step so that the stationary image produced can be studied in detail. Figure 5 shows this second step. The lower illustration is the image formed from part of the hologram shown in Fig. 4. As discussed earlier, two images are formed, one virtual and the other real. The real images of all the particles can be viewed directly on a screen with or without magnification. Since a collimated beam is used, the images all have the same unit magnification in the system diagrammed here. The virtual images of the particles are in the exact original location with respect to each other; the same is true of the real image except this image is pseudoscopic. Thus, measurements can be made of the positions of all the particles in the original sample volume and their cross-sectional shape as "seen" by the incident beam in the recording step. The holographic method is thus a two-step imaging method that

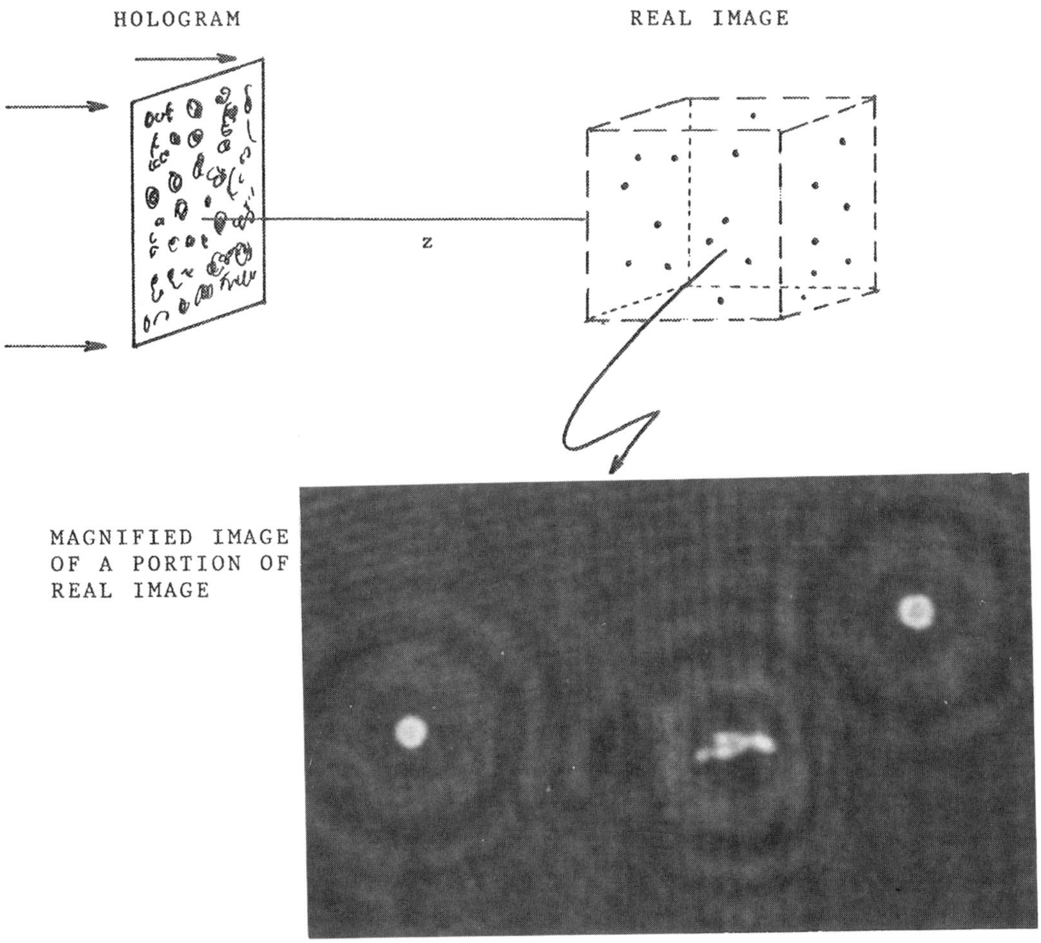

Figure 5 Image formation from an in-line hologram. The lower illustration is the image formed from part of the hologram shown in Fig. 4. (From Ref. 17.)

produces a permanent stationary image of a transient event. The time history of the particles in the volume can be determined by taking a sequence of holograms.

VI. SOME EXAMPLES

First, some individual particle examples will be considered. For a one-dimensional object such as a fiber, a hair, or a thin wire, etc., the cross-sectional shape is rectangular of width $2a$ and the term $\tilde{A}$ in Eq. (4) will have the form

$$\tilde{A}\left(\frac{x}{\lambda z}\right) = 2a \operatorname{sinc}\left(\frac{kax}{z}\right) \tag{11}$$

An in-line far-field holographic record of a wire of width $2a = 160$ microns is shown in Fig. 6b together with a profile of the intensity distribution in Fig. 6a. This hologram was made with a value of $z = 60$ cm and an argon-ion laser was used to provide the illumination, $\lambda = 0.5145$

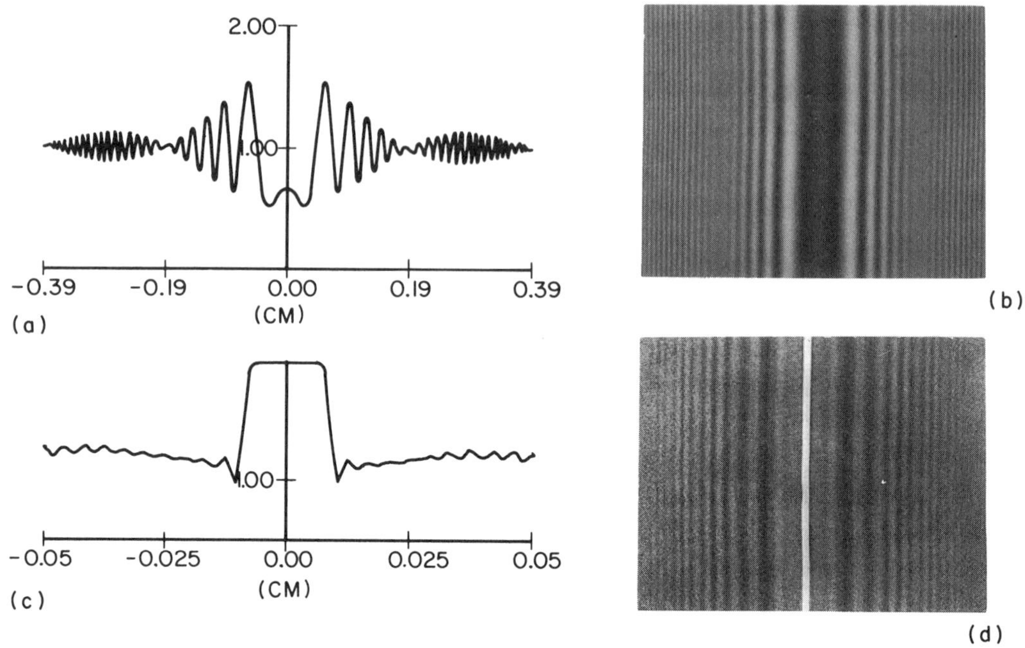

Figure 6 Example of the hologram of a wire of width 160 microns; (a) plot of intensity profile of the hologram; (b) the hologram; (c) plot of the image from the hologram; and, (d) photograph of the real image. (From Refs. 15 and 18.)

micron (after Tyler and Thompson [15], see also Thompson [16,18]). The expression representing this profile is

$$I(x) = 1 - \frac{4a}{\sqrt{\lambda z}} \cos\left(\frac{\pi x^2}{\lambda z} - \frac{\pi}{4}\right)\left[\frac{\text{sinc } 2kax}{z}\right] + \frac{4a^2}{\lambda z}\left[\frac{\text{sinc } kax}{z}\right]^2 \tag{12}$$

The image formed from this hologram is shown in Fig. 6d together with an enlarged plot of the intensity profile in Fig. 6c. The required image is the bright line image in the center. This image is surrounded by another hologram formed at an effective distance of 120 cm as described earlier. The relative advantage of the far-field method is apparent here in that the light associated with the other image is contained in the background hologram that does not interfere with the image. In near-field holograms this is not the case.

Many real objects have a circular cross section; for example, liquid droplets in air, or gaseous bubbles in a liquid. The cross-sectional geometry of diameter $2a$ produces a two-dimensional Fourier transform that has circular symmetry; thus

$$\tilde{A}\left(\frac{r}{\lambda z}\right) = \pi a^2\left[\frac{2J_1\,(kar/z)}{kar/z}\right] \tag{13}$$

where r is the radial coordinate and J_1 denotes a Bessel function of the first kind and order 1. For this example, the intensity in the hologram has the form

$$I(r) = 1 - \frac{2\pi a^2}{\lambda z} \sin\left(\frac{\pi r^2}{\lambda z}\right)\left[\frac{2J_1(kar/z)}{kar/z}\right] + \frac{\pi^2 a^4}{\lambda^2 z^2}\left[\frac{2J_1(kar/z)}{kar/z}\right]^2 \tag{14}$$

Figure 7 shows such a hologram, the radially symmetric sine term is readily apparent as is the envelope.

Particles with irregular cross section produce related results. While it is possible to calculate $\tilde{A}(\ )$ for any particular geometry, a closed-form expression similar to those shown above cannot be given. Figure 8 shows a number of images of other shapes; Fig. 8a shows images formed from a hologram of irregular-shaped carborundum particles, the smallest of which are a few microns in diameter and the larger ones some 20 to 30 times bigger (Thompson [16]). Figure 8b shows the image of a single pollen grain that measures about 20 microns in nominal diameter. Most pollen grains are not smooth spheres but have irregular surfaces. The small protuberances

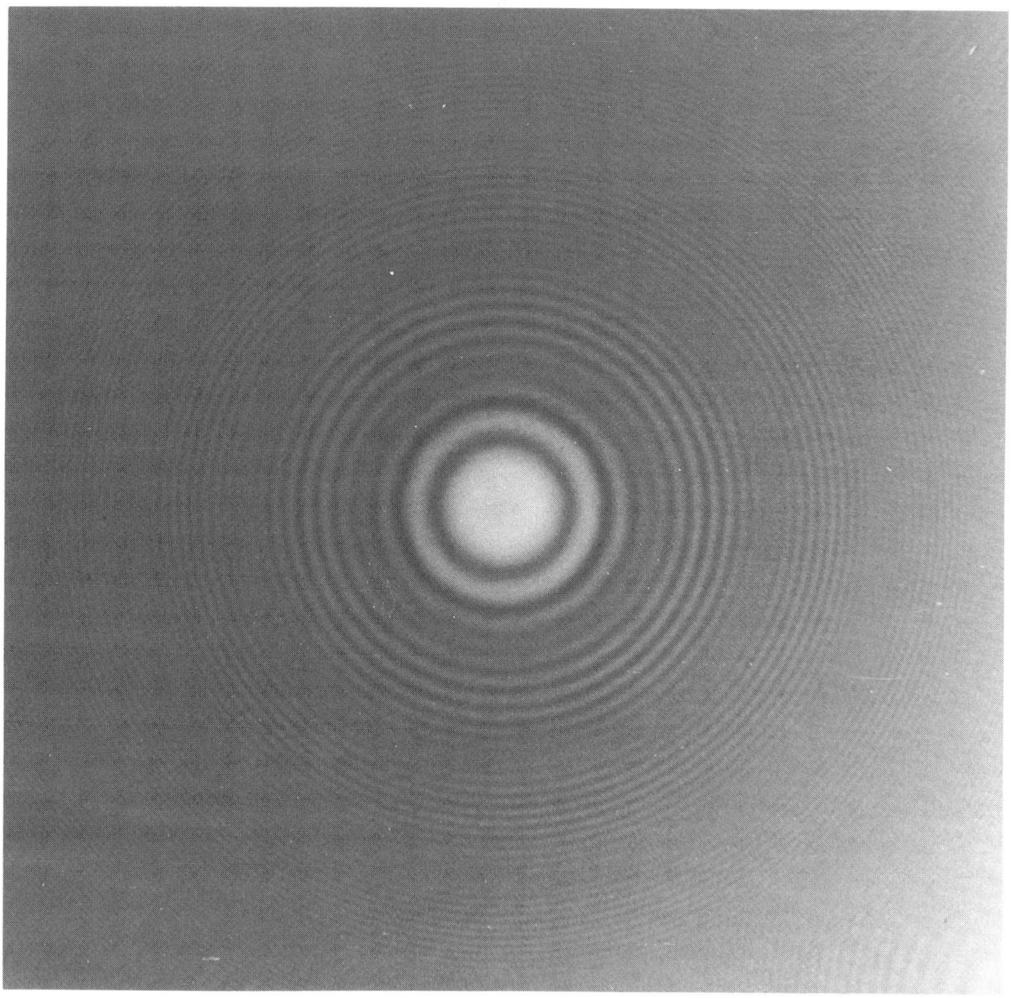

Figure 7 Hologram of a particle with a circular cross section. (From Ref. 19.)

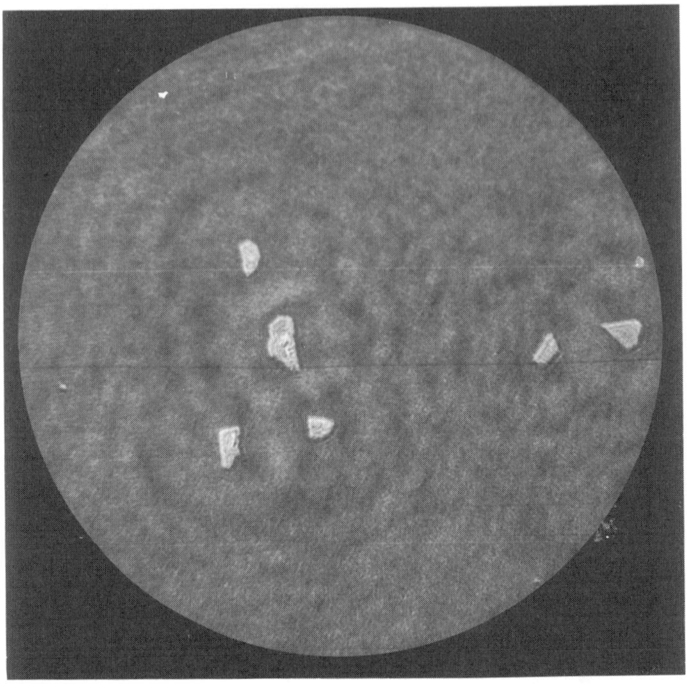

(a)

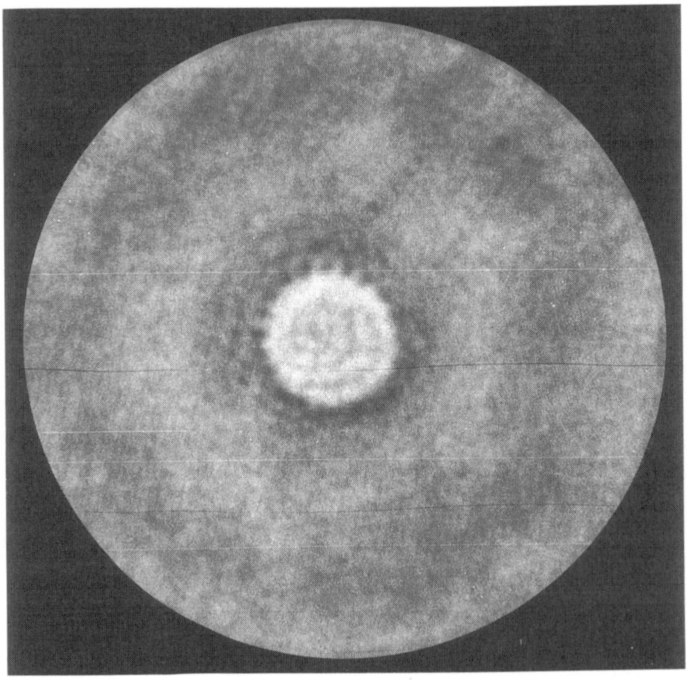

(b)

Figure 8 Images of irregular shaped object formed from a hologram; (a) carborundum particles (after Thompson); (b) a single pollen grain. (From Ref. 19.)

on the surface of the pollen particles are about 1 micron across and protrude about 2 microns above the surface.

VII. SYSTEM PARAMETERS

Designing holographic systems requires consideration of a number of parameters. Ultimate systems performance will depend upon these parameters—the most important of which is particle resolution. In the end, the size resolution will depend on the wavelength of the illumination.

A. Far-Field Condition

In recording the hologram it is best to meet the far-field condition that is worth redefining as an equality of the form

$$z = N\frac{(2a)^2}{\lambda} \tag{15}$$

where N can be chosen appropriately (the factor N is usually called the number of far-field distances), and $2a$ is the dimensional size of the particle. The system will work from N being a small multiple of 2 or 3 up to N values of several hundred. If N is too small then the pattern will not be a far-field pattern and the image formed from the hologram will have some distortion from the other image. When N is very large the contrast of the pattern that has to be recorded gets very low and techniques are used to enhance the contrast on recording.

This discussion does not mean that quite effective near-field holograms cannot be recorded and used, they clearly can; but there will be some possible degradation in the image. Often, if the size range of a sample volume is large then inevitably some of the larger particles will be in the near field when the hologram is recorded. Figure 9 shows a portion of a hologram of a field of particles that have circular cross section; some eight full or partial individual holograms can be seen. The one in the middle at the top is clearly a far-field hologram as is the partial hologram in the upper half of the right-hand side. The others are either just into the near field or well in the near field, for example, the hologram at the center is actually of the largest particle and is quite near field and the size could be measured approximately using the technique mentioned earlier of treating this record as an out-of-focus image. It will be further noted that the near-field patterns vary in detail because of the Fresnel diffraction component that is associated with the particle. Thus $\tilde{A}$ in Eq. (4) needs to be replaced with the appropriate Fresnel pattern function which varies with distance. $\tilde{A}$ in Eq. (4) for the far-field does not vary in functional form with position but only changes in scale.

B. Fringe Visibility

Getting the best possible recording of the hologram is clearly important; a poor hologram will, without doubt, give a poor image! In recording the hologram it is the visibility of the fringe pattern that is all important since these fringes actually carry the information about the particle's cross-sectional shape, i.e., the envelope function on the fringes.

The intensity distribution in the hologram for a circular cross section particle is described in Eq. (14). This equation may be recast by making some substitutions from Eq. (15). Thus,

$$I(r) = 1 - C_r \sin\left(\frac{\pi r^2}{\lambda z}\right)\left[\frac{2J_1\,(kar/z)}{kar/z}\right] + C_r^2\left[\frac{2J_1\,(kar/z)}{kar/z}\right]^2 \tag{16}$$

where $C_r = \pi/4N$.

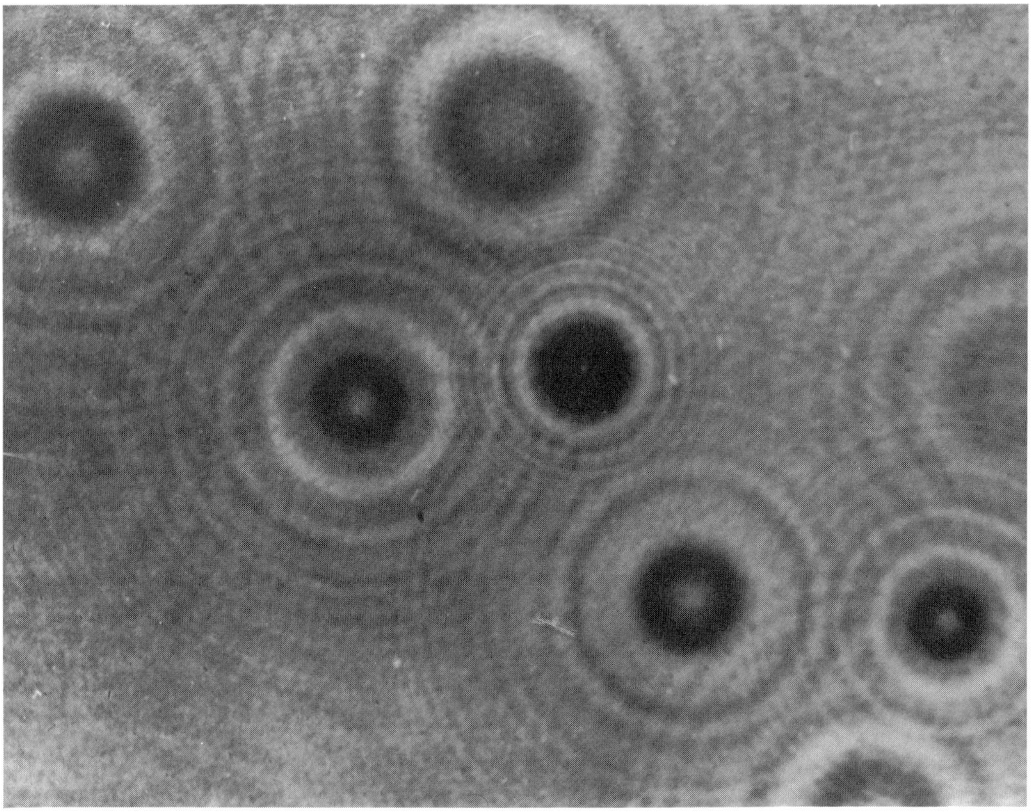

Figure 9 In-line near and far-field hologram of particles with circular cross sections. (From Ref. 20.)

The visibility of the fringes as a function of position, $V(r)$, is

$$V(r) = \frac{I_{\max} - I_{\min}}{I_{\max} - I_{\min}} \tag{17}$$

and

$$I_{\max}(r) = 1 + 2C_r\left[\frac{2J_1\,(kar/z)}{kar/z}\right] + C_r^2\left[\frac{2J_1\,(kar/z)}{kar/z}\right]^2$$

$$I_{\min}(r) = 1 - 2C_r\left[\frac{2J_1\,(kar/z)}{kar/z}\right] + C_r^2\left[\frac{2J_1\,(kar/z)}{kar/z}\right]^2$$

Hence, the visibility as a function of the radial coordinate r is

$$V(r) = \frac{2C_r\left[2J_1(kar/z)/kar/z\right]}{1 + C_r^2\left[2J_1(kar/z)/kar/z\right]^2} \tag{18}$$

or

$$V(\vartheta) = \frac{2C_r D(\vartheta)}{1 + C_r^2(D(\vartheta))^2} \tag{19}$$

where $D(\vartheta)$ is the term in brackets of Eq. (18).

Figure 10 shows the visibility plotted as a function of ϑ for this case with varying values of N (after Dunn and Thompson [20]). The visibility clearly decreases with increasing values of N. However, there are reports of quite large values of N being used (several hundred) where much care is taken to enhance the contrast during the actual photographic recording by using special detection and processing tricks. A standard method for recording diffraction phenomena in general is to overexpose and then underdevelop by using regular developer for a shorter time or by using a weaker developer for the standard time. A more detailed discussion of fringe visibility and resolution can be found in the literature [20,21].

The exposure level and the overall contrast of the fringes is determined by the background intensity and the ratio of the background intensity to the diffracted intensity. A number of techniques have been developed that can change the ratio (see, e.g., Molesini et al. [22], Özkul et al. [23]). Other methods have modified the illuminating beam by using say a Gaussian profile (Vikram and Billet [24]). However, this is only advantageous when the beam profile can be centered on the particle.

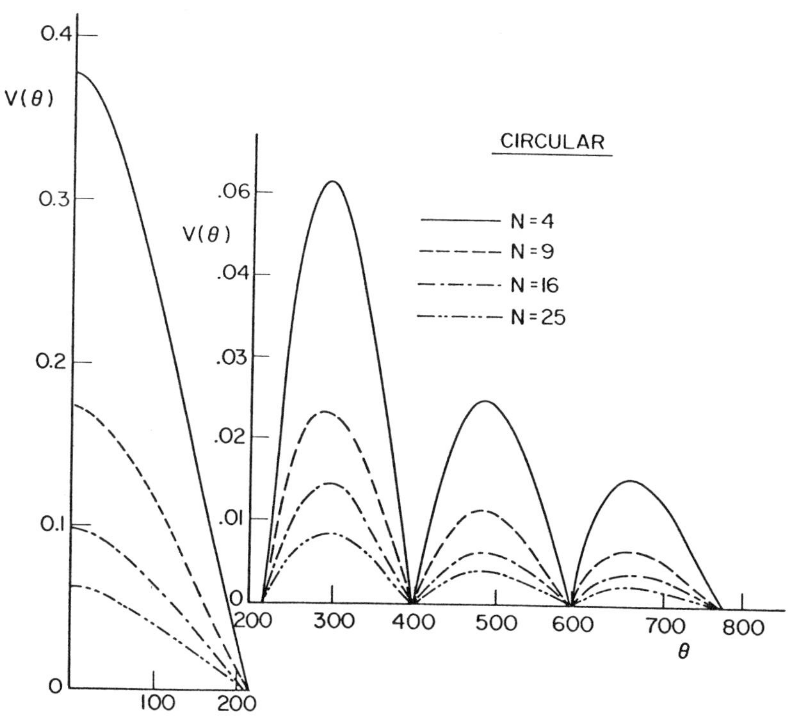

Figure 10 The visibility of the fringes in a far-field hologram of a particle of circular cross section. The visibility is determined along a radial coordinate for various values of N (Dunn and Thompson [20]).

C. Independent Reference Beams

Some of the issues discussed above arise because the background beam is the same beam that illuminates the particles. Thus, under some circumstances it is worthwhile to use an entirely separate reference beam and to use that beam in the off-axis mode, so that the diffracted beam and the reference beam are propagating at an angle with respect to each other. The problems with the overlapping light associated with the two images formed from the resulting hologram is, of course, overcome.

The basic concept of the off-axis method is illustrated in Fig. 11. Collimated illumination is again used but part of the beam is reflected by the beam splitter and then reflected from a mirror to reach the recording plane to interfere with the beam that is diffracted by the particle. The angle of the separate reference beam is chosen to make sure that the two images formed from the hologram are completely separated in the reconstruction; i.e., the beam associated with one image does not propagate through the plane containing the other image. In the case depicted here the object is still transilluminated and, hence, the usual in-line hologram is formed as well as the off-axis hologram. Thus, actually there will be four images, but this time the in-line images are not used. It will also be noted in this figure that recording is carried out in the far-field although this is much less important with the off-axis reference beam method. Very high resolution film is required since the fine fringes formed by the angle between the beam diffracted by the object and the reference beam are, of necessity, of higher frequency than the fringes in the in-line system.

The versatility of the off-axis system also allows the sample to be nontransmitting and, hence, light reflected or scattered from the particles can be used to form the hologram. This is particularly important for dense sprays.

Early work using this method with a pulsed laser was successfully completed by Brooks et al. [25] and applied to particulate measurements by Wuerker et al. in 1968 [26] and by Trolinger et al. [27]. Excellent reviews of these techniques are available in the literature (see, e.g., Trolinger et al. [7,28]). It will be recalled that the very first concept of the idea of holographic particle measurements was using a Q-switched pulsed ruby laser. The image formation (or reconstruction as it is often called) is usually carried out with a He-Ne laser.

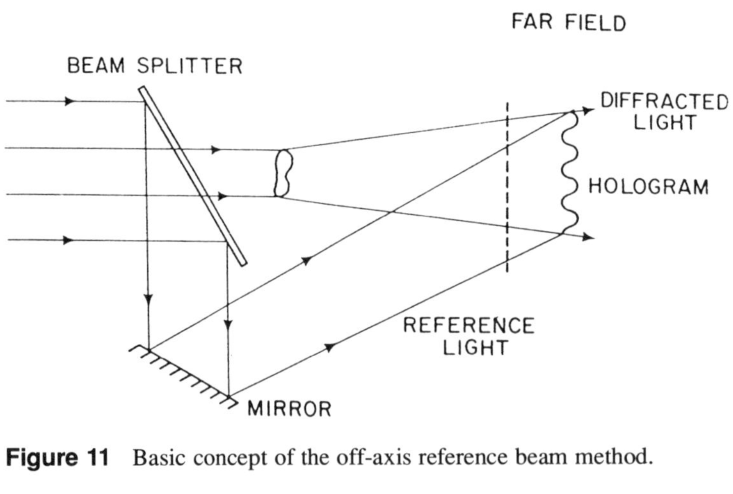

Figure 11 Basic concept of the off-axis reference beam method.

Because of the closeness of the wavelength of the ruby laser and the He-Ne laser only minimal aberration occurs.

D. Relay Optics

Collimated illumination is usually used in these holographic systems since this gives uniform magnification throughout the image volume. This magnification is unity if no auxiliary optics are used. Magnification can be introduced by using an auxiliary lens or optical system. This has the advantage of allowing holograms to be recorded at more remote distances from the actual sample volume. Demagnification can also be used if larger particles over a long sample depth are to be recorded. The early work on the recording of naturally occurring fog droplets used magnification method to ease the resolution requirements on the recording film (Thompson et al. [29]).

E. Field of View; Depth of Field

The field of view will be limited by the magnification used and by the effective aperture of the relay system if used. Eventually, the field will be limited by the dimensions of the recording material. It is important that the central maximum and at least two sidelobes of the diffraction pattern associated with the particle are recorded; the extent of this pattern must be recognized when recording. The depth-of-field could be limited by the temporal coherence of the illumination but again it is more likely that the recording material resolution and dimensions limit the depth-of-field. The depth-of-field can be as expressed in terms of multiples of the parameter $2a^2/\lambda$ (see Eq. (15)) and can range from 50 to 100 far-field distances, although larger depth-of-fields have been achieved under special circumstances.

Figure 12 shows part of a hologram of a volume of droplets. It will be clear that the droplets exist in many planes. The broad hologram in the center is of a droplet at a long distance from the recording plane. The holograms (a group of three) just to the left of this hologram are at much closer distance. From the fringe pattern it can be determined that there is at least a factor of 16 between the distances involved. The hologram in the middle of the right-hand edge shows a factor of 25 between its distance and the first hologram discussed.

F. Resolution

Many of the parameters discussed above are related to the final resolution of the image formed in these holographic two-step image-forming systems. In the end the resolution will be determined by the highest spatial frequency of fringes that can be recorded [29] and stored on the recording material. This in turn is determined by the resolution limit of the recording material. Careful attention to the various parameters discussed here can result in resolutions up to the limit set by the wavelength of the illumination.

Detection of particles, as opposed to measurement of particles, requires some different considerations. Beyond the resolution limit the "image" of an unresolved particle will be the overall impulse response of the entire system. Submicron particles have been detected and counted under these conditions (Thompson and Zinky [30]). A system with an impulse response width of 3 microns was used to detect particles almost an order of magnitude smaller than this, i.e., 0.365-micron-diameter latex spheres.

Figure 13a shows the hologram of a field of the 0.365-micron-diameter latex spheres [30]. This hologram was made by illuminating a latex hydrosol on a microscope cover glass with light

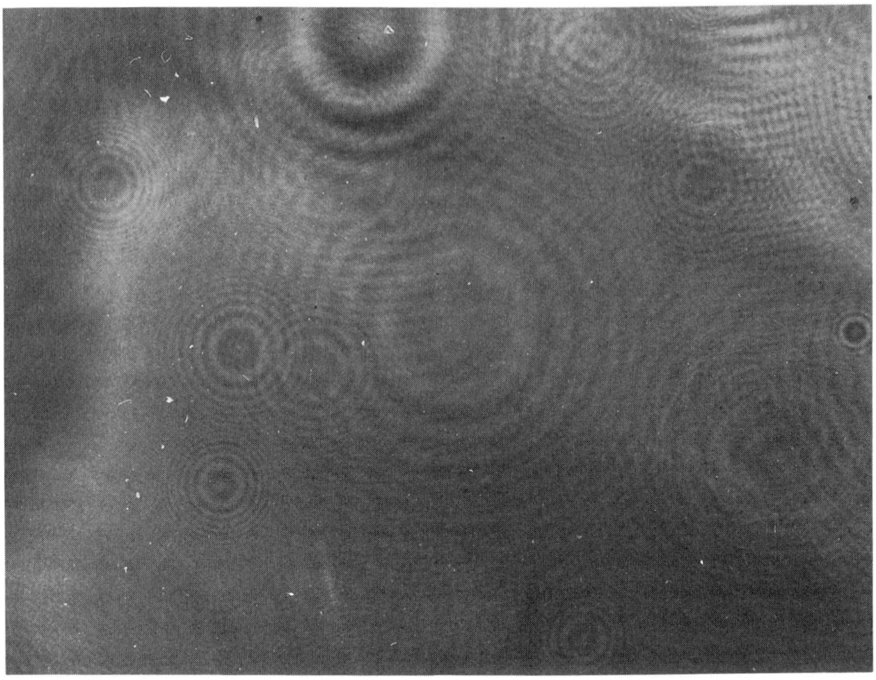

Figure 12 Hologram of particles in a volume at significantly different distances.

from a mercury arc suitably filtered through a pinhole to get sufficient spatial coherence, and through a color filter to select the green line. A simple lens relay system was used to relay an image of the hydrosol before the hologram was made. The "images" of individual particles in Fig. 13b are actually impulse responses of the relay optical system except for the large object in the upper left-hand corner. A count can be made of the particles in this plane, although when several particles are grouped together it is more difficult. Such "images" also exist in other planes as described in the original paper on this work [30].

To push the resolution to smaller sizes than that possible with visible light, investigations have been carried out using electron-beam holography (Tonomura [31,32]). The images are reconstructed optically. X-ray holography of particles has been explored successfully in a series of experiments by Howells et al. and by Jacobsen et al. (see, e.g., Refs. 33 and 34).

G. Positional Resolution

Apart from the resolution of the particle cross section there is also the question of positional resolution within the sample volume. The lateral resolution is quite high and depends directly upon the particle resolution. On the other hand, the resolution in the longitudinal direction is much poorer and is about an order of magnitude poorer. It must be remembered that this is a two-step coherent imaging system. Thus, if the resolution in object space of an individual particle is R_o, then the resolution in longitudinal position is equivalent to the depth of field of the coherent imaging system which is $2R_o^2/\lambda$. Hence, for a system performing close to the limit, let $R_o = 2\lambda$, then the positional depth resolution is appropriately 8λ.

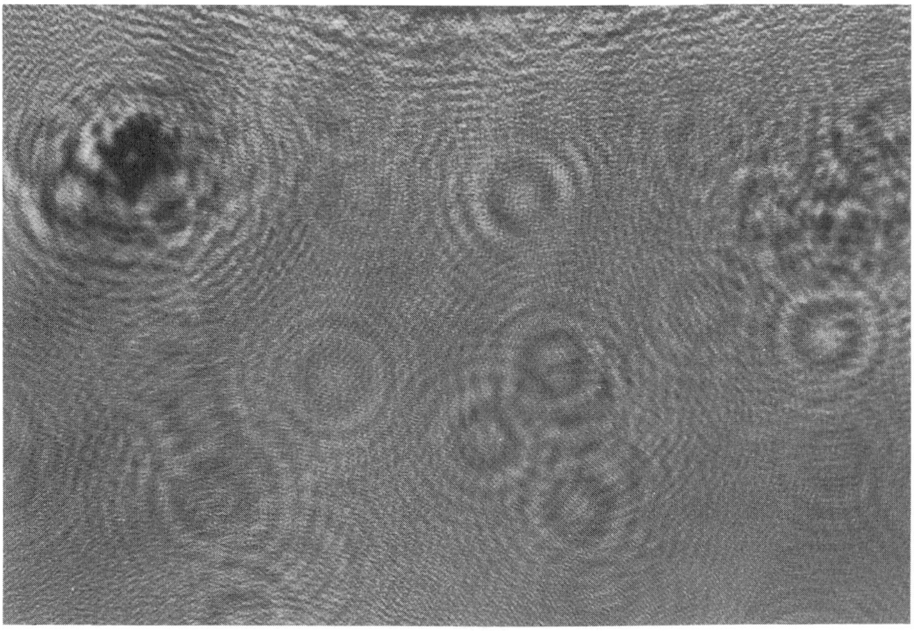

(a)

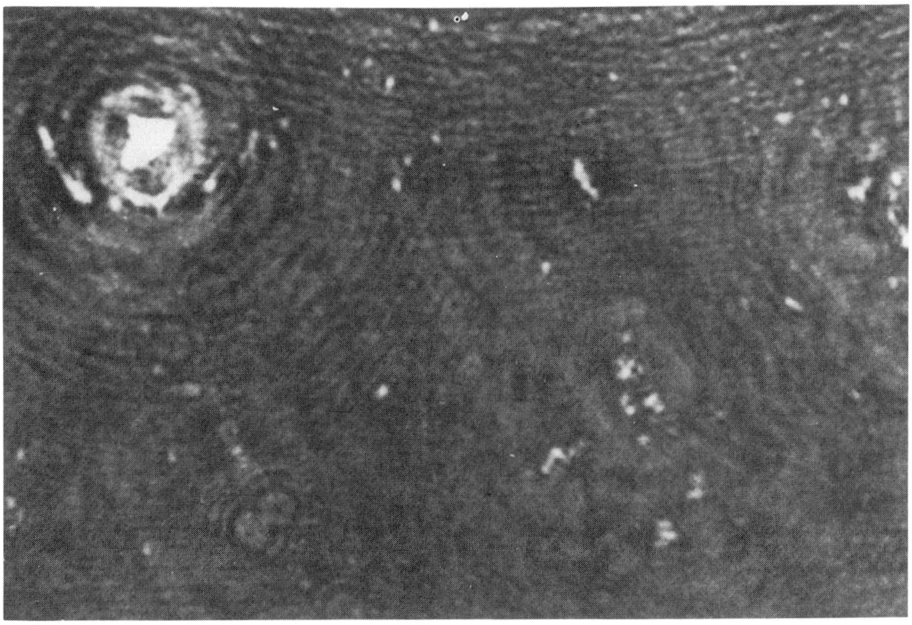

(b)

Figure 13 Holographic method applied to the detection of submicron particles; (a) hologram of 0.365-micron-diameter latex beads in a hydrosol; (b) "image" of one plane formed from the hologram.

VIII. HOLOGRAPHIC SYSTEMS FOR PARTICLE MEASUREMENT

There have been a wide variety of systems built for recording holograms of particle fields using both the in-line far-field, off-axis far-field and near-field systems depending upon the application. All these systems have been custom built for the particular character of the particular problem being investigated. There really is no universal system because the parameters vary over a wide range of particle size, velocity, concentration, depth of sample, ambient conditions of temperature and pressure, circumstances of the measurement including the field conditions including vibration, humidity and portability and finally the need for installation within existing systems.

Once the hologram has been recorded, the image needs to be formed and interrogated and some automation needs to be included in this second system. Again, most of these systems have been customized based on the particular application, the recording material used and its format, the depth of the sample volume measured, the magnification required in the readout system, and whether the system was in-line or off-axis.

A. Holographic Recording Systems

Figure 14 shows a straightforward optical system for recording the far-field in-line hologram of sample volumes of dynamics particulates or droplets. The source is a Q-switched ruby laser and the beam after spatial filtering illuminated the sample volume with a collimated beam. A relay lens produces an image of the sample volume at a more convenient location. It will be noted that the recording plane can be in a variety of positions including actually within the image. In this later case, one plane recorded would be an image plane and adjacent planes would be near-field planes and then others far-field. This is the type of system used in the original fog studies and in explosively generated aerosol work. It is worth noting that in some early work Knox [35] recorded in-line holograms of living plankton organisms by inserting the photographic plate directly into an aquarium containing the plankton.

Figure 15 shows a more versatile type of system used for recording in-line far-field holograms. The illumination is provided by a Q-switched laser with spatial filtering to clean up the beam with the 200-micron pinhole being the final element of this subsystem. Lens L_2 produces a collimated coherent beam to illuminate the sample volume. A two-lens unit-magnification system is used to relay an image of the sample volume to a more convenient location and a hologram is then made behind that image plane. Film recording is carried out with a 35-mm format. The two-lens relay system is designed so that the separation between the lens is the sum of their focal lengths; this system has the advantage that the image formed has the

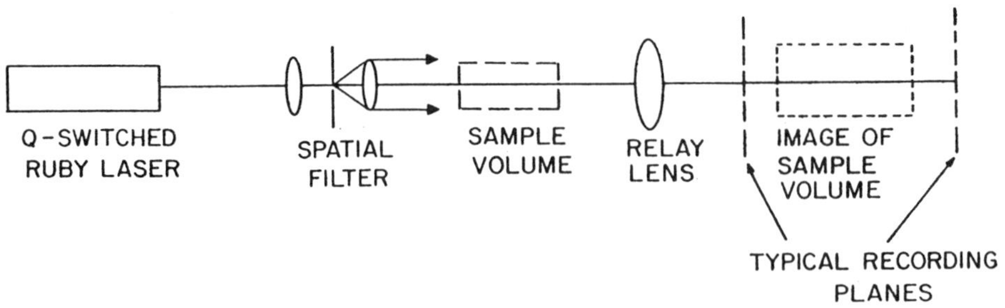

Figure 14 Typical holographic recording system with single-lens relay optics.

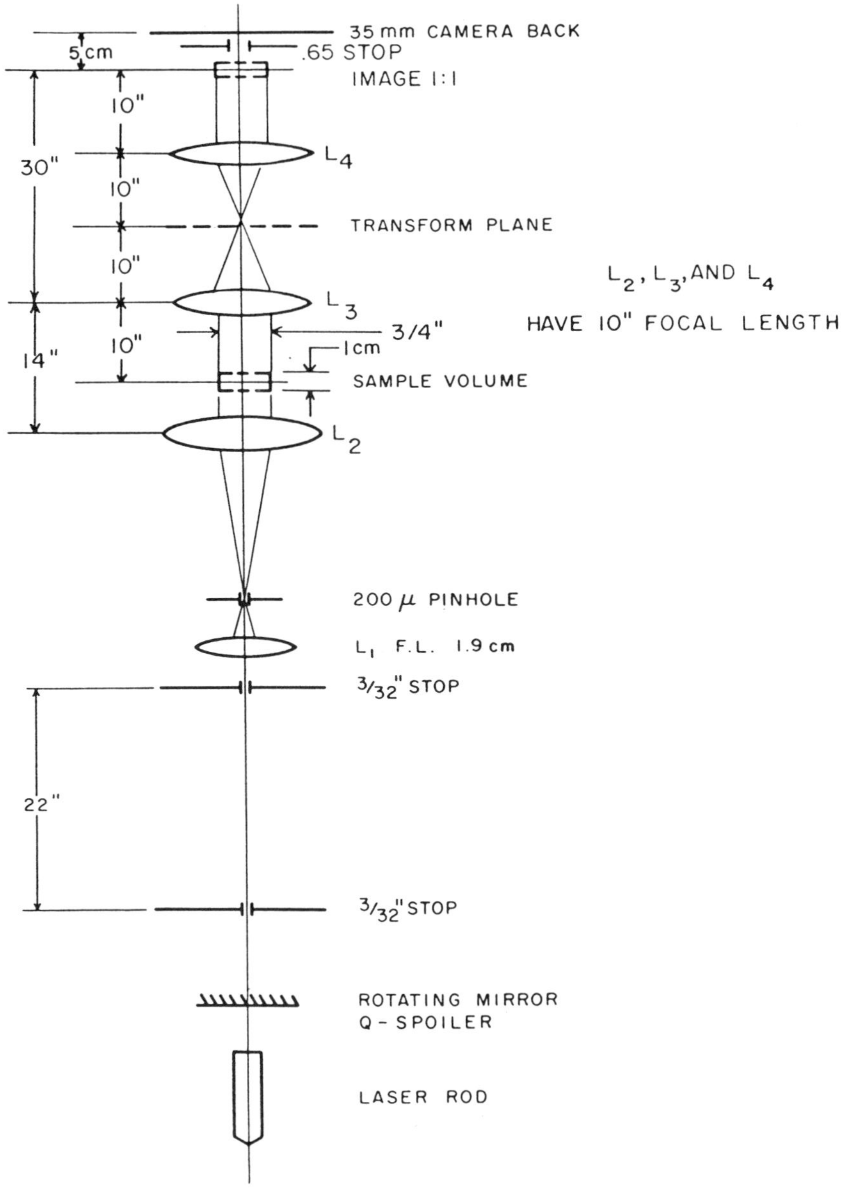

Figure 15 Holographic recording system with two-lens relay optics and spatial frequency filtering.

same magnification independent of the position of the object. Secondly, this system has an intermediate plane in which the transform of the input can be accessed if required. Note, in this plane the individual Fraunhofer diffraction pattern associated with each particle all are formed on the optical axis and overlap together with the focused undiffracted beam. Spatial frequency filtering can be placed in the plane to change, for example, the ratio of the background light to the diffracted light. Another example of the types of system that have been built is shown in Fig. 16. These systems were built by Trolinger (see Ref. 36) as airborne systems. In the upper

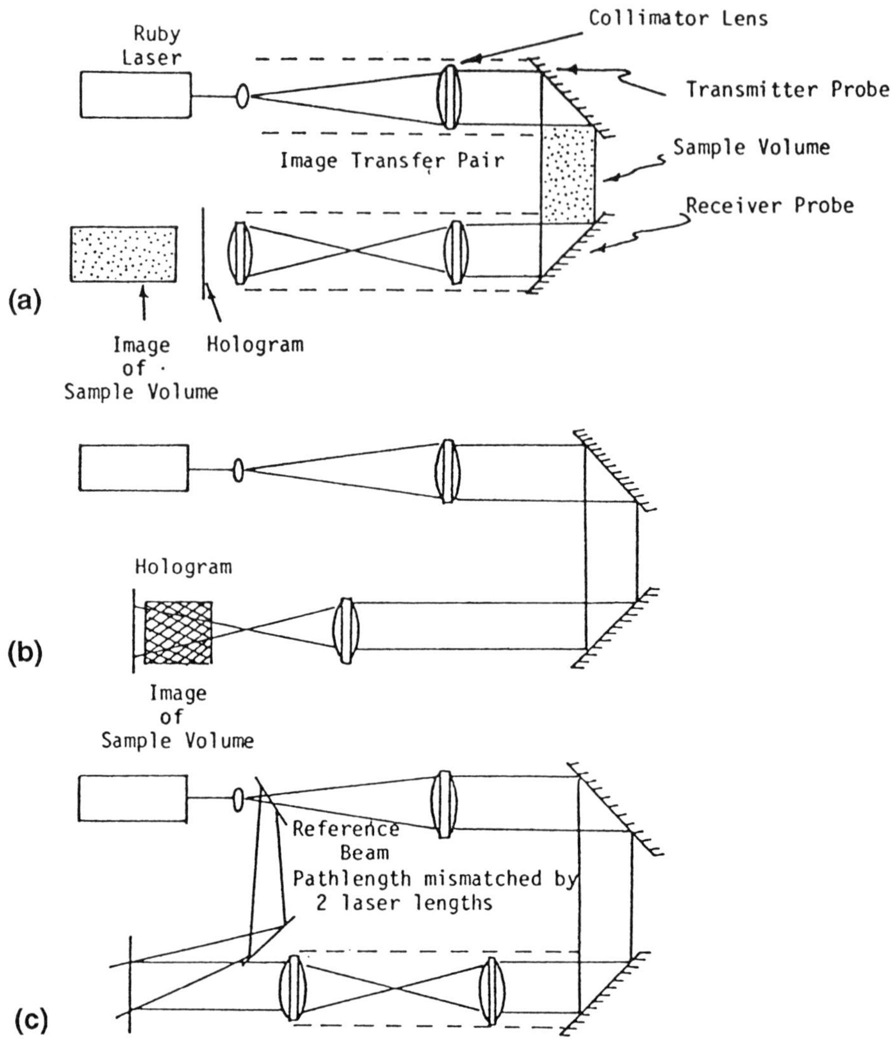

Figure 16 Some airborne holographic recording systems; (a) in-line two-lens relay; (b) in-line single-lens relay; and (c) off-axis configuration. (From Ref. 36.)

part of this figure an in-line system is shown using a two-lens relay optical system. Note that because of the application and the installation configuration this system is folded. The second system has a single relay lens and the final system is an off-axis configuration. These systems were used in a radome of a WB57 weather reconnaissance aircraft using sample volumes of 0.3 l and an optical resolution of about 20 microns [36]. Cloud sampling produced images of droplets and ice crystals. Other installations were in a Cessna aircraft.

B. Imaging System

The hologram provides a permanent record of the particles in the sample volume and may in fact need no calibration. A stationary image can then be formed from the hologram and studied in detail. Figure 17 shows the type of system used. A continuous wave laser is required and a He-Ne laser is often chosen if the original laser was a ruby laser. Other studies have used multiwatt argon lasers with conventional shutters to obtain the hologram and then the same argon laser can be used to produce the image. Collimated illumination of the hologram is used and the holographic record is held on a traveling stage that has motion along the axis as well as in the plane. The reconstructed real image is formed and investigated. The usual method is to examine the image volume with a TV camera with auxiliary optics and then to view the image plane by plane on a TV screen. The sample is scanned through by moving the hologram. The cross-sectional size, shape and *x, y, z* location of each particle in the sample volume can be recorded. Figure 18 shows one of these types of systems.

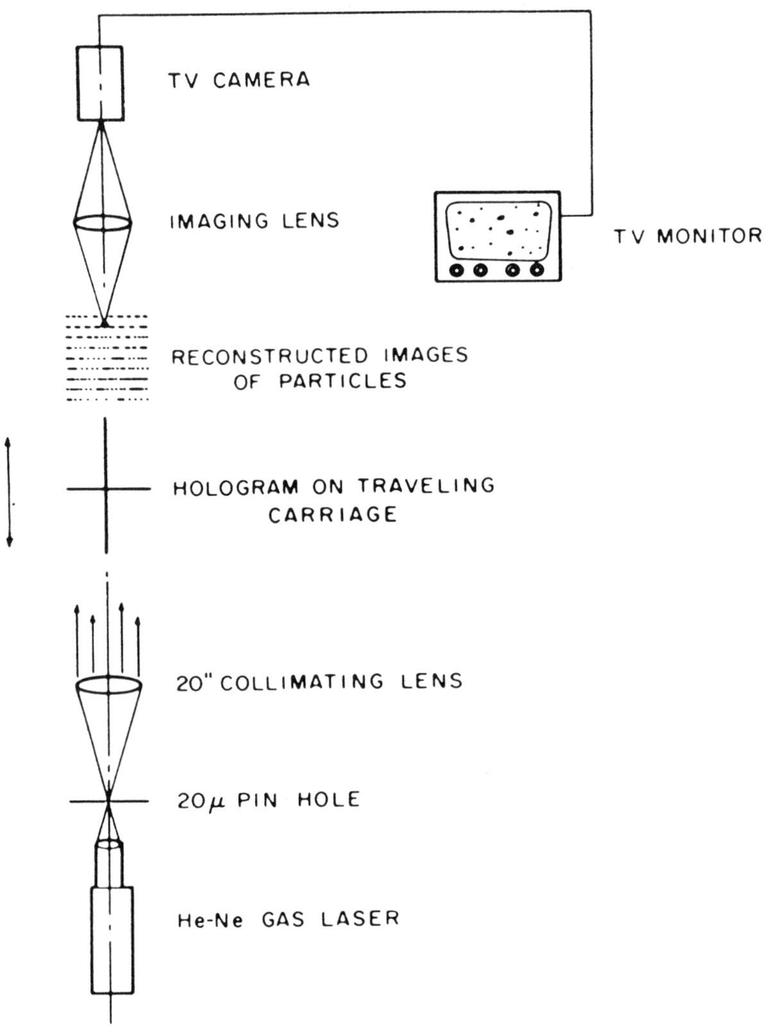

Figure 17 Typical system for forming an image from the hologram.

Figure 18 Photograph of a system for "reconstructing" the individual particle images from the hologram.

IX. SOME ILLUSTRATIVE EXAMPLES

The list of applications of holography to particle size measurement is quite lengthy and it would not be particularly appropriate to go through all these here. Some results have already been shown in the early figures. For example, the original records when shadowgraphs were attempted are the earliest results. Figure 1a shows a far-field in-line hologram of a wire but the large bead in the middle is a very near field in-line hologram. The other illustration in Fig. 1 are all far-field holograms. Figure 4 shows the first hologram of a particle field taken deliberately. This illustration contains mainly far-field holograms but some of the larger objects have produced near-field holograms (see, for example, the particle of diameter 746 microns at the upper left hand side of this holographic record). The hologram (far-field) of a wire is shown in Fig. 6b and a circular cross-sectional particle in Fig. 7. Figure 9 again illustrates both near and far-field holograms of water droplets, and Fig. 12 shows individual holograms at a wide variety of distances. Holograms of submicron, and hence unresolvable, particles are illustrated in Fig. 13a.

Images from some of these holograms are shown in Figs. 5, 6d, 8a,b, and 13b.

As a further interesting example of a hologram of a dynamic sample of water droplets in air, the illustration in Fig. 19 is one of the many frames from the original system used for fog droplet measurements—the problem that got all this started. The sample depth in the system was seven centimeters and the cross section was 1 cm^2. Obviously, there is a range in position of these particles as can be seen from the individual holograms. There is also a significant range in size; some droplets are obviously in the far-field but there are some large drops very much in the near field. The largest drop in the near field is seen in the middle at the bottom of the frame. Histograms of particle size distributions were obtained from these holograms with the smallest block centered at 5 microns. Since the positional information about each droplet in the same volume is known, a check can be carried out to see if there is any biasing in the sampling of the volume. Size distribution can be obtained by layers through the volume and then statistical tests made to make sure that they all come from the same sample. Interestingly, the total water content

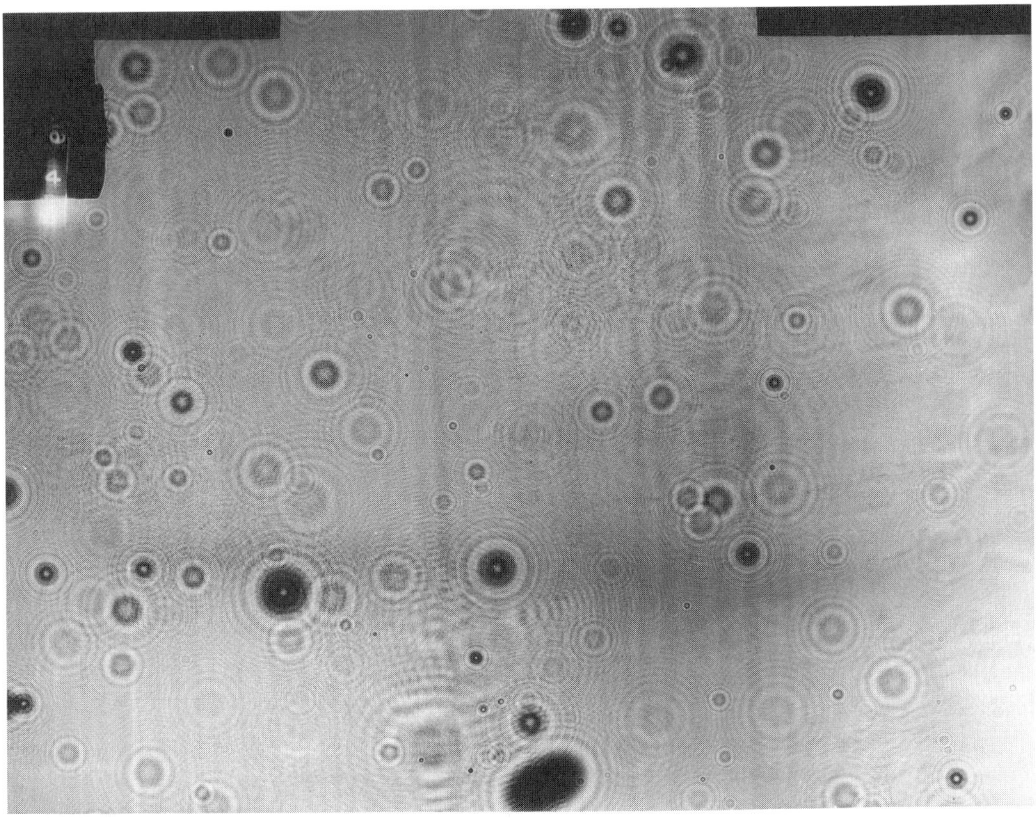

Figure 19 Hologram of a sample of water droplets. (From Ref. 37.)

of the fog can be obtained by integration under the distribution curve. Measurements made this way agreed well with conventional measurements made at the same time.

Figure 20 shows an example of results obtained for both roughly spherical objects (a) and irregular shaped particulate matter in (b). The roughly spherical particles are pollen particles, and a number are in focus in this image formed from the hologram. Some significant clumping can be seen but the individual pollen grains can be easily identified in the image. These grains are 20–30 microns in diameter so good resolution is obtained. The irregularly shaped particles are carborundum with the small pieces being about 20 microns in size. These irregularly shaped particles have good edge definition. It is worth noting that small spherical droplets are the most difficult to get with good resolution because of the low contrast of the holographic fringes caused by the envelope function (see the earlier discussion on fringe contrast and also the more detailed discussion referenced earlier [20,21]). Nevertheless, resolution is down at the very few micron range in these examples.

As discussed earlier, the limit of resolution is the wavelength of the light used to illuminate the sample volume. In practice this means about 1 micron for holographic systems using visible light. It is important to note that with careful technique this result can be achieved. Figure 21 shows in the upper part of the photograph part of a hologram of a large number of 20-micron ragweed pollen particles; each of these individual holograms is in the far-field. There is, of course, considerable overlap of the individual holograms but this is of no concern. An image

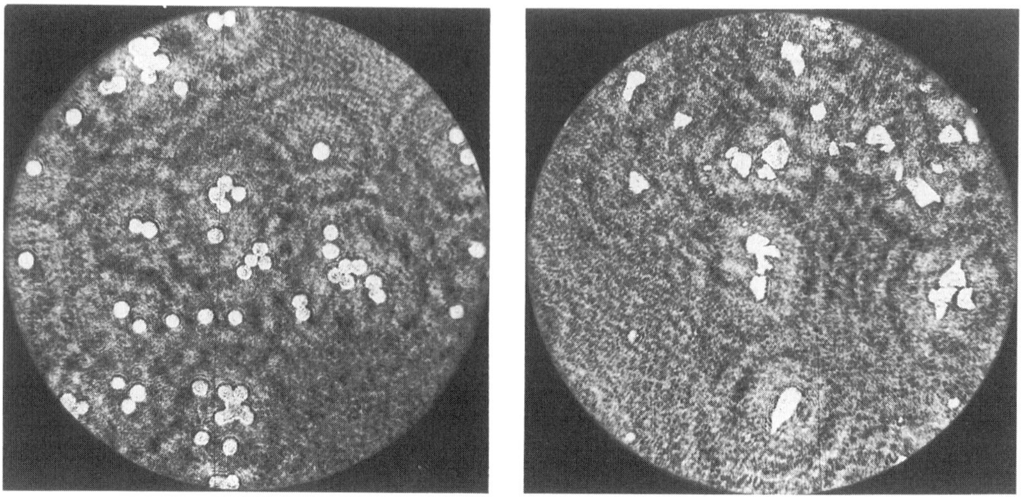

(a) (b)

Figure 20 Typical images of particles formed from an in-line far-field hologram; (a) image of pollen particles of approximately circular cross section; (b) image of irregular shaped carborundum particles. (From Ref. 17.)

from a few of these individual holograms is shown in b, and shows two clumps of particles, one of five particles stuck together and the other a group of 8. The resolution is excellent (see also Fig. 8b). These pollen particles have a fairly detailed structure as can be seen in the electron microscope photograph in Fig. 21c. The surface has small spicules protruding from it which causes them to clump together. This surface structure is just at the resolution limit of the holographic system but they can be seen in Fig. 21b and particularly in the individual image of Fig. 8b.

Holographic particle and droplet measurement methods make a very good verification technique for methods of generating controlled droplet size distribution. Rotary atomizers, for example, provide a large quantity of droplets of a vary narrow size distribution. These devices often consist of a spinning plate or cup onto which a constant flow of liquid is fed; this liquid is thrown to the edge and then off the edge as ligaments and droplets form from the end of the ligaments. Figure 22 shows one of these results (Dunn and Walls [38]). This image is formed from an in-line hologram, the ligaments are clearly seen as are the droplets that are produced at the end of the ligaments. These droplets are flung off. Interestingly, the ligament at the left hand side shows a droplet being formed at the end and two droplets that have already been thrown off. Satellite droplets can be seen behind the droplet that last left the ligament. There are two smaller satellite droplets that are visible. The plate was spinning at 8000 rpm in this example and the main droplet size is 50 microns in diameter. Dunn and Walls [39] used these methods to examine agricultural spray techniques.

Another example comes from a study of ampule contaminants (Crane et al. [40]). A topic of considerable interest in the pharmaceutical industry. In this study a method of verification of a more indirect method was sought. The drug to be tested is contained within cylindrical ampules

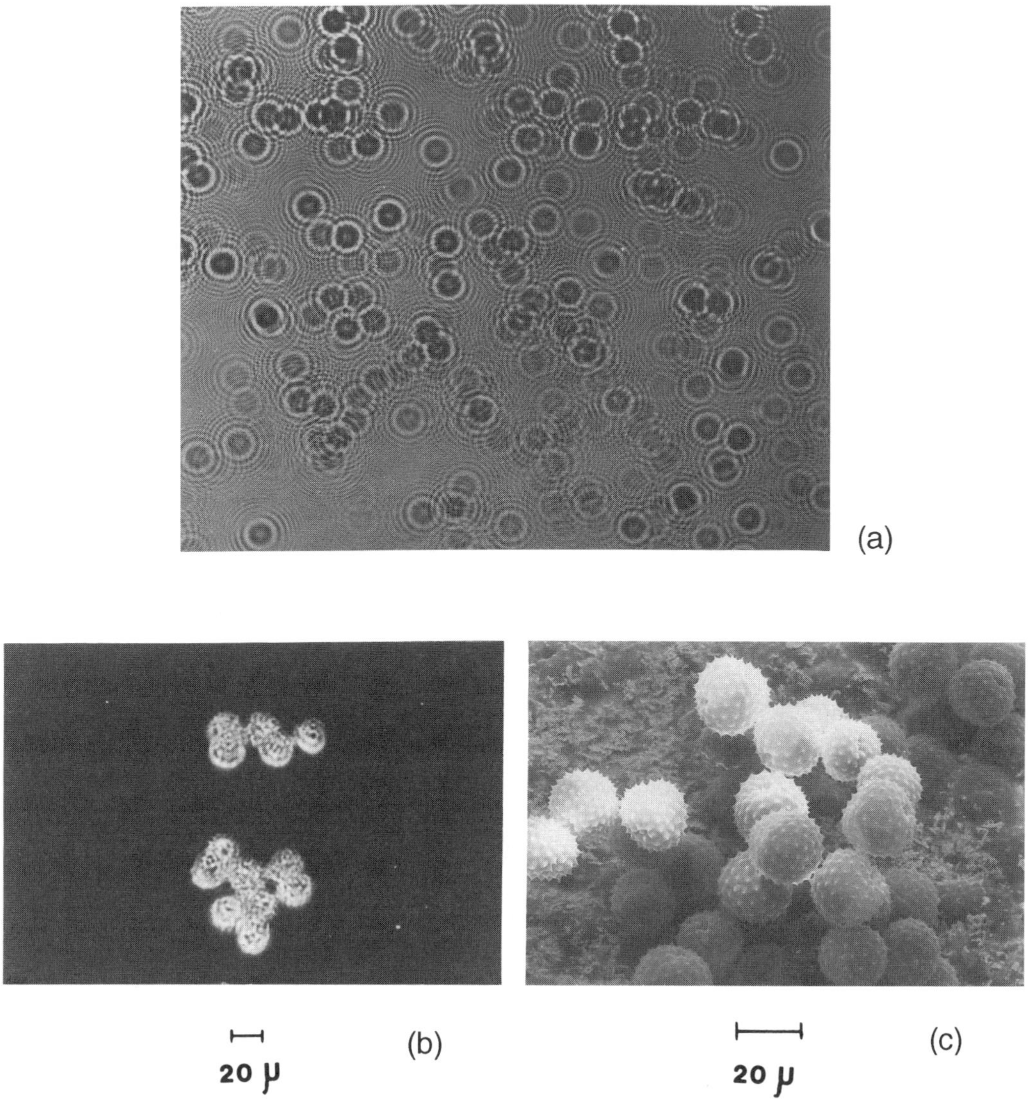

Figure 21 Far-field holograph with high resolution; (a) hologram of a large number of ragweed pollen particles; (b) image of a few of these pollen particles formed from a; (c) an electron microscope photograph of ragweed pollen particles. (From Ref. 5.)

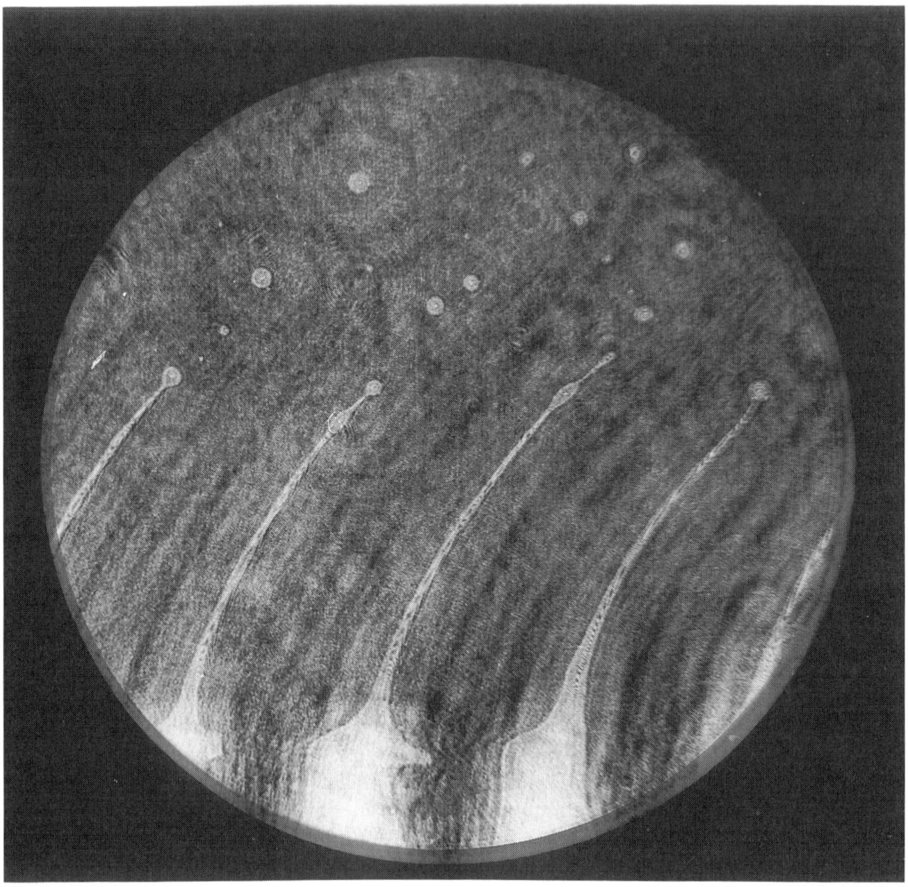

Figure 22 Image formed from an in-line holographic study of spray produced by a rotary atomizer. Complete ligaments are formed and can be seen breaking up into 50-micron-diameter primary droplets with smaller satellites. (From Ref. 5.)

which are sealed. The walls of the ampule and the fact that the ampule is cylindrical can produce significant distortion, hence, the ampule is immersed in an index matching bath before the contents are illuminated and the hologram produced. In addition, it is necessary to spin the ampules at the start of the test to make sure that any contaminants are in suspension. Figure 23 shows the image produced from these holograms; they are of individual contaminants. Figure 23a shows a ringlike contaminant that turns out to be particularly interesting. After the drug has been placed in the ampule the neck is sealed by heating it. Thus, if any of the liquid has splashed into this region during filling the liquid, it would be heated sufficient to cause evaporation of the liquid in any drop that has adhered to the side of the ampule. Upon evaporation a small ring of solid material that was dissolved in the liquid is left behind.

The discussion so far has been devoted to particulates that were generally opaque or where the light transmitted by the particle is relatively small. Water droplets are, of course, transparent and there is a small component of transmitted light that can be taken in to account (see Prikryl and Vest [41] Lu and Meng [42]). It should be noted that the hologram records the field

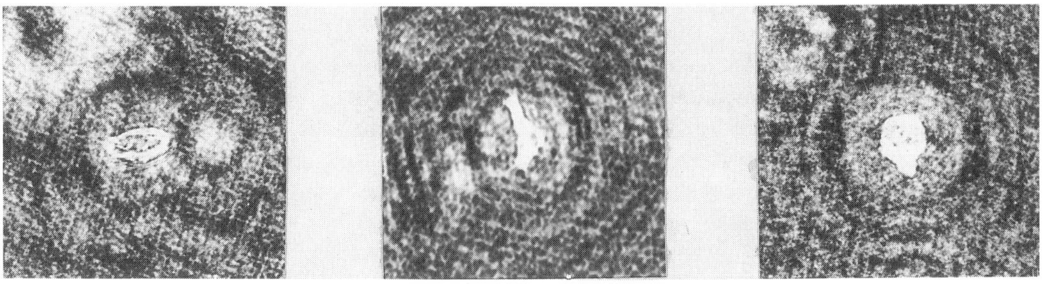

(a) (b) (c)

Figure 23 Images of contaminants contained in sealed ampules of drugs. (From Ref. 17.)

diffracted by the particle as well as any transmitted field. Thus, in general the actual optical field, in amplitude and phase, that occurs right after the particle is reconstructed from the hologram. The intensity is then recorded. Lomas [43] showed that when an image of a transparent tapered fiber was formed from the hologram, a set of fringes were observed that provide information about the taper. The effects observed in particles which are essentially flat transparent plates have been discussed by Cartwright et al. [19].

Figure 24 shows the image formed from a hologram of a field of 400-micron-diameter microballoons. These are thin-walled glass hollow spheres. Excellent images are formed but some effects of the transmitted component are apparent.

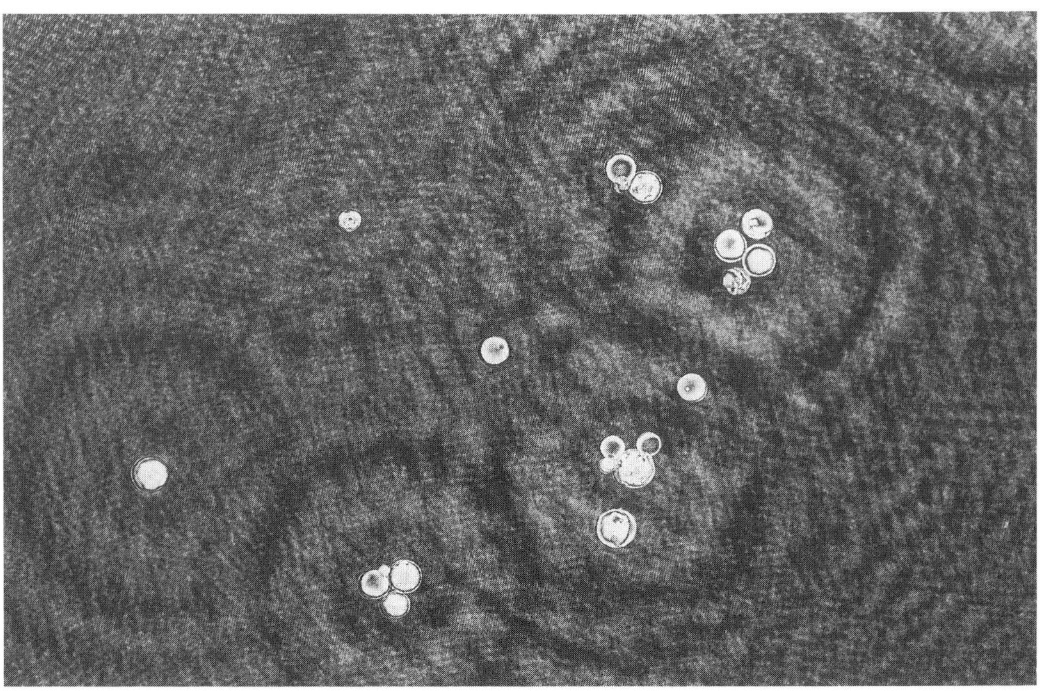

Figure 24 Images of 400-micron microballoons formed from an in-line hologram.

X. AUTOMATED READOUT

The measurement of a particle size distribution from the image formed from a hologram can be quite tedious even with semiautomated output techniques. Hence, it is not surprising that there have been a number of important approaches to fully automating the analysis. The methods generally fall into two categories, the first automating the measurement of the real images produced from the hologram, and the second to interrogate the hologram itself directly.

A. Image Analysis

A number of image analysis methods have been applied with some considerable success (see, e.g., Payne et al. [44]). This and other related methods take the output from the image plane as recorded with a TV or video camera, digitize the output and conduct a computer analysis so that image processing can be accomplished (Bexton et al. [45], Witherow [46], Haussmann and Lauterborn [47]).

B. Hologram Analysis

The second approach to automation of the output is to operate on the hologram directly. One method is to digitally decode the hologram and with a computer calculate the required parameters of the particle sample (see Onural and Scott [48]).

A different approach is to illuminate the hologram with a collimated beam of coherent light and then to form its optical Fourier transform by means of a lens. The required transform is formed in the focal plane of the lens and each particle forms its own transform (Fraunhofer diffraction pattern). Thus, this is equivalent to forming the Fourier transform (composite diffraction pattern) of the original sample volume. Since there are a number of systems that produce particle size distributions on line by this technique, they can clearly be adapted to the holographic record. The advantage is that the method is automated and a permanent record is still stored in the hologram for further study. This idea of using this so-called total diffraction method with the in-line hologram as the input was pioneered by Ewan [49,50]. Other workers have successfully applied this method for the measurement of size distribution (see, e.g., Hess and Trolinger [51]).

XI. VELOCIMETRY

The application of holographic methods to the study of the velocity of particles in a three-dimensional sample was recognized quite early by Menzel et al. [52] and discussed in more detail in a subsequent publication (Menzel and Shofner [53]).

There are several possible techniques. The first is to have the exposure time for recording the hologram to be longer than that needed to freeze the motion of the particle. The hologram will then be "smeared" slightly and, hence, so will the image. Dotson [54] used the change in edge definition to measure the distance moved and, hence, determine the velocity. A less sophisticated method is to increase the smear and then measure the length of the smear (see, e.g., Crane et al. [55]). If the particle is of known size, such as in the case in seeded flow, then it is possible to carry out a deconvolution to determine how far the particle has moved.

A better method is to record two separate holograms, separated in time, but on the same recording material (i.e., same photographic film or plate) so that the registration is automatic. A typical double-exposure hologram is shown in Fig. 25; since the holograms are recorded at different times, the two holograms do not interfere with each other. If the time interval between exposures is known, then the change in position of individual particles in the field can be converted to an individual velocity. There is, of course, an ambiguity as to which individual

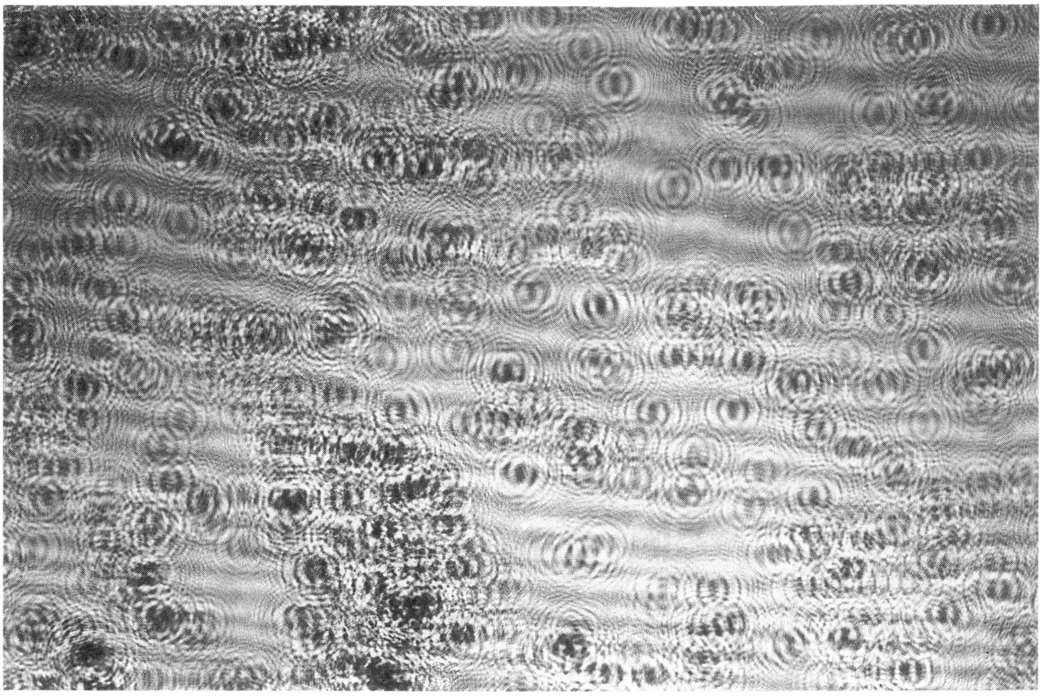

Figure 25 A doubly exposed hologram of field of moving particles. (From Ref. 55.)

hologram is on which exposure, this results in a 180° ambiguity in the direction of motion. This method has been used by Fourney et al. [56] and by Royer [57], who applied the method in an interesting way to submicron particle velocities. If the holograms are recorded on separate films then the 180° ambiguity is removed, but the registration problem has to be faced.

More than two holograms recorded on the same film can produce variable results and, of course, multiple particle images (see, e.g., Crane et al. [55]). If the time interval between exposures is varied then the 180° ambiguity discussed above can be avoided.

The direct method of analysis is simply to form an image of the particle fields from the multiple exposure hologram and then measure the distance moved for each particle—both the in-plane components and the out-of-plane components. The resolution of the in-plane component is a great deal better than in the out-of-plane component. This suggests that the two exposures should be obtained in two different directions simultaneously. That is, each record will be of a double-exposure hologram. If they are at right angles, for example, then the resolution is the same in each direction. The registration problem again must be solved.

This direct technique is obviously quite tedious and a more rapid and automated method is required such as the Fourier method developed by Ewan (1979). In Ewan's [58] method the Fourier transform of the double-exposure hologram is formed in the same way as discussed earlier for the single-exposure hologram. For explanation, we consider the effect of a single-velocity component. Thus, the double-exposure hologram represents two recordings of the same particle field physically displaced. The transform will then show cosine-squared fringes in intensity, the argument of that cosine term is kbv/f, where $2b$ is the separation of the two samples of particles and v is the coordinate in the Fournier plane. Figure 26a shows such a result (Crane et al. [55]). This ideal single case produces a set of fringes of a single frequency and, hence, a unique spacing. When other velocity components are present there will be other sets of fringe

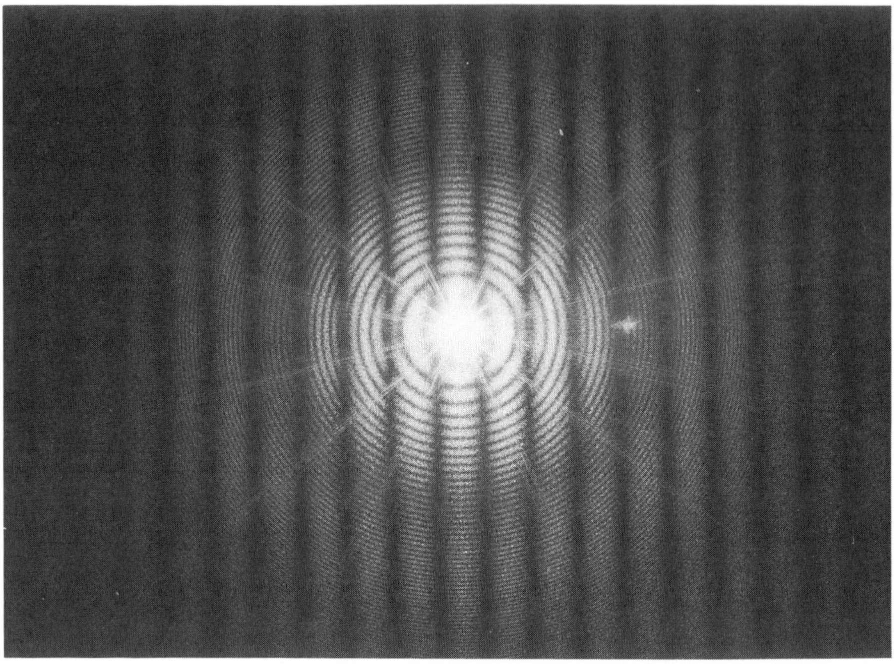

(a)

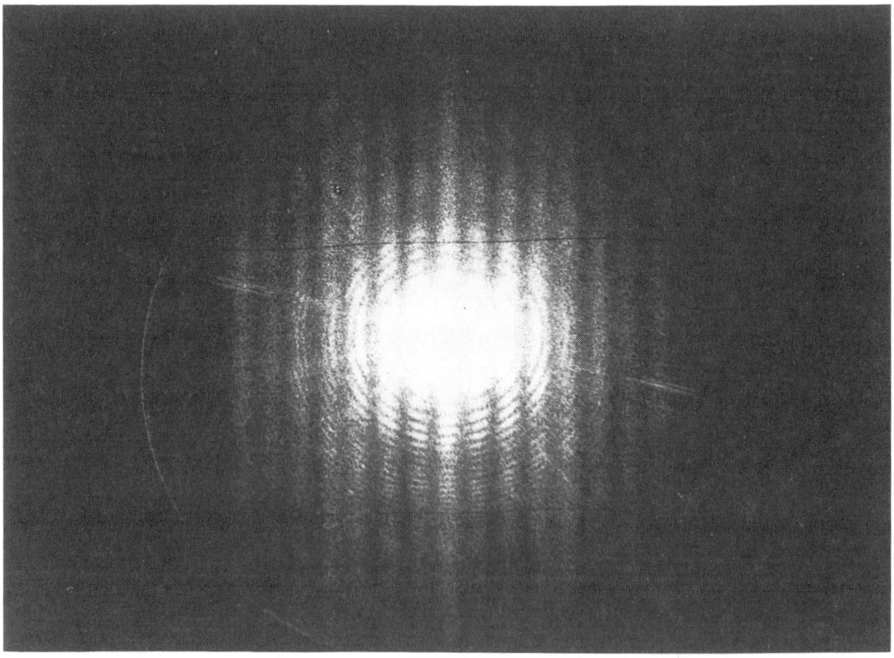

(b)

Figure 26 Fourier transform of doubly exposed holograms of moving particle fields; (a) single velocity components; (b) distribution of velocity components all in the same direction. (From Refs. 55 and 59.)

with different spacing and in different orientations. The sum of all these fringes is the required pattern that is formed in the Fourier plane of the transform lens. This composite pattern can be detected and the original velocity distribution determined. Figure 26b shows the pattern formed from a double-exposure hologram when there is a velocity distribution, but the motion is all in the same direction as in a flow field (Malyak and Thompson [59]).

The steps necessary for the process of data acquisition and processing are reported in some detail by Malyak [59], and this process does need to be followed for accurate results. The first step is to record the required double-exposure hologram and know the time interval between the exposures. The second step is to produce the Fourier transform of this record. Figure 27 shows a test of this process that details the necessary date analysis steps that need to be taken. The object used in this example and test is a random array of 40 μ diameter opaque disks such that

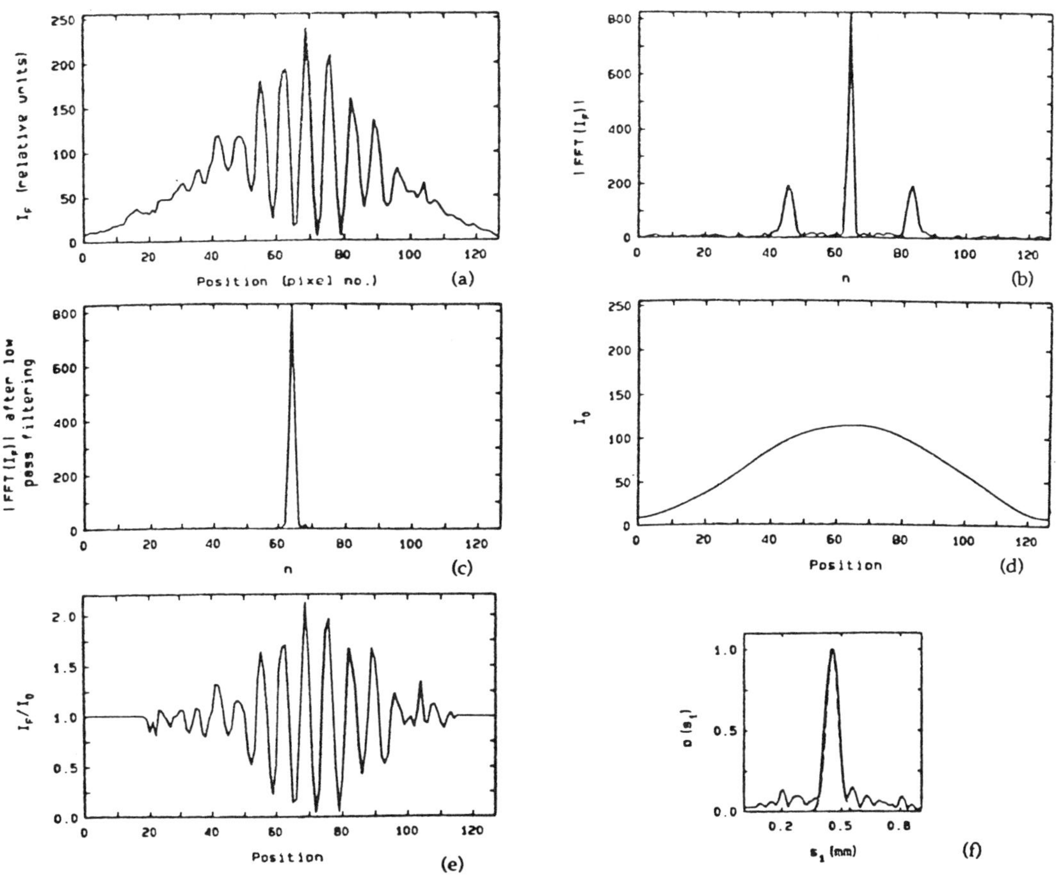

Figure 27 Particle velocity distribution measurement. (a) Plot of the intensity profile in the transform plane of a far-field hologram of a random array of opaque disks. The separation between disk pairs is characterized by a Gaussian probability density function (pdf); $\mu = 460$ μm and $\sigma = 30$ μm. (b) Magnitude of the FFT of the intensity profile vs. the discrete transform variable η. (c) Plot of $|FFT(I_F)|$ vs. η after low-pass filtering. (d) The envelope function I_O is obtained by computing the inverse FFT of the data in (c). (e) Plot of I_F/I_O vs. position; random fluctuations at the left and right portions of the curve were removed by setting I_F/I_O to unity. (f) Plot of the computed (—) density function and the theoretical (—) density function vs. displacement. (From Ref. 59.)

the separation between the disk pairs is characterized by a Gaussian probability density function centered at a displacement parameter of 460 μm and with a half-width of 30 μm. There are 5000 disk pairs in this data set. In Figure 27a the intensity profile that is recorded is displayed and the fringe pattern is readily apparent but are not cosine fringes since there is a distribution of displacements present. Once this recording has been detected a fast Fourier transform (FFT) of this profile is calculated and is displayed in Figure 27b, the peaks are located at the mean displacement and the distribution at those locations is related to the probability distribution. This calculated FFT is then low-pass filtered to obtain the distribution of Fig. 27c and its inverse FFT is shown in Fig. 27d, which is the required envelope function for processing the record in Fig. 27a. This envelope function is then removed to produce the curve I_F/I_O shown in Fig. 27e. In addition, the outer portions of this curve have been set to the mean value (unity in this case) to reduce noise. Finally, this record is retransformed to produce the probability density function that is shown here in Fig. 27f together with the known input function. This method has been tested in a variety of circumstances including particulates falling in water.

XII. CONCLUSIONS

This summary discussion of the holographic methods of particles and droplet diagnostics is only an introduction to a quite sizable literature that continues to grow month by month. It has not been possible to acknowledge the important contributions of many of the workers in the field and for that I apologize, but do refer the reader again to the volume of selected reprints (Vikram [3]) and also to the bibliography that it contains. Suffice it to say, in summary, that holography is a valuable technique in particle diagnostics for size, shape, position and velocity measurements.

Finally, a word about nomenclature. In the early days (the 1960s) the term "Fraunhofer hologram" was used to distinguish this particular type of hologram that gives excellent results. The method works because of the special character of being in the Fraunhofer region of the diffraction by the particle as opposed to the Fresnel region. The special properties of being in the Fraunhofer region that have been discussed in this chapter had not been recognized in earlier studies of holography (i.e., pre-1963). The condition for being in the Fraunhofer region is the so-called far-field or far-zone condition, hence, the more recent and more descriptive terminology of far-field holography. This later term seems to be the one most used today, and systems are described as in-line far-field or off-axis far-field (near-field, i.e., Fresnel diffraction field can be similarly described). It should be noted that this terminology refers to the near-field or far-field of the particle cross section not of the total field containing the particles. A few workers have described these types of holograms as Gabor holograms since the original holographic systems were in-line. Gabor, of course, did not distinguish the Fresnel and Fraunhofer regions.

ACKNOWLEDGMENT

In writing this article I do acknowledge the many contributions of my own co-workers as well as colleagues around the world—there are well over 100 authors represented in the Milestone volume [3]! As one who has spent a long time, on and off, in this field, I congratulate them on their contributions and insights.

REFERENCES

1. Thompson, B. J., Selected topics in holographic particle diagnostics, Workshop on Particle Image Velocimetry, Heriot-Watt University, Edinburgh, Scotland, 1991.
2. Vikram, C. S., *Particle Field Holography,* Cambridge University Press, 1992.

3. Vikram, C. S., *Selected Papers on Holographic Particle Diagnostics* (B. J. Thompson, ed.), SPIE Milestone Series, vol. 21, 1990.
4. Thompson, B. J., Holographic particle sizing techniques—current status, *J. Phys. E.: Sci. Instrum. 7*(10):781–788 (1974).
5. Thompson, B. J., and Dunn, P., Advances in far-field holography; theory and applications, *Recent Advances in Holography,* Proc. SPIE 215, 1980, pp. 102–111.
6. Crane, J., Dunn, P., Malyak, P. H., and Thompson, B. J., Particle velocity and size measurement using holographic and optical processing methods, Proc. 15th Int. Cong. on High Speed Photography and Photonics, Proc. SPIE, 348, 1983, pp. 634–642.
7. Trolinger, J. D., Particle and flow field holography—a critical survey, in *Holography* (L. Huff, ed.), Proc. SPIE 532, 1985, pp. 40–62.
8. Lee, Y. J., and Kim, J. H., A review of holography applications in multi-phase flow visualization studies, *J. Fluids Eng. 108*(3):279–288 (1986).
9. Thompson, B. J., Holographic methods for particle size and velocity measurements, *Proc. SPIE 1136*:308–326 (1989).
10. Thompson, B. J., and Ward, J. H., Particle sizing—the first direct use of holography, *Sci. Res. 1*:37–40 (1966).
11. Silverman, B. A., private communication, 1963.
12. Thompson, B. J., and Malyak, P. H., Accuracy of measurement in coherent imaging of particulates in a three-dimensional sample, *Proc. SPIE 573*:12–20 (1985).
13. Parrent, G. B., and Thompson, B. J., On the Fraunhofer (far-field) diffraction patterns of opaque and transparent objects with coherent background, *Opt. Acta 11*(3):183–193 (1964).
14. Thompson, B. J., Diffraction by opaque and transparent particles, *J. SPIE 2*:43–46 (1964–64).
15. Tyler, G. A., and Thompson, B. J., Fraunhofer holography applied to particle size analysis: a reassessment, *Opt. Acta 23*(9):658–700 (1976).
16. Thompson, B. J., Principles and applications of optical holography, *Applied Optics and Optical Engineering,* vol. VI (R. Kingslake and B. J. Thompson, eds.), Academic Press, New York, 1980, pp. 357–391.
17. Thompson, B. J., Holographic methods of dynamic particulate measurements—current status, *Proc. SPIE 348*:626–633 (1982).
18. Thompson, B. J., Fraunhofer holograms, *Handbook of Optical Holography* (H. J. Caulfield, ed.), Academic Press, 1979, pp. 157–163.
19. Cartwright, S. L., Dunn, P. and Thompson, B. J., Particle sizing using far-field holography: new developments, *Opt. Eng. 19*(5):727–733 (1980).
20. Dunn, P., and Thompson, B. J., Object shape, fringe visibility and resolution in far-field holography, *Opt. Eng. 21*(2):327–332 (1982).
21. Özkul, C., Far-field diffraction patterns of an opaque object with rectangular cross-section, *Opt. Eng. 23*(6):821 (1984).
22. Molesini, G., Bertani, D., and Cetica, M., In-line holography with interference filters as fourier processors, *Opt. Acta 29*(4):479–484 (1982).
23. Özkul, C., Allano, D., and Tinité, M., Filtering effects in far-field in-line holography, *Opt. Eng. 25*(10):1142–1148 (1986).
24. Vikram, C. S., and Billet, M. L., Gaussian beam effects in far-field in-line holography, *Appl. Opt. 22*(18):2830–2835 (1983).
25. Brooks, R. E., Heflinger, L. O., Wuerker, R. F., and Briones, R. A., Holographic photography of high speed phenomena with conventional Q-switched ruby lasers, *Appl. Phys. Lett. 7*(4):92–94.
26. Wuerker, R. F., Heflinger, L. O., and Zivi, S. M., Holographic interferometry and pulsed laser photography, *Proc. SPIE 15*:97–104 (1968).
27. Trolinger, J. D., Betz, R. A., and Farmer, W. M., Holographic techniques for the study of dynamic particles, *Appl. Opt. 8*(5):957–961 (1969).
28. Trolinger, J. D., Lal, R. B., Batra, A. K., and McIntosh, D., Particle image velocimetry experiments for the IML-I space flight, *Proc. SPIE 1557*:98–109 (1991).

29. Thompson, B. J., Ward, J. H., and Zinky, W. R., Application of hologram techniques for particle size analysis, *Appl. Opt. 6*(3):519–526 (1967).
30. Thompson, B. J., and Zinky, W. R., Holographic detection of submicron particles, *Appl. Opt. 7*(12):2426–2428 (1968).
31. Tonomura, A., Fukuhara, A., Watanabe, H., and Kanoda, T., Optical reconstruction of image from Fraunhofer electron hologram, *Jap. J. Appl. Phys. 7*(3):295 (1968).
32. Tonomura, A., Electron holography, *J. Inst. Electron. Inform. Comm. Eng. 73*(10):1075–1076 (1990).
33. Howells, M. R., Iarocci, M. A., and Kirz, J., Experiments in x-ray holographic microscopy using synchrotron radiation, *J. Opt. Soc. Am. A* 3(12):2171–2178 (1986).
34. Jacobsen, C., Howells, M., Kirz, J. and Rothman, S., X-ray holographic microscopy using photoresists, *J. Opt. Soc. Am. A 7*(10):1847–1861 (1990).
35. Knox, C., Holographic microscopy as a technique for recording dynamic microscopic subjects, *Science 153*(3739):989–990 (1966).
36. Trolinger, J. D., Particle field holography, *Opt. Eng. 14*(5):383–392 (1975).
37. Thompson, B. J., Fraunhofer holography, *Proc. SPIE 15*:25–39 (1968).
38. Dunn, P., and Walls, J. M., An introduction to in-line holography and its applications, Symposium Controlled Droplet Application, Reading University, 1978, pp. 23–34.
39. Dunn, P., and Walls, J. M., Droplet size distribution from an agricultural spray using in-line holography, *Particle-Size Analysis* (M. J. Grove, ed.), Heydon, London, 1977, pp. 196–201.
40. Crane, J. S., Dunn, P., Thompson, B. J., Knapp, J. Z., and Zeiss, J. C., Far-field holography of ampule contaminants, *Appl. Opt. 21*(14):2548–2553 (1982).
41. Prikryl, I., and Vest, C. M., Holographic imaging of semi-transparent droplets or particles, *Appl. Opt. 21*(14):2541–2547 (1982).
42. Lu, Z. K., and Meng, H., Holography of semi-transparent spherical droplets, *Proc. SPIE 599*:19–25 (1986).
43. Lomas, G. M., Interference phenomena in Fraunhofer holograms and in their reconstruction of tapered glass fibers, *Appl. Opt. 8*(10):2037–2041 (1969).
44. Payne, P. R., Carder, K. L., and Steward, R. G., Image analysis techniques for holograms of dynamic oceanic particles, *Appl. Opt. 23*(2):204–210 (1986).
45. Bexon, R., Gibbs, J., and Bishop, G. D., Automatic assessment of aerosol holograms, *J. Aerosol Sci. 7*(5):397–407 (1976).
46. Witherow, W. K., A high resolution holographic particle sizing system, *Opt. Eng. 18*(3):249–255 (1979).
47. Haussmann, G., and Lauterborn, W., Determination of size and position of fast moving gas bubbles in liquids by digital 3-D image processing of hologram reconstructions, *Appl. Opt. 19*(20):3529–3535 (1980).
48. Onural, L., and Scott, P. D., Digital decoding of in-line holograms, *Opt. Eng. 26*(11):1124–1132 (1987).
49. Ewan, B. C. R., Fraunhofer plane analysis of particle field holograms, *Appl. Opt. 19*(8):1368–1372 (1980).
50. Ewan, B. C. R., Swithenbank, J., and Sorusbay, C., Measurement of transient spray size distributions, *Opt. Eng. 25*(5):620–625 (1984).
51. Hess, C. F., and Trolinger, J. D., Particle field holography data reduction by Fourier transform analysis, *Opt. Eng. 24*(3):470–474 (1985).
52. Menzel, R., Russell, T. G., and Shofner, F. M., Recording fluid velocity fields holographically, *Proc. SPIE 15*:167–170 (1968).
53. Menzel, R., and Shofner, F. M., An investigation of Fraunhofer holography for velocimetry applications, *Appl. Opt. 9*(9):2073–2079 (1970).
54. Dotson, W. P., A theoretical calculation of edge smear in far-field holography, *Proc. Symp. Engineering Applications of Holography* (R. Aprahamian, ed.), SPIE, 1972, pp. 113–122.

55. Crane, J., Dunn, P., Malyak, P. H., and Thompson, B. J., Particle velocity and size measurements using holographic and optical processing methods, Proc. 15th International Congress on High Speed Photography and Photonics, Proc. SPIE 348, 1983, pp. 634–642.
56. Fourney, M. E., Matkin, J. H., and Waggoner, A. P., Aerosol size and velocity determination via holography, *Rev. Sci. Instr. 40*(2):205–213 (1969).
57. Royer, H., Holographic velocimetry of submicron particles, *Opt. Comm. 20*(1):73–75 (1977).
58. Ewan, B. C. R., Holographic particle velocity measurement in the Fraunhofer plane, *Appl. Opt 18*(5):623–626 (1979).
59. Malyak, P. H., and Thompson, B. J., *Proc. SPIE 573*:2–11 (1985); *Opt. Eng. 23*(5):567–576 (1984); see also Malyak, P. H., Ph.D. thesis, University of Rochester, 1985.

8

Instrumental Analysis of Particulate Matter

Thomas A. Barber

Baxter Healthcare Corporation, Round Lake, Illinois

I. INTRODUCTION

Although light-microscopic methods often provide the simplest and most cost-effective way to perform particle identification, light-microscopic analysis of particulate matter has a number of limitations. If particles are to be analyzed by polarized light microscopy, they must have a unique morphology, unique optical properties, or be transparent in order to be identified. Another serious general limitation of this method is that its power is dependent upon the skill of the microscopist. Analysts with this skill are few in number and have become increasingly more difficult to find. In this consideration, the use of microscopy as a stand-alone method of particle identification has decreased, but it remains a valuable component in the broader analytical approach applied in most laboratories. Thus, although the author makes a distinction between "microscopic" and "instrumental" (i.e., nonmicroscopic) means of particle analysis, the analyst must continue to rely on the microscope as an initial stage in a progression of analyses that will involve more complex and expensive instrumentation. For both polarized light microscopy and nonpolarized transmitted light microscopy, magnification is limited to approximately 1000×.

Particle identification is currently more dependent on instrumental methods of analysis than on microscopy. The purpose of this chapter is to discuss the general principles of the most useful instrumental techniques and to provide appropriate references. In particular, the papers by Borchert et al. [1] and Aldrich [2] provide excellent summary overviews of instrumental compositional analysis of particulate matter.

Just as the analytical chemist depends heavily upon X-ray spectrometry, infrared spectroscopy, and mass spectrometry of bulk materials, those interested in the identification of particles frequently apply X-ray analysis in combination with electron microscopy, microinfrared techniques, and micro methods of mass spectrometry. Transmission electron microscopy (TEM) and scanning electron microscopy (SEM) provide powerful means of examining particles at high magnification. Crystallography is useful for particle identification in both X-ray and electron modes. Other more specialized techniques are also available.

Particulate matter differs to some extent from the traditional trace analytical sample in that it is often heterogeneously dispersed in various matrices and requires isolation prior to analysis. The path of analysis most frequently applied in particle identification represents a stepwise progression from low to high magnification and simple to complex techniques [2].

Portions of this chapter were adapted from *Pharmaceutical Particulate Matter* by Thomas Barber, 1993. Reprinted with the kind permission of Interpharm Press

I must emphasize that any and all subjective judgments, opinions, and viewpoints expressed in this chapter are my own and not necessarily those of Baxter Healthcare Corporation or any organization representing manufacturers of medical devices or injectable products.

Cost and time requirements are always key considerations to be carefully compared to the value of the analytical result. As a matter of principle, light microscopy and visual inspection of a sample should invariably be conducted before any instrumental analysis is begun. In fact, light microscopy should be applied as a preliminary step in any endeavor directed at particle identification. Table 1 provides a listing of analytical techniques applied in analyses of particulate matter. The order of the list generally represents a progression from the simpler techniques to the more complex; it is to be expected that many particles of interest will be identified by the initial application of microscopy, and that some few will prove complex in nature so that a comprehensive application of a number of analytical techniques will be required.

A consideration of the information in Table 1 leads the reader to two interesting conclusions. First, although a wide range of analytical techniques is presented, all are based on the simple expedient of how the analyte particle or particles interact with some sort of energizing radiation. This latter may be of low energy, such as infrared radiation or may consist of much higher energy electromagnetic radiation, such as X-rays. Secondly, the modes of analyses listed may be generally categorized into one of three types of methods based on whether they yield atomic (elemental), molecular (structural), or morphologic data. The primary data interpretation applied with each type of information is visual (i.e., relying on the analyst's scrutiny of spectra of different types or of an image) to make a simplistic "same or different" judgment. Irrespective of the complexity of the problem presented by the analysis of unknown particulate matter, the light microscope must remain a trusted and proven ally of the analyst.

Table 1 Techniques for Particle Identification

Technique
Visual inspection
Low-magnification (stereomicroscopic) examination
Compound light-microscopic examination (LM)
Polarized light microscopy (PLM)
• physical properties
• microchemistry
• crystal morphology
• thermal analysis
Scanning electron microscopy (SEM)
• surface morphology
• imaging of ultra fine (< 1 μm) particles
• electron energy loss spectrometry
• electron diffraction
X-ray spectrometry (EDXS, WDXS)
• elemental composition
Infrared microspectrophotometry (FTIR)
• structural information
Mass spectrometry (LAMMA, SIMS)
• mass-specific molecular data
X-ray crystallography
• crystal structure-specific molecular data
Transmission electron microscopy (TEM)
• electron energy loss spectrometry (EELS)
• electron diffraction (crystallography)

II. COLLECTION AND ISOLATION OF PARTICLES

Prior to analysis, particles of interest must often be isolated as single particles or as suitably homogeneous populations for presentation to an analytical instrument. If the particles occur as a homogeneous population, the task is fairly simple and involves bulk manipulations, such as their collection on a filter substrate and subsequent removal using a microspatula or probe. In some cases it may be necessary to physically separate particles of interest from their neighbors using micromanipulation techniques. This discussion assumes that the sample preparation discussed will be carried out in a suitably clean area (typically in a laminar flow hood) that provides class 100, filtered high-efficiency particulate air (HEPA).

Ideally, the analyst will be able to personally collect the particulate matter samples for analysis. If the problem is point-generated particulate matter observed in a specific location in a manufacturing process, the best course is to have the analyst go to the production area to collect the samples. If components of a solution container or device are involved, the parts submitted for analysis should be collected using cleaned forceps, and be placed in containers (bags, bottles, jars) that have been tested for cleanliness. Importantly, polyethylene bags and some other widely used containers with a significant inherent particle burden should not be used. Minimally, telephoned instructions on how to obtain the samples should be given to those collecting the samples. In the author's experience, analytical samples collected by filtration and submitted by another laboratory are liable to be contaminated, and may be suspect on other grounds, such as sample identification. Since a single touch of a finger can deposit thousands of skin cells on a sample, it is essential that particle-free gloves are used when samples are handled. Despite precautions, a particle laboratory expects to receive samples collected with various degrees of care, and with experience an analyst learns to make the best of samples collected under less-than-ideal conditions.

If the particles of interest are suspended in a liquid, filtration is most frequently the method of choice to isolate them. Filters used for particle analysis are usually of two types: (1) cast film filters that are spongiform in structure and are made by casting a polymer film containing volatile solvents and pore-forming liquids on a smooth surface [3]; and (2) track-etched or nucleation track filters produced by etching a thin polycarbonate film that has been exposed to gamma radiation [4]. The latter type are often referred to generically as "Nuclepore" filters, based on the name of the original vendor. The generic term *membrane,* which may be used in reference to either of these types of filters, refers to their thinness and their generally fragile nature. The book by Brock [5] is an invaluable basic text on filtration that discusses the principles of manufacture of microporous filters of all types, as well as filter function and application.

The cast film filters that are most commonly used for particle collection in the laboratory are of a porous, open structure with only a small fraction of the membrane volume occupied by the polymer substrate (Figs. 1a,b). In a representative filter of this type, 80–85% of the membrane is open space (void volume). This openness allows for a relatively high fluid flow rate in comparison to what can be achieved with the track-etched membranes. In the days before SEM, it was thought that filters of this type contained cylindrical pores of defined size. Examination of such filters using scanning electron microscopes reveals a random structure, characterized by a polymer network surrounding spherical or oval voids. Depending on the type of cast film filter considered, the polymer structure surrounding the voids of the filter may be penetrated by various numbers of pores or consist of fibrous strands.

Kesting [3] appropriately described cast film filters as elaborate systems consisting of

> . . . open-celled foams (i.e., vacuoles with breached walls). Fastening the cellular network together [are] long hose and chain-like ribs that spread out in three dimensions.

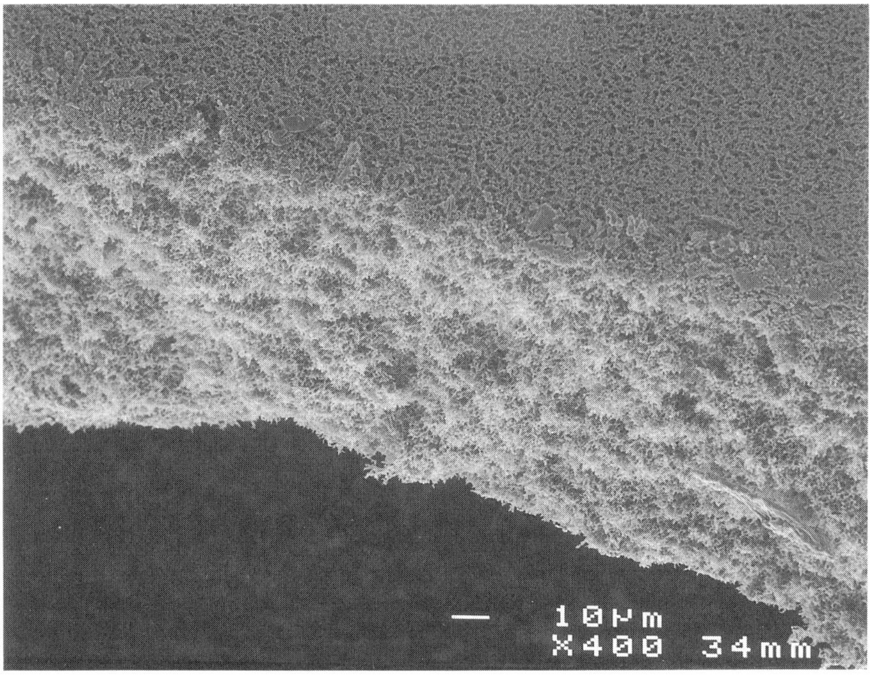

Figure 1a Cast film filter at ×400.

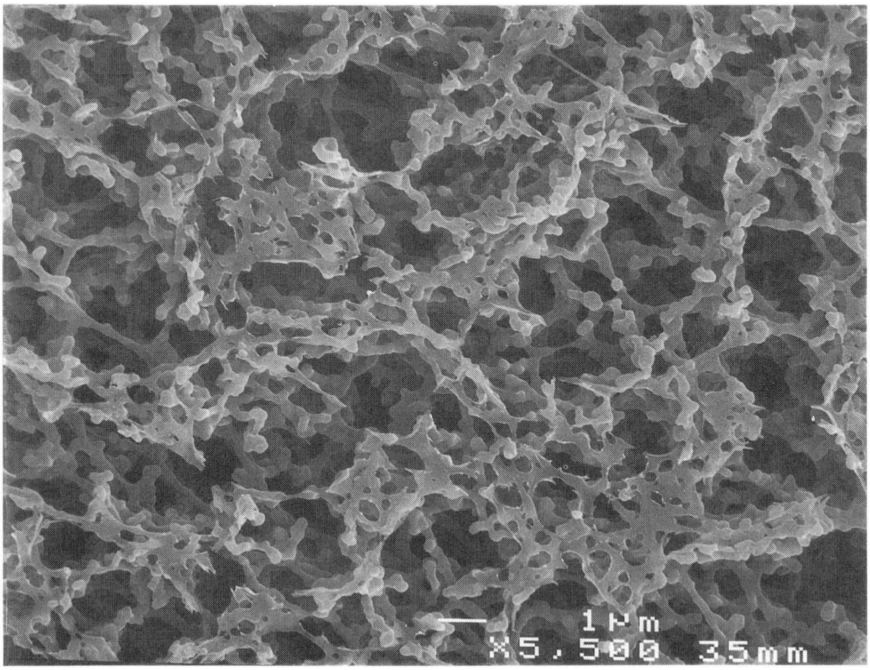

Figure 1b Cast film filter at ×5500.

The term *interconnected voids* have also been used to describe the closely adjacent spherical cells of cast film filters. The route followed by a particle in passing through a filter of this type is determined by the flow path in which it is entrained. This flow path is typically complex and changes direction and velocity numerous times between the upper and lower surfaces of the filter: Thus, the term *tortuous path* filters has also been applied. Although cast film filters are very thin (100 to 150 μm), the filter presents to a particle a relatively thick complex structure, and the surface may differ from the structure within the depth of the filter. Importantly, many particles removed by this type of filter are trapped within the depth of the filter rather than on the surface.

The spongiform structure of cast film filters contrasts significantly with that of the nucleation track membranes that are made by chemically etching polycarbonate films subjected to localized damage by nuclear radiation [4]. A Nuclepore track-etched membrane filter is shown in Figs. 2a,b. This type of filter is characterized by cylindrical pores of uniform diameter that cause the filter to function as a sieve. Nuclepore filters are well-suited for use when particles must be trapped on the filter surface, as is the case in many analytical applications. However, as can be seen in the figure, this filter also has less open pore area and consequently lower liquid flow rates than cast film filters; this makes Nuclepore filters unsuitable for some general applications. They also tend to "blind" or become blocked more easily than cast film filters, and their extreme thinness (10 to 15 μm), flexibility, and sensitivity to static electrical potential make handling difficult. Despite these disadvantages, they are extremely useful in collecting particles for further analysis, since their smooth surface allows particles to be easily removed for transfer to the supports used in analytical instruments, such as scanning microscopes or X-ray microprobes. This type of filter is also more durable than cast film filters and is not as likely to be damaged as particles are removed.

Airborne particles as well as those suspended in liquid may be collected by filtration. Particles filtered from air or gases may also be isolated and identified using techniques applied to solution-borne particulate matter. Particles may be collected from the air in controlled environment areas by the use of a vacuum pump attached to an appropriate filtration funnel. A 47-mm filter is generally a good choice if large volumes of air or air with a high particle burden are to be sampled.

Techniques for isolation and collection of individual particles for analysis involve the use of specialized microinstruments to physically pick up and transfer particles from a filter or other collection substrate, such as a microslide or sample support for an analytical instrument. This technique is not new; Louis Pasteur used the microscope to differentiate the optical isomers of tartaric acid. By methodically segregating individual crystals of two morphotypes, Pasteur was able to resolve the optically inactive mixture into the two optically active isomers. This method of isolating individual particles under the microscope using a microneedle, probe, forceps, or another instrument is often essential to identification of particles of interest. Such "particle picking" requires a hood or other HEPA-filtered enclosure, a clean work area, a steady hand, a stereomicroscope, and tools such as those shown in Fig. 3. Needle probes and glass fiber brushes are used to pick up single particles, while fine forceps are useful for larger particles. Drawn plastic micropipettes can be used to pick up single particles in a drop of fluid. The technique for isolating single particles is generally simple, but it is only through considerable practice that one acquires the necessary skill for handling particles as small as 10 μm. The techniques, necessary tools, and procedures for particle manipulation are discussed in the papers by Teetsov [6] and DeLuca and Bodapatti [7], and in the text by McCrone et al. [8]. The various volumes of McCrone's *Particle Atlas* [9] (now available on CD Rom) serve as comprehensive references both to analytical methodology and particle morphology. Needle

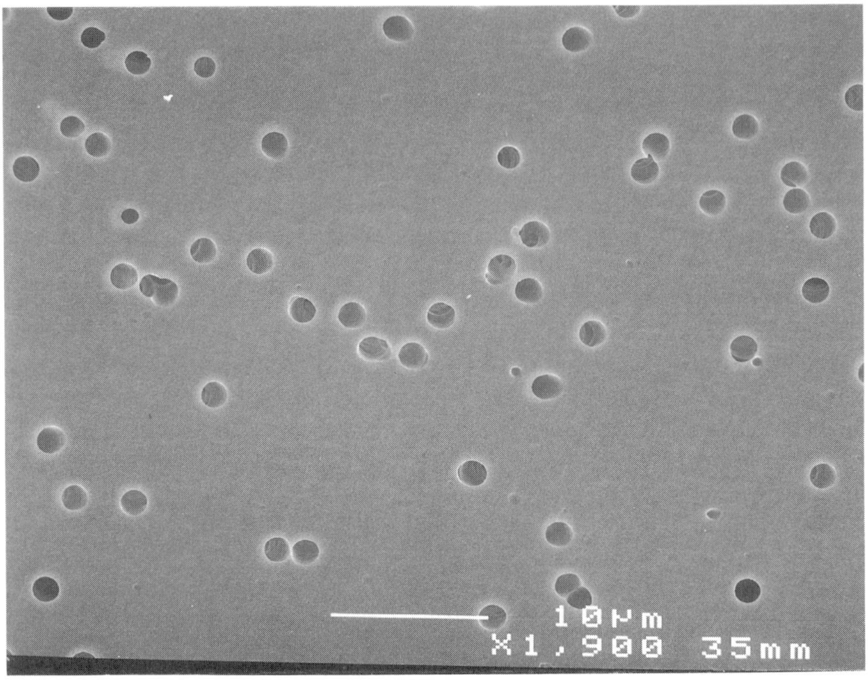

Figure 2a Track-etched membrane filter surface at ×1900.

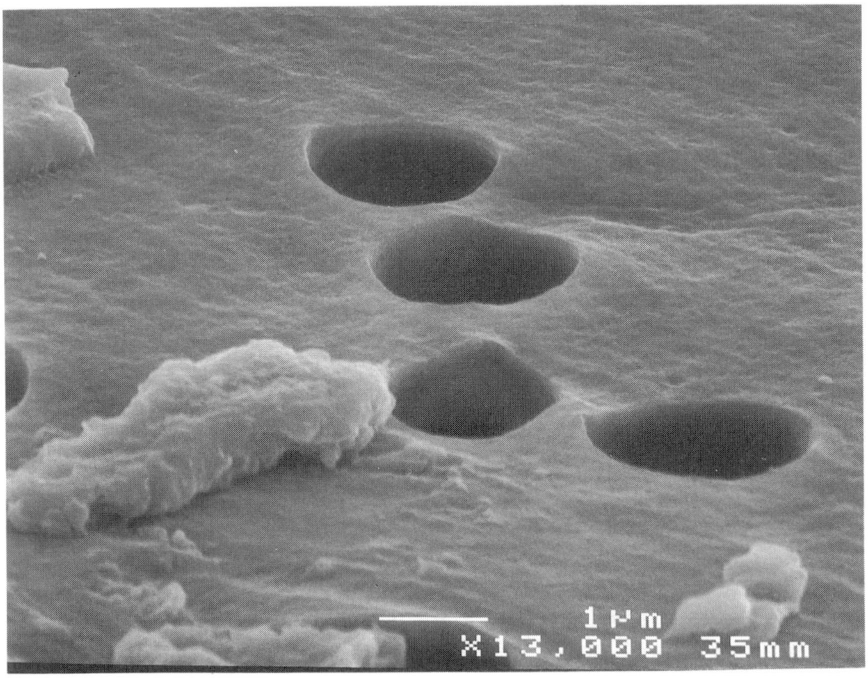

Figure 2b Filter surface at ×13000 (retained particles on surface).

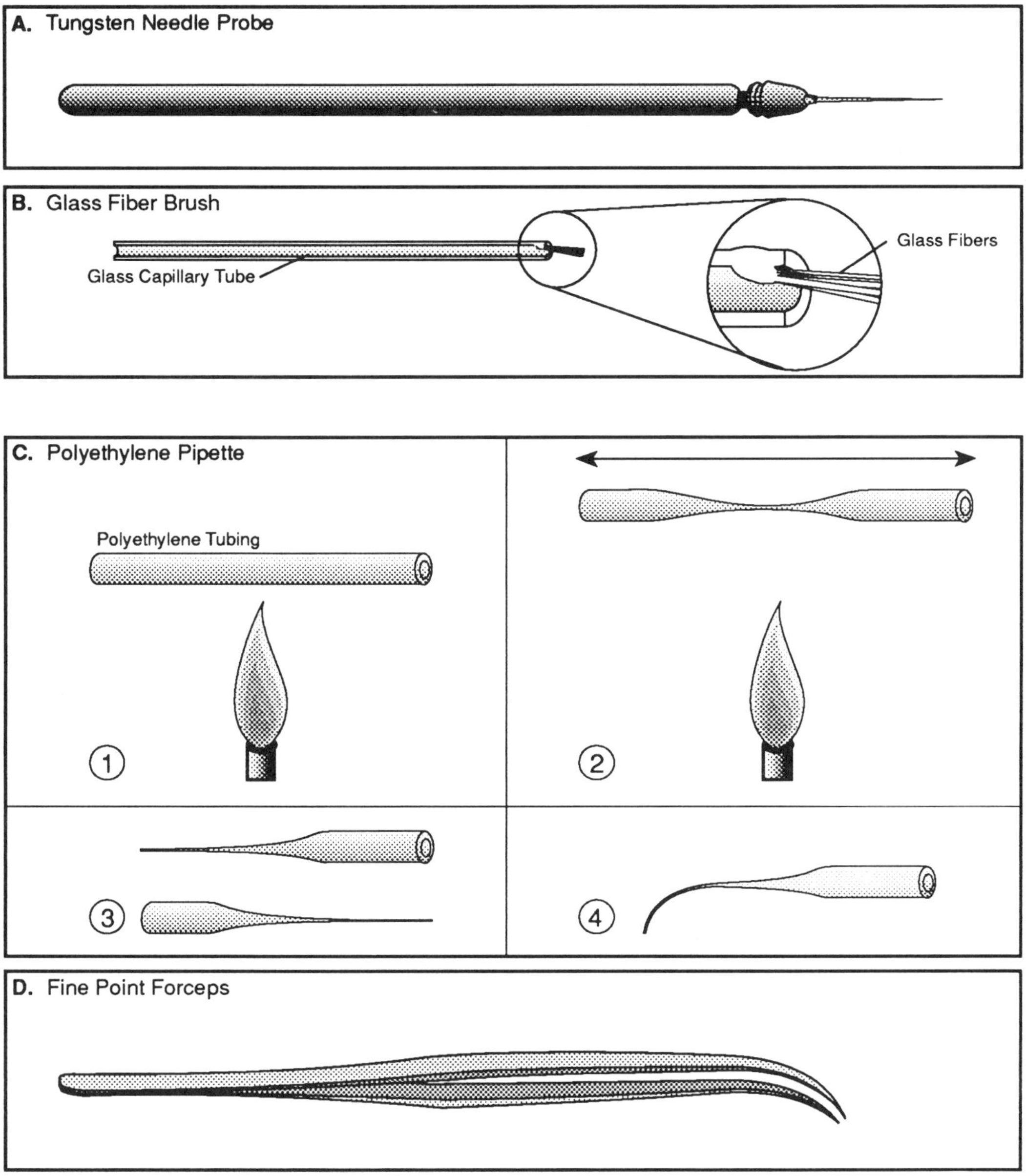

Figure 3 Tools for particle isolation.

probes and fine forceps specifically designed for particle isolation are available from McCrone Associates, Chicago, Illinois.

Particles for instrumental analysis should be isolated using as little manipulation as possible, since each handling step risks loss of the particle of interest and the addition of chemical or particulate contaminants. Typically, one particle is handled at a time. If the analyst is studying a membrane filter sample by reflected light under a stereomicroscope, specific particles may be removed by simply touching them with a finely pointed tungsten needle. Adherence of particles to the needle may be enhanced by wetting the tip in a 1% solution of Aroclor in a solvent such as acetone, and allowing the solvent to evaporate. The coated tip of such a needle with a tip size approximating that of the particle to be isolated is then used to pick up one or several particles under the stereomicroscope. The particles may then be transferred to small drops of mountant, such as collodion or Formvar plastic in amyl acetate, on a clean microscope slide. After adding a cover slip, the sample may then be examined with a polarizing microscope. Using a similar technique, particles may be isolated for transfer to a carbon or beryllium planchette for examination by SEM or X-ray microprobe, or isolated on a salt plate for analysis by infrared microspectroscopy.

Fine-pointed tungsten needle probes are a primary tool for particle isolation. Their fine points allow collection of subvisible particles and their handling and transfer under a microscope. In general, three different needle tip sizes are required. McCrone's #1 needle has a blunt tip about 40 μm in diameter and is very durable. The #2 needle has a finer tip (about 25 μm in diameter), and the #3 needle has a tip diameter of about 1 μm, which is most frequently used for the manipulation and transfer of single particles. Needle tips quickly become dull with use and require frequent resharpening by chemical etching over a hot stick of sodium nitrite. The chemical etching is produced as a result of the exothermic reaction between tungsten and sodium nitrite to form sodium tungstate. To etch the tip of a tungsten needle, the sodium nitrite stick and a tungsten needle are held over a Bunsen burner. When the tip of the sodium nitrite stick begins to melt and the needle is red hot, the end of the needle is drawn lightly through the melted sodium nitrite. Usually a single stroke is sufficient to resharpen the needle. Following sharpening, the needle is then washed with water, dried, and the sharpness of the tip checked using a microscope.

Flexible polyethylene micropipettes are used to deliver tiny (~100 μm diameter) droplets of liquid such as collodion or Formvar solutions to entrap single particles on the surface of a microscope slide. These micropipettes are stable in many organic solvents, and they will not scratch soft surfaces. Polyethylene micropipettes are made by heating and drawing fine-diameter polyethylene tubing. First, a 10- to 20-mm length of polyethylene tubing with a 3- to 5-mm inside diameter is rotated over the flame of an alcohol lamp, about 5 to 10 cm above the flame. When the tubing becomes transparent, it is gently stretched with continued rotation. As the tubing becomes thinner, it is moved away from the flame, and the thinned portion is drawn until it becomes threadlike. After the tubing cools, it may be cut into halves, trimmed to the desired lumen diameter and length, and the tip beveled as desired by cutting with a razor blade. A drop of Formvar solution placed over a particle of interest will spread into a thin film that immobilizes the particle. The flake of plastic containing the particle can then be detached from the slide for further analysis or for storage. The plastic can be readily removed using a drop of solvent if this is necessary.

If the particles have been isolated from solutions by filtration, the filters are first examined using a compound microscope with incident or reflected illumination. Following the initial observation of types of particles on the filter, particles of interest are transferred to glass slides. For this procedure, a glass slide is positioned on the stage of a stereomicroscope and a circle drawn on the underside of the slide with a fine felt-tip marker to indicate the location to which the isolated particles will be transferred. The filter is then placed on the stereomicroscope stage, and particles of interest on the membrane filter surface are selected for isolation. Particles to be analyzed are picked up with a #2 or #3 tungsten needle and placed on the microscope slide under the stereomicroscope. After the desired number of particles has been isolated, the slide is transferred back to the compound microscope for the determination of the optical properties of the particles. If particles on a slide are not all of the desired type, they may be further segregated. At this point, the particles can be mounted in a 1% solution of colloidion or Formvar in amyl acetate and retained or transferred for instrumental analysis.

In lieu of this procedure, if instrumental analysis is to be performed without further microscope work, the particle of interest may simply be transferred to a drop of collodion or Formvar mountant on a suitable specimen support. If a Nuclepore membrane is used, a small drop of collodion in amyl acetate may be transferred to the filter on the tip of a tungsten needle and spread over the particle. When the collodion dries, the thin film that contains the particles may then be removed from the filter and transferred to a glass slide. To complete mounting the particle, a small drop of amyl acetate is placed on a glass slide next to the collodion flake containing the particles. The collodion will be softened by the amyl acetate, and the particle will be anchored to the slide. Using this procedure, the particle is embedded in a transparent film that allows it to be examined microscopically by transmitted polarized light. If necessary, the collodion film can be dissolved with amyl acetate and the particle removed for transfer to the appropriate support for instrumental analysis.

The value of the skill for particle isolation using physical techniques cannot be overemphasized. The present sensitivity limits of microanalytical instruments allow nanogram qualities of material to be analyzed. For example, if a single 10- to 30-μm crystalline particle can be isolated and mounted on the tip of a 10-μm glass fiber, one can obtain an X-ray diffraction pattern. If a clean 10-μm particle can be mounted on a polished beryllium or carbon stub, it can be examined by SEM or quantitatively analyzed with X-ray spectrometry. Secondary ion mass spectrometry (SIMS) is useful with particles of similar size. Micro Fourier transform infrared (FTIR) spectrophotometers can obtain spectra with sharply resolved absorbances from single particles 10 to 20 μm in size.

Experts in X-ray diffraction, electron microscopy, or microprobe analysis are usually not adept at handling single particles of a size approaching the sensitivity of their instruments, and depend on particle analysts with this specialized technical skill to acquire samples for their instruments.

There are several helpful hints regarding particle isolation for the beginner. These include the following:

Never take your eyes off the particle during a manipulation.

Remove the particle from a needle only by "washing" it off into a microdrop of liquid in which the particle will be trapped.

Beware of static electricity.

Ensure that the path the needle will take in transferring the particle on a microscope slide or specimen mount is unobstructed.

An important principle to observe is that any particle to be isolated for analysis by instrumental means should be characterized by light microscopy prior to isolation. The mounted particle can then be recognized as the original particle and not another particle collected by accident.

III. ELECTRON MICROSCOPY

In TEM a sample is analyzed by detection of the transmitted electron beam. In order to obtain information, the TEM's 50–100 kilo electron volt (keV) electrons must pass through the sample. In SEM the electron beam is scanned across the sample in a raster pattern, and the variation in the detected signal provides contrast information that is used in image formation. The resulting image reveals the surface features of the sample, giving a pseudo three-dimensional image. A scanning microscope typically employs a lower energy electron beam (1–30 keV) than a transmission instrument. A variety of analytical modes can be incorporated into an electron microscope system to be used for chemical analysis. These include energy dispersive X-ray spectrometry (EDXS), wavelength dispersive X-ray spectrometry (WDXS), and selected area election diffraction (SAED).

The electron microscope does not eliminate the need for optical microscopy, but rather provides an additional method to be used in particle identification and analysis with its own unique advantages. Although it is a powerful technique, electron microscopy should not be considered to be the ultimate method of morphologic analysis, but rather as one of an analyst's tools to gather information about a specific particle sample. Optical microscopy, especially low-magnification stereomicroscopy, must be used as a first step in all analyses by SEM or TEM.

Scanning electron microscopes and transmission electron microscopes, as well as some other analytical instruments, are based on the use of electromagnetic lenses to focus the electrons that serve as their illuminating radiation. Simple electromagnetic solenoids (hollow electromagnets) may be constructed by coiling a conductor around a hollow nonferrous core. Such a helix or solenoid (Fig. 4) unites the magnetic field of each coil of the conductor and has a north and south pole and an internal and an external magnetic field. This type of solenoid forms a simple, weak electromagnetic lens. Around and within any such cylindrical electromagnet, the field exists as concentric cylinders of magnetic flux lines. The central field in a solenoid will be composed of flux lines of a density proportional (within limits) to the coil current. The field is uniform in the center of the solenoid, but at its ends the lines of force curve and diverge, forming closed loops along the outer surface. Thus, the flux density is greater in the center of the solenoid than at other points. The lines of force are parallel in the center of the solenoid, and a permanent magnet placed in the center of the coil will orient itself perpendicular to the turns of wire.

An electromagnetic lens is, in reality, a powerful solenoid. The path of an electron entering the central lens opening at some angle to the axis of symmetry is defined by (1) resolution of a vector acting perpendicular to the lens field flux lines and (2) one that acts parallel to the field. The path will then be a straight line if the solenoid is not energized. If current is applied to the windings, the path becomes a summation of the linear and arcuate vectors—that is, a spiral helical path. This principle results in the focusing action of electromagnetic lenses in an electron

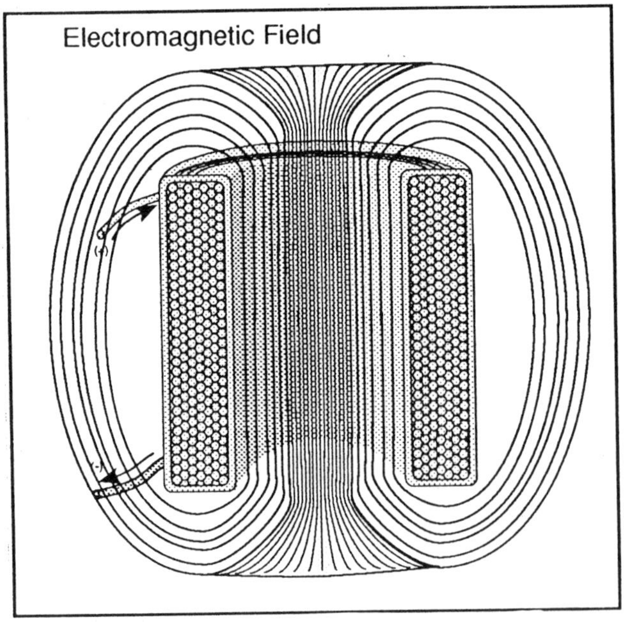

Figure 4 Electromagnetic solenoid.

microscope (Fig. 5). Regardless of their angle of entry, electrons entering the field at a given point complete the same number of turns as they pass from one end of the solenoid field to the other and are brought to focus at a single point behind the lens. As a result, the point at which an electron strikes an image plane will be precisely comparable to its relative position in the specimen plane before the lens. The path of an electron after leaving the lens field (behind the lens) and passing into an area where the lens influence is removed, will again be a straight line at some angle to the optical axis determined by the original angle (Θ) at which it entered the lens field. Electrons thus deflected and focused can form a diffuse magnified image on a behind-the-lens image plane. The end result is a lens system for electrons that mimics the focusing action of biconvex glass lenses for light.

The condenser lenses of an electron microscope allow a demagnified spot of illumination to be focused on the specimen. As in the light microscope, the role of these lenses is to allow optimum illumination by focusing the electron beam on the specimen. They are weak electromagnetic lenses with focal lengths measured in millimeters rather than the micrometer focal length of the objective. Instead of focusing a substage condenser, as with the light microscope, one simply changes beam focal conditions by varying lens current. The changeable focal length of the condenser lens in the transmission electron microscope makes

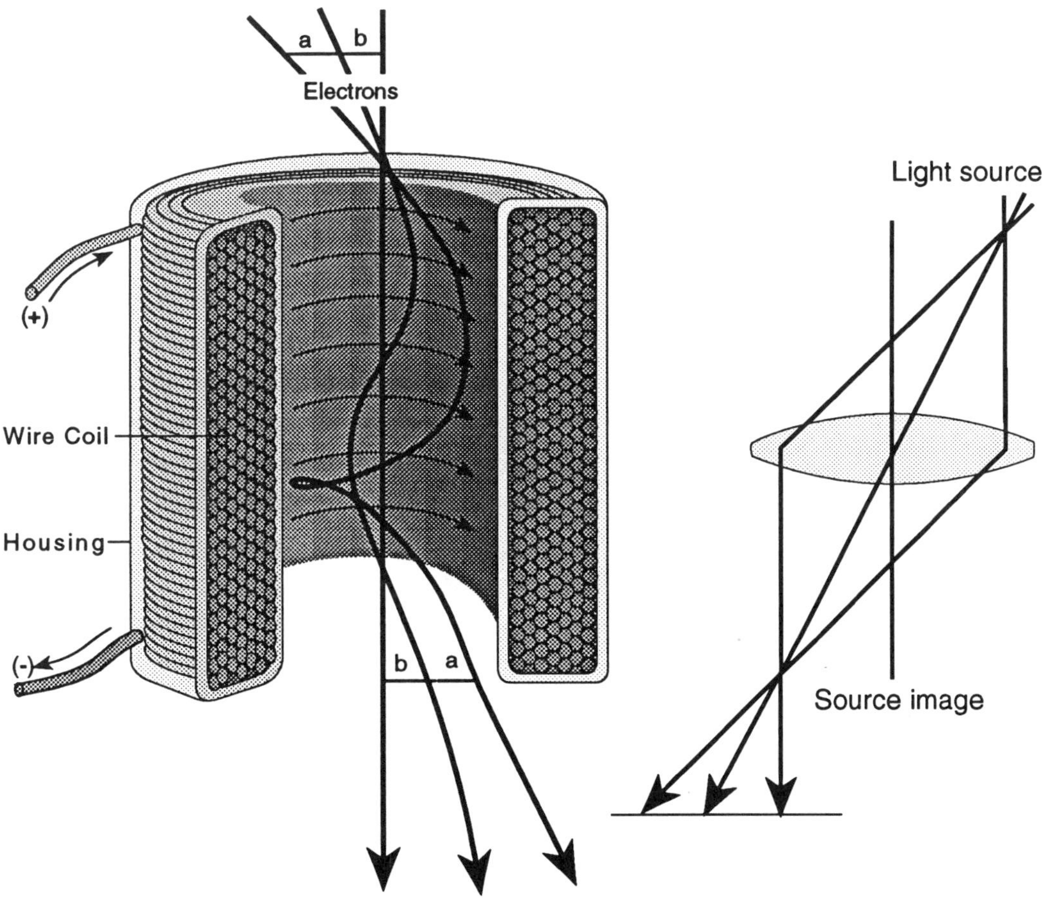

Figure 5 Electromagnetic lens construction and electron focusing.

possible the simultaneous control of both the aperture angle of illumination and the beam intensity (electrons/unit area/time), and it makes the beam more coherent than that obtained from the electron source (gun).

IV. TRANSMISSION ELECTRON MICROSCOPY

The function of electron microscopes is covered in the basic text by Wischnitzer [10]. Figure 6 is a diagrammatic comparison of transmission light and transmission electron microscopes illustrating the similarity between the two instruments. Each contains a condenser lens to focus illuminating radiation on the object being observed. In the light microscope this lens and other lenses are made of glass or a similar transparent material. Glass is opaque to electrons, and in the electron microscope the lenses are cylindrical electromagnets. The focal length of a glass lens is fixed, and focusing the light microscope is achieved by moving the objective lens an appropriate distance from the object. In the electron microscope, the focal length of the lenses is dependent on the amount of current passed through the lens coil. As a result, the lenses of the electron microscope are not moved in relation to the object, but instead the lens current is changed to attain focus. The image of the electron microscope is viewed by allowing the imaging electrons to impact a fluorescent screen; atomic electrons in the screen phosphor excited by the beam electrons emit a green-yellow light visible to the human eye as the electrons return to a ground state.

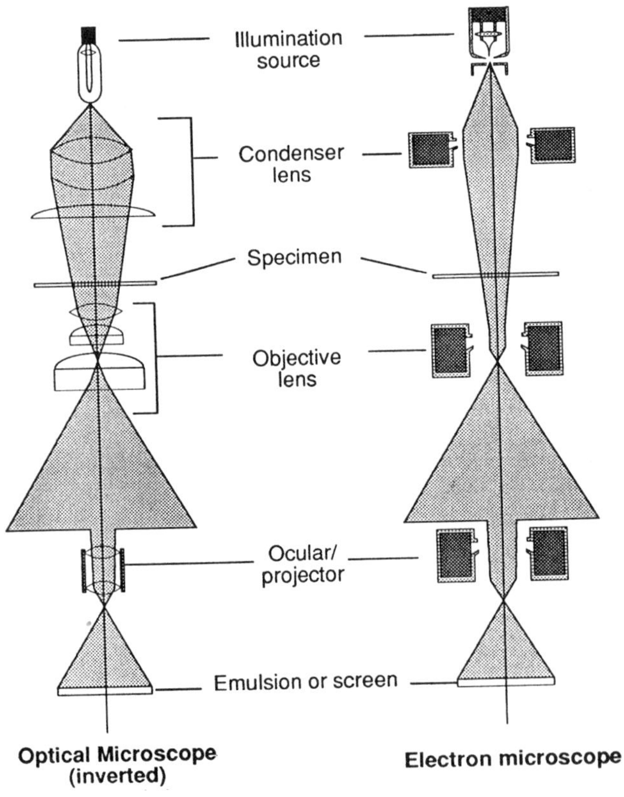

Figure 6 Transmission light and electron microscopes.

Resolution is an important consideration in electron microscopy. If the point-to-point resolution of a light microscope is 0.5 μm, two specimen points closer together than 0.5 μm will appear as a single object. Conversely, when the two points are separated by a distance greater than 0.5 μm, they will be detected separately. Four principal factors determine the resolution attainable with the light microscope. These are diffraction effects, chromatic aberration, spherical aberration, and astigmatism. In electron microscopes, these factors also apply, but resolution is most critically limited by spherical aberrations in the objective lens. The great resolving power of the TEM results from the short wavelength of electron illumination. In 1920, de Broglie postulated that any moving particle could be assigned a wavelength according to the relation

$$\lambda = \frac{h}{mv}$$

where h = Planck's constant
m = mass of the particle
v = velocity of the particle

For an electron accelerated across some voltage potential difference v, this becomes

$$\lambda = \frac{12.3}{v}$$

For 100-keV electrons, the wavelength is about 4.0×10^{-3} nanometers (nm) or about 10^{-5} times the wavelength of ultraviolet light. The resolution achievable with this wavelength of illumination is approximately 0.25 nm. This is an approximate 10^3 increase in resolution over that achievable with a light microscope.

In both transmission light and electron microscopes the image is formed by interaction of the specimen with the illuminating radiation. In the light microscope, most contrast effects result from the differential absorption of light by different areas of the specimen. Additionally, as light passes the edges of objects in the specimen, diffraction patterns result from the interference of diffracted and undiffracted light, and these diffraction patterns are recombined by the lens system in the final image. In the TEM, contrast is provided by electron scattering due to charge effects as the electrons pass through the object. Scattered electrons are removed from the beam and as a result, regions of the object that are electron dense are represented on the screen as dark areas.

The imaging lens system of a TEM consists of an objective lens, two or three intermediate lenses, and a projector lens. The final lens that forms an image on the fluorescent screen is the projector lens, and those lenses between the projector lens and the objective are called intermediate lenses. The objective lens forms the initial magnified image of the specimen; in so doing, it takes as its object field some portion of the specimen and projects a magnified image of this specimen field in the front focal plane or object space of the lens beneath it. The final image projected on the fluorescent screen of the instrument results from compound magnification of the objective and projector lenses, and will consist of an intricate contrast pattern resulting from the differential electron scattering properties of the specimen. An objective aperture placed beneath the specimen removes electrons that are scattered from the beam at an angle greater than the optimum acceptance angle for the lens below the objective.

The objective lens is the most critical optical component of the transmission microscope. In order to obtain a necessarily high magnification at a short image distance, the specimen must be situated close to the focal plane of the objective. The initial magnification provided by the objective lens approximates 100× in most instruments. The objective lens determines the resolving power of the instrument since resolution in the initial magnified image cannot be improved by other lenses. Usually the magnetic field of the objective lens is located eccentrically in relation to the axial extension of the lens coil, so that the field is close to the end of the coil where the specimen is introduced.

Because of the weak penetration of electrons, TEM is only effective in revealing the internal structure of particles that are less than 50 to 100 nm thick. Unless higher accelerating voltages are used, thicker particles are revealed only in silhouette. Three-dimensional information for thick particles can sometimes be obtainable by shadowing the particles with evaporated metal atoms (gold, platinum, palladium) at a low angle with respect to the sample support surface.

V. SCANNING ELECTRON MICROSCOPY

The electron-optical column of a scanning instrument, like that of a transmission microscope, consists of an electron source (a conventional heated filament or field emission gun), and a series of electromagnetic lenses that focus and demagnify the filament image onto the specimen. The electron gun and gun chamber assemblies are also similar to that of a transmission instrument. The column of the scanning microscope is shorter than that of the transmission instrument; there is no need for imaging lenses, and the entire column functions as a multiple condenser lens system that demagnifies the source image into a small intense "spot" of electrons that is scanned over the specimen.

The image produced on the screen of a transmission microscope is generated by a continuous flow of electric current that varies in density over the screen area in response to specimen electron-scattering qualities. The illumination in the transmission instrument is produced with a static beam spot that remains in one position on the specimen. In a scanning electron microscope, the specimen remains in place and the beam is scanned in a rigidly controlled pattern of lines (raster) on the specimen surface. As the raster is scanned on the specimen surface, the beam spot swells for a certain time interval at each of a large number of points on the specimen surface, and excites from each spot a secondary electron signal. The image is formed on a cathode ray tube (CRT) as a result of the variation in intensity and angular orientation of secondary electrons emitted at different sample points (Fig. 7). The pattern of secondary emission and numbers of secondaries generated will depend both on surface contour and emissivity. A conductive, dry electron emissive specimen surface is a necessity for conventional (high voltage) scanning microscopes that operate at beam potentials of 5 to 35 keV.

The mode of image formation most often applied in SEM results from collection of the secondary electrons generated by the interaction of primary beam electrons with the specimen surface. Signal collection is accomplished by the use of a phosphor or aluminum-coated scintillator that converts the electron signal to photons that impinge on a photomultiplier tube (PMT) as shown in Fig. 8. The amplified signal from the PMT is used to intensity modulate the beam of a visual CRT and form a television-mode image. The instrument is somewhat similar in operation to a stereo dissecting microscope that forms an image using light reflected from the specimen surface; it may also be likened to a closed-circuit TV system that provides for high magnification of the imaged field.

The specimen surface is located at the back focal point of the objective (second or third) condenser lens and is scanned in a raster pattern by a small, intense, symmetrical beam spot that is generally 5 to 20 nm in diameter. In the secondary electron imaging mode, a secondary electron signal is generated at the point of beam impact as the 5–20 keV electrons (primaries) excite or eject secondary electrons of 2–5 eV potential from the specimen surface. The secondaries "see" the 50–100 V positive potential on the screen anode of the collector and are attracted to it as shown in Fig. 8. Once secondaries are in close proximity to the screen anode,

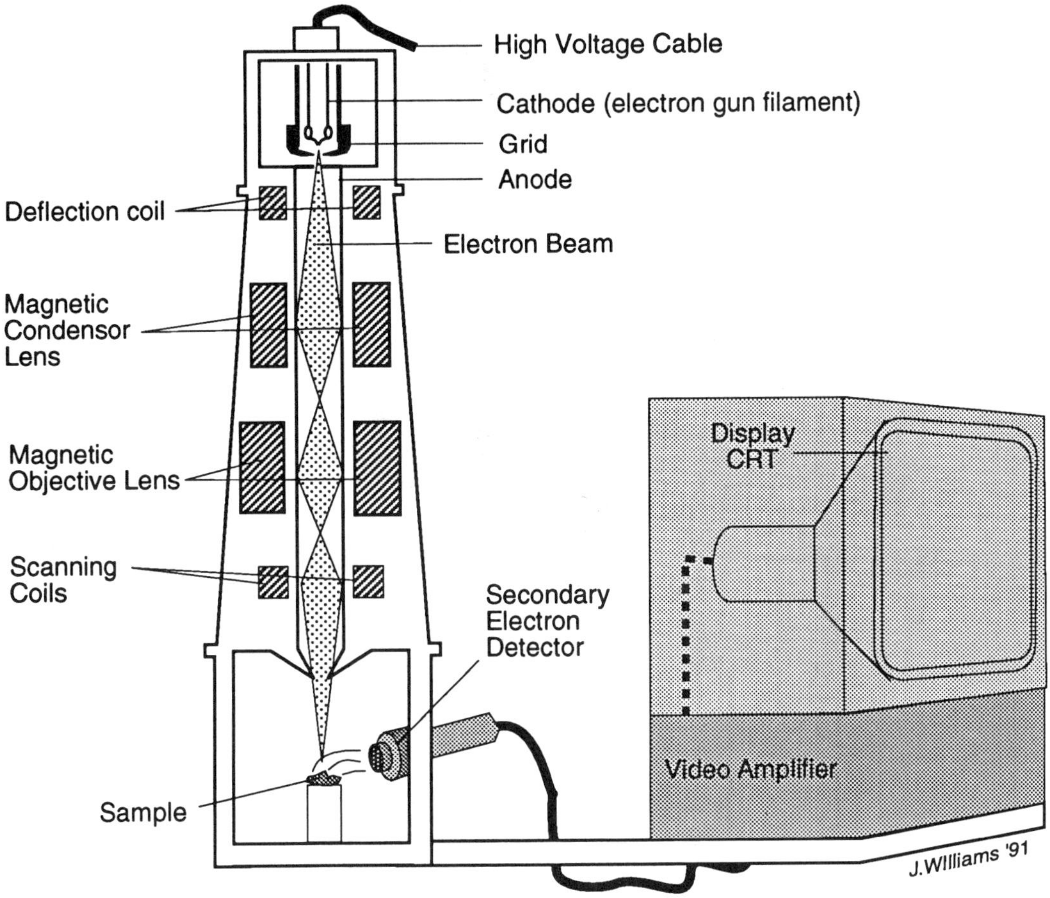

Figure 7 Scanning electron microscope.

they are attracted by the much higher positive potential on the scintillator disc itself (usually 10,000–12,000 V positive) and accelerated across the space between the screen anode and the scintillator. They then strike the scintillator, a disklike or hemispherical structure of clear plastic coated with a fine phosphor. As electrons impact the surface of the scintillator, photons are ejected from the opposite side of the phosphor into a photomultiplier tube that generates the current used to intensity modulate the imaging beam on the CRT monitor.

The resulting video image is composed of a series of spot images (picture points) obtained by scanning the beam spot over the specimen surface and the CRT synchronously. Both are scanned in a precisely controlled raster pattern so that a series of imaging picture points is formed representing a series of secondary electron-emitting specimen points. The degree of synchronization is such that the beam position on the CRT corresponds precisely with the position of the primary beam spot in the raster on the specimen surface; points on the object and in the image are struck by the microscope electron beam and the CRT electron beam at the same instant. Secondary electrons collected at a given raster point do not actually form an image, but rather intensity-modulate the CRT image at the corresponding picture point. There is no need to focus the secondary electrons because those collected at any given point in time come from the exact

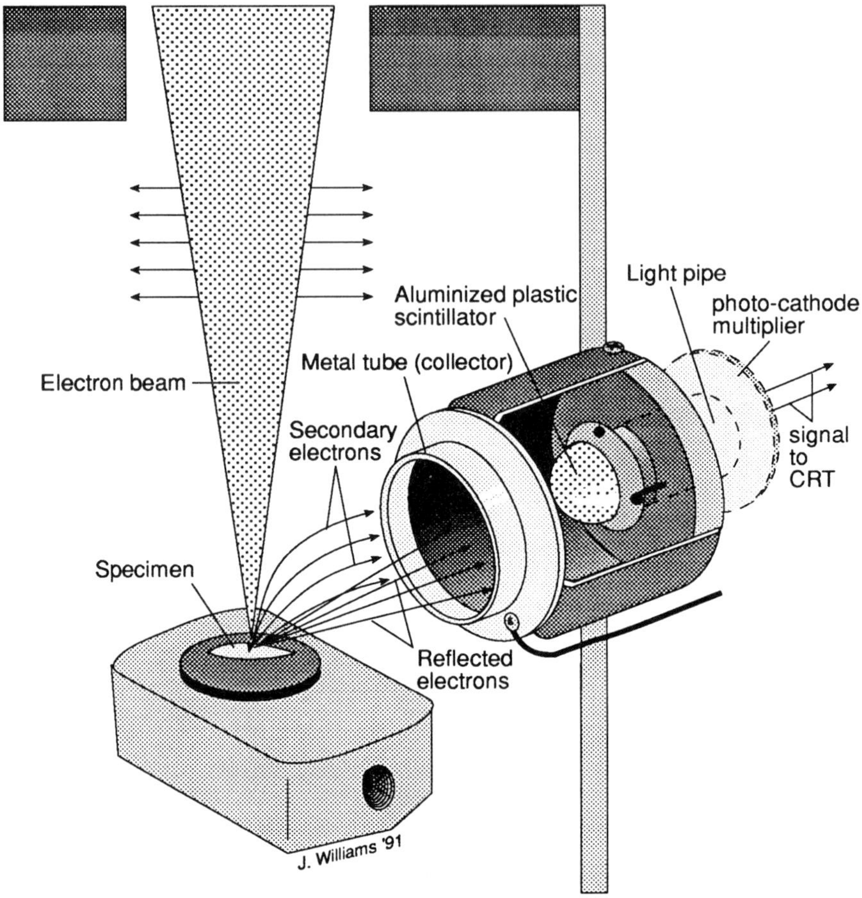

Figure 8 Secondary electron collection in a scanning electron microscope.

point where the probe has entered the specimen. Magnification with the scanning microscope depends upon the ratio of the screen raster size (approximately 4 in. × 5 in.) to that of the raster scanned on the specimen.

When the electrons in the probe strike a particle, other types of radiation in addition to secondary electrons are generated. These additional forms of radiation can be detected and collected to obtain morphologic or compositional data regarding the sample (Fig. 9). As electrons penetrate the specimen they lose energy and slow down; they may also be scattered through large angles. Finally, some electrons may reemerge from the specimen surface and travel in a direction unrelated to that of the incident beam. In principle, any signal can be used to produce an image if it is related to a given point on the specimen surface. Primary (beam) electrons reflected with relatively little loss in energy are not appreciably deflected by a few hundred volts potential difference, and will not enter the secondary collector unless it lies directly in their path. However, these backscattered electrons can generate secondary electrons at surfaces other than the specimen that will degrade the final signal. The probability that secondary electrons will escape from the sample decreases exponentially as their point of generation moves away from the surface layer; this depth is 0.5 to 5 nm for metals, and somewhat greater for insulators.

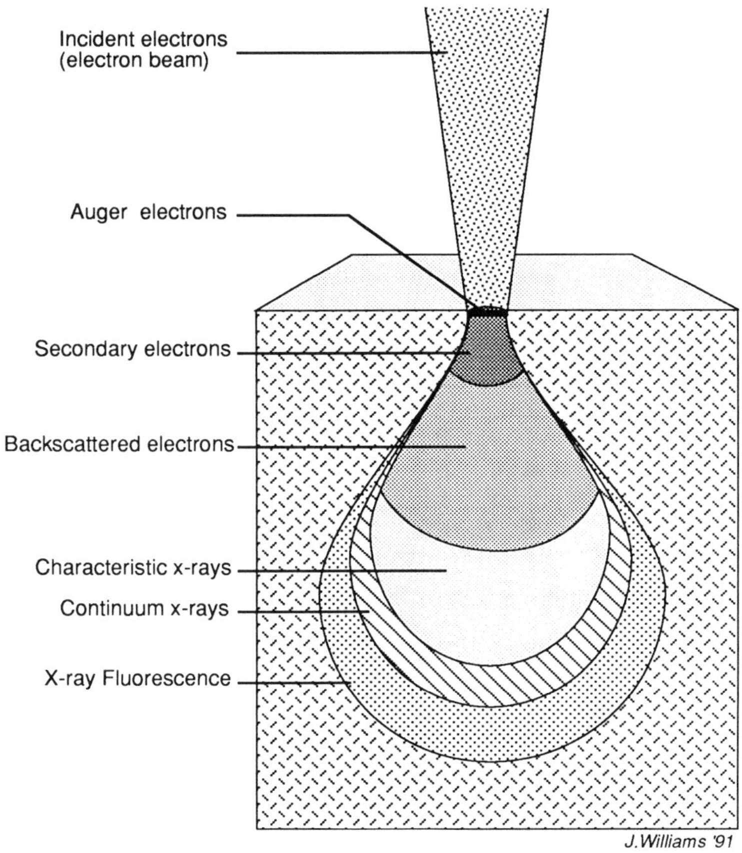

Figure 9 Signal generation in a scanning microscope.

VI. X-RAY SPECTROSCOPY

As the electron beam impinges on the specimen surfaces, X-rays are emitted that are characteristic of the material from which they are excited. X-rays with an approximate wavelength range of 0.1 to 100 Å are produced as the rapidly moving electrons decelerate and convert their energy of motion into X-ray quanta. The wavelength of the emitted radiation will depend on the energy of the electrons.

Under the usual conditions of X-ray excitation, most of the electrons are not brought to a full stop by a single collision, and a continuum of radiation is formed as the electrons continue to decelerate and generate lower energy X-rays. The minimum wavelength of this white radiation (λ) is determined by the instrument accelerating voltage (V) and may be calculated from the relation

$$\lambda_{\text{min}} = \frac{12.30}{V_{\text{acc}}}$$

Thus, higher-energy X-rays are formed in the initial collision of the beam with a particle. The greatest intensity of emission will occur at a somewhat longer wavelength. As the accelerating voltage is increased, not only are the cut-off and peak intensity moved to shorter, more penetrating wavelengths, but also the total emission intensity will increase even though the electron current remains the same.

The ability to perform chemical analysis of particles with a scanning microscope is due to the existence of a simple and unique relation between the wavelength (λ) of the characteristic X-rays emitted from an element and its atomic number (Z). The origin of characteristic X-rays is illustrated in Fig. 10. If an electron beam incident on a sample has sufficient energy to exceed the excitation potential (bonding energy) of a core electron, then that core electron can be ejected from its parent atom leaving an orbital vacancy behind. The atom is then in an excited state, and orbital vacancies formed are filled by electronic relaxation from outer orbitals accompanied by emission of an X-ray photon with discrete energy levels corresponding to the difference between the orbital energy levels.

If beam electrons have energies of 10 keV or greater they can remove electrons from the innermost (K) shell of atoms of $Z = 25$ or greater. A vacancy in the K shell is filled by an electron from the next higher shell (L) or the one above that (M). These vacancies are filled by electrons from outer shells, with the result that X-rays of decreasing energy are produced. Each decrease in potential energy levels, as electrons move from a higher to a lower levels, appears as an X-ray, and as the energies of the shells are well defined, each transition gives a line or peak at a single energy level. The major transition peaks seen in energy dispersive spectra are

Ka1, Ka2 L→K
Kb1, Kb2 M→K

Once created, the X-ray photon is either emitted from the sample or its energy may be transferred to another electron (called an Auger electron), which is ejected from the atom. High-energy X-rays may also fluoresce lower-energy X-rays. The fraction of electronic relaxations that result in X-ray photon emission is called the fluorescent yield; this depends both on the atomic number of the element and the orbital where the ionization occurs. The characteristic X-ray spectral lines are denoted by the symbols *K, L,* and *M,* which relate to the values of the principal quantum numbers ($n = 1$, 2, or 3) for the atom. The polychromatic continuum (white) radiation, or bremsstrahlung, emitted in addition to the characteristic elemental X-rays when the electron beam strikes a specimen, is the major source of the background which limits the sensitivity of X-ray spectrometry.

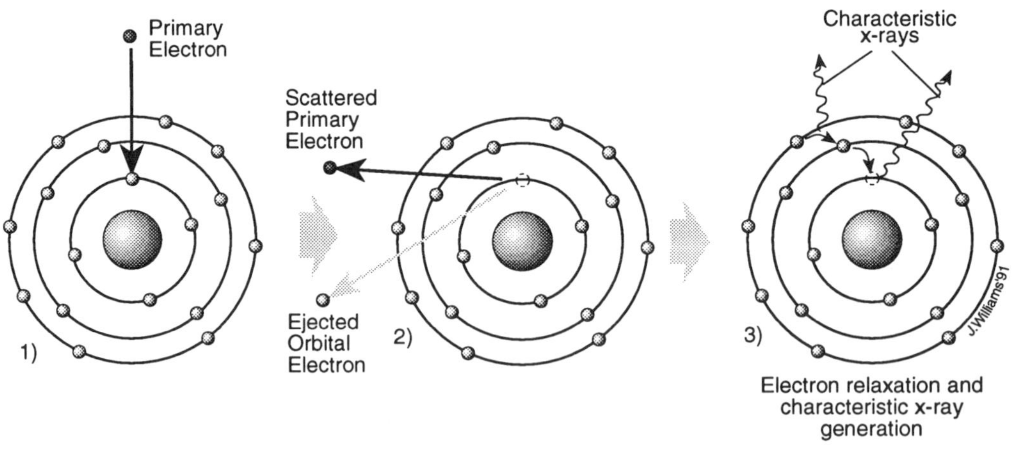

Figure 10 Generation of characteristic X-rays.

The collection of elemental X-rays emitted by a specimen material is the basis of the application of scanning or transmission microscopes in combination with X-ray spectrometry. The characteristic X-rays emitted are used to obtain an elemental analysis of the specimen area being scanned. If the beam is scanned in synchrony with a CRT and the X-ray signal used to modulate the brightness of the CRT spot, the instrument can be used to give a "dot map" image containing information about the distribution of particular chemical elements within the specimen. Simultaneous collection of X-ray and electron signals can be used to form split screen images for combined morphologic and compositional analysis.

Solid-state energy dispersive X-ray spectrometers measure X-ray energy directly, producing a spectrum of counts versus energy. As each X-ray enters the detector, it produces a shower of photoelectrons that strike silicon atoms and raise bound electrons to the conduction band. Each such event requires 3.8 V of energy, so that the total number of electrons freed is linearly proportional to the energy of the entering X-ray. The charge produced by each X-ray is collected by a bias voltage before the next X-ray enters the detector. In order to prevent loss of the electrons during this collection process, imperfections or dislocations in the silicon lattice that might trap electrons must be eliminated. This is done by diffusing in (drifting) lithium atoms. The energy dispersive X-ray detector is operated at liquid nitrogen temperatures to reduce its dark current and thermal noise, and to permit the accurate collection of very low levels of current generated by X-rays striking the collector.

Energy dispersive X-ray spectrometry allows the detection of X-rays from the entire energy spectrum at one time, as opposed to wavelength dispersive analysis (WDXS), which only examines a single X-ray line at a time. This constitutes both an advantage and a disadvantage, since a single line cannot be isolated for critical analysis as with a crystal spectrometer. A distinct advantage of solid-state detectors used for EDXS is that they are nearly 100% efficient over a wide energy range. This results in a proportionality between counts per second and the intensity of the X-ray line at a given energy, thus allowing the comparison of ratios of different lines as a means of identifying the elements in the sample.

Whereas energy dispersive spectrometers simultaneously collect X-rays over a wide range of energy levels, the electron microprobe (wavelength dispersive) microanalyzer uses crystal X-ray spectrometers to identify and count X-rays based on their wavelength. The energy of an X-ray photon that corresponds to the transition energy between orbital shells in the atom can be expressed as an equivalent wavelength (λ) according to the same formula earlier applied to electrons:

$$\lambda = \frac{h}{mv}$$

where h = Planck's constant
m = mass of the particle
v = velocity of the particle

Since X-rays are subject to diffraction and reflection phenomena, these principles can be used to separate X-ray lines of different energies. Crystal WDXS spectrometers are based on the principle shown in Fig. 11.

While engaged in experimental studies in the early part of this century, Sir Lawrence Bragg (1890–1971) noted the similarity of diffraction to ordinary reflection and deduced a simple equation treating collective reflection from planes in the lattice. To derive Bragg's equation, we consider an X-ray beam incident on a pair of parallel planes P_1 and P_2 with interplanar spacing d. The parallel incident rays I_1 and I_2 make an angle θ with these planes. The X-rays reflected at O and C will have the same wavelength as those of the incident beam. For that particular

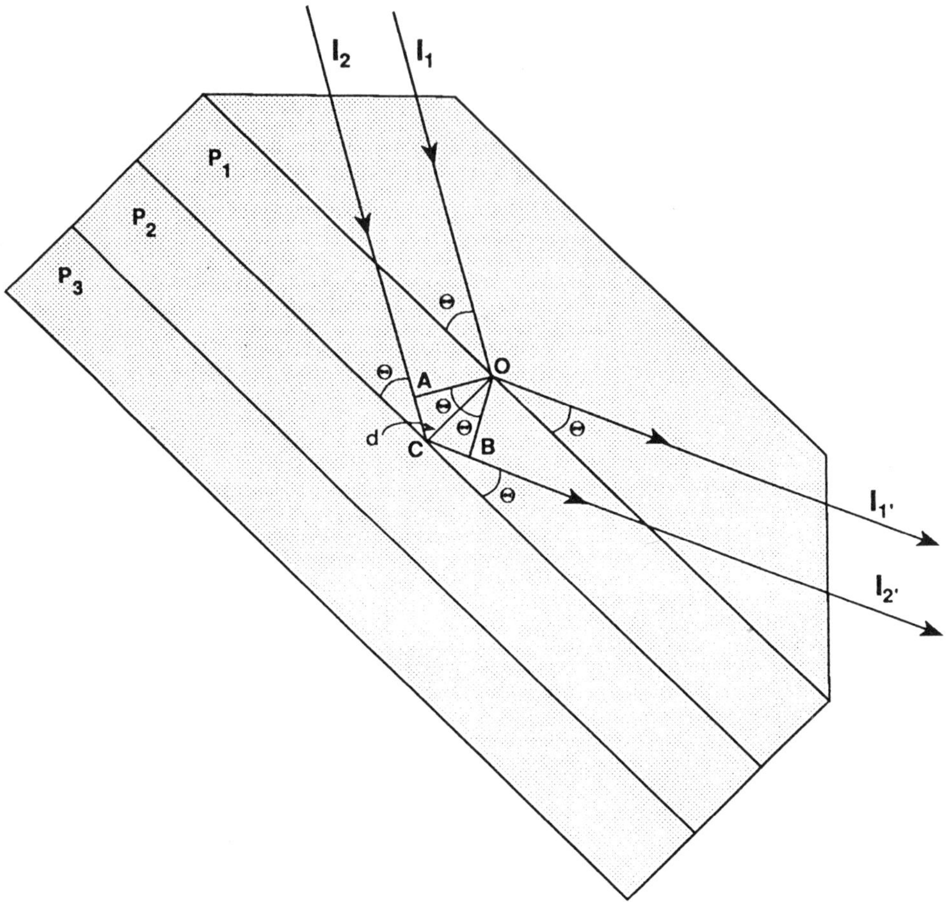

Figure 11 Reflection of X-rays at crystal lattice planes.

direction where the parallel secondary rays 1′ and 2′ emerge at angle θ as reflected from the planes, a diffracted beam of maximum intensity will result if the electromagnetic waves of these rays are in phase and reinforce one another.

By dropping perpendiculars from O to A and B, respectively, it is evident that

$$\angle AOC = BOC = \theta$$

Hence $AC = BC$, and waves in ray I_2' will be in phase (crest to crest) with those in I_1' if $AC + CB = 2AC$ is an integral number of wavelengths (λ). This is expressed by the equality

$$2AC = n\lambda$$

where n is an integer. By definition $AC/d \sin\theta$ and, by substitution,

$$d \sin\theta = n\lambda$$
$$\text{or } n_\lambda = 2d \sin\theta$$

where n is an integer, λ is wavelength of the X-rays, d is interplanar spacings, and θ is diffraction angle. This latter wavelength:diffraction relationship at successive crystalline planes is Bragg's law.

The application of this principle in a basic WDXS spectrometer design is shown in Fig. 12. A crystal of defined interplanar spacing is moved in an arc on the spectrometer focal circle. The crystal at various positions is oriented with its diffracting planes normal to a line connecting the center of the crystal and point *O*, which is on the focal circle and equidistant from the source (*S*) and detector (*D*) of the spectrometer. As the crystal is scanned through a range of positions on the focal circle, Bragg reflection of monochromatic X-rays is achieved for X-rays of differing wavelengths.

The principal limitation of this type of spectrometer is that the geometrical requirements of X-ray focusing restrict the solid angle of X-rays from the specimen to be analyzed. The small solid angle covered by a typical high-resolution, fully focusing spectrometer combined with absorption losses in the diffracting crystals produces a total efficiency of only about 10^{-4}. Despite this drawback, the sensitivity of crystal spectrometers is greater than EDXS with regard to detection limits [11].

The quantitation of electron energy levels after specimen interactions can also serve as a means of particle identification, since the number of backscattered electrons or the loss of energy from the incident electron beam by inelastic collisions is characteristic of the composition of the sample. Using this mechanism, EELS spectroscopy analyzes the decrease in energy of the incident electron beam of a transmission microscope in passing through a thin specimen. The EELS analysis is accomplished with detectors that selectively collect electrons with energies at a specific value via a magnetic deflection detector, in which the magnetic field can be adjusted so that only electrons with a specific velocity and energy will be deflected to the detector. Varying the magnetic field thus allows a collection of the transmitted electron energy spectrum. (The principles of EELS are discussed in a later section.) Auger electron spectroscopy is useful in the study of surface composition of a sample. Auger electrons are produced by the internal conversion of a characteristic X-ray, and occur with energies of up to about 4 keV. Electrons with these energies are very effectively absorbed by inelastic collisions and, as a result, only

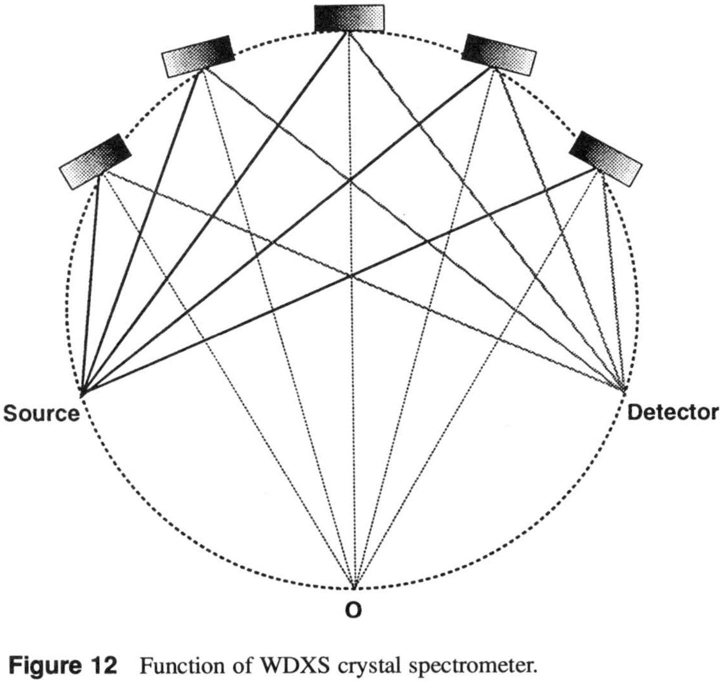

Figure 12 Function of WDXS crystal spectrometer.

electrons emitted from a 1- to 2-nm surface layer are detected; this makes Auger electron spectroscopy well-suited to studying surface characteristics of particles.

VII. ELECTRON MICROSCOPY AND X-RAY SPECTROMETRY FOR PARTICLE ANALYSIS

It has long been recognized that X-ray spectrometry in combination with SEM or TEM is invaluable to the particle analyst [7,12,13]. Particles less than 0.1 μm in size may be examined at high resolution and simultaneously analyzed for 1% or greater concentrations of elements of $Z \geq 6$ (carbon); this analysis represents detection of subpicogram quantities of elements. One of the greatest advantages of the method is that elemental composition and morphological data are gathered simultaneously, and interpretation of the acquired line spectrum can be accomplished by an analyst who has very limited training. The use of X-ray spectrometry is widely practiced both in general [14] and in pharmaceutical particle analysis [15–17].

If they are composed of a conducting material, particle samples may be directly viewed in a scanning electron microscope without application of a conductive coating. For viewing and X-ray spectroscopic analysis, such conducting particles are simply attached to carbon specimen supports of appropriate size with some adhesive material, such as acrylic cement, and placed in the instrument. Particles can also be mounted using two-sided plastic tape. Using this method, particles to be analyzed are picked up from the substrate with one side of the tape or dusted onto the tape surface, and then the tape is attached to the specimen support. The ease of sample preparation for SEM contrasts with the more involved procedures required for TEM. Specimens for transmission microscopic analysis must be sufficiently thin to allow electron penetration through their whole volume. This requirement often makes the preparation for TEM involved and time consuming.

Particulate matter specimens to be analyzed in a TEM are most commonly placed on a cast thin-film support attached to a metallic or nonmetallic grid. Consideration must be given in grid selection to the type of analysis to be performed (morphologic or compositional) and the thickness of film that may be employed, because these factors determine the size of the grid openings used and the composition of the grid. Thermally evaporated films, such as silicon, require smaller grid openings, and plastic or carbon grids may be used for X-ray analysis. Supporting films are usually made of plastic or carbon, though other materials may be used. Suitable solutions for film production may be made up of 2% collodion in amyl acetate or 0.5% Formvar plastic in ethylene dichloride. Several methods may be used to produce films. The most common method involves stripping a collodion or Formvar film from a microscope slide. A microscope slide is first cleaned with a detergent solution and polished with a soft cloth without rinsing, to facilitate later stripping of the film. The slide is then dipped into the Formvar or collodion solution and allowed to drain dry. A razor blade is used to cut through the film at the periphery of the slide, and the film is then floated off onto a water surface. Grids are placed on the surface of the film with a forceps, and the film with the attached grids is picked up by touching it from above with a strip of masking tape or newspaper. A simpler method of producing a TEM preparation from a powder sample is to place a few milligrams of the powder on a microscope slide, add a drop of the suspension of Formvar in a suitable solvent, and rub the powder out on the slide with a glass rod. If prefilmed grids are used, a drop of the particulate matter suspended in water may simply be placed on the filmed grid and allowed to dry. All methods of bulk particle mounting for either TEM or SEM have two undesirable features: (1) they may lead to particle agglomeration, and (2) the degree of dispersion may vary with particle size/shape. Both of these artifacts will result in a sample that will not truly represent the original material.

When particle identification is the primary objective, it will often be desirable to select and mount individual particles for analysis. As discussed earlier, a stereomicroscope is used for the selection of the particle, which is then picked up with the tip of a tungsten needle and adhered to a carbon or beryllium support with a suitable adhesive. After a particle has been mounted, it must typically be provided with a conductive coating. Nonconducting materials or materials that are poor conductors will accumulate a negative charge from the electron beam in typical high-voltage scanning microscopes. The accumulation of charge (charging) produces a potential that causes a deflection of low-energy electrons and increases the artifactual emission of secondary electrons from the sample surface. This results in sporadic variations in the emissivity of the sample, poor resolution, possible radiation damage to the sample, and X-ray emission from areas adjacent to particles of interest. In general, as the energy of the electron beam increases, so does charging. This is overcome by coating the particle surface with a conducting material, such as gold or gold-palladium, to provide a conductive path for the removal of the built-up charge on the sample surface. The use of low-voltage scanning electron microscopy (LVSEM) allows visualization of uncoated particles at accelerating voltages of a few hundred volts. The use of EDXS has become invaluable to the particle analyst over the past 15 years. The 1988 Kevex handbook [18] is an excellent introduction and basic text on this form of analysis. In qualitative analyses, a spectrum on which particle identification can be based often requires less than a minute to acquire from a submicron particle. State-of-the-art thin window detectors allow detection of carbon and oxygen with excellent resolution and sensitivity. The light element detection capability proves very valuable in analyzing many of the particles that are isolated from pharmaceutical injectable solutions.

The first step in qualitative analysis of an unknown sample using EDXS is to determine the major X-ray lines that are collected. If the critical excitation energy for the element involved is less than the beam energy, then all of the characteristic lines of that element should be excited. If a strong line for a given element is detected (e.g., a Ka line), then the next step is to look for another line that serves to confirm the first, such as the Kb. If the element in question is actually present in the particle, the second line should be found, and the ratio of its intensity to that of the principal line should be within 10% of published data. If the second line is absent, it strongly suggests that the element is not present and that the primary line is due to some other element. If the confirming line is present, but the intensity differs by more than 10% from the published value, peak overlap may be occurring. If so, additional characteristic lines of the elements should be verified to resolve the discrepancy. To some extent, the analyst must always be skeptical of the line identifications assigned by instrument software. Observation of peak shape and presence of confirming lines should routinely be made.

Wavelength dispersive X-ray analysis (WDXS) can also be used for particle identification. This subject is well covered in the book by Heinrich [10]. The resolution and sensitivity of this method are typically more than an order of magnitude better than energy dispersive systems. In addition to an inherent higher peak-to-background ratio, wavelength dispersive analysis allows the identification of elements present at lower concentrations (as low as 0.1%). While the procedure for qualitative analysis is generally the same as with energy dispersive systems, the higher resolution of a wavelength dispersive system increases the number of lines that must be evaluated. Historically, the relative complexity of operation of crystal spectrometers and the greater time requirements for WDXS caused many particle analysts to avoid the use of these instruments. The EDXS method is far simpler and quicker to apply. Further, the expertise required for routine use of WDXS in analyzing particles historically took substantially more time to acquire than most particle analysts felt was justified. These detractions to the use of WDXS have been almost totally eliminated by current-generation instruments. Imaging capability of current-generation multiple spectrometer microprobes are equal to those of conventional high-voltage scanning microscopes. In instruments of contemporary design, multiple computer-driven spectrometers can be applied simultaneously to scan for elements over a wide Z range. Despite the advances in WDXS technology, it is doubtful that the additional cost of wavelength dispersive capability is justified in most cases when an EDXS instrument is available.

X-ray spectrometry is most commonly applied in a qualitative analysis mode for particle identification. The X-ray spectra of some contaminant particles encountered in pharmaceutical products are shown in Table 2. In this application, the presence of specific elements and their proportional concentrations are frequently sufficient to allow the composition and source of a particle to be rapidly identified. If elemental composition is not diagnostic, comparison standards can be used in the same fashion as for microscopy and other forms of analysis.

For qualitative analysis, particles may be transferred directly to carbon or beryllium stubs (planchettes). The latter material is toxic and care should be taken to avoid abrading the stubs or generating particles that may become airborne. Sections of filter material with adherent particles of interest may be attached directly to carbon supports using carbon paste. Paste is superior to paint for many applications since it does not run and is not drawn into filters by capillarity. Samples for EDXS or WDXS analysis are routinely coated with carbon by sputtering or vacuum evaporation; if a section of filter material with particles is being analyzed, care must

Table 2 X-ray Spectra of Contaminant Particles from Medical Products

Material	Elemental composition[a]
Acoustic tile fabric	<u>Ca</u>, Si, Fe, Mg, Al
Glass fibers	<u>Si</u>, Ca, Fe, Mg
Calcium phosphate	<u>Ca</u>, P
Kaolin	<u>Si</u>, Al, Ti, Fe
Gypsum	<u>Ca</u>, S, Fe
Skin cells	<u>S</u>, Ca, Na, K, P
Barium sulfate	<u>Ba</u>, S, Fe
Calcium carbonate	<u>Ca</u>, C, O
Titanium dioxide	<u>Ti</u>, O
Sand	<u>Si</u>, O, Fe
Abrasive powder	<u>Al</u>, S, Fe, Ba
Rubber (chlorobutyl)	<u>Cl</u>, S, Ca, Si
Plant fragments	<u>P</u>, S, Ci, K, Na, Fe
Rust (stainless steel)	<u>Fe</u>, Ni, Cr, O
Wear particles (homogenizer/bearings)	<u>Ca</u>, Fe, Zn, Al
Dried blood	<u>S</u>, Ca, Na, P, K, Ca, Fe
Paper (bond)	<u>Ti</u>, C, Ca, O, Cl, S
Calcium stearate	<u>Ca</u>, C
Calcium oxalate	<u>Ca</u>, C
Paint (white)	<u>Ti</u>, Si, Al Fe
Paint (brown)	<u>Fe</u>, S, Ti
Face powder (cosmetic)	<u>Si</u>, Mg, Ti, Fe, Zn
Drug degradant (cephalosporin)	<u>S</u>, C
Amino acid residue	<u>S</u>, C
Teflon	<u>F</u>, C
Cardboard	<u>C</u>, Ca, O, Cl, S

[a]Predominant elements underlined; others appear in order of relative decreasing concentration.

be taken that the carbon adhesive makes contact with the evaporated carbon film to avoid a charging result due to the high beam currents used.

If particles are transferred directly to stubs, a circle should be drawn around them (except on beryllium stubs) with a tungsten needle to facilitate the viewing of their location in the scanning microscope. Particles may be anchored on the stubs by microdrops of Formvar or collodion or by carbon paint. The analyst should view the prepared stubs under a compound microscope, and record the location of particles prior to the X-ray analysis. The particles are also viewed by SEM to verify their location at magnifications of 100× to 250×. If the particles are lost on the stub, the isolation procedure must be repeated.

The quantitative analysis of particles is significantly more difficult. Today the quantitative analysis of bulk samples by either type of X-ray spectrometry is straightforward, applying known correction factors and software-based calculations of elemental concentrations. When particles are analyzed, the correction for beam penetration and fluorescence become more complex. Difficulties in qualitative analyses of particles arise primarily from the fact that the dimensions of a particle may be less than the range of the beam electrons, and that the X-ray absorption effects and fluorescence will differ from that of a bulk sample. Hence, the classical bulk correction factors for atomic number X-ray absorption and fluorescence effects (ZAF) do not apply.

Several methods for the calculation of correction factors for particle size effects have been described. These methods are useful only if computer-based calculations can be used for various steps in correcting the X-ray data to obtain elemental concentrations, as the corrections become exceedingly complex. The NBS publication by Heinrich [20] referenced will provide a useful overview. Particle standards provide a valid way of comparing the unknown sample to a standard with the same or similar size and shape, so that particle size effects can be minimized. Particle standards can be made by pulverizing bulk materials, but it may not be possible to obtain the appropriate particle size and shape suitable for a particular sample. Particle population analysis for compositional characteristics can also be performed by EDXS [21].

Most particles analyzed also have rough or irregular surfaces that interfere, to greater or lesser extents, with X-ray collection geometry. Despite the theoretical complexities of quantitative analysis, good results can be obtained if particles to be analyzed are mounted on thin films of Formvar plastic on carbon grids and irradiated with a raster frame slightly larger than the particle. The thin-film support will minimize background (substrate) effects, and the use of a raster scan (instead of point excitation) will tend to average out density differences within the particle.

VIII. INFRARED MICROSPECTROPHOTOMETRY

In the opinion of some, the most important tool of the organic analytical chemist is the infrared spectrophotometer. The addition of a microscope accessory to a current generation, state-of-the-art Fourier transform infrared (FTIR) spectrophotometer allows the analytical capabilities of this instrument to be applied to a wide range of particulate matter [22,23]. Contaminant particles can sometimes be identified by microscopic examination of their morphological characteristics, but when unique morphology is lacking, infrared spectral data may become critically important for particle identification [24–25]. Infrared microspectrophotometry is widely used in the identification of contaminant particles in the pharmaceutical, aerospace, and microelectronics industries.

In most of today's particle analysis laboratories, the micro-FTIR spectrophotometer powerfully complements the use of the SEM with EDXS capability [26]. In combination with SEM-EDXS, this instrument has revolutionized the chemical analysis of particulate matter. The inorganic elemental constituency of an unknown particle can readily be obtained by EDXS; the FTIR microspectrophotometer can obtain well-defined infrared absorption spectra from particles of 10 μm or smaller in size. Thus, the combination of the two instruments provides for identification, or at least the structural elucidation and characterization of the large majority of particles encountered in pharmaceutical research or problem-solving investigations.

Microscopic infrared analysis requires an all-reflecting (Cassegranian optical) microscope attached to an infrared spectrometer with interferometric rather than dispersive detection. Interferometric detection allows the advantages of Fourier transformation to be realized and avoids overheating the sample with an intense beam of infrared radiation. Infrared microscopes are usually used in a transmission mode, requiring the transfer of a sample to an appropriate substrate such as BaF_2, KCl, or NaCl. Reflectance optics, available on some commercial instruments, allow a double-pass transmission measurement to be made from particulate matter, provided the sample is in close contact with a reflective surface and is not too thick. With particles, as with bulk samples, thickness and shape can have detrimental effects on infrared spectra. Fibers and irregularly shaped particles can create focusing and scattering effects leading to baseline distortions and very inaccurate absorbance measurements. These problems can be of such magnitude that it is not possible to perform spectral subtraction, which is a valuable method of interpreting spectra of mixtures. Flattening the sample manually or in a KBr cell is necessary in some cases.

In general, micro-FTIR is very sensitive and is capable of recording usable spectra for subnanogram sample quantities. As with bulk samples, the particle infrared spectrum is useful because it provides functional group information, and it has a fingerprint region that is specific for a given molecule. Comprehensive spectral libraries are available with most commercial infrared data systems. The most common type of microsampling accessory uses a reflecting infrared microscope that includes a visible illumination system for the visual examination of the samples, and a dedicated on-axis, small area, solid-state mercury-cadmium-tellurium (MCT) detector (Fig. 13). The small area detector reduces the detector-induced noise, thus yielding higher sensitivity. The sampling size can range upwards from less than 10 mm in size [27,28].

The sensitivity of the microsampling technique in transmission studies is illustrated by Figs. 14a,b. These figures show a $CaHPO_4$ particle and the spectrum obtained from the particle, which allows comparison to reference spectra. The signal-to-noise ratio of the spectrum from the particle compares favorably to that of the bulk spectrum down to 1000 cm^{-1}.

Microsampling accessories can be adapted for use in the near-infrared region by replacing the MCT detector with an InSb detector. Overtone and combination bands and electronic transitions can also be studied using this method. When one studies the transmission spectra in the mid-infrared, the sample thickness must be cut down to the 10- to 20-μm range in some cases by sectioning the sample with a microtome. Thicker samples can be used in the near-infrared.

The application of micro-FTIR analysis is inherently simple. The particle of interest is isolated by a conventional technique (i.e., by microneedle) and transferred to a circular salt plate that is placed on the *x-y* stage of the instrument. The particle is then centered in the microscope field and delimited by an adjustable aperture restricting the area through which the infrared (IR) beam passes. Once the particle is selected, Cassegranian optics are swung into place on the microscope turret, and the beam passes from the optical bench of the spectrophotometer through the particle and into the solid state detector at the top of the microscope. For dichroic studies, an infrared transmitting polarizer (gold wire grid supported on KRS-5) may be inserted in the optical train above the eyepiece of the microscope.

Micro-Raman spectroscopy may be a useful adjunct to micro-FTIR [29] for some samples. This methodology is based on elastic scattering by the sample molecules. Infrared absorption requires a change in the molecular dipole moment, whereas Raman scattering requires a change

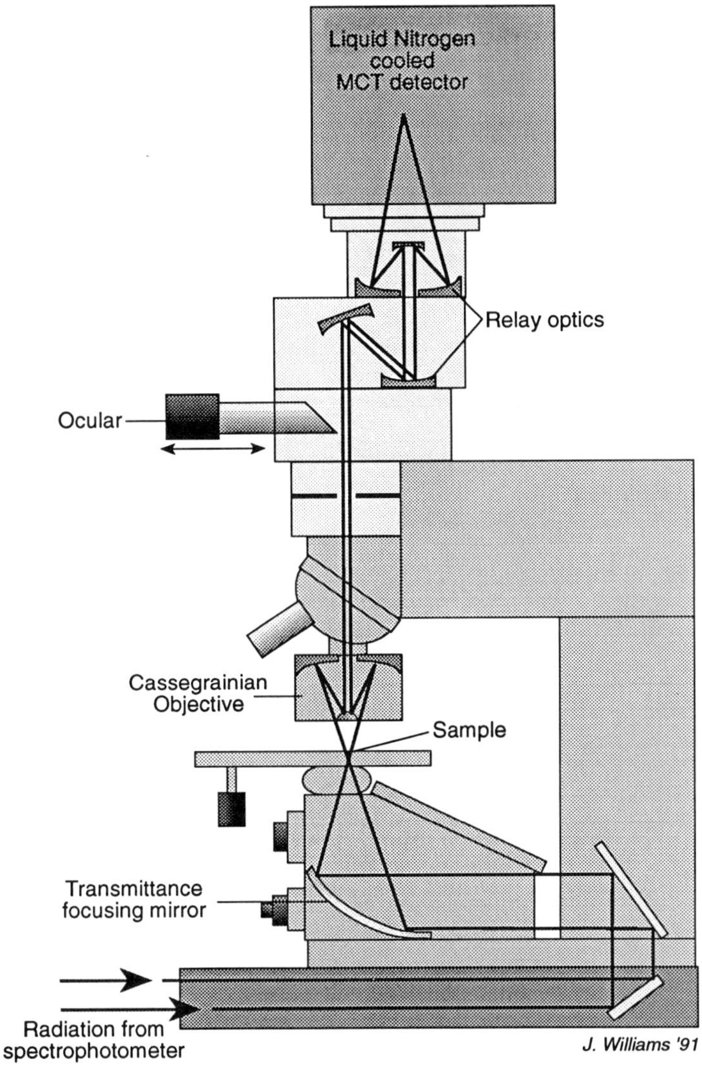

Figure 13 Infrared microspectrophotometer with solid-state detector.

in the molecular polarizability. If the molecules of a material are symmetrical, the fundamental vibrations will sometimes be Raman active rather than IR active. If the molecules of the material studied have only a slight degree of symmetry, IR absorption and Raman scattering are complementary. In general, vibrations of the molecule involving polar bonds are strong in IR spectroscopy, and those involving nonpolar bonds are strong in Raman. A laser beam is an ideal source for Raman spectrometry of particles because of its monochromaticity, its intensity, and its degree of polarization. Micro-Raman spectroscopy may be applied in the analysis of discrete, fine particles less than 1.0 μm in linear dimension as reported by Rosasco et al., in 1975 [29]. The single particles analyzed included thorium oxide (from 0.9 to about 40 μm), crystals of sodium nitrate and ammonium nitrate (0.7 to 20 μm), and polystyrene latex spheres (about 5 μm diameter).

In a laser micro-Raman spectrometer the sample to be analyzed is mounted on a microscope

Figure 14a $CaHPO_4 \cdot 2H_2O$ crystal rosette.

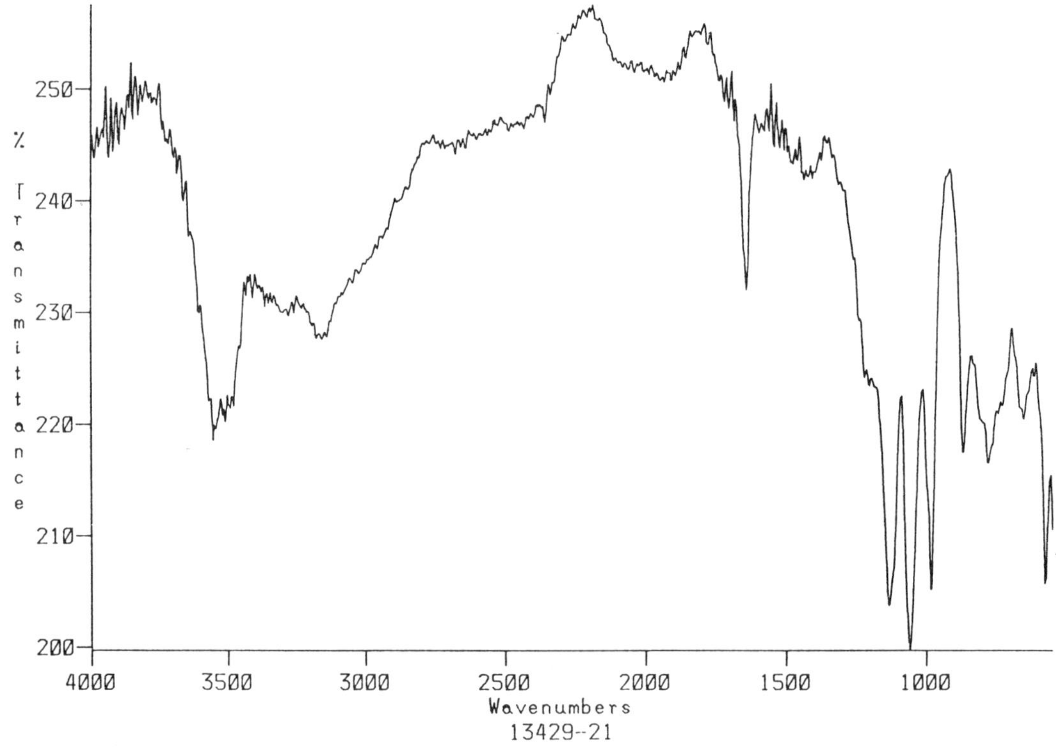

Figure 14b Micro FTIR spectrum from a single $CaHPO_4 \cdot 2H_2O$ crystal.

slide and illuminated by a laser beam. Molecular analysis of a specific particle within the sample area under microscopic examination can be accomplished using its Raman spectrum in a monochannel mode. In the image mode, scanning the spectrometer at a characteristic frequency of a specific compound produces only the image of the particle having this chemical structure (similar to an X-ray dot map). To study the relatively weak Raman lines, light emerging from the microscope is processed in a spectrometer capable of isolating a line of interest for analysis. The light of the frequency selected from different origins in the field of view forms an image on the screen. The image formed can be viewed on a video monitor, much in the manner of a scanning microscopic image. The picture generated on the video monitor will show only the image of the material that has the principal line for which the spectrometer is set. Thus, this analytical method allows the qualitative localization of organic, inorganic, and biological samples on a microscopic scale. It can also be applied to the microanalysis and selective micrography of particular particles within a mixed sample of different particle types.

IX. MASS SPECTROMETRY

Mass spectrometry (MS) is, on occasion, extremely useful in characterizing or identifying particulate matter. Modern instruments provide high mass resolution and accuracy in mass determination for both atomic ions and complex organic species. There are a number of ways of obtaining mass spectra from particles. If a conventional instrument is to be used, probably the best is direct insertion on a wire probe and desorption by heating, chemical ionization, or fast atom bombardment. This sampling method is useful for many types of samples and allows relatively small amounts of particulate matter to be analyzed in the high-resolution, double-focusing mass spectrometers typically used for bulk materials. Mass spectrometry is a very sensitive means of analysis capable of analyzing subnanogram quantities of particles. As with all analytical methods, there are some disadvantages. Mass spectrometry exhibits greater sensitivity for volatile compounds, so that particles of volatile material or those with volatile components are forward in the analysis. For this reason, it is important to check the integrity of the sample before and after analysis to ensure that the spectrum seen is representative of the bulk of the sample and not simply a volatile component. Some particles are difficult or impossible to ionize, thus preventing their analysis by this method.

The laser microprobe mass analyzer (LAMMA, Fig. 15) meets many of the requirements for chemical analysis of particles, including delimitation of the ionizing radiation to the area of a single particle if necessary. The instrument illustrated consists of a laser ion excitation mechanism attached to a conventional time-of-flight mass spectrometer. The particle to be analyzed is visualized using the microscope accessory, and targeted with the pilot laser. Ions are produced by a short, powerful burst from the Nd-YAG laser. In the case of techniques such as direct insertion, MS is primarily limited to particles greater than 10 μm in size, but the size range amenable to LAMMA analysis starts at approximately 0.2 μm.

In the LAMMA, the microprobe objectives and other diffracting elements are made from UV transparent material. The specimen is located beneath a thin (0.15 mm) quartz slide that serves both as a vacuum seal and optical window. This configuration permits the use of a wide range of light microscope objectives, including 100× immersion objectives. Visualization of the specimen is possible by either incident or transmitted light, including UV microscopy at 260 nm. This not only improves the imaging resolution of the microscope, but also provides for better optical classification of an inhomogeneous particle sample. During analysis, the specimen can only be seen by incident illumination because the optical condenser underneath the specimen has to be exchanged for an electrostatic ion lens. To select the area of interest, the vacuum flange mounting the specimen is manually shifted in the x- and y-direction via two micrometer screw

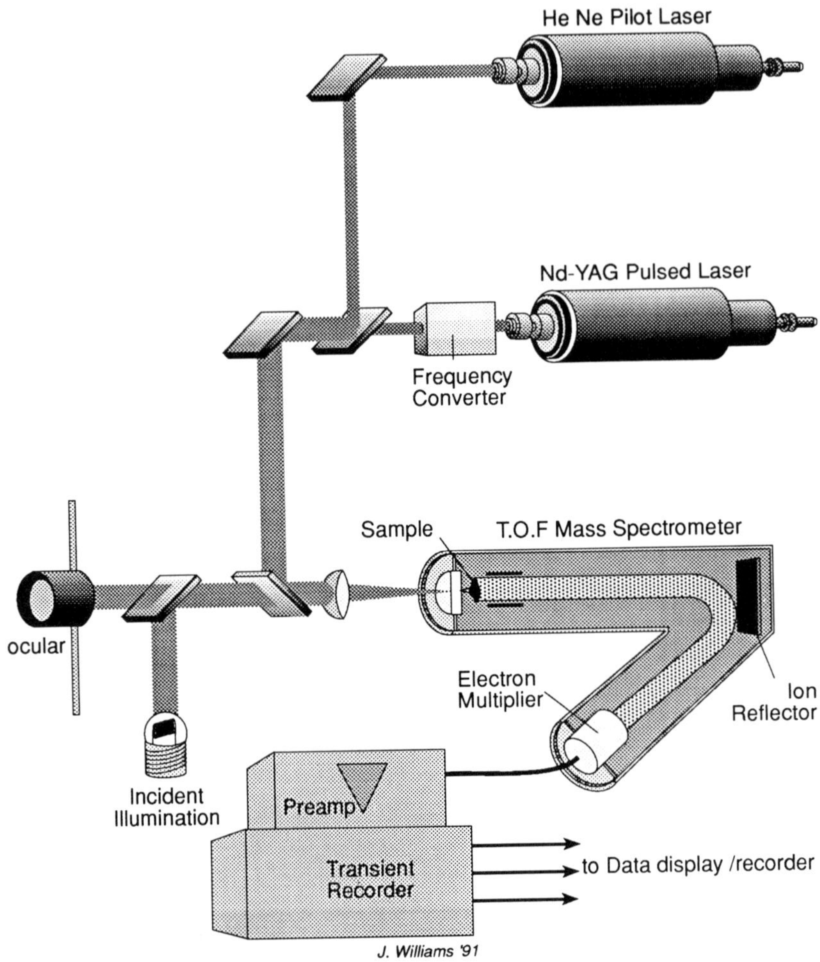

Figure 15 LAMMA instrument.

drives, and focused in the *z*-direction by a screw drive acting on the microscope. The area of interest is indicated by focusing a red target (He-Ne) laser into the microscope field of view. The typical diameter of the analyzed area is about 0.5 μm with a 100× objective, and 1 μm with a 30× objective. The pulsed YAG laser is then used to ionize the sample.

The time-of-flight spectrometer consists of an accelerating and focusing electrostatic lens underneath the specimen, a field-free drift tube 1.20 m in length, an ion reflector at the end of this tube reflecting the ions to an angle of nearly 180°, and a second drift tube about 60 cm in length. An open secondary electron multiplier is used as the ion detector. The ion reflector compensates for the spread of initial ion energy (0 to 50 eV).

When laser light is focused on solid-state material, the prevalent energy transfer process depends strongly on irradiance (i.e., the power density in the focus) and may cover classical absorption at the low irradiance end of a large variety of nonlinear optical processes occurring at irradiance $\geq 10^6$ to 10^7 W/cm^2. The amount of energy deposited per unit volume determines the occurrence and the amount of physical changes within the irradiated material, such as heating, phase transition processes, ionization, and thermal degradation.

The primary disadvantages of the LAMMA technique are (1) the high laser beam energy that may result in uncontrolled ionization or carbonization of an organic sample rather than a controlled generation of ions; and (2) the rather poor mass resolution of the time-of-flight spectrometer. The results obtained with LAMMA show a strong dependence on sample preparation and the geometry of the sample as well as of the laser beam. In particle analysis, there will typically be a predominance of volume evaporation. There is an upper limit of particle diameter beyond which the particle can no longer be totally evaporated by laser desorption [30]. This limit is around 1 μm if one relies on the laser beam geometry at the focal point. The upper limit at which signal intensities are proportional to the volume of the particle was found to be in the range of 0.6 to 0.8 μm by Surkyn and Adams [31] employing glass microspheres. Kaufmann and Weiser [32] found a limit of 1.2 μm for a 32× objective, and 3.0 μm for a 100× objective in a size dispersive specimen of NaCl microcrystallites. If particle size exceeds these upper limits, selective or only partial evaporation and the eventual redeposition of material must be taken into account. Further disadvantages are saturation effects on the multiplier for signals related to leading constituents, and peak deformations due to relatively broad time and energy spreads of the produced ions. Despite all of these negative points, the instrument's sensitivity allows for the detection of trace contaminants in single particles, and analysis time is fast enough to obtain statistically meaningful data in a comparatively short time period.

A short time after its introduction into particle and aerosol research, the LAMMA technique demonstrated its capability to provide unique insights into the chemical nature of particles not readily attainable by alternative techniques [33]. However, it is not widely used at present.

X. SECONDARY ION MASS SPECTROMETRY (SIMS)

A SIMS instrument consists of a mass spectrometer coupled to a highly sophisticated excitation mechanism. In SIMS, particles or surfaces to be analyzed are excited by a focused collimated beam of high-energy ions (most commonly $^{16}O^-$, $^{32}O^+$, $^{40}Ar^+$ or $^{133}Cs^+$). The primary ions acquire their high-energy levels by acceleration across a potential of 5 to 20 kV in the "gun" of the instrument. The most commonly used ions for analysis are $^{32}Cs^+$ (produced in cold cathode field emission sources) and $^{133}Cs^+$ (produced by thermionic emission). Secondary electron images may be obtained by using an electron multiplier to detect secondary electrons emitted from the sample under ion bombardment; these images may be used to localize the area for ion analysis. Secondary ion images are obtained by adjusting the mass spectrometer to detect only the element of interest and using this signal to modulate a CRT. The primary ion beam can be used to perform bulk or surface mass analyses on isolated particles less than 10 μm in size. The instruments currently available differ both in column design and mechanism of ion detection; an instrument based on a quadrupole mass spectrometer is shown in Fig. 16. The SIMS technique constitutes a very useful nonroutine method of particle analysis. The reference by Gavrilovic [34] provides an excellent overview of the method.

Two principal designs of SIMS instruments (ion microscope and the ion microprobe) are used today. Within the ion microscope as developed by Castaing et al. [35], a 10- to 300-μm ion beam strikes the sample and a magnified secondary ion image is produced through ion transfer optics. A portion of this image can be selectively isolated, examined, and analyzed in detail for chemical composition, isotopic ratios, or other information. The spatial resolution of an ion microscope is limited to about 1 μm, while the mass resolution depends upon the mass spectrometer design. The ion microprobe, first described by Liebl [36], is based on an electrostatic ion optics system that focuses a primary ion source into a 2-μm ion beam. This primary ion beam is rastered over either a large or small area to generate secondary ions from the specific sample area of interest. The secondary ions are analyzed using a mass

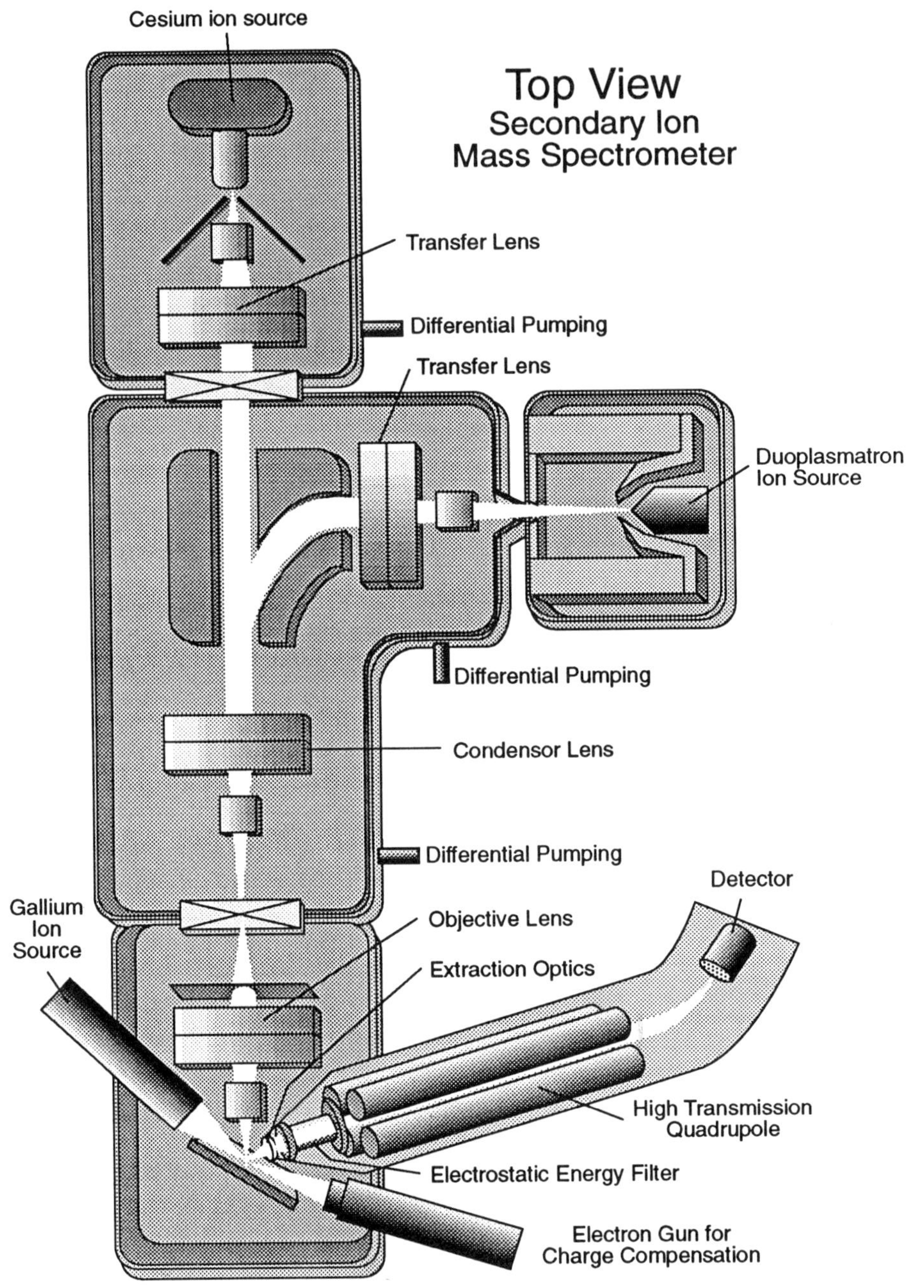

Figure 16 SIMS instrument with quadrapole mass spectrometer.

spectrometer. The microprobe system is capable of nearly 10-fold higher resolution than the microscope design.

Each of these instruments have advantages and disadvantages. Ideally, when performing an analysis on a small particle, one would like to use a magnified secondary ion image for the location of small particles. In performing surface analysis, the quality of light optics is not only important but essential, since when the ion optics system is used to visualize the area of analysis the surface layer of interest may be removed before the analysis can be performed. Based on the quality of light optics, the ion microprobe has advantages for both surface and particle analysis [37,38].

The reaction that occurs in the plasma cloud around a solid particle as it is sputtered away by the ion beam is complex and not entirely understood (Fig. 17). Most of the sputtering occurs within 5 to 10 μm of the surface of the particle. Neutral atoms, positive and negative ions, and electrons are all created in the small plasma cloud above and around the small particles being bombarded. The condensation of some of the positive and negative ions formed occurs in the immediate vicinity of the particles. This phenomenon may be important to the conduct of analyses of small particles, since it provides additional material for analysis [34].

Some of the ions generated on the sample surface are neutralized by collision with electrons, and other ions combine with each other or with the ions from the primary beam. Energies within the primary beam impact area are greater than 1000 eV, and the energies of the chemical bond of a compound on a solid surface are well below that level. In addition, the plasma temperature may be 10,000 to 12,000 K. Therefore, the breakdown of chemical compounds on the sample surface and within the plasma cloud is almost complete. When organic compounds are analyzed, their molecules are broken down to relatively small fragments, most of them having mass to charge (*m*/*e*) ratios below 50. Many short-lived molecular fragments (such as CH^+, CH_2^+, CH_3^+) may be detected.

As the sputtering process proceeds, the sample is gradually eroded away with the erosion rate of the small particle being dependent upon numerous factors including

Incident energy of the primary ions
Type and polarity of primary ions

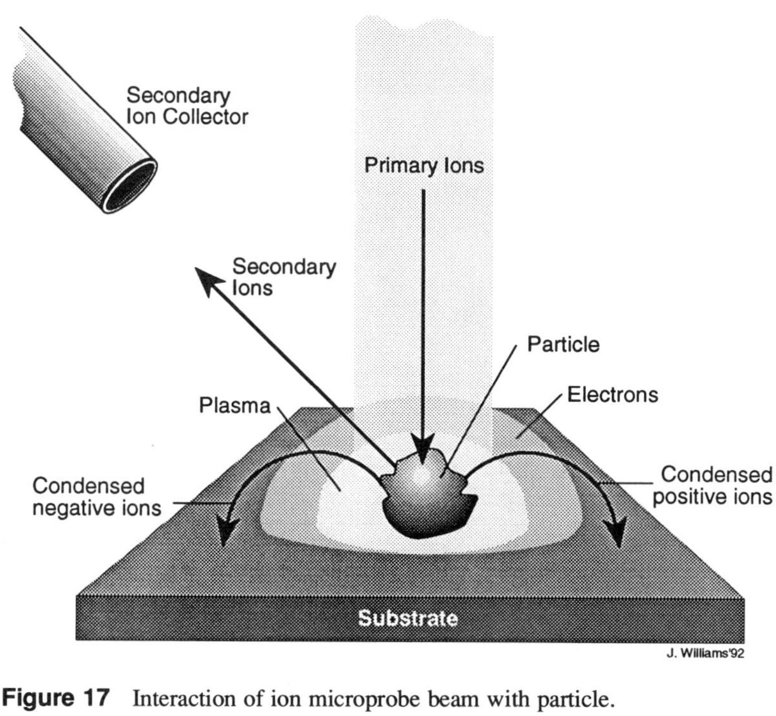

Figure 17 Interaction of ion microprobe beam with particle.

Density of the ions in the beam (ions/μm)
Shape and size of the particle and its surface morphology
Chemical composition of the sample

Although SIMS analysis of small individual particles is ultimately restricted by particle size and the total number of atoms available, such analyses are highly practical. Assuming a 1% secondary ion yield and a 10% collection transmission for the mass spectrometer system, the number of counts that a particle will yield at 10% consumption is such that even a submicrometer-size particle can provide a sufficient count rate for the analysis of a number of elements in a second, and the signal for an associated mass peak may reach over a million counts per second. Thus, for the mass analysis of fine particles, SIMS constitutes an extremely sensitive analytical method; in some cases reaching to the parts per billion range. This sensitivity cannot be achieved in the surface analyses of small particles, due to the small amounts of material that must be analyzed.

Methods for preparing particles for SIMS analyses are critical. Particles exposed to high-energy ion bombardment tend to dissipate some of the absorbed energy by becoming electrically charged and heated, which tends to make the particle move about on its substrate. Thus, some common methods used in mounting very small particles for other microanalysis techniques cannot be used on particles for SIMS when surface analysis is desired. In the surface analyses of small particles, the first several monolayers are of concern, so most of the sample must be left uncontaminated and intact by the mounting technique. For surface analyses, particles larger than 10 μm are usually placed in a "tacky" collodion film. With this technique, the collodion does not flow and coat the upper part of the particle. In spite of all of its limitations, SIMS is a viable analytical technique for determining the surface chemical composition of particles down into the submicrometer size range. For the detection of minute quantities of specific elements on or in small particles, SIMS is very effective, and it may be the most sensitive surface analytical method available for many applications.

XI. CRYSTAL MORPHOLOGY AND CRYSTALLOGRAPHY

Crystallographic chemical analysis has a limited application in the analysis of particulate matter, but in some (few) analyses the definitive identification of particles may be obtained by no other means. Some compounds have identical chemical composition, but a distinctive crystal morphology. For example, diamond and graphite are identical chemically, but they can be differentiated by X-ray diffraction. Titanium dioxide is available commercially in the three crystal forms of rutile, anatase, and brookite. Quartz and crystobalite are different forms of silica. Calcite and aragonite are both calcium carbonate. All of these can be differentiated based on crystal structure. Many compounds show one or more metastable crystalline forms, each with distinctive properties and uses. Characteristic diffraction patterns for use in identification can be obtained from either single crystals or a powder mixture. If the X-ray diffraction data for a given compound have been determined and tabulated, then subsequent samples of that compound can be identified by comparing the measured data with the tabulation. The American Society of Testing and Materials (ASTM) data file reference [39] lists powder diffraction data for over 20,000 compounds.

The principle applied in X-ray crystallography involves the use of X-ray irradiation to produce a highly specific diffraction pattern from a crystal of the material of interest. The text by Stout and Jensen [40] provides both introductory and sample-specific technical information on this technique. If a particle can be obtained in pure crystalline form, either through its

Figure 18 Two-dimensional array of atoms in a crystal.

formation as a precipitate or by selective dissolution and recrystallization, this mode of analysis will generally provide unequivocal identification. Development of computer databases in recent years has reduced the labor associated with the determination of crystal structures by diffraction analysis. Electron radiation can also be used to produce characteristic diffraction patterns as described in the following sections. Today X-ray crystallography, and to a lesser extent electron diffraction, are powerful tools for particle analysis. The single essential criterion that a particle must meet to be analyzed by this means is that it can be obtained in the form of a single pure, or essentially pure crystal. The interpretation of X-ray and electron diffraction patterns is a complex undertaking, even with the development of comprehensive computer databases. Thus, the application of these modes of analyses to particle characterization will almost invariably require the assistance of a specialist.

The reader is directed to three references for an introduction to crystal morphology and diffraction analysis: McCrone et al. [8,9] and Kay [41]. As an introduction to crystal structures, it is convenient first to consider the presentation of crystal structures as a two-dimensional point array. The nature of such an array is illustrated in Fig. 18 by regular repetition of a simple symbol (in this case, a comma).

This array of commas forms a two-dimensional lattice or grid system. Such a grid may be organized in a number of different ways by the addition of lines representing band structures. Figure 19 shows three ways of representing the array in Fig. 18.

In each diagrammatic representation, the grid lines are drawn at equal intervals corresponding to the uniform separation of atoms in the crystal. Crystals of pure material exist as highly uniform, three-dimensional arrays of atoms in the form of lattice planes. The grid systems shown in Figs. 18 and 19 may be extended into three-dimensional arrays of lattice planes. In X-ray crystallography the reflection of the X-rays in a collimated beam at the successive planes of such a three-dimensional crystal lattice gives rise to X-ray diffraction patterns uniquely descriptive of

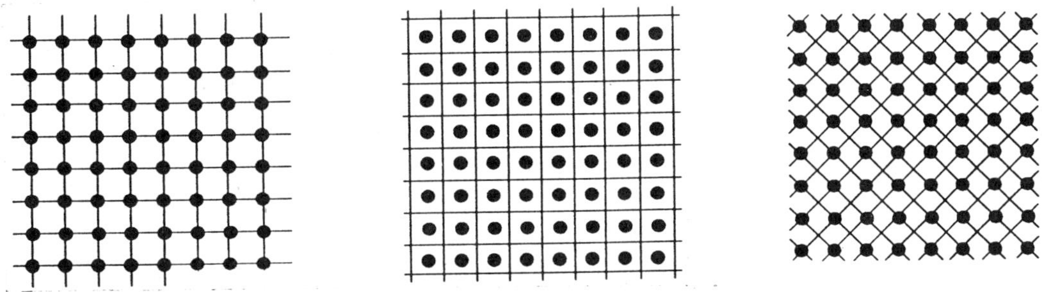

Figure 19 Three different grid systems referred to in array of Fig. 18.

specific chemical compounds; these patterns provide a fingerprint of a crystal that may be used in its identification.

Performance of X-ray (or electron) diffraction studies relates to basic crystal structure. It is the regular arrangement of atoms in space that constitutes the distinguishing feature of the crystalline state. These regular arrays define the crystal structures and give rise to the internal and external symmetry of the crystal. This distinct symmetry of different crystal structures leads to the concept of the crystal lattice, which is a three-dimensional array of atoms at points in space with identical surroundings.

While there are many different crystal structures based on interatomic spacing of component atoms, the various angular arrangements of the component atoms result in only six crystal systems. Atomic arrangements in the different crystal systems are described below and shown in Fig. 20 based on the orientation of the crystallographic axes:

Cubic	Equal spacing of atoms along three mutually perpendicular axes.
Tetragonal	Equal spacing of atoms on two mutually perpendicular axes. Normal to the plane thus formed but the interatomic spacing is different.
Orthorhombic	Unequal spacing of atoms along three mutually perpendicular axes.
Hexagonal	Equal spacing of atoms along three axes 120° apart in the same plane. In a plane normal to the plane thus formed but the spacing differs.
Monoclinic	Unequal spacing of atoms along two mutually oblique axes. Normal to the plane formed by the two axes: the spacing of atoms is different.
Triclinic	Unequal spacing of atoms along three mutually oblique axes.

The atomic arrangement in solid materials is obviously very important since it explains (1) why some solids are crystalline, and (2) why the macrostructure (shape) of crystals of different materials may be significantly different. The notations usually used for naming crystal faces are called Miller indices and are expressed in notation as $\{hkl\}$. These notations are based on the above-described assignment of axes and on an expression of the intercepts of the face on the three axes (the hexagonal system has four). The axes are taken in order: a_1, a_2, a_3 (cubic); a_1, a_2, c (tetragonal); a_1, a_2, a_3, c (hexagonal); a, b, c (orthorhombic, monoclinic and triclinic). In a consideration of axial arrangement, crystals are oriented by convention so that a_1 or a lies front (positive) to back (negative), a_2 or b lies right (positive) to left (negative), and a_3 or c lies top (positive) to bottom (negative). Simple examples of the concept of axes are provided by the axes of the cubic system (Fig. 21).

Thus, the lattice of a crystal may be described with reference to three coordinate axes along which the lattice points have spacings a, b, and c. The angles between the axes are denoted by α, β, and γ. The volume defined by the parallelepiped of sides a, b, and c is termed the primitive or basic unit cell when a, b, and c are the smallest distances that can be chosen. The lattice points then occur only at the corners of the crystal cell or basic unit structure. More complex unit cells may in addition have lattice points at the center of the cell (body centered) or at the centers of the cell faces (face centered).

The simplest cubic lattice is shown in Fig. 22. It has one atom per unit cell with $a = b = c$ and $\alpha = \beta = \gamma = 90°$. In all there are 14 possible lattices distributed between the six crystal systems.

For the purpose of crystallographic analysis, it is more useful to think of the planes that can be drawn through lattice points rather than the lattice points themselves. Taking any point O as the origin of three axes OX, OY, and OZ drawn parallel to the three principal directions in the lattice, a set of parallel planes can be specified by the intercepts on the three axes, measured in

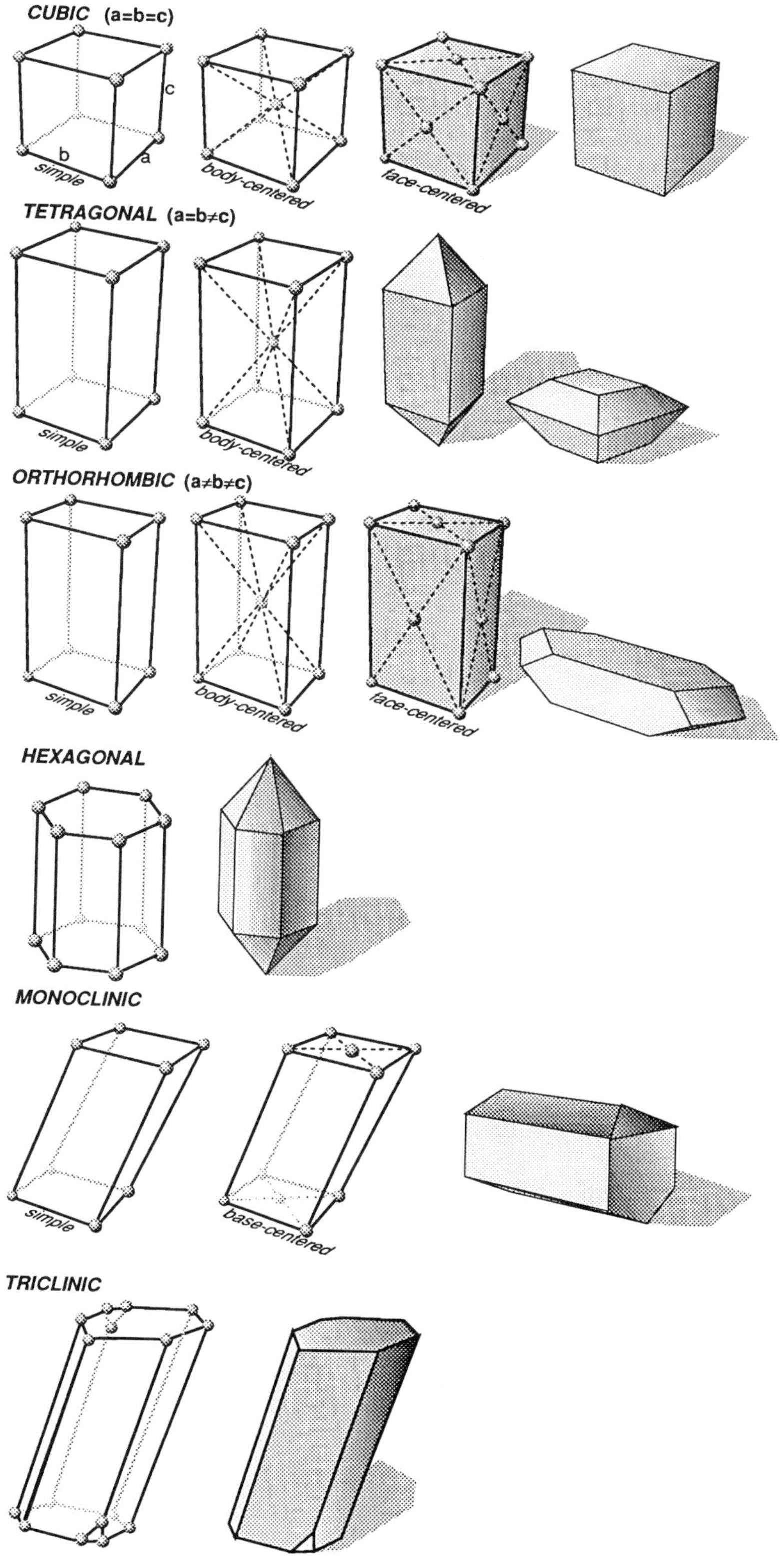

Figure 20 The six crystal systems.

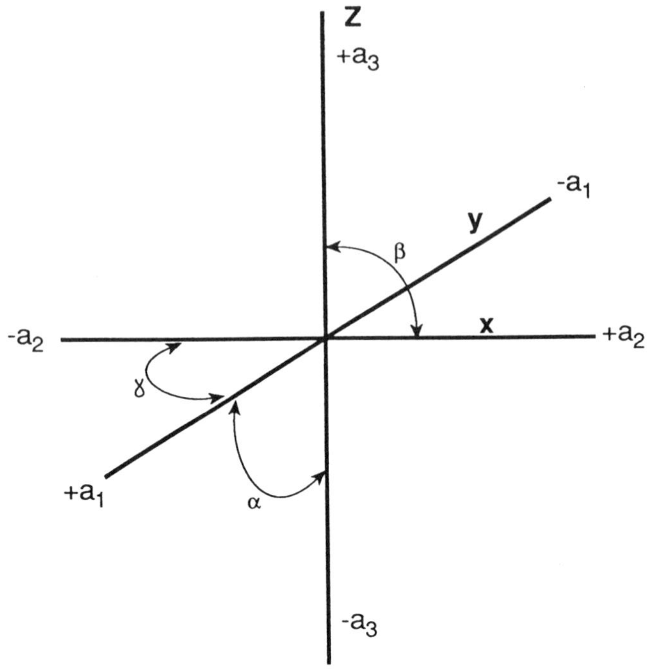

Figure 21 Crystallographic axes in the cubic system.

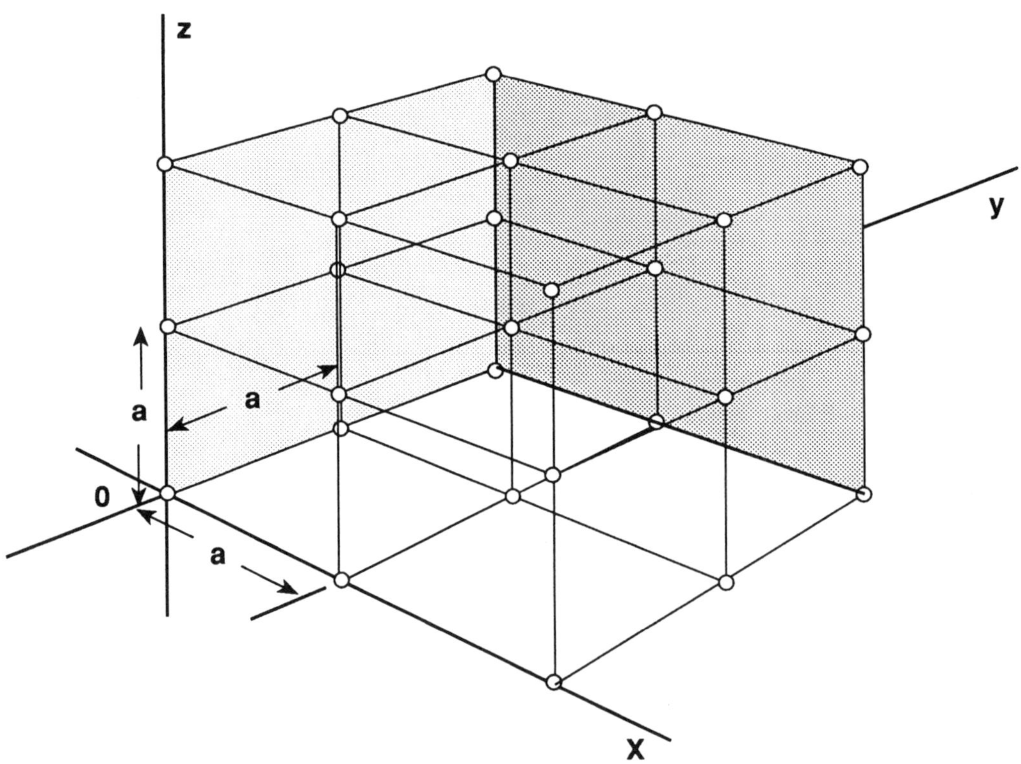

Figure 22 Cubic crystalline lattice.

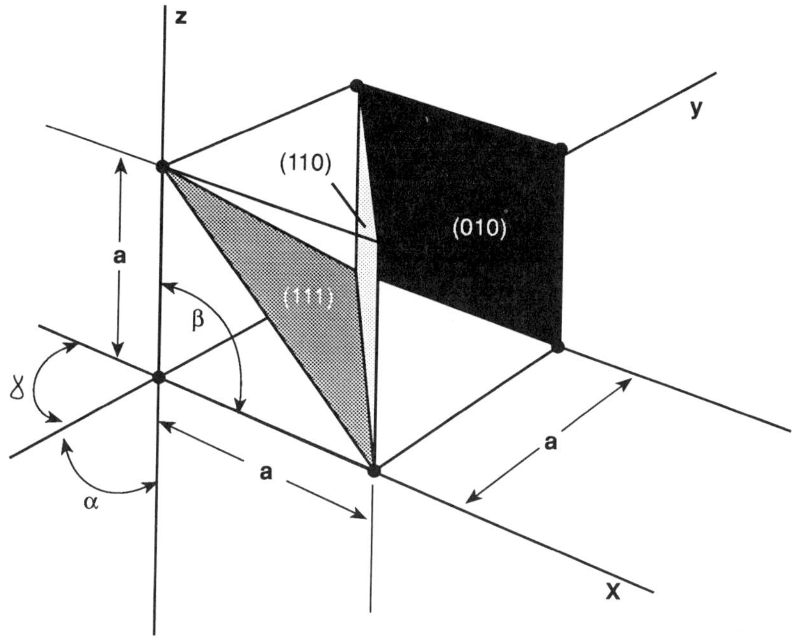

Figure 23 Cubic unit cell showing (010), (110), and (111) planes.

terms of the unit-cell spacings. In practice it is more convenient to use the reciprocals of the intercepts reduced to the smallest possible whole numbers (the Miller indices). For example, the (321) planes would be parallel to a plane making intercepts of $a/3$, $b/2$, c along the axes. The (010), (110), and (111) planes in a cubic system are shown in Fig. 23.

XII. X-RAY DIFFRACTION

Two basic physical phenomena related to the interaction of radiation with the crystal result in distinctive X-ray or electron diffraction patterns. These interactions are (1) reflection of radiation at successive crystal planes, and (2) diffraction of radiation at individual atomic or molecular positions. Whereas reflection consists of the bending back of a ray path at the surface of an object, diffraction is the bending or spreading of ray paths into the area behind an object or opaque barrier of which they have passed.

In the earlier discussion of the Bragg equation and reflection of radiation at successive crystalline planes (Fig. 11) we discussed constructive two-dimensional interferences of the reflected rays. The presentation dealt with diffraction based on collective reflection at single parallel planes. The diffractive phenomena that occur when X-rays interact with a three-dimensional crystal lattice is more complex, both on the basis of multiple diffractive interactions and the third dimension that is added. Despite the added complexity, the Bragg criteria for constructive interference also provide the basis for an understanding of the three-dimensional diffractive phenomena.

Thus, to understand the characteristic ring and spot patterns produced by X-ray diffraction, one must consider the possibilities for three-dimensional diffraction of the radiation at individual atoms or molecules of a crystalline lattice. Figure 24 shows diffraction or scattering of a plane wave at an atom A through the formation of spherical wavelets traveling at an angle 2θ to the original direction of wave motion.

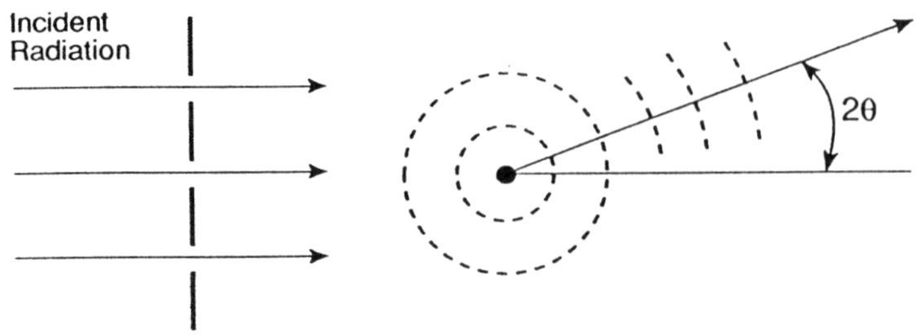

Figure 24 Diffraction of radiation at scattering center.

Before considering diffraction by a regular three-dimensional array of atoms, the principles involved may be discussed by analogy with the diffraction of monochromatic light by a grating (Fig. 25). In this diagram there is a set of slits consisting of plates that are of infinite length perpendicular to the page and have spacing α. A screen is placed at a distance from the grating and the radiation is observed on screen *AB*. If the waves scattered from openings 1 and 2 are in phase, that is their path difference *PD* is an integral multiple of their wavelength *A, X* will be bright. Two in-phase waves are shown in Fig. 25 arriving at the screen. At other points on the screen the path difference may be such that the waves are out of phase, and the total intensity is either very low or zero. When this interference process is considered over the whole screen, alternate bright and dark lines are obtained running perpendicular to the paper with an approximate intensity distribution as shown in the figure. As α increases, the intensity of the fringes decreases, because the efficiency for scattering through large angles is lower. In the case of a three-dimensional crystal, a similar path difference consideration will show that the diffraction of the monochromatic electron beam by the regularly spaced three-dimensional array of atoms gives an interference pattern of beams instead of lines, so that the pattern may consist of dots instead of lines or rings.

In X-ray diffraction analysis, the essential requirements are for a collimated X-ray beam that one can cause to be incident on a crystal of interest (a single, hopefully pure crystal of the

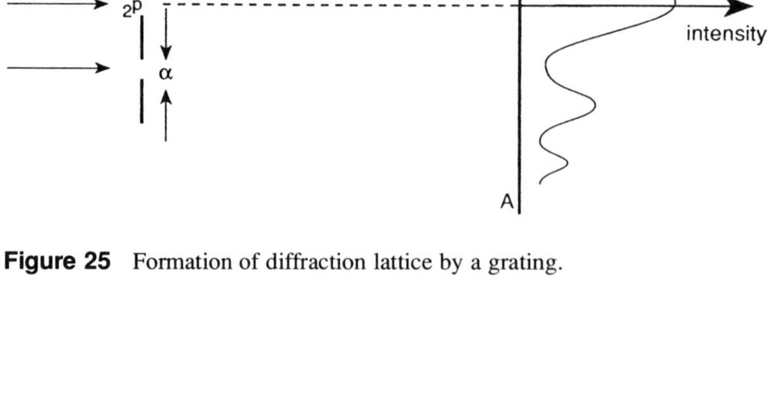

Figure 25 Formation of diffraction lattice by a grating.

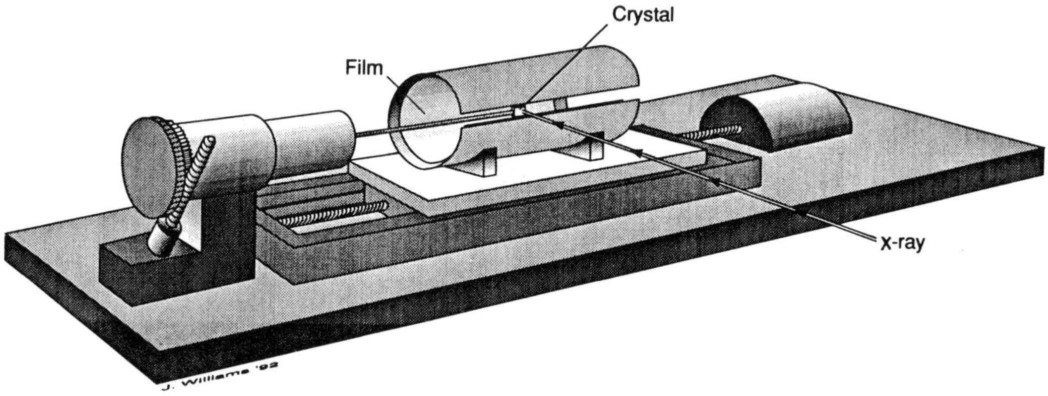

Figure 26 X-ray diffraction camera.

unknown), and a means of projecting/recording the diffraction patterns under rigid geometric conditions. An X-ray diffraction camera is shown in Fig. 26. Diffraction patterns from the unknown crystal are formed and recorded on the photographic film in the cylindrical film holder enclosing the sample.

The relationship of the diffraction angle, θ, to the measured arc length, *S* from the exposed film is shown in Fig. 27. From this the following expression can be written

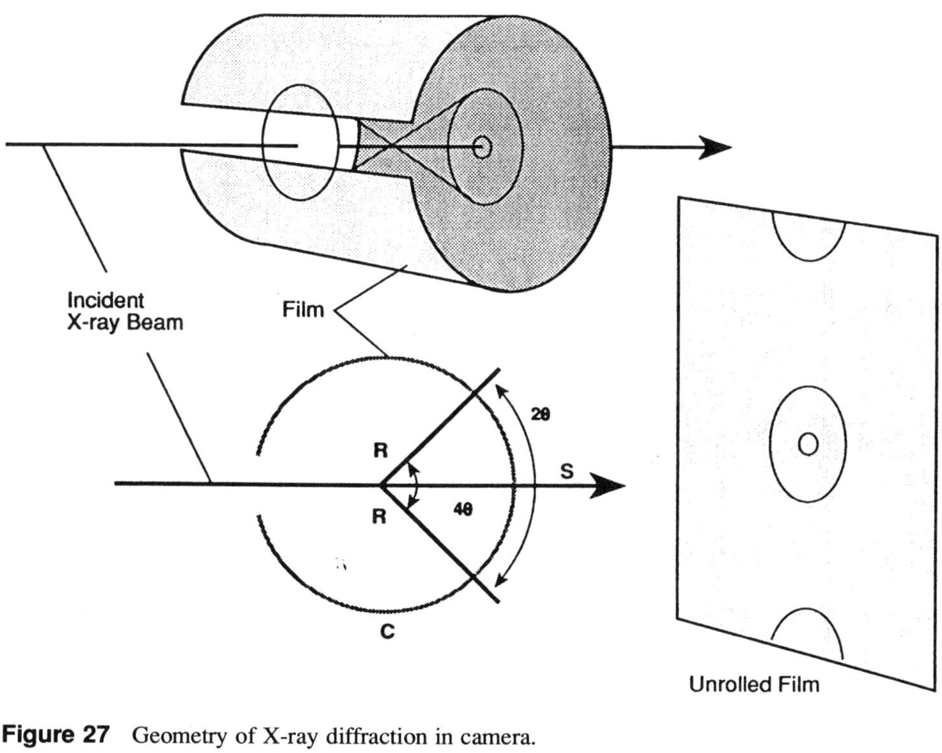

Figure 27 Geometry of X-ray diffraction in camera.

$$4\theta = \frac{S}{R} \text{ radians}$$

where S is the distance of the arc in millimeters and R is the radius of the camera.

Converting to degrees, the expression can be written

$$\theta = \frac{45S}{R\pi} \text{ degrees}$$

With a camera diameter of 28.65 mm, 1 mm on the film corresponds to a 2θ angle of 4°. With any diameter other than 28.65 mm, it is necessary to determine the exact camera diameter before proceeding with the analysis of the sample.

From the values of the distances of the diffraction lines, (S), the value of 2θ is calculated by the above formula. From the values of 2θ, one can calculate the values of interplanar d-spacings by using the Bragg equation. In fact, for a particular X-ray diffraction apparatus of known X-ray wavelength, one can compile a table of interplanar d-spacings as a function of the values of 2θ. In addition to the values of d-spacings, it is necessary to estimate or measure the relative intensity of the diffraction lines corresponding to these d-spacings. Many diffraction patterns resulting from single crystal analysis have been cataloged for use in identification of unknowns [42]. In the absence of such data files, diffraction patterns may be collected from known comparison standards as a means of identification. While spectral information such as IR absorbances or X-ray lines may be equivocal, diffraction patterns provide a unique signature for a given compound.

This camera system X-ray background can be reduced by (1) beam collimation; (2) placing the sample under vacuum; (3) decreasing the air path (using a smaller diameter camera); and (4) reducing the scatter due to the sample mount by using a smaller sample support (e.g., a fine glass fiber). In a typical setup for crystal analysis, the exposure to X-ray radiation in the diffraction camera typically takes place under vacuum maintained in a polymethyl methacrylate vacuum box that is fitted around the X-ray camera. The energy output of the X-ray source is adjusted to an appropriate range (such as 25 to 35 kV and 15 to 25 mA), using a Cu–K X-ray source with a nickel filter, and the film is exposed for 14–18 h. Once the film is developed, the diffraction lines of the X-ray pattern are measured and converted to diffraction angles and d-spacings using the Bragg equation. The relative intensity of the diffraction lines and the d-spacings of the unknown material are then compared to those reported for substances that are suspected to be identical to the unknown sample.

A specialized small-angle X-ray diffractiometer and diffraction camera, especially useful in studies of powders and simple particles, was recently introduced by the Molecular Data Corporation of Cleveland, Ohio. This instrument is applicable in both powder and single-crystal studies with only minor changes in the diffractiometer, and it is worthy of evaluation if purchase of an instrument of this type is contemplated.

The first step of the X-ray diffraction analysis is to obtain the particulate matter of interest in a pure form as a single crystal. It is then mounted in a diffraction camera to produce a diffraction pattern for comparison to reference patterns or to those produced by known standards [43,44]. Obtaining a well-formed crystal of the unknown material may turn out to be far more difficult than the actual analysis. It is at this point that the particle analyst should obtain assistance from a crystallographer or chemist with appropriate training.

If the particulate matter consists of crystals that originated as a slowly forming precipitate without mechanical disturbance in a solution, a single crystal may simply be isolated for analysis. If, in the far more likely case, the analyte exists as poorly defined crystals or as an amorphous material intermixed with miscellaneous extraneous particles, dissolution and recrystallization are required. This generally entails collection of a significant mass (several milligrams) of the material and recrystallizing it to form a single crystal 0.1 to 0.3 mm in size. One of the simplest methods of growing crystals is to allow a saturated or nearly saturated solution to stand under conditions whereby the solvent evaporates slowly. The best crystals are usually produced when the solution is free of mechanical vibration and is allowed to evaporate without disturbance. Some solutions, however, show a marked tendency to become supersaturated, and when crystallization is finally induced it is usually so rapid that only microscopic crystals result. Gross supersaturating can often be prevented by mild mechanical shock or by seeding the solution with a few crystal fragments of the solute. This is obviously a delicate procedure.

A second method commonly employed to grow crystals from solution is slow cooling. Cooling rates from elevated temperatures for solutions contained in laboratory vessels of ordinary size are often too rapid to produce anything but microscopic crystals. The rate of cooling may be slowed somewhat by placing the solution-containing vessel in a large container of hot water, or by surrounding it with sufficient insulating material. The whole process may also be carried out in a Dewar vessel. Extremely slow rates of cooling can be obtained in a well-insulated temperature-controlled oven by reducing the temperature in increments over a period of time.

If difficulty is experienced in growing satisfactory crystals from single solvents by either of the above methods, mixtures of two or more solvents may be attempted. Changing the nature of the solvent often has a pronounced effect on the habit and size of crystals grown. Solvent properties that may influence crystal growth, such as density, dielectric constant, viscosity, solvation, and solubility can be varied in a controlled way over a wide range and adjusted to fit the needs of a particular analysis.

One method that can produce good single crystals with milligram amounts of particulate matter involves crystallization from a solvent by vapor diffusion [40]. Figure 28 illustrates this procedure. A solution of the particulate matter in a primary solvent is placed in a small Erlenmeyer flask or in a test tube. A second solvent is placed in the closed beaker surrounding the flask. The secondary solvent is chosen on the basis of its ability to form a mixed solvent system with the primary solvent in which the material to be crystallized is of lower solubility than in the primary solvent alone. If the secondary solvent is more volatile, it is possible for the volume of solution to increase during crystallization, thereby avoiding the surface "crusts" of precipitated material that often form when concentration occurs during evaporation or diffusion.

Fine glass fibers are used to support a crystalline particle sample for X-ray diffraction in a powder camera [7,45,46]. The use of fine glass fibers (20 to 50 μm in diameter) can help minimize the background on the X-ray film. First a thin glass fiber approximately 0.5 mm in diameter is drawn out over a flame. To mount particles on the drawn glass fiber, the tip is dipped in a small drop of Duco cement, and particles are collected under the stereoscope by touching them with the drop of cement on the end of the fiber. The brass pin and fiber are then transferred to the camera for analysis.

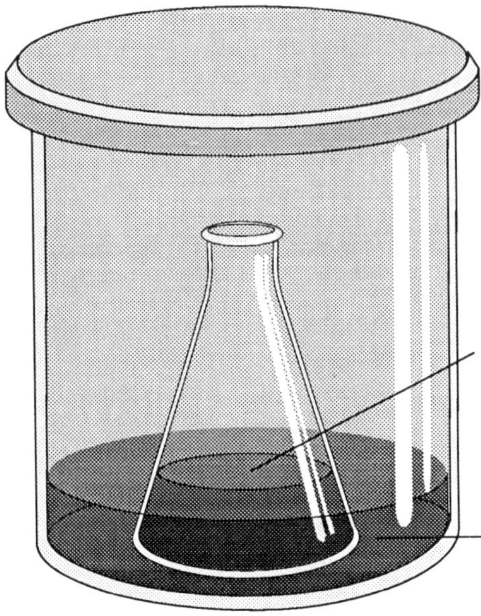

Figure 28 Crystallization by vapor diffusion.

Analysis of particulate matter using X-ray crystallography is most frequently the analytical mode of last resort, due to the fact that most analyses can be completed using more commonly applied instrumentation, such as EDXS and/or FTIR.

With regard to pharmaceutical particulate material, this means of analysis is most frequently applied in the identification of drug degradant materials that may be present in raw materials or that may form as a result of processing (i.e., freezing, lyophilization, or recrystallization). An example of this application is provided by the author's experience with novel degradants, which may be present in penicillin powders. Oxacillin, particularly, has historically been shown to contain a number of degradant materials that have required the use of nuclear magnetic resonance, mass spectrometry, and X-ray crystallography for identification. In one case, degradant problems were first observed as high instrumental and microscopic particle counts in aqueous solutions of sodium oxacillin. The moiety responsible, a thietan-2-one, was observed to form by precipitation in aqueous solutions of the drugs held at room temperature, and particles increased in number with increasing time of storage.

The morphology of the particles was distinctly crystalline, suggesting the possibility of its dissolution in suitable solvents and recrystallization in pure form for crystallographic analyses. Collection of the particulate matter from multiple units on nuclepore membranes and its extraction into solution based on differential solubility experiments allowed isolation of quantities of the material sufficient for a number of instrumental analyses. Following failure of PLM, EDXS, FTIR, nuclear magnetic resonance, and mass spectrometry to provide an identification, analysis of well-formed crystals of the material yielded crystallographic data necessary for unequivocal identification and eventual elimination from the product (Fig. 29).

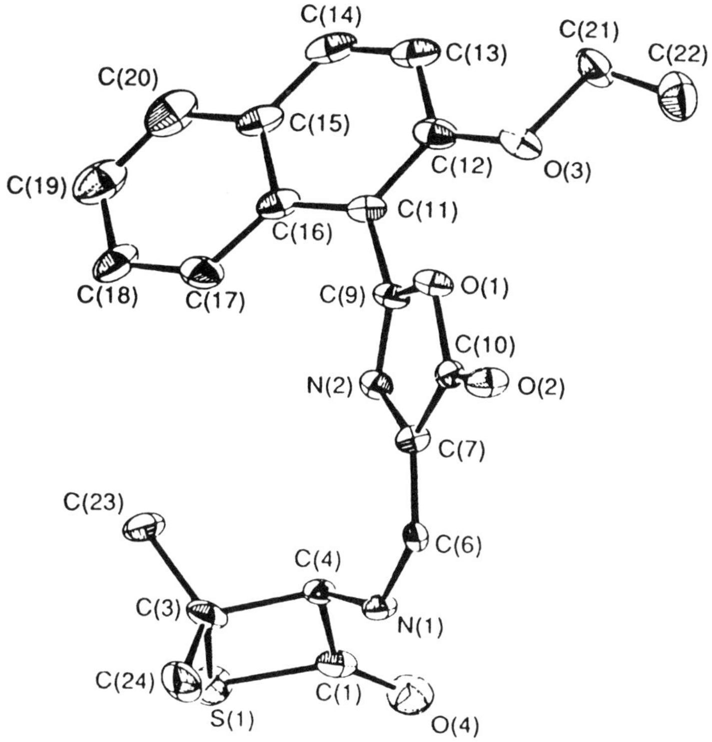

Figure 29 Computer-generated molecular structure of thietan-2-one degradant from oxacillin solution based on X-ray crystallography (crystal grown from acetonitrile).

XIII. ELECTRON DIFFRACTION

The diffraction phenomena by which electrons, light, X-rays, and other forms of electromagnetic radiation are affected are similar. The separation of diffracted X-rays into ring or dot patterns as discussed in the previous section is similar to the analytical principle used to obtain characteristic electron diffraction patterns in the transmission electron microscope. As is the case with X-rays, electron diffraction patterns can be used for structural identification of isolated particles or homogeneous populations. This method, although very useful in specialized cases, is not currently of wide application; though it was once the subject of a good deal of historical interest. The following discussion of electron diffraction to some extent constitutes a continuation of that for X-ray crystallography.

When a beam of electrons is incident on the top surface of a thin crystalline electron microscope specimen, specific diffracted beams arise at the bottom exit surface of the specimen. Although each individual atom in the crystal scatters the incident beam, the scattered wavelets will only be in phase (reinforced) in particular crystallographic directions. This results in the production of diffraction patterns akin to those formed in X-ray diffraction. The mode of analysis is most often referred to as selected area electron diffraction (SAED) since the area of the sample analyzed must be carefully delimited. As with X-rays, this scattering interaction may be discussed in terms of the phase relationships between the scattered waves from each atom in the crystal. As was also discussed with reference to X-ray diffraction, electron diffraction is a somewhat specialized form of analysis where the particle analyst with general training is advised to seek out an expert. Probably the best introductory text is that of Edington [47]. Other general information references are Hirsch et al. [48], Rymer [49], and Ball [50]. Interpretation of electron diffraction patterns is covered in the books by Andrews et al. [51] and Henry et al. [52].

Figure 30 shows the ray paths in the TEM under diffraction conditions. Electron diffraction patterns, like those formed by X-ray diffraction, will be formed according to the Bragg relation, $n\lambda = 2d \sin \theta$ (where n is the diffraction order, λ the wavelength of the radiation, d the spacing between the diffracting planes of atoms, and θ the Bragg diffraction angle). The diffracted beams come to a focus in the objective lens back focal plane. To observe the diffraction pattern, it is necessary for the intermediate lens to focus the objective back focal plane instead of the first intermediate image formed by the objective lens. If an aperture is now inserted in the plane of the first intermediate image, it will pass electrons from only a small area of the sample, hence the designation SAED.

The diameter of the selected area is determined by dividing the aperture diameter by the magnification of the objective lens. For example, an objective lens magnification of 50 and a 50-μm-diameter aperture give a 1-μm selected area diameter. To properly prepare the microscope for SAED requires that the aperture should coincide exactly with the first intermediate image plane of the objective. In practice, the position of the selected area aperture is fixed. The intermediate lens is first focused on the selected area aperture, and the objective lens current is then varied to bring the object into focus in the same plane as is the selected area aperture. The objective aperture is then removed to permit transmission of higher-order diffracted beams, and the strength of the intermediate lens is reduced until the objective back focal plane (i.e., the diffraction pattern) is clearly in focus.

The earlier treatment of X-ray diffraction introduced the important point that strong

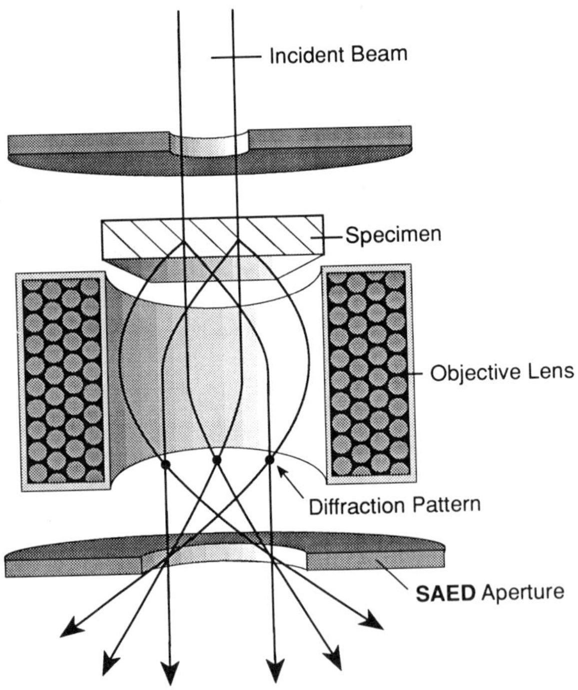

Figure 30 Electron diffraction in TEM.

diffracted beams arise because scattered wavelets are in phase in particular directions in the crystal (that is the path difference is an integral number of wavelengths). Figure 31 shows how diffraction occurs with a thin electron microscope specimen 100 to 300 nm thick. As with X-rays, the incident beam is made up of plane waves in phase and oriented at an angle relative to two (*hkl*) crystal planes (I and II in the figure). If the two waves are reflected by these crystal planes at an angle θ, two situations may occur: (1) The two waves may be in phase, in which case reinforcement will occur and a strong reflected beam will be present; or (2) the waves may be out of phase (that is, they will interfere) and the result will either be zero or a very weak reflected beam. Thus, Bragg's law applies to the (*hkl*) lattice spacing of the crystal for the conditions of electron diffractions as well as for X-ray diffraction.

In electron microscopic diffraction, focusing the objective lens in the same plane as the selected area aperture results in a fixed focal length for the lens, which will be invariant unless the excitation voltage is changed. Thus, for any one excitation voltage, the diffraction pattern will have a defined magnification and so can be standardized. To relate this to an X-ray diffraction camera, we can say that we have fixed the camera diameter. In electron diffraction, this measurement is referred to as standardization of the camera length *L*. In Fig. 32, we can see the relationship between *L* and the interplanar spacing of the crystal lattice.

If *R* is the radius of a diffraction ring on the final visual or recorded image, then

$$\frac{R}{L} = \tan 2\theta$$

But, since ~ is very small (less than 3°), tan 2θ ~ 2θ (radians),

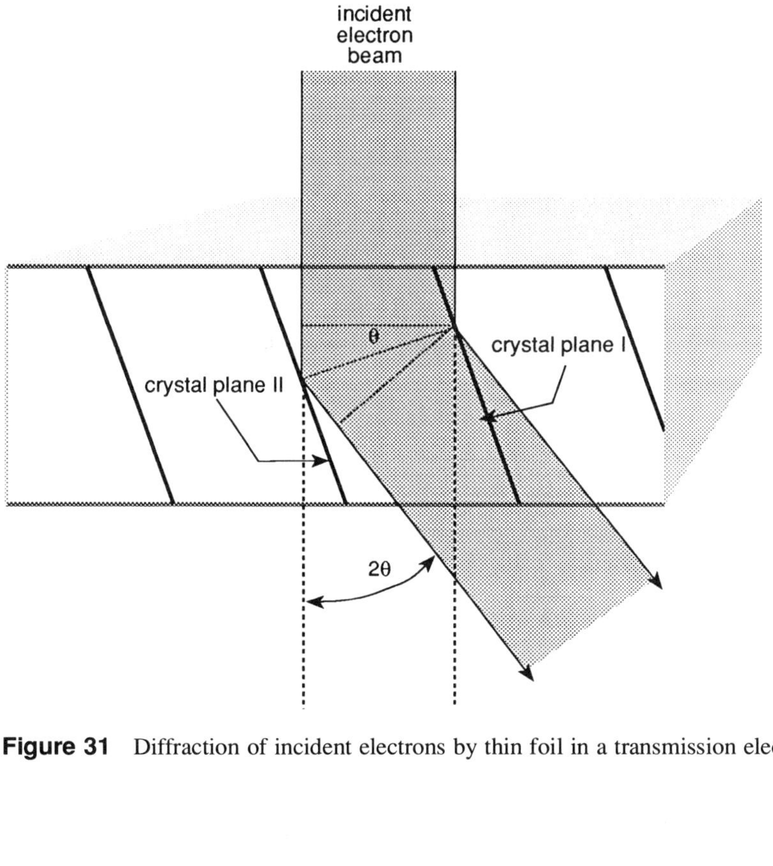

Figure 31 Diffraction of incident electrons by thin foil in a transmission electron microscope.

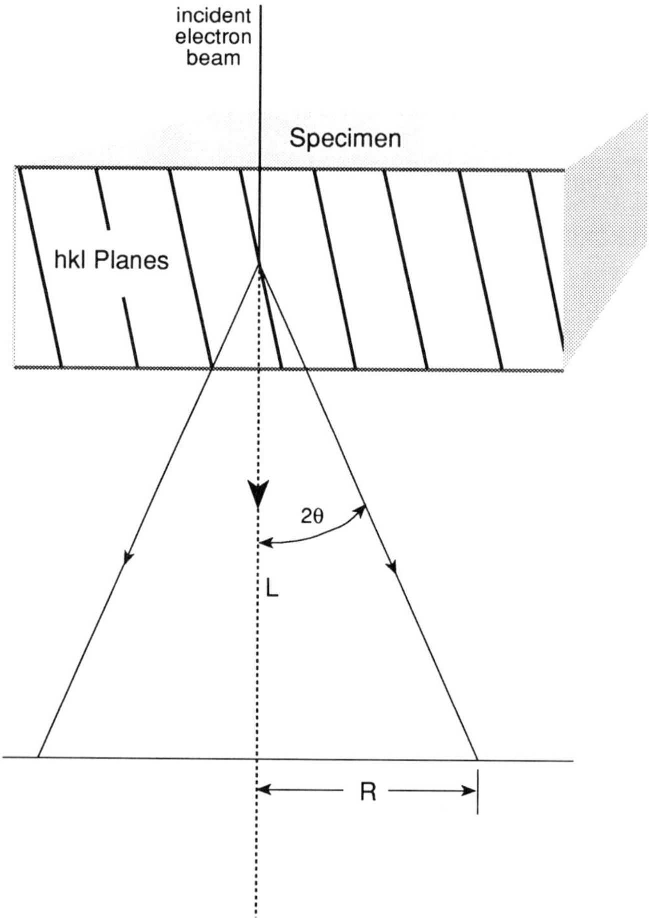

Figure 32 Formation of electron diffraction pattern.

$$\frac{R}{L} = \tan 2\theta$$

And from Bragg's relationship, $nA = 2d \sin \theta$, where again

$$2 \sin \theta \approx 2\theta l$$

Then

$$2\theta = \frac{n\lambda}{d} \text{ (}\theta\text{small)}$$

Therefore,

$$dR = n\lambda L$$

In this equation, the integer n is the order of the diffracted I beam, and A and L are constant for one instrument at a given accelerating voltage. In practice, n is generally regarded as unity since a higher-order diffraction beam can be regarded as the first order diffraction beam if we divide the original interatomic spacing by n. For example $d_{200} = d_{100}/2$. Thus, the latter equation can be changed to

$$dR = \lambda L = K$$

Here $dR = K$, but may also be written as

$$dD = K$$

Since K is a constant, changing from the radius of a ring R to the diameter D will only introduce a new constant ($K = 2K'$). Also, it is much more practical to measure the diameter rather than the radius with its ill-defined center position. K, then, is the instrument factor—effectively a constant. By measuring the diameter of a particular diffraction ring corresponding to a known spacing d for a known substance, K is obtained. In turn, using K and measuring D on any subsequent pattern, we can calculate the interatomic spacing for that compound as

$$d = \frac{K}{D}$$

In effect, this shows that there will be a strong diffracted beam on the exit side of the crystal only if there is a set of crystal planes oriented at a critical angle (θ) relative to the incoming beam. It is possible to obtain a series of diffraction rings from a polycrystalline sample whose diameters can be measured, whose d values can be calculated, and whose intensities can be estimated or measured. These data identify the compounds present by comparing the diffraction data with the standard diffraction data contained in the ASTM index in exactly the same way as is done for X-ray diffraction.

Diffraction rings are formed if there are a large number of individual tiny crystals present in the selected sample area. A single crystal gives a spot pattern instead of a ring pattern. The rings given by a polycrystalline sample are, of course, the accumulated pattern of spots from many individual crystals. As is the case with X-rays, the single-crystal pattern must be interpreted by reference to the reciprocal lattice of the crystal. The reciprocal lattice is a mathematical device designed to facilitate the interpretation of diffraction patterns. As the name implies, it is a set of lattice points that bear a reciprocal relationship to the main crystal lattice planes. Each reciprocal lattice point lies on the normal (passes through the origin) of a set of parallel planes in the crystal lattice. Its distance from the origin is inversely proportional to the interatomic spacings of the planes (Fig. 33). The reciprocal lattice projection for one crystal plane of a face-centered-cubic crystal is shown in Fig. 34.

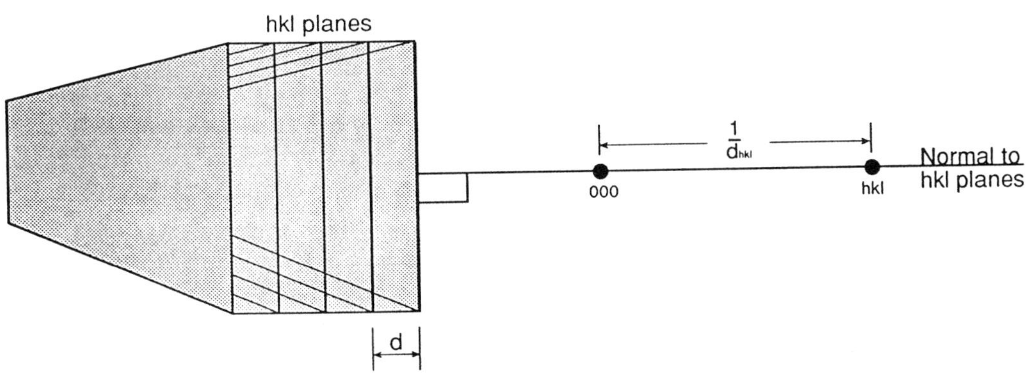

Figure 33 Formation of reciprocal lattice.

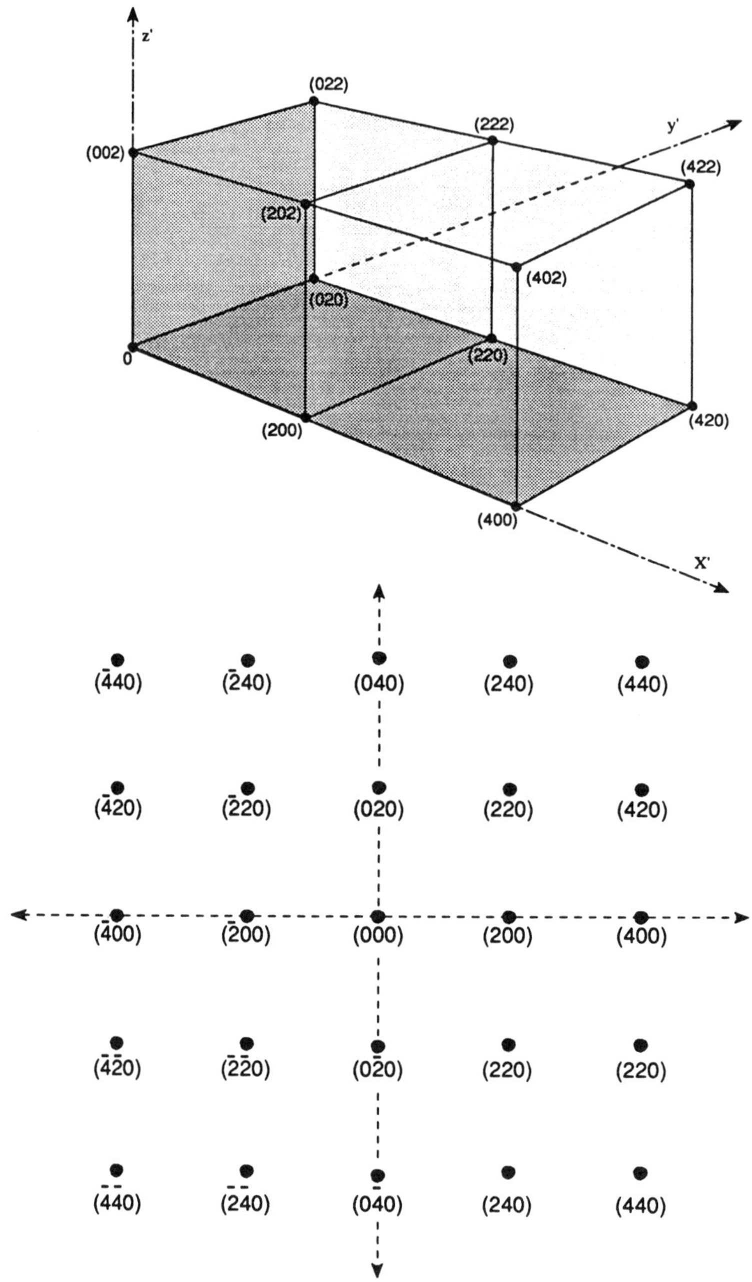

Figure 34 Face-centered-cubic crystal and corresponding reciprocal lattice plane.

The diffraction pattern obtained from a single crystal represents a projection of a single plane of the reciprocal lattice. By taking measurements of this pattern, it is possible to derive constants for the reciprocal lattice; these in turn are related to the unit-cell parameters, *a, b,* and *c*. Thus, it is possible to derive crystallographic constants to be used in identifying single particles. The third direction of the crystal is not included in most single-crystal patterns. As shown in Fig. 35,

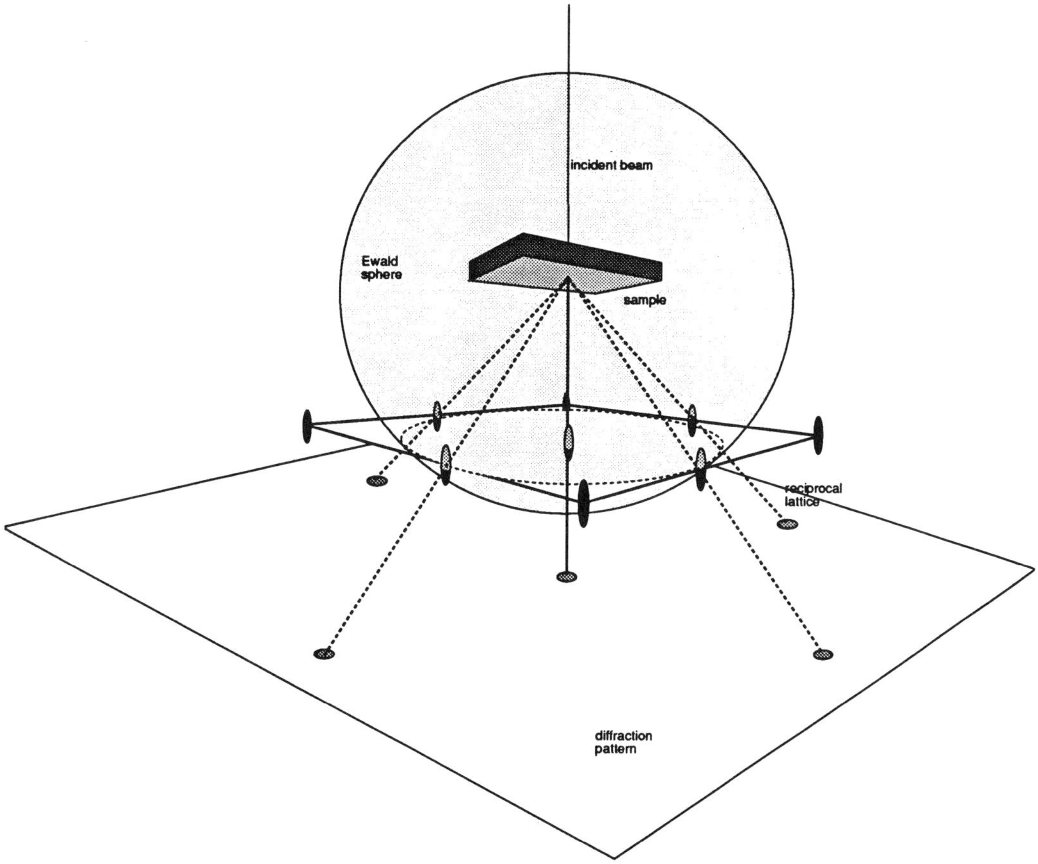

Figure 35 Geometry of a Ewald sphere.

the Ewald sphere is a device that defines which points on the reciprocal lattice will diffract. When an arc of the sphere touches or passes close to a reciprocal lattice point, that point will form a spot in the diffraction pattern (Fig. 36).

One use of ring patterns is in the identification of phases using extraction replicas. These diffraction patterns also arise from very fine, grain-size polycrystalline material such as physically or chemically deposited vapor on electrodeposited thin foils. For a given beam direction, a number of precipitate particles within the area illuminated by the beam will be oriented to satisfy the Bragg law for all allowed reflecting planes, with each individual particle producing a beam reflected from a particular (*hkl*) plane, such that the angle between it and the incident beam is 2θ, satisfying the Bragg law. For randomly oriented particles and a specific reflecting plane (*hkl*) all of these beams will lie in a cone with apex angle 4θ centered on the incident beam direction to produce a ring in the diffraction pattern.

Current applications for electron diffraction in particle analysis typically relate to the rapid identification of particles too small to be conveniently dealt with using a scanning electron microscope with EDXS. One example is the fine-diameter asbestos microfibers that may occur in air or ground water. The EPA methodology [53] for definitive identification of these particles (EPA 100.2) is based on electron diffraction patterns (Fig. 37).

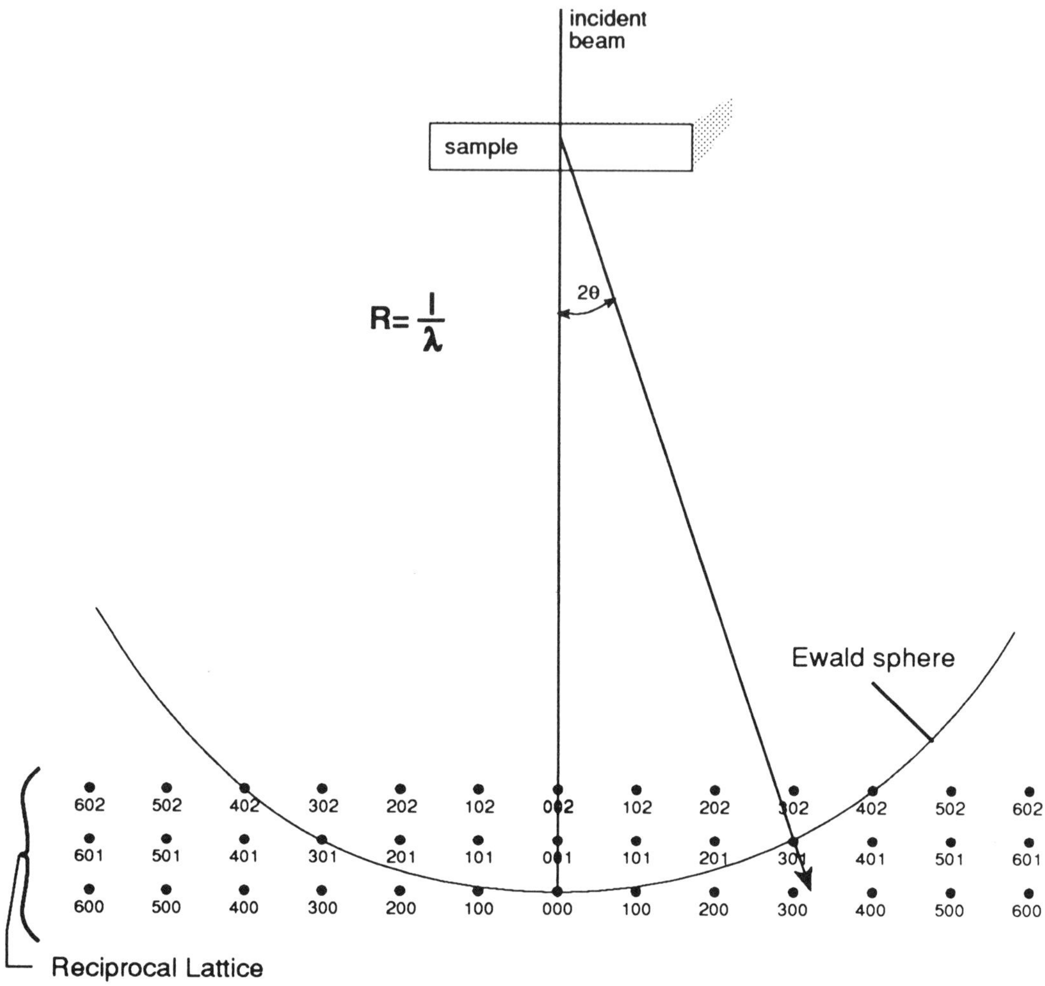

Figure 36 Ewald sphere and reciprocal lattice (dot pattern) formation.

XIV. ELECTRON ENERGY LOSS SPECTROMETRY (EELS)

A fast (80- to 100-kV) electron passing through a thin film consisting of fine particles dispersed on an electron-transparent support loses energy to the material through inelastic scatter [54,55]. The excited states produced by this energy loss decay by emitting the transferred energy in the form of an X-ray, a visible photon, Auger electron, heat, or other radiation. Various spectroscopies utilizing these signals, some discussed earlier, have emerged in the last four decades. Analysis by EELS occupies a special position among these spectroscopies, because it deals with the primary excitation and therefore registers each excitation event independent of any deexcitation mechanism.

Three energy regions in energy loss spectra can be distinguished. Energy losses of a few milli electron volts (meV) to a few hundred meV are predominantly due to vibrational excitations (phonons), but they can only be studied with energy resolution significantly better than is typical of most EELS systems. Collective excitations (plasmons), intraband, and interband transitions

Figure 37 Ring pattern of chrysotile asbestos microfibers produced by electron diffraction. (Courtesy of James Di Orio.)

cause energy losses between a few electron volts to about 40 eV. At energy losses above around 40 eV, one finds the inner shell excitations. These occur at energies

$$\Delta E \geq E_F - E_B$$

where ΔE is the energy loss, E_F is the Fermi level energy, and E_B is the binding energy of the inner shell (always negative). The corresponding feature in an energy loss spectrum is an inner shell loss edge, the threshold energy of which usually agrees within a few electron volts with the known ionization energy for the appropriate electron shell of the atom. Thus, EELS of inner shell losses provides a convenient method of identifying atoms of different types in a thin-film sample.

The various inner shell edges that can arise are labeled following standard spectroscopic notation. Not all the transitions listed occur in every atom. For instance, in aluminum, only the K and L shells are completely filled, and these give rise to the K and L edges. The M shell in aluminum is partially filled, and electrons from this shell give rise to the plasmon loss at 15 eV and other low-energy features. Further, electrons occupying certain shells such as M_1 and N_1 almost never give rise to recognizable features in an EELs spectrum, and the energies for the innermost shell transitions in heavy elements are high (tens of kilo electron volts) and inaccessible to EELS. However, in general there is always at least one strong edge in the usable energy region 100 to 3000 eV for every element, and for many elements there are edges in the optimum EELS energy region which spans roughly 200 to 1000 eV.

Any EELS system has three basic components: (1) an illumination system consisting of an electron gun and one or more electron lenses, (2) a specimen chamber, and (3) a spectrometer, usually preceded by and sometimes followed by one or more imaging lenses. Within these broad constraints, a very large number of designs have been developed in the last 40 years. Among all these systems the single magnetic-sector energy analyzer placed at the end of the electron optical column is predominant. This design has a number of advantages. Being based on magnetic electron separation, it does not require high voltage. This makes possible a compact design even for primary voltages of 200 kV and above. When interfaced to an electron microscope, this type of instrument can be used in standard microscope operating modes to analyze either selected specimen areas and selected scattering angles or both without interfering with the microscope operation. Performance can be optimized so that both energy resolution and collection efficiency are high. Finally, it can be interfaced to an electron microscope with only minimal modifications. All EELS systems typically incorporate a minicomputer-based multichannel analyzer (MCA) for the acquisition, display, and on-line processing of energy loss spectra.

Electron energy loss spectrometers have traditionally employed serial detectors, in which the energy spectrum output is built up by scanning electrons of various energy levels across a single detector. In 1986, the Gatan Corporation introduced the parallel-detection electron energy loss spectrometer (PEELS) [56,57]. The PEELS employs a photodiode array of 1024 channels in place of the single detector, and delivers about a 500-fold increase in operating efficiency compared to serial spectrometers. A diagram of a PEELS instrument is shown in Fig. 38.

In the PEELS instrument, the electron beam, which has passed through a thin sample in a TEM, passes through the entrance focusing coil assembly and is dispersed into an energy spectrum by the magnetic prism. The doubly focusing prism has a 10-cm bending radius. Its dispersion is 1.8 μm/eV at 100 kV primary voltage. The spectrum is magnified by four quadrapole lenses and projected onto a single-crystal YAG scintillator that is fiber-optically coupled to a photodiode array of 1024 channels. The magnification factor is adjustable from 7× to 300× at 100 kV, giving a usable range of dispersions from 2 eV/channel to 0.05 eV/channel. Spectra detected by the photodiode array are read out by a computer over a 16-bit parallel interface.

The application of EELS in analyses are somewhat limited by the specific requirements of the analysis (i.e., the particulate matter of interest must be of sufficient thinness or of sufficiently small size so that a significant portion of the electron beam penetrates the sample). Compared with other spectroscopies such as EDXS, the PEELS minimum detectable limit is lower for practically all the elements. The PEELS is also highly competitive on the

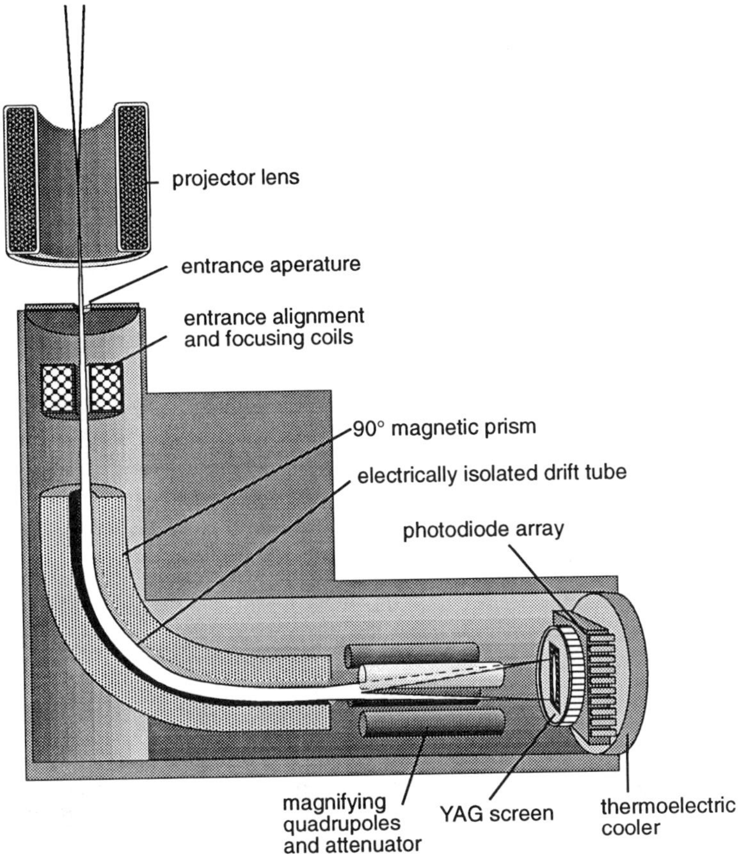

Figure 38 Main components of the PEELS spectrometer.

minimum detectable mass fraction (MDF). For elements of $Z < 30$ that have inner shell loss edges between about 80 and 700 eV, the PEELS signal-to-noise ratio (S/N) obtainable for small concentrations of the element is typically better than the EDXS S/N (approximately equal to 100 ppb for some elements). Particulate matter samples are typically mounted on carbon-filmed grids for examination.

There are several levels of usage for EELS data [58,59]. The simplest level is the identification of the elements present in a thin-film microscopic sample or a fine particle dispersion (also see the Gatan Atlas [60]). Next, the composition of a sample can be determined by comparing the strength of its inner shell losses either to calculated cross sections or to spectra obtained from comparison standards of known composition under identical experimental conditions. The valence of an atomic species can be determined from the precise energy at which its edge occurs (chemical shift). The low loss spectrum contains information on the physical properties of the material, such as its optical reflectivity and its thickness. Finally, EELS can be combined with the many other capabilities of the electron microscope, and this opens up an additional range of experimental possibilities. Thus, energy-filtered images or diffraction patterns can be recorded for improved contrast, or elemental distribution maps may be collected. Performing EELS under channeling conditions makes it possible to selectively probe different sites in a complex crystal, and it is also possible to explore the bonding and near-neighbor distances and numbers along specific crystallographic directions in a crystal by selectively analyzing energy loss events in which the ejected inner shell electron traveled along that direction.

A typical application of EELS is represented by the specimen shown in Fig. 39 and the spectra of Figs. 40a,b. The matrix Si_3N_4 material has an included microprecipitate impurity of TiC. The dimension of the inclusion dictates electron diffraction or EELS in a transmission microscope as the method of choice. In the matrix material the EELS spectrum shows peaks corresponding to the absorption edges for the Si*L* shell and N*K* shell. In the precipitate, peaks corresponding to the absorption edge of the Ti*L* shell and the C*K* shell are present, allowing identification of the particulate impurity.

Combined EELS-SAED at this level of imaging and analytical resolution require the use of an analytical transmission electron microscope, such as the JEOL JEM 2000FX. This instrument's five-lens illumination system, which includes a combination of condenser objective lens and condenser minilens, allows free selection of beams (spot size 2 nm to 1 μm diameter) ranging from a beam with a large convergent angle used in convergent beam electron diffraction to a coherent beam with an extremely small illumination angle used in high-resolution image observation. The beam diameter, illumination angle, incident angle, and axis alignment conditions—all suited for the observation mode—can be stored in the instrument's internal computer in order that given analytical parameters can be duplicated.

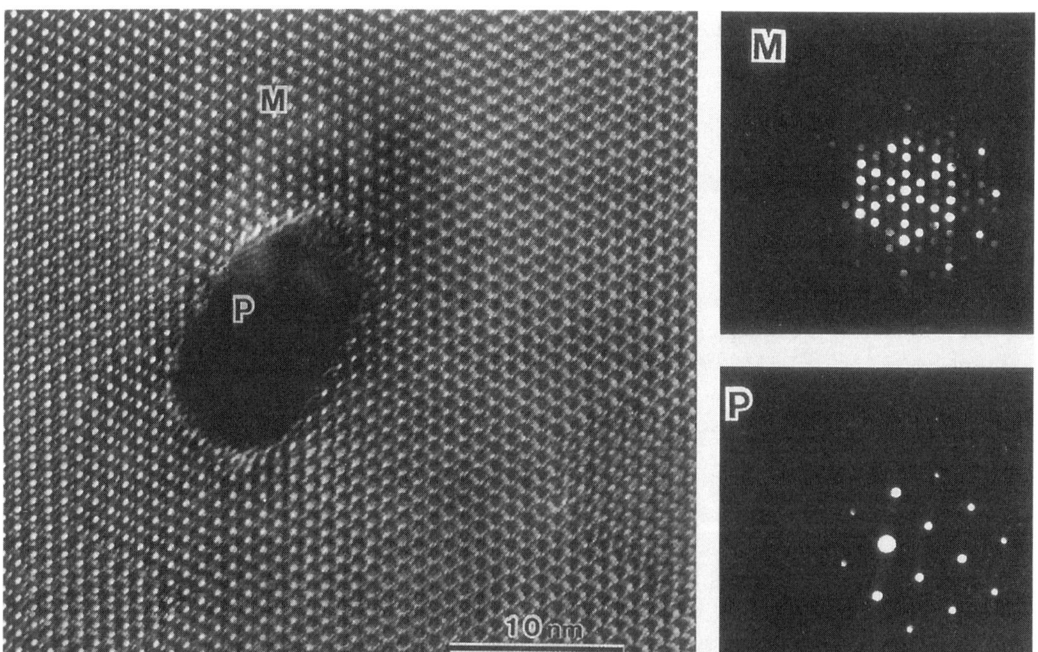

Figure 39 Transmission micrograph showing TiC microprecipitate (P) in Si_3N_4 matrix (M), and corresponding SAED microbeam dot patterns. (Courtesy of JEOL Ltd.)

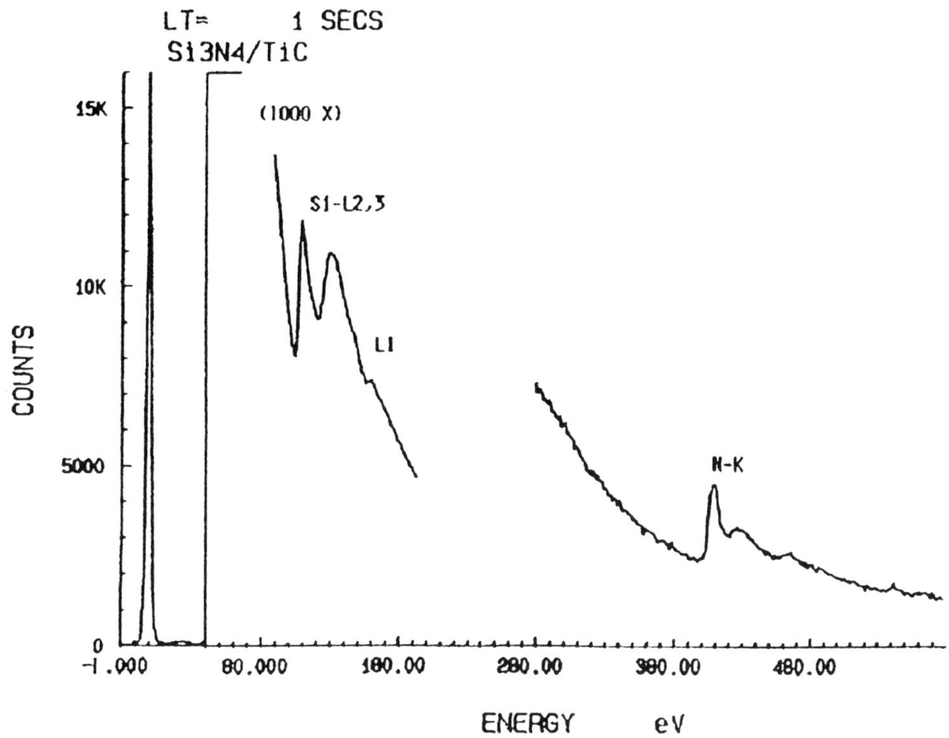

Figure 40a Matrix EELS spectrum.

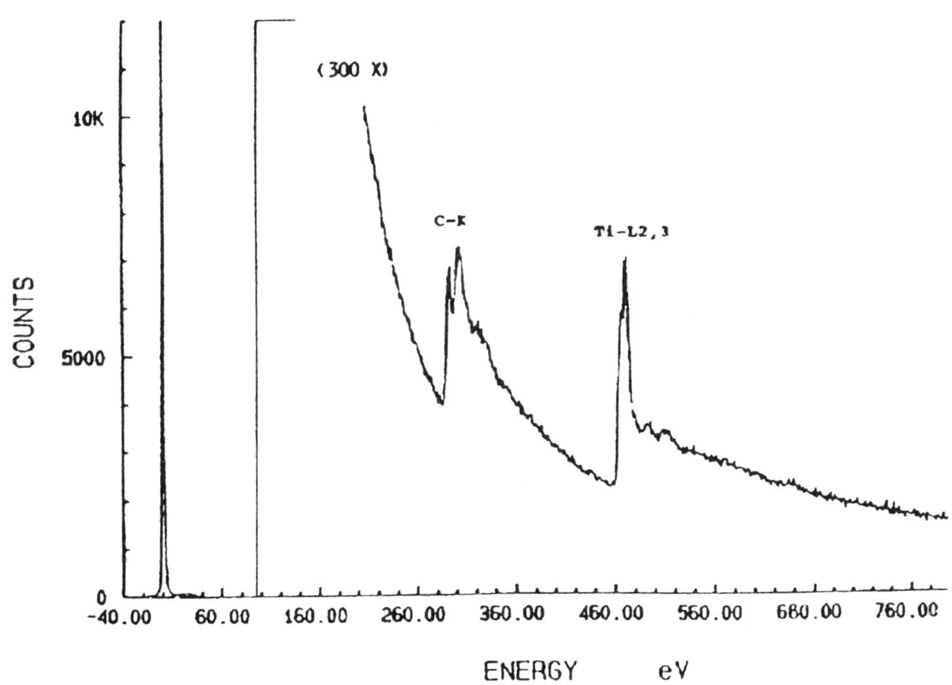

Figure 40b Precipitate EELS spectrum.

XV. SUMMARY

Instrumental techniques discussed in this chapter represent methodologies with a narrower range of applicability than microscopy and require a higher degree of specialization on the part of the operator. All currently receive some level of use in research and technical activities in which unknown particles must be identified. Although many instrumental techniques can yield information on the composition of particles, there is a wide variation in their sample requirements and capabilities. Characteristics of an analytical technique important in the analysis of particles include high spatial and imaging resolution, simultaneous multielement analysis, good sensitivity, applicability to microsamples, ease of sample preparation, and the ability to focus the exciting radiation on a single particle or a narrowly delimited area. Classical atomic absorption and emission techniques are poorly suited to the analysis of particulate matter samples, and dissolution of particulate matter to yield solution samples negates the concentrating effect achieved during particle isolation. The use of EDXS as an atomic spectroscopic technique in combination with micro-FTIR to gather molecular data represents a powerful combined approach for the basic laboratory. Mass spectrometry with laser or conventional means of excitation is useful in a smaller number of cases, and X-ray crystallography is another powerful technique with a limited range of applications.

The method chosen for the instrumental analysis of particles is determined by the objectives of a specific study and the type of data required to meet those objectives. For example, if the elemental composition of individual submicron particles is required, then the most suitable initial means of analysis will be electron microscopy and X-ray spectrometry. The selection of an analytical approach has significant impact on sampling and particle collection. General considerations important in the selection of a mode of analysis involve whether or not it is labor intensive, whether there are interferences in the analysis for which corrections must be applied, how contamination of samples will be avoided, and (importantly) the cost of the total analytical procedure and the magnitude of this cost in relation to the findings of the analysis. The light microscope remains a cost- and time-effective component of any analysis of particle composition.

TRADEMARKS

The following trademarks used in this chapter are the property of the companies indicated. A brief product description is included; additional product information may be obtained from the vendor.

Gatan®. Gatan Inc., 6678 Owens Drive, Pleasanton, CA 94588.

Nuclepore®. Track-etched Membranes, Costar Corp., Cambridge, MA 02140.

ASTM®. American Society for Testing and Materials, 1916 Race Street, Philadelphia, PA 19103.

JEOL®. Japan Eletron Optics Lab (USA), 11 Dearborn Rd., Peabody, MA 01960.

JEM 2000FX®. Analytical electron microscope, Japan Eletron Optics Lab (USA), 11 Dearborn Rd., Peabody, MA 01960.

MiniLens™. Auxiliary electron lens system, Japan Eletron Optics Lab (USA), 11 Dearborn Rd., Peabody, MA 01960.

PEELS™. Parallel detection electron energy loss spectremeter, Gatan Inc., 6678 Owens Drive, Pleasanton, CA 94588.

Microforge™. X-ray Sample Mounting Accessory, McCrone Associates Inc., 850 Pasqurincblic Dr., Westmount, IL 60559.

Duco®. Plastic cement, DuPont Chemicals, Wilmington, DE 19898.

Formvar®. Polyvinyl formal plastic, Monsanto Co., St. Louis, MO 63167.

Teflon®. Fluorocarbon resin, DuPont Chemicals, Wilmington, DE 19898.

REFERENCES

1. S. J. Borchert, A. Abe, S. D. Aldrich, L. E. Fox, J. E. Freeman, and R. D. White, *J. Parent. Sci. Technol. 40*:212–239 (1986).
2. D. S. Aldrich, Identification of parenteral particles, Proceedings of PDA International Conference on Particle Detection, Metrology and Control, Arlington, VA, 1990, pp. 269–292.
3. R. E. Kesting, *Synthetic Polymeric Membranes,* McGraw-Hill, New York, 1971.
4. R. L. Fleischer, P. B. Price, and E. M. Symes, *Science, 143*:249–250 (1964).
5. T. D. Brock, *Membrane Filtration: A User's Guide and Reference Manual,* Sci. Tech. Inc., Madison, WI, 1983, p. 381.
6. A. S. Teetsov, *Microscope 25*:103–113 (1977).
7. P. P. DeLuca, and S. Bodapatti, Guidelines for the identification of particles in parenterals, *FDA Guidelines,* July 3, 1980.
8. W. C. McCrone, L. B. McCrone, and J. G. Delly, *Polarized Light Microscopy,* Ann Arbor Science, Ann Arbor, MI, 1979.
9. W. C. McCrone, L. B. McCrone, and J. G. Delly, *The Particle Atlas,* 2nd ed., Ann Arbor Science, Ann Arbor, MI, 1973.
10. S. Wischnitzer, *Introduction to Electron Microscopy,* 3rd ed., Pergamon Press, New York, 1981.
11. N. H. Farlow, *Anal. Chem. 29*:883–885 (1957).
12. O. Winding, and B. Holms, *Am. J. Hosp. Pharm. 33*:1154–1159 (1976).
13. S. Bodapatti, L. D. Butler, S. Im, and P. P. DeLuca, *J. Pharm. Sci. 69*:608–610 (1980).
14. M. E. Anderson, *Microbeam Anal. 17*:197–201 (1982).
15. R. O. Bollinger, A. M. Preuss, P. R. McClain, and W. T. Hill, Jr., *Am. J. Hosp. Pharm. 35*:312–317 (1978).
16. R. G. Draftz, and J. Graf, *Bull. Parent. Drug Assoc. 28*:35–52 (1974).
17. P. J. Oles, *Microscope 26*:41–48 (1987).
18. *Energy Dispersive X-ray Microanalysis—An Introduction,* Kevex Corporation, San Carlos, CA, 1988.
19. K. F. Heinrich, *Electron Beam X-ray Micro-Analysis,* Van Nostrand Reinhold, New York, 1981, p. 578.
20. K. F. Henrich (ed.), *Characterization of Particles,* NBS (NIST) Spec. Pub. 533, 1980, p. 189.
21. J. E. Post, and P. R. Buseck, *Environ. Sci. Technol. 18*(1):35–42 (1984).
22. P. L. Long, J. E. Katon, and A. S. Bonanno, *Appl. Spectrosc. 42*:313–317 (1988).
23. R. G. Meserschmidt, and M. A. Harthcock, *Infrared Microspectroscopy Theory and Applications,* Marcel Dekker, New York, 1987, p. 451.
24. R. J. Cotter, *Anal. Chem. Acta,* 45–49 (1987).
25. M. E. Lacy, *Inst. Envir. Sci. 28*:185–188 (1982).
26. H. H. Humecki and R. Z. Muggli, *Microbeam Anal. 17*:243–246 (1982).
27. R. Cournoyer, J. C. Shearer, and D. H. Anderson, *Chem. 49*:2275–2277 (1977).
28. C. J. Curry, M. J. Whitehouse, and J. M. Chalmers, *Appl. Spectrosc. 39*:174–180 (1985).
29. G. J. Rosasco, E. S. Etz, and W. A. Cassett, *Appl. Spectrosc. 29*:396–412 (1975).
30. R. J. Conzemius and J. M. Capellan, *Int. J. Mass Spectrom. Ion Phys. 34*:197 (1980).
31. P. Surkyn, and F. Adams, *J. Trace Microprobe Tech. 1*:79 (1982).
32. R. Kaufmann, and P. Wieser, Laser microprobe mass analysis (LAMMA) in particle analysis, *Characterization of Particles,* (K. F. J. Heinrich, ed.), U. S. Government Printing Office, Washington, DC, NBS Spec. Publ., 1980, p. 5533.
33. R. F. Kruger and B. Scheuler, *Adv. Mass Spectrom. 718*:8 (1981).
34. J. Gavrilovic, Surface analysis of small individual particles by secondary ion mass spectroscopy,

Particle Characterization in Technology, Vol. I: *Applications in Microanalysis* (J. K. Beddow, eds.), CRC Press, Boca Raton, FL, 1984, p. 246.

35. R. B. Castaing, B. Jouffrey, and G. Slodzian, *Acad. Sci.* 122 (1960).
36. H. Liebl, *J. Appl. Phys. 38*:5277 (1967).
37. A. Erike, W. Sichtermann, and A. Benninhhoven, *Org. Mass Spectrom. 15*:289 (1980).
38. J. R. McHugh, *Methods of Surface Analysis, in SIMS* (S. P. Wolsky and A. W. Czanderna, eds.), Elsevier, Amsterdam, 1975.
39. "X-ray Powder Data File," Joint Committee on Powder Diffraction Standards, American Society for Testing and Materials, Philadelphia, PA, 1981.
40. G. H. Stout and L. H. Jensen, *X-ray Structure Determinations,* Macmillan, New York, 1968.
41. D. Kaye, *Techniques for Electron Microscopy,* Blackwell Scientific, Oxford, England, 1965.
42. M. J. Buerger, *IUCR world list of crystallographic computer programs,* 2nd ed., A. Oosterhoek's Vitgevers Mij., NY, Utrecht, Netherlands, in *Crystal Structure Analysis* (D P. Shoemaker, ed.), Wiley, New York, 1960.
43. M. J. Buerger, *Crystal-Structure Analysis,* Wiley, New York, 1960, pp. 152–241.
44. R. W. James, *The Optical Principles of the Diffraction of X-rays,* Cornell University Press, New York, 1965, Chaps. 3, 5.
45. F. D. Bloss, *An Introduction to the Methods of Optical Crystallography,* Holt, Rinehart and Winston, New York, 1961.
46. W. C. McCrone, J. A. Brown, and I. A. Stewart, *The Particle Atlas,* 2nd ed., Vol. 5, Ann Arbor Science, Ann Arbor, MI, 1980.
47. J. W. Edington, *Practical Electron Microscopy in Materials Science,* N. V. Phillips' Gloeilampenfabneken, Eindhoven, 1975.
48. P. B. Hirsch, A. Howie, R. B. Nicholson, D. W. Pashley, and M. J. Whelan, *Electron Microscopy of Thin Crystals,* Butterworths, London, 1967.
49. T. B. Rymer, *Electron Diffraction,* Methuen, London, 1970.
50. C. J. Ball, *An Introduction to the Theory of Diffraction,* Pergamon Press, Oxford, 1971.
51. K. W. Andrews, D. J. Dyson, and S. R. Cowan, *Interpretation of Electron Diffraction Patterns,* Hilger, London, 1971.
52. M. F. N. Henry, H. Lipson, and W. A. Wooster, *Interpretation of X-ray Photographs,* Macmillan, New York, 1951.
53. Method for Detection and Analysis of Asbestos Fibers in Drinking Water, *EPA Analytical Method 100.2,* Environmental Protection Agency, Washington, DC, 1993.
54. M. Isaacson, All you wanted to known about ELS and were afraid to ask, Proceedings of 11th Annual SEM Symposium, SEM Inc., Chicago, IL, Vol. 1, 1978, pp. 763–776.
55. D. C. Joy, *Introduction to Analytical Electron Microscopy* (Hren, Goldstein, and Joy, eds.), Plenum Press, New York, 1979, pp. 223–244.
56. O. L. Krivanek, *Ultramicroscopy 28*:118–121 (1989).
57. O. L. Krivanek and M. K. Kundman, Improved detection limits with parallel-detection EELS, Proceedings of 49th Annual Meeting of the Electron Microscopy Society of America (G. W. Bailey, ed.), San Francisco Press, 1991.
58. O. L. Krivanek, M. M. Disko, J. Taftø, and J. C. H. Spence, *Ultramicroscopy 9*:249–254 (1982).
59. O. L. Krivanek, A. J. Gubbens, and N. Delby, *Microsc. Microanal. Microstruct 2*:315–332 (1991).
60. C. C. Ahn and O. L. Krivanek, *EELS Atlas,* Gatan, Inc., Warrendale, PA, 1983.

9

Evaluation and Validation of Nondestructive Particle Inspection Methods and Systems

Julius Z. Knapp

Research & Development Associates, Inc., Somerset, New Jersey

Lee R. Abramson

Statistical Consultant, Washington, D.C.

I. INTRODUCTION

Improvement of productivity in pharmaceutical manufacturing is the driving force behind the expanded use of automated inspection for particle contamination in parenteral containers. To facilitate this expansion, simple accurate methodology to implement a good manufacturing practice (GMP) acceptable validation of these systems is essential. The validation demonstration required for an automated system performing nondestructive inspection of injectable products for contaminating particles must pass one test. This test is that the inspection security achieved by the alternative system is at least equivalent to the manual effectiveness test. The burden of false rejects of usable material must be substantially less than the value of the cost reduction and the quality improvement achieved.

The human performance yardstick of particle inspection security, first defined in U.S.P. XIV, continues unchanged into U.S.P. XXIII [1]. It is ". . . and that every container whose contents show evidence of contamination with visible foreign material be rejected." The performance that must be demonstrated is equivalence to the security obtained by the present manual nondestructive inspection of sealed containers of injectable fluids for particle contamination.

The root problem in this validation is that the comparison rests on variable human behavior in the execution of a task that has many subtle difficulties. To begin a cGMP acceptable validation of a semiautomated or fully automated inspection system, the first requirement is a statistically replicable measurement of the performance of a manual inspection for contaminating particles. Methodology with which to compare the performance of the semi- or fully automated inspection system to the calibrated performance of manual inspection is then the second validation requirement. With these two problems solved, definition of statistically defined inspection performance measures remained as the final roadblock.

The first successful step [2] through this maze of difficulties, published in 1980, was followed by eight other publications by these workers [3–10] in the 10 years that followed. Today, a statistically replicable validation demonstration of equivalent inspection performance for a new inspection technique or automated inspection system is a manageable laboratory task.

The initial papers reported the experimental results of actual injectable product inspections. The data reported was analyzed with a new method that resulted in statistically replicable evaluations of the inspection security in human inspections. The application of this new methodology commenced with the recognition that there is variability in any task performed by a human being. The variability in any individual response is such that accurate measurements and comparisons must be based on statistical measures of performance such as the mean value of a number of repetitive actions such as the accept/reject decision. Statistical methodology was described in the 1980 paper [2] that resulted in simple inspection security evaluations.

Accurate comparisons of validation experiment results are best made when the particle contamination in the inspected containers is standardized. Particle standardization must also accommodate the wide variety of particle densities and refractive indexes to be truly representative of production line rejects. A bootstrap approach to this problem utilizing inspection line rejects, the Working Standard Test Method was described in 1980 and is at present a de facto worldwide standard.

Following establishment of a statistically replicable human performance yardstick, the evaluation of semi- and fully automated system performance becomes possible. With the passage of time, links have been established to varied disciplines. These disciplines provided the tools that have helped shape the validation procedures described in this chapter. With them the improved, cost effective, statistically replicable demonstrations of equivalent security and economic effectiveness have been achieved.

The conditions under which manual inspection is performed can now be selected, measured and accurately described. The problem of extracting statistically replicable data from measurement of the responses of human beings is well described in behavioral experiments. To achieve statistical replicability for human responses, two requirements must be satisfied. First, it is necessary to devise and maintain carefully structured experimental conditions. Second, sufficient data must be generated to permit the employment of statistical comparison measures.

The information base necessary to accomplish a GMP validation of alternative manual, semiautomated and fully automated inspection systems for the inspection of contaminating particles also applies to all the visual and optical inspection requirements of quality maintenance in the pharmaceutical industry. This information base has been assembled and organized for accessibility in the following pages.

Easy-to-apply methods combined with simple calculations are described in this chapter. These procedures and calculations make available the precision of statistical methodology in evaluating and improving the cost effectiveness of the dollars allocated to visual and optical inspection. An essential topic in this information is the use of multiple inspections. Why are multiple inspections an important topic? At the present time there are no automatic particle inspection systems that can deliver the desired combination of inspection security and an acceptable false reject rate in a single inspection. The choice of inspection strategy to deal with the functionality of the inspection system will be seen to set a limit to both the inspection security and the false reject rate that result.

The objective measurements of inspection security and discrimination described in this chapter evolved in the successful automation of particle inspection at Schering-Plough Corporation.

Within the recent past, the accept/reject decision resulting from visual or optical inspection of parenteral products has been critically examined and documented [2–10]. This examination commenced with investigations into the results achieved by visual inspections of particulate contamination. The investigation concluded with a report that the results obtained were supported by an extensive biophysics literature describing human vision experiments.

The most significant fact that emerged in this investigation was that manual visual inspection for quality attributes of parenteral products is a probabilistic process. This means that, in addition to recognizing absolutely good and absolutely bad quality, there is now recognized a continuous variation between these quality extremes. When the rule-of-thumb industry quality categories, "good-good," "bad-bad," and "gray" are defined as segments of the continuous quality variation curve, a foundation for accurate comparisons, and thus, evaluations is established. The determination of two simple numerics, based on these definitions, completely describes the inspection security achieved and the effectiveness of the accept/reject decision in discriminating acceptable from unacceptable quality units.

With the knowledge that the accept/reject decision is probabilistic, it is clear that matching reject rates without reference to individual container rejection probabilities is an adequate procedure for the evaluation or comparison of visual inspection results. Accepting the probabilistic nature of the accept/reject decision, experiments can be designed to measure both the inspection security and the discrimination achieved to any degree of precision required. For GMP validations, 0.05 significance is judged adequate and is employed.

The effective design of manual inspection procedures and the comparative evaluation of automated inspection devices against the human performance yardstick require acquaintance with a miscellany of topics, including:

1. Light intensity selection
2. Type of lighting
3. Contrast
4. Resolution
5. Magnification
6. Velocity of effects
7. Topics in probability and statistics.

At the present time, parenteral processes, components, quality standards, and regulatory guidelines are changing. A point of view is essential from which to evaluate the impact of current and new information on particulates and revised particulate inspection requirements. The view that will be presented is that the delivery of adequate particulate contamination quality is necessary to the well-being of the patient and the economic well-being of the pharmaceutical industry. This adequate quality level is achieved at progressively lower cost with the aid of newer techniques, current technology, and validated, nondestructive inspection systems. To assist in achieving this viewpoint, a brief, hopefully balanced presentation of relatively new information follows.

II. ANALYTICAL AND VISUAL PHARMACEUTICAL PARTICULATE STANDARDS

The progression from U.S.P. XXII to U.S.P. XXIII, effective January 1, 1995 brought with it new and tighter standards for particle contamination in injectable fluids. It also brought with it a response to the measurement accuracy problems that occur when methods with different accuracy capabilities are used to measure the same variable. Table 1 compares the particle contamination standards of U.S.P. XXIII with its predecessor, XXII.

For both large- and small-volume injectables, there has been an evolutionary change in contamination limits. This change is based on the progressive quality improvement that has occurred in the United States due to incremental improvement in the technology and methodology of parenteral production in the last decade. One big, optimistic step has also been taken: the extension of the use of light extinction particle counter quality measurements to large-volume parenteral products. The extension of particle counter measurements to large-volume products implies that the two materials that give false particle counts in particle counter assays are now in substantial control. The false counts are contributed by plasticizer droplets extracted from primary plastic packaging or from droplets of silicone oil that have been washed free from the lubricated stopper surface used with glass and plastic primary packaging.

Table 1 Comparison of U.S.P. XXII and U.S.P. XXIII Particle Contamination Standards for Injectable Fluids

U.S.P. Revision	XXII		XXIII		
Particle range specified	> 10 μm	> 25 μm	> 10 μm	> 25 μm	Method
SVI	10,000	1,000	6,000	600	Light extinction
LVI			25*	3*	particle counter
LVI	50*	5*	12*	2*	Membrane/microscope

*Asterisked limits are particles per ml; all others listed are per container.

When a contaminating particle assay problem is suspected, a manufacturer can fall back on the more costly and time-consuming improved microscope assay.

The magnitude of the measurement interference introduced by these droplets in particle counter assays compared to those of the membrane/microscrope assay can be estimated from the quality limits listed in Table 1. An estimate of the magnitude of this interference is provided by comparing the ratio of particle counts > 10 μm and those > 25 μm specified for large-volume injections in U.S.P. XXIII for the membrane/microscope method compared to the limits for particle counter data. For particles > 10 μm, the ratio is approximately 200%. For particles > 25 μm, the ratio is 150%.

The latest revision of objective particle standards by the U.S.P. has not affected the U.S.P. reliance on the use of human vision and judgment in the visual and optical inspection for particle contamination in injectable fluids. How to select, adjust and validate semi- and fully automated inspection devices against the human performance standard established in U.S.P. XIV is maintained unchanged in U.S.P. XXIII. It states "... that every container whose contents show evidence of contamination with visible foreign material be rejected."

The information and data presented in this chapter is concerned with the evaluation and validation of particle inspection methods and machines against the U.S.P. human inspection standard.

III. VISIBLE PARTICLE SOURCES

The maintenance of particle contamination quality standards in parenteral production is a demonstration of the maintenance of total product quality. The sources of visible particle contamination commences with a complete listing of all primary packaging components.

1. *Glass Container.* The freedom from particulate contamination of the glass container depends on the initial cleanliness of the container as it is received from the manufacturer and on the washing, sterilizing and holding methods employed before production line use. The use of plastic-wrapped particle-free skids of fresh glassware has done much to improve the initial condition of the containers. The initial condition of the glassware is a summary of all handling prior to washing and sterilizing. Just as sterility statements commence with the initial condition of the sterilized product, final cleanliness depends not only on the cleaning process but also on the initial received condition of the container. Particle contamination testing of received and ready-to-use glassware is a conservative control measure.

2. *Product.* Visible particulates can result from the interactions of the injectable product with trace compounds in the stopper or in the primary container. The temperatures reached in terminal sterilization can accelerate both visible and subvisible particle production.

3. *Staff, Air, Water.* The technology available to maintain high-quality, particle-free air and

water is now commonplace. Staff minimization at sensitive production steps, laminar flow areas within the clean room and barriered production steps are all used to control particle contamination.

Stopper particulates are a vitally important contamination source and will be discussed separately.

IV. VIAL STOPPERS: A MAJOR PARTICULATE SOURCE

Widely used pharmaceutical stoppers are molded in batches (heats) that can be as high as 1024 stoppers. These stoppers are molded in individual cavities that are interconnected by excess molding rubber following vulcanization. Following vulcanization, the individual stoppers are die-cut from the molding sheet. Stripping of the stopper heat from the forming cavities is assisted by the use of a mold-release liquid that is sprayed onto each die prior to vulcanization. Of all the components used in the manufacture of an injectable product, stopper cleanliness is the most difficult to achieve on a consistent basis. Particle contamination testing of received and sterile rubber stoppers is a conservative control measure.

Papers by J. M. Garvan and B. W. Gunner in 1963 [11] and in 1964 [12] called attention to the harmful effects of particulates due to rubber stoppers. From those early papers to the recent past there have been many advances in the forming and cleaning of pharmaceutical stoppers. In 1983, W. Whyte [13] reported that ". . . the most cost effective way of increasing the cleanliness of products in a pharmaceutical clean area is to improve the quality of the containers and closures." The use of widely-available automated washing equipment with economically-effective water and air treatment systems and modern continuous sterilizers can and do effectively eliminate container cleanliness from the problem list when adequately employed. At the 1983 annual Parenteral Drug Association (PDA) Meeting, Mr. Whyte reported an extension of his previous work: "Both rubber closures and bottles are a source of contamination but the most significant source of contamination is caused by autoclaving the closures" [14].

Whyte's attention was fixed on the initial inspection that is employed after manufacture. When quality variation over product shelf life is considered, the problem increases.

The data in Fig. 1 are from an accelerated stability study of stopper particulate shedding in which the performance of standard bare butyl stoppers was compared to a commercial Teflon-faced stopper and to developmental butyl stoppers with polypropylene facing. The butyl compound was designed and optimized for the pH 3.5 solution used. For each stopper category shown here, particulation increased during the test period. The rate of particulation for the bare butyl stopper in these test conditions resulted in a 10% reject rate increase in the 12 days of test. Other unpublished data indicate that in these circumstances, the reject rate with natural rubber stoppers is a multiple of that shown for butyl rubber. Both the Teflon- and polypropylene-faced stoppers exhibit improved performance over that of the bare butyl stopper [15].

Figure 2 is a demonstration of what happens when the stopper compound is incompatible with the solution (this butyl compound, designed for low pH use, was tested here at a pH of 9.0). A 100% reject rate can be quickly achieved as shown here. The rate of increase to a final limiting asymptote of the plastic-covered stoppers is attributed to the particulates embedded in the plastic face during the molding process [12].

The reject rates reported in Figs. 1 and 2 were taken on an automated particulate inspection system developed in Schering. To assess inspection system performance prior to its use in the stopper evaluation experiments, the day-to-day variability of a stable test group on the Schering Automated Inspection System was determined and is shown in Fig. 3. The inspection variability of a 250-vial reject-seeded test group is recorded in the 19-day span of measurements shown. The average rejection rate is 31.2%; its 95% confidence interval is ±1.66%. This test group is

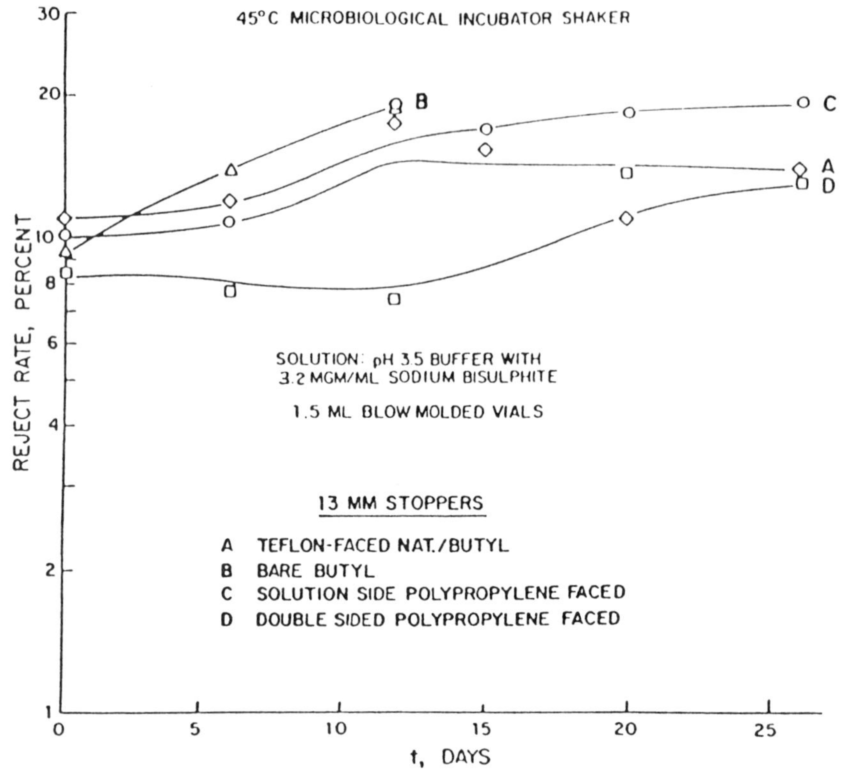

Figure 1 Stopper particulate as shown in an accelerated test of stopper/solution compatibility in 1.5-ml vials.

approximately 18 months old. The stability of the accept/reject decision was judged adequate for the projected use.

The inspection characteristics of the automated inspection machine employed for these tests was matched to a validated inspector group at Schering P.R., Manati, Puerto Rico. The match obtained is shown in Fig. 4.

Figure 4 is a computer summary of the manual/machine inspection comparison. The inspection system was adjusted to match the performance of a group of 10 validated inspectors with 10 repeated inspections per inspector at the Schering Manati, Puerto Rico, plant. This performance was obtained with the Working Standard Test Method by matching the inspection security attained by the manual inspection team as measured by the Reject Zone Efficiency (RZE) and the RAG, the undesired rejected rate in the Accept and Gray Zones. The manual inspection used the Schering test booth, which employed 225 foot-candles, a 3-diopter lens, and a white/black background in a paced inspection.

The information above and below the histogram in this figure summarizes experimental results. The correlation coefficient is 0.793; there are 38 containers with rejection probabilities greater than 0.70 and therefore defined to be in the Reject Zone. Thirty machine runs were used to obtain 95% confidence limits for the inspection security comparison. MRZR1 = 0 and ARZR1 = 17 indicate that none of the manually identified Reject Zone containers was found 100% of the time manually and that 17 of the containers were found 100% of the time in the

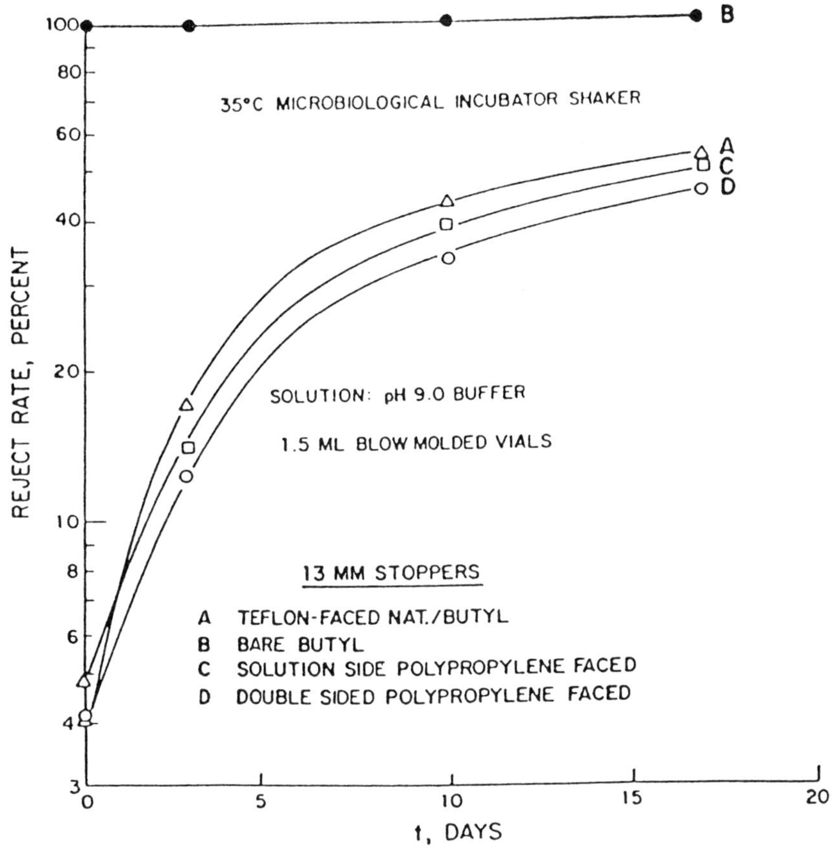

Figure 2 Stopper particulate contribution as shown in an accelerated test as high pH of four stopper types ranging from incompatible to compatible stoppers.

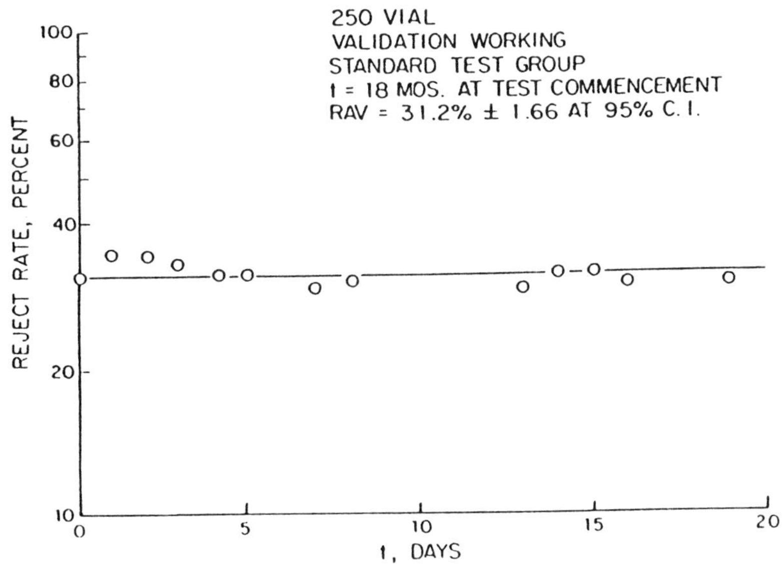

Figure 3 Stability of automated particulate inspection system matched to validated Schering manual inspectors.

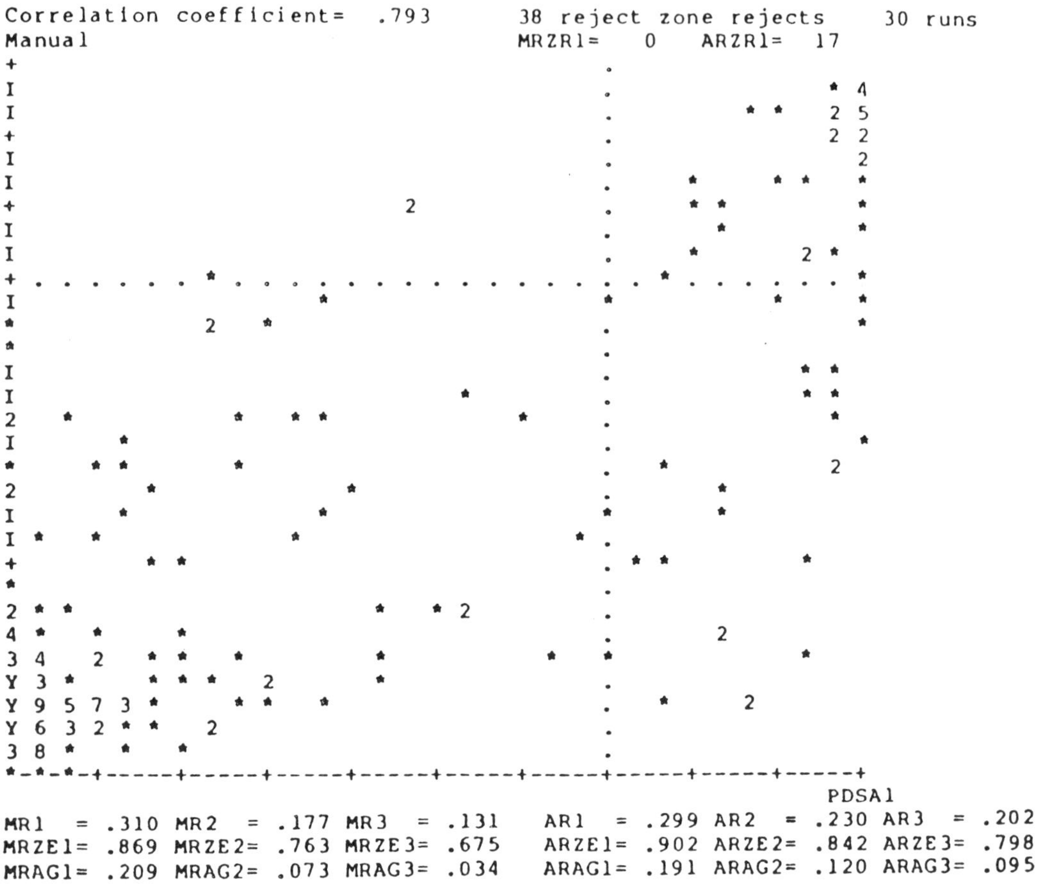

Figure 4 Correlagram of man/machine match between automated inspection system used in stopper experiments and Schering's validated inspectors.

automated system. The three lines under the correlagram list the manual results on the left and the automated inspection results on the right side for three sequential inspections.

V. THE PROBABILISTIC INSPECTION

The Human Inspection Standard for Visible Particulates

The U.S.P. standard of performance for nondestructive tests of particulate contamination in sealed containers is that of the skilled, validated, human inspector. The performance of the skilled manual inspector has therefore been used in the validation procedures that will be described as the basic performance yardstick.

Good manufacturing practice requires a well-documented description of the sequence and conditions of this standard manual inspection. When the inspection sequence is examined, two types of this standard manual inspection can be identified: the single-inspection and a two-inspection sequence. In the single-inspection procedure, those containers that are rejected are destroyed. This procedure is shown in Fig. 5.

The two-inspection sequence consists of a first inspection followed by a cull inspection. In the cull inspection, the containers rejected in the first inspection are reexamined and those accepted join the initially accepted containers in the accepted stock. The sequence of actions that are described above is shown in Fig. 6. Those using the first-and-cull inspection sequence do so on the basis of the cost effectiveness of the sequential inspection. Improving the quality of the first inspection to the degree that the desired product quality is maintained reduces the reject rate of good material. This assertion will be examined using the methods that are described in this chapter. The two-inspection sequence with product accepted after each inspection is called an "Accept in Two" inspection.

Accurate, reproducible results from the visual inspection for contaminant particles and defects in the primary container must satisfy three prerequisites. These are control of the conditions under which the inspection is executed, control of the time allocated to the inspection and definition of the sequence of inspection actions performed. Variability in these details increases the variability of the inspection process.

Figure 7A shows a typical inspection booth used at Schering to obtain the experimental accept/reject data reported. The booth was equipped with an open, white-enameled luminaire yielding a 14° crosswise light shielding to avoid direct glare into the inspector's eyes. Two 20-W 24-inch T12 Cool White (4100 K) fluorescent lamps that maintained the initial light intensity of 225 foot-candles approximately 8 inches from the lamp surface for 1500 hours were used in the fixture. A two-lamp ballast for the lighting fixture was a design prerequisite. The use of the two-lamp ballast results in a flicker rate of 240 per second and reduces the observed flicker by one-third. Although flicker is diminished with the use of a two-

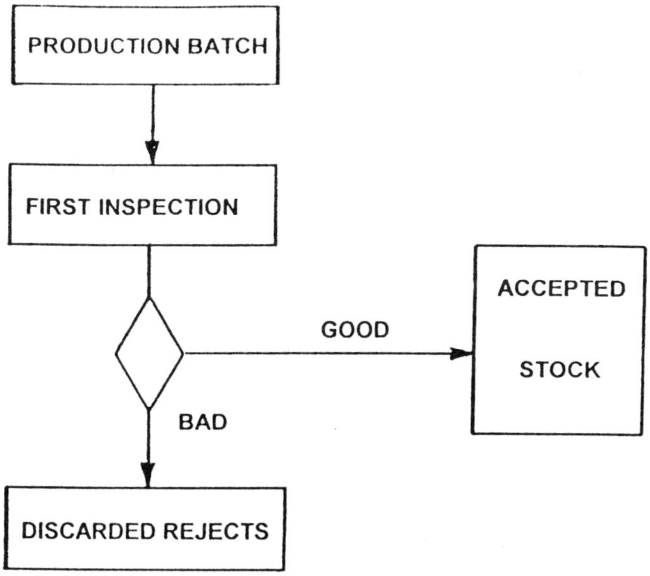

Figure 5 Operational sequence of the standard manual inspection in which product is inspected once and is either accepted for stock if judged good or discarded if rejected (M1-A/1).

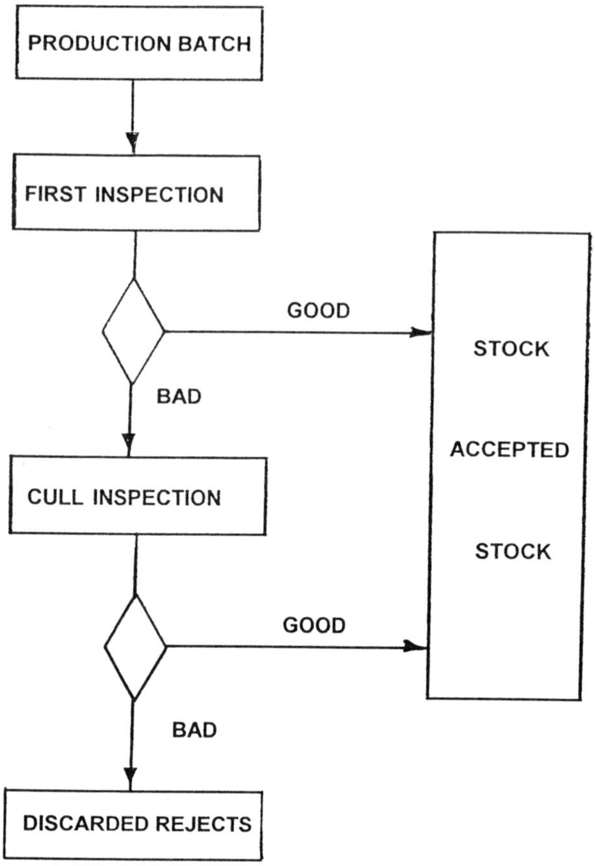

Figure 6 Operational sequence of two sequential manual inspections, a first inspection followed by a cull inspection in which material judged good in either inspection is accepted for stock.

lamp ballast, the strobing of movement remains and is also tiring. The use of a single-lamp or individual ballasts on two lamps results in a flicker rate of 120 per second. The increased amount of flicker and its lower flicker rate increases the strobing effect and increases inspector fatigue.

Recent advances in lighting technology now make possible a better light source for an inspection booth. A flicker-free fluorescent light source with light intensity that is a better match for the difficult task of inspecting for particles is now available. Flicker-free operation is a result of operation using high-frequency excitation sources (20 and 25 kilohertz ballast sources) with the improved fluorescent lamps. The new 24-inch F40 BX/SPX 41/RS Daylight fluorescent lamp provides approximately 20% more initial light intensity than the dual 20-W lamp it replaces. It is rated to maintain its initial brightness for 20,000 hours of operation with a –10% loss of the initial light intensity. A more uniform and more intensely-lighted inspection area also results when the new lamp is used in an appropriate luminaire.

Control of the inspection time for each container is set with a control timer, a pacer [16] that lights a pilot light in the easy vision area of the inspector. The inspection period commences with a touch to a foot switch that initiates the inspection period. At the completion of the inspection period the inspector has three options: accept, reject, or in the event of a question, repeat the inspection period. The use of the repeated inspection cycle diminishes rapidly as inspectors become proficient. A selectable white/black background was used to optimize contrast. The test containers were positioned 8 inches below the center of the lamps in a region yielding approximately constant illuminance of 225 foot-candles. (An increase to approximately 350 foot-candles following the latest I. E. S. task lighting recommendations results when the new lighting technology is used.) The inspectors used a 5-inch 3-diopter magnifying lens to lower the minimum detectable particle size from 100 to 65 μm, thus providing a margin between this production inspection and any user inspection. Figure 7B is the M.W. Technologies Model MIH-1, a modern commercial version of the manual inspection booth used at Schering.

The cumulative effect of a 5-minute "eyebreak" each hour plus 15-min breaks at midmorning and in midafternoon was found to be beneficial in maintaining inspector efficiency at a good level throughout the work day.

The inspection for contaminating particles requires undivided and undisturbed attention by an inspector operating within carefully planned and documented test conditions. Extraneous actions of any type in the area surrounding the inspectors must be minimized to avoid any attention lapse in the inspection. The ambient light in the inspection room should be approximately half of the inspection light intensity.

Since the manual inspection procedure is affected by inspector fatigue, booth and chair design, incoming-material- and inspected-material-container placement, and lens height adjustability have all been carefully reviewed.

In addition to the physical requirements described above, the inspector's operational rate, management, and motivation are also significant elements in the results achieved.

At Schering, experience has shown that a training period of 12–14 weeks was required to ensure that the proper combination of social responsibility, manual dexterity, and functional intelligence level were available. Even with the best training and test inspection environment, when the inspection results of selected inspectors were compared, the results appeared unusable. The degree of frustration achieved can be summarized in a Venn diagram; the nature of the difficulty is shown in the next figure. The problem with the U. S. P. human-capability-based visible particle quality standard was demonstrated in a 1949 FDA court action [17] against Bristol Laboratories. The FDA expert, while on the witness stand, could not securely distinguish his rejected containers from a group that included randomly selected production line material. The FDA case was dismissed.

The failure of the FDA expert witness was studied in an often-repeated experiment summarized in the Venn diagram of Figure 8. Experienced inspectors, in repeated inspections of a test group, rejected only a fraction of the "bad" test group reliably, even though the average number of containers rejected was accurately maintained. The erratic performance of the FDA expert was echoed many times in PDA and PMA efforts to standardize human particulate inspections. It was believed that the variable inspection performance was due to human variability. When human performance variability was eliminated by RCA with its construction of the first generation of automated inspection machines in the early 1950s, the results were just as bad. "Visible" inspection quality, Uhlir [18] concluded in validation trials of automated inspection machines in West Germany in 1983, was a philosophic concept in the heart and mind of the observer. Any attempt to utilize reject rate matching as the sole index of performance equivalence is of limited value. In the work that follows it will be seen that the Venn diagram results are both reasonable and understandable.

Figure 7A Schering standard manual inspection booth showing physical arrangement used in 1970.

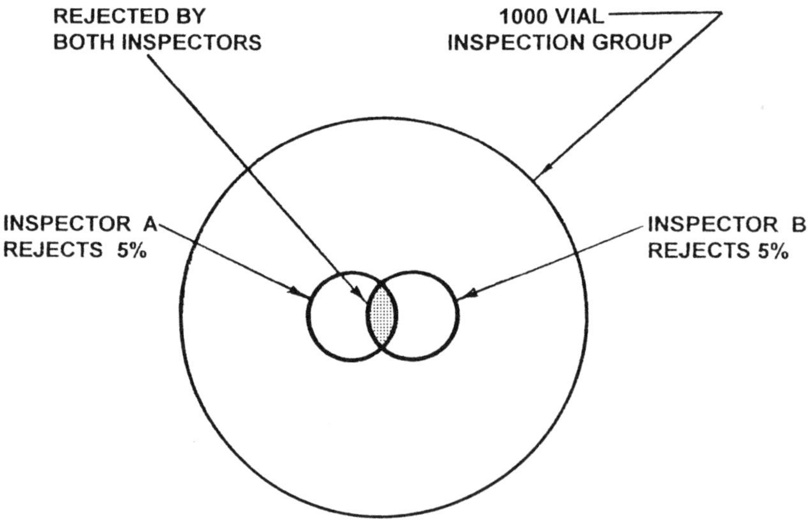

Figure 8 Venn diagram of two-inspector particulate inspection showing lack of agreement in containers rejected.

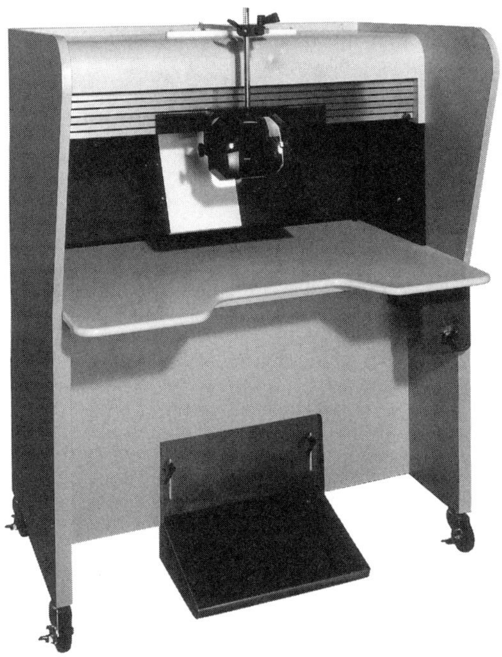

Figure 7B Current commercial version of Schering inspection booth, Model MIH-1 produced by M. W. Technologies.

VI. A PROBABILISTIC EVALUATION OF THE VISUAL INSPECTION FOR PARTICULATES

At the 1979 PDA Annual Meeting, a new approach toward the analysis of visual inspection results was presented [2]. The accept/reject record of *each* container in multiple trials under standard conditions was recorded. Figure 9 is representative of the results obtained from an inspection of a 250-vial test group. Each of six inspectors inspected the test group 12 times under standard conditions. The standard conditions included the use of the inspection booth just described. In the booth, light intensity, the background contrast and the time spent examining each container are defined. The results obtained helped solve the mystery.

Since the performance of the group of inspectors defines the inspection process, the test results for all inspectors are merged. The total of 72 inspections for each container yielded a container rejection test score ranging from 0 to 72. The rejection test score stated as a decimal fraction is an estimate of the experimental rejection probability for the container. The error of this estimate is sufficiently low so that it will be treated as a true probability. A data summary for this experiment is shown in Fig. 8. The lower part of Fig. 9 is a histogram showing the distribution of these containers as a function of the observed reject rate.

The only containers whose inspection results could be securely predicted were those that were absolutely clean and those with such gross particulates that they were securely rejected in any inspection. The absolutely clean containers were never rejected and those with gross defects were rejected in 100% of all inspections. In between these two quality extremes, the repeated inspections showed that all other containers were rejected with varying probability. This analysis which has been repeated many times, characterizes the inspection for contaminating particles as probabilistic in nature. *Simply to state that a particle has been observed is an inadequate summary of essential information.*

AVERAGE FOR ALL INSPECTORS 72 inspections

Ampuls	Rejection Probability									
1	.11	.47	.06	.09	.08	.01	.04	.91	.01	.04
11	.02	.02	.04	.04	.01	.06	.01	.04	.93	.80
21	.04	.11	.02	.04	.01	.02	.01	.36	.06	.65
31	.04	.93	.70	.04	.80	.01	.18	.05	.40	.02
41	.02	.01	.11	.02	.22	.05	.80	.94	.01	.98
51	.05	.01	.31	.93	.08	.25	.01	.02	.04	.12
61	.06	.01	.40	.05	.05	.01	.86	.02	.93	.08
71	.02	.02	.09	.05	.06	.52	.93	.02	.06	.81
81	.05	.09	.08	.91	.09	.02	.04	.02	.05	.04
91	.05	.02	.30	.02	.01	.91	.04	.05	.86	.38
101	.93	.05	.02	.91	.02	.06	.66	.09	.05	.12
111	.66	.08	.34	.08	.88	.06	.05	.91	.05	.04
121	.06	.36	.95	.01	.90	.02	.04	.02	.08	.01
131	.79	.02	.06	.02	.11	1.00	.08	.06	.05	.22
141	.08	.05	.05	.94	.50	.76	.13	.05	.05	.77
151	.97	.44	.86	.90	.05	.08	.72	.05	.04	.09
161	.13	.81	.05	.04	.04	.87	.05	.01	.01	.18
171	.81	.08	.08	.09	.01	.05	.01	.01	.01	.04
181	.04	.06	.05	.04	.02	.04	0.00	.04	.98	.09
191	.97	.12	.16	.75	.76	.02	.09	.44	.06	.90
201	.98	.05	.05	.98	.45	.04	.05	.94	.90	.29
211	.84	.04	.11	.01	.72	.02	.94	.95	.04	.25
221	.02	.04	.02	.02	.25	.04	.97	.04	.33	.11
231	.01	.04	.69	.04	.31	.91	.06	.02	.02	.04
241	.04	.05	.01	.04	.22	.88	.11	.93	.04	.02

PR	FREQ.	PERCENT	
0.0	156	62.40	*******************************
.1	15	6.00	***
.2	7	2.80	*
.3	8	3.20	*
.4	6	2.40	*
.5	2	.80	
.6	4	1.60	
.7	8	3.20	*
.8	13	5.20	**
.9	30	12.00	******
1.0	1	.40	

AVERAGE PROBABILITY OF REJECTION = 26.40%

Figure 9 Objective assessment of rejection rate for each container in a 250-vial test group by a group of six inspectors, each inspector inspecting the test group 12 times. A histogram of rejection probability and the average probability of rejection for the entire test group are also shown.

VII. STATISTICAL BASIS FOR INSPECTION VALIDATIONS

Of major importance in the evaluation of particulate contamination is the conclusion that secure quality determinations can be based only on *groups,* and not on individual containers. Tracking the source of this concept led to a 1693 paper by Edmund Halley (of Halley Comet fame) in the *Transactions of the Royal Society of England* [19]. The title of his paper in this issue is: "An Estimate of the Mortality of Mankind Drawn from Curious Tables of Births and Funerals

AGE	0	10	20	40	60	80
WHITE						
MALES	8.2	0.1	1.3	2.6	14.6	79.5
FEMALES	6.2	0.1	0.4	1.1	8.3	48.9
ALL OTHER						
MALES	6.5	0.1	2.5	5.6	21.7	83.6
FEMALES	3.5	0.1	0.6	2.5	12.2	54.5

Figure 10 Mortality probability per 1000. (From 1990 U.S. census.)

at the City of Breslau; with an Attempt to Ascertain the Price of Annuities on Lives." This is the concept source not only for the treatment of the reject rate of parenteral container groups but of the people data upon which life insurance tables are based.

Figure 10 is a modern version of Halley's data: it was extracted from the 1990 U.S. census. Life insurance rates are successfully based on these group mortality averages. Except for special cases, no one to date has been able to accurately predict the longevity of a single person or the accept/reject decision for a single container in a single inspection.

VIII. EFFECT OF CONTAINER CLARITY ON VISUAL PARTICULATE INSPECTION

The data reviewed thus far are based on the examination of clear glass containers. The relative effect of translucent containers on the detection of particulates within them was examined briefly in the following experiment, which is summarized in Table 2.

In addition to the test conditions listed in this slide, light at 2000 lumens was delivered

Table 2 Description of Experiment to Illustrate the Effect of Translucent Container Walls on Inspection Security

Test conditions	
Sample quantity	1 typical syringe of each material
Inspection time	10 seconds
Number of inspections	5
Number of inspectors	1
Magnification	6×
Particulate material	Precision polystyrene DVB spheres
Number of particulates in each sample	Cloud of 50 or more
Sample solution	2 ml of distilled water

The results in detection probability are considered optimistic because the number of particulates in each sample and the magnification is very high.

Table 3 Transparent/Translucent Container Effect on Particulate Detection in 2-ml Syringes

Material	Clarity	Detection probability 80 μm	40 μm	19 μm	10 μm
Glass	TP	E	E	G	F
TPX (crazed)	TP	E	G	G	F
TPX (new)	TP	E	E	G	F
PP J004	TL	F	F	B	Z
PP 6513	TL	F	F	B	Z
PP DN6005	TL	F	F	B	Z
PP DN6006	TL	F	F	B	Z
PP PB397-1	TL	F	F	B	Z

TP = transparent; TL = translucent; E = excellent—particulates visible during entire inspection; G = good—seen more than half of inspection time; F = fair—seen less than half of inspection time but more than 10% of inspection time; B = bad—seen less than 10% of inspection time; Z = zero—never seen.

through two rectangular-orifice fiber optics. The fiber-optic rectangles were approximately 20% of syringe diameter in ensure achievement of maximum contrast for the particulate clouds contained in the test syringes. The results of the experimental procedure shown in Table 2 are summarized in Table 3.

Using the glass syringe inspection line as our comparison standard, it is fully matched by the data shown in line 3 with a TPX syringe having a newly manufactured surface. In line 2 are the inspection results for a TPX syringe with a slightly worn (crazed) surface. The crazed surface reduces visibility of the 40-μm spheres from excellent (always seen) to good (seen more than half the inspection time).

By comparison, the five lower lines for polypropylene syringe inspection results show detectability for 40- and 80-μm spheres to be equal in detection probability to 10-μm spheres in clear transparent containers. The translucent polypropylene syringe walls curtail the effectiveness of the visual inspection.

This concludes a cursory examination of the effect of the container on inspection results. All further comments are restricted to the inspection of solutions in transparent containers.

IX. REJECT RATE ANALYSIS BY HISTOGRAM SHOWING DISTRIBUTION OF VISUAL PARTICULATE REJECTS

The essential technique step that made possible the data summarized in the solid black line histogram shown in Fig. 11 was the identification of each container in the test group. Since each container was individually identified, an analysis of its accept/reject history in multiple inspections became possible. This accept/reject history permitted the segregation of the containers into rejection probability groups as shown in this figure. Senior inspectors with unlimited time were able to establish a size distribution from very small to large particles paralleling the detection probability progression from 0 to 1.0. This summary is the foundation on which the reported objective evaluation of nondestructive inspection systems rests.

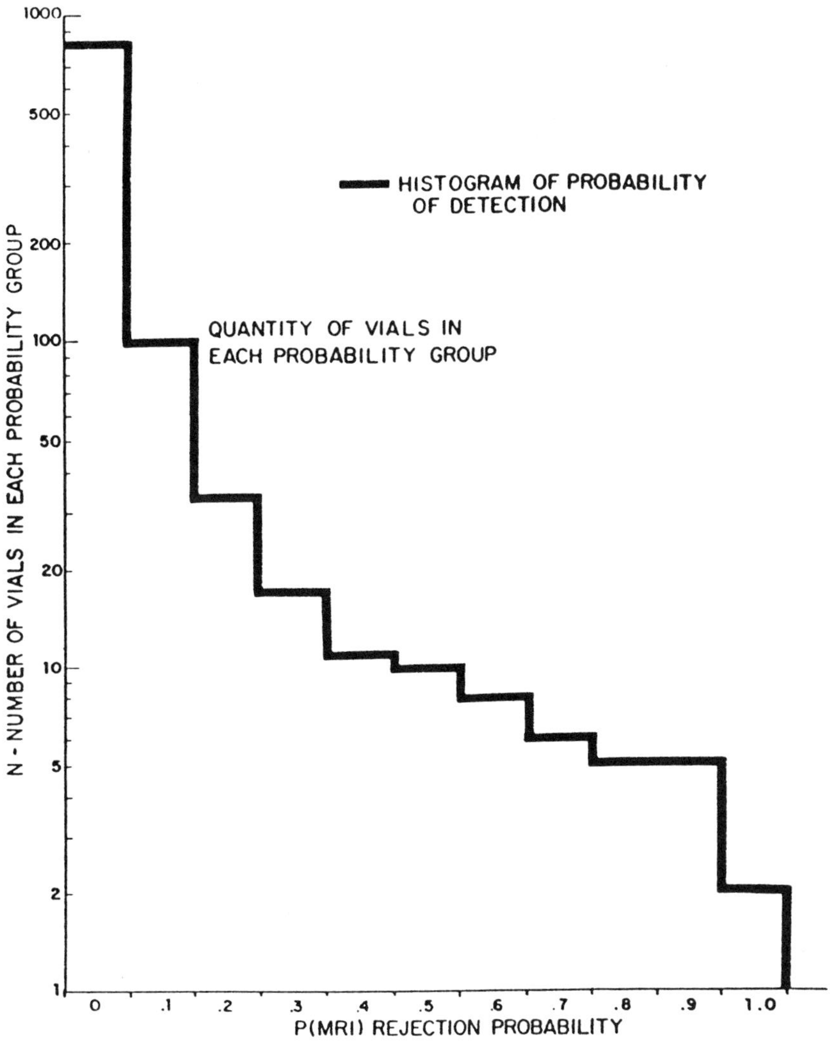

Figure 11 Histogram of detection probability for a 1000-vial, randomly selected test group. The Schering standard, 17-second, paced inspection was employed.

X. VISUAL INSPECTION REJECT RATE UNMASKED AS A MORTALITY TABLE EFFECT

The first result of this histogrammic analysis was an understanding of the lack of agreement between inspectors shown in the Venn diagram. This lack of agreement is a variant of the life insurance mortality table problem. Given the population in each probability segment and knowing the probability of rejection, the average reject rate for the segment can be computed. This computation can accurately specify quantity rejected, but not which units will be rejected in a single inspection except for those absolutely clean units and those with gross particulates. It is also clear that given two inspectors with different sets of rejection probabilities, reject rates can be matched without any evaluation of the relative inspection security and discrimination each inspector achieves. To achieve a replicable analysis of the results of visual particulate inspection, methodology must be formulated to apply the life insurance mortality table concept to visual particle inspection results. This can be achieved by evaluating the reject rate of containers within defined quality limits. This methodology will now be described.

XI. RELATIONSHIPS OF REJECTION PROBABILITY AND REJECTS IN A NORMAL PARENTERAL PRODUCTION BATCH

The average number of vials rejected in a single inspection in each probability group is the product of the rejection probability for that group by the number of containers within it. This data is summarized by the dashed-line histogram in Fig. 12. The average rejection rate for a single inspection is then the sum of the contributions from each segment of the distribution.

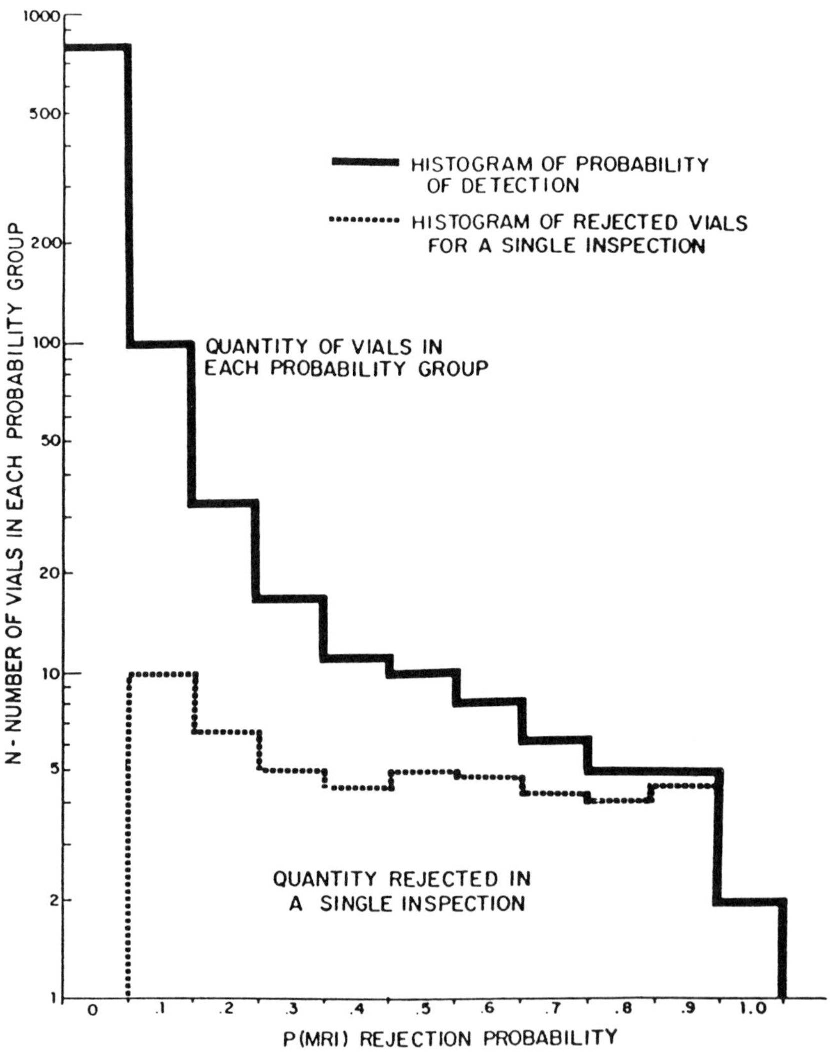

Figure 12 Histograms of detection probability and of the quantity rejected in a single inspection for a 1000-vial, randomly selected test group.

XII. EFFECT OF A SECOND OR CULL INSPECTION ON FINAL REJECT DISTRIBUTION IN A NORMAL PARENTERAL PRODUCTION BATCH

Figure 13 illustrates the effect of a cull operation, i.e., a reinspection of initial rejects upon the initial reject rate. For two sequential inspections, the effective probability in each segment of the distribution is the square of the empirically determined single inspection rejection probability. The rejection data for two sequential inspections is shown as the dotted-line histogram in this figure. The reject rate contribution from the low rejection probability end of the distribution has been reduced. Gross particulates at the high end of the scale are little affected. For three sequential inspections, the effective rejection probability is the cube of the single-inspection probability, etc. Each successive culling inspection reduces the quantity previously rejected.

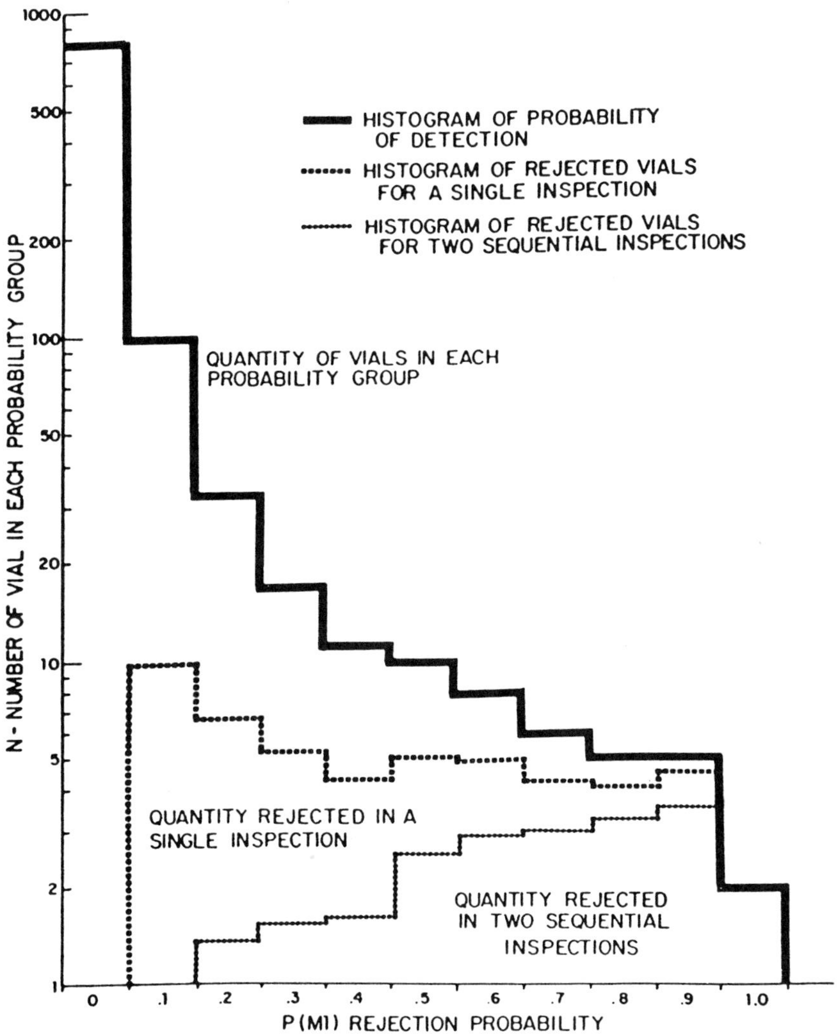

Figure 13 Composite histograms showing detection probability, quantity rejected in a first inspection, and quantity rejected after a second or cull reinspection of rejects.

A functional definition of what should be rejected is required. Such a definition based on prevailing industry practices has been proposed.

XIII. TRANSFORMATION OF RULE-OF-THUMB-QUALITY CATEGORIES INTO PROBABILISTICALLY DEFINED REGIONS

Three rule-of-thumb-quality categories have seen wide industry usage. They are good-good, gray, and bad-bad. These three categories can be transformed into numerically defined probability of detection regions. Recognizing the accepted use of a first inspection followed by a cull inspection, we can define three zones.

An *Accept Zone* is defined as one with all containers that have less than 1 chance in 10 of being rejected in two sequential inspections, i.e., a rejection probability of 0.10 or lower.

A *Reject Zone* is defined as one with all containers that have at least 1 chance in 2 of being rejected in two sequential trials, i.e., a rejection probability of 0.50 or greater.

The *Gray Zone* is defined as one where the containers could be accepted but that can be sacrificed to ensure the security of the Reject Zone rejections. The Gray Zone, therefore, exists in the rejection probability region greater than 10% and less than 50% in two sequential inspections. For convenience, single inspection probabilities will be used in the following discussion. The corresponding single-inspection rejection probability limits for the three zones

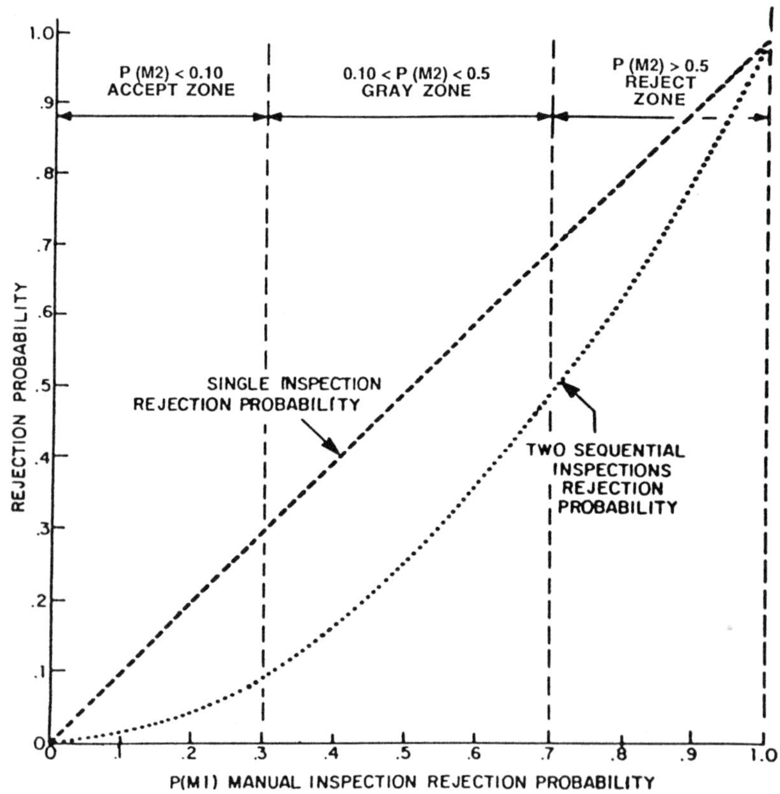

Figure 14 Conversion of good-good, gray, and bad-bad quality into Accept, Gray, and Reject Zones probabilistically defined for one and two inspections.

are as follows: Accept Zone, between 0 and 0.3; Reject Zone, from 0.7 to 1.0; and Gray Zone, from a rejection probability limit greater than 0.3 to, but not including, 0.7.

In Fig. 14, the relative rejection probabilities of a single and two sequential inspections are shown. The downward bow of the rejection probability curve following two sequential inspections shows that there could be a serious impairment to the security of the inspection as a side effect of the discrimination improvement at the low rejection probability end of the quality spectrum. This effect is much reduced with real data because of the preponderance of good material produced in the well-controlled parenteral manufacturing environment.

The three zone boundaries just defined have been graphically transformed to the histogram examined in Fig. 15. In any parenteral manufacturing operating within acceptable process limits, Reject Zone rejects are a small fraction of the entire batch. The effect of this distribution on inspection results is clearly shown in this figure. In the Accept and Gray Zones, a large quantity of usable material is recovered with the reinspection technique. In the Reject Zone, some loss of the desired rejections occurs. A numerical index with which inspection security can be evaluated is an essential for progress.

Since each zone is now numerically defined, the total population in each zone can be easily

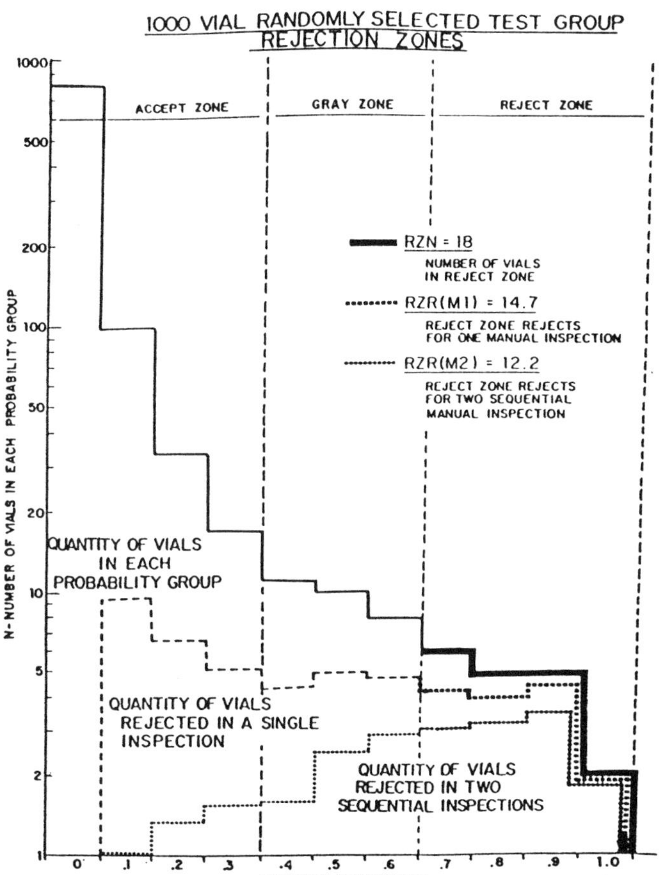

Figure 15 Application of Accept, Gray, and Reject Zone boundaries to the 1000-vial, randomly selected, manual inspection data. Reject Zone population and the quantity of Reject Zone rejects in one and two serial inspections are also shown.

determined. The Reject Zone population is symbolized as RZN, an acronym for Reject Zone number, and for this distribution is 18 containers. The average reject rate of the containers in the Reject Zone after a single inspection is computed as the sum of the contributions from each rejection probability segment in the zone.

Shown here is the Reject Zone reject rate for one and two sequential inspections. This Reject Zone reject rate is symbolized as RZR followed by parentheses to indicate that it corresponds to the first or second manual inspection; i.e., RZR (1) = 14.7, RZR (2) = 12.2. With this population and the average reject rate for one and two inspections established, the relative security of inspection can be evaluated numerically.

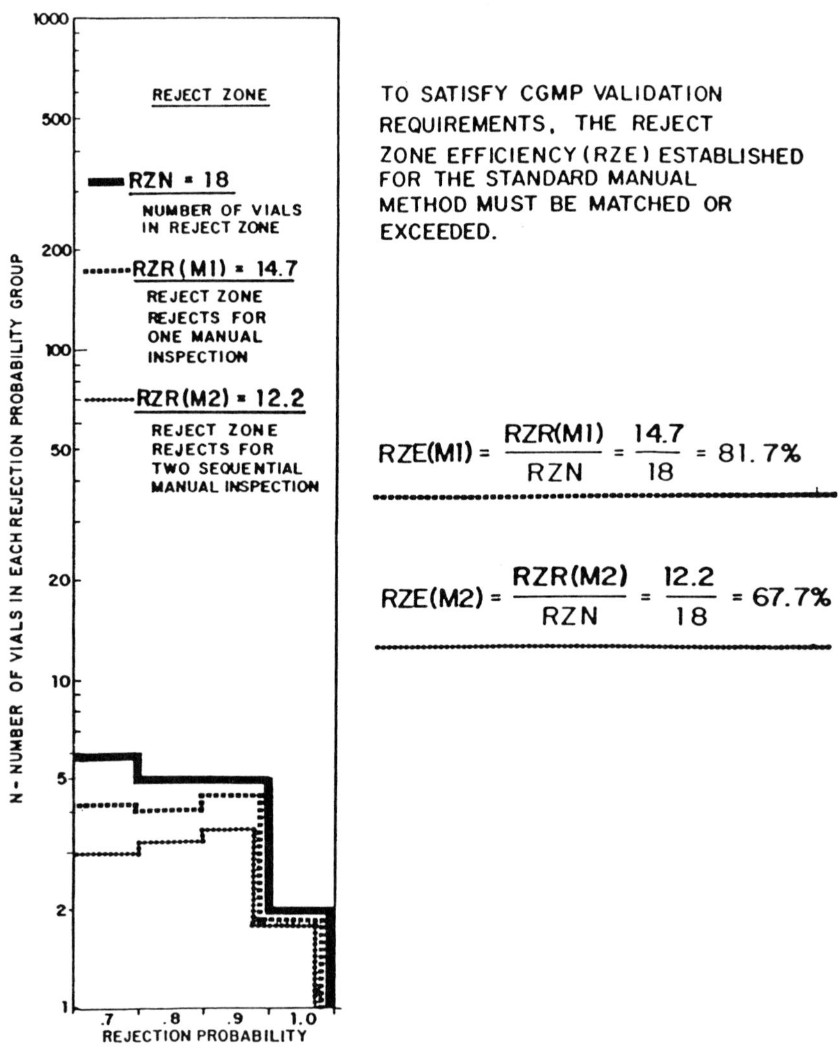

Figure 16 The Reject Zone portion of the 1000 randomly selected vial test group is separated to show the calculation of the GMP-required equivalent inspection security demonstration. This is shown to be the Reject Zone Efficiency parameter, which is defined as the ratio of the number of Reject Zone containers rejected to the total Reject Zone population.

XIV. INSPECTION SECURITY EVALUATED WITH THE REJECT ZONE EFFICIENCY (RZE) PARAMETER

With both the Reject Zone population and the average reject rate established for each inspection, a relative efficiency factor can be calculated for each sequential inspection. This is the ratio of the number of containers rejected in the zone to the total population within it. This ratio is the average rejection probability for all containers in the Reject Zone and is symbolized as RZE, the Reject Zone Efficiency. In Fig. 16 these definitions have been applied to the Reject Zone portion of the inspection data discussed.

$$\mathrm{RZE(M1)} = \frac{\mathrm{RZR(M1)}}{\mathrm{RZN}} = \frac{14.7}{18} = 81.7\% \tag{1}$$

The corresponding efficiency estimate for two sequential inspection is defined as

$$\mathrm{RZE(M2)} = \frac{\mathrm{RZR(M2)}}{\mathrm{RZN}} = \frac{12.2}{18} = 67.7\% \tag{2}$$

Matching or exceeding this objective inspection security measure, within an acceptable confidence interval, should be the only requirement for validation of any alternate inspection technique or process. The inspection security achieved by manual inspection, for the containers identified as occurring in the Reject Zone, is the comparison standard that must be matched or exceeded.

XV. PRODUCT INSPECTION AND PRODUCT QUALITY

From the analysis, it is clear that the final delivered parenteral product quality is not determined by particulate inspection alone. The produced quality leaves an unmistakable imprint.

The current GMPs place responsibility on the user to establish equivalence between a proposed and an existing technique or process. The methodology just described provides a simple numeric evaluation of an existing manual inspection technique. To satisfy current GMP validation requirements, the Reject Zone Efficiency established for the standard manual method must be matched or exceeded by an alternative method. Through a shift in emphasis from overall reject rates to the average reject rate within defined rejection probability limits that bound Accept, Gray, and Reject Zones, a statistically replicable analysis has become available. With this analysis, the security achieved in the particulate inspection of parenteral products can be evaluated with any degree of precision required. The challenge posed by the apparent paradox that resulted when particulate inspection replicability was evaluated with a Venn diagram has been resolved.

Figure 17 summarizes the concepts and definitions presented so far and lists the additional information for real-world applications of this methodology. For the Reject Zone, the single-inspection rejection probability bounds are from 0.70 to 1.0, the number of containers in the Reject Zone is RZN, the number of containers rejected in a single inspection is RZR, and the security parameter RZE is the ratio of RZR to RZN. The ideal RZE is 1.0; this means that every container in the Reject Zone is rejected with a probability of 1.0 in every inspection. In actual practice, RZEs range from 0.80 to 0.95 for a single inspection.

	ACCEPT	GRAY	REJECT
ZONE REJECTION PROBABILITY	$0.0 \leq p \leq 0.3$	$0.3 < p < 0.7$	$0.7 \leq p \leq 1.0$
POPULATION	AZN AGZN = AZN + GZN	GZN	RZN
AVERAGE REJECTED IN A SINGLE INSPECTION	AGZN-R		RZN-R
AVERAGE ZONE REJECTION PROBABILITY	$\text{RAG}^{**} = \dfrac{\text{AGZN-R}}{\text{AGZN}}$		$\text{RZE}^{**} = \dfrac{\text{RZN-R}}{\text{RZN}}$

Figure 17 Basic definitions and relationships required for replicable validation of probabilistic particle inspection systems. RZE is a measure of the inspection security obtained, RAG is a measure of the false reject rate. Note: The asterisks indicate those two parameters that measure the effectiveness and the excess economic cost of the inspection procedure.

XVI. ACCEPT/REJECT DISCRIMINATION EVALUATED WITH THE REJECT RATE IN THE ACCEPT AND GRAY ZONE (RAG) PARAMETER

In addition to the security of the inspection, its discrimination measured as the undesired reject rate in the Accept and Gray Zones must also be evaluated.

Since we can inventory the containers in the Accept and Gray Zones and determine their rejection rate in sequential inspections, we can define a quantity similar in character to the RZE just discussed. This is RAG. It is the false reject rate of containers in the Accept and Gray Zones and is computed as the ratio of the average quantity of Accept and Gray Zone rejects, AGZN-R, to the number of containers in both these areas, AGZN. The ideal AGZN-R is 0; ranges from 1 to 10% are usually found in the production environment. The relationship is expressed as

$$\text{RAG} = \frac{\text{AGZN-R}}{\text{AGZN}} \tag{3}$$

These two parameters RZE and RAG effectively characterize any manual particulate inspection and thus provide a convenient comparison basis when newer manual inspection techniques are compared with the presently accepted procedure or with semiautomated and fully automated inspection systems. They represent the average reject rate for all containers in either the Reject Zone or in the combined Accept and Gray Zones.

One core concept underlies most of the information presented to date: *The probability with which the observed particulates can be detected is essential information.* The application of this core concept transforms visual inspection of particulates from a craft to a measurable science. Using the life insurance mortality tables as a model, rejection probability zones have been defined with which the security and the selectivity of the accept/reject decision can be evaluated to any required degree of precision.

The conclusions presented so far are based on empirical parenteral inspection data only. In a search for more generality and greater insight, the psychophysics vision literature was investigated.

XVII. THE SCIENCE OF VISUAL INSPECTION

A. Biophysical Vision Data Support for the Concept of a Probabilistic Visual Inspection

Scientific support for the observed probabilistic inspection is readily found in the extensive psychophysics literature. This literature has established that objective description of visual inspection has four essential elements:

1. The capability of the viewer
2. The size of the target
3. The total background illumination
4. The contrast of the target against its background.

The data in Fig. 18 are due to Crozier [20], who was concerned with the establishment of detectability thresholds. The methodology this worker employs is typical of psychophysics. The objective brightness, or luminance, of a target of fixed size in fixed position was varied while its background luminance was held constant. This graph shows that the percent of correct responses is a function of the target contrast. The data show that the variation of target contrast, measured by the log of the luminance, can vary detectability from 1 to 99%. Since this graph is

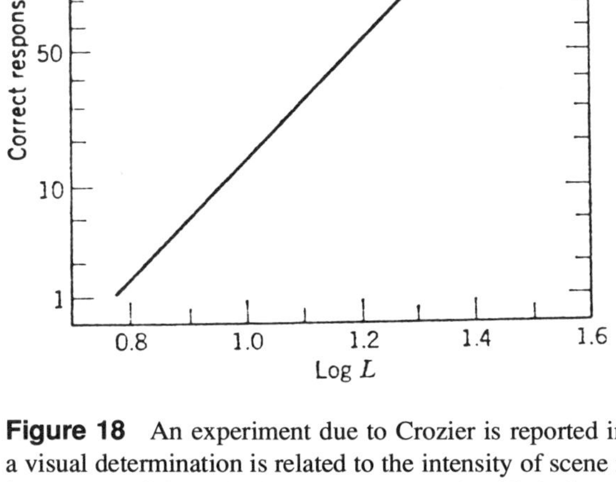

Figure 18 An experiment due to Crozier is reported in which the relationship between the accuracy of a visual determination is related to the intensity of scene illumination. This biophysical experiment is cited in support of the Schering-developed probabilistic inspection data on "percent frequency of seeing" for various values of log L, target luminance.

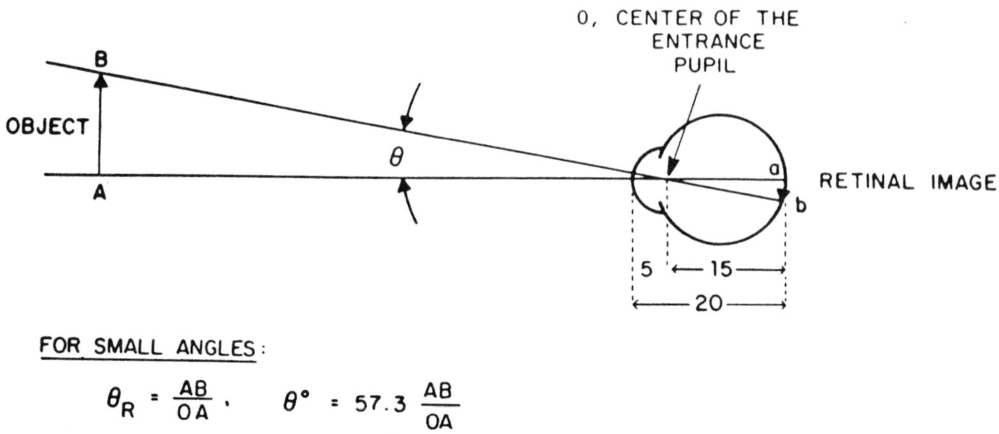

Figure 19 A section of the reduced Laurance eye illustrates the significance of visual angle measurement.

based on the responses of a single person, averages based on the responses of a number of individuals are required for any practical work.

B. Visual Angle: A Basic Vision Measurement Parameter

Much of the biophysics literature utilizes the concept of "visual angle" to measure the subjective perception of size. Since this measurement is used in some of the following material, it is described in Fig. 19.

Visual angle is a basic variable in the evaluation of relative size. It is depicted on the Laurance reduced eye [21] as the angle subtended by a particle referenced to the center of the entrance pupil, as shown in Fig. 19. This angle is simply expressed in radian measure as shown in Eq. (4) and is readily converted to degrees as in Eq. (5):

$$\theta_R = \frac{AB}{OA} \tag{4}$$

$$\theta^\circ = 57.3\frac{AB}{AO} \tag{5}$$

XVIII. EFFECT ON THE 50% DETECTABILITY RESPONSE OF A TARGET WHEN CONTRAST AND BACKGROUND LUMINANCE ARE VARIED IN TIME-LIMITED TRIALS

The data reported in Fig. 20 are due to Blackwell [22,23] and are based on the averaged response of thousands of observations to obtain a detection probability of 50%. These have been selected to illustrate the data available in the psychophysics literature. The experimental design was such that subjects knew exactly where the target might appear. The observer was required to detect the presence of a circular target located at the center of a field of uniform luminance. Single presentations of the target were made at various contrast levels, and the relation between detection accuracy and target contrast was determined. A family of curves for 50% detection accuracy at an exposure time of 0.3 seconds was then compiled. For the usual conditions of

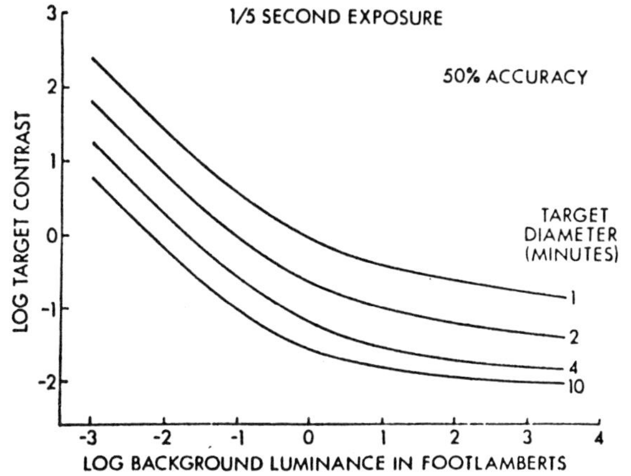

Figure 20 In further support of the finding that visual inspection is probabilistic, the work of Blackwell is reported. Blackwell developed 50% accuracy curves for target sizes between 1 ft. and 10 ft. for various conditions of background illumination and contrast of target. The target diameter at 20 cm close vision distance is equal to 58 μm per minute of visual angle.

particulate inspection (i.e., unknown target location), the contrast must be increased eightfold to achieve the same 50% detection probability. The selection of a 0.3 seconds viewing period is based on using approximately the average fixation of the moving eye in reading or scanning tasks. Using visual angle, as diagrammed in Fig. 19, a 58 μm particle at 20 cm near vision distance occupies one minute of arc. The 2-, 4- and 10-minute target angles at this near vision distance measure multiples of the 58-μm particle. The 1-ft. target diameter, using 20-cm near vision point, translates to a 58-μm particle. The 2-, 4-, 10-, and 60-ft targets can then be interpreted as multiples of 58 μm. From the data in this figure, it can be seen that as size decreases, contrast requirements for detection become increasingly more stringent. Some additional comments are in order: This discussion concerns itself with nonspecular objects only. When specular reflections, such as those that can be observed from glass fragments, are considered, detectability increases sharply.

XIX. REVIEW OF BIOPHYSICS VISION EXPERIMENTS SUPPORTING AND EXTENDING PARTICULATE INSPECTION EXPERIMENTS RESULTS

In practical problems more than 50% detection probability is required. The relationship between contrast and detection security must then be adjusted as shown in Fig. 21, also taken from Blackwell's data [24,25]. Here it can be seen that relative target contrast below 5% is undetectable by the human eye. The achievement of 99% detectability is seen to require a contrast ratio sixfold greater than that required for minimum detectability.

Figures 22 and 23 explore aspects of the same problem: the physical size that can be detected in a fixed 0.1-seconds time interval as contrast and subjective brightness are varied is examined in Fig. 22. Given a fixed target size, how long it takes to detect it for the same variation of contrast and subjective brightness is examined in Fig. 23.

The work reported by Blackwell in Fig. 22 has more direct bearing on particulate detection than much of the other information reviewed. The 1-ft target in this figure translates to a 73-μm particle at a 25-cm near focus or 58 μm at 20-cm near-focus distance. To achieve detectability

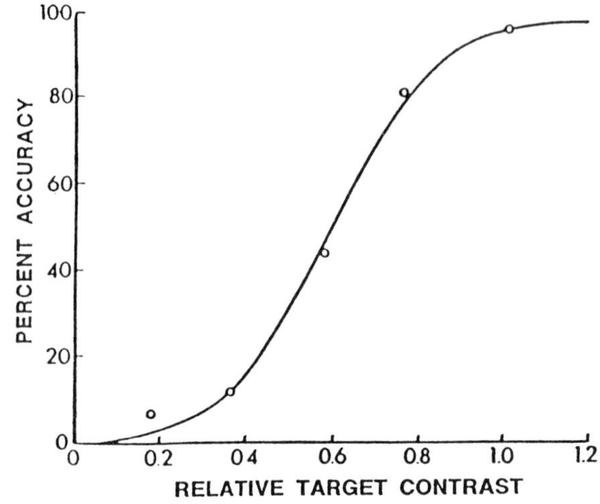

Figure 21 Blackwell also developed a relationship between visual accuracy requirements and target contrast.

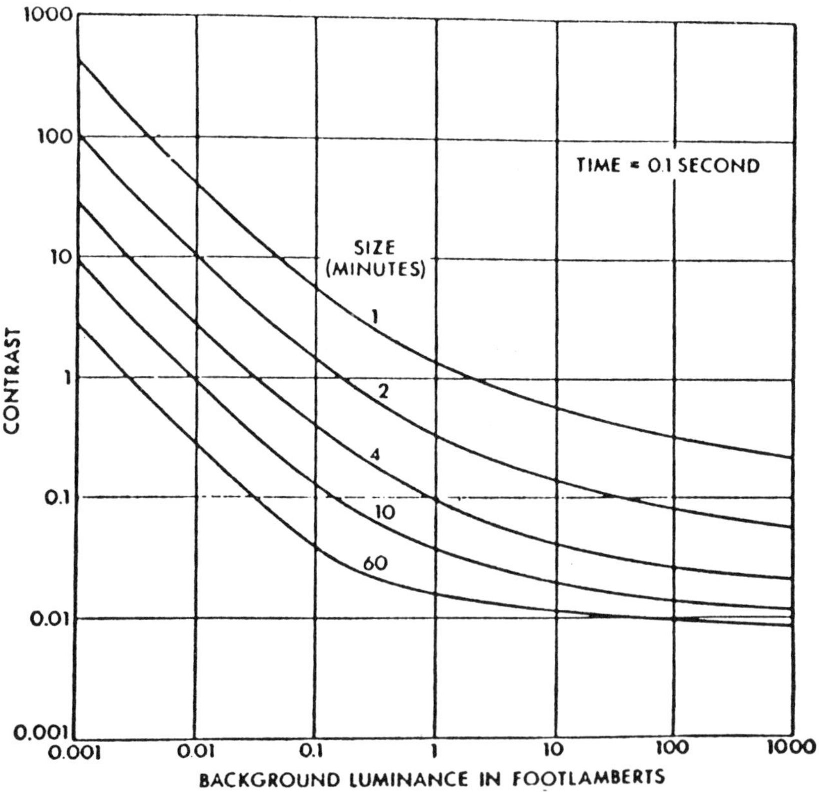

Figure 22 The relationship between accurate visual responses, target size, target contrast, and background luminance is shown as mapped by Blackwell for a fixed, 100-millisecond experiment period.

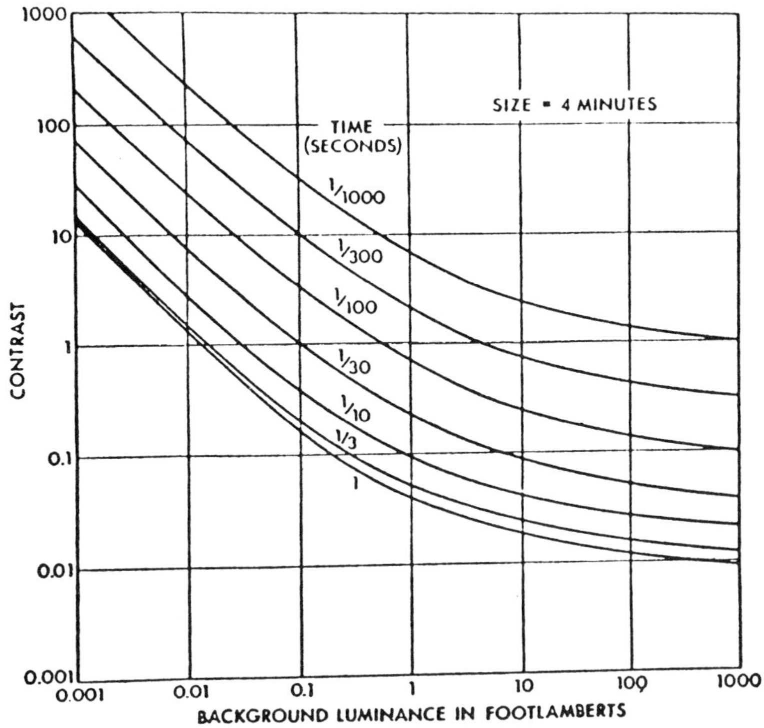

Figure 23 Another aspect of visual capability mapped by Blackwell is the time required for a correct visual response for a target of fixed size as target contrast an background luminance are varied.

for particles of 1 ft and below at the 50% accuracy level reported upon requires background luminance well above the 100 foot-lambert mark.

Blackwell also studied the time required to "see" an object of 4 ft with 50% accuracy for various contrast ratios and background [25] objective brightness or luminance; the data in Fig. 23 summarize his findings. The 4-ft object at 20-cm near-vision distance is approximately a 230-μm-sized particle. What is clear in this data is that the rate at which an object can be detected is affected by both contrast and background luminance. When the time required to detect an object exceeds the eye scan rate of 2.0 seconds per pause, the probability of detection is seriously diminished. Particle size, contrast, and background luminance are seen to be the determining factors in the detection of particulates.

Another investigator whose results are of special importance in manual inspection for particulates is Guth. Guth investigated the reliability of detecting a standard 17% reflectance gray object against both a high-reflectance background and a dark, low-reflectance background [26]. His conclusion presented in Fig. 24 is that for a dark object, a light, high-reflectance background, even at the highest illumination values examined, provides improved detectability. When a white, highly reflective object is substituted, the high- and low-reflectance labels should be interchanged. Guth's conclusion applied to particulate detection optimization for both manual and machine methods says that one lighting and background scheme are not optimum for the full range of particulates encountered. Since the very basis of manual inspection for particulates is the minimum detectable size, the data in Fig. 25 are of special interest.

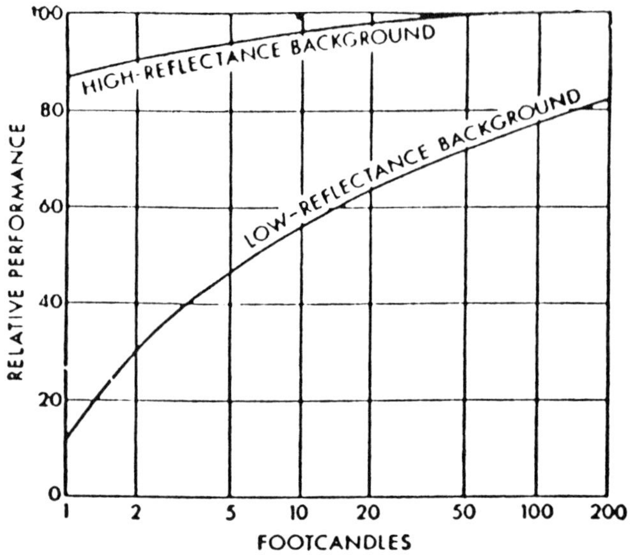

Figure 24 Work by Guth established, as shown here, the relative visual accuracy obtained when a standard 17% reflectance gray card was employed against low- and high-reflectance backgrounds.

The data in Fig. 25 are from Lythgoe's work [27] and were taken for a high-contrast circular image against a background of varying objective brightness or luminance. The left-hand scale, visual acuity, is the reciprocal of the visual angle scale presented at the right. The horizontal axis is shown in log luminance units. At a background luminance of 110 foot-lamberts, 90% of maximum visual acuity can be observed. Using the 25-cm near-distance viewing, this translates to a 31-μm particle. Increasing the background luminance to 1400 foot-lamberts changes the

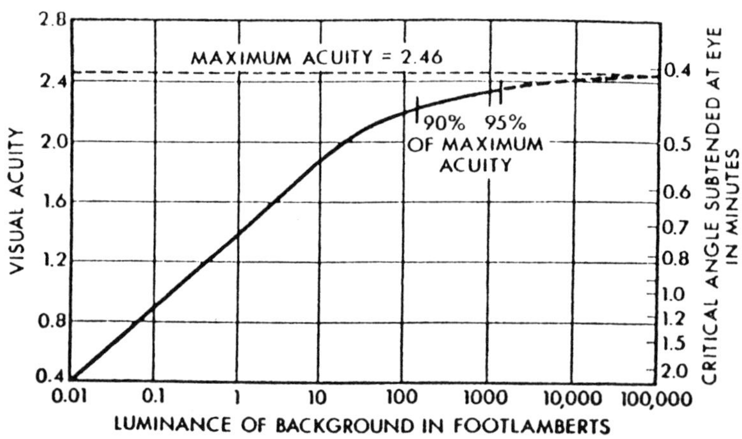

Figure 25 Lythgoe's work in determining visual acuity relationships with high-contrast targets as background luminance is varied is shown here. For the conditions of the experiment, and extrapolated minimum detectable target of 24 μm is possible with a background luminance of 100,000 foot-lamberts.

minimum detectable size to 25 μm when ideal black/white contrast is viewed. The asymptotic limit for visual acuity shown in this figure is 23.6, which is approached at a background luminance of 100,000 foot-lamberts. This substantially smaller value than the 58.6 limit reported by Blackwell is due to the difference in background luminance employed. This brief selection from the psychophysics literature can be no more than an indication of the available material. It has been selected to illustrate the concept that the manual inspection for particulates can be better understood using the well-developed methodology of psychophysics.

XX. IES LIGHTING RECOMMENDATIONS FOR VARIOUS VISUAL TASKS

Since the amount of light used for inspection is an important variable, Fig. 26 reports the Illumination Engineering Society's (IES) judgment and experience on this topic.

This figure is an extract of the 1981 Illumination Engineering Society Lighting Handbook Reference Volume [28]. It can be seen that the selection of the 225-foot-candle light level as in the Schering inspection booth for inspection of particulates is, even by today's standards, an acceptable choice.

In this last edition of the *Lighting Handbook*, it was recognized that a range of recommended lighting conditions was a more realistic response than the implied precision of previous single-point recommendations. It is appropriate to comment that the Illumination Engineering Society *Handbook* recommends that background to workpiece illumination be maintained with a 3 : 1 range. Utilization of the well-developed procedures described in illumination engineering practice can do much to standardize the conditions employed for manual inspection.

VISUAL TASK	RECOMMENDED RANGE OF ILLUMINANCE, FOOT-CANDLES		
	MIN	MID	HIGH
HIGH CONTRAST OR LARGE SIZE	20	30	50
MEDIUM CONTRAST OR SMALL SIZE	50	75	100
LOW CONTRAST OR VERY SMALL SIZE	100	150	200
LOW CONTRAST AND VERY SMALL SIZE OVER PROLONGED PERIOD	200	300	500
EXACTING TASK OVER VERY PROLONGED PERIOD	500	750	1000
VERY SPECIAL VISUAL TASK OF EXTREMELY LOW CONTRAST AND SMALL SIZE	1000	1500	2000

Extract: IES Lighting Handbook 1981 Reference Volume
Section A-3, Table 1

Figure 26 An extract from the Illumination Engineering Society recommendations for illuminance ranges for visual tasks in the parenteral container inspection range.

XXI. INSPECTOR SELECTION AND VALIDATION

A Review of Inspector Selection Criteria

In the absence of objective particulate standards, the Working Standard Test Method has evolved to facilitate comparisons between two inspection methodologies. This method, as presently used, relies on the statistically replicable performance of a group of well-trained manual inspectors as a comparison standard. A prerequisite for the successful use of the Working Standard Test Method is validation of individual inspector performance and assurance that the group as a whole meets some minimum, achievable standard.

A brief summary of the work in which these goals were achieved follows.

When this work began, there was no accepted hypothesis on what should be measured to differentiate between good and undesirable inspectors, so a wide-ranging test program was required. With this in mind, the entire Schering staff of 23 inspectors was evaluated to obtain closely defined performance. The review of each inspector's performance was based on a comparison of the results of 10 inspections per inspector against the 230-trial inspection data shown in Table 4.

Table 4 lists the RZE and RAVI (average reject rate) values for all 23 inspectors for three sequential inspections. From these data, the sequential RAGs for each inspector were computed.

Table 4 Reject Efficiency and Average Reject Rate for 23 Inspectors, 10 Inspections on Seeded Test Group of 250 Vials, Three Sets of Inspections

	RZE			RAV		
Inspector	1	2	3	1	2	3
1	0.96	0.93	0.90	0.285	0.239	0.219
2	0.99	0.97	0.96	0.284	0.233	0.210
3	0.96	0.93	0.90	0.355	0.238	0.199
4	0.93	0.87	0.82	0.295	0.220	0.190
5	0.65	0.48	0.38	0.140	0.083	0.061
6	0.94	0.89	0.85	0.242	0.186	0.163
7	0.89	0.80	0.73	0.200	0.149	0.125
8	0.89	0.83	0.78	0.218	0.171	0.149
9	0.76	0.59	0.47	0.419	0.216	0.130
10	0.82	0.70	0.61	0.226	0.141	0.108
11	0.86	0.76	0.69	0.243	0.167	0.137
12	0.70	0.53	0.42	0.182	0.099	0.069
13	0.99	0.98	0.97	0.478	0.354	0.310
14	0.74	0.59	0.50	0.157	0.102	0.079
15	0.95	0.91	0.88	0.265	0.208	0.184
16	0.71	0.58	0.51	0.137	0.095	0.080
17	0.86	0.76	0.68	0.282	0.170	0.129
18	0.96	0.92	0.89	0.310	0.232	0.202
19	0.57	0.40	0.30	0.151	0.085	0.063
20	0.90	0.82	0.76	0.304	0.192	0.152
21	0.94	0.88	0.84	0.493	0.328	0.256
22	0.95	0.91	0.87	0.278	0.234	0.209
23	0.79	0.65	0.56	0.331	0.173	0.117

The question of inspector performance uniformity was first addressed with the histogram of Fig. 27, which summarizes the first inspection reject rates achieved.

In Fig. 27, each inspector is represented by a unique number to preserve privacy. In this histogram, five inspectors (5,12,14,16,19) can be grouped and described as a low-rejection group. Four inspectors (3,9,13,21) can be grouped and described as a high-rejection group. The remaining 14 inspectors can be described as a midgroup whose rejection rate for this seeded test group ranges from 20 to 35%. Since the number of Reject Zone vials is a substantial proportion of the reported reject rates in this experiment, the reject rate in the Accept and Gray Zones was also investigated.

An essential characteristic for the selected inspectors was their discrimination—their ability to make effective accept/reject decisions. This was investigated using their calculated RAG scores, the undesired reject rate in the Accept and Gray Zones as shown in Fig. 28.

In Fig. 28, the high-rejection group of four has increased to five and appears to be more clearly distinguished from the rest. The low group, except for No. 16, appears to have merged with the midgroup. Number 16 has the distinction of wasting the smallest quantity of acceptable material. Since the security of inspection is the overriding consideration in parenteral drug manufacturing, the first inspection reject zone efficiency performance for all inspectors has also been examined.

When the inspection security ranking of the 23 inspectors was examined in the histogram of Fig. 29, only three inspectors were in the mean group. Two of the inspectors could be eliminated as ineffective. At this point, all the data examined could not be summarized into an effective selection technique. No single result or combination of results appeared to yield a conclusive selection system. Investigation was then extended to subjective size variation with and without the use of a magnifying lens. For those cases where no magnifying lens was used, investigation revealed that the differences in close focusing capability of the eye translate into differences in particulate size detection thresholds, all other factors being equal.

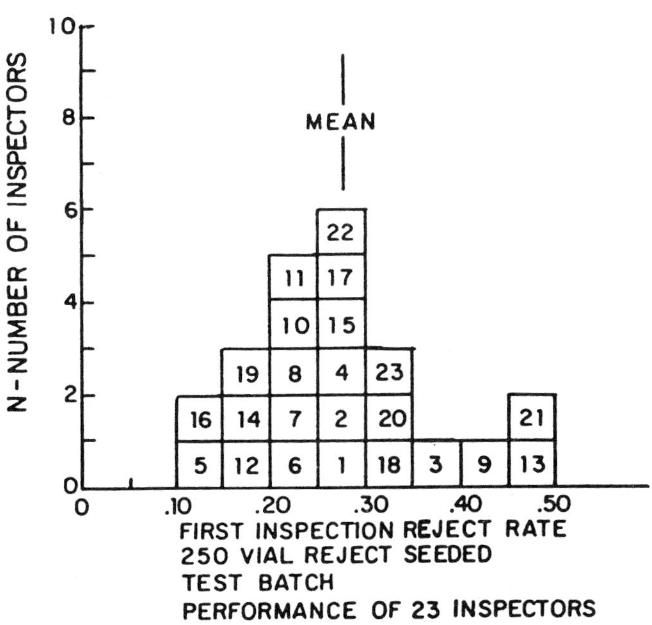

Figure 27 Histogrammic summary of the first inspection reject rates for 23 inspectors.

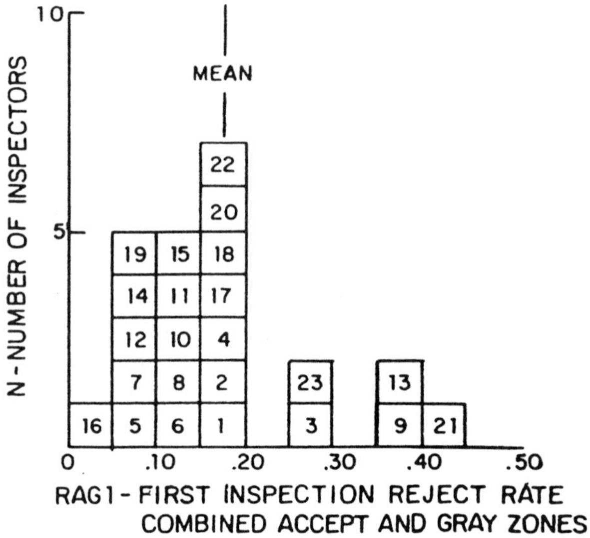

Figure 28 Histogrammic summary of the undesired reject rate in the Accept and Gray Zones (RAG) for 23 inspectors.

The data in Fig. 30 summarize the variability of perceived size for a 50-μm particle in a container at optimum close vision distance for a group of 19 inspectors without any magnifying lens. Inspector number in this and next figure is unfortunately incompatible with the sequence numbers used previously. The horizontal axis of the graph represents the visual angle for this particle positioned at the near focus point—that is, the closest distance from the eye at which sharp focus can be comfortably obtained.

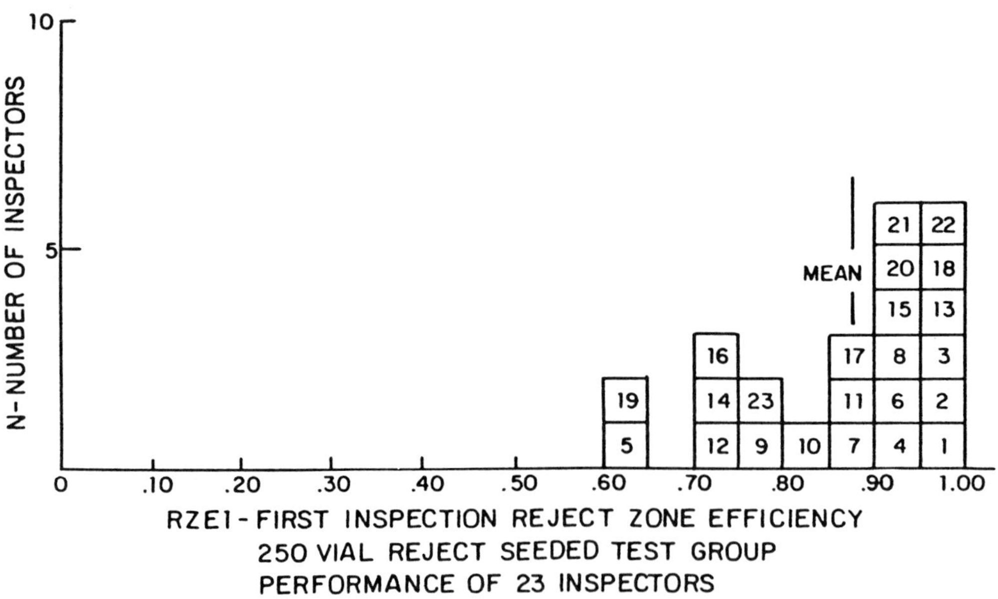

Figure 29 Histogrammic summary of the Reject Zone Efficiency (RZE) for 23 inspectors.

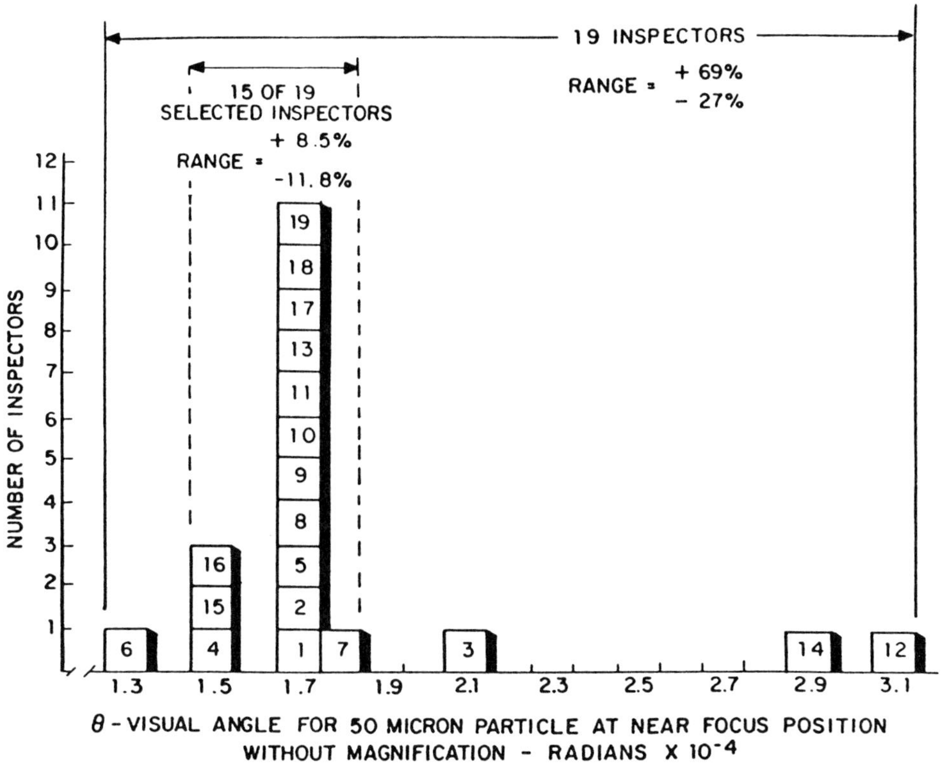

Figure 30 Visual angle variation in 19 spectors at near-focus position for a 50-μm particle without magnification.

The range of variation for the full group is 95% around the mean. A selection of 15 of the 19 inspectors reduced the range of variation to 20.3%, an approximate fivefold improvement. Since the use of the 3-diopter magnifying lens is considered an essential requirement in the inspection for particulates, subjective size perception with the use of a 3-diopter lens was also investigated.

XXII. PARTICULATE INSPECTION WITH A SIMPLE MAGNIFYING LENS

Simple magnifying lenses have been customarily employed in parenteral manufacturing to improve the accuracy with which manual inspections are performed. A brief discussion and summary of magnification relationships follows. For a more detailed discussion, Kingslake is recommended [29]. When the particle is at the focal point of the magnifier, its image is in focus to an unaccommodated eye independent of the eye's distance from the lens. In this case, the magnifying power (M. P.) of a lens is shown in Eq. (6) as the product of the lens power in diopters, D, multiplied by the close focus distance of the eye, V, in meters:

$$\text{M.P.} = VD \tag{6}$$

When an average eye close-focusing distance of 0.25 m is assumed, the widely used 3-diopter, 5-inch-diameter lens yields an M. P. of 0.75, a 25% reduction in size. When this 3-diopter lens is used with an accommodated eye, the magnifying power is expressed as follows

$$\text{M.P.} = 1 + VD \tag{7}$$

For the 3-diopter lens and the 0.25-cm close-focus distance assumed in the previous example, the resulting magnifying power is 1.75. *When this lens is used without adequate specification of the eye and object position, the variability in magnification can be* 1.75/0.75 *or* 233%. The unaccommodated eye position yielding image reduction is objectively a less demanding position in which to spend a working day.

To investigate the effect of a selected magnifying lens on the performance of the group of inspectors previously discussed, the more general form

$$\text{M.P.} = \frac{VF}{Fa + E(F - a)} \tag{8}$$

is used, where

a = object-to-lens distance
E = eye-to-lens distance
F = focal length of lens
V = nearest distance of distinct vision.

The visual angle after magnification for the same group of 19 inspectors was calculated and the results plotted in Fig. 31.

The effect of magnification can be seen in the increase of the average visual angle for the group of 0.182 to 0.243 m, an average magnification of 1.5. The unselected inspector variation is approximately 47% around the mean. Again, a selection of 15 of the 19 inspectors reduces the

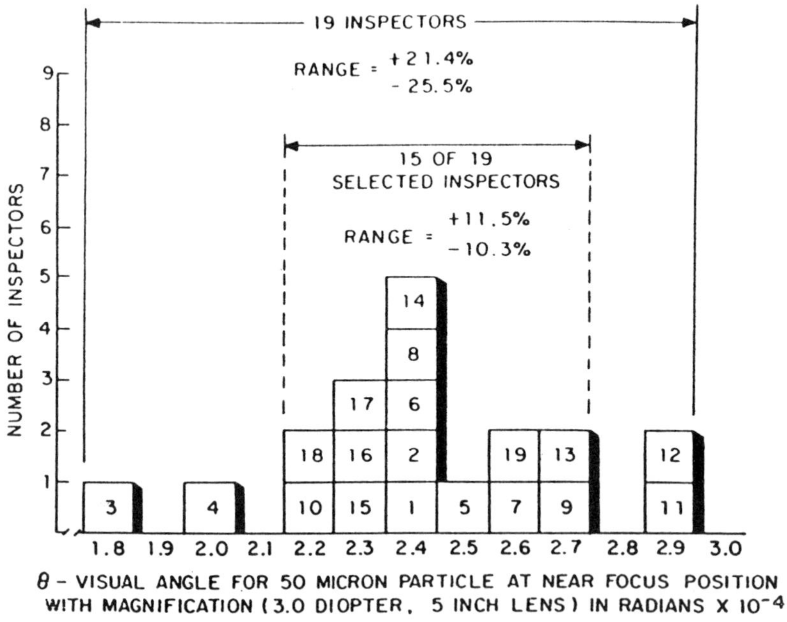

Figure 31 Visual angle variation in 19 inspectors for a 50-μm particle when a 5-inch, 3-diopter magnifying lens is employed.

calculated size variability to 21.5%, quite similar to that of the nonmagnified inspection. It should be noted, however, that the inspectors selected are not the same as in the preceding discussion, only 2 of the group of 19 being excluded from both mean groups. The average magnification reported here is 1.5; this should be compared with the 1.75 calculated with the fixed eye distance equation shown earlier.

In summary, the full library of static and dynamic tests taken individually could not be utilized to select inspectors.

XXIII. BOOTSTRAP INSPECTOR SELECTION UNDER ACTUAL INSPECTION CONDITIONS USING RZE AND RAG

Experience has shown that inspector selection based on the static criteria reviewed above is less desirable than a dynamic evaluation under actual work conditions. Two independent parameters are available with which to summarize the significant dynamic characteristics of an inspection procedure or an inspection device: These are RZE, the Reject Zone Efficiency, and RAG, the undesired reject rate in the Accept and Gray Zones. RZE and RAG values for each inspector can be used as plotting coordinates to achieve a simultaneous statement of the security and discrimination each inspector achieves. The results of an evaluation of 23 inspectors has been plotted in Fig. 32. In this figure, each inspector's performance and variability is succinctly displayed. The mean performance in RAG and RZE is symbolized with a dot; the 95% confidence limit variability is symbolized by the rectangle enclosing each mean data point. This graphical analysis clearly reveals that the group of 23 inspectors is separable into three subgroups: a low group of five inspectors, a midgroup of 13 inspectors, and a high group of five inspectors.

The low group of five inspectors is characterized by low undesired reject rate, approximately one-quarter of midgroup, and low Reject Zone Efficiency, approximately one-third less than the midgroup. The high group of five inspectors has an average undesired reject rate approximately twice that of the midgroup and an average Reject Zone Efficiency approximately equal to that of the midgroup. The performance of the high group may, in part, be attributable to eye examination standards that set minimum limits for visual acuity but set no upper limits.

Close inspection of the midgroup indicates that 2 of the 13 inspectors may be taken as outlying data points: numbers 7 and 10. The 13-member midgroup and the 11-member selected midgroup have been utilized to make the consensus reject zone decisions required. Their comparative performance is summarized in Table 5. This table summarizes the performance comparison of the 23-member total group, the 13-member subgroup, and the 11-member subgroup with the two outliers removed. The RZN increases from the 37 established by the total group to 50 units and is identical for both the groups of 13 and 11 inspectors. This table shows that the average reject rate for all groups is, in practical terms, identical. Some upward improvement in Reject Zone Efficiency can be noted. To appraise this improvement in commensurable terms, the RZE data for the 23-member inspection group, which is based on a Reject Zone population of 37, has been reevaluated in terms of the higher Reject Zone population both of the other groups recognize. This reevaluation is shown as 23* in Table 5. Since the 11 selected midgroup inspectors provide the most discriminating inspection data, the performance of the other groups will be compared with it. The RZE1 ratio of the group of 13 to the group of 11 selected mid-inspectors is 0.98. Using the RZE1 of the more discriminating 11 selected mid-inspectors as the comparison standard, the ratio of the corrected RZE achieved by the group of 23*

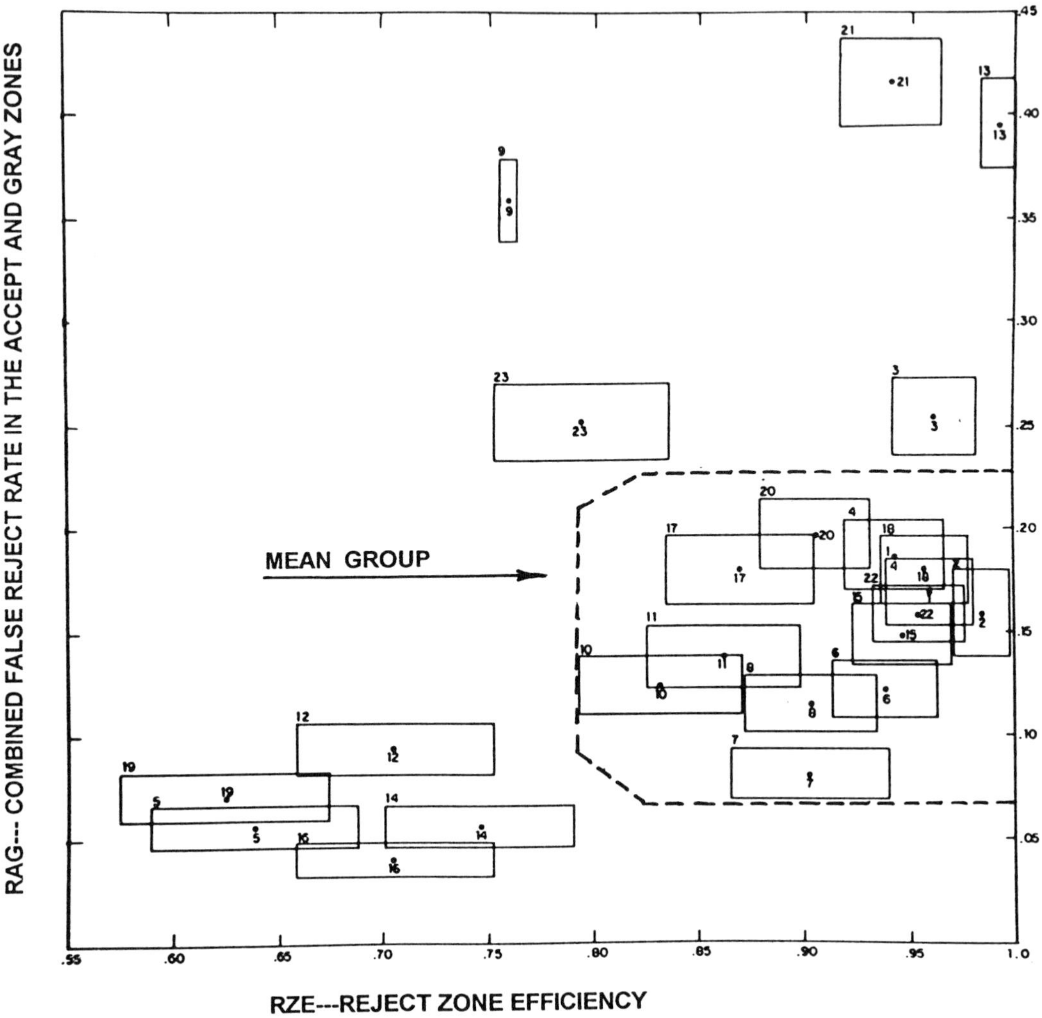

Figure 32 RZE/RAG plane use to select and validate inspectors under dynamic conditions. Each inspector is identified, mean values are shown as dots, 95% confidence limits are shown as rectangles around each mean value.

inspectors is 0.84 of the best achieved. The validated inspection group shown within the dotted line provided the requisite balance between security and discrimination. This group furnished the manual inspection data that were used in the container selection. It is clear that inspector selection based on dynamic testing improved the accuracy and reliability of the standard manual inspection. As a result of this dynamic selection process, 13 of the 23 inspectors were validated.

This table can also by used to evaluate the historical position that two sequential inspections yield an acceptable balance of security and undesired reject rate, RAG. Three sequential inspections performed by human inspection will be seen to yield unacceptable performance. The

Table 5 Performance Comparison of 23-Inspector Total Group, 13-Inspector Midgroup, and 11-Inspector Selected Midgroup

Inspectors in group	RAV 1	RAV 2	RAV 3	RAG 1	RAG 2	RAG 3
23	0.273	0.188	0.154	0.171	0.087	0.057
13	0.264	0.182	0.150	0.112	0.035	0.016
11	0.273	0.192	0.161	0.118	0.040	0.020

Inspectors in group	RZE 1	RZE 2	RZE 3	RZR 1	RZR 2	RZR 3	RZN
23	0.860	0.770	0.710	31.8	28.5	26.3	37
13	0.873	0.770	0.685	43.1	38.5	34.3	50
11	0.891	0.800	0.723	44.6	40.0	36.2	50
23*	0.747	0.640	0.585	37.4	32.0	29.2	50

*Equivalent security and selectivity obtained when the inspection standards set by the selected groups are used.
RAG = average undesired reject rate; RAV = average reject rate; RZE = Reject Zone Efficiency; RZN = number of vials within the Reject Zone; RZR = average number of Reject Zone vials rejected.

lower 95% confidence limit of RZE can be taken as an indicator of the relative security achieved in the inspection. Similarly, the mean reject rate in the Accept and Gray Zones can be used as a relative indicator of the economic desirability of the procedure. A ratio of the lower 95% confidence bounds for RZE2 to RZE1 is approximately 0.9. A ratio of corresponding RAG2 to RAG1 is approximately 0.5. A reduction of 10% in Reject Zone Efficiency resulted in a 50% reduction in undesired rejects. The further recovery of 17% Accept and Gray Zone rejects for any further decline in inspection security has been judged undesirable.

The selection of the RZE/RAG plot to validate inspector performance appears to successfully generate a joint evaluation of both the inspectors' security and discrimination. Some tests of both the generality of the conclusions reached and of the stability of selection were performed.

XXIV. DURABILITY OF RZE AND RAG DETERMINATIONS FOR A VALIDATED INSPECTOR GROUP

The inspection data compared in Fig. 33 span inspections that occurred over a seven-month period and cover two container types: 2-ml blow-molded vials and ampules. The mean Reject Zone Efficiency is 0.868 with a range of 0.026. The mean RAG is 0.36 with a range of 0.064. The small variation in RZE is interpreted as maintenance of plant quality standards within very close limits. The close agreement between ampule and vial data suggests that a good inspector remains a good inspector independent of container types. The greater variability in RAG might be traceable to the composition of the vials-1 reject-seeded test group as shown in Fig. 33.

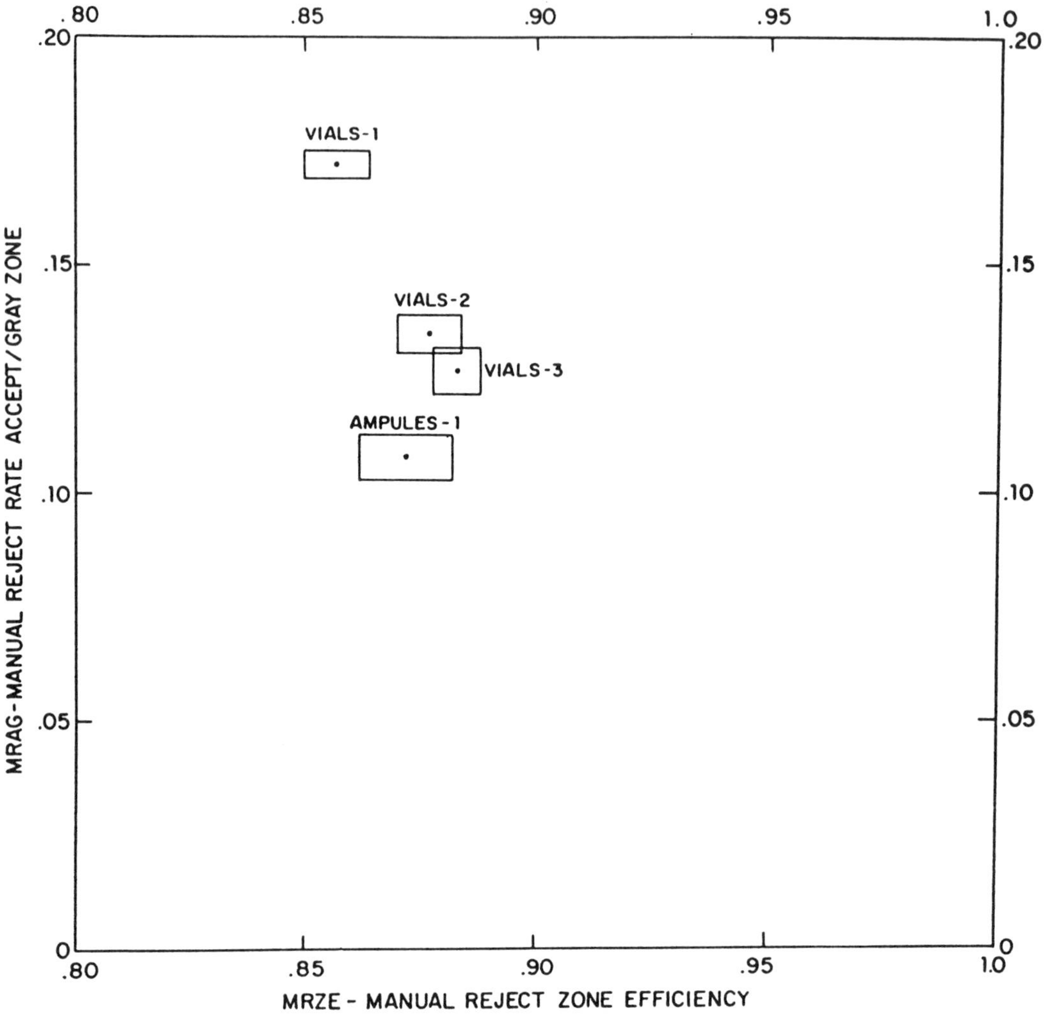

Figure 33 The stability of the RZE and RAG scores for a group of six randomly selected inspectors from the validated group was evaluated over a seven month period with both ampules and vials, with the results shown here. The scatter in RAG results is believed to originate in test group differences.

XXV. DURABILITY OF INDIVIDUAL AND GROUP RZE AND RAG PARAMETERS

Summarized in Fig. 34 is the work of six inspectors from the validated mean group with a reject-seeded test group of 293 ampules of 2 ml used in the PDSA validation. When each inspector's performance is compared against that shown in the 23-inspector comparison, drift in relative performance can be established. However, the mean response for this group taken as a whole is seen to exhibit very desirable characteristics: high RZE, 0.872, and low RAG, 0.102. The replicability of both parameters is indicated by the 95% confidence limits shown graphically by the small rectangle around the mean group values. It is felt that the combined use of these

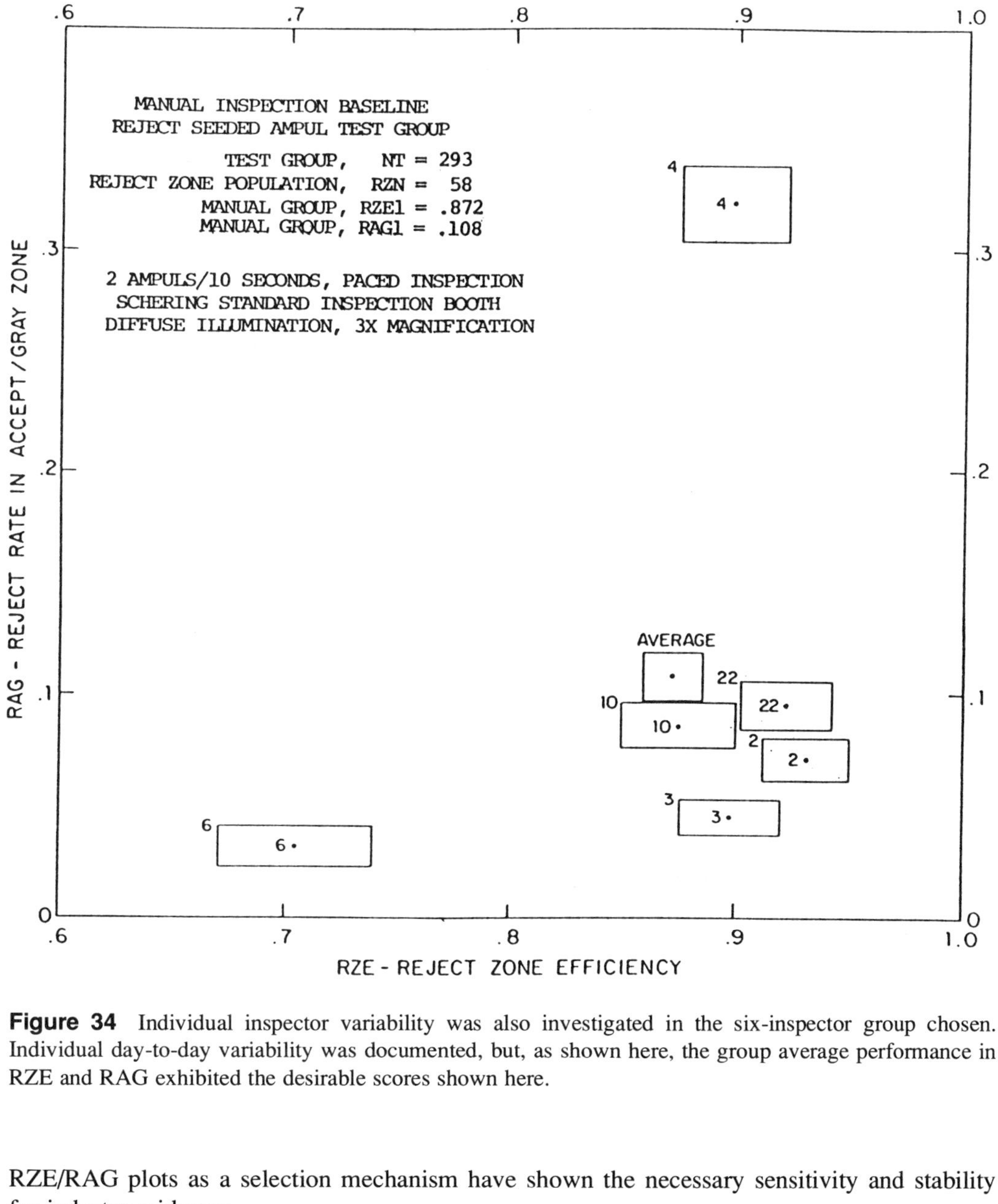

Figure 34 Individual inspector variability was also investigated in the six-inspector group chosen. Individual day-to-day variability was documented, but, as shown here, the group average performance in RZE and RAG exhibited the desirable scores shown here.

RZE/RAG plots as a selection mechanism have shown the necessary sensitivity and stability for industry-wide use.

XXVI. VIDEO INSPECTION: SYSTEM DESCRIPTION, INSPECTION STRATEGY AND QUALITY GOALS

Figure 35 is a photo of a semiautomatic video inspection system dating from the early 1970s. This system is still in use as a research tool and is used to identify particle types. At the time this system was validated, small-volume injectable standards were still being debated. Consequently, as a defensive measure, this system was designed to reject the particle content in 2 ml of the

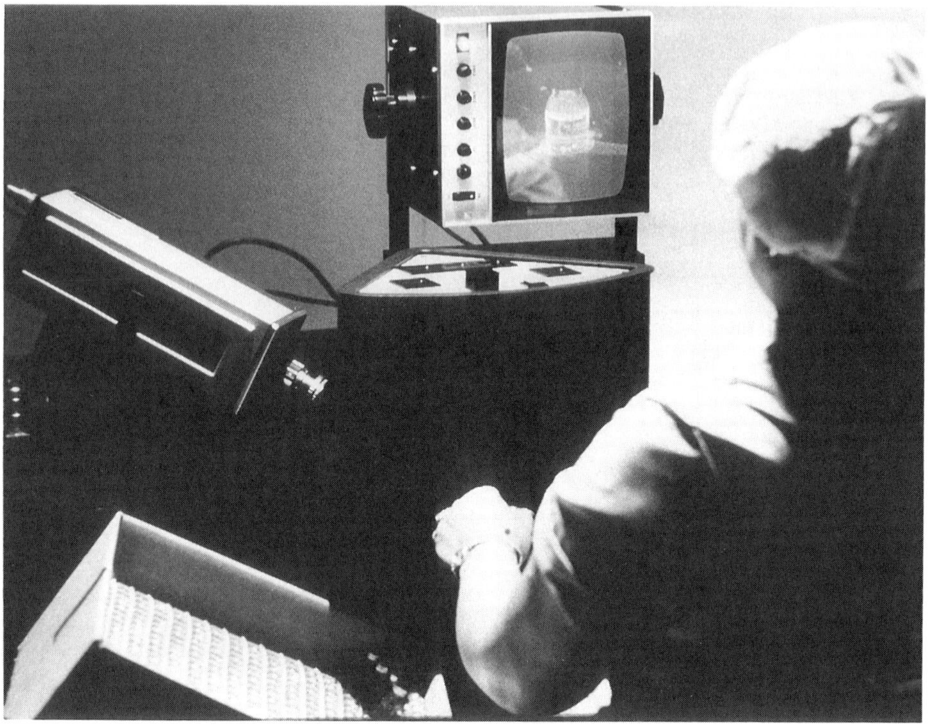

Figure 35 Photo of the video-assisted inspection system in production use. An enlarged view of the forward-scatter illuminated vial is seen on the monitor screen. The camera/monitor characteristics were manipulated to yield a go–no-go characteristic.

U.S.P. XIX large-volume injectable standard. The inspection limit for particles was 100 particles greater than 10 μm and no more than 10 particles greater than 25 μm in each container. Concerning "visible" particles, the zero-defect U.S.P. quality standard was observed. This semiautomated system and its successor, a fully automated system, were found to inspect to a quality well beyond the capability of the manual inspectors. Adhering to the GMP philosophy, the quality resulting from the inspections of both systems employed the inspection strategy shown in Fig. 36.

Figure 36 shows that the small-volume injectable batch was inspected in three sequential inspections. The rejects of each preceding inspection were reinspected in the second and third inspection shown. This sequence has been called an "Accept in Three" inspection.

In the photo of Fig. 35 can be seen the enlarged image of the 2-ml blow-molded vial filled with garamycin being inspected. The semiautomatic system eliminated the need for a rigid head position during the inspection to optimize the magnification achieved. It also reduced the manipulations required of the inspector during the inspection. The system assisted the human decision maker by substituting accurately repeatable machine motions for the subtle manipulations previously employed. It also improved particle detectability by using high-intensity forward-scattered light that produced high-contrast images. Although the light source brightness was well beyond human comfort range, the video camera monitor combination reduced the container light intensity to human comfort levels. The video camera/monitor characteristics were

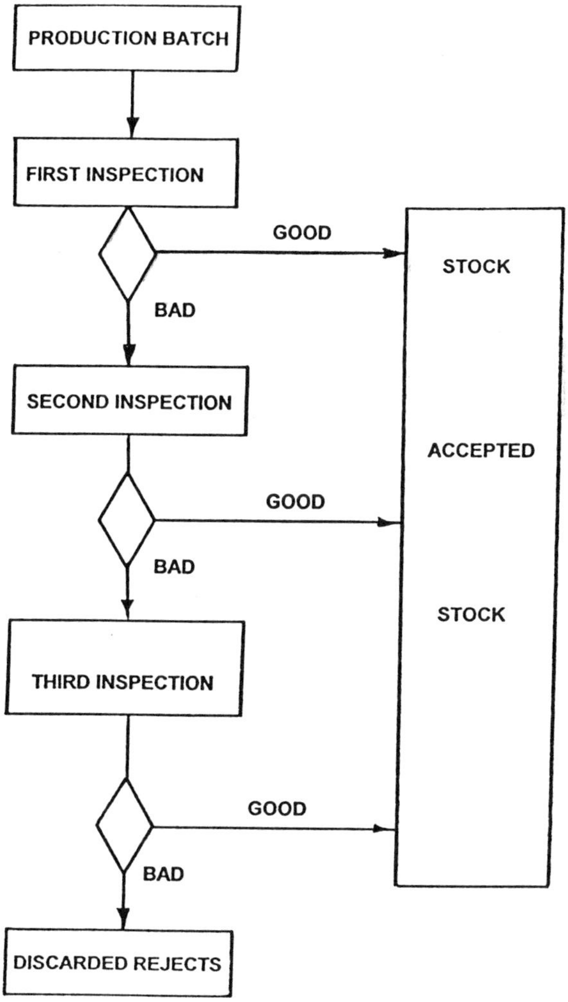

Figure 36 The automated three-inspection sequence (A3) in which product that is judged good is accepted after each of the three inspections is described as an automated "Accept in Three" (A/3) inspection (A3-A/3).

manipulated to achieve a go–no-go inspection characteristic by suppressing the signals of particles below the range of interest. If a particle was detected, the container was rejected by the inspector.

XXVII. COMPARISON OF DIFFERENT PROBABILISTIC INSPECTION TECHNIQUES

So far, our discussion has centered on the characterization of the manual inspection. We will now examine this semiautomatic video inspection system from the same viewpoint. In this examination, the core problem that arises in the comparison or validation of any alternative particulate detection system, either manually assisted as in the video system or fully automatic, will be defined. Methodology to achieve the comparison or validation will be extended from that used to evaluate the manual standard inspection.

The early work with the semiautomated video inspection established the fact that it, too, was a probabilistic inspection. Since then, all particulate nondestructive testing of parenteral products has been shown to be probabilistic in nature. The height of each peg in Fig. 37 is a measure of the number of containers within each video rejection probability group and can be evaluated against the calibrated background. The inner jacket surrounding each peg represents the quantity rejected in a single inspection; the outer jacket height represents the rejection quantity in two successive inspections. To validate the newer video system against the manual benchmark system, the history of each container in *both* inspection systems was studied.

In Fig. 38 is a summary of the manual rejection history for the working standard test group that was used to evaluate the video method. The distribution of rejection probabilities for the working standard test group of containers was found to be as individual as a fingerprint in each system evaluated. The jackets are used as in the earlier figure, the inner jacket indicating rejects in a single inspection, the outer jacket indicating rejects in two successive inspections. Both methods, manual and video, yielded data in the form of one-dimensional probability

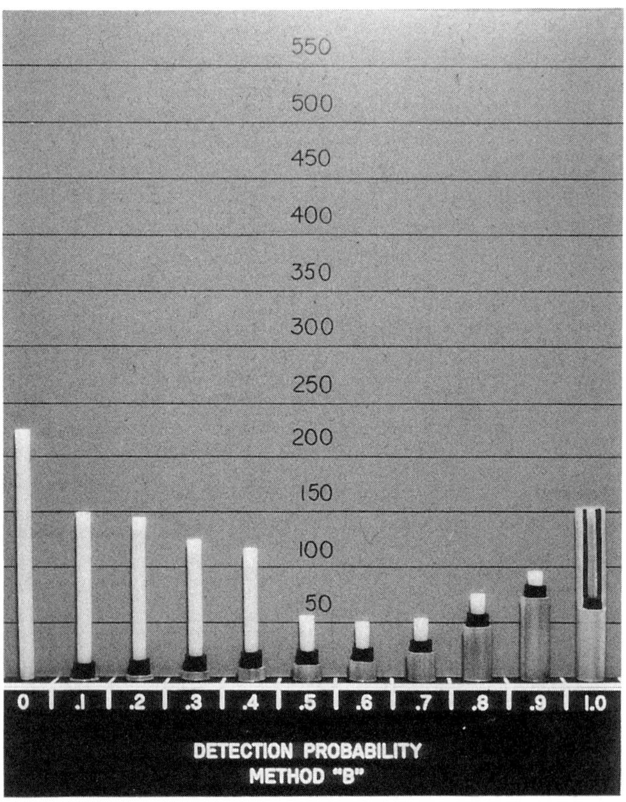

Figure 37 Pegboard histogram of video inspection results showing that it too is a probabilistic inspection. The peg height is proportional to the number of containers in each group; the inner jacket height is proportional to first-inspection rejects, the outer jacket height is proportional to second-inspection rejects.

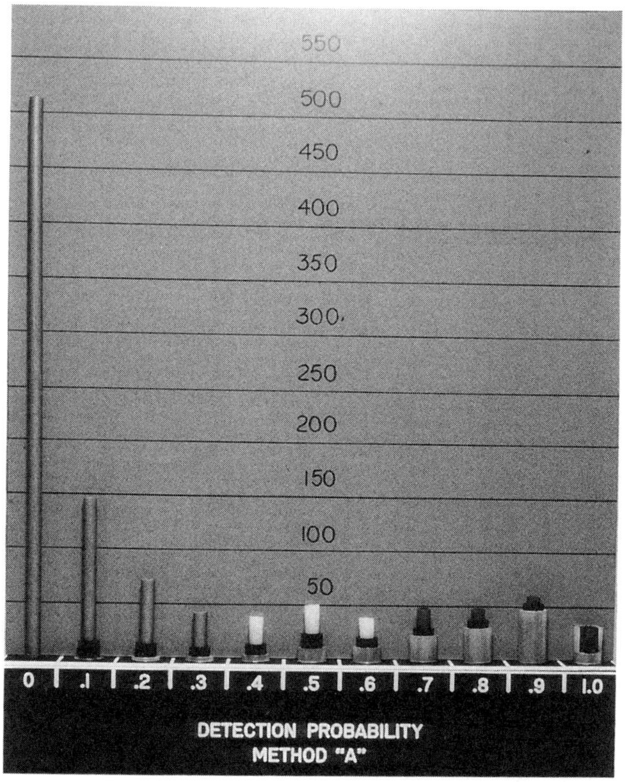

Figure 38 Pegboard histogram summary of manual inspection results on the working standard test group used with the video inspection system. The pegboard symbols are the same as those used earlier. Core height is population, inner jacket height is first-inspection rejects, outer jacket is second-inspection rejects.

histograms as shown. The GMPs clearly state that the newer method must be shown to be as efficacious and secure as the one in present use. This could only be interpreted to mean that the accepted manual method was the benchmark for comparison. Methodology for establishment of an analytical relationship between two probabilistic inspection methods initially appeared to be an intractable problem.

XXVIII. TWO-DIMENSIONAL REJECTION PROBABILITY DISTRIBUTION

The solution was visualized through the use of a pegboard in which each square was labeled with rejection probability in both systems, as shown in Fig. 39. The symbols used in Fig. 39 are those used in the other pegboard models. If, in fact, two systems were identical, then the results would be as shown in this figure. Here, the checkerboard shows that each probability group identified has identical rejection probability in both systems. What, then, of the actual comparison of manual versus video system inspection? Reality was somewhat more complex, as Fig. 40 shows.

In Fig. 40 the results of the standard manual inspection for particulates is shown as method A; the video method labeled method B. The peg heights indicate the number of

Figure 39 Two-dimensional pegboard illustrating a hypothetical pair of probabilistic inspection systems with identical performance characteristics.

containers, with the joint detection probability shown for each square. The light gray pegs at checkerboard left in the 0, 0.1, 0.2, and 0.3 detection probability columns represent all containers determined to be in the Accept Zone by the standard manual inspection. The dark gray pegs at the right end of the method A side represent the containers determined to be in the Reject Zone by the standard manual inspection. The white pegs in the center of the method A side represent the group of the containers in the Gray Zone.

The height of each peg is a record of the number of containers found to have the joint rejection probability for the selected square. For improved visualization, the background calibration used here is logarithmic. The most important results seen in this display are that each probability group in one system can fragment into a complete distribution in the other. No matter how fragmented, each square of the two-dimensional distribution can be evaluated in terms of the manually-defined zonal groupings. With the relationship between the two systems mapped out as shown, the determination of equivalent Reject Zone Efficiency can be computed simply.

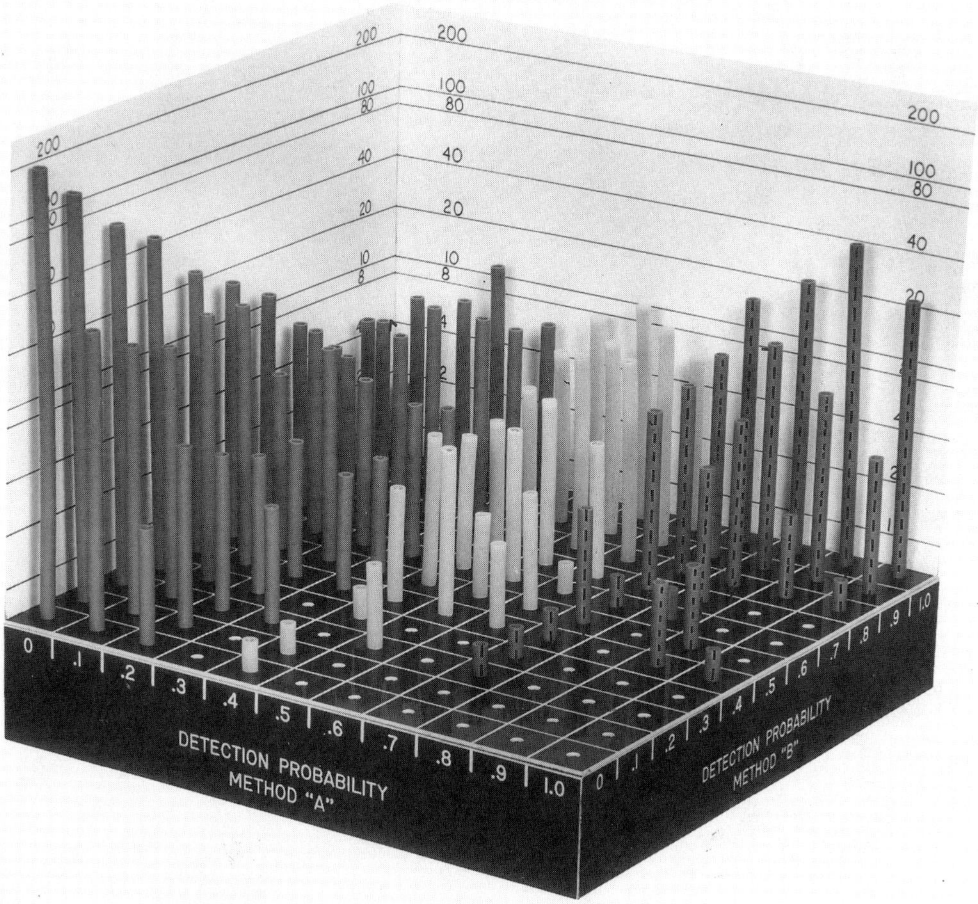

Figure 40 Three-dimensional pegboard correlagram showing the relationship between detection results of a manual system, labeled method A, an alternative video inspection system labeled method B.

XXIX. APPLICATION OF THE MANUAL INSPECTION PERFORMANCE YARDSTICK

In the pegboard photo of Fig. 41 only the reject zone is shown. Since the manual particulate inspection is the performance yardstick, the video-assisted inspection must be evaluated in equivalent manual terms. In this pegboard photo of the manually-defined Reject Zone, containers seen in four manual inspection probability groupings from 0.7 to 1.0 are distributed in 24 video rejection probability groupings as shown in Table 6. The manual equivalent reject zone rejections for the video-assisted inspection system are determined in each column of manually-defined rejection probability from the sum of all the individual products of the population in a square by its video rejection probability. From this example, there are 49 containers that are, on average, rejected in the manual standard system with a probability of 0.70:

$$R(\mathrm{M1}) = 0.70(49) = 34.3 \tag{9}$$

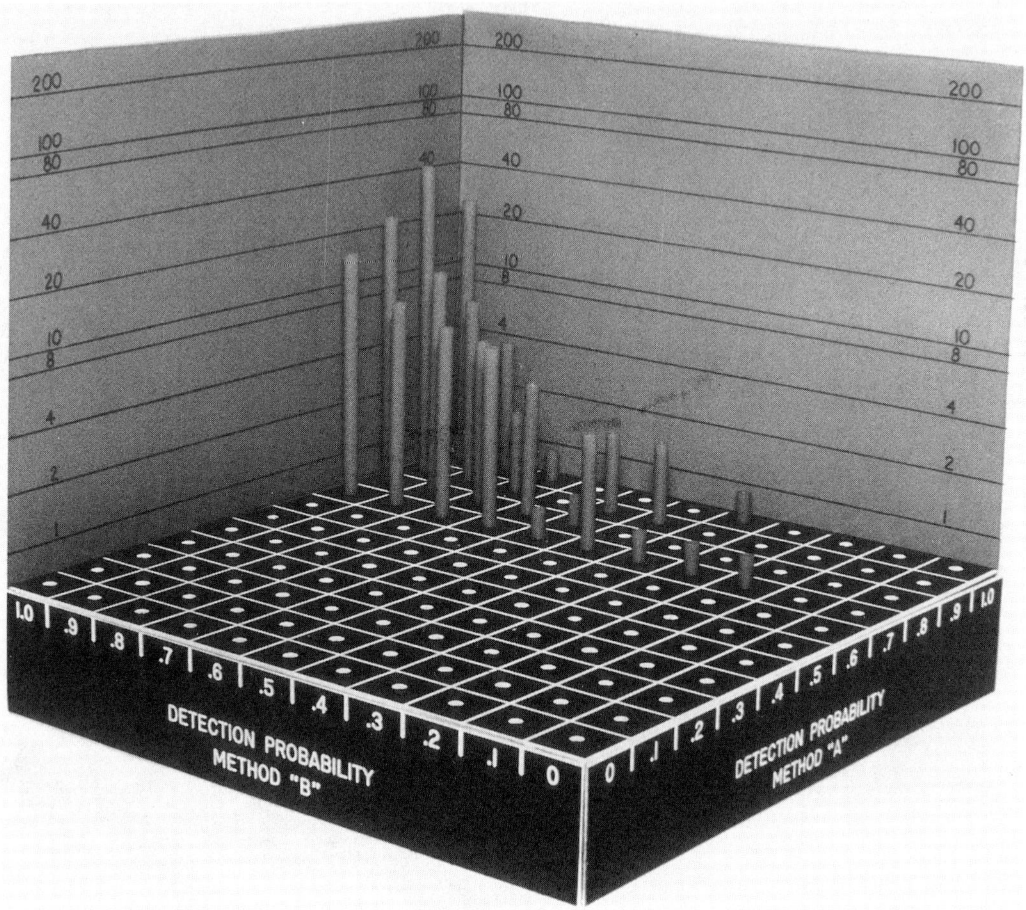

Figure 41 A view of the reject zone portion of the pegboard correlagram. The equivalent of four manual probability groups (0.7, 0.8, 0.9, 1.0) or buckets is seen to be distributed in 24 TV system buckets.

Table 6 Relationship Between Four Manual Probability Buckets and 24 Video Probability Buckets

P(M1) manual rejection probability	P(V1) video rejection probability
0.7	9
0.8	5
0.9	6
1.0	4
Four probability buckets	24 probability buckets

To evaluate the equivalent rejection capability in the video system at the 0.70 manual rejection probability segment requires a square-by-square performance evaluation in the manually-defined 0.7 rejection probability column as shown:

$$R(\mathrm{V1})@(M = 0.70) = 0.2(1) + 0.3(1) + 0.4(1) + 0.5(3) + 0.6(1) + 0.7(7) + 0.8(9) + 0.9(10) + 1.0(17) \tag{10}$$

$$R(\mathrm{V1})@(M = 0.70) = 40.3 \tag{11}$$

From this we conclude that of the 49-container Reject Zone group that the manual method rejects with a probability of 0.70, 34.3 containers are on average rejected by the manual system. In the video system, the equivalent rejection quantity is 40.3, demonstrating an improvement in the security of the inspection.

While the pegboard display is dramatic, a more convenient way to review this kind of two-dimensional probability data is in the format shown in the histogram in Table 7. The data organized to show the number of containers at each rejection probability at the extreme left and the bottom. Adjacent to each population number is its rejection probability in the system described. The number at the row and column intersection of any two probabilities is the number of containers with that joint set of rejection probabilities in both systems. The manually identified Accept and Gray Zones are grouped together; the Reject Zone and the video detection probability region greater than 0.70 are both delineated to focus attention on the inspection security achieved.

The viewpoint presented here has been that the achievement of inspection security equivalent to the present manual method is determined by a match of the Reject Zone Efficiency ratings. The precisely defined results of a standard manual method are used as the basis for a determination of the relative security attained by any proposal manual method or mechanism. A comparison of the effect of two manual inspections versus three video inspections is shown in Table 8. In the data shown, the RZE for A1 and A2, the first and second semiautomated inspections, are much higher than the standard manual scores, shown at M1 and M2, 84% and 71.6%. Adding a third TV inspection adjusted the semiautomated inspection RZE to 80.7%, with corresponding benefits in RAG, the undesired reject rate, and RAV, the average reject rate.

Table 7 Early Correlagram Comparing the Performance of the Video Inspection System to the Results of the Standard Manual Inspection of a 1100-Vial Reject-Seeded Test Group

N(V)	*R*(V1)											
155	1.0	2	1	3	7	9	12	7	17	25	46	26
96	0.9	6	9	10	8	7	9	12	10	13	8	4
76	0.8	5	8	11	11	5	9	10	8	6	2	1
52	0.7	8	6	9	4	4	6	4	7	4	0	0
49	0.6	14	10	6	5	4	5	1	1	1	2	0
56	0.5	20	7	1	3	5	2	3	3	0	2	0
67	0.4	27	20	4	3	3	6	2	1	0	0	1
76	0.3	49	21	4	0	1	0	0	1	0	0	0
97	0.2	69	17	5	3	0	2	0	1	0	0	0
150	0.1	121	21	7	0	1	0	0	0	0	0	0
226	0	192	30	3	0	1	0	0	0	0	0	0
	R(M1)	0	0.1	0.2	0.3	0.4	0.5	0.6	0.7	0.8	0.9	1.0
	N(M)	513	150	73	44	40	51	39	49	49	60	32
		Accept and Gray Zones							Reject Zone			

Table 8 Comparison of Semiautomated Video and Manual Inspection

	Standard manual inspection		Semiautomated video inspection		
	M1	M2	A1	A2	A3
RZE (%)	84.0	71.6	91.1	85.2	80.7
RAG (%)	11.8	4.6	15.9	5.4	2.3
RAV (%)	24.3	16.1	43.6	32.9	27.7

NT = 1100; RZN = 190; AGN = 910.

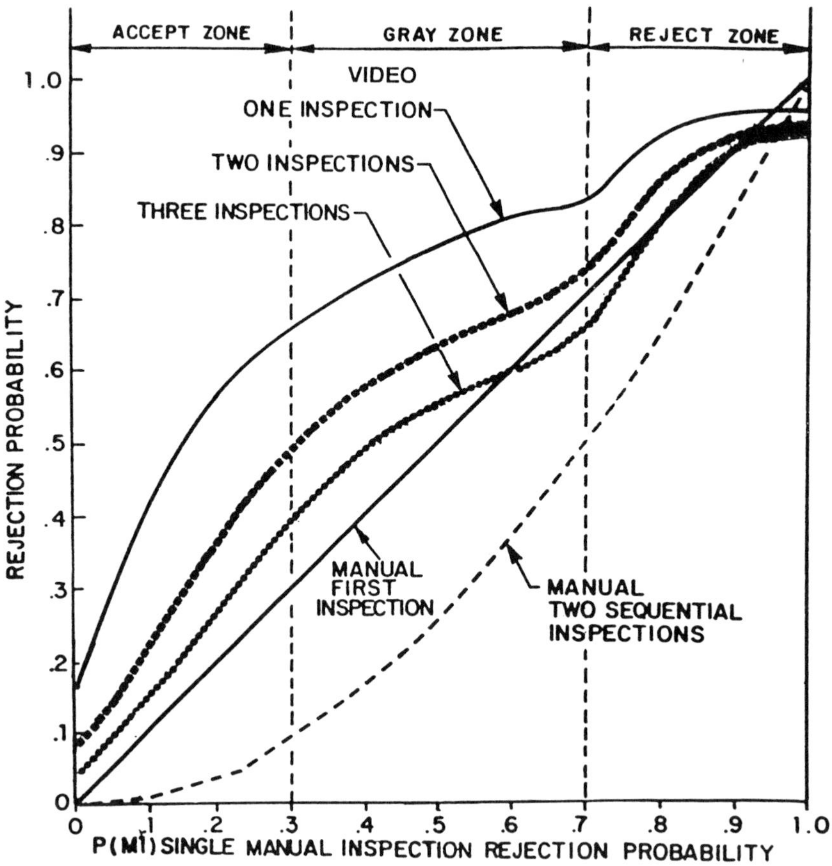

Figure 42 Graphical comparison of the inspection results of a 1000 vial reject-seeded test group after one and two manual inspections and one, two and three video inspection system results. The third video inspection is shown to give more secure results than the second manual inspection through most of the control range.

Table 9 Summary of Effect of TV System Use

Total vials inspected	> 90 million
Machines delivered	28
Reduced inspection time	17 s to 3 s
Labor savings	40 inspectors

Point-by-point computation has been used to determine the relative rejection probabilities for the video system compared to manual performance in Fig. 42. It can be seen that in the Reject Zone, the third video inspection is quite close to manual first inspection rejection probability and is significantly more secure than the use of two sequential manual inspections. The less effective discrimination for the video system in most of the Accept and Gray Zones compared to manual inspection is also clearly visible. When cost/performance was examined, the increase in throughput from 1000/day to 1000/h justified the higher reject rate resulting from video use.

In the course of the video inspection program, the concepts and methodology to evaluate fully automated inspection systems were generated. The bottom-line view of this automation project is shown in Table 9.

In addition to the savings shown, approximately $1 million savings of construction costs more than compensated for the capital investment in the inspection systems. The greatest achievement, however, was confirmation of the concept that the correct use of automation could simultaneously improve quality and reduce costs.

XXX. EXTENSION OF PROBABILISTIC INSPECTION METHODOLOGY

The primary focus of the work at Schering was support of the automated particle contamination inspection machines that had been developed here. These inspection systems were in use to ensure the contamination-free quality of the antibiotic that was being shipped. An indication that the utility of this validation methodology was not limited to the Schering inspection systems which were controlled by the probability of *rejecting* Reject Zone reject containers was the publication of an Eisai Automated Inspection System validation by Borchert and his colleagues [30]. The controlling probability of the Eisai inspection system was that of *accepting* Reject Zone rejects.

An extensive review of the difference in inspection results between these two methods follows.

The Capability and Inspection Strategy of Nondestructive Serial Inspection Systems for Liquid-Borne Particles

Much of the earlier work in the validation of visual and automated inspection [2–8] accomplished at Schering used the results of two sequential human inspections, an initial inspection followed by a culling inspection as the validation performance benchmark. This manual inspection system is described as an "Accept in Two" system. As shown in Fig. 6, containers judged good in either inspection were accepted for stock. The automated inspection system, described in the flowgraph of Fig. 36, used three sequential inspections in which material accepted after each inspection was accepted for stock.

The previous validation methodology for visible particle inspection systems which my co-workers at Schering and I described must be seen in perspective to evaluate its present limitations. The work was accomplished to support the growing production requirements for a life-saving drug: garamycin. As such, it was focused on the need to obtain and document, beyond any possible question, that the inspection security achieved with the cost-reducing automation that had been constructed was at least as good as the quality that was and is a Schering corporate trademark.

The maintenance of quality equal to or better than that delivered with the manual inspection process was imperative. To ensure a conservative product quality match, the inspection security for the different inspection strategies and techniques was compared *on a worst-case basis.* The immediate practical problem that had to be resolved was that the inspection security achieved varied with the number of inspections employed. For the manual inspection and the automated systems evaluated, most of the material inspected was accepted in the first inspection. A fraction, generally less than 25%, was inspected twice. The final inspection processed less than 10% of the initial batch. Within the variability of the batches produced, there was no practical way to determine the proportion of the batch that would pass through the full inspection sequence. The decision reached was that validation must be based on the worst-case security that would have been achieved if the entire batch went through the full inspection sequence described in each flowgraph.

In 1990, this previous work was revised and extended [10] to include Japanese, Italian and German inspection systems that are in current use. This extension was required because the basic inspection strategy, used in most of the automated inspection machines now available, is one in which the controlling probability is that of *accepting* Reject Zone containers. The controlling probability in our previous papers [2–9] is that of *rejecting* Reject Zone containers. This difference stems from the inherent detection capability of each system and is seen to have an important effect on the accept/reject decision and thus on both the true and false reject rates that are obtained. Where the inspected quality is higher than the manual standard, repeated "accept" inspections reduce the level. Where the inspected quality is not equal the manual standard, Repeated "reject" inspections will improve the inspected quality to acceptable levels.

XXXI. MULTIPLE INSPECTIONS

Why are multiple inspections an important topic? *At the present time there are no automatic particulate inspection systems that can deliver the desired combination of inspection security and an acceptable false reject rate in a single inspection.* Recognizing this problem, our earlier publications included convenient tables to determine the number of repeated Reject Zone container inspections needed to validate multiple-inspection automated systems within 0.05 significance limits. The tables were prepared for the planned ways in which the *serial accept* systems could be used. The controlling probability on which these tables are based is the average probability of *rejecting* any Reject Zone container in each of the serial inspections. The controlling probability in the validation of systems that employ the "Reject in Two" inspection strategy is the average probability of *accepting* any Reject Zone container in each of the planned inspections. An extension both of the earlier analysis and the published tables are required to make available the convenience and security of this visible particulate inspection validation methodology for the "Reject in Two" inspection strategy. In addition to the differences in validation methodology, it will be seen that both the total reject rate and the false reject rate are affected by the strategy employed.

From this analysis, it is clear that the final delivered parenteral product quality is not determined by particulate inspection alone. Particle inspection whether manual, semiautomated or fully automated is not an ideal inspection that can eliminate every rejectable container. The produced quality leaves an unmistakable imprint.

GMP philosophy places full responsibility on the user to establish equivalence between a proposed and an existing technique or process. The methodology just described provides a simple numeric evaluation of an existing manual technique. The individually identified test containers can now be separated into subgroups; in each subgroup the number of containers multiplied by the probability of rejection established for the subgroup determined the average number of rejects from it. Using this fact, quality categories corresponding to the earlier rule-of-thumb categories good-good, gray and bad-bad were established within numerical probability boundaries. The earlier categories were respectively renamed the Accept Zone, the Gray Zone and the Reject Zone. These boundaries were not only used as the basis for secure validation of semi- and fully automatic inspection systems, they also provided the necessary tools with which the economic effectiveness of the inspection process could be evaluated. The unchallenged quality definitions in our first inspection paper proposed that any container rejected less than 30% in a single visual inspection was in the Accept Zone; any container rejected 70% or more was in the Reject Zone. The containers that were rejected more than 30% and less than 70% were defined to be in the Gray Zone. Gray Zone containers can be sacrificed to securely reject the bad-bad containers in the Reject Zone. Those Gray Zone containers that are not sacrificed are considered good material and stocked. With these definitions in mind, a determination of the reject rate in the Accept and Gray Zones identifies the false reject rate.

To satisfy GMP validation requirements, the Reject Zone Efficiency established for the standard manual method must be matched or exceeded by an alternative method. Through a shift in emphasis from overall reject rates to the average reject rate within defined rejection probability limits that bound Accept, Gray, and Reject Zones, a statistically replicable analysis has become available. With this analysis, the security achieved in the particulate inspection of parenteral products can be evaluated with any degree of precision required. The challenge posed by the apparent paradox that resulted when particulate inspection replicability was evaluated with a Venn diagram has been resolved.

By defining an *average* rejection probability for a group of containers, repeatable inspection results became possible. Validation of an inspection system for visible particulates is transformed into a requirement that the average rejection rate for all the containers that the human inspection identified as "must rejects," i.e., were in the Reject Zone, had to be matched by the new device within an acceptable confidence interval. In addition to inspection security, the excess costs imposed by the false rejection was evaluated by determining the rejection ratio of Accept and Gray Zone containers. Evaluating both the inspection security and the false reject rate of a proposed device provides an objective analysis of its merits.

Figure 17 (see page 318) summarizes the concepts and definitions presented so far and lists the additional information for real-world applications of this methodology. For the Reject Zone, the single inspection probability bounds are from 0.70 to 1.0, the number of containers in the Reject Zone is RZN, the number of containers rejected in a single inspection is RZR, and the security parameter RZE is the ratio of RZR to RZN. The ideal RZE is 1.00; this means that every container in the Reject Zone is rejected with a probability of 1.00 in every inspection. In actual practice, RZE values range from 0.80 to 0.95 for a single inspection.

XXXII. PARTICLE INSPECTION VALIDATION REQUIREMENTS

The standard particulate manual inspection and all of the automated particulate inspection systems now in use share a common problem. The accept/reject decision is not a simple yes/no affair. It is a graded response ranging from 0% to 100% depending on the particular inspection characteristic employed. To validate such a system requires multiple trials and the services of a statistician or, consultation of the specially prepared statistical tables discussed below.

Four facts are required to plan a validation trial:

1. The RZE($M1$), capability of the manual inspection
2. The type of manual inspection (single or two pass)
3. The number of containers in the test group determined by the standard manual inspection to be in the Reject Zone
4. An estimate of the controlling probability of the automated inspection system.

XXXIII. STATISTICAL DESIGN OF VALIDATION DEMONSTRATIONS

GMP validations require a demonstration that a proposed alternative system or automated inspection device is at least as good as the manual inspection system against which it is compared.

The probabilistic nature of the accept/reject particulate inspection decision dictates a validation based on the mean result of many accept/reject decisions. To plan for a cost-effective validation requires an estimate of the number of manual and alternate or machine inspections that will satisfy the significance level requirements for an acceptable "demonstration of equivalent performance." The repetitive nature of the statistical design for these validation experiments led to the concept that the required statistical information could be supplied in table form, thus minimizing the need for statistical consultation as the experimental conditions changed. This concept was implemented by designing a standard validation experiment for all the presently available inspection strategies over the usable range of RZEs between 0 and 1.0 as described below.

The statistical problem posed in a GMP validation demonstration is a test of two hypotheses. The null hypothesis, H_0, is that the RZE of the alternate or automated system is equal to or better than the manual standard inspection system. The alternative hypothesis, H_1, is that the manual system RZE is better than the alternative or automated system being considered. Both hypotheses can be tested simultaneously when the minimum inspection security difference that we wish to detect between them, ΔRZE, is defined. The selection of ΔRZE equal to 0.05 as a practical validation choice continues in this chapter.

This validation choice means that the demonstration of an automated RZE within ± 0.025 RZE units of the manual RZE is a demonstration of equivalent performance. This concept is illustrated in Fig. 43.

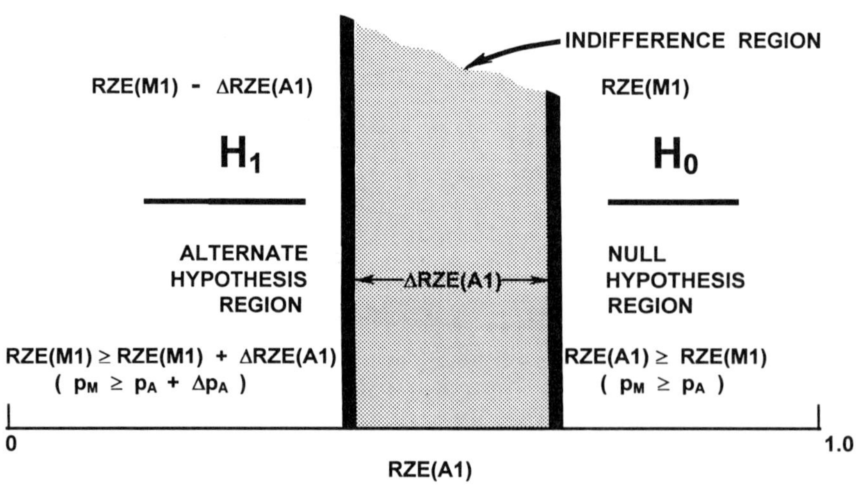

Figure 43 Graphical summary of the hypothesis testing formulation.

XXXIV. INSPECTION REQUIREMENTS FOR GMP VALIDATION OF PARTICLE INSPECTION SYSTEMS OR METHODS

The present RZNI tables for validation of non-destructive inspection systems in terms of human capability ("visible particles") have been extended to include all current manual, automated and semiautomated inspection strategies. Drawing on the accumulated experience of many users, the nomenclature has been simplified, clarified and is listed in Table 10. New tables have been added to assist in the design of validation demonstrations in which validation costs and time are held to a minimum.

For validation demonstrations in which the throughput of the automated system is considerably higher, two sets of tables have been calculated. The first set has been calculated for a balanced combination of manual and alternate system inspection requirements that minimizes the number of manual inspections without disregard for automated inspection time. The second set has been calculated to reduce manual inspection requirements to practical minimums. The combinational validation requirements for all the alternatives discussed are listed below.

The validation procedure described is applicable to the comparison of the effectiveness of an individual inspector, inspector groups and alternative manual procedures, as well as auto-

Table 10 Nomenclature Used in the Discussion of Validation Demonstrations Using Probabilistic Inspection Data

1. Economic effectiveness of the inspection	
AGZN	quantity of containers in the Accept and Gray Zones
AGZA	quantity of containers accepted in the Accept and Gray Zones
AGZR	quantity of containers rejected in the Accept and Gray Zones
RAG	false reject rate in the Accept and Gray Zones
RAV	average reject rate for inspected batch
2. Inspection security	
D	constant used to relate manual and machine inspections required for validation
I	number of inspections of a test group
K	constant used to study the validation inspection requirements
RZA	average acceptance probability of the containers in the Reject Zone
RZE	average probability of rejection of the containers in the Reject Zone
	Note: RZA and RZE are also used with a descriptive suffix enclosed in parentheses. Within the parentheses is the type of the inspection, *M* or *A* for manual or automatic, the number of the inspection and the inspection strategy. The RZE suffix for a two-inspection automated system using a "Reject in Two" inspection strategy is (A2-R/2). The parentheses following a two-inspection manual sequence in which product is accepted for stock in each inspection is (M2-A/2)
RZ$\overline{\text{R}}$	quantity of Reject Zone containers *accepted* in a single inspection
RZR	quantity of Reject Zone containers *rejected* in a single inspection
3. Statistics	
α, β, d, $G(d)$	constants used to define the ±0.025 RZE unit indifferent zone with t_α, t_β which the capability of the tested system is judged to be equivalent in performance to the benchmark system employed
H_0	The null hypothesis states that the RZE of the automated system is larger than the RZE of the manual system.
H_1	The alternative hypothesis states that the manual system RZE is larger than the sum of the RZE for the automated system and the indifference margin.

mated or semiautomated inspection systems. In each case, the number of Reject Zone container inspections required for a 0.05 significance level is related to the controlling probability of the first inspection. This controlling probability is established by the strategy, the inspection sequence and the capability of the inspection system employed. For all inspection strategies that utilize serial accepts, the controlling probability is that of *rejecting* Reject Zone rejects. These include the standard single inspection and the "Accept in Two" or "Accept in Three" inspection strategies. Use of the tables proceeds as follows: the number of Reject Zone container inspections required for the validation demonstration are listed by method and estimated RZE. The number of Reject Zone container inspections is determined by the strategy employed and by the security achieved for each method. The tables list a number of Reject Zone container inspections by the Reject Zone Efficiency or the Reject Zone acceptance rate for each method employed. When an "accept" strategy is employed, the number of Reject Zone container inspections is listed against RZE(M1) for the manual inspection and RZE(A1) for the alternate or automated inspection. Separate tables are provided for each method employed. In these strategies, containers accepted as good in each inspection are added to stock.

For those inspection strategies which use serial rejects and are validated against a single manual inspection, RZE(M1) is still used to determine the number of Reject Zone containers inspected manually. RZE(A1) is replaced by the probability of *accepting* Reject Zone containers in the first of the repeated inspections, RZA(A1). In this type of inspection, a container must be judged good in each of the sequential inspections to be accepted as good for stock and shipment.

XXXV. RZNI TABLES: THEIR PAST

The validation methodology previously described uses a single, well-defined, statistically replicable number to fully characterize the relative security achieved by the manual standard inspection and that of the proposed alternate system or device. The single number now in use is the probability controlling the first inspection in both methods.

To use current terminology, the validation methodology and the statistical tables required when a system or device uses the "Reject in Two" inspection strategy are now harmonized with the preceding treatment of the "accept" inspection strategy. This harmonization will facilitate the planning of validation trials for any automated systems which use *either* inspection strategy.

The RZNI tables were and are grouped in inspection validation pairs. In each set of tables for a validation pair, an RZNI number specified an equal number of inspections of the Reject Zone containers by both methods. The RZNI number in the validation pair table was accessed at the intersection of the controlling probability for the first inspection of each method. The calculation selected for the tables provided a total number of inspections that approximated the theoretical minimum total. The demonstration of equivalent inspection security is based on a demonstration of equal effectiveness in rejecting the Reject Zone containers. To demonstrate statistically replicable performance, a number of test group inspections is required. As has been noted, to maintain inspector performance equivalent to production conditions, an upper limit of 25% Reject Zone containers in the test group should be maintained.

For a particular validation demonstration, the number of test group inspections required, I, is determined by dividing the total number of container inspections specified in the RZNI tables by the number of containers in the validation test group ($I = \text{RZNI/RZN}$). The effect of this computational approach can be reviewed in Fig. 44.

The curves in Fig. 44 are based on the results obtained when a single manual inspection is used to validate a "Reject in Two" automated inspection system. The data for this graph begins with an estimate of the controlling probability in the first inspection. For the benchmark manual

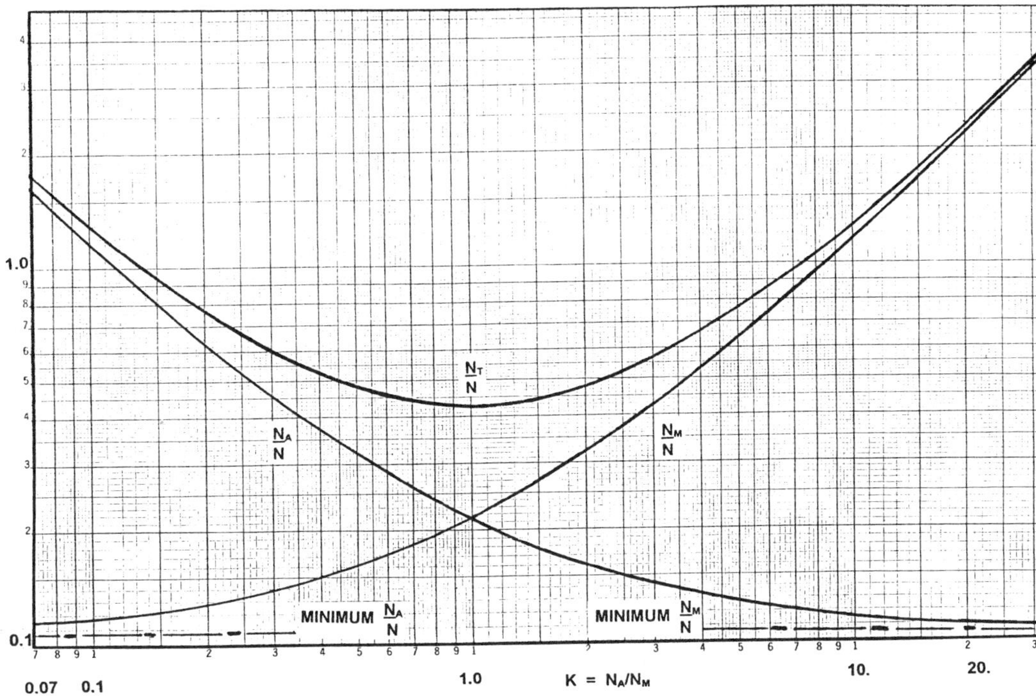

Figure 44 Automatic, manual and total inspection replications for RZE(M1) = RZE(A2-R/2) = 0.88, $N = 4295$ for 0.05 significance validations. All automated and manual vertical inspection replication pairs provide equivalent results.

method, this is the Reject Zone efficiency, the average rejection probability for all Reject Zone containers. The controlling probability for the "Reject in Two" inspection system is the probability of accepting Reject Zone rejects. The final inspection security achieved in both systems has been adjusted to be identical: RZE(M1) is equal to RZE(A2); both are equal to 0.88. When an annual number of inspections for both systems is selected, $N_M = N_A = N_{MA}$, 927 Reject Zone container inspections are required for the 0.05 significance level validation required.

This figure illustrates the fact that there are many possible combinations of manual and alternative or automated inspection that will result in the selected 0.05 significance level validations. These combinations are shown as vertical pairs of N_M/N and N_A/N at the same value of N_A/N_M. The ratio of the total number of inspections at the selected inspection conditions is shown as N_T/N and is located at the same value of N_A/N_M. N_T/N, the ratio of the total number of reject zone container inspections required for validation, is seen to pass through a broad minimum. The selected operating point for this validation demonstration is at the intersection of the N_M/N and N_M/N curves at the point where N_A/N_M is equal to 1. An equal number of inspections for each method is the basis for the RZNI Tables and is labeled N_{MA}. Operation at or beyond the N_{MA} value conserves the amount of manual inspections required.

The intersection point at which the number of manual Reject Zsone container inspections is equal to the number of automated system inspections, N_{MA}, is seen to be in the vicinity of the broad minimum of the ratio of total inspections required for 0.05 significance system validation. The particular closeness of the selected operating point to the theoretical minimum total number of inspections required shown here is a happenstance of the RZE values employed.

RZNI Tables: Their Present

The augmented statistical tables in this chapter now offer convenient multiple choices for the ratio of manual to automated reject zone container inspections that will result in 0.05 significance level results. An index of the tables is found in Table 11, auxiliary Tables 12 and 13 provide adjustment capability for the ratio of manual to automated inspections to maintain the significance level desired. The RZNI Tables follow Table 13.

In Fig. 44 the variation of automated, manual and total inspections are all curvilinear. An improvement in estimating convenience results when the ratio of automated inspections has a linear relationship with the selected normalizing parameter. This is achieved by choosing the normalizing parameter, D, equal to the ratio of the number of automated inspections to the product of the statistical constants for the inspection type, A, and for the significance level desired, N. These definitions mean that D must be positive and greater than 1. Operation with D approaching 1 results in the number of manual inspections becoming unmanageably large. With this method, operation to achieve a controlled ratio of manual to automated inspections as shown in Fig. 45 can be calculated simply. The effect of operation at N_{MA} or at a multiple of N_{MA} is immediately apparent.

When inspection costs are approximately equal, using an equal number of manual and alternate inspections yields satisfactory results. When high-speed automated inspection systems are validated, the use of a minimized number of manual inspections can reduce the time and cost required for validation. Experience has shown that more control of the relationship between the number of manual and machine inspections is needed to enable a planned minimization of manual inspection labor. To maintain the significance level desired for the validation, a compensating increase in the number of automated inspections is required.

Fig. 45 is a simplified result of the reexamination of this problem. A consideration in using this analysis is the selection of an operating margin to ensure that D remains safely greater than 1. Since high rates of manual Reject Zone container reinspections are undesirable, this limitation is easy to accept. Another problem to be considered is an optimum selection of the ratio of manual to alternative inspections. As the ratio of manual inspections decreases toward its asymptotic value, the number of alternative inspections compensates by increasing to maintain the desired significance level. The approach to this problem that was adopted was to select operating margins 5%, 10%, and 20% above the limiting value of the manual inspection requirement.

In Fig. 45, the normalized ratio of the total number of inspections goes through a minimum just as in the previous analysis. The total inspection minimum occurs at the point where the number of manual and automated inspections are equal. The equal inspection number can be calculated simply and is symbolized as N_{MA}. From the equal inspection ratio point, the manual inspection ratio decreases asymptotically to a limit of 1.0 and the automated inspection ratio increases without limit. Any vertical pair of inspection ratios at the same value of D can be

Table 11 RZNI Table Index and Application Notes

Tables	RZE/RZA range	Inspection sequence
RZNI-1.0 to RZNI-1.2	0.70–1.00	M1/A1
RZNI-2.0 to RZNI-2.9	0.00–1.00	M1/A2, M2/A1
RZNI-3.0 to RZNI-3.9	0.00–1.00	M2/A2
RZNI-4.0 to RZNI-4.2	0.70–1.00	M2/A3

Table 12 Statistical Estimation Parameters, M for Manual and A for the Alternative or Automated Method That Is to Be Validated

Single inspection		Two serial inspections		Two serial inspections		Two serial inspections		Three serial inspections	
RZE	M	RZE/RZA	M or A	RZE/RZA	M or A	RZE/RZA	M or A	RZE	A
0.70	0.2100	0.01	xxxx	0.34	0.1037	0.70	0.4116	0.70	0.4538
0.71	0.2059	0.02	xxxx	0.35	0.1115	0.71	0.4151	0.71	0.4709
0.72	0.2016	0.03	0.0001	0.36	0.1194	0.72	0.4180	0.72	0.4876
0.73	0.1971	0.04	0.0002	0.37	0.1276	0.73	0.4201	0.73	0.5038
0.74	0.1924	0.05	0.0005	0.38	0.1361	0.74	0.4215	0.74	0.5193
0.75	0.1875	0.06	0.0008	0.39	0.1447	0.75	0.4219	0.75	0.5340
0.76	0.1824	0.07	0.0013	0.40	0.1536	0.76	0.4215	0.76	0.5477
0.77	0.1771	0.08	0.0018	0.41	0.1626	0.77	0.4200	0.77	0.5603
0.78	0.1716	0.09	0.0027	0.42	0.1719	0.78	0.4176	0.78	0.5716
0.79	0.1659	0.10	0.0036	0.43	0.1812	0.79	0.4141	0.79	0.5815
0.80	0.1600	0.11	0.0047	0.44	0.1908	0.80	0.4096	0.80	0.5899
0.81	0.1539	0.12	0.0061	0.45	0.2005	0.81	0.4039	0.81	0.5962
0.82	0.1476	0.13	0.0076	0.46	0.2102	0.82	0.3970	0.82	0.6006
0.83	0.1411	0.14	0.0095	0.47	0.2201	0.83	0.3888	0.83	0.6027
0.84	0.1344	0.15	0.0114	0.48	0.2300	0.84	0.3793	0.84	0.6022
0.85	0.1275	0.16	0.0137	0.49	0.2400	0.85	0.3684	0.85	0.5990
0.86	0.8600	0.17	0.0163	0.50	0.2500	0.86	0.3562	0.86	0.5927
0.87	0.1131	0.18	0.0192	0.51	0.2600	0.87	0.3425	0.87	0.5832
0.88	0.1056	0.19	0.0222	0.52	0.2700	0.88	0.3271	0.88	0.5700
0.89	0.0979	0.20	0.0256	0.53	0.2799	0.89	0.3102	0.89	0.5528
0.90	0.0900	0.21	0.0292	0.54	0.2898	0.90	0.2916	0.90	0.5314
0.91	0.0819	0.22	0.0333	0.55	0.2995	0.91	0.2713	0.91	0.5054
0.92	0.0736	0.23	0.0374	0.56	0.3091	0.92	0.2491	0.92	0.4746
0.93	0.0651	0.24	0.0420	0.57	0.3185	0.93	0.2252	0.93	0.4383
0.94	0.0564	0.25	0.0469	0.58	0.3278	0.94	0.1994	0.94	0.3963
0.95	0.0475	0.26	0.0520	0.59	0.3368	0.95	0.1715	0.95	0.3482
0.96	0.0384	0.27	0.0575	0.60	0.3456	0.96	0.1415	0.96	0.2935
0.97	0.0291	0.28	0.0632	0.61	0.3541	0.97	0.1095	0.97	0.2318
0.98	0.0196	0.29	0.0693	0.62	0.3622	0.98	0.0753	0.98	0.1627
0.99	0.0099	0.30	0.0757	0.63	0.3701	0.99	0.0388	0.99	0.0856
1.00	0	0.31	0.0822	0.64	0.3775	1.00	0	1.00	0
		0.32	0.0892	0.65	0.3845				
		0.33	0.0963	0.66	0.3910				
				0.67	0.3970				
				0.68	0.4025				
				0.69	0.4074				

The parameters are grouped by the sequence of inspections employed and by the controlling probability of the first inspection. The ratio of M/A is used to evaluate the economic effectiveness of the selected pair of manual and alternative inspections in decreasing the number of manual inspections employed.

Table 13 Manual Inspection Reduction Options

		Margin over N_M limit					
		20% $D = 6$		10% $D = 11$		5% $D = 21$	
M/A	$N_{MA}/NA = 1 + M/A$	N_M/N_{MA}	N_A/N_{MA}	N_M/N_{MA}	N_A/N_{MA}	N_M/N_{MA}	N_A/N_{MA}
0.10	1.10	0.11	5.45	0.10	10.01	0.10	19.09
0.20	1.20	0.20	5.00	0.18	9.17	0.18	17.50
0.25	1.25	0.24	4.80	0.22	8.80	0.21	16.80
0.30	1.30	0.28	4.61	0.25	8.46	0.24	16.15
0.40	1.40	0.34	4.29	0.31	7.86	0.30	15.00
0.50	1.50	0.40	4.00	0.37	7.37	0.35	14.00
0.60	1.60	0.45	3.75	0.41	6.88	0.39	13.13
0.70	1.70	0.49	3.53	0.45	6.47	0.43	12.35
0.80	1.80	0.53	3.33	0.49	6.11	0.47	11.66
0.90	1.90	0.57	3.16	0.52	5.79	0.50	11.05
1.00	2.00	0.60	3.00	0.55	5.50	0.53	10.50
1.10	2.10	0.63	2.86	0.58	5.24	0.55	10.00
1.20	2.20	0.65	2.73	0.60	5.00	0.55	9.55
1.30	2.30	0.68	2.61	0.62	4.78	0.59	9.13
1.40	2.40	0.70	2.50	0.64	4.58	0.61	8.75
1.50	2.50	0.72	2.40	0.66	4.40	0.63	8.40
1.60	2.60	0.74	2.31	0.68	4.23	0.65	8.08
1.70	2.70	0.76	2.22	0.69	4.07	0.66	7.78
1.80	2.80	0.77	2.14	0.71	3.93	0.68	7.50
1.90	2.90	0.79	2.07	0.72	3.79	0.69	7.24
2.00	3.00	0.80	2.00	0.73	3.67	0.70	7.00
2.10	3.10	0.81	1.94	0.75	3.55	0.71	6.77
2.20	3.20	0.83	1.88	0.76	3.44	0.72	6.56
2.30	3.30	0.84	1.82	0.77	3.33	0.73	6.36
2.40	3.40	0.85	1.77	0.78	3.24	0.74	6.18
2.50	3.50	0.86	1.74	0.79	3.19	0.75	6.00
2.60	3.60	0.87	1.67	0.79	3.06	0.76	5.83
2.70	3.70	0.88	1.62	0.80	2.97	0.77	5.68
2.80	3.80	0.88	1.58	0.81	2.89	0.77	5.53
2.90	3.90	0.89	1.54	0.82	2.82	0.78	5.38
3.00	4.00	0.90	1.50	0.83	2.75	0.79	5.25
3.10	4.10	0.91	1.46	0.83	2.68	0.79	5.12
3.20	4.20	0.91	1.43	0.84	2.62	0.80	5.00
3.30	4.30	0.92	1.40	0.84	2.56	0.81	4.88
3.40	4.40	0.93	1.36	0.85	2.50	0.81	4.77
4.00	5.00	0.96	1.20	0.88	2.20	0.84	4.20
4.10	5.10	0.96	1.18	0.88	2.16	0.84	4.12
4.20	5.20	0.97	1.15	0.89	2.12	0.85	4.04
4.30	5.30	0.97	1.13	0.89	2.08	0.85	3.96
4.40	5.40	0.98	1.11	0.90	2.04	0.86	3.89
4.50	5.50	0.98	1.09	0.90	2.00	0.86	3.82
4.60	5.60	0.99	1.07	0.90	1.96	0.86	3.75
4.70	5.70	0.99	1.05	0.91	1.93	0.87	3.68
4.80	5.80	0.99	1.03	0.91	1.90	0.87	3.62
4.90	5.90	1.00	1.02	0.91	1.86	0.87	3.56
5.00	6.00	1.00	1.02	0.92	1.83	0.88	3.50

M and A are the statistical model parameters for the manual and alternative or automated inspection systems. The second column lists the normalized ratio of N_{MA}, the crossover number of reject container inspections where N_M is equal to N_A. Each adjacent pair shows the manual inspection reduction ratio and the alternative or automated ratio referenced to the N_{MA} equal inspection number listed in the RZNI tables for both inspections.

RZNI TABLES

RZNI - 1.0 M1/A1 M: 0.70 - 1.00, A: 0.70 - 0.79

RZE(M1)	RZE(A1) 0.70	0.71	0.72	0.73	0.74	0.75	0.76	0.77	0.78	0.79
0.70	1818	1800	1782	1764	1745	1726	1706	1686	1665	1644
0.71	1800	1782	1764	1745	1726	1706	1686	1665	1644	1623
0.72	1782	1764	1745	1726	1706	1686	1665	1644	1623	1601
0.73	1764	1745	1726	1706	1686	1665	1644	1623	1601	1579
0.74	1745	1726	1706	1686	1665	1644	1623	1601	1579	1556
0.75	1726	1706	1686	1665	1644	1623	1601	1579	1556	1533
0.76	1706	1686	1665	1644	1623	1601	1579	1556	1533	1509
0.77	1686	1665	1644	1623	1601	1579	1556	1533	1509	1485
0.78	1665	1644	1623	1601	1579	1556	1533	1509	1485	1461
0.79	1644	1623	1601	1579	1556	1533	1509	1485	1461	1436
0.80	1623	1601	1579	1556	1533	1509	1485	1461	1436	1411
0.81	1601	1579	1556	1533	1509	1485	1461	1436	1411	1385
0.82	1579	1556	1533	1509	1485	1461	1436	1411	1385	1359
0.83	1556	1533	1509	1485	1461	1436	1411	1385	1359	1332
0.84	1533	1509	1485	1461	1436	1411	1385	1359	1332	1305
0.85	1509	1485	1461	1436	1411	1385	1359	1332	1305	1278
0.86	1485	1461	1436	1411	1385	1359	1332	1305	1278	1250
0.87	1461	1436	1411	1385	1359	1332	1305	1278	1250	1221
0.88	1436	1411	1385	1359	1332	1305	1278	1250	1221	1193
0.89	1411	1385	1359	1332	1305	1278	1250	1221	1193	1163
0.90	1385	1359	1332	1305	1278	1250	1221	1193	1163	1134
0.91	1359	1332	1305	1278	1250	1221	1193	1163	1134	1104
0.92	1332	1305	1278	1250	1221	1193	1163	1134	1104	1073
0.93	1305	1278	1250	1221	1193	1163	1134	1104	1073	1042
0.94	1278	1250	1221	1193	1163	1134	1104	1073	1042	1011
0.95	1250	1221	1193	1163	1134	1104	1073	1042	1011	979
0.96	1221	1193	1163	1134	1104	1073	1042	1011	979	947
0.97	1193	1163	1134	1104	1073	1042	1011	979	947	914
0.98	1163	1134	1104	1073	1042	1011	979	947	914	881
0.99	1134	1104	1073	1042	1011	979	947	914	881	847
1.00	1104	1073	1042	1011	979	947	914	881	847	813

RZNI - 1.1 M1/A1 M: 0.70 - 1.00, A: 0.80 - 0.89

RZE(M1)	RZE(A1) 0.80	0.81	0.82	0.83	0.84	0.85	0.86	0.87	0.88	0.89
0.70	1623	1601	1579	1556	1533	1509	1485	1461	1436	1411
0.71	1601	1579	1556	1533	1509	1485	1461	1436	1411	1385
0.72	1579	1556	1533	1509	1485	1461	1436	1411	1385	1359
0.73	1556	1533	1509	1485	1461	1436	1411	1385	1359	1332
0.74	1533	1509	1485	1461	1436	1411	1385	1359	1332	1305
0.75	1509	1485	1461	1436	1411	1385	1359	1332	1305	1278
0.76	1485	1461	1436	1411	1385	1359	1332	1305	1278	1250
0.77	1461	1436	1411	1385	1359	1332	1305	1278	1250	1221
0.78	1436	1411	1385	1359	1332	1305	1278	1250	1221	1193
0.79	1411	1385	1359	1332	1305	1278	1250	1221	1193	1163
0.80	1385	1359	1332	1305	1278	1250	1221	1193	1163	1134
0.81	1359	1332	1305	1278	1250	1221	1193	1163	1134	1104
0.82	1332	1305	1278	1250	1221	1193	1163	1134	1104	1073
0.83	1305	1278	1250	1221	1193	1163	1134	1104	1073	1042
0.84	1278	1250	1221	1193	1163	1134	1104	1073	1042	1011
0.85	1250	1221	1193	1163	1134	1104	1073	1042	1011	979
0.86	1221	1193	1163	1134	1104	1073	1042	1011	979	947
0.87	1193	1163	1134	1104	1073	1042	1011	979	947	914
0.88	1163	1134	1104	1073	1042	1011	979	947	914	881
0.89	1134	1104	1073	1042	1011	979	947	914	881	847
0.90	1104	1073	1042	1011	979	947	914	881	847	813
0.91	1073	1042	1011	979	947	914	881	847	813	779
0.92	1042	1011	979	947	914	881	847	813	779	744
0.93	1011	979	947	914	881	847	813	779	744	709
0.94	979	947	914	881	847	813	779	744	709	673
0.95	947	914	881	847	813	779	744	709	673	637
0.96	914	881	847	813	779	744	709	673	637	601
0.97	881	847	813	779	744	709	673	637	601	564
0.98	847	813	779	744	709	673	637	601	564	526
0.99	813	779	744	709	673	637	601	564	526	488
1.00	779	744	709	673	637	601	564	526	488	450

RZNI - 1.2 M1/A1 M: 0.70 - 1.00, A: 0.90 - 1.00

RZE(M1)	RZE(A1) 0.90	0.91	0.92	0.93	0.94	0.95	0.96	0.97	0.98	0.99	1.00
0.70	1385	1359	1332	1305	1278	1250	1221	1193	1163	1134	1104
0.71	1359	1332	1305	1278	1250	1221	1193	1163	1134	1104	1073
0.72	1332	1305	1278	1250	1221	1193	1163	1134	1104	1073	1042
0.73	1305	1278	1250	1221	1193	1163	1134	1104	1073	1042	1011
0.74	1278	1250	1221	1193	1163	1134	1104	1073	1042	1011	979
0.75	1250	1221	1193	1163	1134	1104	1073	1042	1011	979	947
0.76	1221	1193	1163	1134	1104	1073	1042	1011	979	947	914
0.77	1193	1163	1134	1104	1073	1042	1011	979	947	914	881
0.78	1163	1134	1104	1073	1042	1011	979	947	914	881	847
0.79	1134	1104	1073	1042	1011	979	947	914	881	847	813
0.80	1104	1073	1042	1011	979	947	914	881	847	813	779
0.81	1073	1042	1011	979	947	914	881	847	813	779	744
0.82	1042	1011	979	947	914	881	847	813	779	744	709
0.83	1011	979	947	914	881	847	813	779	744	709	673
0.84	979	947	914	881	847	813	779	744	709	673	637
0.85	947	914	881	847	813	779	744	709	673	637	601
0.86	914	881	847	813	779	744	709	673	637	601	564
0.87	881	847	813	779	744	709	673	637	601	564	526
0.88	847	813	779	744	709	673	637	601	564	526	488
0.89	813	779	744	709	673	637	601	564	526	488	450
0.90	779	744	709	673	637	601	564	526	488	450	411
0.91	744	709	673	637	601	564	526	488	450	411	372
0.92	709	673	637	601	564	526	488	450	411	372	332
0.93	673	637	601	564	526	488	450	411	372	332	292
0.94	637	601	564	526	488	450	411	372	332	292	252
0.95	601	564	526	488	450	411	372	332	292	252	211
0.96	564	526	488	450	411	372	332	292	252	211	170
0.97	526	488	450	411	372	332	292	252	211	170	128
0.98	488	450	411	372	332	292	252	211	170	128	86
0.99	450	411	372	332	292	252	211	170	128	86	43
1.00	411	372	332	292	252	211	170	128	86	43	0

RZNI - 2.0 M1/A2 M: 0.70 - 1.00, A: 0.00 - 0.09

RZE(M1)	RZE(A1)/RZA(A1) 0.00	0.01	0.02	0.03	0.04	0.05	0.06	0.07	0.08	0.09
0.70	909	909	909	910	910	911	913	915	917	921
0.71	891	891	892	892	893	894	895	897	900	903
0.72	873	873	873	873	874	875	876	878	881	884
0.73	853	853	854	854	854	855	857	859	862	865
0.74	833	833	833	833	834	835	837	839	841	845
0.75	812	812	812	812	813	814	815	817	820	823
0.76	790	790	790	790	791	792	793	795	798	801
0.77	767	767	767	767	768	769	770	772	775	778
0.78	743	743	743	743	744	745	746	748	751	754
0.79	718	718	718	719	719	720	722	724	726	730
0.80	693	693	693	693	694	695	696	698	701	704
0.81	666	666	666	667	667	668	670	672	674	678
0.82	639	639	639	640	640	641	643	645	647	651
0.83	611	611	611	611	612	613	614	616	619	622
0.84	582	582	582	582	583	584	585	587	590	593
0.85	552	552	552	552	553	554	556	558	560	564
0.86	521	521	521	522	522	523	525	527	529	533
0.87	490	490	490	490	491	492	493	495	498	501
0.88	457	457	457	458	458	459	461	463	465	469
0.89	424	424	424	424	425	426	427	429	432	435
0.90	390	390	390	390	391	392	393	395	398	401
0.91	355	355	355	355	356	357	358	360	363	366
0.92	319	319	319	319	320	321	322	324	327	330
0.93	282	282	282	282	283	284	285	287	290	293
0.94	244	244	244	245	245	246	248	250	252	256
0.95	206	206	206	206	207	208	209	211	214	217
0.96	166	166	166	167	167	168	170	172	174	178
0.97	126	126	126	126	127	128	130	132	134	137
0.98	85	85	85	85	86	87	88	90	93	96
0.99	43	43	43	43	44	45	46	48	51	54
1.00	0	0	0	0	1	2	4	6	8	11

RZNI - 2.1 M1/A2 M: 0.70 - 1.00, A: 0.10 - 0.19

RZE(M1)	RZE(A1)/RZA(A1) 0.10	0.11	0.12	0.13	0.14	0.15	0.16	0.17	0.18	0.19
0.70	925	930	936	942	950	959	969	980	992	1005
0.71	907	912	918	925	932	941	951	962	974	988
0.72	888	893	899	906	914	923	932	943	956	969
0.73	869	874	880	886	894	903	913	924	936	950
0.74	849	854	859	866	874	883	893	904	916	929
0.75	827	832	838	845	853	861	871	882	895	908
0.76	805	810	816	823	831	839	849	860	873	886
0.77	782	787	793	800	808	816	826	837	850	863
0.78	759	763	769	776	784	793	803	814	826	839
0.79	734	739	745	751	759	768	778	789	801	815
0.80	708	713	719	726	734	742	752	763	776	789
0.81	682	687	693	699	707	716	726	737	749	763
0.82	655	660	665	672	680	689	699	710	722	735
0.83	626	631	637	644	652	661	670	682	694	707
0.84	597	602	608	615	623	632	641	653	665	678
0.85	568	573	578	585	593	602	612	623	635	648
0.86	537	542	548	554	562	571	581	592	604	618
0.87	505	510	516	523	531	539	549	560	573	586
0.88	473	478	484	490	498	507	517	528	540	553
0.89	439	444	450	457	465	474	483	494	507	520
0.90	405	410	416	423	431	439	449	460	472	486
0.91	370	375	381	388	395	404	414	425	437	451
0.92	334	339	345	352	360	368	378	389	401	415
0.93	297	302	308	315	323	332	341	352	365	378
0.94	260	265	271	277	285	294	304	315	327	340
0.95	221	226	232	239	247	255	265	276	288	302
0.96	182	187	193	199	207	216	226	237	249	262
0.97	142	147	152	159	167	176	186	197	209	222
0.98	100	105	111	118	126	135	144	155	168	181
0.99	58	63	69	76	84	93	102	113	126	139
1.00	16	21	26	33	41	50	60	71	83	96

RZNI - 2.2 M1/A2 M: 0.70 - 1.00, A: 0.20 - 0.29

RZE(M1)	RZE(A1)/RZA(A1) 0.20	0.21	0.22	0.23	0.24	0.25	0.26	0.27	0.28	0.29
0.70	1020	1036	1053	1071	1091	1112	1134	1158	1183	1209
0.71	1002	1018	1035	1054	1073	1094	1117	1140	1165	1191
0.72	984	1000	1017	1035	1055	1076	1098	1122	1147	1173
0.73	964	980	997	1016	1035	1056	1079	1102	1127	1153
0.74	944	960	977	995	1015	1036	1058	1082	1107	1133
0.75	923	939	956	974	994	1015	1037	1061	1086	1112
0.76	901	916	934	952	972	993	1015	1039	1063	1090
0.77	878	893	911	929	949	970	992	1016	1041	1067
0.78	854	870	887	905	925	946	968	992	1017	1043
0.79	829	845	862	881	900	921	944	967	992	1018
0.80	804	819	837	855	875	896	918	942	966	993
0.81	777	793	810	829	848	869	892	915	940	966
0.82	750	766	783	801	821	842	864	888	913	939
0.83	722	738	755	773	793	814	836	860	885	911
0.84	693	709	726	744	764	785	807	831	856	882
0.85	663	679	696	714	734	755	777	801	826	852
0.86	632	648	665	684	703	724	747	770	795	821
0.87	601	616	634	652	672	693	715	739	763	790
0.88	568	584	601	619	639	660	682	706	731	757
0.89	535	551	568	586	606	627	649	673	698	724
0.90	501	516	534	552	572	593	615	639	663	690
0.91	465	481	498	517	537	558	580	603	628	654
0.92	430	445	462	481	501	522	544	568	592	619
0.93	393	409	426	444	464	485	507	531	556	582
0.94	355	371	388	406	426	447	469	493	518	544
0.95	316	332	349	368	388	409	431	455	479	506
0.96	277	293	310	329	348	369	392	415	440	466
0.97	237	253	270	288	308	329	351	375	400	426
0.98	196	212	229	247	267	288	310	334	359	385
0.99	154	170	187	205	225	246	268	292	317	343
1.00	111	127	144	162	182	203	225	249	274	300

RZNI - 2.3 M1/A2 M: 0.70 - 1.00, A: 0.30 - 0.39

RZE(M1)	RZE(A1)/RZA(A1) 0.30	0.31	0.32	0.33	0.34	0.35	0.36	0.37	0.38	0.39
0.70	1237	1265	1295	1326	1358	1392	1426	1462	1498	1536
0.71	1219	1247	1277	1308	1341	1374	1409	1444	1481	1518
0.72	1200	1229	1259	1290	1322	1356	1390	1426	1462	1500
0.73	1181	1209	1239	1270	1303	1336	1371	1406	1443	1480
0.74	1160	1189	1219	1250	1282	1316	1350	1386	1422	1460
0.75	1139	1168	1198	1229	1261	1294	1329	1364	1401	1438
0.76	1117	1146	1176	1207	1239	1272	1307	1342	1379	1416
0.77	1094	1123	1153	1184	1216	1249	1284	1319	1356	1393
0.78	1070	1099	1129	1160	1192	1226	1260	1296	1332	1370
0.79	1046	1074	1104	1135	1168	1201	1235	1271	1307	1345
0.80	1020	1049	1079	1110	1142	1175	1210	1245	1282	1319
0.81	994	1022	1052	1083	1116	1149	1183	1219	1256	1293
0.82	966	995	1025	1056	1088	1122	1156	1192	1228	1266
0.83	938	967	997	1028	1060	1094	1128	1164	1200	1238
0.84	909	938	968	999	1031	1065	1099	1135	1171	1209
0.85	879	908	938	969	1001	1035	1069	1105	1141	1179
0.86	849	877	907	938	971	1004	1038	1074	1110	1148
0.87	817	846	876	907	939	972	1007	1042	1079	1116
0.88	785	813	843	874	906	940	974	1010	1046	1084
0.89	751	780	810	841	873	907	941	977	1013	1051
0.90	717	746	776	807	839	872	907	942	979	1016
0.91	682	711	740	772	804	837	872	907	944	981
0.92	646	675	705	736	768	801	836	871	908	945
0.93	609	638	668	699	731	765	799	835	871	909
0.94	572	600	630	661	693	727	761	797	833	871
0.95	533	562	592	623	655	688	723	758	795	832
0.96	494	522	552	583	616	649	683	719	755	793
0.97	453	482	512	543	575	609	643	679	715	753
0.98	412	441	471	502	534	568	602	638	674	712
0.99	370	399	429	460	492	526	560	596	632	670
1.00	327	356	386	417	449	483	517	553	589	627

RZNI - 2.4 M1/A2 M: 0.70 - 1.00, A: 0.40 - 0.49

RZE(M1)	RZE(A1)/RZA(A1) 0.40	0.41	0.42	0.43	0.44	0.45	0.46	0.47	0.48	0.49
0.70	1574	1613	1653	1694	1735	1777	1820	1862	1905	1948
0.71	1557	1596	1636	1676	1718	1759	1802	1844	1887	1931
0.72	1538	1577	1617	1658	1699	1741	1783	1826	1869	1912
0.73	1518	1558	1598	1638	1680	1721	1764	1806	1849	1893
0.74	1498	1537	1577	1618	1659	1701	1743	1786	1829	1872
0.75	1477	1516	1556	1597	1638	1680	1722	1765	1808	1851
0.76	1455	1494	1534	1575	1616	1658	1700	1743	1786	1829
0.77	1432	1471	1511	1552	1593	1635	1677	1720	1763	1806
0.78	1408	1447	1487	1528	1569	1611	1653	1696	1739	1782
0.79	1383	1423	1462	1503	1544	1586	1629	1671	1714	1757
0.80	1358	1397	1437	1478	1519	1561	1603	1646	1689	1732
0.81	1331	1371	1411	1451	1492	1534	1577	1619	1662	1705
0.82	1304	1343	1383	1424	1465	1507	1549	1592	1635	1678
0.83	1276	1315	1355	1396	1437	1479	1521	1564	1607	1650
0.84	1247	1286	1326	1367	1408	1450	1492	1535	1578	1621
0.85	1217	1256	1296	1337	1378	1420	1462	1505	1548	1591
0.86	1186	1226	1265	1306	1347	1389	1432	1474	1517	1560
0.87	1155	1194	1234	1275	1316	1358	1400	1443	1486	1529
0.88	1122	1161	1201	1242	1283	1325	1367	1410	1453	1496
0.89	1089	1128	1168	1209	1250	1292	1334	1377	1420	1463
0.90	1055	1094	1134	1175	1216	1258	1300	1343	1386	1429
0.91	1020	1059	1099	1139	1181	1223	1265	1308	1351	1394
0.92	984	1023	1063	1104	1145	1187	1229	1272	1315	1358
0.93	947	986	1026	1067	1108	1150	1192	1235	1278	1321
0.94	909	948	988	1029	1070	1112	1154	1197	1240	1283
0.95	871	910	950	991	1032	1074	1116	1159	1202	1245
0.96	831	870	910	951	992	1034	1077	1119	1162	1205
0.97	791	830	870	911	952	994	1036	1079	1122	1165
0.98	750	789	829	870	911	953	995	1038	1081	1124
0.99	708	747	787	828	869	911	953	996	1039	1082
1.00	665	704	744	785	826	868	910	953	996	1039

RZNI - 2.5 M1/A2

M: 0.70 - 1.00, A: 0.50 - 0.59

RZE(M1)	RZE(A1)/RZA(A1) 0.50	0.51	0.52	0.53	0.54	0.55	0.56	0.57	0.58	0.59
0.70	1992	2035	2078	2121	2164	2206	2247	2288	2328	2368
0.71	1974	2017	2060	2103	2146	2188	2230	2271	2311	2350
0.72	1955	1999	2042	2085	2127	2169	2211	2252	2292	2331
0.73	1936	1979	2022	2065	2108	2150	2192	2233	2273	2312
0.74	1915	1959	2002	2045	2087	2130	2171	2212	2252	2291
0.75	1894	1937	1981	2024	2066	2108	2150	2191	2231	2270
0.76	1872	1915	1959	2002	2044	2086	2128	2169	2209	2248
0.77	1849	1892	1936	1979	2021	2063	2105	2146	2186	2225
0.78	1825	1869	1912	1955	1997	2040	2081	2122	2162	2201
0.79	1801	1844	1887	1930	1973	2015	2057	2097	2137	2177
0.80	1775	1818	1862	1905	1947	1989	2031	2072	2112	2151
0.81	1749	1792	1835	1878	1921	1963	2005	2045	2086	2125
0.82	1721	1765	1808	1851	1893	1936	1977	2018	2058	2097
0.83	1693	1737	1780	1823	1865	1908	1949	1990	2030	2069
0.84	1664	1708	1751	1794	1836	1879	1920	1961	2001	2040
0.85	1634	1678	1721	1764	1806	1849	1890	1931	1971	2010
0.86	1604	1647	1690	1733	1776	1818	1860	1900	1940	1980
0.87	1572	1615	1659	1701	1744	1786	1828	1869	1909	1948
0.88	1540	1583	1626	1669	1712	1754	1795	1836	1876	1916
0.89	1506	1550	1593	1636	1678	1720	1762	1803	1843	1882
0.90	1472	1515	1559	1601	1644	1686	1728	1769	1809	1848
0.91	1437	1480	1523	1566	1609	1651	1693	1734	1774	1813
0.92	1401	1444	1488	1530	1573	1615	1657	1698	1738	1777
0.93	1364	1408	1451	1494	1536	1578	1620	1661	1701	1740
0.94	1327	1370	1413	1456	1499	1541	1582	1623	1663	1703
0.95	1288	1331	1375	1417	1460	1502	1544	1585	1625	1664
0.96	1249	1292	1335	1378	1421	1463	1504	1545	1585	1625
0.97	1208	1252	1295	1338	1380	1423	1464	1505	1545	1584
0.98	1167	1211	1254	1297	1339	1381	1423	1464	1504	1543
0.99	1125	1169	1212	1255	1297	1339	1381	1422	1462	1501
1.00	1082	1126	1169	1212	1254	1297	1338	1379	1419	1458

RZNI - 2.6 M1/A2

M: 0.70 - 1.00, A: 0.60 - 0.69

RZE(M1)	RZE(A1)/RZA(A1) 0.60	0.61	0.62	0.63	0.64	0.65	0.66	0.67	0.68	0.69
0.70	2406	2442	2478	2511	2544	2574	2602	2628	2652	2673
0.71	2388	2425	2460	2494	2526	2556	2584	2610	2634	2655
0.72	2369	2406	2441	2475	2507	2537	2566	2592	2615	2637
0.73	2350	2386	2422	2456	2488	2518	2546	2572	2596	2617
0.74	2329	2366	2401	2435	2467	2498	2526	2552	2576	2597
0.75	2308	2345	2380	2414	2446	2476	2505	2531	2554	2575
0.76	2286	2323	2358	2392	2424	2454	2483	2509	2532	2553
0.77	2263	2300	2335	2369	2401	2431	2460	2486	2509	2530
0.78	2239	2276	2311	2345	2377	2408	2436	2462	2486	2507
0.79	2215	2251	2287	2321	2353	2383	2411	2437	2461	2482
0.80	2189	2226	2261	2295	2327	2357	2386	2412	2435	2456
0.81	2163	2199	2235	2269	2301	2331	2359	2385	2409	2430
0.82	2135	2172	2208	2241	2273	2304	2332	2358	2382	2403
0.83	2107	2144	2179	2213	2245	2276	2304	2330	2353	2375
0.84	2078	2115	2150	2184	2216	2247	2275	2301	2324	2346
0.85	2048	2085	2120	2154	2186	2217	2245	2271	2295	2316
0.86	2018	2054	2090	2124	2156	2186	2214	2240	2264	2285
0.87	1986	2023	2058	2092	2124	2154	2183	2209	2232	2253
0.88	1954	1990	2026	2059	2092	2122	2150	2176	2200	2221
0.89	1920	1957	1992	2026	2058	2089	2117	2143	2166	2188
0.90	1886	1923	1958	1992	2024	2054	2083	2109	2132	2153
0.91	1851	1888	1923	1957	1989	2019	2047	2073	2097	2118
0.92	1815	1852	1887	1921	1953	1983	2012	2038	2061	2082
0.93	1778	1815	1850	1884	1916	1946	1975	2001	2024	2046
0.94	1741	1777	1813	1846	1879	1909	1937	1963	1987	2008
0.95	1702	1739	1774	1808	1840	1870	1899	1925	1948	1969
0.96	1663	1699	1735	1769	1801	1831	1859	1885	1909	1930
0.97	1622	1659	1694	1728	1760	1791	1819	1845	1869	1890
0.98	1581	1618	1653	1687	1719	1749	1778	1804	1827	1849
0.99	1539	1576	1611	1645	1677	1708	1736	1762	1785	1807
1.00	1496	1533	1568	1602	1634	1665	1693	1719	1743	1764

RZNI - 2.7 M1/A2 M: 0.70 - 1.00, A: 0.70 - 0.79

RZE(M1)	RZE(A1)/RZA(A1) 0.70	0.71	0.72	0.73	0.74	0.75	0.76	0.77	0.78	0.79
0.70	2691	2707	2719	2728	2734	2736	2734	2728	2717	2702
0.71	2674	2689	2701	2711	2716	2718	2716	2710	2700	2685
0.72	2655	2670	2683	2692	2698	2699	2697	2691	2681	2666
0.73	2635	2651	2663	2672	2678	2680	2678	2672	2661	2647
0.74	2615	2631	2643	2652	2658	2660	2658	2652	2641	2626
0.75	2594	2609	2622	2631	2636	2638	2636	2630	2620	2605
0.76	2572	2587	2600	2609	2614	2616	2614	2608	2598	2583
0.77	2549	2564	2577	2586	2591	2593	2591	2585	2575	2560
0.78	2525	2541	2553	2562	2568	2570	2568	2561	2551	2536
0.79	2500	2516	2528	2537	2543	2545	2543	2537	2526	2511
0.80	2475	2490	2503	2512	2517	2519	2517	2511	2501	2486
0.81	2448	2464	2476	2485	2491	2493	2491	2485	2474	2459
0.82	2421	2437	2449	2458	2464	2466	2464	2458	2447	2432
0.83	2393	2408	2421	2430	2436	2437	2435	2429	2419	2404
0.84	2364	2379	2392	2401	2407	2408	2406	2400	2390	2375
0.85	2334	2350	2362	2371	2377	2379	2377	2371	2360	2345
0.86	2303	2319	2331	2340	2346	2348	2346	2340	2329	2314
0.87	2272	2287	2300	2309	2314	2316	2314	2308	2298	2283
0.88	2239	2255	2267	2276	2282	2284	2282	2276	2265	2250
0.89	2206	2221	2234	2243	2249	2250	2248	2242	2232	2217
0.90	2172	2187	2200	2209	2214	2216	2214	2208	2198	2183
0.91	2137	2152	2165	2174	2179	2181	2179	2173	2163	2148
0.92	2101	2116	2129	2138	2143	2145	2143	2137	2127	2112
0.93	2064	2079	2092	2101	2107	2108	2106	2100	2090	2075
0.94	2026	2042	2054	2063	2069	2071	2069	2063	2052	2037
0.95	1988	2003	2016	2025	2030	2032	2030	2024	2014	1999
0.96	1948	1964	1976	1985	1991	1993	1991	1985	1974	1959
0.97	1908	1924	1936	1945	1951	1953	1951	1944	1934	1919
0.98	1867	1882	1895	1904	1910	1911	1909	1903	1893	1878
0.99	1825	1840	1853	1862	1868	1869	1867	1861	1851	1836
1.00	1782	1798	1810	1819	1825	1827	1825	1818	1808	1793

RZNI - 2.8 M1/A2 M: 0.70 - 1.00, A: 0.80 - 0.89

RZE(M1)	RZE(A1)/RZA(A1) 0.80	0.81	0.82	0.83	0.84	0.85	0.86	0.87	0.88	0.89
0.70	2683	2658	2628	2593	2552	2505	2451	2392	2325	2252
0.71	2665	2640	2610	2575	2534	2487	2434	2374	2308	2234
0.72	2646	2622	2592	2556	2515	2468	2415	2355	2289	2216
0.73	2627	2602	2572	2537	2496	2449	2396	2336	2270	2196
0.74	2606	2582	2552	2516	2475	2428	2375	2316	2249	2176
0.75	2585	2561	2531	2495	2454	2407	2354	2294	2228	2155
0.76	2563	2538	2509	2473	2432	2385	2332	2272	2206	2133
0.77	2540	2515	2486	2450	2409	2362	2309	2249	2183	2110
0.78	2516	2492	2462	2426	2385	2338	2285	2226	2159	2086
0.79	2492	2467	2437	2402	2361	2314	2260	2201	2135	2061
0.80	2466	2441	2412	2376	2335	2288	2235	2175	2109	2036
0.81	2440	2415	2385	2350	2309	2262	2209	2149	2083	2009
0.82	2412	2388	2358	2322	2281	2234	2181	2122	2055	1982
0.83	2384	2360	2330	2294	2253	2206	2153	2093	2027	1954
0.84	2355	2331	2301	2265	2224	2177	2124	2064	1998	1925
0.85	2325	2301	2271	2235	2194	2147	2094	2035	1968	1895
0.86	2295	2270	2240	2205	2164	2117	2063	2004	1938	1864
0.87	2263	2238	2208	2173	2132	2085	2032	1972	1906	1833
0.88	2231	2206	2176	2141	2100	2053	1999	1940	1873	1800
0.89	2197	2173	2143	2107	2066	2019	1966	1906	1840	1767
0.90	2163	2138	2108	2073	2032	1985	1932	1872	1806	1733
0.91	2128	2103	2073	2038	1997	1950	1897	1837	1771	1698
0.92	2092	2067	2037	2002	1961	1914	1861	1801	1735	1662
0.93	2055	2031	2001	1965	1924	1877	1824	1764	1698	1625
0.94	2018	1993	1963	1928	1887	1840	1786	1727	1660	1587
0.95	1979	1954	1924	1889	1848	1801	1748	1688	1622	1549
0.96	1940	1915	1885	1850	1809	1762	1708	1649	1583	1509
0.97	1899	1875	1845	1809	1768	1721	1668	1609	1542	1469
0.98	1858	1834	1804	1768	1727	1680	1627	1567	1501	1428
0.99	1816	1792	1762	1726	1685	1638	1585	1525	1459	1386
1.00	1773	1749	1719	1683	1642	1595	1542	1483	1416	1343

RZNI - 2.9 M1/A2 M: 0.70 - 1.00, A: 0.90 - 1.00

RZE(M1)	RZE(A1)/RZA(A1) 0.90	0.91	0.92	0.93	0.94	0.95	0.96	0.97	0.98	0.99	1.00
0.70	2172	2137	2101	2064	2026	1988	1948	1908	1867	1825	1782
0.71	2187	2152	2116	2079	2042	2003	1964	1924	1882	1840	1798
0.72	2200	2165	2129	2092	2054	2016	1976	1936	1895	1853	1810
0.73	2209	2174	2138	2101	2063	2025	1985	1945	1904	1862	1819
0.74	2214	2179	2143	2107	2069	2030	1991	1951	1910	1868	1825
0.75	2216	2181	2145	2108	2071	2032	1993	1953	1911	1869	1827
0.76	2214	2179	2143	2106	2069	2030	1991	1951	1909	1867	1825
0.77	2208	2173	2137	2100	2063	2024	1985	1944	1903	1861	1818
0.78	2198	2163	2127	2090	2052	2014	1974	1934	1893	1851	1808
0.79	2183	2148	2112	2075	2037	1999	1959	1919	1878	1836	1793
0.80	2163	2128	2092	2055	2018	1979	1940	1899	1858	1816	1773
0.81	2138	2103	2067	2031	1993	1954	1915	1875	1834	1792	1749
0.82	2108	2073	2037	2001	1963	1924	1885	1845	1804	1762	1719
0.83	2073	2038	2002	1965	1928	1889	1850	1809	1768	1726	1683
0.84	2032	1997	1961	1924	1887	1848	1809	1768	1727	1685	1642
0.85	1985	1950	1914	1877	1840	1801	1762	1721	1680	1638	1595
0.86	1932	1897	1861	1824	1786	1748	1708	1668	1627	1585	1542
0.87	1872	1837	1801	1764	1727	1688	1649	1609	1567	1525	1483
0.88	1806	1771	1735	1698	1660	1622	1583	1542	1501	1459	1416
0.89	1733	1698	1662	1625	1587	1549	1509	1469	1428	1386	1343
0.90	1652	1617	1581	1544	1507	1468	1429	1389	1347	1305	1263
0.91	1564	1529	1493	1456	1419	1380	1341	1301	1259	1217	1175
0.92	1469	1433	1398	1361	1323	1285	1245	1205	1164	1122	1079
0.93	1365	1330	1294	1257	1219	1181	1141	1101	1060	1018	975
0.94	1253	1218	1182	1145	1107	1069	1029	989	948	906	863
0.95	1132	1097	1061	1024	987	948	909	868	827	785	742
0.96	1003	967	932	895	857	819	779	739	698	656	613
0.97	864	829	793	756	718	680	640	600	559	517	474
0.98	716	681	645	608	570	532	492	452	411	369	326
0.99	558	523	487	450	412	374	334	294	253	211	168
1.00	390	355	319	282	244	206	166	126	85	43	0

RZNI - 3.0 M2/A2 M: 0.70 - 1.00, A: 0.00 - 0.09

RZE(M1)	RZE(A1)/RZA(A1) 0.00	0.01	0.02	0.03	0.04	0.05	0.06	0.07	0.08	0.09
0.70	1782	1782	1782	1783	1783	1784	1786	1788	1790	1794
0.71	1798	1798	1798	1798	1799	1800	1801	1803	1806	1809
0.72	1810	1810	1810	1810	1811	1812	1813	1815	1818	1821
0.73	1819	1819	1819	1819	1820	1821	1823	1825	1827	1831
0.74	1825	1825	1825	1825	1826	1827	1828	1830	1833	1836
0.75	1827	1827	1827	1827	1828	1829	1830	1832	1835	1838
0.76	1825	1825	1825	1825	1826	1827	1828	1830	1833	1836
0.77	1818	1818	1819	1819	1820	1821	1822	1824	1827	1830
0.78	1808	1808	1808	1809	1809	1810	1812	1814	1816	1820
0.79	1793	1793	1793	1794	1794	1795	1797	1799	1801	1805
0.80	1773	1773	1774	1774	1774	1775	1777	1779	1782	1785
0.81	1749	1749	1749	1749	1750	1751	1752	1754	1757	1760
0.82	1719	1719	1719	1719	1720	1721	1722	1724	1727	1730
0.83	1683	1683	1684	1684	1684	1685	1687	1689	1692	1695
0.84	1642	1642	1642	1643	1643	1644	1646	1648	1651	1654
0.85	1595	1595	1595	1596	1596	1597	1599	1601	1604	1607
0.86	1542	1542	1542	1543	1543	1544	1546	1548	1550	1554
0.87	1483	1483	1483	1483	1484	1485	1486	1488	1491	1494
0.88	1416	1416	1416	1417	1417	1418	1420	1422	1424	1428
0.89	1343	1343	1343	1343	1344	1345	1346	1349	1351	1354
0.90	1263	1263	1263	1263	1264	1265	1266	1268	1271	1274
0.91	1175	1175	1175	1175	1176	1177	1178	1180	1183	1186
0.92	1079	1079	1079	1079	1080	1081	1082	1084	1087	1090
0.93	975	975	975	976	976	977	979	981	983	987
0.94	863	863	863	864	864	865	867	869	871	875
0.95	742	742	743	743	743	744	746	748	751	754
0.96	613	613	613	613	614	615	616	618	621	624
0.97	474	474	474	475	475	476	478	480	482	486
0.98	326	326	326	326	327	328	330	332	334	337
0.99	168	168	168	168	169	170	172	174	176	180
1.00	0	0	0	0	1	2	4	6	8	11

RZNI - 3.1 M2/A2 M: 0.70 - 1.00, A: 0.10 - 0.19

RZE(M1)	RZE(A1)/RZA(A1) 0.10	0.11	0.12	0.13	0.14	0.15	0.16	0.17	0.18	0.19
0.70	1798	1803	1808	1815	1823	1832	1842	1853	1865	1878
0.71	1813	1818	1824	1831	1838	1847	1857	1868	1880	1894
0.72	1826	1830	1836	1843	1851	1860	1870	1881	1893	1906
0.73	1835	1840	1845	1852	1860	1869	1879	1890	1902	1915
0.74	1840	1845	1851	1858	1866	1874	1884	1895	1907	1921
0.75	1842	1847	1853	1860	1867	1876	1886	1897	1909	1923
0.76	1840	1845	1851	1858	1865	1874	1884	1895	1907	1921
0.77	1834	1839	1845	1852	1859	1868	1878	1889	1901	1915
0.78	1824	1829	1834	1841	1849	1858	1868	1879	1891	1904
0.79	1809	1814	1819	1826	1834	1843	1853	1864	1876	1889
0.80	1789	1794	1800	1807	1814	1823	1833	1844	1856	1870
0.81	1764	1769	1775	1782	1790	1798	1808	1819	1832	1845
0.82	1734	1739	1745	1752	1760	1768	1778	1789	1802	1815
0.83	1699	1704	1710	1717	1724	1733	1743	1754	1766	1780
0.84	1658	1663	1669	1675	1683	1692	1702	1713	1725	1739
0.85	1611	1616	1622	1628	1636	1645	1655	1666	1678	1692
0.86	1558	1563	1569	1575	1583	1592	1602	1613	1625	1638
0.87	1498	1503	1509	1516	1523	1532	1542	1553	1565	1579
0.88	1432	1437	1443	1449	1457	1466	1476	1487	1499	1512
0.89	1359	1363	1369	1376	1384	1393	1403	1414	1426	1439
0.90	1278	1283	1289	1296	1303	1312	1322	1333	1345	1359
0.91	1190	1195	1201	1208	1215	1224	1234	1245	1257	1271
0.92	1094	1099	1105	1112	1120	1129	1138	1149	1162	1175
0.93	991	996	1001	1008	1016	1025	1035	1046	1058	1071
0.94	879	884	889	896	904	913	923	934	946	959
0.95	758	763	769	776	783	792	802	813	825	839
0.96	628	633	639	646	654	663	672	684	696	709
0.97	490	495	501	507	515	524	534	545	557	570
0.98	342	347	352	359	367	376	386	397	409	422
0.99	184	189	194	201	209	218	228	239	251	264
1.00	16	21	26	33	41	50	60	71	83	96

RZNI - 3.2 M2/A2 M: 0.70 - 1.00, A: 0.20 - 0.29

RZE(M1)	RZE(A1)/RZA(A1) 0.20	0.21	0.22	0.23	0.24	0.25	0.26	0.27	0.28	0.29
0.70	1893	1909	1926	1944	1964	1985	2007	2031	2056	2082
0.71	1908	1924	1941	1960	1980	2000	2023	2046	2071	2097
0.72	1921	1937	1954	1972	1992	2013	2035	2059	2084	2110
0.73	1930	1946	1963	1981	2001	2022	2044	2068	2093	2119
0.74	1935	1951	1968	1987	2007	2028	2050	2073	2098	2125
0.75	1937	1953	1970	1989	2009	2030	2052	2075	2100	2126
0.76	1935	1951	1968	1987	2007	2028	2050	2073	2098	2124
0.77	1929	1945	1962	1981	2000	2021	2044	2067	2092	2118
0.78	1919	1935	1952	1970	1990	2011	2033	2057	2082	2108
0.79	1904	1920	1937	1955	1975	1996	2018	2042	2067	2093
0.80	1884	1900	1917	1936	1955	1976	1999	2022	2047	2073
0.81	1860	1875	1893	1911	1931	1952	1974	1998	2022	2049
0.82	1830	1845	1863	1881	1901	1922	1944	1968	1993	2019
0.83	1794	1810	1827	1846	1865	1886	1909	1932	1957	1983
0.84	1753	1769	1786	1805	1824	1845	1868	1891	1916	1942
0.85	1706	1722	1739	1758	1777	1798	1821	1844	1869	1895
0.86	1653	1669	1686	1704	1724	1745	1767	1791	1816	1842
0.87	1593	1609	1626	1645	1664	1685	1708	1731	1756	1782
0.88	1527	1543	1560	1578	1598	1619	1641	1665	1690	1716
0.89	1454	1470	1487	1505	1525	1546	1568	1592	1617	1643
0.90	1373	1389	1406	1425	1444	1465	1488	1511	1536	1562
0.91	1285	1301	1318	1337	1357	1378	1400	1423	1448	1474
0.92	1190	1206	1223	1241	1261	1282	1304	1328	1353	1379
0.93	1086	1102	1119	1137	1157	1178	1200	1224	1249	1275
0.94	974	990	1007	1025	1045	1066	1088	1112	1137	1163
0.95	853	869	886	905	924	945	968	991	1016	1042
0.96	724	740	757	775	795	816	838	862	887	913
0.97	585	601	618	636	656	677	699	723	748	774
0.98	437	453	470	488	508	529	551	575	600	626
0.99	279	295	312	330	350	371	393	417	442	468
1.00	111	127	144	162	182	203	225	249	274	300

RZNI - 3.3 M2/A2 M: 0.70 - 1.00, A: 0.30 - 0.39

RZE(M1)	RZE(A1)/RZA(A1)									
	0.30	0.31	0.32	0.33	0.34	0.35	0.36	0.37	0.38	0.39
0.70	2109	2138	2168	2199	2231	2265	2299	2335	2371	2409
0.71	2125	2154	2183	2215	2247	2280	2315	2350	2387	2424
0.72	2137	2166	2196	2227	2259	2293	2327	2363	2399	2437
0.73	2146	2175	2205	2236	2268	2302	2336	2372	2408	2446
0.74	2152	2181	2211	2242	2274	2307	2342	2377	2414	2451
0.75	2154	2183	2212	2244	2276	2309	2344	2379	2416	2453
0.76	2152	2181	2210	2242	2274	2307	2342	2377	2414	2451
0.77	2146	2174	2204	2235	2268	2301	2336	2371	2408	2445
0.78	2135	2164	2194	2225	2257	2291	2325	2361	2397	2435
0.79	2120	2149	2179	2210	2242	2276	2310	2346	2382	2420
0.80	2101	2129	2159	2190	2223	2256	2291	2326	2363	2400
0.81	2076	2105	2135	2166	2198	2231	2266	2301	2338	2375
0.82	2046	2075	2105	2136	2168	2201	2236	2271	2308	2345
0.83	2011	2039	2069	2100	2133	2166	2201	2236	2273	2310
0.84	1970	1998	2028	2059	2092	2125	2159	2195	2232	2269
0.85	1923	1951	1981	2012	2045	2078	2112	2148	2185	2222
0.86	1869	1898	1928	1959	1991	2025	2059	2095	2131	2169
0.87	1810	1839	1868	1900	1932	1965	2000	2035	2072	2109
0.88	1744	1772	1802	1833	1865	1899	1933	1969	2005	2043
0.89	1670	1699	1729	1760	1792	1826	1860	1896	1932	1970
0.90	1590	1619	1648	1680	1712	1745	1780	1815	1852	1889
0.91	1502	1531	1560	1592	1624	1657	1692	1727	1764	1801
0.92	1406	1435	1465	1496	1528	1561	1596	1632	1668	1706
0.93	1302	1331	1361	1392	1424	1458	1492	1528	1564	1602
0.94	1190	1219	1249	1280	1312	1346	1380	1416	1452	1490
0.95	1070	1098	1128	1159	1192	1225	1260	1295	1332	1369
0.96	940	969	999	1030	1062	1096	1130	1166	1202	1240
0.97	801	830	860	891	923	957	991	1027	1063	1101
0.98	653	682	712	743	775	809	843	879	915	953
0.99	495	524	554	585	617	651	685	721	757	795
1.00	327	356	386	417	449	483	517	553	589	627

RZNI - 3.4 M2/A2 M: 0.70 - 1.00, A: 0.40 - 0.49

RZE(M1)	RZE(A1)/RZA(A1)									
	0.40	0.41	0.42	0.43	0.44	0.45	0.46	0.47	0.48	0.49
0.70	2447	2486	2526	2567	2608	2650	2692	2735	2778	2821
0.71	2463	2502	2542	2582	2624	2666	2708	2751	2793	2837
0.72	2475	2514	2554	2595	2636	2678	2720	2763	2806	2849
0.73	2484	2523	2563	2604	2645	2687	2729	2772	2815	2858
0.74	2490	2529	2569	2609	2651	2693	2735	2778	2821	2864
0.75	2492	2531	2571	2611	2653	2695	2737	2780	2822	2866
0.76	2490	2529	2569	2609	2651	2693	2735	2778	2821	2864
0.77	2484	2523	2563	2603	2645	2686	2729	2771	2814	2858
0.78	2473	2512	2552	2593	2634	2676	2718	2761	2804	2847
0.79	2458	2497	2537	2578	2619	2661	2703	2746	2789	2832
0.80	2438	2478	2518	2558	2600	2641	2684	2726	2769	2813
0.81	2414	2453	2493	2534	2575	2617	2659	2702	2745	2788
0.82	2384	2423	2463	2504	2545	2587	2629	2672	2715	2758
0.83	2348	2388	2428	2468	2510	2551	2594	2636	2679	2723
0.84	2307	2347	2387	2427	2468	2510	2553	2595	2638	2681
0.85	2260	2300	2340	2380	2421	2463	2506	2548	2591	2634
0.86	2207	2246	2286	2327	2368	2410	2452	2495	2538	2581
0.87	2148	2187	2227	2267	2309	2351	2393	2436	2478	2522
0.88	2081	2120	2160	2201	2242	2284	2327	2369	2412	2455
0.89	2008	2047	2087	2128	2169	2211	2253	2296	2339	2382
0.90	1928	1967	2007	2047	2089	2130	2173	2215	2258	2302
0.91	1840	1879	1919	1959	2001	2043	2085	2128	2171	2214
0.92	1744	1783	1823	1864	1905	1947	1989	2032	2075	2118
0.93	1640	1679	1719	1760	1801	1843	1885	1928	1971	2014
0.94	1528	1567	1607	1648	1689	1731	1773	1816	1859	1902
0.95	1407	1447	1487	1527	1569	1610	1653	1695	1738	1782
0.96	1278	1317	1357	1398	1439	1481	1523	1566	1609	1652
0.97	1139	1178	1218	1259	1300	1342	1384	1427	1470	1513
0.98	991	1030	1070	1111	1152	1194	1236	1279	1322	1365
0.99	833	872	912	953	994	1036	1078	1121	1164	1207
1.00	665	704	744	785	826	868	910	953	996	1039

RZNI - 3.5 M2/A2 M: 0.70 - 1.00, A: 0.50 - 0.59)

RZE(M1)	RZE(A1)/RZA(A1) 0.50	0.51	0.52	0.53	0.54	0.55	0.56	0.57	0.58	0.59
0.70	2864	2908	2951	2994	3036	3079	3120	3161	3201	3240
0.71	2880	2923	2966	3009	3052	3094	3136	3177	3217	3256
0.72	2892	2936	2979	3022	3064	3107	3148	3189	3229	3268
0.73	2901	2945	2988	3031	3073	3116	3157	3198	3238	3277
0.74	2907	2950	2993	3036	3079	3121	3163	3204	3244	3283
0.75	2909	2952	2995	3038	3081	3123	3165	3206	3246	3285
0.76	2907	2950	2993	3036	3079	3121	3163	3204	3244	3283
0.77	2901	2944	2987	3030	3073	3115	3157	3198	3238	3277
0.78	2890	2934	2977	3020	3062	3105	3146	3187	3227	3266
0.79	2876	2919	2962	3005	3048	3090	3131	3172	3212	3251
0.80	2856	2899	2942	2985	3028	3070	3112	3153	3193	3232
0.81	2831	2874	2918	2961	3003	3045	3087	3128	3168	3207
0.82	2801	2844	2888	2931	2973	3015	3057	3098	3138	3177
0.83	2766	2809	2852	2895	2938	2980	3022	3063	3103	3142
0.84	2725	2768	2811	2854	2897	2939	2981	3021	3062	3101
0.85	2678	2721	2764	2807	2850	2892	2934	2974	3015	3054
0.86	2625	2668	2711	2754	2797	2839	2880	2921	2961	3000
0.87	2565	2608	2651	2694	2737	2779	2821	2862	2902	2941
0.88	2499	2542	2585	2628	2671	2713	2754	2795	2835	2875
0.89	2425	2469	2512	2555	2597	2640	2681	2722	2762	2801
0.90	2345	2388	2431	2474	2517	2559	2601	2642	2682	2721
0.91	2257	2300	2343	2386	2429	2471	2513	2554	2594	2633
0.92	2161	2205	2248	2291	2333	2375	2417	2458	2498	2537
0.93	2058	2101	2144	2187	2230	2272	2313	2354	2394	2433
0.94	1945	1989	2032	2075	2117	2160	2201	2242	2282	2321
0.95	1825	1868	1911	1954	1997	2039	2081	2122	2162	2201
0.96	1695	1739	1782	1825	1867	1909	1951	1992	2032	2071
0.97	1557	1600	1643	1686	1729	1771	1812	1853	1893	1932
0.98	1408	1452	1495	1538	1580	1623	1664	1705	1745	1784
0.99	1250	1294	1337	1380	1422	1465	1506	1547	1587	1626
1.00	1082	1126	1169	1212	1254	1297	1338	1379	1419	1458

RZNI - 3.6 M2/A2 M: 0.70 - 1.00, A: 0.60 - 0.69

RZE(M1)	RZE(A1)/RZA(A1) 0.60	0.61	0.62	0.63	0.64	0.65	0.66	0.67	0.68	0.69
0.70	3278	3315	3350	3384	3416	3447	3475	3501	3525	3546
0.71	3294	3331	3366	3400	3432	3462	3490	3516	3540	3561
0.72	3306	3343	3378	3412	3444	3475	3503	3529	3552	3574
0.73	3315	3352	3387	3421	3453	3484	3512	3538	3562	3583
0.74	3321	3358	3393	3427	3459	3489	3517	3544	3567	3588
0.75	3323	3360	3395	3429	3461	3491	3519	3545	3569	3590
0.76	3321	3358	3393	3427	3459	3489	3517	3543	3567	3588
0.77	3315	3352	3387	3421	3453	3483	3511	3537	3561	3582
0.78	3304	3341	3377	3410	3442	3473	3501	3527	3551	3572
0.79	3289	3326	3362	3395	3427	3458	3486	3512	3536	3557
0.80	3270	3306	3342	3376	3408	3438	3466	3492	3516	3537
0.81	3245	3282	3317	3351	3383	3413	3442	3468	3491	3512
0.82	3215	3252	3287	3321	3353	3383	3412	3438	3461	3482
0.83	3180	3216	3252	3286	3318	3348	3376	3402	3426	3447
0.84	3139	3175	3211	3245	3277	3307	3335	3361	3385	3406
0.85	3092	3128	3164	3198	3230	3260	3288	3314	3338	3359
0.86	3038	3075	3111	3144	3177	3207	3235	3261	3285	3306
0.87	2979	3016	3051	3085	3117	3147	3175	3201	3225	3246
0.88	2913	2949	2985	3018	3051	3081	3109	3135	3159	3180
0.89	2839	2876	2911	2945	2977	3008	3036	3062	3086	3107
0.90	2759	2796	2831	2865	2897	2927	2955	2981	3005	3026
0.91	2671	2708	2743	2777	2809	2839	2867	2893	2917	2938
0.92	2575	2612	2647	2681	2713	2743	2772	2798	2821	2843
0.93	2471	2508	2544	2577	2609	2640	2668	2694	2718	2739
0.94	2359	2396	2431	2465	2497	2528	2556	2582	2606	2627
0.95	2239	2275	2311	2345	2377	2407	2435	2461	2485	2506
0.96	2109	2146	2181	2215	2247	2278	2306	2332	2355	2377
0.97	1970	2007	2043	2076	2109	2139	2167	2193	2217	2238
0.98	1822	1859	1894	1928	1960	1991	2019	2045	2069	2090
0.99	1664	1701	1736	1770	1802	1833	1861	1887	1911	1932
1.00	1496	1533	1568	1602	1634	1665	1693	1719	1743	1764

RZNI - 3.7 M2/A2 M: 0.70 - 1.00, A: 0.70 - 0.79

RZE(M1)	RZE(A1)/RZA(A1) 0.70	0.71	0.72	0.73	0.74	0.75	0.76	0.77	0.78	0.79
0.70	3278	3315	3350	3384	3416	3447	3475	3501	3525	3546
0.71	3294	3331	3366	3400	3432	3462	3490	3516	3540	3561
0.72	3306	3343	3378	3412	3444	3475	3503	3529	3552	3574
0.73	3315	3352	3387	3421	3453	3484	3512	3538	3562	3583
0.74	3321	3358	3393	3427	3459	3489	3517	3544	3567	3588
0.75	3323	3360	3395	3429	3461	3491	3519	3545	3569	3590
0.76	3321	3358	3393	3427	3459	3489	3517	3543	3567	3588
0.77	3315	3352	3387	3421	3453	3483	3511	3537	3561	3582
0.78	3304	3341	3377	3410	3442	3473	3501	3527	3551	3572
0.79	3289	3326	3362	3395	3427	3458	3486	3512	3536	3557
0.80	3270	3306	3342	3376	3408	3438	3466	3492	3516	3537
0.81	3245	3282	3317	3351	3383	3413	3442	3468	3491	3512
0.82	3215	3252	3287	3321	3353	3383	3412	3438	3461	3482
0.83	3180	3216	3252	3286	3318	3348	3376	3402	3426	3447
0.84	3139	3175	3211	3245	3277	3307	3335	3361	3385	3406
0.85	3092	3128	3164	3198	3230	3260	3288	3314	3338	3359
0.86	3038	3075	3111	3144	3177	3207	3235	3261	3285	3306
0.87	2979	3016	3051	3085	3117	3147	3175	3201	3225	3246
0.88	2913	2949	2985	3018	3051	3081	3109	3135	3159	3180
0.89	2839	2876	2911	2945	2977	3008	3036	3062	3086	3107
0.90	2759	2796	2831	2865	2897	2927	2955	2981	3005	3026
0.91	2671	2708	2743	2777	2809	2839	2867	2893	2917	2938
0.92	2575	2612	2647	2681	2713	2743	2772	2798	2821	2843
0.93	2471	2508	2544	2577	2609	2640	2668	2694	2718	2739
0.94	2359	2396	2431	2465	2497	2528	2556	2582	2606	2627
0.95	2239	2275	2311	2345	2377	2407	2435	2461	2485	2506
0.96	2109	2146	2181	2215	2247	2278	2306	2332	2355	2377
0.97	1970	2007	2043	2076	2109	2139	2167	2193	2217	2238
0.98	1822	1859	1894	1928	1960	1991	2019	2045	2069	2090
0.99	1664	1701	1736	1770	1802	1833	1861	1887	1911	1932
1.00	1496	1533	1568	1602	1634	1665	1693	1719	1743	1764

RZNI - 3.8 M2/A2 M: 0.70 - 1.00, A: 0.80 - 0.89

RZE(M1)	RZE(A1)/RZA(A1) 0.80	0.81	0.82	0.83	0.84	0.85	0.86	0.87	0.88	0.89
0.70	3555	3531	3501	3465	3424	3377	3324	3265	3198	3125
0.71	3571	3546	3516	3481	3440	3393	3340	3280	3214	3141
0.72	3583	3559	3529	3493	3452	3405	3352	3292	3226	3153
0.73	3592	3568	3538	3502	3461	3414	3361	3302	3235	3162
0.74	3598	3573	3543	3508	3467	3420	3367	3307	3241	3168
0.75	3600	3575	3545	3510	3469	3422	3369	3309	3243	3170
0.76	3598	3573	3543	3508	3467	3420	3367	3307	3241	3168
0.77	3592	3567	3537	3502	3461	3414	3361	3301	3235	3161
0.78	3581	3557	3527	3491	3450	3403	3350	3291	3224	3151
0.79	3567	3542	3512	3477	3435	3388	3335	3276	3209	3136
0.80	3547	3522	3492	3457	3416	3369	3316	3256	3190	3116
0.81	3522	3497	3467	3432	3391	3344	3291	3231	3165	3092
0.82	3492	3467	3438	3402	3361	3314	3261	3201	3135	3062
0.83	3457	3432	3402	3367	3326	3279	3226	3166	3100	3026
0.84	3416	3391	3361	3326	3285	3238	3185	3125	3059	2985
0.85	3369	3344	3314	3279	3238	3191	3138	3078	3012	2938
0.86	3316	3291	3261	3226	3185	3138	3084	3025	2958	2885
0.87	3256	3231	3201	3166	3125	3078	3025	2965	2899	2826
0.88	3190	3165	3135	3100	3059	3012	2958	2899	2832	2759
0.89	3116	3092	3062	3026	2985	2938	2885	2826	2759	2686
0.90	3036	3011	2981	2946	2905	2858	2805	2745	2679	2605
0.91	2948	2923	2893	2858	2817	2770	2717	2657	2591	2518
0.92	2852	2828	2798	2762	2721	2674	2621	2561	2495	2422
0.93	2749	2724	2694	2659	2617	2570	2517	2458	2391	2318
0.94	2636	2612	2582	2546	2505	2458	2405	2346	2279	2206
0.95	2516	2491	2461	2426	2385	2338	2285	2225	2159	2085
0.96	2386	2362	2332	2296	2255	2208	2155	2095	2029	1956
0.97	2248	2223	2193	2158	2117	2070	2016	1957	1890	1817
0.98	2099	2075	2045	2009	1968	1921	1868	1809	1742	1669
0.99	1941	1917	1887	1851	1810	1763	1710	1651	1584	1511
1.00	1773	1749	1719	1683	1642	1595	1542	1483	1416	1343

RZNI - 3.9 M2/A2 M: 0.70 - 1.00, A: 0.90 - 1.00

RZE(M1)	RZE(A1) /RZA(A1) 0.90	0.91	0.92	0.93	0.94	0.95	0.96	0.97	0.98	0.99	1.00
0.70	3045	2957	2861	2757	2645	2524	2395	2256	2108	1950	1782
0.71	3060	2972	2876	2773	2661	2540	2410	2272	2124	1966	1798
0.72	3072	2984	2889	2785	2673	2552	2423	2284	2136	1978	1810
0.73	3082	2994	2898	2794	2682	2561	2432	2293	2145	1987	1819
0.74	3087	2999	2903	2800	2688	2567	2438	2299	2151	1993	1825
0.75	3089	3001	2905	2802	2690	2569	2439	2301	2153	1995	1827
0.76	3087	2999	2903	2800	2688	2567	2437	2299	2151	1993	1825
0.77	3081	2993	2897	2794	2682	2561	2431	2293	2144	1987	1818
0.78	3071	2983	2887	2783	2671	2550	2421	2282	2134	1976	1808
0.79	3056	2968	2872	2768	2656	2536	2406	2267	2119	1961	1793
0.80	3036	2948	2852	2749	2636	2516	2386	2248	2099	1941	1773
0.81	3011	2923	2828	2724	2612	2491	2362	2223	2075	1917	1749
0.82	2981	2893	2798	2694	2582	2461	2332	2193	2045	1887	1719
0.83	2946	2858	2762	2659	2546	2426	2296	2158	2009	1851	1683
0.84	2905	2817	2721	2617	2505	2385	2255	2117	1968	1810	1642
0.85	2858	2770	2674	2570	2458	2338	2208	2070	1921	1763	1595
0.86	2805	2717	2621	2517	2405	2285	2155	2016	1868	1710	1542
0.87	2745	2657	2561	2458	2346	2225	2095	1957	1809	1651	1483
0.88	2679	2591	2495	2391	2279	2159	2029	1890	1742	1584	1416
0.89	2605	2518	2422	2318	2206	2085	1956	1817	1669	1511	1343
0.90	2525	2437	2341	2238	2126	2005	1875	1737	1589	1431	1263
0.91	2437	2349	2253	2150	2038	1917	1787	1649	1501	1343	1175
0.92	2341	2253	2158	2054	1942	1821	1692	1553	1405	1247	1079
0.93	2238	2150	2054	1950	1838	1718	1588	1449	1301	1143	975
0.94	2126	2038	1942	1838	1726	1605	1476	1337	1189	1031	863
0.95	2005	1917	1821	1718	1605	1485	1355	1217	1068	910	742
0.96	1875	1787	1692	1588	1476	1355	1226	1087	939	781	613
0.97	1737	1649	1553	1449	1337	1217	1087	948	800	642	474
0.98	1589	1501	1405	1301	1189	1068	939	800	652	494	326
0.99	1431	1343	1247	1143	1031	910	781	642	494	336	168
1.00	1263	1175	1079	975	863	742	613	474	326	168	0

RZNI - 4.0 M2/A3 M: 0.70 - 1.00, A: 0.70 - 0.79

RZE(M1)	RZE(A1) 0.70	0.71	0.72	0.73	0.74	0.75	0.76	0.77	0.78	0.79
0.70	3747	3821	3893	3963	4030	4094	4153	4208	4257	4300
0.71	3762	3836	3909	3979	4046	4109	4169	4223	4273	4315
0.72	3775	3849	3921	3991	4058	4122	4181	4236	4285	4328
0.73	3784	3858	3930	4000	4067	4131	4190	4245	4294	4337
0.74	3789	3863	3936	4006	4073	4136	4196	4251	4300	4343
0.75	3791	3865	3938	4008	4075	4138	4198	4252	4302	4344
0.76	3789	3863	3936	4006	4073	4136	4196	4250	4300	4343
0.77	3783	3857	3930	4000	4067	4130	4190	4244	4294	4336
0.78	3773	3847	3919	3989	4056	4120	4179	4234	4283	4326
0.79	3758	3832	3904	3974	4041	4105	4164	4219	4268	4311
0.80	3738	3812	3885	3954	4022	4085	4145	4199	4248	4291
0.81	3713	3788	3860	3930	3997	4060	4120	4175	4224	4267
0.82	3684	3758	3830	3900	3967	4031	4090	4145	4194	4237
0.83	3648	3722	3795	3864	3932	3995	4055	4109	4158	4201
0.84	3607	3681	3753	3823	3890	3954	4014	4068	4117	4160
0.85	3560	3634	3706	3776	3843	3907	3967	4021	4070	4113
0.86	3507	3581	3653	3723	3790	3854	3913	3968	4017	4060
0.87	3447	3521	3594	3664	3731	3794	3854	3908	3958	4000
0.88	3381	3455	3527	3597	3664	3728	3787	3842	3891	3934
0.89	3308	3382	3454	3524	3591	3655	3714	3769	3818	3861
0.90	3227	3301	3374	3444	3511	3574	3634	3688	3738	3780
0.91	3139	3213	3286	3356	3423	3486	3546	3600	3650	3692
0.92	3044	3118	3190	3260	3327	3391	3450	3505	3554	3597
0.93	2940	3014	3086	3156	3223	3287	3346	3401	3450	3493
0.94	2828	2902	2974	3044	3111	3175	3234	3289	3338	3381
0.95	2707	2781	2854	2923	2991	3054	3114	3168	3217	3260
0.96	2578	2652	2724	2794	2861	2925	2984	3039	3088	3131
0.97	2439	2513	2585	2655	2722	2786	2845	2900	2949	2992
0.98	2291	2365	2437	2507	2574	2638	2697	2752	2801	2844
0.99	2133	2207	2279	2349	2416	2480	2539	2594	2643	2686
1.00	1965	2039	2111	2181	2248	2312	2371	2426	2475	2518

RZNI - 4.1 M2/A3 M: 0.70 - 1.00, A: 0.80 - 0.89

RZE(M1)	RZE(A1) 0.80	0.81	0.82	0.83	0.84	0.85	0.86	0.87	0.88	0.89
0.70	4336	4364	4382	4391	4389	4376	4348	4307	4250	4176
0.71	4351	4379	4398	4407	4405	4391	4364	4322	4265	4191
0.72	4364	4391	4410	4419	4417	4403	4376	4335	4278	4203
0.73	4373	4401	4419	4428	4426	4412	4385	4344	4287	4213
0.74	4378	4406	4425	4434	4432	4418	4391	4349	4292	4218
0.75	4380	4408	4427	4436	4434	4420	4393	4351	4294	4220
0.76	4378	4406	4425	4434	4432	4418	4391	4349	4292	4218
0.77	4372	4400	4419	4428	4426	4412	4385	4343	4286	4212
0.78	4362	4390	4408	4417	4415	4402	4374	4333	4276	4202
0.79	4347	4375	4393	4402	4401	4387	4359	4318	4261	4187
0.80	4327	4355	4374	4383	4381	4367	4340	4298	4241	4167
0.81	4302	4330	4349	4358	4356	4342	4315	4274	4216	4142
0.82	4272	4300	4319	4328	4326	4312	4285	4244	4186	4112
0.83	4237	4265	4284	4293	4291	4277	4250	4208	4151	4077
0.84	4196	4224	4243	4252	4250	4236	4209	4167	4110	4036
0.85	4149	4177	4196	4205	4203	4189	4162	4120	4063	3989
0.86	4096	4124	4143	4152	4150	4136	4108	4067	4010	3936
0.87	4036	4064	4083	4092	4090	4076	4049	4007	3950	3876
0.88	3970	3998	4017	4026	4024	4010	3983	3941	3884	3810
0.89	3897	3924	3943	3952	3950	3936	3909	3868	3811	3736
0.90	3816	3844	3863	3872	3870	3856	3829	3787	3730	3656
0.91	3728	3756	3775	3784	3782	3768	3741	3699	3642	3568
0.92	3633	3660	3679	3688	3686	3672	3645	3604	3547	3472
0.93	3529	3557	3575	3584	3583	3569	3541	3500	3443	3369
0.94	3417	3445	3463	3472	3470	3457	3429	3388	3331	3257
0.95	3296	3324	3343	3352	3350	3336	3309	3267	3210	3136
0.96	3167	3194	3213	3222	3220	3206	3179	3138	3081	3006
0.97	3028	3056	3075	3084	3082	3068	3040	2999	2942	2868
0.98	2880	2907	2926	2935	2933	2919	2892	2851	2794	2719
0.99	2722	2750	2768	2777	2775	2761	2734	2693	2636	2562
1.00	2554	2581	2600	2609	2607	2593	2566	2525	2468	2393

RZNI - 4.2 M2/A3 M: 0.70 - 1.00, A: 0.90 - 1.00 RZE(M1)

RZE(M1)	RZE(A1) 0.90	0.91	0.92	0.93	0.94	0.95	0.96	0.97	0.98	0.99	1.00
0.70	4083	3971	3837	3680	3498	3290	3053	2786	2487	2153	1782
0.71	4098	3986	3852	3695	3513	3305	3068	2801	2502	2168	1798
0.72	4111	3998	3864	3708	3526	3318	3081	2814	2514	2181	1810
0.73	4120	4007	3874	3717	3535	3327	3090	2823	2523	2190	1819
0.74	4126	4013	3879	3722	3540	3332	3096	2828	2529	2195	1825
0.75	4127	4015	3881	3724	3542	3334	3097	2830	2531	2197	1827
0.76	4125	4013	3879	3722	3540	3332	3095	2828	2529	2195	1825
0.77	4119	4007	3873	3716	3534	3326	3089	2822	2523	2189	1818
0.78	4109	3997	3863	3706	3524	3316	3079	2812	2513	2179	1808
0.79	4094	3982	3848	3691	3509	3301	3064	2797	2498	2164	1793
0.80	4074	3962	3828	3671	3489	3281	3044	2777	2478	2144	1773
0.81	4050	3937	3803	3646	3465	3256	3020	2753	2453	2119	1749
0.82	4020	3907	3773	3616	3435	3226	2990	2723	2423	2089	1719
0.83	3984	3872	3738	3581	3399	3191	2954	2687	2388	2054	1683
0.84	3943	3831	3697	3540	3358	3150	2913	2646	2347	2013	1642
0.85	3896	3784	3650	3493	3311	3103	2866	2599	2300	1966	1595
0.86	3843	3731	3597	3440	3258	3050	2813	2546	2247	1913	1542
0.87	3783	3671	3537	3380	3198	2990	2753	2486	2187	1853	1483
0.88	3717	3605	3471	3314	3132	2924	2687	2420	2121	1787	1416
0.89	3644	3531	3398	3241	3059	2851	2614	2347	2047	1714	1343
0.90	3563	3451	3317	3160	2978	2770	2533	2266	1967	1633	1263
0.91	3475	3363	3229	3072	2890	2682	2445	2178	1879	1545	1175
0.92	3380	3267	3133	2976	2795	2586	2350	2083	1783	1449	1079
0.93	3276	3164	3030	2873	2691	2483	2246	1979	1680	1346	975
0.94	3164	3052	2918	2761	2579	2371	2134	1867	1568	1234	863
0.95	3043	2931	2797	2640	2458	2250	2013	1746	1447	1113	742
0.96	2914	2801	2667	2510	2329	2120	1884	1617	1317	983	613
0.97	2775	2663	2529	2372	2190	1982	1745	1478	1179	845	474
0.98	2627	2514	2381	2224	2042	1834	1597	1330	1030	697	326
0.99	2469	2357	2223	2066	1884	1676	1439	1172	872	539	168
1.00	2301	2188	2055	1898	1716	1508	1271	1004	704	371	0

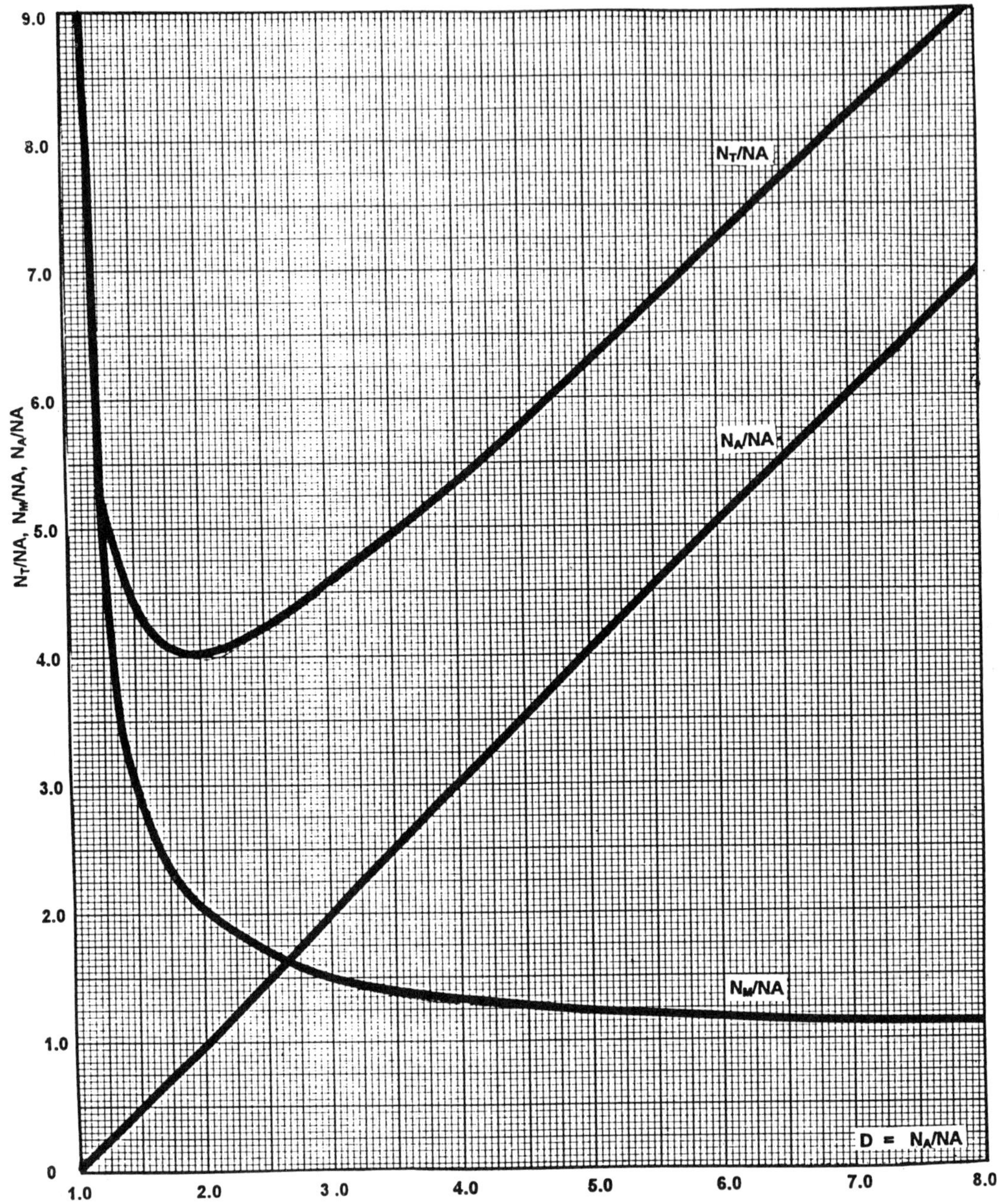

Figure 45 Variation of the number of normalized total, N_T/NA, manual, N_M/NA and automated or alternative inspections, N_A/NA, as D is varied. The intersection of the manual and automated curves defines the N_{MA} inspection value at which the RZNI Tables have been calculated.

selected for a validation demonstration with 0.05 significance level results. The effect of a choice of an operating point that was a multiple of N_{MA} can also be determined simply. The one drawback to the use of the curves shown in Fig. 45 is that they represent one set of a family of curves in which the ratio of the generalized statistical modeling constants is a significant factor. The curves of Fig. 46 capture the full set of relationships required to successfully analyze and control the reduction of the number of manual inspections employed in validation demonstrations.

A reduction in the number of required manual inspections achieved with manipulation of the *D* ratio is possible only when N_A/NA is greater than N_M/NA. The curves in Fig. 46 show that as the ratio of the probability constants *M*/*A* increases, a decrease of manual inspection reduction results. The area to the right of the N_A/NA in this Figure is the region in which a manual inspection reduction can be compensated by an increase in the number of balancing automated inspections.

The normalized curves of the manual inspection ratio, N_M/NA in Fig. 46 begin at a value of $D = 1.25$ and are truncated in their approach to their asymptote at $D = 8$. Most of the experimental inspection data seen occurs at the *M*/*A* span from 1/4 to 2. *M*/*A* experimental data from the widely used "reject in two" automated system is usually greater than 1.

The effect of *M*/*A* ratio variation, the result of selecting of *D* values higher than the one at which N_{MA} occurs, the locus of the minimum total number of inspections and the locus of the N_{MA} value at which the number of inspections by both methods is equal can all be seen in this figure.

To harness the understanding of the data shown here, the RZNI tables are indexed in Table 11 and two new tables have been prepared. These new tables bring new flexibility to the task of adjusting the ratio of manual to automated inspections used in a validation demonstration. The new table numbers are 12 and 13.

These tables are used when minimization of the ratio of manual inspections is to be considered. If manual inspection minimization is not to be considered, the RZNI tables are accessed and used as previously described. When manual inspection minimization is to be considered, Table 12 provides access to the statistical modeling parameters *M* and *A*. These parameters are listed by the number of serial inspections employed and the controlling probability of the first inspection in the sequence employed. Reduction ratios for N_M and the corresponding increase in the requirement for N_A begin with calculation of the *M*/*A* ratio for the planned validation.

Access to Table 13 data is listed by the *M*/*A* ratio and by validation pair. Three manual inspection reduction options are listed by *M*/*A* in Table 12 by the selected margin to the asymptotic value of manual inspection requirements: 20%, 10% and 5%.

The data available from the RZNI tables is the value of N_{MA}. The adjusted numbers of inspections required are determined by multiplication of the RZNI table value by the adjustment constants listed in Table 13. The adjustment constants are paired, the left column in each pair is the adjustment constant to decrease the manual inspection requirement, the right hand column constant increases the number of automated inspections to balance the manual inspection decrease effected.

Four options can be selected with the augmented RZNI tables. Option 1 is the selection of an equal number of inspections for each system. This is an appropriate choice when the cost of each method of inspection is approximately equal.

Options 2, 3 and 4 are those in which a reduction in the amount of manual inspections has been selected. These options differ in the margin to the limiting value of N_M selected and consequently to the amount of manual inspection reduction accomplished. Margins of 5%, 10% and 20% are listed in Table 13. An index of the tables and their contents is shown in Table 11. A more complete treatment of these topics is found in Sections XLII through LII.

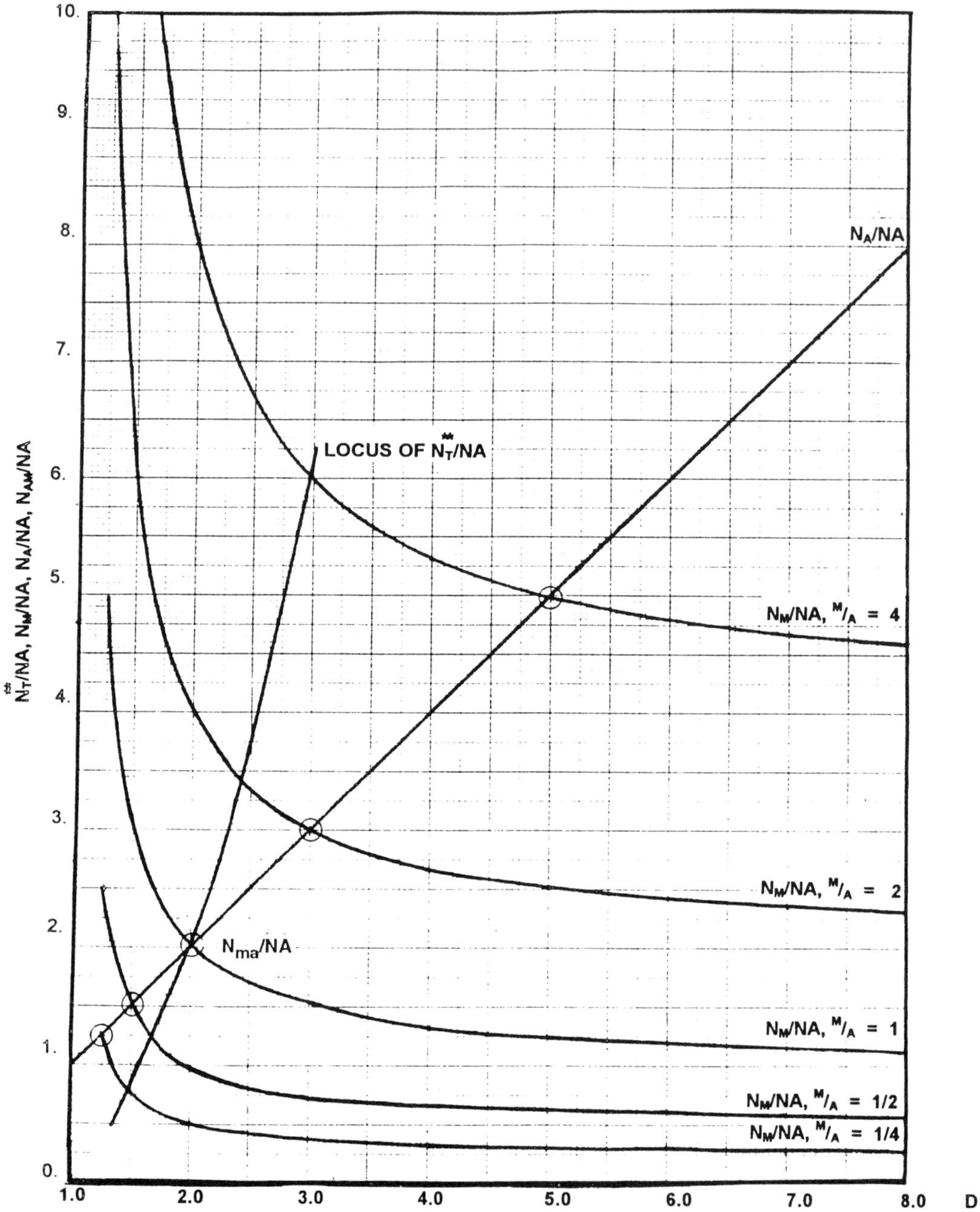

Figure 46 Effect of variation of *D* and the *M*/*A* ratio on the validation parameters. The locus of the minimum ratio of total Reject Zone container inspections, N_T/NA, of the ratio of alternative or automated inspections, N_A/NA, the ratio of manual inspections, N_M/NA and the point at which N_{MA} occurs for each combination of parameters can all be identified. The minimum limit for the N_M/AN ratio is equal to M/A in each case. The circled data points indicate operation with an equal number of manual and automated inspections. The equal inspection condition was used as the basis for calculating the earlier RZNI tables.

XXXVI. SERIAL ACCEPT SYSTEMS AND TECHNIQUES

The publications that resulted from the validation of the automated inspection systems at Schering used the standard two-stage manual inspection as the performance yardstick. Both inspections used a 3-diopter magnifying lens, a white and separate black background and a controlled inspection duration. These manual inspection results were used to determine the effectiveness of one, two or three automated or semiautomated inspection sequences. Both the manual inspections and the machine inspections used the same inspection strategy: Following the initial manual inspection, the rejects were segregated and reinspected. The accepts from either the first or the second inspection went into stock. When the Schering semiautomated and automated inspection systems were evaluated, their heightened sensitivity to the presence of particulates increased the reject rate to a degree that made it necessary to add a third inspection to the two used in the manual sequence. In each case of reinspection only the rejects were reinspected, the accepts from each went into stock. The flowgraph of Fig. 6 shows the manual inspection strategy employed at Schering. The manual inspection is followed by a cull inspection. In the manual inspection followed by a cull inspection, the containers initially rejected are reinspected. The material accepted from either inspection enters stock and the material rejected in the second inspection is destroyed.

The automated "Accept in Three" inspection is flowgraphed in Figure 36 (see page 337). In this inspection sequence the rejects of the first and second inspections are reinspected. The accepts from any inspection are entered into stock.

In any particle inspection, inspection security is evaluated by determining the average rejection probability for all containers that the standard manual inspection evaluated as the "bad-bad" must reject group. The rule-of-thumb "bad-bad" container category, as defined in our earlier papers, has a rejection probability per container equal to or greater than 70% in a single inspection. This average rejection probability in the Reject Zone was labeled the Reject Zone Efficiency (RZE). Serial inspections, correctly conducted, were shown to have independent rejection probabilities. This meant that the RZE achieved in sequential automated inspections were simply related: RZE(A2), the Reject Zone Efficiency after two automated inspections was the square of that determined for the first inspection; RZE(A3), the security achieved after three sequential inspections was the cube of first inspection RZE, $RZE(A1)^3$. Since RZE is less than 1.00, it is evident that repeated inspections with this strategy will reduce the inspection security achieved. In any case, the inspection security resulting from the inspection strategy and sequence selected must be shown to be equal to or greater than the standard manual inspection for a successful validation.

When the Schering automated inspection systems were put on line, it was done by demonstrating equivalent security between the results of the second manual inspection with those of the third automated inspection. Considering the background reviewed there should be no surprise in finding that the published papers do not fully support validation requirements for inspection systems that do not use the inspection strategy described.

XXXVII. SERIAL REJECT SYSTEMS

One Japanese vendor, Eisai, is the source of most of the installed fully automated particle contamination inspection machines in the United States. A second Japanese vendor, Takeda appears to have withdrawn from the U.S. market recently; an Italian concern, Brevetti CEA which has a significant part of the European and Middle Eastern Market is attempting to reenter the U.S. market. Seidenader has shipped some early versions of its own automated inspection

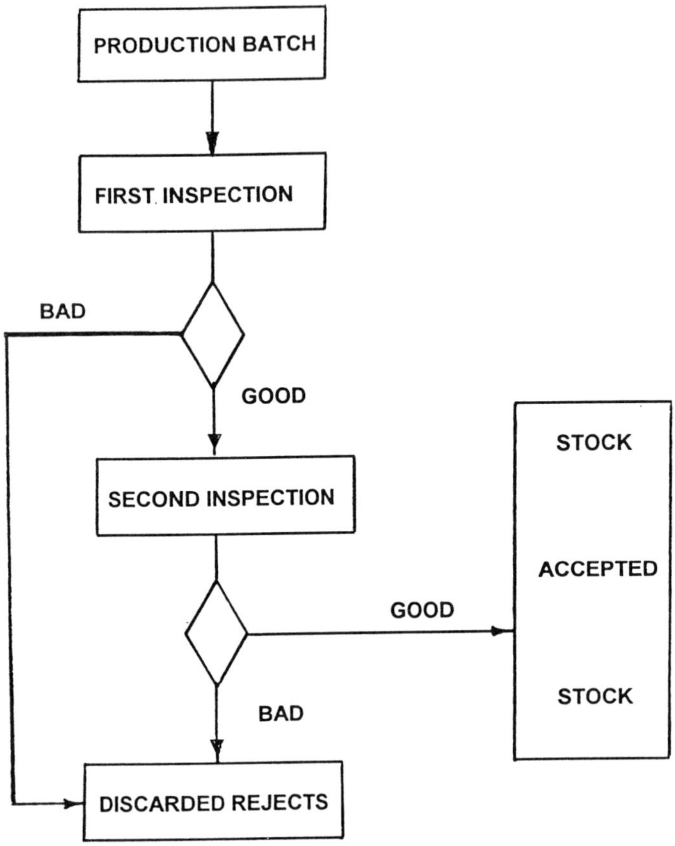

Figure 47 The automated two-inspection sequence in which product is rejected if it is judged bad in either of the inspections (A2-R/2).

system and is participating in operating tests. All use an inspection sequence in which a container is rejected if it does not pass either of two serial inspections. A container which fails the first inspection is judged a reject. Even if a container is accepted in the first inspection it is reinspected a second time. If the container fails either the first or the second inspection it is judged to be a reject and is destroyed. This inspection strategy is described as "Reject in Two." The flowgraph of Fig. 47 illustrates the "Reject in Two" inspection strategy. Prior to 1990, the available publications did *not* describe validation methodology that applies to this inspection strategy. The 1990 report [10] includes a bridge between the earlier papers and present needs.

XXXVIII. THE LINK BETWEEN INSPECTION SECURITY AND INSPECTION STRATEGY

The basic particle detection capability of the automated inspection determines its inspection sequence and strategy. The inspection strategy of the system determines whether the controlling probability is that of accepting or rejecting Reject Zone containers. In every case the sum of the probability of accepting, RZA or rejecting the Reject Zone containers, RZE, is equal to 1:

$$RZE + RZA = 1 \tag{12}$$

When RZE controls, as in serial accept inspections, RZE in the nth inspection, RZE(n) is equal to RZE raised to the nth power:

$$\mathrm{RZE}(n) = (\mathrm{RZE})^n \tag{13}$$

When RZA controls, as in serial reject systems, RZA in the nth inspection, RZA(n) is equal to RZA raised to the nth power:

$$\mathrm{RZA}(n) = (\mathrm{RZA})^n \tag{14}$$

The data available from automated serial rejection systems in the validation mode is used to calculate the average rejection probability. This is programmed for a fixed number of inspection stations for a selected number of inspections. In validation mode, one container is held in each turret position for a fixed number of inspections. Calculating RZE(A2) from the available data is the first step. RZE(A2) is the ratio of the total number of rejections recorded to the total number of container inspections:

$$\mathrm{RZE(A2)} = \frac{\mathrm{RZNR(A2)}}{I \cdot \mathrm{RZN}} \tag{15}$$

RZE(A2) is the Reject Zone Efficiency in the second automated system inspection. RZNR(A2) is the total number of rejects recorded for the Reject Zone containers in 1 inspection of the validation test group. RZN is the quantity of containers that have been identified by the manual inspectors as having a reject rate greater than 0.70.

The relationship between RZE(A2) and the required RZA(A1) is developed as follows:

$$\mathrm{RZA(A2)} = [\mathrm{RZA(A1)}]^2 \tag{16}$$

Using the fact that the sum of RZE(A2) and RZA(A2) is 1.0 and solving for RZA(A1) gives

$$\mathrm{RZA(A1)} = \sqrt{1 - \mathrm{RZE(A2)}} \tag{17}$$

With the estimate of RZA(A1) available, a choice can be made among the options that the revised RZNI tables offer. Following this choice, the number of reject zone container inspections in each system required for a .05 significance level validation trial can be determined.

XXXIX. THE USE OF THE RZNI TABLES, A GUIDED TOUR

For clarity and simplicity, the RZNI tables relate the number of Reject Zone container inspections to an estimate of the controlling probability of the inspection performed. Use of these tables therefore requires an estimate of the type of controlling probability and its magnitude in the first inspection. The inspection sequence and strategy of the automated system determine whether the controlling probability is that of accepting or rejecting Reject Zone containers.

To build familiarity with the 0.05 significance level results available with the RZNI tables, a number of examples will be discussed. The RZNI tables list an equal number of inspections for each method. This methodology was selected since the choice of operating point was in the vicinity of the broad minimum total inspection operating point. Both the equal number of inspections option and the reduced manual inspection calculations will be demonstrated. A more complete analysis and set of worked examples is included in Section III and following.

The following examples include validation calculations for

1. Single manual inspection/single automated inspection
2. Single manual inspection/"Reject in Two" automated inspection
3. Manual and cull inspection/"Reject in Two" automated inspection

4. Manual and cull inspection/"Accept in Three" automated inspection.

In each example, RZE, the final inspection security is equal to 0.80. The examples will use data from a manually identified container test group of 56 Reject Zone containers. The first scenario is the validation of a single-inspection automated system against a standard manual single inspection.

Referring to Table 11 the index to the RZNI tables, Table RZNI-1.0 is listed for single-inspection validations. Using the equal number of inspections option, Table RZNI-1.0 lists 1385 Reject Zone container inspections for an estimated RZE of 0.80 for each method. The number of test group inspections is determined by dividing 1385 by the 56 containers in the Reject Zone test group. The result of 1385 divided by 56 is 24.37 inspections. Rounding up to 25 test group inspections, to ensure the significance level results desired, completes the calculation.

The second scenario is the validation of a two inspection automated system with a reject in two inspection strategy with a single standard manual inspection. The RZE(A2-R/2) is 0.80, the RZE(M1) 0.80. There are 56 Reject Zone containers. Table 11 lists for the determination of the number of inspections of Reject Zone containers for a system using two sequential inspections to be validated with a single baseline inspection.

RZE(A2) matches the 0.80 performance of the manual inspection group. All entries in the RZNI tables are listed against the controlling probability of the first inspection. The probability available from the "Reject in Two" inspection system is that of the second-inspection RZE. Since the controlling probability for the automated system to be validated is that of accepting Reject Zone containers, RZA(A1) must be calculated with Eq. (17):

$$\text{RZA(A1)} = \sqrt{1 - \text{RZE(A2)}} = \sqrt{1 - 0.80} = 0.447 \tag{17}$$

Rounding .447 up to .45, is a conservative estimate of the performance of the automated system (the higher the RZA the less secure is the "Reject in Two" inspection system).

The data for RZNI table use is now available. RZE(M1) is 0.80 and RZA(A1) has been calculated to be 0.45. Table 11 lists RZNI 2 for use with a validation pair of a single manual inspection and an automated system using two sequential inspections, M1/A2 validation demonstrations. The listing for 0.45 RZA(A1) in Table RZNI-2.1 shows that the required number of Reject Zone container inspections to be 1,561. With a 56-containers test group, rounding up the number of required inspections, 28 test group inspections are required.

In the third scenario, a 20% margin over the limit value of the number of manual inspections has been selected using the same given data as the previous example. The RZE(A2-R/2) is 0.80, the RZE(M1) is 0.80; RZN is 56. As calculated in the previous example, RZA(A1) is 0.45. RZNI-2.4 lists 1561 inspections of the Reject Zone containers by both methods.

The first step in evaluation of the use of a reduced number of manual inspections in determination of the values for the statistical modeling constants M and A in Table 12. The M value is determined in the single inspection columns of Table 12 against the manual RZE(M1) of 0.80 and is 0.1600. The A value is determined in the second group of "two serial inspection" columns against an RZA(A1) of 0.45. It is 0.2005. The M/A ratio is 0.3333.

Entering Table 13 with an M/A ratio of 0.33, the selection of a 20% operating margin on N_{MA} results in a manual inspection reduction ratio of 0.28, a manual labor saving of 72%. Multiplying the RZNI-2.4 table value by this factor results in the reduced manual inspection requirement; it is 437.

The multiple increase in the number of required automated inspections is listed adjacent to the manual inspection reduction factor. The increased multiple for automated inspections is 4.29. Multiplying the RZNI-2.4 value of 1561 results in the balancing increase in automated inspections. The balancing number of automated inspections required is 6697.

In the fourth scenario, a first-and-cull manual inspection sequence (in which the accepts in either inspection are accepted for stock) is the basis for validation of an automated inspection system using a two-serial inspection “Reject in Two” inspection strategy. For a realistic comparison, the final inspection security achieved, (M2-A/2) and (A2-R/2), has been adjusted to 0.80 for all systems compared. The corresponding RZE(M1) is calculated as the square root of the RZE after the second inspection and is 0.89 and RZA(A1) is .45 as in the earlier example. RZN, the number of Reject Zone containers, is 56.

The RZNI tables make four validation demonstration designs possible. The first is the method using the RZNI constant to determine an equal number of container inspections by both methods. There are three manual inspection reduction options calculated in order of diminishing margin to the asymptotic limit for N_M : 20%, 10% and 5%. All four options will be implemented in this example. RZNI-3.4 for M2/A2 validations lists 2211 Reject Zone container inspections required for both methods. With the 56-container RZN group the result is 39.5 test group inspections, which is rounded up to 40 by each method.

To access the manual inspection reduction information in Table 12 the statistical modeling constants are required. The M and A constants are listed by the number of sequential inspections employed against the controlling probability of the first inspection in Table 12. For an RZE (M1) equal to 0.89, the value for M is 0.3102. For an RZA (A1) equal to 0.45 the value of A is .2005. The M/A ratio is 0.3102/0.2005 = 1.547. Rounding the calculated ratio down to 1.5, the pairs of Table 13 adjustment constant values for 20%, 10% and 5% margins are listed in the following extract.

		20% $D=6$		10% $D=11$		5% $D=21$	
M/A	$N_{MA}/NA = 1 + M/A$	N_M/N_{MA}	N_A/N_{MA}	N_M/N_{MA}	N_A/N_{MA}	N_M/N_{MA}	N_A/N_{MA}
1.50	2.50	0.72	2.40	0.66	4.40	0.63	8.40

The 40 manual inspections resulting from direct use of the RZNI constant can be reduced to 29, 27 and, in the case of the 5% margin, can be reduced to 26. The price in the increased numbers of automated inspections, using the adjustment constants in the extract, is respectively, 96, 176 and 336 test group inspections. The extra number of test group inspections is therefore 46, 136 and 296 additional over the number resulting from direct use of the RZNI table listing.

All computations have been rounded for conservative results. The optimum choice among these possibilities is affected by so many considerations that deciding among them is a judgment call.

In the fifth and last scenario, a manual inspection followed by a cull inspection is used in a validation trial of a system that uses three sequential inspections in an “Accept in Three” inspection strategy. To match the final results of the manual inspection whose RZE(M1) matches the 0.80 of the earlier example means that an improved manual inspection must be used. The improved manual inspection with an RZE(M2) equal to 0.80 has an RZE(M1) equal to 0.89. There are 56 Reject Zone containers.

Assuming that the “Accept in Three” system is adjusted to match this validation requirement, the RZE(A1-A/3) is calculated to be 0.93. Table 13 shows the three manual inspection options previously explored are available for this validation inspection pair. Direct use of RZNI-4.3 results in a requirement for 3241 Reject Zone container inspections by both methods. Dividing through by the 56-RZN group results in a requirement for 58 test group inspections. The M and A statistical modeling constants, using Table 12 are respectively, 0.2102 and 0.2318. The M/A ratio is 1.34. Rounding down to 1.30, the 20% M/A margin column in Table 13 shows a manual inspection reduction constant of 0.68 at a price of multiplying the number of automated inspections by a factor of 2.61.

We make some general comments in the preceding examples. It can be seen that the number of repeat inspections required increases as the complexity of the inspections increases and the RZE1 decreases or RZA1 increases. One other fact must be stressed: It is essential that the correct combination of automated and manual inspection strategy is specified in order to achieve a correctly validated experimental plan. To accomplish this, the actual internal inspection sequence within the machine must be determined accurately. Failure to select the correct combination of inspection strategies can result in a less accurate validation then desired. It must be stressed that validation or comparison methodology described above is based on the *average* acceptance or rejection probability of groups of containers.

XL. EFFECT OF INSPECTION STRATEGY

The following paper-and-pencil experiments will demonstrate the computations used in particle inspection validation demonstrations. They will also illustrate the effect of inspection strategy upon both the security of the inspection and its economic effectiveness.

In each case a graphic summary of a Reject Zone population has been prepared and is accompanied by a calculation work sheet. To facilitate a focused comparison on the effect of inspection strategy, all examples discussed have been adjusted to an equal RZE of 0.80 in the final inspection. Before any discussion of the inspection security achieved with automated inspection systems, two examples characterizing, first, a standard manual inspection and, second, a manual inspection followed by a cull inspection will be discussed.

The distribution shown in Fig. 48 is a for a single manual inspection using the capability of trained, validated inspectors selected according to the published guidelines [6]. The worksheet for this distribution is shown in Table 14. There are 56 vials in the Reject Zone; the RZE1, the average reject rate for all the Reject Zone containers is 80. This average reject rate is the experimentally established rejection probability for all the containers in the Reject Zone. Characteristic of this single manual inspection is that the peak rejection rate is higher than the average rejection probability shown in the Reject Zone Efficiency. In each probability segment, the rejection contribution is determined by the product of the number of containers in the segment and its rejection probability. It should be emphasized that detection of the specific containers identified by this manual inspection as being in the Reject Zone is the basis for any inspection system validation.

Figure 49 is representative of a standard manual inspection followed by a cull inspection, a reinspection of the initial rejects. The worksheet for this distribution is shown in Table 15. The concept behind the use of this manual two-inspection sequence is that operation of an improved manual inspection can result in both the maintenance of the quality of a single inspection and have improved recovery of Accept and Gray Zone containers. The inspection security following the second inspection is shown to be 0.80. To achieve this result, the inspection security of the first manual inspection must be increased to 0.89. An indication of the improved quality obtained is the movement of the rejection peak to 0.90 from its single-inspection location at 0.80 as well as the increase in RZE(M1). A characteristic of this "Accept in Two" inspection strategy is the decrease in RZE for successive inspections: RZE(M1) is 0.89, RZE(M2) is 0.80. From the rejection probability segment rejection data listed, some deterioration of rejection efficiency is seen for all but the 1.00 rejection probability group. The match of the RZE(M2) to the earlier RZE(M1) demonstrates that the initial inspection security has been maintained and a substantial part of good material has been conserved. When the data here is analyzed, 63% of the test group has been inspected with a rejection probability greater than 90%.

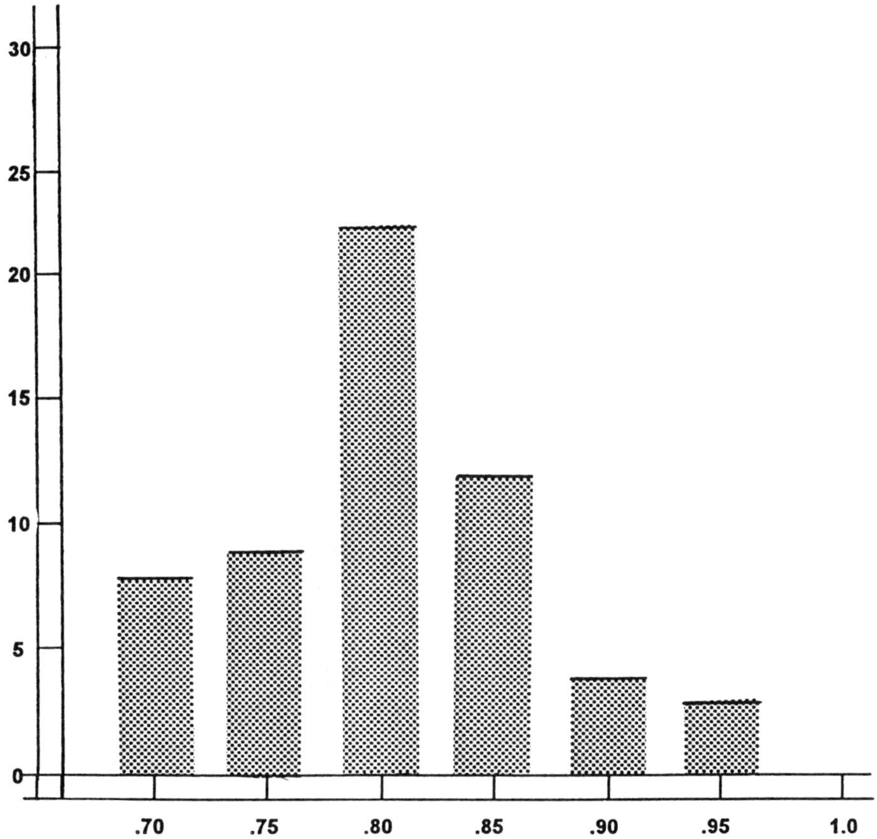

Figure 48 Single manual inspection example. RZE(M1), the inspection security achieved in a single manual inspection. The results have been matched with the initial-and-cull manual inspection, an automated two-stage inspection using a "Reject in Two" inspection strategy and with an "Accept in Three" inspection to show the effect of strategy on the accepted material. See Table 14 for the RZE calculations.

Table 14 Manual Standard First Inspection (M1)
Reject Zone Calculations for an RZE(M1) Equal to 0.80

RZE(M1) Manual first inspection rejection capability	Probability segment quantity	Rejections per probability segment
0.70	8	5.60
0.75	9	6.75
0.80	22	17.60
0.85	11	9.35
0.90	4	3.60
0.95	2	1.90
1.00	0	0.00
Totals	RZN = 56	RZNR(M1) 44.80

$$RZE(M1) = \frac{RZNR(M1)}{RZN} = \frac{44.80}{56} = 0.80$$

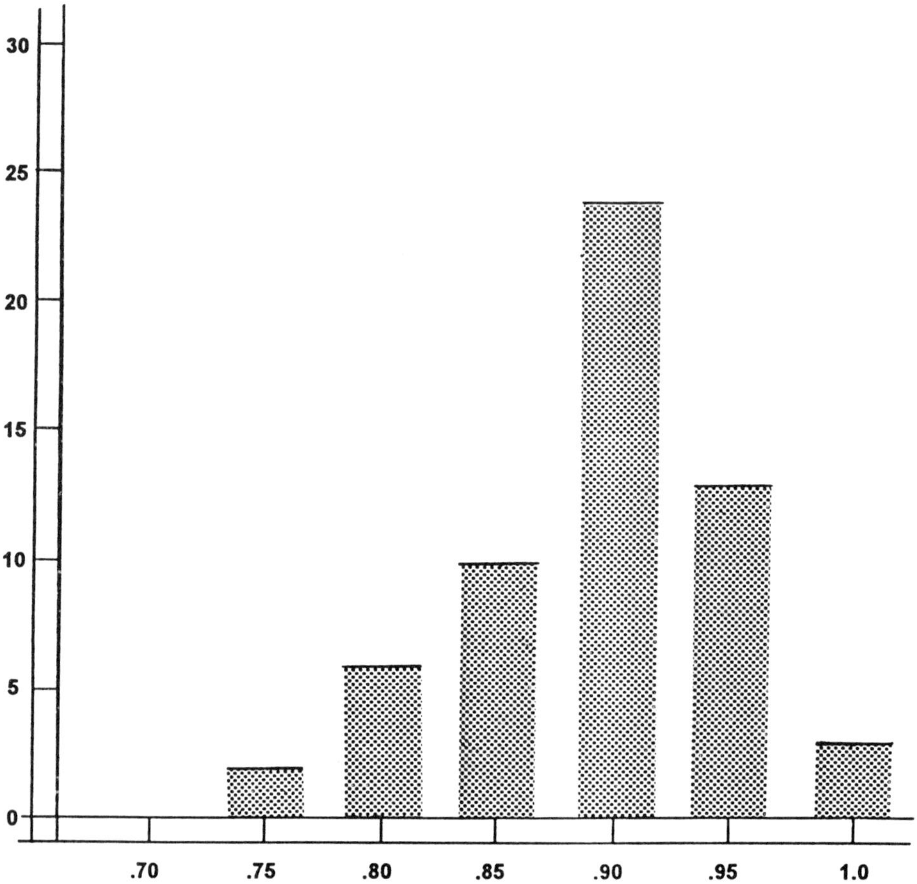

Figure 49 Manual and cull inspection example. RZE(M1-A/2), the security achieved in the first inspection of a manual inspection followed by a cull inspection. The probability of accepting a container is the controlling probability in this inspection. The results following the second inspection have been matched with the results of a single manual inspection, an automated two-stage inspection using a "Reject in Two" inspection strategy and with an "Accept in Three" inspection to show the effect of strategy on the accepted material. See Table 15 for the RZE calculations.

The Reject Zone Efficiency after the *second* inspection is the worst case for inspection security: When this double-inspection sequence is used, this worst-case inspection security must be matched or exceeded in validation trials.

Figure 50 is a representative single-inspection rejection probability distribution for a "Reject in Two" system. The worksheet for this distribution is shown in Table 15. With this type of inspection strategy the controlling probability is that of *accepting* Reject Zone containers, RZA. The controlling probability after two inspections is $RZA(A1)^2$. The data available from the inspection system is the number of container rejections for the number of test group inspections that have been selected. To use the RZNI tables requires determination of the controlling probability for the first inspection. The data listed in Table 16 is a step-by-step computation of RZA(A1) from the segment population number and its rejection probability.

The key piece of required information is that the sum of the Reject Zone Efficiency and the Reject Zone acceptance probability is 1.0. To calculate the RZA for any probability

Table 15 Manual "Accept in Two" Inspection (M2–A/2). Manual Reject Zone Calculations for a Two-Pass Manual Inspection: A First Inspection Followed by a Cull Reinspection of the Initial Rejects for a Matched Final Reject Zone Performance of RZE = 0.80

Manual inspection rejection probability			Rejections per probability segment	
RZE(M1)	RZE(M2)			
First only	First + reject reinspection	Probability segment quantity	RZNR(M1)	RZNR(M2)
0.70	0.4900	0	0	0
0.75	0.5625	1	0.75	0.563
0.80	0.6400	6	4.80	3.840
0.85	0.7225	10	8.50	7.225
0.90	0.8100	24	21.60	19.440
0.95	0.9025	13	12.35	11.733
1.00	1.0000	2	2.00	2.000
		Totals 56	50.00	44.798

$$\text{RZE(M1)} = \frac{\text{RZNR(M1)}}{\text{RZN}} = \frac{50.00}{56} = 0.89$$

$$\text{RZE(M2)} = \frac{\text{RZNR(M2)}}{\text{RZN}} = \frac{44.798}{56} = 0.80$$

segment is accomplished by determining the RZE of the segment and then subtracting this value from 1.0.

As an example of this computation, 28 containers have been determined to have a rejection probability, RZE, of approximately 0.80 following the second inspection. The RZA(A2) is equal to 1.0 – 0.80 = 0.20. The RZA(A1) can then be calculated as the square root of 0.20, which is 0.45. Equations (12) through (17) present this topic in an organized fashion.

The greatest single group occurs at a rejection probability of 0.55. This group, following the reinspection of accepted containers, is rejected with a probability of 0.80. Approximately 7% of the entire group is rejected after the second inspection with a rejection probability greater than 90%; 42% with a rejection probability greater than 80%. The increase from an RZE1 of .56 to an RZE2 of .80 with successive inspections is typical of an inspection strategy that reinspects accepted material. The match of the automated RZE2 to the appropriate manual RZE constitutes validation of this system.

Figure 51 is representative of rejection probability groups in an "Accept in Three" inspection system. The worksheet for this distribution is shown in Table 17. Calculation of the RZE for each probability segment consists of raising the probability for the first inspection of that segment to a power equal to the number of inspections performed. As a case in point, 0.90 RZE(M1) in a segment becomes $(0.90)^2$ or 0.81 in the second inspection and $(0.90)^3$ or 0.729 in the third inspection. This continued slide in inspection security for all particles whose rejection probability is less than 1 is characteristic of serial accept systems.

An important characteristic of this type of inspection is that approximately 50% of the Reject Zone containers are rejected with 100% rejection probability, thus ensuring that in the repeated reinspections of rejects the worst group will be securely rejected. In sequential accept inspection systems $\text{RZE}(n) = \text{RZE}^n$. In the Automated "Accept in Three" (A3-A/3) inspection sequence $\text{RZE(A2)} = [\text{RZE(A1)}]^2$ and $\text{RZE(A3)} = [\text{RZE(A1)}]^3$. Thus if RZE(A1) = 0.93, $\text{RZE(A2)} = (0.93)^2 = 0.89$ and $\text{RZE(A3)} = (0.93)^3 = 0.80$. The greatest portion of this test group,

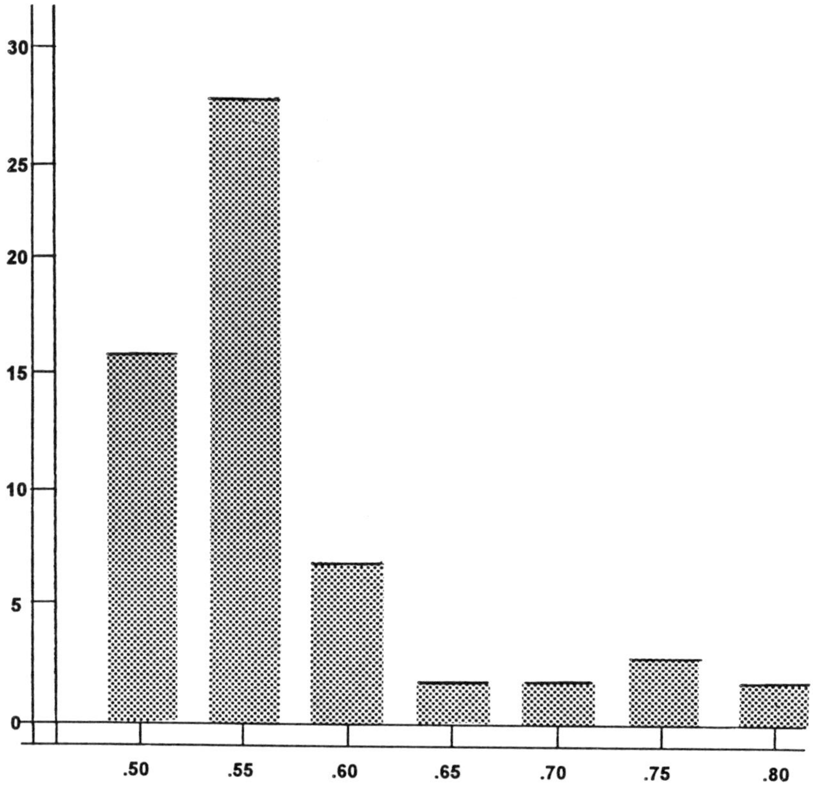

Figure 50 Automated "Reject in Two" inspection example. RZE(A1-R/2) the inspection security following the first inspection of an automated two-stage inspection using a "Reject in Two" inspection strategy. The probability of rejecting a container is the controlling probability in this inspection. The final results have been matched with a single manual inspection, an initial-and-cull manual inspection and with an "Accept in Three" inspection to show the effect of strategy on the accepted material. See Table 16 for the RZE calculations.

or of a production batch under these conditions, will be inspected with an RZE(A1) of 0.93, a reduced quantity will be inspected with an RZE(A2) of 0.89 and the smallest portion of the batch will be inspected with an RZE(A3) of 0.80.

When the validation demonstration for this system uses a two-stage manual inspection, an initial inspection followed by a cull inspection, the worst security of each method is matched. The worst-case security match concept, in validating the "Accept in Three," requires that the RZE(A3) following the third automated inspection must match the security obtained following the second manual inspection as measured with RZE(M2). It can be seen that, while there is a worst-case match, the average product will have been inspected at a much more rigorous standard. In fact, in the automated inspection results shown, approximately 90% of the Reject Zone group has been rejected with a probability of 90%.

Summarizing this discussion, two manual inspections and two different automated inspection strategies have been considered in Figs. 48 and 50. The fact that an increasing number of reinspections of the Reject Zone container group is reported as the inspection becomes more complex (and so more variable) is evident.

The automated inspection system strategies are linked to the inspection capability of each

Table 16 Automated "Reject in Two" Inspection (A2–R/2). Reject Zone Performance of a Two-Inspection "Reject in Two" Strategy Automated System with Matched Final Reject Zone Performance of RZE = 0.80

Automated system acceptance and rejection probability				Qy. per prob.	Segment acceptance and rejections			
RZA(A1)	RZE(A1)	RZA(A2)	RZE(A2)	seg.	RZNR(A1)	RZNA(A1)	RZNR(A2)	RZNA(A2)
0.50	0.50	0.25	0.75	16	8.00	8.00	12.00	4.00
0.45	0.55	0.2025	0.7975	28	15.40	12.60	22.33	5.67
0.40	0.60	0.16	0.84	7	4.20	2.80	5.88	1.12
0.35	0.65	0.1225	0.8725	1	0.65	0.35	0.8775	0.1225
0.30	0.70	0.09	0.91	1	0.70	0.30	0.91	0.09
0.25	0.75	0.0625	0.9375	2	1.50	0.50	1.875	0.125
0.20	0.80	0.04	0.96	1	0.80	0.20	0.96	0.04
Totals				56	31.25	24.75	44.832	11.168

$$\text{RZE(A1–R/2)} = \frac{\text{RZNR(A1)}}{\text{RZN}} = \frac{31.25}{56} = 0.56$$

$$\text{RZA(A1–R/2)} = \frac{\text{RZNA(A1)}}{\text{RZN}} = \frac{24.75}{56} = 0.44$$

$$\text{RZE(A2–R/2)} = \frac{\text{RZNR(A2)}}{\text{RZN}} = \frac{44.832}{56} = 0.80$$

$$\text{RZE(A2–R/2)} = \frac{\text{RZNA(A2)}}{\text{RZN}} = \frac{11.168}{56} = 0.20$$

RZE = Reject Zone efficiency (probability of rejecting Reject Zone containers)
RZA = Reject Zone accepts (probability of accepting Reject Zone containers)
RZNR = Number of Reject Zone rejected containers (parentheses following this number show the sequence number and type of inspection implemented)
RZNA = Number of Reject Zone accepted containers (parentheses following this number show the sequence number and type of inspection implemented)

system. Those systems that do not deliver adequate inspection security in one inspection are forced to reinspect the accepts using a "Reject in Two" strategy. This reinspection of accepted material raises RZE to acceptable levels by effectively lowering the rejection threshold of the system. A side effect of the reduction in the rejection threshold is that both the true and false reject rates are also raised. Those systems whose initial particle detection sensitivity is too high use reinspection of rejects to lower the effective RZE, and thus the reject rate, to economically sustainable levels. The false reject rate of a system must be evaluated to ensure that the inspection of the product is economically effective.

XLI. STRATEGY EFFECTS ON AN IDEAL RESPONSE CURVE

In Fig. 52 the effect of inspection strategy on false reject rate can be evaluated by comparing the results of each strategy against an ideal acceptance curve. The average acceptance probability of the ideal curve in the Accept and Gray Zones, curve 1, is set equal to 100%. The area under the curve in the Reject Zone is zero. To isolate the effect of inspection strategy, a single-pass manual rejection probability curve is taken as the standard; this is curve 4 in this figure. All other inspection strategy curves are derived from the response of curve 4. To evaluate the effect of

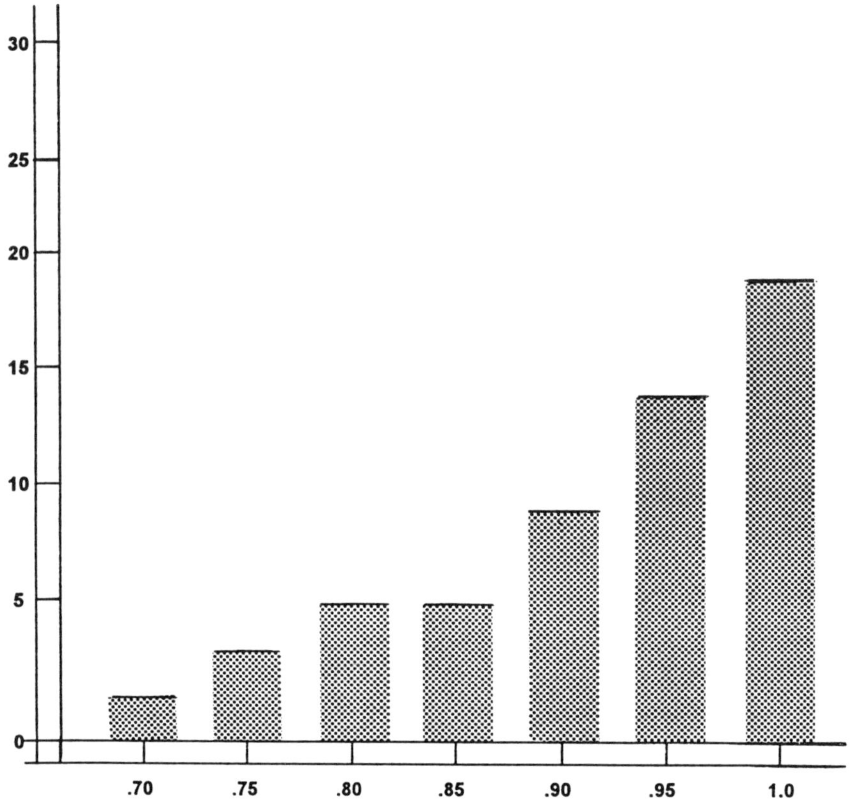

Figure 51 Automated "Accept in Three" inspection example. RZE (A1), the inspection security achieved in a three-inspection "Accept in Three" (A3–A/3) inspection strategy. The probability of accepting a container is the controlling probability in this inspection. The results have been matched with a single manual inspection, with that of an initial-and-cull manual inspection and an automated two-stage inspection using a "Reject in Two" inspection strategy. See Table 17 for the RZE calculations.

Table 17 Automated "Accept in Three" Inspection Calculations (A3–A/3). Reject Zone Performance of a Three-Serial Inspection "Accept in Three" Automated System for Matched Final Reject Zone Performance RZE = 0.80

RZE(A3-A/3) Automated inspection rejection probability				
RZE(A1) First	RZE(A2) Second	RZE(A3) Third	Quantity per probability quantity	Segment rejections RZNR(A3)
0.70	0.490	0.343	1	0.343
0.75	0.562	0.422	3	1.266
0.80	0.640	0.512	5	2.560
0.85	0.723	0.614	5	3.070
0.90	0.810	0.729	9	6.561
0.95	0.903	0.857	14	12.002
1.00	1.000	1.000	19	19.000
		Totals	56	44.802

$$\text{RZE(A3–A/3)} = \frac{\text{RZNR(A3)}}{\text{RZN}} = \frac{44.802}{56} = 0.80$$

inspection strategy in areas under each curve in the Accept and Gray Zone are compared to the area under the ideal curve.

It is clear that curve 2, the "Accept in Three" strategy, at 0.91 approaches closest to the ideal curve. There is a significant acceptance improvement noted when the initial rejects are re-inspected or culled. Reinspection of the rejects increases acceptance in the Accept and Gray Zones to 0.84 of the ideal response compared to the 0.65 achieved in a manual single inspection. The "Reject in Two" strategy shown here as curve 5 has a loss of accepted material approximately as large as the improvement obtained with the culling inspection shown in curve 3: It passes 47% of the Accept and Gray Zone material. It must be noted that when production data is considered, the effects discussed there are magnified.

To evaluate the impact of inspection strategy on performance in the Reject Zone, the result

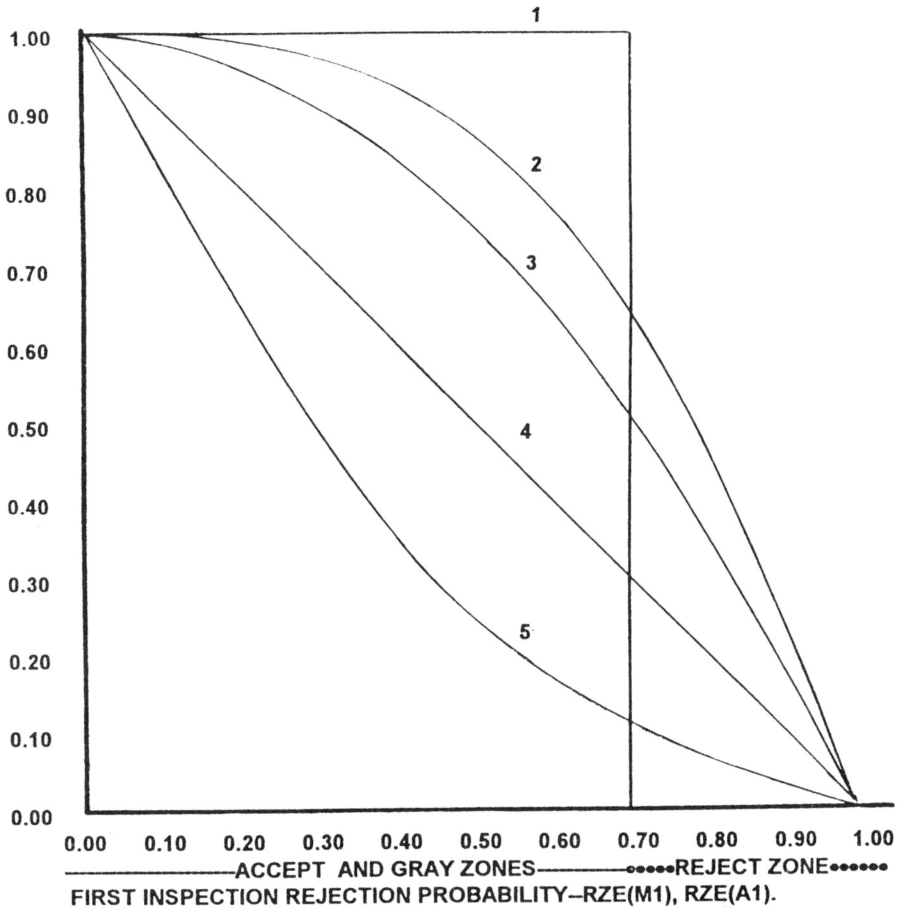

Figure 52 A comparison of the effect of inspection strategy on the acceptance of Accept and Gray Zone containers compared to an ideal container acceptance response. The ideal response is one in which all Accept and Gray Zone containers are accepted and all Reject Zone containers are rejected. Inspection strategies: (1) Ideal acceptance characteristic, 100% for all Accept and Gray Zone containers. (2) "Accept in Three," accept if accepted in any of three sequential inspections. (3) Manual first inspection followed by a cull reinspection of initial rejects. (4) Single manual standard inspection. (5) "Reject in Two," reject if rejected in either of two serial inspections.

of each strategy is compared to that of the manual single inspection in Curve 4. In this comparison, the "Reject in Two" inspection strategy shines: Only 32% of the Reject Zone rejects that are accepted in the manual first inspection are accepted. In the "Accept in Two" manual inspection the Reject Zone accepts are increased 179% compared to the manual first inspection results. In the "Accept in Three" strategy, the Reject Zone accepts are increased 221% compared to the manual first inspection.

The evaluation of these strategies against the ideal performance curve shows that there are potential benefits in the use of an "Accept in Two" or an "Accept in Three" inspection strategy. The choice of such strategy must be supported by an inspection characteristic with greater sensitivity and discrimination than that associated with the manual inspection for particulates. The ideal curve evaluation was undertaken to create a mechanism for comparison of inspection strategies. Any direct attempt at comparative evaluations of the various available inspection systems founders on the lack of comparable evaluation data in the literature.

It is difficult to compare published information concerning Reject Zone and false reject rate performance, since each set of published data has been accomplished with its own working standard test group. In addition, significant differences can exist between the inspector groups that prepared the reject-seeded working standard test sets. With these disclaimers in mind, a comparison of the results obtained with two imported systems and one that was produced in this country for 2-ml containers is shown in Table 18.

One note of caution: Inspection of larger container sizes and container surface defects can result in difficulties. These problems should be approached with caution and stressed in any evaluation or validation.

The serial reject strategy, specifically the "Reject in Two" inspection strategy is the one in greatest use at present in automated inspection systems. This strategy evolved in the inspection of antibiotics and sterile liquids. The labor saving due to the application of automated inspection of these injectable products far outweighed the excess cost of the resulting false rejections. For many products low container factory costs made the excess cost of the rejects an acceptable charge against the labor savings resulting from the automated inspection.

This assumption of acceptable false reject loss must be reviewed when the new generation of costly bio-engineered products are to be inspected for particulate contamination. This review of the effect of different inspection strategies must consider both the inspection security achieved in the inspection and the false reject rate that results from the controlling strategy for each inspection.

In serial reject systems, the controlling probability is that of accepting rejectable containers. As an example, a container with a 10% rejection probability has a 90% acceptance rate in a single inspection. When this container is inspected twice in a serial reject system, the acceptance probability in the second inspection is the first inspection acceptance probability squared.

Table 18 Performance Comparison of Three Automated Inspection Systems Showing the Effect on Performance of "Reject in Two" versus "Accept in Three" Inspection Strategies

Vendor	RZE	RAGZ	Container	Rejection strategy
Japanese import	0.93	0.14	Tubing data only	"Reject in Two"
Italian import	0.91	0.29	Tubing data only	"Reject in Two"
U.S. product	0.96	0.12	Blow-molded and tubing data	"Accept in Three"

RZE measures the inspection security achieved. RAGZ, the false reject rate in the Accept and Gray Zones, evaluates the excess cost of the automated inspection.

The probability of acceptance in the second inspection acceptance is $0.90^2 = 0.81$. The resulting reject rate in the second inspection is 1 minus the second inspection acceptance probability or 19%. For any who reinspect the results of the first two inspections a second time on a "reject in two" inspection system, the acceptance rate after the fourth inspection, that is two passes through the inspection system, is the first inspection acceptance probability raised to the fourth power. $0.90^4 = 0.6561$. The reject rate after the fourth inspection in this serial reject system is $1 - 0.6561 = 0.3439$, 34.39% for a container with a 10% rejection probability that is considered to be a good-good container. As noted, the inspection security improves after each serial reject reinspection.

In serial accept systems, the controlling probability is the probability of rejecting a container. In these systems, the probability of a rejection in two serial inspections is the rejection probability in the first inspection squared. The rejection probability for the container in three serial inspections is the first inspection rejection probability cubed. Considering a container with a 10% rejection probability as in the earlier example, the rejection probability after two inspections is $0.10^2 = 0.01$. The initial 10% reject rate is reduced to 1% as the result of reinspecting the initial rejects. If the inspection security is high enough to sustain three serial reinspections, the container that has an initial 10% rejection probability will have a 0.10^3 rejection probability or 0.10%. As noted, in the serial accept system the inspection security decreases with each successive inspection. There is one exception to this rule: containers whose rejection probability is 1.00 are unaffected by the serial reinspections.

Table 18a systematically examines the effect of the inspection strategy employed on Accept, Gray, and Reject Zone containers, thus providing information on the relative security and the false reject rate of each inspection strategy. This is accomplished by examining the rejection probability of containers ranging from 0.01 to 1.00.

The rejection probability for each container is calculated for a "Reject in Two", R/2, system and an "Accept in Two", A/2, strategy in use as the manual-and-cull inspection strategy and an "Accept in Three" system. Following this set of calculations, the rejection ratios of the "Reject in Two" system to both serial accept systems is calculated. This comparison of rejection ratios shows that use of the serial reject strategy increases the loss of low-rejection-probability Accept Zone containers. That is, the group of containers with rejection probabilities ranging up to 0.30. Where the unit production cost of containers must be considered, this inspection strategy may not be desirable. The successful use of the serial accept inspections strategy, however, is premised on the availability of an inspection system in which the rejection probability of Reject Zone containers is efficiently high.

The serial accept inspection strategy reduces the false reject rate in the Accept and Gray Zones as shown by the listed rejection ratios. In previous trials, the use of forward-scattered light in a serial accept system has resulted in better discrimination, a rejection probability of 1.00 for approximately ½ of all containers identified as Reject Zone containers. Table 18 compares two "Reject in Two" inspection systems with an "Accept in Three" inspection strategy system. The two "Reject in Two" systems listed in this Table have been validated with approximately equal Reject Zone efficiency scores of 0.91–0.93. The 3-pass serial accept system has been validated with a Reject Zone efficiency of 0.96. This measured equivalence in inspection security provides the rationale for using a technique which reduces the rejection rate of usable containers.

Minimization of manual inspection labor in validation demonstrations has always been a matter of concern. To this end, some users of automated inspection systems have attempted to use them in a preliminary screen to concentrate the Reject Zone containers. This was attempted through the use of multiple, serial inspections. Table 18b evaluates the effect of using multiple inspections on "Reject in Two" inspection systems. This Table shows that multiple inspections on "Reject in Two" particle inspection systems does not result in an effective segregation of

Table 18a The Forward Scatter Light System Employed with the Serial Accept Inspection Strategy Reduces the False Reject Rate in the Accept and Gray Zones as Shown by the Listed Rejection Ratios. In Previous Trials, the Use of Forward-Scattered Light Has Resulted in Better Discrimination, a Rejection Probability of 1.00 for Approximately ½ of All Containers Identified as Reject Zone Containers. The Three Inspection Strategies in this Table Have Been Validated with Approximately Equal Reject Zone Efficiency Scores of 0.91–0.93 for the Serial Reject Systems and 0.96 for the 3-Pass Serial Accept System. This Demonstrated Equivalence in Inspection Security Provides the Rationale for Using a Technique Which Reduces the Rejection Rate of Usable Containers

	Container rejection probability				Rejection probability ratio	
	Single	Reject in Two R/2	Accept in Two A/2	Accept in Three A/3	(R/2) : (A/2)	(R/2) : (A/3)
Accept and Gray Zones	0.010	0.0199	0.00010	0.000001	199.0	19900.
	0.025	0.0494	0.00063	0.00002	78.4	2470.
	0.050	0.0975	0.0025	0.000125	39.0	780.
	0.100	0.190	0.010	0.001	19.0	190.
	0.200	0.360	0.040	0.008	9.00	45.0
	0.300	0.510	0.090	0.027	5.6	18.9
	0.400	0.640	0.160	0.064	4.00	10.0
	0.500	0.750	0.250	0.125	3.00	6.00
	0.600	0.840	0.360	0.216	2.33	3.89
Reject Zone	0.700	0.9100	0.4900	0.3430	2.6531	2.6531
	0.800	0.9600	0.6400	0.5120	1.8750	1.8750
	0.900	0.9900	0.8100	0.7290	1.5000	1.3580
	0.950	0.9975	0.9025	0.8574	1.1053	1.1634
	0.975	0.9994	0.9506	0.9269	1.0782	1.0782
	0.990	0.9999	0.9801	0.9703	1.0202	1.0305
	1.000	1.0000	1.0000	1.0000	1.0000	1.0000

Table 18b This Table Illustrates the Progressive Increase of Rejection Probability for Accept and Gray Zone Containers with Repeated Automated Inspections that Use the "Reject in Two" Inspection Strategy Employed by Presently Available Automated Inspection Systems

Number of repeated inspections	Single manual inspection rejection probability—percent								
	1.0	2.0	5.0	10.0	20.0	30.0	40.0	50.0	60.0
	Equivalent manual inspection rejection probability								
1	1.99	3.95	9.75	19.00	36.00	51.00	64.00	75.00	84.00
2	3.94	7.76	18.55	34.39	59.04	75.99	87.04	93.75	97.44
3	5.85	11.41	26.49	46.86	73.79	88.24	95.33	98.44	99.59
4	7.73	14.92	33.66	56.95	83.22	94.24	98.32	99.61	99.93
5	9.56	18.29	40.13	65.13	89.26	97.18	99.40	99.90	99.99
6	11.36	21.53	45.96	71.76	93.13	98.62	99.78	99.98	100.00

Reject Zone containers. On the contrary, an undesirable inclusion of Accept and Gray Zone containers due to the weak accept/reject decision boundary in these automated systems is demonstrated. The rate of inclusion of these undesired containers is seen to increase with each repeated inspection emphasizing the undesirability of this procedure.

XLII. VALIDATION STATISTICS FOR NONDESTRUCTIVE PARTICLE CONTAMINATION TEST METHODS

The probabilistic nature of the accept/reject particulate inspection decision dictates a validation based on the mean result of many accept/reject decisions. The accept/reject decisions with which performance equivalence of an automated inspection system is demonstrated to be equivalent to that of manual inspectors is the accept/reject record compiled in repeated inspections of Reject Zone containers. To plan for a cost-effective validation requires advance knowledge of the number of manual and alternate or machine inspections that will satisfy the significance level requirements for an acceptable "demonstration of equivalent performance." The repetitive nature of the statistical design for these validation experiments led to the concept that the required statistical information could be supplied in table form thus minimizing the need for statistical consultation as the experimental conditions changed.

Methodology for validation of nondestructive inspection systems in terms of human capability ("visible particles") has been extended to include all current manual, automated and semiautomated inspection strategies. Drawing on the accumulated experience of many users, the nomenclature has been simplified, clarified and is listed in Table 10 (see page 349). New tables have been calculated to assist in the design of validation demonstrations in which validation costs and time are held to a minimum.

The RZNI tables that were initially published were limited: they were calculated for manual and automated Reject Zone Efficiencies between 0.70 and 1.0 for serial acceptance strategies. Tables were calculated for the "Accept in Two" and "Accept in Three" systems that were of interest at that time. In this strategy, the controlling probability is that of *rejecting* Reject Zone containers. This controlling probability is evaluated with a single numeric, the Reject Zone Efficiency (RZE). For simplicity in the use of these tables, they were compiled to show the number of total Reject Zone container inspections versus the estimated controlling probability of the first inspection. This was noted as RZE(M1), RZE(M2) for, respectively, a single manual inspection or for a manual inspection followed by a cull reinspection of the initial rejects. The RZE of automated systems were noted as RZE(A1), RZE(A2) and RZE(A3) for one inspection, two and three serial inspections respectively. The first version of the RZNI tables was calculated only for validation needs of the Schering automated particle inspection systems.

In 1990 the earlier work was extended [10] to the many commercial inspection systems in which the controlling probability is that of *accepting* Reject Zone rejects. As in the earlier work, the controlling probability is evaluated with a single numeric, RZA. RZA is the probability of *accepting* Reject Zone containers. The sum of RZE and RZA is one. High RZE capability is translated into low RZA capability. This difference in controlling probability meant that the preceding RZNI tables required downward extension to cope with validation of serial rejection inspection systems.

A characteristic of all present systems in which RZA is the controlling probability is the use of two sequential inspections. These serial rejection systems are therefore characterized as "Reject in Two" inspection systems. The controlling probability of these systems in the first inspection is symbolized RZA(A1). Both the first and the revised RZNI tables were based on an equal number of manual and machine inspections. To provide for minimization of manual validation requirements by substituting increased machine inspection replications additional RZNI tables

Table 19 RZNI Table Index and Application Notes

Tables	RZE/RZA range	Inspection sequence
RZNI-1.0 to RZNI-1.2	0.70–1.00	M1/A1
RZNI-2.0 to RZNI-2.9	0.00–1.00	M1/A2, M2/A1
RZNI-3.0 to RZNI-3.9	0.00–1.00	M2/A2
RZNI-4.0 to RZNI-4.2	0.70–1.00	M2/A3

12 and 13 have been added. The validation choices now available are (1) an equal number of manual and machine inspections, for those occasions in which the inspection cost by both methods is approximately equivalent, and (2) a minimized number of manual inspections balanced by an increased number of machine inspections in three steps of increasing manual inspection minimization.

For those validations in which either inspection strategy could be employed, a combined RZA(A1), RZE(M1) table covering the entire range of rejection probabilities from zero to 1.0 has been prepared. For most uses the convenient tables that have been prepared will prove adequate. The combinational validation requirements for all the alternatives discussed are listed below. These tables list the number of Reject Zone container inspections for a 0.05 significance level validation. Numerical examples of each type of validation demonstration have been included and are analyzed in order of complexity. For those who need other significance levels in their results, a table with the necessary information has been included.

The validation procedure described is applicable to the comparison of the effectiveness of an individual inspector, inspector groups and alternative manual procedures, as well as automated or semiautomated inspection systems. In each case, the number of Reject Zone container inspections required for a 0.05 significance level is related to the controlling probability of the first inspection. This controlling probability is established by the strategy, the inspection sequence and the capability of the inspection system employed. For all inspection strategies that utilize serial accepts, the controlling probability is that of *rejecting* Reject Zone rejects. These include the standard single inspection, and the "Accept in Two" or "Accept in Three" inspection strategies. Use of the tables proceeds as follows: the number of Reject Zone container inspections required for the validation demonstration are individually listed. The number of Reject Zone container inspections is determined by the strategy employed and by the security achieved for each method. The tables list a number of Reject Zone container inspections by the Reject Zone Efficiency or the Reject Zone acceptance rate for each method employed. When an "accept" strategy is employed, the number of Reject Zone container inspections, RZNI, is listed against the controlling probability of the first inspection for each method. This is RZE(M1) for the manual inspection and RZE(A1) for the alternate or automated inspection. Separate tables are provided for each pair of methods employed. In these strategies, containers accepted as good in each inspection are added to stock.

For those inspections strategies which use serial rejects and are validated against a single manual inspection, RZE(M1) is still the baseline method used to determine the number of Reject Zone containers. RZE(A1) is replaced by the probability of accepting Reject Zone containers in the first of the repeated inspections, RZA(A1). In this type of inspection, a container must be judged good in each of the sequential inspections to be accepted as good for stock and shipment.

An index of the tables and their contents is listed in Table 19.

XLIII. REJECT ZONE CONTAINER INSPECTION REPLICATIONS FOR 0.05 SIGNIFICANCE RESULTS

Four validation demonstrations using the serial accept strategy are examined. In this type of inspection, the controlling probability for the alternative inspection is that of *rejecting* Reject Zone rejects. A standard single manual inspection in a validation demonstration of a single machine inspection is first discussed. The basic concepts used in this first example are then extended to the case of a validation demonstration of two sequential manual inspections (Accept in Two) to validate a three-sequential machine inspection (Accept in Three) procedure. This section also includes a two-sequential manual (Accept in Two) inspection used to validate a two-sequential inspection (Accept in Two) machine procedure. For completeness, although not presently usable, a two-sequential (Accept in Two) manual inspection as the basis for a single-inspection machine validation is also included.

The serial reject strategy is, at present, the one most widely used in high-speed automated inspection of injectable compounds. In this inspection strategy, the controlling probability is that of *accepting* Reject Zone containers. Two validation demonstrations of automated inspection systems using the serial reject strategy are discussed. The first example uses a standard single manual inspection; the second example uses a manual inspection followed by a cull inspection as the basis for a similar validation.

XLIV. PROBLEM DEFINITION BASICS

For any batch of containers, let

p_M = true average rejection probability for the manual standard inspection in the Reject Zone.

p_A = true average rejection probability for an alternative or automated inspection in the manual-inspection-defined Reject Zone.

p_{MO} = observed value of p_M, the manual reject rate in the Reject Zone, RZE(M1).

p_{AO} = observed value of p_A, the alternative or automated system acceptance or reject rate in the manually-defined Reject Zone.

For the containers in the batch that are "bad", i.e., have manual rejection probabilities ≥ 0.70 in a single manual inspection, p_{MO} is the observed first inspection Reject Zone Efficiency, (M1). p_{AO} is either the observed acceptance rate of Reject Zone containers in serial rejection machines, RZA(A1), or the observed rejection rate of these containers in serial accept machines, RZE(A1). Whether manual or machine inspection is better depends on the true rejection probabilities p_M and p_A. However, since these are unknown, the decision as to which inspection scheme is best must be made on the basis of the observed first inspection values RZE(M1) and RZA(A1).

The decision technique used is that of statistical hypothesis testing. The null hypothesis is that machine inspection is at least as good as manual inspection. This is written as

$$H_0 : p_A \geq p_M \tag{18}$$

The alternative hypothesis, that manual inspection is better than machine inspection, is written as

$$H_1 : p_M \geq p_A \tag{19}$$

While this formulation of the hypotheses is an exact equivalent of the verbal formulation in the paragraph above, it is inappropriate for an hypothesis testing procedure. To test two

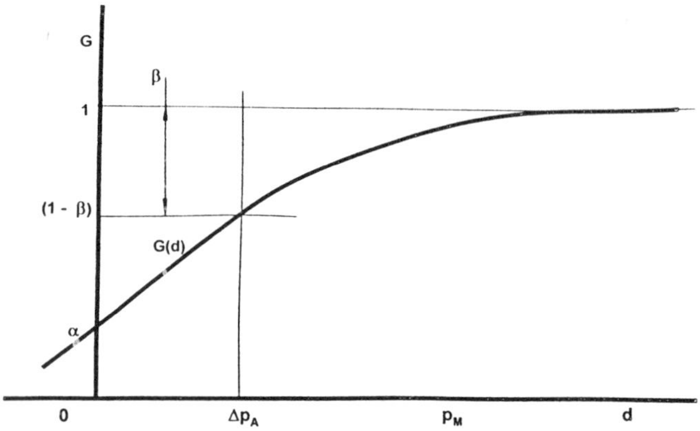

Figure 53 Sketch of a typical power function where $\alpha = G(0)$ and $\beta = 1 - G(\Delta p)$. α is the probability of rejecting H_0 when H_0 is true "on the boundary." β is the probability of accepting H_0 when H_1 is true "on the boundary."

hypotheses, it is necessary that some "distance" exists between them. Accordingly, we revise our formulation as follows:

$$H_0 : p_A \geq p_M \tag{20}$$

$$H_1 : p_M \geq p_A + \Delta p_A \tag{21}$$

Here Δp_A represents the minimum difference between ARZE and MRZE that we wish to detect. The role of Δp_A will be clarified by the power function.

Any test of H_0 versus H_1 is evaluated by its power, which is defined as the probability of rejecting H_0. The power clearly depends on p_A and p_M and especially on the difference between them. We denote this difference by d:

$$d = p_M - p_A \tag{22}$$

For any fixed value of p_M, the power is a function of d and is expressed by the power function:

$$G(d) = \text{Prob}\,\{\text{reject } H_0 \text{ given} \quad p_M - p_A = d\} \tag{23}$$

A sketch of a typical power function is shown in Fig. 53.

The quantities of α and β in Fig. 52 are defined in Eqs. (24) and (25).

$$\alpha = G(0) \tag{24}$$

$$\beta = 1 - G(\Delta p_A) \tag{25}$$

By the definition of power function, α is the probability of rejecting H_0 given $p_A = p_M$. In other words, α is the probability of rejecting H_0 when H_0 is true "on the boundary." Similarly, β is the probability of not rejecting H_0 given $p_M = p_A + \Delta p_A$. In other words, α is the probability of accepting H_0 when H_1 is true "on the boundary." As is evident from the definition of power function, and as indicated by Fig. 53, $G(d)$ is a monotonically increasing function of d in the interval defined in Eq. (26):

$$p_M - 1 \leq d \leq p_M \tag{26}$$

This interval is the full domain of d, since

$$d = p_M - p_A \quad \text{and} \quad 0 \leq p_A \leq 1 \tag{27}$$

From the sketch of Fig. 53, three regions can be defined:

$$G(d) \leq \alpha \quad \text{for} \quad d < 0 \tag{28}$$

$$\alpha < G(d) < 1 - \beta \quad \text{for} \quad 0 < d < p_A \tag{29}$$

$$G(d) \geq 1 - \beta \quad \text{for} \quad d \geq \Delta p_A \tag{30}$$

The condition that $d \leq 0$ is equivalent to $p_A \geq p_M$ or H_0; and the condition that $d \geq p_A$ is equivalent to $p_M \geq (p_A + \Delta p_A)$ or H_1. In other words,

$$\text{Prob}\,\{\text{Reject } H_0 \text{ given } H_0 \text{ true}\} \leq \alpha \tag{31}$$

$$\text{Prob}\,\{\text{Accept } H_0 \text{ given } H_1 \text{ true}\} \leq \beta \tag{32}$$

In the language of statistical hypothesis testing, these two probabilities are, respectively, the probabilities of Type I and Type II errors. They represent the protection against making the wrong decision when using a statistical hypothesis testing procedure.

The indifference region for d is defined by Eq. (29). In other words, when the difference between manual and machine inspection lies between the hypotheses being tested, the probability of rejecting H_0 lies between α and $1 - \beta$. This result is a consequence of the protection afforded when H_0 or H_1 is true. It implies that Δp_A should be chosen so that we are indifferent as to whether H_0 is accepted or rejected when d lies between the hypotheses. Fig. 43 graphically illustrates the concepts discussed (see page 348).

XLV. SINGLE MANUAL VERSUS SINGLE AUTOMATED INSPECTION: (M1/A1)

The above discussion of statistical hypothesis testing has been solely in terms of the true values p_M and p_A and their difference d. To actually perform the test, the observed values, the manual and automated reject zone efficiencies, RZE(M1) and RZE(A1) must be used. As an example, consider a test of a single-inspection manual procedure against a single-inspection automated system. Assume that α, β, and Δp_A have been chosen. Let N_M = total number of manual inspections of the Reject Zone containers, N_A = total number of machine inspections of the Reject Zone containers, R_M = total number of manual rejections, and R_A = total number of machine rejections.

The observed rejection probabilities in the Reject Zone for the manual and alternative inspections respectively are

$$\text{RZE(M1)} = \frac{R_M}{N_M} \tag{33}$$

$$\text{RZE(A1)} = \frac{R_A}{N_A} \tag{34}$$

Given α, β, and Δp_A, the theoretical formulas for the required sample sizes N_M and N_A depend on the true values p_M and p_A. However, the form of the dependence is such that only a rough estimate of the average of p_M and p_A is necessary to obtain an excellent approximation for N_M and N_A. Accordingly, p_O is defined as the average observed container rejection probability using p_M and p_A from Eqs. (33) and (34):

$$p_O = \frac{\text{RZE(M1)} + \text{RZE(A1)}}{2} \tag{35}$$

Then choose N_M and N_A to satisfy

$$\frac{p_O(1 - p_O)}{N_M} + \frac{p_O(1 - p_O)}{N_A} = \left(\frac{\Delta p_A}{t_\alpha + t_\beta}\right)^2 \tag{36}$$

where t is the $100(1 - \alpha)$th percentile of the normal distribution shown in Eq. (37):

$$1 - \alpha = \int_{-\infty}^{t_\alpha} \frac{1}{\sqrt{2\pi}} \exp\left[\frac{-x^2}{2}\right] dx \tag{37}$$

Equation (36) specifies the constraints on the number of inspections required in order to achieve the desired levels of protection when testing the specified hypotheses. Dimensionless analysis shows that the right side of Eq. (36) has units of (number of inspections)$^{-1}$. Equation (36) is rewritten for clarity as

$$\frac{p_O(1 - p_O)}{N_M} + \frac{p_O(1 - p_O)}{N_A} = \frac{1}{N} \tag{36A}$$

Here N equals $[t_\alpha + t_\beta]/\Delta p_A$ which has units of number of inspections.

Note that N does not correspond to any specific sample size. Table 20 lists parameters and values of N for a $\Delta p_A = 0.05$ and significance levels from 0.01 to 0.10. From Eq. 36 there are many pairs (N_M, N_A) which can be used. The particular choice can be made for convenience. However, a minimum total number of inspections will result when N_M is chosen equal to N_A.

If no prior estimate of p_O is available, a preliminary sample can be taken and the observed rejection probability can be used. If p_M and p_A are larger than one-half, a conservative procedure is to choose p_O between one-half and their average, as in Eq. (37):

$$\frac{1}{2} < p_O < \frac{1}{2}(p_M + p_A) \tag{38}$$

Any p_O satisfying Eq. (38) will yield overestimates of N_M and N_A when used in Eq. (40). This will guarantee achieving the desired protection as defined by a choice of p_O and Δp_A.

Table 20 Parameters and Constants for Four Levels of Significance in Validation of Nondestructive Particulate Inspection Systems

Significance α, β	t_α, t_β	$\frac{1}{N} = \left[\frac{\Delta p_A}{t_\alpha + t_\beta}\right]^2$	RZNI	RZNI/4329
0.01	2.326	0.0001155	8656	1.9996
0.02	2.055	0.0001480	6757	1.5609
0.05	1.645	0.0002310	4329	1.0000
0.10	1.282	0.0003803	2630	0.6075

Note the use of RZNI rather than the abbreviated form used for convenience in notation. $\beta = 1 - G(p_A)$, t_α, t_β, is the $100(1 - \alpha)$th percentile of the normal distribution, $\Delta p_A = 0.05$; $\Delta p_A = \Delta\text{RZE(A)}, \Delta\text{RZA(A)}$.

Once N_M and N_A are chosen, the required inspections can be carried out and RZE(M1) and RZE(A1) can be calculated. The test procedure is then as follows: Accept H_0 if, and only if,

$$\text{RZE(A1)} - \text{RZE(M1)} \geq C, \quad \text{where } C \text{ is defined in Eq. (40)} \tag{39}$$

$$C = -\left(\frac{t_\alpha}{t_\alpha + t_\beta}\right)^{\Delta p_A} \tag{40}$$

For the important special case where $\alpha = \beta$ the test becomes accept H_0 if, and only if Eq. (41) is satisfied:

$$\text{ARZE} - \text{MRZE} \geq -\frac{\Delta p_A}{2} \tag{41}$$

As an example, suppose that $\alpha = \beta = 0.05$ and that $\Delta p = 0.05$. In other words, we want a 5% or smaller chance of making the wrong decision, i. e., that either the machine is better than manual inspection (H_0 is true) or the true manual rejection probability is at least 5 percentage points larger than the true machine rejection probability (H_1 is true). From tables of the normal distribution, $t_\alpha = t_\beta = 1.645$. With these assumptions, $N = 4329$ as shown in Table 21. Suppose further that the Reject Zone Efficiency of both the inspectors and the machine are about 80%. Substituting $p_O = 0.8$ and $N = 4329$ into Eq. 36 we get

$$\frac{0.8(0.2)}{N_M} + \frac{0.8(0.2)}{N_A} = \frac{1}{4329} \tag{42}$$

If equal numbers of machine and manual inspections are desired, then $N = N_A = N_{MA} = 1385$ and the total number of inspections is 2771.

For the case where RZN, the number of Reject Zone containers is 50, the total number of inspections required is rounded up to 28. Four inspectors each performing seven inspections could be used. Once the inspections have been made and RZE(M1) and RZE(A1) calculated, C can be determined from Eq. (40) $C = -.025$: the test is to accept H_0 if and only if RZE(A1) ≥ RZE(A1) – 0.025.

In other words, we accept the hypothesis that machine inspection is at least as good as manual inspection whenever the observed machine rejection efficiency is at least as large as the manual rejection efficiency less 2.5 percentage points. If, for instance, RZE(M1) = 0.80 and RZE(A1) = 0.78 we would accept H_0.

The discussion above shows how to compare two systems that perform a single inspection whether they are performed manually or by machine.

XLVI. TWO SEQUENTIAL MANUAL INSPECTIONS VERSUS THREE SEQUENTIAL ALTERNATIVE OR AUTOMATED INSPECTIONS USING AN "ACCEPT IN THREE" INSPECTION STRATEGY [(M2)vs(A3-A/3)]

To compare the security of a two-sequential-manual-inspection system with a three-sequential machine inspection, operating with an "Accept in Three" is desired, the results above are modified as follows. In the two-sequential-inspection manual procedure, the true average rejection probability is $(p_M)^2$ and under the three-sequential-inspection machine procedure, the true average machine rejection probability is $(p_A)^3$. Accordingly, we test the null hypothesis of Eq. (43) versus the alternative hypothesis of Eq. (44):

$$H_0 : (p_A)^3 > (p_M)^2 \tag{43}$$

$$H_1 : (p_M)^2 > (p_A)^3 + \Delta p_A \tag{44}$$

For this case, the formula for the sample sizes N_M and N_A requires preliminary estimates of both p_M and p_A. Denote these by p_{MP} and p_{AP}. Then, for a given α, β and Δp_A, N_M and N_A must satisfy

$$\frac{4p_{MP}^3[1 - p_{MP}]}{N_M} + \frac{9p_{AP}^5[1 - p_{AP}]}{N_A} = \frac{1}{N} \tag{45}$$

When the number of manual and machine inspections are selected to be equal, $N_M = N_A = N_{MA}$, we get

$$N_{MA} = N[4p_{MP}^3(1 - p_{MP}) + 9p_{AP}^5(1 - p_{AP})] \tag{46}$$

As in the M1/A1 case, this choice of N_M and N_A minimizes the total number of inspections.

Once N_M and N_A are chosen in Eq. (45), the required inspections can be carried out and the observed rejection probabilities RZE(M2) and RZE(A3) can be calculated as follows:

$$\text{RZE(M2)} = [\text{RZE(M1)}]^2 \tag{47}$$

$$\text{RZE(A3)} = [\text{RZE(A1)}]^3 \tag{48}$$

Here RZE(M1) and RZE(A1) are the first inspection results for each method.

As in the comparison of two single-inspection methods, the test procedure is as follows: Accept H_0 if, and only if, Eq. (49) is satisfied:

$$\text{RZE(A3)} - \text{RZE(M2)} \geq C \tag{49}$$

where

$$C = -\left(\frac{t_\alpha}{t_\alpha + t_\beta}\right)\Delta p_A \tag{40}$$

As an example, consider a situation similar to the comparison of the single-manual and single-automated inspection developed in Eq. (41) for the case where $\alpha = \beta = 0.05$ and $C = -\Delta p_A/2$ to again achieve a 5% or smaller chance of making the wrong validation decision. The exception for this case is that we assume $p_{MP} = 0.80$ and $p_{AP} = 0.86$. This choice of values results in equal Reject Zone Efficiencies following the completion of the two-sequential-manual and the three-sequential-automated inspections: 0.64 is obtained by substituting the selected numerical values into Eqs. (47) and (49).

As before, we select a 0.05 significance level. From Table 21, the summary listing of applicable equations, the required sample sizes must satisfy Eq. (45) as shown:

$$\frac{0.410}{N_M} + \frac{0.593}{N_A} = \frac{1}{4329} \tag{45A}$$

If $N_M = N_A = N_{MA}$, then $N_{MA} = 4342$. For the example discussed previously with 50 containers in the Reject Zone, the total number of inspections required becomes 88. If the manual inspection is carried out with 11 inspectors, each inspector is required to perform eight inspections of the test batch. The increase in the number of required inspections from the 1 × 1 case is due to the increased inspection variability in rejection efficiencies.

Table 21 Summary of Formulas Used in Validation Demonstrations of Alternative Particle Contamination Inspection Methods Against the Human Inspector Visibility Standard

Comparison	Probability function
1. M1/A1	$\frac{p_{MP}(1-p_{MP})}{N_M} + \frac{p_{AP}(1-p_{AP})}{N_A} = \frac{1}{N}$
2. M1/A2	$\frac{p_{MP}(1-p_{MP})}{N_M} + \frac{4p_{AP}^3(1-p_{AP})}{N_A} = \frac{1}{N}$
3. M2/A1	$\frac{4p_{MP}^3(1-p_{MP})}{N_M} + \frac{p_{AP}(1-p_{AP})}{N_A} = \frac{1}{N}$
4. M2/A2	$\frac{4p_{MP}^3(1-p_{MP})}{N_M} + \frac{4p_{AP}^3(1-p_{AP})}{N_A} = \frac{1}{N}$
5. M2/A3	$\frac{4p_{MP}^3(1-p_{MP})}{N_M} + \frac{9p_{AP}^5(1-p_{AP})}{N_A} = \frac{1}{N}$

XLVII. TWO MANUAL VERSUS TWO ALTERNATIVE OR AUTOMATED INSPECTIONS WHERE BOTH USE AN "ACCEPT IN TWO" INSPECTION STRATEGY [(M2-A/2)vs(A2-A/2)]

Another useful example is that of a system using two sequential manual inspections, an "Accept in Two" procedure against a similar automated two-inspection system. In this case the true average rejection probability for both methods is p_M^2 and p_A^2 and the null hypothesis is

$$H_0 : p_A \geq p_M \tag{50}$$

The alternative hypothesis is

$$H_1 : p_M^2 \geq p_A^2 + \Delta p_A \tag{51}$$

As in the case for the three sequential inspections versus the two sequential inspections, the formula for the sample sizes N_M and N_A requires preliminary estimates of p_M and p_A denoted by p_{MP} and p_{AP}. Then, as before, for a given α, β and Δp_A, N_M and N_A must satisfy

$$\frac{4p_{MP}^3(1-p_{MP})}{N_M} + \frac{4p_{AP}^3(1-p_{AP})}{N_A} = \frac{1}{N} \tag{52}$$

when $N_M = N_A = N_{MA}$, we have

$$N_{MA} = N[4p_{MP}^3(1-p_{MP}) + 4p_{AP}^3(1-p_{AP})] \tag{53}$$

Once N_M and N_A are determined, the required inspections can be carried out and the observed rejection probabilities RZE(M1) and RZE(A1) can be computed. As in the previous comparisons, the test procedure is as follows. Accept H_0 if, and only if, Eq. (54) is satisfied:

$$\text{RZE(M2)} - \text{RZE(A2)} \geq C \tag{54}$$

When α is selected equal to β, C is equal to $-\Delta p_A/2$. Continuing the example from the other two cases, assume that: $p_{MP} = 0.80$ and $p_{AP} = 0.80$ so that, $p_{MP}^2 = p_{AP}^2 = 0.64$. Thus manual and machine inspection have the same rejection efficiency. Then for $\alpha = \beta = 0.05$ and $\Delta p_A = 0.05$, and, $N = 4329$ substituted into Eq. (52) results in

$$\frac{0.410}{N_M} + \frac{0.410}{N_A} = \frac{1}{4329} \tag{52A}$$

For $N_M = N_A = N_{MA}$, the number of inspections for each method is 3550. For 50 containers in the Reject Zone, the total number of inspections required becomes 72. With nine inspectors, each inspector is required to perform eight inspections of each Reject Zone container.

XLVIII. TWO SEQUENTIAL MANUAL INSPECTIONS ("ACCEPT IN TWO") VERSUS A SINGLE ALTERNATIVE OR AUTOMATED SYSTEM [(M2-A/2)vs(A1)]

At the present time there are no automated or semiautomated machines that have the selectivity to deliver an economically acceptable single inspection equivalent in quality to that achieved in a manual inspection followed by a cull inspection. This validation procedure is included for completeness. In this case the true average rejection probabilities for the manual and automated methods are, respectively, p_M^2 and p_A. The null hypothesis for this case is

$$H_0 : p_A \geq p_M^2 \tag{55}$$

The alternative hypothesis is

$$H_1 : p_M^2 \geq p_A + \Delta p_A \tag{56}$$

As in the cases previously discussed, the formula for the sample sizes N_M and N_A requires preliminary estimates of p_M and p_A, denoted by p_{MP} and p_{AP}. Then as before, for a selected significance level, N_M and N_A must now satisfy

$$\frac{4p_{MP}^3(1 - p_{MP})}{N_M} + \frac{p_{AP}(1 - p_{AP})}{N_A} = \frac{1}{N} \tag{57}$$

when $N_M = N_A = N_{MA}$ we have

$$N_{MA} = N[4p_{MP}^3(1 - p_{MP}) + p_{AP}(1 - p_{AP})] \tag{58}$$

Again, as in the cases previously discussed, once N_M and N_A are determined, the required inspections can be carried out and the observed rejection probabilities RZE(M1) and RZE(A1) can be computed. As in the previous comparisons, the test procedure is as follows. Accept H_0 if, and only if

$$\text{RZE(A1)} - \text{RZE(M2)} \geq C \tag{59}$$

When α is selected equal to β, C is equal to $-\Delta p_A/2$.

Continuing the example from the other two cases, assume that for $\alpha = \beta = 0.05$ and $\Delta p_A = 0.05$; from Table 1, $N = 4329$. Given $p_{MP} = 0.85$ and $p_{AP} = 0.70$ completes the information needed with which to calculate N_M and N_A. Substituting these values into Eq. (57) results in

$$\frac{0.368}{N_M} + \frac{0.210}{N_A} = \frac{1}{4329} \tag{57A}$$

Setting the number of manual container inspections equal to the number of automated device inspections, $N_M = N_A = N_{MA}$ results in N_{MA} equal to 2502 for the protection level selected.

The preceding examples illustrate procedures for making validation comparisons between a single inspection and a system employing two sequential inspections. A manual two-sequential inspection ("Accept in Two") versus an automated three-sequential-inspection system ("Accept in Three") is also discussed. In each case the controlling probability is that of *rejecting* Reject Zone containers. To satisfy cGMP requirements each of the validations discussed is based on a demonstration of equivalence to manual inspection results.

The "Reject in Two" inspection strategy is used in the majority of all automated inspection systems for injectable products. The analysis of the number of inspections of Reject Zone containers required to demonstrate statistically replicable validations of these systems is developed in a manner similar to those previously discussed. There is one major difference: the controlling probability is that of accepting the Reject Zone containers.

XLIX. "REJECT IN TWO" INSPECTION STRATEGY: ONE MANUAL INSPECTION VERSUS TWO SEQUENTIAL ALTERNATIVE OR AUTOMATED INSPECTIONS [(M1)vs(A2-R/2)]

When a single manual inspection is used as the basis for validating an automated two-sequential "Reject in Two" inspection system, an estimate of the true average inspection probability for each strategy is required. For this case, the true manual inspection probability is p_M and the true automated device inspection probability is $(p_A)^2$. Accordingly, we test the null hypothesis of Eq. (60) versus the alternative hypothesis of Eq. (61):

$$H_0 : (p_A)^2 \geq p_M \tag{60}$$

$$H_1 : p_M \geq (p_A)^2 + \Delta p_A \tag{61}$$

For this case, the formula for the sample sizes N_M and N_A requires preliminary estimates of both p_M and p_A which are again denoted as p_{MP} and p_{AP}. Assuming α, β and Δp_A equal to 0.05 as before results in a value of N in Table 1 of 4329. N and N_A must satisfy

$$\frac{p_{MP}(1 - p_{MP})}{N_M} + \frac{4p_{AP}^3(1 - p_{AP})}{N_A} = \frac{1}{4329} \tag{62}$$

The following validation example uses Reject Zone Efficiencies that are more likely to be encountered in practice. The Reject Zone Efficiency for a manual inspection team is determined to be 0.75. The data available from the automated inspection device is RZE(A2) of 0.76. The controlling probability for the "Reject in Two" inspection strategy is that of accepting Reject Zone containers. This is calculated from

$$\text{RZA(A1)} = \sqrt{1 - \text{RZE(A2)}} \tag{17}$$

It should be recalled that RZA(A1), the probability of accepting Reject Zone containers in the first inspection is a decimal fraction. The sum of RZA(A1) and RZE(A1) is unity:

$$\text{RZE(A1)} = 1 - \text{RZA(A1)} \tag{17A}$$

RZA(A1) in this example is calculated to be 0.3162. Using equal numbers of inspections, $N_M = N_A = N_{MA}$, the number of Reject Zone container inspections is calculated as

$$N_{MA} = 4329[p_{MP}(1 - p_{MP}) + 4p_{AP}^3(1 - p_{AP})] \tag{63}$$

Substituting the data given into Eq. (63) results in a determination that 1850 Reject Zone container inspections are required for a 0.05 significance level validation. With a 50-container Reject Zone container group, four inspectors would inspect the test group containers 10 times each for the validation demonstration. It is to be noted that if the machine results were interpreted as a single inspection to be matched against a single manual inspection, only 32 test group inspections would have been selected. The 0.05 significance level desired for this validation would therefore not have been obtained.

L. "REJECT IN TWO" INSPECTION STRATEGY: MANUAL INSPECTION AND CULL INSPECTION VERSUS TWO SERIAL MACHINE INSPECTIONS [(M2-A/2)vs(A2-R/2)]

The statistical relationships for two sequential manual inspections to validate a system using two automated sequential inspections in an "Accept in Two" strategy has been previously developed: $\alpha = \beta = 0.05$ and $\Delta p_A = 0.05$ have been previously examined. The controlling probability for the manual inspection remains RZE(M1) which is the average *rejection* probability for the manually determined Reject Zone containers. The controlling probability for the "Reject in Two" automated inspection is the average probability for *accepting* Reject Zone containers. As in the preceding example, the data available from the automated device is RZE(A2), the Reject Zone Efficiency at the end of the second machine inspection. The number of Reject Zone container inspections and the number listed in the RZNI tables are separately determined for each type of inspection system. The number of Reject Zone container inspections for each type of inspection system is listed by the controlling probability of the first inspection, RZE(M1) and RZA(A1) respectively. RZA(A1) is computed with Eq. (17) as shown above; an example follows.

A manual inspection team is determined to have RZE2 of 0.70, the automated inspection system as in the previous example has an RZE2 of 0.76. To use the tables, the first inspection controlling probabilities for each method are required. The manual Reject Zone Efficiency, RZE(M1) is the square root of the RZE(M2) value given. After rounding, this is 0.84. The automated device RZA(A1) is determined with Eq. (17) as in the preceding example and is 0.55. Use of Eq. (52) yields 2897 Reject Zone container inspections by each method to achieve the desired 0.05 significance level validation. With a 50-container Reject Zone inspection group, 58 inspections by each method are required.

LI. SUMMARY OF THE STATISTICAL FORMULAS FOR VALIDATION DEMONSTRATIONS

Table 20 lists RZNI requirements for significance levels from 0.01 to 0.10. The RZNI requirements have been normalized to the requirements for a significance level of 0.05. Using Table 21 which summarizes the formulas used to calculate Reject Zone container inspections for these validation demonstrations, Reject Zone container inspections for other significance level validations can be easily calculated with the right-hand column of Table 20.

LII. GENERALIZED INSPECTION MINIMIZATION

A general treatment of the relationship between the minimum total number of reject zone container inspections and the corresponding number of manual and automated inspections will now be presented. The expression for the probability constants for the manual and automated containers inspections shown in Table 21 are generalized as *M* and *A* respectively. Following the form of the five equations listed in Table 21, the generalized equation is

$$\frac{M}{N_M} + \frac{A}{N_A} = \frac{1}{N} \tag{64}$$

In Eq. (64) N must be a positive number. The minimum values of N_M and N_A are, respectively, MN and AN. The total number of manual and automated container inspections, N_T, is

$$N_M + N_A = N_T \tag{65}$$

The ratio of automated to manual container inspections is denoted by K:

$$K = \frac{N_A}{N_M} \tag{66}$$

Substituting for N_A and then N_M their values from Eq. (66) into Eq. (64) and then simplifying, results in Eq. (67) and Eq. (68) respectively:

$$\frac{N_M}{N} = \frac{A + KM}{K} \tag{67}$$

$$\frac{N_A}{N} = A + KM \tag{68}$$

Adding these equations yields the total number of normalized container inspections:

$$\frac{N_T}{N} = [A + KM]\left[\frac{K + 1}{K}\right] \tag{69}$$

Differentiating Eq. (69) with respect to K to determine a minimum and then simplifying gives Eqs. (70)–(72). These equations express the minimum value of the total number of inspections, N_T, and the number of manual and automated inspections that occur at that point.

$$\frac{N_M^{**}}{N} = M\left(1 + \sqrt{\frac{A}{M}}\right) \tag{70}$$

$$\frac{N_A^{**}}{N} = A\left(1 + \sqrt{\frac{M}{A}}\right) \tag{71}$$

$$\frac{N_T^{**}}{N} = \left(\sqrt{M} + \sqrt{A}\right)^2 \tag{72}$$

Equations (70)–(72) use the generalized statistical parameters defined in Eq. (64). These equations show that the minimum total number of inspections and the number of manual and automated inspections that occur at that operating point all vary with the ratio of the manual and automated statistical constants. These equations have been used in the curves shown in Fig. 44 (see page 351).

The data from which Fig. 44 has been prepared incorporates the results of a standard manual inspection followed by a "Reject in Two" automated system inspection. The manual RZE(M1) is 0.85; the automated system RZE from the second of the two sequential inspections, RZE(A2) is 0.90. The RZA(A1) is calculated to be 0.3162.

Figure 44 graphically illustrated the existence of a minimum in the number of total inspections and the corresponding variation of N_A and N_M, the number of automated and manual inspections to achieve validation of a "Reject in Two" inspection system with 0.05 significance. The intersection of the N_M and N_A curves is $K = 1$ and defines the point at which N_M is equal to N_A. This point, at which there is an equal number of inspections in both methods, was used in all previous versions of the RZNI tables.

Figure 44 was derived from a specific validation demonstration of an automated "Reject in Two" inspection system against a single manual inspection. The matched RZE in this demonstration was 0.88, 356 Reject Zone container reinspections are required by each method for a 0.05 significance level validation.

The computations for Fig. 44 commence with performance data estimates for each of the systems involved. The possible use of additional automated inspections to reduce the amount of manual inspection labor and time required for a validation demonstration requires a different approach. An analysis that achieves this goal will now be developed and demonstrated.

A concern in any validation demonstration is control of the projected costs. To achieve a cost-effective inspection system validation, the relative costs of container inspection on the systems to be compared must be reviewed. Once the relative costs are determined, selection of the relative number of inspections by the standard manual inspection and the alternate manual, semiautomated or fully automated inspection system can be determined.

LIII. MINIMIZATION OF THE TOTAL NUMBER OF REJECT ZONE CONTAINER INSPECTION REQUIREMENTS

Where a manual system is to be compared to another manual system or an alternate inspection system with comparable throughput, minimization of the total number of inspections is appropriate. When the system to be validated is a high-speed automated inspection system with substantially lower container inspection costs than in the baseline human inspection system, minimization of the total number of manual inspections is desirable. Equation (64A) is a modification of Eq. (64) to clarify the relationships of both types of inspection to the RZNI significance level constant:

$$\frac{1}{N_M/M} + \frac{1}{N_A/A} = \frac{1}{N} \tag{64A}$$

Rewritten in this manner, Eq. (64A) is seen to be a sum of reciprocals in which N, N_M/M and N_A/A are all positive numbers. For physical realizability, N_M/M and N_A/A must be at least equal to N. If either term approaches N in value, the balancing increase in the second term approaches infinity.

The minimum number of manual inspections for physical realizability is determined by considering N_A in Eq. (64A) to be so large that the value of the N_A term approaches zero. Under these conditions, the minimum N_M/N ratio, $N_M{}^*/N$ is determined from

$$\frac{N_M^*}{N} = M \tag{64B}$$

The minimum value for the N_A/N ratio, $N_A{}^*/N$ is determined from

$$\frac{N_A^*}{N} = A \tag{64C}$$

Both of these asymptotic ratio limits can be seen in Fig. 44.

Assuming that both N_A/A and N_M/M are within their limiting values, an increase in N_A/A can be calculated to offset a reduction in N_M/M. In Eq. (73) a multiple value of N is assigned to N_A/A.

$$\frac{N_A}{A} = DN \tag{73}$$

Substituting for N_A/A in Eq. (64A) its value from Eq. (73) and then solving for N_M/N gives

$$\frac{N_M}{N} = M\left(\frac{D}{D - 1}\right) \tag{74}$$

The total number of inspections, N_T, is the sum of the number of manual and automated inspections as shown in Eq. (65). This total number of inspections in terms of D is obtained by substituting Eq. (73) and Eq. (74) into Eq. (65). The simplified results are

$$\frac{N_T}{N} = AD + \frac{MD}{D - 1} \tag{75}$$

In the sum of the reciprocals shown in Eq. (75) as D approaches 1, the value of the alternative inspection fraction term approaches that of N_T/N and the value of the manual fraction term approaches infinity. Dividing both sides of Eq. (75) by A results in the normalized form

$$\frac{N_T}{(NA)} = D + \frac{(M/A)D}{D - 1} \tag{75A}$$

When $M = A$, Eq. (75A) reduces to

$$\frac{N_T}{(NA)} = D + \frac{D}{D - 1} \tag{75B}$$

In this application D is limited to real positive values. The limitation on the range of D imposed by the form of Eq. (75B) is examined in Fig. 45. The $N_A/(NA)$ contribution is seen to vary linearly. The contribution of $N_M/(NA)$ is seen to be negative in the region below 1 and to go through infinity at 1. As D increases indefinitely, the value of $N_M/(NA)$ approaches 1. When the effect of M/A ratio is included, the minimum value of $N_M/(NA)$ approaches M/A. An operating margin on $N_M/(NA)$ is established by setting a minimum value of D. For an operating margin of 20%, 10% and 5% the value of D is respectively 6, 11 and 21. The normalized total of Reject Zone container inspections, $N_T/(NA)$ is seen to go through a sharp minimum at $D = 2$ and to increase sharply below as D approaches 1. The intersection of the curves for $N_A/(NA)$ and $N_M/(NA)$ define D_{MA}, the point where the number of Reject Zone container inspections are equal, $N_{MA}/(NA)$. This is the number of Reject Zone container inspections listed in the RZNI tables. Selection of a pair of N_M and N_A values for a validation demonstration is accomplished by selecting a value of D. Figure 45 shows that, when minimization of manual inspection labor is to be considered, a value of D equal to or greater than the value at which N_{MA} results is required.

The K parameter used in the first analysis of the inspection ratio problem and D are linearly related. The minimum value of N_T/N remains as calculated in Eq. (72). The normalized form of Eq. (72) in Eq. (72A) is provided by dividing each side of the equation by A:

$$\frac{N_T^{**}}{(NA)} = \frac{\left(\sqrt{M} + \sqrt{A}\right)^2}{A} \tag{72A}$$

Normalized expressions for the ratio of manual and automated Reject Zone container inspections at the minimum total number of inspections result when each side of Eq. (70) and (71) are divided by A. The results are

$$\frac{N_M^{**}}{(NA)} = \frac{M}{A}\left(1 + \sqrt{\frac{A}{M}}\right) \tag{70A}$$

$$\frac{N_A^{**}}{(NA)} = 1 + \sqrt{\frac{M}{A}} \tag{71A}$$

The value of D at which the minimum total number of inspections occurs is determined by differentiating N_T/N with respect to D in Eq. (75) and solving for D^{**}. The simplified results are

$$D^{**} = 1 + \sqrt{\frac{M}{A}} \tag{76}$$

In practical validations A and M are defined by the strategy and efficiency of the automated system being validated and the benchmark capability of the manual inspection system. Selection of operating conditions to minimize the amount of manual inspection labor for a validation requires additional information on the way that variation of M and A affect the test conditions. A graphical analysis is helpful in this investigation. The development of this graphical analysis is simplified by the use of normalized expressions for $N_M/(NA)$, $N_A/(NA)$ as well as the normalized expression for $N_T/(NA)$ in Eq. (75).

Equation (73A) is a normalized expression for N_A:

$$\frac{N_A}{(NA)} = D \tag{73A}$$

The normalized form of $N_M/(NA)$ in Eq. (77) is provided by dividing each side of Eq. (74) by A:

$$\frac{N_M}{(NA)} = \frac{(M/A)D}{D - 1} \tag{77}$$

The value of D and D_{MA} at the crossover point where the number of manual and automated inspections is equal, N_{MA}, is the value at which the RZNI tables have been calculated. This point is determined by equating the expressions for manual and automated inspections in Eqs. (73A) and (76). The simplified results are

$$D_{MA} = 1 + \frac{M}{A} \tag{78}$$

When $M = A$, $D_{MA} = 2$. Equation (73) defined D equal to $N_A/(NA)$. D_{MA} in Eq. (78) similarly defines $N_{MA}/(NA)$ as the equal inspection ratio for both methods as

$$\frac{N_{MA}}{(NA)} = 1 + \frac{M}{A} \tag{78A}$$

The ratio N_M/N_{MA} is calculated as the ratio of Eq. (73A) to Eq. (78A). Similarly, N_A/N_{MA} is calculated as the ratio of Eq. (76A) to Eq. (78A). The results shown in Eqs. (79) and (80) can be graphed to provide a visual analysis of the effect of the variation in D on the ratio of manual and automated inspections used.

$$\frac{N_M}{N_{MA}} = \frac{\frac{M}{A}}{1 + \frac{M}{A}} \frac{D}{D - 1} \tag{79}$$

$$\frac{N_A}{N_{MA}} = \frac{D}{1 + \frac{M}{A}} \tag{80}$$

Equations (79) and (80) both show that the ratio of M/A is an essential part of the possibility of any planned reduction in manual inspection requirements. Equation (79A) expresses the effect of the M/A ratio upon the N_M/N_{MA} determination.

$$f\left(\frac{M}{A}\right) = \frac{\frac{M}{A}}{1 + \frac{M}{A}} \tag{79A}$$

Substituting the value of D_{MA} into Eq. (76) evaluates N_T/N at the crossover point. The results are

$$\left(\frac{N_T}{N_A}\right)_{MA} = 2\left(1 + \frac{M}{A}\right) \tag{81}$$

A. Validation Choices

Figures 44–46 all show that there are many pairs of manual and automated inspection replications that can be used to obtain the same significance level validation demonstration. For normal validation demonstrations in which the inspection cost of both methods is approximately equal and adequate time is available, an equal number of inspections is usually selected. For validation demonstrations of high-speed automated inspection systems, selection of a reduced number of manual inspections can result in faster, lower-cost validations. As these three figures show, selection of a reduced manual inspection validation requires a compensating increase in the number of automated inspections that are used. These figures also show that for optimum results in minimizing manual inspection, operation at a D value that is equal to or higher than D_{MA} is required.

B. Validation Ratios

Table 12 summarizes the statistical estimation parameter values M and A by the number of sequential inspections employed and by the controlling probability of the first inspection. These parameters, included in the equations of Table 21 are

$$p_1 = p_1(1 - p_1) \tag{82}$$

$$p_2 = 4p_2^3(1 - p_2) \tag{83}$$

$$p_3 = 9p_3^5(1 - p_3) \tag{84}$$

where the subscript indicates the number of sequential inspections.

The crossover point and the minimum total number of inspections are seen in Fig. 46 to coincide at $D = 2$. Also shown in this figure is the locus of the minimum total number of inspections of $N_T/(NA)$, as D and M/A vary. Use of the minimum total number of reject container inspections is beneficial when fractional values of M/A are encountered. The convenience and utility offered by use of the crossover number of inspections for M/A ratios greater than 1 is evident.

Reduction of manual inspection with a resulting increase in alternative or automated inspections is accomplished by selection of a value for D greater than D_{MA}. The relative benefit of an increase of D to a D' greater than D_{MA} encounters an asymptotic limit, the limiting value of $N_M/(NA)$ is M/A. The effect of D/D_{MA} ratios that provide 5%, 10% and 20% margins to the asymptotic limit of $N_M/(NA)$ can be examined by substitution in Eq. (76). The effect of these margins is examined in Table 13.

The relative effect of using the RZNI data directly, compared to the use of a reduced number of manual inspections, can be gauged against N_{MA} in Table 13. In each data group the ratio of N_M/N_{MA} is a direct measure of the possible improvement, and the ratio of N_A/N_{MA} a direct evaluation of the increased number of alternative inspections required to offset the reduction in manual inspection requirements.

LIV. RELATIONSHIP OF PRESENT AND EARLIER RZNI TABLES

The RZNI tables are calculated for an equal number of manual and alternative inspections, N_{MA}. The data in the published tables is accessed in a two-dimensional listing against the controlling probability of the first inspection in each method. New flexibility is provided by the capability to determine a reduction in manual inspections and a compensating increase in the number of alternative or automated inspections that are employed in the validation demonstration. The use of Tables 12 and 13 inject flexibility into the design of validation demonstrations by facilitating the shift of inspection time and labor cost from the manual to the automated inspection system. In the most widely used parameter range for the validation of "Reject in Two" automated systems, reductions of 30% to 50% in the number of manual reinspections of Reject Zone containers are now possible.

A. Higher Accuracy Inspection Ratio Adjustment

The manual to automated inspection ratio adjustment available from Table 13 will be found adequate for most use. For those occasions when more adjustment accuracy is required, the following sequence of calculations is used.

1. Calculate the M/A ratio.
2. Calculate $N_{MA}/(NA)$, the equation inspection value from

$$\frac{N_{MA}}{(NA)} = 1 + \frac{M}{A} \tag{78A}$$

3. Calculate the effect of the M/A ratio on the inspection ratio adjustment with $f(M/A)$ using

$$f\left(\frac{M}{A}\right) = \frac{M/A}{1 + M/A} \tag{79A}$$

4. Calculate $N_M/(NA)$ for 20%, 10% and 5% margins by multiplying the results of step 2 by 1.2, 1.1 and 1.05, respectively.
5. Calculate the balancing increase in $N_A/(NA)$ from

$$\frac{N_A}{(NA)} = \frac{D}{1 + M/A} \tag{85}$$

For the 20%, 10% and 5% operating margin on D substitute 6, 11 and 21, respectively, to determine the balancing increase in $N_A/(NA)$.

B. RZNI Tables: An Overview

The RZNI tables have been calculated for all current automated and semiautomated inspection strategies. The combinational validation demonstration requirements covered are listed below.

Tables	RZE/RZA range	Inspection sequence
RZNI-1.0 to RZNI-1.2	0.70–1.00	M1/A1
RZNI-2.0 to RZNI-2.9	0.00–1.00	M1/A2, M2/A1
RZNI-3.0 to RZNI-3.9	0.00–1.00	M2/A2
RZNI-4.0 to RZNI-4.2	0.70–1.00	M2/A3

All RZNI tables have been caluclated for a number of Reject Zone container inspections to obtain 0.05 significance level validations of the combinations of alternative or automated inspection techniques listed. Both the manual single inspection and the manual first-and-cull inspection are available as the human inspection performance benchmarks. The RZNI tables are labeled for use with the controlling probability of the first inspection: RZE(A1) where the controlling probability is that of *rejecting* Reject Zone containers and RZA(A1) where the controlling probability is that of *accepting* Reject Zone containers. In the shorthand notation used for validation pairs, M refers to manual inspections, A to alternative or automatic system inspections, the numeric following M or A describes the number of sequential inspections employed.

Cost-effective validation demonstrations for systems of comparable inspection cost are performed using an equal number of inspections for both methods. This option results from the use of the number of Reject Zone container inspections listed in the RZNI tables.

When high-speed automated inspection systems, whose inspection costs can be substantially lower than manual inspection are to be validated, capability to respond to this new situation is important. Flexibility to choose other pairs of the number of manual and automated Reject Zone container reinspections can result in cost and time savings for the validation demonstration. The savings depend on the relative inspection security and the number of sequential inspections used in both inspection methods. In each case in which the number of manual reinspections is reduced, the number of alternative inspections is increased to obtain the required significance level for the validation demonstration. The following set of worked examples will extend these remarks.

LV. EXAMPLES OF RZNI TABLE APPLICATIONS

Example 1. Single manual inspection versus single automated inspection, (M1) vs (A1):

$$\text{RZE(M1)} = \text{RZE(A1)} = 0.80, \quad \text{RZN} = 50$$

Selecting the equal number of inspection option, RZNI-1.3 lists 1385 Reject Zone container reinspections by both methods. RZNI-1.0 shows that N_{MA}, the number of Reject Zone inspections required for each method is 1385. Dividing RZNI by 50 Reject Zone containers, the number of test group inspections is equal to 27.7, which after rounding is 28. This number of test group inspections required for validation can be satisfied with four inspectors each inspecting the test group seven times.

Example 2. Two sequential manual inspections with an RZE(M2) equal to 0.80 versus two sequential alternative inspections using "Accept in Two" inspection strategy with RZE(A1-A/2) equal to 0.92: (M2-A/2) vs (A2-A/2). The validation demonstration is based on the worst-class inspection security achieved by each method, the inspection security after the second inspection. Table entry, however, is based on the RZE after the first inspection. Since the "accept" strategy is used, the first inspection RZE for each method is the square root of the second inspection Reject Zone security. The values for both methods are, respectively, 0.89 and 0.9650 Reject Zone containers that are available in the test sample group.

$$\text{RZE(M1)} = 0.87, \quad \text{RZE(A1)} = 0.92, \quad \text{RZN} = 50$$

Both inspection methods have closely comparable costs, therefore the equal number of inspection option is selected and RZNI-3.8 is used.

$$N_{MA} = 2{,}561$$

With the 50-container Reject Zone test group, dividing RZNI by 50 results in 51.22 which is rounded up a total of 52 inspections of the test group for this validation demonstration.

Example 3. Two sequential manual inspections using an "Accept in Two" inspection strategy versus a single-inspection automated system: (M2-A2) vs (A1).

$$\text{RZE(M2)} = 0.82, \quad \text{RZE(A1)} = 0.85, \quad \text{RZN} = 50$$

For RZNI table entry, RZE(M1) is required; RZE(M1) is the square root of RZE(M2) and is 0.906. Estimating RZE(M1) conservatively, it is rounded down to 0.90. Selecting the equal number of inspections option, RZNI-2.8 is used (Note: RZNI-2.8 is used with reversed titles for this determination). RZNI is located at the intersection of 0.90 in the RZE(M1) column and 0.85 in the RZE(A1) column. RZNI is 2108. Dividing RZNI by an RZN of 50 determines N_{MA}. The calculated N_{MA} is 42.16, which is rounded to 42 test group inspections.

Example 4. A single manual inspection is used to validate an automated system that uses a "Reject in Two" inspection strategy: (M1) vs. (A2-R/2).

$$\text{RZE(M1)} = 0.84, \quad \text{RZE(A2)} = 0.88, \quad \text{RZN} = 50$$

The controlling probability for a "Reject in Two" inspection strategy is the probability of accepting Reject Zone rejects. RZA(A1), the probability of accepting Reject Zone rejects in the first inspection, is required for use of the tables.

$$\text{RZA(A1)} = \sqrt{1 - \text{RZE(A2)}} = \sqrt{1 - .88} = \sqrt{.12} = .3464 \tag{17}$$

Note: For conservative rounding, calculated values of RZE are rounded down, calculated values of RZA are rounded up. The RZNI table entry data is

$$\text{RZE(M1)} = 0.84, \quad \text{RZA(A1)} = 0.35$$

RZNI-2.3 is used; RZNI = 1,094.

$$N_{MA} = 1065$$

Dividing RZNI by the 50 RZN containers in the test group results in 21.3 inspections, which is rounded up to 22 inspections by each method.

To investigate a reduction in the number of manual reinspections required, the first step is the determination of the statistical estimation parameters for each inspection method in Table 12. The value of M is determined in the single inspection column against an RZE(M1) of 0.84; it is 0.1344. The A parameter is determined under the heading "two serial inspections" against the RZA(A1) of 0.35; it is 0.1115. The M/A ratio is 1.21. Table 13 offers three reduction options. Using the listed M/A ratio of 1.20, the 20% margin on N_M provides a reduction to 65% of the number of manual inspections compared to direct use of the RZNI tables. This reduction in manual inspections requires a compensating increase to 273% of the number of automated inspections. When a margin of 10% on N_M is selected, a reduction of 60% of the RZNI table value correlates with a compensating automated inspection increase of 500% over the inspections that would have resulted from direct use of the RZNI tables. The selection of the 5% margin on $N_M/(NA)$ reduces the number of manual inspections to 55% of the N_{MA} value listed in the tables but requires an increase in the number of automated inspections by a factor of 9.55. Which option is best will depend on varied circumstances. The important conclusion is that such a flexibility now exists.

Example 5. Two sequential manual inspections are used to validate an automated system that uses a "Reject in Two" inspection strategy: (M2-A/2) vs. (A2-R/2). The RZE(M2) = 0.84 and the RZE(A2) = 0.88 as in the previous example. RZE(M1) is the square root of 0.84 and is, after rounding down, 0.91. RZA(A1) = 0.35, after rounding up as in the previous example. RZN = 50. RZNI-2.2 shows an RZNI of 1657, division by the 50-container RZN group results in 33.14 which is rounded up to 34 inspections. The two-serial inspection heading in Table 12 lists $M = 0.2713$ for an RZE(M1) = 0.91 and an $A = 0.1115$ for an RZA(A1) = 0.35. The M/A ratio is 2.43. Using the 2.40 M/A ratio and a 5% margin for $N_M/(NA)$ a reduction to 74% of the original RZNI value is shown for an increased multiple of 6.18 automated inspections.

Example 6. Two sequential manual inspections ("Accept in Two") with an RZE(M1) of 0.85 are used to validate a three sequential inspection ("Accept in Three") automated inspection system with an RZE(A1) of 0.96: (M2-A/2) vs. (A3-A/3). RZN = 50. RZNI-4.3 lists an RZNI of 2866. The calculated results after dividing through by the 50-container RZN is 57.32 which is rounded up to 58 test group inspections. M and A, from Table 12 are, respectively, 0.3684 and 0.2935. The M/A ratio is 1.26.

Using the 10% $N_M/(NA)$ margin option in Table 13 for an M/A ratio of 1.20 shows a reduction in manual inspections of 60% of the RZNI listed value with a fivefold increase in the number of automated inspections required. A reduction from 58 manual inspections of the test batch down to 35 has been realized with a corresponding increase in automated inspections to 290.

A. Manual Inspection Reduction Choices

The selection of the number of manual Reject Zone container inspections for validation of automated inspection systems now has four possible choices. The first is use of the unmodified RZNI table values. This choice can be made when the effective cost of both methods is approximately equal and there is little time pressure.

The remaining three choices are in the amount of manual inspection reduction that should be made. This choice involves not only the margin over the minimum $N_M/(NA)$, but required also the return available for the increased number of automated inspections that are required as the reduction in manual inspections approaches its asymptotic limit. With so many variables present, recourse to a judgment call is reasonable.

B. Objective Standards

A major obstacle that must be overcome before equivalent inspection security is attained is that there is no 1:1 relationship between the accept/reject decision results of the human eye/brain combination and the variety of automated nondestructive or destructive particulate tests now used. The automated inspection systems currently available evaluate the indirect "optical signature" of a particle against stored acceptability criteria. Nondestructive physical measurement of particulates in sealed containers is required to improve our understanding and use of the present manual inspection techniques and to provide validation support methodology for high-speed, automated inspection systems. Toward this end, the potential of far-field holography to achieve direct, nondestructive physical measurement of particulates in sealed, sterile containers appears attractive. With a reconstructed hologram, the capability to insert a microscope into the volume of a sealed parenteral product is brought into being. The present limitations of depth-of-focus imposed by microscope or camera lens systems are by-passed. Holography can be used to calibrate actual production line rejects within intact containers. The intact containers with typical production line rejects, particulates whose size, shape, and quantity were physically determined, could then be used as transfer standards. These transfer standards could be used to

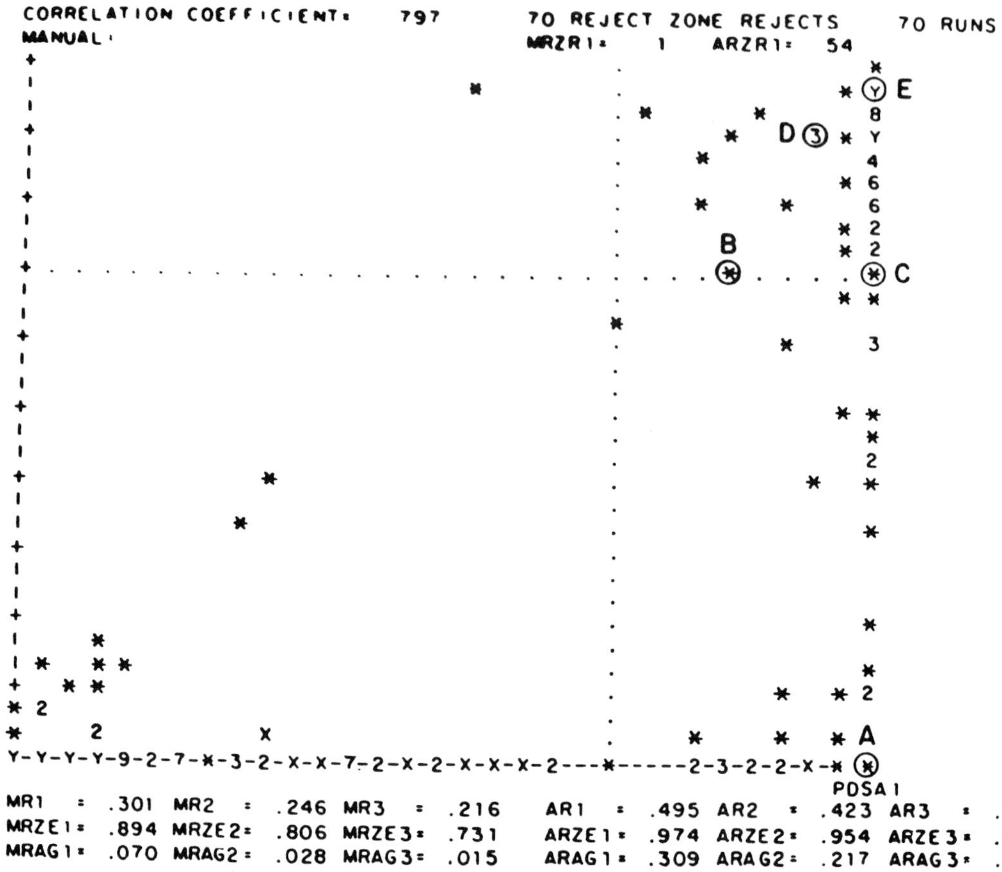

Figure 54 Computer-generated correlogram showing manual and automated system rejection probabilities for five production samples selected for hologrammic analysis. MR1, MR2, MR3, AR1, AR2, and AR3 are shorthand labels for the average reject rate following the first, second and third manual and automated inspection respectively. Similarly MRZE1, MRZE2, MRZE3, ARZE1, ARZE2, ARZE3 are the Reject Zone Efficiencies for the manual and automated systems for the first, second and third inspection respectively. MRAG1, MRAG2, MRAG3, ARAG1, ARAG2 and ARAG3 identify the false reject rate in the Accept and Gray Zones for the manual and automated systems after first, second, and third inspection respectively.

efficiently shape the accept/reject decision of high-speed inspection machines as an alternative to the use of the visual probabilistic standard. The work discussed is the result of a joint project between R&D Engineering at Schering and the Institute of Optics at the University of Rochester.

In the project, far-field transmission holography was used for the inventory and measurement of particulates in sealed, sterile containers. The physical count and measurement of particulates thus achieved was related to the probabilistic rejection data that resulted from Working Standard Test Method validation of a high-speed automated ampule inspection system whose development at Schering has been reported in detail in the *Journal of the Parenteral Drug Association.* The manual inspection data used as the basis of the validation were the result of the standard Schering particulate inspection using a 3-diopter, 5-inch magnifying lens, a 17-second paced inspection with a white-and-black background and a 225-foot-candle illumination. With this technique, direct determination of the detectability of particulates as a function of size and shape was achieved. The Reject Zone Efficiency (RZE) and the undesired reject rate in the Accept and Gray Zones (RAG) were, with holography, translated into counts of physical shapes.

To ensure that the samples selected were of maximum interest, the computer-generated correlagram shown in Fig. 54 was utilized. This computer-generated correlgram is a convenient way of representing the comparison of two probabilistic inspection systems. The manual inspection probability axis is vertical; the PDS/A inspection rejection probability axis is horizontal. The two dotted lines indicate 70% rejection probability limits for each system. A single container at a data point is shown by an asterisk; quantities from 2 to 9 at a data point are shown by the numeric. A "Y" at a data point indicates a quantity of containers greater than nine. Just below the data heading, we can read that there are 70 Reject Zone rejects and that the manual inspectors detect only one of them with 100% probability. The automated system, by comparison, detects 54 of the 70 with 100% probability. Comparing the security of the second manual inspection versus the third automated inspection, we compare RZE(M2) to RZE(A3-A/3) which is listed directly below the histogram. The security achieved in two manual inspections is seen to be 0.806; the automated system yielded a higher degree of security as shown by the 0.936 score, 13% higher for the automated system. The inspection discrimination as measured by RAG(M2) versus RAG(A3) is seen to be 0.028 versus 0.187. This indicates that the undesired reject rate on the automated machine when measured with the reject-seeded test group is approximately sixfold higher. Since in normal production the rejects are a small fraction of the containers produced, this ratio has small economic impact.

LVI. HOLOGRAMMIC EVALUATION OF AUTOMATED AND VISUAL INSPECTION PERFORMANCE

Five data points on this correlagram were selected for hologrammic investigation to help establish a link between physical measurements and the rejection probability data available. They were shown as A through E in the correlagram of Fig. 54 and their relative manual and automated detection probabilities are listed in Table 22. It should be noted that the manual detection data include the use of a 3-diopter lens with an average magnification factor of 1.5.

Figure 55 is a photograph of the Schering Optics Laboratory setup to generate ampule holograms. The windowed mechanism topped with the cylindrical motor seen in the center of this photo is the unit that spins the ampule in an optical index-matched liquid cell to generate particulate movement. The cell, in large measure, neutralizes the container's optical effects, thus making possible the hologram. The spin motor is a low-rotary-inertia servo motor used to

Table 22 Relative Detectability of Five Selected Samples in Both Manual and Automated Inspection Systems

	Detection probability		
Particle	Manual	Automated	Remarks
A	0%	100%	Subvisible
B	70%	100%	Reject Zone boundary
C	70%	84%	Reject Zone boundary
D	90%	90%	"Bad-bad" reject
E	97%	100%	"Bad-bad" reject

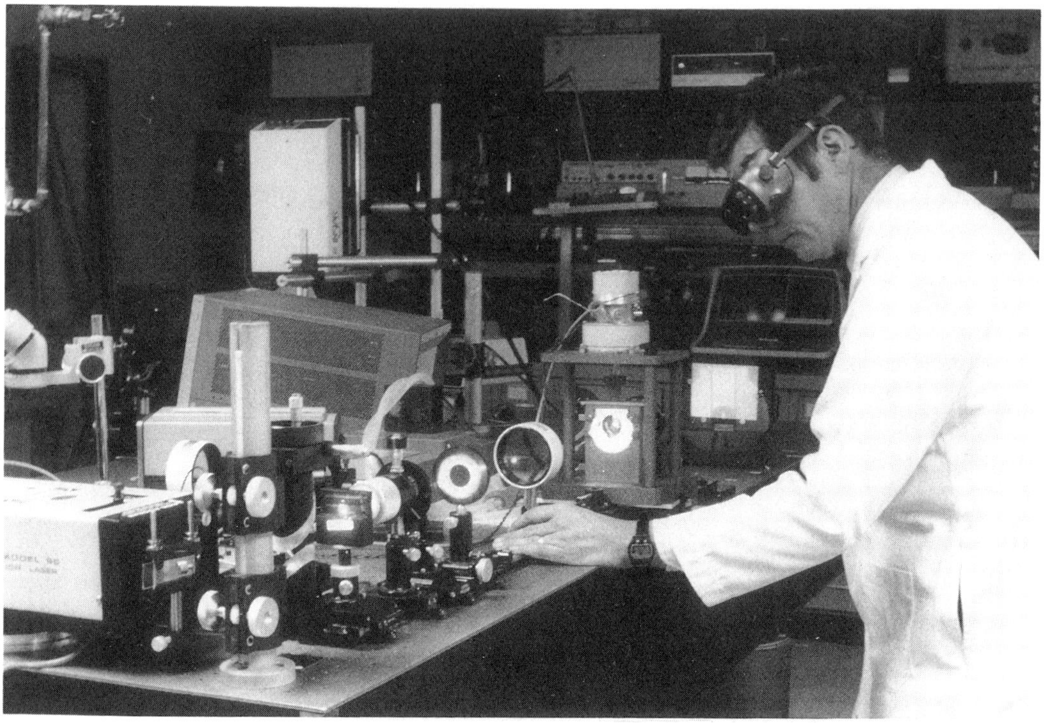

Figure 55 Schering Optics Laboratory setup to generate ampule holograms.

achieve the fast starts/stops required to generate the needed particulate movement. At the far right is a ground glass in the position of the holographic plate. The inverted image of the ampule from meniscus through ampule base is clearly seen on the ground glass, which substitutes here for the 4-inch by 5-inch holographic plate. The hologram will be seen to capture the entire contents of the ampule with all particulates recorded. The entire ampule volume can be sharply focused in the reconstructed hologram. Particulates down to 5 μm in size can be located and accurately measured in the reconstruction.

The details of hologram formation are more readily seen in the block diagram of Fig. 56.

Coherent light from an argon ion laser passes through a collimator/spatial filter assembly before entering a liquid cell with optically flat end windows in which the ampule is held, spun, and quickly stopped. The recording lens and film placement for the hologram are seen at the extreme right. The advantages of the hologram become apparent as we proceed through the block diagram in Fig. 57 which describes reconstruction of the hologram image.

Figure 57 shows that in reconstruction the hologram is illuminated using the same laser and collimator/spatial filter assembly as in recording. The reconstructed volume appears behind the hologram plate. As shown here, two choices of magnification were considered: 1:1 imaging, which yields particulates in real size and shape, or 5:1 magnification. The use of fivefold magnification minimized film resolution requirements and facilitated secure measurements to at least 5 μm. The fivefold magnification in the *XY* plane bears with it squared magnification (25-fold) in the *Z*-axis direction, i.e., into the container depth. This should be acceptable if most particulates are roughly spherical in shape. Particulate search in the reconstructed volume was carried out be transversing the volume with a bare vidicon, a TV camera tube whose output was displayed on a TV monitor as shown. For clarity, the TV camera in this block diagram has been shown external

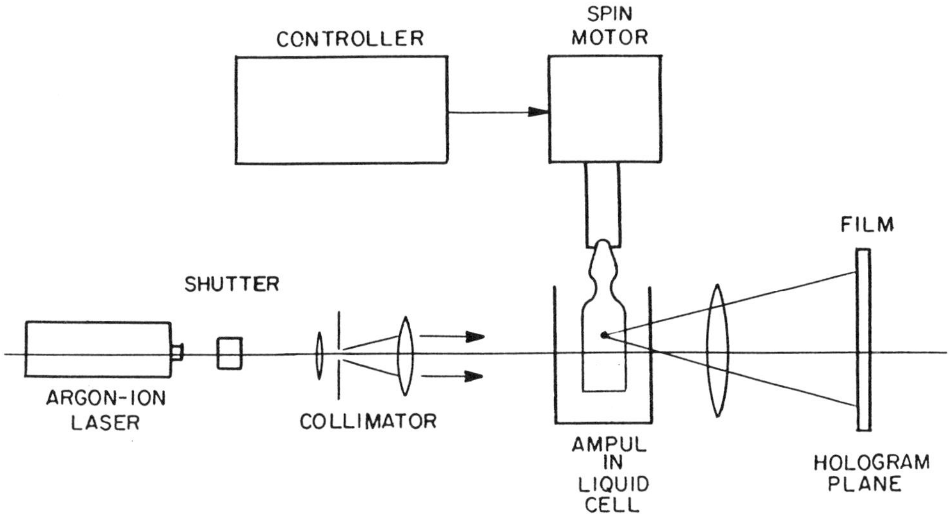

Figure 56 Schematic description of hologram formation.

to the reconstruction volume. Total particulate magnification in this system is 150. Fivefold magnification is due to the hologram and 30-fold to the vidicon/monitor display. The position of sharpest focus and thus true magnified size for a particle is relatively easy to determine.

Figure 58 is a photo of the TV monitor that was used to view hologram sections displaying the contents of an ampule with 100-μm standard spheres. Two sharp and one partially imaged 100-μm standard are seen in this hologram section; also seen is the lens-less imaging vidicon. The spot at the lower right is a chance reflection of an indicator light onto the CRT face.

Figures 59 and 60 are photographs of the calibration of the holographic system with precision latex spheres. These and all other holographic photos are at constant 150× magnification.

The precision 20-μm spheres in the hologram seen in Fig. 58 measure 3 mm on the

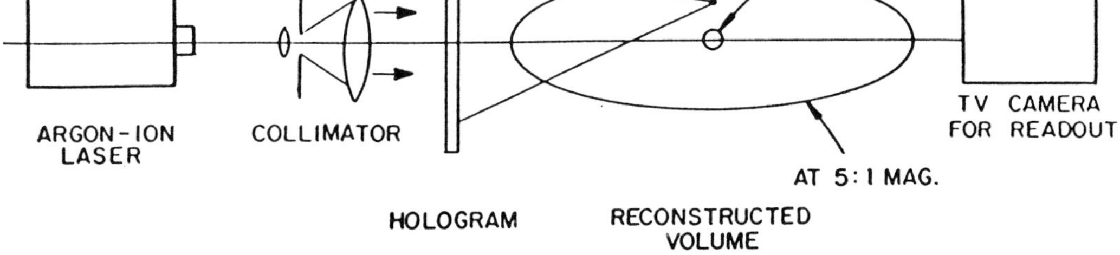

Figure 57 Schematic description of hologram reconstruction.

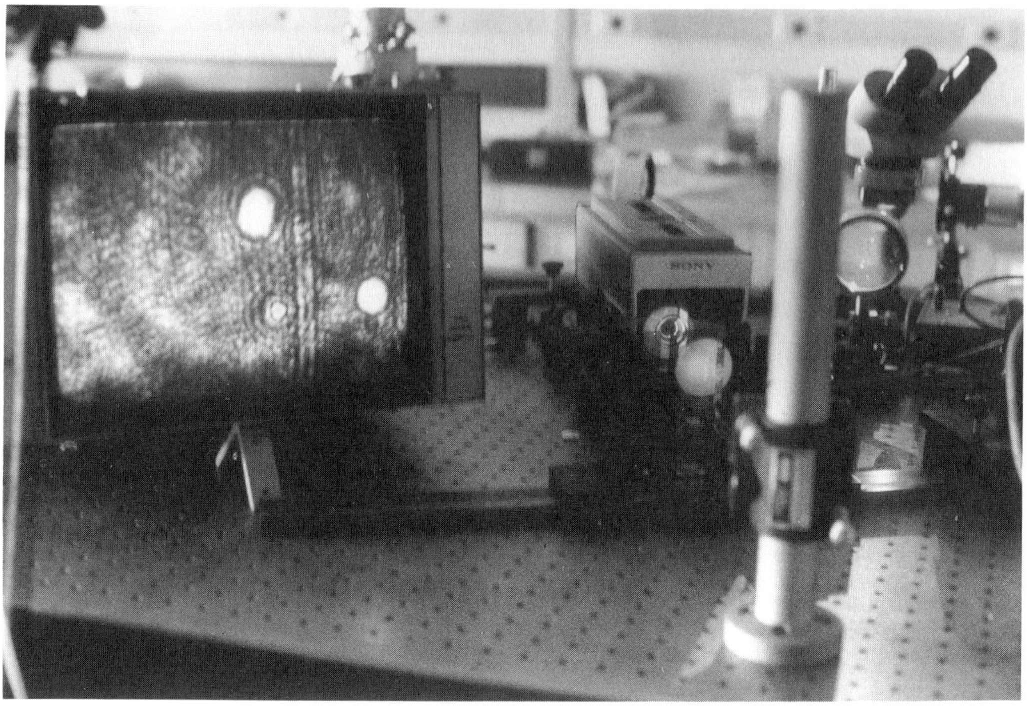

Figure 58 Video monitor used to display hologram sections. On the screen are two sharp and one partially imaged 100-μm latex spheres.

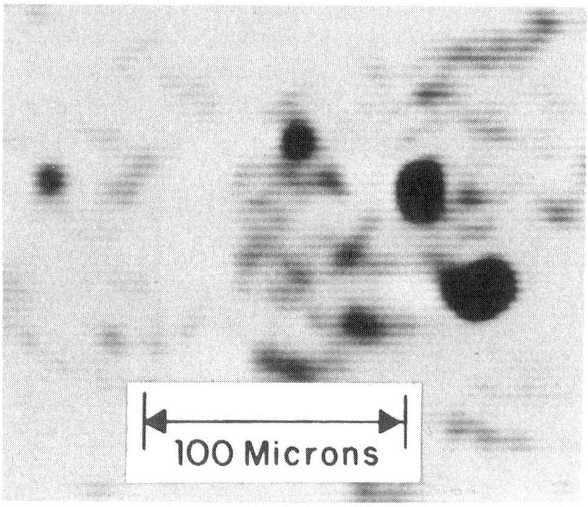

Figure 59 Calibration of the system was achieved with precision polystyrene spheres. Shown here are two "20-μm spheres." Also seen is some contamination at the 5-μm level. Net system magnification is 150.

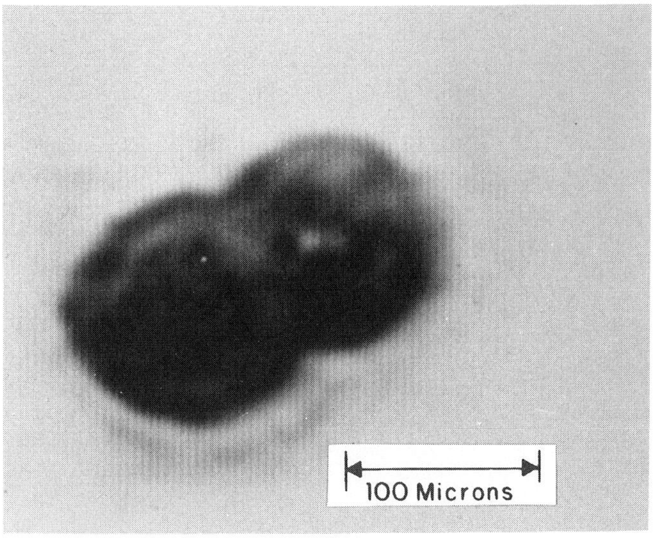

Figure 60 The system was also calibrated with the 100-μm spheres shown here. System linearity and magnification factor of 150 were confirmed. Selected for display is a hologram section showing merged 100-μm spheres.

TV screen and occupy 10 TV lines. The noise level is below one TV line. Accurate measurement to at least 5 μm is possible.

For interest, Fig. 59 has also been selected to show that occasionally the "standard" spheres are not quite standard and that some particulate contamination at the 5-μm size is within the container.

The hologram section in the photo of Fig. 60 shows the adhesion of two 100-μm calibration spheres. The 100-μm diameter translates to a video display of 15 mm and confirms the fact that this is a linear magnification system. Surface details of the spheres can be seen in this photo, which provides support for the view that an image of the particle under analysis contains substantially more information than light-beam interruption measurements or conductive channel modulation measurements make available.

The next group of five figures are from holograms of the selected joint rejection probability containers listed in Table 22 and shown circled on the correlagram of Fig. 54. These particles are all production rejects, not ideal spheres. The reported dimensions are the maximum particle dimensions.

A. Particle A: Manual Detection 0%, Automated Detection 100%

The low-contrast particle shown in Fig. 61 is 35 μm in diameter and appears to be at or just below the threshold of detectability in the standard manual inspection described earlier. The manual detection probability for this particle is zero. This is the first direct experimental evidence under actual inspection conditions of the limits of visual inspection. This particulate has been shown to have a detection probability of 100% in the correlagram of Fig. 59, in which the automated ampule inspection system developed at Schering was evaluated. (Note: This automated system is produced commercially as the PD-100 Automated Particulate Detection System Electro-Nucleonics, Inc.)

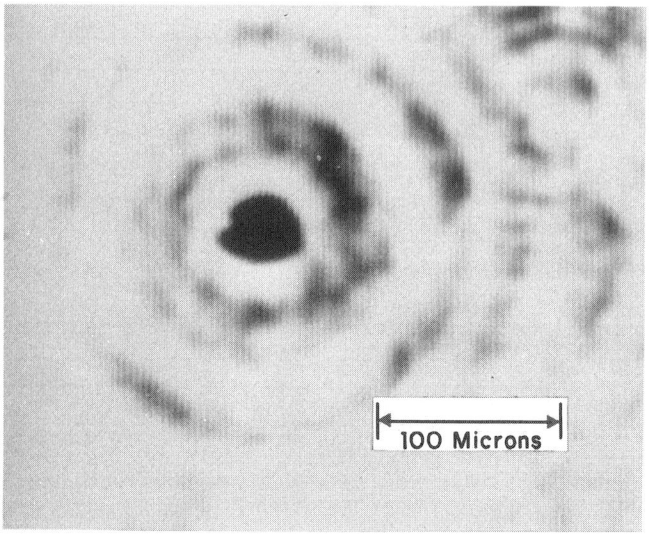

Figure 61 Hologram section of particle A. This particle has a manual detection probability of zero in the Schering standard inspection and an automated detection probability of 100%. The particle measures 35 μm in diameter.

Since an average 1.5× magnification factor results from the 3-diopter lens use, this would establish 50 μm as the equivalent *threshold* detectability limit for low-contrast particulates without the use of the magnifying lens in a similar 225-foot-candle illumination environment. This is supported by a recent review of the biophysical literature.

B. Particle B: Manual Detection 70%, Automated Detection 100%

Figure 62 is typical of the group of spherical particulates that have a 70% manual and 100% automated system detection probability. The 70% manual detection probability is of major significance, since it defines the boundary of the Reject Zone described in six PDA papers. The particulate is approximately spherical and has a 65-μm diameter. Since this result was obtained with 1.5× magnification, the equivalent 70% detection probability sphere without magnification in the same lighting environment would be 100 μm in diameter. The manual and automatic rejection probability is strongly affected by both the hydrodynamic and optical characteristics of the particulate.

C. Particle C: Manual Detection 70%, Automated Detection 84%

Figure 63 shows a chip whose dimensions are 150 × 25 μm by 20 μm thick. The rejection probability for this chip is 70% manual and 84% on the automated ampule inspection system previously discussed. Since this is a chip with marked specular reflection, its rejection probability is essentially controlled by chance orientation to the inspector or to the inspection means.

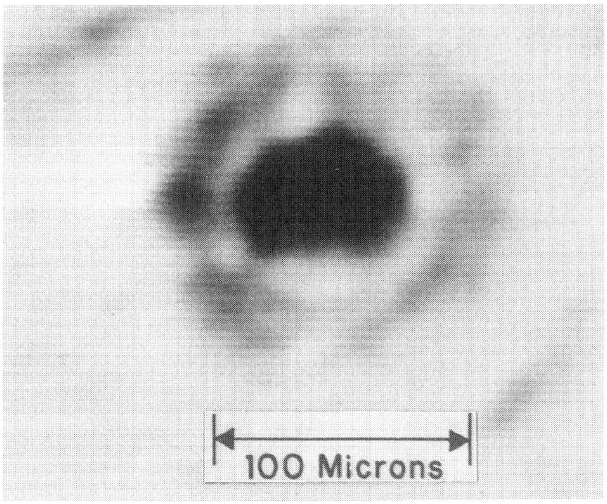

Figure 62 Hologram section of particle B. This particle has a manual detection probability of 70% in the Schering standard inspection and an automated detection probability of 100%. The particle has a diameter of 65 μm.

D. Particle D: Manual Detection 90%, Automated Detection 90%

Particle D, shown in the hologram section photo in Fig. 64 is representative of the effect on detectability of dark and colored particles.

Colored and black particles are generally picked up with high security in manual inspec-

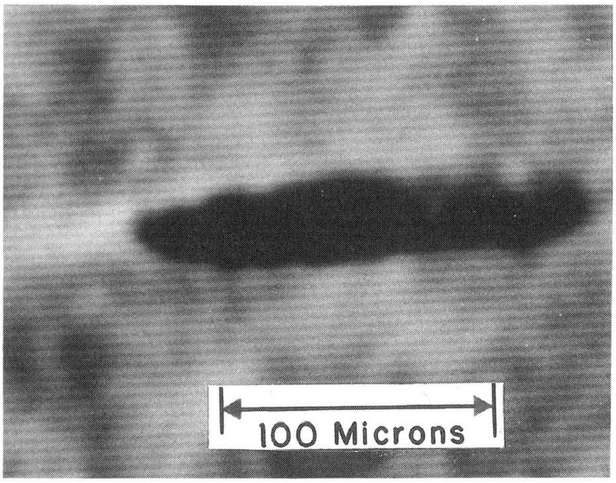

Figure 63 Hologram section of particle C. This particle has a manual detection probability of 70% in the Schering standard inspection and an automated detection probability of 84%. This particulate is a chip 150 × 25 μm by 20 μm thick and has marked specular reflection. Detection is determined by chance orientation to either the inspector or the inspection system.

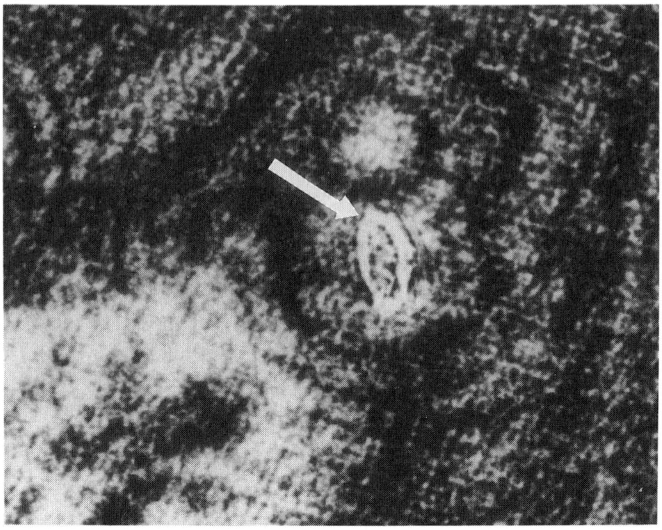

Figure 64 Hologram section of particle D. This particle has a manual detection probability of 90% in both the Schering standard inspection and in the automated inspection system. The particle is a 50-μm charred ring resulting from a droplet of antibiotic spatter.

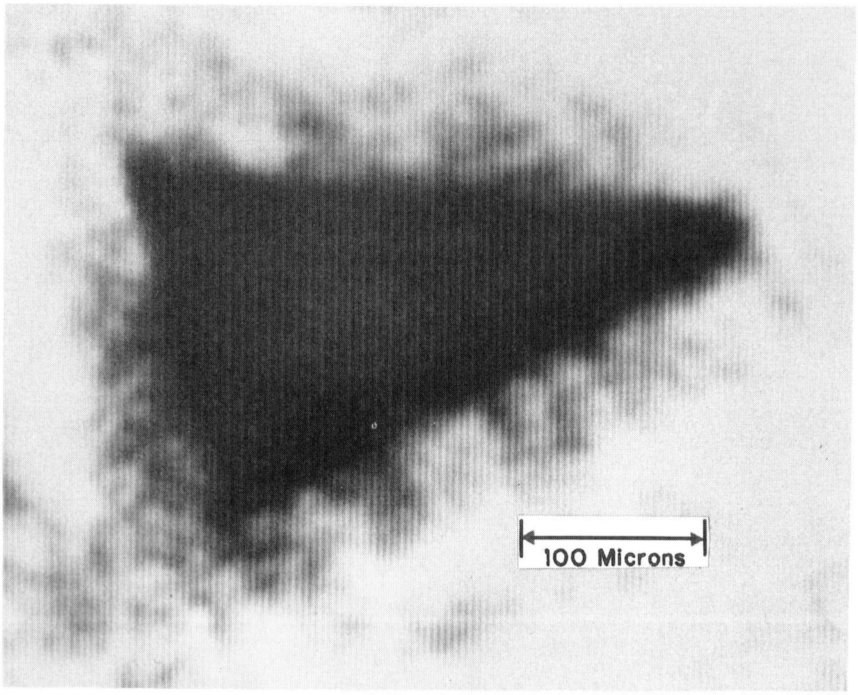

Figure 65 Hologram section of particle E. This arrowhead particulate is 160 × 260 μm. The incidence rate for this particulate, which was found in a normal production batch, is so low that its control in any sampling inspection scheme is virtually impossible.

tions. The 50-μm donut shown in Fig. 64 has a 90% manual detection probability and 90% on the automated system. It is the result of an antibiotic splatter droplet in the ampule neck area that has been charred and has flaked off into the solution.

E. Particle E: Manual Detection 97%, Automated Detection 100%

Figure 65 shows an arrowhead-shaped particulate that was found in a normal batch produced in a modern parenteral manufacturing facility. Its dimensions are 160×260 μm. The incidence of this type of defect is low, some tenths of a percent. Since this is a visible Reject Zone particulate, most pharmaceutical manufacturers would agree that its removal from the batch is essential. Its manual detection probability is 0.97 and its automated rejection probability is 1.00. The rate of incidence for this particulate is so low that its detection in any sampling scheme is improbable.

LVII. VISIBILITY REGION BOUNDARIES FOR ACTUAL PARTICULATES IN PARENTERAL PRODUCTS

The holography data just reviewed can be used to separate low-contrast spherical particulates for unassisted vision inspections into three categories: subvisible, proportionately visible, and visible. The first category, subvisible, is for spheroids equal to or less than 50 μm in diameter. The second category, proportionately visible, extends from 50 μm to an estimated 200 μm. Reject Zone particulates commence at 100-μm-diameter spheroids, which are visible with 70% probability in a single inspection. Fully visible spheroids from this table are at least 200 μm or 0.005 inch in diameter. These visibility zones are summarized in Table 23. With this table in view, it is possible to focus attention on particulate contamination sources in these size ranges, thus achieving improvements in general plant product quality. With the continued requirement for zero-visible particulates as expressed in USP XXIII a perspective of closer particulate control in parenteral products appears desirable.

Table 23 Visibility Regions

	Diameter (M)	
Visibility region	Unassisted vision	3-Diopter lens
Subvisible	$50 \leq P$	$35 \leq P$
Proportionately visible	$50 < P < 200$	$35 < P < 130$
70% Manual rejection (Reject Zone boundary)	100	65
Visible	$P \geq 200$	$P \geq 130$

Note: Low-contrast spheroidal particulates with 225-foot-candle illumination and white-and-black background particle dimensions listed are those of particles in rejected production line containers. Since the particles are not ideal spheres, the maximum particle dimension has been reported.

LVIII. ACCURACY LIMITATIONS IMPOSED BY SAMPLING INSPECTION PROCEDURES

When the inspection of parenteral products is reviewed, occasionally a view is expressed that a sampling inspection scheme is good enough to maintain the parenteral production process within control limits. The security attained with sampling schemes should be carefully examined when visible contamination with very low incidence rates is concerned.

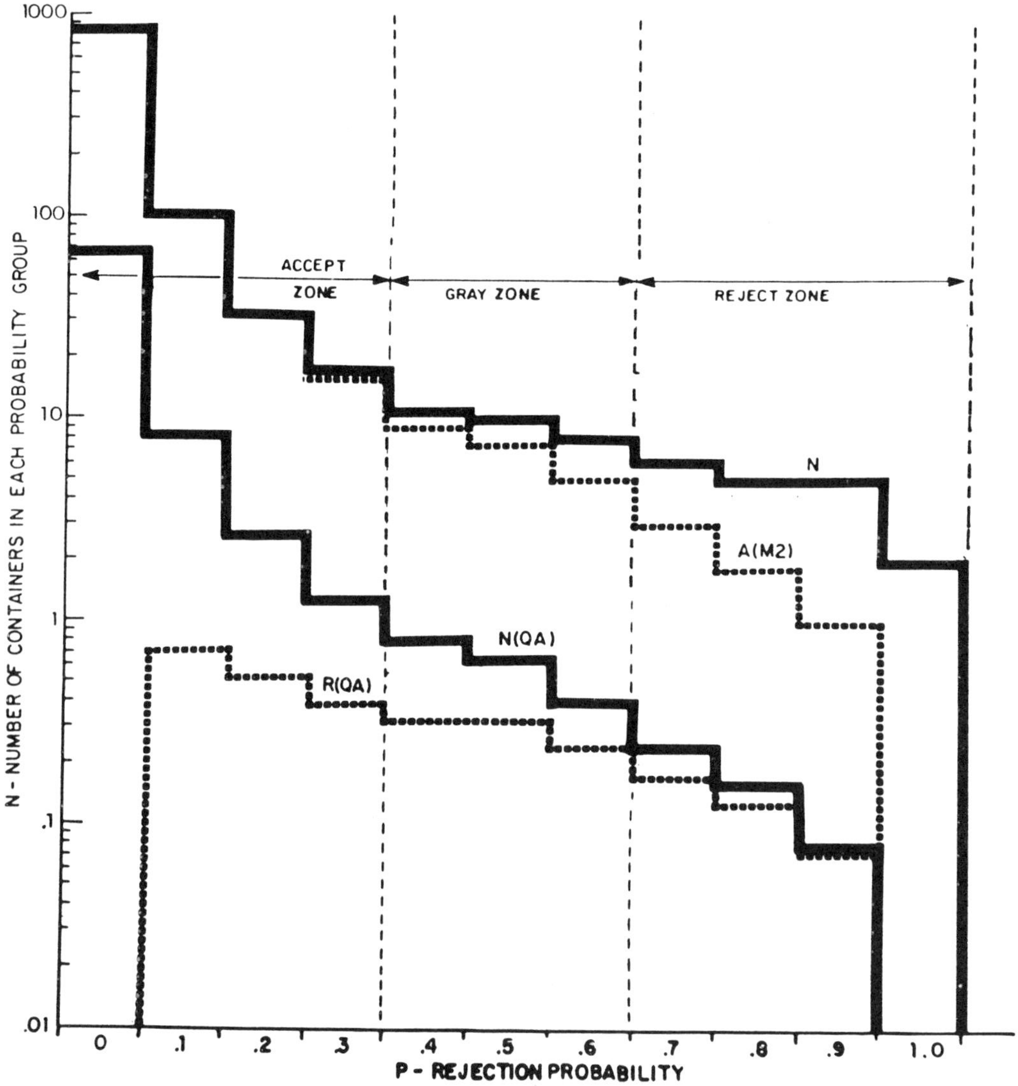

Figure 66 Histograms representing a synthetic 1000-vial sample that has been realistically modeled. The upper curve shows population distribution, *N*. The distribution of accepted material (following a first and then a cull inspection) is shown as *A*(M2). *N*(QA) represents an 80-container, randomly selected quality assurance sample (selected as per MIL SPEC 105D). *R*(QA) represents containers rejected in the QA inspection. The lack of tight process control is indicated by the fact that the number of Reject Zone rejects has to increase *10-fold* to trigger a QA rejection.

Table 24 Sampling Requirements for Evaluation of Low-Occurrence-Rate Serious Particulate Contamination

Percent serious rejects	QA Aliquot for 95% confidence limits for stated accuracy of determination		
	±100%	±50%	±25%
1	800	1600	3200
2	400	800	1600
4	200	400	800

The QA Aliquots shown are very much smaller than the batch size.

In Fig. 66 is a synthetic example with which the utility of sampling schemes in parenteral quality assurance programs is challenged. In the upper two curves, this histogram shows the population distribution of a 1000-container randomly-selected test group, *N*, and the distribution of the accepted material after an initial inspection followed by a cull inspection, *A*(M2).

The lower two curves show the distribution of containers in an 80-sample quality assurance aliquot and the rejects that result after inspection. In this simplified presentation, the Reject Zone rejects would have to increase 10-fold to attain the reject trigger point of 4. With control data as insensitive as that resulting from this MIL SPEC 105D sampling scheme shown, maintenance of parenteral process control becomes much more difficult.

A different view of the problem is shown in Table 24. In Table 24, a destructive sampling scheme using an analytical device is considered and the accuracy of determination is examined. This table shows that when serious rejects are no more than 1% of a batch, a sample of 800 containers will result in an empirical determination from 0 to 2%. When the QA aliquot is increased to 1600 containers, the empirical determination will range from 0.50% to 1.50%. When 3200 containers are selected for this evaluation, the empirical results range from 0.75% to 1.25%. This rate of sacrifice to ensure usable data, when destructive testing is considered, lies beyond the bounds of acceptable economics. It can be concluded that 100% nondestructive tests are and will continue to be used in parenteral manufacturing in the foreseeable future to ensure that the manufacturing process remains within control limits and thus minimizes the number of crises encountered.

LIX. UHLIR'S MODEL OF PARTICULATE INSPECTION

This discussion of a probabilistic validation methodology would be incomplete without mention of an alternative validation methodology due to Uhlir, who developed his analysis working with the Bayer inspection system and described it in the West German pharmaceutical industry press. Uhlir also concluded that the inspection for particulates was probabilistic [29,31]. He attempted to evaluate systems based on the work chart shown in Fig. 67. His focus was on containers that were repeatedly good in sequential inspections, on containers that were repeatedly bad in sequential inspections, and on containers that "wavered," whose current inspection differed in

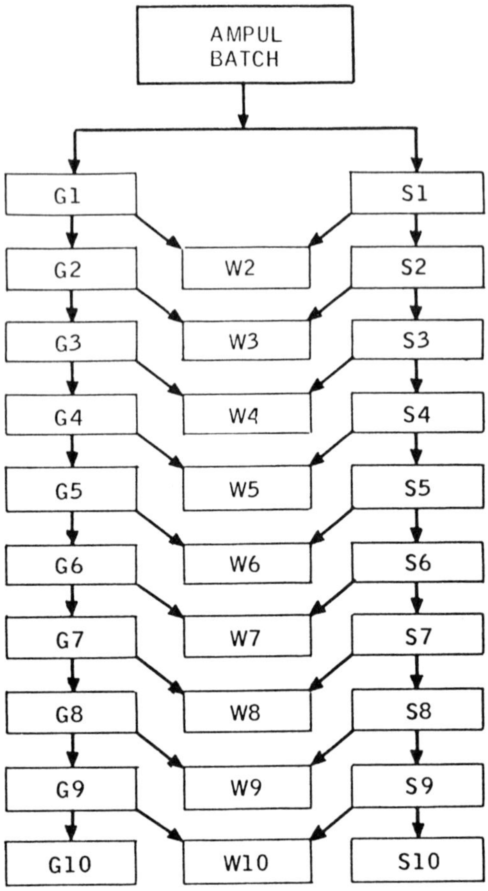

Figure 67 Inspection work scheme for Uhlir's probabilistic inspection model separates containers into good, bad, and "wavering" categories for those containers with differing inspection results in sequential inspections.

results from the preceding one. Uhlir generated various parameters with which to compare the results of the test scheme shown.

Klein and Reuter [32] applied the Uhlir developed methodology in a validation demonstration and reported extensive experimental results in 1978. Uhlir's method was unchallenged from its appearance in 1973 until 1980. Applying the more complete probabilistic inspection model now available to Uhlir's evaluation method, the results shown in Table 25 were obtained.

Uhlir's methodology was challenged with three synthetic samples. The first, labeled "Manual," is a data set in which there is an ideal 1:1 comparison between manual and machine inspection results. For this case, acceptable results are obtained with either the Uhlir or the probabilistic Working Standard Test Method. Systems I and II were constructed to have identical Uhlir coefficients and range from good to unusable performance as shown in this summary. In terms of the two-dimensional probabilistic mode, Uhlir utilized two unrelated one-dimensional

Table 25 Comparison of System I and System II with the Standard Manual Inspection Using Probabilistic Inspection Model

	RZE1	RAG1	RAV1
Manual	0.85	0.30	0.50
System I	0.95	0.20	0.47
System II	0.53	0.39	0.47

System I and System II have identical Uhlir performance parameters and greatly different performance.

probability distributions: manual and machine. This approach has been shown to have limited utility for the problems in view. A more complete treatment of this topic can be found in Ref. 5.

This completes an overview of information and methodology that can be utilized for the evaluation and validation of automated inspection systems; none of the available devices is inexpensive. With the information that has been discussed, the move from a present manual or semiautomated particulate inspection system into the automated inspection world can be accomplished with efficiency and safety. Is the move worthwhile? Figure 68 summarizes the results of automation at Schering over a 13-year period. As Fig. 68 shows, there are marked benefits to automation. An increase from 1000 inspections per day to 10,000 per hour achieved in 1976 was accompanied by a proportionate reduction in inspection cost. The reduction was accompanied by a simultaneous 70% reduction in reject rate. This reduction can be traced to two sources: (1) a more discriminating inspection and (2) using the replicable data available from the investigation system. Production management was able to reliably test changes to their process to achieve the higher reliability shown.

A practical, cost-efficient validation procedure will now be described.

YEAR	INSPECTION RATE	TYPE	TRAINING PERIOD	NORMALIZED REJECT RATE
1968	1000/DAY	MANUAL	14 WEEKS	1.0
1970	1000/H	SEMIAUTO, VIDEO	2 WEEKS	1.2
1976	10,000/H	PDS/V AUTOMATIC	0	0.3
1981	3600/H	PDS/A AUTOMATIC	0	0.3

Figure 68 Summary of the Schering automated inspection machine project, supporting the view that the use of modern technology can improve quality and reduce costs.

LX. A PRACTICAL COST-EFFECTIVE INSPECTION VALIDATION PROCEDURE

The validation procedure described below is an application of the probabilistic working standard methodology developed by Knapp and Kushner in a sequence of five papers [1–5] published in *Journal of the Parenteral Drug Association* in issues from January–February 1980 through March–April 1982. This procedure has been organized to permit selection of the best combination of operating conditions and number of sequential inspections for a probabilistic particulate inspection technique or device. This selection will yield a statistically replicable match of security and economic effectiveness of the results of the inspection system in present use.

For a realistic demonstration of the simplicity and ease of this methodology, data from an actual validation is used below.

A. Validation Procedure

This is a self-contained, step-by-step guide to the use of the Working Standard Test Method for validation of inspectors, new inspection techniques, and mechanisms. This method utilizes the probabilistic inspection model to calculate objective performance criteria for a comparison of alternative procedures. The Working Standard Test Method utilizes a mortality table concept. Considering average rejection results within statistically defined Accept, Gray, and Reject Zones, any degree of evaluation precision required can be achieved. For validation, the reliability of the inspection in rejecting significant particulates is the only performance criterion of significance. This parameter, the RZE, is an estimate of the average probability of rejecting vials that the manual standard method identified as Reject Zone vials. The economic effectiveness of the inspection procedure is evaluated by the magnitude of the undesired reject rate, RAG, in the Accept and Gray Zones. These two parameters, RZE and RAG, completely characterize any manual or machine inspection. The data required for this procedure is the accept/reject record in a representative test group of serially numbered containers. The following sections will describe each step required in the selection of validated inspectors and for an efficient, successful validation.

B. Manual Inspection Basics

The following discussion assumes a prior inspection of about ¼ of the total inspection time for container integrity and acceptability.

Data comparability for particle contamination requires a well-defined inspection environment and sequence of inspection actions. The specification of these prerequisites must be based on an understanding of the biophysical effect of light intensity, contaminating particle contrast, the duration of the inspection and the detection characteristic of the human eye/brain combination.

The visual angle for high resolution is approximately 2°. This defines a 1-mm zone of high resolution at the 25-cm close vision distance. The 150° binocular movement detection zone, in contrast, provides a zone 48 cm wide at the 25-cm close vision distance. The human eye/brain combination utilizes movement detection to localize the position of the particle within the high-resolution zone of the eye. To inspect effectively, a liquid-filled container is moved to impart motion to any particles within the liquid. For an interval after the container stops moving, the suspended particles will continue to move. It is this interval of particle movement after the container movement comes to a stop in which particle detection occurs. The amount of movement should be vigorous enough to impart motion to the contaminating particles but should be less than the motion at which cavitation bubbles occurs. Cavitation prolongs and interferes with the inspection process.

The major category of contaminants in injectable liquids are low-contrast particles that are inspected against a black background. The movement of heavy glass or metallic fragments is quickly damped by friction against the container bottom or walls. The first part of the black background inspection period therefore follows immediately after the container movement stops, and is for these heavy fragments which occur near the bottom of the container for about ¼ of the inspection period. The heavy fragment inspection period is followed by the inspection for low-contrast particles against the black background for an additional ¼ inspection period. This is followed by a change to the white background, repeated container movement and an inspection for dark particles. The dark particle inspection should be about ¼ of the total inspection time per container.

C. Inspection Environment

1. The particle inspection can be accomplished in a working volume approximately 42″ wide by 18″ deep by 24″ high with a flat black interior for optimum detection of the majority of the low-contrast particle contaminants encountered. Within this volume a flat white area of approximately 8½ × 11 inches that can be provided by a clean white sheet of paper is required for optimum detection of the dark or black particles. Automating the change of background from black to white avoids the repositioning of the container and the eye focus change that accompanies the change of container position, thus reducing inspector fatigue and improving inspection efficiency.
2. Following the Illumination Engineering Society recommendations, a flicker-free, lighted inspection volume providing an illuminance between 750 and 1,500 foot-candles should be employed. To improve the comparability of the inspection results, lighting uniformity within 10% in the test volume should be employed. To accommodate a reasonable range of inspector size and height variation, the light inspection uniformity should be maintained for a 10-inch height variation centered in the inspection enclosure.
3. Standardization of the inspection time per container improves inspection comparability. The use of a digital timer whose control period is initiated with a foot switch and which lights a small pilot light in the peripheral vision area of the inspector at the completion of the inspection cycle can be employed. If the inspector is doubtful of the accept/reject decision, a second cycle can be initiated. An initial inspection of a single container at a 20 second inspection rate is useful in establishing baseline performance.

D. Inspection Procedure Prerequisites

To help achieve good, unbiased manual inspection data on which all inspection validations are based, the following prerequisites must be satisfied.

1. The workflow of the inspector during the inspection must be arranged for smooth continuity.
2. The selected containers must be presented in randomized sequence in each of the multiple inspections.
3. The serial numbers for vial coding should be small and should not obscure the contents of the container or be read by the inspector.
4. The accept/reject data should not be recorded by the inspector. The use of a clerk for this recording is desirable.
5. Since there is serious degradation of detection capability with fatigue, all data should be taken under identical conditions and, most desirably, at the start of the working day.

6. The occurrence of reject containers in the test batches must be significantly higher than that in a normal production batch to achieve statistically usable groups, but should not exceed 25%. Otherwise, positive reinforcement will occur and the inspectors will perform with more discrimination than in normal inspection procedures.

E. Sample Group

To achieve validation, a broadly representative test group of containers is required that includes containers from the three statistically defined product quality zones, the Accept Zone, the Gray Zone, and the Reject Zone. The size of the test group should be at least 125 containers. The data in this Chapter was based on 250-container test groups serially numbered and assembled from five separate production batches in the following manner:

1. 170 randomly selected uninspected containers
2. 40 containers selected from material rejected in two sequential inspections using the present method
3. 40 samples accepted by the present inspection method but rejected by the Quality Assurance Test procedure.

An alternative container selection procedure that can be employed in start-up situations is based on a sequence of inspections that will separate containers into the desired groups.

1. Sequentially number a group of containers.
2. Inspect the container group 3 times, recording the rejections for each container in each inspection.
3. For the Reject Zone containers select 25 containers from those rejected in each of the three inspections.
4. For the Accept Zone containers select 75 containers from those that were either never rejected or rejected in a single inspection.
5. For the Gray Zone containers select 25 containers from the group that have been rejected in 2 of the three inspections.
6. The container exteriors must be clean before the test commences and must be maintained free of skin oils during the extended handling involved in the test sequence. For this purpose, lint-free gloves are essential.

F. Data Recording

Data sheets should be headed with the necessary descriptive information: date, inspector's name, inspection beginning and ending times. The data sheets should be prenumbered with the serial number sequence assigned to the test batch. The accept/reject record for each serial-numbered vial in each inspection is the data required.

G. Inspector Selection Methodology

The evaluation of inspectors and the comparison of the effectiveness of new manual techniques or automated systems against a current manual standard share an identical methodology. The core concept employed is the average rejection probability in a zone defined by the present manual inspection technique. In this technique, every container rejected with a probability equal to or greater than 0.70 (as determined from the average of multiple inspections) is defined to be in the Reject Zone. The average rejection probability of *all* the containers in the Reject Zone is called the Reject Zone Efficiency, RZE, and is a measure of the security achieved by the inspection.

The Accept Zone is defined to contain those containers with an average rejection probability below 0.30; the Gray Zone lies between the 0.30 limit of the Accept Zone and the 0.70 lower limit of the Reject Zone. The economic effectiveness of the inspection is evaluated by the magnitude of the undesired reject rate in the Accept and Gray Zones, which is noted as RAG. These definitions are summarized in Fig. 17 (see page 318).

In each use, the standard manual inspection defines the quality category of each container in the test group.

In the example following, the effectiveness of a group of inspectors is evaluated individually, then as a group, and the entire inspector group is used to evaluate an automated inspection system. Only basic arithmetic and paper and pencil need be used. The following discussion can also be used as the basis for a simple computer program if desired.

H. Data Organization

Since the particulate inspection is probabilistic, to generate secure comparisons of performance, we must deal with the mean performance generated by many replications. For the data volume that is generated, early consideration of the data recording forms is essential. Figure 69 is a copy of the data generated in the set of 10 inspections of the working standard test set by inspector 1.

The basic organization of the data in Fig. 69 is used throughout. Each container is identified and its rejection probability in sequential trials is determined. While validated inspectors were used, the rough histogram and average rejection probability shown at the bottom of this figure are useful as performance indicators.

Since we require characterization of the inspection group as a working block in the manufacturing process, the data for each inspector is combined as shown in Fig. 70. This figure is then a summary of the results of 100 inspections performed by 10 inspectors. This set of data will yield the RZE inspection security parameter and RAG, the undesired reject rate in the Accept and Gray Zones. The probabilistic inspection data resulting from the automated inspection system of the working standard test group is shown in Fig. 71.

I. Inspector Selection

In order to minimize the effect of inspectors with widely varying performance, it is necessary to select an average group of inspectors on which to base method comparisons. This initial inspector validation and selection is a boot-strap process. The merged performance of the entire experienced inspector staff is used to define both the baseline inspection process and the group of containers in the Accept, Gray and Reject Zones as described in Fig. 17 (see page 318). In a five-inspector test group, 10 inspections of the container sample test group by each inspector are usually adequate for a statistically acceptable analysis of the accept/reject decision results within 95% confidence limits. Following the separation of the containers into the three zones, the performance of each inspector can be evaluated against the average performance of the entire inspector group. The two important characteristics of an inspector's performance are inspection security (RZE) and discrimination (RAG). The criteria for inspector selection are based on a balance of these two parameters, which is achieved through the use of the RZE-RAG plane in the step-by-step procedure described below.

INSPECTOR 1 10 inspections

Ampuls	Number of times rejected								
1	0	0	0	0	6	10	1	0	0	1
11	6	10	2	1	1	1	7	4	10	0
21	1	0	1	1	3	0	3	10	2	0
31	1	1	10	10	10	8	10	1	1	0
41	0	0	6	1	7	8	3	10	10	9
51	3	10	10	10	0	0	10	10	9	0
61	9	9	1	2	1	0	3	6	4	2
71	0	1	10	0	0	0	1	0	1	3
81	8	2	6	0	2	1	6	8	0	1
91	0	0	8	10	10	0	0	1	1	1
101	6	5	0	1	0	6	2	1	0	0
111	0	1	2	0	1	1	0	10	9	0
121	0	1	0	1	0	1	7	0	9	0
131	10	1	0	0	1	1	1	0	0	8
141	4	6	10	0	0	0	0	4	1	0
151	10	4	0	0	1	9	9	1	2	0
161	2	2	6	0	9	6	1	0	0	0
171	0	1	0	0	1	1	9	0	4	3
181	0	0	0	0	5	2	10	9	0	0
191	0	0	6	0	0	10	1	1	0	0
201	0	0	0	10	10	2	1	2	0	5
211	0	0	7	9	2	0	1	0	0	0
221	10	0	0	0	0	6	0	9	0	3
231	0	0	0	1	0	1	1	2	0	5
241	9	3	1	1	6	2	10	0	1	0

PR	FREQ.	PERCENT	
0.0	96	38.40	********************
.1	52	20.80	**********
.2	17	6.80	***
.3	9	3.60	*
.4	6	2.40	*
.5	4	1.60	
.6	14	5.60	**
.7	4	1.60	
.8	6	2.40	*
.9	14	5.60	**
1.0	28	11.20	*****

AVERAGE PROBABILITY OF REJECTION = 28.92%

Figure 69 A single inspector data sheet showing the accept/reject record of each serial-numbered ampule in a test group of 250. Also shown is the average reject rate and a distribution of the rejects.

J. Data Evaluation

Step 1. Generate Inspector Data Sheets: Each inspector in the group inspects the container sample test group 10 times. The accept/reject test results for each container inspection are recorded as shown in Fig. 69.

Step 2. Summarize Inspector Data: The summary of the 10 inspection results for each inspector are recorded on separate sheets to provide for step-by-step checking capability in forms similar to Fig. 69.

Step 3. Generate Batch Summary Sheet: The total number of rejections for each container by all inspectors is calculated and the empirical rejection probability is entered on the form

AVERAGE FOR ALL INSPECTORS 100 inspections

Ampuls	Rejection Probability									
1	.06	.08	.11	.02	.37	.90	.04	.13	.04	.19
11	.52	.75	.08	.13	.13	.10	.73	.45	.75	.12
21	.20	.11	.08	.10	.34	.13	.43	.79	.14	.08
31	.31	.25	.98	.96	.68	.65	.94	.08	.11	.09
41	.20	.16	.48	.07	.66	.65	.65	.94	.97	.80
51	.43	.73	.95	.83	.11	.16	.97	.94	.69	.11
61	.79	.64	.14	.12	.14	.16	.15	.37	.12	.06
71	.19	.11	.98	.05	.09	.08	.40	.19	.20	.48
81	.70	.32	.55	.09	.39	.10	.58	.88	.09	.21
91	.04	.05	.67	.85	.94	.13	.06	.17	.14	.07
101	.22	.56	.14	.08	.13	.18	.11	.19	.09	.07
111	.09	.26	.10	.18	.09	.19	.05	.93	.83	.46
121	.12	.08	.11	.02	.16	.14	.50	.10	.69	.14
131	.97	.14	.08	.50	.18	.09	.11	.11	.33	.52
141	.41	.37	.52	.11	.14	.14	.12	.61	.17	.13
151	.95	.17	.09	.13	.07	.81	.76	.16	.35	.05
161	.26	.10	.58	.14	.81	.52	.16	.10	.17	.15
171	.12	.10	.11	.08	.46	.26	.86	.11	.23	.25
181	.09	.12	.09	.13	.44	.35	.85	.85	.03	.10
191	.04	.07	.46	.07	.09	.91	.08	.04	.13	.14
201	.11	.08	.10	.95	.91	.26	.17	.33	.07	.44
211	.12	.08	.32	.52	.06	.12	.16	.11	.12	.50
221	.55	.14	.09	.12	.09	.41	.07	.74	.25	.36
231	.10	.19	.06	.24	.10	.18	.20	.15	.13	.21
241	.89	.29	.08	.19	.45	.20	.94	.12	.12	.10

PR	FREQ.	PERCENT	
0.0	53	21.20	**********
.1	91	36.40	******************
.2	18	7.20	***
.3	13	5.20	**
.4	14	5.60	**
.5	13	5.20	**
.6	10	4.00	**
.7	9	3.60	*
.8	11	4.40	**
.9	18	7.20	***
1.0	0	0.00	

AVERAGE PROBABILITY OF REJECTION = 30.97%

Figure 70 Grouped inspection data resulting from a total of 100 manual inspections of each ampule in the test group. Also shown is the average reject rate and a distribution of the rejects.

shown in Fig. 70, the batch summary sheet. The empirical rejection probability is determined by the percentage of rejections in the total number of inspections for each container.

Step 4. Generate Inspector/Inspection Team Evaluation Form: Transfer the inspector group rejection probability for each container from the batch summary sheet to a set of forms as shown in Figs. 71 and 72, which list all container serial numbers.

Starting at the left in these forms is a complete serial number sequence for all containers used in the study. The next column, the group rejection probability, would be entered from the batch summary sheets. The next column requires a clear-cut decision. Rejection probabilities equal to or greater than 0.70 are entered over the diagonal, rejection probabilities below 0.70 are entered below the diagonal. *The number of entries above the diagonal is equal*

Batch 6/83 Test 1	RZN 38	N_m = 100 Date 9/12/83
Quantity NT 250	AGN 212	N_a = 30 Page 1 of 11
Description	Inspector Validation	

Container Serial Number	Group Rej. Prob.	Group Zone RZ	Group Zone AGE	Inspector 1 RZ	Inspector 1 AGE	Inspector 2 RZ	Inspector 2 AGE
12	.75	.75		1.0		1.0	
13	.08		.08		.20		.00
14	.13		.13		.10		.10
15	.13		.13		.10		.00
16	.10		.10		.10		.00
17	.73	.73		.70		1.0	
18	.45		.45		.40		.90
19	.75	.75		1.0		1.0	
20	.12		.12		.00		.10
21	.20		.20		.10		.30
22	.11		.11		.00		.00
23	.08		.08		.10		.00
24	.10		.10		.10		.00
	6.19	3.13	3.06	3.70	2.60	4.00	3.30

Figure 71 Extract from inspector evaluation data. The above example illustrates a partial paper-and-pencil evaluation sheet. (Courtesy of Schering-Plough Corporation, Bloomfield, NJ.)

to RZN, the manual group determination of the number of containers in the Reject Zone. Enter this information in the labeled block. The number of entries below the diagonal is equal to AGN, the number of containers in the Accept and Gray Zones. Enter this also in the labeled block. The sum of the two is, of course, the total of the test group containers.

The remaining columns on the right are for the data from individual inspectors. *The position for entry of each inspector's data is determined by that of the whole group entry. Enter the probability for each inspector from the inspector summary sheet.*

BATCH 6/83 Test 1	RZN 38	N_m 100	Date 9/12/83
QUANTITY NT 250	AGN 212	N_a 30	Page 11 of 11
DESCRIPTION	INSPECTOR VALIDATION		

CONTAINER SERIAL NUMBER	GROUP REJ. PROB.	GROUP ZONE RZ / AGE	INSPECTORS 1	INSPECTORS 2
TOTALS pg. 11	3.38	1.83 / 1.55	1.90 / 1.40	2.00 / 2.40
pg. 10	5.27	.74 / 4.53	.90 / 3.00	.90 / 6.10
pg. 9	6.56	2.77 / 3.79	3.00 / 3.60	3.00 / 4.90
pg. 8	6.06	2.56 / 3.50	2.80 / 1.70	3.00 / 4.70
pg. 7	7.52	3.33 / 4.19	3.70 / 3.00	3.80 / 4.80
pg. 6	6.50	.97 / 5.53	1.00 / 5.10	1.00 / 7.40
pg. 5	5.39	1.76 / 3.63	1.90 / 2.90	1.90 / 3.60
pg. 4	9.06	4.35 / 4.71	4.60 / 3.10	4.60 / 7.20
pg. 3	10.68	6.98 / 3.70	8.80 / 4.10	7.80 / 4.20
pg. 2	10.77	4.61 / 6.16	5.00 / 5.50	5.00 / 7.50
pg. 1	6.19	3.13 / 3.06	3.70 / 2.60	4.00 / 3.30
GR. TOTALS	77.38	33.03 / 44.35	37.30 / 36.00	37.00 / 56.10
m RZE1	.869			
m RAG1	.209			

Figure 72 Extract from inspector evaluation data. Partial summary sheet and example of paper-and-pencil inspector evaluation computations. (Courtesy of Schering-Plough Corporation, Bloomfield, NJ.)

Step 5. Compute the Inspection Group RZE

1. List the individual page subtotals on the last data sheet (as shown in Fig. 72) for all Reject Zone containers.
2. Total all Reject Zone rejection probabilities.
3. Divide the total by RZN to get the group RZE and enter the information in the block provided.

In the worked example the RZN is 38, the total rejection probability for all containers in the

Reject Zone is 33.03: the average rejection probability is 33.03/38 or 0.869 as shown for MRZE1. Similar computations are used to establish individual inspector RZE's.

Step 6. Compute the Inspection Group RAG

1. List the individual page subtotals for the whole group on the last data sheet (as shown in the worked example) for all Accept and Gray Zone containers.
2. Total all Accept and Gray Zone rejection probabilities.
3. Divide the total by AGN, the number of containers in the Accept and Gray Zones, to get the group RAG and enter the information in the block provided.

In the worked example, the sum of all rejection probabilities in the accept and reject zones is 44.35; the number of containers in these zones is 212. RAGZ(M), the discrimination parameter, is the average rejection probability in these two zones and is 44.35/212 = 0.209 as shown. Similar computations for individual inspectors are used to establish their individual inspection security and discrimination.

Step 7. Determine Average Reject Rate: The RAV, the average reject rate, is determined from the average rejection probability for all containers in the test group. This is the sum of all container rejection probabilities shown in the group zone column, divided by the total number of containers in the test group.

In the worked example, the sum of all container rejection probabilities for the entire groups is 77.5. This sum divided by the total of all containers in the test group is RAV, the average reject rate. MRAV1 is therefore 77.38/250 = 0.310.

Step 8. Generate Inspector Selection Data Using Confidence Limit Calculations: The RZE and RAG parameters are based on probabilistic determinations, so calculation of the 95% confidence limits interval for these parameters will establish performance replicability. The confidence interval is dependent on the number of containers in the Reject Zone (RZN), and the number of containers in the combined Accept and Gray Zones (AGN), and the number of inspections employed. The extent of the 95% confidence limits for each of these parameters, assuming a normal distribution, can be computed from

$$\mathrm{CI} = 1.96\sqrt{\frac{\mathrm{RZE}(1 - \mathrm{RZE})}{(\mathrm{RZN})(NI)}} \tag{86}$$

where

RZE = Reject Zone Efficiency
RZN = number of Reject Zone containers
I = number of inspections
CI = confidence interval

Thus, for example, if there are 50 containers in the Reject Zone being inspected 10 times by each manual inspector, for an inspector with an RZE of 0.9, the 95% confidence interval would be computed in the following manner: RZE = 0.90, RZN = 50, NI = 10; hence,

$$\mathrm{CI} = 1.96\sqrt{\frac{(0.9)(0.1)}{(50)(10)}} \qquad \mathrm{CI} = 0.03$$

Thus, for this inspector,

$$0.87 \leq \mathrm{MRZE} \leq 0.93$$

In order to find the range for this inspector along RAG axis in the RAG-RZE plane, the confidence interval for RAG must be computed.

$$\mathrm{CI} = 1.96\sqrt{\frac{\mathrm{RAG}(1 - \mathrm{RAG})}{(\mathrm{AGN})(N)}} \tag{87}$$

where

$$\mathrm{RAG} = \frac{\mathrm{AGR}}{\mathrm{AGN}} \tag{88}$$

and AGR = average quantity rejected in the Accept and Gray Zones, AGN = total number of containers in the Accept and Gray Zones population; N, as above, is 10. For the example we are working with, AGN = 200, AGR = 24, $N = 10$ inspections. Thus,

$$\mathrm{RAG} = \frac{\mathrm{AGR}}{\mathrm{AGN}} = \frac{24}{200} = 0.12$$

$$\mathrm{CI} = 1.96\sqrt{\frac{(0.12)(0.88)}{(200)(10)}} = 0.01$$

$$0.11 \leq \mathrm{RAG} \leq 0.13$$

The inspector's position can now be located in the RZE-RAG plane, at a point RZE = 0.90 and RAG = 0.12 with 95% confidence limits $0.87 \leq \mathrm{RZE} \leq 0.93$, $0.11 \leq \mathrm{RAG} \leq 0.13$. This is approximately the position of inspector number 8 in Fig. 32 and would place this inspector in the mean group.

Step 9. Graphical Selection of Validated Inspectors: The means and 95% confidence intervals for each inspector are plotted as shown in Fig. 32. The inspectors are separable into three groups, which can be described as follows:

Low-security group: Low reject rate in the Reject Zone (low RZE); low rate in the Accept and Gray Zones (low RAG)

Low-discrimination group: High reject rate in the Reject Zone (high RZE); high reject rate in the Accept and Gray Zones (high RAG)

Balanced midgroup: High reject rate in the Reject Zone (high RZE); low reject rate in the Accept and Gray Zones (low RAG)

Once a balanced midgroup of inspectors has been identified, this group, or a subset of it, can be used to inspect and generate standard test groups for validation machine (i.e., automated) inspection. It is now possible to use the balanced midgroup as a validated inspection team.

Since the inspectors used in this study have been through a selection and validation process as shown in Fig. 32 their RZE and RAG scores are tabulated for review in Table 26.

In Table 26 inspectors 4 and 5 display the periodic variability in efficiency that can occur in an inspector group of adult females. The test scores for the entire group are quite satisfactory, and it is the merged average performance that characterizes the security and discrimination achieved in the inspection.

Table 26 Inspector Group and Individual Inspector RZE and RAG Scores for Inspection of the Working Standard Test Group

Inspector	Group	1	2	3	4	5	6	7	8	9	10
RZE	0.869	0.981	0.973	0.868	0.776	0.705	0.802	0.905	0.837	0.918	0.911
RAG	0.209	0.170	0.264	0.132	0.109	0.078	0.127	0.253	0.401	0.311	0.263

K. Example of Direct Validation Approach

When a validated inspector group is available, a more direct approach to validation of an automated particle inspection system is possible. A key element in this alternate approach is a tabulation of the experimental rejection probabilities of each container. An example of such a tabulation for the rejection probabilities of each container in a validation test group is shown in Table 27. An example of this alternate approach using Figs. 73 through 76 follows. It must be stressed that the core of this validation demonstration is equivalence of performance to match that of skilled human vision. These figures are used to systematize comparison and validation of an automated inspection system against the performance of a manual inspection standard inspection. It has been organized to facilitate easy, efficient checking of the manipulations and arithmetic required in system or inspection technique validation. The parameters that will be evaluated are as follows:

RZN Number of containers in the test group with rejection probabilities greater than 0.7
AGZN Number of containers in the test group with rejection probabilities below 0.7

Table 27 Test Group Rejection Probabilities

6-17-83, 13:30, HD and transport slower — File: PRID6TO8DAT
Ampules

1	0.07	0.07	0.03	0.07	0.70	1.00	0.20	0.03	0.03	0.03	0.27	0.97	0.00
14	0.00	0.17	0.00	1.00	0.77	0.93	0.00	0.00	0.03	0.00	0.87	0.33	0.10
27	0.40	1.00	0.00	0.00	0.73	0.50	1.00	1.00	0.37	0.30	0.90	0.03	0.10
40	0.00	0.20	0.00	0.13	0.00	1.00	0.23	0.00	1.00	1.00	0.83	0.83	0.77
53	0.97	0.80	0.00	0.00	1.00	1.00	1.00	0.00	0.83	0.23	0.03	0.00	0.00
66	0.20	0.17	0.13	0.00	0.03	0.10	0.13	1.00	0.03	0.03	0.23	0.00	0.27
79	0.00	1.00	0.23	0.77	0.53	0.03	0.83	0.03	0.97	1.00	0.00	0.03	0.03
92	0.00	0.90	0.80	0.87	0.00	0.03	0.03	0.00	0.17	0.83	0.97	0.03	0.07
105	0.03	0.00	0.10	0.10	0.10	0.00	0.00	0.53	0.03	0.03	0.00	0.03	0.00
118	0.97	1.00	0.10	0.00	0.00	0.30	0.00	0.30	0.07	0.07	0.00	0.70	0.00
131	0.97	0.00	0.00	0.00	0.93	0.10	0.00	0.10	0.17	0.97	0.00	0.37	0.33
144	0.13	0.30	0.23	0.07	0.00	0.20	0.07	1.00	0.00	0.00	0.00	0.03	0.47
157	0.93	0.03	0.03	0.00	0.07	0.00	0.93	0.00	0.47	0.60	0.00	0.10	0.63
170	0.00	0.37	0.13	0.07	0.00	0.00	0.43	0.93	0.77	0.10	0.03	0.00	0.00
183	0.07	0.10	0.13	0.67	0.90	1.00	0.03	0.00	0.03	0.00	0.97	0.00	0.13
196	0.97	0.03	0.13	0.00	0.00	0.00	0.00	0.03	1.00	1.00	0.53	0.00	0.93
109	0.23	0.97	0.03	0.00	0.20	0.00	0.03	0.00	0.43	0.10	0.00	0.37	0.93
222	0.00	0.00	0.03	0.00	0.17	0.03	0.80	0.00	0.10	0.07	0.17	0.03	0.00
235	0.07	0.43	0.00	0.00	0.87	0.83	1.00	0.00	0.00	0.70	0.27	0.00	0.97
248	0.18	0.27	0.00										

NT Total number of containers in the test group is equal to the sum of RZN and AGN

RZE Inspection security parameter is determined as the average rejection probability for all containers in the Reject Zone

RAG Undesired reject rate in the Accept and Gray Zones is determined as the average rejection probability for all containers in the test group in these zones

RAV Average reject rate for the test group is determined as the average rejection probability for all containers in the test group

In Figs. 73–76 the first column on the left is a listing in serial order of all containers in the working standard test group. The second and third columns are, respectively, the manual rejection probability for each container and the automated inspection device rejection probability for the small container.

The analysis starts with the manual inspection data and consists of an either-or decision. If the rejection probability for a container is equal to or greater than 0.70, it is entered in the M1 column of the RZE computation block. The total number of entries made is equal to RZN, the number of containers in the Reject Zone; this value is entered in the titled area on the data sheet. The population of the Reject Zone, RZN, for this set of data was determined to be 38 containers, the number of containers in the test batch with rejection probabilities equal to or greater than 0.70 which are listed in the M1 column of the RZE computation.

The next step in the analysis is the RZE computation. For this, transfer the rejection probability from the automated inspection column into the A1 column of the RZE data block for each manual entry whose rejection probability is 0.70 or higher.

Totals for each category per data page, as shown in the figures, are recommended to improve error checking capability.

Figure 74 is a section of the eleventh and final sheet in the automated inspection system RZE computation. In addition to the page 11 totals, the subtotals from the previous pages have been entered here and the sums indicated. In Fig. 75 is illustrated the final step in the manual and automated inspection RZE computation. The RZE was defined as the average rejection probability for all containers identified to be in the manually determined Reject Zone. This average is determined from the rejection probability sums in column 2 of Fig. 75 divided by RZN (which is 38) for the manual system. For automated system, RZE is determined from the sum of rejection probabilities in column 3 of Fig. 75 divided by the manually defined RZN, which is 38. For the data shown in Fig. 74, the manual first inspection RZE, MRZE1 is 0.869; the automated system single-inspection RZE, ARZE1, is 0.902. Since these comparisons are based on the number of trials required for 95% confidence limits, the GMP demonstration of equivalent performance for validation of the automated inspection system is judged complete.

There is, however, additional information required in the real world: RAV, the average reject rate, and RAG, the false reject rate in the Accept and Gray Zones. The computation of the average reject rate is indicated in Fig. 75.

The second and third column totals in Fig. 74 are the sums of all the individual container rejection probabilities for the entire batch for first, the manual method, and second, the automated method. Since RAV is the *average* of all container rejection probabilities in the test group, this computation is carried out in the lower-left section of this figure. As shown, this is simply the sum of all rejection probabilities for each method divided by the total number of containers in the working standard test group, 250. As shown, the RAV(M) is 31.0 and the RAV(A) is 29.9%. The next computation, for RAG, is quite similar: It determines the average reject rate in a numerically defined rejection probability zone. Figures 75 and 76 are utilized to obtain RAG, which measures the discrimination achieved in the accept/reject decision, and thus,

Batch 6/83 Test 1	RZN 38	N_m = 100	Date 9/12/83
Quantity NT 250	AGN 212	N_a = 30	Page 1 of 11
Description	PD 100 Validation		

Container Serial Number	Rejection Probability		RZE Computation	
	Manual	Auto	M1	A1
012	.75	.97	.75	.97
013	.08	0.00		
014	.13	0.00		
015	.03	.17		
016	.10	0.00		
017	.73	1.00	.73	1.00
018	.45	.77		
019	.75	.93	.75	.93
020	.12	0.00		
021	.20	0.00		
022	.11	.03		
023	.08	0.00		
024	.10	.87		
TOTALS	6.19	7.24	3.13	3.90

Figure 73 Extract from first sheet of a form suitable for validation of new inspection techniques and devices by evaluating the false reject rate in each system. (Courtesy of Schering-Plough Corporation, Bloomfield, NJ.)

the economic effectiveness of the inspection. As shown in Fig. 75, if the rejection probability for a container is less than 0.70, it is entered into the M1 column of the RAG computation block, the manual-first-inspection rejection probability column. For each entry into the M1 column of the RAG computation block, a corresponding entry from the automated inspection data block is made into the A1 column. The total number of entries made into the RAG computation block is equal to AGN, the Accept and Gray Zone population. The AGN in this example is 212. This

Batch 6/83 Test 1	RZN 38	N_m = 100	Date 9/12/83
Quantity NT 250	AGN 212	N_a = 30	Page 11 of 11
Description	PD 100 Validation		

Container Serial Number	Rejection Probability		RZE Computation			
	Manual	Auto	M1	M2	A1	A2
TOTALS pg. 11	3.38	3.39	1.83		1.97	
pg. 10	5.27	5.43	.74		.80	
pg. 9	6.56	7.18	2.77		2.97	
pg. 8	6.06	6.39	2.56		2.83	
pg. 7	7.52	5.53	3.33		2.87	
pg. 6	6.50	5.51	.97		.97	
pg. 5	5.39	5.12	1.76		1.97	
pg. 4	9.06	9.58	4.35		3.90	
pg. 3	10.56	10.05	6.89		7.20	
pg. 2	10.77	9.32	4.61		4.90	
pg. 1	6.19	7.24	3.13		3.90	
GR. TOTALS	77.38	74.74	33.03		34.28	
MRAV = $\frac{77.4}{250}$ = 31.0			mRZE1 =	33.03/38	ARZE1 =	34.28/38
ARAV = $\frac{74.7}{250}$ = 29.9			=	.869	=	.902

Figure 74 Extract from summary sheet showing the computation of average reject rate and Reject Zone Efficiency for a standard manual inspection and for an automated inspection system. These provide the basis for establishing the false reject rates of the two systems. (Courtesy of Schering-Plough Corporation, Bloomfield, NJ.)

sum is entered into its labeled position on the data sheet. The sum of AGN and RZN (previously determined) is equal to NT, the number of containers in the test group, and is checked against the data sheet entry.

Figure 76 is a section of the eleventh and final sheet in the RAG computation. The sum of all manual rejection probabilities in the Accept and Gray Zones is seen to be 44.35. The average

Batch 6/83 Test 1	RZN 38	N_m = 100	Date 9/12/83	
Quantity NT 250	AGN 212	N_a = 30	Page 1 of 11	
Description	PD 100 Validation			
Container Serial Number	Rejection Probability		RAG Computation	
	Manual	Auto	M1	A1
012	.75	.97		
013	.08	0.00	.08	0.00
014	.13	0.00	.13	0.00
015	.13	.17	.13	.17
016	.10	0.00	.10	0.00
017	.73	1.00		
018	.45	.77	.45	.77
019	.75	.93		
020	.12	0.00	.12	0.00
021	.20	0.00	.20	0.00
022	.11	.03	.11	.03
023	.08	0.00	.08	0.00
024	.10	.87	.10	.87
TOTALS	6.19	7.24	3.06	3.34

Figure 75 Extract from the first page in a computation of inspection selectivity for both a manual standard system and for an automated device. Selectivity is evaluated by determinations of the undesired reject rate in the Accept and Gray Zones. (Courtesy of Schering-Plough Corporation, Bloomfield, NJ.)

rejection probability in these zones is then 44.35 divided by 212, the population of these zones. The MRAG1, the manual first inspection RAG, is, as shown in this figure 0.209.

In a similar fashion, the total of all rejection probabilities for the automated system for all containers in the manually-identified Accept and Gray Zones is 40.49. This sum is divided by 212, the Accept and Gray Zone population AGN. For this example the ARAG1, the automated

Batch 6/83 Test 1	RZN 38	N_m = 100	Date 9/12/83
Quantity NT 250	AGN 212	N_a = 30	Page 11 of 11
Description.	PD 100 Validation		

Container Serial Number	Rejection Probability		RAG Computation			
	Manual	Auto	M1	M2	A1	A2
TOTALS pg. 11	3.38	3.39	1.55		1.42	
pg. 10	5.27	5.43	4.53		4.63	
pg. 9	6.56	7.18	3.79		4.21	
pg. 8	6.06	6.39	3.50		3.56	
pg. 7	7.52	5.53	4.19		2.66	
pg. 6	6.50	5.51	5.53		4.54	
pg. 5	5.39	5.12	3.63		3.18	
pg. 4	9.06	9.58	4.71		5.68	
pg. 3	10.68	10.05	3.70		2.85	
pg. 2	10.77	9.32	6.16		4.42	
pg. 1	6.19	7.24	3.06		3.34	
GR TOTALS	77.38	74.74	44.35		40.49	
			mRAG1 =	44.35/212	ARAG1 =	40.49/212
			=	.209	=	.191

Figure 76 Extract from the summary sheet in a comparative evaluation of inspection selectivity for both a manual and an automated inspection system. Sample calculations are shown. (Courtesy of Schering-Plough Corporation, Bloomfield, NJ.)

first inspection RAG, is 0.191. Comparing these two scores, we deduce that the automated system has somewhat superior discrimination compared to the manual standard inspection.

To conclude this validation example, we have seen simple effective methodology yield three objective parameters with which to evaluate automated inspection systems as well as alternative manual procedures. When containers with physically calibrated particulates become generally available, the methodology described can be utilized directly.

It should be noted that the working standard test group is organized to generate a uniform distribution of rejection probability particulates ranging from 0% to 100%. This is done to achieve efficient inspection security and inspection discrimination evaluations. The test results from this specially constructed test group are not directly useful in estimating the average reject rate with normal production material. In the well-controlled parenteral manufacturing facility, the particulate rejects encountered are low in incidence and randomly distributed throughout the production batch. To generate realistic estimates of production floor reject rates, a randomly selected test group should be assembled and inspected by both the standard manual method and the alternative technique or inspection system.

LXI. INSPECTION SYSTEM VALIDATION WITH OBJECTIVE PHYSICAL CRITERIA

The initial "bootstrap" system validations used an assortment of accept and reject containers culled from five normal production lines. The basis for selection of the final 250-container test group was to achieve assurance that the kinds of particles that could be encountered would be represented in this test group. This approach gave assurance that the characterization of experimental manual rejection efficiency and the response of the automated or semiautomated inspection system were evaluated realistically.

The containers with particles collected in this manner are truly random. Comparison of Reject Zone Efficiencies with these randomized containers has shown small variation. The false reject rate in the Accept and Gray Zones exhibit much wider variability. Because of this variation, it is difficult to compare these factors from inspection group to inspection group and from machine to machine.

In the 14 years since first publication of the Working Standard Test Method, there has been enough accumulated experience to support an objective particle contamination transfer standard. Work on an artificial transfer standard commenced with Borchert and co-workers in 1986 [33]. A very promising direction for these objective standards was reported by Shabushnig, Melchore and co-workers at the 1994 P.D.A. annual meeting [34]. The advent of standard test containers with one particle per container holds out the hope of fully-controllable laboratory testing for all types of inspection systems. For this perspective to be fully realized, the particles in the test containers must be representative of the full range of contaminating particle sizes and densities that are encountered in practice. When this is accomplished, these standard containers will provide a new and significant capability to the pharmaceutical industry. With these precision standards, accurate comparisons of the performance of manual inspectors and automated devices can become a customary validation exercise.

Before the distant goal can be realized, the use of these containers seeded with standard low-density particles, and validated by the inspection of many inspectors, can be used to facilitate "preflight" testing of particle inspection systems on a convenient daily basis.

LXII. INSPECTION SYSTEMS: PRACTICAL CONSIDERATIONS

A. Inspection Security and Its Particle Detection Probability Limit

In all presently available particle inspection systems, the random position of the contaminating particle at the commencement of the inspection period is a basic limit to the effectiveness of the contaminating particle inspection. This basic inspection-security limit results from the fact that detection of the particle can only occur in a relatively small portion of the liquid volume within the container. For those inspection systems that use forward-scattered lighting, the detection zone lies within a slice of the container volume parallel to the viewing lens and is about 1 mm deep. For those systems using light extinction detection to determine the presence of a particle, the detection band is about 1 mm wide and is perpendicular to the viewing lens.

Particle detection in each system depends on the transfer of orbital motion to the particle by spinning the container and stopping the spin abruptly. For particle detection to occur, the continuation of the orbital motion of the particle within the container after the end of container spin must bring the particle through the detection zone. Since the position of the particle is random with respect to the detection zone, some proportion of all contaminating particles will have exited the detection zone before the inspection period begins. To detect these "just missed" particles requires a prolonged inspection period that limits inspection speed.

When the reject zone efficiency of a single inspection in either of these systems is evaluated against a standard manual inspection, the single inspection is shown to be distinctly inferior to the standard manual inspection. The standard manual inspection has been shown to have an RZE of at least 0.80 for trained inspectors; the single automated inspection has an RZE of 0.55 to 0.65. This lack of an effective single inspection is, in part, compensated by reinspecting the results of the first inspection and requiring that all containers that are accepted in either of the sequential inspections be accepted for stock. When this second inspection is properly programmed, the Reject Zone Efficiency of the inspection can increase up to 0.92.

The reinspection of accepted material to raise the Reject Zone Efficiency of the process brings with it an additional cost. The cost is an increased loss of Accept and Gray Zone containers compared to the discrimination of the manual standard inspection.

B. Effect of the Light Source

In any detection system, the signals generated by moving particles are fundamentally determined by the type and quantity of light present. The stability and adjustment accuracy of the light source is therefore an important consideration in any automated inspection system. Constant light intensity is so essential that automatic alarm and shutdown for out-of-bounds operation are a fundamental requirement. This is especially so for modern, high-speed inspection systems. With constant light intensity ensured, particulates of identical optical density or reflectivity will generate accurate size-graded signals. Provision for objective light-source measurement to ensure system operation within desired limits is a system operating condition that should be documented. As significant as the quantity of light and its stability is the quality of the lighting method and the contrast achieved.

Forward-scattering lighting improves the net signal-to-noise ratio for small, low contrast, low optical density particles; direct light is more effective for larger, opaque or dark particles. The signal created in a forward-scattered lighting system is a detector current increase whose pulse height is determined by the size and reflectivity of the particle. In direct light systems the signal created by a particle is a decrease in detector current. The effect of these lighting types is similar to that achieved with the white/black background change in manual particulate inspection. Some commercially available systems use both types of lighting either in combination or in two sequential inspections. The proportionality of the pulse height signal to particle size is different for both types of lighting.

An opaque particle signal created by a direct light source results in a reduced detector current. A low-optical-density particle in a forward-scattered light system increases the detector current. At the present time there are no commercially available inspection systems that can detect both types of signals without interference and signal loss.

C. Detectors and Detection Limits

Since the signal generated by the moving particulate is proportional to its projected size on the photosensitive detector, it is clear that with conventional detectors (CCD's, etc.), the minimum particle size that can be detected is determined by the size of the container imaged. The larger the container, the less sensitive the system is to small particles. If a system with a conventional detector is selected, testing should commence with the largest container and the smallest particulates that are of control interest. A detection system that maintains a constant particulate size threshold is to be preferred. In this type of system, the number of photodetectors used is proportional to the container-size inspected. The limit on particulate size detectability is the signal-to-noise ratio achieved for the full system.

The noise due to the optics and electronics is usually much less than that generated by the interfering signals created by light reflections from meniscus movements. The complex meniscus surface that results from the combination of container movement and spin reflects the light source onto the walls and container bottom. The movement signals so created are indistinguishable from particulate signals. This troublesome signal is enhanced by a vertical light source entering the container bottom; it can be minimized, but not eliminated, by the use of vertical narrow-slit lighting, which provides light entry perpendicular to the container sides.

Other interfering signals are those created by cavitation bubble movement. These cavitation bubbles are created by either shock impact or by excessive spin prior to inspection. During the brief inspection period, the cavitation bubble signals cannot be separated from particulate signals, and therefore contribute to the false reject rate. An additional noise signal source can result from the combination of container or detector vibration and either glare spots or defects on the container itself.

D. Container Rotation

In place of the subtle hand motions employed in manual inspection, automated systems employ container rotation followed by an abrupt stop to generate rotation of the container's liquid contents. To achieve replicable inspections in the brief time available, particle velocity must be controlled with good accuracy.

The achievement of adequate particulate speed requires rapid deceleration of the container from an adequately high velocity to transfer movement to the contained liquid. A practical limit is imposed, however, by cavitation. Cavitation creates bubbles, typically 10 to 15 μm in diameter. Since many cavitation bubbles are generated simultaneously, their summed detection signal will appear as that due to large, rejectable particles. During practical inspection time limits, the movement of these bubbles cannot be distinguished from particulate contamination, thus resulting in false rejects. This observation must be followed by the conclusion that cavitation in spin or handling must be avoided at all costs.

E. Movement Detection Considerations

To avoid the many false rejections that would occur due to optical defects in the containers inspected, some form of movement detection scheme is usually employed. These movement detection schemes have many of the functional characteristics of a frog's eye. They are blind to stationary images; when a moving image is detected, the image is compared with a stored set of descriptions to determine whether a response should occur. This blindness to stationary images eliminates system response to container markings and flaws.

Since a moving image system has been selected as appropriate, its disabilities and limitations must be recognized and defensive actions instituted. For those movements parallel to the plane in which the image is viewed, maximum detection sensitivity is zero. The orbital path of particulates imposed by spin varies the particle path cyclically from being parallel to the detection plane to one in which the path is perpendicular to the plane. This disability places a critical limit on the inspection speed of single-axis detection devices. This cyclical blindness can be eliminated by the use of two simultaneous viewing axes to achieve orthogonal viewing when high operating speeds and maximum inspection security are required.

For detection to occur in a movement detection scheme, the particle must traverse the scanned volume. Only those particles moving within the liquid can be detected. To assure that the quality of the container system is adequately tested, a rinsedown by inversion, or similar technique, of any particulates in the neck of the container or on the stopper is a prerequisite before inspection commences. To optimize inspection speed, the scanned volume must approach 100% of the container volume. When less than the full container volume is scanned, inspection time must increase to allow the particle trajectory to intersect the volume scanned. When this condition is not satisfied, an additional multiplicative probability enters the scene. In the limit, for an inspection of adequately short duration, this probability is simply the decimal fraction of the total volume scanned. From this discussion, it is clear that inspection security is enhanced and inspection time minimized when the full container volume is scanned.

Intuitively, when less than the full container volume is scanned, an increase in orbital speed of the particle will minimize inspection time.

F. Detector Dynamic Range Requirements

Even a cursory review of parenteral ampules and vials will show that in either manual or automated inspection systems, glare from markings, flaws or surface curvature is frequently encountered. The retention of detection sensitivity under glare conditions is a prerequisite for secure inspections. If the required wide dynamic range cannot be achieved throughout the entire image area, then detector saturation in the glare areas effectively reduces the security of the inspection.

To achieve reliable glass particle detection requires movement detection in or close to glare points originating from container bottom markings or optical flaws. For reliable glass fragment detection, inspection must commence as close to spin cycle end as possible, because glass particle movements are quickly damped out by drag against the container bottom. This requirement imposes a 250-ms maximum delay between end of spin and inspection start. For this type of particulate, a scan of the full container bottom improves detectability. Conversely, when the full bottom is not scanned, glass detection reliability is decreased. The container meniscus is another region in which glare signals must be considered. With the minimization of Kraft fiber in parenteral container packaging, the total amount of packaging particulates encountered has been reduced. However, the change to plastic packaging materials has brought with it an increased incidence in floating particulates, which should be matched by increased inspection of the meniscus surface. Thus, dynamic range in the meniscus area is desirable.

G. Automated Inspection Validation and Revalidation

It must be appreciated that an adequate inspection of each container size requires the solution of a specific complex of mechanical, physical and optical requirements. A demonstration of inspection adequacy, i.e., validation, is therefore a prerequisite when container sizes are changed. Validation will also be required when the container size is constant but the color and/or viscosity of the solution are varied.

The period between revalidations is determined by the rate at which machine performance will vary from adjusted conditions. For those devices that do not utilize feedback adjustment of the significant variables, i.e., light intensity, spin speed and the duration of the inspection period, 30-day intervals are appropriate. For systems with feedback adjustment, quarterly revalidation periods should be adequate.

H. Particle Inspection Perspectives

Two limitations to the security of the contaminating particle inspection process have been identified. The first limitation is the result of the fractional volume that is scanned for particles in each system. Unless the orbital path of the particle passes through the detection volume, the particle cannot be detected. In the limit, the maximum inspection security achieved cannot be greater than the proportion of the container volume that has been examined. Due to the imaging limitations, the scanned container volume is a 1-mm-deep section in forward-scattered systems. In light extinction inspection systems, the scanned container volume is a section 1 mm wide.

As noted, in each system a bias exists for the type of particle that offers optimum detection. Forward-scattered detection systems favor the detection of white or light-colored particles. The particle signal in such a system is an increase of detector current. Light extinction detection systems favor the detection of dark or opaque contaminating particles. In such a system, the particle signal is a decrease of detector current. Can either type of particle be disregarded? An answer based on experience is that balanced sensitivity to both kinds of particles is a safe choice. Experimental data shows that when either forward-scatter or light-extinction particle detection systems are pushed to obtain sensitive detection for the less-easily-detected particle, the system reject rate escalates. The opposite physical effect of the particle signal in both of the lighting schemes bars the use of the same detector area for a noninterfering detection of both signals. The random position of the particle within the container limits the utility of inspections in which successive positions are dedicated to each type of lighting.

A recent patent [35] appears to have found a way to increase imaging depth to encompass containers up to 32 mm in depth and to bypass the signal polarity problem by allocating separate detector areas to both types of signals. A major benefit that is anticipated from this new patent is an improved accept/reject discrimination which should decrease the false reject rate presently experienced.

LXIII. APPLICATION OF PROBABILISTIC INSPECTION METHODOLOGY TO STATIONARY DEFECT INSPECTION

Although some tentative exploration of the use of video inspection for primary container defects was initiated, the first published application of the probabilistic inspection methodology for this purpose is due to Rothrock [36]. Since that report, the continued increase in the use of lyophilized products for injectable preparations has steadily increased the need for a convenient, accurate evaluation and validation of automated inspection systems performing inspections of lyophilized cake preparations.

Since the defects are stationary, a booth inspection by a senior inspector with unlimited time can be employed to sort containers into four categories: critical defects, major defects, minor defects and defect-free. With the use of the good lighting conditions discussed and possibly magnification with a 3-diopter lens, the sorting can be very accurate. Structuring a test group using 20 critical defects, 40 major defects, 60 minor defects and 130 defect-clean containers provides a broad enough range of inspection probabilities to provide a well-calibrated inspection test set. Selection of a smaller group that may not be representative of the range or severity of the defects that must be analyzed can limit the effectiveness of any analysis that can take place.

With these stationary defects, a high RZE value can be expected when the test group is inspected under test booth conditions. The high RZE value should result in a validated test group of containers in approximately 10 repeated inspections. With the validated test group prepared, validation trials of an automated inspection machine can take place in the normal fashion. Considering the possible fragility of some of the lyophilized cakes that have been seen, minimization of the repeated manual inspection by selecting maximum machine inspections for the validation demonstration appears desirable ($D = 8$).

LXIV. COMMERCIAL INSPECTION SYSTEMS

The nine-year period since publication of "Particulate Detection in Sealed Containers" in Chapter 14 in *Filtration in the Pharmaceutical Industry* edited by Theodore H. Meltzer and published by Marcel Dekker, has been marked by innovation and many changes. Of the four automated particle inspection systems offered nine years ago (Brevetti, Eisai, Lasko and Takeda) only two (Brevetti and Eisai) are actively offered in the United States. The lion's share of all automated particle inspection systems in the United States have been installed by Eisai. In these nine years, Eisai has positioned themselves as a reliable and responsible partner for the many U.S. companies who rely on Eisai particle inspection systems to ship their product. In those few circumstances where extraordinary levels of support were critically required, the support was immediately forthcoming even when that meant moving people and equipment from Japan to the United States. Past history alone, however, provides a limited guide for future events. Technological and economic forces are at present combining to transform the pharmaceutical manufacturing environment. There is a present perception that the pace of rationalization and integration of production activities must accelerate as an essential defensive reaction.

In response to these market forces, Eisai has expanded their original particle inspection systems, designed for ampules and vials, into a complete set of finished container product inspection systems. This has been accomplished by integrating commercially available image analysis system into their well-designed container-handling systems.

Within the last two years has there been a serious attempt by Brevetti to reenter the United States market. This move by Brevetti, paralleling the Eisai market direction, has been based on the Brevetti perception that image analysis systems are now well-tested, commercially-available packages. These sturdy systems are offered as adjuncts to the particle inspection systems that they have been actively marketing in Europe and the Near East.

This use of generalized image analysis has found a special niche in the inspection of lyophilized products. The Brevetti and Eisai concept is that the addition of image analysis systems to their particle inspection systems extends their product inspection capability in a contiguous application area. It also concentrates inspection activity which is essential to the security of product delivery and accomplishes this goal with a reduced requirement for inspectors and floor space.

A new entry in the race for a portion of the automated inspection machine market has been recently fielded by Seidenader. Seidenader has an established position in the sale of semiautomated inspection systems and within the recent past has been selling empty glassware inspection systems. The extension into automated inspection systems has been helped by the recent expansion of interest in the automated inspection of the static images that are inspected in lyophilized products. Seidenader now has available particle inspection systems for ampules, vials and syringes.

A listing of vendors of parenteral inspection products, devices and systems is summarized in Table 28. In addition to the previous focus on fully automated particle inspection systems, the

Table 28 Inspection Booths, Semiautomated and Fully Automated Systems

Brevetti CEA Represented in USA by: IMA North America, Inc. 418 Meadow Street Fairfield, CT 06430	SA,A	203/331-0331	FAX 203/384-2555
Dabrico, Inc. 1244 Grinnell Road P.O. Box 2163 Kankakee, IL 60901	SA	815/939-0580	FAX 815/939-7798
Driam USA 185 Davis Chapel Road P.O. Box 4216 Spartanburg, SC 29305	SA	803/579-7850	FAX 803/579-7852
Eisai, U.S.A., Inc. 300 Frank W. Burr Blvd. Teaneck, NJ 07666-6741	A	201/692-0999	FAX 201/692-1972
F. P. Developments Inc. 333 Blue Anchor Rd. P.O. Box 2163 Berlin, NJ 08009	SA	609/768-3003	FAX 609/767-4176
Lakso Division Package Machinery Co. 44 Mead Street Leominster, MA 01453	A	508/537-8534	FAX 508/840-0730
M.W. Technologies, Inc. 71 Midland Ave. Elmwood Park, NJ 07407	B,SA	201/797-0700	FAX 201/797-0745
Seidenader Equipment, Inc. 25 Hanover Road Florham Park, NJ 07932	SA,A	201/301-9800	FAX 800/342-6910

B = inspection booth; SA = semiautomated inspection system; A = fully automated inspection system.

list of vendors has been extended to include manual inspection booths and semiautomated inspection systems. Tables 29–32 provide a sampling of system capability and price ranges to expect from these vendors.

ACKNOWLEDGMENT

This is to acknowledge the vision and leadership of Schering-Plough Corporation, who believed that better quality and lower price could be simultaneously achieved and, in particular, to those who encouraged the project from which these insights emerged: W. H. Conzen, S. Finzi, A. Gibson, E. B. Hershberg, N. Kirsch, and W. Pace. To the colleagues whose work and support helped bring this project to a successful completion: Paul M. Bjorndal, Sam DiFabritis, Walter Ivanoff, Harold K. Kushner, Kay R. Messina, Diane M. Pisano, Domenic Roselli, Rush Slivjanovsky, Henry R. Sochon, David Willard, and John C. Zeiss.

Table 29 Brevetti CEA Inspection System Listing (Represented by IMA North America, Inc.)

ATM model	Particle detection container size range	Inspection rate (containers/hour)
18S	Up to 19 mm	9,000
18D	Up to 19 mm	18,000
32S	32 mm	9,000
32D	32 mm	18,000
60S	69 mm	3,600
60D	69 mm	7,200
	Lyophilized inspection systems	
18C	Up to 19 mm	9,000
18C-D	Up to 19 mm	18,000
32C	Up to 32 mm	9,000
32C-D	Up to 32 mm	18,000
60C	Up to 60 mm	3,600
60C-D	Up to 60 mm	7,200
STM model	Syringe inspection systems	
15S	Up to 15 mm	7,000
15D	Up to 15 mm	14,000
15-3X8	Up to 15 mm	24,000
	Ampule leak testing systems, LTM	
Jupiter	Up to 24 mm	18,000

Notes: All models can be made for both in-line and off-line use. Models listed are complete particulate inspection systems for clear and amber containers with clear or colored solutions. Particle detection uses the CCD imaging technology developed at Brevetti with a variety of lighting options selected by the customer. The fixed container imaging size means that particle size resolution must be tested in the largest container to be processed. A container volume check is optional on the Brevetti system.
Prices: Prices start at $175,000 but can approach $1,000,000 per system depending on the speed of the machine and the degree of container defect inspection.

Table 30 Eisai U.S.A., Inc. Inspection System Listing

Particle inspection systems		
AIM model	Container size range	Inspection rate (containers/hour)
360	Up to 27 mm	4,000
287	Up to 18 mm	7,000
587-1	Up to 33 mm	5,500
287-2	Up to 23 mm	14,000
587-2	Up to 33 mm	11,000
787-1	Up to 43 mm	4,000
787-2	Up to 43 mm	8,000
288	Up to 23 mm	18,000
588	Up to 33 mm	13,000
294	Up to 18 mm	24,000
1088m	Up to 43 mm	15,000
1088w	Up to 55 mm	10,000
1088v	Up to 95 mm	5,000
287-2/c	Up to 18 mm	14,000
294/c	Up to 18 mm	18,000
294/s	Up to 18 mm	18,000
594/s	Up to 27 mm	12,000
Lyophilized product inspection system		
795	Up to 40 mm	9,000

c and s suffix additions describe cartridge and syringe inspection systems.
Particle inspection in Eisai systems uses the patented lighting and detection system that can be expanded to maintain a constant particle size detection threshold as container size increases. For container defects and defects in lyophilized cake products, Eisai functions as a systems integrator. The lyophilized product inspection system has been designed and priced at $650,000 as an integrated system.
All models can be made for both in-line or off-line use. AIM models are complete particulate inspection systems for clear and amber containers with clear or colored solutions. Some suspensions and oily products can be inspected with these systems but this capability should be verified by testing. All systems are designed to accommodate from 5 to 10 camera positions for container inspection. The specific details of each machine are determined by the intended application. Prices start at U.S. $185,000.00 but can approach $1,000,000.00 depending on the degree of each container defect inspection required.

Table 31 Seidenader Equipment, Inc.

Semiautomated inspection systems	
V 90 AVSB/60	Container size range 0.5–1000 ml
	Infusion bottles, vials, ampules, prefilled syringes
	The inspection rate is determined by the product quality level selected. The cost of a single-inspector system ranges from \$70,000 to \$100,000, dual-inspector systems range in price from \$120,000 to \$150,000.
Automated inspection systems for prefilled syringes	
SYI	Container size range 0.5–20 ml
	Inspection rate up to 18,000/h
	System price range with 5 to 9 camera positions is \$700,000 to \$800,000
Automated inspection systems for lyophilized products	
LI 30	Container size range: ampules 1–5 ml
	Vial diameter 30 mm maximum
	Inspection rate up to 18,000/h. System price range with 5 to 9 camera positions is \$600,000 to \$750,000
Integrated automated container and content inspection system	
VI 30	Container size range: vial diameter 30 mm maximum
	Inspection rate up to 18,000/h. System price range with 5 to 9 camera positions is \$800,000 to \$900,000.

Table 32 Booths and Semiautomated Inspection Systems

Dabrico, Inc.

Syringe inspection system

Dabrico's syringe inspection system has automated removal and replacement after manual inspection of flanged syringes from Becton Dickenson Hypak System tubs. Syringe size capability ranges from 2 to 50 ml. As an alternative, following inspection, syringes can also be transferred to an additional operation.

Vial inspection system

Vial inspection systems cover the range from 2 to 500 ml in configurations for from 1 to 5 inspectors. A dual-tray loader system with automated switchover to the next tray when the first one is filled is available as an accessory.

Single inspection stations range from $45,000 to $65,000 depending on the lights, counters and other accessories selected.

Driam U.S.A.

The Visomat IV inspection system has been designed to cover ranges of container and/or syringe sizes from 1-ml ampules to 1000-ml bottles. Within the designed size range, a change of star wheel and transport belt can be accomplished in ½ to 1 h. Price for a single inspection station ranges from $79,000 to $120,000 depending on the lights, counters and accessories selected. A two-lane inspection system ranges in price from $156,000 to $240,000. The system has been designed to accommodate a rotary feed input and a large buffer table on the output side. A dual-tray loader system with automated switchover to the next tray when the first one is filled is available as an accessory.

F. P. Developments, Inc.

Varied configurations for standard packaging areas, clean rooms and hazardous products. Container size range: 2-ml tubing vials to 500-ml molded vials. F.P. developed, patented and has available a special patented star wheel reject system. Prices start at $45,000 for a single station with a manual reject system.

M. W. Technologies, Inc.

Booths

MiH-dx Tabletop booth

MiH-1 Free-standing booth

The booths are equipped with a 3-diopter rectangular magnifying lens. A variety of container lighting schemes against a selectable black-and-white background is provided. Pacing timers are available as an option.

Prices are in the $1,000.00 to $9,000.00 range.

Semiautomated inspection systems

Model	Maximum container diameter (inch)
VIS 1000	3.0
SENTINEL 2000	2.4

Single-sided semiautomated inspection systems are in the $35,000 to $150,000 price range. Double-sided semiautomated inspection systems are in the $100,000 to $250,000 price range.

REFERENCES

1. The United States Pharmacopeia: XXIII revision, United States Pharmacopeial Convention, Inc. 1995.
2. Knapp, J. Z., and Kushner, H. K., Generalized methodology for evaluation of parenteral inspection procedures, *Bull. Parenter. Drug Assoc. 34*: 14 (1980).
3. Knapp, J. Z., and Kushner, H. K., Implementation and automation of a particulate detection system for parenteral products, *Bull. Parenter. Drug Assoc. 34*: 369 (1980).
4. Knapp, J. Z., Kushner, H. K., and Abramson, L. R., Automated particulate detection for ampules using the probabilistic particulate detection model, *J. Parenter. Sci. Technol. 35* (1): 21–35 (1981).
5. Knapp, J. Z., Kushner, H. K., and Abramson, L. R., Particulate inspection of parenteral products: an assessment, *J. Parenter. Sci. Technol. 35*: 176 (1981).
6. Knapp, J. Z., and Kushner, H. K., Particulate inspection of parenteral products: from biophysics to automation, *J. Parenter. Sci. Technol. 36*: 121 (1983).
7. Knapp, J. Z., Zeiss, J. C., Thompson, B. J., Crane, J. S., and Dunn, P., Inventory and measurement of particles in sealed sterile containers, *J. Parenter. Sci. Technol. 37*: 170 (1983).
8. Knapp, J. Z., Detection and measurement of particles in sealed containers, in *Filtration in the Pharmaceutical Industry* (Theodore H. Meltzer, ed.), Marcel Dekker, New York, 1986.
9. Knapp, J. Z., Process control by non-destructive testing, International Conference on Liquid-Borne Particle Inspection and Metrology, May 11–13, 1987, Arlington, VA.
10. Knapp, J. Z., and Abramson, L. R., Automated particulate inspection systems: strategies and implications, *J. Parenter. Sci. Technol. 44* (2): 74–107 (1990).
11. Garvan, J. M., and Gunner, B. W., Intravenous fluids: a solution containing such particles must not be used, *Med. J. Austral. 2*: 140–145 (1963).
12. Garvan, J. M., and Gunner, B. W., The harmful effects of particles in intravenous fluids, *Med. J. Austral. 2*: 1–6 (1964).
13. Whyte, W., The influence of clean room design on product contamination, *J. Pharm. Sci. Technol. 37* (4): 138–144 (1983).
14. Whyte, W., A multicentered investigation of clean air requirements for terminally sterilized pharmaceuticals, *Pharm. Sci. Technol. 38* (3): 103–108 (1984).
15. Knapp, J. Z., Dull, H. F., Bjorndal, P. M., Brackl, S. F., and Yuen, P. A wide pH range stopper for improved particulate quality in parenteral solutions, *J. Parenter. Sci. Technol. 38*: 128 (1984).
16. Sokol, M., and Kirsch, N., The pacing of mechanized pharmaceutical production operations, *Bull. Parenteral Drug Assoc. 18*:13–19 (1964).
17. U.S. vs. Bristol Labs, Inc., Federal Food Drug and Cosmetics Act: Judicial and Administrative Record, 1938–1949, p. 310–314, Kleinfeld, Vincent A., and Dunn, Charles Wesley.
18. Uhlir, A., Testing of ampuls for undissolved impurities. Part 1: evaluation method , *Pharm. Ind. 35* (6): 356–360 (1973).
19. Halley, Edmund, An estimate of the degrees of the mortality of mankind, drawn from curious tables of the births and funerals at the city of Breslaw; with an attempt to ascertain the price of annuities upon lives, *Philos. Trans. 18*: (1693).
20. Crozier, W. J., On the visibility of radiation at the human fovea, *J. Gen. Physiol. 34*: 87 (1950).
21. Field, J., ed., *Handbook of Physiology*, Sec. 1, Vol. 1, (1959) pp. 647–670.
22. Blackwell, H. R., Contrast thresholds of the human eye, *J. Opt. Soc. Am. 36*: 624 (1946).
23. Blackwell, H. R., Brightness discrimination data for the specification of quantity and quality of illumination, *Illum. Eng. 48*: 602 (1952).
24. Blackwell, H. R., Use of performance data to specify quality and quality of illumination, *Illum. Eng. 50*: 286 (1955).
25. Blackwell, H. R., Development and use of a quantitative method for specification of interior illumination levels on the basis of performance data, *Illum. Eng. 54*: 317 (1959).
26. Guth, S. K., Lighting Research, *Am. Ind. Hyg. Assoc. J., 23*: 359 (1962).
27. Lythgoe, R. J., Measurement of visual acuity, Special Report No. 173, Medical Research Council, H. M. Stationery Office, London, UK (1932).
28. I. E. S. *Lighting Handbook*, reference volume Section A-3, Table 1, 1981.

29. Kingslake, R., ed., *Applied Optics and Optical Engineering*, Vol. 1, Academic Press, New York, 1965, pp. 233–235.
30. Borchert, S. J., Gaines, R., and Kraska, J. R., Validation of the Eisai visual inspection system, in Proc. Europ. Conf. on Visible and Subvisible Particles in Parenteral Products, European Organization for Quality Control, Conference on Visible and Subvisible Particles in Parenteral Products, Symp. Proc. 224–260, 34, 1985.
31. Uhlir, A., Testing of ampules for undissolved impurities. Part 2: Evaluation method, *Pharm. Ind. 36* (7): 582–588 (1973).
32. Klein, H. J., and Reuter, E. W., Automatic electronic inspection device for the detection of particles in ampules, *Pharm. Ind. 40* (12): 1357–1366 (1978).
33. Borchert, S. J., Maxwell, R. J., Davison, R. L., and Aldrich, D. S., Standard particulate sets for visual inspection systems: Their preparation, evaluation, and applications, *J. Pharm. Sci. Technol.* 40(6): 265–276 (1986).
34. Shabushnig, J. G., Melchore, J. A., Gieger, M., Chrai, S., and Gerger, M. E., A proposed working standard for the validation of particulate inspection in sterile solutions, P. D. A. Annual Meeting, Philadelphia, PA. 1994.
35. Knapp, J. Z., Light flux determination of particle contamination, U.S. Patent 5,365,343, November 15, 1994.
36. Rothrock, C. H., Gaines, R., and Greer, T., Evaluating different inspection parameters, *J. Parenter. Sci. Technol. 37* (1): 64 (1983).

10

Validation of Counting Accuracy in Single-Particle Counters: Application of a New Coincidence Model

Julius Z. Knapp

Research & Development Associates, Inc., Somerset, New Jersey

Lee R. Abramson

Statistical Consultant, Washington, D.C.

I. INTRODUCTION

Light extinction particle counters are well-behaved instruments fully capable of the workhorse task of making accurate, routine single particle contamination measurements in injectable products. In principle, any particle counter instrument now in use can deliver accurate single-particle counting data.

Single-particle count accuracy requires operation within particle size and count limits estimated with the new particle-triggered Poisson counter model both within and below the U.S.P. XXIII <788> [1] size limits. Operation within these limits will ensure single-particle count accuracy without false counts or undercounts. These count limits vary with particle size and are determined by the capability of the counter system in use. In practice, selection of a counter with sufficient capability to provide the desired accuracy without constant dilution is an important consideration. When particle concentration exceeds the selected count accuracy contour, dilution and a repeat of the assay provides a practical solution.

An essential requirement for the achievement of accuracy in any measurement is the existence of an accurate, well-defined standard. Accurate particle size standards have been available for some time. With them, particle size calibrations can be routinely transferred to a user's particle counter. Neither accurate count standards nor a validated model of the counting performance of a particle counter have been available. Their absence introduces problems in the establishment of accurate concentration limits for the evaluation of the high SVI particle concentrations that are acceptable in U.S.P. XXIII <788>.

In response to this problem, the U.S.P. has supported the development of an ad hoc qualitative particle count standard and has included an instrumental count accuracy specification in U.S.P. XXIII. This counter accuracy specification has been defined in terms of a limiting acceptable particle concentration for 10-μm particles at which there is a 10% ratio of coincident occurrences. Since this count accuracy specification cannot be measured directly, it must be calculated from experimental results using a model of the counting process. The model of particle counting in the literature is the geometric Poisson model due to Jaenicke [9] and extended by Lieberman [10]. Recent papers [23–25,28] have shown that calculations based on this model do not agree with experimental data. The accuracy specification relies on counter vendor statements without specification of the statistical model or methodology to be employed.

Table 1 Glossary

Symbol	Units	Definition
A	—	Arbitrary constant in curve-fitting equation
A_F	$cm^2 \times 10^{-6}$	Area of detector flow channel perpendicular to direction of sample stream
B_H	μm	Height of the scan beam
c	counts/ml	Detected counts per milliliter
c/n	—	Count efficiency ratio
$\overline{c}$	—	Particle count loss per milliliter due to the coincident count of multiples
c/n	—	Count loss ratio
C^+	counts/ml	Total of coincident counts per milliliter
C_I	counts/ml	Initial, maximum, counts in a dilution experiment
C_0	counts/ml	Background counts, counts at infinite dilution
C'	counts/ml	Counts from the U.S.P. XXIII specified –10% undercount concentration
C_{MAX}	counts/ml	Maximum count that a detector can record, counts/ml
$C(1)/N$	—	Single-particle count ratio
$\overline{C}(1)/N$	—	Single-particle count loss ratio
$\overline{C}_B(1)/n$	—	Bounded single-particle count loss ratio
$C_B^+(1)/n$	—	Bounded multiple-particle group count ratio
$C3$	—	Count limit for a particle size at a selected count error
D_P	μm	Diameter of test particle
D_{PE}	μm	Effective diameter of test particle including the analysis threshold selected
ε_C	—	Count error
f	—	Constant stepwise dilution factor
f_T	—	Total dilution factor between the concentration for a –1% to the noise limited +1% count
k	—	Arbitrary constant in curve-fitting equation
K^+	—	Ratio of all multiple event counts to single particle
n_T	—	Total number of particles in a suspension
N_E	—	Effective detector volumes per milliliter
N_{E3}	—	Effective detector volumes per milliliter for a selected particle size
N_I	—	The initial concentration of particles/ml in a dilution experiment
N'	—	U.S.P. XXIII specified concentration of 10-μm particles at which there is a 10% undercount error for a detector
N''	—	Concentration of 10-μm particles for a 1% undercount error
N_3	—	Concentration for a selected count error
s	—	The sequential step number of a dilution experiment
S	—	The number of steps in a dilution experiment Note: S^* indicates approximate value
V	ml	Unspecified volume
V_A	ml	Sample aliquot volume in a dilution experiment
V_{DE}	$cm^3 \times 10^{-6}$	The effective volume of a detector including the effect of particle size
V_{DG}	$cm^3 \times 10^{-6}$	The geometric volume of a detector based on its static dimensions
V_0	ml	The initial volume of a dilution experiment

In retrospect, the use of particle counters for particle contamination assays was validated in a period of crisis for the low particle concentrations found in LVI product samples. The low-concentration samples minimized the effect of counting model inadequacies. Counting accuracy at the relatively high concentrations found in small-volume injectables was not addressed. Validation of counting accuracy at high particle concentrations required two missing components: a validated model of the counting process and a computation of particle concentration that included the interaction of the particle size with the height of the scan beam.

In many research problems, the initial problem statement is deceptively simple. Complexity grows as the solution to the problem is approached. In the case of particle counting accuracy, the reverse has been the case. Achieving accurate, comparable particle count data can become a commonplace laboratory occurrence. The conclusions reached are simple enough to be of use during the course of a particle counting assay.

The control of particle contamination in injectable products is an essential part of the production process and its evaluation is considered a key index of injectable product quality. To achieve and maintain accuracy in the counting of contaminating particles, the basic counting capability of these particle counters must be clearly understood. This understanding includes the information required to evaluate the functional capability of a particle counter for a given degree of accuracy in particle counting. It also includes methodology to avoid the present undercounts and introduction of false counts during a U.S.P. <788> assay. These count errors are masked by the replicable performance of the particle counter. Replicability alone however is a poor substitute for assay accuracy.

Considering the importance of particle contamination to all concerned with the research, production and quality assurance of injectable products, the essential content of this chapter is communicated in narrative style. Graphs and tables have been included that can be understood without recourse to mathematical details. Two appendixes are included. The first details a particle counter validation protocol, and the second is a straightforward presentation of the essential mathematical details keyed to the narrative section of the chapter. Definitions and special symbols are listed in Table 1.

II. INTENT AND REALIZATION OF PARTICLE COUNTING ACCURACY

Before the results of the counts recorded by particle counters can be evaluated, the intent of these U.S.P. contaminating particle assays must be determined. Rereading of the new U.S.P. XXIII <788> as well as earlier U.S.P. revisions shows that the intent of these contaminating particle assays is to set objective particle contamination quality standards for injectable products based on the number of single particles in a selected sample volume.

The selection of objective particle contamination quality standards was based on two considerations. These considerations were patient insult and the accuracy required of a U.S.P. assay. With these considerations, 10 μm was selected as the minimum size for particle contamination quality standards based on the perceived accuracy of the membrane/microscope contaminating particle assay technique. With patient insult considerations in mind, a much lower acceptable particle contamination level was set for particles greater than 25 μm.

With these measurement intentions in mind, count accuracy itself is a necessary but insufficient criterion with which to make judgment concerning the suitability of electronic particle counters for U.S.P. <788> contaminating particle assays. Critical examination of the relationship of the counts recorded to the concentration of particles counted is essential if the intent of the contaminating particle assay is to be achieved.

There are three prerequisites that must be satisfied before the present confusion surrounding particle counter counting capability can be dissipated. The first is a validated model of the counting process. This validated model will provide verified and verifiable counter performance data. Such a model can provide the secure foundation upon which standard, industry-wide, particle counter specifications and performance operating limits can be based.

The second requirement is for test methodology that will accurately relate particle counting performance to the validated model of the counting process in particle counters. This relationship will facilitate the determination of operating limits within which accurate counting data will result.

The third essential requirement is the adoption and implementation of operating limits that will give the end user the accuracy that is an essential part of any quality assurance procedure. These three prerequisites will result in accurate, correlated U.S.P. <788> particle contamination measurements independent of vendor and instrument operator.

This chapter consolidates the information required to satisfy these three requirements necessary to achieve single particle counting accuracy. It also includes a discussion of some basic limitations on both the functioning of particle counters and the meaning of the data recorded.

III. VALUE JUDGMENTS

There have been many occasions when the utility of an advance in technology is so clearly visible that it is pressed into service immediately. Neat, all encompassing explanations of how the advance works and discovery of its limitations are eventually made and are dutifully documented in the scientific literature. The use of the Coulter, the light extinction and the light scattering particle counters are prime examples of this phenomenon. This chapter analyzes count accuracy in light extinction particle counters.

These counters offer the capability to register counts of suspended particles as the particles pass through a sensing zone. They replace tedious, eye-straining microscope sizing and counting assays by highly trained personnel with an automated device that can be operated by a technician. These devices have shown good replicability; their convenience has tended to obscure the inherent limitations of these counters. The methodology described in this chapter can add measurement accuracy to the list of the benefits that these counters have made available.

IV. LIMITATIONS

This chapter does not address the replicability or stability of particle sizing or particle size resolution. It examines and analyzes the accuracy and replicability achieved in the counting of sized particles by electronic particle counters. For effective use of the data from these counters, some comments on particle "sizing" are needed. The original approach to particle sizing in the U.S.P. used the maximum particle dimension. As it became clear [2] that the reliability of the membrane/microscope measurement of these low contrast particles was too low to result in measurement accuracy, a shift to the replicable results of light extinction particle counter measurements took place. A result of the change in measurement method was a change from a direct measurement of the maximum length of the contaminating particle to an indirect measurement of an associated particle characteristic. The associated particle characteristic evaluated in light extinction particle counters is the equivalent spherical volume of a particle. This measurement is based on its projected two-dimensional shadow. The best accuracy results when an ideal spherical particle is measured at a ratio of the calibration point of a liquid medium in which there is a match to the refractive index ratio of the calibrating particle/liquid combination. As a result of the shift to light extinction detector particle sizing, the greater the divergence from the assumed spherical shape, the greater the measurement variability becomes [3]. Fibers, spicules and platelets are among those particles with the least measured dimensional accuracy in light extinction particle counters.

V. IMPLICIT AND EXPLICIT PARTICLE "SIZE" MEASUREMENTS

The measurement of particle "size" with light extinction particle counters is not in agreement with particle "size" measurements by a conductimetric detector such as the Coulter counter or with the membrane/microscope method. There is no scientific law that requires that these measurements be correlated. The equivalent of this comment is that inferred measurements of particle "size" by measurements of particle volume, particle weight and by the measurements of the projected cross section of a particle are not in agreement: of course.

Since all particle shapes are homogenized into "equivalent" spheres in light extinction particle count data, these counters are not suitable for referee-grade particle analysis. The size recorded for any nonspherical shape depends on its orientation as it passes through the scanning beam. The largest measurement errors occur for particles whose shapes vary most from the ideal spherical shapes, such as for platelets and fibers. The increased measurement variability of nonspherical shapes is due to the variation in the three-dimensional orientation of the particle as it passes through the scan beam of the detector. From the randomized orientation with which a particle enters the sample pickup tube through the traverse of the scan beam in the sensitive volume of the detector there is no mechanism to orient nonspherical particles into some preferred position. A sampling of the effect of random particle orientation on particle size measurements in light extinction particle counters [3] is shown in Tables 2 and 3. Table 2 examines the variation in the measured spherical diameter of a light extinction particle counter for two different flat platelike shapes. The square and triangular or arrowhead shapes selected were of uniform thickness between 5 and 20 μm thick. The maximum linear dimension of these particles ranged from 35.4 to 200 μm. The equivalent spherical measurements of the arrowhead shapes ranged from 36% to 56% of the maximum linear dimension. The equivalent spherical measurements of the square shapes ranged from 36% to 80% of the maximum linear dimension due to the random position of the particle as it traverses the scanning beam. Comparison of the light extinction equivalent diameters with the maximum lineal dimension of the selected particles affirms the opinion that these counters are not capable of referee quality particle assays.

The expanding use of electronic particle counters for routine quality assurance measure-

Table 2 A Comparison of Light Extinction Equivalent Sphere Diameter Measurements with Maximum Particle Dimensions

Particle shape	Dimensions			Light extinction equivalent diameter		Maximum linear dimension
	t	L_1, L_2	L_3	D_{MIN}	D_{MAX}	
Arrowhead	5	50	25	17.8	27.8	50
	15	100	50	35.7	55.5	100
	20	200	100	71.4	111.0	200
Square	5	25	—	12.6	28.2	35.4
	10	50	—	25.2	56.4	70.7
	20	100	—	50.5	112.8	141.4

L_1, L_2 are the equal sides of a triangular chip or the sides of a square chip. L_3 is the length of the base of the triangular chip; all dimensions are in micrometers.

Table 3 A Comparison of Light Extinction Equivalent Sphere Diameter Measurements with Fiber Dimensions

Fiber length/diameter ratio	Fiber dimensions		Light extinction equivalent diameter		Maximum linear dimension
	D_F	L_F	D_{MIN}	D_{MAX}	
5:1	5	25	5	12.6	25
	10	50	10	25.2	50
	20	100	20	50.5	100
10:1	5	50	5	17.8	50
	10	100	10	35.7	100
	20	200	20	71.4	200
20:1	5	100	5	25.8	100
	10	200	10	50.5	200
	20	400	20	17.4	400

All dimensions are in micrometers. D_F and L_F are the diameter and length of the fiber.

ments has involved a step away from the use of the "maximum particle size dimension" in contaminating particle assays. The greater the departure from an ideal spherical shape, the greater the loss of dimensional accuracy. The particle shape whose shape is most divergent from a sphere is a fiber. A rule-of-thumb definition of a fiber shape is a particle whose length is at least five times greater than its diameter. A 5-μm-diameter fiber whose length was 25 μm was judged to be equally as dangerous as a spherical particle whose diameter was equal to 25 μm. Similarly a thin platelet shape whose thickness was 5 μm and whose length was equal to 25 μm was also judged to be equal in patient insult to a 25-μm-diameter particle.

Table 3 compares the results of light extinction particle counter measurement of a range of fibers with the actual fiber length. Since the projected image of a fiber is least comparable to a sphere, the range of differences between the measured to the actual fiber length shown in this table, from approximately 1/2 to 1/20 is not unexpected. The measurement disparity increases as the ratio of fiber length to fiber diameter increases. The data in this table are a persuasive argument for periodic visual examination and identification of the kinds of particulate contamination found in injectable products.

VI. PARTICLE "SIZE" CALIBRATION

The light extinction particle counter is a measurement transfer device. For ideal spheres, suspended in a liquid medium that has the same refractive index ratio as the calibrating spheres in the calibrating liquid employed, there are two accuracy limits. They are the resolution of the measurement instrument and the accuracy of the particle standard employed. The availability of accurate particle size standards make possible the secure transfer of spherical calibration data from the designer's laboratory to the ultimate user. Between calibration points, particle size is determined by interpolation using an algorithm selected by the instrument vendor. In recent times the advent of laser diode particle counters, with their improved durability, has brought with it a series of changes that must be considered. These changes include decreased scan beam heights which improve the concentration limits for accurate particle counting operation. The use of these light sources has also resulted in departures from the previously assumed continuous square-law particle sizing characteristic throughout the pharmaceutical measurement range.

A consequence of the calibration method is that maximum instrument accuracy results at the instrument calibration points. In view of the specification of 10 μm as the minimum particle size that is of concern and the increased penalty for particles greater than 25 μm in size, prudence dictates that these, or any future breakpoints, coincide with calibration points. To ensure the best sizing accuracy at the critical <788> 10- and 25-μm particle size points, U.S.P. XXIII specifies that both of these particle sizes coincide with calibration points.

VII. REPLICABILITY AND COMPARABILITY OF PARTICLE COUNTER DATA

The central focus of this chapter is on the effect of particle counting occurrences within the sensitive volume of the detector and on its electronic control system. The unbiased functioning of the detector requires an unbiased sample stream to and through the detector. The requirements for an unbiased particle sample stream have been defined. The requirements are agitation of the sample stream to prevent sedimentation and maintenance of a sample stream velocity high enough to select and deliver an unbiased sample aliquot to the sensitive volume of the detector. A brief overview of the problem follows.

The sensitive volume of the detector is positioned vertically over the suspension being evaluated. Suction is employed to draw a liquid volume from the sample container through a relatively short tube into and through the sensitive detector volume. This type of sample selection and transport is in widespread use for particle counter applications. Liquid velocity in the sampling tube must be high enough to transport the largest and densest particle that should be evaluated.

VIII. U.S.P. PARTICLE CONTAMINATION STANDARDS AND THEIR MEASUREMENT

The 1966 National Symposium on "Safety of Large Volume Parenteral Solutions," organized by the Food and Drug Administration [5], was a defining moment in the development of particle contamination problems, patient safety and particle contamination measurements. One conclusion reached in 1966, that has been supported in all U.S.P. revisions since that time, is that patient insult increases sharply with the particle "size." This reasoning led to a particle sizing methodology in which the greatest dimension was used to classify contaminating particles.

The initial selection of acceptable particle contamination limits was based on an appraisal of acute patient insult by particle contamination in injectable fluids, the then current manufacturing technology and the perceived accuracy of the available particle contamination measurement. With the approximate 1-μm sizing accuracy limit imposed by the available microscope objectives, a choice of 10 μm for the lower size limit to be controlled could be measured with a 10% error. For a large volume injectable container, a contamination limit of 50 particles per milliliter or 50,000 per liter was selected. A 10-fold reduction of the acceptable contamination level was selected for larger particles, those equal to or greater than 25 μm, resulting in a limit of 5000 particles in this size range per container. (These limits have been revised in U.S.P. XXIII to 12 particles ≥ 10 μm and 2 particles ≥ 25 μm per milliliter.)

The driving engine behind the selection of particle size limits was the belief that filtration though an appropriate membrane and evaluation of the trapped particulates with a microscope was an appropriate referee analysis method. The belief in the membrane/microscope evaluation of particles was challenged with measurements of actual production particles. It soon became apparent that the low contrast of most of the contaminating particles seriously reduced their detectability and thus the accuracy of the particle assay. This low-contrast detection problem was "solved." The use of a grazing light source improved the detectability of low-contrast particulates. The "solution" resulted in roughly comparable particle contamination data.

In the period from 1977 to 1980 the acceptance of terminally sterilized dextrose solutions by the membrane/microscope particle contamination assay became a critical problem. Terminal sterilization of these solutions produced a 5-HMF polymer, a natural dextrose byproduct. This polymer appears as a gelatinous or thin filmy mass whose thickness approached the wavelength of visible light. The thin polymer film in the dextrose sample, deposited on the filtration membrane under the influence of the 15 PSI vacuum filtration, can appear as a balled object, a rolled-up cylinder, or can be stretched out flat showing its full surface area on the membrane surface. The filmy material appeared as a thin "stain" in sufficient quantity and size to block acceptance of this critical product under the particle contamination limit requirement. After detailed examination and review, these thin "stains" were judged to be an interference to the particle contamination measurement rather than a contaminating particle. Fortunately, particle contamination measurements with light extinction particle counters were demonstrated to be relatively immune to this thin film.

In June 1980, the PMA LVP Committee presented a report from Walter C. McCrone Associates, Inc., which identified the filmy material as an amorphous mass rather than a discrete particle. In the 1980 July–August issue of the *U.S.P. Pharmacopeial Forum,* an interim notice was published which allowed the analyst to ignore the appearance of the 5-HMF films. This note suggested the use of a light blockage electronic particle to distinguish between particles and the 5-HMF filmy material. This recommendation was based on the fact that the poor detectability of the filmy dextrose breakdown polymer effectively eliminated its interference with the U.S.P. contaminating particle assay. Considering the importance of the problem to all concerned, the PMA sponsored a collaborative study of the light extinction particle counter made by HIAC/Royco. This study was completed and submitted to the U.S.P. in May 1983. The crisis in acceptance of terminally sterilized dextrose was resolved by the cooperative validation of particle counters counting small aliquots of large-volume injectables tested at low particle concentrations. The acceptance limits were 50 particles per milliliter equal to or greater than 10 μm of which 5 were equal to or greater than 25 μm. In retrospect, the use of particle counters for particle contamination assays was validated for the low particle concentrations found in LVI products. This "validation" was a narrow gauge response to the 1980 crisis of accepting terminally sterilized dextrose. In this narrow gauge validation, implemented in a crisis atmosphere, instrumental specifications and limitations were, understandably, ill understood and scantily addressed.

In 1985 U.S.P. XXI listed a single vendor's light extinction particle counter as an accepted measurement device for small-volume injectable <788> particle contamination evaluations. It soon became apparent that the data from identical models could not be correlated. In parallel with the emergence of data comparability problems with a single vendor, the Pharmacopeia became aware that designation of a single commercial instrument for a required assay was untenable. Consequently, on November 15, 1990, the Pharmacopeia deleted the single-instrument source designation. The acceptance of multiple vendors for these instruments without an accepted referee methodology increased the importance of resolving the problems of data replicability. The achievement of accuracy and replicability of counter data is still a future goal.

In 1990, Munson as FDA spokesman [2], reported ±50% particle assay errors using the membrane/microscope particle assay. These results were obtained by the same person on the same preparation with the same microscope and lighting system. The low accuracy of the membrane/microscope method of particle contamination measurements limited its analysis and enforcement value. Data from the membrane/microscope method of particle contamination analysis was usable only when it exceeded the error margin. In the report cited, the FDA spokesman challenged the P.D.A. Conference at which the report was presented saying, "Compare this (the ±50% FDA measurement error) to the estimated error for the electronic particle counting method of 15 to 20%." This mirage of reduced error was shown to be a result of the interpretation of counter data replicability as count accuracy. The replicability of particle counter data combined with its ease of use were arguments so seductive that a goal of increased application of the electronic counter to all contaminating particle assays was initiated.

The widely variable results seen in the use of the membrane/microscope assay for SVI batch release assays, reported by Munson, fueled a search for alternatives. The demonstrated replicability of particle assay data recorded with these counters made them appear a desirable alternative for both SVI and LVI particle contamination assays. Validated usage for high particle concentrations was assumed but never substantiated. Lacking resources and budget, the FDA neither initiated the essential measurements nor supported them in Academia. The continuing role of the U.S.P. was that of an interested observer.

IX. ACCURACY LIMITATIONS

A. Refractive Index Effects

An additional problem that has been defined and demonstrated experimentally in these counters is the possibility of undersizing or, in an extreme case, of missing particles which should be counted. Complete cancellation of the particle signal can occur when the refractive index of the particle and the liquid coincide as shown in the graph of Akers 1972 reported by Russell [6] in Fig. 1.

In a more recent and more general paper by Knollenberg and Gallant [7] undersizing and

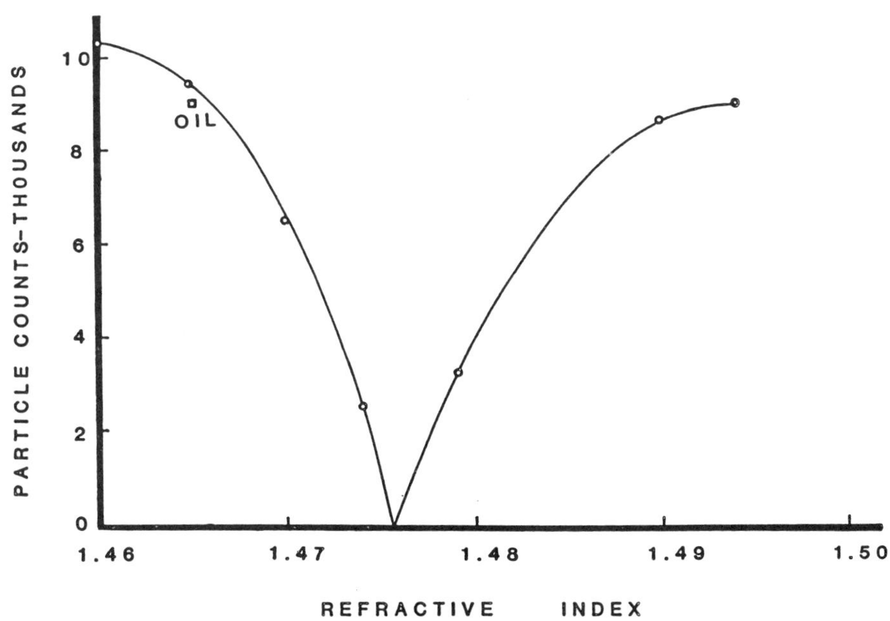

Figure 1 An experiment by Akers demonstrated not only complete cancellation of counting in light obscuration detection systems when the refractive indices of particle and liquid are matched but significant errors as the match point was approached.

ultimately signal cancellation can occur when the refractive index ratio of the liquid approaches that of the contaminating particle. A graph of the Knollenberg and Gallant results is shown in Fig. 2. Fortunately, this effect has not been reported to be a problem in pharmaceutical measurements.

B. Immiscible Droplet Interference

In addition to the accuracy problem introduced by non-spherical contaminating particles, immiscible droplets suspended in the product being assayed introduce an additional measurement variable. These immiscible droplets are counted as variable-sized contaminating particles. The variation in the recorded size of these droplets results from the mechanical agitation of the container prior to measurement and the physical characteristics of the suspending liquid and the immiscible droplet.

X. AGITATION AND TRANSPORT

A similar critical review of the particle transport system is essential. The maintenance of particle transport velocity through the detector within the operating range determined by the instrument designer is an essential requirement for the maintenance of sizing accuracy. Without unbiased selection and transport of the full range of randomly occurring particle sizes, shapes and densities from the test container to the sensitive volume of the detector, count accuracy cannot

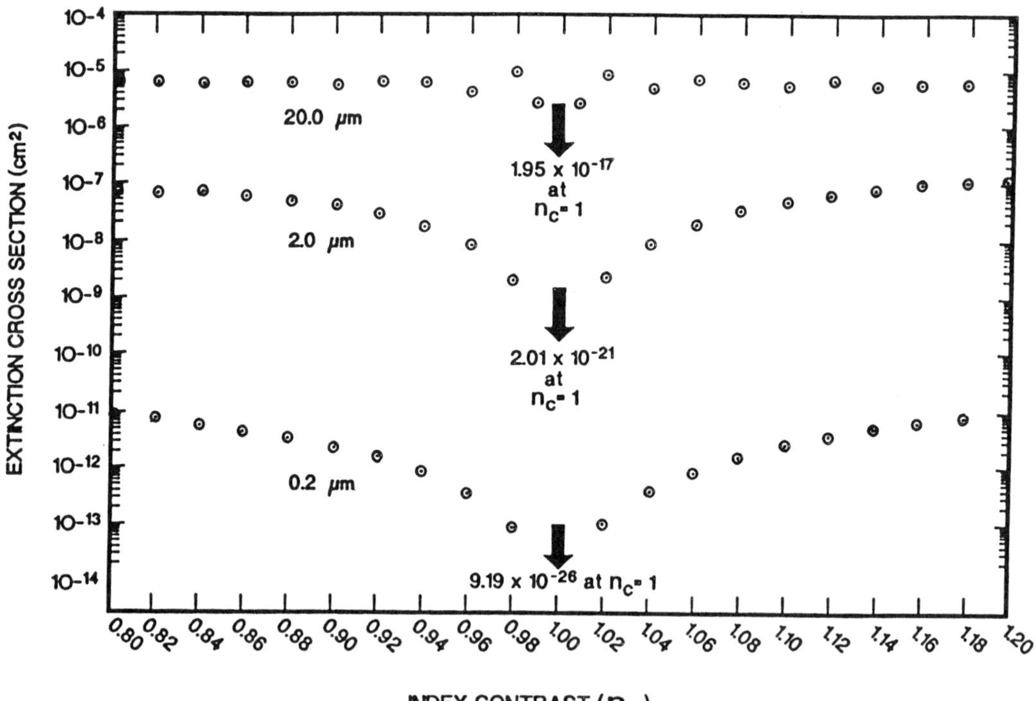

Figure 2 The extinction cross section of spherical particles is shown as a function of the index contrast ratio, the ratio of the refractive index of the particle to the refractive index of the suspending medium, $n_c = n_p/n_m$. The detectability of the particle is effectively zero when the refractive index ratio is equal to one. (From Ref. 7.)

be achieved. The specification of container shape, material, method and speeds of agitation enter into adequate randomization of the sample volumes measured. Accurate particle counts cannot be expected unless the randomized samples are delivered within the pickup zone of the sampling tube. Since counting error is determined by particle concentration, inadequately controlled sedimentation modifies the particle concentration sampled and is a multiplier of counting error [8]. The final hydrodynamic barrier to the delivery of unbiased samples to the detector is liquid velocity within the sampling tube. Unbiased delivery of the full range of particle sizes and densities to the detector through the sampling tube is a prerequisite for counting accuracy.

There is no interrelationship between agitation of the sample volume and an acceptable sample tube flow velocity. While agitation can minimize the flow velocity required for adequate particle sampling, ideal agitation, by itself, cannot deliver particle samples against the force of gravity.

Calculation of an upward flow velocity at which the force of gravity for a particle is balanced, the neutral buoyancy flow velocity, can be estimated with Stokes' law. Table 4 lists the settling velocity in still 20°C waters of a range of particle densities that can be encountered as contaminating particles. The selected list includes latex, borosilicate glass and stainless steel spherical particles. This table lists the terminal settling velocities for spheres of these materials in a size range from 10 to 200 μm.

The flow velocity that will result in neutral buoyancy of a spherical particle within the flow stream of the tubular probe is a limit calculation. Any flow velocity equal to or less than the neutral buoyancy flow velocity results in nondelivery of the particle to the sensitive detector volume and thus a failure of the counter to detect and count that particle. The computed velocities can be used to estimate minimum volume flow velocity for unbiased particle transport to the detector. When the 50-μm stainless steel particle is used as a limit test, 22 ml per minute is the neutral buoyancy volume flow velocity in a 3½-mm-diameter sampling tube.

Additional flow velocity multiples are required when the full range of particle contaminant shapes are considered. Another consideration for adequate, unbiased delivery of particle distributions to the sensitive volume of the detector is the excess flow velocity required to generate an adequate particle pickup zone in the sample volume surrounding the input end of the probe.

The table of Stokes' law settling velocity is applicable to another source of bias in particle contamination measurements. The U.S.P. <788> contaminating particle assays involve sample pooling and manipulation. If sedimentation occurs during the period of sample manipulation, the bias introduced will affect the particle contamination analysis in an unretrievable fashion. Trasen [4] concluded that, "In order to keep sampling errors at a minimum it is important to

Table 4 Stokes' Law Computation of the Terminal Settling Speed for Spheres from 10 to 200 μm in 20° Still Water

Particle diameter (μm)	Density	Latex 1.05	Borosilicate glass 2.57	Type 303 s. steel 7.86
		Terminal settling speed (mm/s)		
10		0.0029	0.088	0.38
25		0.018	0.54	2.38
50		0.072	2.18	9.50
100		0.29	8.17	38.00
200		1.15	34.82	152.00

sample directly from the final packaged container and to sample the entire volume." Trasen's remarks were made in a discussion of the membrane/microscope technique. Kushner [8] demonstrated the effect of sedimentation on data accuracy in an experiment in which a succession of 5-ml samples were drawn and analyzed from continuously stirred 500-ml LVI containers. The detector employed was a HIAC/Royco HR-60HC operated at a flow rate of 10 ml per minute. The sampling tube was lowered after each sample to ensure correct operation. The experiment was terminated after approximately 350 ml were sampled due to air entrapment. The containers were seeded with AC fine test dust. The AC fine test dust ranging in size from 3 to 60 μm was selected on the basis that it was more representative of the random particulates than latex particles. To analyze the effect of sampling variability due to sedimentation, all of the 5-ml analyses were summed and then expressed as a proportion of the total detected in each 5-ml sample for the entire container. The Fig. 3 analysis shows that even with these small particles, over 100% count variability resulted from the joint effect of particle size and the particle pickup zone.

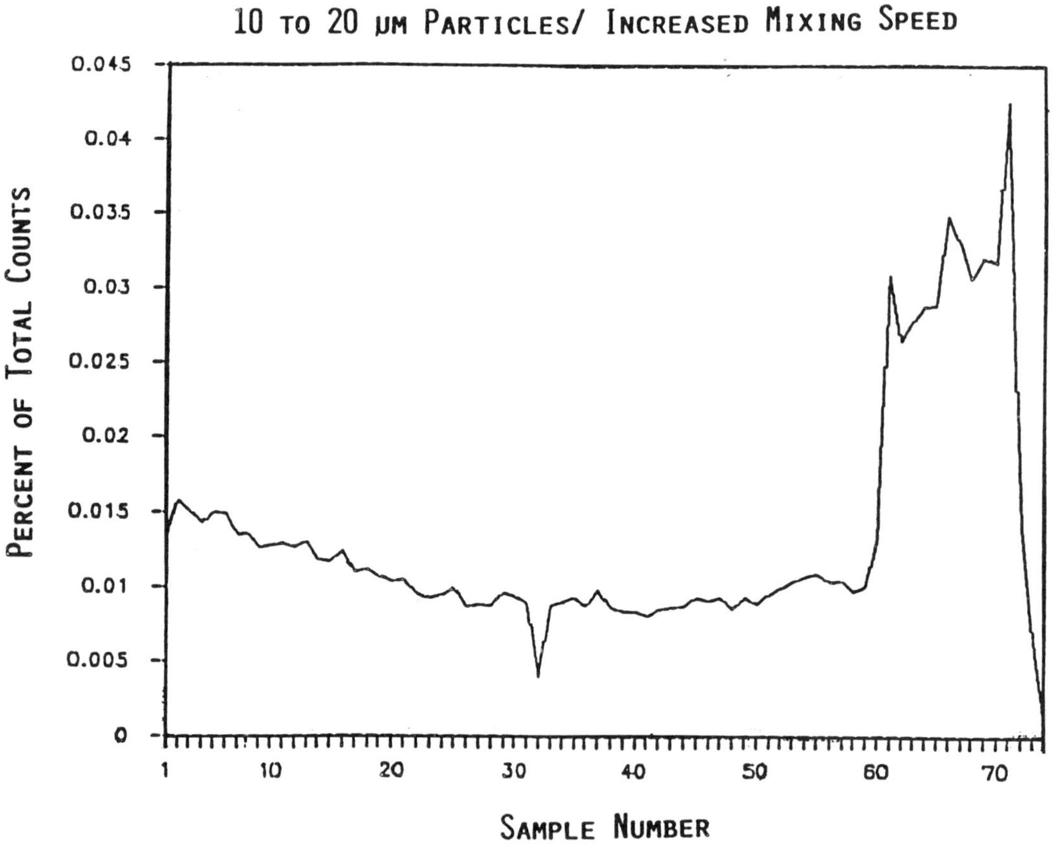

Figure 3 Demonstration of sampling error effects. The effect of sedimentation during the course of <788> measurements is demonstrated in an experiment with AC fine test dust. The agitation employed is due to a Teflon-cased magnet with external drive at twice the normal speed employed for particulate analysis. The data shown is for particle sizes 10–20 μm in diameter. Since control of the coincidence error is based on computations in which the distribution of particles is described by the *average* number of particles per unit solution volume. It is clear that control of sedimentation is a prerequisite to data accuracy.

These hydrodynamic problems require theoretical analysis and empirical confirmation of critical operating conditions. These problems have, as yet, not been adequately addressed. Inadequate consideration of these hydrodynamic requirements can nullify the effectiveness of any particle counter system.

XI. PARTICLE COUNTING

Although particle size standards are now a well-established resource, the parallel development of accurate particle count standards and an accurate model of the counting process has been much slower in arriving. There is a generally accepted view that the sizing and counting of high particle concentration levels affect both the sizing and counting of the particle suspension. The result of such high particle concentration measurements is seen to be a reduction in the total number of particles counted and an increase in the recorded particle size distribution. This result has been labelled the "coincidence effect."

Liquid-borne particle counters that are presently available have one limitation in common: they can accurately count or size only one particle at a time [3,8–26]. The accuracy of the relationship between the instrumental count of multiple particles and the actual number of particles in a suspension is a concern in many disciplines. This is a significant problem for some pharmaceutical and hydraulic measurements since contaminating particle measurements may be made at relatively high particle concentrations. For parenterals, the release for sale of injectable pharmaceutical products is conditioned upon a quality standard specifying size and count limits for any included particles [1]. Without accurate measurements to provide the necessary medical and economic basis for improvements in parenteral products, progress is hampered. The number and size of contaminating particles in the manufacture and sale of parenteral products has economic, legal and medical consequences. Particle measurement inaccuracy is therefore a concern to the entire parenteral community. Improvement in the accuracy of this measurement requires access to methodology with which instrument counts can be accurately related to the number of particles in suspension. Access to the data provided by this methodology can be used to validate an analytical model of the counting process in which instrument capability and operating limits can both be considered. Ideally, the results obtained with this analytical model will make accurate, replicable measurements possible.

The selection of an analytical model that is accurately representative of the realities encountered in particle counting is imperative if a goal of correctable particle count measurements is to be reached. In the absence of a verified analytical method, reliance upon an empirical method of determining count linearity becomes an urgent necessity. There has been no tested model or method with which to accurately estimate this loss of counts or the introduction of false counts into the U.S.P. <788> batch release assay. The chapter describes a practical method to relate the number of particles present in a suspension to instrument counts recorded.

A search of the particle-counting literature shows continuous concern with the relationship between the number of particles in a suspension, the particle size distribution and the counts recorded from them. From the pioneering Wales and Wilson paper [14] in 1961 to the present, there has been a continual progression of estimates of the relationship between the number of particles counted to the actual number of particles in a suspension. One common concept appears in most of these papers, the concept that the counting model is determined by the interaction of the events within the physical or geometric volume of the detector and the electronic circuitry employed. One common conclusion is that there is a size shift in the reported particle size distribution from smaller to larger particles, and a reduction in the total number of particles counted. The recorded size shift is caused by the simultaneous appearance, or coincidence, of particles in the sensing volume, and the consequent addition of their signals into a spurious larger particle. The reduction in the number of particles counted is also a result of their coincident appearance in the sensing volume of the detector, since the particle has been counted as part of a "coincident" particle group. A common problem in most of the papers that have treated the coincidence count loss error has been lack of experimental challenge to the published conclusions.

The 1972 paper by Jaenicke [9] made available the current Poisson model of count loss due to the "particle signal coincidence." Jaenicke's analysis was based on aerosol data: an appraisal of this paper reaches the conclusion that it is suggestive but not definitive. Jaenicke's interest was in the measurement of aerosol droplet distributions in air. He reported that increasing droplet concentration resulted in a decrease of particle counts. He also reported that high particle concentrations resulted in a distortion of the measured particle size distribution. This distortion increased the reported numbers of large particles and simultaneously decreased the reported number of small particles present. A quotation from Jaenicke's paper is appropriate. He said, "Since the subject of coincidence has not been adequately treated in the past . . . , we include a detailed discussion here."

Jaenicke proposed a static Poisson statistical model to describe the experimental loss of counts. In this paper Jaenicke also reported an attempt at experimental verification of his theoretical analysis with 80-μm polystyrene microspheres. The five data points he reported appear to follow the response for the selected model but are too scanty to be more suggestive.

In 1975 Lieberman [10], using the static geometric analysis model proposed by Jaenicke, reported similar but more extensive experimental results for liquid borne particles. Lieberman used concentrations of AC fine test dust to compare the performance of liquid-borne particle detectors. He verified the inverse relationship between detector volume and the linear count range of particle detectors. Working with "standard" and "high concentration detectors" he demonstrated an improvement in the linear counting range when detector volume was decreased. Lieberman also reported a test of the theoretical static Poisson model proposed by Jaenicke. He compared eight liquid-borne particle concentration data points with the theoretical curve coincidence count loss proposed by Jaenicke. Lieberman's data, while more extensive than that due to Jaenicke, falls short of definitive verification of the static Poisson model for coincidence count loss. The four lower concentration data points appear to follow theoretical curve. The counts recorded for the four highest concentration points appear to drop off faster than the theoretical estimate.

Willeke and Liu summarized the capability of optical particle counters for aerosol measurements in 1976 [11]. In this publication, the undercount loss estimate due to particle coincidence is based on the static Poisson model proposed by Jaenicke. While the discussion and application of the Poisson model are more extensive than reported by either Jaenicke or Lieberman, no experimental work to support the reported analysis is included.

Jaenicke's analysis, and consequently Lieberman's analysis, of coincidence count loss, used a static Poisson model of the counting process. A first-step assumption in Jaenicke's work was that the ratio of the detector volume to that of the particles being measured was so large that the particle size could be disregarded. This assumption simplified the calculation of the effect of particle concentration upon coincidence. The only remaining variable was the ratio of particles to detector volumes in the test volume. The Jaenicke-Poisson model divided the test volume into a number of unit volumes. Each unit volume was equal to the physical or geometric volume of the detector. The parameter that determined that amount of count loss was the ratio of particles to detector volumes. The theory predicted that at low concentration each particle would be accurately counted. As concentration increased, the count-to-particle ratio would decrease continuously until some limiting value related to the detector volume was reached.

The use of an infinite series in a Jaenicke analysis of count loss limited its use in laboratory measurements. In 1988, Knapp and DeLuca [3] showed that ratios of the terms in this series resulted in simple expressions that could be used to estimate the counting error and the coincident shift of particle size during the course of a measurement. The Knapp and DeLuca work suffered from the same defect that characterized most of the papers on coincidence count loss; it was not supported by experimental results.

In contrast to the particle sizing problem, concentration limits for counting accuracy rested on essentially unverified vendor claims of performance. As a stop-gap measure, the U.S.P. recently introduced a qualitative count "standard" to provide an interim answer to the need for a functional test of counting capability.

The lack of methodology to objectively evaluate the counting performance of single particles resulted in variability in counting performance of identically specified instruments. This problem was investigated by Gordon Oxborrow, an F.D.A. measurement specialist. At the 1987 International Conference on Liquid Borne Particle Inspection and Metrology, Oxborrow [13] reported an average variability of 429% for the contamination measurements of four marketed small-volume injectable products measured with five commercial particle counters.

The Parenteral Drug Association Research Committee, chaired by Dr. Michael Korczynski, funded an investigation of Knapp and DeLuca into the measurement variability and the lack of correlation between particle concentration measurement methods. These included the manual membrane/microscope method, the light extinction and conductimetric particle counters. Their review was submitted in August 1987 and appeared as *Journal of Parenteral Science and Technology* Technical Report No. 9, a supplement to the 1988 January–February issue [3].

The counting error issue was addressed in this report. The count error analysis commenced by assuming that the Jaenicke [9] Poisson model was correct. The report included the observation that using this model, a doubles occurrence was the most probable coincident event. By taking ratios of terms in the Poisson series solution some simple calculations became possible during the course of an experiment. This paper offered an approximate estimate of the undercount error. The undercount error was simply half of the ratio of particles to the geometric detector volume: a ratio of 0.10 particle/detector volume indicated a count error of 5%.

Two years later, Toshinobu Aoyama [20] at the 1990 International Conference on Particle Detection Metrology and Control compared data from two identically specified particle counters and concluded that "particle counting data could be compared but could not be correlated." At this conference Knapp and Abramson [12] reported on their investigation of particle counter undercounting for high particle concentrations. Their results were applicable to the concentration limits of contaminating particles specified in U.S.P. XXII <788> quality determinations of small-volume injectable solutions. The data reported showed that the size of a particle affects the concentration limit for count accuracy in a detector. It raised serious questions concerning the suitability of the use of the Poisson statistics in the accepted Jaenicke model for counting loss analysis.

It is clear that the present methods of measuring particle contamination limits can result in both falsely rejected batches and in undercounted measurements at the maximum particle acceptance limits for injectable products specified by the Pharmacopeia. It does not address the accuracy, resolution and stability of particle sizing that are essential to the evaluation of contaminating particles.

Counting contaminating particles in an injectable solution is a complex, dynamic process whose problems are not all well defined or well understood at this late date in their accepted use. This chapter describes a solution of the coincidence counting error problem. The counting error problem includes particle undercounts in the 10- to 25-μm particle size range and the contribution of false counts of coincident particle signals. These false counts can occur in two regions, particles less than 10 μm in size can be counted as greater than 10 μm by coincident occurrences. Similarly, groups of coincident particles between 10 and 25 μm in diameter can be counted as single particles greater than 25 μm in diameter.

In light extinction particle counters two mechanisms for this change in particle size distribution have been identified. The first is the simultaneous appearance of multiple particles in the path of the scanning beam. In this case, the particle size recorded is evaluated from the sum of the particle areas projected upon the detector. The second mechanism is the random chaining of particle signals, the arrival of a sequence of particle signals without a sufficient interval to allow the counter to interpret each particle signal as a separate entity. In this case, only one particle size is recorded for each chain of particles. The recorded particle size signal can be equal to the largest particle in the chain or it can be equal to a sum of particle signals that have occurred at some point in the particle chain.

Estimates have been made of the loss of counts for particles greater than 10 μm due to their coincidental appearance in the scanned volume of a detector as a result of concentration. As particle size decreases there is a rapid increase in the random concentration. The coincidence concentration effect, given the high population of particles below the U.S.P. <788> control threshold can also be the source of false count introduction into the mandated U.S.P. <788> batch release measurements. The envelope of the <788> acceptance limits for injectable product contaminating particle size distribution can be used as a guide to the exponential nature of the underlying particle distribution. Delly [17] has reported that in some products 81% to 96% of all particles are below 10 μm in size. Selection of a high measurement threshold does not eliminate this effect, it simply hides the origin of the false counts. *Coincident signals greater than 10 μm traceable to groups of smaller particles cannot be distinguished from single particles equal to or greater than 10 μm in size.*

Two conditions must be satisfied to achieve accurate counts of particle contamination independent of vendor and laboratory usage:

1. Standard specification and evaluation of the counting capability of a counter
2. Operation within a specified particle size concentration limit both below and within the <788> particle size range for the accuracy desired.

The methodology needed to satisfy both conditions is described below. Control of particle contamination is such a basic part of parenteral manufacturing and quality control that any advance in its measurement should be communicated across disciplinary boundaries. This chapter has been organized with this requirement in mind. The details of present and new particle counting theories have been included as appendixes.

XII. THE EFFECT OF CONCENTRATION ON PARTICLE COUNTING DATA

The Coulter, light extinction and light scattering counters all share a common limitation. They are single-particle counters. They are designed to accurately count low concentrations of particles. Concentrations so low that there is a very small probability that a simultaneous particle grouping will appear in the sensitive detector volume and be counted as a single, larger particle. Although the particle "size" standards are now a commonplace, particle counting accuracy and particle counting standards are still troublesome topics.

XIII. CONSIDERATIONS IN PARTICLE COUNTING

Experimentation with the counting capacity of particle counters using mono-sized PSL microspheres has confirmed the view that counting accuracy depends on particle concentration.

A typical-count-to-particle concentration relationship for a well-designed system used within and beyond its design limits is shown in Fig. 3. The calculations on which this figure is based are discussed in detail in Sec. LIII, page 523.

Figure 4 shows that counter operation can be described in four regions defined by increasing particle concentration. The first region, where particle concentration is low, is one in which background counts affect the accuracy of the recorded count. In the second region, the linear region, background counts are negligible and the number of counts observed increases linearly with increasing particle concentration. In the third region, there is a continuing loss of counts with increasing particle concentration and finally, in the fourth region, there is system failure due to overload. System performance in the overload region can depart widely from model predictions and can result in a display of zero counts for high concentrations.

When a pulse height analyzer is connected to the output of the detection circuitry of a particle counter, the linear output shown in Fig. 5A and the logarithmic record of Fig. 5B record the detection of single- and multiple-particle occurrences in the sensitive volume of the detector. Attempts to validate a statistical model based upon the use of the relative peak heights shown in these figures have been unsuccessful. The merging of the peaks and the particulate tail near zero, visible in this figure, contributed to this lack of success.

Visible in Fig. 5 is the physical basis for the observed reduction of the ratio of counts to particles as particle concentration increases. It is the appearance of integral multiple-particle size signals clearly seen in both the linear and log displays of this figure. In the log display the continuation to higher order particle size multiples is more clearly visible. Consideration of statistical analyses suited to the counting data encountered in particle counters points to a Poisson analysis. It must be understood, however, that the need for a Poisson analysis is not limited to the Poisson count accuracy analysis described in the literature. The selected analysis

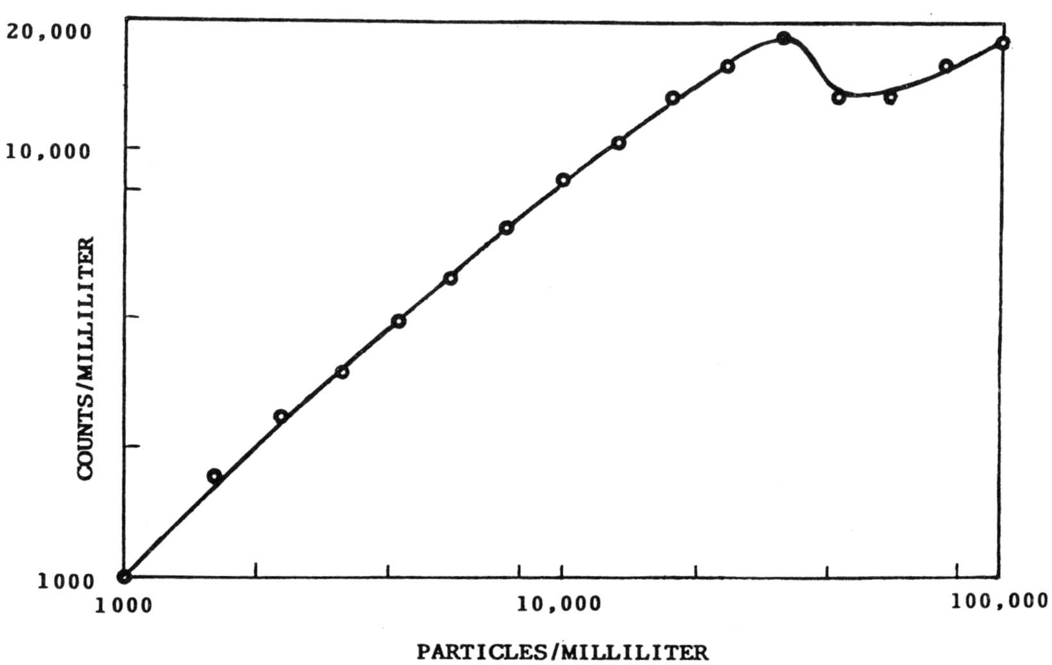

Figure 4 Count response to an increasing concentration of 32.2 μm particles. Initial linearity is followed by increasing count loss and then system collapse under extreme overload.

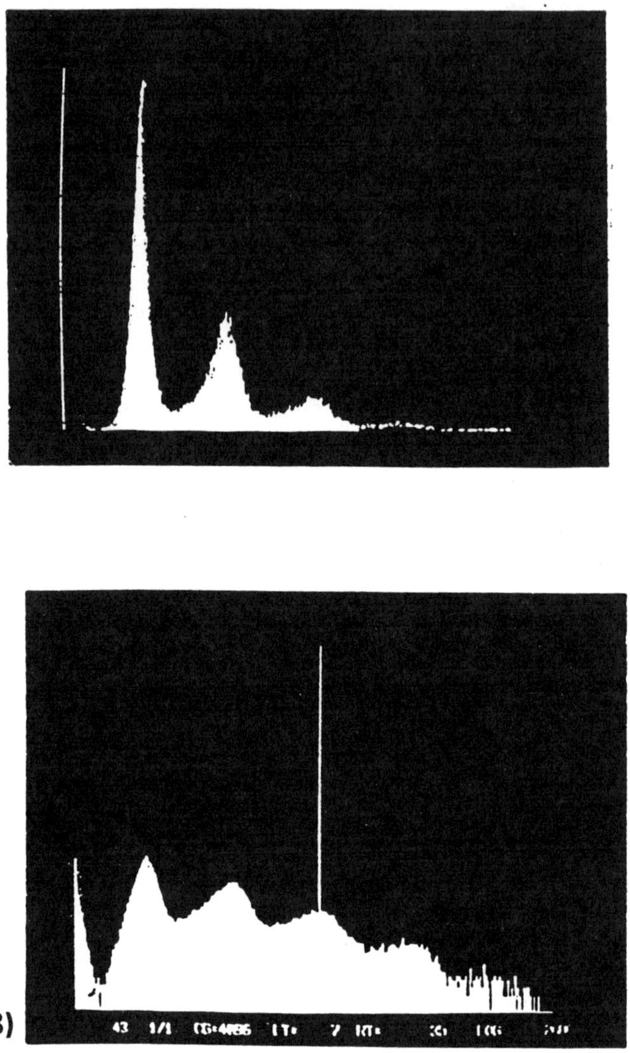

(A)

(B)

Figure 5 A pulse height analysis of the size and count of signals from a light extinction particle detector counting a high concentration of monosized particles. The horizontal axis is particle size; the vertical axis is particle concentration. This recording shows the number of single particles counted and the occurrence of two, three and four multiple coincident particle events. The particle size scale in both recordings is linear. (A) Linear analysis of particle concentration; (B) Particle concentration in log form.

underestimates the number of particles present with increasing severity as particle concentration increases.

A clear visualization of the accepted Poisson statistical model with which counting errors were estimated could be pictured as a box packed with sugar cubes. The sugar for this box of sugar cubes contains some grains of sand. Depending on the ratio of sand grains to the total number of these sugar cubes, the number of cubes that have zero, one, two, three or other multiple sand grains can be estimated with this Poisson statistical model.

When the time course of actual particle signals was examined, it became apparent that groups of particle signals occurred randomly throughout the measurement period. These grouped particle signals were separated in the selected model. The defect visualized in Fig. 6 impairs the count loss calculations based on this model. The effect of the arbitrary division of particle events is modeled in this figure by the segmented, continuous platform of detector volumes supporting the particle bars. This type of Poisson counter analysis has been labeled geometric Poisson analysis.

The overhead bars represent an alternative application of the Poisson analysis in which the measurement of a detector volume commences at the detection threshold of a particle. This model of particle counting events has been named the triggered Poisson model to distinguish it from the earlier geometric Poisson model. Both models are equally effective with the events shown in Fig. 6A; the triggered model performs better with the particle events shown in Fig. 6B. With the new model, a coincident event occurs when one or more particles enter the sensitive volume before the final exit of the initiating particle. An extended comparison of the effectiveness of these two models follows.

The use of arbitrarily positioned detector volumes in the geometric analysis, cannot adequately describe the random chaining of particle occurrences as shown in Fig. 6B. With this flaw in concept, underestimates of particle counting error result. Each case in which the geometric Poisson model provides an accurate estimate describes a static situation. In particle counters, however, there is an interaction between the functioning of the detector and the size

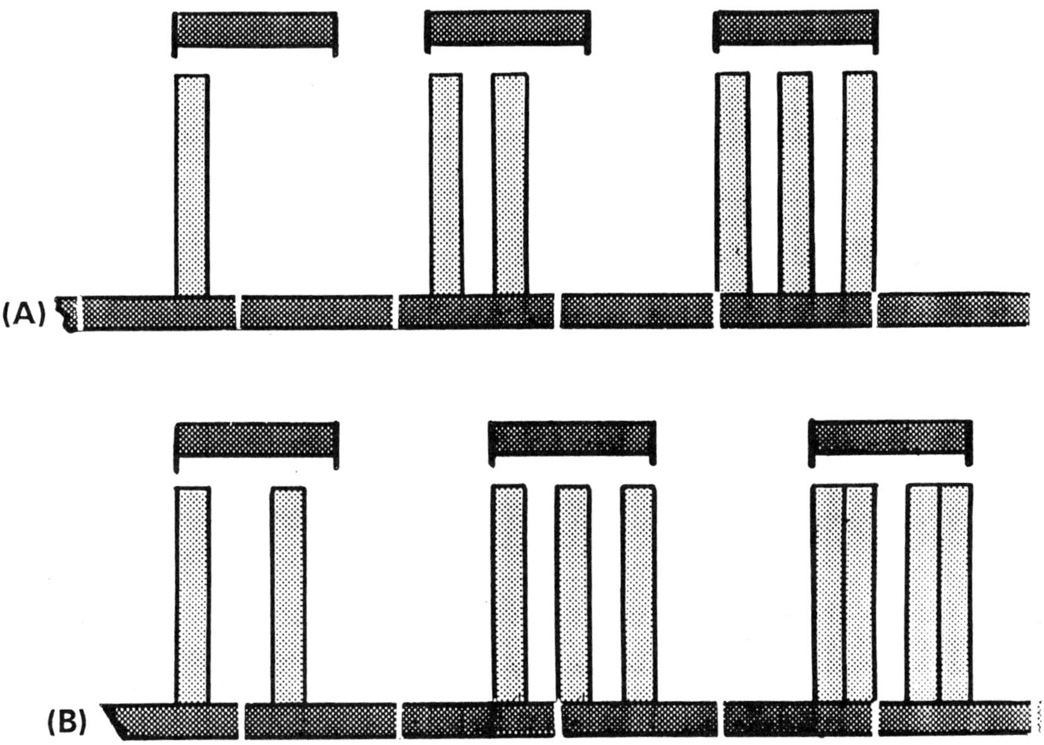

Figure 6 Comparison of the relative effectiveness of the geometric versus the particle-triggered Poisson counting efficiency models. The arbitrary division of particle events by the geometric model is shown by the segmented platform of detector volumes supporting the particle bars. The overhead bars represent the detector volumes of the particle-triggered model. Both are equally effective in the sequence of particles shown in (3A). The particle-triggered model performs better in (3B).

of the particle being measured. This interaction affects the determination of the concentration of particles in the test volume.

An important consideration emerges from the comparison of these detection models. This consideration is independent of the final choice of an optimum statistical counting model. All Poisson models of the counting process estimate counting error based on particle concentration during measurement. When particle concentration is incorrect, count accuracy estimation cannot be achieved. The use of the incorrect concentration and the current Poisson model of the counting process lead to unrealistic concentration performance claims.

The method of calculating particle concentration in present use is based on the assumption that particle volume was so much smaller than the detector volume that it could be ignored. This type of particle concentration estimate is now called the geometric concentration.

At the 1990 Conference on Liquid Borne Particle Detection, Measurement and Control, Knapp and Abramson [12] presented a review of the particle counting process from the viewpoint of a complete systems analysis. This analysis provided a joint framework for many particle counter errors. In addition to the systems analysis of particle counters, an experimental challenge of the commonly accepted Poisson model of the particle counting process was described.

The combined concentration and particle size dependent counting error are shown in Figs. 7 and 8. The data in Fig. 7 were taken with a detector whose geometric or physical volume is 7.5×10^{-6} cm^3 that has a window height of 50 μm. Concentration data for 5- and 19.6-μm PSL particles was taken with the detector operated at the manufacturer's specified flow rate. The data in Fig. 8 were taken with a detector whose geometric or physical volume is 50.0×10^{-6} cm^3

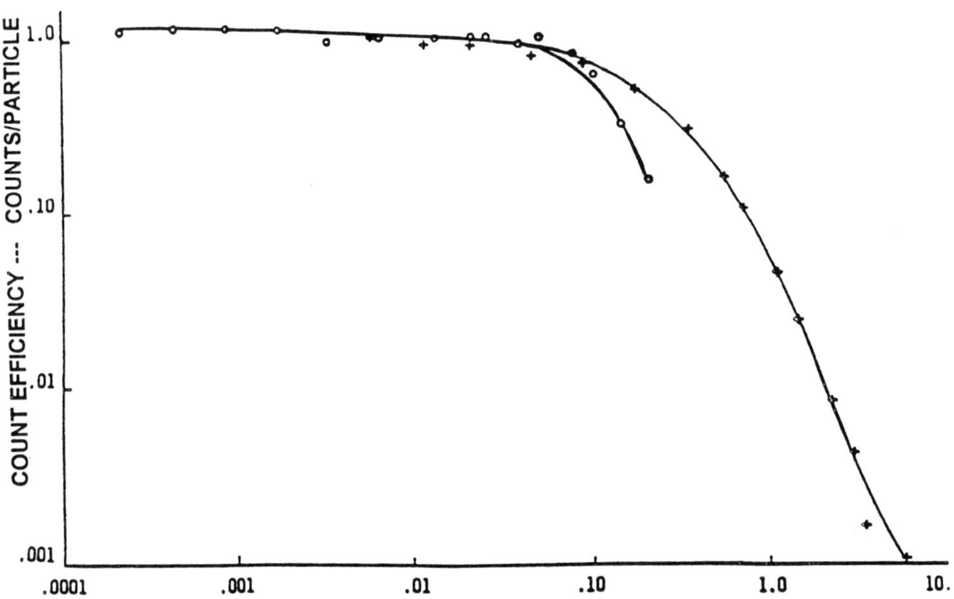

Figure 7 Count efficiency for a Climet detector with a white light source whose physical or geometric volume is 7.5×10^{-6} cm^3 and whose window height is 50 μm. The detector was operated at the manufacturer's specified flow rate of 7.5 ml/min. Two experimental runs are reported. The first with 5.002-μm particles shown as pluses, and the second with 19.582-μm standard spheres, shown as open circles.

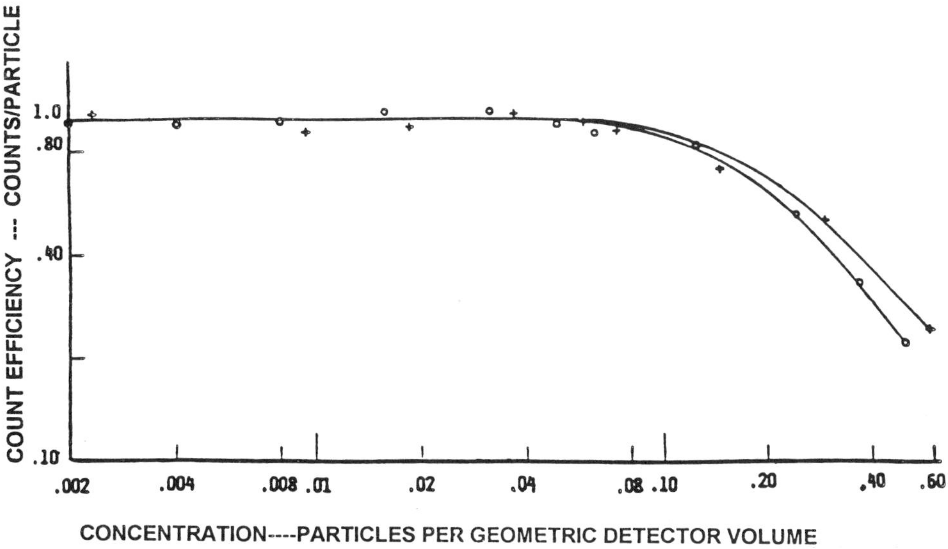

Figure 8 Count efficiency for a Climet detector with a white light source whose physical or geometric volume is 50.0×10^{-6} cm^3 and whose window height is 250 μm. The detector was operated at the manufacturer's specified flow rate of 40.0 ml/min. Two experimental runs are reported. The first with 19.582-μm particles shown as pluses, and the second with 99.0-μm standard spheres shown as open circles.

that has a window height of 250 μm. Concentration data for 19.6- and 99.0-μm PSL particles was taken with the detector operated at the manufacturer's specified flow rate.

These figures support the earlier conclusion that a reduction in detector volume increases the particle concentration that can be accurately counted. Neither the original Poisson model, due to Jaenicke, nor the improved Knapp and Abramson Poisson model adequately described the rapid drop in counting efficiency, the ratio of counts recorded to particles counted, as the concentration increased. Another result that could not be explained with the accepted coincidence analysis was the interaction of the detector volume with the diameter of the particle on the concentration-dependent counting error. The authors commented on their data: "The test data is insufficient for precise results. It is sufficient to conclude that neither of the models discussed accurately represents the loss of counting efficiency. An effect not previously predicted is that the loss of counting efficiency is a function of particle size—the larger the particle the earlier the onset of counting loss."

The size dependency of the coincidence effect proved to be a critical finding. Jaenicke's simplifying assumption that the effect of particle size upon the measurement could be disregarded was invalid. The increase in the effective height of the detector increases the effective volume of the detector. The increase of detector volume with measured particle size directly affects the concentration ratio of particles per detector volume. This ratio of particles per effective detector volume is the critical parameter in the estimation of the effect of concentration on the particle coincidence effect.

A review of the 1990 experimental results showed that the proposed Poisson model did not respond to the random groups of particles that occur as particle concentration increases. A new Poisson model of the particle counting process more closely akin to the actual particle detection mechanism was formulated and tested. In this new model, the calculation of a unit detector volume commences only after a particle signal crosses the selected analysis threshold. With this approach, the separation of particle counts by arbitrary unit detector volumes is minimized.

The next step toward an understanding of the counting process in particle counters was presented at the Joint PDA/IES Conference [22], in February 1992. At this conference, a reformulated Poisson analysis more accurately related to the actual particle-counting process was described and evaluated with experimental data. The critical steps forward in this paper are the recognition of the dynamic effect of the particle groups upon counting accuracy and the crucial effect of the interaction of the size of the particle being measured with the accuracy of the results obtained. The paper reported that the relative size of the particle and the scan beam affect the duration of the measurement and thus the estimate of particle concentration. The mechanism of this interdependence is pictured in Fig. 9.

In Figure 9A the geometric or physical cross section of the scanning beam height is shown

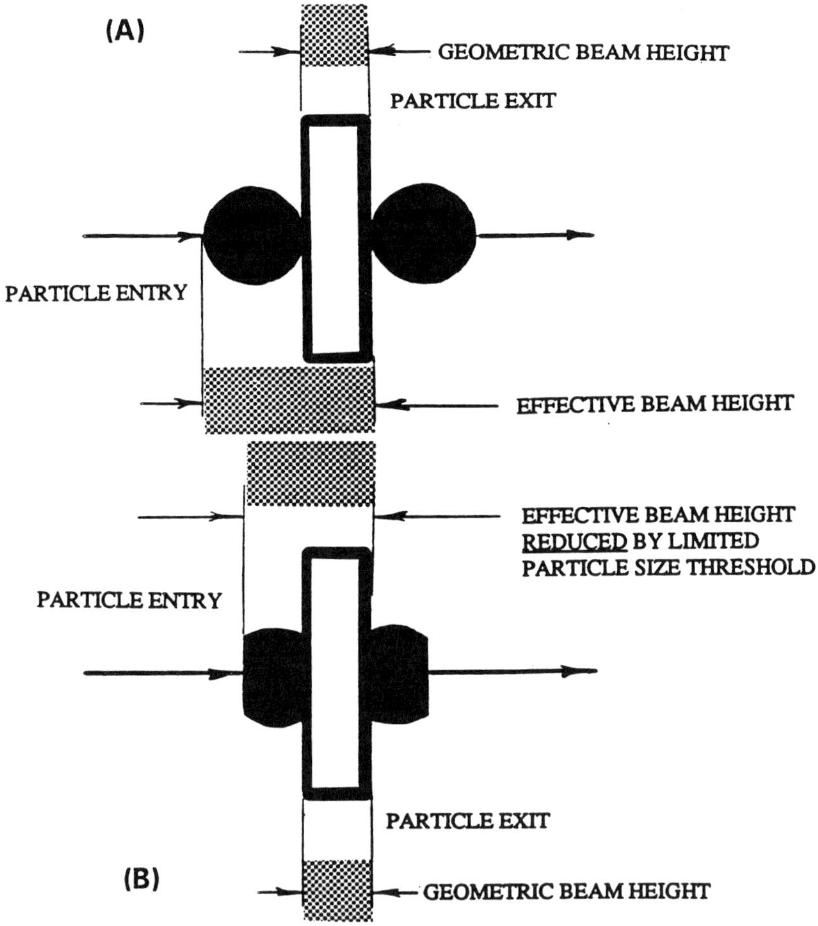

Figure 9 Interaction between particle diameter, geometric window height and the threshold selected sensitivity in determining the effective scan window height and thus the detector volume. (A) Effect of a threshold set for a much smaller particle. (B) Compression of particle length that occurs with threshold discrimination.

as an upright rectangle. The geometric detector volume is the product of beam height × window length × detector width. The particle signal commences when the particle intercepts the scanning beam. The signal continues until the particle exits the scanning beam. The effective duration of the particle signal is determined by the sum of scanning beam height and the diameter of the particle. This increase in the effective scanning beam height increases the effective volume of the detector. From this discussion it can be concluded that the condition necessary for particle counting to be independent of particle size is that the particle diameter must be negligible with respect to the scanning beam height. This requirement cannot be met with commercially available particle counters. The effect of this conclusion is to make the detector volume dependent upon the size of the particle to be measured. This requirement can be satisfied by considering the largest particle size to be measured. Establishing concentration accuracy limits for the largest particle results in smaller errors for all smaller particles.

Figure 9B illustrates signal suppression by the analysis threshold. The area of the signal is reduced by two particle areas whose diameters are equal to the analysis threshold. This suppression does not affect particle size measurement, since the amount of suppression is added back to the sensed particle signal. Count loss due to concentration is, however, determined by the threshold reduced particle width.

When the physical volume of the detector is used to calculate the number of particles per detector volume, the particle counting error is underestimated. The effective volume of the detector, determined by the relative size of the particle, the scanning beam and the response of the entire measurement system, must be employed for accurate results. The concept that the duration of the particle signal modifies the effective volume of the detector transforms the calculation of particle concentrations. Assessment of particle concentration is the basis for any estimate of the ratio of counting efficiency. Counting efficiency is calculated as the ratio of counts recorded to particles counted.

A conclusion that results from a review of the 1992 Knapp-Abramson paper [23–25] is that a clearly defined set of operating limits relating counting accuracy to particle concentration is essential. For a practical instrument user, this will result in a maximum predetermined counting error limit for any combination of particle sizes within the bounds specified by the manufacturer. This selection, based upon the largest particle size designated, will result in at least equivalent accuracy for any other combination of particle sizes measured by the instrument. This specification can be supported solely by empirical methods or with improved generality by an analytical model of the particle-counting process.

A practical method to relate the number of particles present in a suspension to the instrument counts recorded is required. Methodology to achieve this relationship was formulated and tested with latex spheres. Four possible models were defined. The four models are based on combinations of two different methods of calculating particle concentration and two different Poisson models with which counting error can be estimated. A simple method of ensuring count accuracy is proposed.

Two flaws have been described in the previously accepted analysis of the relationship of counts by single-particle counters to the number of particles being counted. The first is the lack of capability to accurately estimate the effect of groups of particles on the recorded count. The second flaw results from an ineffective computation of the concentration of particles being counted. In any statistical model that can be employed, the effect of particle concentration upon counting accuracy depends on a correct determination of the particle concentration present. Even if the geometric statistical analysis of the particle-count-to-concentration ratio were correct, the wrong concentration measure would result in ineffective count error estimates.

XIV. A COMPARISON OF GEOMETRIC AND EFFECTIVE PARTICLE CONCENTRATION

The first step in any evaluation of the counting accuracy of particle counters is a determination of the concentration of particles to be counted. The concentration is estimated as a ratio of particles to detector volumes in the test suspension.

To simplify the calculations in count accuracy determinations, a 1-ml test volume has been selected. This choice results in the often used particles/ml ratio without the need for intervening calculations. The concentration measure that results from this choice is simply the ratio of the number of particles per milliliter, n, to the number of detector volumes per milliliter. N_G is used for the number of geometric or physical detector volumes per milliliter; and N_E is used for the number of effective detector volumes per milliliter. The effective detector volume is calculated with the detector height increased by the diameter of the particle measured. The symbols for these ratios are respectively n/N_G and n/N_E. The notation used for these symbols employs uppercase letters for constants and lowercase letters for the variable quantities.

A graphical comparison of the effect of geometric and effective detector volumes on the calculation of the ratio of particles to detector volumes, n/N_G and n/N_E respectively, is shown in Fig. 10. In this figure the selected analysis uses the full particle area; the effective detector volume is seen to increase with increasing particle size. To maintain a constant particle-counting error as particle size increases, the number of particles in suspension must be decreased. The dependence of counting error on particle concentration makes the determination of the appropriate detector volume an important issue.

Figure 10 compares the effect of particle size on particle concentration calculations for two detectors. The physical or geometric volume of the first detector is 10×10^{-6} cm^3; its window height is 20 μm and the number of geometric detector volumes per milliliter is 100,000. The geometric volume of the second detector is 25×10^{-6} cm^3; its window height is 50 μm and the number of geometric detector volumes per milliliter is 40,000. The calculated concentration in particles per geometric detector volume is independent of particle size. This independence is shown as an invariant straight line as particle size is varied.

Assuming 4000 particles per milliliter, the unchanging concentration for the smaller detector is 4,000/100,000 or 0.04 particle/detector volume without regard to particle size. For the larger detector, this 4000-particle/ml concentration is a concentration of 4,000/40,000 or 0.10 particle/detector volume and is also independent of particle size.

The independence of concentration from particle size is seen as the two straight lines in this figure. The first straight line is at a concentration of 0.04 particle/detector volume for the smaller detector and the second at 0.10 particle/detector volume for the larger detector. The two curves in this figure show particle concentration increasing with increasing particle size. For a 15-μm particle, the effective concentration for the larger detector is 0.13 particle per detector volume as shown by the upper curve. The lower curve, for the smaller detector, shows a corresponding increase of concentration as particle size increases. For 4000 of the 15-μm particles, the effective concentration is 0.07 particle per effective detector volume. Independence of geometric concentration from particle size is seen as a limiting case, applicable to the detection of particle sizes much smaller than the scan window height.

To maintain a constant particle-counting error as particle size increases, the number of particles in suspension must be decreased. The linkage between particle concentration and counting error makes the determination of the appropriate detector volume to be employed an important issue. The effect of this difference on particle counting accuracy will be demonstrated with experimental data.

Two different methods to calculate the concentration of particles in a suspension have

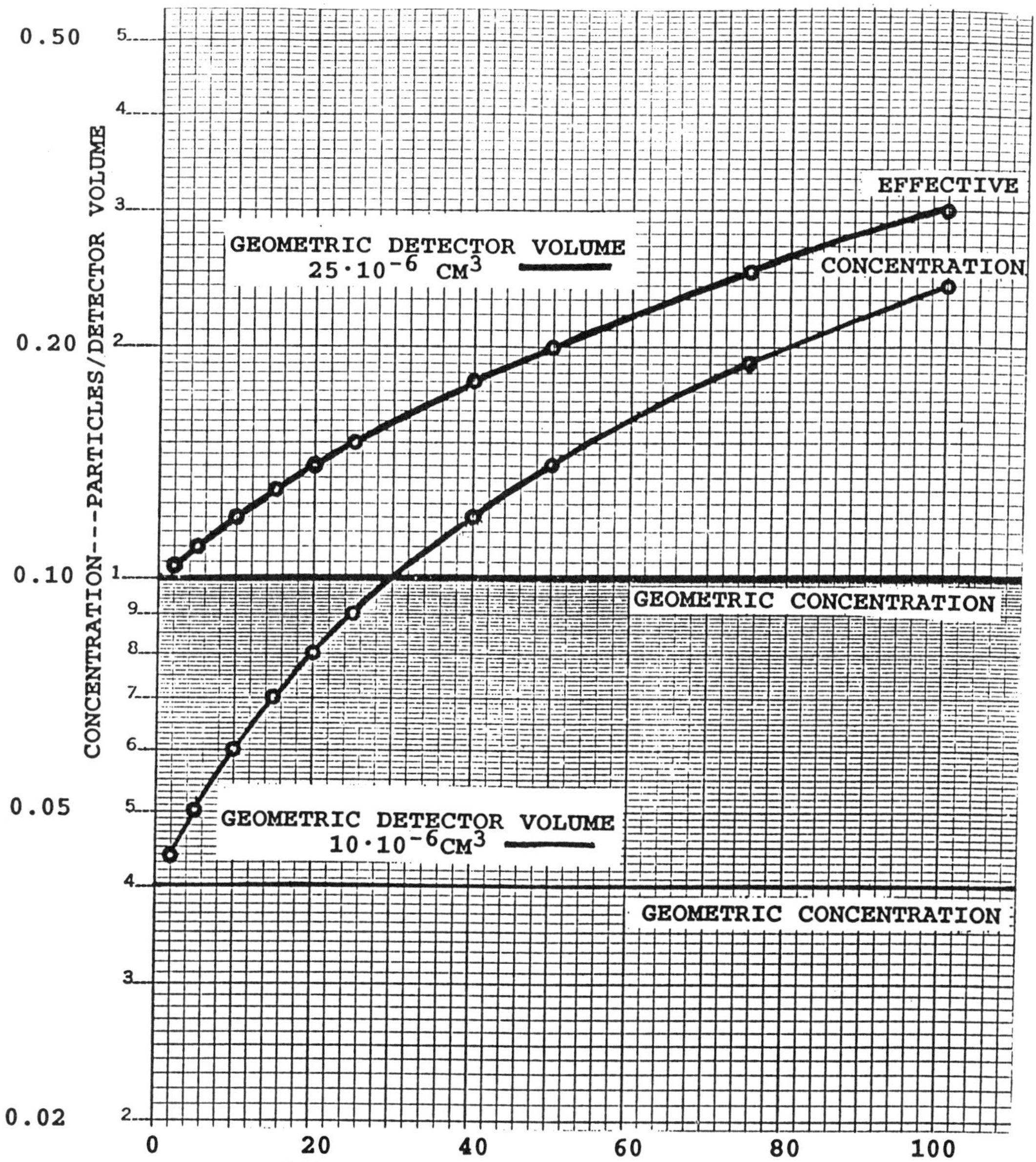

Figure 10 Comparison of the geometric and effective estimates of a concentration of 4000 particles/ml from 2 to 100 μm in diameter for two detectors whose volumes are 25 and 10×10^{-6} cm^3. Window heights are 25 and 20 μm respectively. Count loss in any Poisson model is determined by particle concentration. Counting accuracy cannot be achieved without concentration accuracy.

been described. The first method, that in present use, is based on the geometric volume of the detector, V_{DG}. Its concentration parameter is symbolized as n/N_G and is invariant with particle size. This concentration parameter is applicable to the measurement of particles whose diameter is negligible with respect to the scan window height.

The second method is based on the effective detector volume, V_{DE}. Its concentration parameter is n/N_E. The effective detector volume and the effective concentration parameter both increase with increasing particle size. For the same number of particles in suspension, calculations based on n/N_E will indicate a greater concentration of particles than those based on n/N_G. The effective concentration parameter is applicable to those cases where the ratio of particle size to scan beam height results in a nonnegligible fraction.

The selection of a concentration parameter, n/N_G or n/N_E, is independent of the selection of one of the two statistical models of counting error that will be discussed below. The two models used are first, the widely used geometric Poisson model and second, a triggered Poisson model introduced in 1992 [22]. (See Appendix 2, Secs. A–E, pp. 534–559.)

XV. MODELS OF THE COUNTING PROCESS FOR ESTIMATION OF COINCIDENCE

A. The Geometric Poisson Model due to Jaenicke

To provide continuity with the current literature on coincidence error, the following analysis of the Jaenicke geometric Poisson model uses the geometric detector volume, V_{DG} and its concentration n/N_G. The curves of Fig. 11 have been prepared using the geometric concentration assumption that the particle volume is negligible compared to the detector volume.

Ideal counting performance is identified with a linear asymptote labeled counts/ml = particles/ml. In this model of the counting process, the test volume is arbitrarily divided into

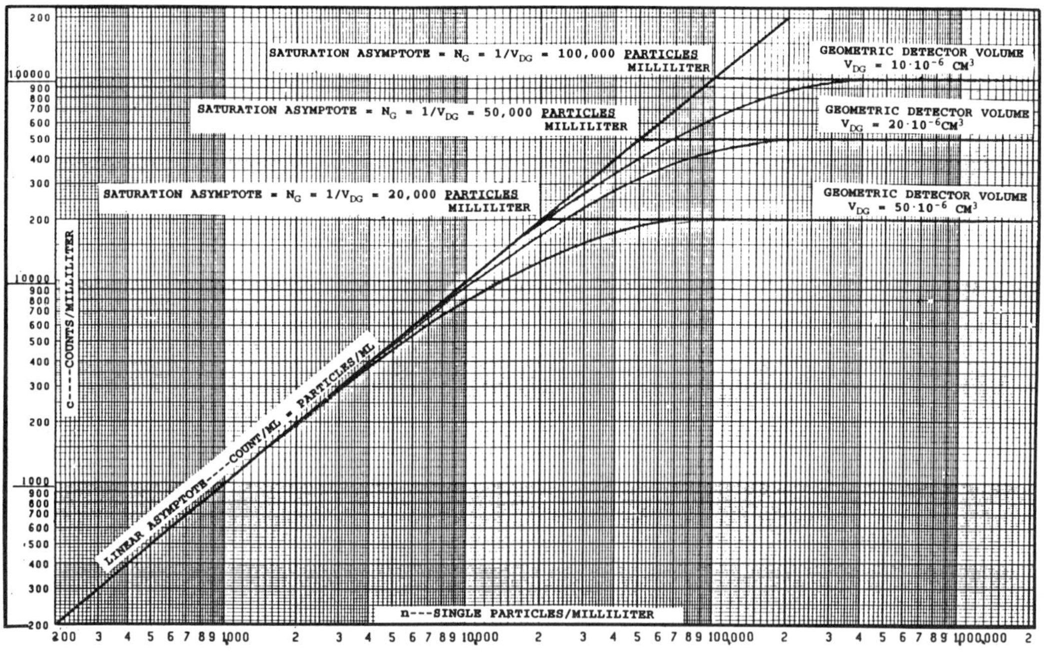

Figure 11 Counts versus single particles in suspension for the geometric Poisson model of three detectors whose volumes are 10, 20 and 50×10^{-6} cm^3 respectively, showing monotonic increase of counts versus concentration to a saturation limit. The corresponding geometric detector volumes/ml are 100,000, 50,000 and 20,000. A peak of the counts versus concentration curve has been common experience.

detector volume cells. The ratio of particles to geometric detector volumes in each milliliter of test volume is used to estimate the particle concentration. For a realistic appraisal of current and older detectors, detector volumes of 10, 20, and 50×10^{-6} cm^3 have been selected.

The estimated counts for monosized particles in suspension using the geometric Poisson counting model are shown in the three curves in Fig. 11. These three curves increase monotonically to a maximum count limit equal to N_G, the number of geometric detector volumes per milliliter. This maximum count value is maintained as particle concentration increases indefinitely. The three curves shown support the Jaenicke/Lieberman conclusions that a reduction of detector volume increases the particle concentration that may be accurately counted and that a decrease of detector volume increases the linear counting span of a detector (see Appendix 2 Sec. A pp. 534–536).

It has been common laboratory experience that as particle concentration increases, a maximum count is experienced. The maximum count value is followed by increasing loss of particle counts with further increases of concentration. Until now this drop in counts has been felt to be solely due to circuit limitations. Since counter performance in this overload region is of limited practical value, this indication of the inadequacy of the geometric Poisson model has been overlooked.

The undercount error, ε_C, for smaller concentration values in the geometric model is half the geometric particle concentration. (see Appendix 2 Sec. B p. 540)

B. The Particle-Triggered Poisson Model

An alternative to the Jaenicke geometric Poisson model, more closely related to the actual functioning of the particle detector, is the triggered Poisson model of the counting process.

The effective count efficiency, c/n, for the particle-triggered Poisson model is determined by the exponential ratio of particles to the number of effective detector volumes:

$$\frac{c}{n} = e^{-n/N_E} \tag{32}$$

Figure 12 summarizes the response of three detectors whose volume are 10, 20 and 50×10^{-6} cm^3 as particle concentration is increased. Three important detector factors and their relationship to the effective detector volume can be seen in Fig. 12. The usable linear counting span, the maximum unambiguous count and the maximum concentration that can be unambiguously counted all increase as detector volume decreases.

The maximum count capability of a particle detector is determined by the number of effective detector volumes per milliliter:

$$C_{MAX} = N_E e^{-1} = 0.3679 N_E \tag{33}$$

The number of single particles per milliliter in suspension, $n(C_{MAX})$ at the maximum particle count, is equal to the number of effective detector volumes per milliliter:

$$n(C_{MAX}) = N_E \tag{33A}$$

Equations (33) and (33A) state that the maximum number of counts per milliliter reported by a counter will be 36.79% of the number of effective detector volumes per milliliter for the detector.

For the 10, 20 and 50×10^{-6} cm^3 detectors examined, the maximum concentrations that can be unambiguously counted are respectively 36,788, 18,394 and 7894 particles/ml. For these

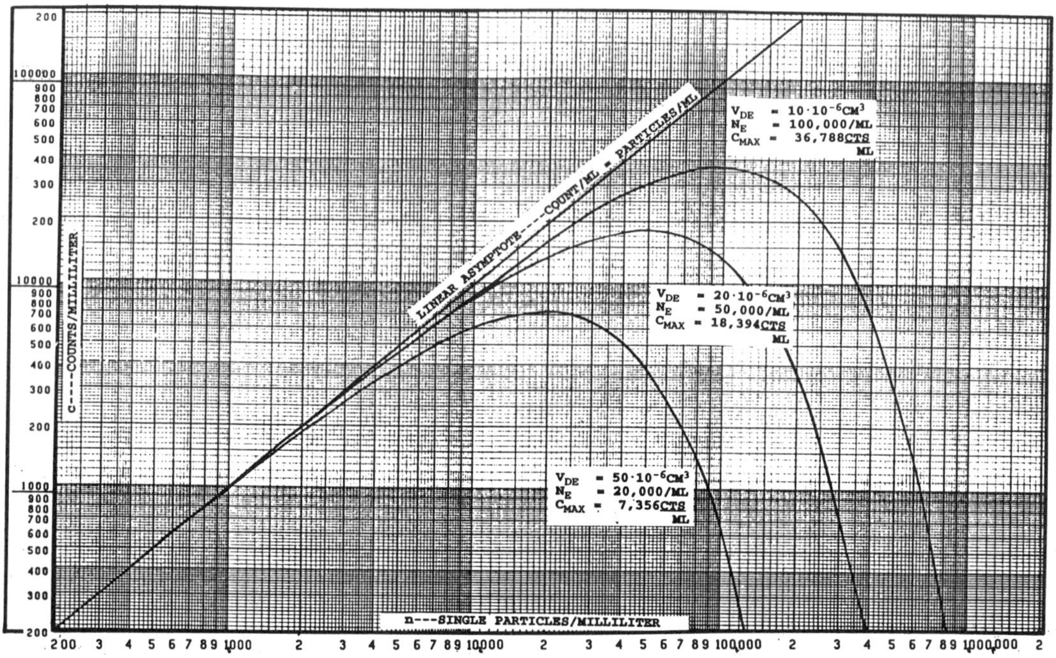

Figure 12 Counts per milliliter versus the concentration of particles/milliliter for detectors with effective detector volumes of 10, 20 and 50 × 10^{-6} cm^3 showing the increase of linear counting span and peak count capability as detector volume is decreased. The peak count for each detector occurs at N_E, the number of effective detector volumes/milliliter, $N_E = 1/V_{DE}$. The magnitude of each peak count is $0.3679N_E$.

maximum recorded counts the undercount error in each case is 63.2%. The danger of ambiguous results when inappropriately selected detectors are employed for U.S.P. <788> determinations is evident. For current particle counters this danger should be nonexistent.

In this model, the computation of the detector volume commences only after a particle signal is detected. With this mode of operation, the detector volume calculation is synchronized to the random appearance of the particle stream. The functioning of the model accurately parallels the functioning of the light extinction particle counter.

As in the geometric Poisson model, decreasing detector volume is accompanied by an increase of the linear counting region. In contrast to the earlier Poisson model, there is a peak in the count characteristic against concentration. The count peak is then followed by a precipitous decline in count capability. The location of the counting peak and its maximum value are well defined.

The curve of Fig. 13 shows the rate of change of the particle-triggered Poisson count linearity curves in Fig. 12. The rate of change in the important low-concentration region is double that of the count error for the same concentration.

Tables 5 and 6 summarize some convenient numerical results. Table 5 compares the estimated ratio of various coincident occurences to the number of single particles counted for the geometric and particle-triggered Poisson models of the counting process. Table 6 compares the counting efficiency, the counting error and the low-concentration approximation of the counting error for each Poisson model.

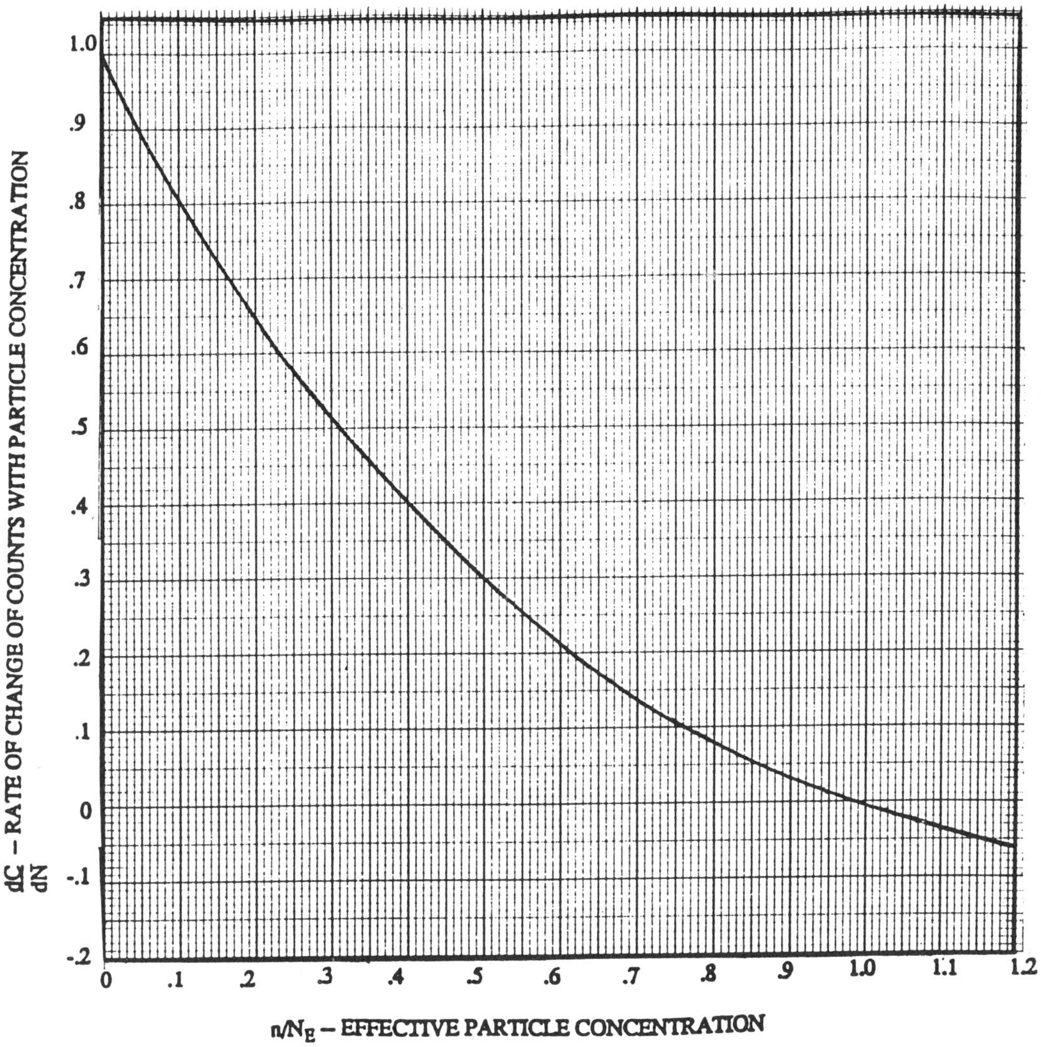

Figure 13 Rate of change of counts with change of effective concentration. The rate of change is approximately twice the count error for low particle concentrations.

The simplified undercount error for low concentrations of particles in the particle-triggered Poisson model is equal to the effective particle concentration:

$$\varepsilon_C = \frac{n}{N_E} \tag{31}$$

The undercount error estimate for the particle-triggered Poisson model is double that of the geometric Poisson model; this doubling is due to the choice of statistical model alone. The increase in counting error, due to the use of the effective detector volume, can be evaluated only for a specific combination of detector and particle size.

A numerical comparison of the count error estimate using both Poisson models can now be

Table 5 Comparison of Geometric and Triggered Estimates of the Double, Triple and Quadruple Coincident Counts to Single Particles in Suspension in Terms of Geometric Particle Concentration, n/N_G, and Effective Particle Concentration, n/N_E, Units

		Poisson model	
Symbol	Ratio	Geometric	Triggered
$C(1)$	—	ne^{-n/N_G}	ce^{-n/N_E}
$\frac{C(2)}{C(1)}$	$\frac{\text{Doubles}}{\text{Singles}}$	$\frac{n/N_G}{2}$	$1 - e^{-n/N_E}$
$\frac{C(3)}{C(1)}$	$\frac{\text{Triples}}{\text{Singles}}$	$\frac{(n/N_G)^2}{6}$	$(1 - e^{-n/N_E})^2$
$\frac{C(4)}{C(1)}$	$\frac{\text{Quadruples}}{\text{Singles}}$	$\frac{(n/N_G)^3}{24}$	$(1 - e^{-n/N_E})^3$

made. A relative particle concentration of $0.05N_G$ for the geometric Poisson model is assured. The low-concentration count error estimation for the geometric Poisson model is $1/2n/N_G$ or 2.5%. For the case in which the diameter of a particle is equal to the beam height, N_E is 0.5 N_G. The relative effective concentration, n/N_E, is then the ratio of $0.05/0.5 = 0.1$. The low-concentration error calculation using the particle-triggered Poisson model is then 10%, fourfold higher than the geometric Poisson model at the same concentration. When the particle diameter is half the beam height in the preceding example, the count error ratio is threefold higher. Table 6 also includes an equation for the estimation of the total number of coincident events for each model. This equation can be employed to estimate the injection of false counts into the U.S.P. <788> batch release particle measurement from particles below 10 μm in diameter in high concentrations (see Appendix 1 Sec. C p. 528).

Table 6 Comparison of Geometric and Triggered Estimates of Counting Efficiency and Counting Error as a Function of Particle Concentration Calculated as a Ratio of the Geometric ($\lambda = n/N_G$) and Effective ($n = n/N_E$) Number of Detector Volumes per Milliliter

		Poisson model	
Symbol	Function	Geometric	Triggered
c/n	Counting efficiency	$\frac{1 - e^{-n/N_G}}{N/N_G}$	e^{-n/N_E}
$1 - c/n$	Counting error	$\frac{e^{-n/N_G} + n/N_G - 1}{n/N_G}$	$1 - e^{-n/N_E}$
$1 - c/n$	Counting error: Low concentration approximation	$\frac{n/N_G}{2}$	n/N_E
C^+	Sum of all coincident particle occurrences	$N_G[1 - e^{-n/N_G} - (n/N_G)e^{-n/N_G}]$	$ne^{-n/N_E}(1 - e^{-n/N_E})$

The small error approximation for the triggered Poisson model is twice that of the geometric Poisson model.

XVI. LOGIC AND EMPIRICAL RESULTS

A physical measurement problem has been described. Experience from the Renaissance to the present has been that there are limitations to the productivity of unsupported logic and assertions in the solution of physical problems. Until logically constructed theories are validated by experiment there can be no defensible choice between them. Accepting this premise, counting experiments were designed and, with the use of borrowed facilities and equipment, were carried out.

In the discussion of the experimental design and data which follows, the primary concern has been the mechanics of particle counting. To focus solely on the mechanics of particle counting, the experimental conditions were selected to eliminate any variability that could be introduced by nonuniformity in the combined effect of energy constancy in the beam or any variability in responsivity across the photodetector's surface. The combination of these two variables is best considered in particle resolution measurements. A simplifying assumption in the following discussion of geometric and effective estimates of particle concentration is that the counter is operating at the flow rate for which it has been calibrated leaving the discussion of flow rate effects for a more complete treatment elsewhere.

Until logically constructed theories are validated by experiment there can be no defensible choice between them. Experimentation was used to determine which of five logical possibilities best described the relationship of the number of particles in solution to the number of counts recorded will now be described. These five are the original geometric Poisson model with particle concentration determined by (1) the geometric detector volume, V_{DG}, or (2) by the effective detector volume, V_{DE}. Alternatively, the triggered Poisson model with particle concentration calculated by either (3) the smaller geometric detector volume, or (4) the larger effective detector volume. The fifth possibility was, of course, that none of the proposed models fit the data adequately and that unsupported empirical calibrations will be used.

XVII. EQUIPMENT

The Micro Liquid Particle Spectrometer, Particle Measurement Systems, Inc. Model mLPS-16, a laser-based particle counter with software settable thresholds, was employed for all measurements. Detector dimensions in micrometers are $L \cdot W \cdot B_H = 1000 \cdot 500 \cdot 20$. The physical or geometric window height of this detector is 20 μm. The sensor is specified for use between 2 and 150 μm. A Hamilton programmable diluter, Model 960, was used. A low-profile, variable-speed external magnetic field stirrer with a Teflon-cased stirring bar was used to maintain particles suspended during measurements. Stirring speeds that produced a meniscus well approximately 5 mm deep were used; vortex formation was avoided.

XVIII. COUNT TO PARTICLE RATIO DETERMINATION

Maintenance of accuracy in the sequential dilutions that are required can be achieved either gravimetrically or by the use of separate stepper-motor-controlled diluters for both the selection of the measurement aliquot and the addition of the dilution aliquot. To avoid clumping of the standard polystyrene particles in the sequential dilutions employed, the diluent should be prepared with 0.1% of a compatible surfactant.

When using stepper-motor-driven syringes for these experiments, the use of full-scale volumes for both the sample and dilution aliquots are recommended; the use of 50% of the syringe volume limits dilution errors to twice that of the calibrated full-scale volume per step.

With a 10-ml 1% syringe used in a 10-step sequential dilution protocol even a 1% error per step results in a volume error of 1 ml. Adequate agitation of the sample container and time delays between sequential dilutions are required to achieve uniformly randomized suspensions prior to each measurement. Care must be exercised to maintain liquid temperatures below, and pressures above, that at which cavitation will occur in fluid entry into the diluter syringe.

The determination of the ratio of counts to particles in suspension is an essential prerequisite for the validation of particle counter systems. To optimize experimental focus solely on counting phenomena, other potential error sources were excluded by selection of a standard sphere size and analysis threshold. An accurately dimensioned 19.5-μm PSL sphere was used with a 10-μm analysis threshold. These choices eliminated any potential variability from the low-occurrence-rate undersized particles and from any potential variability of the size analysis threshold.

The use of a single container in which all dilutions are made and the use of a single dilution factor throughout the count accuracy protocol optimizes the possibility of a successful determination. Methodology to achieve this will now be reviewed. A summary of the relationships employed in achieving the experimental results included in this chapter follows.

The experimental data reported in this chapter were obtained to challenge and characterize the counting capability of single-particle counters. The experimental data consisted of the counts recorded at each dilution step and the net dilution for that step.

Stepwise dilutions were employed to achieve a particle concentration region in which the concentration of particles per milliliter of solution was accurately related to the counts recorded. The end points selected were dilution steps that produced equal deviations from the best straight line fit of the concentration/count data. The high-concentration end point was selected for a 1% coincident undercount; the low-concentration end point was determined by the concentration at which background counts increased the counts recorded by 1%. The initial particle concentration was then extrapolated from the accurate linear count region. The actual particle concentration for each dilution step was determined from the initial concentration with the dilution factor employed for that step. To achieve the desired accuracy, a total of at least five dilution steps was planned in the accurate count/dilution region. The relative standard deviation for counting data described by a Poisson distribution is $\pm\sqrt{c}/c$. To minimize the effect of count variability upon the determination of the initial particle concentration, a 1% relative standard deviation is desirable.

The usual use of the methodology to be described below will be to determine a particle concentration range in which a predetermined maximum error to counting linearity can be expected. The determination of a particle concentration range for light extinction particle counters in which acceptable count accuracy will be available is significantly less work than the experiments that are reported here. In these experiments, the starting point was an arbitrary high concentration. Using the established count replication capability of these light extinction particle counters, a test volume is adjusted to a concentration recommended by the instrument manufacturer. The dilution factor, also recommended by the manufacturer, will then reduce the procedure described to ordinary laboratory practice.

Figure 14 is an experimental curve of counts versus particle concentration for a dilution experiment with a ratio of 1:6. This figure shows the commencement of curvature at both the high- and low-concentration ends of the curve. Dilution factors and counts in the experimental data recorded can vary over three and four decades. To visualize the experimental results, log-log graphs have been employed. The curve of Fig. 14 has been truncated at both the high- and low-count ends where there is observable deviation from a unit slope line. Neither the low end points with visible noise error nor the high end points with visible coincidence effect error are used in the solution of the linear equation. Simple graphical extrapolation of the data points that

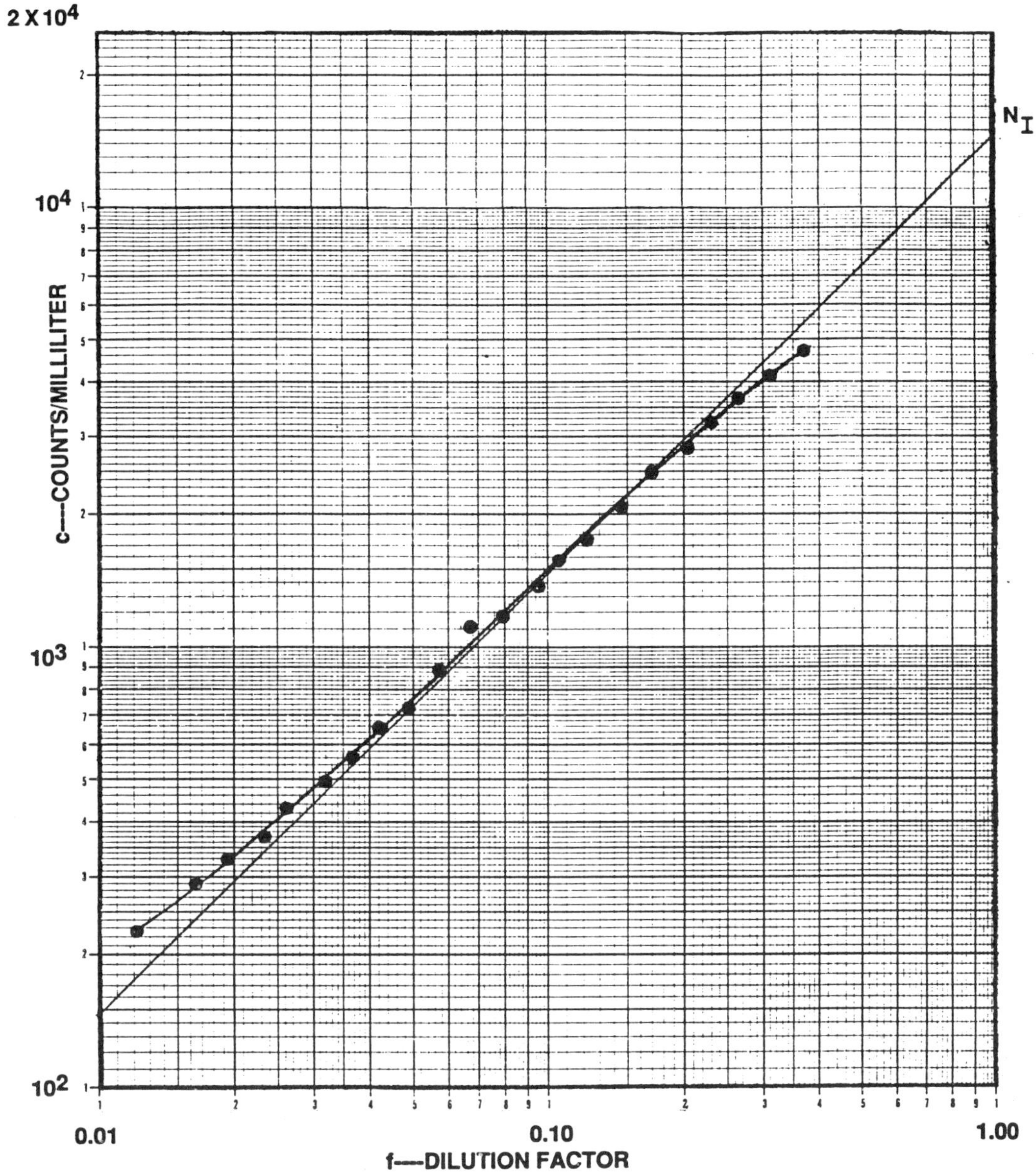

Figure 14 Rough graphical solution of the initial concentration of N_I, of 19.5 μm PSL particles/ml by sequential 1:6 dilutions. A linear regression analysis, including the background count at zero particles, results in a more accurate fit of the experimental data. N_I, by linear regression analysis was 14,739 particles/ml. Criteria for curve selection included minimized offset error and best correlation to a linear equation. Note: Only data points on a line of linear slope in a log-log plane fit a linear solution.

lie on a line of unit slope is used to get an estimate of the initial concentration of single particles. For more accurate results, a least squares line fitted to the selected data points is utilized.

Some additional cautions should be mentioned: many straight lines can be drawn on a log-log plot. The only straight line that yields a linear equation relating particle counts to the number of particles present is one with unit slope; all other straight lines describe exponentially related variables. The accuracy with which the initial particle concentration value is determined is affected by the size of the dilution steps employed. A two-step procedure is helpful; a rough determination of approximate values is first made. Following this first step, selection of a reduced starting concentration is made. With the reduced initial concentration, finer dilution steps now become practical. The combined result of both changes is an improved, more accurate, determination of the initial concentration and the experimental detector constants.

XIX. EXPERIMENTAL PROCEDURE

Prior to any dilution experiments, particle suspension uniformity was evaluated. Twenty successive counts were made of a suspension under uniform agitation. Two definite outliers were eliminated; the balance of the data was within the calculated 95% confidence limits. When adequate particle suspension was achieved, replicability of counts recorded was very good and independent of the loss of counts due to the coincidence error. Two concentrations of 19.5 μm were continuously diluted. A concentration of 14,739 particles/ml. was diluted 1:6. A stock solution of 350 ml was used with 60 ml withdrawn in 10-ml aliquots for analysis: the first 10-ml aliquot was discarded. In this experiment the data employed is the average particle concentration per milliliter. The counts on which the averages are based are therefore 50-fold larger. Following sample withdrawal, 60 ml of high-purity water with 0.1% surfactant were added. The suspension was continuously agitated with a Teflon-cased magnet driven by an external field. Agitation speed was maintained under the speed at which cavitation would result. All aliquots were delivered with stepper motor syringe units at full syringe volumes. Eighteen serial dilutions were made, with the lowest four concentrations being discarded for departure from linearity.

A second concentration of 4349 particles/ml. was diluted in 1:9 in 28 steps. A stock solution of 300 ml was used. 30 ml was withdrawn in six 5-ml aliquots: the first 5-ml aliquot was discarded. Total counts for each dilution are therefore 25 times the data listed. Agitation and dilution were similar to that employed in the preceding case. Twenty-six serial dilutions were made, with the 9 lowest concentration points being discarded due to departure from linearity.

All measurements were made with a lower analysis threshold of 10 μm to ensure relative freedom from count variability due to any outliers in the microsphere samples employed or to any measurement variability that could be traceable to the counter.

XX. DATA AND CALCULATIONS

The data points within the linear region in each dilution experiment were estimated with "best fit" unit slope lines. Figure 14 is a typical result. Following data point selection, the initial concentration of particles in each experiment was then determined by linear regression. The use of linear regression permitted inclusion of the zero concentration point, i.e., the background count point which is suppressed in the graphical estimate. This inclusion provides an accurate anchor for the data at a point where the data is weakest. Also included in Table 7 is N_E. N_E, the number of effective detector volumes/ml, is a result obtained from the count efficiency relationship using particle concentration data from the nonlinear count data points. N_E is the single number that best provides for the comparison of the count accuracy capability of any particle counter.

The relationship between counts and the number of single particles in the initial solution

Table 7 Linear Regression Coefficients and Correlation Coefficients for the Determination of the Initial 19.5 μm Single Particle Concentration in Two Dilution Experiments

		Ratio	
Dilution experiment		9:1	6:1
Slope = initial concentration	N_I	4342.9	14,374.8
Offset = background count	C_0	4.2	−1.5
Correlation coefficient	r	.99981	.99991
Detector volumes per milliliter	N_E	60,600	60,600

The 6:1 dilution experiment provides a full view of the count/particle relationship. The 9:1 dilution experiment provides finer detail in the linear region. The lower analysis threshold was set to 10 μm to minimize the effect of background counts.

was determined with the five-point procedure given in Sec. XXII. This is an experimental procedure applicable to the test of any particle counter model.

XXI. DISCUSSION OF EXPERIMENTAL RESULTS

Four model combinations were considered. The two statistical models are the geometric Poisson and the particle-triggered Poisson models. Each counting model could employ either the geometric or the effective concentration measure. As with many other scientific problems, selection of the pair that best describes experimental particle counting data cannot be deduced by logic alone; the selection must be based on experimental data. Figure 15 summarizes the data required for this determination. The four model combinations are compared with data from the dilution experiment results partially reported in Fig. 14.

The dilution experiments shown in Fig. 14 and 15 were used to establish a relationship between counts recorded and the number of particles represented by the recorded counts. Six 5-ml aliquots were used at each stepwise dilution in this experiment: the first aliquot was discarded and the average count for the next 25 ml was employed to ensure good replicability. The count linearity results shown in Figs. 14 and 15 are the result of empirical observations.

In Fig. 15, data from two dilution experiments with 19.5-μm PSL particles has been superimposed on the set of four theoretical curves. The best agreement appears to be reached with that of the triggered Poisson model using the effective detector volume to determine particle concentration. The experimental difficulties encountered are exemplified in the transient error of the 1.6 dilution points between 2000 and 6000 particles/ml that resulted from the movement of tubing employed.

Analysis of the count versus dilution data was accomplished in two steps. In the first step, "best fit" unit slope lines, as shown in Fig. 14, were used to make an initial selection of data points in the linear count region. In the second step, a solution to the data points was fitted with linear regression. An important criterion for the selection of a linear regression solution was a close match between the calculated background noise count and the experimental background noise count.

The linear regression calculations included correlation coefficients to determine the goodness of fit to the linear relationship assumption; the calculations are summarized in Table 7. The correlation coefficients are 0.99981 and 0.99991 respectively thus supporting the assumption of a linear relationship. The linear regression determinations of the initial particle concentration, N_M, was 4342.9 and 14,374.8 respectively. N_E was 60,600 particles/ml for both dilution

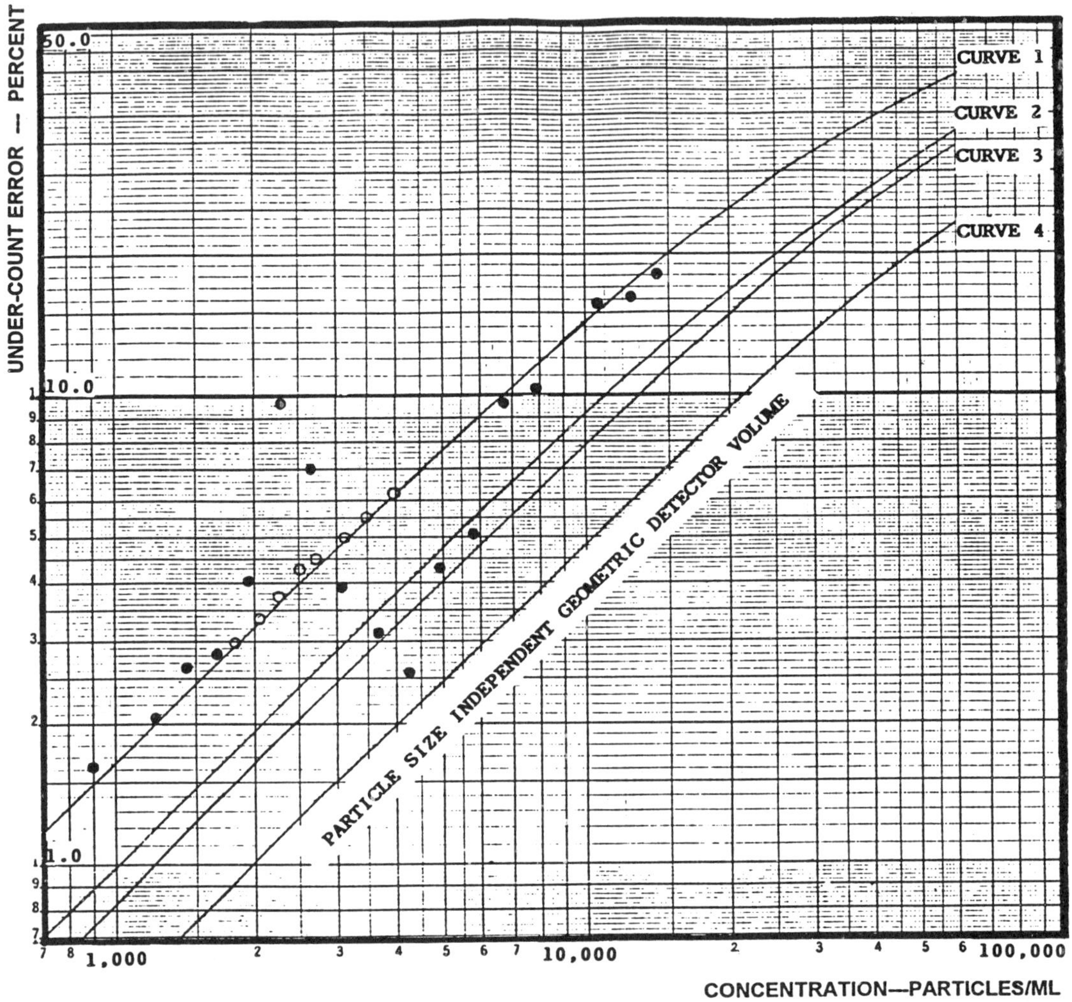

Figure 15 Evaluation of four Poisson count models with sequential dilution data using 19.5-μm PSL spheres. Curves 1 and 2 use the triggered Poisson model. Curves 3 and 4 use the geometric Poisson model. Particle concentration for curves 1 and 3 use the effective detector volume. The geometric detector volume is used to calculate concentration for curves 2 and 4. Open circles represent a 1:9 dilution with N_I of 4349; solid circles represent a 1:6 dilution with N_I of 14,739. All data was taken with a PMS laser sourced detector whose geometric detector volume was rated to be 10×10^{-6} cm^3.

experiments. The experimental effective detector volume, V_{DE}, calculated as the reciprocal of the number of detector volumes per ml, is 1/60,600, and is 16.50×10^{-6} cm^3. The corresponding effective particle diameter with a 10-μm analysis threshold is 13.00 μm.

Based on the selection of a 10-μm analysis threshold, the detected cross-sectional area of the 19.5-μm particle is reduced by the area of two 10-μm particles (see Fig. 9B and Eq. (10A)). The calculated effective diameter of the reduced particle area is 13.43 μm.

The calculated V_{DE} based on the reduced area signal is 16.71×10^{-6} cm^3; the number of detector volumes per milliliter for the reduced area particle is 59,834. The difference between the calculated and experimental values of the effective particle diameters respectively, 13.43 and 13.00 μm is 3.2%. This magnitude of difference could be traceable to experimental error.

The match of experimental data to theoretical curve 1 in Fig. 15 validates the selection of the particle-triggered Poisson model with particle concentration based on the effective detector volume. The validated relationship between particle counts and particle concentration shown in Eq. (32) describes the count efficiency of light extinction particle counters. Count efficiency is determined by the exponential ratio of particles to the number of effective detector volumes:

$$\frac{c}{n} = e^{-n/N_E} \tag{32}$$

Equation (32) is a normalized equation applicable to all single particle counters. This exponential form has been well studied in many disciplines. In this form count linearity is easily quantified. For an exponential value of 1 particle per 100 effective detector volumes, an exponential ratio of 0.01, the undercount error is 0.01 or 1% (see Appendix 2 Sec. C.1 pp. 542–543).

XXII. APPLICATION SUMMARY: THE PARTICLE TRIGGERED POISSON MODEL USING THE EFFECTIVE DETECTOR VOLUME

N_E is determined in a five-step procedure.

Step 1. The initial particle concentration is determined by extrapolation from the accurate count/concentration region. For result accuracy, the initial high particle concentration is selected to be one particle per two effective detector volumes, i. e., $n/N_E = 1/2$. Using an estimated value of N_E based on physical detector dimensions and the selected particle size to start, the initial high particle concentration is calculated. This concentration is then successively diluted until the count region in which counts are accurately related to particles in suspension has been reached. The accurate count region is estimated to occur from an upper limit of –1% undercount error as estimated with Eq. (32) down to a concentration at which the random background counts are 1% of the concentration measured. The selection of a particle concentration which is 100 times the level of the background noise counts recorded while the counter is processing well-filtered diluent provides a measurement range with balanced error at both ends.

Step 2. The initial particle concentration is determined by linear regression using the background count as the particle count at infinite dilution and four dilution data points at and below the concentration at which $n/N_E = 0.01$.

Step 3. The particle concentration at each dilution point is calculated using the stepwise dilution factor and the initial particle concentration.

Step 4. The count efficiency, c/n, the ratio of counts recorded to the number of particles in suspension at each data point is calculated.

Step 5. N_E is calculated from the high concentration data points by substituting the calculated ratio of counts to particles in suspension, c/n, and the number of particles in the suspension for that dilution point into Eq. (29B).

The experimental determination of N_E, the effective detector volume completely characterizes the counting response of the device for any concentration the selected particle size. The alternative to the use of the validated model is the information that can be extracted from the experimental count linearity data. Since the experimental count data is clearly nonlinear, accurate estimation of the effect of particle concentration for any except experimental data points is limited.

XXIII. TECHNIQUE NOTES

Although it is generally assumed that the laboratory techniques to reliably achieve the results discussed in this chapter are common knowledge, it has become apparent that this "common knowledge" is common only to those who specialize in particle measurements. Accordingly, the following Technique Notes have been added for those who might have need of them.

1. Test glassware
   a. Clean glassware is a particle counting prerequisite. The use of new, uncrazed, glassware surfaces minimize the difficulty encountered in achieving adequately clean glassware for particle counting tests. The use of containers reserved for particle counting trials is recommended. The test glassware should be jet-cleaned and maintained under constant overflow conditions to reduce the surface particle burden. A continuous overflow of approximately 100 ml per minute of 0.22 μm deionized filtered water in a 5-liter holding tank will clean and maintain the test glassware in good condition.
   b. Glassware cleaning
   A 25 p.s.i. water source through a disposable 0.4 μm filter connected to a 1/16 orifice can be used to dislodge particles from container surfaces. A final rinse with surfactant diluted water should be used.
   Note. When an ultrasonic bath is used to clean glassware, the cleanliness of the glassware is limited to the cleanliness of the water in the bath. If a bath without a filtered recirculation loop is used, a sequence of water changes should be used.
2. Background count determination
   The background count shall be determined at three sigma below the nominal particle size as determined under system specifications using prepared diluent. The flow rate and volume for the clean diluent shall be at the maximum aliquot volume used during the count accuracy determination.
3. Diluent addition
   To avoid cavitation during the addition of diluent, in the course of the count accuracy determination, the diluent stream should be slowly added at the wall of the sample container. This essential technique step should be practiced with diluent stock prior to the count linearity determination.

XXIV. APPLICATION OF THE PARTICLE-TRIGGERED POISSON MODEL: USING THE EFFECTIVE DETECTOR VOLUME

The limitation to characterization of the count/concentration characteristic of a single particle size described above is traceable to Eqs. (5) and (7). Equation 5 defines the effective detector volume as a function of the geometric detector volume and the scan beam height. N_E is the reciprocal of V_{DE} and is the number of effective detector volumes per milliliter, 1 V_{DE}.

$$V_{DE} = V_{DG}\left(1 + \frac{D_P}{B_H}\right) \tag{5}$$

Equation 5 has two unknowns: V_{DG}, the physical detector volume, and B_H, the experimental value of beam height. The solution of Eq. (5) is accomplished with simultaneous equations using data from dilution experiments with two particle sizes.

XXV. COUNT EFFICIENCY OF SINGLE-PARTICLE COUNTERS

A validated model of the counting process and an effective particle concentration measure are now available. Two core concepts have been described with which evaluation and control of coincidence error in single-particle counters can be accurately quantified. These concepts are: the particle-triggered Poisson model, a new and more accurate statistical model of the particle counting process, and secondly, an effective concentration measure that includes the effect of particle size on the counting capability of a detector.

Use of these two concepts make it possible to calculate particle detector count efficiency capability from experimental data of the coincidence effect. From the data analyzed it can be seen that no single-particle size test can effectively evaluate detector performance. The use of the theory and methodology described can help realize the intent of the U.S.P. <788> SVI particle contamination assay.

When progress is measured against the need for accurate, correlated particle counting independent of instrument fabricator and operating laboratory, the next requirement can be readily identified. There is a need to achieve generalization of the results so that the tools now available to describe the count accuracy of an individual light extinction particle counter can be applied to all similar instruments. This next step is essential for validated comparability of particle counter data independent of vendors, model specifications and commercial claims. This capability can ensure a pharmaceutical manufacturer that particle contamination data from all plants and all instruments are, in fact, accurately comparable.

For a laboratory, even one operating with minimal resources, a choice is now possible. The choice is between continued reliance on the manufacturer's specifications for a production model of a particle counter and the simple experimental evaluation of the operating characteristics of a specific instrument. The determination of a specific instrument's characteristics extends the user's margin of control.

Concentration dependent count accuracy is a durable result of detector design. Following an initial determination upon the receipt of an instrument, repeated determinations should be made only after changes in the light beam or sensor of an instrument occur. With semiconducting laser light sources, the life of a light source can well be the major portion of the instruments useful life.

Cost-effective digital technology has made available counters with accurate particle size calibration and sample volume capability. When these counters are operated within experimentally determined particle concentration limits, accurate comparability of particle count data will become an accepted fact. Until now the accuracy, replicability and transferability of electronic particle counting measurements has been sufficiently questionable for them to be considered almost on the level of qualitative measurements. The first step forward is the development of a practical, accurate coincidence effect measurement procedure that is usable in even a modestly equipped laboratory.

The relationship of count loss due to coincidence was separated from its associated problems by the increased accuracy and stability of particle sizing and the absolute accuracy of standard particles. This was accomplished by the choice of the experimental conditions described.

Methodology with which to quantitate the effect of coincidence upon count accuracy was described and will now be reduced to an operational protocol.

Course 902, taught at the spring 1993 P.D.A. meeting, demonstrated the practicality of the procedure that is described in the model validation section of this chapter. Experimental determination of the maximum concentration for a selected count accuracy limit is accomplished in a sequential dilution procedure. This result will apply to both the determination of undercount error and to the evaluation of conditions resulting in the introduction of false counts into the U.S.P. <788> assay. The experimentally derived test limits should be applied to both the selection of a particle counter and the determination of the operating limits within which it is best used.

Accurate counting error determinations were made with the latest particle counters of two leading manufacturers. The "laboratory" was a hotel room in downtown Philadelphia with untreated "downtown quality" air in a room with the usual contaminating carpet on the floor. Casual clothing, worn without masks or talc-free gloves, was used by the 28 participants. Blank determinations of 0.5 particle/ml, equal to or greater than 5 μm, was demonstrated in these circumstances. Teaming those in the class who had particle measurement experience with the neophytes, the technique to achieve blanks with one particle or less, equal to or greater than 10 μm, per milliliter was successfully used as a laboratory problem. The practicality of this procedure making use of the essential laboratory techniques has been established. In hindsight, knowledge of the essential techniques required for successful particle assays outweighed the requirement for the full panoply of a particle measurements laboratory.

The official activation of the U.S.P. XXIII has changed the operational requirements for particle contamination measurements. Accuracy, acceptance limits and methods for U.S.P. <788> contaminating particle assays published in the XXII Revision are refined in U.S.P. XXII. In both revisions, although different numerical values and methods are employed, particle contamination limits remain constant for all SVI container volumes. The effect of this quality standard is high particle concentration acceptance limits in the smallest SVI container sizes. The effect of these high concentrations is to introduce both undercount errors and false counts into U.S.P. <788> SVI contaminating particle assays.

XXVI. APPLICATION OF THE NEW COUNTING MODEL AND CONCENTRATION MEASURE

Without a validated model of the counting process and an accurate measure of particle concentration, the counting capability of particle counters rested upon variable commercial claims of performance based, in part, on an inadequate theoretical analysis. An alternative is now available. It is a well-defined model of the counting process for single-particle counters.

Figure 12 summarized an evaluation of the counting capability of three detectors with effective detector volumes, V_{DE}, of 10, 20 and 50×10^{-6} cm^3. The independent variable in this figure is particle concentration per milliliter, n. The dependent variable of the Fig. 12 graph is counts/milliliter, c. All curves in this figure show an initial linear count region. In this region, the count of particles is accurately related to the concentration of particles present. This figure supports the widely held belief that counter accuracy diminishes as particle concentration increases. These curves also support the conclusion that the smaller the detector volume, the greater the linear measurement span of a detector. The initial linear counting span in each detector is followed by a region of decreasing sensitivity as particle concentration increases until a maximum count is reached. Following the maximum count peak, a further increase of particle concentration results in a reduction of counts until system collapse occurs. In modern instruments warning signals are employed to alert the operator that the detector capacity has been exceeded.

The count accuracy regions just described have all been observed experimentally and described in the literature. There is a significant difference between these earlier reports and the curves in Fig. 12. The Fig. 12 curves have been calculated from the deduced analytical relationship between the number of particles counted and the number of counts recorded. In Fig. 1 a clearly defined maximum count is visible. The concentration for the maximum count, the magnitude of the count and the count error can now be analytically evaluated. The observed count replicability of these counters is clearly independent of the concentration related count error shown in this figure. Replicability alone is not a useful measure of particle counter counting accuracy.

The result of this analytical evaluation shows that the count peak occurs at a concentration, n, of N_E particles per milliliter. The count efficiency at this maximum count point is e^{-1}, which is numerically 36.7%. The undercount error at this point is $1 - 0.367 = 0.633$ or 63.3%. The procedure section (see Appendix 1 Sec. A–N pp. 527–534.) of this chapter contains methodology with which to experimentally determine N_E for any single-particle detector. The value of N_E is specific to the measurement of a selected particle diameter by a selected detector. It can be used to completely describe the full effect of particle coincidence on the counting performance of a detector for the selected particle size (see Appendix 2 Sec. C.2 p. 543).

The undercount error of each detector due to concentration effects can be evaluated by taking the difference between the ideal linear response curve and the calculated response of one of the detectors shown. The result would be another set of three curves of count error per milliliter versus concentration in particles per milliliter. These two sets of curves would be specific to the three detectors examined. Another drawback to the use of a graphical analysis such as used in Fig. 12 for experimental data is the difficulty in determining accurate concentration limits for a selected count accuracy.

The most important conclusion that can be drawn from Fig. 12 is that an analytical model for the counting process exists. That both instrument specifications and operating conditions can be defined and that counter accuracy independent of designer and measurement laboratory can be obtained. The analytical expression for the relationship between counts and particles facilitates evaluation of the coincidence undercount error and thus counting accuracy determination for these particle counters. A practical limit to application of these curves is the background noise count, C_0. This count is evaluated with well-filtered diluent under normal operating conditions. C_0 ranging from 0.1 to 1.0 counts per milliliter is the controlling error for low concentration measurements. The coincidence error is the controlling error at high particle concentration measurements. Of equal importance, the particle-triggered Poisson model provides the means with which the introduction of false counts into U.S.P. <788> particle assays can be understood and controlled.

XXVII. LIGHT EXTINCTION PARTICLE COUNTERS AND U.S.P. XXIII

The simplicity of the relationship between particle concentration and the counts recorded by a detector has been used to reduce the fog of uncertainty that has enveloped particle count measurements. This relationship has made possible the universal response graph of Fig. 16 which shows the count accuracy capability of all single-particle counters. Figure 16 includes both count efficiency, the ratio of counts recorded to single particles counted, and count error. Count error is defined as 1 minus the count efficiency. The variation of count error and count efficiency with particle concentration is demonstrated in this figure. The effective particle concentration measure used here is the ratio of particles to the effective detector volume. The effective detector volume includes the effect of particle size upon the measurement.

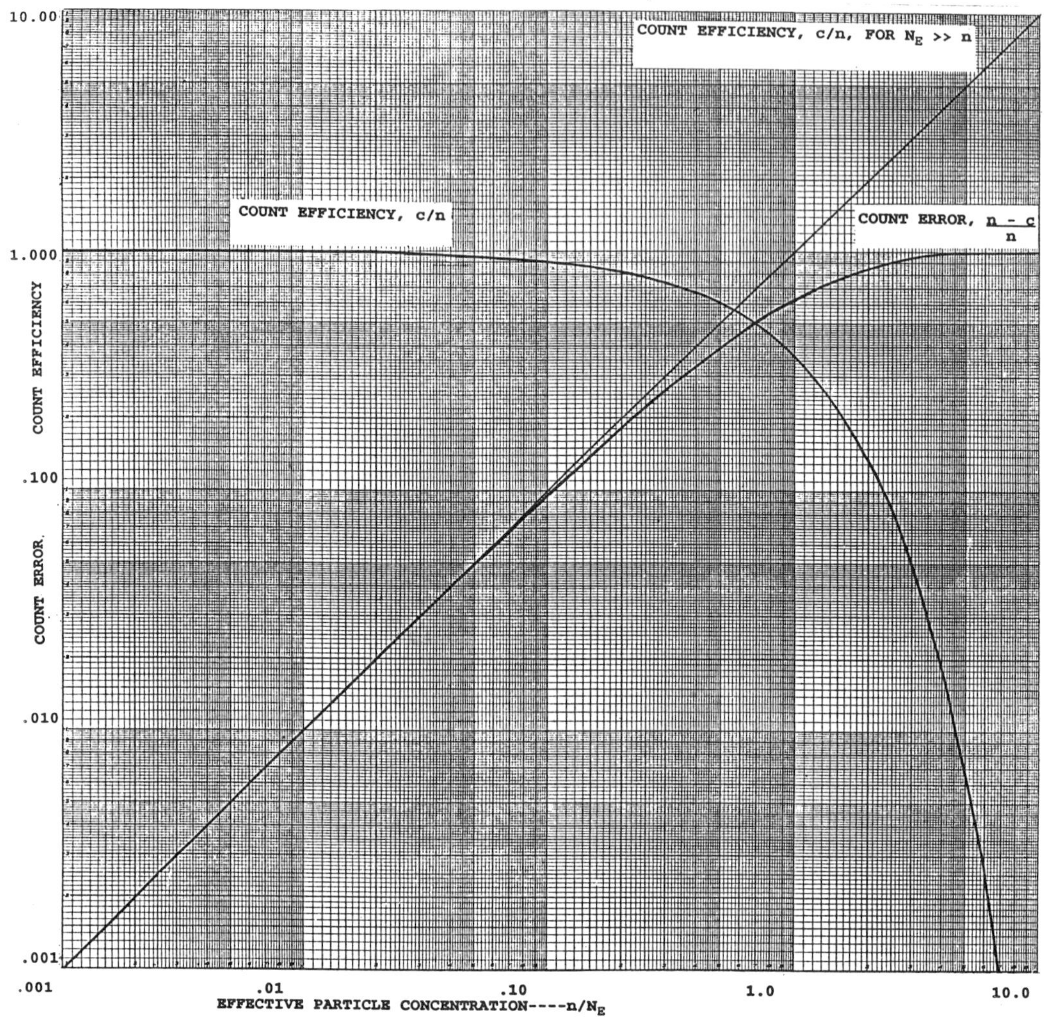

Figure 16 Universal curves for single-particle counters showing count efficiency, c/n, and count error, $(n - c)/n$, versus the effective particle concentration ratio, n/N_E, the ratio of particles to effective detector volumes. The exponential nature of these curves and the lack of a validated model contributed to the wide variability in the results of earlier analyses.

All models of the counting process that use Poisson statistics to estimate the effect of particle concentration use a ratio of particles to detector volumes in the test volume. Without accurate evaluation of this particle concentration ratio, there is no possibility that an accurate Poisson estimate of counting accuracy can be made. The following discussion summarizes the relevant material that is required to transform U.S.P. XXIII undercount error specification into the normalized parameters used in Fig. 16.

To effectively present the ideas that follow, special terminology is essential. A Glossary with a listing of all special terms has been prepared; it includes definitions and measurement units.

Equation (32) defines the experimentally validated single-particle counter model based on the particle-triggered Poisson model, it describes count efficiency as exponential in nature. In this relationship count efficiency can approach linearity but is never ideally linear, c can approach, but is never equal to, n. This equation, uses the effective detector volume to define the relationship between particle counts and the number of particles counted. The selection of 1 ml as the standard test volume has simplified notation and the calculations employed. It makes the results immediately available in per milliliter quantities. The ratio of counts per single particles counted describes a fundamental characteristic of these single-particle counters, the count efficiency. It will be seen that the use of this single, experimentally determined, parameter can be used to evaluate and control both coincident undercounts and false counts in U.S.P. <788> contaminating particle assays.

$$\frac{c}{n} = e^{-n/N_E} \tag{32}$$

The variables of Eq. (32) include counts/ml, c, particles/ml, n, and a new constant, N_E, the number of effective detector volumes per milliliter. In Eq. (32) the exponential ratio n/N_E evaluates the effective particle concentration of the measurement. This ratio has the dimensions of particles per detector volume. This is a result of the selection of a standard 1-ml test volume. The N_E constant in this equation is specific to a detector in the counting of any concentration of a selected particle size.

N_E determined by the volume defined by the physical or geometric dimensions of the flow path of the detector and the effective height of the detected light beam. The effective height of the light beam is the sum of the particle diameter being measured and the physical height of the light beam. Equation (8) expresses this relationship in a convenient form:

$$N_E = \frac{1}{V_{DG}(1 + D_P/B_H)} \tag{8}$$

N_E, the number of effective detector volumes per milliliter, is calculated from sequential dilution experiment data. It provides an operational measurement of the count accuracy capability of any single-particle detector.

A direct relationship between the effective detector volume, V_{DE}, and its predecessor the geometric detector volume, V_{DG}, is obtained by substituting for N_E its value $1/V_{DE}$ from Eq. (3). The results, after simplification, are

$$V_{DE} = V_{DG}\left(1 + \frac{D_P}{B_H}\right) \tag{5}$$

V_{DG} = geometric detector volume
D_P = particle diameter
B_H = scan beam height
N_E = number of effective detector volumes per milliliter

Solving Eq. (5) for V_{DG} gives

$$V_{DG} = \frac{V_{DE}}{1 + D_P/B_H} \tag{5A}$$

V_{DE}, the effective detector volume is larger than V_{DG}, the physical or geometric detector volume. This difference in detector volume is a result of the interaction of the scan beam height and the diameter of the particle being measured. The height of the detector, B_H, physically the height of the illumination beam employed, is increased by the diameter of the particle, D_P, being measured. As a consequence of this difference, concentration estimates based on V_{DE} are greater than those based on V_{DG}. This difference results in greater undercount estimates for the same particle concentration when the validated effective detector volume measure, V_{DE}, is used.

An important use for Eq. (32) is experimental evaluation of N_E, the constant that characterizes the counting capability of a detector. Before the availability of Eq. (32) estimates of the counting capability of particle counters were individual to the experimental technique or estimation procedure of a particular manufacturer or based on an erroneous Poisson analysis. The possibility of experimental evaluation of the counting capability of a detector is

$$N_E = \frac{-n}{\text{Ln}\,(c/n)} \tag{32A}$$

Equation (32A) can be evaluated with particle counters, standard particles and readily available laboratory equipment.

The number of multiple occurrences at each particle size varies with concentration. The effect of the multiple occurrences on the counts recorded is summarized in Table 13 (see page 515). This table shows that the coincidence effect of sub 10-μm particles is the injection of false counts within assay limits. For particles ranging in size from 10–12.5 μm the predominant coincidence effect is an undercount error. For particles from 12.5–17.6777 μm both false counts and undercounts can result based on the concentration of particles counted. From 17.6777–25 μm the predominant (and potentially more serious effect) is the generation of false particle counts > 25 μm.

To examine the effect of coincidence error with realizable, measured particle distributions, the effect of coincidence on two particle sizes was calculated. The measured particle distribution for both particle sizes was assumed to be 5%. The $n = 4$ entry for the asterisked 11.5-μm particle size shows the effect of a coincident occurrence, up to a 4× limit. Up to this limit, the resulting false multiple particle count will remain under 25 μm thus producing undercounts. The asterisked 21.7391-μm particle size, with a similar distribution, indicates that in the absence of coincidence error a $+3\sigma$ retention of single particle counts will remain within the 25 μm control limit. Double coincidence occurrence with this particle size will result in a 30.744-μm particle size peak.

XXVIII. UNIVERSAL CURVE FOR LIGHT EXTINCTION PARTICLE COUNTERS

In the universal response graph of single-particle counters shown in Fig. 16, the independent variable is the effective particle concentration measure, n/N_E, the ratio of particles to detector volumes in a 1-ml volume. The dependent variable in the universal single-particle counter coincidence effect graph shown in Fig. 16 is count efficiency, the ratio of counts recorded to the number of single particles counted, c/n. The results of Fig. 16 are independent of the type of excitation employed in the detector and of the means employed to detect the particle signal. This universal graph for single-particle counters applies to light extinction, forward scatter and conductimetric particle counters.

The first part of this chapter showed that particle detector specifications based on earlier Poisson model and static or geometric concepts of particle concentration were erroneous. The results of Fig. 16 serve to illuminate the difficulty previously encountered in comparing experimental determinations of count accuracy. Arbitrary selection of three or four data points for the analysis of count accuracy without prior knowledge of the relationship between the count efficiency and count error functions can only result in an erroneous analysis of the experiment.

With the count error curve shown in Fig. 16, accurate selection of instrument specifications and operating limits to ensure accurate performance are obtainable with single-particle counters. The limits lie between the boundaries of undercount, imposed by the coincidence effect, and overcount imposed by the operational noise pulses that can be evaluated with particle-free diluent.

The error curve in Fig. 16 shows the linearity of count error with concentration for low-particle concentrations. Combining count confidence limits with control of the coincident count errors shown in this figure, the accuracy of particle counting tests can now approach normal pharmacopeial accuracy limits.

Figure 16 represents an advance in understanding and thus controlling particle measurements: it defines a calculable relationship between the concentration of particles and the counts recorded from them. It is, however, a limited advance. One experimental curve is required to describe the effect of concentration on the counts recorded for a single particle size in a specific detector.

The relationship between the number of particles suspended per milliliter, n, and the number of counts recorded per milliliter, c, can now be restated as a count efficiency relationship. This is the ratio of the counts recorded per milliliter to particles present per milliliter. This count efficiency measure describes the particle count to particle concentration relationship for the class of all single-particle counters. For the user this means that a maximum count limit at each particle size can be calculated within which a selected count accuracy can be maintained. The count efficiency and count loss of all single-particle counters as a class are described graphically in Fig. 16 (see Appendix 2 Secs. C–E, pp. 541–548).

With the data in Fig. 16, the limitations within which particle counters must be operated to achieve a selected counting error can be clearly visualized. At low concentration the error is seen to increase linearly as concentration increases. In fact, for concentrations up to 10% of N_E, the count error and the effective concentration are approximately equal. At the 10% concentration point the count error is 9.52%. As concentration is reduced, the undercount error more closely approaches the calculated effective concentration. At an effective concentration of 1%, the undercount error is 0.995%.

XXIX. GENERALITY OF THE CONCLUSIONS BASED ON THE DATA REPORTED

The conclusions reported were based on experimental data using electronic particle counters from three leading manufacturers. The data analyzed was obtained with both white-light- and laser-diode-based counters. The experimental conditions selected effectively eliminated any effect of sizing variability on the experimental data. The data recording range in the experiments reduced the effect of the noise pulses on the counting experiment to 1% or less by using minimum counts of 100 particles per ml. The statistical counting variability should also be limited by using a minimum 10,000 particle counts at the lowest dilution step. The efficiency of the counting process and the extent of the linear counting range capability of a detector increased as detector volume increased. Halving the effective detector volume doubled the linear count range; decreasing the effective detector volume to one-third tripled the linear count range, etc. All of the detectors examined shared a common characteristic. The exponential relationship between count efficiency and effective particle concentration for each detector was constant within experimental error for the entire dilution experiment. This observation can be used to test for the presence of horizontal coincidence, since the presence of horizontal coincidence will result in the variation of N_E as concentration increases.

One application of the error analysis in Fig. 16 is the determination of constant error contours in graphs of particle size versus particle concentration. This kind of analysis provides a good visual summary of detector capability.

XXX. PARTICLE CONCENTRATION FOR CONSTANT ERROR CONTOURS

Figures 17–20 visualize the effect of particle size variation on the concentration limited counting ability of four typical detectors. One reservation should be noted: these values are based on tests of a single detector in each category and therefore may not be representative of the performance of all similar units. These figures map the 5%, 10% and 20% undercount contours of four representative detectors. Figures 17 and 18 map performance of older white light detectors; Figs. 19 and 20 are maps of modern laser diode dectectors. To evaluate the practical effect of the data in these figures against realisitic SVI maximum particle concentration measurement requirements, the 2-ml container was selected. Justification for this choice is the fact that the preponderant bulk of SVI preparations is in 2-ml or larger containers.

Figure 17 shows the count accuracy performance of a good Climet white light detector. The specified geometric design volume of this detector, V_{DG}, is 7.5×10^{-6} cm^3; its design beam height, B_H is 50 μm. This detector was listed for operation up to a limit of 22,000 2-μm particles per milliliter. It was operated at its design flow rate of 10 ml/min. The experimental results for this detector are respectively $V_{DG} = 20.7 \times 10^{-6}$ cm^3 and $B_H = 28.5$ μm. Close agreement between specified and measured results was not obtained in any detector evaluated. All undercount contours in these four figures are smooth and continuous. Although this is an older detector design, its 10% undercount at 10 μm, 3768 counts/ml, is a specification incorporated in U.S.P. XXIII <788>. The count of the U.S.P. 4000-particle/ml 15-μm working standard will result in a 11.9% undercount.

Figure 18 shows the count accuracy performance of a good, high-flow-rate, Climet white light detector. This detector was not specified for use at high particle concentrations. The following evaluation of this high-flow detector will show the effect of the 4000-particle/ml U.S.P. count standard test.

The geometric design volume of this detector, V_{DG}, is 50×10^{-6} cm^3, its design beam height, B_H, is 250 μm. The detector was specified for operation up to a limit of 2000 5-μm particles/ml. It was operated at its design flow rate of 100 ml/min. The experimental results for this detector are, respectively, $V_{DG} = 109 \times 10^{-6}$ cm^3 and $B_H = 264$ μm. The reservation noted earlier is repeated here: these values are based on tests of a single detector. The contours for all undercount errors are smooth and continuous but are at substantially lower concentration levels than for the earlier smaller volume detector. Any attempt to use this detector for the SVI acceptance testing of 2-ml containers at the previous concentration limits in U.S.P. XXII or in the present XXIII <788> will result in gross error. The 10% undercount contour intersects a concentration of 942 10-μm particles/ml. The actual undercount error calculated for a 4000-particle/ml concentration of 15-μm particles is 36.5%, well beyond the ±18.75% acceptance limits.

Figure 19 shows the count accuracy performance of a modern PMS laser diode sourced detector. The geometric design volume of this detector, V_{DG}, is 10×10^{-6} cm^3, its design beam height, B_H, is 20 μm. The detector was specified for operation up to a limit of 12,000 10-μm particles/ml. It was operated at its design flow rate of 20 ml/min. The experimental results for this detector are respectively, $V_{DG} = 14 \times 10^{-6}$ cm^3 and $B_H = 39.7$ μm. To repeat a previous observation: for no detector evaluated was there close agreement between specified and measured dimensions and results. The reservation noted earlier is repeated here: these values are based on the test of a single detector. The contours for all undercount errors are smooth and continuous and are at substantially higher concentration levels than for the earlier white light detectors examined. The 10-μm 10% undercount contour intersects a concentration of 6012 particles/ml. The undercount error for the U.S.P. 4000 15-μm particle/ml reference count standard is 7.43%, well within the acceptable test variation.

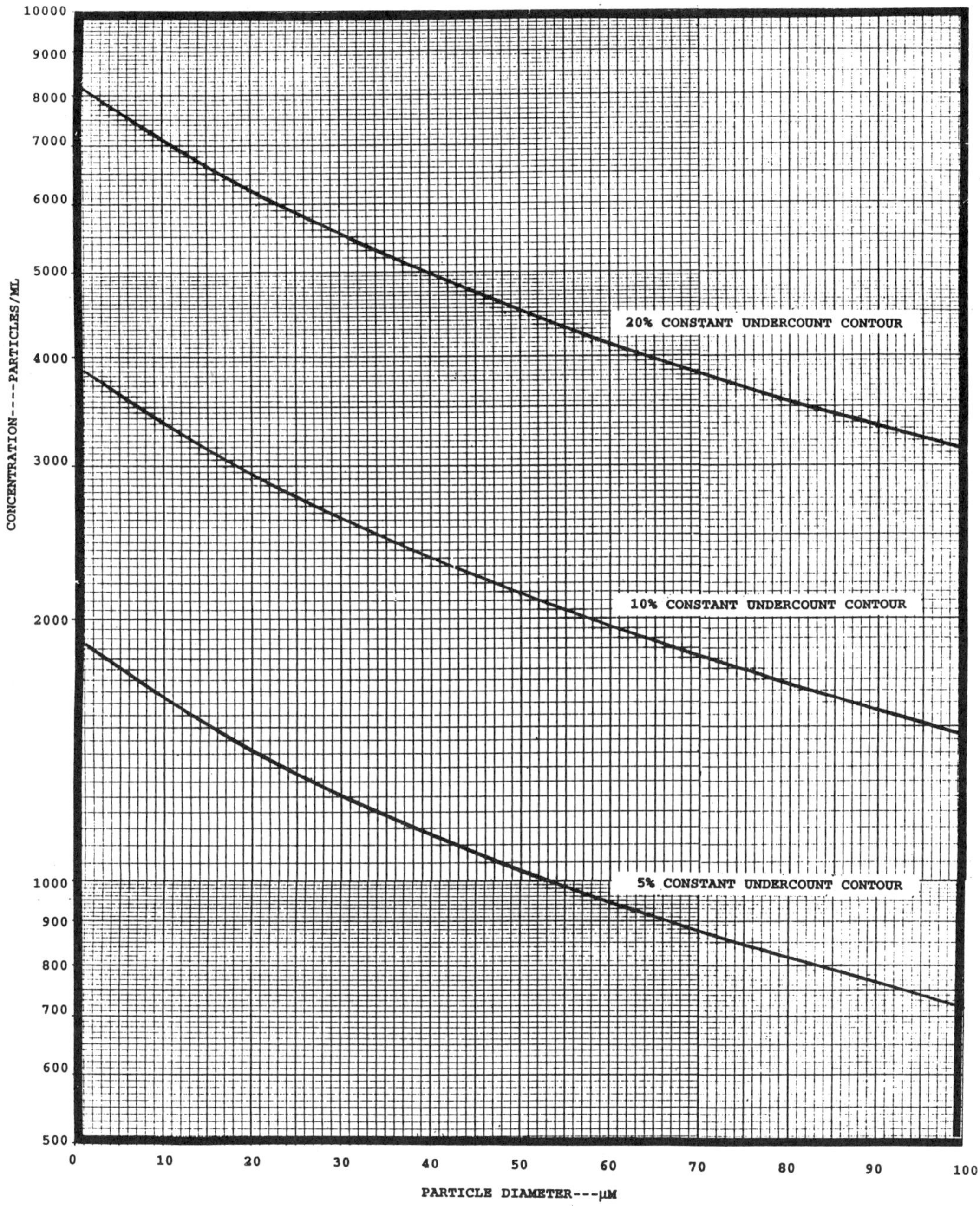

Figure 17 Constant undercount contours for a Climet detector that uses a white light excitation source and has a specified geometric detector volume, V_{DG}, of 7.5×10^{-6} cm^3.

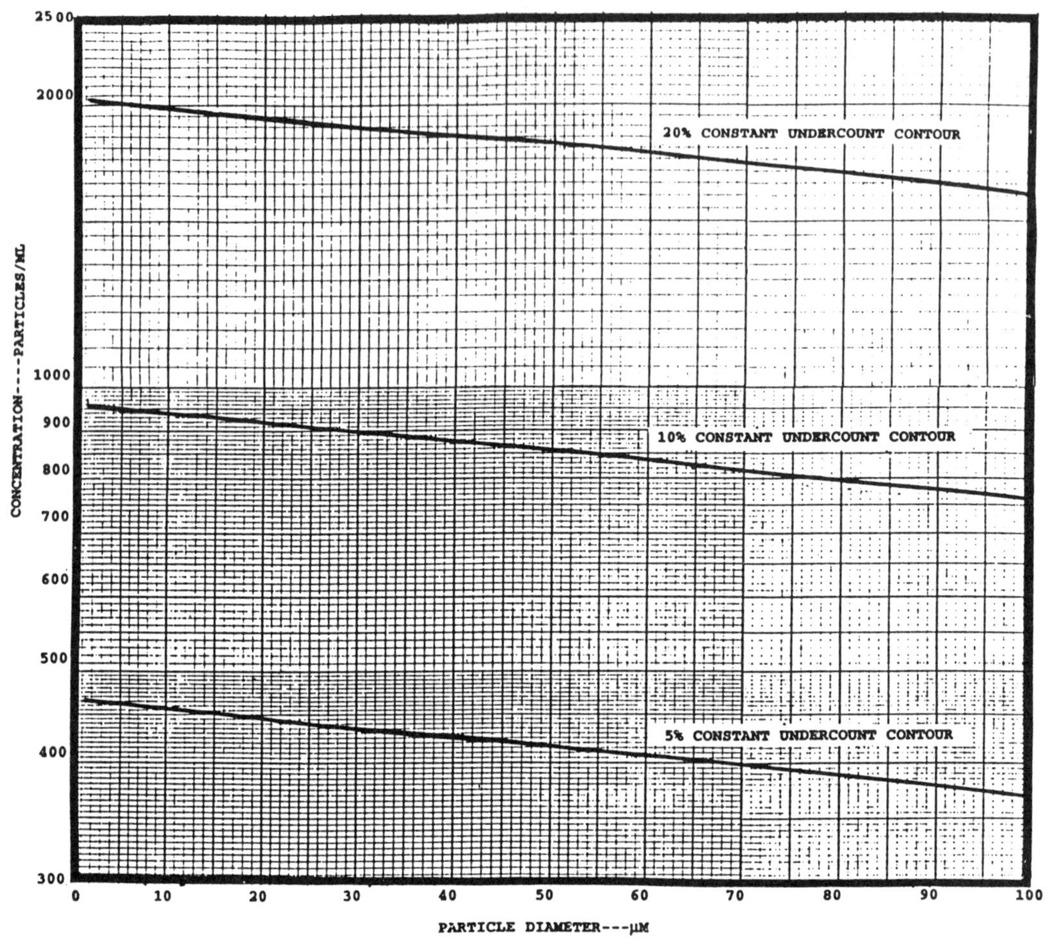

Figure 18 Constant undercount contours for a Climet detector that uses a white light excitation source and has a specified geometric detector volume, V_{DG} of 50×10^{-6} cm^3, and a beam height of 250 µm. The experimental results are $V_{DG} = 109 \times 10^{-6}$ cm^3 and $B_H = 377$ µm.

Figure 20 shows the count accuracy performance of another modern laser diode sourced HIAC/Royco detector. The geometric design volume of this detector, V_{DG}, is 4.5×10^{-6} cm^3, its design beam height, B_H, is 30 µm. The detector was specified for operation up to a limit of 18,000 10-µm particles/ml. It was operated at its design flow rate of 20 ml/min. The experimental results for this detector are respectively $V_{DG} = 7.4 \times 10^{-6}$ cm^3 and $B_H = 42.7$ µm. As with the other detectors evaluated, there was no close agreement between specified and measured results. The reservation noted earlier is repeated here: these values are based on a test of two detectors. The contours for all undercount errors are smooth and continuous and are at substantially higher concentration levels than for both the earlier white light detectors and the first laser diode detector examined. The 10-µm 10% undercount contour intersects a concentration of 11,537 particles/ml. The 15-µm U.S.P. reference count standard test of 4000 particles/ml will have 3.92% undercount error. Calculations show that the count of 4000 100-µm particles with this detector will have a 9.1% undercount error.

An important consideration in the preparation of Figs. 17–20 is that they communicate count

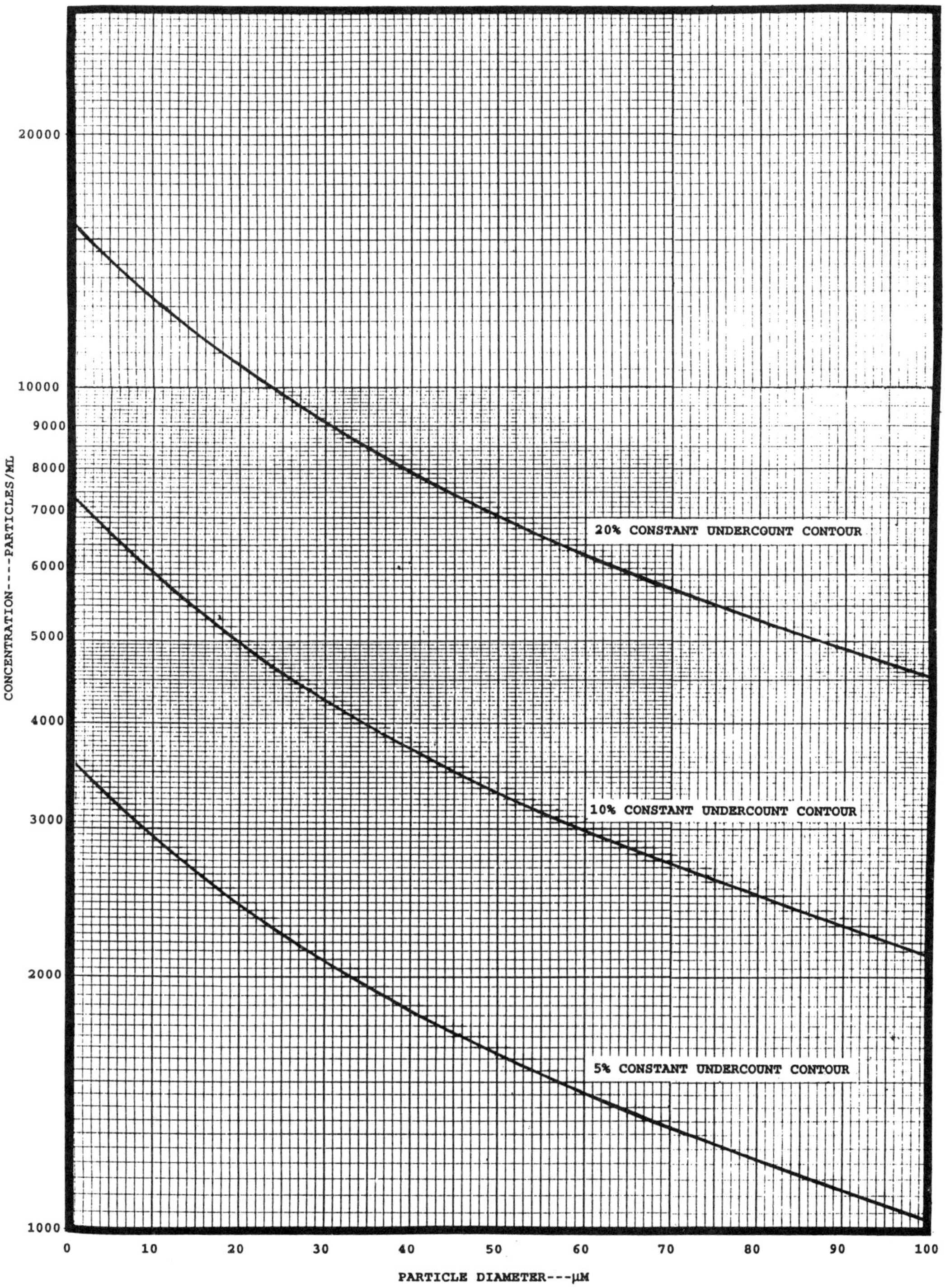

Figure 19 Constant undercount contours for a PMS detector that uses a laser diode excitation source and has a specified geometric detector volume, V_{DG}, of 10×10^{-6} cm^3.

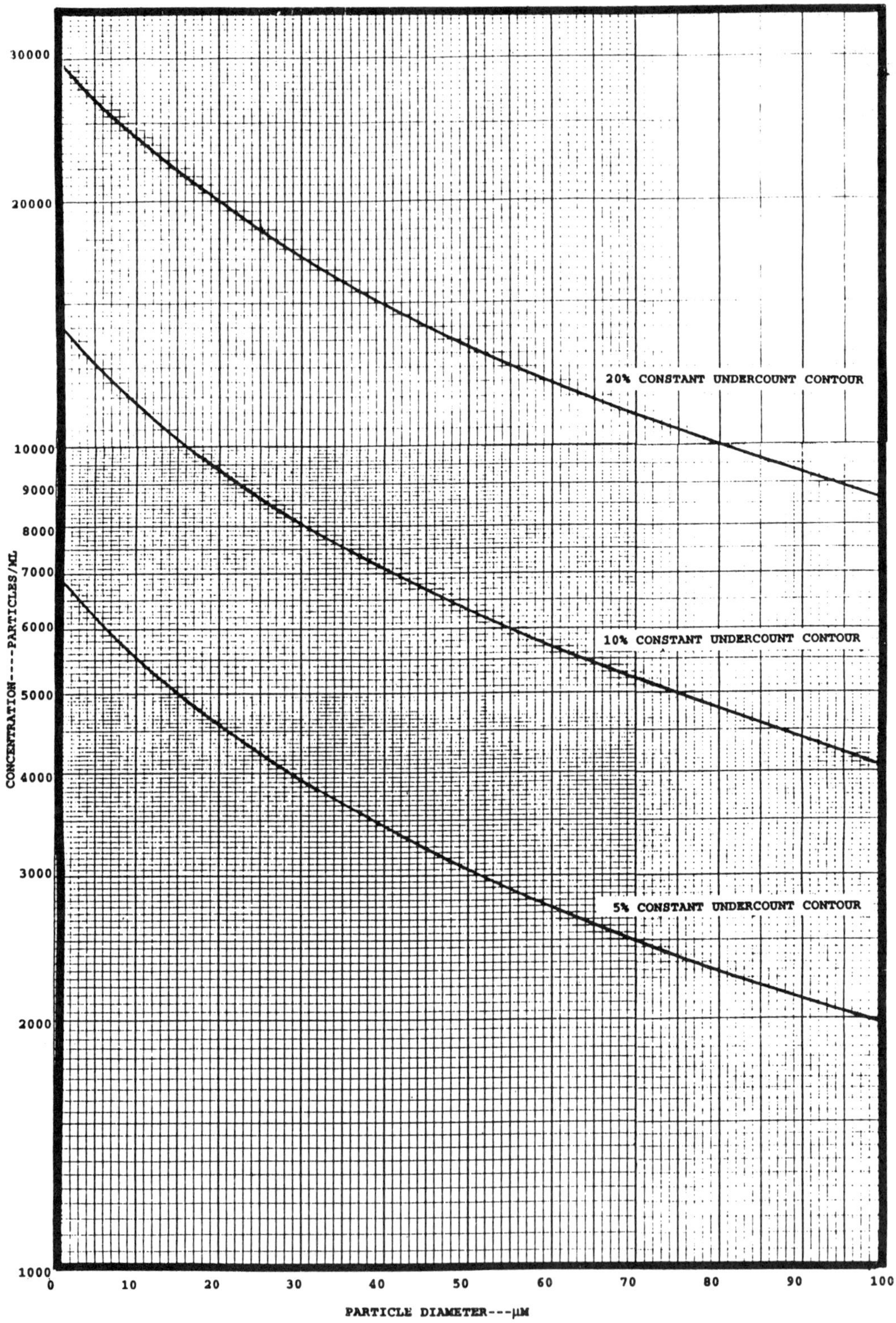

Figure 20 Five and 10% undercount contours for a HIAC/Royco detector that uses a laser diode excitation source and has a specified geometric detector volume, V_{DG}, of 4.5×10^{-6} cm^3 and beam height of 30 µm. The experimental results are $V_{DG} = 7.4 \times 10^{-6}$ cm^3 and $B_H = 42.7$ µm.

accuracy relationships without using new concepts. Although the new particle-triggered Poisson model and effective detector volume have been used in the preparation of these four figures the results displayed can be understood without them.

XXXI. MULTISIZED PARTICLE COINCIDENCE EFFECT CALCULATIONS

An important limitation in the relationship of effective particle concentration and count efficiency is that it describes only a single particle size concentration error curve. To achieve a general solution for the coincidence effects on count accuracy for all particles within the measurement range of a detector, the two unknowns in Eq. (5) must be evaluated experimentally. Equation (5) is modified with bullets in Eq. (5A) to mark the two unknowns.

$$V_{\mathrm{DE}} = \dot{V}_{\mathrm{DG}}\left(1 + \frac{D_{\mathrm{P}}}{\dot{B}_{\mathrm{H}}}\right) \tag{5A}$$

The unknowns are the beam height, B_{H}, and the geometric detector volume, V_{DG}. Solution of Eq. (5A) for these two unknowns requires two dilution experiments with two different particle sizes. With B_{H} and V_{DG} both known, a general solution for the effect of concentration on the counting accuracy of a detector throughout its operational range is possible. Table 8 lists the specified and experimental values of these parameters for the four detectors discussed above.

In current practice, the concentration limited counting accuracy of single-particle detectors is based on a combination of specified physical detector dimensions and static measurements of the white light or laser source employed. The interaction of the energy of the light source with the particle to create the observed signal has not been previously described. At present, particle counting accuracy limits are determined with differing experimental methodologies and variable accuracy. To provide a demonstration of the methodology described and the accuracy that can be expected with the published specifications, four commercial particle counter systems were evaluated using the stepwise dilution methodology described in the chapter. The capabilities of all counters investigated were each found to be overstated as shown in Table 8.

In Table 9, the experimental parameters of Table 8 have been used to calculate the N_{E} values for the four detectors listed here as the measured particle size varies from 2 to 100 μm. The

Table 8 Comparison of Specified and Experimental Value of the Beam Height, B_{H}, and the Geometric or Physical Detector Volume, V_{DG}, for Two White Light and Two Laser Diode Particle Detectors

Detector number	Light type	Geometric volume ($\mathrm{cm}^3 \times 10^{-6}$) Specified	Experimental	Beam height (μm) Specified	Experimental	Vendor
1	White	7.5	20.7	50	28.5	Climet
2	White	50.	109.0	250	377.0	Climet
3	Laser	10.	14.0	20	39.7	PMS
4	Laser	4.5	7.4	30	42.7	HIAC/Royco

Given these two experimental parameters the effective detector volume, V_{DE}, can be calculated for any particle size within the range of a particle detector. N_{E}, the number of effective detector volumes per milliliter for a specific particle size is then calculated as the reciprocal of V_{DE}.

Table 9 Variation of N_E, the Experimental Detector and Particle Size Specific Constant, as Particle Size Varies from 2 to 100 μm

	Geometric detector volume ($cm^3 \times 10^{-6}$) (values from commercial literature)			
	7.5(W)	50(W)	10(L)	4.5(L)
Particle size (μm)	Experimental N_E effective detector volumes per milliliter			
2	45,141	9,126	68,003	129,089
5	41,09	9,054	63,439	120,970
5.773	29,815	9,036	62,360	119,041
7.071	38,706	9,005	60,630	115,936
10.0	35,761	8,937	57,057	109,493
15.0	31,651	8,823	51,841	100,000
17.678	30,859	8,763	49,422	95,569
20.0	28,388	8,712	47,499	92,030
25	25,735	8,604	43,829	85,233
30.0	23,535	8,498	40,685	79,371
40.0	20,099	8,294	35,580	69,774
50.0	17,539	8,100	31,613	62,247
60.0	15,557	7,915	28,442	56,186
70.0	13,978	7,738	25,850	51,200
80.0	12,690	7,568	23,690	47,027
99.0	10,799	7,266	20,445	40,722

Calculation of the undercount for any detector is determined with Eq. (13). The undercount, in particles/ml, can be estimated from the product of the fractional undercount and the N_E value shown. (W) and (L) refer to the employment of a white light or laser excitation source. 7.5 and 50 are Climet designs, 10 and 4.5 are PMS and HIAC/Royco respectively.

information in this table can be used to determine the undercount error for any particle size for any concentration within the capability of these four detectors.

XXXII. VERTICAL AND HORIZONTAL COINCIDENCE

Figure 21 compares the impact of vertical and horizontal coincidence. The additive coincidence illustrated in the pulse height analyzer photos of Figs. 5A and 5B is described as vertical coincidence. The particle flow pattern illustrated in frames 1 through 4 of Figure 21 is indicative of detector design and electronics that minimize or eliminate bridging incidents. The sequential frames 1 and 3 show the vertical or additive particle size coincidence error. The sensed particle signal is determined by the sum of the three uniform particles in frame 1 and the four particle signals of the coincidence occurrence group in frame 4 as the group passes through the scan beam. Frames 2 and 4 are completely blank, allowing time for the particle counter to reset its processing cycle to receive the next particle signal. The time required for counter reset using present solid state electronics is a fraction of the full frame shown in this pictorial.

The second coincidence mechanism is due to random particle chains. In these random

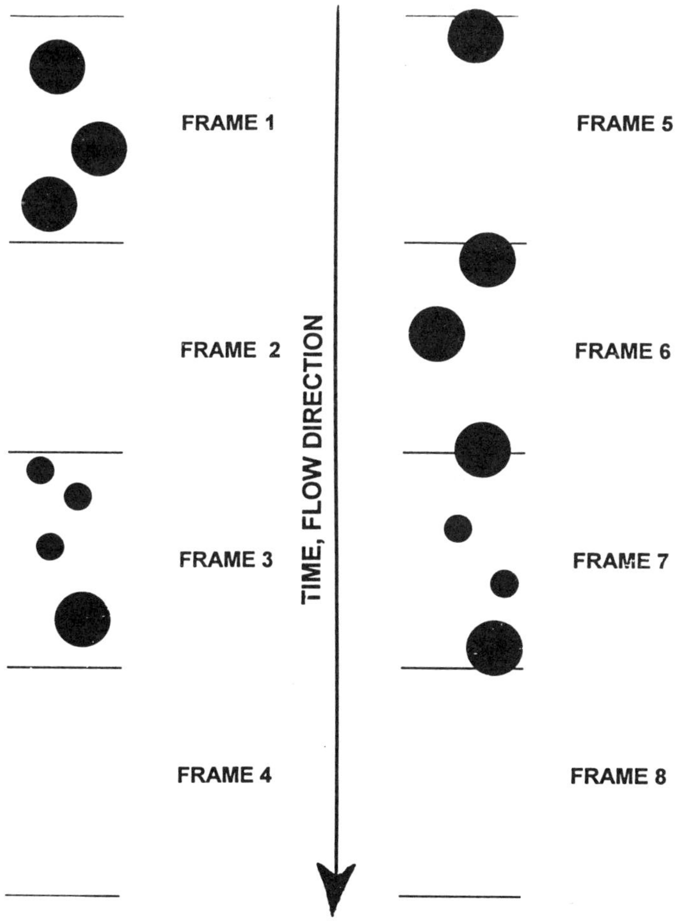

Figure 21 Frames 1 and 3 show vertical coincidence. In each frame a single particle is recorded whose size is equal to the sum of the cross sections of all particles present. Frames 5 through 8 show the effect of horizontal coincidence that results from the lack of counter reset between particle signals. One particle count is reported for the sequence of these four frames, its size is calculated as the equivalent of the sum of the cross section of the two larger particles in frame 6.

particle chains the signals are so closely spaced that the electronic reset signal which initiates the next sizing and counting cycle is suppressed. In a dilution experiment with uniform particles, the count from a horizontal coincidence particle chain can either be of one particle or a vertical coincidental occurrence within the chain. The reset signal level, usually selected to be 10% below the level of the analysis threshold selected, is not reached due to the spacing of the particles in the fluid stream. The effect of the blockage of the reset signal is that each particle chain bridges together a number of detector volumes and produces a single count. In a random particle assay, the count can be from a vertical coincidence group or from the largest particle in a horizontal coincidence chain. Frames 5 through 8 in Fig. 6 represent a random particle chain in which the blockage of particle counter reset results in the recording of a single double-occurrence count for all four frames.

The effective detector volume for the four interconnected frames is four times the unbridged detector volume. The result of the bridging shown is a reduction to one-fourth of the expected linear measurement range and an increased count loss beyond that predicted by the particle-triggered Poisson model. The second type of coincidence count loss is deduced by its effect on Eq. (44) or Eq. (44V) its verbal equivalent, the count balance equation.

If horizontal coincidence was a factor in the dilution experiment data recorded, the exponential relationship of effective concentration to count efficiency could not have the constant characteristic that has been documented. The result of the bridging shown would be evidenced as a progressive change in the exponential relationship between concentration and count efficiency as the effective concentration increased. While horizontal coincidence effects cannot be dismissed out of hand, it should not be a factor with present solid-state signal processing and currently available detectors. In the experimental data recorded, a single exponential relationship between count efficiency and effective concentration was seen in a concentration ratio of 20,000:1. Accordingly, the balance of this chapter uses the particle-triggered Poisson model and the effective detector volume as a tool to evaluate the effect of vertical particle coincidence on particle counting accuracy. The tools with which to achieve replicable, accurate contaminating particle assays with light extinction particle counters are now available.

Technological progress has brought with it gain-bandwidth improvement in the instrumentation amplifiers used in particle counters. This improvement minimizes or eliminates the need to consider "electronic" or horizontal coincidence. These terms were required when the bandwith-limited pulse response of the amplifiers then available blocked the amplification of closely spaced signals. This limitation has been happily obsoleted with current solid-state amplifiers. Accordingly, the application of the particle-triggered Poisson model in this chapter uses effective detector volume as a tool with which to achieve replicable, accurate contaminating particle assays using light extinction particle counters.

XXXIII. THEORETICAL CONSIDERATIONS

In the preceding discussion, the generally held view that the particle undercount error increases as particle concentration increases was confirmed. The experimental data confirmed a usable particle counting accuracy region within two concentration limits. At the low-concentration end of this region, the overcount counting error due to the background count is negligible with respect to the concentration of particles being counted. At the high-concentration end of this region, the acceptable undercount counting error is used to define the maximum concentration that should be measured. Within this region, using a linear approximation at low concentrations, the undercount error is numerically equal to the particle concentration when the particle concentration is measured at the ratio of particles to effective detector volumes.

There are two significant differences between the Knapp-Abramson work and preceding theory. The two differences are the statistical model selected for the analysis and the method of calculating particle concentration. Correction of each of these differences provides a multiplier for previous estimates of count loss with increasing concentration. In addition, the new statistical model predicts significantly greater interference from concentrations of particles below the 10-μm U.S.P. controlled particle contamination region. The new model also predicts interference from counts within the 10- to 25-μm span into false counts of particles greater than 25 μm at concentration levels that have been assumed to be free of measurement difficulty.

The earlier application of Poisson statistics to counting loss used a model better suited to analysis of static occurrences. An example of such a model is the evaluation of the occurrence rate of sand grains in sugar cubes in a box of contaminated sugar cubes. That model cannot accurately estimate counting errors when sequential randomly spaced groups of particles occur.

Prior work considered the volume of the particle to be negligible with respect to the detector volume. This conclusion was the basis for an analysis that considered the calculation of particle concentration to be independent of particle size. However, when the projected area of the particle is not negligible with respect to the size of the scan beam employed, the size of the particle does affect the estimated results. The new effective detector volume is the product of the flow cross section and the effective detector height. The effective detector height is the sum of the particle diameter and the scan beam height. As long as a signal is generated by the presence of part of a particle in the beam of the detector, greater than the selected threshold, the counter system will not reset to report another particle. A consequence of this interaction is the variation of the effective detector volume with the size of the particle being measured. This variation in the effective volume of the detector and thus the calculated concentration of particles with the particle size being measured significantly affects undercount estimates.

The joint effect of both the new Poisson model and the new measurement of concentration on particle size measurements is illustrated in the following discussion. For particles within the linear error range of the new Poisson statistical model, the error estimate for any particle size is doubled over that of the earlier static or geometric model. As particle size increases to that of the scan beam height, the effective detector volume increases to twice that of the physical or geometric detector volume; the two effects are multiplicative. For a particle whose diameter is equal to the scan beam height the joint result of the new Poisson model and new concentration calculation is to require a fourfold reduction of concentration to maintain the previously estimated error limit. Insistence on maintenance of the previously selected particle concentration quadruples the previous error estimate.

Advances: Two normalized equations have been realized that describe the effect of concentration on the counting capability of all single particle counters. Horizontal and vertical coincidence have been investigated. The conclusion reached was that vertical coincidence was the major concern, and with present technology, is probably the only coincidence error concern.

Limitations: No distinction is made between single-particle and multiple-occurrence particle counts. A remaining area of concern in the application of light extinction particle counters is the relationship between the number of counts recorded and the number of single particles that have been counted.

XXXIV. INTENT AND REALIZATION OF COUNTING ACCURACY: PARTICLE COUNTS VERSUS COUNTS RECORDED

Before the results of the counts recorded by particle counters can be evaluated, the intent of these contaminating particle assays must be determined. Rereading of the new U.S.P. <788> [1] as well as those in earlier U.S.P. revisions shows that the intent of these particle assays is to set acceptance limits for injectable products based on the number of single contaminating particles in a sample or in a container. The particle contamination limits above 10 μm and above 25 μm were selected with both patient insult and perceived measurement reproducibility in mind.

Reviewing this measurement intention, count accuracy itself is a necessary but not sufficient analysis end point. The tools with which to realize the U.S.P. quality goal by evaluating the number of single particles counted can be formulated from the series that define the particle-triggered Poisson relationships.

XXXV. COINCIDENCE ERROR REVISITED

The new particle-triggered Poisson model [23–25], described and used in this chapter, focuses on an examination of the counts recorded from each of the particle size groups counted. A common characteristic of all terms in the Poisson series is that each term represents an integral multiple of signal for a single particle. This type of additive coincidence signal can be observed and investigated with a pulse height analyzer as shown in the linear pulse height analyzer display of Fig. 5A or the equivalent data shown in a logarithmic display in Fig. 5B.

Following the development of the universal count accuracy curve for the light extinction particle counters, an investigation into the complement to count efficiency, the loss of count efficiency began. The first step forward was the seemingly trivial observation that the sum of the count efficiency ratio and the count loss ratio was equal to 1. A verbal equivalent to Appendix Eq. (38) is

$$1 = [\text{Count efficiency ratio}] + [\text{Count loss ratio}] \tag{38V}$$

The particle-triggered Poisson model in combination with count efficiency and count loss accurately describe the single particle counting capability of all light extinction particle counters. The count loss ratio evaluates the ratio of single-particle counts that have been merged into false, larger particle sizes.

This particle-triggered Poisson model description is implemented with constants and relationships that can be validated experimentally in modestly equipped pharmaceutical laboratories. The single-particle count capability of light extinction particle counters is developed by expressing the count efficiency ratio as the sum of the single-particle count efficiency ratio and the count efficiency ratio of all multiple-particle occurrence groups. The single-particle count efficiency ratio is available in the particle-triggered Poisson series. This series, whose applicability was validated in the Part 1 coincidence paper [23], is written as a sum of the count efficiency ratios of all particle occurrence count groups.

The probability of occurrence of each particle group is estimated with the particle-triggered Poisson series expressed in terms of count efficiency. Equation (39AV) is a symbolic equation summarizing this Poisson series. It states that the overall count efficiency is the sum of the count efficiency ratio of the occurrence probability terms for each particle group in multiplicity order.

$$\begin{bmatrix}\text{Overall count}\\ \text{efficiency ratio}\end{bmatrix} = \begin{bmatrix}\text{Count efficiency}\\ \text{ratio of single-}\\ \text{particle groups}\end{bmatrix} + \begin{bmatrix}\text{Count efficiency}\\ \text{ratio of two-}\\ \text{particle groups}\end{bmatrix} + \begin{bmatrix}\text{Count efficiency}\\ \text{ratio of three-}\\ \text{particle groups}\end{bmatrix} + \cdots \tag{39AV}$$

The actual particle-triggered Poisson series in Eq. (39V) is an infinite sum of products:

$$\begin{aligned}\begin{bmatrix}\text{Count}\\ \text{efficiency}\\ \text{ratio}\end{bmatrix} = \begin{bmatrix}\text{Count}\\ \text{efficiency}\\ \text{ratio}\end{bmatrix}^2 &+ \begin{bmatrix}\text{Count}\\ \text{efficiency}\\ \text{ratio}\end{bmatrix}^2 \begin{bmatrix}1 - \begin{matrix}\text{Count}\\ \text{efficiency}\\ \text{ratio}\end{matrix}\end{bmatrix}\\ &+ \begin{bmatrix}\text{Count}\\ \text{efficiency}\\ \text{ratio}\end{bmatrix}^2 \begin{bmatrix}1 - \begin{matrix}\text{Count}\\ \text{efficiency}\\ \text{ratio}\end{matrix}\end{bmatrix} + \cdots\end{aligned} \tag{39V}$$

Each right-hand term in Eq. (39V) represents the occurrence probability of the particle multiplicity group represented in Eq. (39AV). Equation (39V) provides the means with which to achieve control of count accuracy in single-particle light extinction counters.

The first right-hand term in the particle-triggered Poisson series is the occurrence probability of single-particle groups. This occurrence probability is the overall count efficiency squared. This observation is the key to understanding and controlling single-particle light extinction counter counting accuracy. The relationship is expressed in a verbal statement of Eq. (40):

$$\begin{bmatrix}\text{Count efficiency ratio of} \\ \text{single-particle groups}\end{bmatrix} = \begin{bmatrix}\text{Overall count} \\ \text{efficiency ratio}\end{bmatrix}^2 \quad (40V)$$

In Eq. (39BV) the series in Eq. (39V) is rewritten as a sum of the count efficiency of single-particle groups and the count efficiency ratio of all multiple-particle groups:

$$\begin{bmatrix}\text{Overall count} \\ \text{efficiency ratio}\end{bmatrix} = \begin{bmatrix}\text{Count efficiency} \\ \text{ratio of single-} \\ \text{particle groups}\end{bmatrix} + \begin{bmatrix}\text{Count efficiency} \\ \text{ratio of multiple-} \\ \text{particle groups}\end{bmatrix} \quad (39BV)$$

The substraction of count efficiency ratio of single-particle groups from the overall count efficiency ratio results in the count efficiency ratio of all multiple-count groups. The results of this manipulation of Eq. (39BV) are shown in Eq. (39CV):

$$\begin{bmatrix}\text{Count efficiency} \\ \text{ratio of multiple-} \\ \text{particle groups}\end{bmatrix} = \begin{bmatrix}\text{Overall count} \\ \text{efficiency ratio}\end{bmatrix} - \begin{bmatrix}\text{Count efficiency} \\ \text{ratio of single-} \\ \text{particle groups}\end{bmatrix} \quad (39CV)$$

Replacing the overall count efficiency in Eq. (38V) by its equivalent value, the sum of the single-particle count efficiency ratio and the multiple-particle count efficiency ratio shown in Eq. (39CV) results in the particle ratio count balance equation.

$$1 = \begin{bmatrix}\text{Singles count} \\ \text{efficiency ratio}\end{bmatrix} + \begin{bmatrix}\text{Multiples count} \\ \text{efficiency ratio}\end{bmatrix} + [\text{Count loss ratio}] \quad (44V)$$

Equation (44V) shows that single-particle count loss ratio has two sources: the count of particle clusters and the loss of counts within the particle clusters that were not recorded. The lost single-particle count ratio is calculated as the sum of the multiple-count ratio and the count loss ratio. This single-particle count loss ratio is expressed in

$$\begin{bmatrix}\text{Single-particle} \\ \text{count loss ratio}\end{bmatrix} = \begin{bmatrix}\text{multiple-count} \\ \text{efficiency ratio}\end{bmatrix} + [\text{count loss ratio}] \quad (44BV)$$

A more convenient expression for the calculation of the single-particle count loss is available by subtracting the single-particle count efficiency ratio from 1 in the count balance equation:

$$[\text{Single-particle count loss ratio}] = [1 - (\text{Single-particle count efficiency ratio})^2] \quad (44CV)$$

Equation (44CV) records a major step forward in understanding and achieving light extinction particle counter accuracy. The single-particle count loss is seen to be equal to 1 minus the square of the overall count efficiency. Equation (44CV) defines the methodology required to achieve the single-particle counting accuracy in light extinction particle counters. Maintenance of a combined size and count limit for the concentration to be sized and counted accurately is the prerequisite that must be satisfied.

XXXVI. UNIVERSAL SINGLE-PARTICLE COUNT ACCURACY AND COUNT LOSS CURVES

The count loss ratio, c/n, and the multiple count efficiency ratio, C^+ shown in Eq. (44BV) are individually graphed in Fig. 22. These two components of the single-particle count loss ratio are approximately equal up to the point where the single-particle count loss amounts to 20%. Clearly, neither component of single-particle count loss can be disregarded if single-particle count accuracy is to be obtained. Neglecting the count loss contribution of either coincidence component results in a 50% underestimation of the single-particle count error. It is important to note that the single-particle count error for all particle sizes large than 10 μm will be worse.

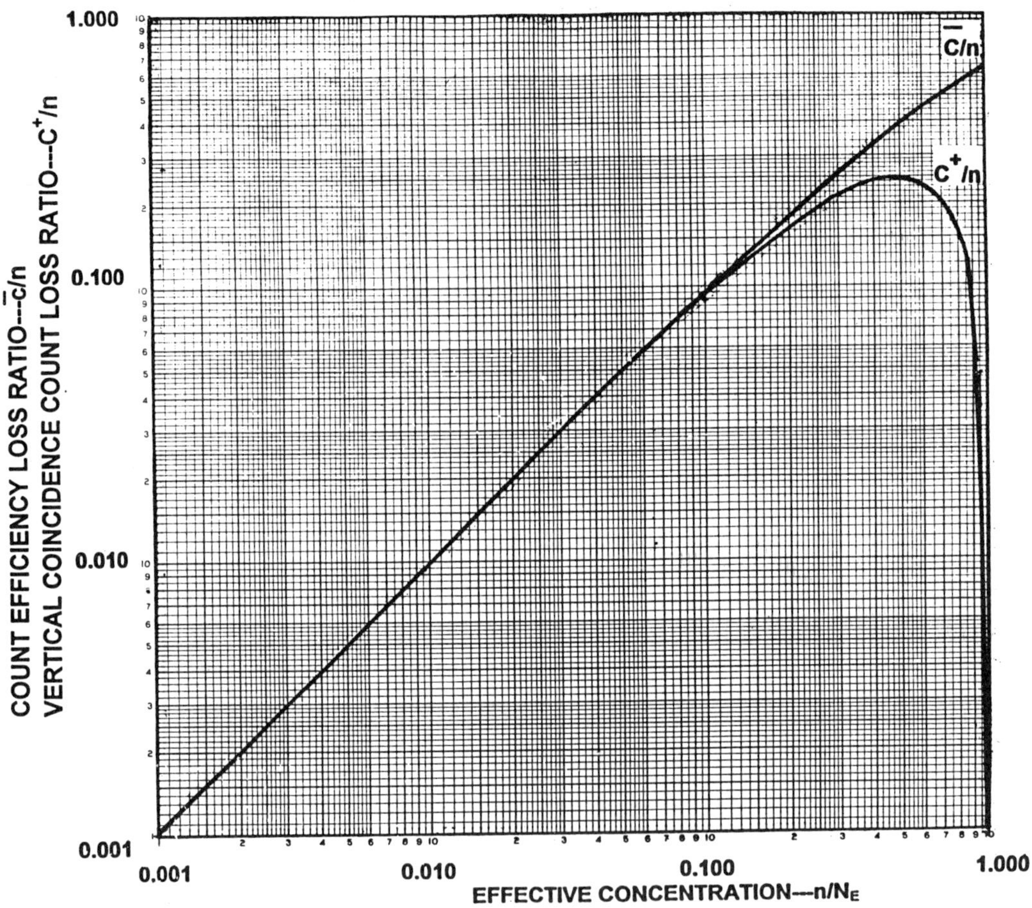

Figure 22 This graph demonstrates the variation of the two components of the vertical coincidence count loss ratio as particle concentration increases. The two particle count loss ratio components are C^+/n, the multiple count loss ratio and $\bar{c}/n$, the count efficiency loss ratio. Both components of the single-particle count loss are seen to be approximately equal up to 10%. The total coincidence count loss ratio is the sum of both losses: neglecting either one reduces the estimate of count loss by 50%.

XXXVII. CONTAMINATING PARTICLES: COUNTS RECORDED VERSUS NUMBER COUNTED

The intent of U.S.P. <788> is to consider contamination measurements evaluated by counts of single particles. The universal count efficiency and count error curves shown in Fig. 23 represent a significant but not final step toward achieving a description of particle counters. In retrospect, the previous graphical analysis of count accuracy must be changed to reflect the fact that these are raw count curves. The universal single-particle count efficiency and count error curves in Fig. 23 provide the capability required for U.S.P. <788> to define single-particle count accuracy and count error. These universal curves, employing the effective particle concentration measure n/N_E, apply to all single-particle counters.

Figure 23 is a set of universal curves for the class of all single-particle counters which

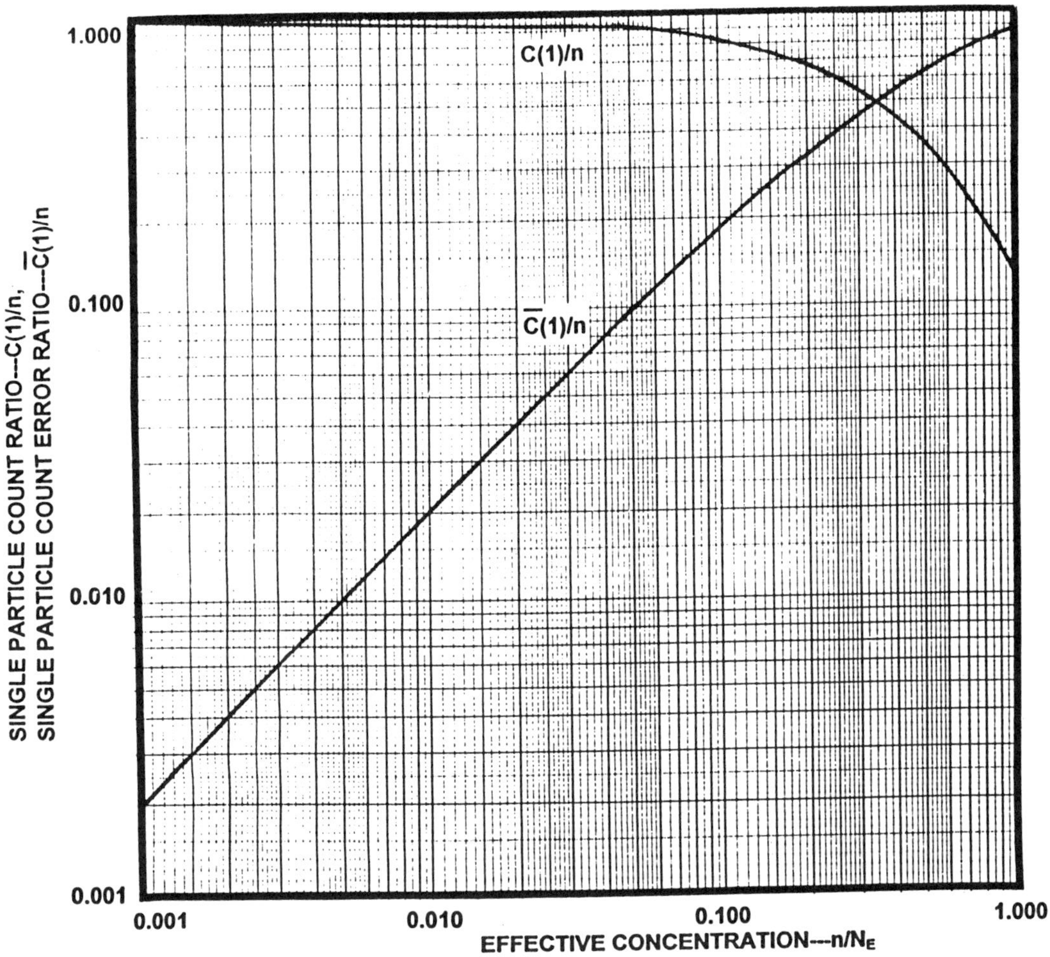

Figure 23 The universal single-particle count efficiency and count error curves for light extinction particle counters are shown. Since the intent of the U.S.P. <788> is to define particle contamination limits in terms of the number of single particles present in a sample, it can now be implemented with acceptable accuracy and replicability.

Table 10 Comparison of Maximum Count and Maximum Single-Particle Group Count Parameters

	c Counts/ml	$C(1)$ Single-particle counts/ml
Maximum concentration n_{MAX}	N_E	$\frac{N_E}{2}$
Maximum count C_{MAX}	$e^{-1} N_E = 0.3679N_E$	$\frac{e^{-1}}{2} N_E = 0.1839N_E$
Count efficiency at maximum $\frac{c}{n}$	$e^{-1} = 0.3679$	$e^{-1} = 0.3679$
Count error at maximum ε_C	$(1 - e^{-1}) = 0.6321$	$(1 - e^{-1}) = 0.6321$

includes the light extinction particle counters specified for the evaluation of random contaminating particles in injectable products. The two curves shown in this figure are the ratio of single-particle group counts to the total number of particles in suspension, $C(1)/n$ and the single-particle count loss ratio $\bar{c}/n$ (see Appendix 2 Sec. E, pp. 546–559).

As in the earlier universal curve of raw counts versus effective concentration, occurrence of a maximum count is observed. The maximum count occurs at the concentration at which there is a ratio of one-half particle per effective detector volume. The maximum count at this concentration is $0.1839N_E$. The peak single-particle count and the concentration at which it occurs are half the raw count values. The single-particle undercount error at the peak count is the same value as for raw counts, 63.2%. These curves provide the information required to operate light extinction particle counters within selected accuracy limits. The selected count accuracy limits are defined by a concentration and particle size contour. These results are compared to those obtained in the raw count analysis in Table 10. Table 10 shows that both the concentration of peak count occurrence and the height of the peak count are both halved when compared to the results obtained in the raw count analysis (see Appendix 2 Sec. E.2, p. 548).

The procedure section of this chapter, Appendix 1 Sec. A–N, pp. 527–534, contains methodology with which to experimentally determine N_E for any single-particle detector. The value of N_E is specific to the measurement of a selected particle diameter by a selected detector. It can be used to completely describe the full effect of particle coincidence on the counting performance of a detector for the selected particle size.

XXXVIII. COUNT EFFICIENCY TOOLS FOR FALSE COUNT AND UNDERCOUNT ANALYSIS

The basic count efficiency relationship in Eq. (32), $c/n = e^{-n/N_E}$, and the series shown in Eq. (39AV) provide the means with which to understand and control the occurrence of false counts and undercounts in particle-counting experiments. A summary of useful relationships and convenient computational forms is shown in Table 11. Table 11 shows the symbols, expressions and verbal explanations for each of the relationships listed.

The tools with which to analyze the effect of coincidence on the counts recorded by light extinction particle counters now exist.

Table 11 The Use of Count Efficiency in Determining Changes in Measured Sizes of a Particle Distribution and the Counts That Have Been Lost as a Function of Particle Concentration

Eq.	Function	Symbol	Expression
40	Single-particle counts / Single particles	$C(1)/n$	$= (c/n)^2$
40B	Double-event counts / Single particles	$C(2)/n$	$= (c/n)^2(1 - c/n)$
40C	Triple-event counts / Single particles	$C(3)/n$	$= (c/n)^2(1 - c/n)^2$
40D	Quadruple-event counts / Single particles	$C(4)/n$	$= (c/n)^2(1 - c/n)^3$
41	All multiple-event counts / Single particles	C^+/n	$= (c/n)(1 - c)$
42A	3 × & higher-event counts / Single particles	$C^+(3)/n$	$= (c/n)(1 - c/n)^2$
38A	Undercount total / Single particles	$\overline{c}/n$	$= 1 - c/n$
44C	Single-particle undercount / Single particles	$\overline{C}(1)/n$	$= 1 - (c/n)^2$

See Appendix 2, Sec. D, pp. 544–548, for detailed examples. All computations are per milliliter.

XXXIX. APPLICATION TO THE U.S.P. <788> CONTAMINATING PARTICLE ASSAY

This section provides an overview of the problem addressed, a more complete treatment follows in Appendix 2, Derivations and Details, beginning on page 534.

U.S.P. XXII, p. 1814, Test Apparatus, Sensor Concentration Limits, defines the limiting concentration that can be assayed as that at which the coincident counts amount to 10% of the single 10-μm particles present.

Determination of the coincident particle group occurrence rate is made with an indirect methodology: the primary measurement is that of count efficiency. The occurrence rate of multiple-particle groups is calculated from the measured count efficiency using one of the published Poisson models and one of the concentration measures published.

Accordingly, an analysis of the U.S.P. <788> particle counter count accuracy specification using the earlier Jaenicke geometric Poisson model and a relative concentration measure that omits the particle/detector interaction is compared with an analysis using the particle-triggered-model of the counting process and the effective concentration measure. An example of the counting errors introduced by the use of the U.S.P. XXIII <788> SVI concentration limit calculated with the Jaenicke geometric Poisson model and the relative concentration measure is analyzed. The example uses the particle-triggered Poisson model, the effective concentration measure and a commercially available laser-sourced detector. The detector selected was the #3 detector listed in Table 8 and experimentally evaluated in the earlier "coincidence" papers. It had a geometric detector volume, V_{DG} of 14×10^{-6} cm^3. The measured beam height was 39.7 μm.

In both analyses, the coincident particle group is equated to 10% of the particle count groups counted. This is expressed as

$$0.10\begin{bmatrix}\text{Overall count}\\ \text{efficiency ratio}\end{bmatrix} = \begin{bmatrix}\text{Multiple-particle group}\\ \text{count efficiency ratio}\end{bmatrix} \tag{46V}$$

Substituting for the multiple-particle group count efficiency ratio its value from Eq. (39CV) which is repeated here for convenience, results in, after simplification,

$$\begin{bmatrix}\text{Count efficiency}\\ \text{ratio of multiple-}\\ \text{particle groups}\end{bmatrix} = \begin{bmatrix}\text{Overall count}\\ \text{efficiency ratio}\end{bmatrix} - \begin{bmatrix}\text{Count efficiency}\\ \text{ratio of single-}\\ \text{particle groups}\end{bmatrix} \tag{39CV}$$

$$0.90\begin{bmatrix}\text{Overall count}\\ \text{efficiency ratio}\end{bmatrix} = \begin{bmatrix}\text{Count efficiency ratio}\\ \text{of single-particle groups}\end{bmatrix} \tag{46AV}$$

The algebraic equivalent of Eq. 46AV is

$$0.90\frac{c}{n} = \frac{C(1)}{n} \tag{46B}$$

Although the symbolic equation is identical for both models, the algebraic realizations will of course differ.

XL. APPLICATION OF THE JAENICKE GEOMETRIC POISSON MODEL

The solution of the Jaenicke geometric Poisson model for the relative concentration of 10-μm particles, the ratio of n particles to N_G detector volumes per milliliter at which there is a 10% ratio of coincident occurrences involves two exponential terms. The solution of this equation is addressed in Appendix 2 Sec. E.4, pp. 549–550, and is a relative particle concentration of 0.2073 correct to four significant figures. The calculated Jaenicke geometric Poisson model count efficiency at this relative concentration is 0.9032. The model-based raw count loss is 1 minus the count efficiency and is 1 – 0.9032 or 9.68%.

The model-based limiting concentration of 10-μm particles for the #3 detector described above is 14,807 particles per milliliter. This result is based on the Jaenicke geometric Poisson model and a concentration measure unaffected by particle size.

XLI. APPLICATION OF THE PARTICLE-TRIGGERED POISSON MODEL

To use the particle-triggered Poisson model with particle concentration evaluated using the effective concentration measure, a specific detector design is required. Using the #3 laser-sourced detector described above, results in a count efficiency ratio of 0.90 at the U.S.P. XXIII <788> coincidence ratio as shown in Eq. (46B). Detailed computations for this example are included in Appendix 2 Sec. C, pp. 541–544.

$$\frac{c}{n} = 0.90 \tag{46B}$$

Using the effective detector volume to determine particle concentration, the particle-triggered Poisson loss ratio of single-particle counts is determined with Eq. (40V):

$$
\begin{aligned}
\text{[Single-particle count loss]} &= [1 - (\text{Single-particle count efficiency ratio})^2] \\
&= [1 - (0.9)^2 = 0.19] \qquad (40V) \\
\text{[Single-particle count loss ratio]} &= 0.19
\end{aligned}
$$

The single-particle count loss for the same detector using this analysis is 19%: this is approximately double the 9.68% count loss obtained with the Jaenicke geometric Poisson analysis for the same 10% coincident occurrence ratio.

The 10-μm concentration calculated with the particle-triggered Poisson model to comply with the U.S.P. XXIII <788> SVI coincidence ratio specification is 6014 particles/ml.

XLII. THE JAENICKE GEOMETRIC POISSON MODEL PARTICLE CONCENTRATION LIMIT EVALUATED WITH THE PARTICLE-TRIGGERED POISSON MODEL AND THE EFFECTIVE PARTICLE CONCENTRATION MEASURE

The U.S.P. coincidence ratio limit concentration calculated with the Jaenicke geometric Poisson model for the same detector was 14,807 10-μm particles/ml. The count efficiency for this concentration in the particle-triggered Poisson model is 0.7714. The 14,807 particles per milliliter result in 11,422 counts per milliliter. The ratio of single-particle counts in the particle-triggered model is calculated as the square of the count efficiency. Squaring 0.7714 results in a single-particle count ratio of 0.5951. The single-particle count loss ratio is 1 – 0.5951 or 0.4049, a 40.49% error. It is important to note that the single-particle count error for all particle sizes larger than 10 μm will be worse.

XLIII. CONVERSION OF U.S.P. XXII <788> COUNTER ACCURACY SPECIFICATION FOR ACCESS TO THE PARTICLE-TRIGGERED MODEL RESULTS

N' and N'_E are respectively the concentration and the number of effective particle volumes per ml that satisfy the U.S.P. XXIII <788> coincidence ratio specification. N'_E, the number of detector volumes per milliliter for 10% coincidence ratio error is calculated with Eq. (14). A summary of all symbols used is listed in Table 1 on page 452.

$$N'_E = 9.4912N' \qquad (14)$$

The conversion constant is 9.4912. The use of this conversion constant transforms the U.S.P. XXII <788> SVI coincidence ratio specification into a measurable and calculable validation procedure that can be accomplished by even a modestly equipped pharmaceutical laboratory. A major consideration in the use of this conversion constant for a concentration of 10-μm particles is that its accuracy relies on the results vendor statements based on unspecified measurement methods.

A methodical examination of single-particle count ratio error versus the count efficiency of operation is shown in Table 12. The data in this table show that a count efficiency of 0.9747 is required to achieve a 5% single-particle error rate. The single-particle count balance expressed in Eq. (10) provides the means to resolve counts of single particles from the total counts recorded. In Fig. 22, the count loss ratio from multiple-particle group occurrences and the count loss ratio due to the uncounted particles within each cluster have been separately graphed. The

Table 12 Effect of Count Efficiency Variation from 0.99 to 0.70 on the Single-Particle Count Ratio and Single-Particle Loss Ratio

Count efficiency c/n	Single-particle count ratio $C(1)/n$	Single-particle loss ratio $\overline{C}(1)/n$
0.99	0.98	0.02
0.9747	0.95	0.05***
0.95	0.90	0.10
0.90	0.81	0.19•••
0.85	0.72	0.28
0.80	0.64	0.36
0.75	0.56	0.44
0.70	0.49	0.51

Note: At 90% count efficiency, the single particle count loss is 19%. The count loss is reduced to 5% at a count efficiency of 0.9747.
***5.00% (at .9747 count efficiency)
•••19.0% (at U.S.P. specified .9091)

total coincidence count loss is the sum of the individual loss ratios. Within a 20% total error limit, the contribution of both loss ratios is seen to be approximately equal. Neglecting the count loss contribution of either coincidence component results in a 50% underestimation of the single-particle count error.

XLIV. COINCIDENT OCCURRENCES AND THE U.S.P. <788> PARTICLE ASSAY

The effect of coincident occurrences on the count accuracy of the U.S.P. <788> contaminating particle assay is determined by both particle concentration and particle size. Coincident occurrences that can affect the accuracy of the U.S.P. <788> particle assay count can originate both within the 10- to 25-μm particle size control region and from the steeply rising concentration of particles immediately below the 10-μm control threshold. The probability of coincidence error-injection into the U.S.P. <788> particle assay arises from both the maximum acceptable particle concentrations defined in the assay and from the uncontrolled particle concentrations immediately below the 10-μm lower boundary.

A summary of these effects is shown in Table 13. The number of multiple occurrences at each particle size that can occur with concentration and their effect on the counts recorded is shown here. This table shows that the result of the coincidence of sub-10-μm particles is the injection of false counts within the assay limits. For particles ranging in size from 10 to 12.5 μm, the predominant coincidence effect is an undercount error. For particles from 12.5 to 17.6777 μm, both false counts and undercounts can result based on the concentration of particles counted. From 17.6777 to 25 μm, the predominant (and potentially more serious effect) is the generation of false particle counts ≥ 25 μm.

To examine the effect of coincidence error with realizable, measured particle distributions, the effect of coincidence on two particle sizes was calculated. The measured particle distribution for both particle sizes was assumed to be 5%. The $n = 4$ entry for the asterisked 11.5-μm particle size shows that up to a 4× coincident occurrence, the resulting false multiple-particle count will remain under 25-μm thus producing undercounts. The asterisked 21.7391-μm particle size, with

Table 13 Effect of Multiple Occurrences on the U.S.P. <788> Contaminating Particle Assay from Below and from Within Controlled Particle Size Range Must Be Considered to Achieve Accurate Count Results

Particle diameter range (μm)	Multiple occurrence *n*	Effect
5.7735 < 7.0711	3	False counts > 10 μm
7.0711 < 10.0000	2	False counts > 10 μm
10.0000	6	U.S.P. limit zone
10.2062 > 11.1803	5	
11.1803 > 12.5000	4	
11.5000*	4	Undercounts 10 μm $< D_P <$ 25 μm
12.5000 > 14.4338	3	
14.4338 > 17.6777	2	False counts > 25 μm
17.6777 > 25.0000	1	
21.7391*	1	
25.0000	1	U.S.P. limit change

a similar distribution, indicates that in the absence of coincidence error a $+3\sigma$ retention of single-particle counts will remain within the 25-μm control limit. Double coincidence occurrence with this particle size will result in a 30.744-μm particle size peak.

XLV. SENSITIVITY ANALYSIS OF U.S.P. <788> SVI CONTAMINATING PARTICLE ASSAY TO COINCIDENT PARTICLE INTERACTIONS

The effect of the coincidence-magnified particle signals on particle contamination assays has been obscured by the focus on undercount error. Figure 24 summarizes the eight major coincidence effects that can occur in the U.S.P. <788> assay as a result of the particle concentration being measured (see Appendix 2 Sec. F, pp. 559–566).

Analysis of the interaction problem begins by dividing the particle sizes that should be considered into five zones. The geometric mean particle size in each zone, shown in parentheses, will be used for performance comparisons. Zone 1 is the particle size group that generates from one-third to one-half of the signal from 10-μm particles. It is the source of false counts greater than 10 μm from triple and higher coincident occurrences.

The particle size group in Zone II commences with particles whose signals range from just below the 10-μm size to ½ the 10–μm size. All coincident occurrences in this zone will register false U.S.P. particle assay counts.

The Zone III particle size group ranges from 10-μm particles to a size that produces less than one-half the signal of a 25-μm particle. Depending on the particle size and the capability of the detector, false U.S.P. <788> SVI assay counts greater than 25-μm and undercounts can both originate in this zone. The undercounts are caused by the count loss due to horizontal coincidence plus that due to the count loss from multiple-occurrence events. Accordingly, Zone III is shown as the origin of three undercount loops and a false count interaction loop. The false count loop shows the possibility of false counts greater than 25 μm originating in Zone III.

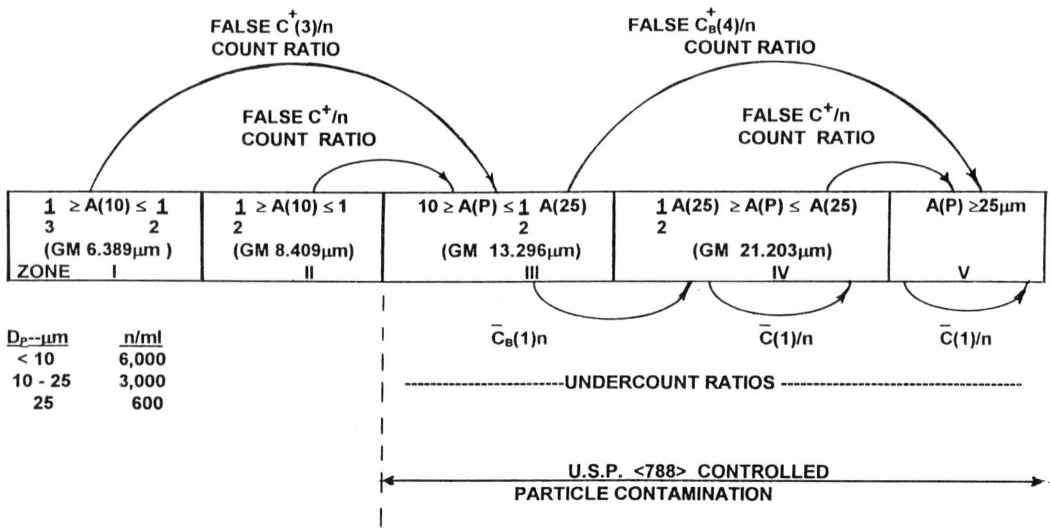

Figure 24 Sensitivity analysis of a light extinction particle detector to coincidence errors in the U.S.P. <788> SVI assay of contaminating particles in a 2-ml container. The errors are based on counts of single particles.

Zone IV commences with particles whose signal is at least half that of a 25-µm particle. Any multiple coincident occurrence in this zone will create false counts in Zone V.

Zone V begins at the 25-µm particle size and extends up to the visible particle zone. The count loss in Zones IV and V consists of both horizontal and vertical multiple occurrences: their effect is shown as self-loops, i.e., loops contained within the same particle size zone.

The zone descriptions, limits and their effect on the U.S.P. <788> SVI contaminating particle assay are summarized in Table 14. Table 15 summarizes the test particle sizes and the N_E values for the detectors used in the sensitivity test.

Table 14 Particle Sizes and Zones for Test of U.S.P. <788> SVI Contaminating Particle Assay Sensitivity to False Counts and Undercounts

	Particle signal range			
Zone	Minimum (µm)	Maximum (µm)	Geometric mean	U.S.P. <788> SVI
I	$\geq \frac{10}{3}$	$\frac{10}{2}$	6.389	Unspecified
II	$\geq \frac{10}{2}$	< 10	8.409	Unspecified
III	≥ 10	$< \frac{25}{2}$	13.296	6000 per SVI container
IV	$\geq \frac{25}{2}$	< 25	21.203	
V	≥ 25			600 per SVI container

In each size range the geometric mean particle size in the zone is used for the sensitivity tests.

Table 15 N_E for Each of the Particle Sizes Used in the Coincidence Sensitivity Test for Both Detectors Employed

Zone	Particle signal geometric mean D_P (μm)	N_E Detector volumes per ml	
		Laser detector	White light detector
I	6.389	61,526	39,463
II	8.409	58,944	37,303
III	13.296	53,508	32,941
IV	21.203	46,561	27,700
V	≥ 25.0	43,829	25,735

XLVI. SVI CHALLENGE OF SENSITIVITY ANALYSIS

To demonstrate the use of this loop analysis, two detectors in current use, were selected. The first is a new, well designed detector with laser excitation. It has high-concentration counting accuracy and a low pressure drop across the detector flow path. The low pressure drop design minimizes the possibility of cavitation-induced false backgrounds. The experimentally determined characteristics for this detector are listed as No. 3 in Table 8 (p. 501).

The experimentally determined characteristics for the second detector selected are listed as No. 1 in Table 8. This is an older, well regarded white light detector. A comparison of the sensitivity to coincidence false counts and undercounts in a U.S.P. <788> SVI contaminating particle assay of the same 2-ml SVI container with both of these detectors follows. This comparison supports the lack of comparability for present particle count data.

The assumed particle contamination in this 2-ml container is based on the new U.S.P. XXIII SVI limits. Any example based on the U.S.P. XXII SVI limits is too extreme to be useful. The contaminating particle concentration is set at the upper U.S.P. XXIII SVI limit of 6000 particles per container or 3000 particles per milliliter ≥ 10 μm. This limit includes a maximum of 600 particles ≥ 25 μm. For this example, a concentration of 6000 particles/ml is assumed in the two unspecified zones below the 10-μm control threshold. In each zone, coincidence calculations use the geometric mean particle size. The effect of the coincidence error as a percent of the USP acceptance level is shown adjacent to the loop affected. (See Appendix 2 Sec. F, pp. 559–566 for calculation details.)

XLVII. NEW LASER DETECTOR SENSITIVITY ANALYSIS

Figure 25 shows the calculated sensitivity of the new PMS laser detector to particle counting interference in the U.S.P. <788> SVI particle assay. The experimental geometric detector volume and scan beam height were evaluated with the protocol included in the *PDA Journal* [2] and are listed in the legend for this figure. The effects of the coincident interactions on the assay results are noted on each loop. The selection of the geometric mean particle size, 13.296 μm, in Zone III for this example eliminates the need for a self-loop in the 10.0- to 12.5-μm span.

The loop from Zone I to Zone III is a false count loop for triple or higher coincident occurrences. At the assumed 6000 particle/ml particle contamination level this loop contributes 94 false counts per container greater than 10 μm or approximately 1.6% of the allowable total. These false counts can be considered a minor annoyance. The Zone II to Zone III loop, contributes 1049 false counts ≥ 10 μm, 17.5% of the allowable total U.S.P. <788> SVI limits per

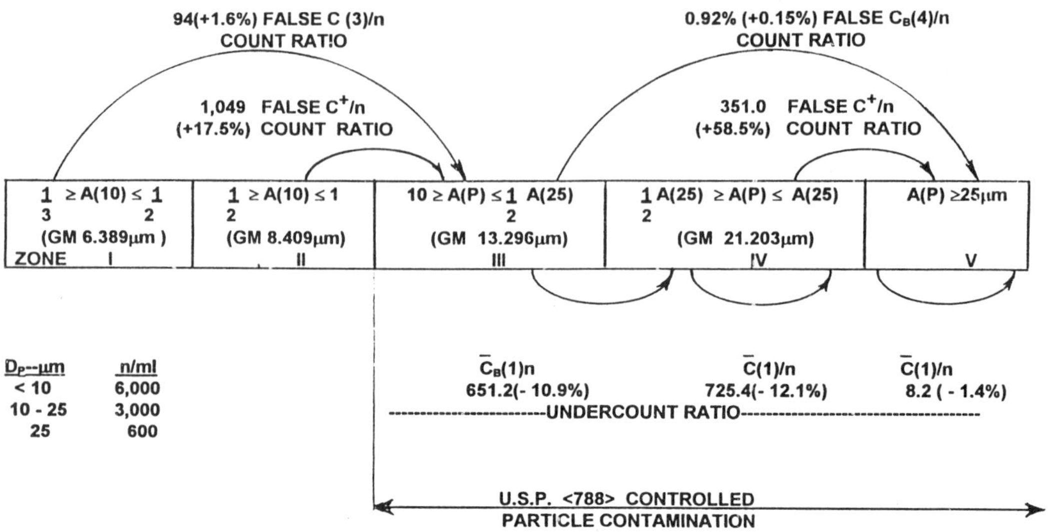

Figure 25 Sensitivity analysis of a new PMS laser-sourced detector to coincidence errors in the U.S.P. <788> SVI assay of contaminating particles in a 2-ml container. $V_{DG} = 14.0 \times 10^{-6}$ cm^3, $B_H = 39.7$ µm.

container, based on the same 6000 particles/ml contaminating particle concentration in the unspecified half 10-µm to full 10-µm size group.

Zone III, with particles ranging from 10 µm to those with half the area of a 25-µm particle, is seen to be a source of both false counts ≥ 25 µm and of undercounts. The undercounts have two sources. They are due to horizontal coincidence count loss and to the portion of the coincidence count loss that remains within the 10- to 25-µm particle size span. The Zone III to Zone IV loop of this laser-sourced detector included 652.1 single-particle undercounts, –10.9% of the allowable 6000 per container within the 10–25 µm span. The Zone III to Zone V loop shows 0.92 false counts ≥ 25 µm, 0.15% of the 600 particles per container that are the acceptable limit.

The single-particle self-loop in Zone IV is due to the count loss ratio and results in an undercount of 725.4, –12.1% of the maximum allowable contamination. In the Zone IV to Zone V loop all coincident occurrences result in false counts > 25 µm. Counting 3000 particles/ml in this 2-ml container this detector records 351.0 false counts >25 µm per 2 ml container, 58.5% of the 600 particle limit.

In Zone V, the count of 600 25-µm particles by this detector is accomplished with distinction. The new laser source detector responds with an undercount of 8.2 particles, –1.4% of the acceptable total.

XLVIII. OLDER WHITE LIGHT DETECTOR SENSITIVITY ANALYSIS

The sensitivity to coincidence interference of the U.S.P. <788> assay data evaluated by the Climet white light detector is shown in Fig. 26. The experimentally determined detector constants [26,27] for the older white light detector are shown in the legend for this figure. There is an approximate doubling in each false count or undercount category that was tabulated for the new laser source detector. The false count contribution from the Zone I to Zone III loop is again negligible. It is 205 counts, 3.4% of the allowable total. The false counts in the Zone II to Zone III loop, 1518, amount to 25.3% of the allowable counts ≥ 10 µm. This error cannot be lightly disregarded.

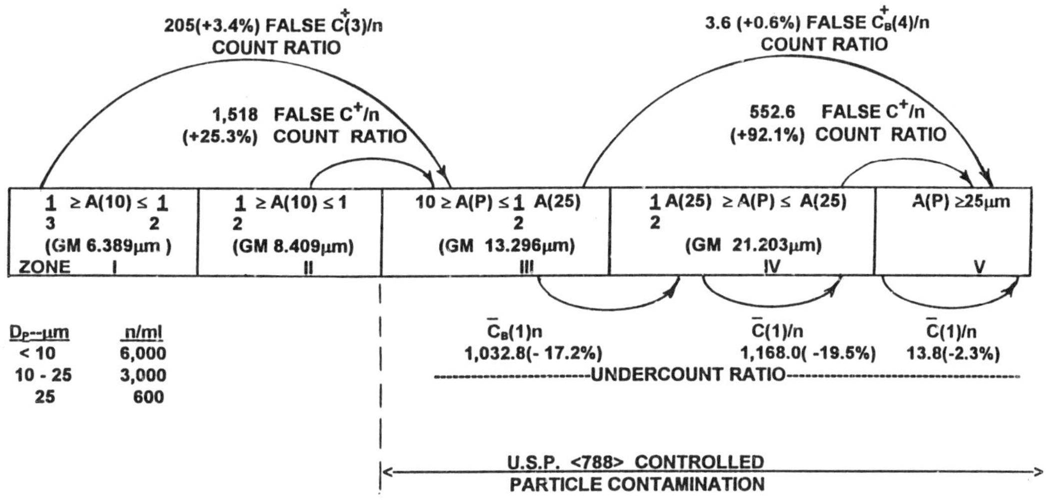

Figure 26 Sensitivity analysis of an older Climet white-light-sourced detector to coincidence errors in the U.S.P. <788> SVI assay of contaminating particles in a 2-ml container. $V_{DG} = 20.7 \times 10^{-6}$ cm^3, $B_H = 28.5$ μm.

The undercount shown in the Zone III to Zone IV loop is 1032.8 counts, –17.2% of the allowable total ≥ 10 μm. There is a negligible false count contribution in the Zone III to Zone V loop of 3.6 counts, 0.6% of the maximum allowable particle count ≥ 25 μm.

The Zone IV self-loop shows an undercount of 1168.0 single particles, –19.5% of the total allowable contamination. The Zone IV to Zone V loop contributes 552.6 false counts ≥ 25 μm, 92.1% of the allowable 600 particle per container maximum.

The Zone V self-loop shows an undercount of 13.8 single particles due to the count loss ratio. This single particle count loss is calculated for a concentration of 600 25-μm particles, –2.3% of the allowable total. This final piece of data from the sensitivity review brings an important note of optimism that count accuracy is possible even with older commercially available particle counters.

XLIX. SENSITIVITY ANALYSIS SUMMARY OF RESULTS

Zone I. The injection of false counts into the U.S.P. <788> SVI particle assay from particles whose light extinction signal is between one-half and one-third of the 10-μm lower control boundary ranges from 3.4% for the older white-light detector shown, down to 1.6% for the laser-sourced detector for the numerical values selected. These false counts do not inject serious errors.

Zone II. False counts from this zone, from particles whose signals lie between one-half to just below the 10-μm particle signal, range from +25.3% for the older white-light detector, down to +17.5% for the newer laser-sourced detector. The count errors in Zone I and Zone II are removable errors. Specification of control limits for this presently unspecified particle size region can control this error.

Zone III. Particle signals in this region range from 10 μm to just less than half the signal of a 25-μm particle. For the present SVI control limits, false counts greater than 25 μm from either detector lie below 0.60% and can be neglected. Single-particle undercounts range from –17.2% for the older white-light detector, down to 10.9% for the newer laser-sourced detector. Control of these errors can be accomplished with sample dilution for both detectors.

Zone IV. This region commences with particle signals that are half that of a 25-μm particle and ends with the particles just below the 25-μm particle size limit. This region is seen to be the source of both undercounts and false counts greater than 25 μm. The undercounts, due to the count loss ratio, calculated on the basis of 3000 particles per milliliter, range from –19.5% for the older white-light detector down to 12.1% for the new laser-sourced detector. The effect of false counts greater than 25 μm from this particle size region, for both detectors, is the most significant count error in the contaminating particle essay. The false count error of particles greater than 25 μm for the older white-light detector is +92.1%; for the laser-sourced detector this error is reduced to 58.5%. Both types of count error are correctable without the advent of new technology. Both types of count error can be corrected by sample dilution.

Zone V. The single-particle undercount errors in this zone, –2.3% for the older white light detector and 1.4% for the newer laser sourced detector, are indicators that the technology in presently available counter can deliver the replicability and count accuracy that are essential to any assay.

L. SINGLE-PARTICLE COUNT ACCURACY CONTOURS FOR THE LASER-SOURCED DETECTOR

The peak count/accuracy contours shown in Fig. 27 are for the PMS laser detector whose characteristics were used in the preceding example. The scan beam height and the geometric detector

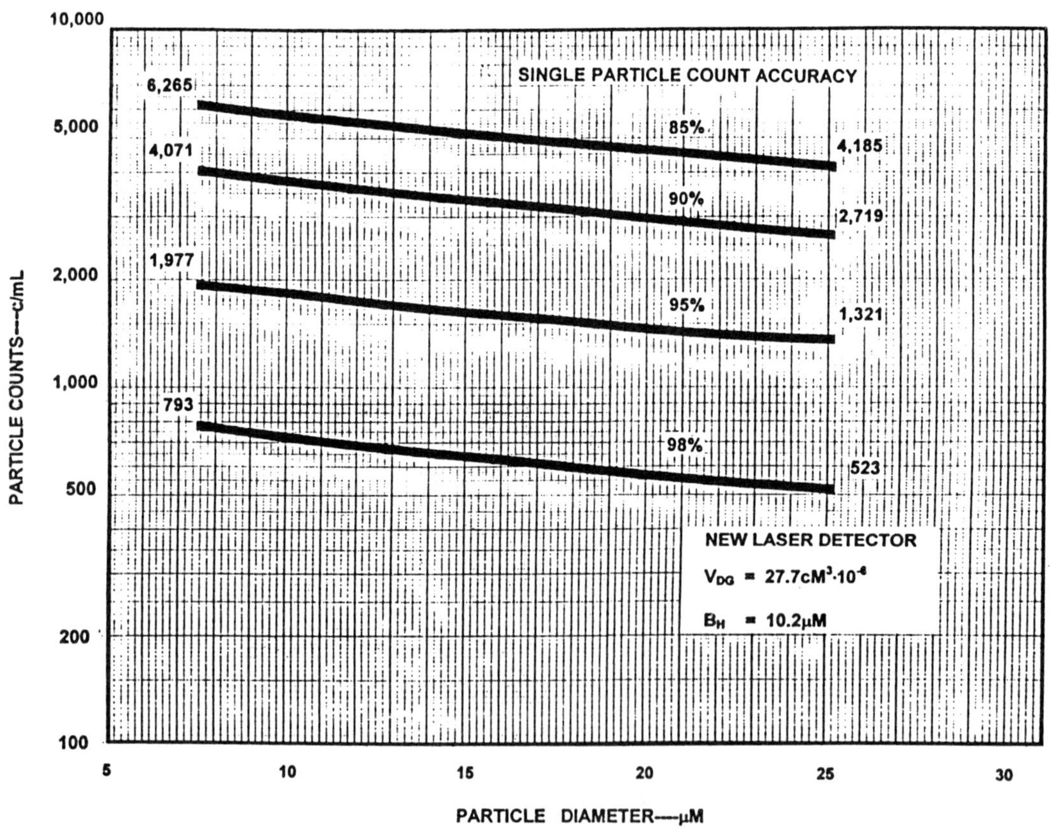

Figure 27 Constant single-particle count accuracy contours.

volume were evaluated with the protocol described in the *PDA Journal* text of this paper. The dilution experiments were conducted with 9.87 and 25.7 μm particles corrected for a 3-μm analysis threshold.

The count accuracy contours for 7.5- to 25-μm particles is proof that practical count accuracy performance is available. The four contours range from a high of 98% accuracy through 95%, 90% and 85% count accuracy contours.

To limit single-particle counting error to 2%, the maximum count ranges from a high of 793 counts/ml at 7.5 μm to a low of 523 counts/ml or 1046 per container for 25-μm particles. This limit is sufficient for many marketed injectable products. To maintain 5% single-particle count accuracy, the count limit ranges from 1977 counts/ml at 7.5 μm to a low of 1321 counts/ml, 2642 per container at 25 μm, a contamination range that should encompass most injectable products. The graph also shows 90% and 85% count accuracy limits.

The experimental results shown here are from a single instrument and, as such, should be considered a tentative evaluation until supported by other independent measurements. These response characteristics are not a brief for a particular instrument. They are a demonstration that count accuracy is possible when the peak count is controlled during an assay. These count accuracy curves also demonstrate that the evaluation of particle counter accuracy is within the reach of even modestly equipped laboratories.

LI. PARTICLE SIZE STANDARD CONSIDERATIONS FOR COUNTING ACCURACY IN U.S.P. <788> SVI PARTICLE ASSAYS

The specification and operation of single-particle counters as practical measurement devices is, in part, determined by their concentration limited counting accuracy. Extending the accuracy will affect both specification and operation of the counts employed for U.S.P. contaminating particle assays.

The continuing problem of undefined count accuracy in light extinction particle counters has been addressed with two action steps in U.S.P. XXIII. A suspension of 4000 15-μm particles/ml is used as an ad hoc counter functionality test in the particle count reference standard. The selection of the 15-μm particle appears to result from the understanding that test results with a single-particle size should be representative of actual use conditions in which a broad distribution of particle sizes can be encountered.

The second action taken has been to require vendors of single-particle counters that are intended for U.S.P. XXIII <788> particle assays to publish the concentration limit for a 10% undercount error with 10-μm particles for each instrument. This selection will replace the wide variation of specifications that previously existed in marketed particle counters and is a marked step forward. The requirement for commercial specification of counting accuracy performance based on undefined methodology is, unfortunately, an overoptimistic response to a difficult measurement problem.

The importance of functional specifications and test methods to ensure that specified functionality is available for use cannot be overstated. The measurements required in U.S.P. <788> are of random distributions of randomly-sized contaminating particles. This type of measurement eliminates any possibility of postmeasurement correction. An associated problem is the lack of an accurate characterization of the distribution of random particles that can be encountered. Varied comments on an envelope for the distribution of all random particles can be found in the literature. Considering the wide range of sources, products, and manufacturing environments in which these particles occur, accurate numerical estimation of the concentration in a single product in its manufacturing environment may not be possible.

LII. U.S.P. <788> COUNT ERROR ANALYSIS

Validation of the particle-triggered Poisson model, using the effective detector volume as the basis for concentration estimates, provided an analytical model of the particle counting process. The use of normalized parameters: c/n, count efficiency, and n/N_E, the ratio of the particles to effective detector volumes, made possible the generation of the universal single-particle count accuracy and count error curves of Fig. 23. These universal curves succinctly summarize count error and count accuracy for all single particle counters, regardless of excitation means employed. The count efficiency parameter will now be used to analyze the U.S.P. <788> count error problems.

Use of the count efficiency parameter made the analysis of count loss due to the increase of particle concentration a simple, credible task. In the following section, the count efficiency parameter will be used to analyze the effect of multiple-particle coincident groupings on particle contamination measurements. This analysis addresses the effect on particle counts of the shift in particle size measurements due to coincidence. This shift in the size of particles counted is the mechanism that produces the observed undercount error. The missing counts will be seen to be a source of false counts and undercounts in the U.S.P. <788> batch release particle assay. Particles in the uncontrolled size region below 10 μm through double and triple coincident occurrences can, depending on the detector employed and the concentration of particles measured, be a matter of concern. Similarly, through coincident occurrences, the particle size span between 10 and 25 μm can contribute to the observed undercount error or to the false count of particles greater than 25 μm in diameter where the contamination penalty is 10 times greater.

The eight equations listed in Table 11 (p. 511) summarize most of the information required to analyze the particle counts displaced and lost due to the coincidence effect. Of particular importance is the fact that these relationships use a theoretically derived, experimentally evaluated count efficiency parameter.

The first equation in Table 11, Eq. (40), is the ratio of single-particle counts to the number of counts recorded. This ratio is shown to be the square of the count efficiency for the particle concentration being counted. At a count efficiency of 95%, 9.75% of the particles counted will be counted as multiple coincident events. Decreasing count efficiency to 90% raises the number of single particles counted as multiple coincident occurrences to 19%.

Further decreases to 85% and 80% count efficiencies raise the incidence of single particles lost to multiple event occurrences to 27.75% and 36% respectively. The disposition of these lost counts is examined with the expressions for double, triple and quadruple coincident events shown in Eqs. (40B)–(40D) of this table.

Equation 41 lists the ratio of all multiple-event counts to single particles counted. This ratio is a convenient expression with which to investigate the effect of multiple-particle groupings of smaller particles on measurements of the mandated size categories of <788>. The variation of C^+/n has been graphed in Fig. 18 (p. 498) to show its approximate linearity in the range of count efficiency from 0.95 to 1.00. Even at a count efficiency of 95%, the generation of multiple counts is 4.75% of the concentration of single particles. The curve of C^+/n Fig. 18 has a maximum at n/N_E equal to 0.5 and goes to zero at n/N_E equal to 1. The count loss curve goes from zero to an asymptotic limit of 1.0.

Equation (42A), the ratio of triple and higher multiple-event counts to single particles, supplements the investigation of the effect of triple and higher particle groupings on smaller particles in the mandated size categories of <788>. This expression is of value when double-event counts do not affect the mandated size categories of <788>, but the inclusion of triple and higher multiple events improve the accuracy of the analysis. Equation (13) is used to calculate the total undercount of the single particles in suspension as a result of the coincidence effect.

Application of these equations to the analysis of the U.S.P. <788> SVI count error problems follows. The false count problem will be addressed first and will then be followed by the undercount problem.

LIII. EVALUATION OF THE COINCIDENCE EFFECT BY SEQUENTIAL DILUTION MEASUREMENTS: A MODEL EXPERIMENT

Coincidence count loss determinations for single-particle counters measure particle concentration effects in sequential dilution experiments. These experiments provide the data with which the injection of false counts and undercounts can be evaluated and controlled. The experiments commence with a high particle concentration that shows excellent count replicability and high count error. The counts recorded for each dilution step are compared to the number of particles for that step, as determined by the initial concentration and the dilution factor employed. Although different techniques are employed to evaluate the initial particle concentration, all use extrapolation of count data from a region of very low counting error to the initial particle concentration, N_I. The determination accuracy of the initial particle concentration is the foundation for any evaluation of the counting capability of a particle detector. This experimental determination of N_I, without the availability of an analytical relationship between counts and the particles in suspension being counted, can approach 99% of the actual value using the methodology described in Appendix 2 Sec. E.9.(a) (pp. 555–558). This negative 1% bias is analyzed in Appendix 2 Sec. E.9.(b) (pp. 558–559). A correction for this negative bias is shown in the worked example in the Appendix. For most purposes, the determination and use of an N_I value within 99% of its actual value provides acceptable results and will now be described.

The experimental technique for the determination of the initial particle suspension suffers from two defects. The first defect is that the critical data points are generated with the most dilute suspensions. These dilute suspensions are affected by the accumulation of any previous experimental errors. The second defect results from the data analysis employed. Since the dilution data spans nearly three decades, log-log plots and a graphical analysis of the data is usually employed. The graphical analysis required the determination of a "best straight-line fit" to the recorded count versus concentration data. This best straight-line fit is an estimate of the linear count asymptote of the detector and is used to determine the maximum initial particle suspension, N_I, by graphical extrapolation.

Determining the best straight-line fit on the log-log plot of data is a more complex operation than it appears to be at first. Although many straight lines can be drawn on a log-log graph, only a line of unity slope can yield the required linear relationship between particle counts and the dilution factor employed. All other straight lines represent exponential relationships of the experimental variables. The required unity slope line will be parallel to one drawn through the upper right vertices of the log-log squares. The combination of experimental error and the slow departure of the count data from linearity makes the graphical selection of the data points to be included in the determination of N_I a difficult judgment call.

The use of the log-log graph also results in the loss of the count at infinite dilution. This is the background count point, C_0, the concentration of contaminant counts in the surfactant-treated, filtered diluent. C_0 is the count/ml recorded with clean diluent in the operating conditions used to generate experimental data.

The addition of this background count point to the data analyzed provides an important anchor for the data. The control of the number of particles counted at the most dilute data points must also be considered if accurate results are to be obtained.

To include the background count data point, C_0, requires the use of linear regression to determine the best straight-line fit to the count/concentration data. This statistical technique for finding the best straight-line fit to data has marked similarity to the graphical technique. In the graphical technique, a straight line is moved until the visible deviation of all plotted points from the straight line have been minimized. In the statistical technique, the best straight-line fit is chosen as the line that minimizes the sum of the squares of the deviations from the straight line to all the data points. It will be seen that the use of linear regression results in more accurate determinations of N_I. Linear regression capability is now available on a wide variety of inexpensive hand-held scientific calculators and is considered a practical device for this important calculation.

The existence of a validated model of the particle-counting process makes possible a detailed examination of the calculations and relationships in a dilution experiment. The "data" from such a dilution experiment is listed in Table 16. For realistic data and results, this experiment is based on the parameters of detector 3 listed in Table 8. The experiment is designed to achieve a set of dilution data points between approximately equal and opposite errors from which to determine the initial particle concentration, N_I, by extrapolation. The reason for the use of extrapolation is evident in examination of Table 16. This table shows that the counts of the high particle concentrations near the initial concentration do not respond accurately to the 0.6 dilution factor employed. The counts recorded at each step of the dilution experiment have been calculated with Eq. (32).

At the high concentration end of the linear span used in the extrapolation, the coincidence effect results in a negative error. The target for this maximum error is –1%.

Table 16 Model-Based Dilution Experiment

1	2	3	4	5	6	7
			Counts/ml			
Step s	Dilution factor $f = (1 - d)$	Particles/ml	$C_0 = 0$	$C_0 = 1$	Effective concentration n/N_E	Count error $(n - c)/n$
0	1.0000	34,114.9	18,761.8		0.5979	0.4500
1	0.6000	20,469.0	14,298.6		0.3587	0.3015
2	0.3600	12,281.4	9,903.0		0.2152	0.1937
3	0.2160	7,368.83	6,476.0		0.1291	0.1212
4	0.1296	4,421.30	4,091.6		0.07749	0.07456
5	0.07776	2,652.78	2,532.3		0.04649	0.04542
6	0.04666	1,591.67	1,547.9		0.02790	0.02750
7	0.02799	955.00	939.15	940.15	0.01648	0.01555
8	0.016796	573.00	567.27	568.27	0.009960	0.008401
9	0.010078	343.72	341.66	342.66	0.006006	0.003083
10	0.006047	206.23	205.49	206.49	0.003619	+0.001260
11	0.003628	123.74	123.01	124.01	0.002173	+0.005246
12	0.002177	74.24	74.15	75.15	0.001303	+0.012258

The low count error dilution steps selected for the extrapolation to N_I are shown in the boxed area. The initial concentration has been increased by 1% from the linear regression computed value to compensate for the selection of the concentration at step 8 as .99 of the linear dilution value. The detector on which this table is based is laser diode based and has the following experimental constants: V_{DE} for 10 μm is 14.0×10^{-6} cm^3 and the beam height, B_H, is 39.7 μm. The assumed C_0 of 1 count/ml has been added wherever its affect is greater than 0.1%. All count errors appearing in the table are negative except those preceded by a plus sign.

Table 17 Linear Regression and Other Results for the Model Sequential Dilution Experiment of Table 15

Slope		33,775.4 particles/ml
Constant = background count	C_0	1.595 counts/ml
Estimated intercept at $f = 1.00000$	N_I^*	33,776.8 particles
Corrected initial suspension	N_I	34,114.6 particles/ml
Correlation coefficient	r	.999996
Experimental initial counts	C_I	18,761.8 counts/ml
Effective detector volumes per milliliter	N_E	57,057

The computed value of N_I is 1% less than its original value. The –1% bias is analyzed in Appendix 2. It results from the exponential relationship of the count efficiency which approaches but never attains ideal linearity.

At the low end of the dilution experiment, the lowest dilution count data used, approximately 100 counts/ml, the positive noise error target is approximately +1%. The choice of equal and opposite error end points for the experiment optimizes the accuracy of the analysis.

Columns 1, 2, and 3 are respectively the dilution step, the cumulative dilution factor and the ideal counts recorded. A background noise count of 1 count per milliliter has been added to the ideal count whenever the net effect was larger than 0.1%.

Column 6 in Table 16 has been calculated with the N_E shown in the linear regression results tabulated in Table 17. Column 7, the count error has been determined as the difference between an ideal dilution curve and the counts plus noise data in column 5. The linear dilution data points used in the extrapolated solution for N_I are the boxed steps 8 through 11 and the C_0 point at infinite dilution.

The linear regression solution is shown in Table 17 which also includes the calculated value of N_E. In an actual experiment, the selected solution would include a C_0 value that approximated the experimental conditions (see Appendix 2 Sec. E.9 (a) and (b) pp. 554–558).

LIV. ACCURACY LIMITS TO THE DETERMINATION OF THE INITIAL PARTICLE CONCENTRATION

The accuracy of the sequential dilution methodology in the determination of N_E rests upon the use of an extrapolation with which N_I, the initial particle concentrations is evaluated. The technique uses data points in the region where both positive errors due to noise pulse injection and negative errors due to the particle coincidence effect are limited to 1% or less to determine the initial particle concentration from data points in a high coincidence error region.

The accuracy of the determination was challenged with a model-based dilution experiment that is described in Appendix 2. The conclusion reached was that there was a 1% undercount error at the highest concentration point used to determine N_I. This 1% undercount error also affects the value of N_I following the procedure described in this chapter. (Appendix 2 Sec. B.9(b), p. 558.)

A Practical Look to the Future

All presently used light extinction particle counters can deliver accurate single-particle count results when a peak count/count accuracy contour is established and maintained in operation.

The use of the count efficiency parameter defined in the particle-triggered Poisson counter model provides simple, useful solutions to high concentration counting problems of single-particle light extinction counters. It can provide a well-defined link between the U.S.P. XXIII

<788> count accuracy specification for light extinction particle counters to an accurately measurable parameter.

This chapter extends and combines the concepts and methods of the three coincidence paper sequence [23–25] with the demonstrated replicability of these particle counters to describe an efficient experimental methodology. Universal graphs of both count accuracy and single-particle counting accuracy using count efficiency and a normalized particle concentration measure are now available. Accurate, instrument-independent counts at the relatively high particle concentration acceptance levels that are defined for small SVI containers in the U.S.P. <788> contaminating particle assay are now possible.

Control of the identified undercount and false count coincidence errors in the U.S.P. <788> SVI contaminating particle assay is now practicable. It is obtained by operation with a concentration limit determined by a selected coincidence count error contour from 5.773 to 25 μm as illustrated in Figs. 3 through 6. When the particle count at any point in this defined particle size span exceeds the selected concentration and count accuracy contour, dilution is utilized to obtain the count accuracy desired.

Four probable counting error sources have been identified in the particle size spectrum. Of the four, three introduce false counts into the U.S.P. <788> contaminating particle assay; the fourth results in undercounts in the 10- to 25-μm particle size span. Controlling the concentration of particles from 5.773 to 10 μm controls the introduction of false counts > 10 μm. The condition for false count control from 5.773 to 7.071 μm is less stringent than that for control in the 7.071- to 10-μm span. Controlling the false count of particles greater than 25 μm in U.S.P. <788> SVI assays has been shown to ensure control of the undercounting error in the 10–25 μm size span of these assays. The four errors are produced by operation of single-particle counters beyond their accurate, concentration-dependent, measurement span. The use of a single-count error contour for the full particle size span that must be controlled is a practical conclusion.

The count accuracy contour of the detector is calculated with the count efficiency of a selected detector. The detector's count efficiency is determined by two characteristics: the geometric or physical volume of the detector and the effective height of the scan beam. These are durable characteristics determined by the physical dimensions of the detector, the light source and the detector employed. They are evaluated with sequential dilution experiments performed with two calibrating particle sizes. The nominal 10- and 25-μm particle sizes can be used for this purpose.

In retrospect, the use of particle counters for particle contamination assays was validated for the low-particle-concentration samples of the LVI products evaluated. This "validation" was a narrow gauge response to the 1980 crisis of accepting terminally sterilized dextrose. In this narrow gauge validation, implemented in a crisis atmosphere, instrumental specifications and limitations were ill understood and scantily addressed. The counting accuracy of these counters at the relatively high particle concentrations that are acceptable in the U.S.P. <788> SVI assay was not defined.

The selection of 10- or 15-μm particle sizes for single-point concentration tests and particle counter specifications is seen to be inadequate when the count accuracy issue is addressed. A more appropriate concentration limit determination must be based on the full concentration accuracy contour required by the U.S.P. <788> SVI high-particle-concentration measurement. This requirement can be satisfied with sequential dilution experiments using two different particle sizes, preferably at 10- and 25-μm particle sizes since measurements at these particle sizes will combine with calibration requirements.

The alternative to the search for new, more efficient particle counters or improved methods for specifying their selection and operating limits is well documented. Kaye [26] says:

Early in the days of the development of stream counters [i.e., single-particle counters in the language of this chapter] such as the Coulter Counter it was appreciated that the only safe way to avoid coincidence errors in the fineparticle characterization study was to continuously dilute the concentration of the suspension being used in the stream counter until the measured fineparticle size distribution function was independent of the further dilution of the suspension.

A dilution step in the routine use of the U.S.P. <788> SVI particle assay is a practical, though burdensome, requirement that could be considered. Recent work, reported in this paper, has demonstrated the practicality of multiple-step sequential dilution experiments performed in a grossly contaminated environment. The addition of a dilution step to maintain count accuracy in the U.S.P. <788> SVI procedure is believed to be a practical immediate step rather than reliance upon some future development that would increase the accurate count capability of these counters. Adequate specification and operating limit definition for single particle counters used in the U.S.P. <788> particle assays is certainly the more efficient choice.

Particle counter evaluations of contamination in parenteral solutions result in relative rather than absolute data. The relative data produced has been shown to be suitable for the control of particulate contamination in parenteral solutions. The need for maximum particle length measurements and for a referee method in the evaluation of particle contamination in parenteral solutions remains.

ACKNOWLEDGMENTS

Sincere thanks are tendered to the management of Particle Measurement Systems, Inc. and those of Climet Instruments Co. where the greatest part of the laboratory data reported in this chapter was accomplished. The detailed measurements were made with the help of Jeannie West at Climet and Paul Hepokoski at Particle Measurement Systems. Heartfelt thanks are extended to Wallace Russell for his encouragement, and to Al Lieberman for his substantial part in shaping the coincidence effect evaluation procedure.

APPENDIX 1: STEP-BY-STEP EVALUATION PROCEDURE FOR THE COINCIDENCE ERROR

A. Scope

This procedure defines the effect of concentration on the coincidence errors of single-particle counters. These errors include both undercounts and false counts that can affect U.S.P. <788> particle assay results. The procedure is designed for volumetric OPCs used for contamination testing of small-volume parenterals as described in U.S.P. XXII <788> and its near future revision U.S.P. XXIII. With world pharmaceutical standards in mind, the subvisible particle size control range is from approximately 2 μm to those particles in excess of 100 μm. The upper concentration limit for this procedure ranges up to 6000 particles per milliliter for the 1-ml container, the smallest SVI container in widespread use. Coincidence count loss determination for particles at the maximum sensitivity of the OPC, is not part of the procedure. Operation within defined count limits over the operating range of the OPC is the basis for accurate count determinations.

B. Limitations

Particle-counting errors occur when the OPC reports more or fewer particles than are actually present in the suspension. The errors can result from incorrect OPC design or construction, from incorrect OPC operation or from incorrect sample handling procedures. Accurate determination of the count efficiency, the ratio of counts to single particles that have been counted, for OPCs must commence with the recognition that the relationship between the number of single particles counted and the resultant count is exponential. The ratio of counts/ml to particles/ml at low particle concentrations approaches but never becomes an ideal linear relationship. Special care must be exercised to obtain accurate results.

With some OPCs, some background electronic noise pulses are large enough to be recorded as counts at the smallest particle size threshold of the counter. These noise pulses will most affect the accuracy of the low concentration, lower size range measurements. While the effect of noise pulses on counting accuracy in higher particle concentration approaches but never becomes an ideal linear relationship. Special care must be exercised to obtain accurate results.

C. Definition of Terms and Symbols Specific to This Procedure

1. **Coincidence.** Signal coincidence describes the mechanism that reduces the count of particles present in a suspension by counting random groups of particles as single, larger, counting events. Coincidence reduces the number of particles counted and distorts the distribution of particles being measured toward larger sizes.
2. **Effective detector volume**. In all presently available particle counters used for the evaluation of particle contamination in pharmaceutical products, the diameter of the particle measured is a nonnegligible fraction of the scan beam height. For this case, the effective detector volume is larger than the physical or geometric detector volume of the detector. The effective height of the detector is the sum of the scan beam height and the diameter of the particle being measured.
3. **Measurement units**. All parameters are defined per milliliter of test volume.
4. **Effective particle concentration**. The effective particle concentration is measured as the ratio of particles/milliliter, n, to the number of effective detector volumes/milliliter, N_E.
5. **Other symbols employed**
   f = stepwise dilution factor
   f_T = total dilution factor between the –1% undercount concentration and the +1% noise-limited count
   N' = concentration at which a –10% count error occurs when 10-μm particles are counted by a detector
   N'' = concentration of 10-μm particles at which a 1% undercount error occurs
   N_I = initial particle concentration in a dilution experiment
   s = sequential step number of a dilution experiment
   S = number of steps in a dilution experiment

Note: An asterisked quantity indicates an approximate value.

D. Summary of Procedure to Determine Counting Accuracy

1. Sensor calibration, carried out at the manufacturer's recommended flow rate. This calibration must precede the count accuracy procedure.
2. The background count, C_0, of the filtered, surfactant-treated, clean water which is used as a diluent is determined. This the count at infinite dilution. The diluent shall contain no more than 1 particle/ml 10 μm. The volume of diluent for this determination shall be no less than 500 ml.

3. The choice of experimental parameters provides four data points in the low count error concentration range. These four data points with C_0, the background count, contribute to the extrapolated determination of the initial concentration, N_I. At each dilution point, six 10-ml aliquots are taken. The first aliquot is discarded and the data of the five remaining aliquots are averaged. The accurate determination of the concentration of particles per milliliter at each dilution step rests on the accuracy with which the initial suspension is determined and the stepwise dilutions are performed.
4. The initial concentration of particles in the undiluted master suspension is selected to have approximately a 5.2-fold increased concentration over that specified for the 10% undercount concentration, N', required in the U.S.P. XXIII. Accurate experimental determination of N_E, the number of effective detector volumes per milliliter is calculated from the initial dilution steps. At these initial steps particle concentration is high and high undercount error is observed. Accuracy in the determination of N_E rests upon the accuracy with which the concentration of the initial particle suspension is determined and the accuracy of the dilution sequence employed.
5. The sequential dilution experiment is continued from the initial concentration until the background noise count, C_0/ml, is 1% of recorded counts. A nominal value for this limit is 100 counts/ml.

E. Significance

OPCs are calibrated with suspensions of well defined particles of known concentration and sizes. However, OPCs are used with liquids whose contamination is not known. If the concentration of particulates exceeds the accurate counting limit of the detector or if the flow rate through OPC is not within the specified range, then control of the coincidence undercount error limit will be degraded. Errors in reported particle size range and/or concentration will occur with no indication of the extent of error. No correction after measurement is possible due to the random nature of the particles being measured. This procedure will aid in defining the operating range of the OPC where acceptable counting and sizing accuracy can be expected.

F. Interferences

1. Improperly cleaned containers used to prepare dilutions and to contain samples will release particles that will result in counting and analysis errors.
2. Inadequately dispersed samples during any measurements will result in wide variations in particle concentration and thus wide variations in data from sequential aliquots. Depending upon the nature of the suspension being counted, particle sedimentation and fluid dynamic forces during sample flow may also cause counting errors. For this reason, this procedure requires continuous stirring of the sample during handling and measurement.
3. Air bubbles resulting from container spin or impact cavitation and immiscible liquid droplets result in erroneous particle counts.

G. Apparatus and Materials

1. A balance with load capability of at least one kilogram and 0.1 g accuracy or a volumetric delivery system that can maintain 1% accuracy.
2. Sample containers from which samples will be fed to the OPC should be approximately 250 ml; for aliquot addition approximately 100-ml containers, with covers, shall be used. The use of narrow mouth containers minimizes the entry of airborne particulates. New containers are desirable since the surface will be uncrazed and thus easier to clean.
3. An external field-driven magnetic stirrer with Teflon-coated stirring bars shall be used.

4. The OPC system shall include a liquid feeder with flow rate control and volumetric measurement.
5. Monodisperse latex particles with nominal size of 10 μm and relative standard deviation no greater than 5% shall be used.
6. Nionic surfactant for use in diluting latex particle suspensions. Triton X-100 (Rohm & Haas), Tween 20 (Atlas Chemical Co.) or equivalent materials have been found satisfactory.
7. Prepare a mixture of approximately 2 liters of clean water with .01% surfactant. Filter the mixture through a 47-mm-diameter filter with a pore size less than .50 μm. Test the mixture for particle contamination. The mixture is acceptable for use if it contains no more than 1 particle/ml > 10 μm.

H. Calibration

Verify that the calibration of the OPC is current and is in accordance with ASTM F658-87 [13] for particle accuracy. Verify that the flow rate used during sizing calibration is set within the manufacturer's recommended range. This flow rate shall be used for all subsequent measurements.

I. Coincidence Count Error Determination Protocol Parameters

1. Using the instrument manufacturer's sensor description, the specified concentration limit, N', for 10% undercount error (a 90% count efficiency particle concentration) calculate the nominal 1% undercount concentration, N'' as .10 N'.

   $$N'' = .10N'$$

   To optimize extrapolation accuracy in the calculation of N_I, a low counting error span with equal and opposite error end points is selected. The maximum particle concentration data point used is limited to the concentration, N'', for a –1% count error. The lowest concentration data point for the extrapolation is determined by the concentration at which 1 count/ml will result in a +1% count error. This choice results in a low concentration end point of 100 counts/ml.
2. Determine the total dilution factor f_T between the –1% count concentration and the background count limited concentration that yields a +1% count.

   $$f_T = \frac{100}{N''}$$

3. Three dilution steps between N'' and 100 counts/ml, when anchored by the data point at infinite dilution C_0, usually provide adequate data for an accurate extrapolation of N_I. The approximate dilution for each step f_S, is calculated as the 1/3 power of f_T.

   $$f_S = \left(\frac{100}{N''}\right)^{1/3}$$

   For convenience in making all the sequential dilutions required, f_S is rounded to f, a convenient decimal or common fraction. Maintain the rounded dilution factor f, calculated in 9.3 throughout the entire dilution sequence.
4. Calculate the approximate value of the initial suspension N_I^* from N', the U.S.P. XXIII specified concentration for a 10% undercount.

   $$N_I^* = 5.2N'$$

5. The approximate number of dilution steps, S^*, is determined from the ratio of N'' to $5.2N'$.

$$f^{S^*} = \frac{N''}{5.2N'}$$

6. Round S^* to the closest integral value, S, and calculate the initial particle concentration, N_I.

$$N_I = \frac{N''}{f^S}$$

7. Calculate the estimated value of N_E^* from the U.S.P. XXIII specified concentration, N', for a 10% undercount.

$$N_E^* = 9.491222N'$$

8. Calculate the initial counts/ml, C_I, resulting from a concentration of N_I particles/ml using the particle-triggered Poisson model.

$$C_I = N_I e^{-N_I/N_E}$$

9. Select the aliquot volume V_A. The aliquot volume is selected to be large enough to provide separation from the preceding particle concentration and to provide an accurate sampling of the suspended particles. The aliquot volume selected should provide an acceptable 95% confidence interval for the smallest count planned. Six 10-ml aliquot volumes have been found to yield accurate results. The first 10-ml aliquot is discarded and the average of the counts for the remaining 50 ml is recorded.
10. Select the sample volume, V_0.

$$f = 1 - \frac{V_A}{V_0}$$

Note: To avoid cavitation, maintain a recommended minimum volume after sampling of 60 ml.

J. Coincidence Count Loss Protocol

1. Validate cleanliness of the containers to be used by rinsing sample components with clean diluent and passing a series of 10-ml samples through the OPC. Discard the data from the first sample and verify that the average is < 1 particle/ml > 10 µm.
2. Tare the selected clean container with the stirring bar and cover until used.
3. Prepare the master suspension of the 10-µm monodisperse PSL particles at a concentration of N_I particles per milliliter as calculated in 9.6 by adjusting counts/ml to the value calculated in 9.7. Adjust the master suspension within ±5% of the calculated counts/ml for the N_I desired.
4. Set the OPC data system within the dynamic range of the instrument to record data in the cumulative mode. Set particle size thresholds including 2, 5, 10, 25 and 50 µm. Smaller and larger size thresholds can also be used. At no time should the smallest size threshold be set so that the particle count for the diluent indicates more than 1 particle per milliliter greater than 10 µm. The count data required is the differential count above 7.5 µm to ensure inclusion of the full tail of the distribution. The sequential dilution experiment is continued from the initial concentration until the background noise count, C_0/ml, is 1% of recorded counts. A nominal value for this limit is 100 counts/ml. This analysis selection is made to ensure inclusion of the lower tail of the 10-µm particle distribution. When the

analysis threshold is greater than 10% of the measured particle, the particle diameter D_{PE} is less than D_P.

$$D_{PE} = \sqrt{D_P^2 - 2D_T^2}$$

5. Place the sample container in the OPC feeder assembly. Set the stirring speed so that a 5-mm meniscus minimum is seen at the liquid surface but no air is drawn into the liquid. Set the OPC sampler to pass liquid to the OPC at the flow rate for which the OPC was calibrated. Pass six 10-ml sample aliquots through the OPC sensor and record the data from the last five samples. Select 10 μm as the lower particle size threshold that will include 99.9% of the 15-μm particle population. Calculate the average particle dispersed, mix the suspension for 1 min before taking an aliquot.
6. Sample the reestablished test volume using six 10-ml aliquots. The remaining volume is $V_0 - 60$ ml.
7. For dilution experiments using gravimetric volume determinations, transfer 60 ml of the prepared clean diluent in a clean tared container to the test volume after the completion of the counter sampling cycle. Cover the container with clean aluminum foil until use. Weigh the container after the transfer of the 60 ml and adjust the total liquid volume transferred. Calculate all concentration factors from the recorded gravimetric data of the sample container only. To ensure that the suspension has been uniformly dispersed, mix the suspension for 1 min before taking an aliquot.

   Note: For samples that include 1-ml Tare volumes, decrease the test aliquot to 9 ml.
8. For volumetrically controlled experiments, add six 10-ml precision syringes of surfactant treated clean water to return to an approximate volume of V_0 ml. Record the test volume to .1 ml. This dilution corresponds to a ratio of d as determined by weight of the total sample plus diluent. To ensure that the suspension has been uniformly dispersed, mix the suspension for 1 min before taking an aliquot.
9. Carry out the procedures of steps 6 and 7 or 8 until the concentration is approximately 100 counts/ml recording indicated average concentration > 10 μm after each dilution.

K. Count Accuracy Data Analysis

1. Review the data. Prepare a data table (e.g., similar in format to Table 16) showing the sequence of dilution steps, the dilution factor and the average counts/ml at each step. Use either step 2 or 3 to select and edit the linear data points for the determination of N_I.
2. From the data table make a log-log graph of the number of counts versus dilution factor. Select the log-log plot data points for numeric analysis that lie on a unit slope line and that appear to be in the linear count region. Edit erroneous data points.
3. Test the linear count data points using C_0 and each point in turn in the linear count region. c is the cumulative particle count > 10 μm, f is the dilution factor for the recorded count, and N_I is the initial particle concentration. Edit erroneous data points.

$$c = C_0 + fN_I$$

4. Prepare a data table of the selected linear response data and use a linear regression analysis to include the C_0 data anchor point.
5. Select the linear regression solution using those data points in the calculation that result in the minimum zero offset to select the optimum value for the initial concentration of particles, N_I.
6. For each dilution data point calculate the concentration of single particles in suspension as the product of the initial concentration and the corresponding dilution factor, $n = N_I f^S$.

7. Determine the count error for each dilution data point.

$$\varepsilon_C = \frac{n - c}{n}$$

8. Determine the 10-μm particle concentration limit for a 10% under-count error by interpolation. Linear interpolation results in an approximate solution since the relationship between counts and particles in suspension is exponential.
9. For accurate determination of the 10% undercount error limit, determine the experimental value of N_E, the number of effective detector volumes/ml. Use data points that depart from the linear dilution equation by at least 30% to ensure accuracy.

$$N_E = \frac{n}{\text{Ln}\,(c/n)}$$

10. Calculate N' the model based on concentration for a 10% undercount limit with Eq. (13) using the experimentally determined value of N_E.

$$N' = N_E \text{ Ln} \cdot 90 \tag{13}$$

11. Calculate C' the model based count at which there is a 10% undercount error.

$$C' = N'(1 - \varepsilon_C)$$

L. Experimental Determination of Geometric Detector Volume and Effective Beam Height

1. To determine the coincidence count loss for the full particle size design range of a detector, V_{DG} and B_H for the detector must be evaluated. B_H is evaluated following two determinations of N_E the number of effective detector volumes for two particle sizes, N_{E1} and N_{E2}.

$$B_H = \frac{(N_{E2}/N_{E1})(D_{P2} - D_{P1})}{1 - N_{E2}/N_{E1}}$$

2. Using the calculated value of B_H from step 1, calculate V_{DG}, the geometric detector volume using paired values of V_{DE} and D_P.

$$V_{DG} = \frac{V_{DE}}{1 + D_P/B_E}$$

M. Calculation of a Selected Count Accuracy Contour for a Range of Particle Sizes

1. Calculate V_{DE3}, the effective detector volume for the selected particle size.

$$V_{DE3} = V_{DG}(1 + D_P/B_H)$$

2. Calculate N_{E3} the number of effective detector volumes per milliliter for the selected arbitrary particle as the reciprocal of V_{DE3}.

$$N_{E3} = \frac{1}{V_{DE3}}$$

3. Calculate N_3, the concentration limit for the selected particle size and count error.

$$N_3 = N_{E3} \, \mathrm{Ln}\, (c/n)$$

4. Calculate the particle count which results from an N_3 concentration. The count efficiency is 1 minus the count error.

$$C_3 = N_3(1 - \varepsilon_C)$$

5. Calculate the 10% error contour for the particle size range, set $1 - \varepsilon_C$ equal to 0.90 for each particle size calculated.
6. Plot the calculated count accuracy contour. To operate with the selected count accuracy, maintain peak counts within the count accuracy contour selected on the basis of false counts and undercounts in the U.S.P. <788> SVI contaminating particle assay.

N. Calculation of a Selected Single-Particle Count Accuracy for a Range of Particle Sizes

Calculate the count efficiency for the single particle count error selected or select the corresponding count efficiency from the following table substituting 14.1 for 13.5 and plot the results as in step 6 of Sec. M. The symbol $\overline{C}(1)/n$ represents the normalized single particle count loss.

$$\frac{\overline{C}(1)}{n} = 1 - \left(\frac{c}{n}\right)^2$$

For single-particle count errors of 20%, 10%, 5% and 2% the respective count efficiencies are

$\overline{C}(1)/n$	0.20	0.10	0.05	0.02
c/n	0.8944	0.9487	0.9747	0.9899

APPENDIX 2: DERIVATIONS AND DETAILS

A. Measures of Particle Concentration

1. Geometric Detector Volumes

The first step in any application of a model to evaluate the counting accuracy of particle counters is a determination of the concentration of particles to be counted. The basic premise for any discussion of the geometric particle concentration is that the volume of the particle is so small compared to the volume of the detector that it can be ignored. The concentration is estimated as a ratio of particles to physical or geometric detector volumes in the test suspension. For the geometric Poisson model this ratio is designated λ. In past practice this has begun with the calculation of the physical or geometric volume of the detector employed, V_{DG}, as the product of the detector length and width and the height of the scanning beam, B_H, as in Eq. (1). (Note: Uppercase letters are used for constants, lowercase letters are used for variable quantities.)

$$V_{DG} = LWB_H \tag{1}$$

The ratio of the total number of particles in a test volume, N_T, to geometric detector volumes, V_{DG}, in this arbitrary volume, V is λ.

$$\lambda = \frac{N_T}{V/V_{DG}} \tag{2}$$

A ratio of particle concentration in particles per milliliter to the number of detector volumes per milliliter is more convenient to manipulate. In all further discussion the test volume, V, is restricted to 1 ml. N_T, the number of particles in the test volume becomes the particle concentration per milliliter, n. The results are

$$\lambda = \frac{n \text{ particles/ml}}{(1/V_{DG})(\text{detector volumes})/\text{ml}} \tag{2A}$$

Simplifying Eq. (2A) results in

$$\lambda = nV_{DG} \tag{2B}$$

Carrying through the indicated division of the fraction $1/V_{DG}$ in Eq. (2A) results in the number of detector volumes per milliliter, N_G:

$$N_G = \frac{1}{V_{DG}} \tag{3}$$

Replacing $1/V_{DG}$ in Eq. (2A) by its value in Eq. (3) gives

$$\lambda = \frac{n}{N_G} \tag{2C}$$

In Eq. (2C) the concentration measure lambda is restated as a ratio of particles per milliliter to detector volumes per milliter. The denominator is also the number of particles per milliliter for a one-to-one ratio of particles to the geometrically defined detector volume, V_{DG}.

2. Effective Detector Volumes

As shown in Fig. 6A, the effective height of the scanning beam, B_H, is the sum of the beam height and the diameter of the particle being measured. The resulting increase in detector volume is described in Eqs. (4) and (4A).

$$V_{DE} = LW(B_H + D_P) \tag{4}$$

Factoring out B_H, Eq. (4) can be written as

$$V_{DE} = LWB_H(1 + D_P/B_H) \tag{4A}$$

Substituting for the product of detector length, width and height of the scanning beam the physical or geometric volume, V_{DG}, as defined in Eq. (1), into Eq. (4B) results in Eq. (5). In Eq. (5), the increase in the effective volume of the detector is related to the ratio of the particle diameter and the height of the scanning beam.

$$V_{DE} = V_{DG}(1 + D_P/B_H) \tag{5}$$

Equation five is then solved for V_{DG}:

$$V_{DG} = \frac{V_{DE}}{1 + D_P/B_H} \tag{5A}$$

Equations (5) and (5A) show that $V_{DE} > V_{DG}$.

Reserving λ for the ratio of particles to geometric detector volumes, the ratio of particles to effective volumes, ν, is

$$\nu = \frac{n}{1/V_{DE}} \frac{\text{particles/ml}}{\text{effective detector volumes per ml}} \tag{6}$$

In Eq. (7) as in Eq. (3), carrying through the division indicated in the fraction of $1/V_{NE}$ results in the number of effective detector volumes per milliliter, N_E.

$$N_E = \frac{1}{V_{DE}} \tag{7}$$

Replacing $1/V_{DE}$ in Eq. (6) by its value N_E from Eq. (7) results in Eq. (6B). In Eq. (6B) the ratio of particles per effective detector volumes is conveniently expressed as ν, a ratio of particles per milliliter to dectector volumes per milliliter.

$$\nu = \frac{n}{N_E} \tag{6B}$$

Substituting for V_{DE} its value from Eq. (5) into Eq. (7) results in Eq. (8) which relates the concentration of particles per effective detector volume to both the effective detector volume and to the number of effective detector volumes per milliliter.

$$N_E = \frac{1}{V_{DG}(1 + D_P/B_H)} \tag{8}$$

The effective particle concentration, $n/N_E = \nu$, is compared with $n/NG = \lambda$, the geometric particle concentration with Eq. (8A) and (2B).

$$\nu = nV_{DG}(1 + D_P/B_H) = nV_{DE} \tag{8A}$$

$$\lambda = nV_{DG} \tag{2B}$$

Comparing ν in Eq. (8) to λ in Eq. (2B), V_{DE}, the effective detector volume, is larger than V_{DG}, the geometric detector volume. For the same number of suspended particles, the larger effective detector volume will result in higher calculated concentrations than those based upon the geometric detector volume. Comparing Eqs. (2B) and (8), it follows that $\lambda < \nu$.

3. Scan Beam Height and Effective Detector Volume

$$V_{DE} = \dot{V}_{DG}\left(1 + \frac{D_P}{\dot{B}_H}\right) \tag{5A}$$

$\dot{B}_H$ = Scan beam height
D_P = Particle diameter
V_{DE} = Effective detector volume = $1/N_E$
$\dot{V}_{DG}$ = Geometric detector volume
• = Unknown terms

Equation (5A) has two unknowns, V_{DG}, the functionally measured geometric detector volume, and B_H, the functionally measured scan beam height. Both the geometric detector volume and the scan beam height, in manufacturers' specifications and in prior publications, have been determined by static measurements. Static measurements cannot fully gauge the dynamic interaction of a particle with the variability of flow within the detector and with the variation of the scan beam scatter as it passes through the detection zone. The two unknown constants in

Eq. (5A) are determined with simultaneous equations using data from two dilution experiments with different particle sizes.

4. The Scan Beam Height: Time-of-Flight Measurements

The effective scan beam height can be experimentally determined using a suspension of monosized precision microspheres with either a time or a sequential dilution-based analysis. For a time-based analysis, the pulse length of the signals generated by a stream of separated precision microspheres, t_p, and the mean flow velocity of the sample stream, V, can be used to compute the physical length of the signal d_E, and the effective height as expressed in

$$d_E = \frac{V}{t_P} = B_H + D_P \tag{9}$$

The liquid flow velocity carrying the particle through the detector is laminar. The position of the particle in the flow stream is random. Since the flow velocity varies along the flow contour, the particle pulse velocity will also vary. To achieve measurement accuracy, the mean of multiple observations of the transit pulse width for a standard sphere is required. In this evaluation, d_E, the system determined signal length, is greater than the sum of particle diameter and window height. This is due to the inertial effect of the electronic circuitry. The effects of flow velocity on particle signal generation and measurement accuracy are considered elsewhere.

5. Effect of Threshold Selection

The proportion of the particle diameter utilized by the measurement system depends on the analysis limits selected. When multiple particle sizes are counted with diameter ratios of 10:1 or greater, essentially the total size of the larger particle image is used for control purposes. When a single particle size is being evaluated, it is often convenient to set the analysis limits closer to the size of the particle. When the low threshold is set at half the particle size to be evaluated, particles with one-fourth of the area of the selected size can pass through the detector undetected. The significant result on the particle being measured is that for the system purposes the particle appears truncated. The first and final 25% of its area are invisible to the sizing circuit of the counter. D_{PE}, the effective size of the particle, is therefore determined by the physical size of the particle and the selected analysis limits. This effect is described in Eq. (10) in which $A(P_E)$ is the sensed, effective, particle area. $A(P)$ is the actual particle area and $A(T)$ is the area of the threshold particle size selected.

$$A(P_E) = A(P) - 2A(T) \tag{10}$$

The corresponding relationship between diameters is

$$D_{PE} = \sqrt{D_P^2 - 2D_T^2} \tag{10A}$$

This reduction in sensed diameter does not affect the recorded size of particles since the suppressed areas are added to the detected peak during recording. It does affect the undercount loss since the physical size of the sensed particle is reduced.

When the analysis threshold reduces the sensed area of the particle, the experimental results have been determined for the effective particle diameter, D_{PE}, not the actual particle diameter. The relationship of the effective particle diameter to the experimental determination of coincident count loss is discussed below. When the particle size analysis threshold selected is no more than one-tenth of the calibrating particle diameter, it will affect the experimental results by 1% or less and can be ignored. The selected threshold reduces the sensed particle diameter as shown in Eq. (10A). D_P is the diameter of the calibrating particle, D_T is the diameter of the particle selected as an analysis threshold and D_{PE} is the effective reduced particle diameter which controls the count loss due to increasing concentration.

The effect of a 5-μm analysis threshold on a 19.5-μm particle is

$$D_{PE} = \sqrt{152 - 2.52} = 13.23\ \mu m \tag{10B}$$

The effect of a 5-μm analysis threshold on a 19.5-μm particle is

$$D_{PE} = \sqrt{19.52 - 2.52} = 18.17\ \mu m \tag{10C}$$

The choice of a detection threshold will affect concentration test results by reducing the detected size of particles. If the reduction in particle width caused by an elevated analysis level is not considered, results based on the nominal size of the calibrated particles will be in error. A constant high-analysis threshold will result in an increasingly oversized determination of N_E as the test particle diameter is reduced. An uncorrected computation based on dilution experiment counting data of 15-, 19.5- and 25-μm particles with a 5-μm threshold will result in an oversizing of N_E by 16.7%, 6.8% and 4% respectively. This oversizing of N_E will result in underestimates of the coincidence effect counting errors for the detector.

B. The Geometric Poisson Model

To provide continuity with the current literature on coincidence error, the following analysis of the geometric Poisson model uses the geometric detector volume, V_{DG}, and its concentration parameter, λ. Since this choice is independent of the statistical model employed, the effective detector volume, V_{DE}, and the effective concentration parameter, ν, can be substituted wherever the geometric parameters are employed. The numerical results will, of course, be different.

A key parameter in this analysis is λ, the ratio of particles per geometric detector volume. λ is given by

$$\lambda = \frac{n}{N_G} \tag{2C}$$

The Poisson analysis which follows uses an infinite series whose sum is 1. Symbolically, this Poisson series is described in Eq. (11), where $P(0)$ is the probability of an empty detector volume, $P(1)$ is the probability of a detector volume containing a single particle, $P(2)$ is the probability of a detector volume containing 2 particles, etc.

$$1 = P(0) + P(1) + P(2) + \cdots \tag{11}$$

The actual Poisson series is shown in Eq. (12). For clarity λ rather than the more useful n/N_G notation is employed.

$$1 = e^{-\lambda} + \frac{\lambda e^{-\lambda}}{1!} + \frac{\lambda^2 e^{-\lambda}}{2!} + \cdots \tag{12}$$

To determine the expected number of counts in each occurrence category for each milliliter of volume, the number of detector volumes in this 1-ml test volume, N_G, is multiplied by the probability for each occurrence category. Multiplying each term of Eq. (12) by N_G yields Eq. (12A), the expected number of counts in each count occurrence group category: 0,1,2,3 etc.

$$N_G = N_G e^{-\lambda} + N_G\frac{\lambda e^{-\lambda}}{1!} + N_G\frac{\lambda^2 e^{-\lambda}}{2!} + \cdots \tag{12A}$$

Equation (12A) can be summarized symbolically as

$$N_G = C(0) + C(1) + C(2) + \cdots \tag{12B}$$

The number of counts per milliliter of single particles, $C(1)$, shown in Eq. (12A) and symbolically in (12B) is

$$C(1) = \frac{N_G \lambda e^{-\lambda}}{1!} \tag{13}$$

The expected number of counts per milliliter, c, is determined by subtracting the number of zero-particle detector volumes from the total number of detector volumes:

$$c = N_G - N_G e^{-\lambda} = N_G(1 - e^{-\lambda}) \tag{12C}$$

These results are directly in particle counts per milliliter.

In the further discussion of the geometric model and in the future discussion of the triggered model, ratios of occurrence group counts will clarify the relationships required to understand and control coincidence error. Since a ratio of counts approaches the ratio of expected counts as the number of events counted becomes large, no distinction will be made between counts and expected counts. In the absence of image recognition capability, the counts generated by single- and multiple-monosized-particle occurrences can be generated and studied. The ratios of such counts provide simple expressions for evaluating counting efficiency, counting error and the various components of the coincidence error. The first such ratio to be examined is the counting efficiency.

The counting efficiency of the geometric Poisson model is defined as the ratio c/n. This ratio of counts per milliliter to the number of single particles in a 1-ml test volume is obtained by dividing Eq. (12C) by Eq. (2C) to get

$$\lambda = \frac{n}{N_G} \tag{2C}$$

$$\frac{c}{n} = \frac{1 - e^{-\lambda}}{\lambda} \tag{14}$$

Equation (14) is the relationship between the expected number of counts and the total number of single particles in suspension for the geometric Poisson model. From Eq. (14), the counting efficiency, c/n decreases with increasing particle concentration. Particle concentration is denoted by λ. As lambda becomes very large, the counting efficiency, the ratio of particle counts to particles in solution, c/n, approaches zero. Replacing λ in Eq. (14) by n/N_G and simplifying results in Eq. (14A). Eq. (14A) directly relates count efficiency, c/n, to the number of single particles in suspension for the geometric Poisson model. The exponential n/N_G is the particle concentration in particles per detector volume; N_G is equal to $1/V_{DG}$.

$$\frac{c}{n} = \frac{1 - e^{-n/N_G}}{n/N_G} \tag{14A}$$

For low concentations, a simple count efficiency expression using the first three terms of the exponential series is employed. c/n after simplification, is approximated by

$$\frac{c}{n} = 1 - \frac{\lambda}{2} = 1 - \frac{n/N_G}{2} \tag{16}$$

The counting error of the geometric Poisson model, ε_C, is 1 minus the counting efficiency.

$$\varepsilon_C = \frac{e^{-\lambda} + \lambda - 1}{\lambda} = \frac{e^{-n/N_G} + n/N_G - 1}{n/N_G} \tag{16}$$

For small λ, i.e., low concentrations, the count error ε_c is approximated by

$$\varepsilon_C = \frac{\lambda}{2} = \frac{n/N_G}{2} \tag{16A}$$

Equation (16A) states that for low concentrations, the undercount error estimated with the present geometrical Poisson model is half the particle concentration measured in particles per detector volume or particles per the number of detector volumes per milliliter.

The ratios of the number of multiple occurrences to the number of single-particle counts utilizes data that can be generated using monodisperse microspheres. These ratios provide simple expressions with which the various contributions to the total coincidence error can be evaluated. To develop these ratios, $N(0)$, $N(1)$, $N(2)$, etc., are defined as the number of detector volumes containing 0, 1, 2, etc., particles. These quantities are noted symbolically by the respective terms in Eq. (12B):

$$N_G = C(0) + C(1) + C(2) + \cdots \tag{12B}$$

The ratio of $C(2)/C(1)$ is shown in Eq. (17), the ratio of $C(3)/C(1)$ is shown in Eq. (18), and the ratio of $C(4)/C(1)$ in Eq. (19).

$$\frac{C(2)}{C(1)} = \frac{n/N_G}{2} \tag{17}$$

$$\frac{C(3)}{C(1)} = \frac{(n/N_G)^2}{6} \tag{18}$$

$$\frac{C(4)}{C(1)} = \frac{(n/N_G)^3}{24} \tag{19}$$

Examining these equations shows that the predominant multiple coincidence occurrence is that of a double; the higher-order occurrences are of less significance than those of the doubles.

The total number of all coincident occurrences per milliliter in the geometric Poisson model, C^+, is determined by subtracting from N_G, the total number of detector volumes in the 1-ml test volume in Eq. (12B), the number of zero-particle volumes, $N(0)$, and the number of single-particle volumes, $N(1)$, in the 1-ml test volume. This is shown symbolically in Eq. (20) and finally in Eq. (20A).

$$C^+ = N_G - N(0) - N(1) \tag{20}$$

Expressing λ as n/N_G in Eq. (12C) and substituting the results into Eq. (20) results in the normalized ratio of multiple counts to the number of single particles that have been counted.

$$\frac{C^+}{n} = \frac{1 - e^{-\lambda} - \lambda e^{-\lambda}}{\lambda} \tag{20A}$$

$$\frac{C^+}{n} = \frac{1 - e^{-n/N_G} - n/N_G e^{-n/N_G}}{n/N_G}$$

Equation (20A) provides a convenient means with which to calculate the total of all multiple-occurrence events.

C. The Particle-Triggered Poisson Model

An alternative model with which to evaluate coincidence error is based solely on counted events. This change in focus to particulate occurrences avoids the artificial boundaries imposed by the geometric Poisson model, since closely spaced particle groups are not arbitrarily separated into adjacent detector volumes. In this model, just as in the physical device, the examination of a unit volume commences with the entry of a particle into the sensitive detector volume. With this model, a coincident event occurs when one or more particles enter the sensitive volume before the final exit of the initiating particle.

The graphical comparison of both Poisson models in Figs. 5A and 5B will now be extended with a statistical analysis of multiple-particle occurrences. In the analysis of measures of particle concentration, two alternatives were discussed. The first was based on the geometric volume of the detector and resulted in λ, n/N_G. The second calculation included the effect of particle sizes that were comparable to the height of the scan beam. This calculation resulted in ν, n/N_E, the effective measure of concentration.

The physical relationship of these two measures of concentration clearly indicates that for the same concentration of particles n/N_E is always greater than n/N_G. Counting error estimates based on n/N_G will therefore always underestimate those based on the effective particle concentration, n/N_E. In the following discussion, the effective detector volume and the effective concentration parameter will be used. Since this choice is independent of the statistical model employed, the geometric paramters can, if desired, replace the selected effective parameters. The selection of a method of calculating particle concentration will modify the results obtained with the triggered Poisson model.

The effect of a change in the statistical model to more closely describe the actual particle counting events will now be described. The introduction of the new model doubles the number of ways that counting error can be estimated from the previous two to four. In the absence of experimental comparisons of their effectiveness, no valid relationship between instrument counts and the number of particles in suspension will be followed by an experimental evaluation of the four models that are now being considered.

For clarity, as in the discussion of the Jaenicke geometric Poisson model, ν, rather than the more useful n/N_E, will be used. The probability of having a single, double and triple occurrence particle count in the particle-triggered Poisson model is respectively, $e^{-\nu}$, $(1 - e^{-\nu})e^{-\nu}$, $(1 - e^{-\nu})^2 e^{-\nu}$, etc. Following circulation of early drafts of this paper Harrison [21], using a different analysis of the coincidence count loss problem, confirmed the inadequacy of the Jaenicke model. Harrison's conclusions support the following analysis.

Since the sum of the occurrence probabilities in this sequence is 1, we get

$$1 = e^{-\nu} + (1 - e^{-\nu})e^{-\nu} + (1 - e^{-\nu})^2 e^{-\nu} + \cdots \tag{21}$$

denoting the total number of counts by c, and multiplying each term in Eq. (21) by c, yields Eq. (22). In Eq. (23) the counts are separated into occurrence groups.

$$c = ce^{-\nu} + c(1 - e^{-\nu})e^{-\nu} + c(1 - e^{-\nu})^2 e^{-\nu} + \cdots \tag{22}$$

$$c = C(1) + C(2) + C(3) + \cdots \tag{23}$$

$C(K)$ in Eq. (23) is the number of counts in each particle count category, $K = 1,2,3, \cdots$. The number of single particles counted, $C(1)$, is determined from Eq. (22) and is shown in Eq. (24). Equation (6B) is used in the final simplified form of Eq. (24).

$$C(1) = ce^{-\nu} = ce^{-n/N_E} \tag{24}$$

The relationship between the total number of particles, n, and $C(K)$ is given by

$$n = C(1) + 2C(2) + 3C(3) + \cdots \tag{25}$$

Substituting each group of counted occurrences from Eq. (21) into Eq. (25) results in

$$n = ce^{-\nu}[1 + 2(1 - e^{-\nu}) + 3(1 - e^{-\nu})^2 + \cdots] \tag{26}$$

Since the bracketed series in Eq. (26) is equal to $e^{2\nu}$, we have

$$n = ce^{\nu} \tag{27}$$

Equation (27) relates counts to the number of particles. Solving for the counting efficiency of the particle-triggered Poisson model, c/n, the ratio of counts to single particles is

$$\frac{c}{n} = e^{-\nu} \tag{28}$$

As in Eq. (14) for the geometric model, c/n approaches zero as ν becomes large. For low particle concentrations, as measured by low ν or n/N_E the effective counting efficiency shown in Eq. (28) is approximated as

$$\frac{c}{n} = 1 - \nu = 1 - \frac{n}{N_E} \tag{29}$$

The counting error for the particle-triggered model, ε_C, is calculated as $1 - c/n$:

$$\varepsilon_C = 1 - e^{-\nu} = 1 - e^{-n/N_E} \tag{30}$$

At low concentrations, small ν or n/N_E, the effective counting error is approximated by

$$\varepsilon_C = \nu = \frac{n}{N_E} \tag{31}$$

For low concentrations, as measured by the effective particle concentration ν, the estimated under-count error for the particle-triggered Poisson model is twice that of the geometric Poisson model.

As in the analysis of the geometric Poisson model, Eqs. (29) and (31) provide a basis for examining the linearity of the relationship of counts versus effective particle concentration expressed as the ratio of n/N_E. The relationship of multiple occurrences to single particles in suspension for the particle-triggered Poisson model follows the structure used for the geometric Poisson model. This relationship for both Poisson models is shown in Table 5. Table 6 provides a side-by-side comparison for both counting error and efficiency for any particle concentration factor. The important low-concentration counting error and efficiency equations are also shown. A comparison of the two models must take into account the different values of n/N_G and n/N_E as well as the different functional forms in Tables 5 and 6.

1. Normalized Count Efficiency and the Particle-Triggered Poisson Model

Substituting for ν its value n/N_E into Eq. (28) results in Eq. (32). Equation (32) provides a normalized expression with which the effective count efficiency or linearity can be conveniently estimated for any single-particle counter.

$$\frac{c}{n} = e^{-n/N_E} \tag{32}$$

At a concentration n equal to N_E the ratio of c/n is equal to e^{-1}. Evaluating e^{-1} results in an estimated counting efficiency of 36.7%, an error of 63.3%.

2. Maximum Count Capability of the Particle-Triggered Poisson Model and the Concentration at Which It Occurs

The relationship between the effective detector volume and the maximum concentration that can be unambiguously counted is determined as follows.

$$C_{MAX} = N_E e^{-1} = .3679 N_E \tag{33}$$

$$n(C_{MAX}) = N_E \tag{33A}$$

Equations (33) and (33A) state that the maximum number of counts reported by a counter will be 36.79% of the number of effective detector volumes per milliliter for the detector. The importance and utility of N_E, the number of detector volumes per milliliter is evident in Eqs. (32A), (33) and (33A).

Solving Eq. (32) for Ln(c/n) as in Eq. (32A) expresses the relationship of instrument counts and particles in suspension in a form that can be tested with data.

$$\text{Ln}\left(\frac{c}{n} = \frac{-n}{N_E}\right) \tag{32A}$$

3. The False Count Problem

The false count problem can be addressed with convenient expressions for significant multiple occurrences and an expression for the total number of all multiple occurrences as a function of particle concentration.

The ratio of multiple occurrences to single particle counts is determined by ratios of the corresponding terms in Eq. (22) and (23) as follows:

$$\frac{C(2)}{C(1)} = 1 - e^{-n/N_E} \tag{34}$$

$$\frac{C(3)}{C(1)} = (1 - e^{-n/N_E})^2 \tag{35}$$

$$\frac{C(4)}{C(1)} = (1 - e^{-n/N_E})^3 \tag{36}$$

Using Eq. (32) as an identity, Eqs. (34), (35) and (36) are rewritten in terms of count efficiency:

$$\frac{C(2)}{C(1)} = 1 - \frac{c}{n} \tag{34A}$$

$$\frac{C(3)}{C(1)} = \left(1 - \frac{c}{n}\right)^2 \tag{35A}$$

$$\frac{C(4)}{C(1)} = \left(1 - \frac{c}{n}\right)^3 \tag{36A}$$

Rewriting the count of single particles per milliliter shown in Eq. (24), in terms of count efficiency results in

$$C(1) = 1 - ce^{-n/N_E} \tag{37}$$

$$C(1) = c\left(\frac{c}{n}\right) \tag{37A}$$

Multiplication of the count ratios in Eqs. (34A), (35A) and (36A) by the count of single particles as shown in Eq. (37A) results in a convenient expression for the counts recorded for each multiple occurrence:

$$C(2) = c\left(\frac{c}{n}\right)\left(1 - \frac{c}{n}\right) \tag{34B}$$

$$C(3) = c\left(\frac{c}{n}\right)\left(1 - \frac{c}{n}\right)^2 \tag{35B}$$

$$C(4) = c\left(\frac{c}{n}\right)\left(1 - \frac{c}{n}\right)^3 \tag{36B}$$

Comparing Eqs. (17A), (18A) and (19A) for the geometric Poisson model with Eqs. (34B), (35B) and (36B) for the triggered Poisson model, a difference in the higher-order multiple-occurrence terms is apparent. The rate with which the concentration of higher-order multiple occurrences falls off in the triggered model is one less than the multiplicity order; in the geometric model the falloff is at the multiplicity order. Accordingly, the effect of coincidence error is of greater significance in the particle-triggered Poisson model than in the preceding geometric model published by Jaenicke [9]. With the particle-triggered model of particle counting, the effect of large groups of particles below the region of control interest is a clearly seen threat to accurate analysis.

D. Count Efficiency Applied to False Counts and Undercounts

Four equations are used to translate multiple-occurrence event ratios in terms of the count efficiency of a detector. The first equation, Eq. (32), is the primary count efficiency equation. The second equation is the series that describes the particle-triggered Poisson model in count efficiency terms, Eq. (40). The third equation is the logical equivalent expression of the Poisson series into count efficiency terms, Eq. (40A). Equation (32) defines the count efficiency relationship between particle counts and the number of particles counted by a detector.

$$\frac{c}{n} = e^{-n/N_E} \tag{32}$$

The balancing concept to the efficiency ratio is the count loss ratio. The count loss ratio, $\bar{c}/n$ (c-not on n) for the class of all single-particle light-extinction particle counters is defined in Eq. (39). The sum of the count efficiency ratio and the count loss ratio is equal to 1.

$$1 = \frac{c}{n} + \frac{\bar{c}}{n} \tag{38}$$

Solving Eq. (38) for $\bar{c}/n$ as 1 minus the count efficiency is shown in Eq. (38A). Incorporation of the effect of the count loss ratio will be seen to be essential to resolving the single-particle count accuracy problem at high concentration ratios.

$$\frac{\bar{c}}{n} = 1 - e^{-n/N_E} = 1 - \frac{c}{n} \tag{38A}$$

Using Eq. (32) as an identity, the particle-triggered Poisson series is rewritten as

$$\frac{c}{n} = \left(\frac{c}{n}\right)^2 + \left(\frac{c}{n}\right)^2\left(\frac{1-c}{n}\right) + \left(\frac{c}{n}\right)^2\left(\frac{1-c}{n}\right)^2 + \cdots \tag{39}$$

Equation (39) is restated in symbolic terms as

$$\frac{c}{n} = \frac{C(1)}{n} + \frac{C(2)}{n} + \frac{C(3)}{n} + \frac{C(4)}{n} + \cdots \tag{39A}$$

Each term in Eq. (39) and Eq. (39A) represents the occurrence probability of a multiple-particle group as shown in the logical equivalent statement of Eq. (40A). This 1:1 relationship between the algebraic series and the logical equivalent statement can be used to determine the efficiency ratio of selected groups.

Equation (39A) is rearranged into the count ratio of single-particle groups and the count ratio of all multiple groups in

$$\frac{c}{n} = \frac{C(1)}{n} + \frac{C^+}{n} \tag{39B}$$

C^+/n, the count efficienty ratio of all multiple-occurrence groups is determined by subtracting the count efficiency ratio of single-particle groups, $C(1)/n$ from c/n, the overall count efficiency ratio. The simplified algebraic result is shown in Eq. (41).

The first right-hand term in Eq. (39), $(c/n)^2$ is the efficiency ratio of the number of particles that have been counted as individual occurrences. The count ratio of single-particle groups is

$$\frac{C(1)}{n} = \left(\frac{c}{n}\right)^2 \tag{40}$$

An alternative form for $C(1)/n$ is

$$\frac{C(1)}{n} = e^{-2n/N_E} \tag{40A}$$

The second, third and fourth terms in the count efficiency series of Eq. (39) are $C(2)/n$, $C(3)/n$ and $C(4)/n$. They are designated

$$\frac{C(2)}{n} = \left(\frac{c}{n}\right)^2\left(1 - \frac{c}{n}\right) \tag{40B}$$

$$\frac{C(3)}{n} = \left(\frac{c}{n}\right)^2\left(1 - \frac{c}{n}\right)^2 \tag{40C}$$

$$\frac{C(4)}{n} = \left(\frac{c}{n}\right)^2\left(1 - \frac{c}{n}\right)^3 \tag{40D}$$

Substituting the value of $C(1)/n$ from Eq. (40) into Eq. (39B) and simplifying results in

$$\frac{C^+}{n} = \frac{c}{n}\left(1 - \frac{c}{n}\right) \tag{41}$$

Similarly, the expression for the sum of the triple and higher coincident occurrence terms is developed from Eq. (39) as the remainder after subtraction of the sum of $C(1)$ and $C(2)$ from c in

$$\frac{C^+(3)}{n} = \frac{c}{n} - \frac{C(1)}{n} - \frac{C(2)}{n}$$

$$\frac{C^+(3)}{n} = \frac{c}{n} - \left(\frac{c}{n}\right)^2 - \left(\frac{c}{n}\right)^2\left(1 - \frac{c}{n}\right) \tag{42}$$

After simplification, we obtain

$$\frac{C^+(3)}{n} = \left(\frac{c}{n}\right)\left[1 - \frac{c}{n}\right]^2 \tag{42A}$$

The proportion of the total coincident occurrences represented by each of the multiple occurrences is determined by taking ratios of each of the multiple-order terms, $C(1)/n$, $C(2)/n$, etc., to the sum of all coincident occurrences. For example, the ratio of double coincident occurrences to all coincident occurrences is evaluated with ratios of the corresponding occurrence terms from Eq. (39) as in

$$\frac{C(2)/n}{C^+/n} = \frac{C(2)}{C^+} = \frac{(c/n)^2(1 - c/n)}{(c/n)(1 - c/n)} = \frac{C(2)}{C^+} = \frac{c}{n} \tag{43}$$

E. Single-Particle Counts Versus Particles Counted

The full significance of Eq. (32) was not appreciated until the second "Coincidence" paper was being written [24]. Equation (32) expresses a normalized relationship between particle counts and particle concentration for the class of all light extinction single-particle counters.

This normalized count to particle relationship and the normalized count error are both graphically demonstrated in Fig. 16. The relationships shown in Fig. 16 apply to all light extinction particle counters.

The normalized count efficiency and count loss equations are, however, only a partial answer to the needs of a U.S.P. <788> particle assay. These equations do not differentiate between counts of single particles and the combinational events in particle + coincidence. With this distinction in mind, the term raw counts will be used for particle counter data.

Since the intent in all particle assays has been to define contamination in terms of single-particle counts, the analysis to date is necessary but not sufficient. A count ratio balance equation expressed in terms of single particles counted provides the essential information required. The symbolic count balance shown in Eq. (44) states that the sum of the singles count ratio, the multiple-occurrence count ratio due to vertical coincidence and the ratio of counts lost in the counting of multiple particle groups is equal to 1.

$$1 = \frac{C(1)}{n} + \frac{C^+}{n} + \frac{\overline{c}}{n} \tag{44}$$

Equation (44) is expressed verbally as

$$1 = \begin{bmatrix}\text{Singles count}\\ \text{efficiency ratio}\end{bmatrix} + \begin{bmatrix}\text{Multiples count}\\ \text{efficiency ratio}\end{bmatrix} + \begin{bmatrix}\text{Multiples count}\\ \text{loss ratio}\end{bmatrix} \tag{44V}$$

This symbolic count balance equation is the basis for experimental verification of the counting accuracy of particle counters. It introduces the concept that the sum of the single-particle count ratio, the multiple-particle count ratio and the count loss ratio due to the occurrence of multiple-particle groups is required to fully describe particle counting occurrences. Each multiple-occurrence particle group produces a single count of a particle whose size is determined by the summation of all the coincident particle signals. In a dilution experiment, the particle size counted will be an integral multiple of the calibration particle size used. In a random particle assay, the particle size recorded cannot be resolved into its individual components.

The multiple-occurrence count efficiency ratio and the multiple-occurrence count loss ratio together are the single-particle count loss ratio. This is expressed as repeated in

$$\frac{\overline{C}(1)}{n} = \frac{C^+}{n} + \frac{\overline{c}}{n} \tag{44A}$$

$$\begin{bmatrix}\text{Singles count}\\ \text{loss ratio}\end{bmatrix} = \begin{bmatrix}\text{Multiples count}\\ \text{efficiency ratio}\end{bmatrix} + \begin{bmatrix}\text{Multiples count}\\ \text{loss ratio}\end{bmatrix} \tag{44AV}$$

For calculation convenience, Eq. (44) is rewritten in terms of the single-particle group count ratio:

$$1 = \frac{C(1)}{n} + \frac{\overline{C}(1)}{n} \tag{44B}$$

$$1 = \begin{bmatrix}\text{Singles count}\\ \text{efficiency ratio}\end{bmatrix} + \begin{bmatrix}\text{Singles count}\\ \text{loss ratio}\end{bmatrix} \tag{44BV}$$

Solving for the single particle loss ratio is accomplished in

$$\frac{\overline{C}(1)}{n} = 1 - \frac{C(1)}{n} = 1 - \left(\frac{c}{n}\right)^2 \tag{44C}$$

Verification of Eq. (44) is obtained by substituting for the three right-hand terms their respective values from Eqs. (40), (41) and (38A) as

$$\begin{aligned} 1 &= \left(\frac{c}{n}\right)^2 + \frac{c}{n}\left(1 - \frac{c}{n}\right) + \left(1 - \frac{c}{n}\right) \\ 1 &= 1 \end{aligned} \tag{45}$$

The particle count balance equation has been validated and can now be applied to the development of universal single-particle count accuracy and count error curves.

1. Universal Single-Particle Count Accuracy and Count Error Curves

The two components of the single-particle count loss, described in Eq. (44A), are the count loss ratio and the multiple-particle group ratio. The total undercount error is the sum of the two components. Both components have been graphed in Fig. 22 as functions of the effective concentration measure, n/N_E. The significant factor that can be observed in these curves is that up to a total count loss ratio of 20%, the loss contribution of each component is approximately equal. Neither component can be neglected without introducing serious error.

$$\frac{\overline{C}(1)}{n} = \frac{C^+}{n} + \frac{\overline{c}}{n} \tag{44A}$$

The relationship between the single-particle group count ratio and the effective concentration measure has been described in Eq. (40). The single-particle count loss is conveniently calculated with Eq. (44C).

$$\frac{C(1)}{n} = \left(\frac{c}{n}\right)^2 \tag{40}$$

$$\frac{\overline{C}(1)}{n} = 1 - \frac{C(1)}{n} = 1 - \left(\frac{c}{n}\right)^2 \tag{44C}$$

Equations (40) and (44C) have been graphed in Fig. 23 as functions of the effective concentration measure, n/N_E. Figure 23 describes the universal relationship between the effective particle concentration measure and single-particle counting efficiency and counting error. These curves are valid for the class of all single-particle counters. This specifically includes light extinction particle counters. The requirements for count accuracy of single particles in the U.S.P. <788> contaminating particle assay can now be defined and controlled.

2. Maximum Single-Particle Group Count Capability of the Particle-Triggered Poisson Model and the Concentration at Which It Occurs

The relationship between the effective detector volume and the maximum concentration of single particle counts that can be unambiguously counted is determined as

$$\frac{C(1)}{n} = e^{-2n/N_E} \tag{40A}$$

Differentiating $C(1)$ with respect to n yields $C(1)_{MAX}$, the peak of the single-particle group count versus particle concentration curve as shown in Eq. (40B). The number of single particles in suspension, $n(C_{MAX})$ at the maximum particle count is shown in Eq. (40C).

$$C(1)_{MAX} = N_E \frac{e^{-1}}{2} = \frac{0.3679 N_E}{2} = 0.1839 N_E \tag{40B}$$

$$n(C_{MAX}) = \frac{N_E}{2} \tag{40C}$$

Equation (40B) states that the maximum number of single-particle group counts reported by a counter will be 18.39% of the number of effective detector volumes per milliliter for the detector. Equation (40C) locates this maximum single-particle group count at a concentration of $N_E/2$ particles per milliliter.

Comparing Eqs. (40B) and (40C) with Eqs. (33) and (33A), the maximum concentration of single-particle group counts is half that of the raw count maximum. The count maximum occurs at half the concentration at which the raw count maximum appeared.

3. Application to the U.S.P. <788> Assay to Determine the Concentration of 10-μm Particles for a 10% Coincident Particle Group Occurrence

Determination of the coincident particle group occurrence rate is made with an indirect methodology: the primary measurement is that of count efficiency. The occurrence rate of multiple-particle groups is calculated from the measured count efficiency using one of the published Poisson models. For an actual determination of the result of this coincidence ratio specification the calculated result must also take into account the manner in which the particle concentration has been calculated. A choice must be made between the present use of the geometric concentration measure, n/N_G, which is independent of particle size, or the use of the new effective concentration measure, n/N_E, which has been shown to vary with the size of the particle being counted. The first step in this comparison is the evaluation of the effect of the Poisson model employed.

Accordingly, the U.S.P. <788> particle counter count accuracy specification using the earlier Jaenicke geometric Poisson model is compared with an analysis using the effective concentration measure and the particle-triggered model of the counting process. In both analyses, the coincident particle ratio is equated to 10% of the total count/particle ratio. This is expressed

$$0.10\frac{c}{n} = \frac{C^+}{n} = \frac{c}{n} - \frac{C(1)}{n} \tag{46}$$

The simplified results are

$$0.90\frac{c}{n} = \frac{C(1)}{n} \tag{46A}$$

Simplification of Eq. (46) shows that a ratio of 0.90 of single particles counted is equivalent to the 0.10 ratio of coincident particle groups. Although the logical statement is identical for both models, the algebraic realizations will of course differ.

To demonstrate the multiplicative effect on counting error introduced by neglecting the particle-detector interaction, an example using a recent laser-sourced detector has been included in the following analysis. The experimental data for the detector selected (detector 3 in Table 8) includes V_{DG}, the geometric detector volume, which is 14×10^{-6} cm^3 and the beam height which is 39.7 μm.

4. Application of the Jaenicke Geometric Poisson Model

The symbolic equation, Equation (46A) translates the U.S.P. XXIII <788> SVI coincidence error specification into algebraic statements of the c/n and the $C(1)/n$ ratios. Substituting for c/n its value from Eq. (18A) and for $C(1)/n$ its value from Eq. (16B) into Eq. (46A) results in Eq. (46B).

$$\frac{0.90(1 - e^{-n/N_G})}{n/N_G} = e^{-n/N_G} \tag{46B}$$

Equation (46C) is the solution of Eq. (46B) for the relative concentration factor, n/N_G. Equation (46C) has been solved by iteration and is correct to four significant figures.

$$\frac{n}{N_G} = 0.2073 \tag{46C}$$

The calculated Jaenicke geometric model count efficiency, c/n, at the specified U.S.P. XXIII <788> SVI concentration is determined by substituting the relative count efficiency into Eq. (14A).

$$\frac{c}{n} = \frac{1 - e^{-n/N_G}}{n/N_G} = \frac{1 - e^{-0.2073}}{0.2073} = 0.9032 \tag{14A}$$

The raw count error due to coincident particle occurrence at this concentration is determined by subtracting the count efficiency from 1. The model-based count error is 9.68%.

The geometric detector volume for detector 3 was reported to be 14 μm [4]. V_{DG}, the number of geometric detector volumes per milliliter is the inverse of the geometric detector volume and is 1,000,000 divided by 14 or 71,429. The calculated concentration accuracy limit, N', for this detector is determined by solving Eq. (14) for the U.S.P. XXIII <788> SVI specified coincidence ratio limited concentration. This is accomplished with the n/N_G ratio calculted in Eq. (46C) and the experimental value of detector volume reported in Ref. 4.

$$\frac{N'}{N_G} = \frac{N'}{71{,}429} = 0.2073 \tag{14B}$$

$$N' = 14{,}807 \quad 10\text{-}\mu\text{m particles per milliliter}$$

The solution of Eq. (14B) shows that a U.S.P. XXIII <788> SVI coincidence limited concentration estimate for 10-μm particles based on the use of the Jaenicke geometric Poisson model with the geometric concentration calculation is 14,807 particles per milliliter.

5. Application of the Particle-Triggered Poisson Model

$$0.90\frac{c}{n} = \left(\frac{c}{n}\right)^2 \tag{46}$$

$$\frac{c}{n} = 0.90 \tag{46C}$$

$$\text{Ln}\ 0.90 = \frac{-n}{N_E} \tag{46D}$$

$$\frac{n}{N_E} = 0.1054$$

$$\frac{\overline{C}(1)}{n} = 1 - \frac{C(1)}{n} = 1 - \left(\frac{c}{n}\right)^2 \tag{44C}$$

$$\frac{\overline{C}(1)}{n} = 1 - (0.9)^2 = 0.19$$

The estimated single-particle count loss, $\overline{C}(1)/n$, based on the particle-triggered Poisson model using this analysis is 19%: this is approximately double the 9.70% count loss obtained with the Jaenicke geometric Poisson analysis for the same 10% coincident occurrence ratio.

$$V_{DE} = V_{DG}\left(1 + \frac{D_P}{B_H}\right)$$

$$= 14\left(1 + \frac{10}{39.7}\right) \tag{11}$$

$$V_{DE} = 17.53 \times 10^{-6}\ \text{cm}^3$$

N_E, the number of effective detector volumes per milliliter is calculated as $1/V_{DE}$ and is 57,057.

$$\frac{N'}{57{,}707} = 0.1054 \qquad (1)$$

$$N' = 6082 \quad \text{10-µm particles per milliliter}$$

Operation of this detector at a concentration of 6082 10-µm particles per milliliter satisfies the U.S.P. XXIII <788> SVI 10% coincidence ratio requirement and results in a single-particle count loss of 19% as calculated with Eq. (13). The difference in the results obtained with the two available statistical models merits further examination.

6. Evaluation of Detector Operation by the Particle-Triggered Model at the Coincidence Limited Concentration Calculated with the Jaenicke Geometric Model

Use of the Jaenicke geometric Poisson model calculations with the experimental constants determined for detector 3, resulted in a U.S.P. XXIII <788> SVI concentration limit for a 10% coincidence ratio of 14,807 10-µm particles/ml. An evaluation of the effect of operation with this concentration using the particle-triggered Poisson model follows. N_E for 10-µm particles with detector 3 is unchanged from the previous example and is 57,057. The count efficiency for 14,807 particles/ml is calculated with Eq. (32).

$$\frac{c}{n} = e^{-n/N_E} = e^{-14{,}807/57{,}057} = 0.7714 \qquad (32)$$

The result of Eq. (32) is a raw count efficiency of 0.7714 and a count loss of 0.2286. The single-particle count loss is calculated as

$$\frac{\overline{C}(1)}{n} = 1 - \left(\frac{c}{n}\right)^2$$

$$\frac{\overline{C}(1)}{n} = 1 - (0.7714)^2 = 0.4049 \qquad (44C)$$

Adherence to the U.S.P. XXIII <788> SVI coincidence ratio requirement using the Jaenicke geometric Poisson model results in a single-particle count loss of 40.49% and a coincidence occurrence ratio of 17.63%:

$$\frac{C^+}{n} = \frac{c}{n}\left(1 - \frac{c}{n}\right) = 0.7714(1 - 0.7714) = 0.1763 \qquad (41)$$

7. Conversion of the U.S.P. <788> Coincidence Specification to Count Efficiency Terms

N' and N_E' are respectively, the concentration and the number of effective particle volumes per milliliter that satisfy the U.S.P. XXIII <788> coincidence ratio specification when the particle-triggered Poisson model is used. N_E', the number of detector volumes per milliliter for 10% coincidence ratio error is

$$N_E' = \frac{N'}{-\text{Ln}\,(c/n)} = \frac{N'}{-\text{Ln}\,0.90} = \frac{N'}{0.1054} \qquad (14)$$

$$N_E' = 9.4912N'$$

The conversion constant is 9.4912. The use of this conversion constant makes possible a validation procedure for the U.S.P. XXIII <788> SVI coincidence ratio specification that can be accomplished by even a modestly equipped pharmaceutical laboratory. A major consideration in the use of this conversion constant for a concentration of 10-μm particles is that its accuracy relies on the results of an unspecified statistical model and unspecified measurement methods.

8. Measurement of the Coincidence Effect: Starting with U.S.P. XXIII <788> Counter Specifications, An Example

The calculations shown in Table 16 have been prepared as an example of a particle dilution experiment to determine the relationship between particle concentration and instrument counts. The protocol can be implemented with any precision microsphere whose total distribution, including outliers, can be counted. The first requirement for the coincidence count loss determination is an accurate experimental relationship of particle counts to the number of particles in suspension. The experimental technique to achieve these results is described in the example. The recorded data in this experiment is the dilution factor and the counts per milliliter at each dilution step.

In U.S.P. XXIII, particle counter vendors are required to specify the nominal particle concentration, N', for 10-μm particles at which a 10% undercount, C', occurs. At this concentration, N', the counting efficiency, C'/N', is 90%. This provides the information needed to compute the nominal value of N_E at 10 μm as with Eq. (32A). The simplified results are shown in Eq. (32B).

$$N_E\ (10\ \mu\text{m}) = 9.4912N' \tag{32B}$$

The inclusion of 10 μm stresses that N_E is particle-size specific. The commercially specified detector volume of $10 \times 10^{-6}\text{cm}^3$, whose experimental parameters are listed as detector 3 in Table 8, has a 10% undercount error for 10-μm particles at a concentration of 6011.6 particles/ml. The N' concentration counted with 90% count efficiency is therefore 6011.6. Substitution into Eq. (32C) provides an estimate of N_E as

$$\begin{aligned} N_E(10\ \mu\text{m}) &= (9.4912)(6011.6) \\ &= 57{,}057 \text{ effective detector volumes per milliliter} \end{aligned}$$

Accurate evaluation of N_E is best made at relatively high concentrations where count efficiency is low. These high-concentration low-count-efficiency data points are those with high count error. A practical initial particle concentration, N_I, for this purpose is approximately $0.6N_E$ as

$$N_I \approx 0.6N_E \tag{17}$$

A relationship between N_I and N', the U.S.P. XXIII limit concentration for a 10% undercount error, will now be developed. The approximate value of this initial concentration is

$$N_I^* \approx 5.2N' \tag{17A}$$

An essential first step in the experimental evaluation of N_E is the determination of the particle concentration at each step of the dilution experiment. This commences with an accurate evaluation of N_I, the initial particle concentration. Following this determination, the concentration at each succeeding dilution step is determined from the precision dilution employed. The initial high-concentration low-count-efficiency concentration designated N_I in Eq. (17) is evaluated by extrapolation. N_I is extrapolated from a set of high-count-efficiency low-concentration data points. To minimize extrapolation error in the determination of the initial concentration from data that approaches linearity but never becomes linear, a measurement span within equal and opposite errors is selected. Using a measurement span with ±1% end points is desirable. Of the data points used in the calculation of N_I, the data at the -1% undercount end point, N'', is the highest particle concentration data point used in the extrapolation calculation. This data point is least affected by dilution errors and the background noise count. To ensure an optimum value for the N'', it is the first concentration calculated. Following the calculation of the N'' value, the dilution ratio from it to the noise limited concentration is used to determine the step dilution ratio. The next determination is the calculation for the number of dilution steps from the -1% undercount concentration to the high concentration of N_I required for a good determination of N_E.

Determination of the concentration at the selected end points follows. The critical -1% concentration, N'', is determined from the 0.99 count efficiency concentration:

$$N'' = -N_E \, \text{Ln}\, 0.99 \tag{18}$$

The concentration for a -1% undercount error can also be calculated from the 10% undercount concentration by substituting for $N_E(10\ \mu\text{m})$ its value, $N'/\text{Ln}\ 0.90$, from Eq. (6A) into Eq. (18) as shown in Eq. (18A). The results for the selected detector are shown in Eq. (18B).

$$N'' = N' \, \text{Ln}\, 0.99 = \frac{0.095390N'}{\text{Ln}\, 0.90} \tag{18A}$$

$$\begin{aligned} N'' &= (0.095390)(6011.6) \\ &= 573.45 \end{aligned} \tag{18B}$$

With current detectors, a noise limit of $+1\%$ on recorded counts results in the selection of a 100 counts/ml as a lower data limit. This lower concentration limit is conservative; some marketed detectors provide better performance.

For accuracy in the determination of N_I, three dilution data points from 573.45 particles per milliliter down to the lower limit of 100 particles/ml are desirable. These four data points anchored by the background count point at infinite dilution usually provide an accurate determination of N_I.

The total dilution factor, f_T, that will be obtained in these three steps is

$$f_T = \frac{100}{573.45} = 0.1744 \tag{19}$$

The required dilution per step, f_S, is calculated as the 1/3 power of f_T in Eq. (19A). For practical convenience f_S is round to f in Eq. (19B) and maintained constant throughout the dilution experiment.

$$f_S = \frac{0.17441}{3} = 0.5587 \tag{19A}$$

$$f = 0.6000 \tag{19B}$$

A practical coincidence error limit experiment must minimize the number of dilution steps to maximize the possibility of accurate results. Ensuring an adequate number of counts at the maximum dilution data points is also essential if the relative standard deviations of those important data points are not to reduce the accuracy of the determination of the initial, maximum particle concentration, N_I.

An approximate value for the number of dilution steps above $.01N_E$, S^*, to the approximate maximum initial concentration, N_I^*, is then calculated as

$$f^{S*} = \frac{N''}{5.2N'} \tag{20}$$

The total dilution from $0.6N_E$ to $0.01N_E$ is the ratio $0.01N_E/0.6N_E$, which is 0.01666. Using the step dilution ratio of 0.600 calculated in Eq. (19B), the approximate number of dilution steps above $0.01N_E$, S^*, is calculated with Eq. (20A). S^* is then rounded to S, an integral number of dilution steps in Eq. (20B).

$$\begin{aligned} 0.600^{S*} &= .0166 \\ S^* &= 8.023 \end{aligned} \tag{20A}$$

$$S = 8 \tag{20B}$$

Rounding the number of dilution steps down from 8.023 to 8, results in a total of 11 dilution steps required for this experiment. The desired value of N_I is then calculated as

$$N_I = \frac{0.01N_E}{0.6008} = \frac{573.45}{0.01680} = 34{,}141.7 \text{ particles/ml} \tag{17B}$$

The number of counts/ml, C_I, resulting from the initial concentration, N_I, is calculated with Eq. (1) as

$$C_I = 34{,}141.7e^{-34{,}141.7/57{,}057} = 18{,}767.8 \text{ particles/ml} \tag{1B}$$

The test volume is adjusted to the desired initial particle concentration by matching the expected counts/ml that result from the N_I particle concentration. This sequential dilution experiment commences with a test volume of 150 ml adjusted to a count of 18,768 counts/ml. The demonstrated replicability of single-particle counters assures the presence of 34,141.7 particles/ml in the test suspension.

A choice of 150 ml for the test volume and 60 ml for the diluent aliquot volume is appropriate. These selections provide an accurately representative sample of the suspension and good separation between successive concentrations. One other experimental requirement must be satisfied to achieve accurate data: adequate mixing time following each dilution step. One minute of mixing time has been found adequate for volumes up to 300 ml. Discarding the first 10-ml aliquot ensures accurate separation of the successive concentration measurements.

Table 16 has been prepared to provide a realistic examination of the data and computations used in an evaluation of the counting capability of a detector. Accordingly, this table is based on the detector employed in the experiments described in Ref. 23. The detector has a laser diode source and a specified geometric volume of 10×10^{-6} cm^3. This is detector 3 in Table 8 in which the nominal and experimentally measured parameters are listed.

9. Model-Based Dilution Experiment

Table 16 summarizes a model-based example with which to examine the problems of sequential dilution experiments. The calculations for this table are based on the particle-triggered Poisson model and the effective particle concentration measure determined for the commercially available laser-sourced detector described earlier. The experimental parameters for this detector are listed in Table 8 as detector 3, the laser-sourced detector with experimentally determined geometric or physical volume of 14×10^{-6} cm^3 and a beam height of 39.7 µm. The data for this table assume a nonlimiting analysis threshold and a background of 1 count/milliliter. The background count is added to the ideal, noise-free, count whenever its addition will increase the total by at least .1%.

These calculations are based on an assumed initial particle concentration of 34,114.9 particles/ml of 10-µm particles. A sequential dilution sequence with a dilution factor of 0.600

was employed. The particle concentration at each dilution point was determined from the sequential dilution for that point.

In column 4, counts/ml were calculated for each concentration with Eq. (32):

$$\frac{c}{n} = e^{-n/N_{\rm E}} \tag{32}$$

The counts shown in column 4 are ideal, noise-free counts. Whenever the assumed background count level of 1 count/ml is greater than 0.1% of the noise-free counts shown in column 4, the total of column 4 is increased by 1 count/ml and listed in column 5.

The entries for column 6 were determined as the ratio of the particles per milliliter for the dilution step to 57,057 detector volumes per milliliter, the 10-mm $N_{\rm E}$ value for the selected detector. Count error in column 7 was computed as the ratio $(n - c)/n$. The n and c entries were taken from column 3 and column 4 or 5 as appropriate.

The dilution steps within the boxed area in this table, steps 8–11, have been calculated to be within the ±1% count error limits. The background noise count is the only positive systematic error encountered, the coincident count errors are all negative; i.e., they reduce particle counts.

(a) Analysis of Model-Based Data. Treating the entries of Table 16 as the experimental data from a sequential dilution experiment, the data analysis that is required will now be demonstrated. The data available at the end of the sequential dilution experiment are the sequential dilution step, the net dilution factor for each step and the counts per milliliter recorded at each step. This data is shown, respectively, in columns 1, 2 and 5.

Identification of the low-counting-error steps 8–11 and C_0, the background noise count at infinite dilution are the first steps following completion of the dilution experiment. $N_{\rm I}$, the initial particle concentration can be determined either by graphical extrapolation to the initial concentration at 1.000 from the low-count-error section of the count-versus-dilution curve or, more desirably, by linear regression from the same set of dilution data.

The linear regression solution shown in Table 17 is based on the background count C_0 and the four selected data points from Table 16. The linearity of the determination is evidenced by the 0.999996 correlation coefficient. The calculated value of the initial concentration, $N_{\rm I}$, and the low value of C_0, which is the calculated background count, is in accord with the design of the model experiment.

Having determined $N_{\rm I}$, the concentration of particles at each dilution step is determined with the 0.600 sequential dilution factor for that step. Determination of the particle concentration at each dilution step of the experiment is then paired with the counts/milliliter recorded for that step. The computation of count error as shown in column 7 can now be made. Dilution step 3 is a sample of the Table 16 data available at this point.

Col. 1	Col. 2	Col. 3	Col. 5	Col. 7
Step	Dilution factor	Calculated particles/ml	Recorded counts/ml	Count error $(n - c)/n$
3	0.2160	7,368.83	6,476.0	0.1212

$$\text{Count error} = \frac{n - c}{n} = \frac{7{,}368.83 - 6{,}476.0}{7{,}368.83} = .1212$$

Column 5 is used in this computation since it is the model equivalent of actual data, in that it represents data that includes the background count, C_0.

This experimental determination of the count-to-particle ratio, the ratio of column 5 to column 3, is independent of any model. This independence, without knowledge of the relationship of counts/ml recorded to particles/ml in suspension limits the utility of the analysis at this point. In instrument terminology, the data that has resulted is a straight-line instrument calibration curve. The data is accurate only at the determination point. Estimates of the relationship of counts to particles in suspension at other than the measurement points should be estimated with straight line interpolation. Clearly, straight-line interpolation of the count/particle data is inappropriate.

The model-based results of this determination are used to define the coincidence count loss of the tested detector for any concentration or particle size with which it can be used. The generalized measure of detector counting accuracy is directly usable as a measurement specification to assure particle count correlation independent of vendor or instrument model number.

The experimental determination of N_E, the parameter that is used to evaluate detector counting accuracy over the full operating range of the instrument, commences when the paired counts/ml and particle/ml count and concentration data are available. This parameter facilitates both standard specifications for particle counters and the definition of the operating limits that should be employed to achieve the desired counting accuracy. Equation (32A) is used.

$$N_E = \frac{-n}{\text{Ln}\,(c/n)} \tag{32A}$$

The accuracy of N_E calculated from experimental data is improved when c/n ratios are substantially less than 1. Values of c and n that are too close together increase the sensitivity of the c/n ratio to small variations in the dilution data. Using the data of step 0, shown below, N_E is calculated with Eq. (6). With N_E available, the ratio of particles to effective detector volumes, the ratio n/N_E, for the step 0 concentration follows.

Col. 1	Col. 2	Col. 3	Col. 5	Col. 6	Col. 7
Step	Dilution factor	Calculated particles/ml	Recorded counts/ml	Effective concen.	Count error $(n - c)/n$
0	1.0000	34,114.9	18,761.8	0.59800	0.45004

$$N_E = \frac{34{,}114.9}{\text{Ln}\,(18{,}761.8/34{,}114.9)} = 57{,}057 V_{DE}/\text{ml}$$

$$\frac{n}{N_E} = \frac{34{,}119.9}{57{,}057} = .59800 \frac{\text{Particles}}{\text{Effective detector volume}}$$

Note: The n/N_E ratio is the effective concentration measure.

These examples show the computation sequence with which to analyze the sequential dilution experiment. While the calculation of N_E shown above is based on the data from a single point, an averaged value calculated from several higher concentration points is generally advisable. For a complete solution, the variation of N_E with particle size must be resolved. To simplify the following discussion, the experimentally determined value of N_E, 57,057 is converted to its reciprocal, the effective detector volume, V_{DE}:

$$V_{DE} = \frac{1}{N_E} = \frac{1}{57{,}057} = 17.53 \times 10^{-6}\ \text{cm}^3 \tag{7}$$

The variation of V_{DE} with particle size is defined as

$$V_{DE} = V_{DG}\left(1 + \frac{D_P}{B_H}\right) \tag{5}$$

V_{DE} is known; V_{DG} the geometric or physical detector volume and B_H, the scan beam height are unknown. With two unknowns in a single equation, the data required for a solution must be supplied by additional measurements. The required data are determined with a second sequential dilution experiment that uses a different particle size. The height of the scan beam is determined as

$$V_{DE1} = B_H + D_{P1} \tag{5D}$$

$$V_{DE2} = B_H + D_{P2} \tag{5D}$$

V_{DG} is calculated with a pair of values for particle diameter, B_H and V_{DE}. Experimental data from a sequential dilution experiment with a 15-μm particle with this detector is available and will be used to illustrate the solution for V_{DG} and B_H. N_E for the 15-μm particle was found to be 51,841.2 detector volumes per milliliter. The corresponding value of V_{DE}, using Eq. (7), is 19.29×10^{-6} cm^3. Substituting the available data into Eqs. (5D) and (5E) results in

$$17.53 = V_{DG}(B_H + 10) \tag{47}$$

$$19.29 = V_{DG}(B_H + 25) \tag{47}$$

The simultaneous solution of Eqs. (5F) and (5G) gives $V_{DG} = 14.0 \times 10^{-6}$ cm^3 and $B_H = 39.7$ μm. With these two parameters established, V_{DE} for any other particle size can be computed and converted to the reciprocal N_E for that particle size. As an example, the value of V_{DE} for a 7.14-μm particle is computed with Eq. (5) and converted to its reciprocal, N_E, with Eq. (3).

$$V_{DE} = 14.0\left(1 + \frac{7.14}{39.7}\right) \tag{49}$$

$$V_{DE} = 16.52 \times 10^{-6}\ \text{cm}^3$$

$$N_E = \frac{1}{V_{DE}} = \frac{1}{16.52 \times 10^{-6}\ \text{cm}^3} = 60{,}540.4 \tag{3}$$

N_E, the reciprocal of V_{DE}, is calculated with Eq. (3) and is 60,540.4 effective detector volumes per milliliter.

Comparison of graphical determinations of N_I from the data of Table 15 with the linear regression results listed in Table 16 show major differences. With experience, an error range of ±20%, at best, can be expected in determination of N_I. This is in part due to the lack of the data anchor provided by the background count, C_0, to subtle variations of the best straight-line fit from unity slope and to the effect of the maximum concentration data point used in the solution. The 33,776.8 particle/ml value for the estimated value of N_I shown in Table 16 has been determined by linear regression from a set of data points whose maximum concentration data point is 1% below its true value.

The linear regression solution for the initial particle concentration is affected by the choice of data from a dilution step whose value is approximately 1% below its true value. The result of this offset is a negative bias in the determination of N_I, the initial particle concentration. For most purposes this bias can be ignored. For those who require maximum accuracy, the initial N_I result of the linear regression analysis is multiplied by $1 + n'/N_E$, where $n'/N \leq 0.01$ is the maximum concentration employed in the linear regression analysis. An analysis of this −1% bias follows below. Applying this 1% correction to the 33,776.8 particles per ml calculated value results in 34,114.5 particles per ml which is within 0.4 particles per ml of the N_I on which Table 16 is based.

The model data employed, neglects the random measurement errors that always occur. However, those errors will have greater effect on the graphic extrapolation without the data anchor provided by C_0, the diluent background count.

Even with the use of linear regression calculations, a graphical initial best straight-line fit is beneficial. This initial selection can also be used to edit high-error data points from the computation. The analysis of those data points that result in the minimum zero offset is used to select the optimum value for the initial concentration of particles.

With linear regression available on many hand-held scientific calculators, its use for the determination of the initial particle concentration in these experiments should replace the more variable graphical determination. Although the use of linear regression for the determination of N_I, the initial particle concentration, is advocated, monitoring experiment progress with a log-log plot of counts/ml versus the dilution factor can provide an early indication of experimental problems.

Table 17 lists the experimental counts/ml at Step 0, C_I, and N_I the initial suspension determined by linear regression. The linear regression solution used the low-count-loss dilution steps 8–11 and C_0, the count at infinite dilution. The infinite dilution count is the background count recorded with clean diluent during counter operation. The background count C_0 has been assumed to be 1.0 in Table 16. The results listed in Table 16 demonstrates the linearity and accuracy that can be obtained with a sound choice of data points and the use of linear regression in the extrapolation for N_I.

(b) Analysis of the Bias in Determination of N_I: The Initial Particle Concentration. The accuracy of the serial dilution experiment rests upon the accurate determination of N_I, the initial particle concentration. N_I is calculated from counts recorded in a low-count-error region. The sequence of particle concentrations counted is prepared by serial dilutions from the initial particle concentration using a dilution factor f as

$$n = fN_I \tag{52}$$

Solving Eq. (5) for f results in

$$f = \frac{n}{N_I} \tag{52A}$$

At $f = 1.0$, n is equal to N_I, the original undiluted particle suspension:

$$nf = 1 = N_I \tag{52B}$$

In these dilution experiments, the value of N_I is determined with data from a concentration region where there are low count errors. The hypothesis that the N_I prepared by sequential dilution is indeed the value of N_I recovered from the experimental data will now be considered. Equation (53) expresses the assumption that c is approximately n:

$$c = fN_I \tag{53}$$

The value of N_I calculated with Eq. (32) from the diluted concentration data by replacing n by c is denoted by N_I^*. Substituting for c its value from Eq. (32) gives

$$ne^{-n/N_E} = fN_I^* \tag{53A}$$

Solving Eq. (53A) for f gives

$$f = \left(\frac{n}{N_I^*}\right) e^{-n/N_E} \tag{53B}$$

Restricting n/N_E to values much less than 1, justifies substitution of the first two terms of the series for e^{-n/N_E} as in

$$f = \left(\frac{n}{N_I^*}\right)\left(1 - \frac{n}{N_E}\right) \tag{53C}$$

Evaluating Eq. (53C) at n/N_E equal to .01, f equal to 1 and solving for N_I, in terms of $nf = 1$ results in Eq. (53D). To test the hypothesis that N_I, determined in a dilution sequence, is equal to that calculated from the count data Eq. (53D) is compared to Eq. (52B).

$$nf = 1 = \frac{N_I^*}{0.99} \tag{53D}$$

$$nf = 1 = N_I \tag{52B}$$

Comparison of Eqs. (52B) and (53D) is summarized as

$$N_I^* = .99N_I \tag{54}$$

The estimation of N_I from sequential-dilution partical-count-data is biased by the relative concentration of the highest data point used in its extrapolated determination. Arbitrary selection of the concentration data points for the determination of N_I without the model employed in this analysis can only multiply the determination error.

The preceding analysis can be verified with the model-dilution experiment calculations of Table 16.

F. Sensitivity of the U.S.P. XXIII <788> SVI Contaminating Particle Assay to Coincidence Effects

A 2-ml container with identical particle loading is assumed in the following comparisons of the coincidence effect on two detectors in current usage. Particle concentration below the 10-μm minimum particle size specified in U.S.P. XXIII <788> SVI contaminating particle measurements is assumed to be 6000 particles per milliliter in each calculation for each particle size. Undercounts and false counts within 10 to 25 μm are assumed to be at the U.S.P. <788> SVI limit of 6000 per container or 3000 particles per milliliter. The sensitivity computations are based on U.S.P. XXIII contaminating particle limits.

The experimentally determined beam height and the geometric detector volume for each detector is listed in Table 7. The geometric mean particle sizes that are used in these calculations are listed in Table 13. The N_E values for each of the test particles is listed for each detector in Table 14.

For each of the test particle sizes,

$$N_E = \frac{1}{V_{DG}(1 + D_P/B_H)} \tag{8}$$

1. Boundary Effects, Undercounts and False Counts

The computation of the single-particle count loss ratio in Eq. (44C) and the ratio of all multiple-occurrence counts in Eq. (41) assumes an effectively infinite measurement span. When boundaries are introduced as in the U.S.P. <788> Contaminating Particle Assay, their effect upon computation of undercounts and false counts must be considered when single-particle count loss is to be accurately calculated. The effect on single-particle count loss of coincident occurrences within a boundary is determined by the coincidence-occurrence multiple that can be contained within the selected boundary. Considering ideal particle samples and ideal counters, the ratio of multiple occurrences that can be contained within the 10- to 25-μm particle size boundaries of the U.S.P. <788> particle assays is determined by the size of the particle as shown in Table 13.

As an example, the bounded single-particle undercount ratio, $\overline{C}_B(1)/n$, and the bounded false count ratio, C_B^+/n for the 13.296-μm geometric mean particle size in Zone III of the sensitivity analysis will be analyzed. Consulting Table 13 shows that double and triple multiple occurrences for this particle size can be contained in the U.S.P.-defined contamination span. The total number of single-particle counts lost within the 25-μm boundary is the sum of the horizontal coincidence count loss and the loss in counts due to the single count for each particle grouping in vertical multiple-occurrence events. This count loss is equal to one count for each double-occurrence event and two counts for each triple-occurrence event. The sum of these losses is

$$\frac{\overline{C}_B(1)}{n} = \frac{\overline{c}}{n} + \frac{C(2)}{n} + \frac{2C(3)}{n} \tag{49}$$

Substituting for each of the symbolic terms in Eq. (49) their algebraic equivalent from Eq. (38B) and Eq. (39) gives

$$\frac{\overline{C}_B(1)}{n} = \left(1 - \frac{c}{n}\right) + \left(\frac{c}{n}\right)^2\left(1 - \frac{c}{n}\right) + 2\left(\frac{c}{n}\right)^2\left(1 - \frac{c}{n}\right)^2 \tag{49A}$$

For simpler manipulation, Eq. (49A) is rearranged as

$$\frac{\overline{C}_B(1)}{n} = \left(1 - \frac{c}{n}\right)\left[1 + \left(\frac{c}{n}\right)^2\left(3 - 2\frac{c}{n}\right)\right] \tag{49B}$$

The single-, double- and triple-occurrence particle counts of the 13.296-mm particles were seen to remain below the 25-μm boundary. Each false count, C_B^+/n will therefore originate in a multiple-occurrence event greater than 4. The bounded false count total is determined by subtracting from the count ratio of all vertical multiple occurrences, Eq. (41A), the ratio of double- and triple-coincident occurrences determined directly from Eq. (39). Equation (50) is the symbolic equation; the algebraic equation, Eq. (50A), and its simplified form Eq. (51) follow.

$$\frac{C_B^+}{n} = \frac{C^+}{n} - \left(\frac{C(2)}{n} + \frac{C(3)}{n}\right) \tag{50}$$

$$\frac{C_B^+(1)}{n} = \frac{c}{n}\left(1 - \frac{c}{n}\right) - \left(\frac{c}{n}\right)^2\left(1 - \frac{c}{n}\right) - \left(\frac{c}{n}\right)^2\left(1 - \frac{c}{n}\right)^2 \tag{50A}$$

$$\frac{C_B^+(1)}{n} = \frac{c}{n}\left(1 - \frac{c}{n}\right)^2 \tag{51}$$

2. New Laser Detector Sensitivity Analysis

The experimentally determined laser-sourced detector parameters are a geometric detector volume, V_{DG} of 10.2×10^{-6} $cm^3$and a beam height, B_H, of 27.3 μm.

Zone I. The geometric mean particle size for the span from one-third to one-half of a 10-μm particle signal is 6.389 μm. The concentration of these particles is 6000 per milliliter, 12,000 per

2 ml container. The injection of false counts from this source into counts of particles ≥ 10 μm can occur only with coincident occurrences of three or more particles in a group. The equation to use is

$$\frac{C^+(3)}{n} = \frac{c}{n}\left(1 - \frac{c}{n}\right)^2 \tag{40C}$$

Table 14 lists N_E equal to 61,526 for the 6.389-μm particle size. At an effective concentration, n/N_E, of 6,000/61,526 = 0.09752 the count efficiency is

$$\frac{c}{n} = e^{-n/N_E} = e^{-0.09752} = 0.9071 \tag{32}$$

<u>False count determination.</u> Substitution into Eq. (40C) with this count efficiency determines the ratio of injected false counts.

$$\frac{C^+(3)}{n} = \frac{c}{n}\left(1 - \frac{c}{n}\right)^2 = 0.9071(1 - 0.971)^2 = 0.007831 \tag{40C}$$

$C^+(3)$ is determined by multiplying both sides of the equation by the 6000 particles/ml of the test particle size (6000)(0.007831) = 46.99 per milliliter, or 93.97 per 2-ml container. This represents 94/6000 or 1.6% of the allowable contamination.

Zone II.

<u>False count determination.</u> The geometric mean particle size for the span from half of a 10-μm particle signal to just under 10 μm is 8.409 μm. The concentration of these particles is 6000 per milliliter, or 12,000 per 2-ml container. N_E for this particle size is listed as 58,944 in Table 15. The injection of false counts from this source into counts of particles ≥ 10 μm can occur with any coincident occurrence. The equation to use is

$$\frac{C^+}{n} = \frac{c}{n}\left(1 - \frac{c}{n}\right) \tag{41}$$

The effective concentration measure is 6,000/58,944 = 0.10179. The count efficiency determination is made by substituting into Eq. (32).

$$\frac{c}{n} = e^{-0.10179} = 0.9032 \tag{32}$$

Substituting 0.9032 into Eq. (41) gives

$$\frac{C^+}{n} = 0.9032(1 - 0.9032) = 0.08743 \tag{41}$$

The number of multiple coincident counts is determined by multiplying the C^+/n by the 6000 single particles per milliliter in the suspension. The result is 524.6 particles/ml, or 1049.2 particles per 2-ml container. This false count injection is 1049.2/6000 or 17.5% of the allowable particle count over 10 μm in the 2-ml container.

Zone III. The applicable equations for the bounded count loss in the 10- to 25-μm particle size span and the bounded false count injection into the group of particles ≥ 25 μm are

$$\frac{\overline{C}_B(1)}{n} = \left(1 - \frac{c}{n}\right)\left[1 + \left(\frac{c}{n}\right)^2\left(3 - 2\frac{c}{n}\right)\right] \tag{49B}$$

$$\frac{C_B^+(1)}{n} = \frac{c}{n}\left(1 - \frac{c}{n}\right)^3 \tag{51}$$

The U.S.P. <788> SVI upper limit 6000 particles per container or 3000 particles/ml is used for this example. N_E for the 13.296-μm particles is listed as 53,508 for the new laser detector. The effective particle concentration, n/N_E, is 3000/53,508, which is 0.05607. The count efficiency for these conditions is

$$\frac{c}{n} = e^{-0.05607} = 0.9455 \tag{32}$$

Bounded count loss determination. The bounded lost counts are determined by substituting the calculated count efficiency into Eq. (49B).

$$\begin{aligned} \frac{\overline{C}_B(1)}{n} &= \left(1 - \frac{c}{n}\right)\left[1 + \left(\frac{c}{n}\right)^2\left(3 - 2\frac{c}{n}\right)\right] \\ &= (1 - 0.9455)[1 + (0.9455)^2(3 - 2 \times 0.9455)] \\ &= 0.1085 \end{aligned} \tag{49B}$$

Multiplying by the test particle concentration of 3000 particles/ml results in a bounded count loss of 325.6/ml or 651.2 per container. This is a count loss of 10.85% per container.

In this case the test concentration, N_E and the count efficiency are the same as those employed for the bounded false count calculation. The count efficiency is therefore 0.9455. The bounded false count insertion is determined with Eq. (51).

Bounded false count determination.

$$\begin{aligned} \frac{C_B^+(1)}{n} &= \frac{c}{n}\left(1 - \frac{c}{n}\right)^3 \\ &= 0.9455(1 - 0.9455)^3 \\ &= 0.0001531 \end{aligned} \tag{51}$$

Multiplying through by the 3000 particle/ml test concentration results in 0.46 particles/ml, or 0.92 per container, falsely injected into the count of particles > 25 μm. For this example, this is a negligible 0.15%.

Zone IV. The geometric mean test particle size in this region is 21.203 μm. The applicable equations for the count loss in the 10- to 25-μm particle size span and the false count injection into the group of particles ≥ 25 μm are, respectively,

$$\frac{\overline{C}(1)}{n} = 1 - \left(\frac{c}{n}\right)^2 \tag{44C}$$

$$\frac{C^+}{n} = \frac{c}{n}\left(1 - \frac{c}{n}\right) \tag{51}$$

The significant differences between Zone III and Zone IV is the change in geometric mean particle size to 21.203 μm and the reduced acceptance limit of 600 particles per container, or 300 per milliliter, for particles > 25 μm.

The N_E value for the 21.203-μm particle listed in Table 15 is 46,561. The particle test population is again 6000 per container, or 3000 per milliliter. The effective concentration measure, n/N_E, is 3000/46,561 = 0.06443. The count efficiency is calculated as

$$\frac{c}{n} = e^{-0.06443} = 0.9376 \tag{32}$$

Count loss determination. The bounded lost counts are determined by substituting the calculated count efficiency into Eq. (44C).

$$\frac{\overline{C}(1)}{n} = 1 - (0.9376)^2 = 0.1209 \tag{44C}$$

Multiplying by the test particle concentration of 3000 particles/ml results in a bounded count loss of 362.7/ml, or 725.4 per container. This is a count loss of 12.1% per container.

False count determination. The test concentration, N_E and the count efficiency are the same as those used for the calculation of lost counts. The count efficiency is therefore 0.9376. The false count insertion is

$$\frac{C^+}{n} = \frac{c}{n}\left(1 - \frac{c}{n}\right) = 0.9376(1 - 0.9376) \\ = 0.05851 \tag{41}$$

Multiplying through by the 3000 particle/ml test concentration results in 175.5 particles per ml, or 351.0 per container, falsely injected into the count of particles > 25 μm. This is 351/600 or 58.5% of the U.S.P. <788> SVI maximum allowable number of particles ≥ 25 μm.

Zone V. The particle coincidence effect in Zone V is limited to the loss of single-particle counts calculated as

$$\frac{\overline{C}(1)}{n} = 1 - \left(\frac{c}{n}\right)^2 \tag{44C}$$

The concentration of test particles in this zone is 600 per container, or 300 per milliliter. N_E for 25-μm particles for this detector is listed in Table 14 and is 43,829. The effective concentration measure, n/N_E, is 300/43,829 = 0.006844. The count efficiency is

$$\frac{c}{n} = e^{-0.006844} = 0.9932 \tag{32}$$

Substituting the calculated count efficiency into Eq. (44C) results in the single-particle 25-μm count loss in Zone V.

$$\frac{\overline{C}(1)}{n} = 1 - (0.9932)^2 = 0.136 \tag{44C}$$

Multiplying the single-particle count loss ratio by the test concentration of 300 particles/ml results in the single-particle count loss of 4.08 particles per milliliter, or a loss of 8.16 per container, 1.36% of the allowable total.

3. Older White Light Detector Sensitivity Analysis

The experimentally determined detector parameters for the older white-light-sourced detector are a geometric detector volume, V_{DG}, of 20.7×10^{-6} cm^{3}and a beam height, B_H, of 28.5 μm. N_E values for the test particle sizes are listed in Table 15.

Zone I. The geometric mean particle size for the span from one-third to one-half of a

10-μm particle signal is 6.389 μm. The concentration of these particles is 6000 per milliliter, or 12,000 per 2-ml container.

False count determination. The injection of false counts from this source into counts of particles ≥ 10 μm can occur only with coincident occurrences of three or more particles in a group. The equation to use is

$$\frac{C^+(3)}{n} = \frac{c}{n}\left(1 - \frac{c}{n}\right)^2 \tag{40C}$$

Table 15 lists N_E equal to 39,463 for the 6.389-μm particle size. At an effective concentration, n/N_E, of 6,000/39,463 = 0.1520 the count efficiency is

$$\frac{c}{n} = e^{-n/N_E} = e^{-0.1520} = 0.8590 \tag{32}$$

Substitution into Eq. (40C) with this count efficiency determines the ratio of injected false counts.

$$\frac{C^+(3)}{n} = \frac{c}{n}\left(1 - \frac{c}{n}\right)^2 = 0.8590(1 - 0.8590)^2 \tag{40C}$$

$C^+(3)$ is determined by multiplying both sides of the equation by the 6000 particles/ml of the test particle size (6000)(0.007831) = 102.5 per milliliter, or 205.1 per 2-ml container. This represents 205.1/6,000 or 3.4% of the allowable contamination.

Zone II. The geometric mean particle size for the span from half of a 10-μm particle signal to just under 10 μm is 8.409 μm. The concentration of these particles is 6000 per milliliter, or 12,000 per 2-ml container. N_E for this particle size is listed as 37,303 in Table 14.

False count determination. The injection of false counts from this source into counts of particles ≥ 10 μm can occur with any coincident occurrence. The equation to use is

$$\frac{C^+}{n} = \frac{c}{n}\left(1 - \frac{c}{n}\right) \tag{41}$$

The effective concentration measure is 6000/37,303 = 0.1608. The count efficiency determination is made by substituting into Eq. (32).

$$\frac{c}{n} = e^{-0.1608} = 0.8514 \tag{32}$$

Substituting 0.8514 into Eq. (41) gives

$$\frac{C^+}{n} = 0.8514(1 - 0.8514) = 0.1265 \tag{41}$$

The number of multiple coincident counts is determined by multiplying C^+/n by the 6000 single particles per milliliter in the suspension. The result is 751.9 particles/ml, or 1518.2 particles per 2-ml container. This false count injection is 1518.2/6,000 or 25.3% of the allowable particle count over 10 μm in the 2-ml container.

Zone III. The applicable equations for the bounded count loss in the 10- to 25-μm particle size span and the bounded false count injection into the group of particles ≥ 25 μm are

$$\frac{\overline{C}_B(1)}{n} = \left(1 - \frac{c}{n}\right)\left[1 + \left(\frac{c}{n}\right)^2\left(3 - 2\frac{c}{n}\right)\right] \tag{49B}$$

$$\frac{C_B^+(1)}{n} = \frac{c}{n}\left(1 - \frac{c}{n}\right)^3 \tag{51}$$

The U.S.P. <788> SVI upper limit of 6000 particles per container, or 3000 particles/ml, is used for this example. N_E for the 13.296-μm particles is listed as 32,941 for the new laser detector. The effective particle concentration, n/N_E, is 3000/32,941, which is 0.0911. The count efficiency for these condition is

$$\frac{c}{n} = e^{-0.0911} = 0.9130 \tag{32}$$

Bounded count loss determination. The bounded lost counts are determined by substituting the calculated count efficiency into Eq. (49B).

$$\begin{aligned}\frac{\overline{C}_B(1)}{n} &= \left(1 - \frac{c}{n}\right)\left[1 + \left(\frac{c}{n}\right)^2\left(3 - 2\frac{c}{n}\right)\right] \\ &= (1 - 0.9130)[1 + (0.9130)^2(3 - 2 \times 0.9130)] \\ &= 0.1721\end{aligned} \tag{49B}$$

Multiplying by the test particle concentration of 3000 particles/ml results in a bounded count loss of 516.4/ml, or 1032.8 per container. This is a count loss of 17.21% per container.

Bounded false count determination. In this case, the test concentration and N_E are the same as those used for the calculation of bounded false counts. The count efficiency is therefore 0.9455. The bounded false count insertion is determined as

$$\begin{aligned}\frac{C_B^+(1)}{n} &= \frac{c}{n}\left(1 - \frac{c}{n}\right)^3 = 0.9130(1 - 0.9130)^3 \\ &= 0.0006012\end{aligned} \tag{51}$$

Multiplying through by the 3000 particle/ml test concentration results in 1.80 particles per milliliter or 3.61 per container falsely injected into the count of particles > 25 μm. For this example, this is a negligible 0.60% per container.

Zone IV. The geometric mean test particle size in this region is 21.203 μm. The applicable equations for the bounded count loss in the 10- to 25-μm particle size span and the bounded false count injection into the group of particles ≥ 25 μm are, respectively

$$\frac{\overline{C}(1)}{n} = 1 - \left(\frac{c}{n}\right)^2 \tag{44C}$$

$$\frac{C^+}{n} = \frac{c}{n}\left(1 - \frac{c}{n}\right) \tag{51}$$

The significant differences between Zone III and Zone IV is the change in geometric mean particle size to 21.203 μm and the reduced acceptance limit of 600 particles per container or for particles ≥ 25 μm. This limit is the equivalent of 300 particles per ml for particles ≥ 25 μm.

The N_E value for the 21.203-μm particle listed in Table 15 is 27,700. The particle test population is again 6000 per container or 3000 per milliliter. The effective concentration measure, n/N_E, is 3000/27,700 = 0.1083. The count efficiency is calculated as

$$\frac{c}{n} = e^{-0.1083} = 0.8974 \tag{32}$$

Count loss determination. The bounded lost counts are determined by substituting the calculated count efficiency into

$$\frac{\overline{C}(1)}{n} = 1 - (0.8974)^2 = 0.1947 \tag{44C}$$

Multiplying by the test particle concentration of 3000 particles/ml results in a bounded count loss of 584.0/ml, or 1168.0 per container. This is a count loss of 19.5% per container.

False count insertion. The test concentration, N_E and the count efficiency are the same as those used for the calculation of lost counts. The count efficiency is therefore 0.8974. The false count insertion is

$$\begin{aligned}\frac{C^+}{n} &= \frac{c}{n}\left(1 - \frac{c}{n}\right) = 0.8974(1 - 0.8974) \\ &= 0.09207\end{aligned} \tag{51}$$

Multiplying through by the 3000 particle/ml test concentration results in 276.2 particles per ml, or 552.4 per container, falsely injected into the count of particles > 25 μm. This is 552.4 or 92.1% of the U.S.P. <788> SVI 600 maximum allowable number of particles ≥ 25 μm per container.

Zone V.

Count loss determination. The particle coincidence effect in Zone V is limited to the loss of single-particle counts calculated with

$$\frac{\overline{C}(1)}{n} = 1 - \left(\frac{c}{n}\right)^2 \tag{44C}$$

The concentration of test particles in this zone is 600 per container, or 300 per milliliter. N_E for 25 μm is listed in Table 15 and is 25,735. The effective concentration measure, n/N_E, is 300/25,735 = 0.01166. The count efficiency is

$$\frac{c}{n} = e^{-0.01166} = 0.9884 \tag{32}$$

Substituting the calculated count efficiency into Eq. (44C) results in the single-particle count loss in Zone V.

$$\begin{aligned}\frac{\overline{C}(1)}{n} &= 1 - (0.9884)^2 \\ &= 1 - (0.9884)^2 = 0.0230\end{aligned} \tag{44C}$$

Multiplying the single-particle count loss ratio by the test concentration of 300 particles/ml results in the single-particle count loss per milliliter of 6.91 particles, or a loss of 13.8 per container, 2.3% of the allowable total.

REFERENCES

1. U.S.P. XXII, U.S.P. XXIII, United States Pharmacopeial Convention, Rockville, Md.
2. Munson, T. E., FDA perspectives on particulate matter testing in small volume injections, Proc. Int. Conf. on Liquid-Borne Particle Inspection and Metrology, Arlington, VA, 1987, p. 87.
3. Knapp, J. Z., and DeLuca, P. P., Review of commercially available particle measurement systems, *J. Parenter. Sci. Technol., 42*(1):(1988), suppl.
4. Trasen, B., Analytical techniques for particulate matter in intravenous solutions using membrane filters, *Ann. N.Y. Acad. Sci. 158*:665–673 (1969).
5. *Safety of Large Volume Parenteral Solutions,* National Symposium Proceedings, FDA, 1966.
6. Russell, A. W., Light blockage sensors for particle metrology, Proc. Int. Conf. on Liquid-Borne Particle Inspection and Metrology, Arlington, VA, 1987, p. 507.
7. Knollenberg, R. G., and Gallant, R. C., Refractive in liquid media by light optical extinction, Proc. Int. Conf. Particle Detection, Metrology and Control, 1990, p. 760.
8. Kushner, H. K., Abramson, L., Barber, T. and Knapp, J. Z., Error contribution in particulate determination from the viewpoint of sampling theory, Proc. Int. Conf. on Liquid-Borne Particle Inspection and Metrology, Arlington, VA, 1987.
9. Jaenicke, R., The optical particle counter: cross-sensitivity and coincidence, *Aerosol Sci. 30*:95–111 (1972).
10. Lieberman, A., Flow rate and concentration effects in automatic particle counters, National Conference on Fluid Power, Chicago, IL, 1975.
11. Willeke, Klaus and Liu, B. Y. H., Single particle optical counter: principle and application, *Aerosol Generation, Measurement, Sampling and Analysis,* Academic Press, 1976.
12. Knapp, J. Z., and Abramson, L. R., A systems analysis of light extinction particle detection systems, Proc. Int. Conf. Particle Detection, Metrology and Control, 1990, p. 283.
13. Oxborrow, G., A comparison of particle counters from five manufacturers. Proc. Int. Conf. Liquid-Borne Particle Inspection and Metrology, 1987, p. 482.
14. Wales, M. and Wilson, J. N., Theory of coincidence in Coulter particle counters, *Rev. Sci. Inst. 32*(10):1132 (1961).
15. Princen, L. H., and Kwolek, W. F., Coincidence corrections for particle size determinations with the Coulter counter, *Rev. Sci. Inst. 36*(5):646 (1965).
16. Bader, H., Gordon, H. R., and Brown, O. B., Theory of coincidence counts and simple practical methods of coincidence count correction for optical and resistive pulse counters, *Rev. Sci. Inst. 43*(10):1407–1412 (1972).
17. Delly, John G., Parenteral solutions—problem of sizing particles using USP methodology, Pharmacopeial Forum, Stimuli to the Revision Process, 1980, p. 359.
18. Raasch, J., and Umhauer, H., Errors in determination of particle sizing distribution caused by coincidence in optical particle counters, *Part. Character, 1*:53 (1984).
19. Thau-Ming Cham, Huey-Mee Yu and Loong-Chyau Tung, Counting of microspheres in electrolytes and parenteral solutions, *Drug Dev. Ind. Pharm. 15*(14–16):2441–2454 (1989).
20. Aoyama, Toshinobu, Perspective: particle detection in Japan, Proc. Int. Conf. Particle Detection, Metrology and Control, 1990, p. 760.
21. Harrison, C., Coincidence response of an idealized optical particle counter, HIAC/Royco Division, Pacific Scientific Co., TR91-01, 1991.
22. Knapp, J. Z., and Abramson, L. R., Validation of electronic particle counter counting accuracy, PDA/IES Joint Conf. Cleanrooms and Microenvironments, Arlington, VA, 1992.
23. Knapp, J. Z., and Abramson, L. R., A new coincidence model for single particle counters. I: Theory and experimental verification, *J. Pharm. Sci. Tech. 48*:110–134 (1994).
24. Knapp, J. Z. and Abramson, L. R,, A new coincidence model for single particle counters. II: Advances and applications, *J. Pharm. Sci. Tech. 48*:255–294 (1994).

25. Knapp, J. Z., and Abramson, L. R., A new coincidence model for single particle counters. III: Realization of single particle counter counting accuracy, *PDA J. Pharm. Sci. Tech., 50,* April–May (1996).
26. Kaye, Brian H., Operational protocols for efficient characterization of arrays of deposited fine particles by robotic image analysis systems, Am. Chem. Soc. Conf. Proc. 1991, pp. 354–371.
27. ASTM F658-87. *Practice for Defining Size Calibration Resolution and Counting Accuracy of a Liquidborne Particle Counter Using Near-Monodisperse Special Particulate Material,* American Society for Testing and Materials, Philadelphia, PA.
28. Knapp, J. Z. and Abramson, L. R., Application of the new particle counting model to count accuracy determinations, *PDA J. Pharm. Sci. Tech.*, (in press).

11

Particles in Semiconductor Manufacturing Process Liquids

Donald C. Grant

CT Associates, Inc., Excelsior, Minnesota

Alvin Lieberman

Particle Measuring Systems, Inc., Fremont, California

I. INTRODUCTION

As indicated in Chapter 1 many industrial processes are sensitive to particulate contamination in both process and product liquids. The problems encountered in measurement and control of particles varies with the nature of the process and the product. In general, the difficulty in characterizing and controlling the particulate content of liquids increases inversely with the size and the concentration of the particles. Of all industrial applications requiring particle measurement, control of particulate contamination in semiconductor manufacturing process chemicals presents the most challenges. Both particle concentration and size specifications are well below the levels permitted in any other industry, and forecasts indicate that both parameters will decrease significantly within the next few years [1]. This chapter focuses on the problems encountered in measurement and control of particulate contamination in semiconductor device manufacturing.

II. IDENTIFICATION OF PROBLEM MATERIALS AND PROCESSES

One industrial process which is especially sensitive to particulate contamination is the manufacture of semiconductor microcircuit devices. These devices are made by forming electrical components (transistors, capacitors and resistors) on wafers made of silicon or other materials with semiconductor properties. These electrical components are connected by conductive lines to form memory chips, microprocessors, and application-specific devices like those used to control automobile ignitions, antilock brakes and microwave ovens.

Modern ultralarge-scale integrated (ULSI) devices contain millions of electrical components and have very small features. The width of the lines connecting the components, termed the design rule, is presently smaller than 1 μm. The dielectric layers in the transistors and capacitors are as thin as 100 Å [2]. Both the design rule and dielectric thickness are decreasing with time as the devices become more complex and manufacturing technology improves. These microcircuit features can easily be damaged by particles which can cause open circuits, short circuits and/or nonintegral dielectric films. Particles with diameters larger than one-tenth the design rule can cause "killer defects" capable of damaging devices. The defects caused by particles can

Table 1 Chemicals Commonly Delivered by Chemical Delivery Systems

Acids	Bases	Solvents
Sulfuric acid	Hydrogen peroxide	Isopropyl alcohol (IPA)
Hydrochloric acid	Ammonium hydroxide	Acetone
Hydrofluoric acid	Tetramethyl ammonium hydroxide (TMAH)	*n*-Butyl acetate (NBA)
Buffered oxide etch (BOE)	Sodium hydroxide	Ethylene glycol monomethyl ether acetate (EGMEA)
Phosphoric acid	Potassium hydroxide	Propylene glycol monomethyl ether acetate (PGMEA)
Phosphoric acid/nitric acid/acetic acid		*n*-Methyl pyrrolidinone
Nitric acid		Dimethyl formamide
		Xylene

Table 2 SEMI Suggested Guidelines for Pure Water During Semiconductor Processing

Test parameter	Attainable	Acceptable	Alert	Critical[a]
Residue, ppm	0.1	0.3	0.3	0.5
TOC, ppm	0.020	0.050	0.100	0.400
Particulates, counts/liter	500	1000	2500	5000
Bacteria, counts/100 ml	0	6	10	50
Dissolved silica (SiO_2), ppb	3	5	10	40
Resistivity, megohm-cm	18.3	17.9	17.5	17
Cations, ppb				
Aluminum (Al)	0.2	2.0	5.0	*
Ammonium (NH_4)	0.3	0.3	0.5	*
Chromium (Cr)	0.02	0.1	0.5	*
Copper (Cu)	0.002	0.1	0.5	*
Iron (Fe)	0.02	0.1	0.2	*
Manganese (Mn)	0.05	0.5	1.0	*
Potassium (K)	0.1	0.3	1.0	4.0
Sodium (Na)	0.05	0.2	1.0	5.0
Zinc (Zn)	0.02	0.1	0.5	*
Anions, ppb				
Bromide (Br)	0.1	0.1	0.3	*
Chloride (Cl)	0.05	0.2	0.8	*
Nitrite (NO_2)	0.05	0.1	0.3	*
Nitrate (NO_3)	0.1	0.1	0.5	*
Phosphate (PO_4)	0.2	0.2	0.3	*
Sulfate (SO_4)	0.05	0.3	1.0	*

[a]Asterisk denotes that values are not assignable at this time.
Source: Ref. 4.

decrease device yield and reduce the reliability of finished product. Since design features are presently below 1 μm, particles smaller than 0.1 μm are of concern.

Numerous liquid chemicals are used in semiconductor device manufacturing. Their largest use is in wafer cleaning and photoresist stripping. The chemicals used in these processes can be divided into four main categories: deionized (DI) water, acids, bases and oxidizers, and solvents. DI water has the highest volume usage. Other commonly used chemicals are listed in Table 1 [3]. These chemicals typically have high particle concentrations compared to the gases used in wafer manufacturing and can deposit particles onto wafer surfaces. There is a need to produce cleaner chemicals. Unfortunately, many of these chemicals are extremely hazardous and corrosive, making particle measurement and control difficult.

Because semiconductor microcircuits are very sensitive to particulate contamination during their manufacture, acceptable concentrations of particles in process liquids have been projected. DI water has received the most attention. Guidelines for water purity developed by Semiconductor Equipment and Materials International (SEMI) are shown in Table 2, which indicates that the acceptable maximum particle concentration for DI water in 1992 was 1000/liter measured using microscopy at 1000× [4]. Reliable particle size definition at this magnification is approximately 0.3 μm. These specifications can be met by state-of-the-art systems incorporating many unit processes, including several filtration steps. These systems are capable of delivering large quantities of water containing 0.1 particle/ml (≥ 0.1 μm) [5]. Considerable effort has been expended to achieve systems with this capability.

Although standards for liquids other than water have not yet been developed, projections of required purity have been made by a number of authors. Figure 1 presents a compilation of a

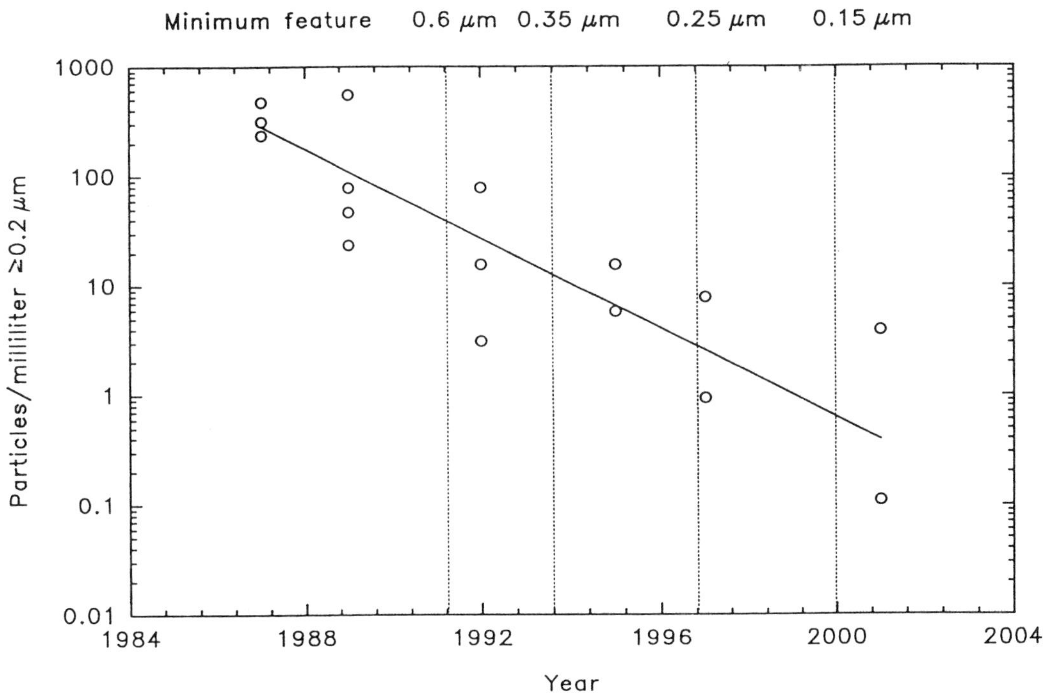

Figure 1 Projected chemical particle purity requirements.

number of these projections along with a mathematical fit. The fit indicates that the maximum allowable particle concentrations in 1993 are 1000/liter (≥ 0.5 μm) and that the allowable concentration is decreasing by a decade approximately every five years [1].

III. LIQUID PROPERTIES OF CONCERN

A. Pertinent Chemical Properties

Many chemical reagents are highly corrosive, especially those used in ULSI device manufacturing. Chemicals such as hydrochloric acid and hydrofluoric acid are extremely aggressive to metals. Due to this aggressiveness they are often stored and used in systems made of plastic, typically Teflon.* Unfortunately, plastics tend to have high permeability to these chemicals and permeating vapors can attack metal components in the surrounding area. Care must be taken to ensure that equipment, including particle counters and other instruments used in the vicinity of these chemicals, are properly protected from vapors.

Hydrofluoric acid, which is commonly used in semiconductor manufacturing, attacks quartz. Particle counter flow cells and optical windows made of quartz or glass rapidly deteriorate if they contact this chemical. Sapphire cells or cells made of other compatible materials must be used in any application using hydrofluoric acid.

B. Pertinent Physical Properties

Some process chemicals have high vapor pressures or high gas solubilities. These chemicals can form gas bubbles when aspirated through particle counter flow cells, where the pressure drop at the rated flow can be as great as 5 psig for liquids with viscosity of 30 centistokes. The bubbles are counted as particles by most particle counting equipment and large errors can result. Bubble formation can be prevented, or at least greatly reduced, by operating the counter at elevated pressure. For example, errors in counting particles in hydrogen peroxide, which is prone to oxygen bubble formation, can be minimized by measuring particle concentrations at pressures in excess of 35 psig [6,7].

A recently developed particle counter, an interferometer, is relatively insensitive to errors caused by bubbles, as described in Chapter 1 [8,9]. This instrument is able to distinguish between particles and bubbles under many operating conditions. An example of its insensitivity to bubbles, compared to darkfield light scattering instruments, is shown in Fig. 2 [10]. This figure shows particle concentrations measured by the two types of counters in hydrogen peroxide as pressure was decreased from 35 to 5 psig. It is known that hydrogen peroxide forms bubbles in this pressure range. Although the particle concentration of the hydrogen peroxide feed did not change, the darkfield scattering instrument measured rapidly increasing concentrations of "particles." The interferometer detected a constant concentration of particles at pressures greater than 10 psig. At 10 psig and above the interferometer was able to distinguish between bubbles and particles.

Optical particle counters use the difference between the refractive index of the particle and that of the surrounding liquid to determine particle size. Counters are usually calibrated using monodisperse polystyrene latex (PSL) spheres in water. If the refractive index of a particle being

*Teflon is a registered trademark of E. I. duPont de Nemours, Inc., Wilmington, DE.

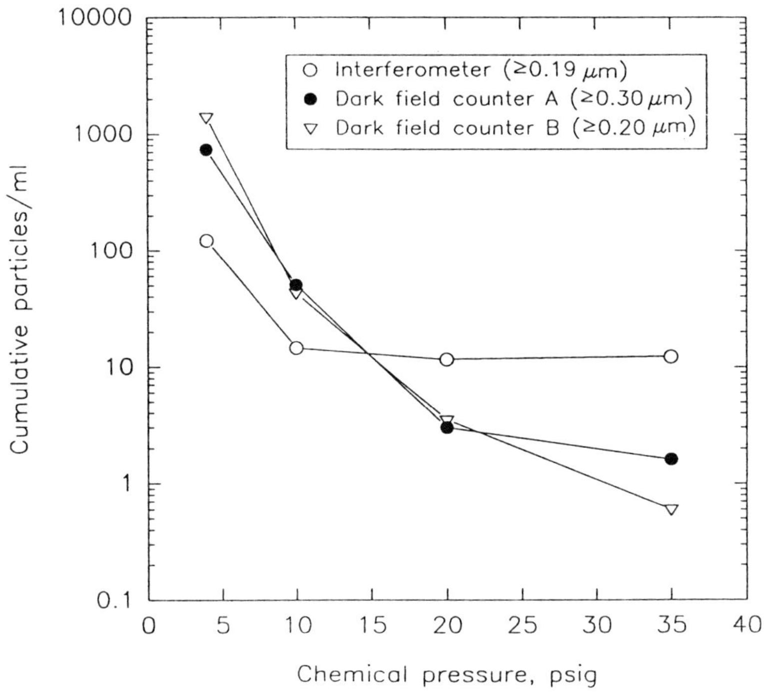

Figure 2 The effect of hydrogen peroxide pressure on measured particle concentrations.

measured differs from that of PSL or if the refractive index of the liquid is different from that of DI water, the particle will be improperly sized. The sizing error becomes largest when the ratio of the refractive index of the particle to that of the liquid, called the index contrast, approaches unity. Because a counter's response to a combination of particle and liquid refractive indices is a function of its optical design, every counter model responds differently. Figure 3 presents the response of one darkfield counter to particle-liquid combinations with different index contrasts [11].

Refractive indices of some liquids and solids commonly used in semiconductor processes are shown in Table 3. Table 4 shows the sizing errors that occur for one darkfield light scattering instrument when particles with refractive index of 1.59 are measured in a variety of liquids. In some cases, large errors result [7].

The sizing errors attributed to differences in index contrast are claimed to be smaller for brightfield instruments, especially if the index contrast is very close to 1. Under some conditions, the sizing error by a darkfield counter is three times that of a brightfield counter [12].

The sizing errors that occur when particles are measured in process fluids cannot be determined unless the refractive indices of both the fluid and the particles are known. The fluid refractive index can easily be determined. However, the refractive indices of the various types of particles present in the fluid are usually unknown. Therefore, the actual size of the particles can only be estimated.

Another fluid property which affects counter performance is viscosity. Many optical cells in particle counters have a very fine flow capillary. If the fluid has a high viscosity, liquid flowing

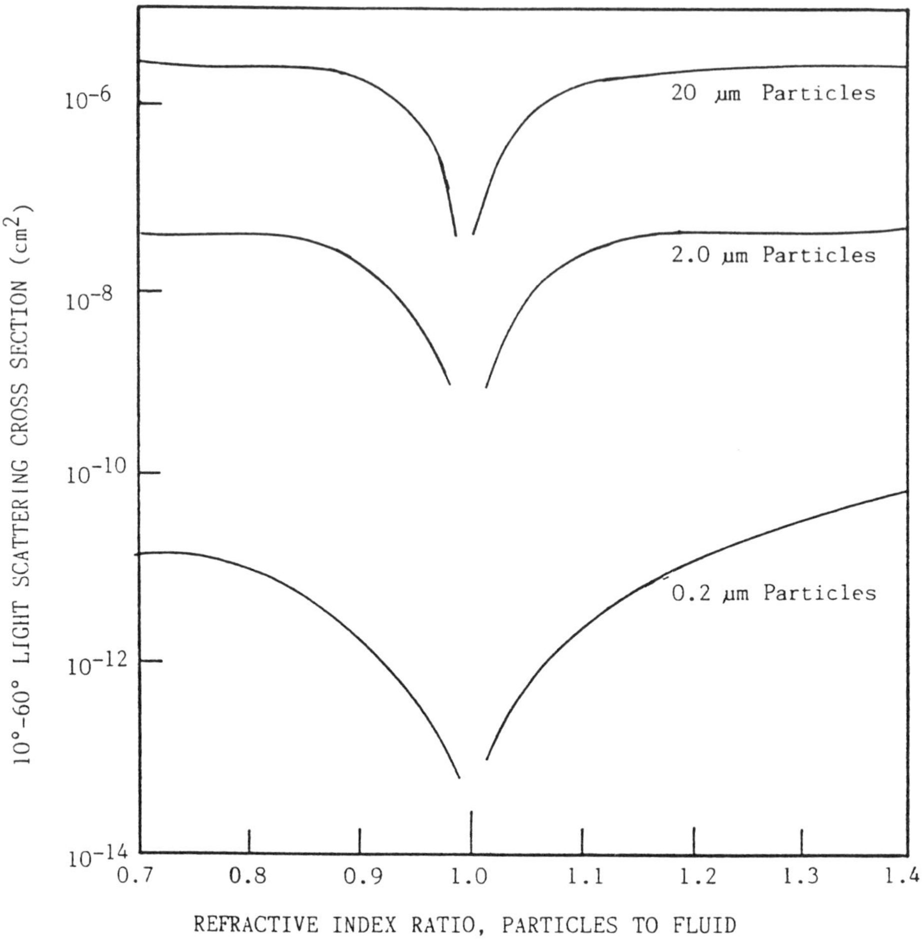

Figure 3 The effect of refractive index contrast on scattering cross section.

through the capillary can create a significant pressure drop. In many counters, it is necessary to reduce the flow rate through the counter to prevent excessive pressure drop. Reducing the flow rate results in reduced statistical accuracy of the measurement, unless excessive sample measurement times are used [13].

The concentration of particles in semiconductor process fluids also influences particle counter accuracy. In chemicals containing very low particle concentrations, instrument background noise can result in large measurement errors. In addition, some instruments require extremely long sampling times to achieve statistically significant data. When long sampling times are used, the ability to observe intermittent particle upsets can be lost. Therefore, it is desirable to use a particle counter with a high sampling rate. Unfortunately, the minimum particle size that an instrument can detect and the instrument sampling rate are strongly related. The smaller the minimum detectable size, the lower the sampling rate [14].

Selection of the best counter for a semiconductor application is not straightforward. A number of factors need to be considered including minimum detectable particle size, sample flow rate, sensitivity to differences in refractive index, background noise and sensitivity to bubbles.

Table 3 Refractive Indices of Commonly Used Semiconductor Materials

Gases[a]	Refractive index
Nitrogen	1.00
Oxygen	1.00
Argon	1.00
Hydrogen	1.00
Hydrochloric acid	1.00
Liquids	**Refractive index**
Hydrofluoric acid (50%)	1.29
Ammonium fluoride (40%)	1.33
Ammonium hydroxide (30%)	1.33
Methanol	1.33
Water	1.33
Acetone	1.36
Hydrogen peroxide (30%)	1.36
Acetic acid (concentrated)	1.37
Methyl ethyl ketone	1.38
2-Propanol	1.38
n-Butyl acetate	1.39
Nitric acid (50%)	1.40
Hydrochloric acid (38%)	1.41
Phosphoric acid (85%)	1.42
Sulfuric acid (96%)	1.46
Trichloroethane	1.48
Trichloroethylene	1.48
Tetrachloroethylene	1.50
Toluene	1.50
Xylene	1.50
Solids	**Refractive index**
Silicon	3.90
Germanium	3.50
Carbon	2.00
Silicon dioxide	1.45
Polystyrene latex (PSL)	1.59

[a]The refractive indices of these gases are nonzero only in the fourth decimal place at STP.

Table 4 Optical Particle Counter Sizing Error Attributable to Index Contrast Differences

Particle diameter (in microns) in H_2O, 28% NH_4OH	Particle diameters (in microns) with equivalent scattering intensity in various solutions			
	49% HF	30% H_2O_2	37% HCl	96% H_2SO_4
0.3	0.28	0.31	0.36	0.40
0.5	0.46	0.52	0.62	0.72
1.0	0.87	1.10	1.80	2.30
5.0	5.20	5.00	6.00	7.90

IV. PARTICULATE CONTAMINATION PROBLEM AREAS

Several steps are taken during chemical production to achieve the low particle concentrations required by the semiconductor industry. Liquid chemicals other than water are manufactured using only clean raw materials, and the chemicals are filtered to ensure cleanliness. They are packaged in bottles or drums specially prepared from selected plastics (typically either ultrahigh molecular weight polyethylene or Teflon) to prevent recontamination of the chemical [15]. Unfortunately, even specially prepared containers can add particles to the chemicals both during filling and transportation to the locations where they will be used. Note that 1 ppb of material with a density of 1 g/cc, if broken into uniform 0.5-μm-diameter spheres, adds 15,000 particles/ml to the chemical. This concentration is far greater than the industry requirements detailed in Fig. 1.

Since bottled chemicals often cannot meet the purity requirements of semiconductor processing, the chemicals can be supplied to the points of use (POUs) by chemical delivery systems. Delivery systems transport chemicals supplied in bulk to the POUs within the facility. These systems are capable of supplying chemical with very low particle concentrations because they supply filtered chemical directly to the POUs without the need for intermediate storage. Systems of this type are capable of meeting present semiconductor particle purity needs [16–19].

A simplified schematic of a chemical delivery system is shown in Fig. 4. The system consists of a chemical delivery module (CDM) and distribution tubing. The purpose of the CDM is to draw chemical from a bulk source, pressurize the chemical, and force it through the delivery tubing to the POUs. The CDM contains pumps, filters and controls to assure the flow of clean chemical to the POUs. All of the wetted parts of the CDM and the distribution tubing are usually made of fluoropolymers with smooth internal surfaces.

The most common use of chemicals in semiconductor processing is for cleaning particulate, metallic, organic and oxide contaminants from wafer surfaces [20,21]. Cleaning is performed many times during the manufacture of microcircuit devices. Each clean usually requires a sequence of chemical solutions to remove different types of contaminants. A commonly used clean is a modified RCA clean consisting of the steps shown in Table 5.

The modified RCA clean shown in Table 5 consists of four chemical solutions, each designed to remove a particular type of contaminant. Organic removal by a combination of sulfuric acid and hydrogen peroxide is the first step because the presence of organic films can render the surface hydrophobic and prevent other cleaning solutions from reaching the surface. A dilute solution of hydrofluoric acid is used to remove the thin layer of native oxide which is always present on wafer surfaces. Inorganic contaminants trapped in the native oxide are also removed during this step, resulting in a surface with very low metallic contamination. Because the oxide removal process tends to add particles to the wafer surface, it is often followed by an ammonium hydroxide/hydrogen peroxide step. This step is effective in both removing particles, presumably by etching the wafer surface from under them, and regrowing a thin oxide film. The oxide "passivates" the surface, making it hydrophilic and less susceptible to organic and particulate contamination. The hydrochloric acid/hydrogen peroxide step removes any residual

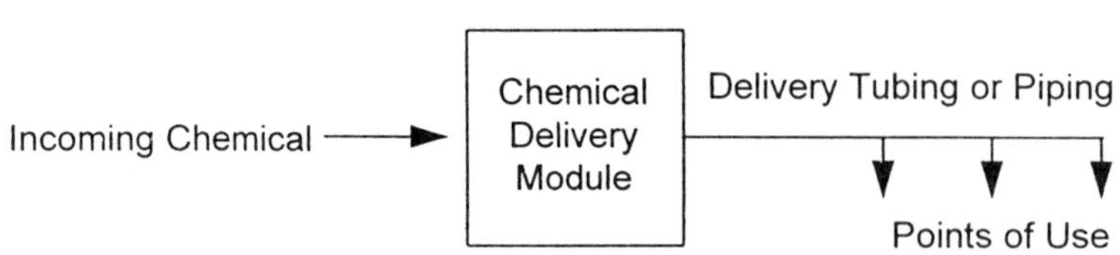

Figure 4 A simplified chemical delivery system schematic.

Table 5 Example of a Wafer Cleaning Process

Clean step	Purpose	Chemicals used	Typical chemical ratio
1	Organic removal	96% Sulfuric acid	4 parts
		30% Hydrogen peroxide	1 part
2	Rinse	DI water	
3	Oxide removal	49% Hydrofluoric acid	1 part
		DI water	100 parts
4	Rinse	DI water	
5	Particle removal	28% Ammonium hydroxide	1 part
	Oxide regrowth	30% Hydrogen peroxide	1 part
		DI water	5 parts
6	Rinse	DI water	
7	Metals removal	37% Hydrochloric acid	1 part
		30% Hydrogen peroxide	1 part
		DI water	5 parts
8	Rinse	DI water	
9	Dry		

metals and prevents them from plating back onto the surface by complex formation. Used together these cleaning solutions remove most wafer contaminants, even when the source and composition of the contaminants are unknown.

Wafer cleaning processes are most commonly performed in either wet sinks or centrifugal spray processors. Wet sinks, also called wet benches or immersion processors, immerse a cassette containing multiple (typically 25 or 50) wafers in a series of chemicals solutions in separate baths. The cassette remains in each chemical bath for a predetermined time, and is rinsed in a DI water bath before being transferred to the next chemical solution. An example of a fully automated wet bench is shown in Fig. 5 [22].

Wet sink processors offer flexibility in choices of cleaning solutions and process times. In addition, baths may be heated or agitated by mechanical movement, ultrasonic energy, or nitrogen bubbling through the solution. Filtered recirculation of chemicals improves bath cleanliness and reduces chemical usage. Recent development of automated processors with robotic handling has improved system performance [23]. Cassetteless transport processors have the advantage that less chemical is carried between baths, therefore contamination is reduced.

The disadvantages of wet sink processors are their high chemical usage and high floor space requirements. Because the contaminant concentrations in the bath increase with time, the chemicals must be monitored for purity and replaced or replenished to ensure the effectiveness of the process. The buildup of particulate contaminants is often controlled by recirculating the chemical through filters using a system like that shown in Fig. 6. Chemical from a bath is drawn off by a pump and passed through a microporous membrane filter. Particulate contaminants are removed by the filter, and the chemical is returned to the bath.

Centrifugal spray processors (Fig. 7) are automated processors that perform an entire cleaning process in a closed chamber [24]. A cassette of wafers on a turntable is rotated past a stationary spray post. Pressurized chemical solutions are sprayed uniformly at the wafers. Used solutions are continuously drained from the bottom of the chamber. The entire cleaning

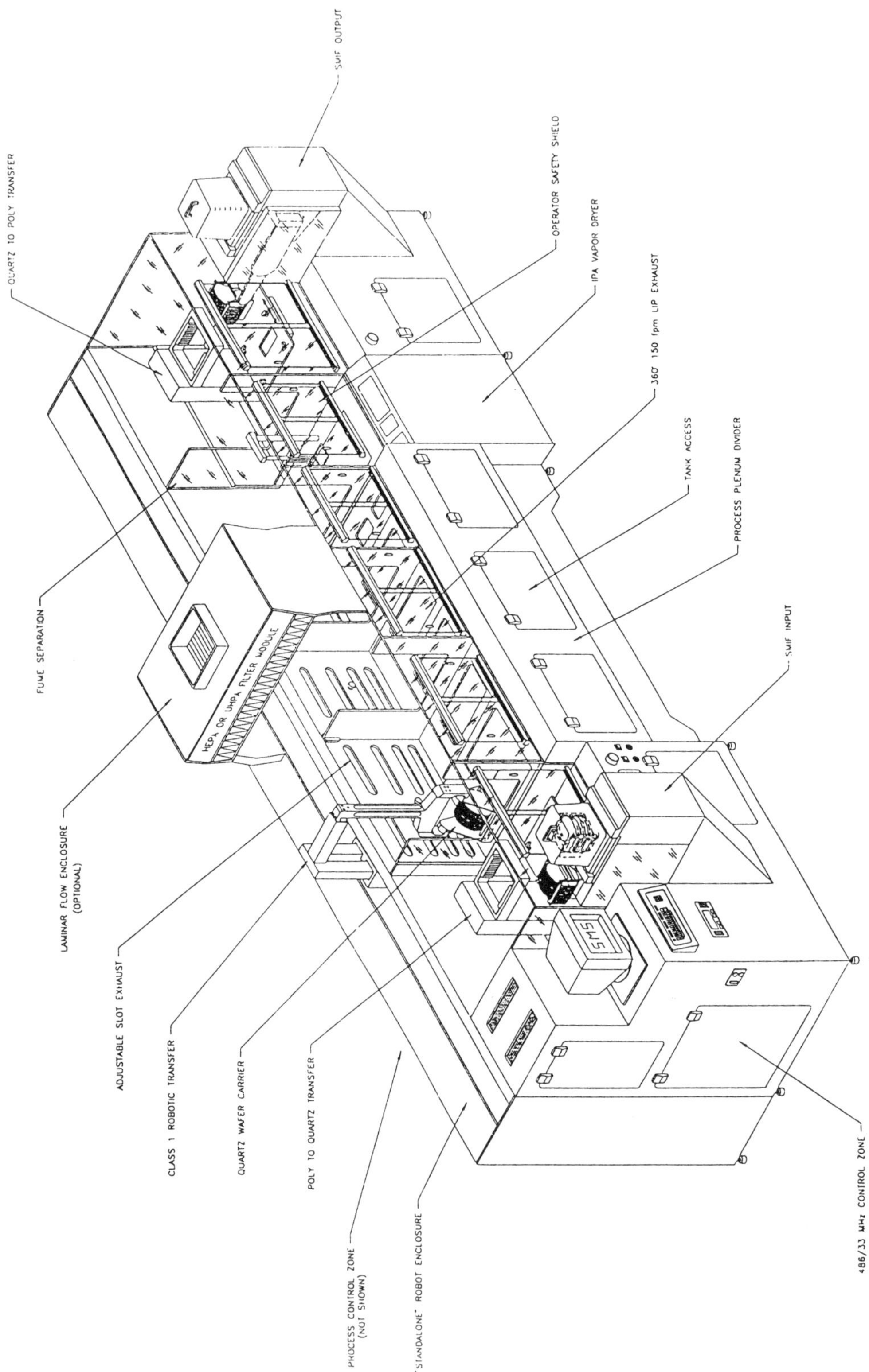

Figure 5 Example of fully automated wet sink cleaning equipment.

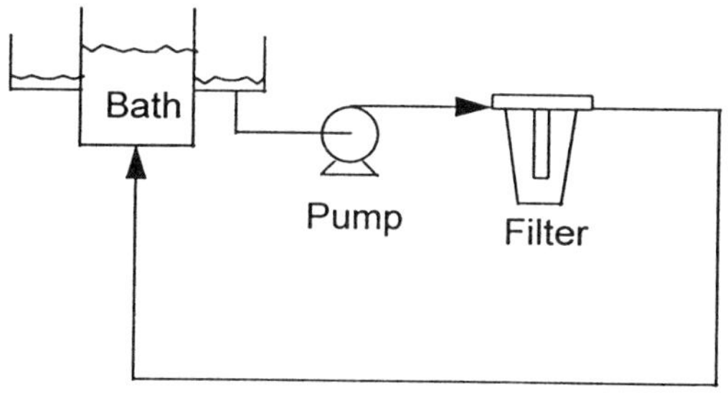

Figure 6 Recirculating etch bath flow schematic.

sequence, including rinsing and drying, is performed without removing the wafers from the processor. Because the chamber is enclosed, chemical cleanliness and operator safety are improved. In addition, wafers see only fresh chemical. Any contaminants, including particles, which are removed from the wafer or generated during the cleaning process, are isolated from the wafer [25].

The filters used to remove particles from semiconductor process fluids are usually all-Teflon microporous membrane cartridges. Microporous membrane filters are used because of their

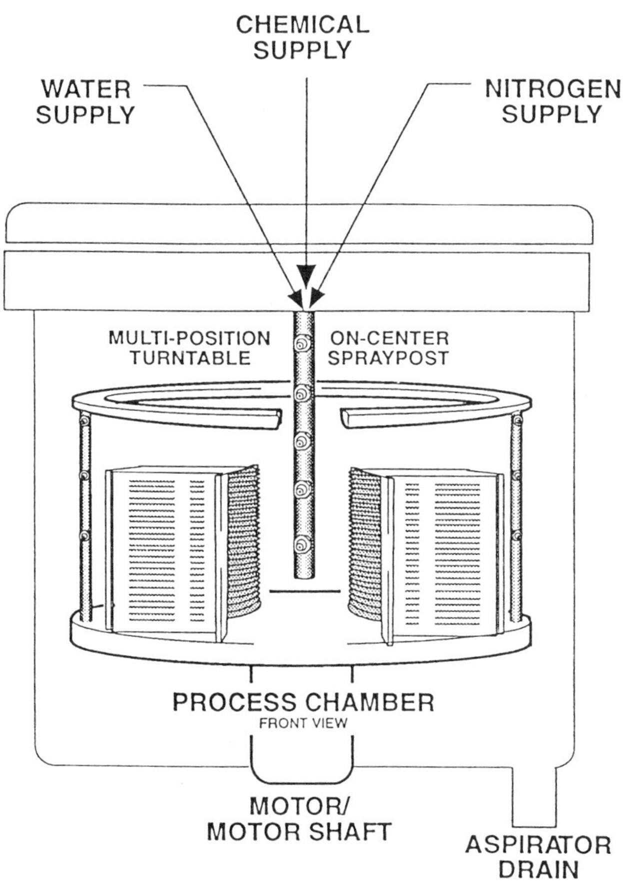

Figure 7 Spray processor schematic.

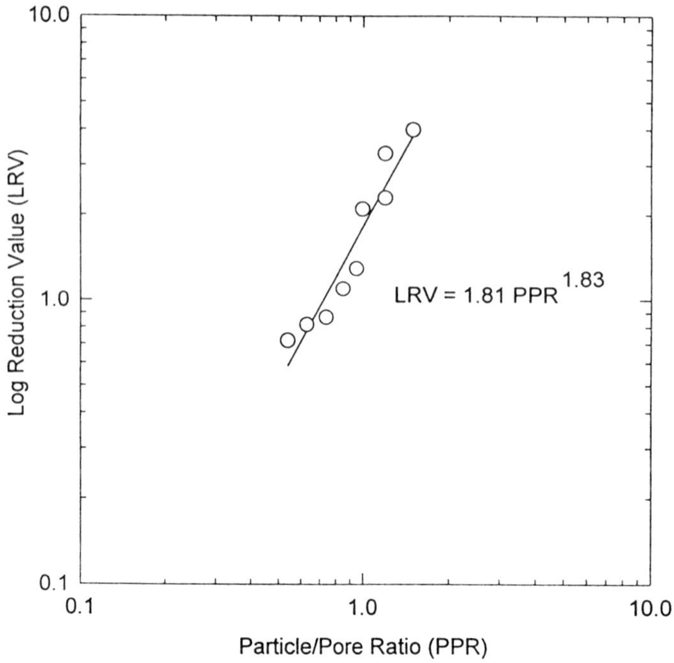

Figure 8 The relationship between filter retention and the ratio of particle diameter to pore diameter (PPR).

ability to retain small particles [26]. Teflon is often the material of choice because of its chemical inertness. Teflon filters are very hydrophobic and must be subjected to a time-consuming prewetting process prior to use in aqueous chemicals. Some chemicals, like concentrated ammonium fluoride, will spontaneously dewet Teflon filters even though they have been prewet. In these cases, materials other than Teflon must be used or the surface of the filter must be modified to make it hydrophilic [27].

The ability of microporous membrane filters to remove particles from liquids is described by its rated pore size. This rating scale is based on the ability of filters to quantitatively remove microorganisms. Quantitative retention is defined as having a log reduction value (LRV) ≥ 10, where LRV is defined as

$$\text{LRV} = \log_{10}\left[\frac{\text{Incoming concentration}}{\text{Filtrate concentration}}\right]$$

The relationship between particle removal and LRV is

Particle removal (%)	LRV
90	1
99	2
99.9	3
99.99999999	10

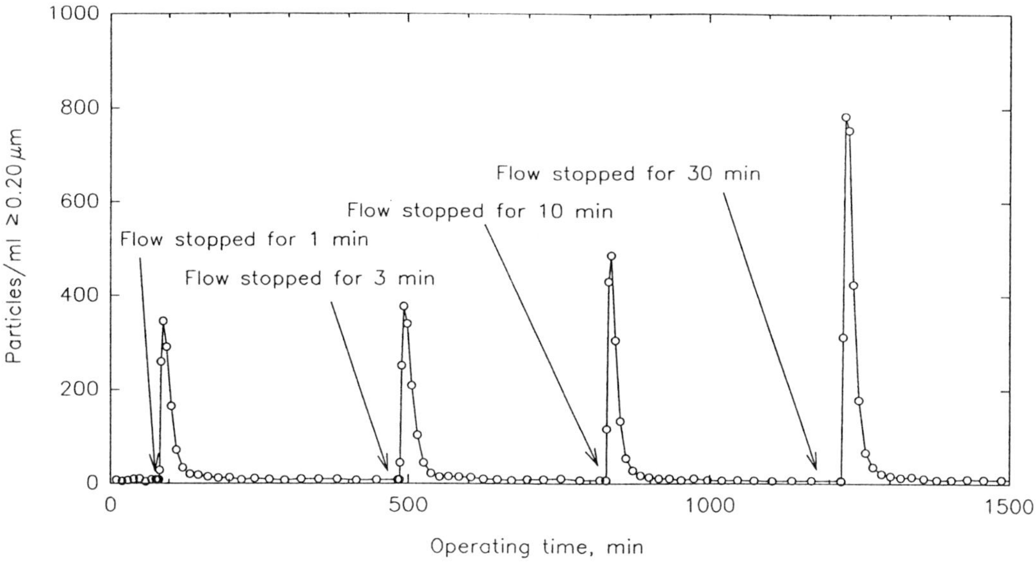

Figure 9 The effect of hydraulic shocks on particle concentrations in sulfuric acid in a system with 0.2-μm filters.

Since the microorganism size ratings were instituted, hard, spherical particles made of polystyrene latex (PSL) have been developed. Testing of filters with these particles has demonstrated that filters remove 99–99.9 of the particles with diameters equivalent to the pore size rating, as shown in Fig. 8. Quantitative retention of PSL particles is obtained when the particle diameter is approximately 2.5 times the rated pore size [28].

Particles downstream of filters can come from sources other than incomplete retention. Filter cartridge materials can shed particles and particles which have been previously captured can be released when filters are subjected to hydraulic shocks. Filter shedding can be the major contributor to particle concentrations downstream of a filter. Hence, filters developed for use in the semiconductor industry are specially prepared to meet its extremely stringent requirements.

Large changes in the flow rate through filters create a sufficient hydraulic shock to cause release of particles. Hydraulic shocks can increase particle concentrations by a factor of 100 or more, as shown in Fig. 9. The magnitude of these particle spikes can be greatly reduced by maintaining a stabilizing flow through the filters at all times. This mode of filter operation, termed Stabilized Distribution* by FSI International, is very effective in reducing particle concentrations, as shown in Fig. 10 [1].

Measurement of particle concentrations in semiconductor process chemicals poses several problems. Because the chemicals usually contain very low particle concentrations, it is difficult to transport samples from the sampling location to the particle counter sensor without influencing the measurement. On-line sampling, in which the chemical is passed directly through the sensor, is the preferred sample collection method. Unfortunately, this method requires a system pressure sufficient to prevent bubble formation in the chemical. Hence, it can be used to measure particles in chemical delivery systems and sometimes in recirculating sink systems downstream of the

---

*Stabilized Distribution is a trademark of FSI International, Chaska, MN.

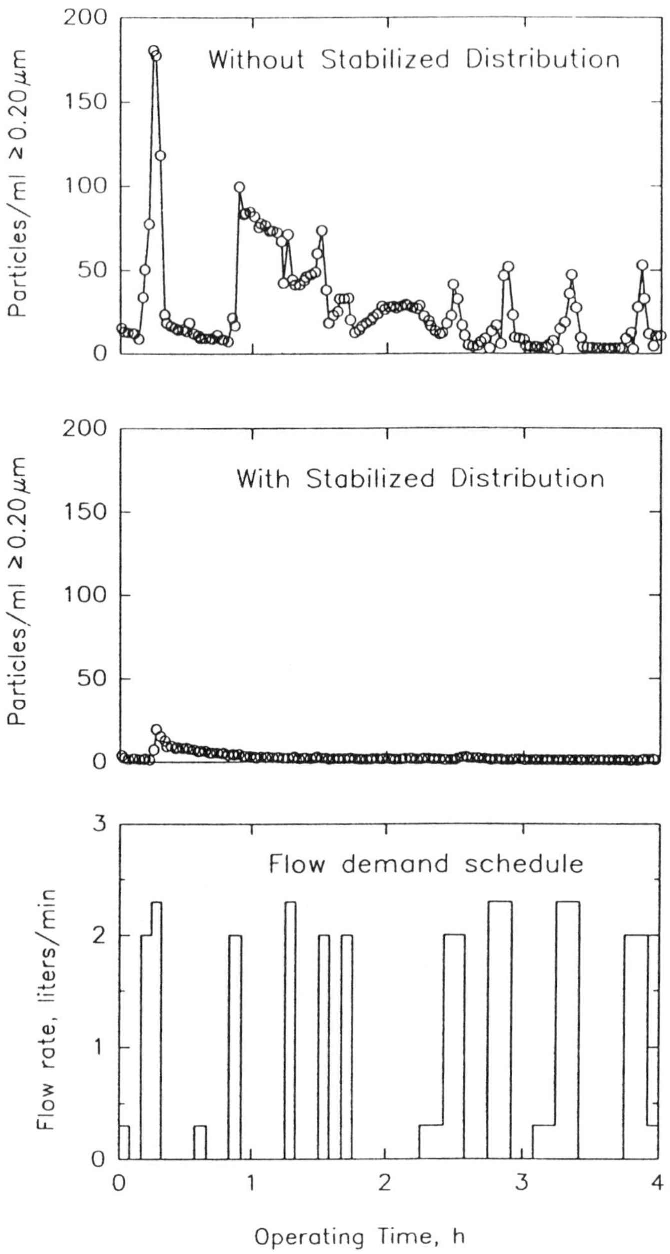

Figure 10 The effect of stabilized distribution on particle concentrations in H_2SO_4 in a system with 0.1-μm filters.

circulating pump. It cannot be used to measure particle concentrations in wafer cleaning baths or in chemical containers.

Particle concentrations in chemical baths and containers can be measured using in situ or batch samplers. Although these devices have been discussed in Chapter 6, some precautions are repeated here. Proper operation of these devices is required to minimize errors due to particle loss or artifact generation in the sample handling process. Sample feeder components must be compatible with all solutions to be measured to prevent degradation by aggressive chemicals. The components of concern include not only structural elements, but such items as seals, O-rings, and valve elements. Where hazardous materials are characterized, a safe handling and disposal protocol should be established before any measurements are made.

Systems using either batch or sidestream sampling must be designed so that the sample is truly representative of the population being sampled. Isokinetic sampling should be used where practical and isoaxial sampling should be used where possible. Sample lines carrying process fluids to the particle counter cannot include long horizontal runs where the liquid flow is laminar, since gravitational losses can occur. If extreme variations in flow occur, the sample line may collect slugs of liquid or foam that may be interpreted as particles.

When batch samples are measured, sample containers must be clean enough that no particles are flaked or abraded from the container walls. When the containers are taken to a remote location for particle measurement or stored for any period of time, the sample liquids must be mixed sufficiently to suspend the particles randomly before batch sampling is initiated. Bubbles produced by mixing should be removed before measuring takes place. The batch sampler's ability to feed a sample of known volume at a known, fixed flow rate to the particle counter should be verified. A tare volume may be required between samples to ensure that particle counts are not inaccurate because of residue from a previous sample.

Spent process chemicals must be disposed of in a manner acceptable to the local environmental pollution control agencies. Often samples from CDMs can be returned to a chemical tank in the system, thus eliminating disposal issues and recovering the valuable material. Other samples can usually sent to drain. Solvents, acids, bases and fluoride-containing liquids require separate drains. Even after recycling, some waste materials can only be disposed of by packing in drums for removal by a qualified hazardous material handling organization. Typical cost for disposal of a single "hazardous material" drum is approximately $500.

V. CONCLUSIONS

Because of the stringent particle purity needs of the semiconductor industry, particle measurement and control present unique and difficult problems. This need for extremely pure chemicals continues to drive the development of particle measurement instruments. Current particle counters can detect particles with diameters as small as 0.050 μm in water and 0.065 μm in chemicals. State-of-the-art chemical delivery systems with Stabilized Distribution are capable of delivering chemicals which meet the projected 1997 semiconductor industry's requirements.

REFERENCES

1. Deal, D. B. and D. C. Grant, Chemical delivery systems: past, present and future, Proceedings of the Electrochemical Society, Pennington, NJ, 1994.
2. Ramakrishna, V. and H. D. Fitch, Strategic road maps (defect learning), *Proceedings of CleanRoom 93*, Witter, Flemington, NJ, 1993, pp. 83–99.
3. Zahka, J. G. and D. C. Grant, Predicting the performance efficiency of membrane filters in process liquids based on their pore-size ratings, *Microcontamination, 9*(12):23–29 (1991).
4. *SEMI Pure Water Guidelines,* Semiconductor Equipment and Materials International, Mountain View, CA, 1992.
5. *Pure Water Specifications and Guidelines,* Balazs Analytical Laboratories, Sunnyvale, CA, 1994.
6. Dillenbeck, K., Advances in particle counting techniques for semiconductor process chemicals, *Microcontamination 5*(2):30–38, 65 (1987).
7. Grant, D. C., Improved methodology for measurement of particle concentrations in semiconductor

process chemicals, *Particles in Gases and Liquids* 1: *Detection, Characterization and Control* (K. L. Mittal, ed) Plenum Press, New York, 1989, pp. 121–134.

8. Blackford, D. and D. C. Grant, Evaluating a novel liquid-borne particle counter based on interferometry, *Microcontamination 11*(2):27–32, 66–67 (1992).
9. Blackford, D. B., J. S. Batchelder, M. Taubenblatt and D. C. Grant, Interferometric detection of particles in semiconductor process chemicals, Proceedings of the 11th ICCCS Meeting, London, 1992.
10. Grant, D. C., Evaluating a novel liquid-borne particle counter based on interferometry: experimental testing, *Microcontamination 11*(3):37–43 (1992).
11. Knollenberg, R. G., The importance of refractive index in evaluation of liquid and surface microcontamination measurements, *J. Environ. Sci. 30*(2):50–58 (1987).
12. Batchelder, J. S. and M. A. Taubenblatt, Evaluating a novel liquid-borne particle counter based on interferometry: comparing detection of nearly-matched particles, *Microcontamination 11*(4):35–40 (1993).
13. Batchelder, J. S. and M. A. Taubenblett, Measurement of the size and refractive index of a small particle using the complex forward-scattered electromagnetic field, *Appl. Opt. 30*(33):4972–4979 (1991).
14. Grant, D. C., Measurement of particle concentrations in central chemical delivery systems, *J. Environ. Sci. 23*(4):32–37 (1990).
15. Hackett, T. B. and K. D. Dillenbeck, Examination of polyethylene as a packaging material for electronic chemicals, Proceedings of Microcontamination 93, Canon Communications, Santa Monica, CA, 1993, pp. 19–28.
16. Grant, D. C. and W. R. Schmidt, Particle performance of a central chemical delivery system, 7th Annual Microelectronics Technical Symposium, Millipore Corporation, Bedford, MA, 1989.
17. Gruver, R., R. Silverman and J. Kehley, Correlation of particulates in process liquids and wafer contamination, Proceedings of the 36th Annual Technical Meeting, Institute of Environmental Sciences, Mt. Prospect, IL, 1990, pp. 312–315.
18. Grant, D. C. and D. P. Min, The effect of operational mode on particle concentrations in bulk chemical delivery systems, Proceedings of the 37th Annual Technical Meeting, Institute of Environmental Sciences, Mt. Prospect, IL, 1991, pp. 796–801.
19. Wang, F. C., D. Charest, J. Campaneria, L. S. Wai, I. Lye, S. K. Yeo, C. K. Sing, K. Y. Tey, D. Kaur, D. C. Grant, D. Smith and P. Palm, Design, certification and verification of technology for delivering sub-ppb, low-particle chemicals to semiconductor cleaning baths in wafer fabs, Proceedings of the 40th Annual Technical Meeting, Institute of Environmental Sciences, Mt. Prospect, IL, 1994.
20. Burkman, D. C., C. A. Peterson, D. B. Deal and D. C. Grant, Wet chemical processes: cleaning processes based on aqueous solutions, *Handbook of Silicon Wafer Cleaning Technology* (W. Kern, ed.), Noyes, Park Ridge, NJ, 1993, pp. 111–146.
21. Kern, W., The evolutions of silicon wafer cleaning technology, Proceedings of the First International Symposium on Cleaning Technology in Semiconductor Device Manufacturing, Vol. 90-9, The Electrochemical Society, Inc., Pennington, NJ, 1990, pp. 3–19. Figure provided by R. Novak, SubMicron Systems, Inc., Wayzata, MN.
22. Menon, V. B. and R. P. Donovan, Reviewing particle control issues associated with silicon wafer cleaning, *Microcontamination 8*(11):29–34, 66 (1990).
23. Burkman, D. C., Optimizing the cleaning procedures for silicon wafers prior to high temperature operations, *Semicond. Int. 4*(7):103–114 (1981).
24. Kern, W., *J. Electrochem. Soc. 137*(6):1887–1892 (1990).
25. Grant, D. C. and M. Heisler, Evaluation of a novel dual PTFE asymmetric membrane cartridge for filtration of sulfuric acid, Proceedings of Microcontamination 93, Canon Communications, Santa Monica, CA, 1993, pp. 60–70.
26. Zahka, J., V. A. Anantharaman, M. Carroll and K. Vakhshoori, Characterization of filters used in recirculated buffered oxide etch baths, Proceedings of the 38th Annual Technical Meeting, Institute of Environmental Sciences, Mt. Prospect, IL, 1992, pp. 309–318.
27. Zahka, J. G. and D. C. Grant, Predicting the performance of membrane filters in process liquids based on their pore size rating, *Microcontamination 9*(12):23–29 (1991).

12

A Systematic Approach for Particle Counting Petroleum Oils

Barry M. Verdegan, Brian W. Schwandt, and Christopher E. Holm

Nelson Industries, Inc., Stoughton, Wisconsin

I. INTRODUCTION

Petroleum-based oils are used as lubricants, fuels, and feedstocks in industry. Particulate contaminants in oil reduce its economic value and render it unsuitable for use in many applications. For example, particles in oil cause wear. Wear rates in engines and hydraulic systems increase with the concentration of particles in their oil [1–3]. Hydraulic components, which use petroleum oil for lubrication and to transfer power, are particularly sensitive to these contaminants. Solid particulate contamination has been estimated to account for 70–85% of hydraulic component wear problems [4]. In aircraft hydraulic oil, the concentration of particles larger than 10 μm should be maintained below 1000/ml [5]. In high-pressure applications, lower concentrations of even finer particles must be maintained. Filtration and separation are used to control particulate contamination. Particle counting is the primary analytical tool for determining filter efficiency and for rating filters. It is also useful for monitoring contamination levels in used oil.

Despite its importance to industry, particle count data have a reputation for being unreliable. Though individual laboratories may generate reliable, reproducible particle count results, it is not unusual for poor agreement to exist between different laboratories or even between different instruments within the same laboratory. There are three reasons for this. First, problems can arise from instrument calibration. Improperly calibrated instruments or instruments calibrated using inappropriate procedures produce unreliable data. Second, particle counter operators may not be properly trained. Particle counting is not a normal part of academic curricula. As a result, substandard analytical techniques may be used and operators may be unable to detect problems. Finally, the particle counter itself is a source of variability. Particle counters are not primary sizing instruments; rather, they allow users to make rapid comparisons of particle concentrations and size distributions of samples to a standard. To do this, a variety of operating principles, light sources, and data processing systems are utilized and variability is introduced into the results. In some cases, the instrument design produces artifacts that aren't obvious except under unusual circumstances. This chapter will focus on optical particle counters, since these are the primary tools of the industry.

Oil samples present challenges not encountered in the analysis of aqueous samples. The viscosity, density and color of oil varies over a broad spectrum. The types of particles being counted are equally diverse, as shown in Fig. 1. These micrographs were obtained using a

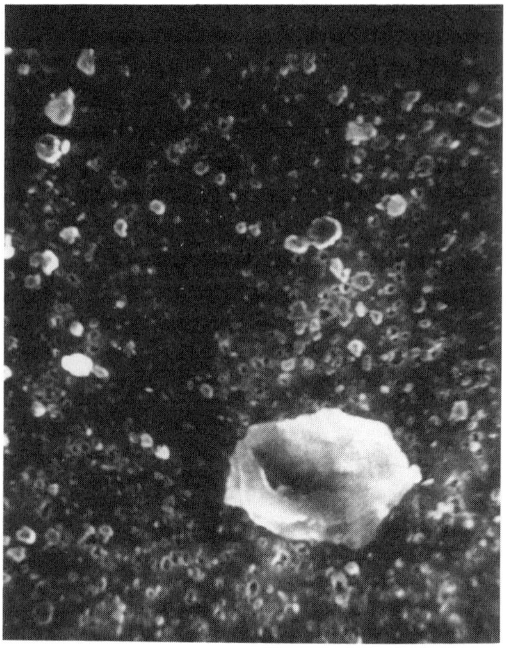

(a)

(b)

Figure 1 Scanning electron micrographs of contaminants from (a) used engine oil from a city bus and (b) AC fine test dust. (Courtesy of Nelson Industries, Inc.)

scanning electron microscope. Contaminants from the used engine oil of a city bus are shown in Fig. 1a. These include large amounts of submicron soot, as well as lesser amounts of wear metal, silicate, and oil additive residue particles. The largest particle in Fig. 1a is 14 μm in length. Figure 1b shows AC fine test dust which is used to evaluate oil filters. These particles are primarily oxides of silicon, aluminum and iron. The largest particle shown in this figure is 10 μm in length. Compared to the used oil sample, these particles tend to be more angular in shape and more large particles are present. Both used oil and AC fine samples are polydispersed (i.e., have a broad particle size distribution) and contain a variety of particles with differing refractive indices, colors, and shapes. As will be discussed, particle sensors are sensitive to the size distribution and optical properties of samples. Over time, the instruments and techniques used for oil analysis have evolved in order to deal with the complex nature of these samples.

As complex as the samples are, an even greater challenge for oil analysis is the manner in which the results are used. Particle count data from different instruments and labs are routinely compared. Oil filter performance and hydraulic system cleanliness levels are evaluated using particle counters. Data from vendors and customers are routinely compared as part of quality control programs. Within the same lab, the results from different instruments are often used to generate a single filter efficiency result. Therefore, agreement among different sources of data is essential. If the results don't agree, large sums of money can be lost due to rejected shipments, trying to reconcile differences, or erroneous filter ratings. For these reasons, the requirements for oil analysis can be more demanding than in aqueous applications.

This chapter presents a basis for developing a systematic methodology for particle count analysis of petroleum oil samples. Since conditions and needs differ from lab to lab, no single methodology is the "best" for all purposes. However, general principles have evolved which facilitate the development of protocols by individual laboratories. Interferences and sources of error are discussed, as well as methods for recognizing their effects and minimizing them. Both instrumental and analytical considerations are addressed. Emphasis is placed on those considerations unique to oil analysis.

II. INSTRUMENTATION

An optical particle counting system consists of a fluid delivery system, particle sensor, and data acquisition and analysis circuitry. Each must be properly designed and operated in order to yield *reproducible data representative of the particle size distribution of the sample* being analyzed. The operational characteristics and limitations of the instrument should be determined before use. These include the operating principle, noise level, flow rate limitations, resolution, coincidence error limits, and calibration curve. Failure to work within the instrument's limitations can result in the generation of misleading data. Therefore, it is important for operators to understand the meaning and significance of each of these characteristics.

A. Operating Principle

Optical particle counters used for oil analysis are single-particle counters using either light extinction or light scattering principles. The term *single-particle counters* refers to the fact that particles are individually sized and counted. This is in contrast to quasi-elastic light scattering or Fraunhofer diffraction systems which analyze populations of particles.

The operating principles of extinction and scattering sensors are illustrated in Fig. 2a and 2b, respectively. In each case, a particle passing in front of the light source perturbs the light impinging on the detector and generates a voltage pulse signal. In light extinction, the perturbation reduces the amount of light reaching the detector. The amount of light blocked increases with particle size. In light scattering, the perturbation is an increase in light intensity due to

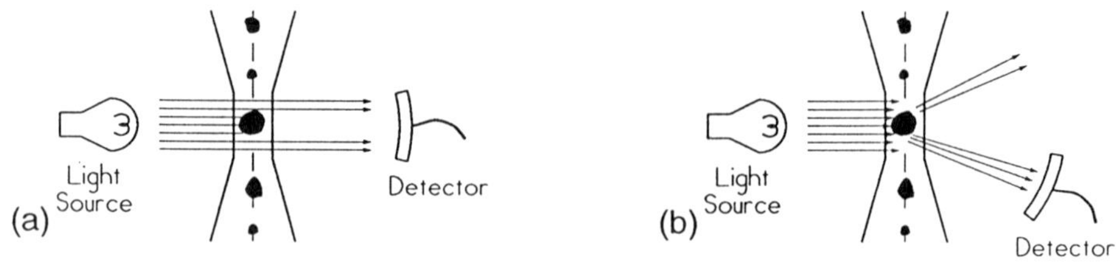

Figure 2 Operating principles for optical particle sensors: (a) light extinction and (b) light scattering.

forward scattering of the light. Scattered light intensity increases with particle size. Since particles simultaneously block and scatter light, sensors do not rely exclusively on a single operating principle. Extinction sensors measure some scattered light, and vice versa. Particle counters are calibrated with regard to size to account for this. This is also why irregularities are often observed in calibration curves. This is especially true at transitional sizes between 1 and 5 μm. Instrument designers minimize contributions from competing processes by selecting appropriate wavelengths of light and light collection angles.

In light extinction, the relationship between pulse height and particle size may range from first to second order depending on the size of the particle [6–7]. The apparent size is also influenced by the particle and fluid refractive indices and light absorbance properties, as well as sensor geometry and the wavelength of light utilized [6–7]. As a result, particles having the same size but differing optical properties are sized differently. In general, extinction sensors are useful for particles larger than 1–2 μm. At smaller sizes, the signal is weak and scattered light makes an important contribution to the signal. However, the size range and versatility of this sensor makes it a common choice for oil analysis.

In light scattering, the relationship between pulse height and size is more complex. Depending on the particle size and the system under consideration, the relationship may range from first to sixth order [7–8]. The same factors influencing apparent particle size in light extinction influence light scattering, but to a greater degree. Highly absorbing (black) and highly reflective particles tend to be sized inaccurately. Thus, scattering sensors should not be used for the analysis of used oil, which contains wear metal particles and soot. For particles smaller than 1–2 μm, light scattering is the only viable option. Scattering sensors can also be used for larger particles, although they tend to utilize the light extinction principle in this range.

B. Electronic Noise Level

Particle counters utilize pulse height analyzers or comparator circuitry to analyze the voltage output of particle sensors. Voltage pulses of appropriate amplitude are counted as particles. However, not all pulses are the result of particles in the sensing zone. Some are the result of electrical noise. Figure 3 shows the output of a typical particle sensor. The single large pulse shown was generated by a spherical latex particle while the numerous smaller pulses are the result of electrical noise. The inability of instruments to distinguish pulses due to noise from those due to particles is a potential source of error, particularly at small particle sizes.

The threshold noise level of an instrument provides an indication of the potential for results to be influenced by electronic noise. It is operationally defined as the maximum threshold voltage or pulse amplitude which yields more than 60 counts per minute in the absence of particles in the sensing zone [9]. The noise level limits the minimum useful size of a particle counter. Sizing and counting errors result when the pulse height due to particles

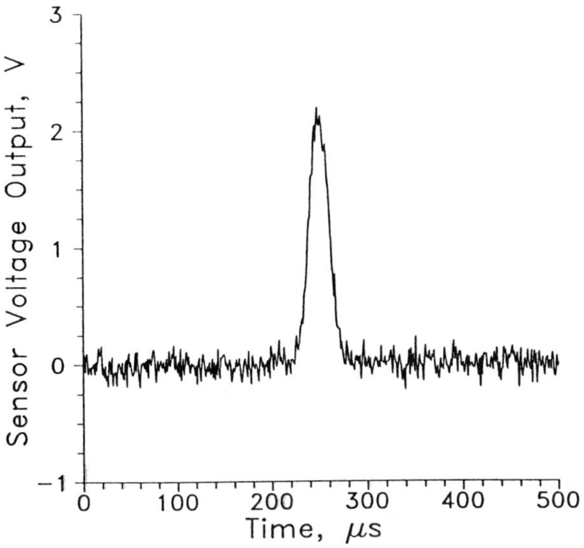

Figure 3 Voltage output for a light extinction particle sensor.

approaches that of instrument noise. At very low threshold voltages (i.e., sizes below an instrument's detection limit), the number of counts due to noise exceeds the contribution due to particles. If data from below the instrument noise level are used for instrument calibration, the resultant calibration curve will be skewed. In order to avoid this undesirable artifact, the National Fluid Power Association (NFPA) defines a particle counter's minimum detection limit at 1.5 times the instrument noise level.

Instrument noise levels have improved steadily, but environmental noise remains a problem. Inadequate electrical shielding or grounding of particle counters, as well as their placement, can result in transient electrical noise. In these cases, the noise level may be elevated above the instrument noise levels or it may vary in a transient manner. Random voltage pulses, larger than the threshold noise level, may be observed. These are not the result of instrument noise, but are artifacts of the instrument's operating environment and design.

Processes which cause power supply fluctuations or which generate stray electrical voltage contribute to environmental noise. Electromagnetic devices, such as electrical generators and motors, operating in the vicinity of a particle counter are a common source of environmental noise. Discharges of static electricity can also contribute to the problem. It is not uncommon for the clean oil used to dilute samples to be located near the particle counter. Static discharges resulting from pumping operations can contribute to environmental noise if the clean oil doesn't contain an antistatic additive and isn't grounded. Improvements in the electrical shielding and grounding, as well as the use of voltage isolators, help to reduce the effects of environmental noise. In extreme cases, the particle counter or problem equipment may need to be relocated.

C. Electronic Saturation and Flow Rate Limitations

Electronic saturation results when particles arrive in the sensing zone so fast that they are indistinguishable as individuals particles, resulting in both sizing and counting errors [10–11]. Therefore, saturation can be a problem at high particle concentrations or high flow rates. Electronic saturation is influenced by a variety of factors, including particle size. Meaningful quantification is difficult, and the use of the results to solve analytical problems is not straightforward. Instead, operationally defined parameters, including the flow rate and concentration limit of an instrument, are more useful. These parameters are sensitive to instrumental factors in addition to saturation and provide meaningful data which can be used to establish analytical protocols.

Flow rate limitations are determined by the speed of the electronics and, in part, by the baseline correction circuitry. In typical sensors, the sensing zone is about 10^{-5} ml in volume and particles are present in it for about 30 μs. Within this short time, the voltage pulse for a particle must develop. If the flow rate is too high, the electronics do not have time to respond and voltage pulses do not develop completely. Particle counters may also misinterpret voltage pulse information if the flow rate is too low. In this case, particles are present too long in the sensing zone. Whether the flow rate is too high or too low, inaccurate particle sizing may result.

Particle counters can operate over a relatively broad range of flow rates, but always should be used at the same flow rate. Calibration curves are sensitive to flow rate, as shown in Fig. 4. As a result, particles are sized inaccurately if the particle counter is used at a different flow rate than that for which it was calibrated. Observed particle concentrations vary with flow rate, especially for polydispersed samples. Therefore, care should be taken to ensure that instruments are operated at the same flow rate for which they were calibrated. In a later section, the effects of flow rate variability are discussed.

Many instruments possess baseline correction circuitry. This circuitry analyzes the sensor output and defines a baseline from which pulse heights are measured. This was developed to allow the instruments to adjust for baseline drift. The baseline may change for a variety of reasons, including differences in the optical properties of the fluid, the presence of high concentrations of very fine particles, and aging of instrument electronics and optics. Baseline correction allows different colored fluids to be analyzed without the need to recalibrate. This is an important feature in used oil analysis. Without it, a different calibration curve would be needed each time a different sample was analyzed. However, baseline correction can also introduce undesirable artifacts which are discussed later.

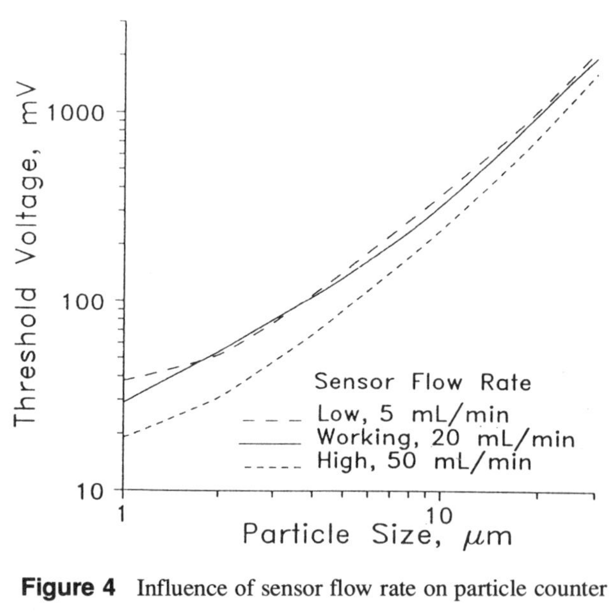

Figure 4 Influence of sensor flow rate on particle counter calibration curves.

D. Resolution

Resolution is a measure of the ability of an instrument to distinguish between similar sized particles. It is determined using well-characterized monodispersed latex particles in a manner described later in this chapter. Monodispersed latex exhibits a near Gaussian particle size distribution. Ideally, the size distribution observed by a particle counter should also be near Gaussian with the same standard deviation. However, the observed standard deviation is actually larger due to contributions from instrument resolution and the calibration technique. Typically, resolution is the largest contribution to the observed standard deviation.

Resolution is, in part, controlled by the quality of lighting in the sensing zone. The more homogeneous the light intensity, the better the resolution [7]. Sensor optics, including the light source and collimating lens, influence illumination. In general, laser and laser diode light sources provide more uniform illumination and have better resolution than white light sources. This is one reason for gradual replacement of white light sensors by laser diode sensors.

Resolution is an indicator of instrument quality and condition. Better agreement in particle count results is obtained among instruments having resolution less than 10% than among those having poorer resolution [9,12]. In the hydraulics industry, it is recommended that instruments possess resolution better than 10% [9,13,14]. The resolution of instruments made prior to the mid-1980s is typically in excess of 10%. Resolution is approaching 5% for instruments made after 1988. To a large degree, this reflects improvements in the optics and electronics. For a particular instrument, a loss of resolution over time is indicative of a need for servicing. Routine resolution checks are recommended in order to identify instrument problems before they become significant.

E. Coincidence Error Limits

Optical particle counters are single particle counters. Ideally, only one particle should be present in the sensing zone at a time. When two or more are present, they are counted as a single particle. Since sensors monitor changes in the total illumination across the sensing zone, they cannot determine if these changes are due to the presence of one or many particles [11]. Error resulting from the presence of multiple particles in the sensing zone is referred to as coincidence error. Coincidence error results in inaccurate sizing and counting of particles [10,11,15]. As a result, apparent particle concentrations may be higher than the actual particle concentrations at large particle sizes, but lower at smaller sizes [11].

The terms "coincidence error limit" and "concentration limit" are used interchangeably to refer to the maximum particle concentration which yields acceptable data. The term "coincidence error limit" is somewhat misleading since it may be determined by either coincidence error or electronic saturation, depending on instrument design, the conditions of use, and particle size distribution of the sample. In the hydraulics industry, it is the concentration such that actual and observed particle concentrations differ by less than 5% [9]. The coincidence error limit of a sensor can be estimated from the following equation:

$$C \approx \frac{1}{10V} \tag{1}$$

where C is the coincidence error limit (particles/ml) and V is the volume of the sensing zone (ml). For reasons which are discussed later, this equation provides only a rough estimate of the limit.

Coincidence error limits estimated using Eq. (1) are relatively accurate for samples containing monodispersed particles such as latex. This is shown in Fig. 5 using the data obtained for samples containing 9.87-μm latex particles. Equation (1) predicts that coincidence error will be

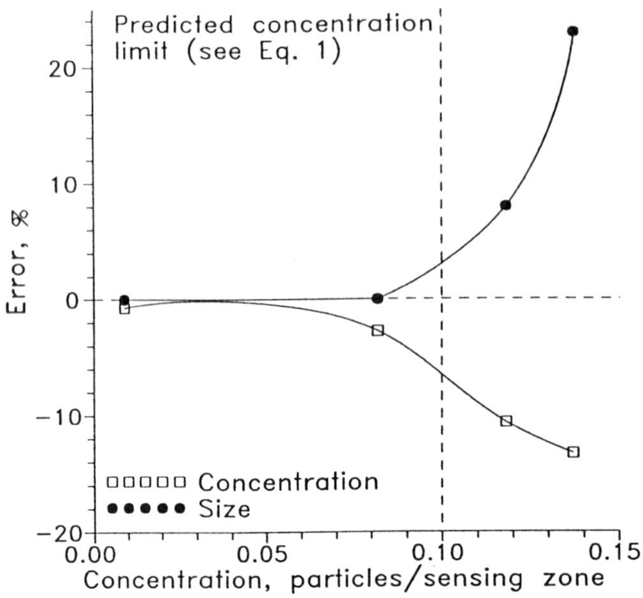

Figure 5 Effect of concentration on counting and sizing errors for monodispersed 10-μm latex particles.

insignificant at concentrations lower than 0.1 particles per sensing zone. As shown, the particle sizing and counting results are relatively accurate (< 5% error) under these conditions. At higher concentrations, the random passage of particles through the sensing zone creates intermittent conditions of optical coincidence and electronic saturation causing particles to be oversized and undercounted. Thus, coincidence error contributes to both sizing and counting errors.

Typically, instrument manuals report concentration limits determined using monodispersed latex particles. However, these are only guidelines since the actual limits may vary considerably. Polydispersed samples tend to have lower limits. The limits determined with monodispersed latex may differ by more than a factor of 3 from the limits for polydispersed samples [12]. In recognition of this, the NFPA empirically defines coincidence error limits using the same AC fine test dust (ACFTD) used for filter testing [9,16]. ACFTD limits are more representative of the limits for actual samples than latex, but for samples containing a high percentage of fine particles, e.g., multipass filter test or used lubricating oil samples, they may be even lower.

III. CALIBRATION

There are two existing standards for the calibration of particle counters used for oil analysis. The first and most widely used standard, ISO (International Standards Organization) 4402, utilizes a standard contaminant, AC fine test dust [16]. The second, ANSI (American National Standards Institute)/(NFPA)T2.9.6R1, utilized monodispersed latex spheres as contaminants [9]. There are inherent shortcomings of each method. This has led to efforts on the part of the NFPA and ISO to develop a hybrid method which combines the best features of these two standards [13,14].

A. AC Fine Test Dust Calibration

The AC fine calibration method, ISO 4402, establishes calibration curves by comparing particle counts from prepared samples of ACFTD to those of a published size distribution [16]. Figure 6 shows a graphic representation of this size distribution. Samples of ACFTD are prepared at

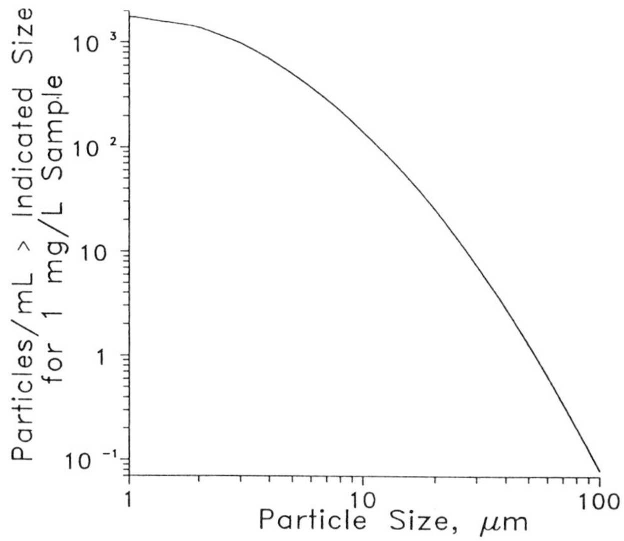

Figure 6 Particle size distribution of AC fine test dust according to ISO 4402.

known concentration and analyzed at arbitrarily chosen instrument threshold voltage settings. The results are normalized with respect to dust concentration and compared to the published distribution in order to determine the particle size corresponding to each threshold setting. This is the basis for constructing a calibration curve.

The strength of ISO 4402 is the similarity between the calibration contaminant and the contaminants found in oil samples. The same dust is used for calibration and filter testing. Further, the calibration dust is similar to silica-type contaminants commonly found in used oil. Thus, the optical differences between the calibration dust and real contaminants are negligible. ISO 4402 also has practical advantages in that it is relatively inexpensive and is not time consuming. The hydraulics, automotive, and aerospace industries have over 20 years of experience with this standard, and there exists wide international acceptance of it.

A fundamental weakness of ISO 4402 is that the published size distribution is known to be inaccurate [17–20] below 10 μm and above 60 μm. ACFTD may contain 10 times the number of particles smaller than 5 μm than is reported. As a result of the inaccuracy in the size distribution, the micrometer, as defined by ISO 4402, is not the same as this internationally defined unit of length. Since industry is increasingly interested in 1–10-μm particles, there is a greater need for accurate calibration at small sizes.

Though ISO 4402 is an internationally accepted standard, the calibration material itself has never been certified by a national or international agency. Historically, there has only been one source of ACFTD and batch-to-batch variations in the dust have been observed. Quality control for particle sizes below ~5 μm is questionable [21]. This was acceptable when the calibration standard was developed over 20 years ago and 10 μm was the smallest size of interest. However, this is no longer the case. Today, ACFTD is no longer manufactured by the original supplier. The new supplier uses different manufacturing processes than the original supplier. This has resulted in decreased batch-to-batch variations, but the size distribution itself has changed somewhat [22]. As a result of this, the Society of Automotive Engineers (SAE) and ISO are in the process of developing new test dust standards, and ACFTD will be replaced by ISO fine test dust [23].

Another problem with ISO 4402 is a lack of internal quality control checks [24]. Thus, procedural errors made in preparing the suspended dust samples or in counting the samples may shift the resultant calibration curves to higher or lower sizes. In a past round-robin evaluation, significantly better agreement in particle count results was found among laboratories which utilized a common source of ACFTD suspensions for calibration than among laboratories which prepared their own suspensions [9]. Since internal quality control checks are lacking, no conclusions can be drawn concerning which data most accurately reflect the true size distribution of samples. However, it is apparent that the preparation of ACFTD calibration samples is a major source of variability [25].

Another issue is that the ACFTD calibration method does not require verification that an instrument is operating properly. In the previous round-robin [9], nearly one-third of the instruments involved were identified as substandard. Poor resolution and defective equipment were the most common instrument problems. This reduces agreement among instruments. Resolution is an important indicator of instrument condition and quality. Deteriorating resolution is an indication that an instrument needs servicing. Some older instruments have inherently poor resolution. For example, one instrument in the round-robin had a resolution in excess of 50%; hence, it was unable to accurately distinguish between 5- and 10-μm particles. Such instruments cannot be identified as substandard using ISO 4402.

ISO 4402 does not adequately determine the operating limitations of an instrument. The coincidence error limit is determined, but it is overestimated [24]. Instrument noise level and working flow rate are not determined at all. This has led some laboratories to operate below the noise level of their instruments in an effort to measure smaller sizes. Obviously, the resultant data reported are meaningless.

A final concern with ISO 4402 is that it is based on the longest chord dimensions of irregularly shaped particles [26,27]. Particle counters don't measure longest chord dimensions. Instead, equivalent diameters defined by the instrument operating principle are determined. This is not a major concern for dusts with a constant particle aspect ratio, but it can be significant in field samples which have variable aspect ratios. Equivalent diameter is a more meaningful dimension than longest chord for such applications.

ISO 4402 is a useful calibration standard which has achieved international acceptance and has been the basis for particle counter calibration in the hydraulics, automotive and aerospace industries for over 20 years. The strength of the procedure is that the optical properties of the calibration material, ACFTD, are essentially the same as the contaminants used in oil filter testing and those found in used oil samples. Despite this, the calibration dust lacks traceability to any national or international agency, and questions have been raised concerning historical batch-to-batch reproducibility. The published size distribution is inaccurate. The procedure lacks internal quality control checks, such as verification that the instrument is operating properly, that good analytical technique is used, and that the instrument is used within its operating limits. Erroneous calibration curves can result from inaccuracies introduced by these sources. As a result of these criticisms, the latex-in-oil calibration method was developed.

B. Latex-in-Oil Calibration

The latex-in-oil calibration method, ANSI/(NFPA)T2.9.6R1, establishes a direct relationship between particle size and threshold voltage setting. In contrast, the ACFTD method utilizes an indirect correlation between particle counts and size. The latex-in-oil method is similar to latex in water techniques [28–29] used for aqueous applications. However, latex in water methods are not applicable to oil analysis since water and oil differ in terms of their optical properties. Most oils absorb light at the wavelengths used by optical particle counters. Further, the index of refraction for petroleum oil is ~1.47, as contrasted with 1.33 for water. Since agreement in particle count data improves as the refractive index of the calibration and sample fluids approach one another [24], water is not a suitable calibration medium for oil analysis. There are also pragmatic reasons for using oil as the calibration fluid. Since oil and water are immiscible, a series of solvents, such as 1,1,2-trichlorotrifluoroethane and 2-propanol, must be used to clean the sensor of one fluid before it can be used for the other. For these reasons, water is not a suitable calibration medium for oil applications.

The latex-in-oil calibration procedure has five parts, including verification that the instrument is working properly and determination of its operating limits. These include

1. Determination of the instrument's threshold noise level and working flow rate
2. Primary size calibration of the instrument
3. Determination of instrument resolution
4. Determination of coincidence error limits
5. Calibration verification

The first step of the procedure is critical but is often overlooked. The noise level is checked to ensure that the instrument is in good working condition. The results are later used to establish the instrument's detection limits. The working flow rate data enable the operator to choose a suitable flow rate for subsequent calibration and sample analysis.

The primary size calibration utilizes monodispersed latex particles suspended in MIL-H-5606 hydraulic oil. The latex must have a coefficient of variation (CV) less than 5% and its supplier must certify that it was sized using NIST (National Institute of Standards and Technology) traceable methods. In the past, the creation of stable latex-in-oil suspensions was problematic. On one hand, mixing aqueous latex suspensions with oil yielded emulsions unsuitable for calibration. On the other hand, polystyrene latex swells and changes size in many organic solvents [30]. Both problems were overcome by adding Aerosol OT (sodium dioctylsulfosuccinate) at high concentrations, i.e., $\approx$12 g/l, to MIL-H-5606 oil. Aerosol OT is a surfactant which causes water and oil to form a microemulsion. The resultant homogeneous latex suspension still contains water droplets, but at sizes too small to interfere with particle counting [31]. Further, latex particles suspended in this manner do not swell and change size over extended periods of time [24,31]. Thus, Aerosol OT allows latex to be used in oil for calibration purposes.

With the latex-in-oil method, the threshold voltage corresponding to a particular latex size is established using a modified differential half-count method. The manner in which this is done depends on whether the counter uses a pulse height analyzer or comparator circuits to analyze the sensor output. A pulse height analyzer is a multichannel analyzer which sorts voltage pulses into discrete channels permanently defined by the electronic hardware. Typical pulse height analyzers collect data into over 1000 different channels, from which the operator chooses those to be monitored. In comparator circuit instruments, channels are nearly independent circuits which are hardware or software adjustable to different threshold voltages. Since each channel is a separate circuit, fewer channels (6–32) are available, but they can be adjusted over the same voltage range as pulse height analyzers.

For instruments with pulse height analyzers, a sample of latex in oil is analyzed in the differential mode of operation. Figure 7 shows a typical graph of this data. The large peak with near Gaussian behavior corresponds to counts from individual latex particles. Outlier counts at lower voltages correspond to fine particulate contamination. Outlier counts at higher voltages result from contamination or latex aggregates. The median threshold voltage of the large peak corresponds to the particle size of the latex. Utilizing this method to determine the threshold voltages corresponding to different sizes of latex, a calibration curve can be defined.

In instruments with comparator circuits, the procedure is similar. In this case, at least three

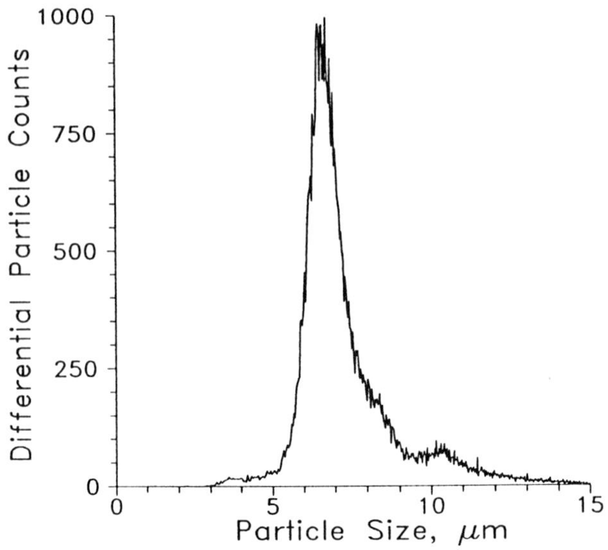

Figure 7 Differential size distribution for 7-μm monodispersed latex particles.

instrument channels are required to relate latex size to threshold setting. One (channel B) is set at a threshold voltage corresponding to the approximate particle size of interest. The other two (channels A and C) are set to correspond to sizes 15% smaller and larger than that of the particle. By assuming that threshold voltage and particle cross-sectional area are directly related, a logarithmic relationship between particle size and voltage with a slope of 2 is expected. This is a reasonable approximation for extinction sensors operating well above the instrument noise level. Based on this assumption, the following relationships among the three channels must be maintained during calibration [9]:

$$V_A = 0.72V_B \tag{2}$$

$$V_C = 1.32V_B \tag{3}$$

where V_i is the threshold voltage of the indicated channel. Samples of latex are analyzed and the channels adjusted while maintaining this relationship until the ratio of the number of counts in channels A and B is as close to one as possible. When this condition is met, the voltage of channel B corresponds to the particle size. As in the previous case, this process is repeated for different sizes of latex, and a calibration curve constructed. There are many situations when the underlying assumption of this procedure is violated and the slope of the calibration curve deviates from 2. For example, the slope often varies with particle size. For scattering sensors, it may be as great as six for small sizes and as small as one at larger sizes. For extinction sensors, the slope may deviate if the sensor output is electronically conditioned or if the sensor is operating near the instrument noise level. In these cases, the calibration curve approximates the true relationship between particle size and voltage.

Instrument resolution is determined by assuming a normal size distribution and analyzing latex using the counter in the cumulative mode. The process is illustrated in Fig. 8. Channel B is set to correspond to the latex particle size. Channels A and C are adjusted until the ratio of counts in channels A, B, and C, respectively, is 5.29:3.15:1. The resultant threshold settings are

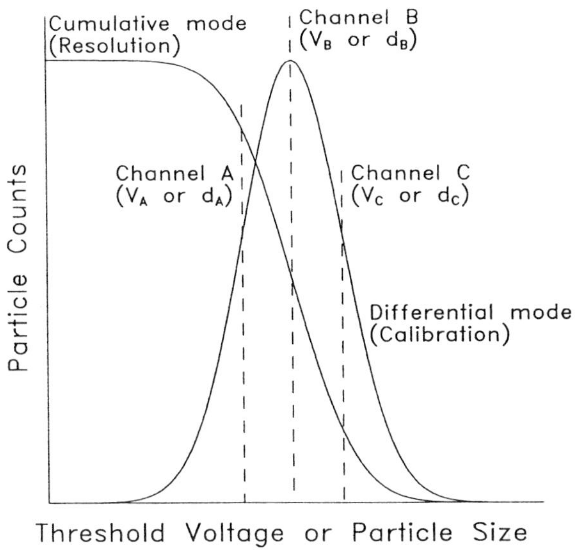

Figure 8 Conceptual illustration of threshold settings for modified half-count method of calibration (differential mode) and for resolution determination (cumulative mode). Refer to text for nomenclature.

then converted to sizes using the calibration curve. The positive and negative side resolutions (R_+ and R_-, respectively) are given by

$$R_- = \sqrt{(d_A - d_B)^2 - s_L^2} \tag{4}$$

$$R_+ = \sqrt{(d_B - d_C)^2 - s_L^2} \tag{5}$$

where d_i refers to sizes corresponding the indicated channels and s_L refers to the standard deviation of the latex size distribution. The positive and negative side resolutions should agree within 5%. If not, this suggests a need to have the particle counter serviced or asymmetry in the latex size distribution [9]. The reason for the asymmetry can be determined by microscopic examination of the latex and appropriate action taken.

With the latex method, the coincidence error limits of the instrument are determined in a manner similar to ISO 4402. In both procedures, different dilutions of a concentrate containing a known concentration of ACFTD are analyzed. In the latex procedure, the particle count results at the smallest particle size of interest are plotted as a function of ACFTD concentration. Results obtained at concentrations less than or equal to 40% of the expected coincidence error limit are used to establish the theoretical relationship between the number of counts and concentration. The limit is defined as the concentration at which the data deviates from the theoretical result by 5%. In ISO 4402, a weighted average is used to determine the limit. Depending on the number and concentrations of samples analyzed, ISO 4402 significantly overestimates the limit [24]. This is avoided with the latex procedure.

The final step of the latex in oil is calibration verification. Prepared samples of ACFTD are analyzed and the results compared to past results for identical samples. This is done for routine verification that the instrument calibration has not changed. The results can also be used to establish a relationship between the latex and ACFTD methods.

The latex-in-oil procedure has a number of advantages over the ACFTD method [24,32]. The latex method is a true size calibration, not a count correlation method like ISO 4402.

As such, it is relatively insensitive to analytical technique. This allows the operator to verify that the instrument is working properly and to establish its operating limits. Quality control checks permit the operator to ensure that good analytical technique is being used. Another advantage is that the calibration material is NIST traceable. As a result, concerns regarding the particle size distribution and variability of the calibration contaminant disappear.

Despite its advantages, problems exist with the latex-in-oil procedure. Significant differences exist between the optical properties of latex and the contaminants normally found in oil. ACFTD and related test dusts are among the most commonly encountered contaminants. ACFTD has a refractive index of about 1.55 with a nonzero imaginary component (i.e., it absorbs light) while latex has a refractive index of 1.59, but no imaginary component (i.e., it is transparent). These differences influence the manner in which particles interact with light and how they are sized [9]. The term *extinction efficiency* is a theoretically defined term used to describe the effect of various factors on the ability of extinction sensors to size particles.

Extinction efficiency is the ratio of the extinction cross section of a particle to its actual cross section. Simplistically, it can be thought of as the ratio of the cross-sectional area of the "shadow" cast by a particle on the detector to its actual cross section. Extinction efficiency is a function of particle size as well as the real and imaginary components of refractive index. When it equals 2, particles are accurately sized. For larger values, particles are oversized while the reverse is true for smaller values.

For a given system, extinction efficiency exhibits near sinusoidal behavior when plotted as a function of particle size [6]. Thus, particles may be undersized or oversized depending on their actual size. As size increases, the extinction efficiency approaches a constant value of 2 and sizing by particle counters becomes more accurate. Sizing also becomes more accurate as the magnitude of the imaginary component of refractive index increases and as the difference in refractive index between the fluid and particle increases. Since latex has no imaginary component and since its refractive index is similar to that of oil, there is enormous potential for sizing inaccuracies during a latex in oil calibration. This manifests itself in terms of the variability in results obtained by different instruments calibrated with this method.

Agreement among particle count results from various instruments is worse with the latex in oil calibration than with the ACFTD calibration [25]. Depending on the size distribution of the sample, differences can exceed a factor of 4. These differences are related to instrumental factors. Instruments made by the same manufacturer tend to show better agreement than among different manufacturers. Agreement is also better between the same model of sensor than among different models. Because the contaminant used for calibration and analysis are the same, this is not observed with the ACFTD method. Since one of the primary uses of particle counters is oil filter evaluation, the problem of agreement among instruments is a major drawback of the latex in oil method.

Several instrumental factors have been identified which influence agreement among sensors. Instrument resolution is one of the biggest sources of variability in particle count data [9]. Better agreement is obtained among instruments with resolution less than 10% than among those with poorer resolution [9,12]. Since poor resolution sensors erroneously size some small particles as larger ones, agreement is sensitive to the size distribution of polydispersed samples. Agreement deteriorates as the fraction of finer particles increases [25]. Another source of variability is the light source used [25]. Poor agreement is obtained among results obtained from white light, laser diode and HeNe laser sensors, but better agreement is achieved when sensors using the same type of light source are considered. Another source of variability is the operating principle [9,25]. Since extinction follows a second-order relationship while scattering is sixth order [7,8], this is not unexpected. The problem should be especially noticeable at small particle sizes. Other instrumental factors, such as sensor geometry and collection angle, also influence agreement among sensors.

The latex-in-oil calibration procedure is superior to ISO 4402 in that it is a true size calibration procedure and is relatively insensitive to analytical technique. Unlike ISO 4402, instrument operation can be verified and the quality of the analytical methods can be confirmed using the latex method. The latex method also provides NIST traceability lacking with ISO 4402. However, the optical properties of latex particles and the contaminants actually analyzed differ considerably. This results in significant differences in the particle count results obtained from different types of instruments. For this reason, it is not recommended that this calibration method be used for oil filter testing [9].

C. Hybrid Calibration Method

The ACFTD and latex methods compliment one another in terms of their strengths. The ACFTD method utilizes calibration materials optically identical to the samples which will be analyzed, while the latex method allows instrument operation to be verified and its working limits determined. However, each procedure has significant shortcomings. For this reason, a hybrid method is being developed by NFPA and ISO [12–14,25].

The hybrid method uses calibration suspensions of ISO medium grade dust [23] (i.e., SAE 5–80-μm test dust) suspended in MIL-H-5606 oil. Since the optical properties of the suspensions are identical to multipass samples, the sizing inaccuracies inherent in the latex-in-oil method are avoided. ISO medium dust rather than ACFTD is used [25] since it contains fewer 0–5-μm particles. This reduces coincidence error and increases the population of larger particles. Thus, a more reliable calibration and better agreement among sensors results [25]. The cumulative number concentration of particles and corresponding coefficient of variation for particle sizes between 0.5 and 80 μm will be certified by NIST. The CV at 5 and 30 μm will not exceed 4% and 7%, respectively. The use of calibration suspensions from a single source avoids past variability resulting from the preparation of suspensions by individual laboratories [9] and from batch-to-batch variability in ACFTD [21]. For reasons of convenience, secondary calibration suspensions prepared by sources other than NIST can be used. This reduces the cost of calibration and ensures that the supply of NIST-certified suspensions is not depleted too rapidly. Sample preparation, statistical quality control and data reporting methods are strictly defined to maintain traceability [13,14].

The first step of the hybrid method involves determination of the threshold noise level and the repeatability of volume measurement by the particle counter. The noise level is checked to ensure that the instrument is working properly. The repeatability of volume measurement plays a crucial role in obtaining reproducible results and agreement among various instruments. Uncertainty in volume measurement during calibration introduces uncertainty in the size determination.

An iterative process is used for primary size calibration. The threshold settings of the instrument are adjusted to cover the size range of interest. The threshold noise level limits the smallest size which can be measured. A working flow rate is chosen which will be used for all calibration and analysis work. A calibration sample is then ultrasonically dispersed, shaken, degassed and counted five times. The data are then statistically evaluated in order to determine their acceptability. The sizes corresponding to the observed particle concentrations are determined by interpolation from NIST certified data and used to construct a tentative calibration curve. The process is then repeated twice, but each time the threshold settings are readjusted to correspond to sizes for which NIST size distribution data exist and the calibration curve refined. The final curve is constructed using data from all three iterations.

Instrument resolution, coincidence error and flow rate limits are determined similar to ANSI/(NFPA)T2.9.6R1 [9]. ISO ultrafine grade dust [23] is used to define coincidence error limits and working flow rate. Coincidence error limits are sensitive to particle size distribution [11] and tend to be lower for field and multipass samples than for ACFTD. Hence, ISO ultrafine yields a more conservative estimate of coincidence error limits than ACFTD, which is used in the latex method. The hybrid method defines a range of working flow rates for which counting errors do not exceed 5%. In contrast, the latex method yielded a single flow rate, but provided no guidance concerning how precisely it needs to be maintained. On the other hand, ISO 4402 does not address flow rate issues at all. Thus, the hybrid method provides more realistic assessment of both flow rate and coincidence error limits than either ISO 4402 or the latex method.

Past experience indicated the need for an instrument specification in order to obtain agreement among different instruments [25]. ISO 4402 contains no instrument specifications, while ANSI/(NFPA)T2.9.6R1 contains only recommendations. The hybrid procedure includes a performance specification requiring instruments to

1. Be able to deliver volumes with a CV $\leq 3\%$
2. Have a resolution $\leq 10\%$ at 10 μm or the smallest size of interest
3. Yield particle counts for ISO ultrafine test dust that agree within $\pm 10\%$ of the mean obtained by an interlaboratory round-robin

It has been estimated that 30–50% of the instruments used in past round-robins could not meet these specifications [9].

Volume measurement uncertainty is the most basic instrument specification. This influences the accuracy of calibration and subsequent sample analysis. If, for example, the CV of volume measurement exceeds 3%, agreement among different instruments should exceed 5%. In the past, some instruments have exhibited CVs exceeding 10%.

Resolution is important as an indicator of instrument quality. Excessive variability is observed when resolution exceeds 10% [9,12,25]. In past work, good agreement could be obtained among instruments utilizing different light sources and operating principles, if their resolution was less than 10% [12]. However, agreement was not obtained among similar instruments when resolution exceeded 10% [9,12].

The ISO ultrafine specification is necessary since agreement is sensitive to sensor design [12,25,32]. Even though agreement is obtained for ISO medium or ACFTD, there is no certainty that it will also be obtained for samples with truncated size distributions, such as multipass test or used oil samples. Agreement tends to be worse for samples enriched with fine particles [25]. The ISO ultrafine specification helps ensure valid comparisons of data from various labs.

Though the hybrid method is not yet a standard calibration method, it is being actively pursued by both NFPA and ISO. Technically, it has significant advantages over the alternatives. Both contaminant and fluid used are optically identical to those commonly found in oil samples. Problems associated with lack of traceability, batch-to-batch variability, and erroneous size distributions are avoided through the use of calibration suspensions with NIST certified size distributions. Instruments are required to meet minimum performance specifications before they can be used. Further, their operating limitations are determined under realistic conditions. Several studies have demonstrated that better agreement is obtained with the hybrid method than with ISO 4402 or ANSI(NFPA)T2.9.6R1 [12,25]. Conservatively, it is estimated that CVs will decrease by a factor of 2 or more with the hybrid procedure.

IV. SOURCES OF ERROR AND VARIABILITY

Particle counters, like other types of analytical equipment, generate data even when used improperly. Therefore, operators must be familiar with possible sources of error in order to obtain meaningful results. They must be capable of recognizing erroneous data, be able to identify their causes, and know how to take corrective action. There are six basic steps in the analysis of oil samples. First, a representative sample must be collected. Second, particles which settled out during sample transport must be resuspended and then aggregates broken up. The sample must then be diluted and particle-counted. Finally, the results must be interpreted. Many sources of variability and error can enter during each step of the analysis. Table 1 summarizes the steps involved in the particle-counting analysis of oil samples and some of the common sources of error and variability associated with each step. This section discusses particle-counting errors, ways of recognizing them, and methods of minimizing their adverse effects.

A. Contamination

Contamination refers to undesirable changes in a sample's particle concentration or size distribution. As a result, particle count data may not be representative and are often irreproducible. Contamination can occur during any phase of sample analysis and is typically the result of carelessness or the use of dirty fluids or equipment. Contamination from dirty equipment will be emphasized in this section. Other sources of contamination are discussed in later sections.

Particle-free sample containers, typically glass or plastic bottles, should be used for oil sample collection and transport. Oil analysis programs rely almost exclusively on samples collected in the field in particle-free bottles and brought back to the laboratory for analysis. Similarly, samples collected in particle-free bottles are often used in filter testing, particularly when high dust concentrations or coarse filters are involved. For fine and silt control filters, on-line particle counting reduces problems stemming from sample handling and contamination. This is discussed in a later section.

Sampling apparatus, containers, and analytical glassware must meet three criteria. They must be compatible with the fluids of interest. They must not alter the sample's particle size distribution. They must be "particle free." Chemical reactions occurring between the sample and container may result in the loss of sample or production of particles. Containers with porous

Table 1 Summary of Analytical Scheme for Oil Samples

Procedure	Sources of variability and error
Sample collection	Contamination
	Sampling
Resuspension of particles	Stirring and mixing
Deagglomeration	Ultrasonic dispersal
Dilution	Volume measurement
Particle counting	Instrumentation
	Calibration
	Diluent effects
	Entrained air bubbles
Interpretation	Contaminant nature
	Immiscible droplets and bubbles
	Coincidence error
	Particle statistics

walls should be avoided since they are difficult to clean and may capture and hold contaminant. In this way, they alter the sample's apparent size distribution. The various steps of the analysis process should be considered when choosing a container. For example, if heating, ultrasonic vibration, or mechanical shaking is to be used, the container must be resistant to these applied forces.

Several types of containers are used for oil sampling, including both wide- and narrow-mouth bottles. Wide-mouth bottles facilitate sampling, but the risks of contamination, improper sealing, and accidental spillage are increased. Borosilicate glass bottles are widely used. However, different grades of glass are available and some release chips or flakes upon ultrasonic cleaning or aging. Alternatively, leakproof, high-density polyethylene bottles with polypropylene screw closures have been used successfully. They are durable and well suited for collecting field samples. The authors have found that these bottles yield reproducible results, are chemically resistant to oil, and new bottles are typically very clean. The quality of both glass and polyethylene bottles varies depending on the source. Bottles should be thoroughly evaluated when a new source of containers is being considered.

The selection of an appropriate container closure is as important as the selection of the container itself. Much time and effort has been squandered trying to identify contamination problems caused by dirty, leaky or chemically incompatible closures. Many types of closures are available for glass bottles. Particle-free Mylar films are often used to line closures and minimize these problems. However, these films are inconvenient to use, subject to leakage, and must be disposed of after use. These problems can be reduced by using suitable caps with flat polytetrafluoroethylene (PTFE)-faced liners. They are virtually particle free, easy to clean, durable, and chemically resistant. However, they sometimes need to be checked for tight sealing.

In general, sample containers and closures, as well as sampling and analytical apparatus, must be cleaned prior to use. Laboratories typically develop cleaning protocols tailored to their individual needs, but some general guidelines are worth discussing. With the exception of disposable plastic pipette tips, new containers, closures and associated apparatus are rarely clean enough for use in particle counting. For this reason, they must be amenable to solvent cleaning.

Cleaning solvents should be chosen with several factors in mind. They must be chemically compatible with the sampling and analytical equipment. They should not be sources of contamination; hence, they should be filtered to an acceptable degree, typically 0.45 μm. Finally, they must be capable of removing contaminants from the walls of containers and other apparatus. Thus, low-viscosity solvents having low surface tension solvents or containing cleaning or wetting agents are preferred. Petroleum naphtha, pentane, petroleum ether, 1,1,2-trichloro-2,2,1-trifluoroethane, 2-propanol, and water have been used in various stages of the cleaning process. However, the use of pentane, ether, and trichlorotrifluoroethane is diminishing and alternative solvents are being sought as a result of safety and environmental concerns.

A variety of cleaning methods are available. The one chosen should be at least as vigorous as the methods used to analyze samples. The use of high-pressure sprays, mechanical shaking and ultrasonic cleaning may be involved. If ultrasonic contaminant dispersion or degassing will be used to analyze samples, ultrasonic cleaning is essential. It is critical that all surfaces which will be exposed to the sample, including closures and screw cap threads, be thoroughly cleaned in a similar manner. Once particle free, sample containers can be capped and stored for later use.

In order to demonstrate the principles involved in cleaning, it is helpful to discuss one method used to clean sample bottles for filter testing. Since containers may contain a variety of soluble and insoluble particles and deposits, as well as oil residue, three solvents, water, 2-propanol and petroleum naphtha, are used. Both organic solvents are filtered through 0.45-μm membranes prior to use. First, water-soluble contaminants are removed with detergent and water in a dishwasher. Then, the containers are filled with 2-propanol and placed in an ultrasonic cleaner for several minutes. Finally, this step is repeated using petroleum naphtha to remove oil soluble residue and nonpolar contaminants adhering to the container. Containers with fewer than 5 particles/ml greater than 5 μm and fewer than 10 particles/ml greater than 1 μm are obtained by this process.

It should never be assumed that cleaning procedures are satisfactory. The results of cleaning procedures should be evaluated statistically in order to determine their acceptability and to identify problems. ISO 3722 provides guidelines for evaluating cleaning processes [33]. This standard evaluates sample containers for chemical compatibility and cleanliness using the same procedures and fluids for routine particle counting. In general, contamination from all sources should be at least two orders of magnitude lower than those expected from the samples to be analyzed [9,33,34]. In situations requiring a higher degree of accuracy or when high dilution ratios (> 50) are used, more rigorous cleanliness standards need to be applied. Cleaning procedures and methods of evaluating them are discussed elsewhere [5].

Pipettes, syringes and volumetric glassware are commonly used for oil sample dilution and can be sources of contamination. These materials can be cleaned with solvents and checked for cleanliness using clean diluent in a manner similar to that used for sample containers and closures. It is good practice to rinse pipettes and syringes with three volumes of the sample before actually measuring out the sample to minimize artifacts arising from residual cleaning fluid in the volumetric apparatus or from contaminant adhering to its sides.

In off-line particle counting, the batch sampler used to force oil through the sensor is often overlooked as a source of contamination. Most batch samplers used for oil applications pressurize a sample chamber to force oil through the sensor. The air and vacuum lines of the batch sampler may be contaminated with dirt, rust and water. Commercially available in-line filters and desiccants should be used to reduce this source of contamination. The sensor and associated tubing should be regularly back-flushed or rinsed with particle-free fluid to dislodge particles trapped in the sensing zone. Erratic results and unexplained reductions in sample flow rate may be indications of a need to back-flush. It is useful to install a three-way valve on the low-pressure side of sensors to facilitate back-flushing. The entire particle counting system should be flushed with clean fluid at the end of each working day and stored with clean fluid in it. Routine cleanliness checks of diluent should be used to ensure that unexpected sources of contamination don't develop.

B. Sampling

Proper sample collection is essential. Improper collection results in unrepresentative samples with altered size distributions. Erroneous interpretation resulting from improper sampling could allow for catastrophic engine failure, premature disposal of oil, or inaccurate filter ratings. In order to obtain representative samples, both sampling location and method of collection must be considered. Oil samples may be collected from reservoirs, such as oil sumps or barrels, or from the lines of an operating system. The techniques involved are discussed in detail elsewhere [5,35–37], but the general principles will be reviewed in this section.

Reservoir samples are the easiest to collect, but are less representative than line samples due to sedimentation, mixing and separation processes occurring within the reservoir. Reservoir samples are commonly used in situations where there is no alternative or where samples are collected infrequently. For example, reservoir sampling is used to spot-check incoming barrels of new oil and to monitor engine condition for city buses.

Sampling from the lines of an operating system is preferred to reservoir sampling in situations where it is important that representative sample be obtained. This is recommended for stationary engine lube oil systems where representative samples are used to establish trends in contamination levels and monitor wear rates. This type of sampling is also used to test and rate filters [38–42]. In this case, samples are used to establish the conditions existing at different times during a test. In fine and silt control filter testing, samples collected from the test lines can be plumbed directly to a particle counter for on-line particle counting [38–40]. This is discussed in a later section. This section focuses on the collection of field samples.

Sample collection is critical in field situations, since months may be required to establish trends. Lost samples can leave irreparable gaps in the data. These gaps are expensive because more time is required to establish trends, and transient conditions leading to equipment failure may occur. For this reason, serious consideration of all relevant factors, including human ones, should be given to sample collection. The advantages of reservoir versus line sampling has already been discussed. However, consideration should also be given to the person who actually collects the samples. These individuals should be trained in sampling procedures appropriate to their situations. In addition, the conditions under which samples are collected need to be thoroughly considered. If one is required to crawl under a truck and reach around a greasy starter and hot exhaust manifold to collect a sample, then contaminated or otherwise unrepresentative samples will result, if samples are collected at all.

The location of the sampling port is critical. The sampling port should be located at a point that is accessible and free of excess grease, dirt and water in order to minimize accidental contamination. Samples should be collected from points of turbulent flow in order to obtain representative samples. If one is interested in the quality of oil lubricating an engine or hydraulic component, samples should be collected downstream of the full-flow filter and upstream of the component. This provides an indication of the level of protection afforded to the equipment by the filter. Some recommend sampling upstream of the filter. They suggest that artificially low contaminant levels are monitored after the filter and that this reduces the value of the data as a diagnostic tool for preventive maintenance programs. However, there is little evidence to support this. Particle counting provides an indication of the concentrations of potentially damaging particles in oil, whereas wear metal data are more indicative of wear rates. Thus, wear metal results are useful as a diagnostic tool. Because wear metal concentrations are not altered by passage through filters as fine as 3 μm [43], samples obtained downstream of a filter can provide useful information about both the wear rate and level of protection.

The accessibility of sampling ports can be a concern for operating systems. Engine blocks typically have ports leading to oil galleries which can be fitted with sampling ports. Figure 9 shows a sample being taken from such a port mounted on an engine block. When sampling is completed, the tubing can be disconnected and the port capped to protect it from external contamination. Hydraulic equipment and laboratory filter test stands can usually be sampled through existing plumbing. Filter housings often have convenient entry points or can be readily adapted to accept sampling ports. In general, long horizontal runs of plumbing should be avoided to reduce particle settling. For similar reasons, the dead volume between the sampling point and the sample dispensing locations should also be minimized through the use of small-diameter tubing.

The sampling apparatus, including sampling port, tubing and valves, should be purged with at least five times the dead volume before a sample is actually collected [35,36]. Ball valves, as opposed to metering valves, should be used to dispense samples. Samples should only be taken with the valve fully opened, as particles can become trapped in a partially opened valve. Care must be taken not to contaminate the inside of the container or closure with dirt, oil drippings, or greasy fingers. After purging, the sample should be collected as the oil continues to flow. Apparatus and techniques for collecting oil samples are described in detail elsewhere [5,35–37].

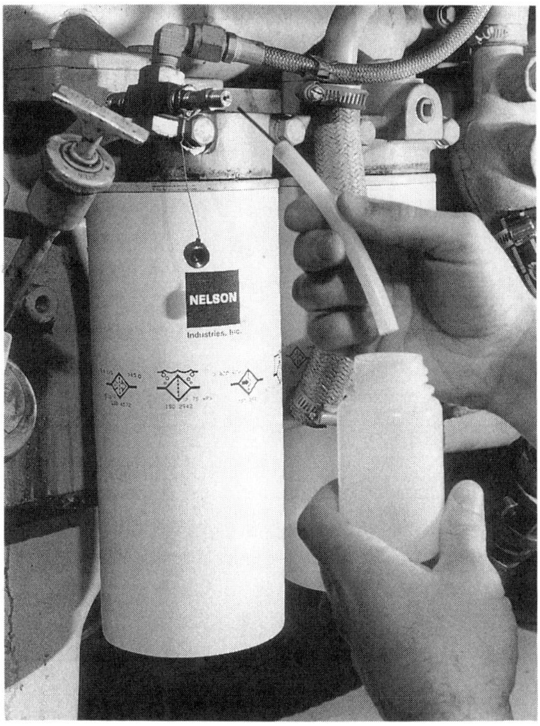

Figure 9 Oil sampling port on a diesel engine. (Courtesy of Nelson Industries, Inc.)

C. Resuspension and Deagglomeration

Deagglomeration and resuspension of contaminants are necessary when oil samples cannot be analyzed immediately. Particle counters are rarely present at sampling sites. The time between sample collection and analysis may range from several hours to several weeks. During this time, particles settle, adhere to container walls, and agglomerate. Therefore, deagglomeration and resuspension are necessary. The importance of resuspension is illustrated in Fig. 10a for typical SAE 5–80-μm test dust and used engine lube oil samples. In each case, the results represent an average of three consecutive particle counts for the indicated sample. The solid lines show data obtained by mechanically shaking the sample bottle between each count. In the case of the dashed lines, no resuspension method was utilized during the 3 min required to complete the analysis. It is significant to note that the data obtained without resuspension tends to be lower than the resuspended sample by more than a factor of about 2. This is a result of settling for particles larger than 10 μm in size. The effects of settling increase with particle size and time. These results show that resuspension of samples before and during particle count analysis is essential if representative results are to be obtained.

Resuspension can be accomplished by hand or mechanical shaking, or by magnetic stirring. The purpose of resuspension is to resuspend contaminants that have settled or are loosely adhering to the container walls and to obtain a homogeneous suspension. Mechanical shaking, using paint or laboratory bench top shakers, is the preferred method [34] since it is more reproducible than hand shaking. Depending on shaker design, samples should be shaken for at least 5 min. Mechanical shaking and other resuspension methods should not be used for

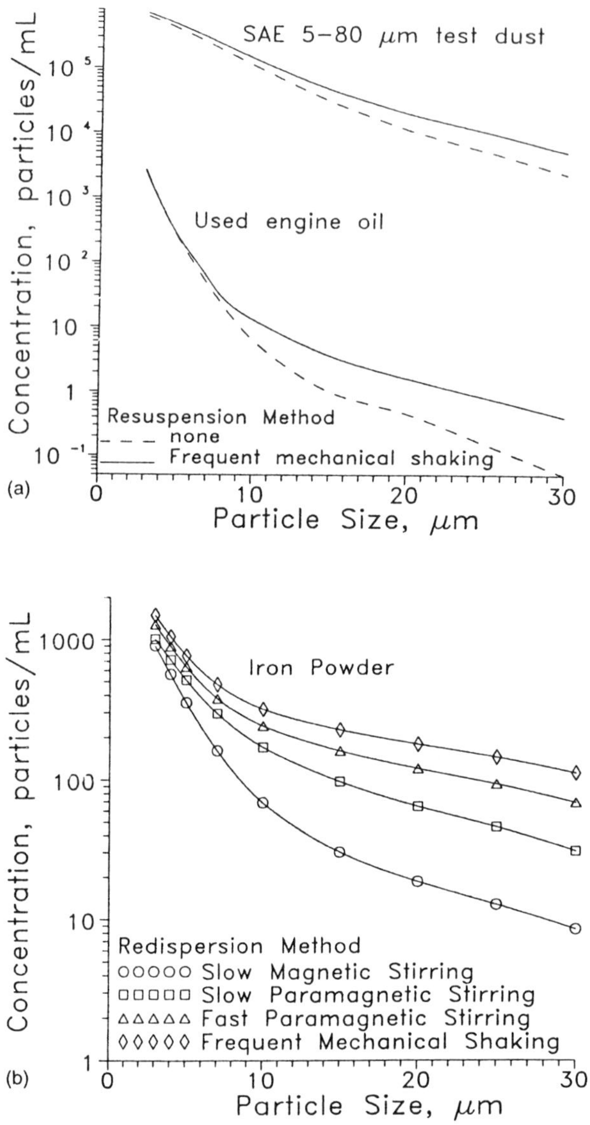

Figure 10 Influence of resuspension on observed particle size distributions: (a) effect of resuspension for SAE 5–80-μm test dust and used engine oil samples, and (b) effect of resuspension method for iron powder sample.

deagglomeration. The forces imparted by shaking are inadequate to break up any but the weakest particle aggregations.

Magnetic stirring is sometimes used for resuspension; however, care must be exercised when using this method. Particle counters equipped with batch samplers often utilize magnetic stir plates. Classification of particles by size and biased particle count data may result if the sample contains magnetic particles. Used lube oil samples typically contain magnetic wear metal particles. ACFTD and related test dusts contain 2.5–5.0% Fe_2O_3 which may be removed by magnetic fields. The effects of magnetic stirring on particle count results is shown in Fig. 10b for oil contaminated with iron powder. When a magnetic stir bar was used to resuspend the particles, the observed concentrations of particles larger than 10 μm were more than five times lower than for samples resuspended using frequent mechanical shaking. Better agreement was obtained when a paramagnetic stir bar was used, but the results were still lower than when mechanical shaking was used. Agreement improved as the mixing speed increased with a paramagnetic stir bar, but sample aeration and bubble formation can interfere with the analysis if excessive speeds are used. Some recommend relatively slow stirring and the use of containers with rounded inside bottoms [11]. If magnetic stirring is used, glass-coated paramagnetic stir bars, not magnetic stir bars, should be used. Frequent mechanical shaking is recommended for samples containing magnetic particles. Samples should not be allowed to stand for more than 1–2 min without resuspension. For batch samplers equipped with magnetic stirrers, it may be necessary to remove the permanent magnet from the unit to avoid problems.

Deagglomeration is accomplished using an ultrasonic cleaner or bath. Contaminant particles can aggregate and change size distribution upon standing. These aggregates are partially disrupted by shear forces upon passage through the sensor [44]. Ultrasonic deagglomeration breaks up aggregates in a reproducible manner. When used immediately prior to analysis, the energy imparted by a fully functional ultrasonic cleaner with a minimum rating of 0.33 W/cm^2 is sufficient for deagglomeration. This minimizes the irreproducible effects of standing and passage through the sensor on particle size distribution. Studies conducted by the authors show that 10 min in an ultrasonic cleaner is needed to fully deagglomerate contaminants in some used lubricating oils. Since the time required varies with oil viscosity and contaminant nature, tests should be conducted to determine an appropriate time.

D. Dilution

For most oil applications, dilution of samples is a necessity to avoid coincidence error and to reduce sample viscosity. In multipass filter testing, contaminant concentrations exceed the coincidence error limits of many particle sensors. Similar problems are encountered in used oil analysis where there is the additional problem that the oil may be too viscous to flow through the sensor at the required flow rate. To address these problems, it is necessary to dilute samples with a clean diluent of suitable viscosity and miscibility characteristics. Since the dilution process adds an additional sample handling step to the analysis, the potential for error exists.

MIL-H-5606 hydraulic oil is the most widely used diluent for oil analysis. It is the standard oil used for hydraulic filter testing, is available worldwide, and is produced to tightly controlled military specifications. MIL-H-5606 is compatible with most petroleum oils, contains few additives, and has a relatively low viscosity of about 17 cP at room temperature. Other diluents, including No. 2 diesel fuel, naphtha, and isooctane, have also been used. In selecting a diluent, it is critical that samples be completely miscible in the diluent, that it be compatible with the sample's additive package and the equipment used, and that it possess optical characteristics similar to the fluid used for particle counter calibration.

Contamination from the diluent is one of the most pervasive problems with dilution. To a large extent, this is controlled through the design and operation of the diluent cleanup and delivery system. A typical system is illustrated in Fig. 11. It consists of an oil sump, pump, cleanup filter, valves and tubing. Each of these is a potential source of contamination. The system should be enclosed and not open to the atmosphere. The sump should be large enough to accommodate several months' supply of diluent, since it is easier to maintain the desired cleanliness levels than to be cleaning new oil continuously. When oil is added to the system, it contains contaminant which must be removed. Pumps, valves, and associated fittings can trap

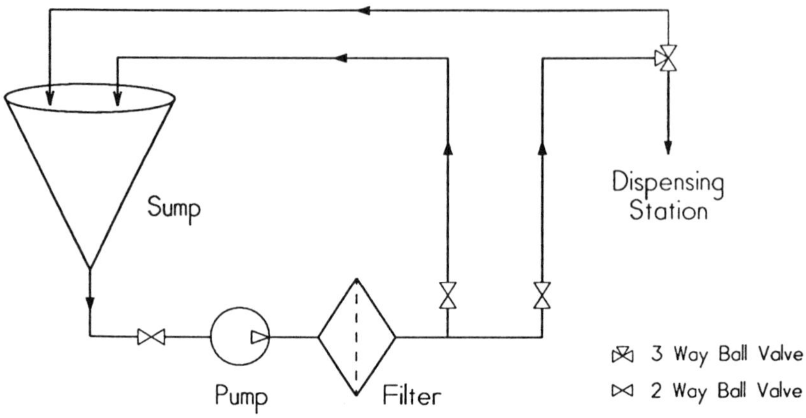

Figure 11 Schematic of diluent cleanup system.

and retain particles for long periods of time. The use of an open impeller centrifugal pump minimizes the pump as a potential source of contaminant. To facilitate cleanup, the pump should be sized such that the oil residence time in the system is less than 5 min. For best results, dilution system components should be made of stainless steel, Teflon, or other inert, easily cleaned material.

At a minimum, the cleanup filter for dilution systems should be capable of removing virtually all particles at the smallest size of interest. However, it should also remove significant quantities of particles smaller than the instrument's detection limit, since these may contribute to coincidence error. For example, a cleanup filter with a filtration ratio greater than 100 at 1 μm and greater than 2 at 0.5 μm is suitable for counting 2-μm particles. The removal of particles smaller than the instrument's detection limit is important. Severe coincidence error problems have been reported, even though insignificant concentrations of particles large enough to be detected were observed. These problems were later attributed to the presence of high concentrations of subcountable particles. Since filters rarely remove 100% of the particles on a single pass, oil should be continuously recirculated through them to obtain the maximum levels of cleanliness. The location of the filter should be as near as possible to the point at which the clean fluid is dispensed. In order to avoid contamination introduced by filter changes, cleanup filters should last at least a year.

The flow of the recirculating diluent should not be interrupted during dispensing. Ball valves can be used to dispense diluent, but it may be necessary to use stainless steel valves that are ultraclean and lubricant free. The silicone lubricants commonly used in these valves are immiscible in petroleum in oil and contaminate the diluent. Ball valves should not be used to control the flow. As an alternative method, short lengths of capillary tubing can be used to obtain the desired flow.

Aside from contamination, the largest source of error in making dilutions is volume measurement. Volumetric glassware is typically used, but *volumetric glassware, e.g., pipettes, flasks, graduated cylinders, etc., are designed to deliver the indicated volume of water.* Since oil differs in terms of its viscosity, density, and surface tension, erroneous and often irreproducible volumes of oil will be delivered. For this reason, use of the mass dilution technique is recommended for oil samples [25,32,45]. In this technique, the mass of sample and diluent used to prepare the diluted sample are determined. The densities of the sample and diluent are determined using a density meter. The volumetric dilution ratio can be calculated from this data.

Only a clean pipette, balance and density meter are required for mass dilution. Thus, contamination from sample handling equipment is minimized. A diluent dispensing/mass dilution system suitable for analyzing large numbers of samples is shown in Fig. 12. Burets for dispensing particle-free diluent, a chamber to protect samples from external contamination, an electronic balance with computer interface, and density meter are shown. Dilution factors in excess of 20,000 can be accurately performed in a single step by this method. In contrast, significant errors are obtained by single-step volumetric methods at dilutions ratios of 100. Multiple-step dilution methods offer a more accurate alternative to single-step volumetric dilution [34], but additional sample handling is required, increasing the chance for error. Since less time is spent cleaning glassware, making multiple dilutions and in data analysis, mass dilution is faster and more efficient than volumetric dilution.

Dilution is used to minimize coincidence error, but at excessive dilutions other problems arise. At high dilutions, the contribution of particles from the diluent, as opposed to the sample, can become important. This can be a significant problem for samples, such as used lube oil samples, which contain high concentrations of fine particles and few large particles. In this case,

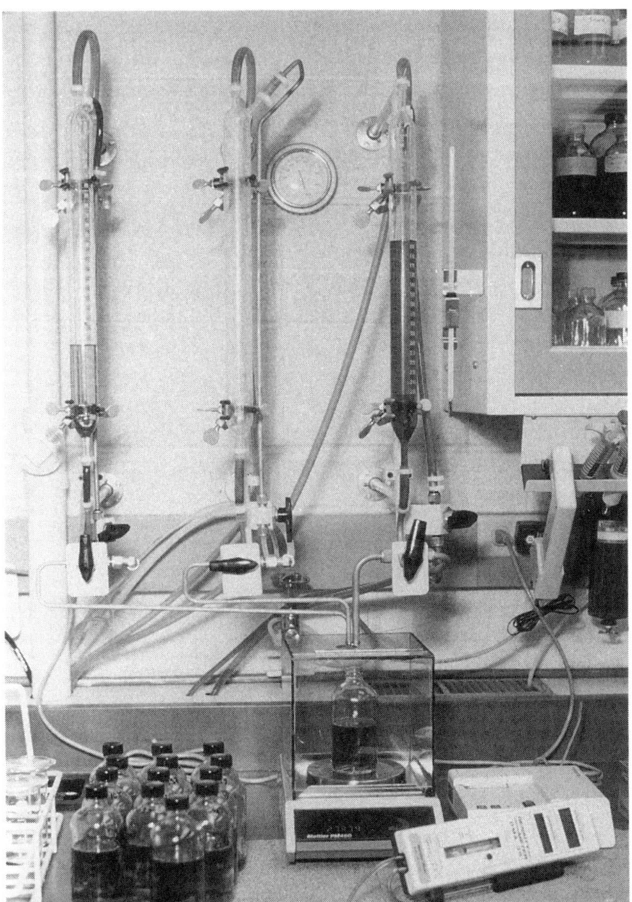

Figure 12 Diluent cleanup and dispensing system for preparing mass dilutions of samples. (Courtesy of Nelson Industries, Inc.)

very few large particles are observed as a result of the high dilution ratios required to minimize coincidence error at the small sizes. If it is assumed that a 1% error due to diluent contamination is acceptable, the following relationship defines the maximum usable dilution ratio, R:

$$R = \frac{101C_d}{101C_d - C_o} \tag{6}$$

where C_o is the observed concentration in the diluted sample and C_d is the contaminant concentration in the diluent at the same size. This relationship is graphically illustrated in Fig. 13 and is only valid for $C_o/C_d < 101$. For a particular C_o/C_d, diluent contamination is insignificant for dilution ratios lower than the corresponding point on the line. At higher ratios, the error due to diluent contamination exceeds 1%. Thus, the line shown in the figure defines the maximum usable dilution ratio as a function of diluent fluid cleanliness levels and the observed particle concentrations in diluted samples.

E. Flow and Volume Measurement and Control

Errors in both flow and rate and volume measurement can occur resulting in erroneous data. Volumetric errors can be on the order of 10% [12]. This produces a 10% error in the observed particle concentration. If such an error occurs during calibration, particle sizing errors are also introduced. The magnitude of the sizing error varies as a function of particle size, but may be as large as 15%. Volume and flow measurement errors may account for a significant portion of the variability observed in past interlaboratory round robin studies.

Batch samplers for automatic optical particle counters typically utilize air pressure to induce oil flow. Alternative methods, including the use of positive displacement and syringe-type pumps to control flow and measure volumes are also available but are less common. For this reason, the discussion of flow and volume errors will focus on air pressure systems.

In air pressure systems, optical level detectors mounted on a graduated cylinder downstream of the sensor electronically signal the counter to start and stop counting. The volume of oil is

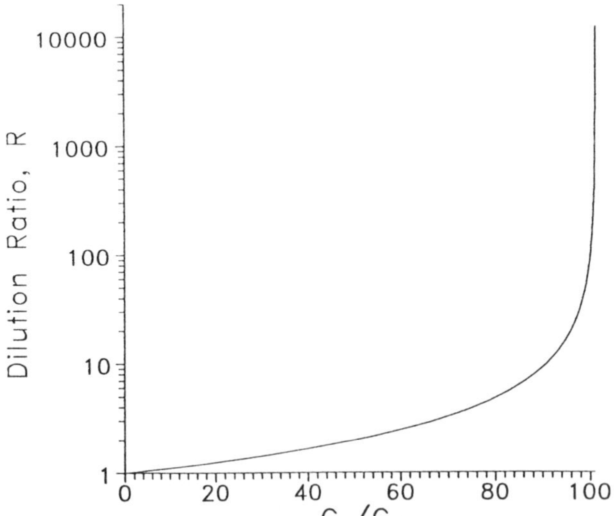

Figure 13 Maximum usable dilution ratio as a function of the ratio of the diluted sample concentration (C_o) to the diluent concentration (C_d).

controlled by the relative positions of the optical level detectors. The flow rate is controlled by varying the applied pressure and restriction.

Transient flow rate errors occur for a variety of reasons and, unless the operator is familiar with them, are difficult to detect. They can occur when counting is initiated before the air pressure stabilizes in the sample chamber. Since the flow rate through the sensor varies during analysis, erroneous results will be obtained. In order to avoid this problem, oil flow can be prevented by a solenoid valve downstream of the sensor, or counting can be delayed, until the pressure stabilizes. Changing viscosity can also cause transient effects. This can occur when samples of differing viscosity are analyzed consecutively without adequately flushing the sensor between samples. Temperature changes also induce viscosity changes.

Changes in restriction across the system can cause flow rate variations. This can occur when particles too large to pass through the sensor or valves become lodged in passageways. A reduction in flow rate or a need for increased pressure to maintain the desired flow rate is often encountered in these instances. However, trapped particles can also prevent particles larger than a certain size from entering the sensor without altering the flow rate. In this situation, reductions in the concentration of particles larger than a certain size can occur. This may be evident when concentration versus size data are plotted or when changes in the data for certain sizes occur between consecutive samples.

Volume measurement errors are one of the least considered sources of error. The authors have measured 10% error in volume measurement for existing systems. The level detectors used to initiate and conclude counting are the source of this problem. The detectors may not trigger in a reproducible manner; hence, variable volumes may be delivered. Changes in the optical properties across the light path of the level detector influence triggering. Oil draining down the sides of the graduate cylinder, turbidity, and changes in sample color or refractive index are among the common causes of irreproducible triggering. Even when level detectors trigger properly, volume errors occur if the detectors are inaccurately placed. The use of relatively-small-diameter graduate cylinders reduces this problem.

F. Coincidence Error and Contaminant Size Distribution

Coincidence error is of special concern in the analysis of oil samples due to the subtle effects of contaminant size distribution. The coincidence error limit of a particle counter varies with particle size distribution, including those particles normally considered too small to count [11]. In routine analyses, only data from a few particle size classes are examined and only a superficial screening of the results for coincidence error is conducted. The quality of the data and resultant interpretation can suffer as a result. In this section, the effects of contaminant size distribution on coincidence error and ways of minimizing these effects are discussed.

The effects of particle size distribution on coincidence error are illustrated in Fig. 14. The particle counter was calibrated according to ISO 4402 and samples of latex and ACFTD were analyzed. For ACFTD, the abscissa, labeled as the actual concentration, was calculated using the ISO 4402 particle size distribution. The coincidence error limit for ACFTD is much lower than predicted by Eq. (1). This is due to the presence of high concentrations of small "subcountable" particles. Individually, these are too small to be detected, but together they produce coincidence errors and are counted as larger particles. Thus, coincidence error limits are much lower for ACFTD samples than for latex, as shown.

Coincidence error can give rise to complex behaviors. This is illustrated in Fig. 15 for 1-μm particles in used engine lubricating oil. At dilution ratios in excess of 2000 (i.e., 1 part sample to 1999 parts of diluent), coincidence error is negligible and the apparent concentration

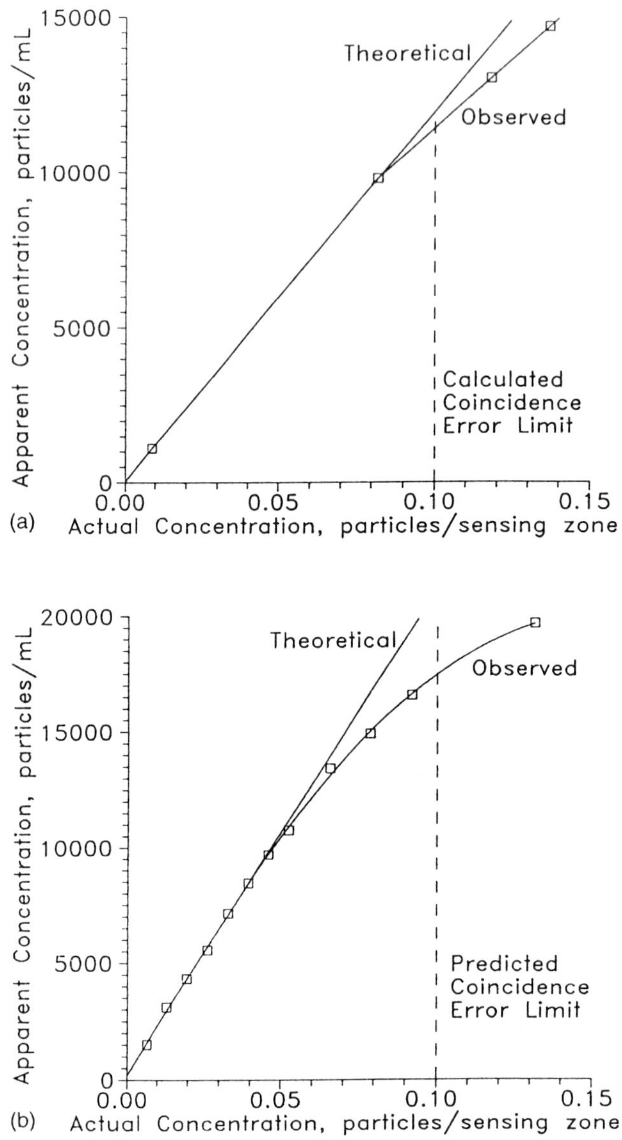

Figure 14 Effect of particle size distribution on coincidence error limits for (a) monodispersed latex and (b) polydispersed AC fine test dust.

does not vary with dilution ratio. At dilution ratios less than 500, classic coincidence error behavior is observed as the apparent concentration increases with dilution ratio. However, the apparent concentration decreases with increasing dilution ratio for dilution ratios between 500 and 2000. This, too, is an artifact of coincidence error. When high concentrations of subcountable particles pass through the sensing zone with larger particles, an apparent increase in the population of large particles is observed, as has been reported previously [11]. As shown in the figure, the effects of coincidence error tend to diminish with increasing particle size. Since there are many more small than large particles, coincidence error is less important at the larger particle sizes.

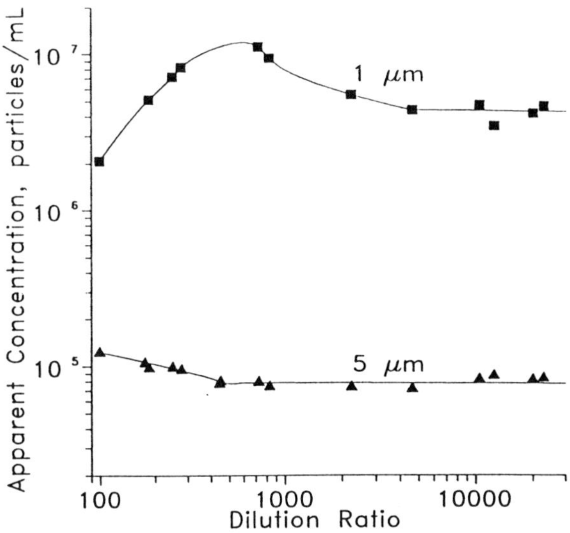

Figure 15 Effect of coincidence error on apparent concentrations for 1- and 5-μm particles in used engine oil.

Both optical and electronic effects influence coincidence artifacts as shown in Fig. 16. The results shown are for samples containing 0.017 particles/sensing zone of 7-μm latex (i.e., 17% of the estimated coincidence error limit) and varying concentrations of the 0–3-μm test dust. Less than 1% of the dust particles were large enough to be detected by the extinction sensor. As shown, the apparent latex size increased with increasing dust concentration. For extinction sensors, high concentrations of fine contaminants, such as 0–3-μm dust, reduce the

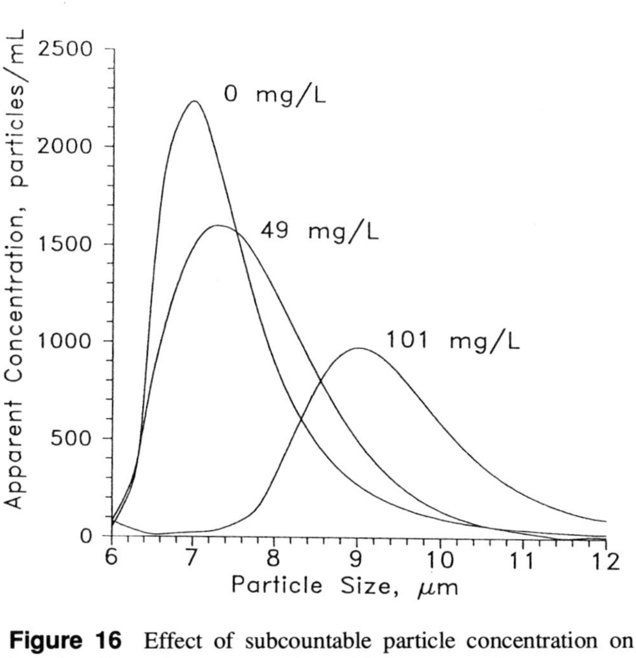

Figure 16 Effect of subcountable particle concentration on apparent latex particle size distribution. Labels refer to the concentration of 0–3-μm test dust.

amount of the light reaching the detector. For scattering sensors, the amount of light reaching the detector increases under the same conditions. In each case, the background noise and apparent particle sizes increase due to the presence of the fine particles [11], as shown. The opposite behavior occurs in scattering sensors when there are high concentrations of particles with large imaginary components to their refractive index, such as black soot particles. In this case, light is absorbed and the apparent particle size is reduced. Figure 16 also shows that the apparent concentration of latex decreased with increasing dust concentration. This, too, can be explained in terms of electronic and optical saturation [11].

In some particle counters, baseline correction circuitry can also yield particle sizing errors under certain conditions. At very high concentrations of fine particles, some will always be present in the sensing zone. These influence the gross optical properties of the fluid. Baseline correction adjusts for them as if they were part of the fluid continuum. At high concentrations of polydispersed particles, a stable baseline may not be established since conditions do not remain stable long enough for an accurate reference to be established. Though this is a concern for field samples, especially if insufficient dilution is used, errors associated with the use of instruments without baseline correction are probably even greater for samples of this type.

Due to size distribution and concentration effects, it is not always obvious that coincidence error and electronic saturation are occurring, particularly when a sample has only been analyzed at a single dilution. The only method to ensure that coincidence error is not occurring is to compare the results from several different dilution ratios. If the normalized results (i.e., the observed particle concentration times the dilution ratio) do not vary with dilution ratio, then coincidence error is insignificant. Other cursory screening tools are also available. Most particle counters have overconcentration alarms to alert operators of these conditions. However, coincidence error and saturation may occur at concentrations well below those required to trigger a warning. For this reason, it is common practice to monitor the apparent concentration at the smallest detectable size. If the apparent concentration exceeds the coincidence error limits, then additional dilutions are made. Though useful, cursory methods are limited in that concentrations may be so high that the apparent concentration may appear to be less than the coincidence error limits. However, the absence of cursory warnings doesn't mean that coincidence error is negligible. One can only ensure this by analyzing several different dilutions of a sample and comparing the normalized results.

G. Contaminant Nature

Oil samples contain many types of particles of varying shapes and optical properties. Silicates enter oil via leakage, contamination and atmospheric sources. Highly reflective, shiny wear metal particles are common in lubricating and hydraulic systems. In engine lubricating oil, soot is present. Transparent water droplets may be found in fuel, lubricating and hydraulic oil, especially in marine, off-road, and agricultural equipment. Sludge resulting from biological growth or oil additive breakdown may also be present. In residual fuel oil, black semisolid asphaltenes, as well as siliceous catalyst particles are present. Occasionally, cellulose, polymer and glass fibers may be present. The diversity in particle size, shape, and substance influences particle-count results.

The optical properties of particles are particularly important. As was discussed previously, particle sizing errors for extinction sensors can result from differences in the imaginary component of a particle's refractive index. For scattering sensors, even larger errors may result. Especially large sizing errors are encountered by scattering sensors for black particles and shiny particles, like wear metal chips, due to their unusual optical properties. As a result, scattering sensors should not be used for used oil analysis.

Fibers and other particles with long aspect ratios present special problems in particle counting. They may plug sensor orifices. Since fibers may be longer than the sensing zone, itself, other contaminants passing through at the same time as a fiber will be incorrectly sized and counted. The apparent size of nonspherical particles, particularly those with large aspect ratios, depends upon their orientation in the sensing zone with respect to the light source [26,27,46].

H. Immiscible Droplets and Bubbles

Optical particle counters detect immiscible droplets and gas bubbles as well as solid particles. Since droplets and bubbles are normally not of interest and since their size distribution is easily changed, they require special attention as a class of contaminants. Air bubbles result from sample agitation occurring during resuspension and other steps of sample analysis. Immiscible droplets, usually water, may be present as a result of leakage or condensation. Special procedures are necessary to identify and deal with droplets and bubbles.

Water droplets increase the chance of coincidence error, skew observed particle size distributions, and yield irreproducible results [31,34,45,47]. Skewed results typical of the presence of water are shown in Fig. 17. The observed size distribution in the presence of water drops often deviates from the log-log^2 relationship typically exhibited by field samples and ACFTD samples. Excess concentrations of particles are counted at sizes corresponding to the average drop size. For oils containing high concentrations of additives or emulsifying agents, the concentrations of small particles are excessive. This behavior is illustrated in Fig. 17 using a diesel-fuel-in-water emulsion. For relatively additive-free oil or oil with extremely high water content, the concentrations of large particles tend to be high. In some cases, a plateau may occur over which the apparent concentration does not change as a function of size. This is shown in Fig. 17 for an engine oil sample. Results from samples containing immiscible droplets or bubbles should be regarded as inaccurate, unless appropriate precautions have been taken.

Bubbles are relatively easy to remove. Oil samples should be degassed immediately prior to particle counting using an ultrasonic cleaner or vacuum. In some cases, bubbles can form

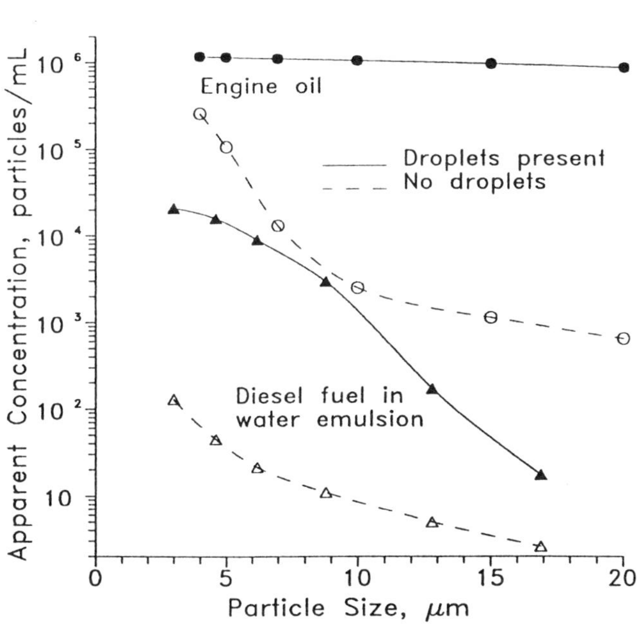

Figure 17 Effect of water on apparent particle size distributions.

within the sensor. This is often a problem for volatile fluids, but it can also occur with some petroleum products. The problem can be avoided by selection of an appropriate nonvolatile diluent or by installation of an adjustable valve downstream of the sensor. In a partially closed position, the valve increases the pressure downstream of the sensor, reducing bubble formation.

Water droplets are difficult to deal with. In petroleum oils, the solubility of water is 100–200 mg/l. Water droplets are present at higher concentrations, even if they are not visible. Oil samples, particularly field samples, should be routinely inspected for free water before particle counting. The appearance of free (settled) water in the bottom of the sample container or a cloudy appearance to the sample indicates the presence of undissolved water. A qualitative test, referred to as a splatter test, is recommended for routine checks for the presence of water in field samples [34]. Microscopy is also useful for qualitative screening. Infrared spectroscopy and Karl-Fisher water analysis techniques quantitatively measure water concentrations in oil, but are time consuming and some oils contain interferences which influence the results. Samples containing excess water should not be analyzed unless precautions are taken to deal with this interference.

The water incorporation technique has been shown to be effective in dealing with water droplets [31,47]. It uses a diluent consisting of MIL-H-5606, naphtha, or similar solvent, containing a minimum of 12 g of Aerosol OT per liter of diluent. This is the same additive used in the latex in oil calibration procedure [9]. At this concentration, Aerosol OT forms reverse micelles. Undissolved water present in the sample is incorporated into the reverse micelles upon dilution. The resultant water-swollen micelles are about 0.002 μm in size [31,47,48], which is undetectable by optical particle counters. Thus, only solid particles will be counted in the diluted samples. The technique is effective for most petroleum oils [31,45,47] over a wide range of water concentrations. With appropriate dilution, samples of used cutting oil containing 95% water have been analyzed [31]. A variation of the technique, referred to as the phase inversion technique, has been used to analyze water samples contaminated with silicone oil droplets [47]. Past work has shown that Aerosol OT, used in this manner, does not significantly alter the observed particle size distribution [31,45]. Therefore, Aerosol OT can be routinely used in diluents, even if no water is present, without adversely affecting the quality of the results.

I. Particle Counting Statistics

Variability in particle-count data results from both systematic and random sources of error. Systematic errors bias particle count results in a particular direction. For example, an improperly calibrated sensor that counts 4-μm particles as 5-μm particles will yield high counts at "5 μm" for polydispersed samples. On the other hand, random errors increase the variability in particle-count results. Because of the random distribution of particles in a sample, some variability inevitably occurs. Variability increases as the concentration of particles decreases. The effects of random errors on particle-count results can be quantified using statistical analysis [49–50].

While particle counting a sample, particles pass through the sensor at random time intervals. The average number of incoming particles over a given time period is described by the Poisson distribution [49]. An important property of this distribution is that the variance, σ^2, has the same magnitude as the mean. Thus, the *CV* for particle-count results that can be expected due to chance alone is

$$CV = \frac{100}{\sqrt{X}} \tag{7}$$

where $\overline{X}$ is the mean number of counts per sample. This equation is the theoretical *CV* that would be expected if the random distribution of particles in the fluid was the only source of variability. In practice, the actual *CV* will be larger because other factors also contribute to variability.

Another source of variability is the measurement of the sample volume. This results from variability in volume and flow measurement devices. This contribution to variability was not considered in previous treatments [49–50]. The overall variance that can be expected from multiple sources of error is the sum of the variances due to each contributing error. If volume measurement and random distribution (Poisson) errors contribute to actual variability, then

$$\sigma^2_{\text{Actual}} = \sigma^2_{\text{Poisson}} + \sigma^2_{\text{Volume}} \tag{8}$$

where the subscripts refer to the source and type of variability. σ^2_{Volume} can be expressed in terms of the *CV* of volume measurement, CV_{Volume}, by the following equation:

$$\sigma^2_{\text{Volume}} = \left(\frac{CV_{\text{Volume}}\overline{X}}{100}\right)^2 \tag{9}$$

and the expected *CV*, CV_{expected}, due to Poisson and volume measurement error is

$$CV_{\text{expected}} = 100\sqrt{\left(\frac{CV_{\text{Volume}}}{100}\right)^2 + \frac{1}{\overline{X}}} \tag{10}$$

Previously, it was recommended that the *CV* in volume measurement not exceed 3% [13–14]. This value was calculated using the preceding equation in order to maintain CV_{expected} below 10% for particle counts greater than about 100. The relationship between the expected *CV* with and without volume measurement error as a function of the observed particle count is shown in Fig. 18. It shows that the CV_{expected} decreases with increasing particle counts. A comparison of

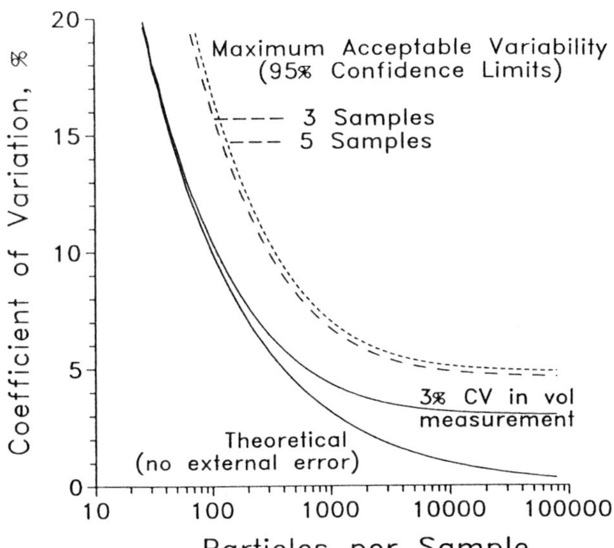

Figure 18 Influence of the numbers of particles counted and other factors on the coefficient of variation in particle count results.

the lines with and without volume error considered shows that Poisson statistics dominate at particle counts fewer than 100. In contrast, volume errors govern variability at high counts. One would expect to improve counting statistics by using lower dilution ratios, but coincidence error is also increased. This suggests that a more effective way to reduce variability is to increase the number of particle counts by analyzing larger volumes of sample.

From the preceding equations, objective guidelines for rejecting results based on excessive variability can be obtained. Multiple counts (three or more) must be made on every sample in order to statistically evaluate the results. The *CV* from these counts can then be used to conduct a χ^2 test. Details concerning the development of this method are covered in statistics textbooks and will not be repeated here. Based on the assumptions previously made, variability is excessive if

$$CV_{\text{sample}} > CV_{\text{expected}}\sqrt{\frac{\chi^2}{N-1}} \tag{11}$$

where CV_{sample} is the *CV* for the sample, CV_{expected} is the *CV* calculated from the preceding equation, *N* is the number of times that the sample was counted, and χ^2 is the chi-squared statistic for the desired confidence level and a number of samples. The results for a particular sample can be rejected if CV_{sample} exceeds the value of the right-hand side of the inequality.

For purposes of illustration, Fig. 18 shows these critical levels as a function of particle counts for $N = 3$ and $N = 5$ assuming CV_{Volume} is 3% and a confidence level of 95%. If the CV_{sample} exceeds the critical values shown for the corresponding numbers of counts, the results should be rejected and the source of the variability determined. It is important to note that the critical levels of CV_{sample} change relatively little as *N* is increased above three. Thus, for routine work each sample should be counted at least three times while for more critical work, such as particle counter calibration, five counts should be used. If only one count is made, little confidence can be placed in the results.

Typical sources of variability include particle settling, air bubbles, volume or flow measurement problems, and sensor plugging. Settling can often be identified by a pattern of decreasing numbers of particles at the larger sizes of consecutively run counts. The problem of air bubbles and immiscible droplets has already been discussed. Typically, it is manifested as an increase in the numbers of particles of a certain size. A plot of the log of the number of counts versus particle size that deviates from a smooth curve is often observed in conjunction with this problem, but such deviations may also indicate partial sensor plugging. Random volume and flow problems are difficult to detect. During calibration, the fluid delivery system should be thoroughly checked out in order to minimize occurrences of this problem. Sensors should be routinely back-flushed to keep flow and optical paths clear of obstructions. A number of other problems can also contribute to excessive CV_{sample}. These are discussed elsewhere [9].

V. BASIC OIL ANALYSIS METHODOLOGY

In order to demonstrate the basic oil analysis methodology, filter test samples from a multipass filter test [38,40–42], will be considered. This type of sample is commonly analyzed in conjunction with hydraulic and lube oil filter testing. Particle count analyses of this type are critical since they provide the basis for filter size ratings and for comparisons of competing products. In comparison to used oil samples, the analysis of these samples is relatively straightforward since a standardized test fluid and contaminant are used and no water is present. The oil is a standard test fluid, MIL-H-5606, with specified chemical and physical properties. The contaminant is AC fine or a related test dust. AC fine is a naturally occurring, polydispersed test dust with controlled particle size distribution consisting primarily of SiO_2, Al_2O_3, and Fe_2O_3 [51]. The major concerns in the analysis of multipass samples involve contamination and coincidence error.

To demonstrate the basic methodology, this section will deal with the analysis of samples collected in particle-free glass bottles. These samples are typically obtained from tests involving filters with beta ratios at 10 μm of less than 50. Such samples may contain up to 60 mg/l of ACFTD; hence, dilution is a necessity to avoid coincidence error. Depending on the work load, samples may stand several days before analysis. In preparation for particle counting, the permanent magnet used as a stirring device for the batch sampler should be removed, since AC fine test dust contains magnetic particles. The instrument must be calibrated and its operating limits determined. The instrument is used in the cumulative mode and threshold settings set to correspond to particle sizes greater than 2, 5, 10, 15, 20, and 30 μm.

Resuspension and deagglomeration are necessary if samples have been allowed to sit for more than a day. However, it is good practice to use resuspension and deagglomeration even if samples have not sat for long periods of time. After several days, a brown layer of test dust ACFTD sediment is often visible at the bottom of sample bottles. This layer consists of the particles and aggregates of greatest interest, i.e., greater than 2 μm. First, the aggregates are broken up by placing the sealed sample in an ultrasonic bath rated at 0.33 W/cm^3 for 30 s. The sample is then shaken and returned to the ultrasonic bath. This process is repeated twice to ensure complete resuspension and deagglomeration of particles. Following this, the sample is then placed on a mechanical shaker for about 5 min to achieve a homogeneous suspension. The time needed for resuspension and deagglomeration will be different for other types of samples and equipment. Once this has been completed, the sample is not allowed to sit undisturbed until the analysis is completed.

Before dilution, diluent cleanliness must be determined. This is done by filling a clean bottle about 75% full with clean diluent (i.e., MIL-H-5606) and obtaining several particle counts of at least 10 ml each. The desired flow rate through the sensor can be established and residual contamination flushed from the particle counter during this process.

The next step is dilution. Dilution is important since AC fine contains high concentrations of particles smaller than 5 μm and since filtration further enriches the sample with regard to fine particles. Often, analysts are able to estimate the required dilution ratio based on past experience. The estimate can be confirmed by analyzing additional samples at dilution ratios differing by factors of 2 and 4. Since the true particle concentration of the sample is unknown, the analysis of three different ratios is needed to confirm that coincidence error is negligible.

Dilutions are made on a mass basis in the following manner. A particle-free bottle is weighed. Using a clean pipette, the required amount of oil sample is added and its mass recorded. The diluent is then added and its mass recorded. The diluted sample is then capped and placed on a mechanical shaker until it is analyzed. The dilution ratio is calculated from the mass of sample and diluent.

To begin counting, the diluted sample is removed from the shaker and degassed. Degassing can be done by placing the sample in an ultrasonic bath for 15 s or by using vacuum. The sample is then placed in the sample chamber, sealed and flow through the sensor initiated. The particle counter should be purged with 5–10 ml of sample. After the flow has stabilized, counting is initiated. At least three consecutive 10-ml particle counts are obtained. The total time elapsed between the time the sample is removed from the shaker until completion of counting should not exceed 3 min. If the time exceeds this, this sample should be removed, redispersed, and degassed.

The final step in the process is data interpretation. The mean, standard deviation and *CV* of the raw data, as well as the normalized particle concentration should be calculated for each size. The normalized concentration is the observed concentration, expressed in terms of particles larger than the indicated size per milliliter times the dilution ratio. The *CV*s should be compared to the corresponding observed number of counts in order to determine if the results are acceptable. In Fig. 18, the relationship between acceptable *CV* and number of counts was shown. If the *CV* exceeds the value shown for a given number of counts, the data should not be accepted. Typically, errors due to particle settling, the presence of air bubbles, flow rate fluctuations, or blockage of the sensor's optical or flow path cause excessively high *CV*s [9,13,14]. These errors must be corrected and the sample reanalyzed before the data can be accepted.

Once acceptable data has been obtained, the normalized results from the various dilutions should be compared to determine the conditions under which coincidence error is negligible. When coincidence error is negligible, similar normalized results should be obtained for consecutive dilutions. If this is not observed, further dilution is required, unless the observed counts are so low that the results can be attributed to "noise" (refer to Fig. 13).

In order to obtain statistically sound data, several different dilutions may be required. It may be necessary to use different dilutions for different particle sizes. At large sizes, the size distribution is truncated by the test filter and very low dilution ratios are required. Meanwhile, the sample is enriched in small sized particles, and high dilution ratios may be required for these sizes. Thus, a series of dilutions may be required to obtain the data necessary for all particle sizes. Different dilutions may also be needed for samples collected upstream and downstream of the test filter. Since filter efficiency changes over the course of a test, it may be necessary to utilize various dilution ratios.

The basic methodology just described is intended for use with multipass filter test samples collected in bottles. For used oil samples and for on-line particle counting, additional considerations are involved. Even with this basic analysis, significant interlaboratory variability has been reported in some studies [9,12,25]. The major sources of variability are associated with instrument calibration and dilution. Reasons for the variability inherent in ISO 4402 [16] and ANSI/(NFPA) T2.9.6R1 [9] have already been discussed. In the case of dilution errors, the problems stem from "shortcuts" taken to speed up analysis and failure to identify conditions of coincidence error.

VI. ON-LINE PARTICLE COUNTING

On-line methods couple a particle counter directly to the sampling and dilution systems. In this way, many sources of error inherent in conventional off-line methods are reduced, as is the

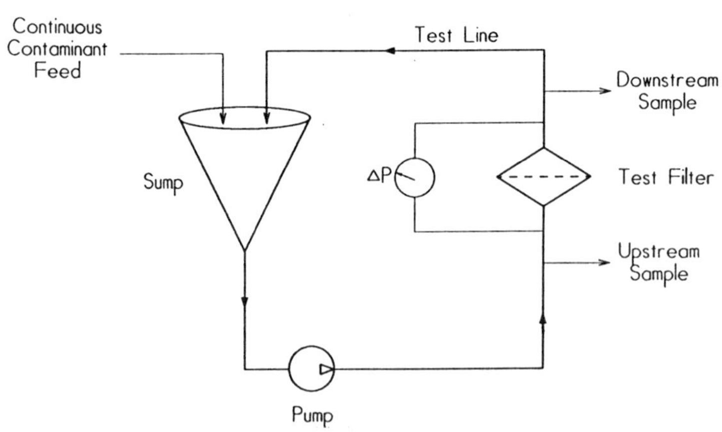

Figure 19 Schematic of multipass filter test stand.

amount of labor involved in particle counting. On-line particle counting is primarily used for testing fine oil filters, although it is becoming important for evaluating oil cleanliness in sensitive equipment. These oil filters are designed to remove particles between 1 and 40 μm in size. In order to discuss on-line particle counting, it is necessary to introduce the filter test procedures with which it is used.

Multipass filter test methods are used to evaluate the performance of hydraulic and lube oil filters [38,40–42]. In the multipass test, oil contaminated with a standard test dust is continuously recirculated through the test filter while upstream and downstream samples are collected and analyzed by particle counting. Throughout the test, additional contaminant is added to the system in slurry form to replace that which has been filtered out. The test rig is schematically illustrated in Fig. 19.

Filter removal efficiency varies as a function of particle size. This is illustrated graphically in Fig. 20 for a typical filter used for mobile hydraulic applications. In this figure, filter performance is expressed in two ways, as a beta (i.e., filtration) ratio and as an efficiency for particles larger than a particular size. In hydraulic applications, the beta ratio [38,40] is used, whereas efficiencies are used in lube oil applications [42]. Beta ratio, β_X is defined by the following equation:

$$\beta_X = \frac{N_U}{N_D} \tag{12}$$

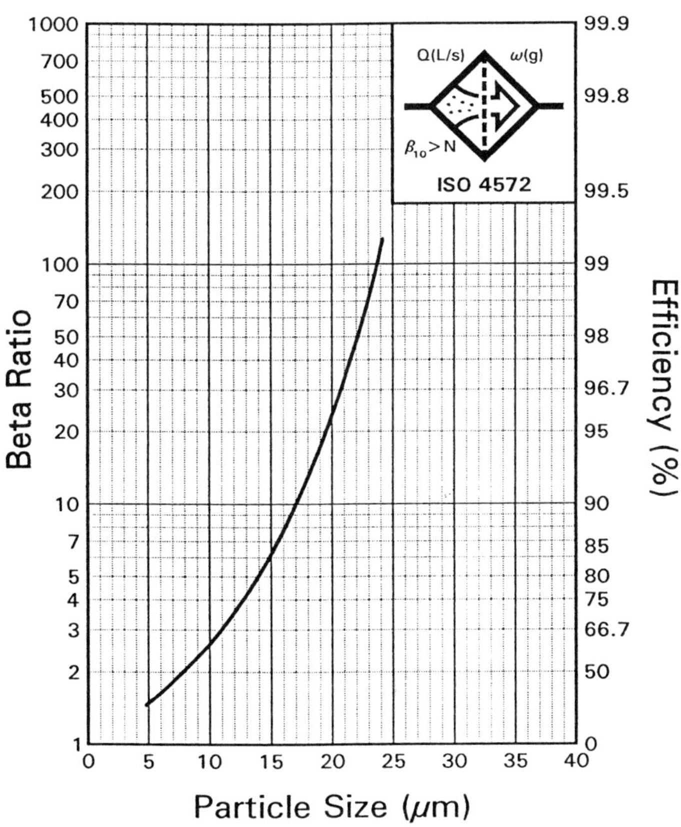

Figure 20 Particle removal as a function of particle size for a typical hydraulic oil filter.

where N_U is the upstream concentration of particles larger than the indicated size, X, and N_D is the downstream concentration of particles larger than X. Percent efficiency, E_X, as defined by the multipass test, is related to beta ratio in the following manner:

$$E_X = 100\left(\frac{\beta_X - 1}{\beta_X}\right) \tag{13}$$

Figure 20 has significant implications for particle counting. It shows that the filter truncates the size distribution of the contaminant. The size distribution becomes enriched in fine particles and depleted with regard to coarse particles. Since agreement between sensors deteriorates as the proportion of fine particles increases [25], the upstream and downstream sensors may be sizing particles differently. This is discussed in detail later.

Coincidence error is a major concern in multipass filter testing. Since contaminant is continually injected into the test stream and since fine particles are removed with relatively low efficiency (refer to Fig. 20), their concentration increases over the course of a test. Depending on the filter being tested, concentrations exceeding 60 mg/l may be present at the end of a test. The magnitude of coincidence error is, in part, governed by the choice of contaminant. Three standard contaminants are used in oil filter testing: AC fine [38,40], AC coarse [41], and SAE 5–80-μm [42] test dusts. Their particle size distributions are shown in Fig. 21. AC fine test dust is used for fine hydraulic filters; AC coarse is used for coarse hydraulic filters; lube oil filters utilize SAE 5–80-μm test dust. As a result of the contaminant size distribution and test conditions, coincidence error is an important concern when AC fine and AC coarse are used. This is of less concern with SAE 5–80-μm dust since it contains fewer fine particles than AC fine.

During the course of a multipass filter test, the contaminant concentration and size distribution changes. This is illustrated in Fig. 22 for a typical hydraulic filter test. The changing conditions imply that the analytical technique (e.g., dilution ratio) must be adjusted. Depending on the sizes of interest, it may be desirable to use results obtained with different dilution ratios in order to obtain significant numbers of particles.

On-line particle counting is finding increasing use. The amount of labor required for on-line

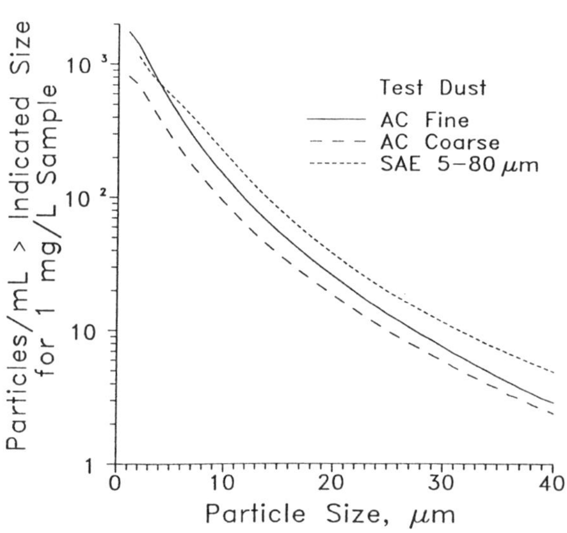

Figure 21 Particle size distributions of multipass filter test contaminants: AC fine [16], AC coarse [41], and SAE 5–80-μm (ISO medium) test dust [25].

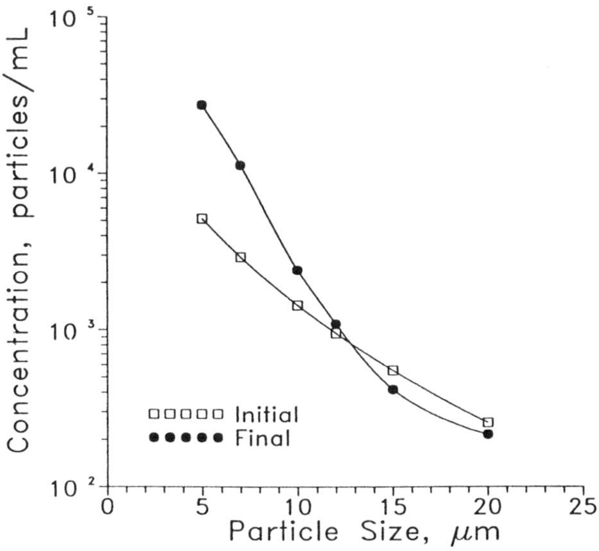

Figure 22 Changes in contaminant size distribution during a multipass filter test.

counting is much less than for off-line counting. With advances in computers, data acquisition and control, it is increasingly more cost-effective to utilize on-line particle counting. For high-efficiency fine filters, on-line particle counting is a necessity since small amounts of contamination can drastically affect the beta ratio results. Because sample handling and associated steps are avoided, more accurate results can be obtained with on-line methods.

On-line particle counting systems are integral parts of some multipass filter test rigs. A schematic of a typical on-line particle counting system is shown in Fig. 23 [39]. Sampling

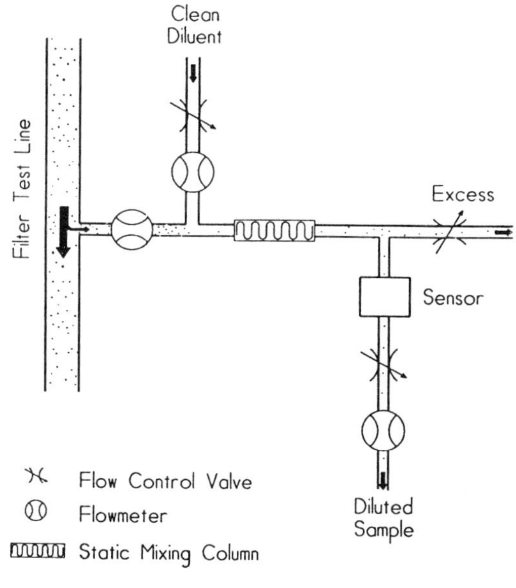

Figure 23 Schematic of on-line particle counting system.

points are located upstream and downstream of the test filter in sections of piping where the fluid is well mixed and turbulent flow conditions exist. The distance between sampling points and sensors should be as short as possible, and long runs of horizontal tubing must be avoided. The flow rate through the sampling lines should also be maximized to prevent particles from settling. Turbulent flow in the sampling, transport and mixing lines is desirable. Particular care is needed if the system is intended to measure particles larger than about 40 μm. A desirable side effect of these precautions is that the time from the test loop to the sensors is minimized.

The dilution system is a critical part of the system, since dilution ratios up to 100 may be needed. Dilution is achieved by mixing sample and diluent flow streams. In order to attain the required accuracy, the diluent must contain fewer than 5 particles/ml greater than 5 μm, unless it can be shown that higher concentrations will not increase the error by more than 1%. The upstream and downstream sample flow rates, as well as the corresponding diluent flow rates, must be accurately measured and controlled. The total flow rate of the combined sample and diluent streams must also be accurately measured and controlled since the results are sensitive to flow rate and sample volume is not measured directly. The sample volume is calculated from the measured flow rate and time required to obtain the data. Since particle concentrations change over the course of a test, the dilution system should be adaptive, i.e., as the concentration of particles increases, the system should automatically adjust the dilution ratio to avoid over-concentration, yet still yield significant numbers of particles. High-precision flowmeters and computer-controlled variable flow valves which do not act as particle traps or sources are necessary to achieve this. As with bottle sampling, it is important that the diluent and sample be well mixed before counting. To ensure this, the combined diluent and sample streams are passed through in-line static mixers.

On-line systems differ from off-line systems in that different sensors are used for the upstream and downstream samples. In order to obtain accurate filter efficiency results, the two sensors must be matched. Care must be taken to ensure that the upstream and downstream sensors agree (within tolerance limits) on the number of particles at each size. This can be difficult to achieve since agreement between sensors is influenced by the size distribution of the contaminant.

On-line counting has significant advantages over off-line counting. The risk of contamination during sample handling is negligible in on-line particle counting, since the particle counter is plumbed directly to a sampling device. Deagglomeration is unnecessary and the passage of the sample stream through static in-line mixer is sufficient to ensure sample homogeneity. As a result, sample preparation, the most time-consuming part of analysis, is eliminated and on-line results are available almost instantly. The latter advantage is important in filter testing and equipment condition monitoring where circumstances change and it may be necessary to take corrective action to avoid costly failures or wasted time. Despite these advantages, calibration and dilution pose critical challenges and the issue of agreement among instruments is especially important.

ANSI/B93.73 [39] utilized a three-step process to calibrate on-line particle counters used for multipass filter testing. First, the particle counter itself must be calibrated, then the calibration is verified on-line, and finally the dilution system is validated. There are three methods used for obtaining on-line particle counter calibration curves. The most basic is to calibrate the on-line particle counter off-line with a batch sampler. This involves disconnecting the counter from the test stand and reconnecting it to a batch sampler. The instrument can then be calibrated according to ISO 4402 [16] in the usual way. A modification of this utilizes a separate calibration loop installed on the test stand. A suspension of AC fine test dust in MIL-H-5606 hydraulic oil is recirculated at the appropriate flow rate through on-line sensors plumbed in series. For this and subsequent steps of the calibration, the multipass test filter housing should be replaced with a short straight section of pipe. With the test rig configured in this manner, the on-line particle counter can then be calibrated on-line according to ISO 4402 [16]. Both of these methods are acceptable primary on-line calibration methods [39].

The third method of calibration is not a primary calibration method but rather is a verification method. It involves setting up the test stand as for a multipass validation test with a section of pipe replacing the test filter. AC fine contaminated MIL-H-5606 containing concentrations well below the coincidence error limit is recirculated through the stand while upstream and downstream particle counts are obtained with the on-line system. No dilution is used. Simultaneously, bottle samples of the recirculating suspension are collected from a point near the on-line sampling points. These samples are then analyzed using a calibrated off-line particle counter with threshold settings corresponding to the sizes of interest. The results from the bottle samples are compared to those from the on-line sensors and appropriate adjustments made to the on-line settings in order to obtain agreement. Since this method relies on a previously calibrated instrument, it is a secondary method and the associated errors are greater than with a primary method. This "match calibration" method is useful for routine verification of instrument calibration since it does not require plumbing changes be made to the test rig.

The second step in calibrating on-line systems [39] utilizes a suspension containing a known concentration of AC fine test dust in MIL-H-5606. The AC fine concentration should be $50 \pm 5\%$ of the sensors' concentration limits. This suspension is prepared in the test stand, recirculated and analyzed using the on-line system without dilution. The results are compared to the corresponding numbers of counts for each size which would be expected for AC fine at the designated concentration. The expected and observed results for both sensors must agree within ±5%. This step constitutes verification of the calibration under the conditions with which it will be used.

The final step in on-line particle counter calibration is to validate the accuracy of the dilution system. A suspension of AC fine test dust in MIL-H-5606 is prepared as in the previous step, but this time the concentration should be $80 \pm 5\%$ of the sensors' concentration limits. The suspension is analyzed using a dilution ratio of 2, the maximum dilution ratio, and two intermediate dilution ratios. The normalized results at all four dilution ratios for both sensors must agree within ±5%. If the final requirement is met, the on-line system can be considered calibrated and validated.

Even though an on-line system has been calibrated and validated, there are limitations that need to be considered in interpreting the results. Agreement between sensors is sensitive to the contaminant size distribution. It has been shown that beta ratios obtained by the same laboratory using different particle-counting systems can vary more than a factor of 2 due to differences in instrument response to changes in size distribution [25]. This is of enormous significance industry-wide, since many different types of instruments are used to analyze filter test samples.

In a single laboratory, identical "matched" sensors are normally used for on-line counting; however, ANSI/B93.73 [39] only compares the results from the two sensors for AC fine test dust. The size distribution becomes enriched in fine particles over the course of a multipass test and the sensors see different size distributions as a result of truncation by the test filter. It is not uncommon for the sensors to respond differently to size distribution changes. This influences the apparent beta ratio. The failure of the sensors to yield similar results for truncated size distribution is, in part, due to sensor resolution effects [12]. As the size distribution becomes steeper, more particles smaller than the size of interest are erroneously counted. Uncertainty in sensor calibration is another reason for poorer agreement.

The effects of uncertainty in calibration is shown in Table 2 for AC fine test dust, as well as upstream and downstream multipass samples. A 5% error in size determination, as was assumed in the table, can produce an 8% error in the observed number of particles for AC fine,

Table 2 Effect of a 5% Particle Sizing Error on Observed Concentrations of 5-μm Particles

Sample	$\frac{N_3}{N_7}$[a]	Relative error in observed particle concentration (# ml^{-1} > 5 μm)
AC Fine test dust	3	8%
Multipass upstream	10	15%
Multipass downstream	23	20%

[a]Ratio of the concentrations of particles larger than 3 μm to those larger than 7 μm.

but a 20% error for a downstream sample. Sizing errors influence agreement between sensors and have a large impact on the calculated beta ratios. The sensor performance specification which is part of the hybrid calibration procedure [13,14] is the best means available for ensuring agreement between sensors and laboratories [13,14,25].

Changing test conditions, which do not occur during calibration and validation, can introduce transient errors. System temperature and volume must be controlled to within 1% while flow rates must be controlled to within 5% during testing [39]. Vibration and electrical noise can introduce artifacts into particle-count data. Valve adjustments, made in response to pressure, flow or concentration changes, can be a significant source of contamination. Valve components can shed particles while being adjusted; hence, adjustments should be avoided during and immediately preceding a particle count. Sensors can become blocked and should be back-flushed regularly with clean diluent during tests in order to clear their optical and flow paths.

Inaccuracies in flow rate measurement and control are a major source of uncertainty and contribute to lack of agreement among sensors. Figure 24 illustrates the two most common arrangements for measuring and controlling flow. Both directly measure the diluent flow rate, but they differ in terms of how the sample flow rate is determined. In Fig. 24a, the sample flow rate is measured directly using a flowmeter placed before the point where it is mixed with diluent. Since the sample flowmeter is located ahead of the sensor, there is a possibility that it could alter the particle size distribution. Care should be taken to avoid this. In Fig. 24b, the sensor flow rate is held constant and the flow rate of the excess diluted sample measured. The sample flow rate can then be calculated. Though the location of the flowmeter precludes its alteration of the size distribution, there are other concerns with this arrangement, as shown in Fig. 25.

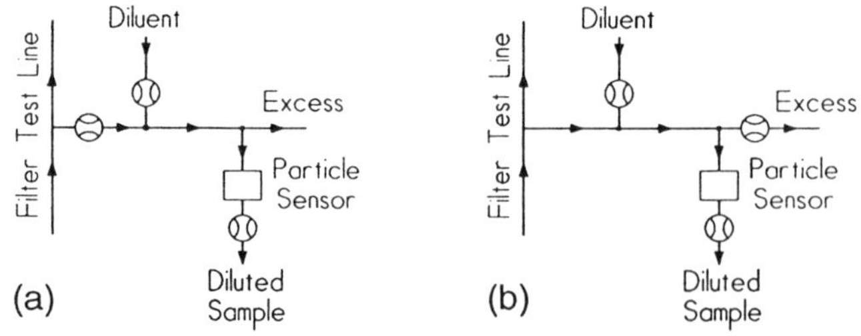

Figure 24 Schematic showing methods of controlling flow rates for on-line particle counting systems: (a) direct measurement of sample flow rate and (b) measurement of excess diluted sample flow rate.

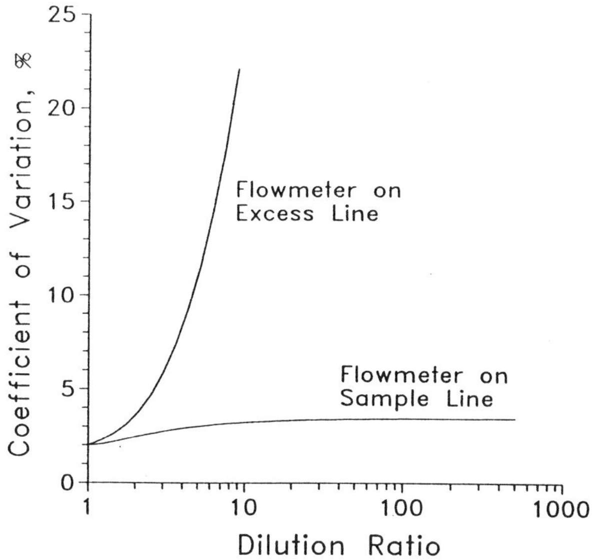

Figure 25 Effect of flow rate control method on coefficient of variation for on-line particle count results: flowmeter on sample line (Fig. 24a) and flowmeter on excess line (Fig. 24b).

Figure 25 shows the effect of flowmeter location on the *CV* in particle counts. Using the arrangement shown in Fig. 24a, the *CV* for the particle-count results never exceeds twice the *CV* of flow measurement, even at high dilution ratios. If the flow measurement guidelines previously discussed are maintained, the net *CV* will be less than 4%. In contrast, *CV*'s for the arrangement shown in Fig. 24b increase rapidly with dilution ratio. Thus, inaccuracies are minimized by measuring flow rates directly if the flowmeters used do not alter the particle size distribution. The upper limit on the dilution ratio which can be used is determined by the contamination level of the diluent relative to the sample, as has previously been discussed. When operating within this limit and by directly measuring sample and diluent flow rate, the effects of dilution ratio on *CV* are negligible.

Particle-counting statistics are important in filter testing and on-line particle counting because the results from two sensors employed to measure different populations are used to calculate beta ratios and efficiencies. The effect of statistical uncertainty on beta ratios as a function of particle size for three filters is shown in Figs. 26a,b. In each figure it is assumed that 100 ml of sample is analyzed. In Fig. 26a, it is assumed that data for all sizes is obtained at the same dilution ratio, i.e., the ratio which eliminates coincidence error for the smallest particle size of interest. This is the normal manner in which on-line particle counters are used. In contrast, Fig. 26b, assumes that several different dilution ratios are used such that the optimal dilution ratio (discussed in the next section) is used for each size.

Figure 26 illustrates a limitation of on-line particle counting. Typically, the dilution ratio is optimized with regard to the smallest particle size for on-line counting. Since a single dilution ratio is used, the uncertainty may become excessive at larger sizes, as shown in Fig. 26a. This uncertainty is the result of the small population of particles being counted. If, instead, several dilution ratios are used to optimize the results for each size, the uncertainty in beta ratio is reduced as shown in Fig. 26b. In general, the dilution ratio used in on-line counting is determined by the characteristics of the filter. However, caution should be used in comparing beta ratios much greater than 100 for the reasons just discussed.

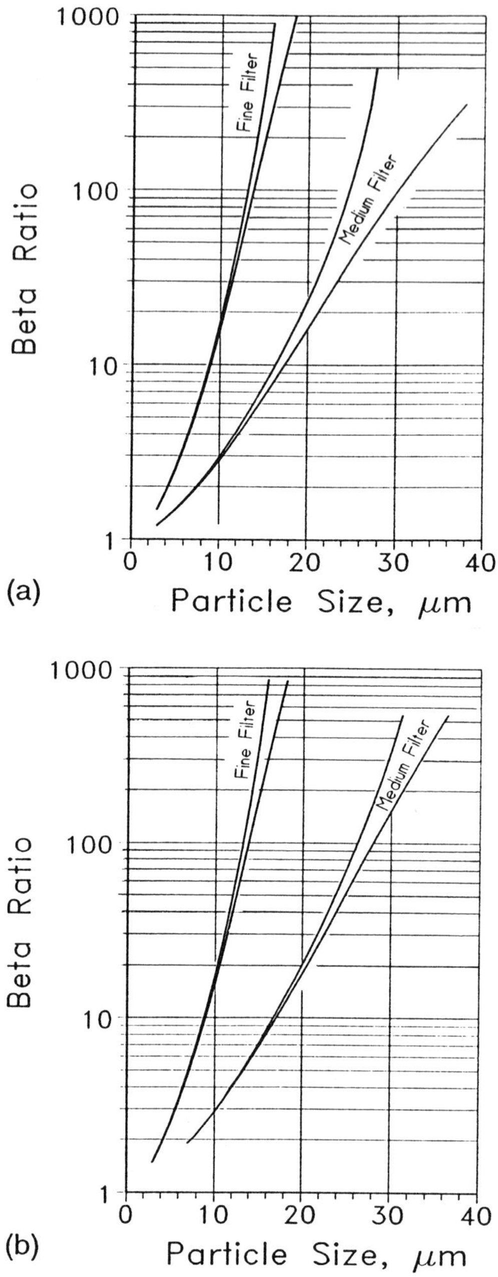

Figure 26 Effect of method of optimizing dilution ratios on beta ratios: (a) optimization for concentration of smallest-sized particles and (b) optimization for each individual particle size.

Methods exist which enable one to decrease the uncertainty associated with filter test and on-line particle counting. One is to increase the sample volume. The effects of increasing sample volume on uncertainty is shown for a 10-μm filter in Fig. 27. As the sample volume is increased from 30 to 500 ml, the *CV*s for beta ratios decrease by about a factor of 4. Another approach is to increase the particle size used for optimizing dilution ratio to a value closer to the rated size

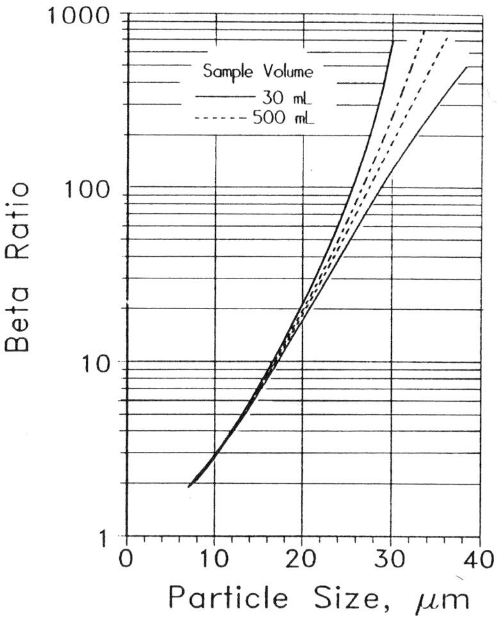

Figure 27 Effect of sample volume on coefficient of variation of filter beta ratios.

of the filter. A lower dilution ratio can then be used and the particle population will increase correspondingly. The result is less uncertainty in the beta ratio results at larger sizes.

On-line particle counting offers a number of advantages over off-line methods. Many of the sources of error associated with sample collection and preparation are avoided, less labor is required for analysis, and the results are available in near real time. Care is required to ensure that on-line dilution and mixing are done properly and that there are no transient effects. Problems obtaining agreement among different sensors can result unless well-matched sensors, i.e., those meeting the previously described performance specifications, are used. Though particle statistics restrict the size range over which useful results are obtained, this is not a serious limitation since the size range of interest is governed by the filter's removal characteristics.

VII. ANALYSIS OF USED OILS

Truck, bus, and off-road vehicle fleets, together with power plants and large production facilities, utilize particle count results in preventive maintenance programs. Since these results reflect contamination levels for those sizes of particles most damaging to equipment, increases in their concentrations are typically accompanied by increased wear rates. Therefore, well-designed oil analysis programs are critical to these operations.

Oil analysis programs utilize samples collected from equipment at regular intervals. Ideally, contamination levels should be monitored for each individual piece of equipment. Significant differences can exist between similar equipment used in comparable environments. The frequency of sample collection is determined by considering the relative costs of equipment failure and of the analysis program. Though some collect a single sample during each oil change, there are advantages to collecting multiple samples. Frequent oil sampling is recommended during the development of an oil analysis program, since components wear, oils degrade and filter performance changes over time. Once baseline data has been established, sampling frequency can be adjusted as needed. With multiple samples, trend analysis of the data can be used proactively to minimize failures rather than respond after they have occurred.

Trend analysis provides an indication of changes occurring within equipment which may otherwise go unnoticed. It can be used to establish oil and filter change intervals and to anticipate or prevent impending failures. Deviations from established trends can be indicative of impending failure or a need for servicing. This allows operators to plan maintenance activities in advance, significantly reducing costs.

Particle counting used oil presents challenging problems to the analyst. Unlike multipass filter test samples, in which the same standard oil and contaminant is always used, a wide variety of oils and contaminants are found in used oil, and each is unique. As a result, the basic methodology previously described must be modified for used oil. These modifications involve sensor selection, deagglomeration, redispersion, and dilution.

The choice of instrumentation is crucial in used oil analysis. Wear metals, silica, soot, salt, water and coolant droplets, fibers, and even insect parts have been found in used engine oils. As a result of differences in their optical properties, each will be sized somewhat differently from one another. This is not a major concern for extinction sensors. However, scattering sensors are especially sensitive to the differences. Shiny particles (e.g., wear metal chips) and black particles (e.g., soot and asphaltenes) are especially prone to be sized differently than other contaminants by scattering sensors. Scattering sensors have been observed to measure about half the concentration reported by extinction sensors for used engine oil [12]. As a result, impending equipment failure could be overlooked as a result. For this reason, extinction sensors, not scattering sensors, should be utilized for used oil analysis.

When choosing a sensor, the geometry of the flow and optical paths merit consideration. Sensors with extremely small orifices and flow paths are susceptible to plugging. Large particles and particles with large aspect ratios (e.g., fibers) which can plug sensors are found in field samples. Since large numbers of particles are often present, the use of sensors with coincidence error limits greater than 10,000 particles/ml is desirable. This can increase productivity since less dilution is needed.

The primary concern in used oil analysis is coincidence error. Contaminant concentration and size distribution vary depending on the application. Figure 28 shows typical particle size distributions for a variety of hydraulic and lube oil applications [43,52]. The ordinate expresses concentrations using both the number concentration of particles and the ISO code [53]. The ISO code uses two numbers separated by a / to describe the size distribution. The numbers refer to the concentration of particles larger than 5 μm and larger than 15 μm, respectively. For example, a sample with 30,000 particles greater than 5 μm/mL and 550 greater than 15 μm/mL would have an ISO code of 22/16. The ISO code is used to facilitate communication of particle-count results for field samples.

From Fig. 28, it is apparent that the contaminant concentrations vary over several orders of magnitude and that dilution is often necessary. Used engine oils are particularly troublesome, as they contain high concentrations of fine particles. This problem may get worse in the future since there is a tendency for lubrication systems to operate at higher pressures. As a result of this and tighter tolerances, finer wear debris is generated and finer filtration is needed to protect engines from abrasive wear than had been the case in the past. At the same time, vehicle manufacturers must decrease engine weight to meet fuel economy and emissions regulations. This necessitates downsizing the filtration systems. The result is an increase in the level of fine contaminants. The coincidence error limits for such samples are significantly lower than for ACFTD, and higher dilution ratios and cleaner diluent are needed. Since the particle size distribution in used engine oils tends to be enriched in fine particles, the most critical aspect of the analysis is to determine

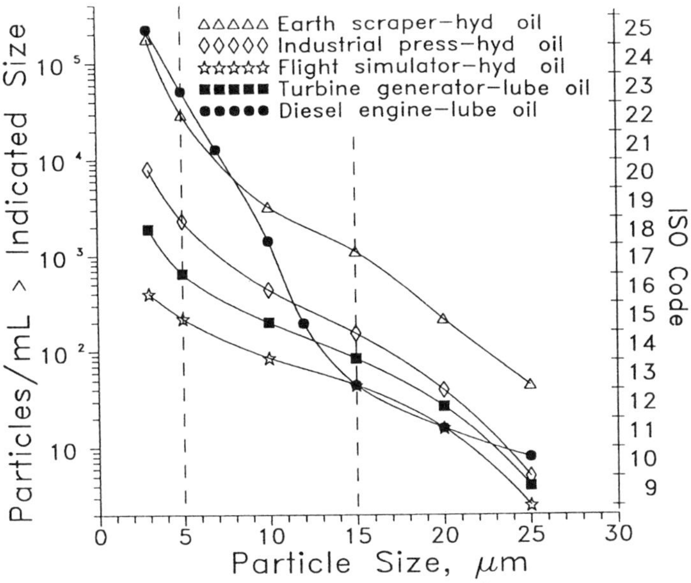

Figure 28 Particle size distributions for oil samples from various applications [43,52].

the proper dilution ratio. As previously discussed, the sample must be diluted enough to eliminate coincidence error without adversely affecting particle statistics (i.e., refer to Fig. 18).

Figure 29 illustrates a technique for determining appropriate dilution ratios for samples with an unknown concentration and particle size distribution. Results from a city bus oil sample are used in this example. As shown, the sample was analyzed at 12 different dilution ratios and the normalized results for each size plotted as a function of dilution ratio. For each size, a plateau is observed over which the normalized results do not vary significantly with dilution ratio.

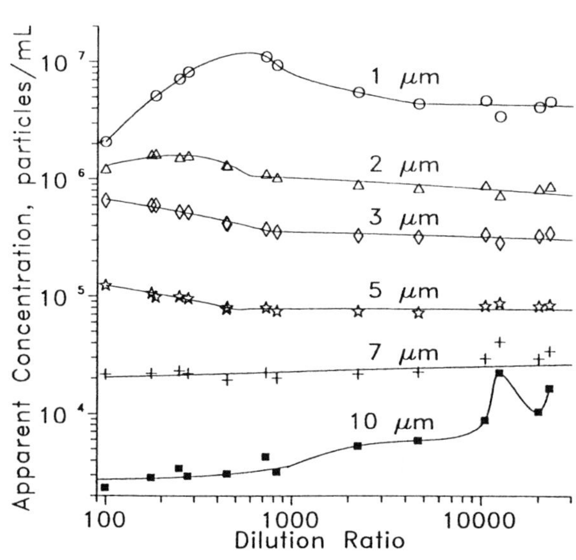

Figure 29 Selection of dilution ratios for used engine oil sample analysis.

Coincidence error occurs at lower dilution ratios, whereas diluent contamination and particle statistics become problems at higher dilution ratios. Results taken from the plateau region for each size are representative of the actual concentration for the indicated size. In order to generate representative particle size distribution data for used oil, several dilution ratios should be analyzed. The behavior shown in Fig. 29 is typical of used oil samples, though the plateau tends to narrow as the size distribution becomes increasingly fine.

The method just described is useful for analyzing samples of unknown size distribution and concentration. Though the procedure seems tedious, analysts become able to make relatively accurate estimates of the necessary dilution ratios and samples can be analyzed quickly. Some laboratories engaged in the analysis of large numbers of samples have evolved "shortcuts" to increase productivity. An unacceptable shortcut involves failure to dilute samples or to use the same, single dilution ratio for all samples. This compromises the quality of the results and can render them meaningless. Results obtained without dilution or using a single dilution ratio should be regarded with skepticism.

Dilution not only minimizes the effects of coincidence error but facilitates analysis in other ways. The viscosity of used oil, especially lube oil, is typically greater than that of the diluent. Extra time may be required for the redispersion, mixing and degassing compared to multipass test samples. If volumetric glassware is used, measurement errors increase with sample viscosity. Mass dilution, as previously described, reduces this error to acceptable levels and facilitates sample handling. When weighing apparatus is coupled to a computer, mass dilution is faster and more accurate than volume dilution.

The presence of water and other immiscible droplets can be a problem in field sample analysis. It complicates data interpretation and gives rise to experimental artifacts. Immiscible droplets are often present as a result of coolant leakage or condensation. This is especially common in marine, agricultural and off-road equipment. For these samples, preanalysis screening of samples for the presence of water or the routine use of Aerosol OT in the diluent is needed to minimize the effects of this interference. For routine work, the latter method is preferred. Using Aerosol OT in the diluent, the need for screening procedures is reduced and potential miscibility problems between the sample and diluent are minimized.

Oil additive packages indirectly influence particle count results. A variety of additives can be found in oil which improve its functional characteristics. Additives tend to be surfactants and include

Dispersants to inhibit sludge formation
Detergents to reduce, prevent or remove deposits
Antioxidants to resist reactions with air and unburned fuel
Antiwear, extreme pressure agents, and friction modifiers which bond to metal surfaces
Pour point depressants to improve pumpability at low temperature
Corrosion inhibitors to coat metal surfaces and repel water
Antifoamants to reduce oxidation and maintain oil pressure
Viscosity index improvers for all-season, multigrade oils.

Dispersants and detergents hinder aggregation and flocculation and clean metal surfaces. This facilitates contaminant deagglomeration but increases the concentration of fine particles. This may result in coincidence error and electronic saturation problems. Corrosion inhibitors and surfactants, in general, reduce the size of water droplets, making it difficult to detect their presence. Antifoamants expedite sample degassing. Viscosity index improvers allow lower-viscosity oil to be used in equipment. This aids degassing and assists mixing of sample and diluent.

In the past, users of oil analysis programs found it necessary to ensure that all results were obtained from the same source. This is the result of the lack of agreement among different labs and instruments as has been previously discussed. Ideally, comparisons of particle-count data should be made only among samples analyzed using similar instruments calibrated in the same manner. Strict comparisons of particle-count results among laboratories is difficult at present due to the varying levels of expertise of the analysts, instrument condition, and inherent limitations in existing calibration standards. These concerns are being addressed in a variety of ways. Instrument problems are becoming less of a concern as older instruments are replaced. NFPA and ISO are developing the hybrid calibration method which promises to resolve many of the problems which reduce agreement among laboratories. Despite the problems with agreement, optical particle counters remain a useful part of preventive maintenance programs for hydraulic and lube oil applications. Within the limitations discussed, particle count data are useful for establishing trends in oil contamination and can be used to develop oil and filter change intervals.

VIII. CONCLUSIONS

Particle counting is a primary tool for evaluating filter performance and monitoring oil contaminant levels in machinery and equipment. Oil analysis of this type is particularly demanding. Oil samples are extremely diverse in terms of their physical and chemical properties. The size distributions of oil samples are often enriched in fine particles. In many cases, results from different laboratories or even different types of instrument are compared. Thus, there are numerous sources of error when particle-counting oil samples.

There have been tremendous advances in the areas of particle-counting instrumentation and methodology in recent years. In the past, agreement among different instruments was poor at best due to instrument limitations. Today, instrument quality is better and, in part due to industrial user group performance specifications, good agreement among diverse types of instruments is possible. Several benefits have resulted from recent advances in calibration. Particle counters can now be calibrated more reliably. The new calibration methods explicitly provide for appropriate statistical evaluation of the results. There has been an associated increase in the level of expertise of particle counter operators. However, advances in methodology are occurring so rapidly that it is difficult for operators to remain current. Techniques, which are described in this chapter, are available to deal with a broad range of problems. The means necessary to deal with the high contaminant concentrations and changing size distributions encountered in the analysis of multipass filter test samples have been developed and are becoming accepted in industry. On-line particle counting facilitates automation of filter testing while improving the caliber of particle-count data. Thus, the quality of filter test results is improving. Similar problems are faced in field sample analysis, as well as the added concern of extreme variability in the nature of the samples. In this case, the required techniques and instrumentation exist, but flexibility on the part of the analyst and laboratory is needed to implement them. Professional analytical laboratories involved in routine sample analysis must decide the relative importance of data quality versus quantity.

ACKNOWLEDGMENTS

The authors gratefully acknowledge the help of T. Lippert, K. Anderson, T. Ptak, J. Wake, P. Steaffens and Carroll Wiseman in this work. We would also like to thank C. Montague (Hiac-Royco) and C. Gustafson (Met One) for their comments and technical assistance with the experimental work.

REFERENCES

1. B. W. Schwandt and B. M. Verdegan, *SAE Tech Paper* #902238 (1990).
2. D. R. Staley, *SAE Tech Paper* 881825 (1985).
3. J. Fodor, *Filtration Separation Jan/Feb* 64 (1982).
4. J. G. Eleftherakis, *Hydraul. Pneumat. Nov* 34 (1992).
5. E. C. Fitch and I. T. Hong, *Contamination Control for the Fluid Power Industry,* HIAC/ROYCO, Silver Springs, MD, 1991.
6. R. G. Knollenberg and R. C. Gallant, Refractive index effects on particle size measurement by optical extinction, Proc. Int. Conf. on Particle Detection, Metrology and Control, Arlington, VA, 1990, pp. 154–182.
7. H. T. Sommer, C. F. Harrison, and C. E. Montague, *Particle Size Analysis* (N. G. Stanley-Wood and R. W. Lines, eds.), Royal Society of Chemistry, Cambridge, U.K., 1992.
8. H. T. Sommer, Performance of monochromatic and white light extinction particle counters, Proc. Int. Conf. on Particle Detection, Metrology and Control, Arlington, VA, 1990, pp. 269–282.
9. ANSI, Hydraulic fluid power—Calibration method for liquid automatic particle counters using latex spheres, ANSI/(NFPA) T2.9.6R1, 1990.
10. J. Z. Knapp and L. R. Abramson, A systems analysis of light extinction particle detection systems, Proc. Int. Conf. on Particle Detection, Metrology and Control, Arlington, VA, 1990, pp. 283–352.
11. A. Lieberman, *Technology and Metrology,* ASTM STP 990 (D. Gupta, ed.), ASTM, 1988.
12. B. M. Verdegan, B. W. Schwandt, and C. E. Holm, Development of a traceable particle counter calibration standard, Proc. 6th World Filtration Congress, Nagoya, Japan, 1993.
13. NFPA, NFPA T2.9.6R2 (General Review Draft), Hydraulic fluid power—NIST traceable calibration method for liquid automatic particle counters, 1993.
14. ISO, TC 131/SC8/WG9 Document N66, August, 1992.
15. J. Raasch and H. Umhauer, *Particle Character. 1*:53 (1984).
16. ISO, Calibration of liquid automatic particle-count instruments—method using air cleaner fine test dust contaminant, ISO 4402, 1977.
17. M. Iwanaga, *BFPR J. 14*(4):323 (1981).
18. P. R. Johnston, *Filtration Separation March/April,* 134 (1976).
19. A. F. Lieberman, *J. Environ. Sci. May/June,* 34 (1988).
20. R. Inoue, *BFPR J. 13*(1):35 (1980).
21. W. Mach, Liquid particle counter spherical calibration pros and cons, Proc. 1992 Int. Fluid Power Applications Conf., Chicago, 1992, pp. 343–350.
22. T. Francis, ISO TC22/SC7/WG3 Document N261, Arizona Test Dust Update, 1992.
23. T. Francis, ISO TC22/SC7/WG3 Document N262, Proposal for ISO Test Dusts Standard, 1992.
24. B. M. Verdegan, J. Stinson, and L. Thibodeau, Accurate methods for particle counting, Proc. 43rd Nat. Conf. on Fluid Power, Chicago, 1988, pp. 299–308.
25. B. M. Verdegan, K. McBroom, B. W. Schwandt, C. E. Holm, and L. Liebmann, Advances in oil filter test methods, Proc. 1992 Int. Fluid Power Applications Conf., Chicago, 1992, pp. 333–342.
26. P. Mardosa, Particle counter physics, *Topics in Contamination in Hydraulic Systems,* SAE, Warrendale, PA, 1979.
27. H. Umhauer, A new device for particle characterization by single particle light extinction measurements, *Particle Size Analysis* (N. G. Stanley-Wood and R. W. Lines, eds.), Royal Society of Chemistry, Cambridge, U.K., 1992, pp. 236–245.
28. ASTM, Defining size calibration, resolution, and counting accuracy of a liquid-borne particle counter using near-monodisperse spherical particulate material, ASTM F658, 1987.
29. SAE, Procedure for calibration and verification of a liquidborne particle counter: an absolute standard, *SAE ARP* 1192, 1987.
30. L. B. Bangs, *Uniform Latex Particles,* Seragen Diagnostics, Inc., Indianapolis, IN, 1985.
31. B. M. Verdegan and L. Thibodeau, *Particulate Sci. Technol. 7*(1):23 (1989).
32. B. M. Verdegan, K. McBroom, and L. Liebmann, *Filtration Separation, July/Aug* 327–331 (1992).

33. ISO, Hydraulic fluid power—fluid sample containers—qualifying and controlling cleaning methods, ISO 3722:1976.
34. ANSI, Method for determining the particulate count of an oil sample from a system (using liquid automatic counters), ANSI/(NFPA) T2.9.11-1989, 1989.
35. ANSI, Method for extracting fluid samples from a reservoir of an operating hydraulic fluid power system (NFPA/T2.9.9-1976) ANSI/B93.44-1978, R1987.
36. ISO, Hydraulic fluid power—particulate contamination analysis—extraction of fluid samples from lines of an operating system, ISO 4021:1977.
37. J. W. Robinson, Jr., Methodology of hydraulic fluid sampling and analysis, Proc. 43rd National Conference on Fluid Power, Chicago, 1988, pp. 293–298.
38. ANSI, Multi-pass method for evaluating the filtration performance of a fine hydraulic fluid power filter element, ANSI/B93.31-1973, R1988.
39. ANSI, Hydraulic fluid power—in-line liquid automatic particle counting systems—method of validation, ANSI/B93.73, 1986.
40. ISO 4572, Multipass test method, 1981.
41. NFPA, Hydraulic fluid power—coarse fluid element—multi-pass method for evaluating filtration performance, NFPA/T3.10.8.18-1977, R1989.
42. SAE, *Full Flow Lubricating Oil Filters Multipass Method for Evaluating Filtration Performance,* SAE J1858, 1988.
43. B. W. Schwandt, B. M. Verdegan, C. E. Holm, S. R. Kelly, M. Bohmann, and D. Harvey, SAE Technical Paper Series No. 930017, 1993.
44. R. J. Akers, A. G. Rushton, I. Sinclair, and J. I. T. Stenhouse, *Particle Size Analysis* (N. G. Stanley-Wood and R. W. Lines, eds.), Royal Society of Chemistry, Cambridge, U.K., 1992.
45. B. M. Verdegan, L. Thibodeau, and S. L. Fallon, SAE Technical Paper Series No. 881824, 1988.
46. H. Umhauer, Enhanced-resolution size analysis of non-spherical particles by means of an extended single particle light extinction measuring technique, Preprints of European Symposium Particle Characterization, Nurnberg, Germany, 1992, pp. 519–534.
47. B. M. Verdegan, S. L. Fallon, and B. W. Schwandt, Immiscible droplet effects on particle counting, Proc. Int. Conf. on Particle Detection, Metrology and Control, Arlington, VA, 1990, pp. 209–229.
48. M. Ueda and A. Z. Schelly, *Langmuir 4*:653 (1988).
49. D. W. Cooper, Particle statistics for contamination control: an introduction, Proc. Int. Conf. on Particle Detection, Metrology and Control, Arlington, VA, 1990, pp. 183–208.
50. L. Bensch, Particle counting statistics, Proceedings of the Fluid Power Research Conference, Stillwater, Oklahoma, 1975, pp. 11-1–11-6.
51. SAE, *Air Cleaner Test Code,* SAE J726, 1990.
52. L. Bensch, A modern review of field contamination levels based on analyses of 25,000 samples, Proc. Int. Fluid Power Applications Conf., Chicago, 1992, pp. 285–298.
53. ISO, Solid Contamination Level Code, ISO 4406 (1987).

13

Sources of Particulate Matter in Pharmaceutical and Medical Products

Thomas A. Barber

Baxter Healthcare Corporation, Round Lake, Illinois

I. INTRODUCTION

Sterility, absence of pyrogens, and low levels of particulate matter are widely considered to be the most critical quality standards against which injectable products must be judged [1]. In recent years, particles in medical devices has also become an issue. This chapter is generally directed at sourcing and eliminating particles from parenteral dosage forms and medical devices. It is not the intent to provide a comprehensive listing of sources of particles in parenterals and medical devices, but to present examples of the most common sources of particles and outline general or philosophical approaches for troubleshooting and particle source elimination that will serve as helpful, basic information. The book by Akers [1] is a valuable detailed reference and classic in this regard, providing a discussion of particulate matter in relation to overall product quality. The treatise by Groves [2] also contains very useful discussions of particulate matter in injectables.

Total particle burden of an injectable product/medical device is determined by the additive particle contribution from the following sources:

The environment
Packaging materials
Solution, formulation, and device components
Product-package interactions
Process related particulate matter

The last four items on this list may be categorized as nonenvironmental particle sources, and their control and elimination is typically more important than the control of environmental particles. The control of particles from these four sources usually requires more diverse control measures than does the control of dispersed environmental (airborne) particles, and an uncontrolled situation with these sources will typically have more serious implications for product quality than will failure to control particles from the general environment.

The following have historically been identified as sources of nonenvironmental particles in the manufacture of devices and solutions:

Portions of this chapter were adapted from *Pharmaceutical Particulate Matter* by Thomas Barber, 1993. Reprinted with the kind permission of Interpharm Press ©1993, all rights reserved.

I must emphasize that any and all subjective judgments, opinions, and viewpoints expressed in this chapter are my own and not necessarily those of Baxter Healthcare Corporation or any organization representing manufacturers of medical devices or injectable products.

Human activity
Container-related and closure-related contaminants
Filling operations
Particles arising at process assembly points
Degradation of drug materials
Product-closure interaction
Extraction of materials from containers
Device components
Device assembly

In the following sections, particle sources and particle generation in injectable products is considered first, followed by a discussion of particles in medical devices. There are similarities in the approaches used for particle control in the two areas. A principle that is unchanged, whether devices or injectables is considered, is allowable versus excessive particle burden. Production of injectable products totally free of extraneous particles is not possible. Some level of particulate matter in product, even in a validated, monitored, current good manufacturing practice (CGMP) must be considered unavoidable by compendial groups and regulatory agencies.

By definition, this allowable (unavoidable) level of particulate matter consists of low numbers of randomly sourced particles of heterogenous composition. Implicit in this definition is that particles originating from a single identifiable source due to non-GMP conditions are proscribed and may reasonably be the subject of enforcement activity. A second principle that is unvarying with respect to both injectables and devices is the componentized or additive nature of the total particle burden. The sum total of particles "routinely" found to be present in the finished product is a summation of particles added at each assembly step. ("Routinely" implies that product failures, such as solution precipitation or disintegration of a component of a device, do not occur). Objectively, then, control of this particle burden will consist of a systematic, structured approach to minimizing particulate matter at each process step and eliminating, whenever possible, particles that might be present on product components such as stoppers, cannulae, or bottles.

II. PARTICLES IN INJECTABLE PRODUCTS

All particles that occur in injectables will come from two sources. The great majority of such particles, both kinds and numbers, originate from sources external to product; this includes container components and particles generated at specific process points and those contributed by filling machinery or human contact. The second category is related to the formulation itself, and takes the form of precipitates, degradants, or particles related to an interaction of the formulation with its container.

Obviously, extraneous particulate matter may enter the product as the result of insufficient cleaning, or originate due to product assembly steps, such as particle generation by a filler. Particles can enter either injectable or device manufacturing processes, or components purchased from a vendor. Formation of particles in the product is a complex phenomenon and can involve degradation, aggregation, agglomeration, precipitation, crystallization, and/or sedimentation reactions. Personnel activity and equipment design and maintenance are critical factors, since point-generated particles can arise from a multitude of process steps. Ideally, nonenvironmental particles are eliminated by process design rather than process troubleshooting. Process design incorporates not only the consideration of process machinery, environmental cleanliness, and filtration, but of container and closure cleanliness as well. Currently available process technology for both small volume injection (SVI) and large volume injection (LVI) product incorporates the potential for generation of product with an extremely low nonviable particle burden.

Ideally, process design for elimination of particles should be undertaken when a facility is on the drawing board, so that exclusion of particulate matter from the product is built into the manufacturing line. Obviously, this opportunity is not often present in real-world situations. Too often we are faced with the problem of reducing particulate matter levels arising from a process that began operation a decade or more ago when higher levels of process-related particles were acceptable. Often progress in process design-based particle control must be made stepwise, as new lines are added, as facilities that will use a new process are built, or as process lines are upgraded.

Particles present in injectable product reflect the quality of product generation from formulation development to filling. Particles may gain access to parenterals during manufacture, or they may form during product storage. To minimize access of extraneous particles to product, particle-controlled assembly procedures with high quality packaging components are employed. Particle formation in finished product often proves more difficult to control than extraneous particles from the process or environment. Particle formation through solution degradation requires studies of component quality, raw material uniformity, and rigorous formulation compatibility/stability testing. Especially when plastic solution containers are used, a high level of knowledge regarding product-container interactions is essential. Some extraneous particles are readily identified; in other instances, a comprehensive approach is required to determine the source. The latter must include an understanding of the visual appearance of the particle, as well as its physical and chemical characteristics. Efforts to exclude objectionable material from the product do not begin with highly controlled filling environments; R & D efforts that indicate the best components, excipients, and packaging are essential to ensure product quality. Particle identification is an integral part of the quality process; it allows location of the source and hence its elimination.

III. CONTAINER-CLOSURE

Containers and closures constitute a substantial source of added particles; this contamination is diverse in composition due to the complex nature of containers and stoppers [3]. The most obvious examples of extraneous materials present in injectables is foreign material introduced into the container before or during filling, or resulting from inadequately cleaned container materials. These are classified as process-related particles, since they originate from the immediate preparation, manufacturing, and filling environments. Glass, metals, biologicals, fibers, and machine fragments are the most common but not the only types of process-derived particles that may be encountered; fragments of packaging materials also fall into this category.

Stoppers used to close vials and bottles may also release significant numbers of particles and extractables, especially when they are steam sterilized. Such soluble extractables may then react with the solution and produce insolubles. The first rubber stoppers for parenterals were made from sulfur-cured natural rubber. This material has a relatively high moisture permeability and can release high concentrations of extractables, absorbs drug materials and preservatives, and contains plant proteins and other undesirable materials. Natural rubber has largely been replaced by bromobutyl and chlorobutyl (halobutyl rubbers). These two rubber formulations are compatible with most solutions, and can be cured without sulfur; they are also less likely to release significant amounts of leachables. Stoppers now available have the silicone resin Teflon, polyethylene, or other materials of low reactivity coated onto their inner surfaces. These coatings, along with improvements in the environmental conditions during stopper manufacture, have resulted in significant reductions in closure-related particle contamination.

Many closures for injectables are lubricated with silicone oils. The reconstitution of lyophilized parenterals may be adversely affected if excessive oil levels are employed. Moreover, high silicone levels can cause: (1) elevated "particle" counts via light extinction methods; (2) false indications of visible "particles" due to Schlieren effects; and (3) occurrence of slowly dissolving lyophilized cake fragments from encapsulation of the cake by the oil. The treatment of stoppers with silicone oil or emulsions must be considered a necessary evil if butyl stoppers are used in automatic stoppering machinery. Alternatives to coating with silicone may be found in proprietary surface treatment of halobutyl rubber stoppers, or through the use of "baked-on" silicone coatings. Some drugs, such as cephalosporins, are reactive with rubber, so that an inert coating such as a fluoropolymer is required on the surfaces of the stoppers that contact the product.

If an elastomer contains low molecular weight ingredients (such as accelerators, plasticizers, etc.), that bloom out of the surface as solid particles during molding, sterilization, or storage, a potential source for particles in the product will always exist. Even after an efficient washing cycle that cleans the surface, low molecular weight ingredients may bloom out after weeks or months, completely canceling out the effect of the previous washing treatment [4,5]. Notorious in this regard is mercaptobenzothiazole (MBT), an accelerator once widely used in stopper formulations. It is of the utmost importance that rubber used in stoppers have a high chemical purity. Decreasing the number of ingredients (i.e., control of physical properties with formulations rather than additives) requires a higher level of polymer technology, has the benefit of decreased particle release, and also guarantees a broader compatibility range. Fillers and rubber fragments contribute other stopper-related particles; filler materials may separate at the surface of rubber parts in some molding processes and be subsequently washed off into the product. Extraneous particles added by packaging or handling of the closure must be controlled. Washing should be utilized to remove only a small number of particles left by the total process, and not be relied on to remove leachables or other stopper-related particles.

Closure-related particle problems may be due to fragments of the closure itself. In one instance in the author's experience, LVI solution product indicated an elevated particle count due to large dark-colored particles 25 to 100 μm, which proved to be rubber closure fragments. Microscopic examination of the closures used in these units showed fragments of closure material on the cut edge of the closure. In addition, pitting and fissures were found in the body of the closure. Nondefective units in adjacent batches contained closures exhibiting none of these defects. The problem was determined to be caused by a combination of a dull cutting die used by the vendor and incompletely cured polymer in the stoppers.

Another potential risk of particle generation related to stoppers occurs due to coating materials present on stoppers as received from the vendor. This involves the chemical combination of materials present as a coating on the stoppers with the dry powder material or dissolved drug. Two occurrences of this problem come to mind. The first was due to the presence of traces of mold release agent on the product contact surface of stoppers that was not completely removed by washing. The second involved the presence of a reactive impurity in the silicone emulsion used to lubricate the stoppers. In both cases, the particles were amorphous or gelatinous materials that formed indistinct spots or smears on membrane filters, and imparted a hazy appearance to the solution. Both materials gave EDXS spectra that contained silicon as the predominant detectable element, allowing investigations to focus on the stoppers as the most likely source of the problem. The mold release agent was identified by a computerized FT-IR spectral file search. The reactive impurity in the silicone emulsion was also identified by FT-IR. In this latter case, the supplier of the silicone emulsion indicated that the reactive impurity was a known contaminant of their material. Stoppers with a large surface area can be expected to generate more particles than smaller surface area stoppers of the same rubber. Lyophilization ("lyo") stoppers, due both to conditions of their use and the unique requirements for surface treatment, have frequently been the source of particle issues with lyophilized SVI product.

Containers for injectable liquid and powder dosage forms are made from glass and a wide variety of polymer materials. Glass containers for pharmaceutical products are typically made from two types of glass [6]; these are borosilicate glass (Type I), or soda lime glass with a surface treatment (Type II). Typically, Type I glass containers are cleaner as received from the vendor and are more readily cleaned by washing than Type II bottles. Contamination of glass containers often occurs during manufacturing or packaging. General cleanliness is most often dependent on the effectiveness of cleaning measures performed immediately prior to filling. Large-volume glass bottles are likely to bear high levels of particles on their inner surfaces prior to washing. Despite the initial higher level of particles, it is sometimes easier to attain a given level of cleanliness with a large glass container than a small one. This partially explains why smaller units contain relatively more particles than do large ones, since as the volume of the container decreases, the area of the stopper and walls of the container to which particles can adhere does not decrease proportionately. This allows more surface area for the release or generation of particles [7,8]. Washing in itself is a complex procedure, and washing to remove a low level of adherent particles may actually result in an elevation of particle numbers rather than a decrease. Drying ovens with forced air ventilation and depyrogenation ovens are notorious sources of particles. The best course is to buy certified clean components from vendors and keep them clean, rather than relying on washing.

Much of particulate matter in glass containers is glass particles or contamination from extraneous sources introduced during manufacture, packaging, and storage, and not subsequently removed by washing prior to filling. The amount of particle concentration in plastic containers depends on the type of plastic and method of container fabrication. Sterilization may alter the particle burden in both glass and plastic containers. Though containers are made of various types of glass and plastic, plastic containers almost invariably give lower particle counts, since they often do not require stoppers and are typically protected in manufacture to decrease the chance of contamination from the environment. In some cases, they are made from tubular extrusions, so that the container surface is never exposed to external process-generated particles or environmental contaminants. A similarly high degree of protection is provided by form-fill-seal processes that use laminated films to fabricate, fill, and seal a solution unit within a Class 100 or better machine enclosure. Blow molded plastic bottles may be filled immediately after they are formed, so there is minimal chance of environmental contamination. The ratio of container surface area to solution volume is often a key determinant of particle burden. When plastic containers are used for large volumes of solution, an extremely low particle burden generally results.

Blow-molded LVI units without elastomeric closures are widely marketed in some countries that do not adhere to the requirements of a major pharmacopoeia. In appearance, these units resemble a large plastic ampule, and in principle can be made without exposing the solution to any significant particle burden. In use, however, any advantages of low particulate matter burden in manufacture are negated by the absence of a closure; the units are entered with a steel hypodermic syringe needle, creating particles and providing a questionable sterile barrier. Further, the molding for any blow-molded unit must be carefully controlled to prevent generation of "char" particles by the mold.

IV. FORMULATION COMPONENTS

Particles occurring in parenteral products may provide negative indications regarding:

The level of drug raw material control
The stability of the formulation (physical stability)
Process design
Process execution
Potential sterility problems (aseptic fill process)

Particle elimination results in an improved manufacturing process and the ultimate release of a greater amount of product at lower cost to the user. Quality with regard to particulate matter is difficult to objectively describe, but some believe that in the case of particles in parenteral solutions, the quality of the manufacturing process is inversely related to the number of particles present in the finished product.

The paper by Speed [9] provides an interesting insight into this issue. Precipitation or crystallization has historically been a source of particles in parenteral solutions; some may result from precipitation of one of the components in the parenteral formulation. Generally, products are not intentionally formulated near or above the solubility limits of the components, but it is a possibility that must not be overlooked. Control of solubility as a function of pH and excipient concentration in preformulation studies provides little protection against drug or excipient precipitation that results from temporary storage at cold temperatures. Sodium phosphate buffers may produce the relatively insoluble $Na_2HPO_4 \cdot 12H_2O$ at refrigerated temperatures [10]. The presence of a contaminating species leached from a closure or inadvertently introduced prior to filling can also result in precipitation; this type of contaminating material always has the potential to interact with solution components to form insoluble compounds.

Crystallization is not uncommon in LVI solutions of concentrated dextrose, dextrans or similar materials formulated near the solubility limits, particularly if the product experiences low temperatures. Pharmacy-produced admixtures frequently develop precipitates from either a marginal formulation in terms of solute concentration or through degradation of admixture components (such as the formation of calcium oxalate from degradation of ascorbic acid). Drug solutions may contain crystallized particles resulting from a small amount of impurity that precipitates upon storage of the drug. Penicillinoic acid SVIs are an example of this phenomenon. In addition to oversaturation conditions, precipitation requires the existence of nucleation sites. Thus, the introduction of nucleation centers, such as contact with a closure surface, a glass defect, or a syringe needle, may cause the sudden formation of large numbers of particles. Extremely small amounts of materials can be responsible for a problem level of precipitates. An example is the combination of trace levels of calcium and aluminum from phosphate-containing pH-adjusting reagents that form a precipitate. Precipitation reactions may also be highly variable between units; it is not uncommon to see units containing readily visible particles and unaffected units from the same batch. Similarly, individual units from the same batch may contain crystal precipitates of distinctly different morphology.

The surfactant nature of many parenteral products can cause unique effects with regard to particle formation. Some drugs are ampiphilic in nature: the drug molecules exhibit both polar (hydrophilic) and nonpolar (hydrophobic) elements. This behavior is characteristic of the general class of detergent molecules. Behaving as detergents, some drugs form micelles in solution; micelles also serve to solubilize hydrophobic molecules. There is a critical concentration above which surfactant added to a solution enters micelles, and the concentration of monomeric surfactant remains essentially constant. This concentration is called the critical micelle concentration (CMC). At surfactant concentrations above the CMC, the solubilization capacity for nonpolar materials generally increases with the surfactant concentration.

Surfactant-emulsified microdroplets are readily counted by light extinction particle counters and may be responsible for compendial count failures. Surfactants, especially when present at concentrations above the CMC, can dramatically influence the solubility of some excipients. The nonpolar environment in the interior of micelles acts as a distinct solution phase in which hydrophobic molecules can dissolve.

Dissolved materials in an SVI may precipitate following dilution into an infusion solution if the dilution brings the surfactant concentration below the level necessary for emulsification of hydrophobic materials [11]. A seemingly insignificant increase in the level of an impurity may lead to the precipitation of another species dissolved in the micellular phase if the impurity alters the solubilizing ability of the micelles. Sodium methylprednisolone succinate is an example of a micellular drug [12]. At high concentrations of the sodium salt, solutions of this material contain a large number of micelles. These can dissolve nonpolar materials, such as the free alcohol form of methylprednisolone. With age, methylprednisolone succinate hydrolyzes to methylprednisolone, a nonpolar compound. This reduces the number of micelle-forming molecules and increases the number of water-insoluble molecules that need to be solubilized. As this reaction proceeds, the equilibrium solubility of the drug may be exceeded, in which case a precipitate results.

Historically, the ≥ 2-μm and ≥ 5-μm size categories that were counted in the British Pharmacopoeia method resulted in specific treatment of some products to achieve count levels necessary to ensure compliance at the ≥ 2 μm size. Gupta et al. [13] provide an interesting commentary on the process of reducing counts at this size in lyophilized formulations. Measures employed included addition of a surfactant to the formulations (to solubilize immiscibles), careful control of pH to control free acid or base moieties, charcoal treatment to remove trace impurities, and use of filters with a 0.1-μm retention rating to remove incompletely dissolved fine particles or precipitates. The solubilizing agent and charcoal treatment were found to be most effective.

In addition to precipitated material or particles related to surfactant activity, some drugs and carbohydrate or amino acid solutions may contain amorphous (noncrystalline) material related to a small amount of degradation of a solute ingredient. Cephalosporin antibiotics frequently contain amorphous or gelatinous particles from this source. The tendency of dextrose solutions to contain amorphous or gelatinous fragments of 5-hydroxymethyl furfural is acknowledged by the United States Pharmacopoeia, and they allow for the assay of dextrose-containing solutions by light extinction counting rather than microscopy. A similar brown amorphous material can occur in amino acid or protein-containing drugs due to tryptophan degradation.

On occasion, complex particulate matter problems arise with drug material requiring the application of a wide range of instrumentation and expertise. In the study of Ashline et al. [14], a degradation product of penicillin drug powder required the use of nuclear magnetic resonance, mass spectrometry, and X-ray crystallography for identification. This problem was first observed as high instrumental and microscopic particle counts in aqueous solutions of sodium nafcillin and sodium oxacillin. The moiety responsible, a thietan-2-one, was observed to form by precipitation in solutions of the drugs held at room temperature, and particles increased in number with increasing storage time.

Precipitation frequently becomes an important source of particulate matter in admixtures. Classic examples are provided by (1) the decomposition of ascorbic acid to give oxalate, which can bind with calcium to form calcium oxalate; and (2) the formation of calcium phosphate $Ca_3(PO_4)_2$ crystals in 3-in-1 admixtures containing Ca and PO_4 ions.

Some particle problems that arise during storage of liquid formulations are related to adsorption phenomena. Materials present in solution may be concentrated at surfaces. This is true for trace impurities or contaminants present in solutions stored in glass vials or plastic containers. While the concentration of a reactive impurity in solutions may be insufficient to result in particle formation, surface concentration of the material may result in generation of particles. In one example of a problem of this type, low concentrations of bismuth were present in a solution that contained bisulfite as a preservative (antioxidant). The combination of the bismuth with sulfur ions to form particles was evidenced by a brown ring that appeared on the surfaces of the glass bottles and polyethylene airway straws at a depth of approximately 1 mm beneath the solution-headspace interface. The brown particles of BiS and BiS_2 formed were identified using EDXS; the metal contaminant was then traced to one of the raw material components of the material using inductively coupled plasma (CP) spectroscopy. Peptide-containing drug materials may also form sulfur-containing precipitates, and under some conditions oxidation of sulfur-containing functional groups of these materials may form particles of elemental sulfur.

V. POWDER DRUG MATERIALS

The reduction or elimination of particles in dry powder dosage forms is a critical and complex undertaking. In powder filling operations, particulate matter present in the bulk raw material used cannot be excluded from the product, and the particle burden of the powder in a final product vial will be directly related to the particle burden of the raw material. Typically, lyophilized or spray-dried powders have a lower contaminant (extraneous) particle burden than crystallized powders, due to the application of filtration prior to drying operations. These preparations may, however, be afflicted with the presence of impurities that form amorphous particles on drying, or form particles that do not dissolve during reconstitution. In the case of extraneous insoluble particles in powder-filled material, elimination of extraneous particles may be simple and straightforward. However, elimination of drug-related soluble impurities that can pass through filters to form particles at a later time may call for high-level problem solving and involved procedures for particle identification.

In the case of crystallized powders that are obtained in bulk and processed by aseptic powder filling, the most important potential sources of particles are the processing equipment used in crystallization and drying and the packaging process for the bulk powder. The situation is made more complex because many powdered materials, such as penicillins, may be available from a number of different vendors who may apply a wide variety of particle control measures. Similarly, a single vendor may employ different processes for the same drug manufactured at different facilities. Some vendors of dry powder drug materials may produce different "grades" of raw material based on the particulate matter content, and supply product with fewer or more extraneous particles to customers who are more or less discriminating.

The best course of action for the manufacturer of powder dosage forms is to routinely perform particulate matter assays on raw material lots prior to filling. Conducted appropriately, inspections of the candidate raw materials will prevent filling of a powder with an unacceptably high particle burden, and the consequent inability to release product. Ideally, three types of compendial assays will be applied to the raw material: visual inspection, light extinction counting, and microscopy. The light extinction assay will serve to detect immiscible materials as well as small (≥ 2 μm to ≥ 10 μm) particles, but it is of little or no value in detecting particles that may be observed in a visual inspection at a later time. Microscopy is of great value due to its provision of data on large particles as well as subvisible ones.

Lyophilized materials should be subjected to acceptance inspections as well. In this case, however, freedom from insoluble particles on dissolution does not necessarily indicate that the material is free of soluble impurities that may form precipitates at a later time. An example of this type of particle problem is provided by the penicillinoic acid thiolactones that may be present in penicillin powders. Extremely low levels of this impurity can form subvisible insolubles over the shelf life of product, and in the worst case may necessitate product recall. The only sure protection against this type of defect is a thorough knowledge of the chemical constituency of the raw material; this knowledge is most easily obtained from the manufacturer of the drug. Of particular concern is a single drug that may be manufactured by a number of different processes. As manufacturers working with generic antibiotics have discovered in the recent past, different synthetic pathways and extraction procedures can yield compounds that are identical chemically, but have significantly different levels of impurities. Generally, one should beware of raw material vendors who do not have a comprehensive knowledge of their product as indicated by process execution and extent of testing. Similarly, reluctance on the vendor's part or inability to supply information in depth regarding trace impurities present in their raw materials may be sufficient reason to seek another supplier.

VI. PRODUCT-CONTAINER INTERACTIONS

The most well-known product-container interactions resulting in particle generation relate to interaction of products with the elastomeric closures of glass containers [15,16]. Cephalosporin drug powder antibiotics can react with uncoated butyl or halobutyl stoppers to generate particles due to extraction of stopper components into the drug. As discussed earlier, the composition of a rubber stopper can be quite complex, and in addition to the elastomer may include some or all of the following ingredients: a vulcanizing agent, a filler, an accelerator, an activator, an antioxidant, a pigment, a plasticizer, and a lubricant. In addition to fragments of the elastomeric formulation, a variety of closure extractables can thus result in haze or particle formation. Some of the extractables may not be soluble in the parenteral solution, and others may react with another component to form an insoluble material. Volatiles from stoppers may also cause haze formation in freeze-dried parenterals, and processing-related materials such as hydrocarbon or silicone-based vacuum pump oils may also appear as particle counts.

The paper by White [17] deals with some of the complex interactions of products and containers in which extremely small quantities of insoluble material can cause particle problems. Copper or other metals found at 1 part per million in solutions may cause precipitates. The solubility of iron is about 5 ppb in high pH (7.8) solutions due to the formation of insoluble iron hydroxide $Fe(OH)_3$. The kinetics of formation of insoluble iron hydroxides is complex, unpredictable, and may occur at a slow rate. Species such as $Fe(OH)^{2+}$, $Fe(OH)^{2+}$, and $Fe_2(OH)_2^{3+}$ are also involved. At sufficiently high pH, the iron is somewhat more soluble due to the formation of higher hydroxides. The presence of silicon is often detected in particles, most often resulting from the silicone oil used as a closure lubricant. The presence of sulfur in particles in SVI is often an indicator of the presence of MBT, a vulcanizing agent used in stoppers. MBT can combine with soluble metal ions present in trace amounts to form a metal-organic complex, or may be observed as isolated particles.

Another example of particulate matter from product-container interactions involves barium sulfate [18]. This material has very limited solubility. Subvisible particles of this material have been identified in a wide range of products. It is important to note that although most barium sulfate particles identified in various solutions are subvisible, their size may vary significantly. The crystals are usually symmetrical, which is typical of slow crystal growth via nucleation reactions. The $BaSO_4$ particles are frequently formed by the reaction of the Ba^{2+} ions that may be extracted from glass ampules or vials with SO_4^{-2} ions that originate from the anion of drugs and/or the bisulfite ion of the antioxidant. One remedy to this problem is to use borosilicate glasses that do not contain barium. Some elastomeric closures contain a barium sulfate filler that may segregate at the stopper surface during molding.

In addition to barium, other soluble extractables, such as aluminum and calcium [19,20], may come from glass. Other sources of these elements include drugs, excipients, and buffers. Besides particles resulting from soluble extractables from glass containers, glass flakes occur in some parenteral products, generally resulting from erosive attack of glass solutions (pH 7.6 and above). Most of this material is highly transparent and very thin, although some may appear as white, fluffy particles. Though most of the flakes are quite thin (< 1 μm), they are often detected under visual inspection conditions due to their reflective nature. A critical consideration is that tubular glass vials and ampules may have a significantly lower resistance to attack by basic solution than do bottles formed by inflation molding. The second stage, forming tubular glass to make vials, results in surfaces more susceptible to breakdown and attack, and may be noted at pH 7.0 to pH 7.4 for some types of glass.

Delamination or glass breakdown due to solution attack of the glass container surface is not nearly so common today as it was 10 years ago. Currently, Type I glass is used by manufacturers when the pH of a solution approaches neutrality; Type I glass is resistant to attack by basic solutions and typically resists attack solutions up to pH 8 effectively. The pharmaceutical manufacturer should be aware, however, that there are different Type I formulations available that will vary in resistance to breakdown, and that the annealing process involved in drawing vials or ampules from tube stock must be carefully controlled so that the resistant properties of the original material are retained.

Plastic containers are not totally free of problems related to solution-container interactions. Particular care must be taken when solutions with detergent properties are filled into polymeric containers. These materials typically contain minor components, such as low molecular weight slip agents or lubricants, that are present on the solution contact surface; these materials may be affected by the surfactant properties of a drug.

One of the most interesting considerations regarding the relatively high levels of particles that can occur in dry powder dosage forms regards the elimination of these particles from the final dosage delivered to the patient by filtration. Terminal in-line filtration is effective in this regard, but is not necessary in many cases and increases the cost of intravenous therapy. Terminally filtered liquid or frozen drug products eliminate the extraneous particle burden of the powder by means of terminal filtration in manufacture. Today, a number of drug delivery systems allow mixing of diluent and powder immediately prior to administration; some are designed for in-line use, and others (open vial access systems) allow the entire contents of the powder vial and stopper to drop into a diluent bag or vial. When these latter systems are used, complete dissolution of the drug powder before use should be ensured, and the drug-diluent solution should be carefully subjected to visual inspection for particulate matter.

VII. PROCESS GENERATED PARTICLES

Process point-generated particles result from any manufacturing step in which packaging components are manipulated or handled, including the formation of the container as well as the fill process. Some typical sources are:

Fill nozzle misalignment
Capper misalignment
Stopper feeder
Vial or ampule turntable
Handling of container or closure

Cutting the neck of blow-molded containers
Metal wear fragments

Additionally, human handling of product containers is liable to add particles. Scientific literature has many references to colorless, transparent, ovoid, flake-like particles that are, in fact, human skin cells. Powdered gloves are a potential source of starch, talc, or calcium carbonate particles. A powder-free glove is the only recommended type to use in product filling areas, and even powder-free gloves are not immune to particle shedding problems. The surface treatments used to remove particles from these gloves may cause deterioration or roughening of the surface. Thus, if powder-free gloves are used for long periods of time without being replaced with new gloves, abrasion of the surface causes particle generation.

Non-environmental particles from these sources must be carefully controlled. Despite filtration, it must be assumed that any contaminant particle present in the solution mix tank of an LVI fill process or in a drug solution that will be lyophilized is a potential contaminant of the final product. Similarly, particles originating in process equipment, such as solution lines or fillers, will have a detrimental effect on product quality if they are not denied access to the final product. Control of particles from nonenvironmental sources is, therefore, dependent both upon control of particulate matter sources related to every process component that touches the product, and final filtration; the latter is often overemphasized at the expense of the former. While process filtration is extremely important, its emphasis too often leads to a philosophic disregard of particle sources within processes occurring downstream of filters that can lead to product contamination. In any consideration of fill process particle elimination, contribution of the fill path and fill equipment must be given careful consideration.

For fill equipment, the following are important considerations:

Existence of pockets or dead-legs in fill line circulation
Number and nature of joints and welds in the fill line
Type of pumps used
Filler characteristics (diaphragm vs. piston, etc.)
Length of the fill line path downstream of line filters and filler

Both the elimination of pumps from the fill whenever possible and the use of gravity or pressurized solution fill have many benefits. Centrifugal pumps are highly efficient, but invariably contribute wear particles (gasket, bearing, and seal materials) to process a line. Due to the high rotational speed and close tolerances at which such pumps operate, slight misalignment of the rotors can generate very high levels of particles through abrasion. Similarly, simple piston-type fillers are an inevitable source of wear particles. Contemporary diaphragm or time-pressure fillers with non-particle-shedding valves are far superior with regard to particle generation.

Minimally, filters should be used between the last piece of equipment in the fill line that has moving parts and the fill nozzle; the length of the flow path downstream of the final filter must be kept at an absolute minimum. This, of course, suggests the use of filling head filters; filling nozzle contact with the neck of an ampule or vial is notorious for the production of glass or metal fragments. Fill processes, including placement of the fill needles or nozzles, are designed to be very uniform. Thus, misalignment of vials or the filler may result in almost every unit in a batch being affected by this defect.

Obviously, filtration is of extreme importance. The treatise by Meltzer [21] provides detailed discussions of all critical aspects of filtration relating to pharmaceutical manufacture. We must keep in mind that filter housings shed particles, and the microporous (spongiform) media of most process filters has the potential for particle generation through fragmentation. Thus, the placement of a filter in a process line does not ensure that particle-free solution will be filled. Filters placed in the fill tubing downstream of the filler may be effective in eliminating particles originating in the fill mechanism, but must also be able to reliably withstand the constant pressure surges of the filler. Rupture of a nozzle filter with release of retained particles and filter fragments can lead to tragically high levels of particles in product.

Filters for pharmaceutical solutions marketed today by major vendors are designed to be extremely efficient in particle removal, to have low particle shedding characteristics, and to be inert to the solutions filtered. Construction of a typical process filter is shown in Fig. 1.

Although solution filtration is often viewed as a simplistic process, it is really quite complex. Particles are removed from liquids by different mechanisms depending on the type of solution being filtered, the material of the filter, the filter retention rating (pore size), and the size of the particles of interest. The sieve retention mode of particle capture occurs when a particle is too large to pass through a filter pore. This type of particle capture is largely independent of filtration conditions, including any reasonable pressure. Sieve retention is also generally independent of the particle challenge level. If a sieving mechanism is considered, regardless of the number of particles contacting the filter, all the particles above a given size will be retained. Additionally, sieving particle retention will be independent of the suspending liquid vehicle with regard to its ionic strength, pH, surface tension, temperature, viscosity, and presence or absence of surfactant.

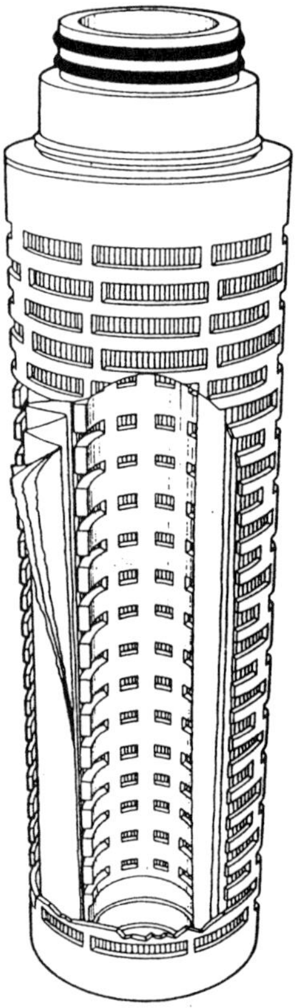

Figure 1 Cartridge filter for application in pharmaceutical manufacture incorporating prefilter and filter media. (Courtesy of Gelman Sciences.)

The membrane-type filters used in parenteral production can be viewed as either depth or screen filters depending on the particle sizes of interest. Generally speaking, the production cartridge filters used for pharmaceutical solutions will affect not only the removal of particles above the retention rating, but also large numbers of smaller particles. Particles below the filter retention rating my be removed by adsorptive contact within the filter matrix; this mode of particle capture is less important than sieving if particles greater than 1 μm in size are a consideration, since the filters used in parenteral solution production will most often have a 0.22-μm or 0.45-μm retention rating. Particles smaller than the retention rating will not necessarily pass through a filter. Generally, both electrostatic and van der Waals forces are active in attracting and adhering particles below the pore size to the filter medium. Most filters possess a negative charge when wet, and negatively charged particles will be more likely to pass through the filter than positive ones. Surface retention on a fluid filter may also result in exclusion of particles smaller than the nominal pore size of the filter.

Filters applied to protect products from extraneous solid particles may contaminate the product filtrate with immiscible contaminants. Phthalates present in a cellulose ester filter have historically been implicated in this regard. In an occurrence reported by Aldrich [3], the problem material was poly(diethyleneglycol) isophthalate (PDEGI) that was not related to the product. The material was traced to a bulk drug isolation filter used to harvest the crystallized drug. The filter was composed of Dacron and contained small quantities of PDEGI. In order to maximize yields, the process involved collection of additional drug from the initial filtrate. Filling of the product through 0.2-μm sterilization filters did not diminish the content of the polymer in the formulation. Due to the high concentration of the drug relative to the polymer and slow formation of the polymer-related particles, the problem was not immediately detected. Further production of the drug without a Dacron isolation filter yielded polymer-free material.

Filling machines are a class of production machinery that has a significant impact on product particle burden. The filling of liquid product into a sterile container is complicated by the flow of air around the mouth of the container during the fill cycle. In the past, filling heads were often set onto bulky supports that controlled the vertical movement of the filling needles. Structures overhanging the filling line disrupted the laminar HEPA airflow and resulted in eddies and turbulence in a portion of the critical area. Typically, in machines of current design, solution metering is done by diaphragm valves or pinch clamps. A design approaching the ideal suspends the filling needles from a slender post that moves up and down or has only small-diameter tubing overhanging the vials being filled. The bulk of a filling machine should be below the opening of the container and below the continuous belt or endless screws that move the bottles.

VIII. SOURCING PARTICULATE MATTER IN INJECTABLES

Isolation and identification of particles are critical steps in problem solving. The identification of the chemical composition of particles may require ingenuity in addition to the use of sophisticated instrumentation. Light optical microscopy is the starting point for most identifications, followed by X-ray analysis, infrared spectroscopy, and microchemical testing. Once the identity of the particles is known, it is usually possible to form a hypothesis regarding their source. This hypothesis must then be tested and confirmed, or modified until a solution to the problem is achieved.

An understanding of the mechanisms of particle formation is often based on simple chemistry. Gross problems generally manifest themselves in either unsatisfactory visual inspection or subvisible particle count results. If a product has consistently shown a reject rate of 1% to 2%, a sudden increase of 10% or higher must result in investigation and corrective action.

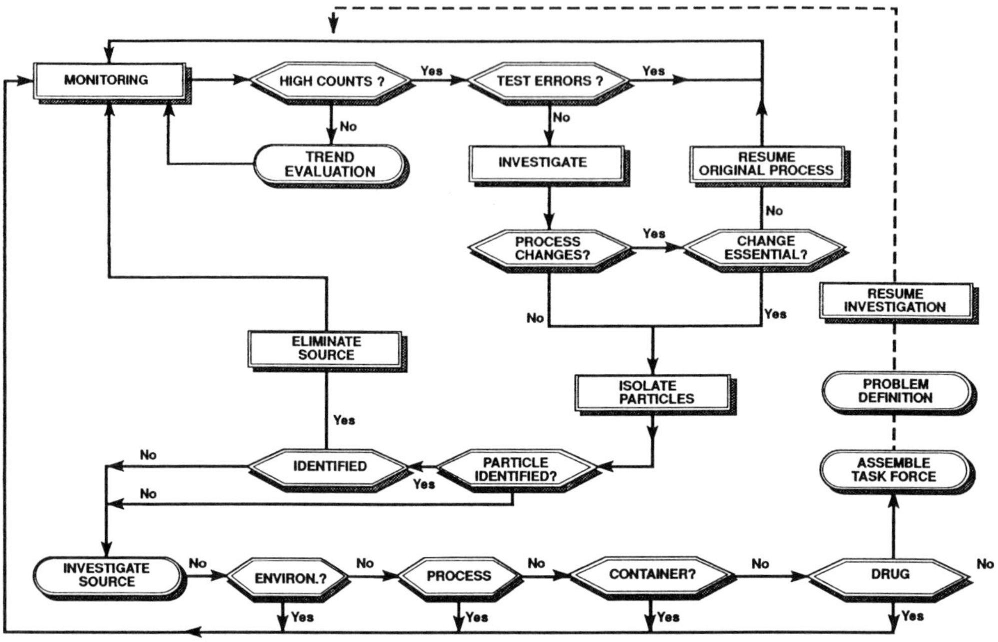

Figure 2 Elimination of particulate matter from process.

Typically, the use of light extinction counting in monitoring product provides a more sensitive indication of problems and is far less variable than human visual inspection. If the number of 10-μm-diameter particles per mL increases from a "historic" 5 to 35, we must assume that something is amiss, even if visual inspection results are not drastically affected. Elevated subvisible particle counts often precede visual inspection problems.

All problem-solving efforts must be well documented. Invariably, the first question to be asked is whether or not the particle inspection or monitoring procedures have changed. Detergents used, substitutions in processing equipment and/or filter vendors, changes in water systems and so forth, often result in particle problems, especially if validation of the change was not performed. The impact of any and all process changes must be evaluated. Even though raw materials and packaging components are released for use after passing established tests, purchasing from a new vendor or process changes by a vendor can still result in particle problems. This source must be considered as a contributor to any problem under investigation.

In most cases, the identity of particles gives a good indication of their source, and thus provides a basis for a rational approach to solving the given problem. Isolation and scrutiny of individual process steps, including samples taken before and after a suspect step of the process, will generally confirm a given source and suggest the corrective action path to be taken. A rational approach to sample collection will invariably narrow down the range of possibilities. With solution products, it is sometimes possible to isolate a causative mechanism by separately evaluating suspected components through substitution of different stoppers, bottles, or control solutions, such as water. As an example, absence of a contaminant in fluid collected into a flask at the filling nozzle suggests container or closure problems. Absence of particles when containers are filled with filtered water or detergent solutions instead of product and then stoppered as usual would suggest that solution-related problems are involved. If the economics of the problem are significant, the use of consultants and/or outside laboratories is usually justified in an attempt to find a quick solution. A flow chart for a generalized attack on a particulate matter problem is shown in Fig. 2 [22].

IX. GENERAL CONSIDERATIONS ON PARTICLES IN INJECTABLES

Implicit in the foregoing discussion is the fact that point-generated extraneous particles are much more important than airborne environmental particles. This is often true, since point-generated or container-related particles are always in the immediate vicinity of the product, and thus have a high probability of gaining access to the final container. We must remember that HEPA filters and directional airflow will not, in fact, remove particles from product. As particles decrease in size, the force required to dislodge them from a surface to which they adhere is increased dramatically due to boundary layer effect, static electrical charge attraction, and van der Waals forces that bind the particles to the substrate.

Laminar airflow itself simply serves to provide directional airflow away from product. However, if large particles become airborne, the laminar flow will serve to shorten their float time in the air and restrict access to product at a remote process point. Larger particles (≥ 10 μm) tend to settle much more rapidly than smaller ones. A 50-μm particle (spherical radius) settles at a rate in the range of 50ft/min. If the particle is elongated or fibrous, this may be decreased 10-fold. In the absence of directional airflow, a particle 1 to 10 μm in size may be suspended indefinitely. Protection of product by HEPA-filtered laminar airflow must be carefully designed and monitored if product is to be effectively protected. It must never be regarded as a cure-all.

It is essential to make containers as clean as possible; it is not practical to depend on cleaning or washing to yield a minimal particle burden. A glass vial is sterile and free from particles at the moment it is made from molten glass. If the cooling process, packing, handling, and transportation are carried out in such a way that no contamination reaches the inside of the vial, there is less concern about washing or sterilization before filling. If vials are handled in a clean environment from the time of manufacture and wrapped in plastic film, it is a much simpler task to clean the vials than if they are packed open in cardboard boxes. Ampules are made from clean tubular glass, and annealed at 500 to 600°C; they contain extremely low numbers of particles. Organic materials, such as fibers and oil droplets, are destroyed in this process. However, no one has yet been able to open an ampule without contaminating the solution with glass particles.

As outlined by Ernerot, monitoring and inspection procedures are very important [23]. If a normal, noncritical visual inspection removes about ⅔ of the units that contain visible particulate matter, by definition one-third are not detected. (Some units will also be rejected even if there are no particles present.) Thus, if the true reject rate is 3%, inspectors will pick up 2% of the true rejects and leave about 1%, some of which will be detected by the customer. Therefore, it is not possible to reach a total quality level by visual inspection. As long as the inspected material has some rate of occurrence of visible particles, human inspectors will always release some items that should have been rejected. Increasing the speed of inspection will result in inspectors rejecting less of a defective product, and with more time they will reject falsely at a higher rate. The same general considerations pertain to machine inspection.

In some situations, washing may be an effective means of decreasing particle burden [24]. The attraction between particles and containers must often be altered with surfactant if contamination is to be effectively removed by washing. Filtered water for injection with a low particle burden should be used as the final rinse for any glass container. Once vials are washed, the transfer of stoppers or vials from the washing area to the sterilizing tunnel through turbulent air must be carefully protected. Whenever possible, containers must be made clean and shipped and stored clean rather than relying on washing to remove contamination. These methods remove one kind of contamination while possibly adding others. The ultrasonic treatment widely practiced in Japan is very useful, but it is time-consuming. The best solution is to have a glass that is as clean as possible when it enters the washing machine. As a generality, the smaller a particle is, the more difficult it is to remove by washing.

In production of aseptically-filled SVI vial product, the outlet from wash tunnels or pass-throughs into the process room is generally less critical than the inlet, since the outlet typically enters a clean area that is Class 100 or better; laminar flow protection between the tunnel and the filling machine remains important. Though it is difficult to avoid a buildup of vials between the tunnel and the filling line, it should be kept at a minimum. Turntables where vials are rotated before filling should be carefully monitored at all times. When clean, sterile glass surfaces are rubbed together they generate particles that may find their way into the vials or result in out-of-limits airborne particle counts.

Stoppers may be manufactured under unclean conditions, and grease and extractables must be washed away [25–26]. As discussed earlier, it may also be necessary to add silicone oil to the rubber surface to achieve "machineability." Here, modern closed-cycle washing machines can remove debris, add silicone, and sterilize in the same operation, and then feed the stoppers aseptically into particle-free nylon bags or directly into a turnkey integrated SVI filling machine. One solution to the SVI particle problem that may enhance solution cleanliness in the future is the use of plastic ampules. Plastic granules, typically polyethylene, are used for form-fill-seal processes. One example of this technique is the Automated Liquid Packaging, Inc. (ALP) bottle-pack system that is currently used for intravenous fluids, irrigation solutions, and eye-drops. This technique approaches an absolutely clean process; it significantly decreases both particle levels and microbial contamination of vials.

As mentioned earlier, personnel continuously and significantly contribute to particle issues; they are the primary contamination source in any "clean area." The benefits of removing personnel from critical areas are seen as cleaner products and lower costs. "Clean" garments dramatically reduce the shedding of particles, but those cleaning the garments may not understand or have forgotten the importance of the procedure and the necessity for cleaning to be done carefully and thoroughly. Fibers from wash or rinse water may contaminate washed garments and then contaminate product. One answer to this problem is disposable garments. The undeniable way of the future is automated and robotic processes that minimize human exposure to product.

The Japanese execution of process design to eliminate particles is interesting. The majority of the LVI solutions sold in Japan are marketed in blow-molded (blow-fill-seal) polyethylene bottles. The processes for blow molding and filling of these units are designed to totally exclude environmental particles. Process-generated particles are low in number due to the specific design of the process machinery and the use of laminar airflow enclosures of the type typically used in Western countries for aseptic filling. The polyethylene containers also incorporate stoppers made from rubber that have extremely low levels of leachables. For product in these containers and in glass, vendors are required to furnish stoppers that are extremely clean, and (in many cases) ready to use.

The Japanese design for filling and product handling machines incorporates minimal potential for particle generation. Filling machines are operated under manufacturing conditions before delivery, and particle generation is monitored. The filling machinery used for LVIs also incorporates a high level of container protection from particles.

X. PARTICLES IN MEDICAL DEVICES

Just as there is no evidence to indicate that the low levels of particles present in injectable products manufactured under CGMP are a hazard to patients, there is no indication that any significant patient health issues are related to the particles from parenteral-type devices. While a very limited number of studies have suggested that, occasionally, specific devices may have high particle burdens [27–29], recent comprehensive studies of Gour [30] and DiPaolo et al. [31] show that a wide range of parenteral-type devices have low solution-contact path particle burdens.

In the Gour study, the majority of a wide range of devices tested met the acceptability criteria imposed by the investigator of a maximum of 25 particles ≥ 10 μm and 5 particles ≥ 10 μm per milliliter of eluent. DiPaolo et al. [31] commented that intravenous administration sets are potential sources of substantially lower particulate matter contamination than are the infused solutions themselves, and that an in-line filter in a set functions effectively not only in removing particles from devices, but from the infusate solutions as well.

The primary concern of both regulatory agencies and manufacturers are devices that have a blood contact path, or solutions that will be infused directly into the patient's venous circulation. This category includes administration sets, blood sets, dialyzers, intravenous catheters, needles, syringes, and other indirectly related products (such as devices for peritoneal dialysis). Examples of devices of this type are

Blood administration sets
Dialyzer, oxygenator-inlet and outlet sets
Hollow fiber dialyzers
Oxygenators
Pump cassettes
Cell elimination filters
Cardiotomy reservoirs
Heat exchangers
Empty plastic containers for IV solutions
IV catheters
Needles
Central catheters
Extracorporeal tubing
Vascular grafts
Extension sets
Pharmacy compounding sets
Solution transfer sets
Autotransfusion devices
Balloon catheters

XI. GENERATION AND CONTROL OF PARTICLES IN DEVICES

Many particles associated with medical devices originate at process assembly points; typically a much smaller number comes from the environment. Each component part assembled into devices typically contributes some variable level of unique particles. Interestingly, because of the distinct sources of particles found in devices, mixed particle populations composed of subvisible 1- to 50-μm particles (representing the inherent burden of components), and small numbers of predominantly process-related larger particles of visible size may occur. Currently manufactured parenteral-type devices span a range of complexity from single administration sets and extension sets to blood oxygenators, extracorporeal circuits, blood warmers, and autotransfusion devices. Large numbers of individual component parts may be included in devices of the latter types. Even simple devices may incorporate parts made from several different types of polymeric material (Table 1).

Specific polymeric materials are selected for an application based on requirements relative to the following criteria: (1) resilience; (2) flexibility; (3) printability; (4) transparency; (5) tensile strength; (6) working pressure; (7) resistance to heat, physical forces, and chemicals; and (8) bonding properties. Cost and resistance to gamma radiation, ethylene oxide or chemical agents that are applied for the purpose of sterilization are also concerns. For some applications, the particle-generating properties of materials may also be a critical concern.

A number of sources of particles in devices may be suggested based on the nature of the components and the assembly processes applied. Medical-grade tubing (silicone, vinyl, polyurethane, polyisoprene or others) may have an inherent particle burden based on its composition or on the process by which it is formed. Additionally, whenever tubing is cut by any means, particles are generated. Injection-molded plastic parts, such as spikes and connectors, may have an associated particle burden originating from fillers or reinforcing agents that may segregate at surfaces, or from minor ingredients that can "bloom" out. The method of collection of injection-molded parts is also critical. As an example, rigid plastic parts with sharp edges that are allowed to drop from a mold into a receiving pan lined with a polyethylene bag can be expected to generate particles from the soft polyethylene. Plastic parts may be washed prior to assembly or may be produced by a "clean" molding process; the latter is invariably a better approach, due to particles being eliminated rather than attempts being made to remove them using washing procedures that are of variable effectiveness.

Elastomeric device components, such as injection sites, diaphragms, and valves, are subject to the same general considerations as stoppers in particle generation. These include blooming or migration of low molecular weight components and surface adherent particles, such as fibers. Filters may be particularly problematic sources of particles; fibrous filters used for blood transfusion invariably shed fibers, and the fragile nature of microporous membrane material makes it liable to generate particles when it is cut or pressed into a housing. Sintered plastic filter

Table 1 Polymeric Components of Medical Devices

Rigid plastic	Flexible (tubing)	Elastomers
Styrene	PVC	Synthetic natural rubber
Impact styrene	Polyethylene	Butyl rubber
Acrylic	Polypropylene	Halobutyl rubber
Clear ABS	Nylon	EPDM
Rigid PVC	Polyurethane	Neoprene
Polycarbonate	Silicone	Silicone
Polysulfone	Teflon	Styrene-butadiene
Polyester	Natural rubber	Nitrile
Nylon	Synthetic natural rubber	Thermoplastic elastomers (TPE)
Polypropylene	Hytrel	
Polymethylpentene		
Cellulosics		
Rigid urethane		
Ektar		
Dacron		

media is prone to shed particles of incompletely sintered material, and cutting hollow fiber filters can generate fragments of the microporous medium. Large plastic parts tend to develop static charges in handling or molding, and can attract and persistently adhere fibrous particles such as paper or cotton. These large particles may create a conspicuous defect if they adhere to the external surfaces of devices. Sometimes metal needles and cannulae bear abrasive particles on their surfaces from sharpening or contain particles within the lumen due to the forming process. Silicone particles (grease or oil) applied to the piston seal of disposable plastic syringes may form agglomerates with particles present in the device, and silicone oil microdroplets may be counted if light extinction counters are applied in enumerating particles in the effluent from a syringe.

Control of particles in the assembly process is critical for devices. The protection of components at each assembly station is important, since a supply of parts is typically held in close proximity to the point at which they will be used. The protection of parts in particle-free bags during handling and storage becomes increasingly critical as more assembly steps are involved, since the potential for the access of process-related particles to the device is increased with the number of assembly steps.

The negative aspects of human contact with devices during manufacture may be more important than when dealing with parenteral products, due to the difficulty of protecting the device solution contact path during assembly and to the higher number of manual operations generally required. We previously discussed particle-generating properties of the feeder bowls widely used in device manufacture. Automatic or robotic assembly of devices is, in general, superior to manual assembly; but machines will also generate particles from their moving parts and from the device components that they handle.

Many of us who work with control and elimination of particles from devices and components are cognizant of the fact that flash or molded parts are a source of particulate matter. Flash from molding processes may be firmly attached to a noncritical area of an affected plastic part, and thus constitute a cosmetic defect rather than a functional or particle issue. If it is released during assembly or during use of a device, however, flash has serious negative implications regarding the process. The same may be said of "angel hair," the long thin filaments of polymer produced by drawing melted plastic during heat-sealing or piercing. Both flash and angel hair particles are relatively easy to source, identify, and control.

Some process machines, such as tubing cutters or vibratory feeder bowls, are obvious potential sources of particles. Other machinery used in manufacturing will also prove capable of generating particles that may gain access to the solution path. Examples include tubing expanders, leak testers, printers, and slitters. Sonic welding or cutting and radio frequency (microwave) sealing may serve to either generate, detach, or disperse particles. Packaging machinery often generates fragments of plastic (Tyvek) or coated paper due to the cutting and drawing processes involved; although sets are most often closed at this point in manufacture, negative customer perceptions may result from this type of particulate matter issue.

It is sometimes necessary to troubleshoot a process to isolate the source of particles or to define individual contributions to the total particle burden. There are generally three steps involved in this type of assessment. The first (essential) step is simply a walk-through inspection observing each process point. The general level of cleanliness of work surfaces is important, as is that of specific areas that contact exposed portions of the device. The presence of paper fibers is usually an indication of a problem, as are unprotected device components exposed to contamination from an operator or by personnel traffic while awaiting assembly. The second step is verifying observations by the use of witness plates that are left in place for some interval in the critical areas and then evaluated microscopically. Collection of particles may also be effected by the use of filter cassettes attached to a vacuum pump, as discussed in a previous chapter.

In this application, the cover of the cassette should be removed so that the maximal surface area of the filter is exposed.

A final step in troubleshooting is to survey the static electric potential. This is crucial in assessing the vulnerability of the components to airborne particles at a given process point. A static potential of thousands of volts is sometimes generated on the surface of large plastic parts during handling or molding. The function of this charge in attracting particles from any great distance is questionable, but static potential will attract particles over short distances and can result in persistent adherence of particles to critical surfaces. Any process step involving plastic sheeting or exposed flat plastic surfaces is suspect; static can be generated by movement of belts, turntables, and rollers or by other movement of plastic surfaces against one another. Static charge may be dissipated by grounding and by ionized air sources used in combination with vacuum nozzles that collect particles loosened by the ionized air. Ionizing grids on the face of HEPA filters are effective in reducing static charges over short distances (less than 1 m). Different means of static dissipation may be evaluated with the assistance of vendors with sufficient knowledge of this type of equipment.

XII. PROCESS DESIGN

Earlier, we discussed the principles of particle control in the manufacture of parenteral products. Since controls regarding particles in devices differ substantially from controls for parenteral products in some specifics, a brief discussion for devices is appropriate. With the wide range of medical devices manufactured, there are a large number of assembly processes and variations of those processes. This situation holds in our present limited consideration of devices having a solution path that contacts infusion solutions or blood. As was the case with parenteral products, minimal particle burden in the final product can be achieved far more cost-effectively through process design than by particle removal operations (such as component washing), rigorous inspection procedures, or an after-the-fact "cleanup" of problem process points.

The simplistic approach to the goal of particle elimination involves the acquisition of clean components and subassemblies from vendors and their clean fabrication into devices. These two steps are equally important. Obtaining clean components will not happen by chance; it requires the vendor's understanding of the customer's requirements and a commitment to this endeavor. Economic factors will obviously be a concern, and it is extremely important that the parties involved accept the fact that cleaner components will result from judicious process design and control, and that a higher level of cleanliness need not be synonymous with higher costs. When particle issues arise with a given device in the field or during manufacture, there is often an unfortunate tendency to focus troubleshooting efforts and resources only on the immediate problem. This approach frequently provides only temporary elimination of the particle source responsible and may result in another problem at a later date at a different process point or with a different vendor. This retrospective approach is not cost effective and results in higher per-unit costs for the finished product.

In many devices, the design of the process for particle elimination is simpler than backtracking to identify, source, and eliminate particles once a problem is detected. Consider the design of a process to assemble a simple administration set consisting of two 1-m lengths of tubing, a "Y" injection site (with integral drip chamber), a spike, and an over-the-needle catheter. There are seven components that will be assembled, and each must be considered separately and carefully in the course of process design.

Most frequently, the manufacturer of the device purchases bulk tubing that will be cut to length on site. Cutting of tubing is notorious for the generation of elongate, crescentic or fibrous particles that mysteriously appear in drip chambers or within in-line filter housings. The cutting mechanism is as critical as the sharpness and durability of the blade used. A regimen for blade changes and careful selection of the cutting machine and blade type (steel or ceramic) is called for; with some materials, laser cutting may have advantages. The bulk tubing used with regard to composition and extrusion should be chosen for a low inherent particle burden. With some tubings, extrusion aids (slip agents) or plasticizers may constitute a significant component of the released particles. Silicone oils are a ubiquitous menace. Therefore, it is very important that the execution of the extrusion process include a consideration of die temperatures, cooling rate, and control of buildup of char or low molecular weight components of the polymer on the die. Bulk tubing rolls cannot be assumed to be immune to particle contamination during storage, handling, and shipping simply because the tubing lumen is closed to the environment. Any adherent material on the tubing surface has the potential for building up at a critical process point and being carried into the lumen of the device, or being transferred to the critical areas of the device on the assemblers' hands (touch contamination).

The other components of the set must be received with vendor assurance of cleanliness based on process control and monitoring during production. The catheter needle tubing should be made by a process that minimizes lumen roughness and particles, and be sharpened by a process that ensures removal of any residual abrasives and/or metal fragments. Assembly of the needle into its hub must be conducted under conditions that provide only minimal access of particles to the hub or needle shaft. Injection molding of the spike should be done under clean-molding conditions that provide for collection of the part and its attachment to a clean-molded drip chamber with maximal control of particles. The rubber chosen for the septum of the "Y" site should give minimal particle generation when penetrated by a needle, and should be free of any additives (modifiers or processing aids) at a level sufficient to produce "bloom" during sterilization or storage. The access of particles to this and other elastomeric parts must be carefully controlled, due to the tendency of these materials to adhere particles with which they come in contact. If the vendor of our "Y" site uses a vibratory feeder bowl to pass the septa singly into the assembly station, this area should receive careful scrutiny as a potential contributor of particles. Feeder bowls invariably require cleaning at periodic intervals to prevent them from becoming reservoirs for particles. A particular concern may be achieving a balance between sufficient surface lubrication of the septa to ensure feeding through machinery, and an excess of a material (such as silicone or slip agent) that will generate particles.

With regard to assembly of the device, specific safeguards related to restricting generation and transport of particles and their contact with the product will be observed. In the absence of automated assembly, personnel should be required to wear smock or jacket overgarments and, depending on the potential for touch contact with product, particle-free gloves. Components awaiting assembly must be protected from particles in the air, both those generated by the process and those generated by or adherent to the surfaces of the containers in which the components are held. Depending on the particle levels generated at the assembly point, ionized air blowdown of the point, and vacuum removal of process-generated particles may be employed. Work surfaces at assembly points should be carefully watched for any buildup of particles.

Even with the effective local measures for cleaning and particle removal, HEPA process-point protection is desirable. Although HEPA airflow is not sufficient to remove particles from a work surface or component, it minimizes the float time of particles, such as fibers, and establishes a flow of clean air onto and around the process point. Modular HEPA enclosures with side curtains provide an economical way to achieve this protection. Periodic cleaning of work surfaces, assembly mandrels, punches, and other equipment with lint-free cloths and a suitable solvent is also effective in reducing work station particle levels.

Vendor monitoring of component particle levels and finished device product by the manufacturer plays an important role in ensuring that a process continues to operate as designed.

Monitoring of particles should not be confused with 100% inspection, whereby some non-destructive analysis is applied to all units manufactured. Monitoring signifies a low level of testing that is applied to ensure that a minimal particle burden is maintained, as assured by a consistent, controlled, and validated GMP process. Given the existence of a consistent, controlled process that has been validated to result in low levels of particles, testing of low numbers of samples suffices to assure that the process is running within limits. Total testing of any device product is ineffectual as a particle control measure; process control is the only approach capable of achieving particle elimination.

As is the case with injectable products, the basic design of any medical device must be carefully executed to prevent particle issues from arising once the complete device is manufactured, or worse yet, when the device is in the field. Selection of materials is particularly important. The glass fiber or other reinforcing agents used in plastic parts, such as connectors, can segregate at surfaces during injection molding processes and be shed into the solution path if molding conditions and the fiber size distribution are not carefully controlled. The use of natural rubber check valve diaphragms that are powdered with talc to prevent sticking must be avoided, so that particles of talc and natural rubber are not shed into the solution path. Pump tubing segments for a peristaltic infusion pump must be of a material chosen for its resistance to spallation and abrasion during pumping. The nonwoven fiber mat in Dacron blood filters must be manufactured so as to ensure that fibers are held in the mat during use. These provide examples of serious potential issues that must be eliminated in the design of a device.

As is the case with drug products, personnel are the largest single source of particles in device manufacture. While it is generally appreciated that people are a significant source of particles in the manufacture of injectables, often less attention is given to protecting product from personnel in production and medical device manufacture. Most of the particles shed by people are skin cells, fragments that have been identified as a source of bacterial contamination, and fibers. These particles are being continually shed by personnel at a rate of 10^5 to 10^7 cells/day. Skin cells, which make up most of the material generated, transport microbes that are present in significant numbers on the human body. Ideally, monitoring should be performed to determine the level of particles generated by workers. Depending on the susceptibility of the process to human contamination, overgarments may be used to filter out the particles that personnel generate, since any exposed area of skin will shed particles. The protection offered by protective garb is only as good as gowning techniques and the aseptic procedures practiced by personnel wearing the garments. Garments may be made of closely woven polyester, Tyvek or (more recently) Gore-Tex. Tyvek garments are available in both disposable and reusable types that control particles shed by a wearer very well; however, they are not "breathable" and may be less comfortable to wear than garments made of other materials.

XIII. SUMMARY

While there are general similarities regarding the sources of particles in injectable products and medical devices, specifics regarding sourcing and elimination may vary. The invariant considerations are as follows: (1) Elimination of particles and particle control must be built into each step and aspect of the process (including product development) for best results. (2) Generation of particles from various sources in the process will be additive and, in this light, minor sources should not be considered acceptable until carefully evaluated; (3) Codevice components must be obtained in a high state of cleanliness and kept clean rather than relying on particle removal. (4) The use of highest quality raw materials for drug or nondrug injectables is warranted if product-related particles are to be avoided. (5) Quality with regard to particulate matter can be built into a drug product only through a thorough understanding of the nature of the drug material and of the container and the interaction of the two. (6) Nonenvironmental (process- or product-related) particles provide the most serious threat of generation of unacceptably high particle burdens.

In both injectable product and medical device manufacture, the control of nonenvironmental particles demands a knowledge of both process and product. Although simple intuition will often suffice, sometimes the use of a comprehensive analytical capability is required. A key factor in most situations will be particle identification. Despite the frequent need for analytical services, those skilled at practical aspects of process particle control can often perform critical problem solving. Whether injectable products or medical devices are considered, particles in the finished product are eliminated by a combination of product design, process design, and the control of environmental and process-generated particles in manufacture. Maintenance of the low levels of particles currently present in the great majority of devices and injectable products produced in the United States, United Kingdom, Europe, and Japan will be achieved by the continuous attention of the manufacturer to eliminating particle sources and the application of GMP, rather than by additional or more stringent regulatory requirements.

TRADEMARKS

The following trademarks used in this chapter are the property of the companies indicated. A brief product description is included; additional product information may be obtained from the vendor.

ALP®. Automated Liquid Packaging, Inc., Woodstock, IL 60173.

Dacron®. Polyester fiber, DuPont Chemicals, Wilmington, DE 19898.

Extar®. Polyester-based polymer, Eastman Kodak Co., Kingsport, TN 37662.

Gelman®. Gelman Sciences Inc., Ann Arbor, MI 48103.

Gore-Tex®. Expanded PTFE, W. L. Gore and Associates, Inc., Newark, DE.

Hytrel®. Polyester elastomer, DuPont Chemicals, Wilmington, DE 19898.

Teflon®. Fluorocarbon resin, DuPont Chemicals, Wilmington, DE 19898.

Tyvek®. Spunbonded polyolefin, DuPont Tyvek, Wilmington, DE 19880.

REFERENCES

1. M. J. Akers, *Parenteral Quality Control: Sterility, Pyrogen, Particulate and Package Integrity Testing,* Marcel Dekker, New York, 1985, p. 253.
2. M. J. Groves, *Parenteral Products,* Hinemann Medical Books, London, 1973.
3. D. S. Aldrich, Particulate formation as a result of packaging, Proceedings of European Conference on Visible and Subvisible Particles in Parenteral Products, European Organization Quality Control, 1985, p. 261–279.
4. S. Nishida, T. Muraki, and S. Ichiyoshi, Ultra-clean pharmaceutical rubber stoppers, Proceedings of European Conference on Visible and Subvisible Particles in Parenteral Products, European Organization Quality Control, 1985, p. 73–86.
5. R. Nishimura, J. Kishimoto, Y. Nishida, Y. Noguchi, and S. Imai, *J. Parent. Drug. Assoc. 33*: 96–103 (1979).
6. P. Abendroth, *J. Parent. Sci. Technol. 39*: 112–113 (1985).
7. H. Green, J. McNelis, and S. Steinman, *J. Parent. Drug. Assoc. 33*: 319–324 (1979).
8. T. J. Roseman, J. A. Brown, and W. W. Scothorn, *J. Pharm. Sci. 65*: 22–25 (1976).
9. W. Speed, Prospective analysis of packaging formulation interactions with regard to particulate

formation and product stability, Proceedings of European Conference on Visible and Subvisible Particles in Parenteral Products, European Organization Quality Control, 1985, pp. 87–146.
10. S. J. Borchert, A. Abe, S. D. Aldrich, L. E. Fox, J. E. Freeman, and R. D. White, *J. Parent. Sci. Technol. 40*: 212–239 (1986).
11. S. H. Yalkowsky, and S. C. Valvani, *Drug Intel. Clin. Pharm. 11*: 417–419 (1977).
12. R. D. White, Mechanism of particulate formation, Proceedings of European Conference on Visible and Subvisible Particles in Parenteral Products, European Organization Quality Control, 1985, pp. 195–223.
13. P. K. Gupta, E. Porenelski, and N. A. Williams, *J. Pharm. Sci. Tech. 48*: 30–37 (1994).
14. K. A. Ashline, R. P. Attrill, E. K. Chess, J. P. Clayton, E. A. Cutmore, J. R. Everett, J. H. Naylor, D. E. Pereira, W. J. Smith, J. W. Tyler, M. L. Veiria, and M. Sabat, *J. Chem. Soc. Perkin Trans. 2* 1559–1566 (1990).
15. P. A. Damme, *Pharm. Act. Helv. 45*: 564–667 (1970).
16. J. W. Danielson, G. S. Oxborrow, and A. M. Placencia, *J. Parent. Sci. Technol. 37 (3)*: 89–92 (1983).
17. R. D. White, Mechanisms of particulate formation, Proceedings of European Conference on Visible and Subvisible Particles in Parenteral Products, European Organization Quality Control, 1985, pp. 195–223.
18. T. Aoyama, and M. Horioka, *Chem. Pharm. Bull. 35*: 1223–1227 (1987).
19. S. J. Borchert, M. Ryan, R. L. Davidson, and W. Speed, *J. Parent Sci. Technol. 42*: 187–194 (1989).
20. S. J. Borchert and S. S. Aldrich, Glass container contributions to parenteral particulates, Proceedings of European Conference on Visible and Subvisible Particles in Parenteral Products, Washington, DC, 1990, pp. 132–154.
21. T. H. Meltzer, *Filtration in the Pharmaceutical Industry,* Marcel Dekker, New York, 1986.
22. L. Ernerot and l. Dahlinder, *Act. Pharm. Sueccica 6*: 401–406 (1969).
23. L. Ernerot, Particulate matter problems during manufactur of parenteral products: a manufacturer's experience, Proceedings of European Conference on Visible and Subvisible Particles in Parenteral Products, European Organization Quality Control, 1985, pp. 48–49.
24. W. K. Hallock, *Bull. Parenteral Drug Assoc. 25*: 65–67 (1971).
25. D. Dolcher, Material and process environmental related particles from elastomeric closures: Proceedings of PDA International Conference on Particle Detection, Metrology and Control, 1990, pp. 103–132.
26. D. Dolcher, *Pharm. J.* pp. 246–267 (1991).
27. E. A. Knopp, F. G. Baumann, D. Pratt, R. Faden, F. P. Catinella, I. M. Nathan, P. X. Adams, J. N. Cunningham, Jr., and F. C. Spencer, *J. Cardiovas. Surg.* 23(6): 470–476 (1982).
28. U. S. Page, J. C. Bigelow, C. R. Carter, and R. L. Swank, *Ann. Thorac. Surg. 9*: 18–22 (1970).
29. J. E. Dimmick, K. E. Bove, J. McAdams, and G. Benzing, *New Eng. J. Med. 292*: 685–687 (1976).
30. L. Gour, *Pharmacopeial Forum* 2506–2522 (May/June 1987).
31. E. R. Di Paolo, B. Hirsch, and A. Pannatier, *Pharm. Weekbl. 12 (5)*: 190–195 (1990).

14

Physiological Consequences of Injected Particulates

Lawrence John Pesko

St. Vincent Medical Center, Toledo, Ohio

A meaningful discussion of particulate sizing should include a scientific explanation of the physiologic effects resulting from the introduction of foreign particles into a patient's circulatory system. Efforts to measure, establish limits, and prevent particulate contamination of injectable products are based on our belief that these "contaminants" are unwanted. They are unwanted because they provide no demonstrable benefit and there is some evidence that they may be harmful. The entire study of particulate matter really began with the work of Garvin and Gunner [1], who quantified and classified particles that were present in intravenous solutions that were manufactured in Australia. In a second paper [2], Garvin and Gunner gave additional information on the method of detection and the nature of these particles. They also presented evidence of the harmful effects of such particles. These observations were based on rabbit experiments and histological evaluation of human lung tissue taken at autopsy (see Figs. 1 and 2). The histopathologic changes in the rabbit consisted of capillary and arterial granulomas resulting in a proliferation of histiocytes and containing one or more foreign body giant cells. Each granuloma contained fragments of cellulose fibers. "Vascular bastosis" was suggested as a specific and descriptive name for these lesions (bast was the material which made up the filter media used to prepare intravenous solutions in Australia). The investigators estimated that for every 500 ml of Australian intravenous fluid injected into each rabbit, 5000 scattered granulomas had developed throughout both lungs. The lesions observed in humans by Garvin and Gunner were similar to those granulomas found in their experiments on rabbits and invariably contained bast fibers. In those cases in which intravenous fluids were given just prior to death, there was a chronic fibrous pneumonitis in which cellulose fibers and crystals were present, as well as vascular lesions in the lungs.

These two studies marked the beginning of a entirely new science which this book addresses. If we were to take a skeptic's view of Garvin and Gunner's studies we might still consider it worthwhile to study the potential effects of particulate matter. Consider the following scenario; let us suppose that the average hospitalized patient receives at least 5 l of intravenous solutions during their hospital stay. We know that approximately 33,500,000 patients are hospitalized in the United States annually. Thus, on average, at least 167,500,000 l of intravenous fluids are administered in the hospitalized setting annually. If we were to additionally consider the administration of intermittent admixtures, directly administered intravenous push medica-

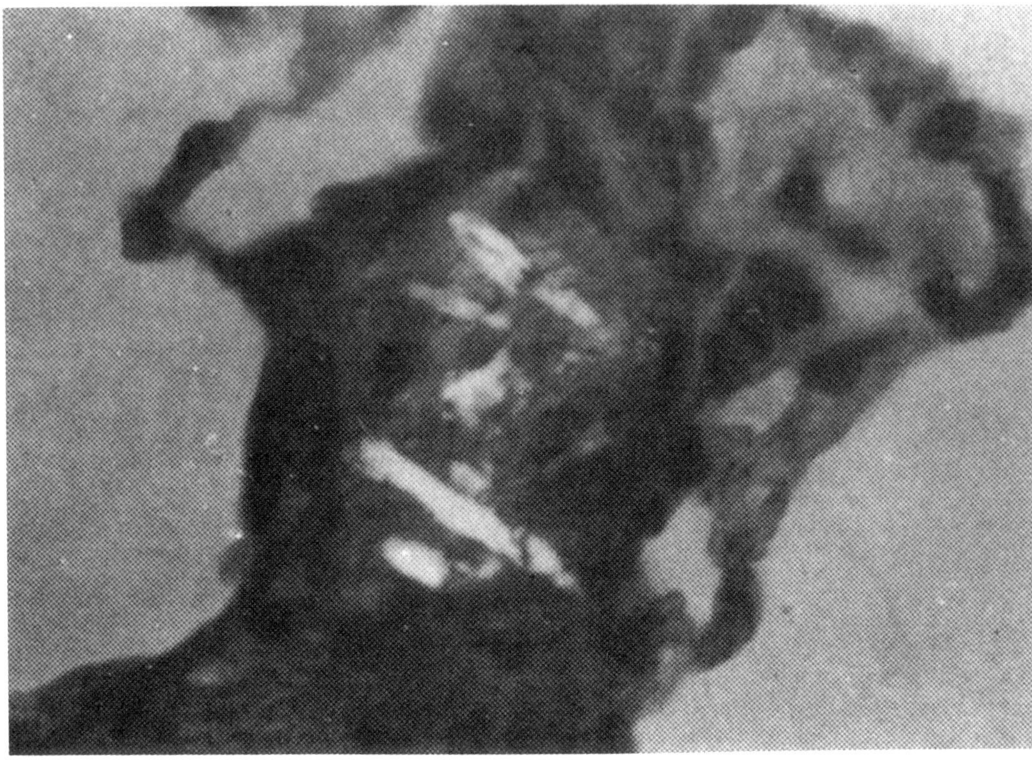

Figure 1 Histological section of experimental rabbit lung showing cellulose fibers as viewed by polarized light. (From Ref. 1.)

tions, and intravenously administered diagnostic agents, we can begin to see the scope of this important route of administering medications to patients. Considering the aforementioned situation, if there is only a remote possibility that injected particulate matter causes some degree of pathology, then it stands to reason that it would be worthy of investigation. Unfortunately, most of the scientific investigation into this question has been incomplete and in many instances the studies have been poorly designed. Despite the paucity of scientifically sound information on this topic, we shall present a summary of the scientific literature that addresses the physiological consequences of injecting particulate matter into the vasculature of patients and in laboratory animal models.

I. ANATOMY AND PHYSIOLOGY OF THE VASCULAR SYSTEM

In order to appreciate and understand the adverse consequences attributed to particulate matter, we need to first discuss the vascular anatomy of the body and how it transports particulate matter throughout the body. Special attention shall be devoted to describing the vasculature of the lungs and the liver, since studies have found these two organs as the major repositories of experimentally injected inert particulate matter.

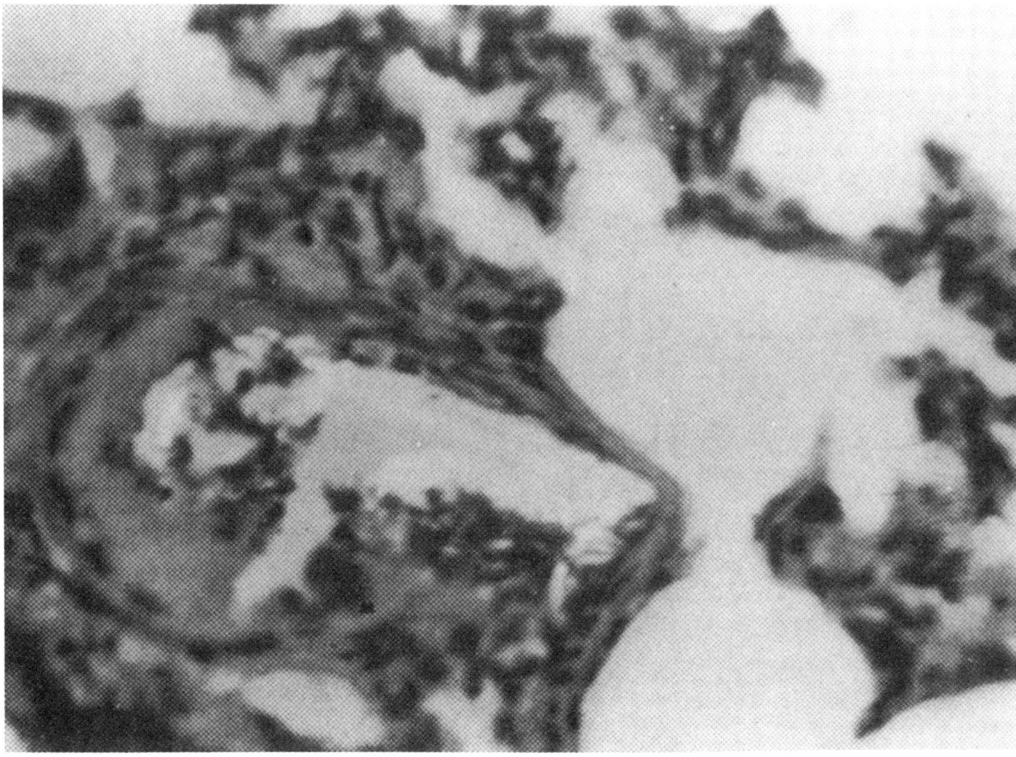

Figure 2 Photomicrograph of a pulmonary granuloma from a 25-day-old infant showing fragments of cellulose fiber as viewed by polarized light. (From Ref. 2.)

A. Overview

In 1628 William Harvey described the action of the heart, veins and arteries as a closed circular system which conveys blood throughout the body. He clearly demonstrated that blood leaves the heart through arteries and returns to it through veins. About 40 years later Malpighi discovered that there was a microscopic link between these two systems. These tiny vessels were called capillaries and are in the tiny vessels that Harvey knew existed between the smallest arterioles of the arterial system and the venules of the venous system (see Fig. 3). For the most part arteries are responsible for delivering oxygenated blood and nutrients to the organs and tissues of the body. Veins are responsible for removing waste materials and deoxygenated blood from these tissues.

The circulatory system can further be subdivided into the pulmonary circulation and the systemic circulation. In higher vertebrates large amounts of oxygen are needed to support cellular respiration. It is equally necessary to dispose of the large amounts of carbon dioxide that are generated by cellular respiration. These needs are achieved through the large capillary mass of the lungs. To achieve a high blood flow rate, blood is pumped by the "two"-pump system of the heart. One pump receives oxygen poor blood from the venous system into the right atrium of the heart and pumps it to the lungs from the right ventricle of the heart. The second pump receives oxygenated blood from the lungs into the left atrium and pumps it to the rest of the body from the left ventricle. Blood entering and leaving the lungs goes through a single

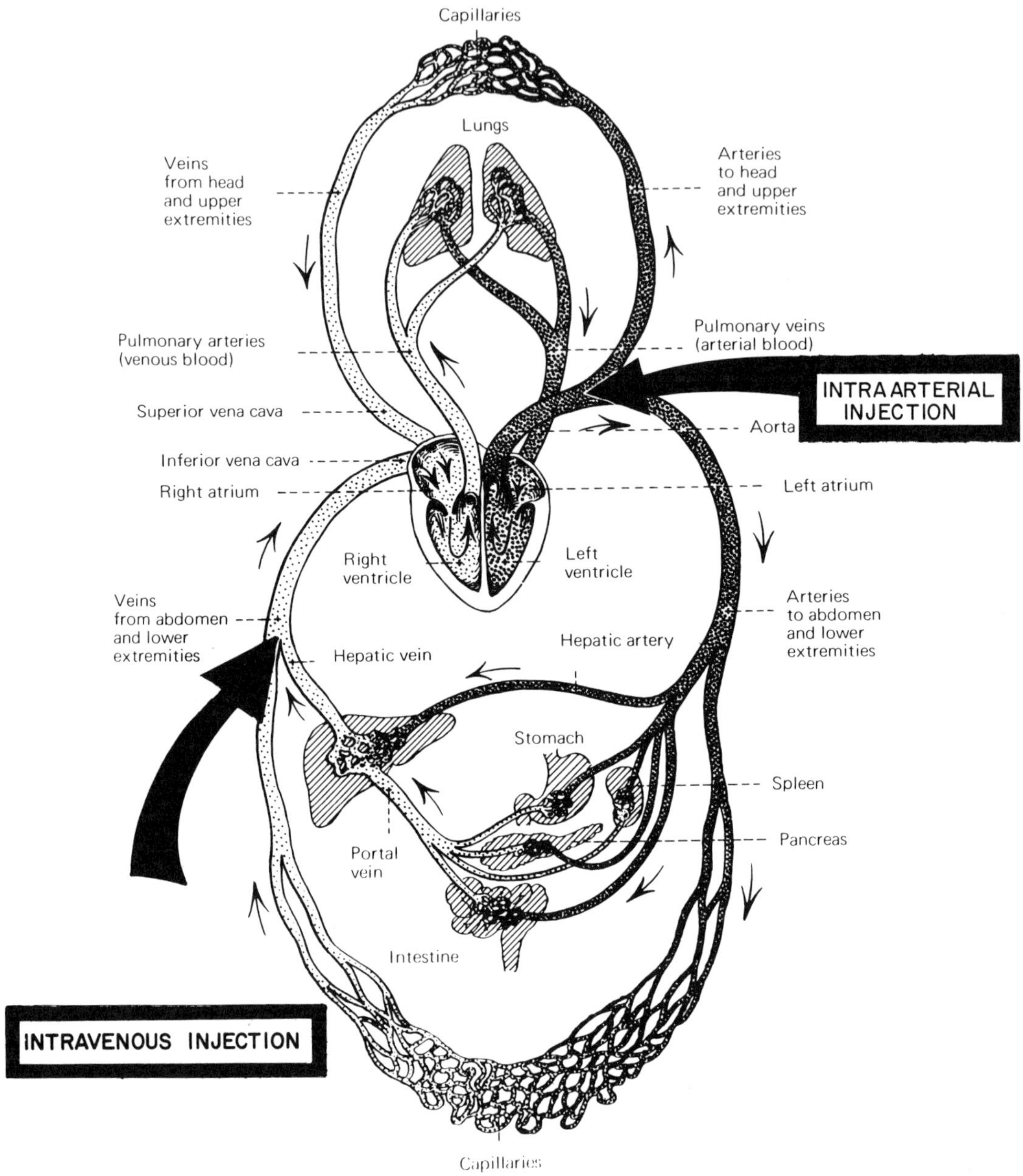

Figure 3 Diagrammatic representation of systemic and pulmonary vascular systems. (From Ref. 29.)

massive network of capillaries. The capillary network is similar to the rest of the body's system for blood leaving the arterial system and being transferred to the venous system. However, a special arrangement exists for the return of blood from the abdominal part of the alimentary canal, the spleen and the pancreas. From the capillaries of these organs, the blood flows into veins which unite to form the portal vein. The portal vein, which is very large passes into the liver where it branches into smaller veins which are capillary like vessels called sinusoids. Therefore, blood which leaves the liver via the hepatic veins has passed through two sets of capillaries as compared to the single set of capillaries that blood from most other organs and tissues pass through.

B. Lymphatic System

The lymphatic system is another system that is extremely important to the effective transport of materials within the body. As blood flows through the capillaries some of its fluid and other substances pass through their thin walls. This fluid makes up the interstitial fluid which surrounds the cellular structures of the body and allows for the exchange of materials resulting from the metabolic activities of the cells. The fluid is drained from the interstitium by reentering the blood capillaries toward the venous end and is returned to the heart. The fluid is also removed by a completely different set of capillaries called lymph capillaries. These capillaries dump into the lymph vessels which make up the lymphatic system. The larger vessels of the lymphatic system eventually discharge their contents into the great veins at the root of the neck. Thus, all fluid which irrigates the tissues is eventually returned to the heart. Not all parts of the body have lymph vessels, they are absent from avascular tissues, bone marrow, muscle and the central nervous system. Lymph vessels are interrupted during their course by lymph nodes which serve as filters for the lymph system. Most of the lymph passes through at least one lymph node and sometimes two nodes before reentering the blood stream.

C. Arterial System

The arterial system is composed of a very elastic vasculature. When the heart contracts, blood is suddenly forced into the aorta and larger branches of the arterial tree. These vessels have a predominance of elastic tissue in their walls and distend with each systole (contraction) of the heart. This distention protects the smaller arteries from the sudden strain of systole, while their elastic nature allows them to relax during diastole (relaxation) which allows the blood to continue to flow down the arterial tree. The expansion/relaxation cycle ensures that the blood will flow in a steady stream to the smaller vessels. Working from within the artery, the three categories of tissue that make up the arteries are the tunica intima, tunica media and the tunica adventitia. The arterioles are the smallest arteries with a mean diameter only slightly larger than that of the capillaries. The mean average diameter of an arteriole is between 100 and 500 microns. The tunica intima consists of simple endothelium with a thin internal elastic lamina. The tunica media consists of circularly deposited visceral muscle fibers arranged in layers which become fewer as the arterioles become smaller. The tunic adventitia in large arterioles may equal the tunica media in thickness, but in the smaller arterioles it becomes very thin and the external elastic lamina disappears.

Capillaries are on average 8 to 10 microns in diameter and have an average length of about 750 microns. Their walls consist of endothelium arranged as pavement cells with sinuous tapering edges which overlap and adhere to adjacent cells. These cells appear to be cemented to the margins of one another to form a continuous tubelike structure. Physiological evidence strongly suggests that capillaries have the ability to contract and to dilate. Contraction may be achieved through the swelling of the endothelial cells which obliterate the lumen of the capillary which is restored when the cells shrink to their normal flattened position.

D. Venous System

In the venous system, blood flows much more slowly and at a reduced pressure. Therefore, it stands to follow that the lumina of the vessels are larger and their walls are thinner then the corresponding arteries. The walls of veins resemble those of the arteries except that they contain less muscular and elastic tissue and contain more white fibrous tissue. The walls of the venous system are so thin and contain so little muscular tissue that they collapse when empty. The three membranous coats that make up the structure of the veins are not always able to be distinguished from one another. At least they do not have as clear a boundary as in the arteries. The tunica intima consists of an endothelial lining, a layer of subendothelial areolar tissue which is less abundant than in the arteries and an outer layer of longitudinally arranged elastic fibers. A distinguishing feature that is peculiar to the inner structure of the vein is a series of valves formed by folds of endothelium strengthened with a small amount of fibrous and elastic tissue. The purpose of the valves is to prevent the backflow of blood to the periphery and to help to support a column of blood in the veins where there is an upward flow. The tunica media is much thinner and contains less muscle and more collagen than the corresponding coat in the arteries. In some veins the muscular tissue is absent. The tunica adventitia consists of white fibrous and elastic tissue. Longitudinally disposed muscle fibers in the outer coat of some large veins allow for changes in length and provides some degree of protection from external pressures.

The venous system commences from a large network of capillaries and form larger trunks. However, as mentioned earlier, the veins of the portal system begin and end in a capillary bed. There are many more veins than there are arteries. Additionally, veins are able to hold a greater capacity of blood and have a tendency to anastomose more freely than do arteries.

Injections of sterile products are most frequently given via the venous route of administration. Veins of the forearm and the dorsal portion of the hand are the most frequently chosen sites for administration of intravenous solutions and injections. One of the veins of the median cubital complex regularly has a large vessel anastomosing with the deep veins. This anastomosis holds the superficial vein in place and will prevent it from slipping away during venipuncture. Other frequently used veins are the cubital, basilic and cephalic veins (see Fig. 4).

E. Lungs

As previously mentioned, the lungs are a vast network of capillaries that anastomose the arterial and venous systems. The purpose of this network is to provide oxygenation of blood and to allow for desaturation of carbon dioxide from the blood. The lungs are particularly important in studying the pathophysiologic effects of injected particles. Most injections are made directly into the venous system. The narrowest point in the venous system is the capillary bed of lungs. These capillaries are typically 8 to 10 microns in diameter. This diameter allows erythrocytes (red blood cells) to easily pass from the poorly oxygenated blood of the pulmonary artery . . . undergo oxygenation and enter the pulmonary vein. It would appear from our previous discussions that the pulmonary capillary bed represents a fine-mesh screen which everything in the venous system must pass through before entering the arterial system. Theoretically, any particle larger than the lumen of the capillary bed might lodge in that capillary bed. Therefore particles larger than 10 microns would not appear in the arterial system. However, this is not the case.There are arteriovenous shunts that are present in the vasculature of the lungs. These shunts allow particles that are larger than the capillary bed to enter into the arterial system, where they may eventually become lodged in another capillary network of some internal organ like the liver, spleen, brain or the eye (see Fig. 5). The system of arterio-venous shunts in the liver, spleen and lungs was first described by Prinzmetal et al. in 1948 [3]. This group injected glass spheres which ranged in size from 10 to 440 microns into rabbits, cats and dogs. Spheres were injected into the portal vein of rabbits and recovered in the lungs. These spheres were 60 to 180 microns in diameter. In dogs, the spheres were injected into the splenic artery and recovered in the liver. These spheres varied in size from 160 to 370 microns. Since injected glass spheres that were 20 to 40 times the average diameter of the lumina of capillaries rapidly passed through

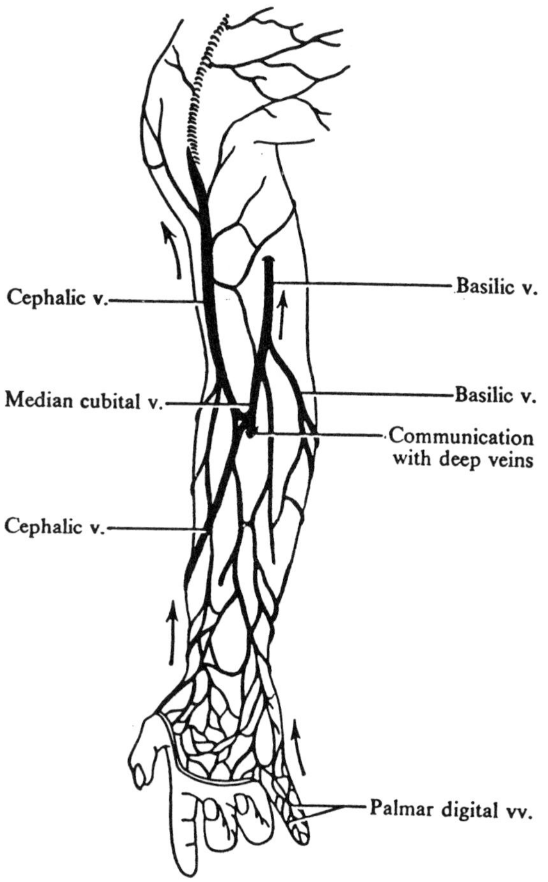

Figure 4 Anterior view of the superficial veins of the upper extremity. (From Ref. 30.)

the liver, spleen and lung of the test animals, the investigators concluded that these organs must have noncapillary channels. Thus demonstrating that these organs do have arteriovenous or noncapillary shunts.

To summarize, the vasculature of the lungs does represent a large filtering system for particulate matter. However, the pulmonary capillary system is not an absolute filter because of the existence of arteriovenous shunts.

F. The Liver

The liver is the largest gland in the body weighing between 1.4 and 1.6 kg in an adult male. Blood is brought to the liver by the portal vein and by the hepatic artery. These vessels bath the liver cells with blood. The hepatic artery enters the liver and ramifies with the portal vein and the hepatic duct. It gives off capsular branches which ramify in the fibrous capsule of glisson. It also gives off capsular branches which extend into the stellate plexuses. The portal vein also enters at the porta and runs through the portal canals dividing into smaller branches as it passes to the hepatic sinusoids. The hepatic sinusoids are large, richly anastomosed, modified capillary beds lying between the cells of the liver. The 1.5 l of blood per minute which enters the liver passes through the sinusoids to the central veins. About two-fifths of this blood is derived from

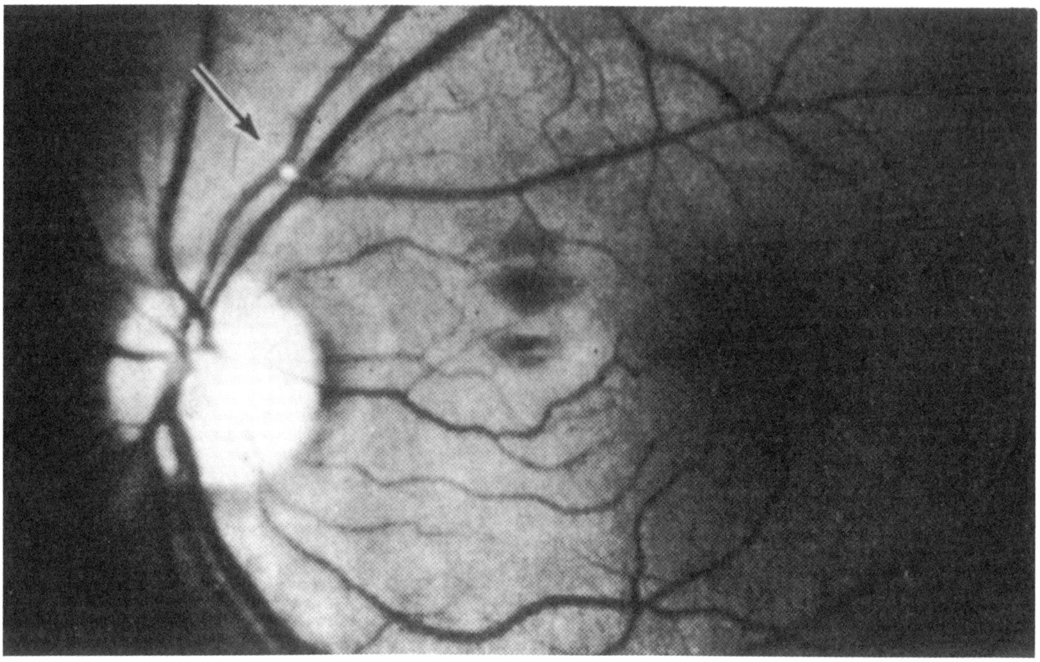

Figure 5 Fundoscopic view of a human retina with an embolus at a bifurcation of the central artery. An incomplete occlusion has resulted. (From Ref. 31.)

the hepatic artery and three-fifths from the portal vein. This dual system of nutrient supply to the liver parenchyma makes the occurrence of infarcts in this organ very rare. The sinusoids are lined with modified endothelium and contain many macrophages attached to their walls. These macrophages play an important part in trapping the very small foreign materials that present themselves to these cells (this includes the very small particulate material which makes its way past the pulmonary capillary bed). In these sinusoids, the endothelial lining is characterized by wide intercellular junctions and large fenestrations which provide gaps for the easy passage of proteins, cells and other microparticulates.

The pulmonary capillary bed can be regarded as the place where the gaseous products of respiration are exchanged; Whereas the hepatic capillary bed is regarded as the place where a vast array of metabolic process occur. Two of these processes involve the removal of foreign material (such as bacteria) and the detoxification of noxious substances.

To review the complex anatomy previously described, let us travel the commonly encountered sanguinous freeway as a foreign particulate that has been introduced through the brachial vein. For argument sake we shall consider ourselves small enough to traverse the entire interstate system. Once we are injected into the brachial vein we will quickly be transported to the axillary vein then to the subclavian vein and through the right brachiocephalic vein and finally into the superior vena cava that will terminate in the right atrium. From the right atrium of the heart we shall be quickly catapulted to the right ventricle where we shall catch the overpass to the pulmonary artery. Once inside the pulmonary artery we will take the left fork and travel down the left pulmonary artery till the artery reaches its final destination down a narrow and tortuous one lane dirt road. The dirt load will anastomose with another dirt road which will eventually open up to a superhighway called the pulmonary artery. The pulmonary artery shall take us to the left atrium where we shall again be catapulted to the left ventricle then to the aorta and we will finally find ourselves on the interstate artery system. Since we are going to spend a few weeks with relatives in liver-land, we decide to take the most direct route there through the abdominal aorta to the celiac artery and then to the common hepatic artery to the right hepatic artery . . . home at last!

II. THE EFFECT OF PARTICLE SIZE ON DISTRIBUTION WITHIN THE BODY

The arteriovenous highway analogy highlights this system as the main mode of transportation for injected particles throughout the body. Our initial assumption was that we were a particle which could easily transverse any highway system. However, we can easily see that if we were a large vehicle (particle) we may be to large to enter certain highway systems and may fall short of our final destination. Another possibility may be that we get off the highway system and need to take the lymphatic ferry. The ferry may eventually get us back on the arteriovenous highway system. There are a number of well-designed and controlled studies that have evaluated the fate of inert particulate matter which is introduced to the vascular system. These studies help us to understand where particles may reside once introduced into the bloodstream.

There are three main factors that influence the pathological effects of particulate matter which is injected into the bloodstream: (1) the site of occlusion and the degree of interruption of the blood supply, (2) the size, shape, and chemical composition of the injected particle, and (3) the host response to the particle.

There have been several studies that have addressed clearance kinetics and the distribution of particles following intravenous administration. One of the first studies involved the use of polystyrene divinyl benzene copolymer (polystyrene DVB) spheres. Schoenberg et al. [4] injected 0.514 to 1.17-microns polystyrene DVB particles intravenously into albino white rabbits at doses up to 1.6×10^{12} particles/kg and found that all the particles were cleared effectively from the bloodstream. More than 90% of the particles were cleared from the blood within the first 60 to 90 min; however, complete clearance had not occurred even after 18 days. Particles were found in the reticuloendothelial (RE) cells of the liver and spleen, but only a small percentage of RE cells contained particles. It was concluded by the authors that only a small percentage of the RE cells participated in phagocytosis since only a small portion contained particles. In a second study by Schoenberg [5], he suggested that the slower phase of particle clearance represented equilibrium between phagocytosis and reentry into the circulation from RE turnover. The larger particles were phagocytized more rapidly than the smaller particle and over time, the particles became redistributed from the liver to the spleen.

Wilkins and Meyer [6] injected polystyrene DVB spheres with a diameter of 1.305 microns and labeled with I^{125} into rats at a dose of 6×10^9 particles/kg. They found that 99% of the spheres were cleared from the blood within 15 min of injection. Examination of the uptake by the various organs revealed that the more positively charged colloids were deposited in the lungs. Since a large accumulation of colloidal material in the lungs is normally due to filtration of flocculated material rather than phagocytosis, the possibility was investigated that the uptake was only temporary. This was indeed that case for small particles. It was also interesting to note that the surfaces of the interior of blood vessels and liver and spleen cell surfaces have a negative potential and, therefore, would be expected to attract positively charged particles and repel negatively charged particles. However, this was not found to be the case. The authors summarized their finds as follows:

Negatively charged particles are taken up to a greater extent by the liver.
Positively charged particles showed an initial accumulation in the lungs and a later accumulation in the spleen.

Singer et al. [7] studied the clearance of polystyrene DVB particles labelled with I^{125} and a mean diameter of 0.23 micron which were injected in mice. He found that 95% of the particles were cleared from the blood within 5 min following injection. However, some residual particles were detectable for as long as six months. The initial distribution of particles was within the lungs and these tended to disappear after 10 to 60 min and show up in the liver and spleen. The immediate agglomeration of spheres were dissociated by polymorphonuclear leukocytes. The phagocytic process was clearly visualized at the cellular level with an electron microscope. These observations revealed coating of particles with a protein-like material, with adhesion to and phagocytosis by the Kupffer cells. Comparison of latex and carbon particles revealed that the clearance curve of carbon differed from that of latex. The rate for carbon removal being inversely proportional to the dose; phagocytosis of carbon was seen to be accomplished by the formation of channels in contrast to the latex particles which were engulfed by peripheral vacuole formation. These authors concluded that the clearance curves neither reflected nor indicated a direct relationship to the actual process of phagocytosis by the reticuloendothelial system cells.

These studies have shown that inert particulate matter that is of a very small diameter may be cleared from the blood stream and phagocytized by the reticuloendothelial system. Madden et al. [8] showed that the transmigration of larger particles can occur if the particles were injected into an artery or the aorta. This was demonstrated by injecting 15-micron microspheres into the aorta of female Wistar rats. Twenty minutes later the microspheres were recovered in the oropharynx, trachea, bronchi, stomach and urine. When the tracheal or bronchial blood supply was interrupted, their appearance was suppressed. Following aortic blood diversion, very few particles were found in the isolated bronchial washings. Following intra-aortic injection, microspheres were readily recovered from the stomach and urine. When either the terminal aorta or renal arteries were ligated, microspheres appeared in the urine. When both were ligated, none appeared, indicating that they passed via both kidneys and cystic arteries. These findings may explain additional routes of elimination of particulate matter from the body.

Most of the aforementioned studies looked at the distribution of particulate matter that was less than 2 microns in diameter. Because their major purpose was to study the distribution of particles, histopathologic slides were rarely performed.

Gesler et al. [9] designed a study to determine the influence of particle size, dosage and time course on the distribution of particles in female Sprague-Dawley rats. The investigators studied polystyrene DVB microspheres with the following size distributions: 0.4, 4, 10, and 40 microns mean diameter. Particle size distribution was determined with a Model B Coulter Counter and a size distribution oscilloscope with adjustable thresholds and gate widths. Preliminary biologic studies of the polystyrene DVB microspheres were conducted to determine acute intravenous toxicity, intradermal irritation, and tissue culture biocompatibility. The previously mentioned four groups of particles were suspended in normal saline and administered to the rats at four different dosage levels for each of the individual groups. The four dosage levels for each group were 10^3, 2×10^4, 4×10^5, and 8×10^6 particles/kg of body weight. The volume of injectable material was kept constant at 10 ml/kg and the amount of material injected was adjusted by varying the concentration of the study concentrations. All injections were made into the tail vein of the rats at a rate of 2 ml/min. The rats were observed carefully following injection and were weighed prior to injection, at weekly intervals thereafter, and again just prior to necropsy. Five rats from each group were killed at 1 h, 1 day, 7 days and 28 days following the injections. Four control groups of 20 rats were injected with a placebo. Blood samples were obtained from each rat via the orbital sinus for determination of total leucocyte count, differential count, hematocrit and hemoglobin concentration. At necropsy, the organs of the cranium, thoracic and abdominal cavities were examined grossly. The brain, lungs, spleen, liver and kidneys were weighed. Portions of these organs and the heart, adrenals, gonads, lymph node and large intestine were prepared for histopathological examination.

The finding of these investigators were as follows:

Thirteen of the 18 rats injected with 8×10^6 particles/kg at the 40-micron size developed labored respiration and prostration immediately following injection. All 13 of the rats died within 3 to 5 min. The remaining five animals in this group showed some signs of hypokinesia following the injection, but fully recovered within 30 min.

No meaningful body weight differences were detected in any of the groups studied.

Total leucocyte count, differential count, hematocrit and hemoglobin concentrations were essentially within normal limits for all treated and control groups.

There were some significant differences in the weights of the individual organs (see Table 1).

No gross pathological changes attributed to the intravenously administered particles were seen.

At the 4-micron size, particles were observed chiefly in the liver and spleen.

With the 10-micron size, particles were predominantly seen in the lung.

The animals injected with the 40-micron-size particles had these particles predominantly in the lungs and a few particles were in the coronary vasculature (see Figs. 6 and 7).

The incidence of particles in the heart and lungs was proportional to the dose.

The deposition of particles in the different organs was related to particle size.

None of the sections that were examined histologically showed signs of an inflammatory reaction even after 28 days.

This study showed that there was a definite relationship of particle disposition based on particle size. Small particles in the range of less than or equal to 4 microns tend to migrate to the liver and spleen. Intermediate particles in the 10-micron range localize in the lung and particles of 40 microns could be found in the lung and heart. The study also demonstrated that this type of test material was inert. Therefore, the granulomatous type reactions previously described in the literature did not present themselves in this study.

A third study of particle disposition in beagle dogs was conducted by Kanke et al. [10] to determine the time dependency of the relocation of small microspheres injected intravenously and intraarterially. The previous two studies had described the distribution of small particles within beagle dogs, however they had not plotted the time course of that distribution. Kanke's group also undertook a more intensive investigation of the histological changes presented by the particles. Kanke injected Ce^{141} radiolabeled polystyrene DVB microspheres with mean diameters of 3.4, 5.4, 7.4, and 11.6 microns into two- and three-year-old female beagle dogs. The dosage of particles were as follows: 3.4 microns: $253–890 \times 10^6$; 5.4 micron: $73–101 \times 10^6$; 7.4 micron: $46–93 \times 10^6$; 11.6 micron: $8.6–10^6$. Intravenous doses were administered through the radial veins of the left and right forelegs. Intraarterial doses were administered via a cannula which resided in the left ventricle. All injections were made over 2 min and the dogs were sacrificed at 20 min, 1 h, 24 h, or 4 weeks following the injection. Images were photographed using a scintillation camera interfaced to a digital computer. Images were recorded at a rate of 12 frames/min for the first 2 min, followed by 2 frames/min for the next 58 min. Delayed static images were recorded at 2 and 24 h, 1, 2 and 4 weeks following microsphere administration. Microsphere distributions were determined externally using computer-generated images integrating the counts with each organ area. The lungs, liver, spleen, heart, kidneys, brain, and tongue were removed from the sacrificed animals and scanned individually to quantify the distribution

Table 1 Summary of Particle Distribution and Incidence in Tissues of Animals Surviving for the Time Indicated (number of animals per group of five in which particles were found in indicated organs)[a]

Particle	Dose	1 h				1 day				7 days							28 days				
size	particles/kg	Lung	Heart	Liver	Spleen	Lung	Heart	Liver	Spleen	Lung	Heart	Kidney	Spleen	Brain	Pancrease	Liver	Lung	Heart	Brain	Liver	Spleen
0.4	10^3	0	0	0	0	0	0	0	0	0	0	0	0	0	0	0	0	0	0	0	0
	2×10^4	0	0	0	0	0	0	0	0	0	0	0	0	0	0	0	0	0	0	0	0
	4×10^5	0	0	0	0	0	0	0	0	0	0	0	0	0	0	0	0	0	0	0	0
	8×10^6	0	0	0	0	0	0	0	0	0	0	0	0	0	0	1[b]	0	0	0	0	1[b]
0.4	10^3	0	0	1	0	0	0	0	0	0	0	0	0	0	0	1	0	0	0	0	0
	2×10^4	0	0	1	2	0	0	0	1	0	0	0	0	0	0	0	1	0	0	0	0
	4×10^5	1	0	0	0	0	0	0	1	0	0	0	1	0	0	0	0	0	0	1	0
	8×10^6	2	1	1	2	0	0	2	1	0	0	0	2	0	0	2	0	0	0	1	1
0.4	10^3	1	0	0	0	0	0	0	0	0	0	0	0	0	0	0	0	0	0	0	0
	2×10^4	2	0	0	0	1	0	0	0	0	0	0	0	0	0	0	1	0	0	0	0
	4×10^5	5	0	0	0	5	0	0	0	3	0	0	0	0	0	0	3	0	0	0	0
	8×10^6	4	1	2	0	5	0	0	0	5	1	1	1	1	1	0	5	0	0	1	2
0.4	10^3	1	0	0	0	1	0	0	0	0	0	0	0	0	0	0	0	0	0	0	0
	2×10^4	2	0	0	0	3	0	0	0	1	0	0	0	0	0	0	1	0	0	0	0
	4×10^5	5	2	1	0	5	1	0	0	5	0	0	0	0	0	0	5	2	0	0	0
	8×10^6 [c]	1	1	0	0	2	1	0	0	1	1	0	0	0	0	0	1	1	0	0	0

[a]The number of surviving rats killed at the specified time periods were 1 h, 1; 1 day, 2; 7 days, 1; 28 days, 1. The particle distribution and incidence in tissues in the 13 animals that died acutely were as follows: lung, 13; heart, 7; liver, 2; kidney, 4; brain, 1.

[b]Possible particles.

[c]Thirteen of 18 animals injected with this dose of this particle size died acutely.

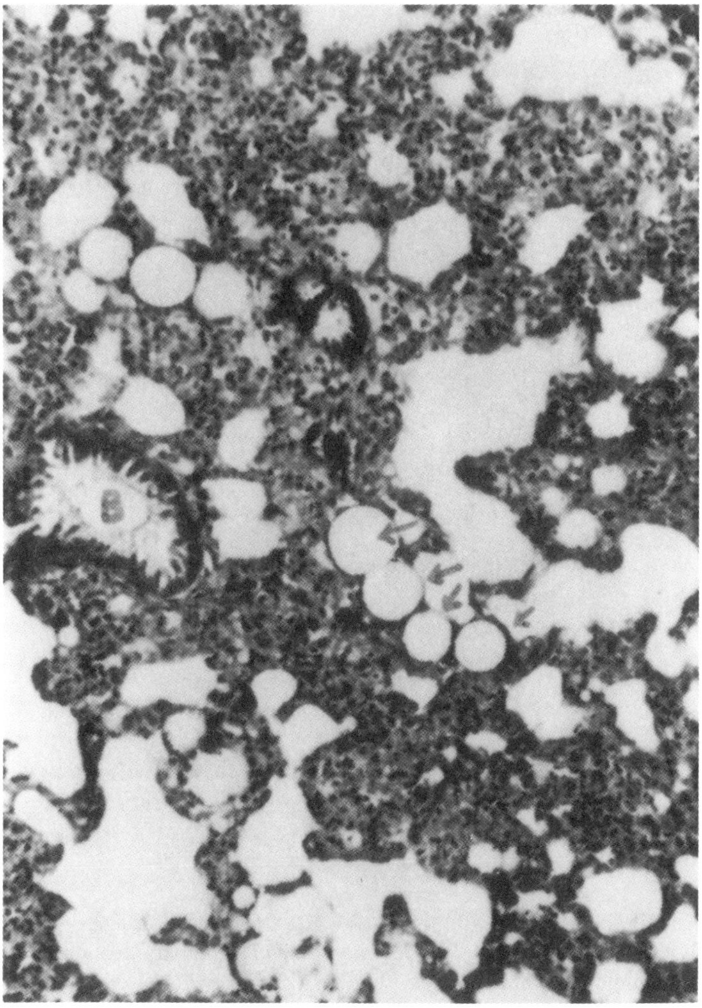

Figure 6 A section of lung from a rat injected with 40-micron particles at a dose of 8×10^6 particles/kg and killed at day 1. The particles are indicated by the arrows and the bronchiole by the B. No inflammation is present. (From Ref. 9.)

of the microspheres at specific times. Following the scanning of the intact organs, the organs were sectioned and a small sample of tissue was removed for histological examination. The remainder of the organ was homogenized and analyzed using a well-type gamma counter. From the results of the study the investigators concluded:

Elimination of microspheres from the circulation following intravascular administration is size dependent, with the larger diameter spheres clearing most rapidly.

Mechanical filtration and prolonged retention of 7-micron and larger spheres in the lungs imply a critical size for passage through the pulmonary vascular bed.

Retention of small microspheres (3 and 5 micron) in the liver and spleen for prolonged periods appears to be due to phagocytosis of the spheres by cells of the reticuloendothelial system.

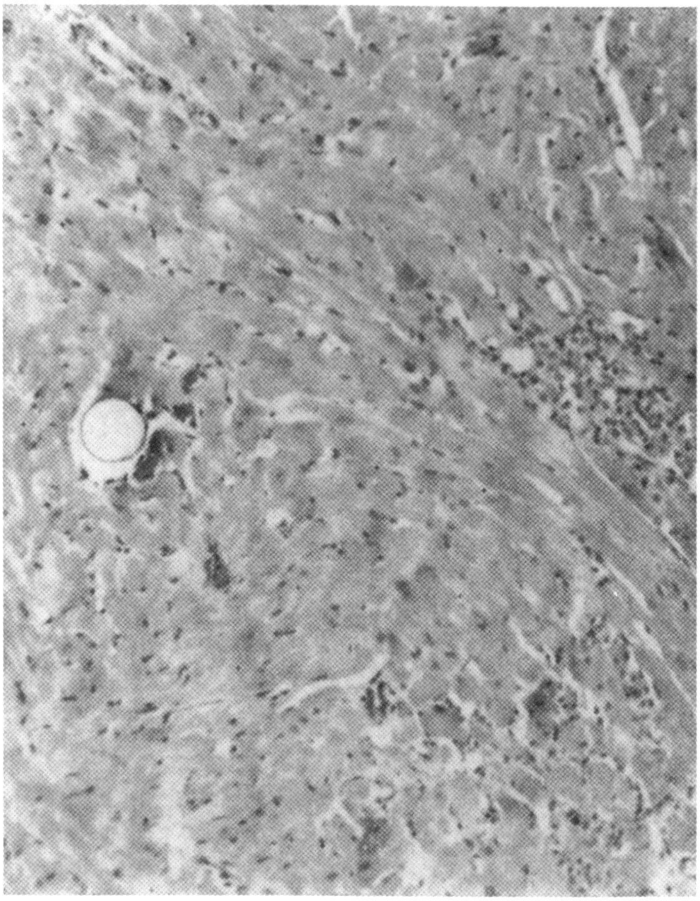

Figure 7 A section of ventricular myocardium from a rat injected with 40-micron particles at a dose of 8×10^6 particles/kg and killed at day 1. This micrograph shows a widely dilated capillary containing a particle. The adjacent myofibers appear slightly compressed but are otherwise intact. (From Ref. 9.)

Small microspheres (3 micron) tend to agglomerate when administered intravascularly, presumably due to protein adherence and a large surface to mass ratio.

No differences in the clearance or distribution of the 3- and 5-micron spheres were observed between the intraarterially and intravenously administered particles.

Large microspheres (12 microns) are not phagocytized by pulmonary macrophages or the reticuloendothelial system within four weeks or intravascular administration as are the smaller spheres.

Two very important unexpected observations were discovered in this investigation. First, there was no difference in the distribution patterns of intravenously and intraarterially administered microspheres in the 3–5 micron range. These spheres either passed through or did not enter muscular or visceral beds and recirculated as if they had been injected intravenously. Second, the large-size microspheres (12 microns) were not phagocytized by the pulmonary macrophages or the reticuloendothelial system. This finding implies that there may be a size limit for phagocytizing inert particles.

In general, the conclusions that are presented in the aforementioned studies are not unexpected if we consider the previous anatomical discussion. One would expect certain size particles to lodge in various capillary beds. However, none of these studies address the effect that particle shape, composition and host response play in determining the physiological consequences of injected foreign particulates. After all, we know that polystyrene DVB spheres do not represent the type of foreign matter that is commonly found in intravenous medications. On the contrary, foreign particulates may be represented by such materials as cellulose fibers, glass, shreds of rubber closure materials, insect parts, various fungi spores, diatoms, metal shavings, and various filter materials.

III. THE BODY'S RESPONSE TO FOREIGN MATERIALS

Before we discuss the various studies that have investigated the effect of foreign particles in the body, it is important for us to understand how the body recognizes a foreign substance and how the body ultimately disposes of the foreign material. Almost any adult person can relate to the effect of foreign material on the body if we consider the sequence of events that are triggered by a wooden splinter that is driven into their hand. After a period of time, the wooden splinter elicits an inflammatory response. As we have seen, large foreign particulate matter will occlude some portion of the vasculature. This foreign material is *not* inert (such as the inert polystyrene DVB microspheres), it will occlude the vasculature and cause an inflammatory response to be initiated. The seriousness of this insult will depend on the degree of the reaction, its location to vital organs, and the organs capacity to compensate for the injury. If the reaction is self-limiting, then the injury may be contained. However, if the reaction is progressive (as in a neoplastic sense) the reaction may prove to be extremely damaging to the body. Another factor might be whether the foreign material is antigenic. If the foreign material is antigenic, this may initiate an allergic response by the host which could be considered another example of a progressive reaction.

A. Inflammation

The basic character of the immediate inflammatory response is almost always the same, regardless of the location or nature of the injurious agent. When most people think about inflammation, they have an infectious etiology in mind. However, as demonstrated by some of the previously cited studies, foreign particles can evoke an inflammatory response as evidenced by the granulomas found by Garvin and Gunner. Moreover, the necrotic byproducts of cellular injury serve as inflammatory stimuli. Although the inflammatory response is almost always the same, the intensity of the reaction is determined by the severity of the injurious stimuli and the reactive capacity of the host. A single transient injury usually evokes a short-lived response. Repeated or persistent injuries usually result in a sustained reaction termed "chronic inflammation." The intensity and duration of the inflammatory response may be viewed as a balance between evoker and responder. In a delibated host, even slight injury may produce serious consequences.

The local signs of inflammation have been classically described as heat, redness, swelling pain and loss of function. All of these characteristics are present in describing phlebitis. Localized heat and redness result from dilatation of the microcirculation in the environs of the injury. The swelling is produced by the escape of fluid from the blood stream into the perivascular tissues. Pain may be caused by the release of bradykinin a direct neural irritator. Dilatation of the microcirculation occurs as a consequence of changes in the permeability of the microvasculature, with resultant tissue swelling. The tissue swelling results from the accumulation of fluid within the extravascular tissue space and from the swelling of individual cells from the leakage that is induced by chemical mediators.

Particulate matter which may be constantly bombarding the intima of the vein has been implicated as a causative agent of phlebitis. We shall undertake a review of these studies later in this chapter.

B. Granulomatous Inflammatory Reactions

Certain types of tissue injury and foreign materials evoke a distincitive pattern of inflammation termed granulomatous inflammation. Particulate matter has been found to elicit this type of reaction in various tissues, most notably the lungs and liver. From a morphologic view, granulomas refer to a 1- to 2-mm nodule which is composed of a collection of modified macrophages or histiocytes that are almost always surrounded by a rim of mononuclear cells, which are principally lymphocytes. In the case of particulate induced granulomas, Langhan's or foreign body giant cells may be present in the granuloma. These modified macrophages give an appearance of an epithelial cell because of their plumpness. The reason for the transformation is thought to be the result of a phagocytized particle which persists within the macrophage and produces this epithelioid transformtion. From a pathologic perspective, tuberculosis is a prototype of granulomatous disease. It is important to remember that the persistence of a particle within the macrophage is the basis for the transformation of granulomas (see Fig. 8).

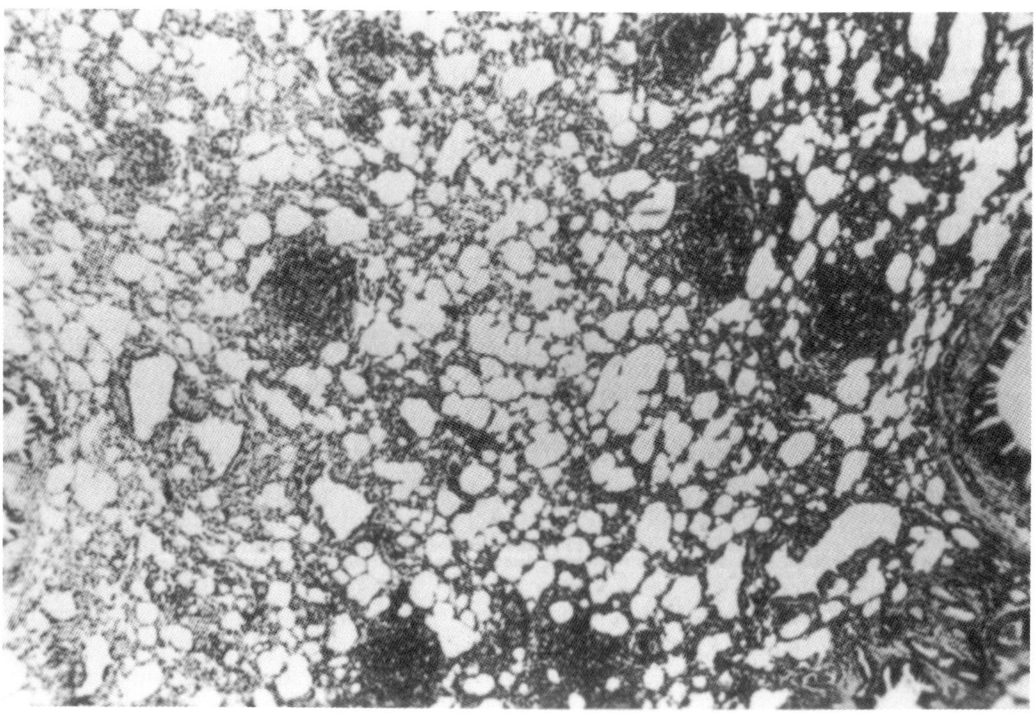

Figure 8 Example of granulomas scattered throughout the lung of a dog receiving 5 mg of ground filter paper five days prior to necropsy. (From Ref. 13.)

C. Reticuloendothelial Reactions to Foreign Materials

The term reticuloendothelial system (RES) was first coined by Aschoff and Kiyono to describe a system of cells found within and outside of typical organs that had as their main characteristic phagocytes of foreign materials (dyes). The cell types that comprise the RES are the intravascular endothelial cells of the liver, lymph node sinuses, splenic sinuses, adrenal capillaries and hypophyseal capillaries, the monocytes and extravascular macrophages and histiocytes.

The function of the RES which is pertinent to our discussion is phagocytosis, storage and clearance of unwanted disease. The major location of phagocytic cells is the spleen and liver, it is in these organs that most phagocytic activity occurs. Additionally, as blood flows through the sinuses of the liver and spleen it slows to a sluggish pace promoting phagocytosis of foreign materials.

The RES also plays a major role in the immune process through the interaction of macrophages and lymphocytes. Immune reactions are generally grouped into two categories: humoral immunity and cellular immunity. In humoral immunity the macrophage plays a duel role; it protects the host against an overload of antigens by catabolizing them, and it presents the lymphocytes with antigenic material, which stimulates them to produce immunoglobulins. In cellular immunity, the macrophage becomes an effector cell. These macrophages respond to foreign materials to form the typical granulomatous inflammation characteristic of delayed hypersensitivity reactions. The ability of the RES to recognize foreign matter is remarkable. When red cells are coated with antibody, they are selectively filtered out of this system even though such cells do not appear morphologically abnormal. In some cases the RES can become overloaded by its phagocytic activity, producing what has been termed *reticuloendothelial blockade.* Such blockade can only occur if the phagocytic overload is continued and heavy, because these cells can digest all biodegradable substances. However, for as long as the blockade lasts, it has potentially serious consequences. By blocking the RES, the body's last line of defence against foreign invaders has been lost. In a real sense, the RES and the lymphatics are the ready reserves called into play when a local inflammatory response is insufficient to deal with the invading substance. Theoretically, foreign particles of a sufficient load could block the system and potentially lead to secondary infections in a debilitated host.

Phagocytosis by RE cells is essentially governed by two broad considerations; the physiologic condition of the host and cells of the RES, and the physical and chemical characteristics of the particle that is introduced to the system. We have already mentioned a few studies (e.g., Schoenberg et al. [4,5] and Singer et al. [7]) that have looked at the role of the RES in removal of foreign material from the body. Most of these studies have used an inert test substance. Very few well controlled studies have examined the effect of antigenic materials.

Wartman, Hudson and Jennings [11] studied the effect of filter paper fibers that were injected into the ear veins of laboratory rabbits. Suspensions of sterile filter paper with fragments of approximately 10 by 80 microns were suspended in normal saline or with human fibrin or rabbit blood. No attempt was made to quantify the number of particles injected into each test animal. Untreated control rabbits were also maintained during the course of study. In about 15% of the control rabbits, small granulomas were found around a few of the end branches of the pulmonary arteries. However, these lesions were much different than those observed in the test animals. There were no differences in the types of lesions produced by the fibers in saline versus the fibers in fibrin or whole blood. All test animals showed widespread impaction or adhesion of emboli with moderate to marked acute arteritis (see Fig. 9). These changes were consistent at intervals of several minutes to seven days following the last injection of test materials. After a week or 10 days the acute lesions became organized. Fibroblasts, a few capillaries and mononuclear phagocytes appeared and the clot retracted. It then became

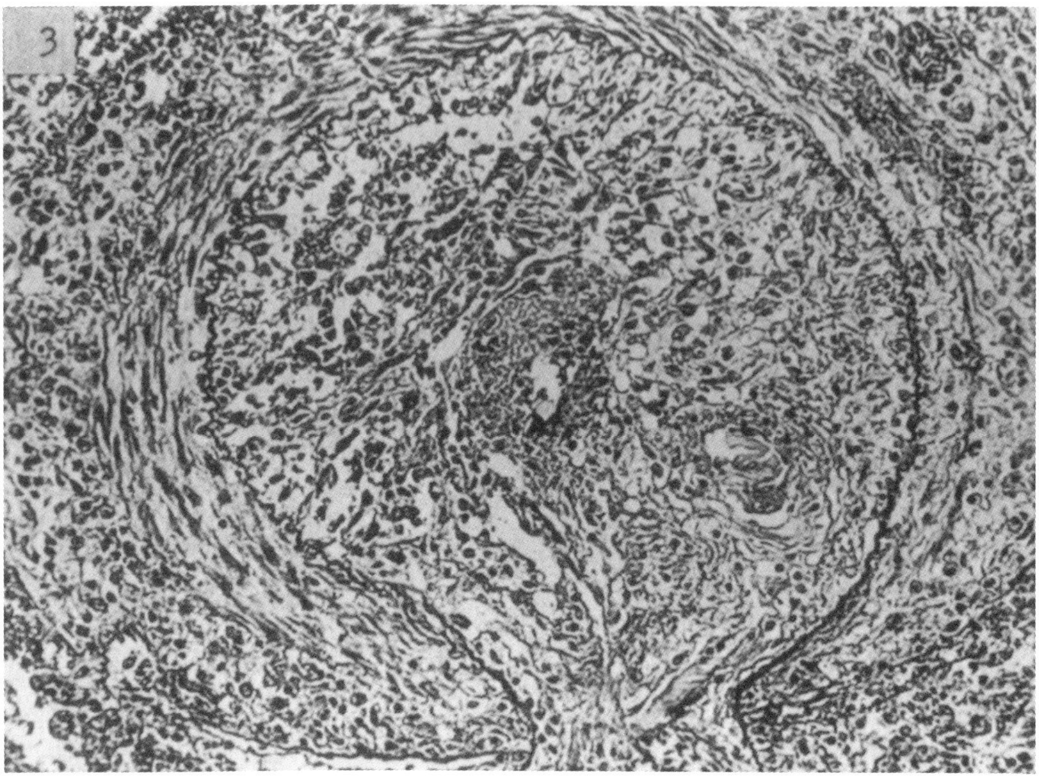

Figure 9 Photomicrograph of the impaction of an embolus in a large artery with organizing acute inflammation. (From Ref. 11.)

endothelialized forming an eccentric plaque on the intima. Within the span of several weeks the process of organization had resulted in fibroelastic scaring of the intima of a considerable number of large and small arterioles.

The filter paper fibers caused an acute inflammatory reaction. Some fibers were engulfed by foreign body giant cells and embedded in the intima in dense plaques of avascular collagenous tissue (see Fig. 10). Other fibers were found in the media and adventitia of the vessels. Fibers were found to be able to pass entirely through the arterial wall into the perivascular space or become lodged within the wall (see Fig. 11). The authors postulated that this could be a possible mechanism for removing large particulate material from the vasculature. Although the mechanism of transport of the fibers was not studied, two possible mechanisms were offered as explanation for the transport of fibers outside the vasculature. Fibers could be pushed out of the vasculature and then phagocytized or the fibers were phagocytized and then carried through the vessel's walls by the phagocyte. The study also demonstrated that there were nodules of lymphoid tissue lying between the artery and the bronchus. These lymphoid nodules varied greatly in size, and in the small nodules it was always possible to identify a lymphatic lying between the artery and the bronchus from which the lymphoid nodule developed. It sees likely that in this situation, after foreign materials clear the artery, it is passed to the lymphatics where depending upon its size is either cleared by the lymphatics or localized. It is also possible that the material could pass from the lymphatic system into the bronchus and be

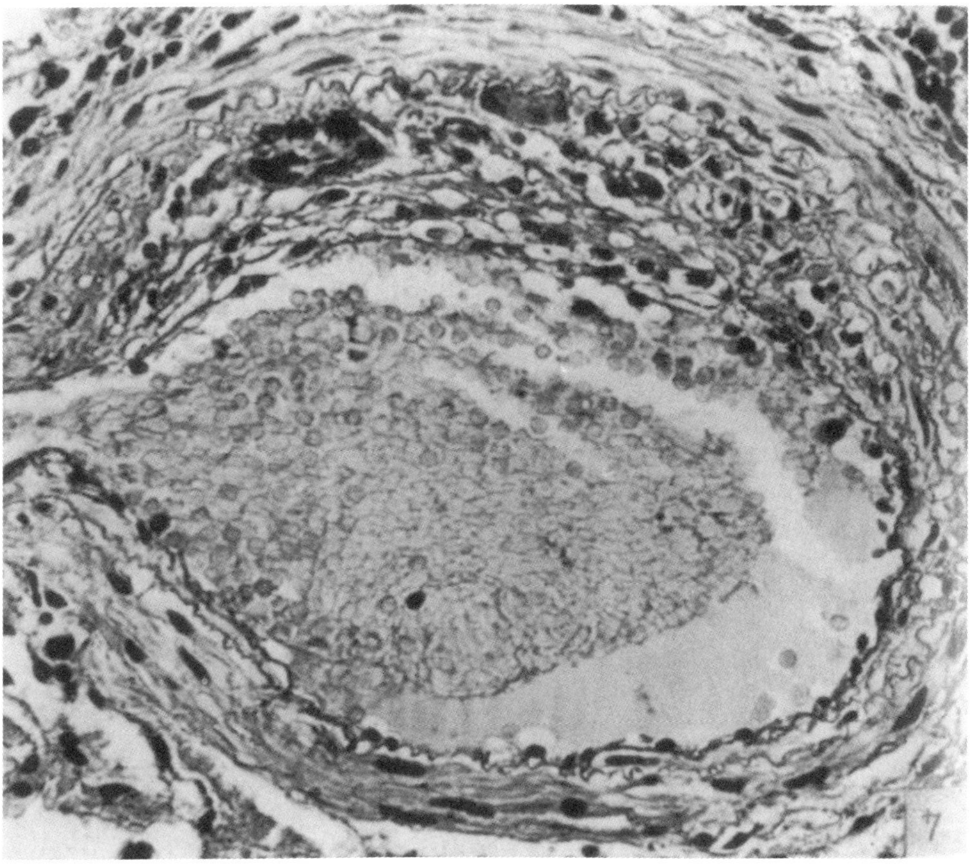

Figure 10 Photomicrograph of chronic inflammation of the intima following injection of minute emboli of filter paper fibers. Note that the giant cell contains the fiber. (From Ref. 11.)

coughed up. However, no evidence was presented by the authors nor any other investigators to support this theory.

A second study, by Stehbens and Florey [12] utilized a variety of different particulate substances and injected these substances into the carotid artery and later into a large ear vein. Observations were made on noninflamed, inflamed and growing blood vessels in the rabbit ear chamber. The borders of the endothelial cells were not outlined by the particulate matter during several hours of observation, although a small amount did adhere to the endothelium. The administration of particulates into the bloodstream led to the agglutination of platelets to which the particles adhered. Thrombi were formed around these platelets which adhered to leukocytes and often entangled some erthyrocytes. The thrombi formed strands in the vasculature which vibrated with blood flow and eventually fragmented forming emboli. Capillaries were frequently blocked by these thrombi and emboli. Some of the particulate matter was phagocytized by leukocytes and appeared inside platelets. Some particles were later seen in macrophages outside the blood stream.

A more recent study was conducted by Gross and Carter [13] of the Food and Drug Administration. In their study they used two types of particulate matter; ground filter paper and

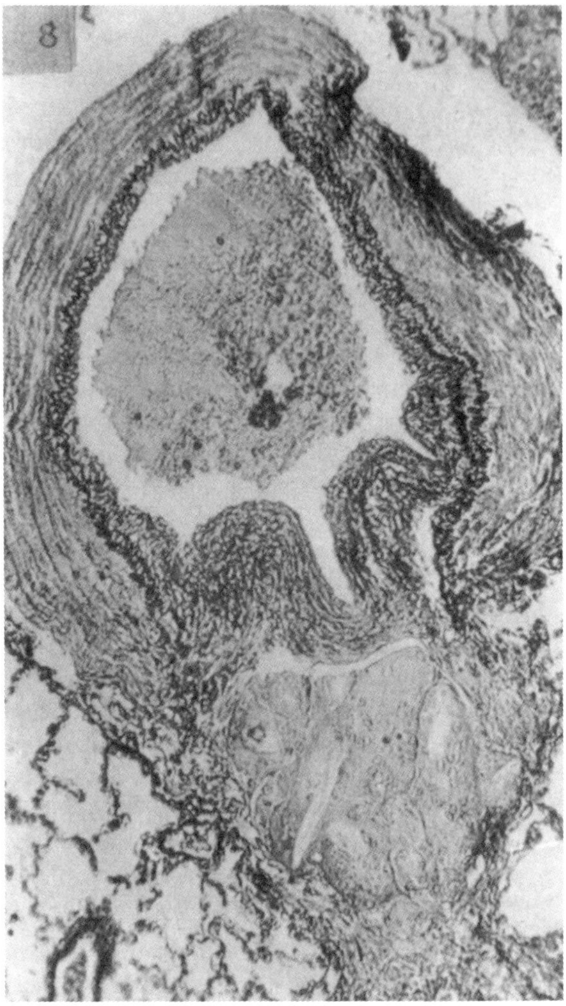

Figure 11 Photomicrograph of the passage of filter paper fibers through the wall of a large muscular artery, with destruction of all coats and replacement fibrosis of intima and media. The fibers are surrounded by a foreign body type of granuloma which is rich in multinucleated giant cells. (From Ref. 11.)

a ground plastic material used in closures for parenteral solutions. The average size of the particles was 74 microns and the number of particles were 8.45×10^5 for the filter paper and 12×10^5 for the plastic. Young adult beagle dogs were selected as test animals for the study which allowed the use of an 18 gauge needle to administer the particulate materials. The larger gauge needle allowed particles of a comparable size to be injected without their aggregation prior to introduction into the circulatory systems of the animals. In all, 20 animals were studied, of which four served as controls. Two of the controls received only normal saline, the remaining two received normal saline and sodium lauryl sulfate (the suspending agent). All animals were injected through the cephalic vein and received the test materials in 10 ml of fluid. Two control animals and four test animals received a second injection of test material either three or eight days following the initial injection. None of the animals demonstrated any ante mortem signs of distress or ill health. The post mortem examinations were similar for all test subjects except for one dog. That dog had received 5 mg of ground plastic material in a single injection five days prior to necropsy. In that animal, the dorsal aspect of the left diaphragmatic lobe of the lung presented a sharply circumscribed area of subpleural nonaeration and induration which was grossly observed. Microscopically, the pathologic picture observed in this and other animals was a nonsupportive chronic granulomatous inflammation of the lung at a stage preceding fibroplasia (see Fig. 12). Most of these lesions were of a focal milliary type scattered throughout the entire lung parenchyma. Their average density throughout the lung, although fairly constant within any given animal, varied among the different animals. In general the level of density reflected the total size of the inoculum. While the densities were not exactly the size of the inocula, the animals which received less particulate matter presented less lesions. The characteristics of the lesions were consistent with those of a classical foreign body reaction.

Fairly large areas of granulomatous reaction were frequently found, and in most cases were determined to be the result of coalescence of two or more adjacent lesions, each thought to have arisen independently (see Fig. 13). Often the air passages in the lung at the bronchiolar level exhibited metaplastic changes in the mucosal epithelium. These changes were consistent of the type of normally found with pulmonary inflammation.

An important finding in this study was that the histologic evaluation of the granulomas did not exhibit any pathognomonic characteristics which distinguished one type of particulate matter from the other. This finding is important because of certain types of infecting agents do present

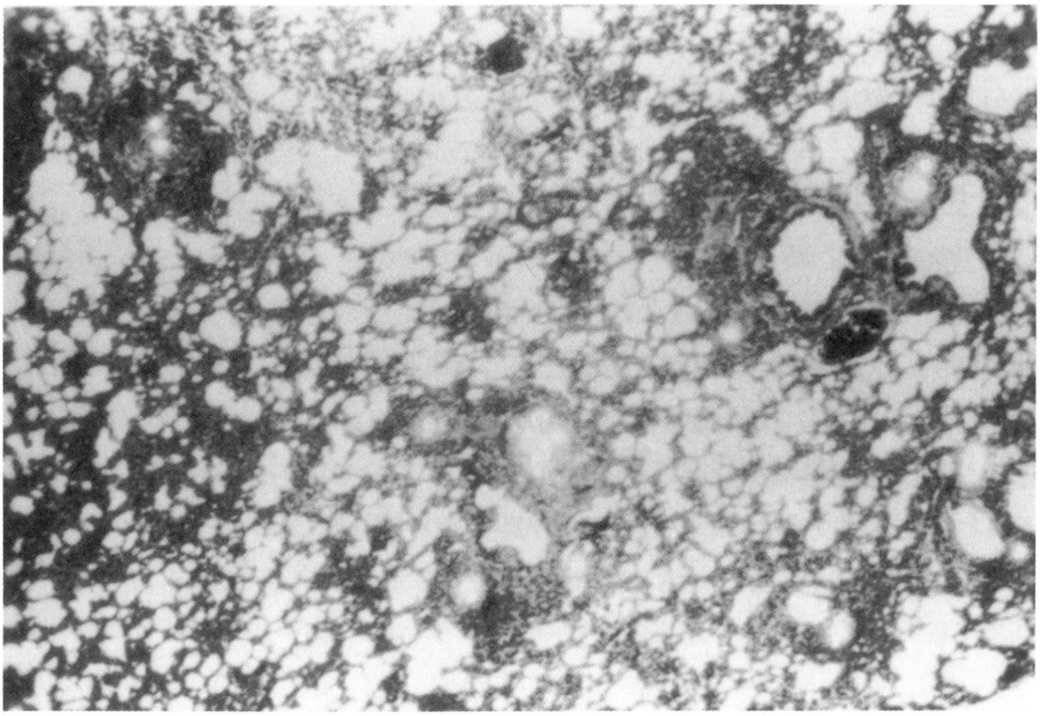

Figure 12 Numerous focal granulomas scattered throughout the lung of a dog which received 5 mg of ground filter paper in a single intravenous dose five days prior to necropsy, as viewed by polarized light. Note that the filter particles can be seen in each granuloma. (From Ref. 13.)

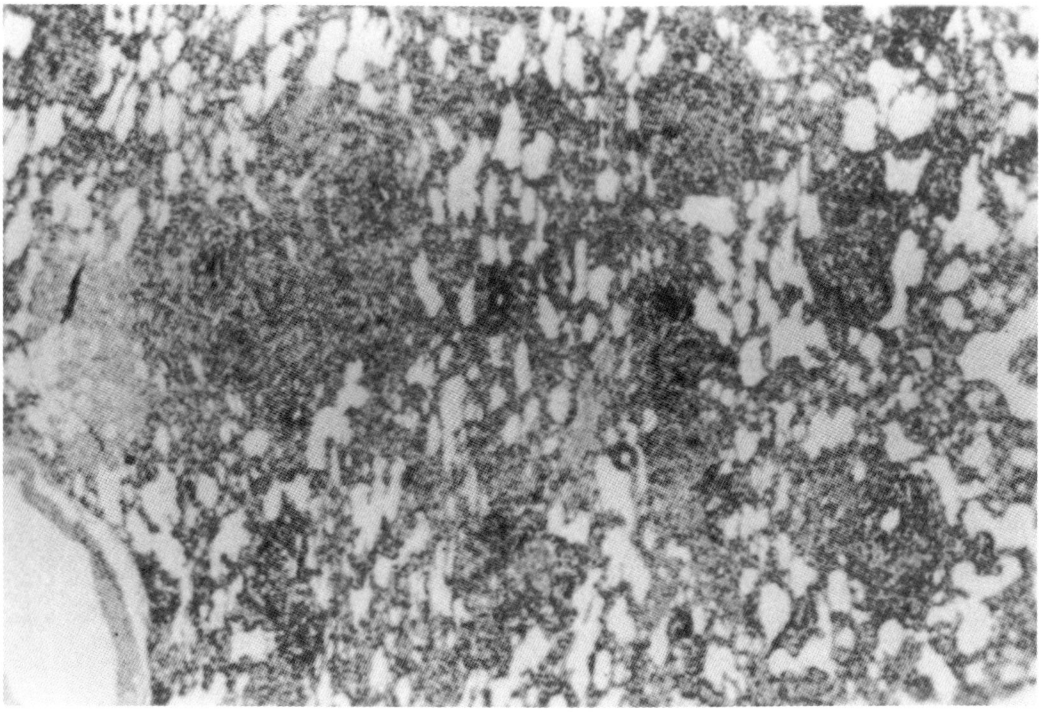

Figure 13 Photomicrograph of the lung of a dog which received 54 mg of ground plastic material in two injections spaced three days apart. Note that an extensive subpleural area of granulomatous inflammation has occurred which renders this tissue nonfunctional. (From Ref. 13.)

different chronic granulomatous processes that are characterized by either their presence or lack of a suppurative process. The authors concluded that if it were not for the actual particles themselves, they would have been unable to determine by morphological examination, if the lesion was caused by the filter material or the plastic material.

D. Phlebitis

The last and most frequently studied reaction to injected particulate matter is infusion phlebitis. Infusion phlebitis is a condition which describes the inflammation of a vein or veins that is the direct result of infusing medications into that vein. Phlebitis is characterized as red, painful, tender, raised areas involving the skin and subcutaneous tissue, usually in linear distribution along the course of visible veins. It can progress from a painful self-limiting condition to more serious life-threatening forms of illness such as thrombophlebitis, pulmonary embolism, suppurative thrombophlebitis and systemic sepsis. Some authors have reported that infusion phlebitis occurs in as many as 70% of all intravenous infusions, depending on the duration of the infusion [14,15]. The cause of infusion phlebitis has been attributed to the composition of the infusion set and catheter [15,17], specific drugs [18,19], pH of the infused solution [20,21], the concentration of the infusion [22], and bacterial migration around the site of the venipuncture [23,24]. The suspicions of Ryan et al., that particulate matter may be the cause of phlebitis, gained considerable interest by investigators in 1973 [14]. Ryan's group demonstrated in a 100-patient clinical evaluation, that a micron 0.45-micron final filter not only protected the patient from receiving particulate matter and bacterial contamination, but dramatically decreased the incidence of the acute infusion phlebitis when compared to a group of nonfilter control patients. DeLuca et al [25] evaluated the effect of final filtration on the incidence of infusion phlebitis in a prospective, double-blind investigation involving 146 postoperative patients. Infusion phlebitis was defined as pain, tenderness, erythema, swelling and a palpable venous cord. Infusion phlebitis was determined to be found if any two of the above criteria were present. All phlebitic determinations were conducted by a physician. The incidence of infusion phlebitis was found to be significantly reduced when an in-line 0.45-micron filter was used. The greatest reduction of infusion phlebitis was in the filter group receiving unbuffered solutions and no set change over the 72 h of therapy. Buffering the infusion solutions or changing the infusion set every 24 h did not affect the incidence of phlebitis. Antibiotic therapy appeared to have a slightly beneficial effect only when in-line filters were employed. Increased white blood cell counts and increased erythrocyte sedimentation rates were observed in the patients who received unfiltered solution. The authors concluded that a 45-micron filter is an effective means of reducing infusion phlebitis and felt that the filter implicated particulate matter as the causative agent of infusion phlebitis. No attempt was made to standardize the type of solutions administered.

Chamberland, Lyons and Brock [26] studied the effects of in-line filtration on the incidence of infusion phlebitis in a commonly hospital setting. Two general surgical units, comparable in patient population, were studied for one month. On one unit an extension set with 0.45-micron filter was attached to every i.v. catheter; no in-line filter was instituted on the second unit. The second unit served as the control unit. The use of filters was not blinded from the person evaluating phlebitis. Filters were replaced every 24 h or sooner if blockage occurred. A total of 49 patients (107 i.v.'s) were studied in the group with filters and 40 patients (84 i.v.'s) in the group without filters. Phlebitis was graded on a scale of 0 to 4. The proportion of patients with phlebitis in the filter group was 86%; in the nonfilter group, 75%. In the filter group, 67% of the catheterizations ended in phlebitis; in the no-filter group, 63%. Upon first catheterization, phlebitis developed in 63% of the filter group and 58% of the no-filter group. None of these differences were significant. There was no significant difference between the two groups in the severity of phlebitis and the incidence of phlebitis following the administration of irritating solutions containing cephalothin sodium or potassium chloride. Based on this study, the authors concluded that the value of in-line filtration in routine clinical condition appeared to be questionable.

A third study, by Allcutt, Lort, and McCollum [27], evaluated the effect of in-line final filtration of intravenous admixtures on phlebitic rates in hospitalized patients. Their study utilized a Pall in-line final filter with a pore size of 0.2 micron. All infusions with a likely duration of 48 h were chosen for the study. The study was conducted over two months during which patients were randomized to the study filter or a suitable placebo filter (lookalike nonfilter). Of the 226 infusions that were randomized, 32 failed within the first 24 h and were excluded leaving 194 of whom 93 had placebo filters and 101 had the final filter. Only 38 infusions with placebo filter survived the 48-h test period, compared to 63 infusions with the filter that survived the test period. Analyzed by life table, the trend for filtered solutions to survive longer than unfiltered solution did not show a power of significance. However, in-line filtration did prolong the phlebitis-free survival of infusions over nonfiltered solutions ($P < 0.01$). These benefits were most marked in the antibiotic solutions that were administered via the drip site. The authors concluded that in-line filtration delays the onset of phlebitis; thus more infusions survive until they are no longer needed.

In a very large double blind study by Falchuk et al. [28], the effect of an in-line 0.22 micron final filter on phlebitic rates were determined for 541 patients. A total of 277 patients received infusions through intravenous sets with a 0.22-micron IVEX-HP filter, and 264 received infusions without filters. Each infusion was evaluated daily for a maximum of three days. The incidence of phlebitis on days 1, 2, and 3 of the study were 14.3%, 31.1%, and 27% for the patients receiving infusions without filters and 6.8%, 9.7%, and 11.3% for patients receiving fluids with filters (*P*, 0.001). Thus, the incidence of phlebitis was reduced by approximately two-thirds in the patients who received infusions through the IVEX-HP filters. The authors concluded that the infusion related to phlebitis was a problem associated with microparticulate matter and could be removed by in-line filtration.

An interesting study was conducted by Dorris et al. [29], which investigated the effect of injecting filtered residue of sodium cephalothin into rabbits and dogs. The investigators reconstituted sodium cephalothin according to the manufacturer's instructions. The reconstituted solution was filtered through a 0.45-micron membrane filter. Some filters were retained with the filtrate, others were placed in 5% dextrose in water USP (D5W) to create an injectable form of particulate material. Filters containing residue were sutured to the buttocks of three rabbits and three rats (after the animals had been shaven). A control filter patch was also applied. The control filter patch contained residue form filtered D5W. After 72 h the filter patches were removed and the underlying skin was inspected for evidence of inflammation. There was no gross evidence inflammation. Samples of skin were removed for histologic examination which revealed only mild inflammation in the papillary dermis in both controls and test animals. However, there were no discernible differences between the two groups. Concentrated solutions of particles that were obtained by sonicating the filtered material in D5W were injected subcutaneously in the buttock of five rabbits. Seventy-two hours after receiving the 0.5-ml injection, the rabbits were observed for gross evidence of inflammation at the injection site. There was no evidence of inflammation at the injection sites of the control or test animals. Histologic examination showed signs of mild inflammation, with rare subepithelial histiocytes and mast cells, but again there was no difference between the test animals and their controls. The third phase of the study tested the effect of injecting 1 ml of resuspended particles administered into the marginal ear vein of five rabbits. Ten rabbits received injections of D5W residue, 6 with D5W filtrate, and 11 rabbits with cephalothin filtrate. Histologic examination of the veins 24 h later revealed venous congestion and perivenous hemorrhage in all subjects. The filtrate of D5W has the lowest incidence of identifiable reaction and the mildest reactions. The residue of D5W yielded a high incidence of mild reactions, but only a tenth of the specimens showed evidence of severe vein-wall damage. The filtrate of sodium cephalothin and the residue of sodium cephalothin caused mild reactions in a high percentage of veins studied, and a substantial number of specimens were found to have severe reactions with damage to the vein wall. Evidence of venous thrombosis was seen only with the injections of the sodium cephalothin filtrate. In the final phase of the study a dog had a mixture of 2 g/L of cephalothin, 20 mEq/L of KCl in 0.2% saline infused for 2 h in 3 veins, 4 h in 12 veins, and 6 h in 12 veins. At the end of the infusion period, each vein was dissected from the point of venipuncture to the saphenofemoral junction and examined histologically. The results of the examination revealed that the 2-h infusion showed no changes, the 4-h infusion showed mild changes of intimal edema. The 6-h infusions showed severe damage of the endothelium as evidenced by focal medial necrosis.

The authors concluded that within the limits of the study, filterable particulate material of sodium cephalothin causes a tissue-specific reaction with venous endothelium but not with cutaneous or subcutaneous tissues.

IV. SUMMARY

Over the past thirty years the evaluation, identification, measurement, and control of particulate matter in parenteral products has become a science in its own right. This science was born from our concerns about the physiologic consequences of introducing particulate matter into the bloodstream of patients receiving parenteral products. These concerns were first raised by Garvin and Gunner more than 30 years ago when they observed multiple granulomas in the lung tissues of patients at autopsy. These investigators also conducted experiments to demonstrate that the granulomas could be reproduced in animal models. A review of the literature on this subject has shown that definite pathophysiologic consequences result from injecting large-size foreign materials into the bloodstream of various animals. It appears that the result of introducing particulate matter into the bloodstream is directly related to three major variables: the size of the particles introduced, the number of particles and the rate at which the particles are introduced, and the chemical composition of the particles introduced. We might suggest a fourth factor could be the "condition" of the host. This "condition" might be defined as the capacity of the host to acclimate to the injurious nature of the foreign insult. This could range from the general health of the host to the size and capacity of various organs to compensate. These factors which effect the biological consequences of introducing foreign particles into the body, have confounded the studies that have been reviewed. There has yet to be performed the perfect study that can definitively answer the question, "What level of particulate contamination can the human body withstand without causing an untoward effect?"

Nonetheless, we clinicians and scientists must continue to weigh the potential dangers of introducing foreign particulates vis-à-vis parenteral products into patients, with our ability to control and eliminate their introduction.

REFERENCES

1. J. M. Garvin and B. W. Gunner, *Med. J. Austral. 2*: 140 (1963).
2. J. M. Garvin and B. W. Gunner, *Med. J. Austral. 2*: 1 (1964).
3. M. Prinzmetal, E. M. Ornitz, Jr., B. Simkin, and H. C. Bergman, *Am. J. Physiol. 152*: 48 (1948).
4. M. D. Schoenberg, P. A. Gilman, V. R. Mumaw, and R. D. Moore, *Brit. J. Exptl. Pathol. 42*: 486 (1948).
5. M. D. Schoenberg, P. A. Gilman, V. R. Mumaw, and R. D. Moore, *Exp. Molec. Pathol. 2*: 126 (1963).
6. D. J. Wilkins and P. A. Myers, *Brit. J. Exptl. Pathol. 47*: 568 (1966).
7. J. M. Singer, L. Adlersberg, E. M. Hoenig, E. Ende, and Y. Tchorsch, *J. Reticuloendothel. Soc. 6*: 561 (1969).
8. R. E. Madden, D. Agostino, and L. Gyure, *Arch. Surg. 101*: 425 (1970).
9. R. M. Gesler, P. J. Garvin, R. U. Robinson, C. R. Thompson, W. R. Gibson, F. C. Wheeler, and R. G. Carlson, *Bull. Parenteral Drug Assoc. 27*: 101 (1973).
10. M. Kanke, G. H. Simmons, D. L. Weiss, B. A. Bivins, and P. P. DeLuca, *J. Pharm. Sci. 69*: 755 (1980).
11. W. B. Wartman, B. Hudson, and R. B. Jennings, *Circulation 4*: 756 (1951).
12. W. E. Stehbens and H. W. Florey, *Quart. J. Expt. Physiol. 45*: 252 (1960).
13. M. A. Gross and C. J. Carter, The pathogenic hazard of particulate matter in solutions for intravenous use, FDA Symposium on Safety of Large Volume Parenteral Solutions, Washington, DC, 1966, pp. 31–37.
14. P. B. Ryan, R. P. Rapp, and P. P. DeLuca, *Bull. Patenter. Drug Assoc. 27*: 1 (1973).
15. G. A. Brown, *Br. J. Clin. Pract. 24*: 197 (1970).
16. M. W. Jones and D. B. Craig, *Can. Anaesth. Soc. J. 19*: 491 (1972).

17. D. L. Maki, D. A. Goldman, and F. S. Rhame, *Ann. Int. Med., 70*: 867 (1973).
18. K. Nordell, L. Mogensen, and O. Nyquist, *Acta Med. Scand. 192*: 263 (1972).
19. J. Carrizosa, M. E. Levison, and D. Kay, *Antimicrobial Agents Chemotherapy 5*: 192 (1974).
20. C. H. Flores-Vega, J. A. Neutzel, and W. A. Knight, *Missouri Med. 67*: 305 (1970).
21. E. W. Fonkalsrud, K. Carpenter, and J. Y. Masuda, *Surg. Gynecol. Obstet. 133*: 253 (1971).
22. E. T. Thomas, W. Evers, and M. B. Racz, *Anesth. Analg. 49*: 150 (1970).
23. H. Smits and L. R. Freedman, *N. Engl. J. Med. 276*: 1229 (1967).
24. D. W. Wilmore and S. J. Dudrick, *Arch. Surg. 98*: 256 (1969).
25. P. P. DeLuca, R. P. Rapp, B. Bivins, H. E. McKean, and W. O. Griffen, *Am. J. Hosp. Pharm. 32*: 1001 (1975).
26. M. E. Chamberland, R. W. Lyons, and S. M. Brock, *Am. J. Hosp. Pharm. 34*: 1068, (1977).
27. D. A. Allcutt, D. Lort, and C. N. McCollum, *Br. J. Surg. 70*: 111 (1983).
28. K. H. Falchuk, L. Peterson, and B. J. McNeil, *N. Engl. J. Med. 312*: 78 (1985).
29. P. P. DeLuca and J. Z. Knapp, Particulate Matter, Pharmaceutical Dosage Forms: Parenteral Dosage Forms (K. E. Avis, H. A. Lieberman and L. Lachman, eds.), Marcel Dekker, New York, 1993, p. 3.
30. B. Pansky and E. L. House, Review of Gross Anatomy, Second Edition, Macmillan Company, New York, 1969, p. 177.
31. A. M. Jonas, Potentially Hazardous Effects of Introducing Particulate Matter into the Vascular System of Man and Animals, FDA Symposium on Safety of Large Volume Parenteral Solutions, Washington, D.C., 1966, pp. 23–28.

15
Diagnostic Applications of Microspheres

Leigh B. Bangs

Bangs Laboratories, Inc., Carmel, Indiana

I. BACKGROUND

Uniform microspheres or latex particles were first made in 1947. For details of their discovery, preparation, and characterization see elsewhere [1]. For several years they were studied primarily by electron microscopy and were regarded as laboratory curiosities. Early uses were as standards—put in with a sample, they served as internal yardsticks.

Microspheres/beads are presently available with diameters ranging > 6 orders of magnitude from ~15 nm to 25 mm (0.015–25,000 μm)! Those from ~0.1 to ~40 μm have the best uniformity; their coefficients of variation are usually < 3%. They are made from various polymers and minerals, with densities ranging from 0.03 to > 2.1 g/ml and they offer a wide choice of surface chemistries. It is now possible to get the microspheres made to individual specifications—with special chemistries [2,3], densities, refractive indices, and all colors of the rainbow plus black, brown, and flourescents. Superparamagnetic microspheres are available in different sizes and surface chemistries, with different levels of magnetite for variable response to a magnet. Uniform silica microspheres, in sizes from ~0.3–5 μm, are made from pure $Si(OC_2H_5)_4$ reacted with water and ammonia.

Microspheres are still used as calibrators for particle size analyzers and flow cytometers, but for the past few decades their main application has been in clinical diagnostics. Microsphere-based *qualitative* (yes/no) diagnostic tests and truly *quantitative* assays are usually based upon the very specific interaction of antigen (Ag) and antibody (Ab). With Ab or Ag adsorbed onto them, these coated or "sensitized" microspheres then act to magnify or amplify the Ag-Ab reaction.

Different microsphere sizes are typically used for different diagnostic test formats: slide agglutination tests use 0.2–0.9 μm; microsphere immunoassays, 0.01–0.3 μm; microsphere capture ELISA, 0.3–0.9 μm (depending on microsphere capture method); strip tests, 0.1–0.4 μm (microsphere size depends on porosity of the strip); and solid-phase immunoassays, > 0.8 μm.

II. CLEANING MICROSPHERES

Most uniform polymer microspheres are made by emulsion polymerization using surfactants. The surfactants (usually negatively charged alkyl sulfonates, sulfates, or carboxylates) become adsorbed on the microsphere surface where they give microspheres a negative charge and increased colloidal stability. Before microspheres can be used in various diagnostic tests and assays, this surfactant must often be removed. In other cases one may want to add more surfactant to ensure colloidal stability of the microsphere dispersion.

Uniform silica microspheres have a natural hydrophilic surface consisting of native hydroxyl groups and no surface-active impurities. They should need no cleanup before use.

Magnetic microspheres may have permanent surface groups, like carboxylic acid, and added surfactant to ensure long-term colloidal stability. The surfactant can be removed by washing.

A. Water

Those working with microspheres often *assume* that the water they have is clean, but often it contains ionic and organic species which can adsorb onto the beads. The microspheres have a hydrophobic surface generally, and a very high specific surface area (~60 m^2/g for 0.1-μm microspheres). They will adsorb organic species from the water, if *any* are present. If the microspheres have been made with good water, they will have only surfactant present. It is important to use a good water source and check to be sure it is *consistently* producing very clean water (low conductivity, zero organics, and no bugs). Otherwise, it is possible to be cleaning the *water* with the *microspheres,* rather than the opposite.

B. Cleaning

The best overall methods for cleaning microspheres are probably ion exchange or cross-flow filtration ("dialysis under pressure" or "filtration without a filtercake"). Various mechanical means may be employed to permit liquid to permeate the filter medium while preventing a microsphere layer build-up.

Seemingly straightforward methods like centrifugation and simple filtration can lead to serious problems like clumping. In addition, as the microspheres get "cleaner", the colloidal stability of the dispersion becomes more fragile. The topic of cleaning microspheres is covered better and in more detail elsewhere [4].

III. MICROSPHERE CHARACTERIZATION

It is advisable to test microspheres at various stages in their processing—before and after cleaning and coating. One can monitor the microspheres monodispersity, colloidal stability, surface charge, and changes in electrokinetic behavior relative to protein coverage.

Size (microscopy or instrumental methods): Monitoring size is one way to ensure that the microspheres are still singly dispersed after every important step of your handling process.

Surface titration: This can be done on clean or coated microspheres to determine lot-to-lot reproducibility, or it may be used to help explain inability to adsorb as much protein after a particular treatment (was cleaning thorough enough?). The capacity of certain lots of COOH-modified microspheres for covalent coupling (active surface COOH groups) can be measured.

Critical coagulation concentration: By titrating with an appropriate salt solution one can pre-determine the colloidal stability of a lot of microspheres to anticipate possible flocculation in a buffer used to coat, couple, or store the microspheres.

Electrokinetics: There are several ways and appropriate instruments to monitor electrokinetic mobility or zeta potential of the microspheres at various steps in cleaning and coating of the microspheres. The most commonly used method measures the speed of motion of individual microspheres in a standard setup. With electrokinetics one can determine whether the microspheres are clean or reproducibly and uniformly coated with protein.

Each of these methods can help to reduce the risk of "flying blind" in the coupling process.

IV. COATING MICROSPHERES

A. Simple Adsorption

Protein adsorbs onto polystyrene (PS) *readily and permanently.* Putting protein, especially polyclonal immunoglobulin G (IgG), on the surface of polystyrene microspheres by simple adsorption is successful > 95% of the time. It is common to underestimate the surface areas involved and the amount of protein that can be adsorbed.

The surface saturation capacity of polystyrene for bovine serum albumin (BSA) and bovine IgG is calculated [5] at ~3 mg BSA/m^2 and ~2.5 mg BIgG/m^2. The surface area/mass ratio for spheres, A/M (m^2/g) = $6/\rho D$, where ρ = density (g/ml) and D = diameter (μm).Then for PS, where ρ = 1.05 g/ml, $A/M = 5.7/D$. Thus, if D = 1 μm, then A/M = 5.7 $m^2/g \approx 6$ m^2/g. Therefore, 1 μm PS microspheres may adsorb ~3 mg/m^2 × ~6 $m^2/g \approx$ 18 mg BSA/g of microspheres. This represents a monolayer of protein and will be about the maximum amount which can be put on the microspheres.

B. Covalent Coupling

If simple adsorption is inadequate or undesirable, other methods of coating microspheres can be considered, such as adsorption of one protein which will capture another molecule of interest or covalent coupling of organic molecules to the microsphere surfaces. This latter procedure is particularly useful for attachment of small molecules or haptens, monoclonal antibodies, antibody fragments, polypeptides, and oligonucleotides.

The above general principles for handling uniform microspheres should provide a basis for developing protocols for reproducible tests and assays with good sensitivity and shelf life.

V. QUALITATIVE AGGLUTINATION (TESTS)

Because of their superior uniformity and stability, polymeric microspheres gradually replaced clay particles, charcoal, killed bacterial cells, and fixed or treated erythrocytes as agglutination markers. Properly sensitized or coated with antibodies, the latex microspheres are a stable colloid of singly dispersed particles, until they become "agglutinated" or clumped by addition of a sample containing the corresponding antigen.

In simple particle agglutination, a positive test results when milky-appearing, Ab-coated particles, uniformly dispersed in a drop of water on glass slide, react with Ag in a drop of sample (whole blood, serum, urine, etc.) to cause particle agglutination. Alternatively, Ag-coated microspheres are agglutinated by a positive sample of Ab (Fig. 1).

In 1956 Dr. Jacques Singer developed the first particle agglutination tests (PATs) or latex agglutination tests for rheumatoid factor [6]. Since then, PAT's have been devised for > 100 infectious diseases, e.g. bacterial, viral, fungal, mycoplasmal, protozoal, rickettsial, and chlamydial, and > 60 other chemical analytes, e. g. human chorionic gonadotropin (pregnancy), rheumatoid factor, C-reactive protein, antistreptolysin O, fibrin degradation products (blood clotting), fecal occult blood (colorectal cancer), myoglobin (heart attack). PATs are rapid and efficient. Since they can be run quickly and simply, they can be used in the clinic or at a bedside so that treatment can commence promptly, before the patient is transferred or discharged [7].

There have also been many new applications in the fields of animal and plant health, law enforcement, and the environment.

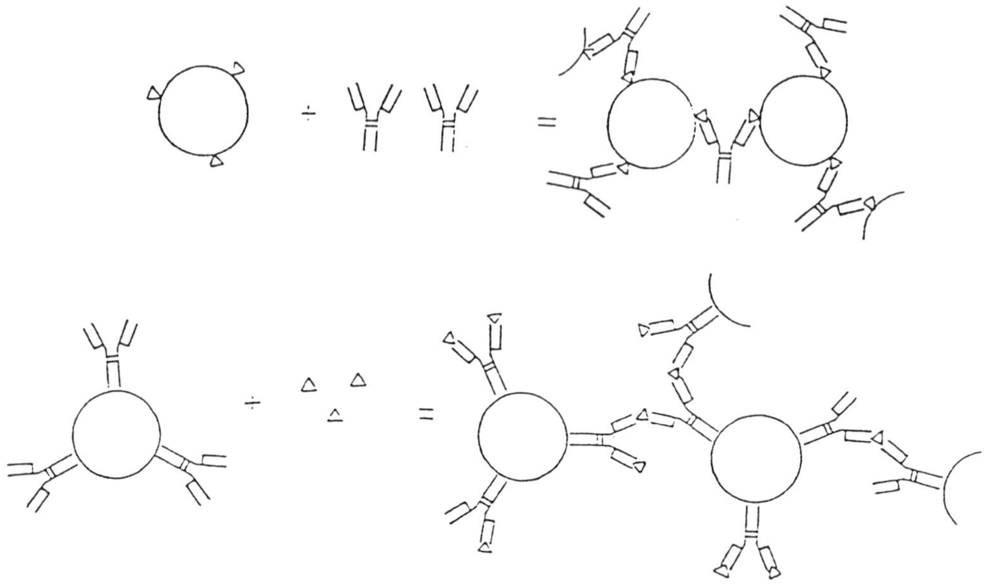

Figure 1 Typical microsphere agglutination tests for antibody (Y) and antigen (Δ).

A. Slide Test Sensitivity

Current PATs generally use 0.2–1 μm diameter microspheres. Calculations show that larger microspheres should theoretically yield more sensitive tests (Table 1). I assume that in order to detect agglutination one must be able to see ~100 clumps of agglutinated microspheres in the field of view on a slide. Each clump must be > 50 μm to be visible. If single microspheres are 1 μm in diameter, then $\sim 50^3 \approx 10^5$ microspheres will be required to make one visible clump. Each microsphere can touch 12 nearest neighbors in tightest packing possible, so a minimum of ~10 bonds may be required per microsphere to hold them all together. Thus, for 10^5 microspheres per clump × 100 clumps × 10 bonds per microsphere, $\sim 10^8$ molecules (10^{-16} mole) of agglutinator (Ag or Ab) would be required to detect agglutination. If MW = 100,000 (say for IgG), then necessary amount would be 10 picograms. If a 10 μl sample is used, then sensitivity would be

Table 1 *Theoretical* Sensitivity of Microsphere Agglutination Test

Assumptions:

- ~100 clumps must be seen to determine agglutination
- Clumps must be ~50 μm in size to be seen by eye
- ~10 bonds required per microsphere to agglutinate
- 10 μl sample used

Microsphere diameter	No. of Microspheres in		Agglutinator required . . .			*Theoretical* sensitivity
	1 clump	100 clumps	Molecules	Moles	wt., if IgG	
0.1 μm	10^8	10^{10}	10^{11}	10^{-13}	10 ng	10 nM
1.0	10^5	10^7	10^8	10^{-16}	10 pg	10 pM
10.0	10^2	10^4	10^5	10^{-19}	10 fg	10 fM

10^{-16} mole/10 × 10^{-6} l = 10^{-11} mole/l = 10 picomoles/l. If microspheres are 0.1 μm, then sensitivity will only be 10 nM. If 10-μm microspheres are used, sensitivity improves by 1000× to 10 fM. At least two limits to the theory are the facts that clumps of larger microspheres may not be strong enough to withstand hydrodynamic forces trying to break them apart, and it will take considerably longer for all the larger microspheres to form clumps.

B. *Active* Slide Agglutination Tests

Earliest tests used liquid reagents made with plain, white microspheres and were run on black glass slides. Sample and latex reagent are mixed, the slide is rocked and rotated to move the liquid in a circular pattern on the slide, and agglutination or clumping, if it occurs, is observed within a few minutes ("milky" becomes *curdled* milk).

Dyed microspheres provide different contrast (dyed microspheres observed against a white background). There are even tests which use antibodies to *three* different antigen groups bound to three different colored microspheres (red, blue, and green). By comparing the color of the combined agglutinated microspheres to a background color, one can decide when salmonella groups are present in the sample.

C. Passive Agglutination Tests (Like Hemagglutination Tests)

Large (>> 1 μm), dense (> 1.05 g/ml), dyed microspheres are used for agglutination tests in 96-well plates [8,9], thus avoiding the variability and limited shelf life of red blood cells. Reactants are mixed in the wells and left undisturbed for 2–60 min while the large heavy microspheres settle out of solution. As in hemagglutination tests, unagglutinated microspheres settle to the bottom and roll to the center of the V-shaped well to form a dark spot there, while agglutinated microspheres settle to form a lacy pattern which covers the bottom of the well.

VI. QUANTITATIVE AGGLUTINATION (ASSAYS)

When the human eye is replaced by a *nephelometer* to measure scattered light directly [10,11] or a *spectrophotometer* to measure change of transmitted or absorbed of light (measure scattered light indirectly) [12,13], it is possible to measure agglutination quantitatively and to develop sensitive microsphere agglutination immunoassays (MAIA) (see Fig. 2). See also Table 2 for an outline of various instrumental methods for quantifying agglutination reactions which are discussed below.

The intensity of light scattered by particles dispersed in water varies with the number of particles, the diameter of the particles, the wavelength of the incident light, the angle of the detector to the incident light, and other parameters. As agglutination starts, single particles first become doublets; thus the number of light scatterers drops dramatically (decreases by half), and the apparent diameters increase rapidly to 2×. After this point the changes in number and diameter are less rapid. These particle immunoassays can be very sensitive, since the change of scattered light intensity is highest at the very beginning of agglutination or at the lowest concentrations of analyte.

Particles which scatter light best have the diameters approximately equal to the wavelength of light being scattered. Therefore, for visible light (λ = 390–760 nm) the best scattering particles have diameters of 390–760 nm (0.39–0.76 μm). Particles outside this range will not function as well. In practice then, one can start with particles < 0.1 μm which are poor scatterers. As they agglutinate, the particles quickly grow to a size where they scatter light much better. Thus, change of scattered light vs. analyte concentration can be the basis for very sensitive endpoint

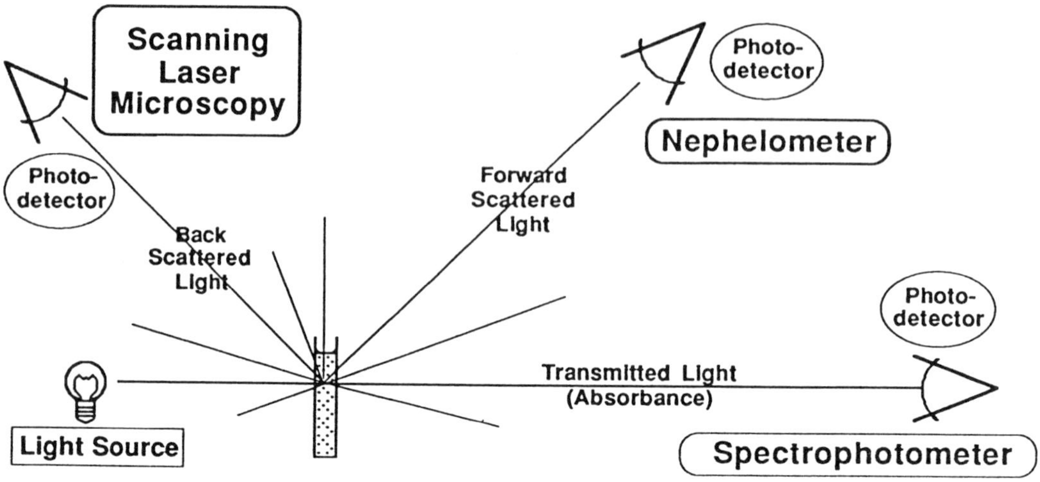

Figure 2 Turbidity from microspheres measured with spectrophotometer or nephelometer.

or rate method immunoassays. UV light requires smaller starting particles (<< 100 nm) and infrared light can use 0.5-μm starting particles. Conversely, one can also begin with particles which scatter well (perhaps 0.5-μm particles) and observe them clumping to sizes where they do not scatter as well, but most assay systems seem to use the principle of small, poorly scattering particles clumping to form big, nicely light scattering clumps.

Modern 96-well plate readers can make spectrophotometric measurements of a complete plate in ~2 s for endpoint or kinetic assays. A lactogen assay measures microsphere agglutination turbidimetrically in a plate reader [14]. The spectrophotometric/nephelometric sensitivity may be as good as 100 pg/ml (1 pg/10 μl) for proteins [15]. Thus, a turbidimetric method using ~0.1-μm microspheres is 10× more sensitive than a typical 1-μm microsphere *visual* PAT (Table 1), and 10,000× more sensitive than a PAT test using 0.1-μm microspheres. Turbidimetry is more sensitive because it detects agglutination when single microspheres first start to agglutinate.

One can use "brighter" polymers, i.e., those with higher refractive indices (n_D), such as

Table 2 Microsphere *Agglutination* Immunoassays: Instrumental Measurement

Microspheres Dispersed
- Turbidimetric (Measure *absorbance* on spectrophotometer or plate reader; endpoint or kinetic)
  Note: Turbidity changes; microspheres do not settle out.
- Nephelometric (measure forward light *scattered*)
- Scanning laser microscopy (back scattered)
- Microsphere counting *Note*: 10–15× greater sensitivity is claimed compared to turbidity
- Image analysis (agglutination pattern) *Note*: Microspheres settle out of suspension and form a pattern on bottom of plates.
- Magnetic moment analysis (magnetic microspheres, measured in magnetometer; magnetic pulse height counting)
- Fluorescence (quenching) analysis (fluorescent microspheres, measured in fluorometer; light quenched by agglutination)

Microspheres caught on a filter
- Agglutination/capture tests and assays (dyed microspheres; color on filter read on reflectometer)

polyvinylnaphthalene, with its $n_D = 1.68$ vs. 1.59 for polystyrene. Such microspheres scatter light better, especially when they agglutinate to the optimum scattering size [16,17].

Some researchers have preferred polybutylmethacrylate ($n_D = 1.43$) microspheres [18]. These less bright microspheres were apparently chosen because they do *not* scatter as well, possibly important in some instruments for lower background scattering (a lower blank value for the calibration curve).

Scanning laser microscopy instruments have been proposed to quantify agglutination [19]. This new method may offer advantages since it uses larger microspheres (> 1 μm) and operates with concentrated microsphere suspensions where agglutination may be faster.

Angular anistropy (or "two-angle light scattering") [20,21] and quasi-elastic light scattering (also called "dynamic light scattering" or "photon correlation spectroscopy") [22] are more powerful techniques which have been investigated and patented for diagnostic applications. These more *computation-intensive* methods *might* deliver better sensitivity and be used after simpler methods have been exhaustively explored.

Other methods for quantifying microspheree immunoassays have been devised. Instruments like *particle counters* which count numbers of single particles or clumps of particles in discrete channels have been used to follow agglutination [23]. Sensitivity is claimed to be 10–15 times better than with turbidimetry.

Passive agglutination can be read by *pattern recognition* for reproducible and semiquantitative assays. An instrument scans the bottom of 96-well plates and detects the difference between a central button of nonagglutinated particles and the typical diffuse, lacy agglutination pattern.

In *fluorescence quenching analysis,* fluorescent particles yield a lower fluorescent output on agglutination. When particles clump together they interfere with and absorb light from each other so less light gets to the detector.

Magnetic microsphere agglutination can be measured with magnetic microspheres by *magnetic moment analysis.* As the microspheres pass through a magnetometer, agglutinated microspheres give a larger signal than single microspheres, so agglutination can be quantified [24].

In another magnetic microsphere application, Ab_1-coated magnetic microspheres mixed with Ab_2-coated plain microspheres and combined with Ag from a sample, then a magnet is used to remove coagglutinated microspheres and the absorbance of the suspension is very different before and after sample addition [25].

VII. AFTER AGGLUTINATION

A. Filter Separation Agglutination Tests (and Assays)

Some test kits have used dyed agglutinated microspheres caught on a filter. Red microspheres coated with Ab were incubated with a sample and poured onto a filter. Single microspheres passed through the filter and no color appeared on the surface. If the sample contained the appropriate Ag, the microspheres agglutinated and the agglutinated clumps were caught upon the filter resulting in a red (or pink) positive color test for the Ag (Fig. 3 is from a Cetus poster paper at an American Society for Microbiology national meeting).

A successful home pregnancy test called First Response uses plain (~1 μm) microspheres and very small (< 50 nm) red gold particles (Fig. 4). *To prepare the test,* the red particles are coated with one antibody (Ab_1) to HCG (human chorionic gonadotrophin, a hormone released by the placenta in pregnant women); the plain microspheres are coated with another antibody (Ab_2); then both kinds of particles mixed and lyophilized. *In use,* both particles are redispersed with a sample of urine; if the sample contains HCG, the particles are coagglutinated, yielding red clumps. The mixture is poured through a filter which catches the red clumps to yield a

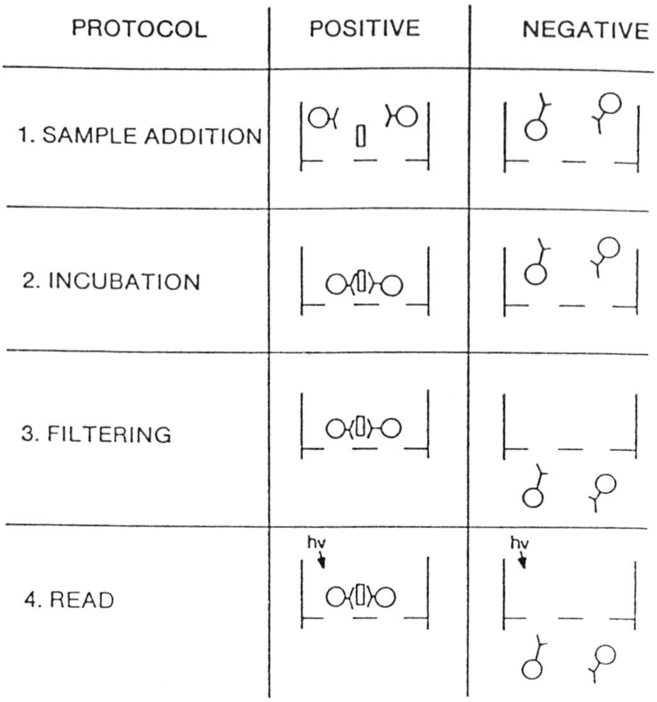

Figure 3 Kodak (Cetus) Sure Cell earliest test format. Diagrammatic illustration of the filter separation agglutination assay. Dye-imbibed latex particles coated with antibodies to group A carbohydrate are mixed with extracted patient sample. In the presence of antigen a cross-reaction between particles occurs, trapping the particles above the filter. In the absence of antigen the particles pass freely through the filter. Color remaining on top of the filter thus indicates a positive reaction. An absence of color indicates a negative reaction.

pink-colored filter. With negative urine unagglutinated red particles pass through the filter and no color develops on its surface.

Other tests use black-dyed antibody-coated microspheres which are mixed with sample and poured onto a strip. If there is no antigen present, then the microspheres migrate up the strip where they are observed. Thus, observed color = "negative color" test result. If the microspheres are agglutinated by the sample, then the clumps are too large to travel up the strip, and no color is observed at the observation point (no color = "positive").

These *test* principles could easily be applied to *assays* where the reflected color intensity (measured perhaps in a dry strip reader?) would correlate with the sample's Ag content.

B. Microsphere Capture Enzyme-Linked Immunosorbent Tests and Assays (ELISAs)

To prepare these tests: Ab is bound to microspheres, the microspheres are caught on a filter and dried (Fig. 5). In use: first, a sample is passed through the filter and any Ag is caught by the Ab on the microspheres. Next, a second Ab-enzyme conjugate is put through the filter; the second Ab is caught by the Ag-Ab on the microspheres to complete the sandwich. Third, enzyme

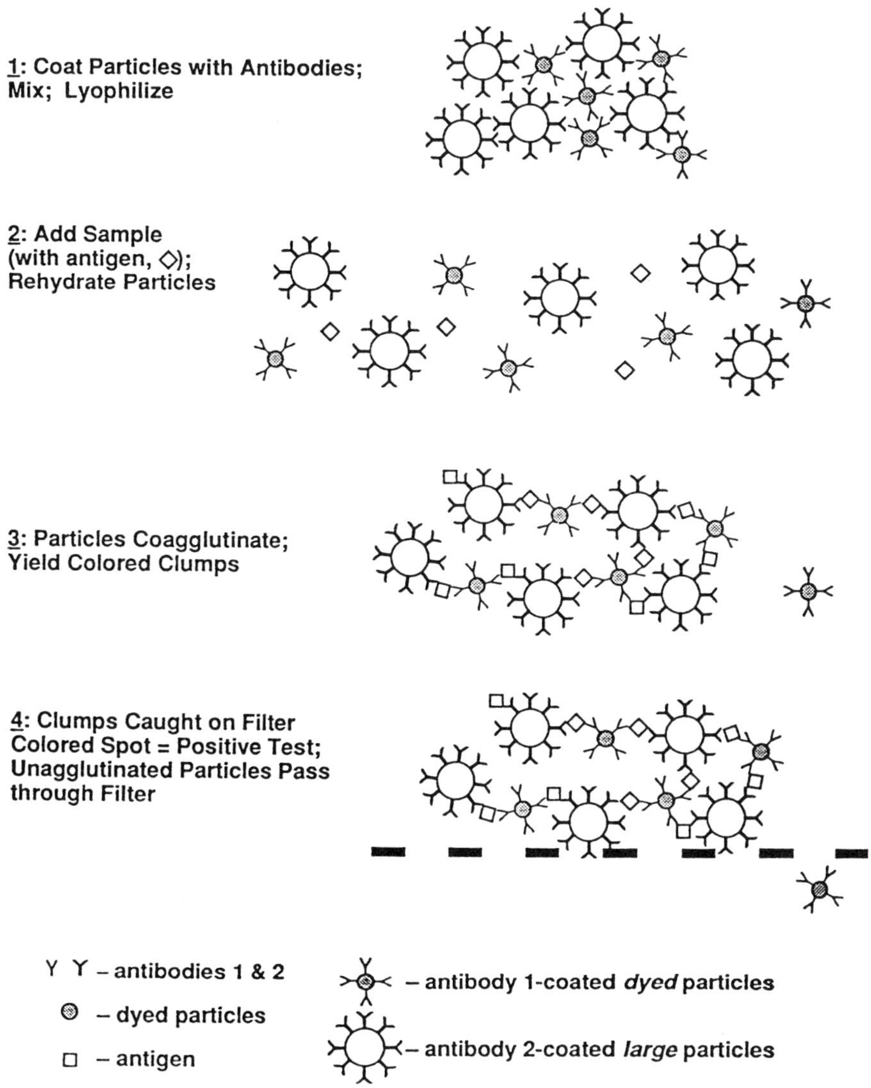

Figure 4 First response (Hygeia/ Carter-Wallace) test format.

substrate is passed through the filter and the enzyme reacts with substrate to create an *insoluble* colored product (on the filter) which is proportional to the amount of Ag caught.

Using this format many different qualitative *tests* have been made for human and *animal* analytes.

Quantitative *assays* have also been made by several companies. With a reflectance meter the colored spots caused by analyte are compared to a blank and a standard spot to yield a true quantitative assay. There is even a "field" instrument (pocket-sized) for measuring plant pests like fungal infestations of soybeans and golf course turf diseases (Neogen, Lansing, MI). A similar hand-held instrument might permit ambulance and bedside quantitative assays in human diagnostics.

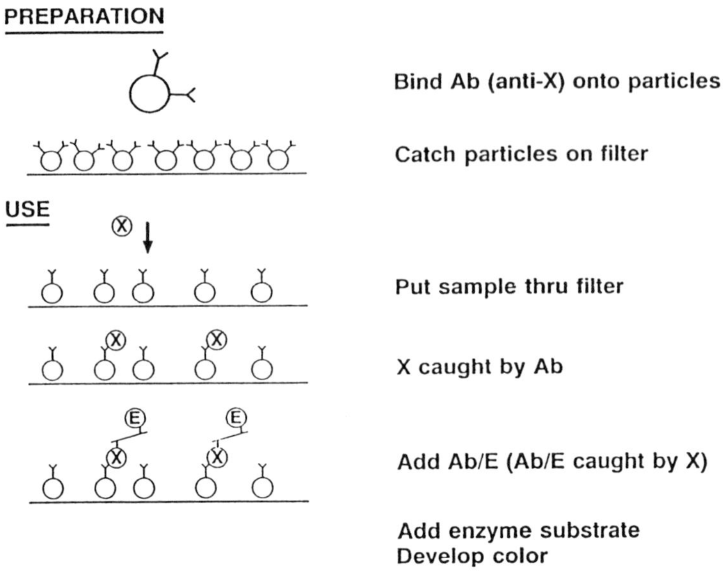

Figure 5 Microsphere capture ELISA test format.

C. Dyed Microsphere Sandwich Test (Also Called One Step Chromatographic, or Strip Tests)

In 1988 a new over-the-counter pregnancy test (Clearblue Easy), developed and patented by Unipath, was introduced. The test uses dyed microspheres in a sandwich format to give a one step test [26] (Figs. 6 and 7). To prepare the test, small dark-blue dyed microspheres (O) are first coated with antibody (Ab_1) to HCG; the microspheres (O-Ab_1) are dried on one part of a nitrocellulose strip; a second antibody (Ab_2) to HCG is immobilized on another section of the strip (Ab_2-1). In use the strip is wetted at one end with urine. As the urine moves by capillary action, it picks up the blue microspheres (O-Ab_1), and carries them downstream; any HCG in the urine reacts with Ab_1 on the microspheres (O-Ab_1-HCG). When the flow reaches the immobilized Ab_2-1, the dyed microspheres with HCG (O-Ab_1-HCG) are captured by Ab_2-I to form a blue line caused by the HCG sandwich (O-Ab_1-HCG-Ab_2-1). The blue line signals a

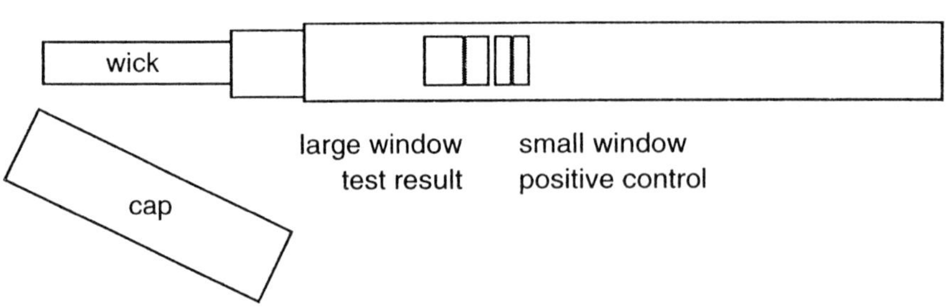

Figure 6 Clearblue Easy home pregnancy test.

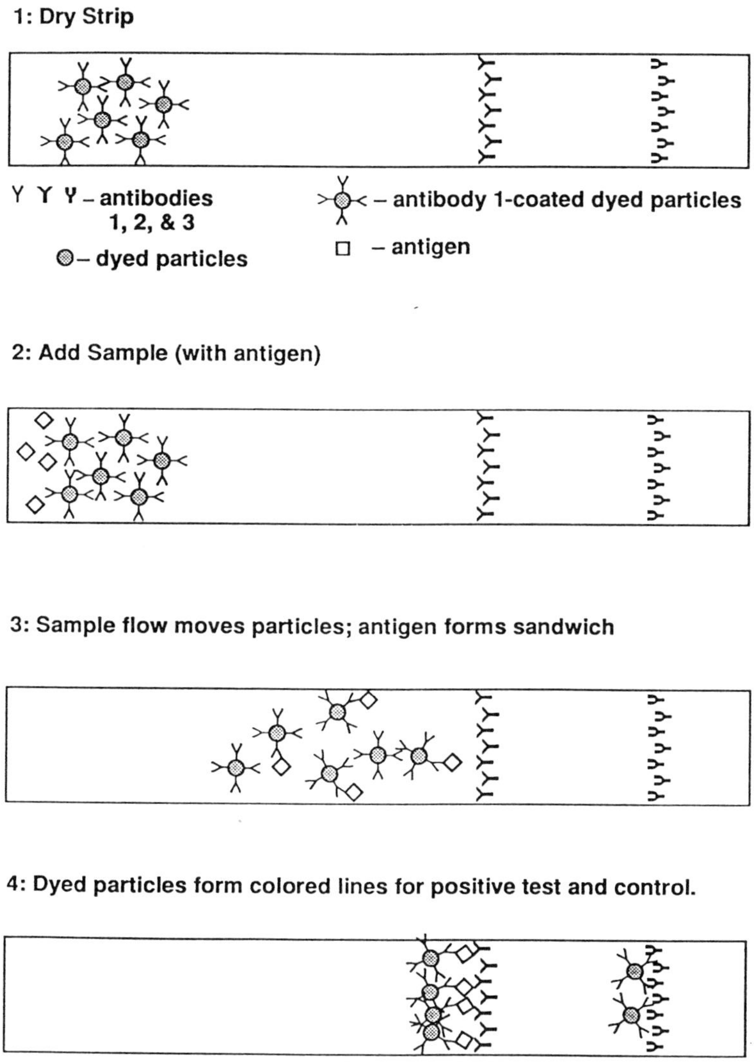

Figure 7 Dyed microsphere chromatographic strip test (Unipath).

positive pregnancy test. Further downstream there is another line of immobilized protein (Ab_3-1) which catches unconjugated ❍-Ab_1 (as ❍-Ab_1-Ab_3-1) (independent of HCG) to form another blue line which acts as a positive procedural control. If the second line does *not* form, the test results are invalid.

Such elegantly simple tests, with a built-in control are well suited to the consumer market. There are also *home* test for LH (luteinizing hormone) and similar *clinical* tests for Strep A, and Chlamydia. Other sources have laboratory single-analyte tests using this chromatographic principle for HCG, common infectious diseases, and DAU (drugs of abuse in urine) (Table 3).

There is even a seven analyte DAU *inhibition* test panel, based on this technology [27,28]. In the first step urine reconstitutes and reacts with a mixture of seven pairs of reagents. Each pair consists of very small gold sol particles conjugated to a drug (❍-drug) + excess

Table 3 Some Dyed Microsphere Chromatographic Strip or Sandwich Tests

Company	Market	Test
Unipath		
	OTC:	HCG: Clearblue Easy (Clearblue One Step in UK)
		LH: Clearplan One Step and Clearplan Easy
	Lab:	HCG: Clearview
		Strep A: Clearview
		Chlamydia: Clearview
Abbott		
	Lab:	HCG: Test Pack Plus
		Strep A: Test Pack Plus
Johnson & Johnson (made by Abbott):		
	OTC:	HCG: Fact Plus
Drug Screening Systems		
	Lab:	Cocaine
		Heroin/opiates
Pacific Biotech/Hybritech		
	Lab:	HCG: $CARDS^+$-OS and Concise
		Strep A: $CARDS^+$-OS and Concise
		Strep B: $CARDS^+$-OS
		Mononucleosis: $CARDS^+$-OS
Biosite		
	Lab:	Panel of seven DAUs: Triage
Quidel		
	OTC:	LH: Conceive

antibody (Ab_1) to that drug. In the second step the urine/reagents mixture migrates on a strip to which seven other antibodies to the seven drugs are bound in different locations, Ab_2-1. If the urine is drug-free, then Ab_1 will bind all ○-drug and none will be available for the second (solid-phase) reaction.

$$\bigcirc\text{-drug} + xsAb_1 + Ab_2\text{-l} \rightarrow \bigcirc\text{-drug-}Ab_1 + Ab_1\text{-l} \ (\textit{no}\ \text{color})$$

If a drug is present in the urine above a certain cut-off level controlled by the amount of excess Ab_1 present, then ○-drug will be free to migrate along a strip to become bound by Ab_2-1. If a colored line is found, then that drug was present in the urine at a level above the threshold level.

$$\bigcirc\text{-drug} + \text{free drug} + xsAb_1 + Ab_2\text{-l} \rightarrow \text{drug-}Ab_1 + \bigcirc\text{-drug-}Ab_2\text{-l} \ (\text{color spot})$$

All of these tests have promise of becoming assays—if the intensity of color formed can be read by a dry strip reader, for example, then a quantitative result can be obtained. In addition, different colors of dyed microspheres could be used to color code different analytes.

D. Solid-Phase Assays

Particles small enough to remain suspended for hours or longer at normal gravity, yet easily centrifuged to provide quick bound/unbound separation have been used for years as solid supports for radioimmunoassays and other newer assays where solid/liquid (bound/unbound) separation is needed (Table 4).

Table 4 Microsphere Use in Solid Phase Assays

Solid-phase /liquid-phase separations
  Separate bound/unbound
  "Read" solids or liquid
Use of microspheres in separation

Microsphere type	Separation method	Assay examples
Large PS (> 0.8 μm)	Centrifugation	LDH-1/Washington University
Smaller PS (< 0.5 μm)	Centrifugal filtration	LDL separation/Genzyme
Silica	Centrifugation	DNA separations
< 1 μm PS/fluor. tag	Filtration	PCFIA particle concentration Fluorometric immunoassay (Fig. 8)
Superparamagnetic	Magnet	(see Figs. 9, 10)

In a test for cardiac specific isoenzyme, LD-1: antibody (D.8.1) is adsorbed on 0.8 μm microspheres; microspheres are mixed with serum; D.8.1 binds LD-2, 3, 4, and 5; microspheres are centrifuged to remove these competing isoenzymes; and free LD-1 is reacted with substrate and measured in a spectrophotometer without interference [29].

A new "Direct LDL [Low Density Lipoprotein] Cholesterol Immunoseparation Reagent Kit" uses ". . . latex beads coated with affinity purified goat polyclonal antisera to specific human apoliproteins, which facilitate the removal of high density lipoprotein (HDL) and very low density lipoprotein (VLDL) in the specimen" [30]. To use, one mixes and incubates serum or plasma with reagent containing Ab-coated microspheres in a separation device. After centrifuging the device (12,000 G's, 5 min) to filter the microspheres from the liquid, one can simply measure LDL cholesterol in the filtrate using conventional enzymatic cholesterol reagent in a spectrophotometric clinical analyzer.

New uniform silica microspheres are useful for nucleic acid separation and with DNA probes and related assays. After addition of chaotropic agents to cell lysates, the DNA or RNA will adsorb onto glass or silica. DNA probes can also be covalently bound to surface-modified silica spheres. The density of the silica microspheres, 2.1 g/ml, makes them easy to centrifuge. The uniform-sized, smooth silica spheres are superior to several commercial products which use silica or "glass milk" (ground glass dispersed in water) to adsorb nucleic acids.

1. PCFIA

In particle concentration fluorescence immunoassay, microspheres coated with one antibody trap a second antibody which traps an antigen or fluorescent-labeled antigen in a competitive binding assay. The microspheres are caught on a filter in the IDEXX (formerly PANDEX) "screen machine." Recently this principle has been applied to an assay for tylosin, a veterinary antibiotic, in animal feeds (Fig. 8) [31].

2. Supermagnetic Microsphere Based Assays

Newer "magnetic" microspheres permit fast and easy separation of solid and liquid phases. Actually *superparamagnetic,* the microspheres respond to a magnet but are not magnets themselves and they retain no residual magnetism after removal of the magnet. These are used in commercial solid phase radioimmunoassays (RIAs), ELISAs, and newer chemiluminescent assays. Immunoradiometric assays (IRMA's) can be designed with magnetic microspheres also [32].

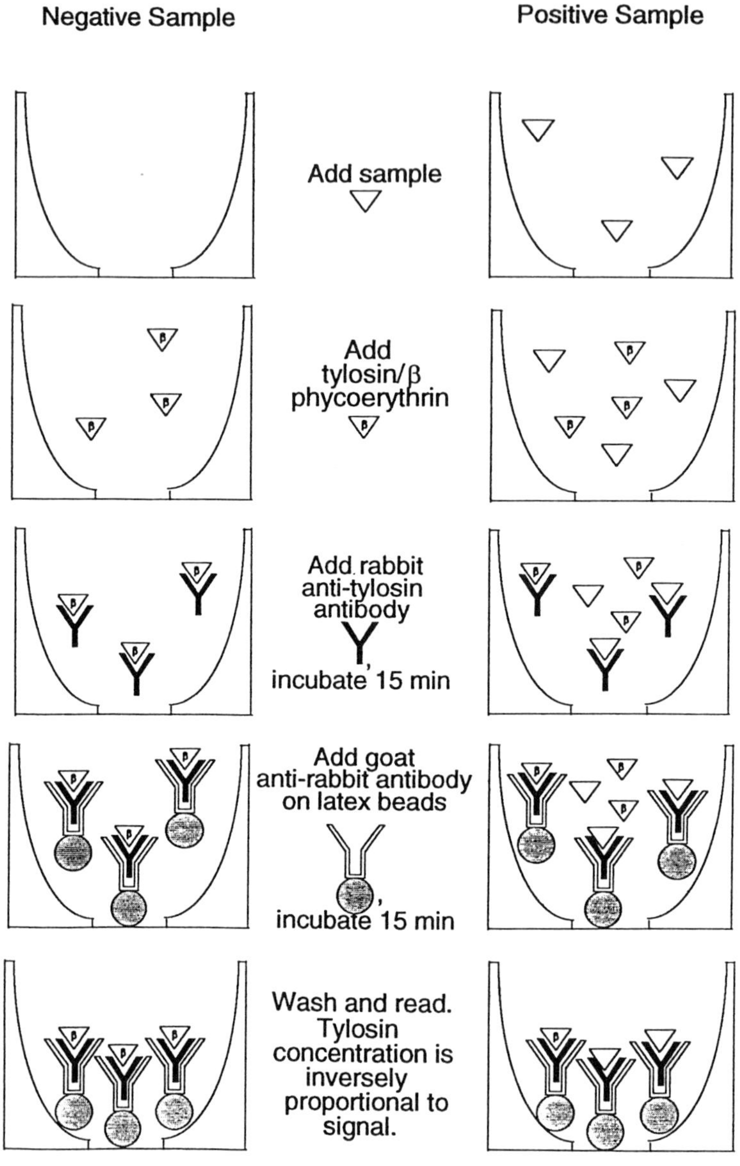

Figure 8 Screen machine fluorometric immunoassay for Tylosin.

The newest of these formats is Sanofi's ACCESS instrument using magnetic microspheres with chemiluminescent assays (first instrument sold in spring of 1993). Over a dozen Sanofi papers and posters have now appeared; one example is their ferritin assays [33] (Fig. 9).

Roger Acey of California State University, Long Beach has developed an interesting format for the aphid-borne plant pathogen citrus tristeza virus (CTV) [34]. In his "Dagwood sandwich" assay he binds sheep antimouse polyclonal antibody (Ab_p) to magnetic microspheres (❍-Ab_p). The Ab_p will bind anti-CTV mouse monoclonal antibody (Ab_m) to form ❍-Ab_p-Ab_m and the Ab_m in turn binds CTV (❍-Ab_p-Ab_m-CTV) from samples of infested trees. Another, biotinylated anti-CTV monoclonal antibody, (Ab_m-B), is bound by the CTV, if present. Streptavidin/alkaline

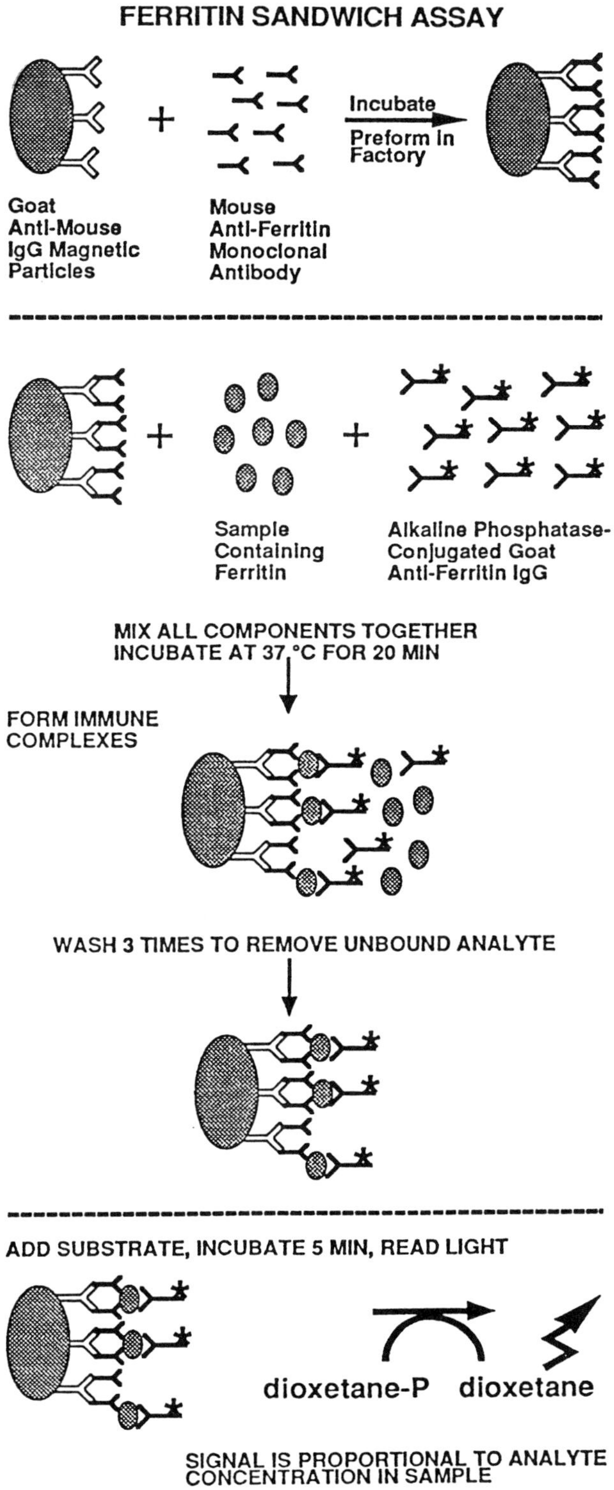

Figure 9 SANOFI magnetic microspheres assay.

phosphatase conjugate (Av-AP) is bound to the biotin (B), to yield the ultimate sandwich: O-Ab_p-Ab_m-CTV-Ab_m-B-Av-AP. After appropriate separation steps pNPP (a colorless dye) is converted to pNP (yellow) by the enzyme. Thus color intensity (absorbance at 415 nm) is directly proportional to CTV present in the sample.

One mRNA isolation procedure uses biotinylated oligo-dT bound to streptavidin coated magnetic particles and hybridized with mRNA (Fig. 10). After magnetic separation, the mRNA can be eluted off the magnetic beads for recovery and further processing [35].

In another DNA hybrid assay (by IGEN), streptavidin-coated magnetic microspheres are used in a human papilloma virus assay based on PCR (polymerase chain reaction), and read by electrochemiluminescence [36].

Other schemes using microspheres in DNA probe assays have been devised. In one, RNA from a sample is first allowed to hybridize with the DNA attached to microspheres. Then enzyme conjugated to an Ab to the DNA-RNA hybrid is added. After appropriate washing and addition of enzyme substrate, the resulting color signals the amount of the specific RNA of interest (Fig. 11).

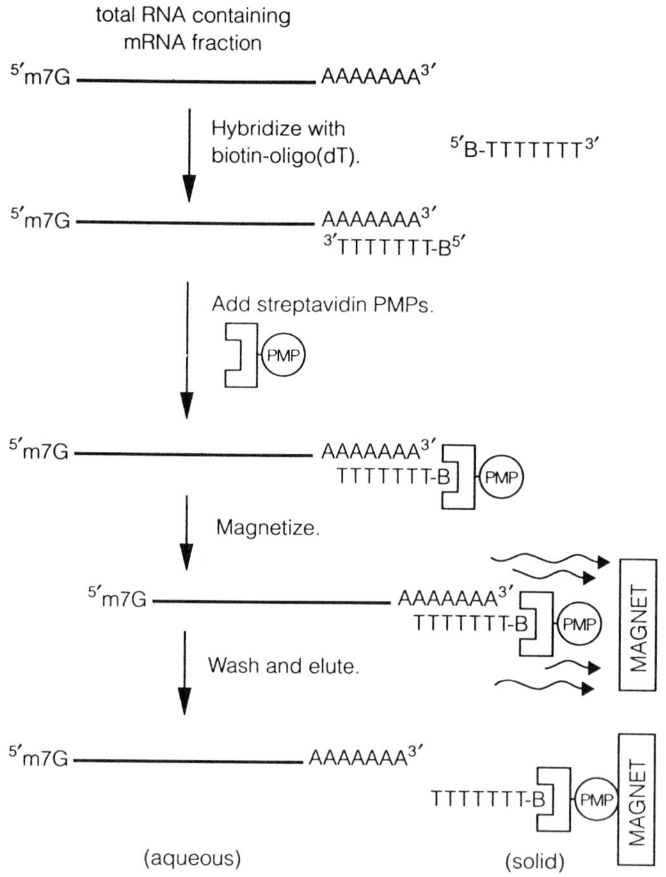

Figure 10 Promega mRNA isolation with superparamagnetic particles.

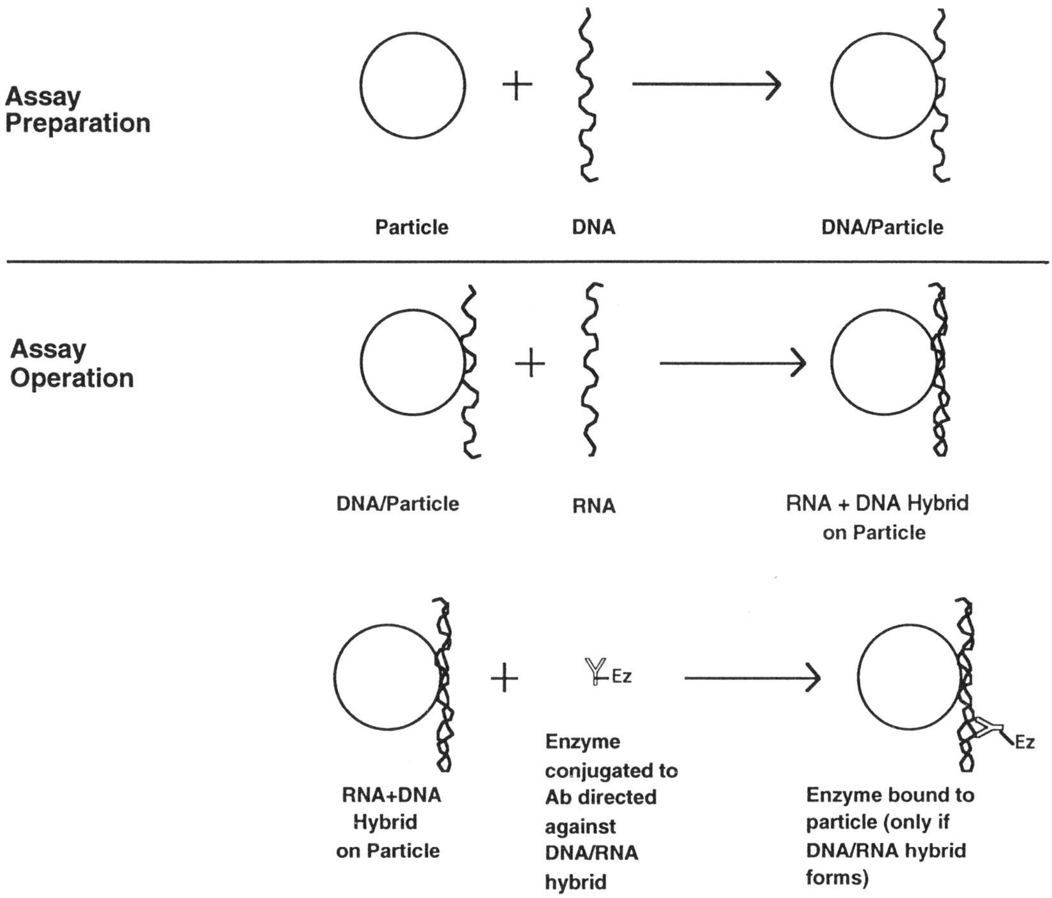

Figure 11 Scheme for DNA probes using microspheres.

VIII. DYED MICROSPHERES AS MARKERS AND STAINS

Since the 1960s (white or colorless) uniform latex microspheres have been used to study phagocytosis and to identify or "mark" phagocytic cells. More recently fluorescent microspheres coated with proteins or pharmacological agents have been used to quantify effects of such agents on the phagocytic process [37].

Appropriately coated plain and dyed microspheres have also been used as cell tags or *exterior* cell labels. Dyed ~5 μm diameter microspheres with appropriate Ab coatings will cluster around certain cells to form rosettes which identify the cells in the light microscope. In other cases the coated microspheres are used to localize cell surface antigens [38].

Dyes are more stable when inside microspheres, and even small microspheres can each hold more than one dye molecule. The ultimate lower limit in microsphere markers is to tag one molecule with one brightly or darkly dyed microsphere.

A decade ago Katz first used very small rhodamine-labeled microspheres as retrograde tracers in neurons—either in vitro or in vivo [39,40,41]. These microspheres travel up the

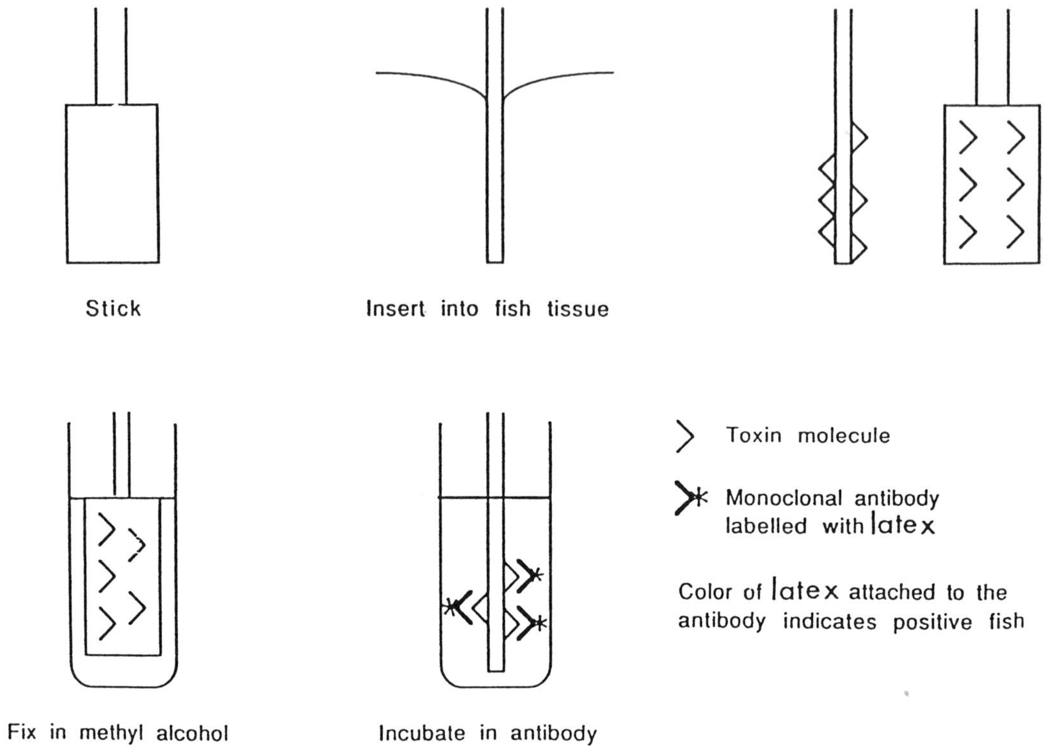

Figure 12 Dyed microspheres ciguatoxin test for fish. (Y. Hokama and Hawaii Chem. Tech.)

neurons and can be used to trace nerve pathways. They may eventually be used to identify accidentally severed nerves to be reattached in reconstructive surgery.

Plain microspheres are used in flow cytometry, (as "zero" or low-level calibrators). Labeled on the surface or internally with fluorescent dyes, they are used to calibrate the instruments for measuring dye-labeled or stained cells.

There is growing interest in "designer" microspheres—dyed with "fluorochromes", "fluorophors" (fluorescent dyes with particular spectral properties), and scintillators. Often only a small amount of these dyes is required to give an intense signal. Stains and flow cytometry are obvious applications of these microspheres, and there will be others.

A fish ciguatoxin test offers a novel use for dyed microspheres as immunomarkers [42]. A roughened paddle is inserted into a cut in the fish. When it is removed some fish flesh adheres to it. Next, the paddle is dipped in a suspension of antibody-coated dyed microspheres and washed. A colored paddle shows a positive test for ciguatoxin (Fig. 12).

Other applications include regional blood flow studies in animals using multicolored 10 and 15 μm microspheres such as those offered by E-ZTrac, Triton Corp., and Molecular Probes [43]. These are replacing radiolabeled particles in this field.

IX. SINGLE MICROSPHERE ASSAYS IN FLOW CYTOMETRY

Originally flow cytometers were rare and were intended only for studying cells; microspheres were used as calibrators, standards and controls. Now Ab- or Ag-coated microspheres can be used in these instruments to study single microsphere interaction with antigens or antibodies and real assays can be run in flow cytometers [44]. Some researchers predict that many immunoassays will soon be done on cells and single microspheres in flow cytometry instruments as use of these instruments becomes more widespread [45].

Perhaps typical of this wave of the future is a highly specific and sensitive microsphere-based fluorescence-activated flow cytometry assay for the measurement of natural latex (NL) protein-specific human immunoglobin E [46]. NL was covalently bound to microspheres. After exposure to serum samples the microspheres were exposed to biotinylated goat antihuman IgE and fluorescein isothiocyanate-avidin. The microspheres were then assessed for fluorescence in a flow cytometer.

X. POSSIBLE FUTURE USE OF MICROSPHERES

The field of biosensors has been "promising" for several years. These promises may soon be delivered and microspheres may be able to help in amplifying the signal from optical- and electronic-based sensors. For example, Battelle has a "biorefractometer" based upon the Mach-Zehnder principle, which depends upon the difference between the two laser light paths, one from a blank and one from an antibody coated surface. Microspheres might be used to amplify this difference. Similarly, watch for opportunities for microspheres in the new evanescent-wave-based sensor technology.

REFERENCES

1. Bangs, L. B., "Uniform Latex Particles," *Amer. Clin. Prod. Rev.*, 7, #1, 22–26 (1988).
2. Kapmeyer, W., "Nephelometric Immunoassay with Shell/Core Particles," *Pure Appl. Chem., 63*/8, 1135–1139 (1991).
3. Maehara, T., Eda, Y., Mitani, K., Matsuzawa, S., "Glycidyl methacrylate-styrene copolymer latex particles for immunologic agglutination tests," *Biomaterials, 11*/3, 122–6 (1990).
4. Bangs, L. B., "Particle-Based Tests and Assays—Pitfalls and Problems in Preparation," *Amer. Clin. Lab. News,* 9, #4, 16–17, May, 1990.
5. Calculated from data by Cantarero, L. A., Butler, J. E., Osborne, J. W., "The Adsorptive Characteristics of Proteins for Polystyrene and Their Significance in Solid-Phase Immunoassays," *Analytical Biochemistry, 105,* 375–382 (1980).
6. Singer, J. M., Plotz, C. M., "The Latex Fixation Test. I. Application to the Serologic Diagnosis of Rheumatoid Arthritis," *Am. J. Med. 21,* 888 (1956).
7. Carney, J., "Rapid Diagnostic Tests Employing Latex Particles," *Anal. Proc., 27,* 99–100 (1990).
8. Proulx, A., Riggin, C. H., "Passive Agglutination with Recombinant ENV Antigen to Detect Antibodies to Human Immunodeficiency Virus," Poster at 4th Int'l AIDS Conference, Montreal, June 1989. Cambridge Bio Tech Corp., Worcester, MA.
9. Fujikawa, H., Igarashi, H.,"Rapid Latex Agglutination Test for Detection of Staphylococcal Entertoxins A to E that Uses High-Density Latex Particles," *Appl. Envir. Microbiol., 54*/10, 2345–2348 (1988).
10. Kapmeyer, W. H., Pauly, H-E, Tuengler, "Automated Nephelometric Immunoassays with Novel Shell/Core Particles," *J. Clin. Lab. Anal., 2,* 76–83 (1988).
11. Delanghe, J. R., Chapelle, J. P., Vanderscheuren, S. C., "Quantitative Nephelometric Assay for Determining Myoglobin Evaluated," *Clin. Chem., 36*/9, 1675–1678 (1990).
12. Medcalf, E. A., Newman, D. J., Gilboa, A., Gorman, E. G., Price, C. P., "A Rapid and Robust Particle-Enhanced Turbidimetric Immunoassay for Serum β_2-Microglobulin," *J. Immunol. Methods, 129,* 97–103 (1990).
13. Medcalf, E. A., Newman, D. J., Gorman, E. G., Price, C. P., "Rapid, Robust Method for Measuring Low Concentrations of Albumin in Urine," *Clin. Chem., 36*/3, 446–449 (1990).

14. Collet-Cassart, D., Limet, J. N., Van Krieken, L., De Hertogh, R., "Turbidimetric Latex Immunoassay of Placental Lactogen on Microtiter Plates," *Clin. Chem., 35*/1, 141–143 (1989).
15. Hager, Hans, Private Communication, November 9, 1990 [Hans J. Hager, Cuddledyke House, The Pingle, Upwell, Wisbech, Cambs., PE14 9DN, UK].
16. Medcalf, E. A., Newman, D. J., Gilboa, A., Gorman, E. G., Price, C. P., "A Rapid and Robust Particle-Enhanced Turbidimetric Immunoassay for Serum β_2-Microglobulin," *J. Immunol. Methods, 129,* 97–103 (1990).
17. Medcalf, E. A., Newman, D. J., Gorman, E. G., Price, C. P., "Rapid, Robust Method for Measuring Low Concentrations of Albumin in Urine," *Clin. Chem., 36*/3, 446–449 (1990).
18. Amiral, J., Miguad, M., "Development & Applications of a New Photometric Method for Fast and Sensitive Immunoassays," *Eur. Clin. Lab., 10*/June, 28 (1991).
19. Putnam, D. L., Private Communication, May 24, 1991 [David L. Putnam, Pacific Technologies, 21806 N. E. 1st, Redmond, WA 98053].
20. Cannell, D. S., Giglio, M., Benedek, G. B., von Schulthess, G. K., Cohen, R. J., "Immunoassay by Light Scattering Intensity Anistropy Measurements," U.S. Pat. 4,174,952 (1979).
21. von Schulthess, G. K., Giglio, M., Cannell, D. S., Benedek, G. B., "Detection of Agglutination Reactions Using Anisotropic Light Scattering: An Immunoassay of High Sensitivity," *Mol. Immunol. 17,* 81–92 (1980).
22. Cohen, R. J., Benedek, G. B., "Immunoassay by Light Scattering Spectroscopy," U.S. Pat. 4,080,264 (1978). See also related U.S. Pat. 4,164,558.
23. Wilkins, T. A., Brouwers, G., Mareschal, J-C., Cambiaso, C. L., "High Sensitivity, Homogeneous Particle-Based Immunoassay for Thyrotropin (Multipact™)," *Clin. Chem., 34*/9, 1749–1752 (1988).
24. Imai, K., Tokinaga, D., Yokogawa, D., "Particle Agglutination Immunoassay Apparatus," U.S. Pat. 4,913,883 (1990).
25. Collet-Cassart, D., Private Communication, November 9, 1990.
26. "Two new babies on the way," *Financial Times (London),* July 7, 1988; Technology Section.
27. "Drugs of Abuse by Immunoassay," *Clin. Lab. Prod. 21*/3, 8–9 (1992).
28. U.S. Pat. 5,028,535 (1991).
29. Vaidya, H. C., Porter, S. E., Landt, Y., Silva, D. P., Dietzler, D. N., Ladenson, J. H., "Quantification of Lactate Dehydrogenase-1 in Serum with Use of an M-Subunit-Specific Monoclonal Antibody," *Clin. Chem. 34*/12, 2410–2414 (1988).
30. Genzyme Product Literature 1001–1018 © 1993.
31. Wicker, A. L., Sweeney, D. J., Mowrey, D. H., Coleman, M. R., Morris, D. K., Brockus, C. L., "Validation of the Particle Concentration Fluorescence Immunoassay of Tylosin," Poster Reprint, AOAC Annual Meeting, Cincinnati, August, 1991; Lilly Research Labs., Greenfield, IN 46140-0708 and International Diagnostic Systems Corp., St. Joseph, MI 49085.
32. Larue, C., Calzolari, C., Léger, J., Léger, J., Pau, B., "Immunoradiometric Assay of Myosin Heavy Chain Fragments in Plasma for Investigation of Myocardial Infarction," *Clin. Chem. 37*/1, 78–82 (1991).
33. Peterson, T., Kapsner, K., Liljander, B., et al., "A Chemilminescent Immunoassay for the Determination of Liver Ferritin," Poster 624 at AACC Meeting 1992, Chicago (Sanofi, Chaska, MN 55318).
34. Acey, R., "Immunology Tutorial," Amer. Soc. Cell Biol. Meeting, December 1991, Boston, sponsored by American Qualex.
35. Smith, C., Ekenberg, S., McCormick, M., "The PolyATtract™ Magnetic mRNA Isolation System: Optimization and Performance," *Promega Notes,* #25, June, 1990, Promega, 2800 Woods Hollow Rd. Madison, WI 53711-5399.
36. Kenten, J., Casadei, J., Link, J., Lupold, S., Farrell, M., Lowke, G., Massey, R., "DNA Hybridization Assays for Polymerase Chain Reaction Products Based on Electrochemiluminescence," Poster Reprint, IGEN, Inc., 1530 E. Jefferson St., Rockville, MD 20852.
37. Falk, L. A., Vogel, S. N., "Comparison of Bone Marrowing Progenitors Responsive to Granulocyte-Macrophage Colony-Stimulating Factor and Macrophage Colony-Stimulating Factor," *J. Leukocyte Biol. 43,* 148 (1988).

38. Reinhardt, J., Balcerzak, S., Lobuglio, A., "Study of the Malaria-Red Cell Relationship with the Use of a New Immunological Marker," *J. Lab. Clin. Med., 78,* 167 (1971).
39. Katz, L. C., Burkhalter, A., Dreyer, W. J., "Fluorescent Latex Microspheres as a Retrograde Neuronal Marker for in vivo and in vitro Studies of Visual Cortex," *Nature,* 310, 498–500 (1984).
40. Callaway, E. M., Katz, L. C., "Emergence and Refinement of Clustered Horizontal Connections in Cat Striate Cortex," *J. Neurosci. Methods, 10,* 1134, (1990).
41. Katz, L. C., Iarovici, D. M., "Green Fluorescent Latex Microspheres; a New Retrograde Tracer," *Neurosci., 34,* 511–520 (1990).
42. Hokama, Y., "Simplified Solid-Phase Immunobead Assay for Detection of Ciguatoxin and Related Polyethers," *J. Clin. Immunoassays, 14*/2, 111–114 (1991).
43. Glenny, R. W., Bernard, S., Brinkley, M., "Validation of Fluorescent-Labeled Microspheres for Measurement of Regional Organ Perfusion," *J. Appl. Physiol., 74,* 2585 (1993).
44. Cook, L., Irving, D., "Microsphere-Based Flow Cytometric Assays," *J. Clin. Immunoassay, 12*/1, 36–39 (1989).
45. McHugh, T. M., "Flow Cytometry and the Application of Microsphere-Based Fluorescence Immunoassays," *Immunochemica, 5*/1, 1–6, (1991), Zymed Labs., Inc.
46. Kwittken, P. L., Pawlowski, N. A., Sweinberg, S. K., Douglas, S. D., Campbell, D. E., "Flow Cytometric Measurement of Immunoglobulin E to Natural Latex Proteins," *Clin. Diagn. Lab. Immunol., 1*/2, 197–201 (1994).

16
Reference Materials in Particle Measurements

Renliang Xu

Coulter Corporation, Miami, Florida

I. INTRODUCTION

Reference materials are one of the most important and indispensable components in analytical science and measurement technology. Reference materials are used as "known" materials for helping develop accurate methods of analysis, as "control" materials analyzed simultaneously with "unknown" materials and as "calibration" materials for calibrating measurement systems, and for ensuring the long-term adequacy and integrity of measurement quality assurance programs. With global demands for quality control and quality assurance in both industrial processes and research and development, and with more and more industries adopting the ISO 9000 Standard Series, in order to meet ever stricter rules and standards, reference materials serve as the guard—in many situations the only guard—to validate correct measurement procedures and results. In particle science and technology, reference materials play a vitally important role in validating particle characterization techniques, in calibrating or checking analysis equipment, and in standardizing quality assurance programs. A partial list of their uses includes calibration of particle characterization systems such as

Particle sizing devices: test sieves, dynamic and static light scattering instruments, laser diffraction instruments
Particle counters: optical single-particle counters, Coulter counters
Microscopes: optical microscopes, electron microscopes
Chromatography apparatus: size exclusion chromatographs, capillary hydrodynamic fractionation systems, field-flow-fractionation devices
Electrophoretic measurements: capillary electrophoresis devices, electrophoretic light scattering instruments
Sedimentation instruments

and as controls for

Scientific research: colloidal systems, microcontamination, pharmaceuticals, as model, carrier or marker particles
Technological studies: light scattering techniques, analytical optical systems, aerosol generation, microporous filtration media

There are many aspects to particle characterization and thus there are many different reference materials for each property, such as surface area or electrophoretic mobility, of a particulate system. However, the main parameter in characterizing particulate systems is size: the mean size and the size distribution. Also, for many applications, measurements of particle concentration (count) are often needed. The content of this chapter is thus focused on the reference materials for particle size. The concept and reality of different reference materials, the primary and the secondary standards, and the traceability of secondary standards to the national standards will be discussed in the second section. The reference materials used in particle sizing, their main properties and proper handling and storage, will be the content of the third section, along with a list of available sources for these reference materials in the United States. A short summary is provided as the conclusion of the chapter.

II. REFERENCE MATERIALS AND TRACEABILITY

What is a reference material? From "Terms and definitions used in connection with reference materials," ISO Guide 30-1992 (E/F): "reference material (RM): Material or substance one or more of whose property values are sufficiently homogeneous and well established to be used for the calibration of an apparatus, the assessment of a measurement method, or for assigning values to materials." The same guide has also classified the best of reference materials as "certified reference material (CRM): Reference material, accompanied by a certificate, one or more of whose property values are certified by a procedure which establishes its traceability to an accurate realization of the unit in which the property values are expressed, and for which each certified value is accompanied by an uncertainty at a stated level of confidence." RMs (and CRMs) may be produced by any firm that uses valid procedures to produce a material with the property value being well established (and certified) by the firm. However, for such RMs or CRMs to be accepted by the analysis community, their property values have to be established according to certain international or national standard reference materials using validated and reliable measurement methods.

Many countries have their own authorities for issuing and maintaining national standards for reference materials. Most of these materials are made internationally available for applications and correlation needs throughout the world. In the United States of America, the national authority for producing and issuing Standard Reference Materials (SRMs) is the National Institute of Standards and Technology (NIST), formerly the National Bureau of Standards (NBS). In Europe, the Community Bureau of Reference (BCR), based in Brussels, issues monodisperse and polydisperse CRMs [1]. NIST prepares and produces SRMs (1) if they would permit users to attain more accurate measurements, or (2) if their production elsewhere would not be economically or technically feasible, or (3) if they would be an industry wide standard for commerce from a neutral source not otherwise available to the public, or (4) if their production by NIST would assume continued availability of a well-characterized material important to science, industry, or government [2]. All certified values for SRMs are obtained by independent, reliable and previously validated measurement methods performed by NIST and by a network of cooperating laboratories. Currently, there are more than 1300 different SRMs available from NIST. The following example briefly describes how a national SRM is produced.

NIST issues a particle size SRM made of polystyrene spheres with a 30-μm nominal diameter (NIST SRM 1961). These microspheres were manufactured using a special procedure developed by Lehigh University and NASA in space aboard the space shuttle *Challenger* during five space shuttle flights in the NASA STS-11 mission. The spheres are extremely uniform in size (with a coefficient of variation equal to only 0.8%) and spherical because they were made under microgravitation in space. Certainly, such production is feasible only by a government agency such as NIST and not by a private firm. These extremely uniform microspheres, which would not be available otherwise, can be an excellent standard material in scientific research and for many industrial applications. The diameters of the spheres were determined by the center distance finding technique, confirmed by metrology electron microscopy, and cross-validated by transmission electron microscopy, Coulter principle and optical microscopy in cooperative determinations by NIST and by other laboratories throughout the nation in coordination with ASTM (American Society for Testing Materials).

The time period for preparing a SRM, from planning, manufacturing, to completing calibration, is typically a few years and the amount of material (a few hundred kilograms in the above instance) is very limited as a national resource. Thus, although the SRMs have precisely determined property values (for example, the SRM 1961 has the mean average diameter 29.64 ± 0.06 μm) they cannot be routinely used as reference materials because of the limited national supply and the high cost. It is also impractical to let NIST calibrate every measuring instrument. In addition, in many situations, the quality of NIST SRMs far exceeds what is really needed for the application. There have to be secondary standards, which are commercially available in large quantities and at low cost, yet with measurement linkage to NIST, that can be used as daily supplies for the ever increasing demands of reference materials.

The questions now are: how good are these secondary standards when compared with the NIST SRMs? Is one going to get the same result if one uses these secondary standards instead of the NIST SRMs to calibrate measurement systems, to develop analytical techniques, or to control the quality of processing or production? Before answering these questions, we need to discuss the popular term *traceability,* which is in fashion in measurement technology nowadays.

In order to make the quality of secondary standards approach the primary standard, more and more vendors who produce these reference materials standardize and control their manufacturing and calibration procedures, thus making their products traceable to the NIST SRMs or NIST analysis methods. Many of them issue a certificate for each batch or bottle of their reference material. Often a certificate contains statements like: "this product is certified for the mean diameter and traceability to NIST," or "the sizes are directly traceable to NIST standards." These statements give users confidence, sometimes serve as vendor protection against litigation, and imply that these reference materials are "good" because they are traceable to the national standard. But what does traceability really mean? What is implied when a material is labeled traceable? Although many users take the concept of traceability for granted, this author believes that it is important to understand the implication of traceability and the proper use of this concept.

Historically, the main applications of traceability began more than three decades ago [2], and are only the contractual agreements between government agencies and their prime contractors, suppliers, and regulatees. Since then, the written concept of traceability in measurement technology has been changing over time. Many discussions are ongoing to develop a better and clearer understanding of this evolving concept and its function in measurement science and technology [3]. In ISO Guide 30:1992 (E/F), *traceability* is defined as: "Property of the result of a measurement or the value of a standard whereby it can be related, with a stated uncertainty, to stated references, usually national or international standards, through an unbroken chain of comparisons." A full discussion of traceability is beyond the scope of the present work. The two important issues, which are at the essence of measurement traceability and standard traceability, are

1. Traceable measurement results should be in an unbroken pathway of calibration to one or more of the following: (a) national standards (in the USA these are the NIST SRMs); (b) fundamental or natural physical constants; (c) ratio type of calibration.

2. A measurement quality assurance system is required to ensure that the accuracy of each measurement is within stated limits of uncertainty.

Since even if an instrument has been accurately calibrated by SRMs, there is still no guarantee that a measurement made with it will be suitably accurate because of operator errors, environmental influence, etc. Only if both of the above conditions are satisfied can true traceability be established. Thus, the generally accepted guidance is: "measurement traceability has been achieved when the reference standards base is specified and when the total measurement uncertainty relative to those standards has been shown to be sufficiently small for the intended purpose [3]." The second portion of this statement is important because it is often neglected when dealing with reference materials.

In particle size measurements, traceable size values imply that these values are determined from one or both of the follwing:

1. A comparative measurement that employs the NIST SRMs as a direct calibrator. An example of such measurement is when using optical microscopy (OM). The sample to be calibrated and the NIST SRM are mixed together. The preparation is observed by OM and photos are taken. Since the magnification rate (G) is the same for both populations, one can use the ratio type of calibration, i.e.,

$$d_{\mathrm{r,sample}} = d_{\mathrm{p,sample}} \frac{d_{\mathrm{r,SRM}}}{d_{\mathrm{p,SRM}}}$$

   with r = real and p = photo. By measuring sizes of the SRM and the sample in the photos, the real sizes for the sample can be obtained. Thus, in addition to using the calibrated magnification rate to obtain the sample size from $d_{\mathrm{r,sample}} = Gd_{\mathrm{p,sample}}$, the sample size can be validated directly to the NIST SRM to avoid possible instrumental artifacts.
2. A calibration using one or more independent reliable measurement methods on instruments that have adequately high metrological quality and the accuracy, range, and resolution for the size values to be determined and have been calibrated or checked either by NIST (seldom the case in practice) or by using the NIST SRMs. The most commonly used methods are TEM (transmission electron microscopy), OM (optical microscopy), PCS (photon correlation spectroscopy), SEM (scanning electron microscopy), and ESZ (electrical sensing zone, also known as the Coulter principle).

In particle size measurement traceability also implies that the stated size and polydispersity values have small enough uncertainty and small enough deviation from the true values that these variances will not affect the intended application purpose. The phrase "small enough," however, is a relative term; it all depends on the purpose of the application. For example, if a reference material is to be used to calibrate an apparatus having a small dynamic range but high precision such as a microscope, an uncertainty of, for example, 6% in the stated size will be too big even if the material is NIST traceable. On the other hand, such material may be suitable as a traceable standard for calibration of a light diffraction instrument that has a dynamic range of thousands to one but lower resolution.

Now let us try to answer the questions posed earlier. In particle size measurement if the intended application for a reference material is for calibrating and/or checking existing sizing instrumentation, or for quality control in industrial particulate processes, a bona fide secondary standard is just as good as the primary standard provided that

1. The secondary reference material comes with a certificate in which the standard's base and

the methods used to determine the property value are specified, and the property value and its associated uncertainty are stated along with a material safety data sheet.
2. The user understands that the uncertainty related to the property value of the secondary standard is small enough relative to the permissible error limit of the application. Recently, many vendors include the main intended uses of the material in the certificate to help users use it more properly.

One important aspect that should not be neglected is the handling and storage of reference materials. The quality and integrity of a reference material are preserved only when proper handling procedures and storage conditions are observed. Most reference materials are stored in sealed containers; once the seal is broken by the user and the container has been opened, special care and laboratory procedure should be taken to avoid any storage condition and contamination that would alter the certified value.

III. REFERENCE PARTICLES FOR PARTICLE SIZE MEASUREMENTS

In particle size measurement the complexity of samples with different shapes requires more than one parameter to truly size a particle, unless it is a sphere or a cube. However, for polydisperse particulate systems, it is neither practical nor possible, and many times even not necessary, to characterize the whole system for the constituent's real dimensions. People often use a volume equivalent spherical diameter to "size" a particle. Although it may not reflect any true dimension, in most cases this equivalent diameter does represent the characteristics of an irregular particle. It is very important to realize that the "size" of a particle and the real dimension of the particle are often two different concepts and have different values. For a spherical particle, the diameter is an accurate measure of its true size. No matter what principle or technology an instrument employs, the basic requirement for a sizing apparatus is to measure a sphere correctly. Thus, to minimize shape effects on the response of analytical systems it is obvious that the best reference materials for sizing instruments should be spherical in shape.

A size reference material serves two purposes in calibrating a particle measurement system: to calibrate the accuracy of the sizing or counting function, and to verify the linearity of the system in sensing particles of various sizes. For accuracy, the material should be monodisperse in size with the size value accurately known; for linearity, the material should be multimodal or polydisperse with a known size distribution. Presently, only NIST issues polydisperse reference materials. Commercial vendors offer only monodisperse spheres as NIST traceable materials. Table 1 lists all major U.S. vendors and their product lines for particle size reference materials along with the calibration methods used for the certified size values.

A. Reference Materials of Monodisperse Sizes

There are two types of materials commonly used as size reference materials: soda-lime glass beads and synthetic polymer latexes. The soda-lime glass beads, which contain trace amounts of metallic elements, are dry spheres with sizes from a few micrometers to a few millimeters. The main applications of the glass beads are for evaluation of test sieves, for calibrating instruments requiring a higher density material (soda-lime glass has specific density of 2.4–2.5 g/ml), such as ultrasonic and sedimentation based analyzers, and for instruments requiring materials having high contrast to incident radiation, such as optical, X-ray and electron beams. These glass beads have been processed to assure a high percentage of perfect spheres. Some carefully prepared reference glass beads contain virtually no nonspherical particles. Besides being used as dry spheres, glass beads can also be suspended in liquid if the measurement has to be

Table 1 Particle Size Reference Materials

Producer	Product name	Size range[a]	Methods[b]	Note[c]
NIST, Standard Reference Materials Program, NIST, Gaithersberg, MD 20899, 301-975-6776	SRM 1003b, 1004a, 1017b, 1018b, 1019a	10–2160,P	SEM, CC, SIEVE	g
	SRM 1690-1692, 1960-1965	0.1–30	CDF, MEM, TEM, CC, OM	
	SRM 659	0.2–10,P	XRS	sn
	SRM 1978	0.3–10,P	XRS	zo
Coulter Corp., 1800 SW 147 Avenue, Miami, FL 33196, 800-523-3713	Microspheres	2–90	CC	
Duke Scientific Corp., 2463 Faber Place, Palo Alto, CA 94303, 800-334-3883	Nanosphere size standards	0.02–0.9	PCS, TEM	
	Monosized polymer microspheres	1–1000	OM, CC	
	Uniform polymer microspheres	5–40	OM, CC	
	Uniform glass microspheres	2–2000	OM	g
	Monosized glass microspheres	250–600	OM	g
Dyno Particles, Norway (distributed by Bangs Laboratories, Inc., 979 Keystone Way, Carmel, IN 46032, 317-844-7176)	Dynospheres calibration kits	0.5–20	OM, CC, PCS	
Interfacial Dynamics Corp., 17300 SW Upper Boones Ferry Road, Suite 120, Portland, OR 97224, 800-323-4810	NIST traceable calibrated microspheres	0.04–2	TEM	

Polymer Laboratories, Inc., 160 Old Farm Road, Amherst, MA 01002, 413-253-9554	NanoCal particle size standards	0.1–0.9	PCS, TEM	
Polysciences, Inc., 400 Valley Road, Warrington, PA 18976, 800-523-2575	Polybead select certified size standards	0.1–2	TEM	
Powder Technology, Inc., P.O. Box 1464, Burnsville, MN 55337, 612-894-8737	Fine test dusts[d]	1–100, P, N	CC	ds
Rhône-Poulenc, France (distributed by Bangs Laboratories, Inc., 979 Keystone Way, Carmel, IN 46032, 317-844-7176)	Estapor calibration standards	0.2–0.8	TEM	
Seradyn, Inc., 1200 Madison Avenue, Indianapolis, IN 46225, 800-428-4007	PRECISION microparticle size standards	0.1–2	TEM, SEM, PCS	
	Large size standards	2–20, N	CC	

[a]All values are in micrometers and are NIST traceable except for the ones marked "N." Several monodisperse materials are available in the range except for the ones marked "P," where the materials are polydisperse with known distributions. For NIST glass beads SRMs, the size range indicates the overall size range covered by these SRMs.

[b]Calibration methods used in determining the property values. The first one in the list is the primary method. CC: Coulter counter (also known as electrical sensing-zone method); CDF: center distance finding; MEM: metrology electron microscopy; OM: optical microscopy; PCS: photon correlation spectroscopy; SEM: scanning electron microscopy; TEM: transmission electron microscopy; XRS: sedimentation with X-ray detection.

[c]All materials are polymer spheres (mostly polystyrene) except g: glass spheres; sn: equiaxed silicon nitride; zo: irregularly shaped zirconium oxide; ds: classified Arizona desert sand.

[d]NIST is in process of developing concentration RMs using polydisperse fine test dust in hydraulic fluid.

performed in liquid phase. A typical application of these glass beads dispersed in electrolyte solution is to calibrate a Coulter counter. These beads are easy to use, to handle, to store and to reuse if they are handled carefully without, for example, being strongly impacted which may break or crack the beads, or being exposed to strong basic environment that will corrode glass.

Synthetic polymer latexes, mostly polystyrene latexes, comprise chains of polymers or copolymers either held together by van der Waals attraction forces to form nonporous, hydrophobic particles or cross-linked to varying degrees to make particles that are more resistant to dissolution by organic solvents. Using well-studied recipes, polymer latexes can be produced by emulsion polymerization, suspension polymerization, or other means to achieve precisely desired diameters from nanometers to millimeters. Especially, when experimental conditions are carefully controlled, these latexes will be fairly uniform in size and spherical in shape with excellent reproducibility from batch to batch. Thus, polymer spheres are the most common materials used for size reference purposes. Besides NIST polymer latex SRMs, many commercial firms manufacture polymer latex spheres of various sizes with different degrees of uniformity. Being calibrated using NIST SRMs or NIST traceable methods for diameter and measurement uncertainty associated with the diameter, these secondary size standards are satisfactory for most industrial and academic applications.

Except for large latexes ($d > 140$ μm, for example) that are dry spheres, polymer latex spheres are dispersed in ultrapure water or in surfactant solution, depending on their synthesis procedure and surface conditions, and in concentrations optimized for ease of dispersion and colloidal stability. These suspensions are usually supplied with 0.1–10% solid concentrations and in many cases may need to be diluted before use.

There are two types of latex spheres, each with different surface conditions. The so-called surfactant-free latexes are synthesized without the use of surface active agents. These latex particles are stabilized in pure water against aggregation by surface charges introduced at the particle surface by various polymer chain terminators used in the polymerization process. Due to the lack of surfactant coverage on the surface these latexes may lose their colloidal stability over time because of the hydrolysis of the sulfate groups to hydroxyl groups. The majority of latex preparations are made by emulsion polymerization using surfactants, usually alkyl sulfonates or alkyl carbonates. These materials adsorb at the particle surface, enhance the charge on the surface and thereby increase particle's colloidal stability and resistance to coagulation and aggregation by electrolyte ions. Most latex spheres are anionic, thus negatively charged with sulfate, carboxyl, or carboxylate modified surface groups. There are also a few positively charged ones with amidine or aldehyde modified surface groups, although these are rarely used as size standards due to their lower stability against negatively charged contaminants in their environment. The variety in surface conditions of latexes may create important differences in their applications in biomedical or clinical diagnostics, but it makes no difference when using these latex spheres as size reference materials once they are stably dispersed without aggregation and their certified size values are maintained.

In order to ensure a correct measurement when using polymer latexes as size standards they must be well dispersed as individual particles. However, latex particles always tend to aggregate due to van der Waals forces because of their large surface areas and small diameters. This tendency is prevented and particles can be stably suspended by the electrostatic repulsion arising from the surface charge groups. Thus, any of the following conditions that could weaken the repulsive force, reduce interparticle distance, and cause instability of the dispersed latex particles should be avoided:

1. High solids concentration
2. Harsh mechanical stirring

3. High electrolyte concentration or high ionic strength. The ion concentration should be as low as possible, especially for small latexes with low surface charge density, and must be lower than the critical coagulation concentration (CCC). The CCC values are a function of particle size and type, surface charge density, and ion type. For univalent ions, the CCC values range from 10–200 mM.
4. Neutralization of surface charge groups, such as the addition of cationic emulsifiers into anionic latex suspension.
5. Addition of polyelectrolytes.
6. Addition of any multivalent cations in anionic latex suspensions (vice versa for anions in cationic latexes) since a very low concentration, often even less than 1 mM, of such ions will induce severe aggregation.
7. Improper pH condition, resulting in surface charge groups that do not remain fully ionized.

The aggregates formed under the above conditions can usually be redispersed by either diluting the suspension, adjusting the pH value, or reducing the ionic strength along with sonication. When an electrolyte has to be used as the diluent, for example in Coulter counter measurements, first dilute the sample with water and then add the electrolyte (all multivalent ions should be avoided as much as possible). Adding sufficient surfactant such as sodium lauryl sulfate (SDS) may help prevent aggregation. Addition of organic solvents should be avoided because some of them dissolve or swell polymer particles, thus changing their sizes. Table 2 lists common physical properties for polystyrene latexes.

Any latex particles larger than about 0.6 μm tend to settle on prolonged standing. In assuring that there are no aggregated latex particles and all particles are uniformly dispersed, manually shaking or inverting the sample bottle a dozen or more times before use is the minimum requirement. Gentle agitation or rotation of a few hours in an appropriate device may be needed. Exposing the bottle to a mild ultrasonic bath for a few minutes before each measurement is highly recommended to redisperse any possible aggregates. The power of the ultrasonic device should be controlled so that it does not overheat or irreversibly flocculate the particles. In the case of very small particles (< 0.1 μm) the sonicated suspension can be briefly centrifuged at high speed (12,000 rpm) to further remove any nonredispersed aggregates from the suspension. For particles 0.6 μm and larger, the completeness of the dispersion or redispersion can be checked using a light microscope.

Reference latex particle suspensions are shipped in sealed containers. A newly prepared or an aseptic suspension can be stored at room temperature for many years. In many commercial polymer latex suspensions trace amounts of sodium azide or other chemicals are added as a biocide to prevent growth of contaminating organisms before the material is packaged. However, once the container is opened, the certified size values of the remaining materials will still

Table 2 Bulk Properties of Polystyrene Latexes

Density	1.05 g/cm^3 at 25°C
Refractive index	1.59 at $\lambda_0 = 590$ nm
Glass transition temperature	80–90°C
Melting temperature	240°C
Dielectric constant	2.53 at 25°C
Solvents	Aromatic hydrocarbons, carbon disulfide, chlorohydrocarbons, cyclohexanone, ethyl acetate, higher ketones, pyridine, THF
Nonsolvents	Acetic acid, acetone, lower alcohols, lower aliphatic hydrocarbons, ether, nitromethane, phenol, water

be valid only under proper storage conditions, including proper temperature and minimal microbial contamination. Polymer latex suspension can be stored in a cold room or refrigerator at +4°C to avoid growth of airborne microorganisms that might be acquired during handling. They should never be frozen, not even accidentally, as this will produce irreversible flocculation or aggregation that will permanently change the particle size. Prolonged storage at high temperature should also be avoided. If after use a prolonged storage is envisioned, the suspension may then be sterilized by one of the following procedures:

1. Pasteurization: 24 h at 70–75°C
2. Gamma radiation: 0.03mrad/h for 24 h

The bottle should be tightly capped to prevent loss of water that changes the solid concentration in case concentration measurement is needed. Latex suspensions older than one year after opening should not be used for primary calibration unless the particle size has been verified.

Monodisperse spherical size reference materials can also be used as counting standards for light and electrical sensing zone counters. Many vendors specify the number of concentration for their standards. Because a NIST SRM for counting is not available, these values are not NIST traceable. The solid content in a suspension of monodisperse spheres can be readily found by weighing a certain amount of homogenized suspension, evaporating to dryness in an oven at temperature < 70°C and then weighing the remaining solids. The number of concentration then can be calculated by a simple algorithm.

B. Reference Materials of Polydisperse Sizes

For calibration of test sieves and other sizing instruments used for size distribution measurements, a polydisperse size reference material with a calibrated size distribution is needed. The linearity, the sensitivity over the entire size range, and the resulting size distribution of an instrument can only be verified by such a polydisperse reference material even though the measurement accuracy can be checked by individual monodisperse size reference materials. Polydisperse reference materials are much more difficult to prepare due to the reason that entire lot of the material has to be homogeneous despite its broad distribution, and that several counting and sizing methods have to be used in order to overcome the bias (or nonlinearity) unavoidably associated with each counting method and to minimize the associated uncertainties with each composition value. Thus, reference materials with certified broad distributions can only be obtained from NIST as the NIST SRMs. They are spherical glass beads, equiaxed silicon nitride, and irregularly-shaped zirconium oxide, all in various size ranges. Figure 1 shows the cumulative size distribution of a polydisperse SRM (NIST SRM 1004a) that can be used to calibrate the linearity of size measurement systems. Any instrument used to obtain size distributions must be able to reproduce the certified distribution of a SRM within allowable error limits, after correction for the shape effects of the nonspherical SRMs. However, a precaution has to be taken when using polydisperse reference materials, such as the NIST SRMs 1003b, 1004a, etc. The entire bottle of the SRM material should be used for measurement, because the calibrated size distribution is based on the entire homogenized sample in the bottle. Otherwise, except when a correct procedure in taking subsamples from the SRM bottle is applied (such as the ones described in [4]) the size distribution obtained from the measurement will never be the same as the certified distribution.

These broad distribution SRMs can be used for calibration or quality control of counting devices, although they can only provide relative mass percentage or volume percentage values. To calibrate absolute counts over a wide size range, a reference material with a known number distribution instead of a percentage distribution has to be used. Because of the difficulties in obtaining batches with the homogeneous and reproducible number distribution, such material is

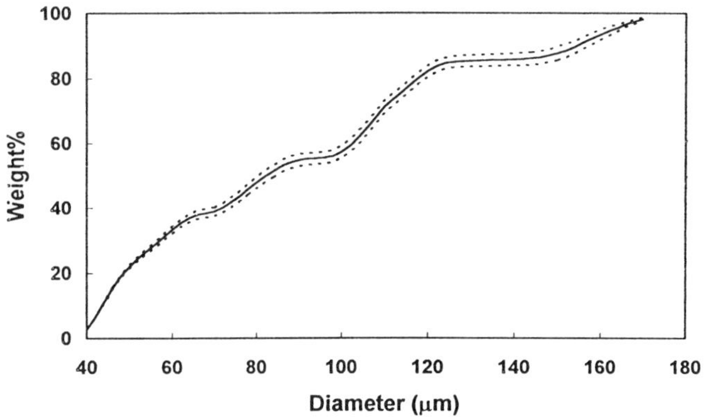

Figure 1 The cumulative size distribution of NIST SRM 1004a. The dashed lines indicate the uncertainty range for the weight percentage values. The size distribution was determined by scanning electron microscopy and by the Coulter principle technique using the Coulter Multisizer IIe.

not available as a certified reference material [5]. However, there is a particulate material that has been proved by the ISO and the American National Standards Institute (ANSI) as the standard for the calibration of particle counters [6]. That material is the classified fine test dust (formerly known as AC FTD from General Motor). Fine test dust comes from Arizona desert sands composed of silicon dioxide and other metal oxides. These particles are nonspherical with a length-to-width aspect ratio of 2.7:1 and closely resemble the type and shape of particles found as contaminants in hydraulic systems. Thus, AC FTD has for several decades been a standard particulate material for use in fluid power contamination control and filter evaluation applications. There is a large natural resource of this material and the number distribution is very close to a log-normal distribution. It has been studied using a wide variety of techniques and yields a reproducible number distribution. Thus, although a shape correction is needed for comparison with spherical materials and therefore it lacks certification for number distribution from a national authority, fine test dust still has served as the main calibration material for optical single-particle counters for years and is recently being developed into an RM by NIST.

IV. SUMMARY

In summary, with industries paying increasing attention to product quality, reference materials are playing an ever important role in today's economic cycle. In particle size measurement even for absolute measurement techniques that, at least theoretically, do not require calibration (such as laser diffraction technology), size reference materials are being used more frequently: by instrument manufacturers to validate designs of both hardware and software, to calibrate instruments, to service customers, and to compete with competitors; and by users to detect operational errors, to increase confidence for the measurement results, and to cross-check system performance. The tendency toward stricter quality assurance poses demands on vendors of reference materials to make their products NIST traceable, with certified size and polydispersity values and associated measurement uncertainty, on instrument manufactures using the primary and secondary size standards in their production and service processes, and on NIST issuing more SRMs not only for size but also for other particle characterization-related reference materials. For example, electrophoretic mobility is an important parameter for colloidal systems, and many scientific research and industrial applications need to know accurate values of electrophoretic mobility. After several years of work, as the latest member of the NIST size

SRMs, NIST now issues an electrophoretic mobility SRM (SRM 1980) that not only fills the blank for such an SRM but also provides the few vendors who have been providing electrophoretic mobility standards the opportunity to verify their stated values*. It is foreseeable that the quality of industrial products thus the quality of our daily life will be continuously improved with the pivotal contributions from reference materials in different fields.

ACKNOWLEDGMENTS

The support from Coulter Corporation is appreciated. The comments and suggestions from Mr. R. Gettings at NIST and Mr. H. Hildebrand at Coulter Corporation, are gratefully acknowledged.

REFERENCES

1. J. P. Mitchell, *Anal. Proc. 29*: 508 (1992).
2. *NIST Standard Reference Materials Catalog,* NIST Special Publication 260, NIST, Maryland, 1995.
3. E. L. Garner and S. D. Rasberry, *J. Test. Eval., 21*: 505 (1993).
4. T. Allen, *Particle Size Measurement,* 4th ed., Chapman & Hall, London, 1990, Chapter 1.
5. A mixture of monodisperse particles can be served as the number distribution standard provided that the number of concentration of each component in the mixture is well calibrated. See S. Aotani, T. Katsuta, and Y. Fukal, "Preparation and Application of Size & Number Standards in Water," Proc. 12th Int. Symp. Contamination Control, Yokohama, Japan, 1994, pp. 469–475.
6. *Hydraulic Fluid Power: Calibration of Automatic-Count Instruments for Particles Suspended in Liquids: Method Using Classified AC Fine Test Dust Contaminant,* ISO 4402: 1991 (E), ISO, Frankfurt, 1991.

*Electrophoretic mobility standards are available from Coulter Corp., Miami, Florida, and from Interfacial Dynamics Corp., Portland, Oregon.

17

Selective Particle Capture with Transparent Sieves in the Evaluation of Contamination in Parenteral Solutions

Julius Z. Knapp

Research & Development Associates, Inc., Somerset, New Jersey

The casual use of the phrase "particle size measurement" has tended to obscure the fact that there are only two techniques that afford a direct measurement of the maximum dimension of a micrometer-sized particle. These techniques are image analysis and sieving. All other "particle size measurement" techniques in present use interpret a signal associated with a selected characteristic of a particle as the "particle size." The applicability of selective mechanical capture of contaminating particles can be defined by examining the capability of the widely used light extinction particle counters for these contamination measurements.

In the United States the convenience, cost effectiveness, and measurement replicability of the light extinction particle counter has made it the primary instrument specified by the U.S.P. to evaluate particle contamination levels in small-volume injectable products. Based on this evaluation, a mandatory determination that the particle contamination is within U.S.P. specified quality limits is required before a batch is released for clinical use. This selection is premised on the view that routine batch release measurements of the contaminating particles in parenteral products are sensitive enough to show changes in the produced quality that can be investigated by other, more accurate, means. The "particle size measurement" of the light extinction particle counter is based on the light extinction coefficient of a particle as it passes through a scan beam. In this type of measurement, the particle counter acts as a measurement transfer device. It compares a stored light extinction calibration curve with an incoming signal to determine the equivalent spherical particle "size" produced by a random particle signal. The calibration curve is based on the response of the instrument to measurements of precision spherical particles. Best accuracy is obtained when the refractive index ratio of the calibrating spherical particle in its suspending liquid matches that of the experimental particle in its suspending liquid. Since a vendor-selected algorithm is employed to determine "particle size" for measurements that do not occur at calibrated points, best accuracy results when the particle size measurement occurs at a calibration point.

An inherent limitation to the measurement accuracy of light extinction particle counters is the variability of particle size determination for nonspherical particles. The variability results from the random orientation of a particle as it crosses through the scan beam. The greater the deviation from the ideal spherical shape, the greater is the variation in the measured size. The greatest variability results in the measurements of particles that are extreme departures from the calibrating spherical shapes. These include fibers, platelike particles, and filmy materials. The maximum dimension of this group of contaminants is grossly undersized in light extinction particle counter measurements.

During product development and in the investigation of deviations from an accepted production quality norm, the data provided by light extinction particle counters can be inadequate. Air bubbles due to cavitation and immiscible droplets are counted as particles when using light extinction particle counters. The measurement of contaminating fibers and platelets is too variable to afford the accurate analytical data that could help to identify the source of the contaminant. Particle identification and a more accurate measurement of particle contamination in a parenteral product is required to guide production in correcting process defects. Other types of contaminating particle analysis must be considered that will provide identification and verify that the count data recorded is free of artifacts.

The belief that the reproducibility of counting results in light extinction particle counters demonstrated their counting accuracy [1] has been disproven [2–4]. The experimental requirements for accurate operation of light extinction particle counters are now firmly established. Considering the limitations of these particle counters, a need exists for a practical, low-cost methodology that can provide accurate particle identification and evaluation of the size and concentration of the contaminating particles.

When image analysis is used, the actual size of a particle is determined from the viewed or recorded particle image and the magnification factor employed. Image analysis measurements are continuous and provide detailed shape information. The concentration of particles that can be counted is a fundamental limitation to the accuracy of image analysis measurement. This limitation is reached at 5% coverage in the field of view and is considered to be a practical maximum for accurate particle sizing measurements [5].

When a selected range of particle sizes is to be evaluated, any particle size present that is not included in the specified measurement range is an interference to the desired measurement. The U.S.P. [6] specifies a maximum acceptable contamination level in injectable products in two size groups: particles greater than 10 μm and particles greater than 25 μm. Due to the perceived patient hazard, the acceptable concentration of particles greater than 25 μm in injectable fluids is held to one-tenth of the acceptable concentration of particles less than 25 μm in diameter. Delly [7] has reported that in some injectable products, particles less than 10 μm in diameter constitute 81% to 96% of all particles present. The presence of these smaller particle sizes constitutes a time-consuming interference in the U.S.P. <788> membrane/microscope measurement of particle contamination in injectable fluids. Measurement interference is also a concern when the contamination levels of injectable preparations are evaluated with electronic particle counters. This measurement interference occurs at a rate determined by both particle size and concentration. At high particle concentrations random coincidental transits of the scanning beam of a counter occur. These random multiparticle occurrences are recorded as a single, larger particle.

When this coincidental particle appearance is uncontrolled during the sizing and counting of contaminating particles, there is a decrease in the total number of particles counted and an increase in the measured size of particles. This apparent increase in particle size can make particles less than 25 μm in size appear to be greater than 25 μm and thus subject to limits 10 times more stringent than is intended.

The 5% coverage limit, for image analysis in the field of view of the microscope, is predicated on the achievement of a usable contrast ratio between the image and the background [5]. When the human eye is the final evaluator of either the direct or the recorded image, the contrast of the particle image against its background is the determining factor in its accurate detection. Biophysical evaluation of human vision by Blackwell [8,9] has shown that a threshold of 5% contrast is the minimum detectable by the human eye. To achieve threshold detection of particles against a stationary background, a contrast ratio of 100% is required for accurate, dependable detection. In the region between threshold detectability and secure detection of a particle, detection is probabilistic and the detection probability is proportional to the contrast of the image against its background. The major group of pharmaceutical particulates consists of low-contrast white particles. Estimates of the proportion of these low-contrast particles run as high as 70% of all particles present. Until the recent appearance of the improved membrane microscope assay in U.S.P. XXIII, most accepted the evaluation of the F.D.A. spokesman [1] that the results of the membrane/microscope assay of injectable contaminants were negligibly comparable. The improved microscope assay results for different observers is now within the outer reaches of usability.

The widespread acceptance of image analysis for the evaluation of parenteral particle contaminants is blocked by practical considerations: cost, the rate of change in imaging technology, and the change in laboratory staffing requirements. In this interregnum of unpredictable length, with no universal methodology, a new device manufactured by Collumated Holes, Inc., of Campbell, California provides access to a cost-effective refereed analysis for pharmaceutical particulates. This new device combines mechanical separation of the analyzed particle size range with optical clarity. This combination of characteristics makes possible the choice of manual or automated microscope counts of the captured particles. Resuspension and counting of the trapped particles by light extinction or conductimetric particle counters is a practical alternative to microscopy. Figure 1 is a photograph of a precision optically transparent glass sieve. The 25-mm total diameter area includes a 1.5-mm solid perimeter and a net active area 23 mm in diameter. Channel diameter is maintained within a tolerance of ± 5%.

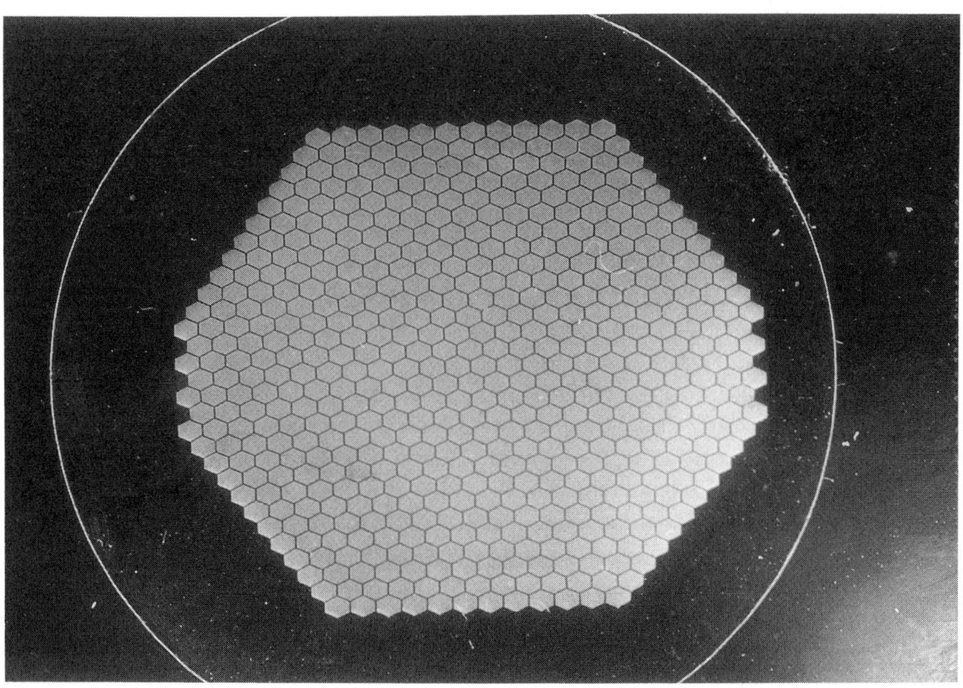

Figure 1 A photograph of a ½-mm-thick, 25-mm-diameter precision transparent glass sieve showing the fused hexagonal building blocks and the 1½-mm solid surround.

Figure 2 An oblique 3000× SEM through the center of a 5-μm precision transparent glass sieve showing the accurate parallelism of the channel walls.

Figures 2–4 are SEMs of 5-, 10-, and 25-μm sieve sections showing the right cylindrical channels and the uniform circularity of the channel ports. The sieves are 0.5 mm thick.

The optical clarity of the sieves is equivalent to that of a microscope slide. From this observation it was concluded that any particle capable of detection with a microscope could be identified, sized, and counted. The combination of the optical clarity of the sieves and the accuracy of their channel dimensions provide both size and quantity data for accurate characterization of parenteral particle contaminants. The accuracy and uniform geometry of these sieves has proven to be of value in varied measurements. One application was as a linear pressure/flow element in a subambient gauging calibration project. Another application was in the acceptance testing of particles used for the production of photographic film base.

In the film base application, the sieves were a critical element in the determination of particle sizes present in a ratio of parts per million. The sieves were also used for outlier determination of particles greater than 50 and 90 μm in diameter in pharmaceutical preparations. The feasibility of sieve use to improve replicability of particle counting data between observers will also be discussed.

The mechanical separation of particles by sieving is probably the oldest particle sizing

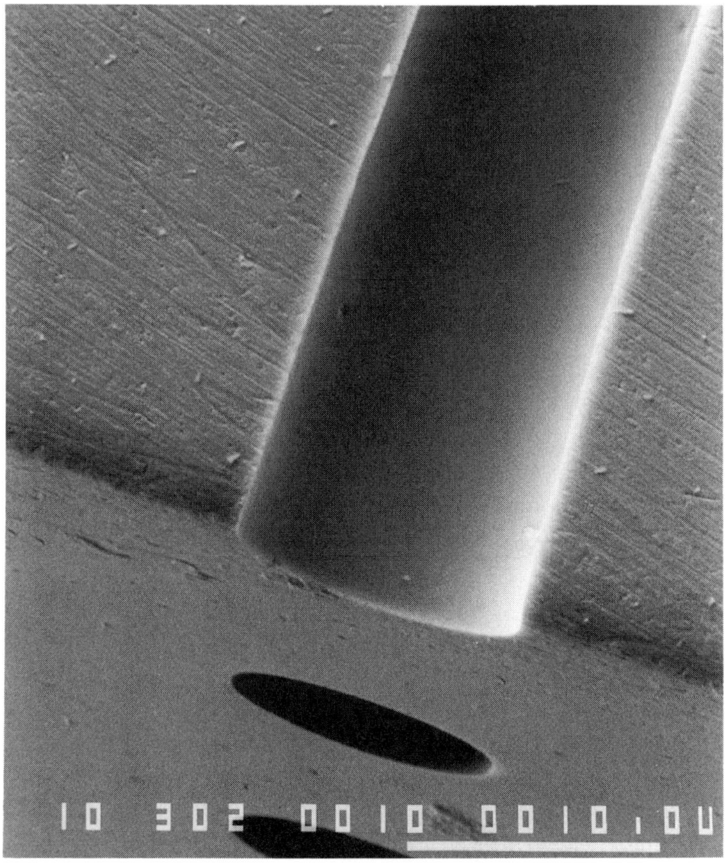

Figure 3 An oblique 3000× SEM through the center of a 10-μm precision transparent glass sieve showing the accurate parallelism of the channel walls. The open area for this sieve is 15%.

technique. Its separation accuracy depends on the three-dimensional uniformity of the channels in the fabricated sieve. In these precision, transparent, glass sieves, channel dimensions approach ideal circular cross section and size uniformity. This combination of characteristics define an important new measurement tool for particle size analysis. As in present application of the Coulter counter, mass:balance checks can be used to determine that the sum of the masses in the separated particle groups is in agreement with the original mass.

For close agreement with theoretical results which are based on calculations with spheres, cylindrical channels with accurately dimensioned circular entry ports are essential. This requirement for analytical accuracy of the results eliminates from consideration the use of depth filter media due to the variable trapping and problematic recovery of particles within the three-dimensional volume of the filter. Figures 5 through 8 are representative of the tortuosity and variability of the channels through a depth filter that contribute to the trapping of particulates. Figures 5 and 6 are 1000× and 3000× SEMs, respectively, of a widely used 0.8-μm polymeric filter that clearly show the range of channel size and the trapping capability of this depth filter. Figures 7 and 8 are similar 1000× and 3000× SEMs of a 0.8-μm silver depth filter. The SEMs in Figs. 7 and 8 also display a wide range of channel sizes and a tortuosity that can entrap particles, thus negating accurate analytical applications.

Figure 4 An oblique 200× SEM through the center of a 25-μm precision transparent glass sieve showing the accurate parallelism of its channel walls and faces. The combination of diamond saw slicing accuracy and the fiber-optic channel structure ensure circular channel ports. The open area for this sieve is 15%.

The use of surface filtration filters prepared with nuclear radiation and acid etching (track etched) provide a step advance over the capability of depth filters. Channel diameters are significantly more uniform and consequently, less particle trapping is experienced. The use of an energetic radiation source followed by acid etching to remove impact areas results in randomly positioned channel entries. To achieve useful flow rates through the sieves, the ratio of channel area to sieve surface area must be held to some reasonable fraction. As a result of the increased exposure, multiple adjacent radioactive particle strikes result. These multiple strike zones produce enlarged holes in a variety of shapes following the etching process. Although an optimistic view is that only 1% of the channel ports are affected examination of sieve microphotographs (Figs. 9, 10 of 0.8-μm polycarbonate surface filters) show that, in reality, the 1% figure can be best described as a target. Figs. 9 and 10 are, respectively, 1000× and 3000× SEMs of a widely-used polycarbonate surface filter sieve. These SEMs indicate that the commercial claims of 1% channel accuracy limits can be overoptimistic. This variation in the produced channel-size limits the separation accuracy that can be achieved with this type of sieve.

An alternative methodology for producing a surface sieve structure is provided by the kind of photolithographic process now widely used in the fabrication of micro-circuit components.

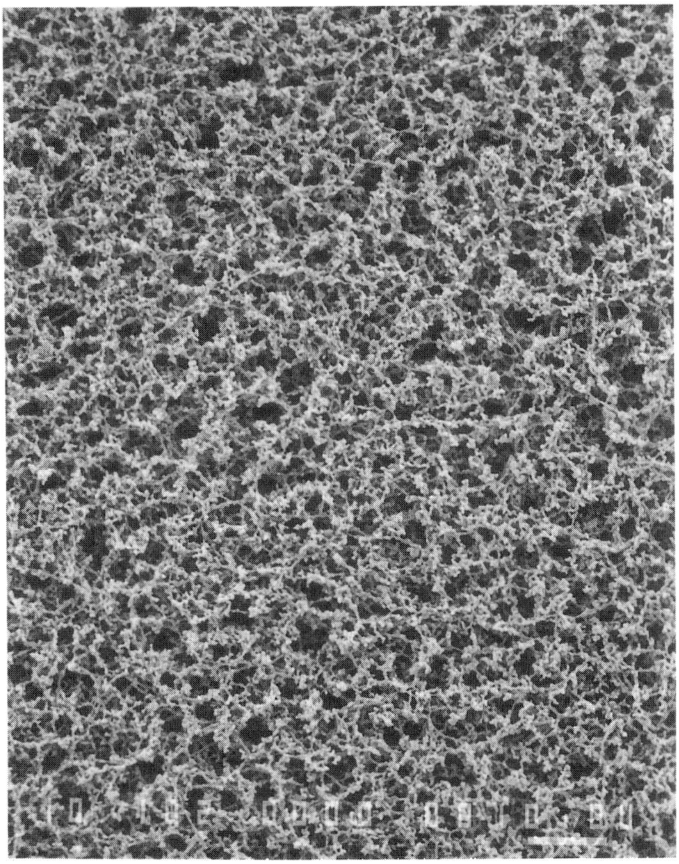

Figure 5 1000× SEM of a 0.8-μm polymeric depth filter showing the range of channel diameters and tortuous channel structure that contribute to particle trapping. These characteristics combine to preclude its use for analytical particle size separations.

This methodology has been successfully used to provide multiple-aperture holographic screens in thin copper foils. In this process, the desired channel entry ports are left bare metal and a protective film is applied to cover the metal surface that is to be retained.

Following etching of the channel ports, the protective film is removed and the surface sieve remains. Experience has shown that the depth of metal that can be successfully etched to produce parallel-walled cylindrical channels is restricted to a metal depth equal to the channel diameter. When the metal depth exceeds the hole diameter, the channel narrows in a rounded conical manner approaching in shape that of an ellipsoid. A practical limit of sieve thickness is twice that of the channel diameter. To achieve this doubling of sieve thickness, a two-sided etching pattern is employed. The difficulties in maintaining registration of the two-sided protective pattern and of maintaining identical etching conditions throughout the etching cycle limits impose high manufacturing costs and result in a fragile device. For the nominal range of channel diameters that are useful in pharmaceutical measurements (5, 10, 25, and 50 μm), film strength for a 25-mm sieve is too low to provide a practical device.

The application of fiber-optic manufacturing technology has recently made possible an optically transparent sieve with very accurate channel port diameters and parallel walls. Use of

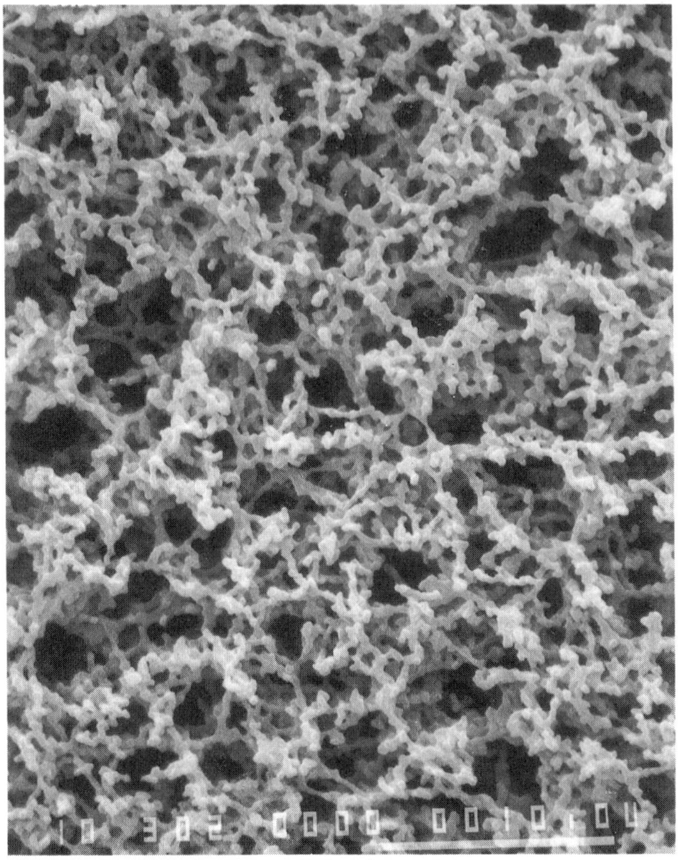

Figure 6 3000× SEM of a 0.8-μm polymeric depth filter showing more clearly the range of channel diameters and tortuous channel structure that contribute to particle trapping. These characteristics combine to preclude its use for analytical particle size separations.

this type of sieve makes possible viewing and counting of trapped particles with any microscope lighting technique. The image contrast and clarity of size-sorted particles against an optically clear work surface realize the target of the U.S.P. <788> microscope/membrane assay for contaminating particles.

In the usual fiber-optical manufacturing sequence, a jacketed glass fiber whose core glass has a higher refractive index than that of the jacket is produced on servomechanism-controlled machinery. To be usable, the optical fiber that is produced must be, and is, maintained within extremely close dimensional tolerances. To produce an accurately dimensioned sieve, the core glass is chosen to be a soluble glass and the outer jacket is a hard optical glass.

Following the fiber drawing operation, the individual fiber strands are laid into molds and then compressed and sintered into small-cross-section hexagonal rods. In the final assembly stage a rod of the desired dimensions is assembled of the individual small-cross-section hex rods and surrounded with a jacket of optical glass fibers. The composite rod is sintered using heat and pressure to become a cylindrical rod approximately a meter in length. Following annealing, the glass rod is sliced into 0.5-mm sections. The slicing is performed with diamond saws that produce plane parallel faces on each sieve. Following sawing, the sieve blanks are polished and finally etched to produce an optically clear transparent sieve. Since the polishing of the wafers

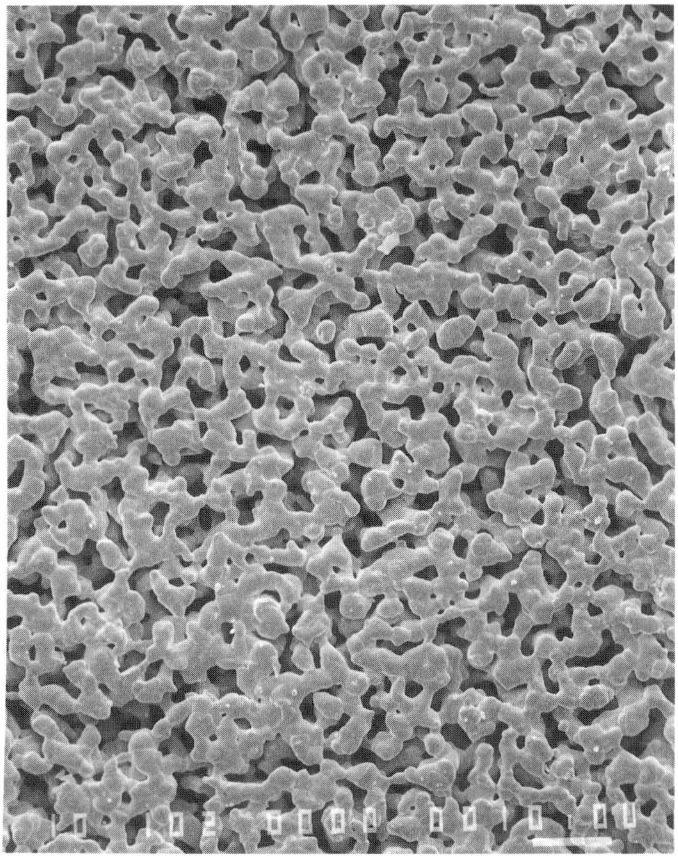

Figure 7 1000× SEM of a 0.8-μm silver depth filter showing the range of channel diameters and tortuous channel structure that contribute to particle trapping. These characteristics combine to preclude its use for analytical particle size separations.

precedes the etching step, the channel ports are fully supported during sawing and polishing steps and thus emerge after polishing as nearly ideal circles.

The optical clarity of the polished sieve permits the use of any available microscope lighting technique to size and identify the sieved particles. Although simple transmitted illumination is usable for many applications, the use of phase contrast and dark-field illumination improves the contrast with which low-contrast particles can be detected. When a 10× objective is used there are 2000 to 5000 accurately dimensioned channels in each field of view. Since the sieved particles are always seen in close proximity to the channel port, accurate sizing by comparison to channel diameters is convenient and possible. Figure 1 shows an entire sieve showing its hexagonal building blocks and the clear glass surround. Figures 2, 3, and 4 are diagonal SEM sections through the center of 5-, 10-, and 25-μm sieves. These show the right cylinder structure of the individual channels and the circularity of the channel ports. Magnification of 3000× has been used for the 5-μm sieve in Fig. 2 and the 10-μm sieve in Fig. 3. The optical flatness of the face of the sieve can be seen in both of these figures. The diagonal section through the 25-μm sieve in the SEM of Fig. 4 is magnified 200×. In Fig. 4, the straight parallel lines of each face of this sieve are testimony to the plane parallelism that has been achieved in the diamond wheel sectioning and the following optical polishing.

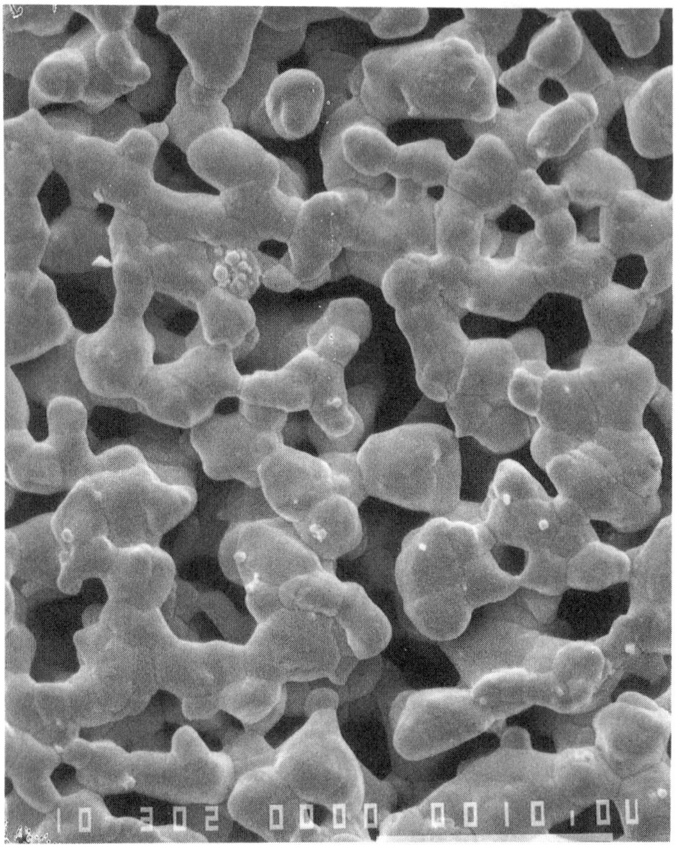

Figure 8 3000× SEM of a 0.8-μm silver depth filter showing more clearly the range of channel diameters and tortuous channel structure that contribute to particle trapping. These characteristics combine to preclude its use for analytical particle size separations.

Channel accuracy is maintained within ±5% for these sieves courtesy of the close tolerances required for usable fiber optics and the sophisticated servo-controlled apparatus with which they are made. The sieves have been manufactured in response to particle separation requirements and are available in 2-, 5-, 10-, 15-, 25-, 50-, and 90-μm sizes [10]. The open area of these sieves is maintained in a range of 15% to 20% except for the 2-μm sieve in which the open area is somewhat greater than 5%.

These glass sieves have been available for approximately four years [11,12] and have been applied to a scattering of problems rather than the U.S.P. particle assay for which they were first envisioned. A major holdback has been the cost of these sieves. Production of the sieves is still at the prototype quantity level at which they were first introduced. A sufficient production level to justify the capital investment required to reduce unit price by 10-fold has not yet appeared. One problem has been the lack of adequate sieve holders and an application methodology flexible enough to facilitate either microscope or particle counter readout of the trapped particles. These lacks are addressed in this chapter.

To obtain accurate single-sieve particle separations, the sieve must be clean and held leak free to the bottom of a clean reservoir during the sieving operation. Following sieving, the sieve must be capable of removal, without contamination, from its holder for viewing under a micro-

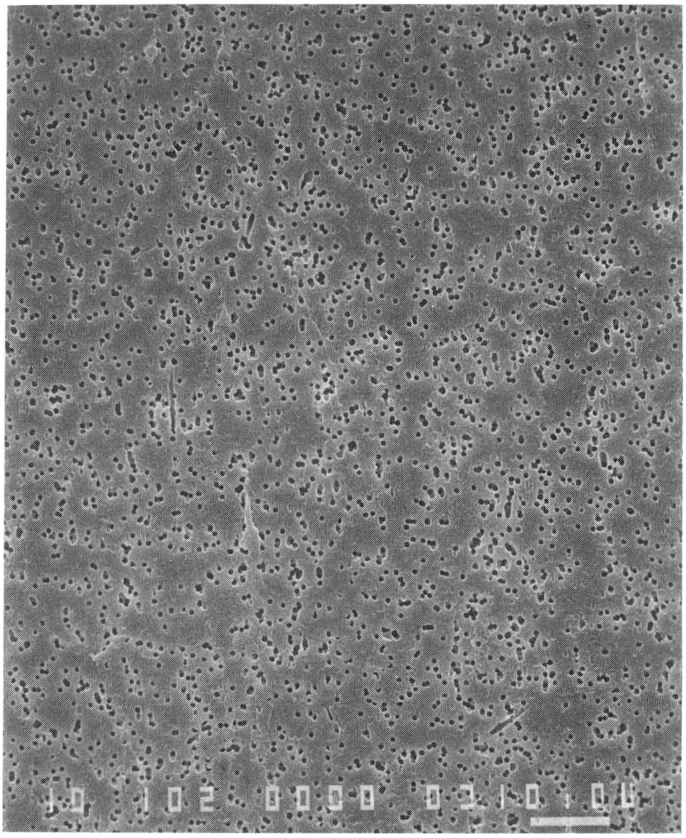

Figure 9 1000× SEM of a 0.8-μm polymeric surface filter showing the range of channel diameters that result from the random strikes of energetic particles. The random targeting characteristic of the energetic source leads to coincidental strikes that occasionally overlap. Coincidental strikes are a concomitant of achieving acceptable levels of flow.

scope. Since the particle-sizing step has been eliminated, analysis has been reduced to the counting of particles that have been sieved from the sampled volume. As an alternative, the trapped particles can be resuspended in an accurate volume of water and counted with a particle counter.

The requirements for accurate results with the sieves listed in priority order are:

1. A clean working area at least the equivalent of a pharmaceutical laboratory
2. Pressurized clean water source
3. Clean parts
4. Particle-contamination-minimized liquid seals
5. Surface smoothness of all parts in contact with the particle suspension
6. Low operating pressure across the sieve.

Contrary to most opinions, a laminar flow hood is not an essential requirement for success in accurate particle measurements. The loosely defined working conditions in requirement 1 have been found satisfactory.

The realization of a jet of pressurized ultraclean water source is, by now, a garden variety problem. Commencing with WFI in a clean pressurizable stainless steel tank, a control valve is

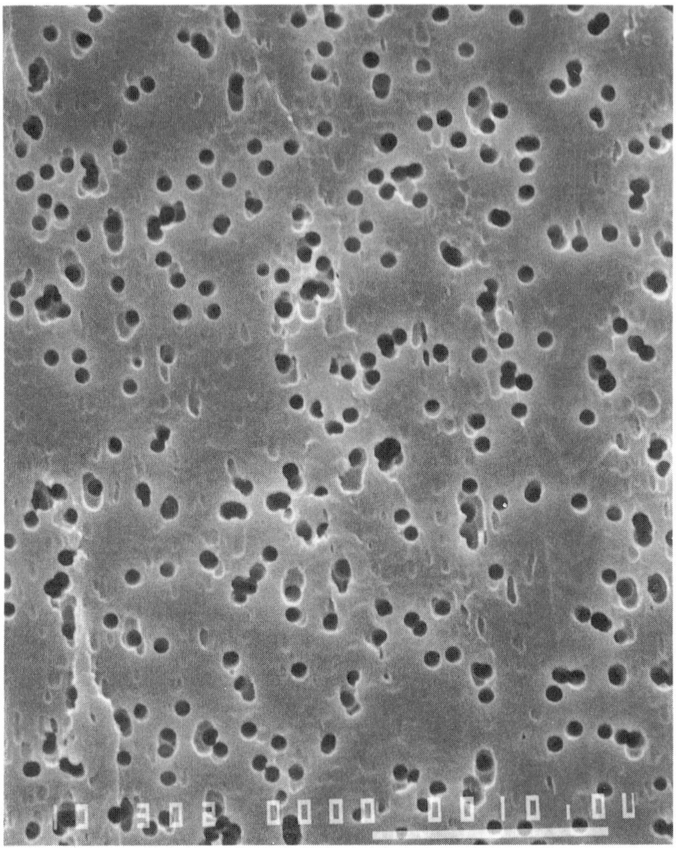

Figure 10 3000× SEM of a 0.8-μm polymeric surface filter showing the range of channel diameters that result from the random strikes of energetic particles. The random targeting characteristic of the energetic source leads to coincidental strikes that occasionally overlap. Coincidental strikes are a concomitant of achieving acceptable levels of flow.

connected via flexible tubing to the tank and to a terminal 0.2-μm filter rigidly connected to a jet nozzle. A methodical evaluation of the liquid pressures and jet velocities required for good glassware cleaning can be found in Stowers [14]. For the pressurized jet cleaning of glassware and sieves that is essential if a low baseline contamination is to be achieved, the 25 P.S.I. pressure limit permissible under OSHA is a good choice. The velocity and force of the jet employed remain as a variable. A stub 14 gauge hypodermic needle (approximately 0.067 in. inside diameter) with rounded edges provides a satisfactory cleaning jet. The requirement for a rigid connection from the filter output to the hypodermic needle is achieved with a disposable hypodermic 0.22-μm filter. The rigid connection between the filter output and the jet nozzle avoids the problem of particle detachment transients that occur when tubing is flexed or agitated.

An understanding of one experimental reality is required if adequately clean parts are to be obtained. Absolute freedom from particle contamination appears to be a dynamically unstable phenomenon. To achieve an acceptable contamination baseline approaching a concentration of zero particles per milliliter for the use of the sieve holder in a test, the total energy expended in cleaning the part must be higher than that which will be employed during the test.

Maximum parts cleanliness for many is associated with use of an ultrasonic bath. Too often the bare assertion that an ultrasonic bath has been used is taken to mean that the component immersed in the bath will be clean. The commonly used low-powered ultrasonic laboratory baths operate around 20kHz. The energy to clean the components immersed in the bath is generated by producing cavitation bubbles in the bath. The implosion of these bubbles produces the desired cleaning action. In the recent past, damage to integrated circuits from the high energy-levels produced by implosion of the cavitation bubbles has been reported. The possibility that container surface damage can occur is a matter of concern. This type of damage is accelerated by the use of harsh cleanup agents. Any surface roughness increases the difficulty of cleaning the parts used in sieving and resuspending the trapped particles. The use of cleaning compounds increases the possibility of container surface damage. The achievement of maximum cleanliness of parts in an ultrasonic bath requires the addition of a recirculating filter system to scavenge the particles detached with the ultrasonic energy since, in the limit, part cleanliness cannot be better than the condition of the water in which it is cleaned. Other considerations that must be addressed include the amount of ultrasonic energy available in the bath and the nonuniform distribution of energy in its volume. An effective demonstration of the variability of ultrasonic energy in the usual general purpose ultrasonic bath can be demonstrated with a single thickness of household aluminum foil. The sheet of aluminum foil is inserted vertically into the bath liquid through the center off the bath. Exposing the foil for about 1 minute to the ultrasonic energy will produce a pattern of punctures in the test sheet physically mapping the relatively small volumes within the bath that can be effectively used for cleaning. In the event that ultrasonic cleaning must be employed, this mapping procedure localizes the bath volumes that should be used.

The effect of both the power available in the ultrasonic bath and its localized distribution within the bath must be addressed by a validation of the reduction in contamination achieved with the bath prior to its use.

For completeness, a new improved ultrasonic cleaning capability that has as yet not made a transition to the general purpose laboratory bath is described below. In the last 10 years there has been mounting interest in a new ultrasonic (megasonic) energy region for cleaning purposes. This ultrasonic energy region is around 800kHz. The cleaning action in this higher frequency ultrasonic region is produced by short sonic pressure waves that are energetic enough to detach particle contamination from a surface but does not posses the destructive potential of the implosive force generated by the cavitation bubbles at 20kHz. Cleaning capability tests were conducted with relatively clean bare silicon wafers. Menon [16] reported that wafer surface cleanliness levels reached down to levels not previously obtainable. Cleaning efficiency better than 90% was achieved with optimized power and cleaning solutions.

As an alternative to ultrasonic cleaning, pressurized jet cleaning is simple and attractive. Cleaning a container commences with positioning it so that the wash liquid can drain out easily. Tilting the container 20°–30° below the horizontal is satisfactory. Use of the terminally filtered water jet described above applied in a continuous spiral, first to the container bottom and then to the container walls, can produce satisfactory results. The use of laboratory cleaning fluids such as Micro or household cleaning fluids such as Sunlight as an initial cleanup step, followed by pressurized jet cleaning has been found to speed the cleaning process. The adequacy of the cleaning efforts should be checked by filling clean containers with terminally filtered water and then checking the contamination baseline with a particle counter. Fortunately, low-particle-shedding "O" rings are an essential requirement in high-performance hydraulic servo systems and are available at relatively low cost. The use of these "O" rings on each side of the glass sieve produces safe, leak-free seals.

An essential requirement for accurate particle size separation with these precision glass sieves is a sieve holder whose surface offers minimal trapping sites for the broad range of particle chemistries that can be encountered. This requirement is basic to the achievement of a sieve holder whose wetted surface can be cleaned to the required degree. It is also basic to the accurate microscope examination of all the trapped particles in the suspension that have been trapped on the sieve. When resuspension and electronic counter analysis is planned, assurance that the trapped particles can be gently resuspended from the sieve-holder surface in a selected volume of liquid is a prerequisite for analytical accuracy. The use of lathe turning and polishing to fabricate the contours and surfaces of the sieve holder produce surface finishes that are at best 10 to 100 times rougher than the best molded or extruded parts. Avoidable surface roughness increases the formidable obstacles that are in the way of validating high-accuracy sieving methodology. Figure 11 is a comparison of the 32-microinch surface roughness achieved in a machining operation versus the specular reflection showing the 0.1-microinch optical smoothness in an extruded plastic surface.

Use of parts molded in well-finished molds results in contoured wetted surfaces for the sieve holders that can be cleaned to the necessary degree. Due to the surface smoothness of these sieve holders any particles left on the walls can be easily rinsed down onto the sieve surface for further analysis. For an effective validation, the sum of the particles trapped on the sieve and the sum of the particles that have been sieved through the sieve provide a count balance equal to the total of the initial particles.

Examination of surface roughness among the various chemically and dimensionally stable plastic piping materials used in high-purity water systems eliminated the use of polypropylene and PVC. The selection of Halar versus Kynar and PVDF was based on comparison of SEM and atomic force micrographs. Figures 12–14 are 1000× SEMs of these three materials at a 20° tilt angle taken with an SSL of 3.0 kV. Visual evaluation of the surface roughness shown in these micrographs lists Kynar (Fig. 12) as the least desirable, then PVDF (Fig. 13), and Halar the smoothest (Fig. 14).

In each of these micrographs, the surface roughness of a 100 × 100 μm area has been investigated. Using maximum surface roughness as a figure of merit, Kynar in Fig. 15 and PVDF in Fig. 16 are approximately equal. The maximum surface roughness of Kynar is 0.92505 μm and that of PVDF in Fig. 17 is 1.082 μm. In Fig. 17 is a similar AFM of a Halar area with maximum surface roughness of 0.24344 μm, a reduction in roughness of approximately fourfold. The atomic force micrograph of Halar in Fig. 17 supports with numerical data the visual evaluation of

Figure 11 Extruded and machined pipe section showing a comparison between a 32-microinch surface finish and the 0.1-microinch finish on the inner surface of an extruded Halar pipe section.

Figure 12 1000× SEM of a Kynar sample taken with a 20° tilt to emphasize surface roughness.

the Fig. 14 SEM of Halar as the best choice available. The evaluations of Kynar and PVDF, which appear to differ in surface roughness in the SEMs of Figs. 12 and 13, are seen to have approximately equivalent surface roughness but differ in the surface characteristic achieved.

Flow through the sieve is initiated with a negative downstream pressure to ensure freedom from pressure source contamination. The use of a minimum negative operating pressure to initiate sieving reduces, but does not eliminate, the impact and adhesion of particles during the sieving operation. The probability of particle attachment to the container walls and onto the sieve in the lands between the channel ports occurs in unidirectional liquid flow even when minimum sieving pressure is employed. Separation probability increases when tangential liquid flow is established during the sieving operation. In lieu of tangential liquid flow during sieving, a postsieving rinse-down can be employed.

To achieve accurate particle size separation, particles attached to the walls are rinsed down onto the sieve surface prior to microscope examination. Any attached fines on the sieve surface can be rinsed through the sieve ports with a capillary-sized clean-water jet stream using a rinse volume of three to five times the original sample volume. The volume of rinse liquid required is a function of the wetted surfaces of the sieve assembly and the type of particulates that are in the suspension. The volume employed should be validated prior to use.

Figure 13 1000× SEM of a GF-PVDF sample taken with a 20° tilt to emphasize surface roughness.

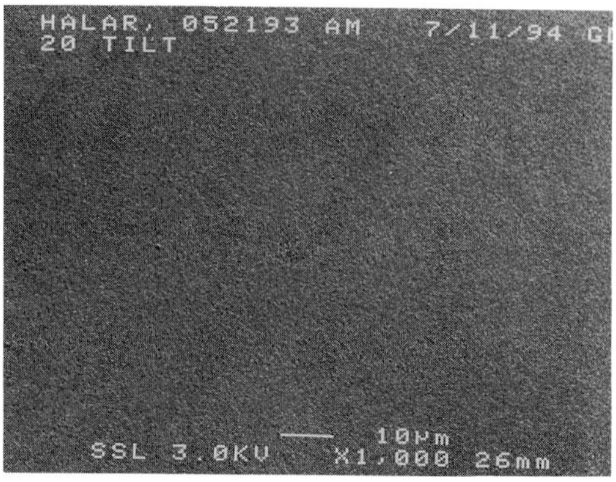

Figure 14 1000× SEM of a Halar sample taken with a 20° tilt to emphasize surface roughness.

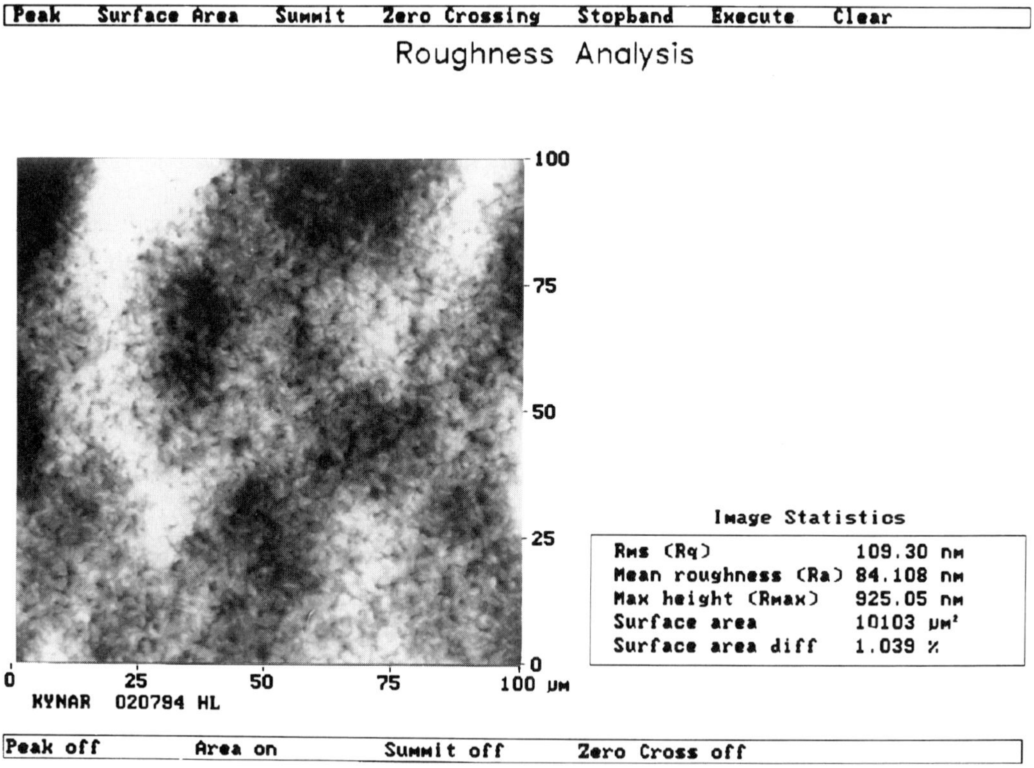

Figure 15 900× AFM of a 100 × 100 μm square Kynar sample taken with a 20° tilt to emphasize surface roughness. The maximum roughness height is 0.92505 μm.

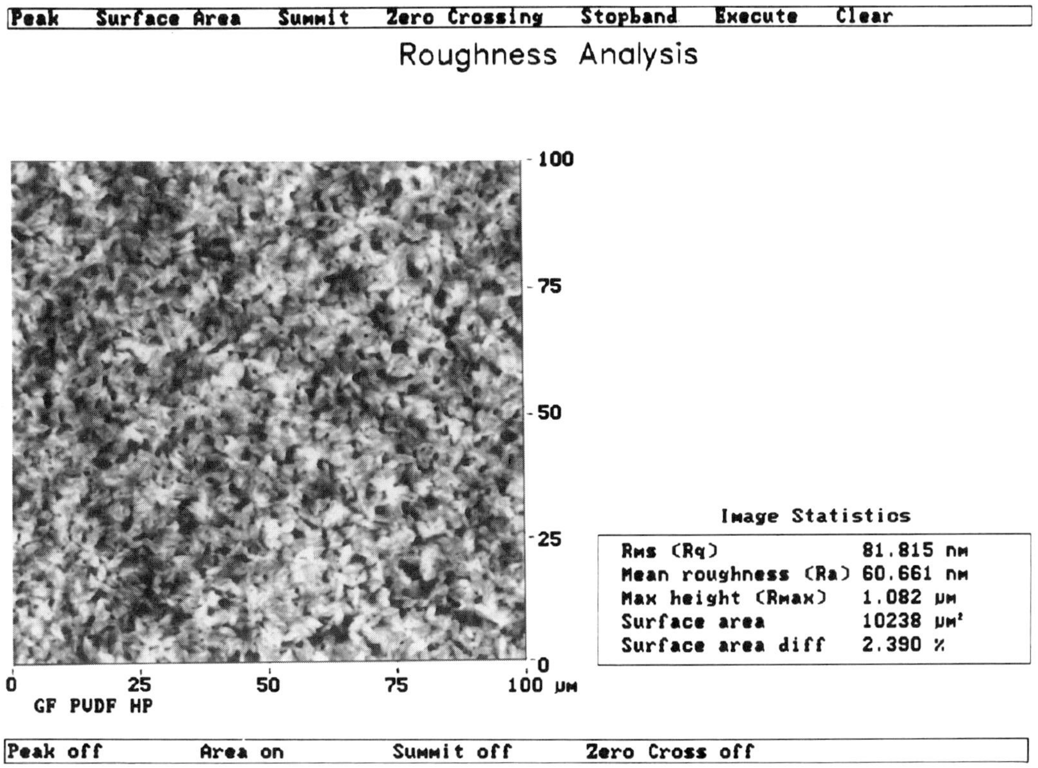

Figure 16 900× AFM of a 100 × 100 μm square GV-PFDF sample taken with a 20° tilt to emphasize surface roughness. The maximum roughness height is 1.082 μm.

Using a 10-ml syringe at full scale on a light extinction particle counter provides the accurate volume and pressure required. A terminally filtered 20-gauge hypodermic needle with rounded end can be used to produce a low-force water jet for rinse-down purposes [15]. The 20-gauge needle produces a jet diameter of 0.024″: this jet stream is strong enough to detach the particles from the sieve surface and to minimize the possible loss of particles by forcing them through the sieves channels. The use of a rigidly connected terminal filter is essential if particle detachment transients are to be avoided. The use of plastic tubing to connect the terminal filter to the jet will introduce transient particle shedding at each movement of the plastic tubing.

Unplanned evaluation of the durability of these sieves occurred in the development of the washing and handling technique. The sieves survived dropping approximately 2 feet onto a stone chemical sink surface and a drop over 3 feet onto a linoleum-covered laboratory floor. With repeated use the capability of the sieve to survive shocks will be reduced. This reduction in shock survival capability is due to surface crazing of the sieve which increases with repeated use. With reasonable care, handling and adequate cleaning, the sieves can be reused indefinitely. The channel length in the sieves maintains particle separation accuracy during its lifetime.

A disassembled single sieve on the sieve removal fixture is shown in Fig. 18. Following sieving, the sieve is disassembled from its holder and transferred to a microscope stage with a soft-surfaced tweezer for visual identification and counting of the trapped particles. Visible is the sieve between its cushioning "O" rings resting on the disassembly fixture. The recess for the sieve is visible just below it; the sieve reservoir and the assembly clamp are visible at the right.

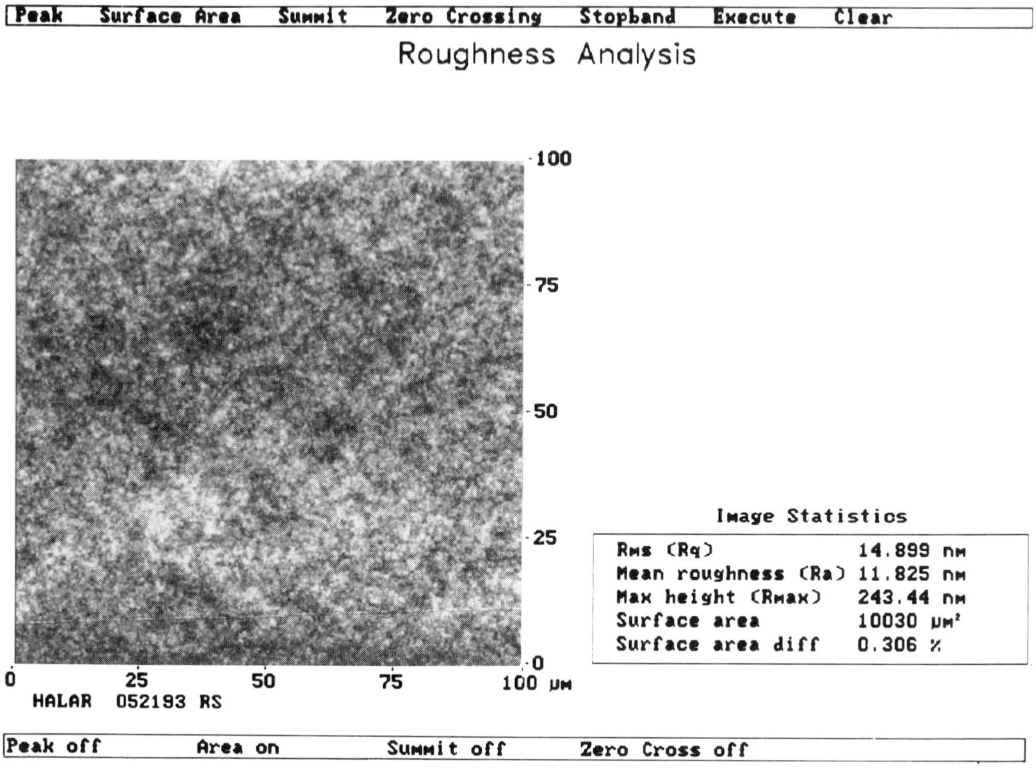

Figure 17 900× AFM of a 100 × 100 μm square Kynar sample taken with a 20° tilt to emphasize surface, roughness. The maximum roughness height is 0.24344 μm, approximately one-fourth the maximum roughness for either Kynar or GV-PFDF.

Figure 18 Glass sieve assembly components and a disassembly fixture.

Figure 19 repeats the group of three parts from Fig. 18 and includes the assembled sieve mounted on a Corning 500-ml sterile disposable filter system receiver (Corning #25942 or any equivalent which uses a 6° taper). Visible in the far left is the sieve-removal post shown in Fig. 18.

Figure 20 shows a two-tandem-sieve assembly that could be used for a U.S.P.<788> contaminating particle assay. The upper sieve is a 25-μm sieve; the lower contains a 10-μm sieve. The Corning disposable filter system receiver provides a stable, inexpensive, readily available base. Following sieving and prior to the separation of the two sieve assemblies, the 25-μm assembly is rinsed down, as described in the manual microscope assay paragraph, to eliminate the smaller particles from the analysis. Resuspension of the trapped particles is then accomplished. Following separation, the 10-μm sieve assembly is similarly rinsed down prior to resuspension and analysis.

Sieving with these glass sieves is accomplished in a matter of seconds. Sieving speed facilitates examination of the full contents of a parenteral container. This is particularly advantageous when an alert level is triggered by the usual quality assurance routines. An early practical experimental result of using a single 10-μm sieve for the U.S.P. contaminating particle analysis is shown in Fig. 21. The results shown here are a 40× visible light micrograph of a 10-μm sieve. The trapped fibers are sheared plastic slivers about 5 μm thick and are respectively 68, 165, and 280 μm in length. Figure 22, also from early work, shows a 10-μm sieve with three trapped particles that should be counted in the 10- to 25-μm range: 13, 20, and 24 μm in maximum dimension, and three particles between 5 and 10 μm in length that remained due to an inadequate rinse-down technique.

Figure 23 shows a 400× examination of an amorphous residue in an out-of-date injectable local anesthetic trapped on a 10-μm sieve. The relatively small amount of this material was detected by sieving the total contents of the 50-ml product containers. This microphotograph was recorded during a three-day evaluation of this sieve at the F.D.A. Minneapolis Methods Development Laboratory.

An alternative to microscope analysis of the trapped particles is resuspension and counting on a light extinction particle counter operated within its defined accuracy limits. The achievement of accurate particle size separations is limited by the extraneous contamination that can result from any sieve disassembly and transfer operation. To minimize contamination, resuspension is performed by a jet rinse-down of the sieve and the wetted surface of the sieve holder, as described above, into a precleaned and tared beaker.

For accurate counting results from the resuspended particles, the volume of suspending

Figure 19 Single-sieve assembly mounted on a vacuum receiver; the individual sieve parts are shown on the right; a disassembly fixture is shown on the left.

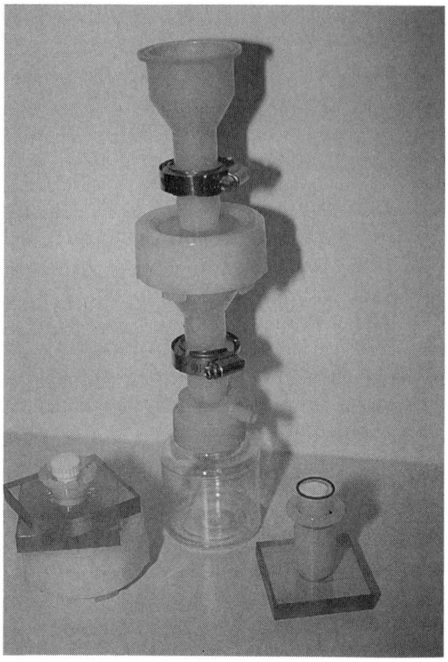

Figure 20 Center is a two-sieve assembly that could be used for a U.S.P. XXIII <788> contaminating particle assay. The individual sieve parts are shown on the right; a disassembly fixture is shown on the left.

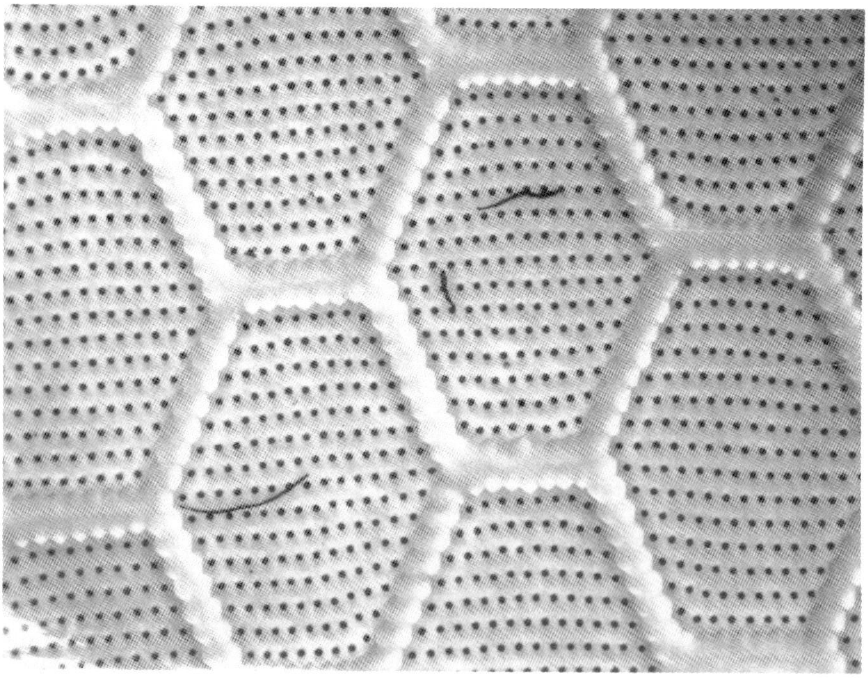

Figure 21 A 40× white light microphotograph of sheared plastic slivers trapped on a 10-μm sieve. The slivers are approximately 5 μm thick and are 68, 165, and 280 μm in length.

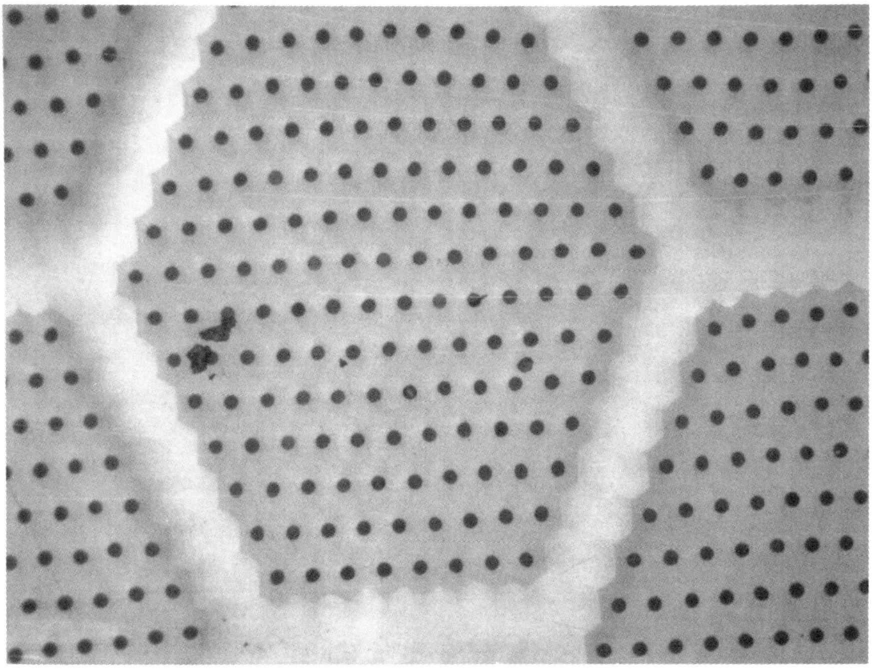

Figure 22 100× microphotograph of particles trapped on a 10-μm sieve. Six particles can be seen, three of which should be counted in the 10–25 μm range (13, 20, and 24 μm in maximum dimension). Three particles between 5 and 10 μm can also be seen and are due to an inadequate washdown technique.

liquid must be accurately established. This is accomplished by accumulating the rinse-down liquid in a tared beaker and then adjusting the initial volume to some simple number. Adequate transfer of the trapped particles is tested following the initial particle transfer by a similar washdown and counting the washdown liquid. For U.S.P. XXIII <788> particle contamination assays, the weight contribution of the random particle contaminants can be neglected and the aliquot is calculated on the basis of the initial gravimetric determination of the water volume. Transferring the trapped particles to a beaker permits the use of the Teflon-cased magnet for stirring and the usual external magnetic field stirring driver.

The choice of washdown and transfer to another container for analysis of the resuspended particles appears to disregard the possible use of the upper surface of the sieve with which the resuspension volume could be defined. From filter theory, a positive downstream pressure, say 10% below the sieve bubble point, should block liquid flow through the sieve. As a result, the addition of an accurate amount of well-filtered water to the sieve holder reservoir provides the volume data with which the total resuspended particle count could be evaluated from a sampled aliquot. Unfortunately, adequate stirring of the resuspension liquid in the present holder has, as yet, not been achieved, leaving this desirable option for future work.

The U.S.P. has consistently used two particle size zones for instrumental evaluation of the degree of contamination of injectable products. The first zone extends from a lower boundary of 10 μm to an upper limit of 25 μm. The second zone commences at 25 μm and continues until the visible particle zone is entered at approximately 100 μm. Concern with patient insult has resulted in a restriction to one-tenth of the acceptable concentration of contaminating particles greater than 25 μm compared to the acceptable concentration of particles within the 10- to

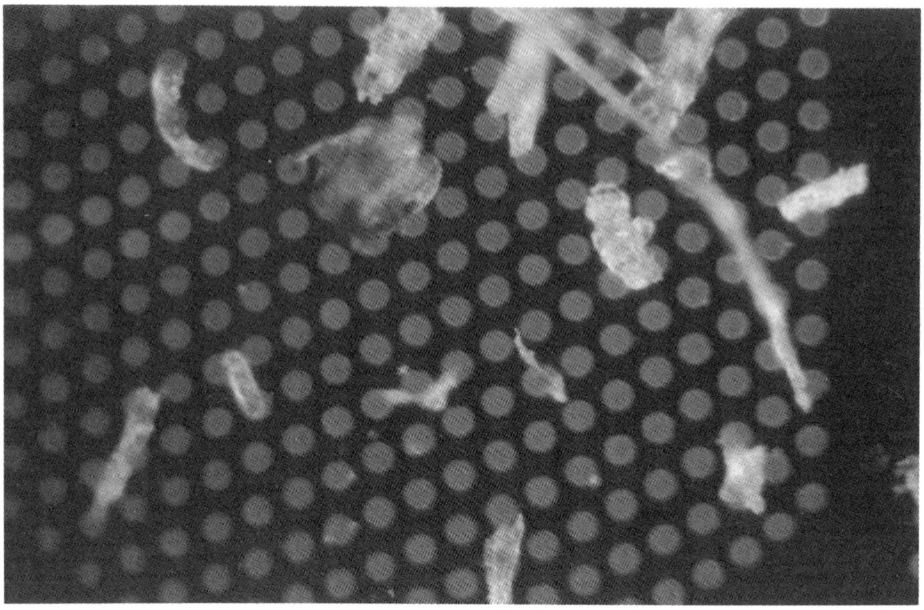

Figure 23 400× microphotograph using phase contrast lighting to investigate product related amorphous particles trapped on a 10-μm sieve. This material was present in a container of out-of-date injectable local anesthetic.

25-μm zone. With these particle size zones defined it would appear self-evident that two transparent glass sieves in tandem are all that is required to achieve a U.S.P. contaminating particle analysis consistent with normal pharmaceutical measures.

Realization of this self-evident referee analysis scheme rests on the use of a tandem arrangement of two sieve assemblies as shown in Fig. 20. The 25-μm sieve is in the upper assembly and the 10-μm sieve is in the lower assembly. The central assembly is a cascade of two precision glass sieves. For the analysis of the U.S.P. XXIII <788> specified particle size ranges in parenteral products, the upper assembly houses a 25-μm sieve, the lower assembly houses a 10-μm sieve. The two foreground assemblies show disassembled sieves awaiting removal and microscopic evaluation.

The two sieve assemblies are mounted on a 500-ml disposable microbiological filter unit base which is used as a receiver for the sieved liquid. The separate mounting of each sieve facilitates an improvement in the accuracy of particle size separation by minimizing retained fines in each sieve assembly. Following completion of the sieving operation, the 25-μm sieve assembly is rinsed prior to separation from the downstream 10-μm sieve assembly. After separation, the downstream 10-μm unit is similarly rinsed prior to any further analysis.

The design and fabrication of these sieves originated as an alternative to the inadequate particle contrast achieved in the lighting specified in the U.S.P. <788> membrane/microscope contaminating particle assay. The inherent advantages offered by these sieves include the following: (1) the improvement of particle contrast for the low-contrast particles normally encountered in parenteral products. Any particle that can be detected on a microscope slide can be securely detected on the optical glass surface of the glass sieve. (2) The mechanical separation of the major portion of the particles less than 10 μm present eliminates introduction of false counts by coincident appearance in the scan beam of the counter or from manual microscope evaluation. (3) The rinse-down also eliminates from consideration the false counting of immiscible droplets whose elasticity facilitate transit through the 10-μm channels.

Resuspension and counting of trapped particulates by particle counters has been shown to result in improved particle count determination accuracy in varied assays. This assay improvement technique has been utilized by Shelton [17] and Webb [18] in the determination of silicon wafer cleanliness and by Allen [13] in the control of the particle size distribution in the manufacture of films. Allen [13] improved the statistical reliability of quality assurance limits for a film product in which control of particles whose occurrence rate was determined to be 40 parts per million particles was required.

Isolation of the upper tail of this distribution facilitated the generation of statistically replicable quality assurance data. The particles were chemically and dimensionally stable in the suspension liquids employed. Referee quality optical sizing, identification and counting of the trapped particles were performed using a microscope. This achievement was made possible by the optical transparency of the sieves. The filter holder employed was machined from Delrin stock.

Automated counting and sizing of the particles with a Coulter counter followed particle dispersion in an accurately determined volume. Allen used high-powered ultrasonic pulses to ensure resuspension of the particles trapped on the walls and on the sieve. Allen remarks that 30-second stirring is generally adequate for resuspension of flocs, weakly bonded aggregates, or BCR calibrating quartz powders. He goes on to say that resuspension of titanium dioxide would require 5 minute (high-powered) ultrasonic probe exposure in a water-cooled container for effective dispersal. Fortunately, most pharmaceutical powder contaminants are in the weakly bonded category.

The possibility of accurately comparable particle counts of contaminating particles between individuals was investigated in a three-day pilot study conducted in the FDA Methods Development Laboratory in Minneapolis, Minnesota. The total contents of 50-ml containers of an out-of-date injectable local anesthetic were sieved through a tandem arrangement of the 25- and 10-μm sieves. As a variable vacuum source was not available, the full house vacuum pressure (25–28 Hg) was used. The sieved particles on each sieve were then counted by two observers. Two measurement sessions were completed.

In the first session it was determined that the use of a phase contrast microscope using the 10× microscope objective and ocular combination specified in U.S.P. <788> for the assay of contaminating particles was less stressful than direct transmitted illumination. Using phase contrast microscopy, the counts for the 25-μm sieve were 1249 ± 5.9%. The number of particles on the 10-μm sieve was too numerous to count. Twelve randomly selected hexes were chosen by each observer. The results were 158 ± 12.7% counts on the counted hexes. To assess the effect of each observer's choice of the hexes to be counted, 12 identical hexes were counted by each observer. The results were 130.5 ± 2.7% counts.

On the second trial day, tandem sieves were again used. The counts for the 25-μm sieve were 356.5 ± 7.7%. Examination of the 10-μm sieve again showed that the results were too numerous to count. To separate data variability introduced by the counting of different selected fields, 12 identical hexes were counted by each observer. One additional refinement was added: any particle touching a hex border would be counted by each observer. The results of this last, better-defined trial were 173.5 ± 0.8% counts. This observation is an indication that simply ensuring that the same particle fields were examined by each observer can reduce the presently observed data accuracy range.

Early verbal reports of ±30% interpersonal variability with the new improved U.S.P. microscope assay show that minimum usable accuracy has been obtained. This achievement is in contrast to the FDA evaluation of the lack of usable accuracy with the preceding membrane/microscope assay. The last experimental determination clearly indicates that the use of good particle contrast in a well-defined procedure can result in test accuracy approaching the usual compendial accuracy range.

The use of electronically recorded video images of the mechanically sorted particles can eliminate the individual variability inherent in particle size judgment. Video images provide a means to bypass the 5% contrast limit of the human eye and provide accurate particle contamination data on which to firmly base process changes that will lead to product improvement.

To date, the major applications of these sieves have included the achievement of secure statistical analysis of the upper tail of particle size distributions and as an accurate flow resistance device for the calibration of subambient pressure flow regimes. Their potential for the accurate determination of the contaminating particle content of injectable products is still to be realized.

ACKNOWLEDGMENT

The author expresses his thanks to T. A. Barber and associates for the Figs. 2 through 10, to Aussimont U.S.A. for Figs. 12 through 18, and to the Minneapolis F.D.A. Methods Development Laboratory for Fig. 23. Thanks are due to Collumated Holes, Inc., of Campbell, California for the precision glass sieves reported upon in this chapter.

REFERENCES

1. T. E. Munson, F.D.A. perspectives on particulate matter testing in small volume injections, Proc. Int. Conf. on Liquid-Borne Particle Inspection and Metrology, 1987, p. 827.
2. J. Z. Knapp and L. R. Abramson, A new coincidence model for single particle counters. Part I: Theory and experimental verification, *J. Pharm. Sci. Tech. 48*(3):110–134 (1994).
3. J. Z. Knapp, A. Lieberman and L. R. Abramson, A new coincidence model for single particle counters. Part II: Advances and applications, *J. Pharm. Sci. Tech. 48*(5):255–292 (1994).
4. J. Z. Knapp and L. R. Abramson, A new coincidence model for single particle counters. Part III: Realization of single particle counting accuracy, *J. Pharm. Sci. Tech.,* in press.
5. B. H. Kaye, Operational protocols for efficient characterization of arrays of fine particles by robotic image analysis systems, *Particle Size Distribution.* II: *Assessment and Characterization* (Theodore Provder, ed.), American Chemical Society Symposium Proceedings, Series 472, 1991, chap. 23.
6. United States Pharmacopeia, XXIII, United States Pharmacopeial Convention, Inc., Rockville, MD, 1995.
7. John G. Delly, Parenteral solutions—problem of sizing particles using USP methodology, *Pharmacop. Forum,* Stimuli to the Revision Process, p. 359, July–August 1980.
8. H. R. Blackwell, Contrast thresholds of the human eye, *J. Opt. Soc. Am. 36*:624 (1946).
9. H. R. Blackwell, Development and use of a quantitative method for specification of interior illumination levels on the basis of performance data, *Illum. Eng., LIV*:317 (1959).
10. Collimated Holes, Inc., 460 Division St., Campbell, CA 95008.
11. J. Z. Knapp, T. A. Barber and R. W. Mead, Transparent precision sieves in micrometer sizes: new answers for old problems, Measurement Science Conference, Disneyland Hotel, Anaheim, CA, 1991.
12. J. Z. Knapp, R. W. Mead, Accurate sorting of micrometer-sized particles, Interphex U.S.A., 1991 Technical Proceedings, pp. 193–202.
13. T. Allen, Application of precision transparent sieves to the determination of low numbers of oversized particles in powders, *Part. Part. Syst. Charact. 9*:252–258 (1992).
14. I. F. Stowers, Advances in Cleaning Metal and Glass Surfaces to Micron Level Cleanliness, *J. Vacuum Sci. and Tech. 15*(2):751–754 (1978).
15. K. Skidmore, Cleaning Techniques for Wafer Surfaces, *Semiconductor International 10*(9):80–87 (1987).
16. V. B. Menon, A. C. Clayton, and R. P. Donovan, Removing Particulate Contaminants from Silicon Wafers, *Microcontamination 7*(6):31–36 (1989).
17. D. Shelton, R. Webb, and J. G. Eleftherakis, Particle Counting Using the Particle Resuspension Method, Proc. Am. Filt. Soc. Produced Water Conf., Houston, January, 1995.
18. R. Webb and J. G. Eleftherakis, Particle Counting Using the Particle Resuspension Method, Filtration News, p. 28, July/August 1995.

18

Particle Size Distribution Analysis Using Light Scattering

Renliang Xu

Coulter Corporation, Miami, Florida

I. INTRODUCTION

Particulate materials exist everywhere. They are the building blocks of modern society and are used everyday in our daily life, industrial production, and academic research. In the United States alone, annual industrial production related to particulate materials was over \$1 trillion of market value in 1993. The characterization of particulate materials is vitally important because its results determine and influence other portions of the life cycle of those materials, such as manufacturing, storage, transportation, and ultimate application. For small particles with dimensions of a millimeter or less, the uniqueness of the huge numbers of particles per unit volume or per unit weight (for example, 1 cm^3 contains only one cubical particle 1 cm long but 1 million particles 100 μm long and 1 trillion particles 1 μm long) make the methods of characterization completely different from the methods used for measurement or characterization of macroscopically chunky or bulky materials.

We can classify the analytical methods used in particle analyses into ensemble methods and nonensemble methods according to whether each measurement detects or senses a material property from an ensemble of particles in the sample. For nonensemble methods, materials have to be separated or fractionated into separate components according to a certain property of the material prior to the measurement. All nonensemble methods are comprised of a separation or fractionation mechanism and, depending on the completeness of the separation or fractionation, the measurement may detect or sense only one particle at a time or a group of particles having the same property value according to how they are separated or fractionated. Two typical methods used to size one particle at a time are the optical particle counter and the Coulter counter. Technologies such as sedimentation, size exclusion chromatography, or field-flow fractionation are really methods of fractionation because additional detection schemes, which are often completely different technologies, are still needed to complete the measurement. Typical ensemble methods for particle size determination are photon correlation spectroscopy and elastic light scattering. There are advantages and disadvantages for both ensemble methods and nonensemble methods. They are often complementary to each other. The advantages of ensemble methods, such as broad dynamic range and fast, nonintrusive measurement, etc., are just the deficiencies of nonensemble methods, namely narrow dynamic range, time-consuming analysis, and sample destruction. Opposed to the disadvantages of ensemble methods, such as low resolution, nonensemble methods have the advantage of high resolution. Thus, the choice of a proper analytical method for particle characterization fully depends on the requirement of the application and accessibility to a suitable analytical technique. Users often have to make a compromise when choosing the best method for their particulate materials.

As a general term, light scattering is related to several branches of physical sciences such as optics, physical chemistry, material sciences, fluid dynamics, and solid physics. There are many technologies that use light scattering phenomena to study various properties of materials. Each of these technologies has its own terminology, expertise, applications, and techniques of measurement, and involves different disciplines in physical science. For measurements performed on the scattering intensity fluctuation as a function of time there are photon migration measurement, transient light scattering, photon correlation spectroscopy, and diffusive-wave light scattering; there are static light scattering and turbidity spectroscopy for measurements performed on time-averaged intensity; and there is phase Doppler analysis for phase analysis of the scattered light. Depending on whether the frequency of the scattered light to be detected is the same as the incident light the scattering may be described as elastic, quasi-elastic, or inelastic light scattering (Raman scattering and Brillouin scattering). For measurements performed under an additional applied field, there are electrophoretic light scattering, electric field light scattering, and force Rayleigh scattering, etc. For measurement of material surfaces or interfaces there is surface Raman scattering. In static light scattering, depending on the relative values of the light wavelength and the material dimension, there are Rayleigh scattering, Mie scattering, and Fraunhofer diffraction. It is certain that even persons with strong science backgrounds will get confused with all of these different specialties and that hardly anyone can be an expert in all light scattering technologies. Thus, before going into a detailed description of light scattering, it is important to clarify the scattering technology and describe its intended applications. Several of the above scattering technologies can be used to measure particle size. In the present text, we will discuss one of the scattering technologies that has gained wide acceptance in the past decade for sizing a broad range of airborne and liquid-borne particulate materials due to recent advancements in laser and computer technology and the subsequent development of commercially viable instruments. That technology is elastic light scattering intensity measurement based on the time-averaged angular intensity pattern scattered from dispersed substances (or particulate materials). For the sake of clarity, hereafter we will call this particular technology light scattering, in order to distinguish it from other scattering technologies that will not be discussed further.

In light scattering measurements one obtains the mean particle size and particle size distribution through a matrix conversion of scattered intensity measurements as a function of scattering angle and wavelength of light based on applicable scattering theory. This is an absolute method, free from the need for calibration or scaling. The measurement is fast and nonintrusive and can cover an extremely broad size range, spanning almost five orders of magnitude from tens of nanometers to a few millimeters.

During the past few years, light scattering has in large extent replaced the conventional sieving method for particles smaller than a few millimeters and taken the place of optical and electron microscopy for particles larger than some tens of nanometers. Due to the immaturity of the technology and lack of computing power, in the early days of light scattering sizing, only the Fraunhofer diffraction theory, which has a simple formulation, was used for calculating particle size even for particle sizes far outside the Fraunhofer theory's applicable range. For that reason, angular light scattering intensity measurement is widely known throughout industry as *laser diffraction*. This term has the advantage in that it distinguishes itself from all other scattering technologies mentioned earlier and has also been accepted by national and international particle characterization communities. However, *laser diffraction* no longer reflects the current state of the technology because presently light scattering is not limited only to simple diffraction effects and also a laser is often not the sole light source in a light scattering measurement. Today, more general approaches to the measurement of scattering intensity are employed when using this technology to size particulate materials, and a light source with continuous wavelength (a white light) is often used to complement the main laser source in order to gain more characteristic information based on the wavelength and polarization dependence of the scattering intensity. Thus, *light scattering* is a more proper term for this technology.

Light scattering is widely used in both academic research and industrial development and quality control. However, except for research articles published in academic journals and texts buried in professional monographs, there is a lack of suitable reading material to allow general users and audiences to systematically understand and master this powerful particle characterization technology [1–3]. Thus, although it is impossible to fully expound the details of light scattering technology in this short chapter, it is the author's intention to describe and introduce light scattering theories and applications in a way that a novice with a college background will be able to understand the fundamentals of light scattering as it is used as a tool in particle characterization. To that end, all derivations of the related equations, as well as many technical details, have been omitted so that readers will learn the essence of the technology without getting lost in the details. For advanced readers, this chapter will serve as a useful reference from which basic formulas of light scattering theory, the present state of the art of light scattering instrumentation, and the sources for further reading may be found. In the following text, theories for the elastic scattering of a single particle, i.e., without consideration of the multiple scattering that occurs when light strikes one particle and is then deflected again by another particle, and applications of light scattering in sizing particulate materials, which can range from solvated macromolecules to suspended particles, are included in Sec. II. Instrumentation for light scattering characterization of airborne and liquid-borne particles is descriptively introduced in Sec. III. Data interpretation is the topic of Sec. IV. A short summary is provided in the last section.

II. LIGHT SCATTERING THEORY

A. Light Scattering and Particles

Light is electromagnetic radiation in the frequency range from approximately 10^{12} Hz (infrared) to 10^{17} Hz (ultraviolet). When light propagates it has the characteristics of both a transverse wave (a light wave) and a particle (a photon). As a wave, one expects wavelike properties such as frequency, wavelength, interference, etc., and as a particle, one expects particle-like properties such as momentum, velocity, position, etc. When light illuminates a material, e. g., a particle, which has a dielectric constant different from unity, depending on the wavelength of the light and the optical properties of the particle, the light will be absorbed, or scattered, or both. In this chapter we discuss the situations where the dimension of the light beam is larger than that of the material. In these cases, descriptions from simple geometric optics, such as reflection and refraction, are no longer sufficient to describe the behavior of light. The absorbed light energy that becomes the excitation energy of the particle will be dissipated mostly through thermal degradation, i.e., converted to heat, or lost through a radiative decay producing fluorescence or phosphorescence depending on the electronic structure of the particle. When light interacts with the electrons bound in the material that reradiate light, scattering is observed. Because most materials exhibit strong absorption in the infrared and ultraviolet regions which greatly reduces scattering intensity, most light scattering measurements are performed using visible light of wavelengths from 350 to 900 nm. The scattering intensity from a unit volume that is illuminated by a unit flux of light is a function of the complex refractive index ratio between the material and its surrounding medium and is, in the regime of Rayleigh scattering, inversely proportional to the fourth order of the light wavelength: i.e., the shorter the wavelength, the stronger the scattering. In fact, the wavelength dependence of scattering power was the first observed scattering phenomenon from nature. The reason that the sky is blue at midday and red at sunrise or sunset is that one sees the scattered sunlight during daytime and sees the transmitted sunlight during dawn and dusk. The gorgeous blue color of the ocean can also be attributed in large extent to the strong scattering of the short-wavelength portion in the sunlight. The deeper the water level, the deeper the blue color becomes because of the reduced scattering from the longer wavelength portion of the sunlight. Utilizing this wavelength dependence, we use red as the color for the stop light and for all traffic control warning signs because red light has the least scattering power in the visible light spectrum, thus allowing the transmitted light to go through the fog, rain, and dust particles and reach the intended detector, in this case the human eye.

"Industrial" particles have dimensions ranging from a few nanometers to a few millimeters, from molecular-sized particle to sand grains, even though the upper end of particle size in industrial applications may extend into the range of centimeters. Within this span, particles may exist in some very different forms. Particles can be synthetic or natural macromolecules in linear forms or in networks, such as proteins, gels, DNAs, and latexes; they can also be ensembles of small inorganic or organic molecules, or even pieces of "space," such as bubbles in a foam. More typically, they may be just minute pieces of bulk materials, such as metal oxides, sugar, pharmaceutical powders, paint, or even the nondairy creamer one puts in coffee. In general, the scattered light intensity of a particle is a function of the following variables: particle dimension, particle refractive index, medium refractive index, light wavelength, and scattering angle. The scattered intensity from a particulate sample is, in addition to the above variables, a function of particle concentration and particle-particle interaction. Some of the variables are constants in a particular experimental setup, such as light wavelength, and some of these variables are constants for a particular sample, such as refractive index. In characterizing particle size using light scattering, one optimizes sample concentration to a proper range so that the sample will scatter enough intensity to enable the measurement to be completed with a desired signal-to-noise ratio, but not to scatter so much as to saturate the detecting system. Sample concentration is also optimized for minimal particle-particle interaction and minimal multiple scattering (i.e., when the light strikes one particle and is then deflected again by another particle) so that the measurement is performed based on elastic single-particle scattering. In addition, in a light scattering measurement one has to make the assumption that the refractive index and density of particles in the sample are uniform, which is true for most particulate systems. Thus, the scattered intensity is only a function of scattering angle, particle shape, and particle size. If the relations between scattering intensity, scattering angle, particle shape, and particle size are known, one is able to resolve size distribution for particles of a particular shape from the measured angular scattering intensity pattern.

Figure 1 is a typical scheme for light scattering geometry in the laboratory frame. In the figure, $\mathbf{K_0}$ is the vector of the incident light (with intensity I_0) of wavelength λ ($= \lambda_0/n_0$, with n_0 being the refractive index of the medium and λ_0 the incident light wavelength in vacuo)

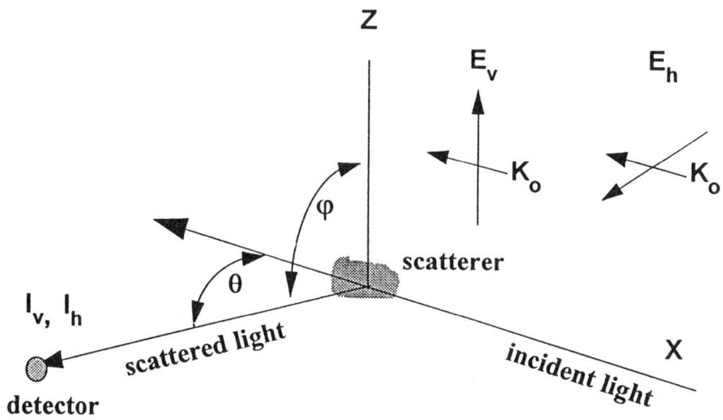

Figure 1 Schematic of light scattering geometry.

propagating along the X direction with the electric vector oriented either perpendicular to the hypothetical tabletop in the Z direction, termed E_v or parallel to the tabletop, termed E_h. The light is scattered in a direction at an angle θ with respect to X and at an angle φ with respect to Z. In most light scattering setups, the angle φ is set to 90° for theoretical simplification and practical convenience, and the angle θ is called the scattering angle. The quantities I_v and I_h are the scattering intensities detected with the electric vector of the light beam oriented in the E_v and E_h directions, respectively. In the above arrangement, the magnitude of momentum-transfer vector K (= $|\mathbf{K}| = |\mathbf{K_0} - \mathbf{K_s}|$) is

$$K = \frac{4\pi}{\lambda} \sin\left(\frac{\theta}{2}\right) \tag{1}$$

B. Scattering Intensity Theories

The formulation of light scattering theory can be traced back to an Arabic physicist early in the eleventh century, although the first systematic studies of scattering effects and the development of explanations took place in the 1860s by Govi and by Tyndall, who independently analyzed suspensions and aerosols. The completion of the scattering intensity theory for spheres was accomplished mainly by Lord Rayleigh, Lorenz, Debye, Mie, and others in the beginning of this century, although it is known to today simply as the Mie theory [4]. For nonspherical or noncubical particles, even in monodisperse systems, there are more than one-dimensional parameters. Although a lot of research works have been published, the applications of light scattering theory in nonspherical particle sizing have been very limited due to the complexity arising from the convolutions of multiple parameters. Thus, we will only list the related references and not discuss these theories in the present text [5–7]. In the remainder of this chapter, we will introduce Mie theory and its two limiting cases, Rayleigh-Debye-Gans theory for small particles and Fraunhofer theory for large particles.

The scattering intensity detected at a far field, i.e., when the distance between the detector and the scatterer is much larger than the dimension of the scatterer, is the result of the scattered electric field from different portions of a scatterer that is homogeneously illuminated by the incident light. Because scattered field from different portions of the scatterer interfere with each other, one has to solve the complete set of Maxwell's equations for the entire scatterer in order to obtain the scattering intensity at any given scattering angle. Rigorous analytical formulas for solving Maxwell's equations for a scatterer exist only for a sphere in the spherical coordinate and an oriented rod of infinite length in the cylindrical coordinate. There are approximate solutions for some regular-shaped scatterers such as ellipsoids and, lately, for irregular-shaped particles [6,7]. For particles with dimensions either smaller or much larger than the incident light wavelength, further approximations can be applied for those having regular shapes: for small ones, instead of rigorous solutions, scattering form factors for particles of a particular shape can be used to describe the interference effect of the scattered light intensity; for large ones, different diffraction formulas are available for nontransparent regular-shaped particles. We will further discuss these two types of treatments in the sections on Rayleigh-Debye-Gans theory and Fraunhofer theory.

1. Mie Theory

For spherical particles the scattering angular pattern is symmetric with respect to the axis of the incident light; i.e., the scattering pattern is the same for the same absolute value of the scattering angle. These patterns form scattering minima and maxima at different locations depending on the properties of a particle. Figure 2 shows the first few scattering lobes for two spherical particles of different sizes.

As shown in Fig. 2, a general characteristic of scattering from spheres is that the location of the first intensity minimum is closer to the axis and the peak intensity is higher for a larger particle (the solid line in Fig. 2) when compared with a smaller particle (the dashed line in Fig. 2). The intensity distribution of the light scattered from a particle may also be displayed as a "radial graph" such as in Fig. 3. In Fig. 3, the particle is located at the center of the graph. The curve connecting the data points is the intensity profile: the greater the distance of a data point from the center, the greater the intensity at that angle.

Because the scattering pattern is axially symmetric, i.e., the intensity curve at the left side (the negative scattering angle) is a mirror image of the right side (the positive scattering angle) in Fig. 2, we need to focus the discussion only on one side of the incident light using all positive angles for convenience.

Mie theory is the rigorous solution for the scattering from a sphere. The derivation of Mie theory can be found in several references and is beyond the scope of this chapter [5]. The analytical formulas of Mie theory can be represented by the following equations:

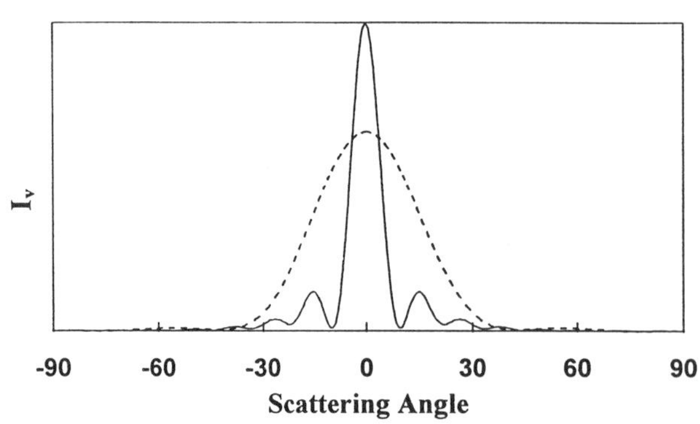

Figure 2 Angular scattering intensity pattern for spheres.

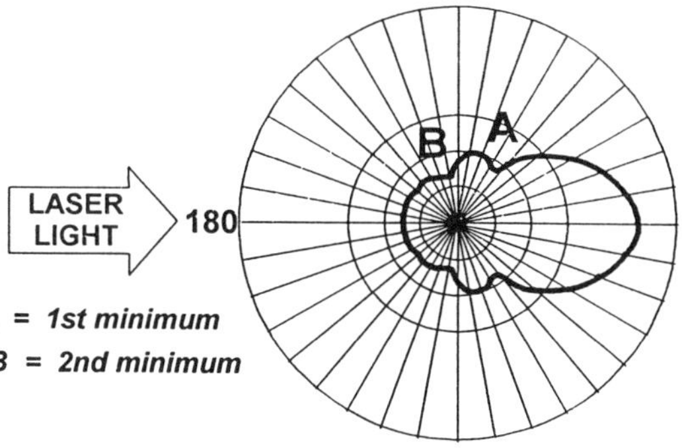

Figure 3 Radial graphic presentation of scattering intensity for a sphere.

$$I_v \propto \left| \sum_{j=1}^{\infty} \frac{2j+1}{j^2+j} \left(a_j \pi_j (\cos\theta) + b_j \tau_j (\cos\theta) \right) \right|^2 \tag{2}$$

$$I_h \propto \left| \sum_{j=1}^{\infty} \frac{2j+1}{j^2+j} \left(a_j \tau_j (\cos\theta) + b_j \pi_j (\cos\theta) \right) \right|^2 \tag{3}$$

where a_j and b_j are functions of α ($= \pi d n_0/\lambda_0$), β ($= \pi d m/\lambda_0$), and m ($= m_1/n_0$) with m_1 being the refractive index of the particle, which may be a complex number if there is absorption, and d being the sphere diameter. The detailed formulations for a_j, b_j, τ_j, and π_j are listed in Appendix I. Equations (2) and (3) are in the form of the spherical Bessel function, which has an oscillating nature.

Figure 4 illustrates in a three-dimensional fashion the scattering intensity I_v of a unit volume from polystyrene latex (PSL) spheres suspended in water, which is a typical model scatterer. The intensity is plotted as a function of scattering angle and sphere diameter. Please note that all three latexes are logarithmically scaled in order to cover a wide dynamic range and to reveal details of the intensity variation. There are two general features in the map. The first is the trend of the scattering intensity variation: the intensity is higher for larger particles at the same scattering angle, and the intensity gets lower as the scattering angle becomes higher for a given size of particle. However, the trend for angular dependence is reduced when sphere size gets smaller. For spheres in the nanometer range the intensity becomes much less angularly dependent until it reaches a flat line for spheres smaller than 50 nm. The second feature is that due to the oscillatory nature of the Bessel function there are many systematic ripples, peaks, and valleys, especially sharp intensity changes from the peak values that center around 0° to the first minimum and to sequential minima and maxima. The pattern is characteristic for a given size particle. As shown in Fig. 4, if both scattering angle and diameter are spaced logarithmically, the first minimum displays a linear decrease in angle as particles get larger. It is from these features that the size information for spherical particles can be resolved through light scattering measurements. The shape of the 3-D map may vary depending on the values of the relative refractive index m and the incident light wavelength, but these two general features always exist. Figure 5

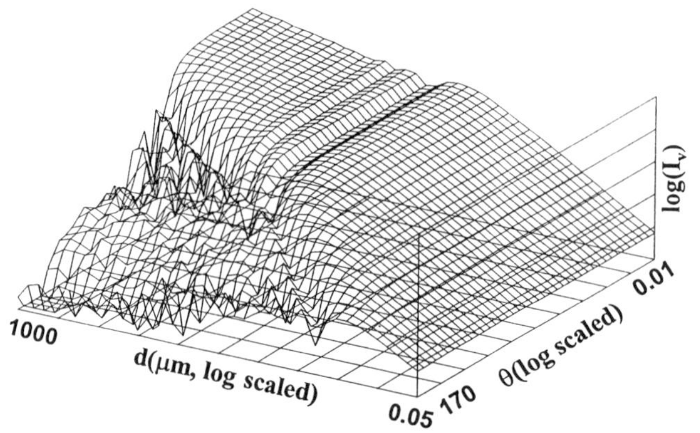

Figure 4 Three-dimensional display of the Mie scattering intensity from a unit volume of spheres with the relative refractive index $m = 1.20 + 0i$. The light wavelength is 750 nm in vacuo and 562 nm (= 750/1.333) in the medium (water).

shows several scattering intensity curves for a 0.5-μm PSL sphere suspended in water and illuminated by light of different wavelengths and different polarizations.

At small angles, typically smaller than 10°, the scattering pattern for spheres is centrally symmetric instead of axially symmetric; i.e., it displays concentric rings in the direction of the incident light. In this range, the intensity will be the same for the same solid angle θ with respect to the incident light, and is not restricted to the angle θ in the scattering plane defined by the tabletop in Fig. 1.

2. Rayleigh-Debye-Gans Theory

For particles much smaller than the incident light wavelength and for particles much larger than the incident wavelength, the solutions to light scattering equations can be simplified, leading to Rayleigh-Debye-Gans theory and Fraunhofer theory, respectively. The oscillation of the scattering intensity as a function of scattering angle and size is mainly due to intraparticle interference. For particles that are much smaller than the light wavelength the intraparticle interference can be neglected or be taken to a first-order approximation in which the interference effect can be

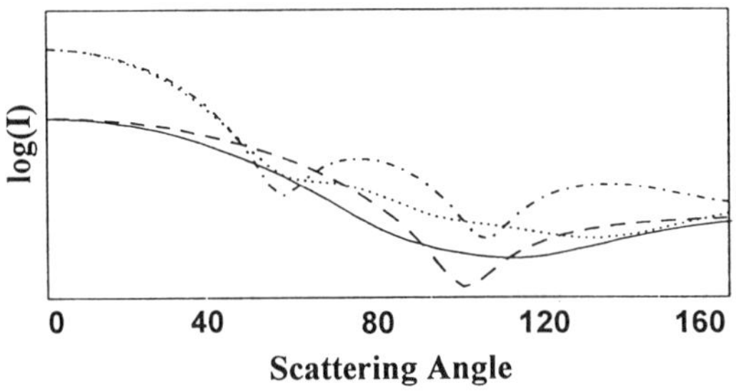

Figure 5 Scattering intensity pattern for a 0.5-μm PSL sphere in water illuminated by different light sources. Solid line: I_h from a diode laser ($\lambda_0 = 780$ nm); dashed line: I_v from the diode laser; dotted line: I_h from an Ar ion laser ($\lambda_0 = 488$ nm); dot-dashed line; I_v from the Ar ion laser.

accounted for by a scattering form factor. The Rayleigh-Debye-Gans theory is applicable in the size range where

$$\frac{2\pi}{\lambda}d(m - 1) << 1 \tag{4}$$

In Eq. (4) d is the longest dimension through the particle. The scattering intensity from an unit scattering volume in the above size range is

$$I(\theta) \propto \frac{1}{\lambda^4}(m^2 - 1)^2 VP(\theta) \tag{5}$$

where V is the volume of the particle and $P(\theta)$ is the scattering form factor. $P(\theta)$ $\left(= (1/n^2)\sum_i\sum_j(\sin Kr_{ij})/Kr_{ij}\right)$ is derived from the consideration that a large particle has n active identical scattering centers, any two of which are separated by a spatial vector $\mathbf{r}_{ij}$ of length r_{ij}. All possible pairwise combinations of these centers within the particle geometry form the basis of the scattering interference pattern. Scattering form factors for many regular shapes have been derived or numerically computed and are available in the literature [5,8,9]. Appendix II lists the form factors for common shapes. The form factors are often only a function of the product of the momentum-transfer vector and the particle dimension. The analytical formulas for the form factors are thus much simpler than the rigorous solutions (if these solutions exit) for a particular shape. In the case of spheres, the complicated 3-D scattering patterns from the Mie theory as shown in Fig. 4 will be simplified into a single variable plot that can be depicted by a 2-D graph. Figure 6 shows the form factors for several common shapes.

The biggest advantage in using form factors is that these functions are independent of the refractive index value of the material. Size and shape are the only variables once the experimental setup is determined. Thus, for small particles, once one knows the particle shape, size

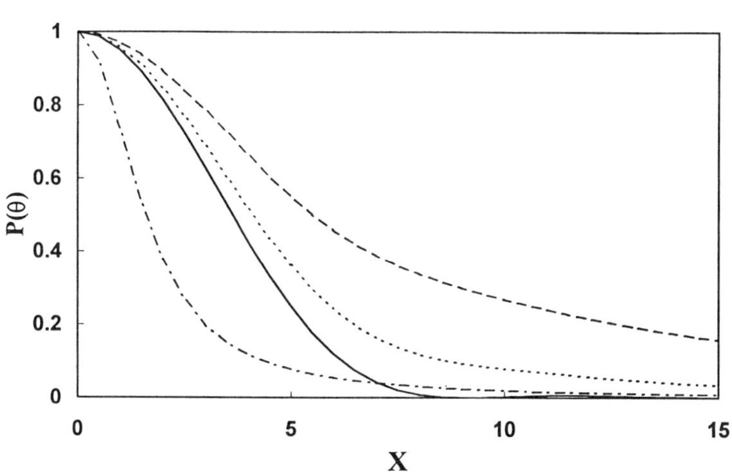

Figure 6 The scattering form factors of several shapes. From the top down: (a) dashed line: randomly oriented circular rods with the length (L) to diameter (d) ratio $L/d = 8$ where $x = 4\pi L \sin(\theta/2)/\lambda$; (b) dotted line: randomly oriented infinitely thin disks where $x = 4\pi d \sin(\theta/2)/\lambda$ with d being the diameter; (c) solid line: spheres where $x = 4\pi d \sin(\theta/2)/\lambda$ with d being the diameter; and (d) dot-dashed line: random coils where $x = 4\pi R_g \sin(\theta/2)\lambda$ with R_g being the radius of gyration.

characterization is readily feasible provided that all particles have the same geometric shape. According to Eq. 4, in the visible light wavelength range the Rayleigh-Debye-Gans theory can be used for particles with sizes up to 100 nm. For example, taking the light wavelength being 750 nm, the upper size limits will be about 100 nm for $m = 1.1$ and 60 nm for $m = 2.0$. This theory has been widely used in the characterization of synthetic and natural macromolecules because most of these molecules that are in coil conformation in solution have sizes less than 1 μm. This theory can also be used to size industrial particles with a uniform shape. However, due to practical limitations, such as polydispersity of shape for industrial particles and weak angular intensity dependence for small particles, such applications are still limited to academic work.

3. Fraunhofer Theory

On the other hand, when particle dimension is much larger than the wavelength of light or the materials are highly absorptive, the edge effect of particles contributes more and more to the total scattered intensity. The interference effect now comes mainly from the contours of the particles, i.e., diffraction produced by the bending of the light at the particle boundary. For large particles, when only diffraction is considered, we can use various diffraction equations for particle sizing. Diffraction can be described by Huygen's principle: each point on a propagating wavefront is an emitter of secondary wavelets. The locus of expanding wavelets forms the propagating wave. Interference between the secondary wavelets gives rise to a fringe pattern which rapidly decreases in intensity with increasing angle from the initial direction of light propagation. The diffraction effects are different for a point source and for a parallel beam, namely Fresnel diffraction for the point source and Fraunhofer diffraction for the parallel beam. In a light scattering measurement, because the light source is far away from scatterers and the optics are usually designed such that the incident beam illuminating the scatterers is homogeneously parallel, only Fraunhofer diffraction will take place. For spheres, the Fraunhofer diffraction is just a simplified form of Eq. (2) with the limiting condition of $d >> \lambda$:

$$I \propto 2\left(\frac{\pi d J_1(\pi d \sin\theta/\lambda)}{\lambda \sin\theta}\right)^2 \tag{6}$$

where J_1 is the Bessel function of the first kind of order unity:

$$J_1(x) = x\sum_{k=1}^{\infty}(-1)^{k+1}\frac{x^{2k-2}}{(k-1)!k!2^{2k-1}} \tag{7}$$

Equation (6) can only be used for particles that are (1) much larger than the wavelength of the light (typically > 30 μm) and nontransparent, i.e., the particles have refractive index values different from that of the medium (typically with the relative refractive index value m being larger than 1.2); or (2) highly absorptive (typically with absorption coefficients higher than 0.5). Equation (6) is derived assuming that a particle scatters as if it were an opaque circular disk of the same projected area positioned normally to the axis of the beam. Because of this assumption, in Fraunhofer theory the refractive index of the material is irrelevant. Although Eq. (6) is valid only for spherical particles, various diffraction equations for other regularly shaped particles are also available [10,11]. Because for large particles where the Fraunhofer theory applies, the scattering intensity is concentrated in the forward direction—typically at angles smaller than 10°, Fraunhofer diffraction is also known as forward scattering. In the Fraunhofer diffraction theory the angle for the first minimum of scattering intensity is simply related to the size by

Table 1 Intensity Distribution of the Fraunhofer Diffraction for Spheres

Intensity ring	Position (θ)	Relative intensity (I/I_0)	Intensity in ring (%)
Central maximum	0.0	1.0	83.8
First minimum	$\sin^{-1}(1.22\lambda/d)$	0.0	
Second maximum	$\sin^{-1}(1.64\lambda/d)$	0.0175	7.2
Second minimum	$\sin^{-1}(2.23\lambda/d)$	0.0	
Third maximum	$\sin^{-1}(2.68\lambda/d)$	0.0042	2.8
Third minimum	$\sin^{-1}(3.24\lambda/d)$	0.0	
Fourth maximum	$\sin^{-1}(3.70\lambda/d)$	0.0016	1.5
Fourth minimum	$\sin^{-1}(4.24\lambda/d)$	0.0	

$$\sin\theta_{\text{1st minimum}} = \frac{1.22\lambda}{d} \tag{8}$$

and most of scattering intensity is concentrated in a very sharp central lobe, as shown in Table 1, which provides a much simpler solution applicable to sizing large particles in a light scattering measurement.

Figure 7 displays a similar 3-D map as Fig. 4, the unit volume scattering intensity as a function of scattering angle and sphere diameter, but uses Eq. (6) instead of Eq. (2). For the regions of large particles and small angles these two figures are identical, which demonstrates that Mie theory can be approximated by Fraunhofer theory.

For particles with smaller m values, even though they may be very big, Eq. (6) will no longer be valid because the effects of light transmitting or refracting through the particles also have to be taken into account. In this anomalous diffraction regime either Mie theory has to be used or, as a first-order approximation, another term, $f(m,\theta)$, accounting for refraction effect has to be added to Eq. (6), yielding [12]

$$I \propto 2\left(\frac{\pi d J_1(\pi d \sin\theta/\lambda)}{\lambda \sin\theta}\right)^2 + \left(\frac{\pi d}{\lambda}\right)^2 f(m,\theta) \tag{9}$$

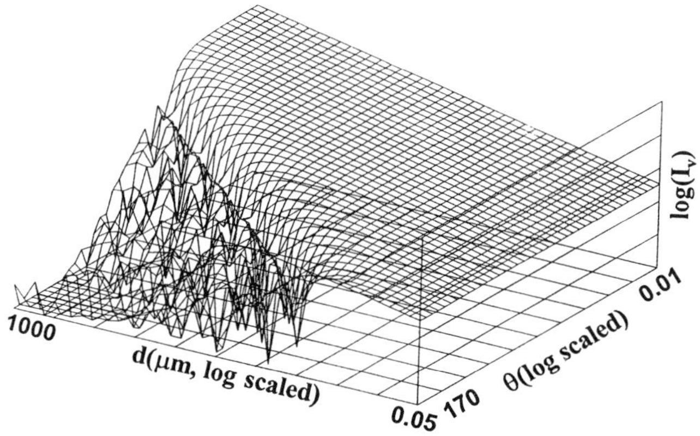

Figure 7 Three-dimensional display of the Fraunhofer scattering intensity pattern from a unit volume of spheres. The light wavelength is 750 nm in vacuo and 562 nm (= 750/1.333) in the medium (water).

The term $f(m,\theta)$ has strong angular dependence that diminishes only when m is larger than approximately 1.2 as shown in Fig. 8. For $m > 1.2$, $f(m,\theta)$ is a constant so that the angular scattering pattern can be treated using the ordinary Fraunhofer theory.

In the above discussion, the assumption is that the incident light is a parallel beam with a plane wavefront and of homogeneous intensity. If the beam intensity is Gaussian or the light is divergent, all formulas listed have to be modified accordingly which will further complicate the problem and lead to the generalized Mie theory [13]. For that reason, one of the important targets in designing the optics for a light scattering instrument is to make the beam intensity homogeneous and to reduce the beam divergence to a minimum. In the above equations and plots only the scattering intensity from an unit volume is computed and displayed. The unit volume scattering pattern is the foundation for particle size distribution analysis. Because particles with a certain size have their own characteristic scattering pattern, the absolute intensity that they contribute to the measured intensity is proportional to the total volume of these particles. By decomposing the measured intensity pattern into individual patterns corresponding to different sizes, the relative amplitudes of these sizes as the size distribution can be obtained. Light scattering is an absolute measurement technology only in the sense that once the experimental setup is correct, calibration or scaling are not necessary in order to obtain the volume (or weight) percentage of each component.

Although the angular scattering pattern is a function of many variables, in a measurement most of these variables are known constants, namely the light wavelength and the refractive indices of both medium and material. Thus, the angular scattering pattern is only a function of the size and shape of the material. If one knows the relations between the scattering intensity, scattering angle, and particle size and shape, then the particle size and shape are not difficult to obtain from measurements of the scattering angular pattern, at least theoretically. Unfortunately, there are no existing formulations for scattering patterns of irregularly shaped particles. Even for regularly shaped particles only limited numbers of formulas are available in the Rayleigh-Debye-Gans region and in the Fraunhofer region. In the region where the dimension of particles is

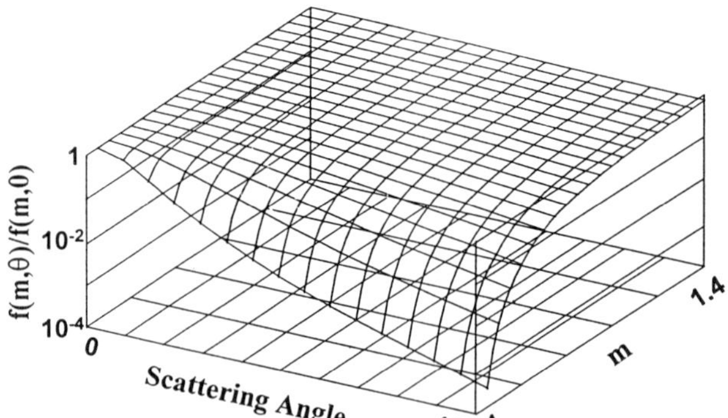

Figure 8 The refraction effect in the anomalous diffraction region where the function

$$f(m,\theta)\left(=4\left(\frac{m}{m^2-1}\right)^4\frac{(m\cos(\theta/2)-1)^3(m-\cos(\theta/2))^3}{\cos(\theta/2)(m^2+1-2m\cos(\theta/2))^2}\left(1+\frac{1}{\cos^4(\theta/2)}\right)\right)$$

shows strong angular dependence.

comparable to the light wavelength, the only formulas available are for spheres (Mie theory) infinitely long circular rods, and ellipsoids. Since most industrial particles closely resemble spheres and the scattering effects from corners and edges of these particles are averaged out due to the tumbling and rotational motion in sample circulation during the measurement, one can apply the spherical model to the most practical systems with one parameter: diameter. Of course, such treatment only yields apparent values that may be different from the real dimension, especially when nonsphericity is severe. This also is the only feasible choice for a commercial instrument that can be used for a broad range of samples.

From Figs. 4 and 7 one can see that the scattering intensity cliffs (large decreases in intensity) shift toward smaller angles with increasing particle size; i.e., larger particles have a slimmer central scattering intensity peak. It is these cliffs, along with the fine structures of the angular pattern, that form the foundation of resolving particle sizes from intensity measurements. However, with decreasing particle size the fine structure in the scattering pattern gets lost and the scattering intensity with respect to angle changes very slowly. Thus, the practical upper limit for using a single wavelength of light to characterize particle size by the angular pattern is based on the smallest scattering angle that the instrument can access. Presently, this upper limit is about 3.5 mm. By applying Eq. (8), the first minimum in the Fraunhofer diffraction for a 3.5-mm-diameter spherical particle is located at $\theta = 0.015°$. Thus, in order to detect and to fit the sharp intensity change within the central lobe, the smallest scattering angle accessible should be at least smaller than 0.01°. Typically, a very long optical bench is needed to reach that range. The upper limit is also partially determined by practical necessity, because the sizing of microparticles with an upper boundary larger than 3.5 mm is uncommon. For particles longer than a few millimeters other analytical methods, such as sieving, may be more suitable. Attention has to be paid when combining or comparing particle size results from two different technologies, such as sieving and light scattering, because each technology has different weighting and sensitivities to the constituents of the sample. The lower limit is determined when the fine structure of the angular pattern diminishes in the detecting angular angle. The value for the lower limit depends on the relative refractive index and the optical setup, which will be discussed in the next subsection.

4. Submicron Size Analysis: Polarization and Wavelength Effects

Currently, light scattering technology is being applied in an ever-widening variety of fields. A lot of samples have particles whose sizes extend into the submicron range, which requires instruments capable of sizing wider ranges of particles. However, when particle size gets smaller, the ratio of particle dimension and light wavelength (d/λ) is reduced. Interference effects are thus reduced and the scattering pattern becomes smoother and less angular dependent. Notice in Fig. 6 how smooth these scattering form factors are when approaching small x values. In this range, the sensitivity of particle size to scattering intensity pattern is greatly reduced causing it to be more and more difficult to obtain correct size values. Obviously, if light of a short wavelength is used, the ratio will be great, and so the lower size limit will be effectively extended. As shown in Fig. 5, for a 0.5-μm PSL there is only a slow change in the scattering intensity if a light of wavelength 780 nm is used. If, however, for the same particle, when a shorter-wavelength light ($\lambda_0 = 488$ nm) is used fine structures exist. On the other hand, scattered light collected at different polarizations also has different fine structures. The increase in size range due to regaining the fine structure is not in a linear relation with the light wavelength. Combining the polarization effects of light scattering with the wavelength

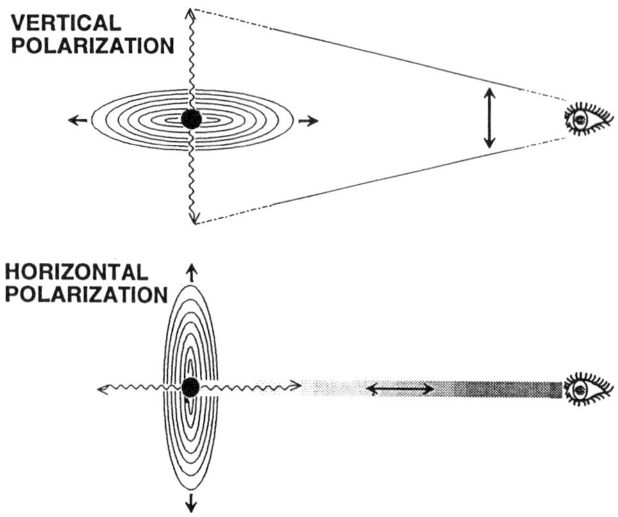

Figure 9 Scattering from different polarizations.

dependence at high angles, we can extend the lower size limit to as low as 50 nm, almost reaching the theoretical limit. This is the patented polarization intensity differential scattering (PIDS) technique [14,15].

The origin of the polarization effect can be understood in the following way. If a small particle is located in a light beam, the oscillating electric field of the light induces an oscillating dipole moment in the particle, i.e., the electrons in the atoms comprising the particle move back and forth relative to the stationary particle. The induced motion of the electrons will be in the direction of oscillation of the electric field and, therefore, perpendicular to the direction of the propagation of the light beam. As a result of the transverse nature of light, the oscillating dipole radiates light in all directions except in the direction of oscillation as visualized in Fig. 9. Thus, referring to Fig. 1, if the detector is facing the direction of oscillation it will receive no scattering from single dipoles. When the light beam is oriented in either E_v or E_h, the detected

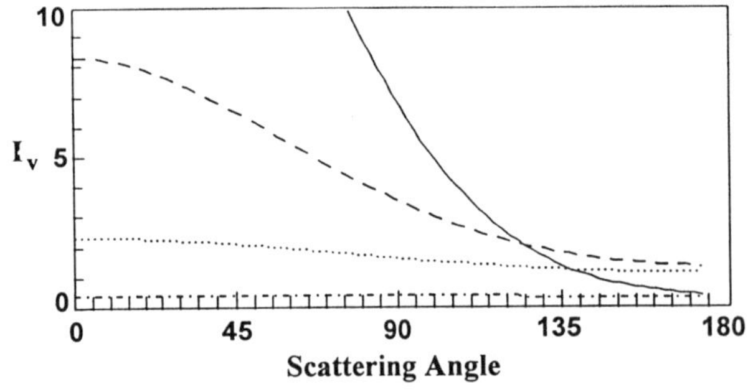

Figure 10 I_v scattering of small PSL in water ($\lambda_0 = 450$ nm). Solid line: $d = 200$ nm; dashed line: $d = 150$ nm; dotted line: $d = 100$ nm; and dot-dashed line: $d = 50$ nm.

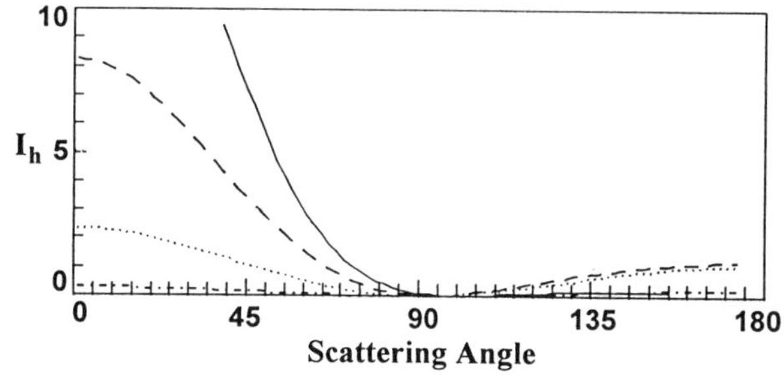

Figure 11 I_h scattering of small PSL in water ($\lambda_0 = 450$ nm). Solid line: $d = 200$ nm; dashed line: $d = 150$ nm; dotted line: $d = 100$ nm; and dot-dashed line: $d = 50$ nm.

scattering intensity I_v and I_h at a given angle will be different. The difference between I_v and I_h ($I_v - I_h$) is termed the PIDS signal.

As the particle size increases the particle will no longer act like a simple dipole and the scattering pattern will become more complex. For small particles the PIDS signal is roughly a quadratic curve centered at 90°. For larger particles, the pattern shifts to smaller angles and secondary peaks appear. Since the PIDS signal is dependent on particle size relative to the light wavelength, valuable information about the particle size distribution can be obtained by measuring the PIDS signal at a variety of wavelengths. In order to visualize the polarization effect in the scattering intensity, let us look the difference in I_v and I_h as a function of diameter and angle for our favorite model system: PSL suspended in water at $\lambda_0 = 450$ nm.

The main difference between the plots in Figs. 10 and 11 is that there is a minimum around 90° in the I_h scattering for each size. Thus, the angular variation of the scattering intensity for small particles can be enhanced if the differential intensity between I_v and I_h ($= I_v - I_h$) is used instead of I_v only as illustrated by Fig. 12.

It has been proven that such an enhancement is significant in the characterization of particles with diameters smaller than 0.5 μm. From Fig. 12, the angular patterns for 100-nm and even for 50-nm particles are recognizable, in addition to the shift in the axis of symmetry. The improvement is from both the wavelength effect and the polarization effect. It has been verified through both theoretical simulation and real experimentation that to accurately size particles smaller than approximately 0.4 μm by scattering intensity without PIDS technique is just unrealistic and cannot be true. Figure 13 is a typical trimodal distribution with the diameter ratios being

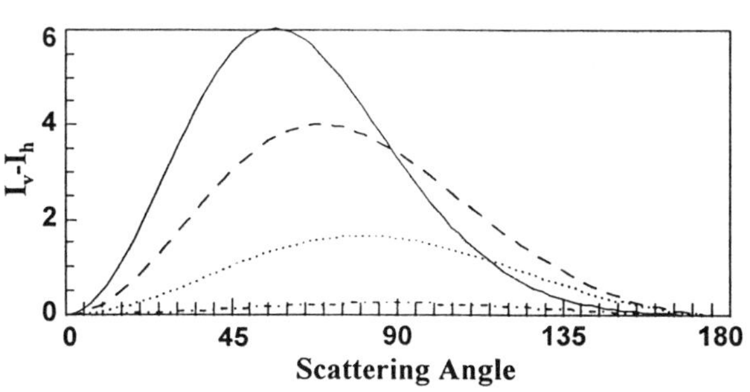

Figure 12 $I_v - I_h$ scattering of small PSL in water ($\lambda_0 = 450$ nm). Solid line: $d = 200$ nm; dashed line: $d = 150$ nm; dotted line: $d = 100$ nm; and dot-dashed line: $d = 50$ nm.

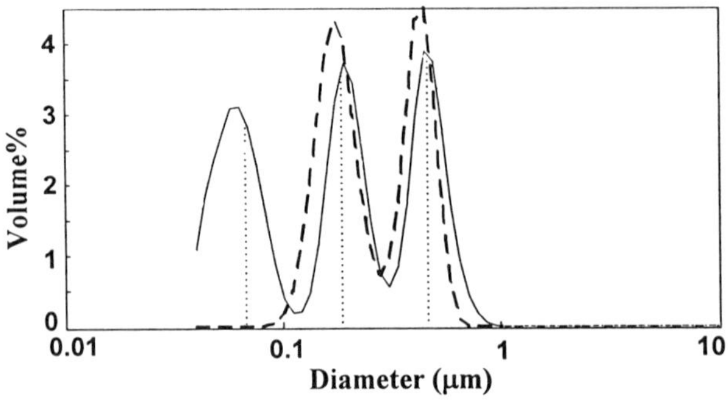

Figure 13 A real trimodal mixture of PSL (1:1:1 volume ratio) measured by Coulter LS 230 light scattering instruments using PIDS technique (solid line) and without using PIDS (dashed line). The dotted lines are the nominal diameter values of the PSL in the mixture from the PSL vendor.

approximately 1:2:4. Without the PIDS technique the smallest component is missed, even using the information gathered from high scattering angles and short wavelengths.

III. INSTRUMENTATION

There are many different avenues in the design and construction of an analyzer for light scattering measurements. Depending on the desired specifications and applications, there are differences in the design of the electronics, optics, mechanics, and computer software used in instrumentation. This is especially true in academic research where unique designs are realized for special research projects. It is neither the desire of this chapter nor is it possible to include all practical designs. In this section, only typical and commonly used designs are described and analyzed.

A. Light Source

There are two types of light sources used in light scattering instruments. One is a white light source, typically a tungsten-halogen lamp, and the other is a laser light source. The former has a broad spectrum range with the wavelength typically ranging from 250 to 3000 nm which can be used when a multiple-wavelength measurement is needed. However, laser light has many advantages in light scattering measurement. Especially valuable are a laser's high stability, long lifetime, monochromatic illumination, and long spatial and time coherence. Thus, although use of a laser as the light source is not essential, it provides definite benefits and is almost always used in modern light scattering instrumentation as the main light source or the sole light source. Table 2 lists the main characteristics of lasers commonly used in particle characterization analyzers.

Recently, along with advancements in semiconductors and fiber optics, diode lasers have replaced gas lasers and ion lasers in more and more instrumentation because of their low cost and compactness. However, as opposed to a helium-neon laser or an argon ion laser, the light from a diode laser is not collimated and has a severe astigmatism. A diode laser's monochromatic light must be conditioned to produce a collimated circular beam with homogeneous flux. The device that is most often used to treat an illuminating source is known as a spatial filter. A conventional spatial filter is a set of optical elements, such as lenses, pinholes, apertures, designed to produce the desired beam. This type of spatial filter is not immune to mechanical movement and is often unstable, large in volume, and expensive. Another type of novel and more

Table 2 Common Lasers Used in Light Scattering Measurements

Laser type	Power (mW)	Wavelength (nm)	Features
Ar ion	30–2000	488, 514.5	Collimated, water cooling needed for high power
He-Ne	1–50	543.5, 594.1, 612.0, 632.8	Collimated
Diode laser	0.1–100	635, 650, 670, 685, 750, 780, 810, 830, 840	Low cost, miniature and interfaceable with fiber optics, elliptical and very divergent output

rugged spatial filters employs a length of single-mode fiber-optic cable coupled to the laser diode together with an optical system using microoptics such as gradient index lenses. Single-mode fibers have very small diameters, typically measured in microns, and only allow the propagation of light waves that are both monochromatic and in phase (coherent). The small diameter of a single-mode fiber creates a near point source of light which is superior in this application to conventional designs. After leaving the fiber, the light is focused and then expanded by a lens system to a homogeneous parallel beam with the beam diameter typically 10–20 mm, which is used to illuminate the sample in the sample chamber.

B. Sample Cell

The main function of the sample-handling module is to expose particles in the sample, without discrimination to their sizes, to the light but not to introduce any undesirable effects such as air bubbles and thermal turbulence. The sample module is typically composed of the sample cell and a circulation system. The circulation system may include certain features such as a circulation pump, heating coils, an ultrasonic probe, and stirring bars to better disperse and circulate the particles. The sample cell design is determined by the type and volume of the sample and suspension medium. It is critical that the width of the sample cell is thin enough to minimize the effects of multiple scattering, that its optical elements, e.g., the windows, do not distort the incident beam, and that by its design the area of illumination contains a subpopulation representative of the sample under analysis.

There are two general categories of dispersed systems and thus two different types of cells. The first type is for dry powders in air dispersion where the dispersing medium is air (refractive index of approximately 1.000). In these systems, complete dispersion on many occasions is not achieved and the particles are seen as is. Typically, the particles that make up the system take the form of agglomerates, aggregates, and/or associations. This is especially true for systems with sizes below 10 μm because there is not enough dispersion force to overcome static charges. The second type of cell is for both dry powders and liquid-borne particles in liquid dispersion where the dispersing medium is either water based or organic based with some additive used to aid and/or to maintain dispersion. In certain situations, particle size distributions obtained from air and liquid dispersions will not give similar results because of different completeness of dispersion.

C. Collecting Optics

Particles in the sample cell scatter light from the incident beam. Upon leaving the sample cell, the scattered light may be guided to the detector plane by Fourier optics or by reverse Fourier optics. Typical schematics of these two optical configurations are shown in Fig. 14. In standard

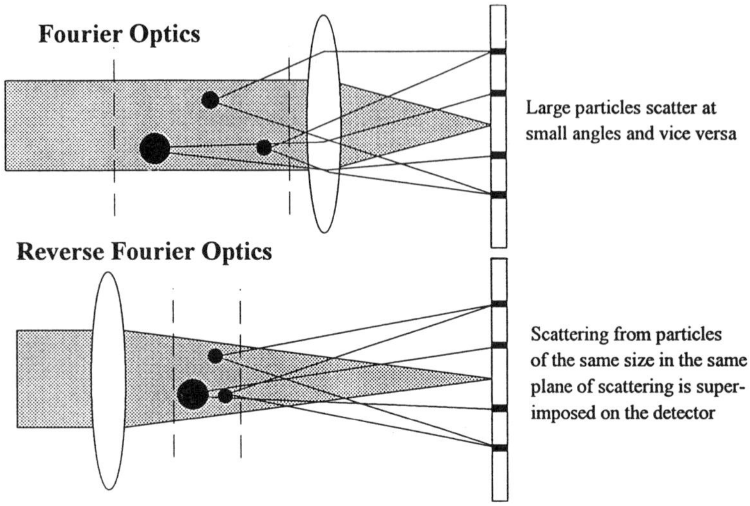

Figure 14 Fourier optics and reverse Fourier optics used in light scattering instrumentation.

Fourier optics, particles pass through an expanded and collimated laser beam in front of a lens in whose focal plane is positioned a photosensitive detector array. The focal length of the lens, the detector array arrangement, and the laser wavelength together determine the particle size range that can be analyzed. No calibration is required. The main incident beam is brought to focus in the detector plane and either reflected out by a mirror or passed through an aperture at the detector center. The angular scattering pattern is recorded by the detector array at the detector plane. This light intensity function is commonly referred to as the primary data, a light flux pattern $f(\theta)$, where flux is defined as light intensity per unit detecting area. An important feature of Fourier optics is that light scattered at a specific angle will be directed by the lens to fall onto a particular detector regardless of the particle's position. Particles may flow or move through any part of the beam, but their diffraction pattern is stationary and proportional in intensity to the total number of particles contributing to the scattering. For a polydisperse sample, the angular pattern may change with time as the number of distribution of particles in the scattering volume varies over time with particles entering and leaving the beam. This temporal variation is usually integrated to give an average true representation of the sample.

In reverse Fourier optics, so called because it reverses the relative positions of the sample and lens, the particles are confined to a plane that is thin compared to the distance from sample plane to detector. The transforming effect is mainly retained, i.e., the scattering from the same angle for all particles will be directed to the same detector regardless of their positions.

As mentioned, the lower size limit in scattering intensity measurement depends on the relative refractive index of the system and the optical setup. Since one cannot change the relative refractive index of the system under investigation, much effort has been spent in improving or optimizing the optical arrangement in order to increase instrument's dynamic size range, in particular, to extend the lower size limit. In practical optical setups, using either standard Fourier optics or reverse Fourier optics, the focusing lens has a finite size so that the scattering angular range from which scattered intensity can be collected by the lens is limited typically to less than 15°–20°. If one wants to size particles by using the characteristic maxima and minima of the intensity, in order to measure smaller particles a wider detecting angle is needed. One obvious way is to increase the dimension of the focusing lens and the size of the detector array. However,

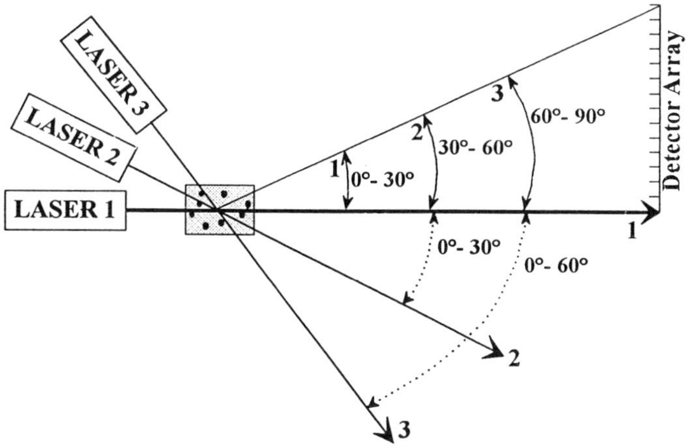

Figure 15 Trilaser optical system.

that is not very practical. Besides the cost of expensive large lenses, the quality of lens grinding and associated aberrations that affect high-angle signals are more severe as the lens diameter gets bigger. Another alternative is to use multiple lasers positioned in different locations so that the same detector array will receive the scattering from the same particle but illuminated by different lasers in different scattering angular ranges as shown in Fig. 15.

However, alignment of multiple lasers, for both light intensity and laser position, is inherently complex and may be unreliable. The sample cell design is also exceedingly complex. In addition, the intensity and polarization fluctuations of the three lasers will greatly affect the uniformity of the angular pattern, leading to possible additional measurement error because the angular pattern comprises three consecutive independent measurements. A better way to achieve wider angular coverage is to use a second lens, and a second detector array placed at a high angle to extend the detecting areas, as shown in Fig. 16.

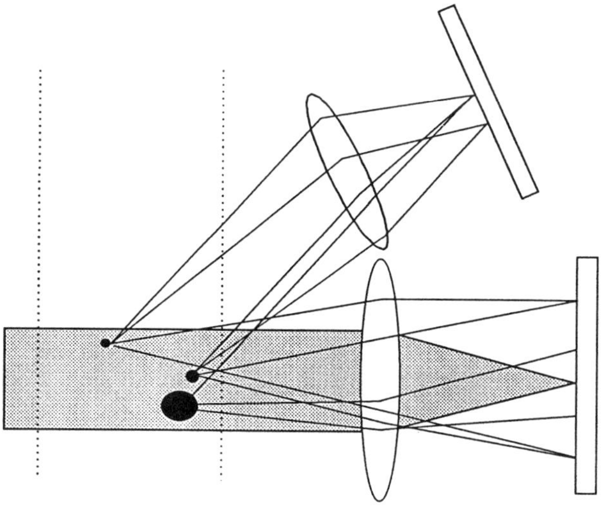

Figure 16 Double-lens optical system.

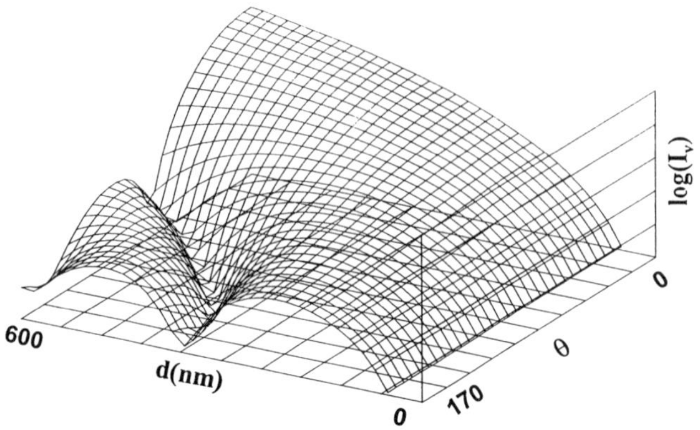

Figure 17 A 3-D display of the Mie scattering of small PSL spheres in water ($\lambda = 750/1.333 = 562$ nm).

The addition of the second lens and second detecting array extends the detecting angle to more than 30° which effectively extends the lower sizing range down to about a few hundred nanometers [16]. Because of the even slower angular variation for smaller particles, further increase of the scattering angle will not be significant in reducing the lower size limit. Figure 17 is a zoomed 3-D map from Fig. 4 that displays the very slow angular variation for small particles. In this case, we can see that once the size is less than around 0.4 μm, the minima and maxima disappear even for scattering angles up to 170°. Simply increasing the detecting angle will not make any more improvement. For further increase of the dynamic range, a different approach has to be used, such as the PIDS technique, which utilizes the wavelength and polarization effects.

In a typical PIDS optics system (Fig. 18), light from a white light source, e.g., a tungsten-halogen lamp, is filtered through filters of several wavelengths at two polarizations (as shown in Fig. 1). The collimated light passing through the sample cell is monitored by a beam monitor. The scattered light is detected by detectors located over a wide angular range, typically from 60° to 145°. Measurement is made at each wavelength and polarization sequentially. The data

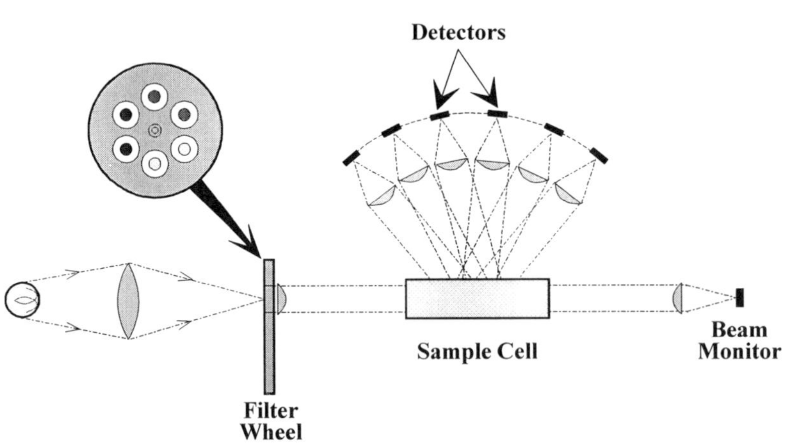

Figure 18 Typical PIDS system that utilizes multiple wavelengths and polarizations to size submicron particles.

collected from different wavelengths and polarizations at all angles provide a light scattering instrument with the capability of sizing submicron particles as small as some tens of nanometers [14,17].

D. Detecting System

A detector array formed by silicon detectors is commonly used as the detecting device in a light scattering measurement because of the wide dynamic range, high sensitivity, and long life of silicon material. The geometry of the detector array depends on the desired angular range, sensitivity, resolution (channel width), and any additional desired features. Due to the nature of the angular scattering pattern, there are two common features in most detecting arrays: (1) the spacing between low-angle detectors is much smaller than that between high-angle detectors in order to keep the same resolution over the entire size range, and (2) the total area of each detector increases with increasing angle in order to maintain similar sensitivity for small particles as that for large ones; for example, in Fig. 19, the total area of detector 10 is 200 times larger than the total area of detector 1. In the forward scattering range there are two commonly used detector array geometries: the concentric ring pattern and the X pattern as shown in Figs. 19 and 20.

The angular position of these detectors is tightly controlled, because even a small error in positioning can result in a substantial distortion in the reported particle size distribution. In comparing these two detector patterns, the elements in the ring pattern have large detecting areas which may result in high sensitivity when the scattering is weak and may be less sensitive

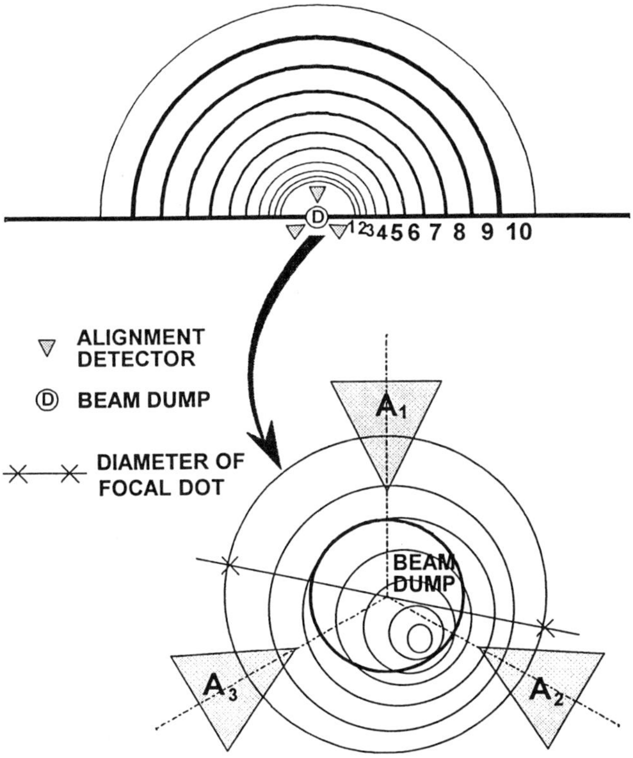

Figure 19 Geometry of the concentric ring pattern detector array (upper portion, only the first 10 rings are shown); and the optical alignment strategy (lower portion).

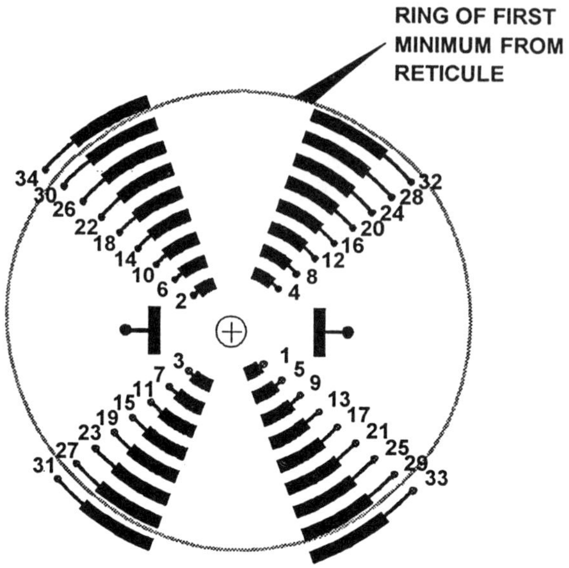

Figure 20 Geometry of the X pattern detector array (only the first 34 detectors are shown).

to particle orientation for non-spherical particles. The X pattern provides several advantages over the concentric ring pattern. For example, it allows for closer spacing of the detector elements than in the ring design. Closer spacing at these low angles enhances the resolution of the instrument in the large end of the size range [18,19].

In the center of the detector array there is either a mirror or a hole which guides the main beam to a beam dump. The beam dump serves two purposes: (1) stray light "noise" is prevented from reaching the detectors; and (2) the optimal amount of sample required for a measurement can be determined from the attenuation of the incident light intensity that is measured and recorded by the beam dump. As sample is introduced to the sample cell, light begins to scatter from its direct path along the optic axis, and the signal in the beam dump decreases. When there is sufficient sample in the sample cell to provide a good signal to the photodetectors, yet not so much as to cause multiple scattering, the sample measurement begins. During the measurement the response from each of the detector elements is amplified by a dedicated amplifier circuit, and this signal is averaged over the duration of the measurement, typically 1 min or longer. When the measurement is completed, these averaged values are written to a data file and converted to a particle size distribution by an analysis algorithm.

Because the scattering intensity changes rapidly with the scattering angle, especially for large particles, a proper arrangement of the detectors at different scattering angles is very important in order to obtain more information from the same number of detectors. As demonstrated in the previous figures, the maxima and minima of the scattering patterns for different size of particles change linearly if the scattering angle is logarithmically scaled. Thus, the rational arrangement for the detectors is to place them at angles that are logarithmically spaced. The detecting area for each detector is also proprietarily optimized to receive scattering energy that will easily yield a stable and correct solution in the matrix conversion. However, the determining factor for obtaining high resolution and correct size distribution is still the number of the detectors employed in the detector array. Even with the best matrix conversion algorithm, without enough number of detectors to collect the necessary information from the scattering pattern, high-resolution results are not achievable. One can never produce more information from less information obtained. Without inputting additional information or providing some restrictions and artifacts, the number of data points in the size distribution should be close to the number of detectors; i.e., the number of detectors determines the resolution in the size distribution. In addition, due to the oscillatory nature of the scattering pattern, if the number of detectors is too small, the fine structures in the scattering pattern will be missed and the ambiguity in the computation will increase resulting in an incorrect size distribution. This is because the computation is based on reported discrete intensity integrated from the detector surface area, not the real continuous angular intensity. Figure 21 depicts the resolution and accuracy in finding the minimum position from a detector array with a large number of detectors as compared with that from a detector array with fewer detectors.

In this example, for the array with a large number of detectors, one finds that the first minimum is located between 0.095° and 0.105°, which corresponds to one detector angular width and represents the instrument resolution. If Eq. (8) is used with $\lambda = 750$ nm, the corresponding size range will be from 552 to 499 μm. The same calculation can be performed for the array with the smaller number of detectors. The minimum now is detected between 0.0875° and 0.1125°. The size range will now be from 599 to 466 μm, a much lower resolution! In Fig. 22, a scattering curve from a mixture of 350-μm/500-μm PSLs in water is recorded using different detector arrays, one with 126 detectors and the other with 32 detectors. It is clear that the signal recorded by using 32 detectors misses much of the fine structure in the scattering pattern when compared with that recorded by 126 detectors. Thus, a correct result resolved from the signal recorded by the 32-detector array will be less likely.

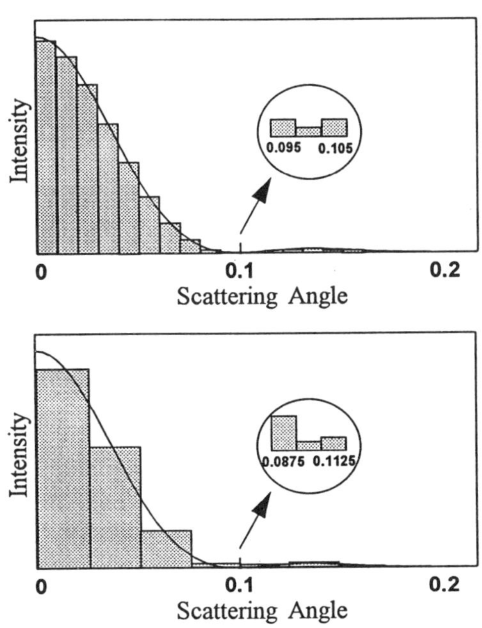

Figure 21 Demonstration of the effect of detector number in size resolution. The upper graph is of a large number of detectors with an angular resolution of 0.01°. The detector array used in the lower graph has an angular resolution of 0.025°.

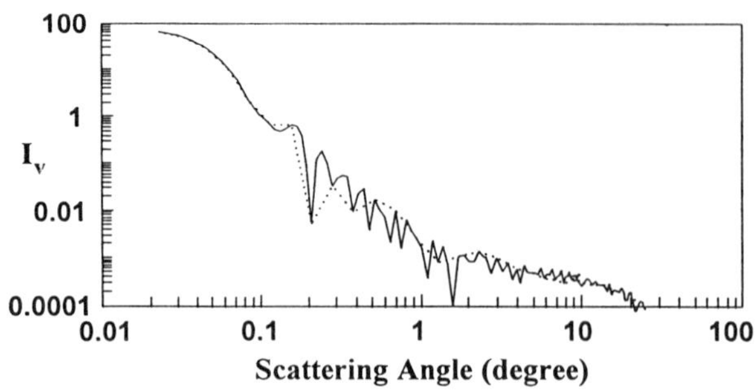

Figure 22 Detected signals from detector array of different detector number. Solid line: signal recorded with 126 detectors; dashed line: signal recorded with 32 detectors.

E. Instrument Calibration

In the common sense of calibration, the procedure may involve measurement of a reference material, whose size has been carefully determined and verified by several techniques, and adjustment of the instrument calibration constants to match the assayed value. However, since light scattering is an absolute measurement technology, it is not required to use reference materials to adjust the instrument, although one may use these materials to verify the validity and status of an instrument. For a light scattering instrument, the calibration is mainly related to the alignment of the optics, i.e., the laser and the detectors. Since in most cases detectors are etched onto a circuit board, they are not individual components. The task is simply to align the laser, the focusing lens, and the detector array.

The precise alignment of the detector array is critical in order to get correct results, especially in the case of large particles. The following simple calculation shows that the error introduced due to misalignment is linearly proportional to the sphere diameter in the Fraunhofer region. Taking the derivative of Eq. (8), one obtains the percentage of size error per degree of angular alignment error ($\lambda = 0.75$ μm):

$$\left|\frac{\Delta d}{d\Delta\theta}\right|\% = \frac{\pi}{180}\sqrt{\left(\frac{d}{1.22\lambda}\right)^2 - 1} \cdot 100\% \approx \frac{d\pi}{180 \cdot 1.22\lambda} \cdot 100\% = 1.9d\% \tag{10}$$

For a 0.01° misalignment, it will produce a 9.5% error for a 500-μm particle, but will produce only a 0.095% error, which is negligible, for a 5-μm particle.

The procedure used in aligning the optical train is to align the focusing lens so that the collimated light is positioned at the center of the detector array. There are two common techniques. One is to use the forward incident beam as the alignment standard as described in the lower portion of Fig. 19.

In this figure, the three triangular detectors are served as the alignment targets. For a circular spot that is focused from the incident light, the perfect alignment will be approached once the three detectors detect the same amount of intensity. This method is based on the assumption that the focal dot is a perfect circle. However, due to imperfections in lenses and light sources, asymmetry of the focal dot is unavoidable and is unique to each instrument since the lenses used in each instrument are different. Thus, although each instrument may achieve a certain degree of reproducibility and precision after compromising for the asymmetry, a sample measured on different instruments using the same sample module and same software may yield different results because the optics in each instrument have their own asymmetry. The typical instrument-to-instrument reproducibility is about ±4% on instruments that are aligned using this technique.

The second method is a combination of an internal dark-field reticule and the X pattern detector array. A dark-field reticule is a single hole or set of holes (a few hundred micrometers in diameter) in flat sheet metal. The holes diffract light just as a large particle of the same size would. By inserting the reticule into the beam one can use the scattering pattern from the hole(s), which is perfectly circular when compared with the focal spot, to perform the alignment (see Fig. 20). If the beam is misplaced from the center, then the intensity from each quadrant of detectors will be distorted differently and the intensity pattern will be a sawtooth curve because of the unique detector sequence. When it is perfectly aligned the recorded intensity will be a smooth curve. Thus, with the built-in reticule and the X pattern, the alignment is easier, quick, and can be fully automated under computer control without tedious manual loading and positioning, which require special skill. By optimizing hardware and software, the alignment takes about 30 s. Instruments aligned with this method typically have a reproducibility of ±1%.

IV. DATA ACQUISITION AND ANALYSIS

A. Matrix Conversion

For monodisperse particles, there is a one-to-one relation between the scattering intensity and the scattering angle for a particular size. For polydisperse samples each particle will contribute its scattering to the angular pattern differently. The intensity detected by a detector located at a scattering angle will be an integration of the scattering from all particles. The scattering pattern then can be represented by the following matrix:

$$\begin{bmatrix} f(\theta_1) \\ \cdot \\ \cdot \\ \cdot \\ f(\theta_m) \end{bmatrix} = \begin{bmatrix} a(\theta_1,d_1) & \ldots & a(\theta_1,d_n) \\ \cdot & & \cdot \\ \cdot & & \cdot \\ \cdot & & \cdot \\ a(\theta_m,d_1) & \cdots & a(\theta_m,d_n) \end{bmatrix} \begin{bmatrix} x(d_1) \\ \cdot \\ \cdot \\ \cdot \\ x(d_n) \end{bmatrix} \tag{11}$$

The left side of the matrix is the detected intensity flux at scattering angles θ_i, i from 1 to m. The flux is from the scattering of all particles with diameters d_j that scatter $a(\theta_i,d_j)$ amount of light per unit volume, j from 1 to n. The matrix $a(\theta_i,d_j)$ is called the scattering kernel, which can be computed from theoretical models according to the theories described in Sec. II. From the measured $f(\theta_i)$ values and the computed $a(\theta_i,d_j)$ values, and by a matrix conversion technique, $x(d_i)$, the particle size distribution, can be resolved. Every instrument has a specified size range that is classified into a certain number of size channels with assigned channel locations and widths, and every instrument has a unique detector array with fixed angular locations and angular coverage for each detector. Thus, a database of $a(\theta_i,d_j)$ with different refractive index values can be built for a particular instrument. The matrix of $a(\theta_i,d_j)$ is often referred to as the optical model.

The matrix conversion is at the heart of the data retrieving procedure in light scattering measurements. Due to the oscillatory nature of the Bessel function, which is the mathematical foundation of both the Mie and Fraunhofer formulations, specialized conversion techniques have to be used. Although there are many matrix conversion and data fitting techniques available in the public domain, the conversion algorithms of commercial instruments are proprietary almost without exception. Besides the optical arrangement, the conversion algorithm is the most important factor in the success of an instrument.

If one has some knowledge about the material, such as whether the size distribution is unimodal or mulitmodal, narrow or broad, the incorporation of such knowledge into the matrix conversion procedure may help to obtain a better result since the fitting can now be finely tuned

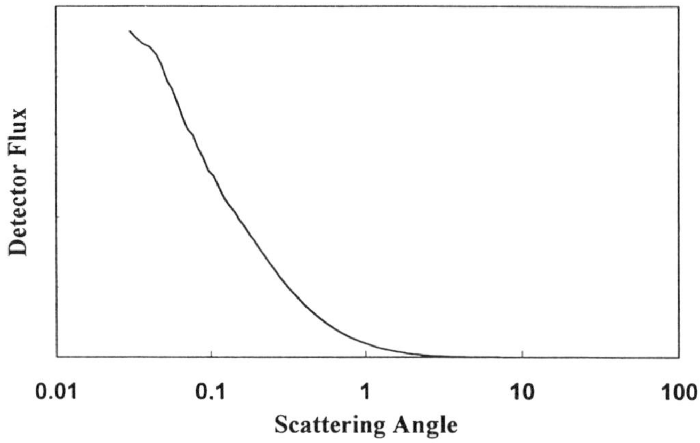

Figure 23 A typical scattering flux pattern measured by Coulter LS 230 light scattering instrument.

accordingly. For example, certain pretuned conditions can be used if the size distribution modality is known, or smoothing techniques can be used to obtain more realistic distributions. However, the drawback of such "prechoices" is that, in most instances, prefixing a distribution shape will bias the result toward the chosen distribution shape; i.e., you will get the result as you want and different choices will lead to different results. For sophisticated conversion algorithms, such "prechoices" have been proven to be unnecessary and occasionally detrimental in obtaining correct results.

Figures 23 and 24 display a typical scattering intensity pattern and the corresponding size distribution obtained from a sample with an extremely broad size range (0.1 μm to 1 mm) by using the matrix conversion method.

Choosing the correct optical method is often a key step in obtaining the correct result from the light scattering measurement. Typical errors that may occur when a wrong optical model is chosen (for example, using the Fraunhofer model for particles smaller than 2 μm) are (1) multiple artificial peaks, (2) excessive peak broadening, and (3) incorrect reporting of mean sizes.

The data analysis procedure for the scattering experiments combining low-angle scattering with high-angle PIDS measurements is similar to the procedure described earlier. The matrix

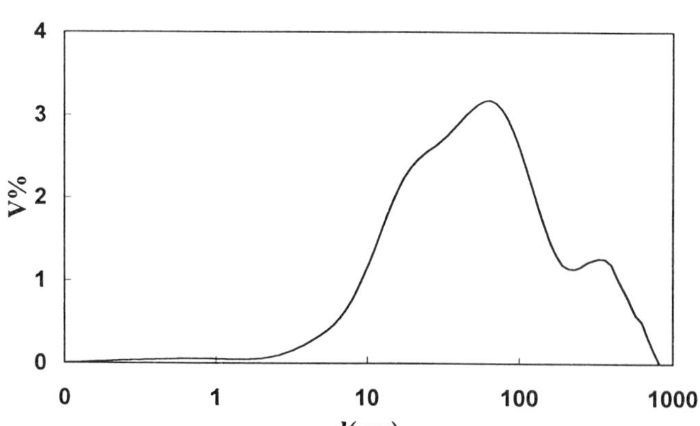

Figure 24 The particle size distribution resolved from the flux in Fig. 22.

form is now extended to include both forward scattering and PIDS information since both forward scattering and PIDS signals are mathematically based on the same theory and physically measured from the same material in the sample chamber. In Eq. (12), $p(\theta_i,\lambda_j)$ is the PIDS signal measured at angle θ_i ($i = 1$ to k) and at wavelength λ_j ($j = 1$ to h). However, since the light sources and the detectors are different for low scattering and PIDS measurement, the signals from them have to be weighted differently before they can be combined into a single matrix. The weighting and scaling constants can be easily obtained from an instrument calibration procedure.

$$\begin{bmatrix} f(\theta_1) \\ \cdot \\ \cdot \\ \cdot \\ f(\theta_m) \\ p(\theta_i,\lambda_1) \\ \cdots \\ p(\theta_k,\lambda_h) \end{bmatrix} = \begin{bmatrix} a(\theta_1,d_1) & \ldots & a(\theta_1,d_n) \\ \cdot & & \cdot \\ \cdot & & \cdot \\ \cdot & & \cdot \\ a(\theta_m,d_1) & \ldots & a(\theta_m,d_n) \\ b(\theta_i,\lambda_1,d_1) & \cdots & b(\theta_i,\lambda_1,d_n) \\ \cdot & & \cdot \\ \cdot & & \cdot \\ \cdot & & \cdot \\ b(\theta_k,\lambda_h,d_1) & \cdots & b(\theta_k,\lambda_h,d_n) \end{bmatrix} \begin{bmatrix} x(d_1) \\ \cdot \\ \cdot \\ \cdot \\ x(d_n) \end{bmatrix} \tag{12}$$

B. Choosing a Proper Optical Model

One of the frequent mistakes in a light scattering measurement is not from the choice of sample module or dispersion procedure but from using an inappropriate optical model in the data analysis. This will often produce an incorrect size distribution even though the measurement is performed correctly and the instrument is functional.

As discussed earlier, Fraunhofer theory is a special case of more comprehensive Mie theory. For spherical, opaque particles whose diameters are much larger than wavelength of the incident light, Mie theory can be simplified. This simplification gives rise to the Fraunhofer calculations. These calculations involve high scattering intensity at very low scattering angles which results from diffraction of the incident light at the edges of the particles. For systems with particles larger than 100 microns, the Fraunhofer optical model gives the same size values as the Mie model. When analyzing systems with particles in the range from 10 to 100 microns, the Fraunhofer optical model usually gives a fair approximation to the Mie model. For particles smaller than 10 μm, the Mie model with correct refractive index values should be used. The Fraunhofer model will give a distorted result that may overemphasize or underemphasize certain portions of a distribution and thus should be avoided. However, when there is no way of knowing or guessing the refractive index of a sample, the Fraunhofer model is the only alternative.

Figure 25 is from a series of light scattering measurements of monodisperse PSLs of various diameters suspended in water. Different size values are obtained when using Mie theory with different refractive values and when using Fraunhofer theory. When compared with the vendor's specified nominal values, Fraunhofer theory could only provide comparable results from $d > 30$ μm. For small particles, even though the refractive index value used for the Mie model may be incorrect, the sizes resolved are still much better than those obtained from Fraunhofer theory.

C. Refractive Index

One variable that distinguishes the scattering pattern of one particle from another particle of the same diameter is the refractive index. The refractive index, also termed index of refraction or refractivity, is a measure of the amount of refraction. It can be a function of wavelength,

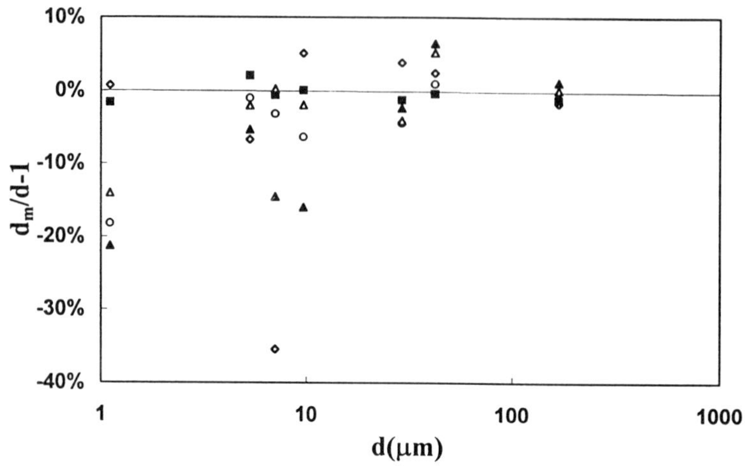

Figure 25 Comparison of particle size (d_m) obtained from measurements of PSL in aqueous suspensions using the Coulter LS 230 particle size analyzer. The relative deviations are based on the vendor-specified nominal diameters d, from NIST, Duke Scientific, and Seradyn. The optical models are generated according to Mie theory by using various m values. Hollow diamonds: $m = 1.05$; filled squares: $m = 1.20$ (the correct m value for polystyrene in water); hollow triangles: $m = 1.5$; and the hollow circles: $m = 2.25$. The filled triangles are the result from Fraunhofer theory computation.

temperature, and pressure. If the material is nonabsorbent and nonmagnetic at any wavelength, then the square of the refractive index is equal to the dielectric constant at that wavelength. For absorbing materials, the complex refractive index $m = n(1 - ik)$ is related to the absorptive index k, where the real term describes the refraction and the imaginary term describes the absorption.

Because the angular scattering patten depends on the relative refractive index between the medium and the particulate material, knowing the refractive indices for both is vital in characterizing particles in the size range where the rigorous Mie formulation has to be used. Fraunhofer theory is independent of refractive index as long as the relative refractive index is not near unity; i.e., the material is not transparent. For transparent materials, even for very large particles, the anomalous diffraction prevents one from using Fraunhofer theory in light scattering.

Refractive index values for the suspending medium, such as air, water and organic solvents, can be found in various handbooks [20]. They are mostly nonabsorbing, so they consist of a real part only. The values listed are often for one temperature at sodium light ($\lambda_0 = 589.3$ nm), although the refractive index values for air and common liquids are less dependent on temperature and wavelength.

Refractive indices of particulate materials can vary depending on the electron structures of the materials. Figure 26 summarizes the range of complex refractive indices of common materials. The plane is drawn in a lin-log form. The real part n is shown on the linear y-axis and the imaginary component (absorption) on the logarithmic x-axis. Although some of these values measured at one or two wavelengths can be found in handbooks, these values usually depend heavily on wavelength, especially for colored materials. Extrapolation, interpolation, or estimation from the listed values to the wavelength used in the light scattering experiment is usually needed. Many of these values can be found in corresponding professional articles and books [20–22]. Typically, for materials such as colored pigments or toners the values of the refractive index at different wavelengths are different and have to be chosen separately.

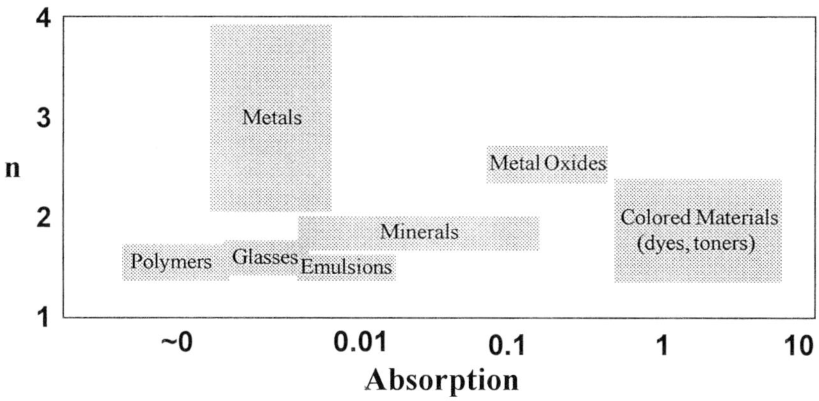

Figure 26 Refractive index values for common materials.

The imaginary part of the refractive index, however, can usually be estimated from the color of the material. For example, when analyzing a blue pigment, one should expect it to have higher absorption for most wavelengths other than blue. Therefore, when selecting the values for the imaginary portion, the value 0.000 should be chosen for a wavelength of 450 nm; and values between 1 and 10 should be given for 600-nm and 900-nm wavelengths.

Keep in mind that in light scattering measurements, only the relative refractive index counts; so for the same glass beads different scattering patterns are expected when they are dispersed in air and when they are dispersed in water. When choosing a refractive index, one should take care to choose the right form of compound being sized. For example, there are five forms of alumina listed in the CRC *Handbook of Physical Chemistry*. Each has its own distinct molecular geometry and color and should be treated differently with regard to its real and imaginary refractive indices. The real part of the refractive index is different for each of these five compounds, ranging from 1.577 to 1.768.

For mixtures, if each particle has a different refractive index, there is no ideal procedure to determine which refractive index value should be used since the measurement is performed on all particles at once. There is a particular instance where correct sizes of the mixture may be obtained. That is when there are two components that have substantial difference in size; for example, if component A is much smaller than component B, then one can use the refractive index value of B to compute the scattering kernel for the low angles and that of A to compute the scattering kernel for the high angles. Because large particles have less scattering in high angles, most of their information can be retrieved from scattering intensity in low angles where small particles contribute less to the overall intensity. The converse is true for small particles at high angles.

V. SUMMARY

In summary, because of the extremely wide dynamic range and fast measurement associated with the technology, elastic light scattering intensity measurement has become one of the most popular methods in sizing particles from nanometers to millimeters—replacing many other existing sizing technologies such as sieving, sedimentation, and microscopy in a broad range of industrial fields (for examples, see Refs. 23–25). The popularity of this technology has led to several national and international collaborative studies of the reproducibility and accuracy of commercial instruments using light scattering technology as compared with other particle characterization methods [26–28]. For narrowly distributed spherical particles the results are very promising in that the mean size and cumulative distributions from different models of light scattering instruments are close to each other and to the values obtained by other methods, and have excellent reproducibility. For nonspherical particles the results from different instruments are scattered, with different degrees of deviation when compared with other methods depending on sample polydispersity and shape. This is possibly due to (1) the approximations used in calculating apparent values (equivalent spherical diameters), for nonspherical particles are theoretically different for different methods; (2) the randomness of particle orientation in the scattering volume may be different when using different sample modules to circulate particles; (3) different measurement sensitivities may exist for particles of various sizes due to the differences in optical design; and (4) the different algorithms used in the size retrieval process. International and national standards are being established to standardize this technology so that results will be less ambiguous, more reproducible, and more accurate.

Although formulations for oriented or randomly oriented nonspherical particles in the Rayleigh-Debye-Gans region and in the Fraunhofer region are available, applications of these theories are still largely based on individual cases and limited to academic research. With further advancements in diode lasers (longer lifetime, collimated beam output and higher power), in optical designs (the use of more fiber optics and better arrangement of detecting optics), and in the data retrieval process (more powerful computers and better algorithms), applications of light scattering will certainly be further expanded to every corner where the characterization of particulate materials is needed.

ACKNOWLEDGMENTS

The support from Coulter Corporation is appreciated. The comments and suggestions from Mr. Hillary Hildebrand of Coulter Corporation are gratefully acknowledged.

APPENDIX I: SCATTERING FUNCTIONS OF A SPHERE

$$I_v \propto \left| \sum_{k=1} \frac{2k+1}{k^2+k} \left(a_k \pi_k(\cos\theta) + b_k \tau_k(\cos\theta) \right) \right|^2 \tag{A.1}$$

$$I_h \propto \left| \sum_{k=1} \frac{2k+1}{k^2+k} \left(a_k \tau_k(\cos\theta) + b_k \pi_k(\cos\theta) \right) \right|^2 \tag{A.2}$$

where

$$\pi_k(\cos\theta) = \frac{2k-1}{k-1} \cos\theta \cdot \pi_{k-1}(\cos\theta) - \frac{k}{k-1} \pi_{k-2}(\cos\theta)$$

$$\pi_0(\cos\theta) = 0, \quad \pi_1(\cos\theta) = 1 \tag{A.3}$$

$$\tau_k(\cos\theta) = \cos\theta \cdot \pi_k(\cos\theta) - (1 - \cos^2\theta)((2k-1)\pi_{k-1}(\cos\theta) + \pi'_{k-2}(\cos\theta))$$

$$\tau_1(\cos\theta) = \cos\theta \tag{A.4}$$

$$\pi'_k(\cos\theta) = (2k-1)\pi_{k-1}(\cos\theta) + \pi'_{k-2}(\cos\theta)$$

$$\pi'_0(\cos\theta) = 0, \quad \pi'_1(\cos\theta) = 0 \tag{A.5}$$

$$a_k = \frac{\Psi_k(\alpha)\left(\eta_k^{(1)}(\beta) - m\eta_k^{(1)}(\alpha)\right)}{\zeta_k(\alpha)\left(\eta_k^{(1)}(\beta) - m\eta_k^{(3)}(\alpha)\right)}, \quad b_k = \frac{\Psi_k(\alpha)\left(\eta_k^{(1)}(\alpha) - m\eta_k^{(1)}(\beta)\right)}{\zeta_k(\alpha)\left(\eta_k^{(3)}(\alpha) - m\eta_k^{(1)}(\beta)\right)} \tag{A.6}$$

$$\Psi_k(\alpha) = \frac{2k - 1}{\alpha}\Psi_{k-1}(\alpha) - \Psi_{k-2}(\alpha)$$

$$\Psi_0(\alpha) = \sin\alpha, \quad \Psi_1(\alpha) = \frac{\sin\alpha}{\alpha} - \cos\alpha \tag{A.7}$$

$$\eta_k^{(1)}(y) = \frac{y^2 + ky\eta_{k-1}^{(1)}(y) - k^2}{ky - y^2\eta_{k-1}^{(1)}(y)}, \quad \eta_0^{(1)}(y) = \cot y, \quad y = \alpha,\beta \tag{A.8}$$

$$\zeta_k(\alpha) = \frac{2k - 1}{\alpha}\zeta_{k-1}(\alpha) - \zeta_{k-2}(\alpha)$$

$$\zeta_0(\alpha) = \sin\alpha + i\cos\alpha, \quad \zeta_1(\alpha) = \frac{\sin\alpha}{\alpha} - \cos\alpha + i\left(\frac{\cos\alpha}{\alpha} + \sin\alpha\right) \tag{A.9}$$

$$\eta_k^3(\alpha) = \frac{\zeta_{k-1}(\alpha)}{\zeta_k(\alpha)} - \frac{k}{\alpha} \tag{A.10}$$

NOTATION

$a(\theta,d)$	element of scattering kernel
$b(\theta,\lambda,d)$	element of PIDS kernel
d	particle diameter
E_v	scattered electric field with vertically polarized light
E_h	scattered electric field with horizontally polarized light
$f(\theta)$	intensity flux at angle θ
I_v	scattered intensity with vertically polarized light
I_h	scattered intensity with horizontally polarized light
J_1	Bessel function of the first kind or order unity
$\mathbf{K}$	momentum-transfer vector
$\mathbf{K}_0$	incident wave vector
$\mathbf{K}_s$	scattered wave vector
m	m_1/n_0
m_1	refractive index of particle
n_0	refractive index of medium
$p(\theta,\lambda)$	PIDS signal at angle θ and wavelength λ
$P(\theta)$	scattering form factor
R_g	radius of gyration of particle
$x(d)$	size distribution
α	$\pi d n_0/\lambda_0$
β	$\pi d m/\lambda_0$
λ_0	wavelength of light in vacuo
λ	wavelength of light in medium
θ	scattering angle
φ	angle between Z-axis and scattered light (Fig. 1)

APPENDIX II: SCATTERING FORM FACTORS FOR RANDOMLY ORIENTED PARTICLES

Shape	Form factor	Variables
Sphere	$\left[\frac{3(\sin x - x\cos x)}{x^3}\right]^2$	$x = Kd/2$, d = diameter
Thin disk	$\frac{2}{x^2}\left(1 - \frac{1}{x}J_1(2x)\right)$	$x = Kd/2$, d = diameter
Thin rod	$\int_0^{2x} \frac{\sin t}{t}\,dt - \frac{\sin^2 x}{x^2}$	$x = Kl/2$, l = length
Random coil	$\frac{2}{x^4}(e^{-x^2} + x^2 - 1)$	$x = KR_g$, R_g = radius of gyration
Concentric spheres (core shell)	$\left[\frac{3(\sin x - x\cos x)}{x^3} + \frac{m_I - m_0}{m_0 - 1}\left(\frac{d_I}{d_O}\right)^3\left(\frac{3(\sin y - y\cos y)}{y^3}\right)\right]^2$	$x = Kd_O/2$, $y = Kd_I/2$ d = diameter m = refractive index O = shell, I = core
Circular rod	$\int_0^{\pi/2} \frac{\pi}{2y\cos\beta}\left[J_{0.5}(y\cos\beta)\,\frac{2J_1(x\sin\beta)}{x\sin\beta}\right]^2 \sin\beta\,d\beta$	$y = Kl/2$, l = length $x = Kd/2$, d = diameter β is angle between axis of rod and bisectrix
Ellipsoid of revolution	$\int_0^{\pi/2}\left[\frac{3(\sin x - x\cos x)}{x^3}\right]^2 \cos\beta\,d\beta,\ x = Ka\sqrt{\cos^2\beta + \left(\frac{b}{a}\sin\beta\right)^2}$	a,b = semiaxes; β is angle between figure axis and bisectrix
Cube	$\frac{2}{\pi}\int_0^{\pi/2}\int_0^1 E^2(x\cos\varepsilon\sin\gamma)E^2(x\sin\varepsilon\sin\gamma)E^2(x\cos\gamma)\,d(\cos\gamma)\,d\varepsilon$	$E(u) = \sin u/u$, $x = Kl/2$, l = edge; γ and ε are the polar angular coordinates of the bisectrix

REFERENCES

1. There are a few monographs that deal with light scattering characterization of macromolecules, see Refs. 2 and 3.
2. M. B. Huglin (ed.), *Light Scattering from Polymer Solutions,* Academic Press, New York, 1972.
3. P. Kratochvil, *Classical Light Scattering from Polymer Solutions,* Elsevier, New York, 1987.
4. G. Mie, *Ann. Phys. 25*:377 (1908).
5. M. Kerker, *The Scattering of Light and Other Electromagnetic Radiation,* Academic Press, New York, 1965.
6. S. Asano and M. Sato, *Appl. Opt. 19*:962 (1980).
7. R. J. Martin, *J. Modern Opt. 40*(12):2467 (1993).
8. C. F. Bohren and D. R. Huffman, *Absorption and Scattering of Light by Small Particles,* Wiley, New York, 1983.
9. R. M. J. Watson and B. R. Jennings, *J. Colloid Interface Sci. 157*:361 (1993).
10. J. Komrska, *Opt. Acta 19*(10):807 (1972).
11. J. Komrska, *Opt. Acta 20*(7):549 (1973).
12. K. A. Kusters, J. G. Wijers, and D. Theones, *Appl. Opt. 30*(33): 4839 (1991).
13. G. Gousebeth, B. Maheu, and G. Gréhan, *J. Opt. Soc. Am. A. 5*(9): 1427 (1988).

14. S. E. Bott and W. H. Hart, *ACS Symposium Series 472: Particle Size Distribution II,* (T. Provder, ed.), American Chemical Society, Washington DC, 1991, p. 106.
15. S. E. Bott and W. H. Hart, U.S. Patents 4953978 (1990); 5104221 (1992).
16. S. E. Bott and W. H. Hart, U.S. Patent 5056918 (1991).
17. G. Row, *Submicron Particle Sizing Using Laser Diffraction Instruments,* Coutler Technical Monograph, Coulter Corp., Miami, FL, 1993.
18. G. Row and S. Bott, *LS Series Reproducibility,* Coulter Technical Monograph, Coulter Corp., Miami, FL, 1993.
19. S. Sugrue and G. Row, *LS Series Resolution,* Coulter Technical Monograph, Coulter Corp., Miami, FL 1992.
20. *CRC Handbook of Physical Chemistry,* 75th ed. CRC Press, Boca Raton, FL, 1995.
21. E. D. Palik (ed.), *Handbook of Optical Constants of Solids,* Academic Press, New York, 1985.
22. I. Fanderlik, *Optical Properties of Glass,* Elsevier, New York, 1983.
23. J. L. Loizeau, D. Arbouille, S. Santiago, and J. P. Vernet, *Sedimentology 41*:353 (1994).
24. F. W. C. den Ouden and T. van Vliet, Special Publication, *R. Soc. Chem. 113*:285 (1993).
25. G. A. Hareland, *J. Cereal Sci. 21*:183 (1994).
26. A. Michoel, N. Jaeger, R. Sneyers, W. D. Wispelaere, E. Geladé, J. Kern, W. Gorter, P. V. Amsterdam, Y. D. Tandt, E. Houtmeyers, and L. V. Cottham, *Part. Part. Syst. Charact. 11*:391 (1994).
27. O. Hayakawa, K. Narahira, and J. Tsubaki, *J. Ceramic Soc. Jpn. 103*(4,5):392, 500 (1995).
28. H. Merkus, O. Bischof, S. Drescher, and B. Scarlett, Precision and accuracy in particle sizing. Round-robin results from sedimentation, laser diffraction and electric sensing zone using BCR 67 and 69, Proceedings of 6th European Symposium of Particle Characterization, Nürnberg, Germany, 1995, p. 427.

19

Industry's Role in the Development of Standards and Controls for Particulate Matter in Small Volume Injectables

John A. Syverson*

Pfizer, Inc., New York, New York

I. INTRODUCTION

In late 1983 at a Parenteral Drug Association meeting in Philadelphia, the United States Pharmacopeia (USP) delivered an address about their plans for official standards to limit particulate matter in small-volume injectables (SVIs). Their announcement that day contained surprising elements. The attention of the audience was caught, however, not so much by the resolve of their intention to move forward quickly, by an ambitious timetable, but more by the selection they made for the referee method of testing. Most had expected that the membrane/microscopic (M/M) procedure, then currently official in the USP for counting and sizing particles in large-volume parenterals (LVPs) would be the basis for the method for SVIs. Instead, the USP broke the news that they had chosen to establish an automated, instrumental means for testing SVIs—an electronic liquid-borne particle counter system that utilized a light-obscuration-based sensor. This was an abrupt and unexpected change having potentially significant implications for manufacturers of small-volume injectable products. Whereas most producer laboratories were familiar with the apparatus and method for USP's M/M procedure, few had rigorously applied, even experimentally, electronic particle counting (EPC) methods to their products.

Questions immediately arose to those affected by this announcement about the instrument, the method, the correlation of results between the M/M and EPC, and the status of validation studies on the proposed method when applied to SVIs. This latter question was pertinent and timely because the Food and Drug Administration (FDA), USP, and industry had all concluded, a short while earlier, that drug testing methods must undergo a systematic, scientific evaluation (validation) before becoming official in order to give reasonable assurance that they would yield acceptable results when performed as directed.

It was soon found that suitable validation studies on the EPC method for SVIs had not been done. Instead, the USP was satisfied that a "qualification" study by an LVP industry group was sufficient to verify its usefulness for SVIs. Thus, one had the ironic development of having a largely untried, unvalidated, and as yct poorly understood analytical method chosen by the

*Retired. John A. Syverson was Vice President, Quality Control Division, Pfizer, Inc. He was chairman of the PMA (now P*h*RMA) SVI Particulate Matter Committee during the period covered by this chapter and also served for a number of years as a member of the association's QC Section Steering Committee.

standards-setting authority of the government, contrary to their own policy, for the assessment of a critical quality factor of SVIs. What followed marked the beginning of more than a decade of uneasy accommodation of industry to this venturesome authority of USP, in a critical area of quality assessment, where only the manufacturers' laboratories had the experience and knowledge needed to deal with the complexities and vagaries of particulate matter control in SVIs. The unifying issue for both sides of the struggle was the single objective of attaining the lowest possible level of extraneous particulates in this important class of drug products.

This chapter will provide a brief overview of the search by industry's laboratory scientists for a more accurate, practical, and universal official standard for evaluating particulate matter in SVIs. Advances were made through the dedicated efforts of individuals and groups who provided USP with critiques on instrumentation and methodology factors, supported by factual information and data collected in collaborative studies and surveys.

II. BACKGROUND

Precedence for drug industry scientists to be called upon to assist government authorities in the establishment of tolerances for quality standards started in the early part of this century. As modern pharmaceutical manufacturing processes evolved with mass production and distribution of drug products, the need to set rigid standards to ensure physical, chemical, and microbiological quality became apparent to government and industry alike.

In the mid-1920s the head of the Division of Drug Control, U.S. Bureau of Chemistry, after finding variation in the strength and quality of some widely used drugs then on the market, met with drug industry scientists to establish a plan for the development of federal tolerance standards for important quality attributes of drug products. He conceived of a mechanism based on constructive collaboration which, without delegating the responsibility for establishing specifications to industry, welcomed their data and well-developed conclusions on methods, standards, and tolerances. This unique arrangement grew in value to the FDA and the Compendia over the years because of the enduring dedication of knowledgeable industry scientists to continuously improve and refine drug quality standards. Both government and industry were aware of the present and future value of a strong and quality disciplined United States pharmaceutical industry.

In the early 1950s, industry was asked by Dr. Frank Wiley of the FDA to help study visual inspection procedures for insoluble particles in ampules. This was one of the early collaborations of government and industry concerning particulate matter in injectables. Twenty years later, USP began a concerted effort to establish limits on particles in SVIs. All interested parties were invited to provide data and information pertinent to the development of a compendial method to detect, size, and count insoluble particulates in SVIs of 100 ml or less. In the early 1970s, the Quality Control Section of the Pharmaceutical Manufacturers Association appointed a committee to undertake this study and thus began a complex, collaborative effort lasting almost a quarter century. In an often contentious arena, the refinement of tests and limits for particulates in SVIs evolved. What follows is a chronological review of highlights and turning points of this important and arduous effort of industry scientists to assist federal regulators to develop effective and practical standards and methods to control particulate matter in SVIs.

III. METHODS DEVELOPMENT

As the work of the committee got underway, there were intensive discussions held by its members on whether particulates in SVIs should be measured by microscopic means or by one of the available electronic systems—light blockage, light scattering, or resistivity. The microscopic method was finally chosen because it was thought to clearly present the best opportunity for accuracy and consistency. Most important was the fact that it had already gained the attention of USP for use in counting particulates in LVPs. Confirmation of its acceptance by regulatory disciplines was noted in a paper out of FDA's National Center for Antibiotics Analysis in September, 1973, authored by Brenda Taliaferro, Thomas Alexander, Margaret Hewlett, and Peter Weiss: "Because of expense considerations and shortcomings of the forced-flow systems (that is, EPC methods) it is quite likely that the 'standard' will have to be the AR-2 (or membrane) filtration and microscopic examination method."

Initial collaborative studies were undertaken by the committee with specially prepared control lot samples of a pollen suspension. Later, blinded samples of actual commercial product were distributed and collaboratively tested. Analysis of results from these studies led to the realization that there was a need for industry/FDA/USP training seminars and workshops. These were held and were very successful in providing a common ground for government and industry technicians to focus on the details of steps in M/M testing. An outgrowth of this effort was the development by the committee of a slide-tape training program which served the purpose of standardizing technique details, leading to a more acceptable degree of uniformity of count results between laboratories on control samples. Over 60 sets of this program were distributed to industry and government laboratories worldwide.

A large M/M survey of marketed products was undertaken shortly after the slide and tape training program was in the hands of collaborating company laboratories. The data resulting from this survey showed logical and expected differences between particulate content of four types of injectable products—aseptic powder fills had the highest counts, solutions in syringes the lowest, while lyophiles and solution in vials were in between. With the completion of this study, the committee provided USP in mid-1983 with a written proposal for an M/M method for consideration as an official, referee method. Four months later, to the disappointment of many, USP announced at the Philadelphia PDA meeting their intention of bypassing the membrane/ microscopic method, and in its place proceeded to establish the light obscuration, EPC procedure as the official method.

Early in 1984, after the USP's plans and justifications for the new method were published in Pharmacopeial Forum (P.F.), the committee studied these in depth and then commented to USP of their concerns about this choice. First, there was disagreement with the proposal to establish the "light obscuration based sensor instrument" (i.e., HIAC/Royco) as the apparatus for the official, referee method for determining particle counts in SVIs. It was granted that the HIAC method could be useful for certain applications, for example, in-process control. But it was expensive, and there were several problems associated with its use which helped to explain why many firms had not adopted it for even routine use with SVIs:

1. Silicone droplets and air bubbles were tallied by the HIAC counter.
2. Factors involving reference standards and calibration were not fully understood.
3. Inaccurate data could be caused by particulate shape and orientation in the fluid.
4. It was impossible to identify particle types.
5. Problems with the sensor resolution of the HIAC were known to exist and needed to be overcome.

Second, from the information provided in P.F. and from what the committee believed to be the facts of the matter, the database used in part to establish shelf-life standards for this class of drugs was very small, not representative, and without consideration of the particulation trends of injectable formulations during aging. In fact, based upon the M/M method survey conducted two years earlier on marketed SVIs of the committee firms, grave concern existed about the possibility that a number of products currently on the market might not meet the parti-

Table 1 PMA Q.C. Survey: Membrane/Microscopic Particle Counts per Individual Container (99.7 Percentile Values)

Product type	Total number of determinations	Number of particles	
		≥ 10 μm[a]	≥ 25 μm[a]
Solutions in vials	845	2900	1200
Solutions in syringes	614	800	400
Aseptic powder fills	580	5300	1800
Lyophilized solids	544	2200	700

[a]Effective linear dimension in micrometers

cle count limits found in the Interpretation section of the proposed HIAC procedure in P.F.: 10,000 ≥ 10 μm and 1000 ≥ 25 μm per container.

The results of the industry survey, from a study involving blinded samples of representative product formulas of committee firms counted by the M/M procedure, were tabulated in Table 1.

The third concern of the Committee had to do with the fact that with this proposal many firms would be forced to purchase expensive equipment which was offered by *only one* U.S. supplier. This put industry and USP in an awkward position of dealing at least currently with only one source of new technology, maintenance, and improvement. Further, once purchased, it was estimated that considerable time could be consumed by a firm before gaining effective use of the apparatus in its applications. The proposal forced industry into unreasonable time pressures to acquire, install, learn, and apply this new method.

Finally, with regard to the proposed test procedure given in the Jan/Feb 1984 P.F. publication, it was considered by the committee to be of doubtful usefulness for the USP to propose a method which had not been carefully evaluated with respect to the specific application in mind (SVIs) by a sizable number of affected laboratories. Although the HIAC had been in use by a few companies, the proposed procedure had not been rigorously evaluated and systematically validated as a referee method for the variety of SVIs included in the USP proposal.

For these four reasons ((1) inherent deficiencies for referee purposes, (2) inappropriate limits—small database, (3) only one supplier of the method's apparatus, and (4) lack of general evaluation and validation leading to questions about the procedure itself), the committee concluded that the proposal was not well founded. Although the EPC method was thought to have some useful applications, the committee had a strong preference for the M/M method for official referee purposes. It stood ready to collaborate with USP in further studies of available particle counting techniques and to participate in a wider study of determining particulate matter levels in SVIs to create a significant database for use in the establishment of the common goal—meaningful and reasonable particulate limits and methods of assessment for small-volume injectable drug products.

During 1984, the USP received a continuing number of comments and recommendations about the EPC method of SVIs after it was published in the Jan/Feb 1984 P.F. It was destined for incorporation into USP XXI in a new chapter, <788> *Particulate Matter in Injections*. USP XXI monographs were to become official in 1985, but because of the constant flow of revisions, the effective date for <788> was delayed. Nevertheless, industrial laboratories were anxious to begin practical studies with the method to gain a solid basis of experience for critical comment that could be provided to USP for necessary improvements.

In early 1985, the committee undertook a definitive collaborative study of the revised and updated USP XXI procedure as published in the Nov/Dec 1984 P.F., with the purpose of challenging the EPC methodology and exploring counting variables and procedure ambiguities. The committee established special arrangements with two firms, Duke Scientific and Ben Venue, to aid in the preparation and subdivision of control lots of 10 ml and 1 ml size fills that were produced from a homogeneous suspension containing gradient sized polystyrene DVB spheres. The samples were sent out to over 30 instrument laboratories, including FDA. Unfortunately, the USP could not enter the study because, by their own choice, they did not have an EPC instrument in their laboratory. Results from responding laboratories, which were instructed to use their then available light obscuration particle count instrumentation, were tabulated and statistically evaluated. Data from some laboratories had to be culled because of extreme outliers. The wide variation in counting results across the 28 instruments completing the study was of significant concern to the committee. Possible causes for the variation, such as the complex array of EPC instrument components of the HIACs used in the study were communicated to USP, along with the statistical interpretation of results. It could be noted that the range of instrument means were of a significantly higher factor than with the earlier M/M aqueous pollen dispersion control lot. The Statistics Committee report pointed out the $\bar{x}$ charts showed a lack of statistical control while the R charts showed good statistical control. They interpreted this to mean that the testing process based on 28 instruments was not in control since there was a wide range among instrument means—from 2- to 14-fold. The ranges of total results for these control lots were so high that they consumed a substantial portion of the range that the USP had proposed for specifications (see table). They concluded that the testing procedure was not a valid method capable of determining acceptability of product using the proposed specifications. The results of this control lot study were reported to USP. Perhaps as a result of this input and others USP received, some important revisions were made by USP in the standard procedure later in 1985.

Phase I Study—28 Instruments November 1985

Lot (ml)	Particle size (µm)	Grand avg.	Range of instrument $\bar{x}$	C.V. of instrument $\bar{x}$
10	≥ 10	4339	2607–6208 (2.4×)	24.8%
	≥ 25	1261	402–2859 (7.1×)	48.4%
1	≥ 10	567	285–1787 (6.3×)	54.7%
	≥ 25	130	37–535 (14.5×)	79.2%

Another survey the committee performed in 1985 at the suggestion of USP was to determine what amount of currently marketed SVIs contacted silicone, that was acknowledged as an interfering substance for the EPC method. It was found that within just the companies represented by the committee, over 397 monographed products, representing 387 million container units annually, were manufactured in packaging systems in which silicone was present. It was estimated that almost 87% of total units produced in the United States were in contact with silicone and therefore subject to silicone interference effects. Later in 1985, a survey was made of actual HIAC count data accumulated by company laboratories on their own products using the then current revision of <788>, though still not official. This was the first multicompany survey of EPC results on marketed product for baseline data. Pooling of units had just been incorporated into the method, but some labs were still using single-unit determinations. This survey was rushed in order to meet USP deadlines, but 17 companies were able to provide data on almost 2000 lots of 280 different products. The most striking result of the survey was that both single and pooled samples gave an overall test failure rate above 3%. A nonconforming lot was defined as one having at least one sample tested exceeding either the ≥ 10 µm or ≥ 25 µm limits. The results of this Phase II 1985 collaborative survey were reported to the USP with the observation that the limits appeared too restrictive for the test.

As a result of all of these studies and surveys during 1985, a number of suggestions for improvement of EPC methodology were provided to USP in December, 1985—17 in fact. Some were accepted and others rejected for stated reasons. The prevailing recommendation of the committee was to replace the troublesome, single-source EPC instrument method by the M/M procedure as the referee method. In deference to USP, however, an alternate two-tier proposal combining both methods was made at that time:

> The electronic particle counting method is the referee method except where there is a known interference from inherent immiscible non-particulate substances associated with the container/closure system in contact with the product. In this case, the Membrane/Microscopic procedure becomes the referee method.

The USP was unwilling to accept this proposal but the effort served to reinforce industry's concerns about EPC methodology:

Interferences of silicone dispersion and air bubbles
Service dependency on single supplier
Calibration/resolution issues
Equipment performance variations
Nonvalidatable status
Nonqualitative results

As 1985 drew to a close, with many items of concern still unresolved, the method development phase of USP's new Chapter <788> test procedure for SVIs reached an important milepost—it became official and effective January 1, 1986.

IV. APPLICATION OF THE EPC METHOD

Despite the lack of success in convincing USP of the unsuitability of the EPC procedure for SVIs, when it became official, the committee pledged to continue both its dialogue with the revision committee and its collaborative surveys and studies that they believed were needed to improve the procedure to better meet criteria consistent with an official, referee method of the USP. Perhaps most important was the fact that the single source instrument specified by the method, the HIAC-Royco, was found to be unreliable and essentially unvalidatable. Specific areas pointed out by laboratories as most needing improvement as official testing started up were (1) instrument calibration, (2) sensor resolution, 3) coincident counting errors, and (4) sample degassing. These difficulties led to poor comparability between similar instruments. The manufacturer of the HIAC instrument was slow to correct the calibration and sensor issues causing frustration in the laboratories as they attempted to comply with the new requirements. The instrument was reported to have a number of defects which the manufacturer was unable to readily correct. The sensors often failed to meet the resolution requirements of <788>, lamp burnout occurred, and master cylinders required frequent replacement in the hydraulic controlled sampling system. Service of instruments during this implementation period was slow, and confidence was low that the failing electromechanical components of the instrument would ever be of the level of quality and reliability desired by the laboratories.

The instrument performance situation began to improve in the latter part of 1986 as industry laboratories continued to report their grievances to USP and HIAC. The Jul/Aug 1986 P.F. included a new footnote in <788> that allowed the use, for the first time, of alternate sources for sensors for the HIAC unit. Replacements with a Russell Laboratory sensor were found to greatly improve resolution results. The Russell bulb was also found to last much longer. This competition was helpful as HIAC subsequently followed with improved versions of their own. This was perhaps a key turning point in the quest for a more reliable instrumentation package for the <788 > EPC test.

In late 1986, the committee decided to undertake a new survey to determine how companies were progressing with the <788> test and what trends, if any, were developing with SVI particulate matter levels industrywide. The intent of the survey was defined as a program to collect and analyze existing particulate matter test data by the HIAC method for a one year period from January 1 to December 31, 1986. The data was to be blinded so that the source of individual test results would be unknown. Pooled and individual unit testing was to be explored and the survey would segregate data by package type for monographed SVI products.

Survey 3, as it was named, was successfully carried out and the data tabulated by an ad hoc statistics committee in 1987. Seventeen companies reported on six types of products providing data on a grand total of 5217 lots. The highest percentage of lots found to be nonconforming were with aseptically filled powder products at 0.4%. Two of the five lots in this category found to be nonconforming were stability samples. Lyophile products and partially filled bags had 0% nonconforming. Overall, a significant improvement was noted in the nonconformance level between Survey 2 (December 1985) of 1.49% versus Survey 3 (January–December 1986) of 0.31%. These results indicated that industry had responded well to the new USP requirement by showing an improvement in their capability to lower or at least maintain low levels of particulate matter in SVIs since the 1985 benchmark Survey 2.

One could speculate that the above improvement trends were due to closer monitoring of the process and materials—bulk product, glassware, and elastomer closures. Nevertheless, the committee remained concerned about the selection of the EPC procedure as the referee method. Though the Survey 3 study turned out to provide rational results, laboratories reported many difficulties with the test and equipment. The committee reported all of the above information to the USP in early 1988 and continued to propose a two-tier approach to <788> allowing recognition of the M/M procedure as qualitatively and quantitatively superior to the HIAC. The M/M method was recognized worldwide, and, in fact, was a *generic* procedure by nature, thus appealing to the international harmonization movement then in its infancy. Along these same lines, the Committee made note to USP in a talk presented at the PDA/IES Conference in February 1990 that overseas in Europe and Japan—the EP, BP, and JP—still did not have standards for particle counts in SVIs. The point was raised that it would then seem appropriate for the USP to develop a common international approach to particle counting. The obvious choice for this would be the microscopic method of the current USP and JP for LVPs. It would provide the basis for a universally acceptable and practical worldwide SVI standard that could be called "progressive" or even "forward looking" harmonization.

Still, as of February 1990, despite the Committee's persistent efforts during the past six years to convince the USP of the wisdom of allowing the membrane/microscopic test to have an official role in the particulate matter assessment of SVIs, there was no affirmative response indicating the likelihood of their acceptance for this purpose. But the outlook was about to change.

V. A PROPOSAL FOR REVISION OF <788>

In April 1990, approximately six and one-half years after the first announcement by USP of their intentions to establish a new particle content limit and EPC test for SVIs, another surprising move on this issue came from USP. The newly appointed Executive Director of the USP wrote to industry groups about several ideas from the USP Revision Subcommittee on Parenterals to make improvements to the official EPC method for SVIs in <788>. The USP Subcommittee wanted to know the views of the PMA Committee on some procedural details and one major and very interesting change. A significant new proposal advanced in this memo was to extend particulate matter assessment in the future to three stages: first, a light obscuration test as currently carried out but with new lower limits; second, if the sample failed the first stage, a chemical test to measure the amount of silicone, DEHP, or other immiscible interferences; and third, if the second stage was passed, a final assessment by an improved microscopic assay (IMA).

Somewhat in advance of the new proposal from USP to improve <788>, it became known that work had been started by the LVP segment of the industry to optimize the microscopic method—still official for the LVP's in the USP. As this work proceeded, the focus of potential changes for improvement narrowed to four: (1) improved lighting, (2) whole unit filtration, (3) improved graticule measurement guides, and (4) adoption of a statistical system for counting. These changes gave promise to provide more accurate M/M results with greater comparability to light obscuration EPC results. If this turned out to be the case, it was obvious why USP was interested in adopting the improved microscopic procedure for the <788> improvement program—the newly proposed three stage test. In fact it was the linchpin component of the new proposal, and the SVI Committee immediately included the IMA development as a major component of a comprehensive collaborative program to respond to USP's request for assistance.

The new three stage testing proposal for SVIs, one that included the M/M procedure that the committee had been recommending for several years, looked very promising on first notice but obviously posed some new questions. First, there was an indication that the EPC limits of the three stage test would be lowered, perhaps significantly, and this would result in higher levels of nonconformance, at least at the first stage. Second, USP was already aware from a previous committee survey reported to them of the prevalent use of silicone in today's injectable container/closure systems. Over 90% of the survey population used silicone for its lubrication properties on glass or stoppers. Furthermore, methodology for the proposed silicone/DEHP/other test at Stage 2 was seen as a definite problem, as no satisfactory quantitative test was available and a limit needed to be established. Third, the "improved" <788> proposal was to include a new EPC instrument qualification section named *Determination of Particle Counting Accuracy.* This was inserted in anticipation of the successful outcome of a USP collaborative study to qualify a particle count reference standard (PCRS). Study results were published in Stimuli, P.F. of Mar/Apr 1990, along with proposed criteria for instrument qualification. (Note: Subsequent statistical and practical analysis by two independent expert groups found the candidate PCRS lot inappropriate due to excessive nonuniformity. A replacement lot was later successfully prepared.)

With regard to the adoption of a revised and "improved" M/M method for use at Stage 3, the USP was informed by the committee that a collaborative study with agreed upon objectives was needed. USP offered the view that any such study of actual product data from an IMA must be carried out on aged file samples in order to give real shelf-life data for referee standard setting purposes. Additionally, within the IMA method it was recognized there were several issues to be settled including apparatus changes, sampling, pooling or individuals, limits and provision for retests. Realizing the scope of their proposed changes, USP requested that the committee develop information in several areas to assist them in their program to improve <788>: (1) to provide data regarding Stage 1 (EPC) limit tightening, (2) provide silicone assays for Stage 2, and (3) study the adoption of the IMA for Stage 3. The committee agreed to undertake these tasks in August 1990 and to report their progress to USP.

In order to prepare for implementation of the IMA, a three-step plan was developed: first, laboratory familiarization and training; second, the establishment of a written procedure for application to SVIs; and third, method qualification through a collaborative study. The first stage, training for technicians, was actively pursued and developed with the assistance of the principal expert involved in the IMA development. Two-day training seminars on the IMA were held at a central location with approximately 10 participants in each of two sessions. The agenda was to cover the procedures in detail for microscope setup, filtration, sizing, and counting, as well as the calculation and recording of data. It was anticipated that this would be a very worthwhile training effort toward the objective of familiarizing committee member laboratory technicians with this new procedure. Carried out in early 1991, 18 participants from 13 companies attended the intensive two-day "hands-on" seminars. The results were excellent. The IMA procedure was well accepted by the trainees and round robin studies on sample membranes that underwent testing during training showed sizing/counting results having reasonable coefficients of variation.

Following completion of the IMA training seminars, the second step toward application of the method, the development of a formalized written procedure, USP style, for testing of SVIs was necessary. Once completed, the committee was prepared to undertake the collaborative study, Survey 5, that was based on determining current capability of recently expired SVI products to meet the requirements of <788>, but with testing performed by a specially designed protocol. The objectives established by the committee for Survey 5 were: (1) to apply and qualify the use of the IMA procedure for SVIs; (2) to compare results from two methods—EPC of <788> and the IMA; (3) to compare results of two sampling plans for the IMA method—a pool of three units versus three individual units. The protocol called for each participant to sample recently expired monographed parenteral products that represented their marketed line. Each laboratory selected up to a maximum of 10 lots. Three tests were run on the samples: (1) the official USP EPC method of <788>; (2) the IMA method on a pool of three samples; (3) the IMA method on three individual samples. The two IMA tests provided data for comparing results per container from pooling versus results per container from averaging three individual counts.

Survey 5 results were reported to the USP in February 1993, with a recommendation that the IMA test be accepted for incorporation into a new <788> procedure for SVI's based on two stages—EPC testing then IMA testing. The silicone testing step (Stage 2) had by now been dropped as a USP requirement. The ultimate scope of Survey 5 covered 154 individual lots of six different parenteral product types. Eighteen laboratories had furnished particle count data and testing information. IMA results from sampling by either pooling or individual test means was essentially equivalent. In analyzing pooled results from both methods based on median values, the EPC method yielded counts 3.6 times higher than the IMA method for ≥ 10 μm and the IMA method yielded counts 3.5 times higher than EPC results for ≥ 25 μm. These differences were not surprising. The overall rate of nonconformance for the EPC method using the current <788> limits of $10{,}000 \geq 10$ μm and $1000 \geq 25$ μm was 0.7% which was only slightly above the normal range of Committee experience. The results of this study clearly indicated, however, that any drastic lowering of the limits, as initially planned by the revision committee, would disrupt the heretofore steady progress of conformance improvement.

VI. CONCLUSION

With regard to methodology alone, the proposal for revision of <788> to include the two stage, EPC/IMA, procedure in place of the single EPC test appeared very satisfactory to the committee. It was the culmination of almost 10 years of prodigious effort by industry laboratories to gain USP's acceptance of having the membrane/microscopic procedure incorporated into the official particulate matter standard for SVIs. The persistence of industry laboratory technicians and the investment of thousands of hours of analytical and statistical evaluations to produce relevant information and facts had made it possible. It was also attributable to the very unique system of open deliberation on drug standards that was first put in place at the beginning of the century by regulatory authorities—USP and FDA—giving industry the opportunity to technically collaborate with them in the mutual and worthy goal of improving the quality of drug products in the United States.

Chronology of Highlight Activities of Industry Committee

Approximate date	Event
1972	Completed methodology survey, decided on M/M
1974	Collaborated on M/M study of aqueous pollen samples
1976–76	Industry conferences with FDA/USP on M/M method
1978–79	Round-robin studies on membrane and product counting
1979–80	Industry Survey 1—collection of all count data available
1980–81	Industry Survey 2—limited data per company—50 sets
1983 (early)	Meetings with USP on various proposals of theirs
1983 (mid)	Industry's SVI M/M procedure sent to USP for publication
1983 (late)	USP's Nov–Dec PF Stimulus to Revision Process for <788>
1984	Study and analysis of HIAC and <788> proposals of USP
1985 (early)	Industry designs/prepares control lot for application of <788>
1985 (late)	Industry sends final report to USP on Control Lot Study
1986	<788> goes into effect
1987	Survey 3—collection/analysis of 1986 HIAC "release" data
1988	USP initiated collaborative study of a Particle Count Ref. Std.
1990	Letter received from USP about their intentions to change <788>
1991 (early)	Survey 4 results (July 89–June 90) on "release" data
1991	Industry carried out technician training on IMA
1991	Industry provided a statistical critique on PCRS report in PF
1992	Survey 5 protocol developed–expired lots survey using EPC and IMA
1993	Industry reports results of Survey 5 and recommends USP implement the two-stage test
1994	Publication in P.F. of two-stage test that specified an improved membrane microscopic assay for Stage 2 testing

Particulate Matter Control of SVI's Collaborative Studies and Surveys of the Industry

Date	Name	What	Why
1974	Initial exploration	Aqueous pollen samples	Explore M/M method
1978	Round-robin	Marketed product	Discover issues of real product
1979	Survey 1	Marketed product	Tabulate *all* current data
1980	Survey 2	Marketed product	Equal data points each comp.
1985	Phase I survey	Count control lot by HIAC	Apply draft <788> procedure
1985	Silicone contact	All marketed product	% product contacting silicone
1985	Phase II survey	Available HIAC data	Collect initial data on HIAC
1987	Survey 3	All 1986 release data	1st year data from <788>
1991	Survey 4	July 89–June 90 release	Determine conformance trend
1992	Survey 5	Recently expired lots-IMA	Qualification of IMA method

Other Important Industry Inputs

Time period	Activity
mid-1970s	Industry sponsored conferences with FDA/USP on methods development
mid-1983	Industry provides USP with formal membrane/ microscopic procedure for publication in PF
late 1990	Industry provides USP with specific comments on new "three stage" revision to <788> with M/M
1991	Industry organized and carried out a two-day technician training session for new IMA procedure
1991	Industry ad hoc statistics committee provided USP with critique on Particle Count Reference Standard report

20

Pharmaceutical Clean Laboratory Operations

Thomas A. Barber

Baxter Healthcare Corporation, Round Lake, Illinois

I. INTRODUCTION

Any laboratory operation carried out for the purpose of counting, characterizing, or identifying particulate matter must be protected from extraneous particles that might introduce error into the data collected. Analyses of hydraulic fluids, fuels, lubricants, and aerospace vehicle components, environmental samples, and materials all have an inherent level of sensitivity to extraneous particles. Obviously, if particle count assays are being performed on any type of product fabricated under extremely clean conditions, even very low numbers of extraneous particles can have serious consequences. While testing pharmaceutical materials according to compendial requirements is much less sensitive to extraneous particles than are analyses related to integrated circuit manufacture, requirements still exist for the control of extraneous particles at levels below a point where data will be adversely affected.

Particle analysis laboratories must be operated under conditions of cleanliness that are adequate for the specific analyses performed. A point implicit in this discussion is that the level of cleanliness in our procedures and in our laboratories should meet but not significantly exceed the requirements of the work being performed. While a sufficiently clean analytical environment is of the utmost importance, higher levels of cleanliness entail additional incremental expense. Since our goal is least-cost generation of our product (data), unnecessarily stringent requirements for laboratory cleanliness will add cost without benefit. The more thoroughly the analyst and laboratory manager understand the requirements of an analysis, the more cost-effectively will adequate cleanliness be attained.

This chapter comprises an introductory discussion of what may become an exceedingly complex subject; specific analyses, particularly those performed on submicron particles, may require extensive modification of the techniques discussed in this chapter. Critical principles essential to the operation of the clean laboratory include

Controlling particles in the general laboratory environment (in the air and on surfaces)
Monitoring of environmental particulate matter in the laboratory
Providing clean solvents and flush solutions
Eliminating extraneous particles in the immediate analytical environment

---

I must emphasize that any and all subjective judgments, opinions, and viewpoints expressed in this chapter are my own and not necessarily those of Baxter Healthcare Corporation or any organization representing manufacturers of medical devices or injectable products.

Adequate cleaning of apparatus to be used in testing
Controlling particles from personnel
Training of all laboratory personnel

This chapter is divided more or less along these lines with subject material addressing each of these components. Though the discussion emphasizes pharmaceutical laboratories, the basic principles discussed are more widely applicable. The paper by DeLuca et al. [1] referenced at the end of this chapter provides a very useful introduction to the subject of clean laboratory operations.

II. ENVIRONMENTAL PARTICLES

For purposes of this discussion, environmental particles are defined as those present as randomly dispersed airborne contaminants in the ambient air of the laboratory. These particles range from the submicron fragments of glass, plastic, metal, or biological material, to large visible fibers of paper or cotton. This particulate matter population is distinct from process- or point-generated particles that may occur at high levels in local areas due to human activity or instrument operations. The control of environmental particles is based on three key mechanisms: (1) continuously supplying clean air to the laboratory area by means of high-efficiency particulate air (HEPA) filters; (2) removing contaminated air through return plenums; and (3) monitoring numbers of airborne particles with optical particle counters (OPCs).

Generally, the most sensitive laboratory analyses with regard to extraneous environmental particles are those involving microscopic particle counting. In these analyses, one typically wants to enumerate heterogeneously sourced contaminant particles associated with a pharmaceutical solution or device, and extraneous particles from the environment may easily be confused with those related to the product. Analyses performed using light extinction counters such as HIAC counters (Pacific Scientific, Silver Springs, MD) are far less sensitive to particles that fall or settle onto the test articles. Light extinction counting (LEC) of many solution products may be performed by simply inserting the instrument probe into the unit prior to analysis; while the unit closure must be carefully removed so that particle generation is minimized, areas with relatively high levels of airborne particles (e.g, 10 to 100 K) may be used for analysis. Some vial products may be sampled using a noncoring needle inserted through the closure of the unit. This type of analysis must be carefully performed to ensure that particles generated by the needle penetration do not affect the result.

There is some divergence of opinion regarding the levels of cleanliness required for work involving the analysis of particle composition using instrumentation such as microprobes or infrared spectrophotometers. If particles to be worked with are large (e.g., ≥ 50 μm) and readily distinguishable on the basis of their morphology, environmental particles may have little detrimental effect on the isolation of the particles of interest. For particles of subvisible size, higher levels of cleanliness are desirable, and Class 1000 environments or Class 100 laminar airflow hoods with HEPA filters are appropriate. In any analytical procedure it must be kept in mind that particles that can settle from the air onto the test articles are not the sole concern; particles that have previously settled onto surfaces or onto the work area, or particles from the analyst's tools or other equipment will also contaminate the analysis if they are transferred to the sample.

III. CONTROL OF PARTICLES IN THE LABORATORY

Control of extraneous particles from all sources is critical to the analysis. Despite air filtration and other safeguards, gross atmospheric contamination and particles from surrounding areas tend to find their way into the laboratory. External contamination (such as combustion particles) may be brought in through the air-conditioning system supplying the laboratory environment with makeup air. Contamination from adjacent areas may enter through doors and pass-throughs. The external contamination from adjacent areas to which work in a laboratory is exposed is critically affected by the type of air filtration used, overpressure gradients, airflow patterns, and human access.

People are the greatest source of contamination in the laboratory due to the fact that we each continually produce and shed particles (both viable and nonviable). Shedding of skin cells generates particles in the 1- to 20-μm range, and exhaled air contains large quantities of particles ranging from submicrometer size to several hundred micrometers. In addition to people, moving contact between surfaces creates contamination. For instance, writing with a pencil on a piece of paper generates an aerosol of fine carbon particles and paper fiber fragments. The movement of a metal part on a bearing surface may generate particulate matter in the form of a fine metallic dust. Wiping a work surface with a cloth may only suspend particles or move them around.

Within the laboratory, air movement results from people walking and from the operation of machinery, fans, and motors. All of these factors tend to cause the air to move with random velocity and direction in the workspace. Particles carried in random currents may easily move from one area to another unless they are controlled by some means. This movement of contamination via random air currents from one part of the laboratory to another is commonly termed cross-contamination. Over time, contamination from all sources will build up in the laboratory area on surfaces and in the air and in a worst case may reach a steady state condition with particles being added as quickly as they are removed. For this reason, it is imperative that work surfaces in the area of microscopes or in areas where filtration will be performed are wiped down before an analyst begins work.

Airborne particles range in size from 0.001 μm to several hundred microns and will settle on containers, filters, and other exposed surfaces at a rate that depends on the size and density of the particle. According to Stokes' law, a spherical particle of 50-μm size with a unit density would take one minute or less to settle 10 f, while a particle in the 1-μm range could take 15 to 20 h to settle on a surface at the same distance.

The monitoring of aerosol particles in clean pharmaceutical laboratory environments is critical to good laboratory practice (GLP) and quality assurance of analytical data. The primary reason for concern with monitoring aerosol particles is that these materials can adhere to surfaces that should be clean or enter the product filter, funnels, clean beakers, etc. Although such particles are generally nonviable organic or inorganic materials, they may transport bacteria and could conceivably be chemically reactive. Particle monitoring to ensure that air quality is maintained is particularly critical to aseptic fill processes and is also important in laboratory operations. Before any method of environmental control of airborne particles in the laboratory is applied, a decision should be made as to how critical the presence of particles is to that specific analytical operation. Allowable size/number of airborne particles at a point within an area depends on the most critical tolerances of the testing being performed at that particular point. Also, the quantity of the particles of a given size that will be generated at laboratory work stations within the area of concern must be considered.

HEPA filters fulfill a number of critical functions in the clean laboratory. They are used to prevent access of particles to the analysis, and also to provide clean airflow that serves to dilute fine particle aerosols generated within the clean area where sensitive analyses are being carried out. Often a HEPA filter is used to generate laminar (unidirectional) airflow that serves to "wash" particles away from a particularly sensitive area.

IV. APPLICABLE STANDARDS

Numerous standards have been generated to deal with particles in controlled environment areas (CEAs) and clean rooms. These are generally applicable to laboratory areas. Probably the two most widely used are Federal Standard (FS) 209D and FS 209E [2] and British Standard (BS) 5295 (Part 1) [3]. These standards specify allowable particle levels at specific sizes for various classifications (Fig. 1). The American standard departs from previous issues of the FS 209 series by providing both metric and English class designations (particles/ft^3).

Such standards are useful since they provide a basis for implementation of uniform

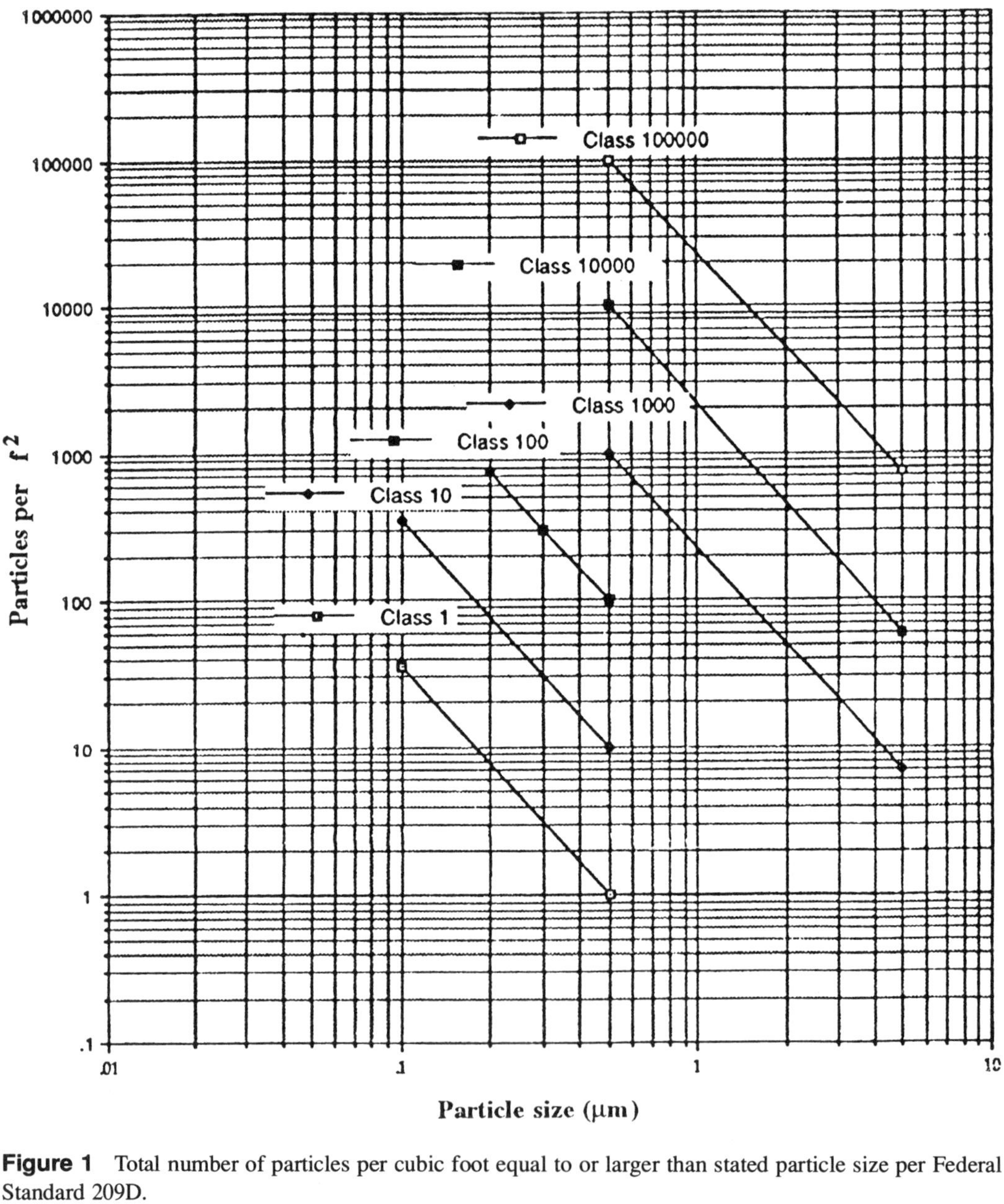

Figure 1 Total number of particles per cubic foot equal to or larger than stated particle size per Federal Standard 209D.

procedures for particle control and monitoring, in both laboratory and production areas. Users of these specifications, however, should avoid two misconceptions. First, both standards imply the existence of particle distributions for various cleanliness classes. Such distributions, unfortunately, are theoretical and have no basis in fact. Monitoring airborne counts at a single size has no implication regarding counts at other sizes. Secondly, both standards specify monitoring schemes for various classifications. The plans specified are useful but not optimal for all areas that will be monitored. With regard to the classifications of laboratory-type CEAs, it is very common to find laboratories designed with two or more different area classifications in a single laboratory. As an example, Class 100 hood or laminar flow areas may be placed at the periphery or even in the center of a Class 1000 or Class 10,000 area. These cleaner areas provide the necessary particle control for the performance of microscopic particle counting or particle isolation; the higher class areas are used for instrumental counting, visual inspection or other less sensitive procedures.

V. PRINCIPLES OF HEPA FILTRATION

HEPA filters are a critical particle control mechanism in pharmaceutical environments [4]. Today, HEPA filter media is made of a mat of glass fibers that are a few microns in diameter held together with a polymeric resin. Despite the fact that the filters are constructed of glass, the HEPA medium is often referred to as "paper." The glass fibers are disposed randomly throughout the depth of the filter medium. In older designs for pharmaceutical use, the glass medium was supported by pleated or corrugated metal dividers (Fig.2); current designs often incorporate pleats or "dimples" for separation (Fig. 3). HEPA filter paper does not have a controlled pore

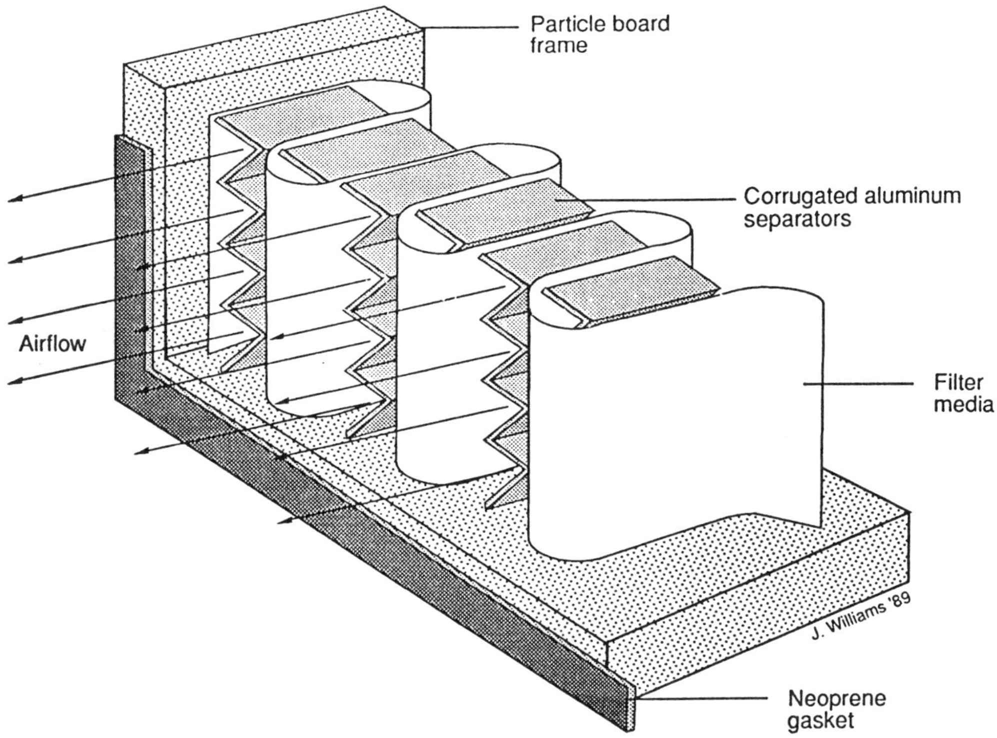

Figure 2 Historical HEPA filter construction.

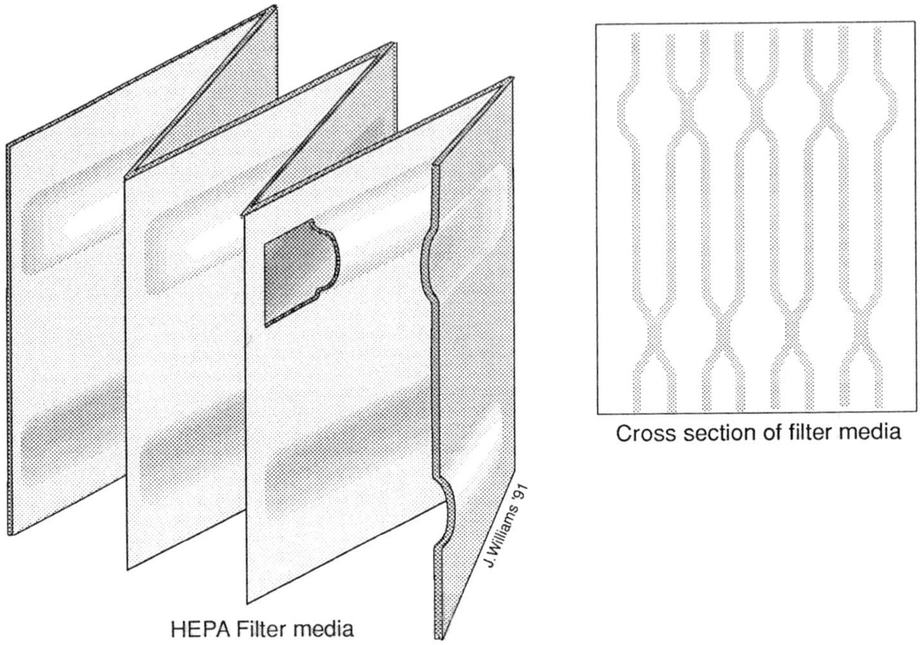

Figure 3 Folded HEPA filter media with "Dimple Pleat" separation of layers. (Flanders Filters, Inc.)

size; spaces between fibers are most often much larger than the particles captured. As an aerosol moves through the filter paper, entrained particles collide with fibers or other particles that are already stuck to the fibers. A strong electrostatic attraction is established when a particle bumps into either a fiber or a particle already adhered, so that a particle once adhered to the fibers of the HEPA filter is unlikely to be dislodged.

All HEPA filters are constructed so that the filter paper is arranged in a large number of parallel pleats. The pleats are narrow and relatively deep and serve to increase the filtration surface area. A typical 2 ft × 4 ft × 6 in. HEPA filter contains 140 to 190 such pleats. The pleats also act to align the flow paths of air molecules moving through the filter. The resistance of the HEPA filter paper to air or gas flow is fairly uniform across the surface, dictating that approximately equal volumes of air move simultaneously through each pleat of the filter. These two factors, the uniform resistance of the filter paper and the large number of parallel pleats, cause the filtered air to converge in generally parallel flow paths 3 to 9 in. downstream of the HEPA filter face. This flow pattern was historically termed laminar flow. Today, the term "unidirectional" is more commonly used. The use of HEPA filters in a laboratory area does not ensure laminar airflow patterns at critical areas. Other parameters, such as airflow rate and placement of equipment (i.e., light fixtures), must be controlled in order to maintain laminar flow (Fig. 4).

Three principal mechanisms (impaction, diffusion, and interception) cause particles to contact a capture point on a fiber of a HEPA filter (Fig. 5). In the process of capture by impaction, large particles with sufficient mass/momentum to leave the airstream as it turns to move around a fiber and contact a capture point. In capture by diffusion, smaller particles without sufficient mass to leave the airstream on their own move about randomly, and this random motion eventually results in their contacting the filter fibers. Larger particles that have deviated from the flow path enough to collide with fibers at a grazing angle are captured by a mechanism

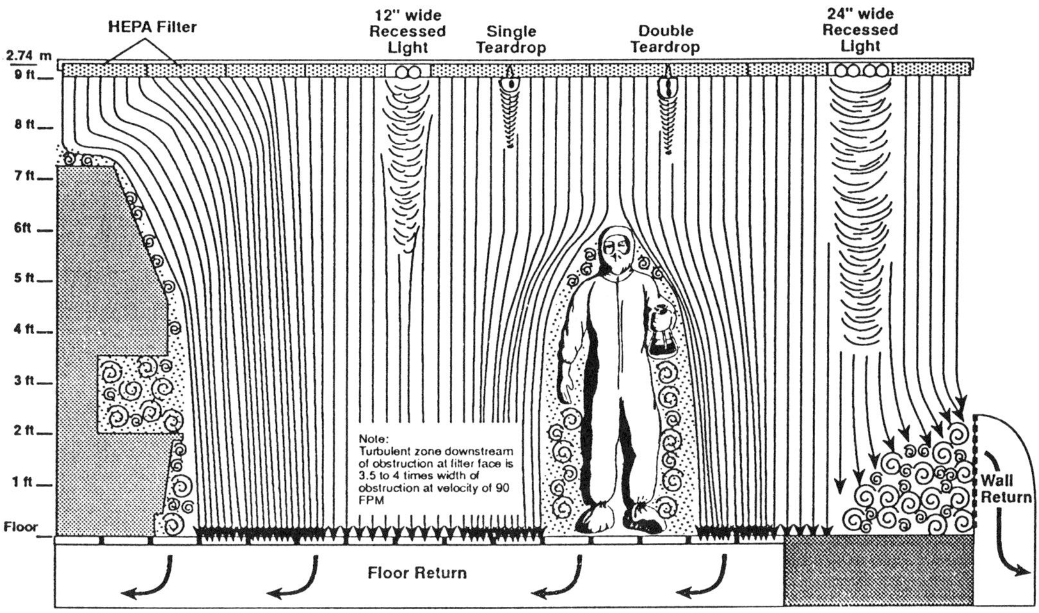

Figure 4 Laminar and turbulent flow in a clean laboratory.

known as interception. A fourth mechanism of filtration, screening or straining, can also occur in a HEPA filter. Straining simply means that the particles lodge in the spaces between fibers. The straining effect becomes important as particles exceed 5 μm in size.

Filtration of small particles by fibrous filters has been the subject of numerous theoretical and experimental studies during the past 30 years. As a result, the dependence of filtration efficiency on particle size is now well established [4]. Larger particles are captured by the interception and inertial impaction mechanisms; smaller particles are collected due to diffusion. There is an intermediate particle size region where both mechanisms are least effective. This is the region where the particle penetration through the filter medium is at a maximum and the retention efficiency of the filter is at a minimum (Fig. 6).

The particle size at which the minimum efficiency occurs will vary depending on the type of filter, retention characteristics of the medium used, and the flow velocity. For fibrous HEPA filters operating at relatively low filtration velocities (75 to 105 fpm), the generally accepted minimum filter efficiency occurs at approximately 0.3 μm. This is the basis of the widely used hot dioctyl phthalate (DOP) test for HEPA filters that makes use of monodisperse DOP particles with a distribution centered around 0.3 μm for testing the filter. The 0.3-μm minimum efficiency figure was established in the 1950s. The more sophisticated equipment available today (i.e., laser particle counters) suggests that 0.12 to 0.15 μm may be the particle size range most difficult to capture. In any case, airflow rate through the filter will impact the filtration efficiency at a given size; a higher flow rate increases capture by impaction and interception and decreases capture by diffusion. A lower flow rate works in the opposite fashion.

Most air cleanliness requirements in pharmaceutical laboratories are well satisfied by average filter efficiencies of 99.95% or greater; areas of greater penetration ("pinhole leaks") may be tolerated as long as the overall penetration does not exceed .05%. This is unlike the case in critical laminar flow systems (clean work stations, clean rooms, downflow hoods) where the HEPA filters are located at the point where the air supply enters a clean room or clean work area.

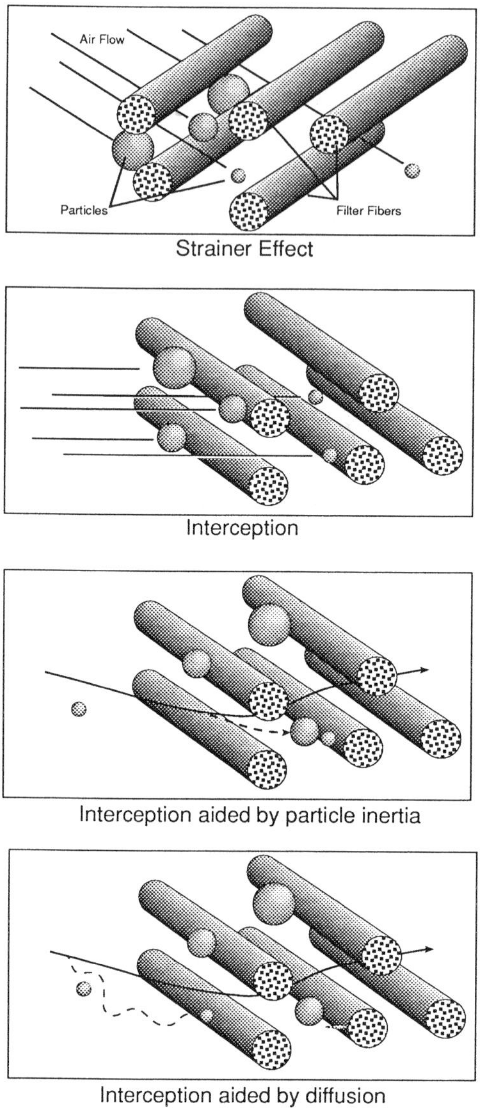

Figure 5 Particle entrapment by media of a HEPA filter.

In this application, particles passing through a pinhole leak are not diluted by the volume of clean air passing through the filter for some distance downstream of the filter face, and it is possible that the product/process requiring particle-free air could be located directly downstream of a pinhole leak. In general, our greatest concern in the clean laboratory will be with particles $\geq 10\ \mu m$ in size. HEPA filtration serves to ensure that particles of this size are not present in the air supplied to the laboratory, and directional airflow will flush particles of all sizes away from work areas. It must be emphasized that HEPA-filtered air will not detach particles from work surfaces: point-generated particles that can contaminate test articles thus must be dealt with by surface cleaning measures, such as wiping the work surfaces down with clean wipes and a solvent such as isopropanol (IPA).

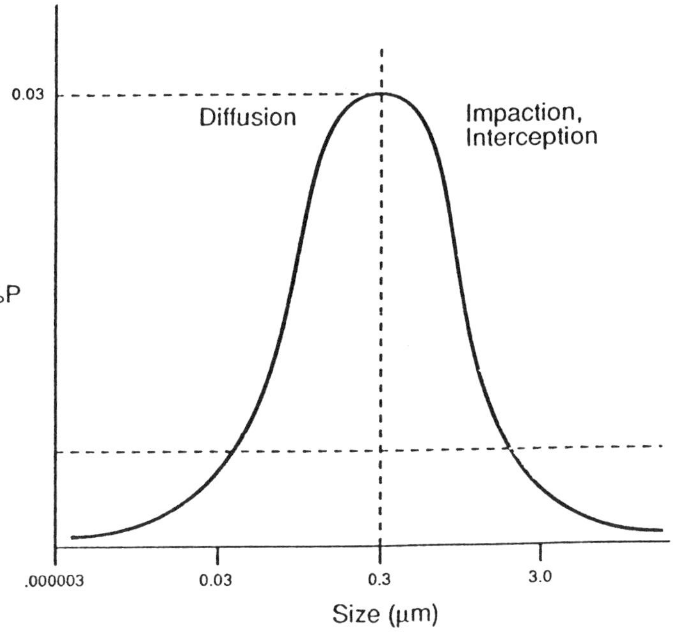

Figure 6 Particle retention curve for HEPA filters.

VI. PARTICLE MONITORING IN THE LABORATORY

The particle monitoring technology available for use in the pharmaceutical laboratory in 1994 is far advanced over that available 5–10 years ago. The optical particle counters that are the key element of airborne particle monitoring have been greatly improved over earlier models [5–8]. Whereas eight years ago the most common practice for airborne particle monitoring in the laboratory was based on portable OPCs and manual monitoring, many laboratories today incorporate remote sampling or remote sensor technology with sophisticated computer-based data collection and reduction.

Since 1980, a variety of new airborne particle counting instruments and systems have been developed. At this time, the most sensitive instruments are capable of counting and sizing 0.05-μm particles in the air at a flow rate of 2.8 l/min or 0.1 cfm [9]. A variety of particle counting instruments are available for measurement of particles larger than 0.1 μm at sample flow rates of 28.3 l/min (1.0 cfm). These instruments are well suited for verifying and monitoring CEAs of Class 10 to Class 100 and lower. For areas rated at Class 1000 and higher, instruments with lesser sensitivity and lower flow rates can be used equally well. Generalized function of a light scattering OPC is shown in Fig. 7. Particles pass through the light beam (sensing volume) one at a time. Intensity of scattered light is measured by a photodetector, and photodetector pulses are analyzed by a pulse height analyzer. Relating the pulse height data to the particle size information yields a response curve for the instrument and allows particles counted at given size thresholds to be quantified.

The accuracy of the size distribution reported by an OPC is dependent upon the response of the counter [10]. The response of the counter is in turn affected by the refractive index and shape of the particles, as well as by the optical design of the instrument (including the light source used). Some understanding of the function of these instruments is critical to the selection of the best instrument for a given application based on its operational parameters. A basic

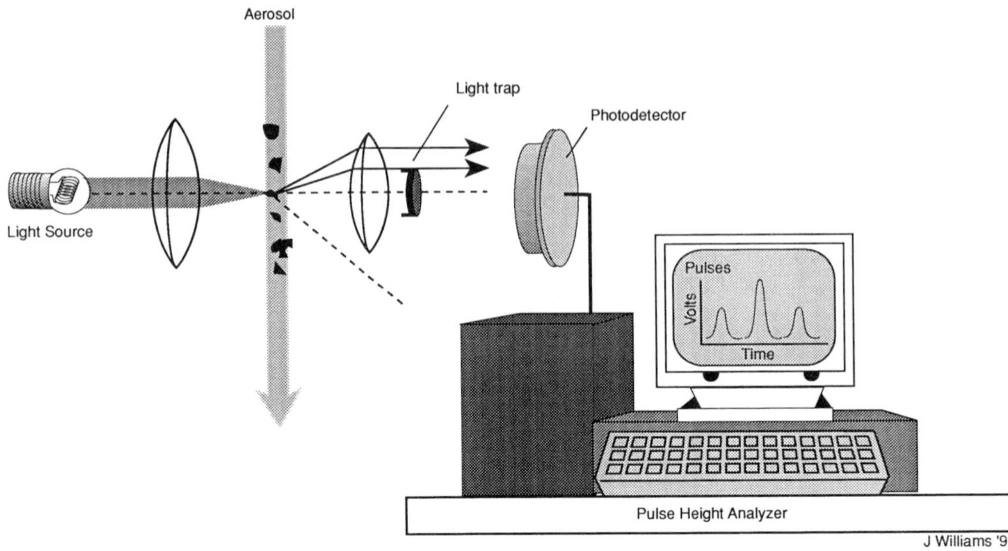

Figure 7 Operational principles of a light scattering optical particle counter.

premise to keep in mind is the following: A counter with greater low end sensitivity (i.e., 0.2 μm) may be expected to give somewhat higher counts at the 0.5 μm size used to define FS 209E classifications than one stated by the manufacturer to have a 0.5 μm low end sensitivity. Despite the statistical advantage of counting at lower sizes in Class 100 or Class 10 areas, the 0.5-μm size is still the most frequently monitored particle size in pharmaceutical laboratories.

The basic, most critical differences between single particle OPCs marketed today involve source illumination (laser or white light) and the geometry (forward, near forward, or wide angle) of scattered light collection [11,12]. A white light sensor with wide-angle collection geometry is shown in Fig. 8. With white light counters, quartz-halogen bulbs are typically used as a light source. Laser counters may use He-Ne hard-seal lasers or a solid-state laser diode. Other factors being equal, sensitivity (i.e., the smallest particle that may be sensed and counted) will be dependent on intensity of the illuminating radiation and its wavelength range. Generally, more intense illumination radiation results in a greater amount of scattered light for particles of a given size. A hard-seal laser in the active cavity mode gives an essentially monochromatic light with an extremely high photon density, and results in the best sensitivity available (0.05 μm). A functional drawing of a contemporary laser-based OPC is shown in Fig. 9. A key design feature is diode array detection of scattered light [13]. The signal-to-noise ratio for light scattered by a particle of given size is increased by this design due to scattering being sensed by a single diode of the array, rather than the much larger single-photodiode detectors some designs use.

Important in this consideration is that counters calibrated by different methods or counters of different designs must be expected to yield different counts for the same particle populations [14,15]. In particular, particle population distribution information from different counters may be expected to show wide variation. The response curve of an OPC is typically nonmonotonic for particles that have different sizes and refractive indices; the shape of the response curve may also differ dependent upon whether the collection optics used are of near-forward, wide-angle, or off-axis design (Fig. 10). At sizes < 1 μm, particles of different sizes with different shapes and refractive indices may, in many instances, give the same pulse height and thus appear to be the same size. While counter pulse height response generally increases with particle

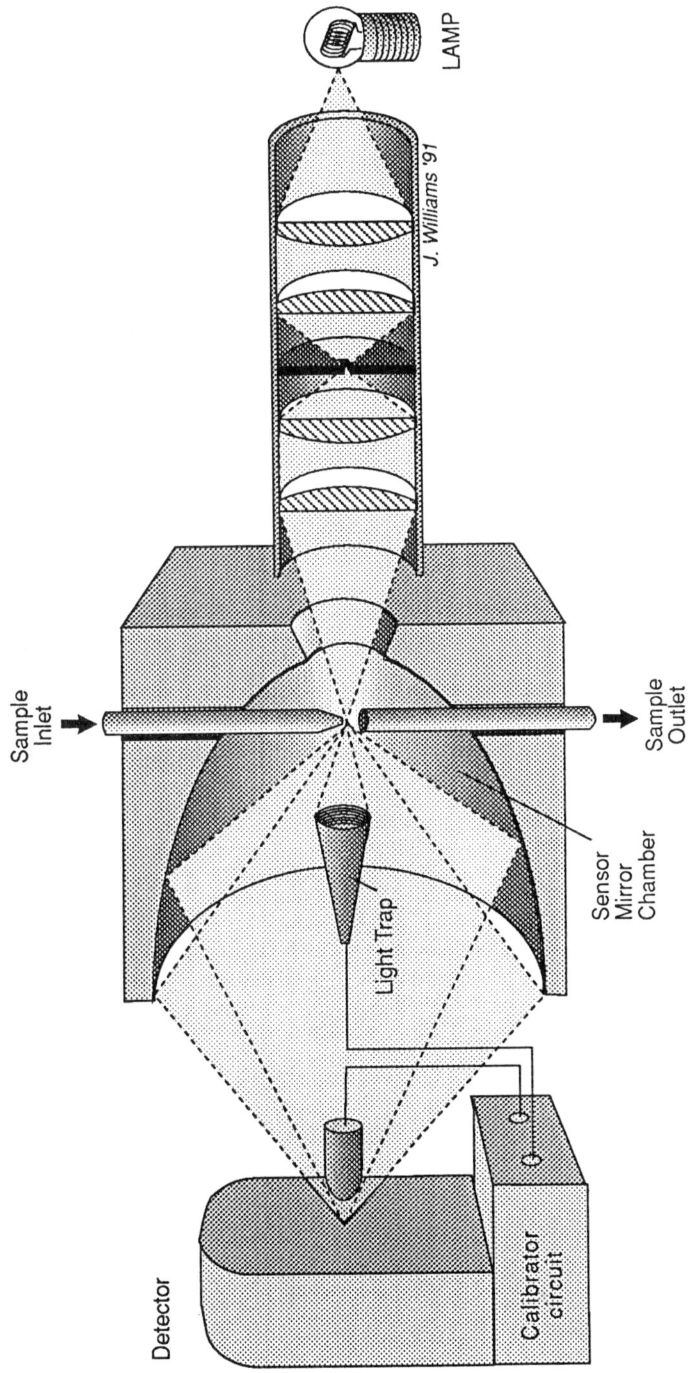

Figure 8 White-light-based OPC with ellipsoidal mirror collection optics. (Climet Instruments.)

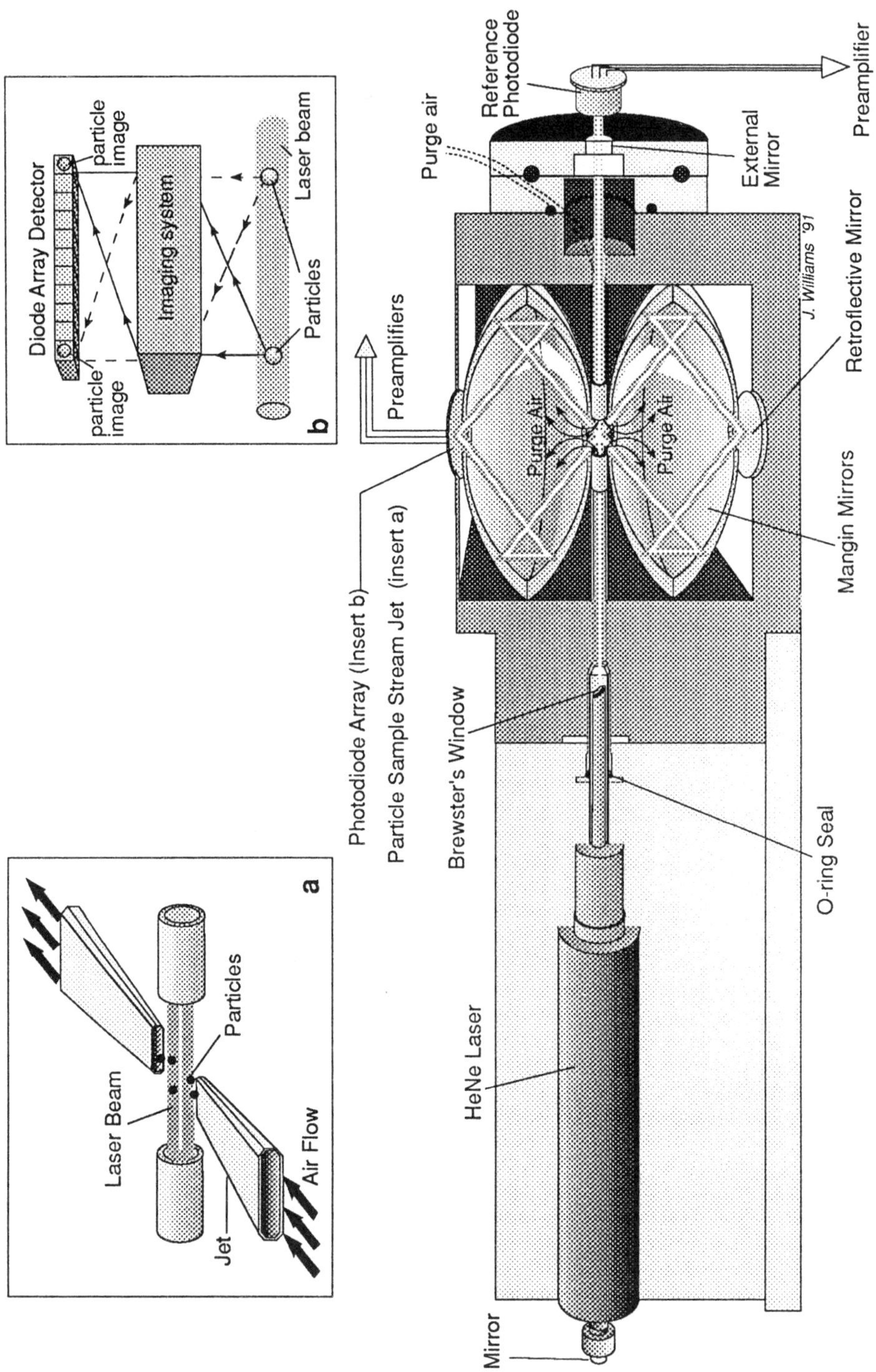

Figure 9 Active cavity laser OPC sensor with strip photodiode detector. (Particle Measuring Systems, Inc.)

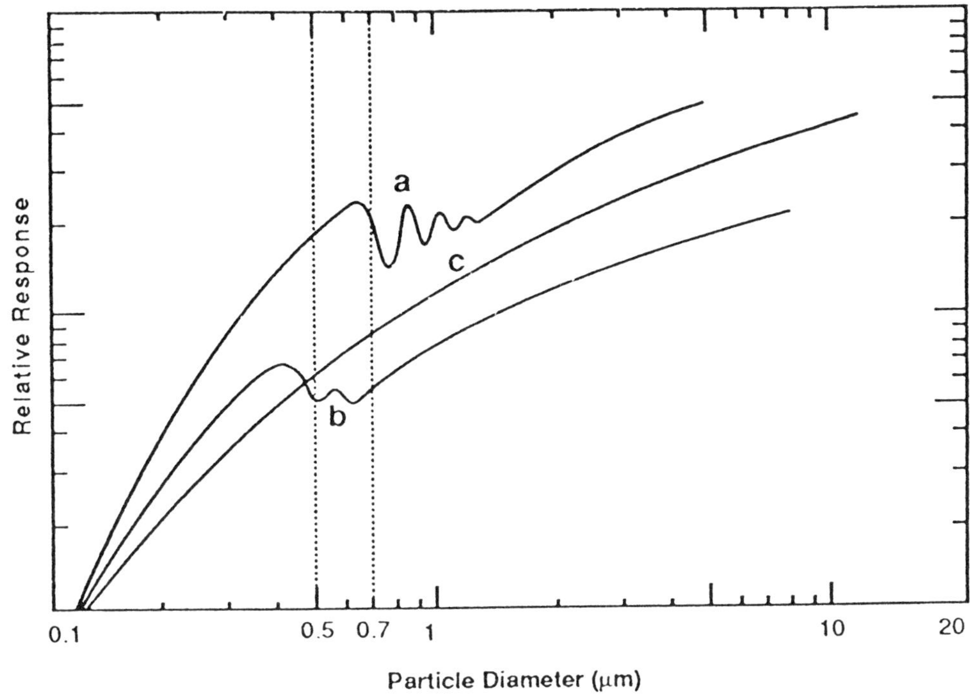

Figure 10 Generalized response curves for counters of different optical design.

diameter, resonances in the response curve may result in very different responses for particles of closely similar sizes [16].

Shape and orientation of nonspherical particles also affect the counter response [17]. If particles have dimensions less than the wavelength of illumination, the measured size may be considered equal to the volume equivalent diameter. For particles with dimensions above the wavelength of illumination, the measured particle size approximates the projected area equivalent diameter for instruments with wide-angle collecting optics. Refractive index differences between particles of the same size may cause an OPC to see the particles differently. The refractive index of a particle is described by a complex number. The real (scattering) and imaginary (absorption) components of this number are both functions of the wavelength. Thus, both components of the refractive index affect the response of an OPC.

One key concern for the laboratory manager in upgrading the OPC technology is the high count levels that may be obtained with some current generation counters versus their older counterparts [17]. The situation may arise where a new counter applied in a Class 100 area indicates out-of-limits conditions, whereas counts with an older model indicate well below 100 particles ≥ 0.5 μm per cubic foot [11]. The higher count with the newer instrument will have no effect on analytical work, but may disrupt operations while the source of the higher counts is investigated. The purchaser is well advised to (1) thoroughly test new units prior to purchase; (2) become familiar with operational principles and calibration techniques; and (3) ensure that the older units to which new ones will be compared are in calibration and are not operating at decreased sensitivity due to their age and long-term use.

One of the chronic problems related to manual airborne particle counting in both laboratory and production areas is that the "snapshot" that the count represents is unlikely to capture random "spikes" or elevated counts that occur sporadically. The easiest way to ensure that outlier data is captured is remote monitoring of critical areas using counters designed for this purpose. Data from these counters may be exported electronically to a computer database, providing a constant record of the count levels in critical areas where analysts are working.

Monitoring of large particles (≥ 10 μm) present in low numbers in controlled environment laboratories requires special techniques. Surface contamination with large particles can be measured by light microscopic counting and sizing, and for surfaces that are not accessible one can use tape sampling methods. There are also video camera-based techniques and scanning laser light scattering devices for this purpose. In electronics manufacturing, these latter devices may be used for evaluation of the product itself, such as silicon wafers or chips.

One of the most practical methods for monitoring larger particles is based upon microscopic evaluation of "witness plates" that have been exposed to the same environment as the product. Although automated devices for reading witness plates are available, the manual light microscopic evaluation of witness plates is the most useful method for pharmaceutical production environments. The technique provides a simple and effective way of measuring large particle fall-out and particles generated at process points.

The technique is simple in application. Witness plate samplers may be constructed using one of the following materials:

Black, double-surface vinyl tape
Agar gel
Filter paper
Polished glass
Aluminum mirrors

In my experience, the first material works best. A one inch square of black double-surfaced vinyl tape is placed in a Petrislide container. The surface is washed with 0.2-μm filtered water from a pressure vessel and dried in a HEPA-protected hood. Following the drying of the Petrislide with the tape, the section is examined microscopically for cleanliness and exposed the area to be tested for 24 h or some other appropriate interval. The number of particles settling on the tape over a given time allows the fall-out rate to be calculated. Multiple plates placed in different process areas can be used to assess the cleanliness of various areas in the laboratory.

VII. NONENVIRONMENTAL PARTICLES

In addition to the detrimental effects of environmental particles, the outcome of a particle assay for pharmaceutical products is critically dependent upon (1) the cleanliness of any glassware or other apparatus that contacts the sample; (2) cleanliness of the fluids or diluents used, and (3) control of particles generated by the analyst or the test procedure. On this basis, standards and technical publications dealing with the particulate matter testing of solutions or fluids specify background particulate matter control measures [18–20]. The paper by DeLuca et al. [1] is also a valuable reference in this regard. These references specify allowable background particulate matter levels and means of denying particulate matter access to the test materials. For some samples, testing may best be conducted in general laboratory environment that is protected by HEPA filters. Other testing demands the protection of a HEPA-filtered laminar flow hood. Even if a hood is not absolutely necessary to perform a specific test, it will serve to provide a clean enclosure for cleaning and storing test equipment and samples, and allows the desirable level of cleanliness to be readily achieved. Filtered water with an appropriately low particle burden must be used as a reference both to verify the cleanliness of the operating environment and as a cleaning fluid; it may also be run between samples and should be available in HEPA-filtered hoods if these are employed.

VIII. FLOORS AND WORK SURFACES

Work surfaces of critical areas must be regularly cleaned to avoid the buildup of point-generated particles. If microscopic particle counting analyses are to be performed, the laminar flow hood work area should be wiped down with a clean wipe and isopropanol prior to beginning work. A number of clean wiping cloths are suitable for this purpose. Synthetic wipes are, in general, significantly less absorbent than cellulosic wipes, but paper or cotton wipes should be avoided unless their particle shedding characteristics are known. Some of the most suitable wiping cloths are blends of natural and man-made fibers. Isopropanol and/or other solvents specified for this role may be used.

Floors are remote from the work surface, but if not cleaned regularly they can become reservoirs of particles that are stirred up by human activity. The best floors for laboratories are of seamless vinyl linoleum, and all corners where the wall joins the floor should be rounded off. Epoxy paint is used for walls and ceiling areas and provides an easily cleaned surface. If vacuum cleaners are used to clean floors or work surfaces, it is imperative that they have well-filtered exhausts. Floors should be mopped with minimal amounts of water and nonshedding mops at least once a week or more frequently as required. If disposable shoe covers are worn and adhesive mats are used at the entrance of the anteroom of the laboratory, buildup of particles "tracked in" on floors is minimized.

IX. CLEANING OF GLASS AND PLASTIC WARE

In the USP <788> test for particles in injections, the most critical factor with regard to nonenvironmental background particulate matter sources is glass or plastic ware used for sample pooling. Laboratory glassware for general use should never be used in the particle counting assay, since any surface scratches present will contribute particles. Clean, dedicated glassware that is stored under HEPA filter protection is far easier to blank (i.e., give lower background counts). One method for cleaning of glassware and other equipment is the method described in the article by DeLuca et al. [1]. It is carried out by first placing the glassware in an ultrasonic bath containing warm detergent solution for 1 min. Both the outside and inside of the glassware are then rinsed, first with hot tap water and then with filtered distilled water before being transferred to the hood. In the hood, the glassware is final rinsed with ultraclean water. The cleaned glassware is then placed in the hood upstream of all other operations and allowed to dry.

The general principle in cleaning is simply the exhaustive removal of adherent particles from glassware, vessels, or apparatus to be used in the test. In lieu of the ultrasonic method, cleaning may be done by soaking glassware, apparatus, or vessels in a detergent solution and pressure rinsing in distilled water. The pressure rinse is performed with a jet of filtered distilled water filtered with a 1.2-μm or lower retention-rated filter. In the final rinse, the water should "sheet" the entire rinse surface of the item being cleaned. A pressure vessel with a filter on the dispenser nozzle (e.g., Filter Jet Solvent Dispenser, Millipore Corp.) also provides an effective means of final rinsing. A useful setup for the final rinse is shown in Fig. 11. A pressure vessel with a filter on the dispenser nozzle may be substituted for the arrangement shown in the figure.

In the latter case, after cleaning in warm detergent solution and then rinsing in filtered distilled water, the final rinse should be performed with filtered water from the pressure vessel in a HEPA filtered hood. The analyst uses a back-and-forth motion to direct the jet spray and works from top to bottom of the object being cleaned. If an item of glassware is needed in a short period of time, a final rinse may be performed with filtered IPA which will dry quickly. Articles washed should be permitted to dry in the hood upstream of any other operation.

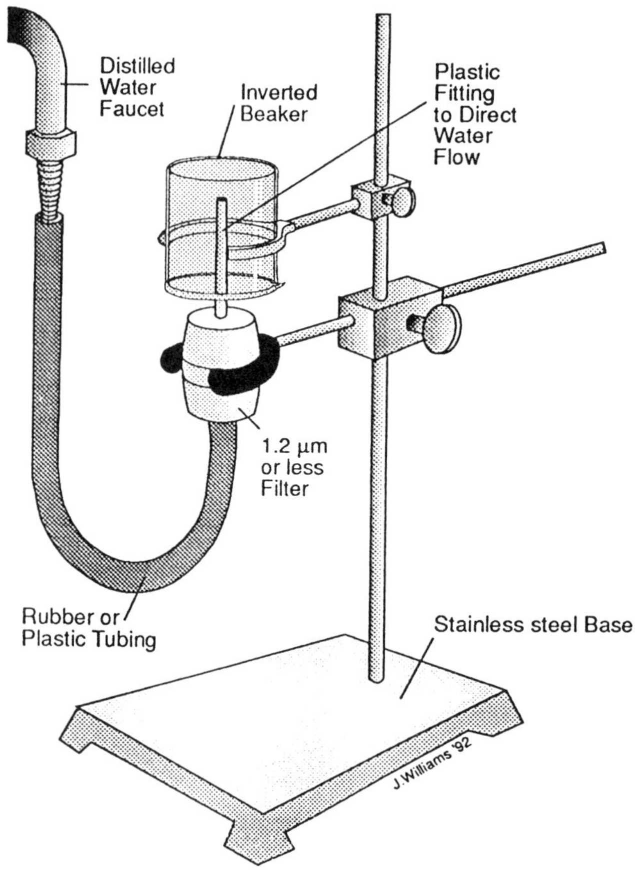

Figure 11 Final rinsing of glassware using filtered distilled water.

Ultrasonic baths for the cleaning of vessels to be used in the analysis of pharmaceutical materials and medical devices is widely practiced and generally effective. The analyst applying this means of cleaning should be aware of the specifics of the technique. First, all ultrasonic baths are not equal in terms of their power to dislodge particles; also the power rating of the bath used does not necessarily determine its effectiveness in cleaning. Some of these devices will not evenly distribute the ultrasound waves throughout the liquid used as a couplant. If this condition pertains, "hot spots" will be observed on the surface of the liquid in the bath, and parts of containers not contacting these points will receive a greatly decreased amount of energy. Ultrasound is frequently applied as a means of removing air or gas bubbles from solutions to be tested. For the majority of the applications in a clean laboratory, a bath of 600 ml to 1 l with a power rating of 75 to 90 W is sufficient.

A word is also in order regarding the effects of sonication in generating particles. Obviously, if a solution sample is to be sonicated in a glass beaker or other vessel prior to the performance of light extinction counting, the beaker must be cleaned using ultrasound before use. Otherwise, particles not removed in a conventional cleaning procedure may be removed during sonication and add spurious counts. If sonication is to be routinely applied in analyses, there is a strong argument for the use of dedicated glassware that has no surface scratches or abraded areas, since glass particles dislodged from damaged areas of the glass can result in artifactual counts.

X. "PARTICLE-FREE" WATER AND SOLVENTS

Since water or other solvents (such as IPA) come into intimate contact with the equipment used in particle testing, the availability of clean fluids is of overwhelming importance to the particle analyst. The need for water with a defined low particle burden is not restricted to laboratories performing particle analysis. Pure laboratory water is needed in virtually all segments of industry, especially pharmaceutical and electronic companies as well as in universities, government agencies, and in hospital laboratories. High-purity water is used for reagent buffers and in sensitive instrumental analyses (such as HPLC, ICP, and GC/MS) as well as in washing and/or preparing biological solutions, such as media for tissue culture.

Vendors of filtration products make purification systems available on a laboratory scale that will produce reagent-grade water with an extremely low particle burden. The systems may either be centrally located for feed to multiple laboratories, or as point-of-use systems for a single laboratory or bench. Virtually any volume and quality specification can be met for today's sophisticated laboratories. The available systems incorporate a wide range of technologies including reverse osmosis, deionization, ultrafiltration, microporous filtration, continuous ion exchange, and carbon adsorption. Millipore's Milli-Q system is an example of this type of system. It produces water that is essentially free of particles $\geq 0.2\ \mu m$ in size and has ppb levels of dissolved solids, organics, silicates, and trace metals. Such systems are efficient and economical to use, but are not a prerequisite for clean water generation in particle analysis laboratories.

"Particle-free" is obviously a misnomer. The water will be essentially free of particles only to the level of filtration (e.g., 0.2 μm, 0.45 μm, etc.). To ensure freedom from particles at the smallest particle size to be counted, filtration should be done at a significantly smaller size. Filtration at 0.45 μm will suffice for both USP and BP analysis of injectables, and 1.2-μm filtration will generally be found adequate for these analyses as well. Filters of the latter retention rating allow a higher rate of solvent flow and are used in solvent dispensers in many laboratories. Efficient, high-surface-area capsule filters such as those marketed by Gelman (NYLAflo), Pall (Ultipor), and Millipore may simply be attached to house distilled water lines to achieve adequate cleanliness for some situations, subject to the cleanliness of the distilled water supply. Depending on the volume of testing to be done, water may be prepared using a recirculating filtration apparatus built in the laboratory (Fig. 12), or are available commercially from the Coulter Corporation in Hialeah, Florida.

In using a cartridge or other type 0.45-μm or 0.22-μm filter to "clean up" distilled water for use in the USP <788> test, a primary concern for a technician to keep in mind is that the first volumes of water flowing through a new filter should not be expected to be free of particles larger than the filter retention rating. Some bits of the filter media will be shed at the beginning of use, and particles will be released from the filter housing and from fittings and tubing downstream of the filter. Particles from the latter must be kept at a minimum. Importantly, whenever a filter is changed or before the first use of filtered water for the day, it is advisable to clean out the filter and lines by passing several liters of water through them. Most stopcock-type valves are capable of releasing significant numbers of particles, so that valves downstream of a filter should be avoided.

Water of adequate cleanliness for most particulate matter analyses directed at contaminant particles is readily obtained by sample filtration at a 0.45-μm or lower retention rating. Most tap water supplies can be used to generate clean water if adequate filtration is applied; the practicality of this source will depend on the use of a filter with sufficient surface area to deal with the significant amounts of organic and mineral matter often found in tap water. In the author's laboratory, upon occasion, we have used production scale (30 in.) cartridge filters to clean tap water for use when the distilled or distilled deionized water supply was disrupted.

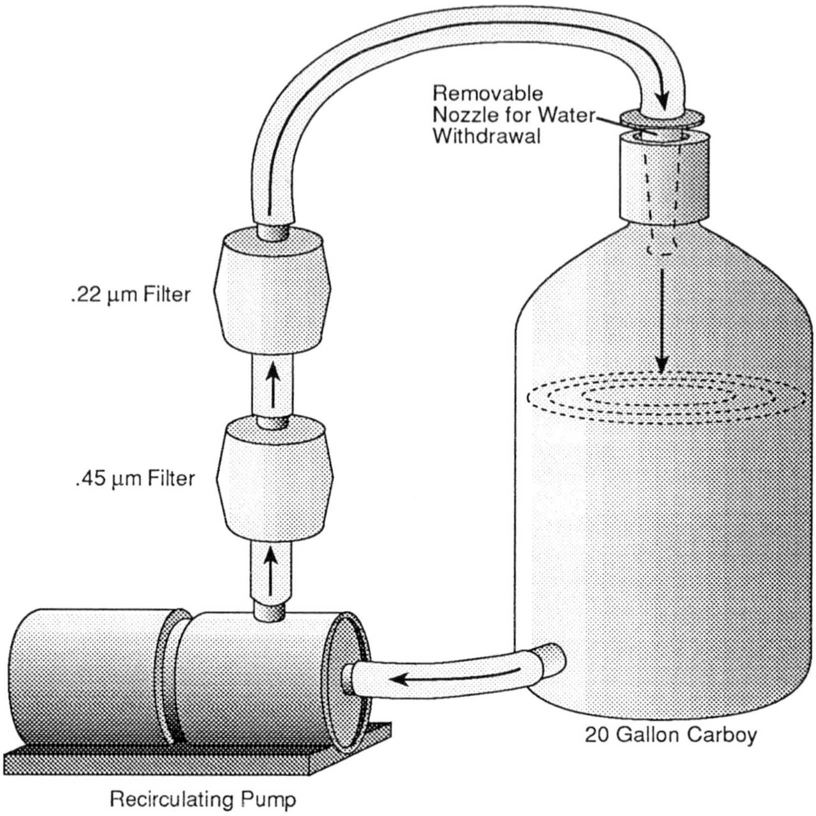

Figure 12 Recirculating filtration apparatus.

The performance of a single compendial (USP) analysis may require 1 to 2 l of water, so that large volumes of clean water may be required for multiple analyses. One procedure used in many laboratories is the attachment of a 0.45-µm or 0.2-µm retention rating capsule filter (Figs. 13, 14) directly to the distilled water supply to obtain water for rinsing glassware, etc. The Pall Ultipor N_{66} capsule filter widely used for this purpose shown in Fig. 13 is a self-contained microfiltration unit featuring a nylon membrane sealed in a durable polypropylene housing. It is used for filtration of small to medium volumes of solvents, reagents, buffers, or laboratory water. It is available in a 0.2-µm pore size, with an effective filtration area of about 2000 cm^2.

Smaller amounts of clean water or other solvents are conveniently dispersed from filtration flasks, pressure cans, or pressure vessels. The solvent flask shown in Fig. 15 is for intermittent application by squeeze-bottle action. These dispensers contain a 25-mm-diameter Millipore filter holder in the outlet of the flask. The outlet of the holder is fitted with a Teflon delivery tube for the purpose of directing the filtered solvent against the surface being washed.

The pressure rinser (Fig. 16) dispenses a pressurized stream of ultraclean liquid at the touch of a finger. It is used to rinse critical mechanical or electronic components, flush filter holders, and to rinse test tubes or laboratory glassware. The 1-quart capacity vessel can be pressurized with a hand pump, gas cylinder, or laboratory air supply. The pressure rinser is equipped with a 25-mm filter holder with a Teflon encapsulated O-ring.

Larger pressure vessels (Fig. 17) are used with "gun-type" solvent dispensers (Fig. 18) for rinsing filter funnels and other apparatus. These steel pressure vessels were originally designed

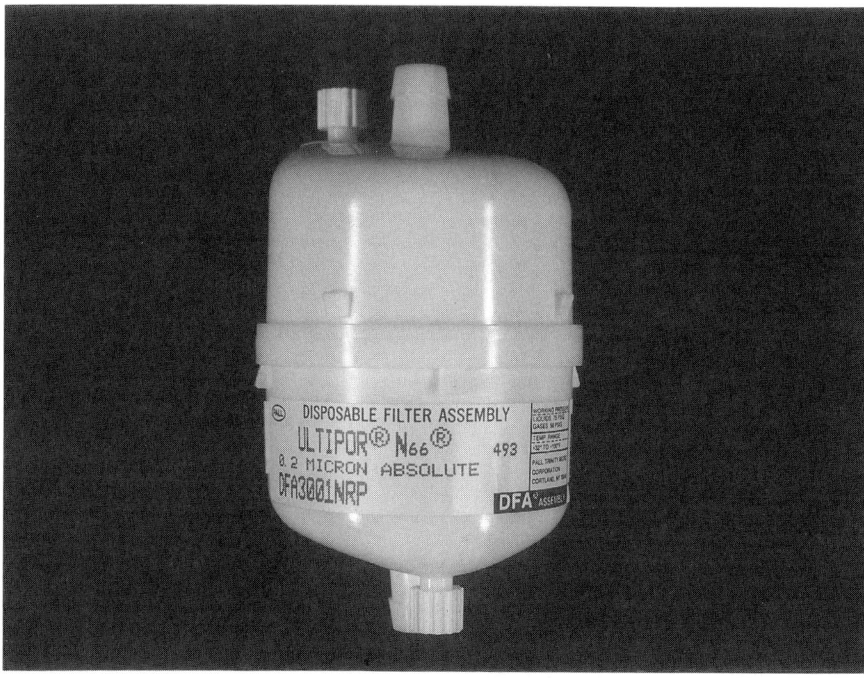

Figure 13 Pall Ultipor N_{66} capsule filter. (Pall Corporation.)

to process small batches of liquid for sterilization and for clarification. They are provided with pressure gauges, safety relief valves, flow shutoff valves, and hose barb inlet/outlet connections that accept 10-mm (3⁄8 in.) ID tubing. They are ASME coded and available in capacities of 3.8 l, 11 l, and 19 l.

Millipore's Filter Jet solvent dispenser is convenient for one-hand use and can apply a spray

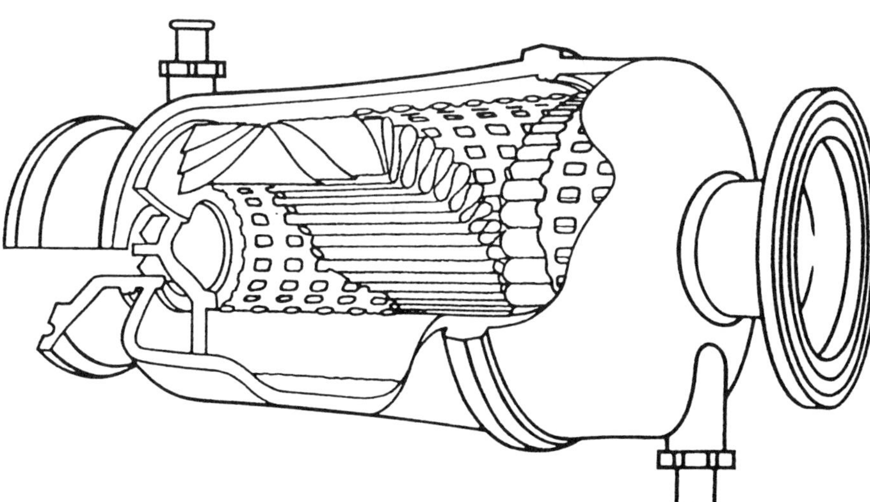

Figure 14 Structure of capsule filter. (Courtesy of Gelman Sciences, Inc.)

Figure 15 Filtration solvent filtering dispenser. (Millipore Corporation.)

or a concentrated jet of ultracleaned solvent or rinse solution to surfaces requiring washing. Solvent can be supplied from a pressure vessel of appropriate volume. A 25-mm filter of desired retention rating is used in the nozzle of the dispenser "gun."

XI. LAMINAR AIR FLOW HOODS

As mentioned earlier, a HEPA-filtered hood (Fig. 19) or similar enclosure is another key item of equipment that facilitates cleaning of articles used in test and sample preparation. Such laminar flow enclosures are available from a number of vendors of clean room supplies. The minimum useful size is 48 in. wide and 24 in. deep. The laminar airflow should be 90 + 15 fpm to provide adequate velocity to prevent particles from the general environment from entering. A shield of clear acrylic plastic sheeting that may be lowered when the hood is not in use is helpful in keeping the work area clean. The hood should contain or provide ready access to pressure vessels, filtered water, glassware, and other items to be used in the test. A sink is an essential feature of the hood. In addition to making it easier to attain the desired level of cleanliness in an analysis, the confines of the hood provide an enclosed work area in which an analyst can focus on the work without distractions or concerns regarding activity in the immediate vicinity that might generate unwanted particles.

Figure 16 Pressure rinser. (Gelman Sciences, Inc.)

XII. FILTRATION OF LIQUID SAMPLES

The filtration process is also critical to precise enumeration of particles and reproducibility of a microscopic assay regardless of whether injectables, medical devices, or other articles are being analyzed. Cleaning of the filtration equipment may be conducted generally as described in the article by DeLuca [1].

All of the steps in a filtration procedure for analysis of pharmaceutical or medical products are best conducted wearing rinsed, particle-free vinyl gloves in a Class 100 laminar airflow hood. For the whole volume filtration of large volume injection (LVI) product, a 100-ml or greater capacity filtration funnel such as the Gelman Sciences, Inc. 25-mm polysulfone filtration funnel with 200-ml twist lock coupling funnel (Gelman Sciences, Inc. #4203) may be used (Fig. 20). To assist in obtaining an even particle distribution, a metal diffuser screen (Millipore stainless support screen, catalog number XX30 025 10) should also be used beneath the filter membrane. In the filtration procedure, the ultraclean filtration funnel is first assembled by placing the metal diffuser on top of the filtration base. A membrane filter previously cleaned with a jet of filtered water is then centered on top of the diffuser using clean forceps to hold the filter edge, and then the filter funnel is attached to the base. The funnel is attached to a shatterproof vacuum flask and the flask to the vacuum source. The neck of the unit to be sampled is rinsed, the unit is gently inverted 20 times, the stopper is removed, and the desired sample volume is emptied into the funnel. To pool a number of small volume injection (SVI) vials, the vials are first emptied into an ultracleaned beaker, and then the beaker is poured into the filter funnel. Vacuum is applied to draw the solution through the filter. If the volume to be filtered is greater than the capacity of the filter funnel, additions of solution are made when the volume in the funnel is approximately 100 ml.

Figure 17 Stainless steel 3.8-l pressure vessel. (Gelman Sciences, Inc.)

Filtration apparatus and accessory items of equipment used in solution filtration are shown in Fig. 20. Stainless steel forceps of the type suitable for handling membrane filters are extremely important to the clean conduct of the filtration analysis. Filtration funnels made of polysulfone or some other readily cleanable material should be used. The forceps allow handling of fragile microporous membrane filters without damaging them, and their smooth surfaces are suitable for thorough cleaning. Many of the laboratories performing microscopic particle counting per USP <788> use polysulfone filter funnel assemblies of the type shown. The smooth, nonshedding surfaces of the funnel and integral filter holder result in low background counts. Petrislides or similar membrane filter containers (Fig. 21) are invaluable as a means of handling, storing, or filing analytical membranes; they seal tightly to protect the membrane and are transparent, so that low magnification viewing may be carried out with the cover in place.

XIII. CONTROL OF BACKGROUND PARTICLE COUNTS

Low background (blank) particle counts are essential to the accuracy and reproducibility of a microscopic assay. Unfortunately, the blank procedure for testing injectable solutions as written in <788> of USP XXII cannot be followed because it is so poorly written. Thus, the burden is

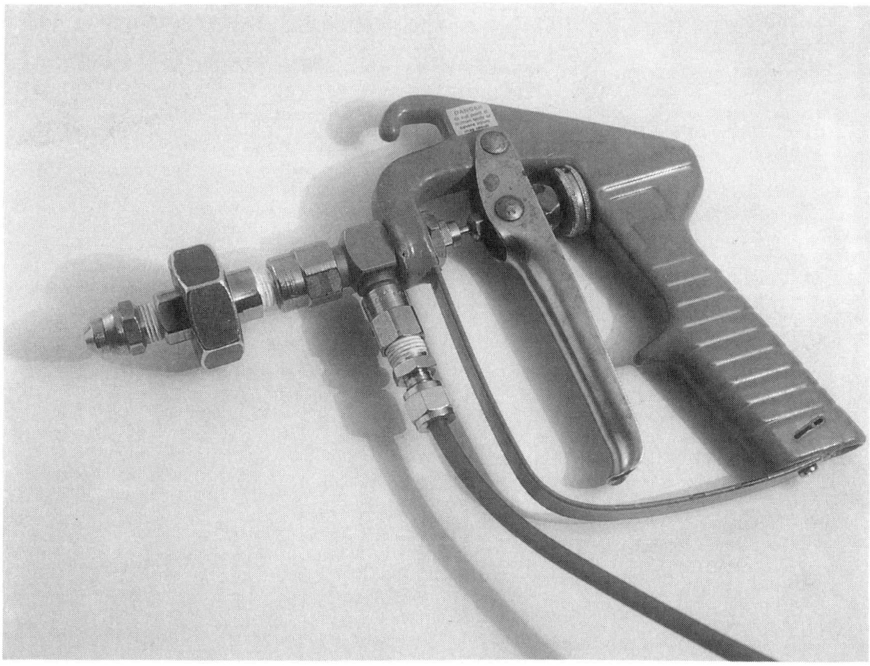

Figure 18 Filter jet solvent dispenser "gun." (Millipore Corporation.)

placed on those performing the test to develop a rational and technically valid approach resulting in a blank as good as or cleaner than the test specified in the requirement. This difficulty will be eliminated in the revised <788> which will appear in USP XXIII. For LEC testing of SVI, a blank will entail filling a cleaned vessel to be used to collect a 10-unit solution pool with 50-ml volume of filtered water used for final rinsing of the glassware before the test is conducted. Then, importantly, a test volume for the blank is drawn that is equal to the sample volume. If two 5-ml aliquots will be counted for record, then two 5-ml aliquots should be used for the blank. If the glassware has been properly cleaned and the water is of appropriate cleanliness, then the combined 10-ml blank volume should have not more than (NMT) 10 particles ≥ 10 μm in size, and NMT 2 particles ≥ 25 μm. For the microscopic assay, a 50-ml volume of filtered water is passed through the cleaned filtration setup, and NMT 20 particles ≥ 10 μm are allowed.

If blank criteria are met, the test system is satisfactory for proceeding with sample determinations. If blank criteria are not met, either the environment is not suitable or the filtered water and/or glassware have not been properly prepared. In this event, the preparatory steps must be repeated until the criteria are met before proceeding with a sample analysis. Some laboratories consider the USP allowable blanks to be too high and substitute a lower allowable count standard. It is important to remember that background particles constitute an unpredictable contribution to counts for record.

A significant contribution to background counts can be attributed to the manipulation associated with opening the containers of injectable material. This is unfortunate but unavoidable. The contribution from opening the container can be assessed by filling blank vials, ampules, etc., with filtered water, and then counting the numbers of particles added by the opening procedure. Ideally the blank vials should be filled under production conditions, and the solution should be removed from containers tested in a fashion that mimics the use of the product. As an

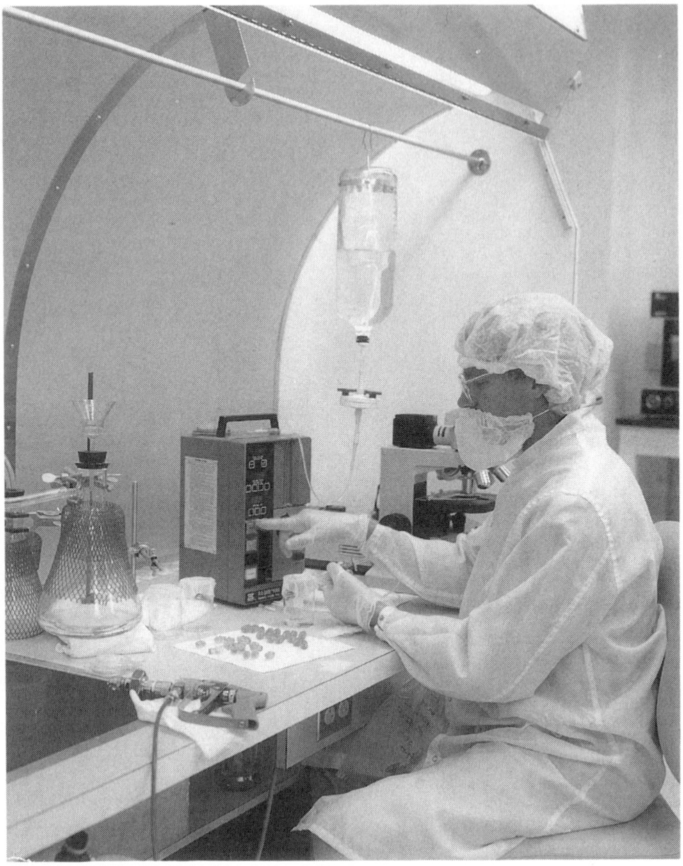

Figure 19 Laminar airflow hood work area.

alternative to removal of the unit closure, vials may be emptied using a cleaned and blanked syringe with a noncoring needle. The caution is that each component used in the test must be blanked for background particle burden.

Since a wide variety of dosage forms may be analyzed by the USP <788> test, the user may find it necessary to modify the blank procedure prescribed in the compendium so that the test process is appropriately represented. Any such modified test procedure should result in a blank test result as good as or better than that specified in the <788> requirement and must be performed in addition to the blank test as specified in <788>. The complexity of blanking procedures for the USP compendial analyses was significantly increased by the way the USP XXII <788> procedure for light extinction counting was written. The instructions for obtaining a blank were incomplete and poorly written, so that some interpretation by an FDA auditor was necessary. This led to adverse audit findings with no real basis. The best course that could be followed by an analyst was to make an honest effort to perform the blank as directed with minimal necessary modifications, and to work for a lower blank count than specified in the compendium. The entire <788> requirement has been rewritten for USP XXIII. The new requirement, written with significant input from manufacturers of injectable products, will eliminate many of the problems, including the blank procedure, that have existed since 1986 with the blank requirement and other sections of <788>.

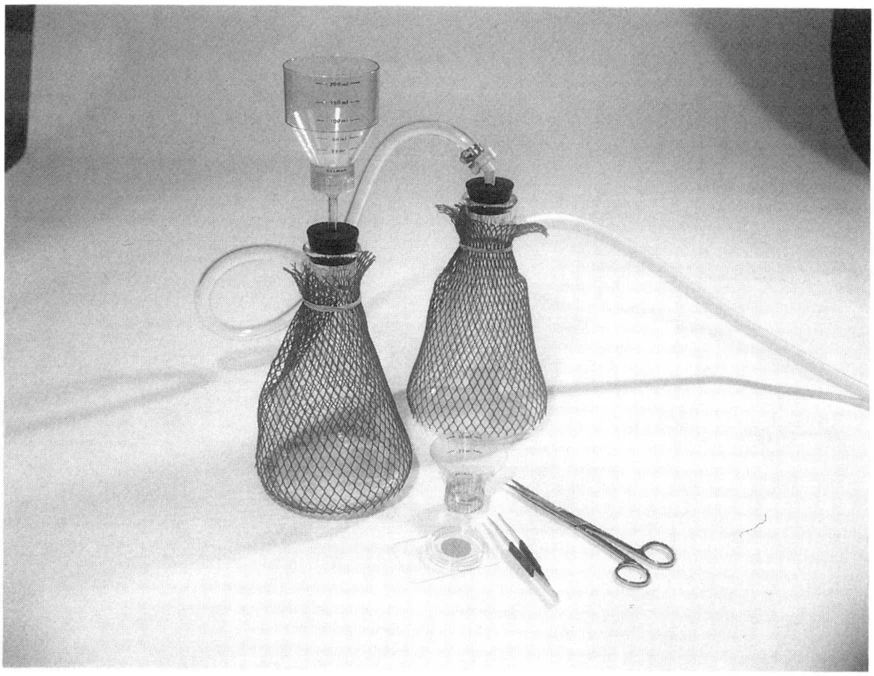

Figure 20 Filtration apparatus with 25 mm Gelman polysulfone filter funnel.

XIV. COMPENDIAL PARTICLE ANALYSIS

Compendial particle analyses are performed to determine whether or not an injectable product meets pharmacopeial requirements. While the principles of protecting these analyses from extraneous particles are the same as those applied for other types of testing, particular care must be given with the instructions provided in the applicable compendium regarding how the analysis is to be performed. Regulatory auditors have the generally reasonable expectation of seeing product assays performed "by the book" when data are being generated. In the US, general chapter <788> of the USP must be followed as closely as possible. As described in the preceding section, there is a specific difficulty related to the blank analysis prescribed in USP <788>. Deviation from the prescribed <788> methodology, for whatever reason, may be the subject of an adverse inspectional observation (a 483). A rule of thumb to follow is that steps may be added to an analysis if desired to achieve higher levels of cleanliness, but those steps should be in addition to, not to the exclusion of those prescribed. Other instructions, such as the requirement that some portions of the microscopic analyses be conducted in a Class 100 hood, must also be followed. Although not specified in the requirement, data should be available to attest to the integrity of HEPA filters in hoods and airborne counts at the appropriate levels.

Although less sensitive to contamination artifact than microscopic testing, LEC of injectables per <788> must also be carefully conducted to prevent extraneous particles from inflicting a bias on count data. In addition to cleaning glassware, another critical step involves the point at which containers are entered and the contents pooled in beakers for testing. If the product to be sampled is a small volume dry powder vial, the removal of the stopper for reconstitution or fluid removal will result in some particles being shed from the stopper. If ampules are not opened carefully, large numbers of glass particles may contribute to the

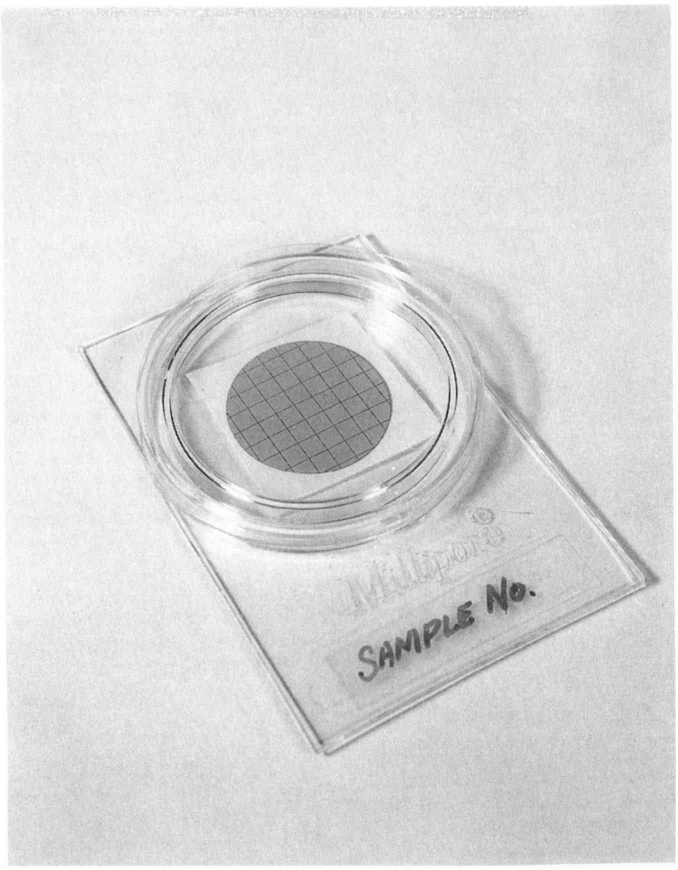

Figure 21 Millipore Petrislide membrane filter holder.

analytical result. Some laboratories have developed a technique for entering stoppered containers using pencil point needles that generate extremely low levels of particles and for opening ampules with an oxypropane torch.

General procedures that may be used when testing different types of injectable product containers are discussed below by product type. Before adopting a procedure for laboratory use, the suggested method should be compared to that in USP XXIII and, if necessary, modified so that compliance is assured. Before sampling any liquid product, it is the standard practice in many laboratories to visually examine each unit for the presence of particles against white and black backgrounds. This simple inspection is not intended to critically detect particles of visible size, but to ensure that the solution appears normal with regard to clarity and color.

A. Ampules

The outer surface of each ampule should be cleaned with warm detergent solution and rinsed before transferring to the hood. After scoring the glass, chips are rinsed from the outer surface of the ampule and the top of the ampule is broken off with a pulling motion. The solution may then be withdrawn with a plastic syringe and needle that have been thoroughly cleaned and rinsed with ultraclean water. If a syringe is used, a suitable blank must be incorporated. A better procedure for opening ampules involves the use of an oxypropane needle point torch to flame a hole in the ampule through which the solution my be withdrawn.

B. Solutions Packaged in Vials and Bottles

The outer surface of each vial is cleaned with a warm detergent solution and rinsed before transferring to the hood. The tear-off seal and inner seal from the test vial are then removed and the outer surface is again cleaned with a lint-free cloth moistened with IPA. The rubber closure and the glass-closure junction are rinsed with a jet of filtered distilled water and allowed to dry. The test solution is then inverted 20 times to resuspend the particles; then the closure is removed with the aid of forceps. The counter probe may then be inserted in the bottle or the contents poured into a beaker (pooled). Gloves should be worn by the analyst.

C. Dry Powder Packaged in Vials

The outer surface is cleaned with detergent as directed above. The tear-off and inner seal are then removed from the test vial and the outer surface of each vial is cleaned with a lint-free cloth moistened with IPA. The glass and the point of closure junction are rinsed with ultraclean water and allowed to dry. The closure is then removed with a forceps and placed upside down within the hood. The powder is dissolved with ultraclean water or other appropriate diluents (e.g., particle-free 5% dextrose, 0.9% sodium chloride solution) at appropriate concentrations recommended by the manufacturer. The closure is replaced and the vial inverted 20 times or until all the powder dissolves. The closure is then removed and the solution pooled in the collection vessel.

D. Powder in Vials with Diluent in a Separate Ampule

Both containers are cleaned with detergent solution and rinsed with water before transfer to the hood. The tear-off seal and inner seal are removed from the test vial and the outer surface of the vial and ampule of diluent are wiped with a lint-free cloth moistened with IPA. The containers are then rinsed with ultraclean water. The ampule is scored lightly with a file or diamond scribe at the breaking point, and then the vial and ampule are rinsed and dried. After breaking the ampule, the diluent is withdrawn with a blanked plastic syringe and needle; a blanked needle filter is then put in place on the syringe. After opening the vial as described earlier, the diluent is injected into the vial. The closure is then replaced and the vial agitated until all powder dissolves. Ten samples are then pooled as required for testing.

E. Prefilled Syringes

The containers are cleaned with detergent as for ampules and vials, and the outer surface of the syringe and syringe parts are wiped with a lint-free cloth moistened with IPA and rinsed after transfer into the hood. The outer surface of the test articles are then rinsed with ultraclean water and let dry. The syringe parts are then assembled according to the manufacture's instructions. The syringe is inverted three times to resuspend the particles. The needle sheath is then removed and the solution expressed into the collection vessel.

XV. ISOLATION OF PARTICLES FOR IDENTIFICATION

A wide variety of particle isolation and collection methods is available to the pharmaceutical particle analyst. The most commonly used analytical path involves the primary isolation of particles from solution by filtration onto a microporous filter followed by secondary isolation of the particle(s) of interest using a fine-pointed probe or forceps. Once the particle is isolated, it may be analyzed by polarizing light microscopy, X-ray spectrometry, or by other instrumental means. Specialized methods are also available for the collection of particles from surfaces, containers, process points, garments, and atmosphere.

For purposes of identification, some particle samples can be examined directly without further isolation. However, if the particles of interest are imbedded in a matrix or must be removed from a mixture of retentate on a filter, manipulation is often required. Techniques for isolation and collection of individual particles for analysis involves the use of specialized microinstruments to physically pick up and transfer particles from a filter or other collection substrate to a microslide or sample support for an analytical instrument. The paper by Teetsov [21] provides an excellent description of how particles may be isolated for analytical procedures, and additional information is provided in the chapter on instrumental analysis of particulate matter.

The requisite cleanliness for particle isolation is typically not as great when particles of visible size are isolated. Often larger particles to be isolated are of unique size or morphology, and are likely to be confused with contaminants. When smaller particles (10 to 25 μm) are to be isolated, the analyst is best advised to take thorough precautions against contamination similar to those employed in particle counting, including working in a HEPA-filtered hood, and wearing gloves. Experience is a great teacher in particle isolation endeavors; the analyst learns to employ safeguards against contamination that vary with the size and type of particle to be isolated.

XVI. PERSONNEL

The importance of personnel as a source of particles in the clean laboratory cannot be overemphasized. People in the laboratory can serve as sources of almost limitless contamination. The referenced paper [22] and others by Dixon are valuable procedural guides.

The human body is continuously shedding particles from fingerprints (skin, oil), skin flakes, dandruff, hair, perspiration residue, and cosmetics. Exhaled air carries with it moisture from the lungs as well as particulate material. Skin flakes are transported by air and can settle on exposed surfaces; they may fragment into smaller particles dispersed across a surface, or they become airborne in the surrounding environment. Skin flakes and oil are also transferred to products indirectly through contact with contaminated tooling, instruments, and surfaces. The threat of human-based contamination cannot be overstated.

Generally, all clean facilities maintain rules of behavior to limit and control the effects of human contamination; the clean laboratory is no exception. Activities such as running, smoking, eating, drinking, etc., are not acceptable due to the amount of contamination generated by these activities. The type of clothing worn under clean room garb is important; heavily-linting materials such as wool, mohair and velvet are banned due to the increased contamination potential. Cosmetics, lotions, hair treatment products, colognes, after-shaves, and perfumes are chemical and particle contaminants. Testing has demonstrated that cosmetics shed in the clean room contain a variety of contaminants [23].

Clean room clothing is designed to reduce the contamination from personnel, including their normal street clothes. No overgarment system, by itself, will ensure product protection from human contamination unless proper gowning techniques, gown storage, laundering, handling, etc., are strictly monitored and followed. Clean room garments have performance properties that enable them to minimize the transmission of human contamination and not generate additional contamination from the garment itself. In the clean laboratory, garment properties must always be evaluated with regard to the degree of control necessary. Here again "overkill" or the application of unnecessarily stringent precautions should be avoided. Disposable hair covers and shoe covers along with polyester smocks provide adequate personnel protection for the majority of pharmaceutical particle analyses. The smocks should have snaps at both the cuffs and the (high) collar. Again, the protection offered by garb is only as good as gowning techniques and aseptic procedures practiced by personnel. Garments may be made of tightly woven polyester, Tyvek, or (more recently) Gore-Tex. Tyvek garments are available in both disposable and reusable types and feature extremely good control of particles shed by a worker. However, they are not "breathable" and are thus somewhat less comfortable to wear than garments made from either materials.

Ideally, the analyst will wear gloves while performing any procedure in which there is exposure of the solution being tested (e.g., sample pooling). Touching the mouth of a sample vial or a cleaned vessel with an ungloved finger can deposit thousands of skin cells that will result in a high level of background count artifact. The best gloves for this purpose are nonpowdered vinyl gloves that are essentially particle-free as purchased. Any type of glove used should be rinsed under filtered distilled water after donning to remove the surface adherent particles present due to manufacture of the glove or the donning procedure. If gloves will not be worn, hand-cleaning agents that leave the hands dry (such as alcohol-containing foams) should be avoided, since they enhance shedding of skin cells.

Most of the particles shed by personnel are skin cells identified as a source of bacterial contamination, and fibers. There is an extremely high level of sensitivity to human contamination in our industry and within regulatory agencies. The viable and nonviable particles shed from people have been described as a "human dust" that is continually being shed at a rate of 10^5 to 10^7 cells/day [24]. No monitoring program for aseptic fill areas is complete without provisions for personnel testing [24]. Monitoring may also be practical for some critical laboratory operations. The goal is simply to keep contamination at a minimum and under control. If results for an individual exceed limits or if a trend is indicated, the person may be counseled or required to take a review course in gowning and aseptic practices.

The importance of training personnel for clean laboratory operations cannot be overemphasized. Training not only in analytical techniques is necessary, but also in the philosophy of laboratory operation. The training process is not given simply to impart knowledge, but through education and involvement on a personal level to alter behavior [22]. Employees working in clean facilities cannot work effectively without a clear understanding of basic clean room technology. Too often training programs overemphasize analytical procedures for clean laboratory work, and omit sufficient background training. Clean room concepts must be described and explained. Training by managers and supervisors should be followed by some level of testing that measures the student's retention and understanding of the concepts taught.

Ongoing support is a critical factor in sustaining motivation and dedication. It is achieved by encouraging people and rewarding their efforts and accomplishments. Recognition and reward are two often neglected aspects of empowering people in this area of particulate matter control [22].

Time must be allowed for contamination control. Technicians and analysts must understand that they are allowed the time to devote to contamination control efforts. Cleaning of surfaces, equipment, and work stations in the laboratory must be viewed as part of the work process rather than a separate function. While increases in productivity and quality have resulted from advances in clean room design, and use of HEPA filters provided improved areas for particle testing, the concepts of contamination control were often poorly understood by laboratory managers. The assumption has been made that contamination can be eliminated by HEPA filters and unidirectional airflow. This lack of understanding the total concept based on HEPA filters, particle counters, clean rooms, garments, and training promoted errors and did not produce the hoped-for results. An understanding of the principles of clean laboratory operations on the part of the manager and the analyst is a most critical element of success.

XVII. LABORATORY DESIGN

The design of a clean laboratory is similar in ways to the design of a CEA for production activities, but will differ with regard to some important considerations. A vestibule or anteroom that isolates the clean laboratory from uncontrolled areas is imperative. Suitable overpressure gradients must be employed. The constraints of GLP and FDA regulations do not provide specifics for the design of laboratory areas. Design of the laboratory is often based almost solely on economics and the need to control particles at different levels in various areas of the laboratory at a level sufficient to protect the analyses being conducted. This is a reasonable principle; cost without benefits due to constructing a cleaner facility than necessary should always be avoided. The design and planning of this type of facility should involve both personnel knowledgeable in clean room design, and individuals with an in-depth understanding of clean laboratory operations and the requirements for analyses to be conducted in the laboratory.

Generally, laboratory design will incorporate different cleanliness "zones" [25] suitable for (1) particle isolation for instrumental analysis [identification]; (2) microscopic particle counting; and (3) particle counting with instrumental particle counters. The "zone" philosophy has a significant impact on working techniques and costs. The use of protective clothing and maintenance are somewhat different for a "zone" design versus a total clean laboratory design. Small "enclosed controlled spaces" will increasingly be used in laboratories, and new types of clean spaces (such as plastic isolators) may be applied in laboratories as well as production areas. This change is akin to that in industrial sectors where contamination control remains important in the face of increasing demands for quality and economy. The microelectroncis industry, which has long been famous for its very large, vertical laminar airflow clean rooms, is changing working methods and utilizing either open-zoned areas or employing curtains or other flexible barriers in laboratories. The concept of minienvironments has been around for years, but only recently has it become more accepted as a clean laboratory approach. Essentially, a clean room may be created in a larger CEA or general laboratory area to isolate more critical analytical procedures from the overall environment. One type of approach to a mixed environment laboratory is shown in Figs. 22a, b.

The main laboratory area is designed to provide Class 1000 to 10,000 conditions. The ceiling HEPA-filtered airflow is directed downward onto the benchtop work surfaces and removed by a floor-level plenum at the base of one wall. The HEPA hoods at the opposite wall provide Class 10 to 100 conditions for microscopic particle counting, particle isolation, and filtration. Airflow is directed from the hoods toward the plenum at the opposite wall. A suspended bank of six HEPA filters provides a larger Class 100 benchtop work area where medical devices/components may be analyzed. The general environment air of the laboratory (1000 to 2000 particles/ft^3 at 0.5 μm) is of more than sufficient cleanliness for instrumental particle counting. Instrument rooms can be built at either side of the main laboratory and separated by vinyl walls to provide an area for particle identification analyses. These areas can also be maintained at Class 10 K or better. Though much less expensive to build and maintain than a stand-alone Class 100 room enclosing all laboratory activities, a laboratory of this type is more than most smaller pharmaceutical companies will need or be able to afford.

At other end of the spectrum of possible designs would be a mobile, hanging curtain HEPA enclosure that could maintain Class 1000 conditions over a 12 ft × 12 ft area (Fig. 23). A separate hood in the enclosure would provide an area for microscopic counting and other analytical activities requiring a higher level of cleanliness; a benchtop area would provide an adequately

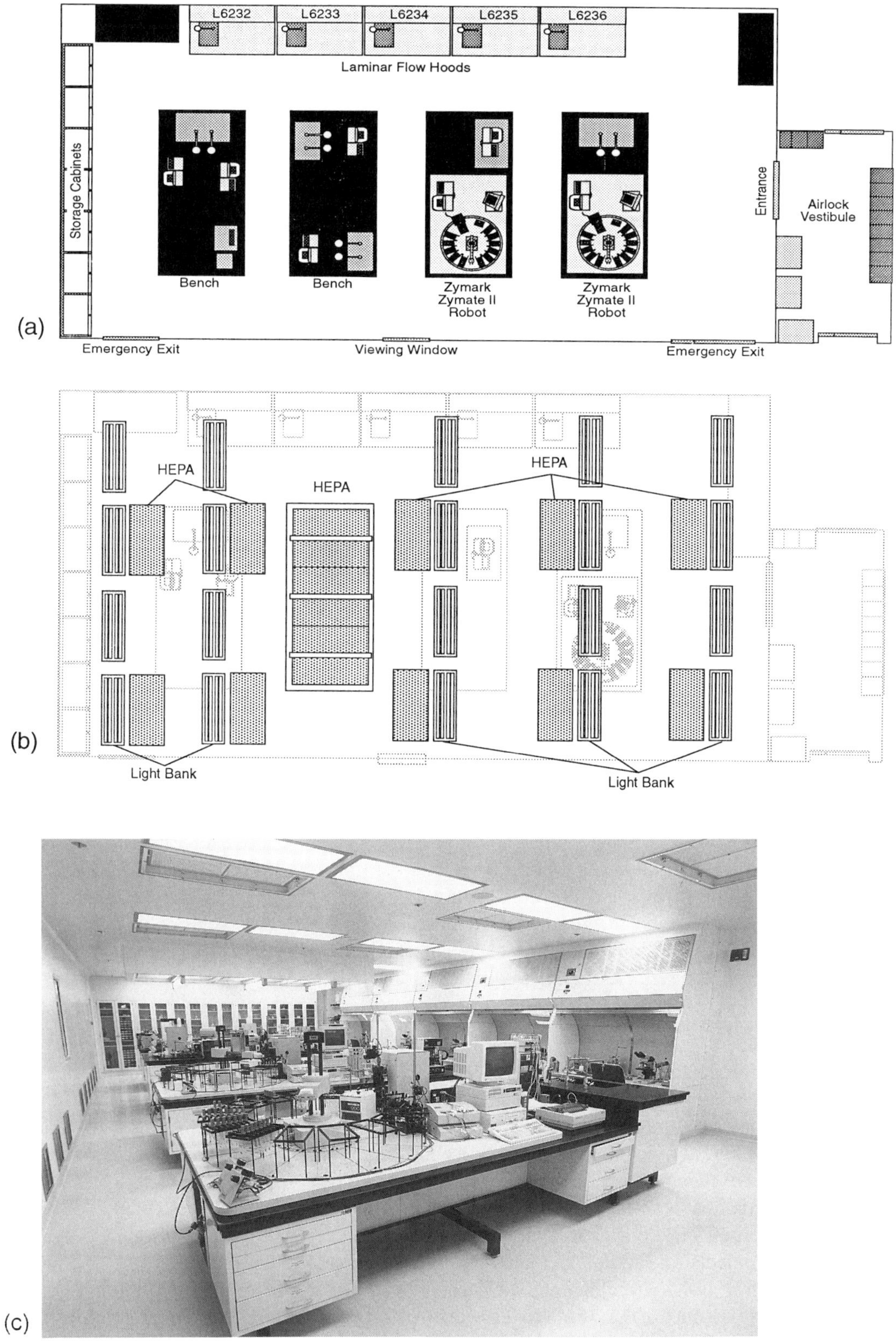

Figure 22 Mixed environment clean laboratory design. (a) Floor plan. (b) Vertical (from above) view showing HEPA coverage. (c) Photograph of lab layout and instrumentation.

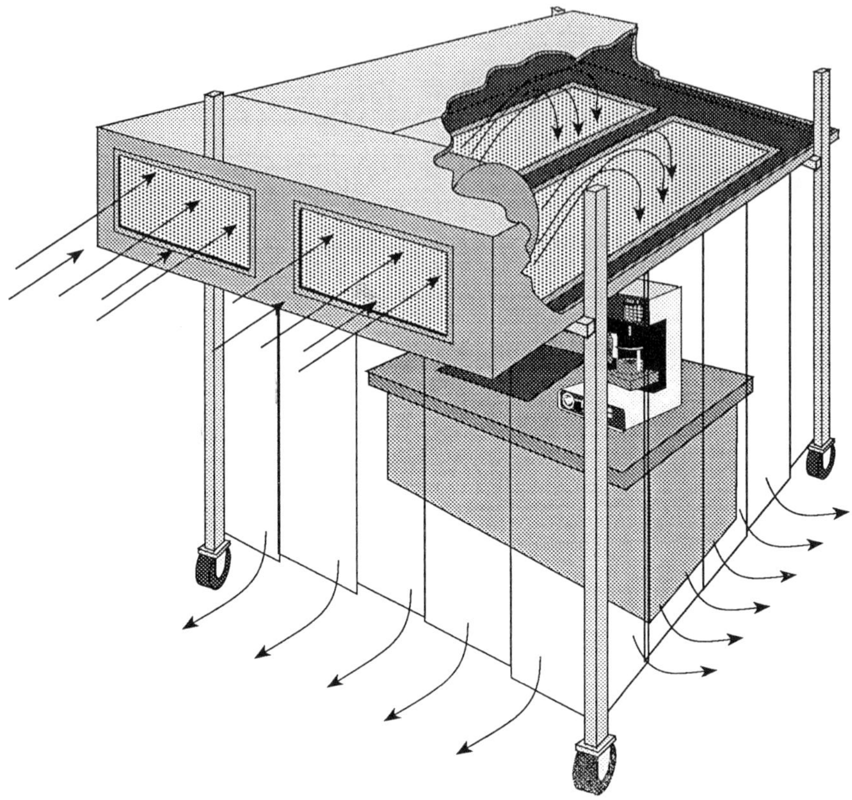

Figure 23 Mobile laminar flow HEPA enclosure.

clean area for a light extinction counter and a polarizing microscope. Enclosures of this type may, in theory, be used to house a microlaboratory in a high classification production environment, but this application could allow contaminants access each time a person entered the module. For the application of this modular technology in high classifications in unprotected areas is facilitated by the simple addition of an anteroom for gowning, which serves to protect the laboratory from the less clean area in which it is located.

XVIII. SUMMARY

Implicit in the foregoing discussion is the premise that any laboratory operation carried out for the purpose of counting, characterizing, or identifying particulate matter must be protected from extraneous particles that might introduce error sources into the data collected. Another important point is that the level of cleanliness in our procedures and in our laboratories should meet but not significantly exceed the requirements of the work being performed. While cleanliness of the analytical environment is of the utmost importance, higher levels of cleanliness entail additional significant expense. Since our goal is least-cost generation of our product (data), unnecessarily stringent requirements for laboratory cleanliness will add cost without benefit.

HEPA control of the environmental air particle burden in the laboratory and monitoring of airborne particle counts ≥ 0.5 μm are key elements in laboratory operations. It must be remembered that neither of these measures have any significant effect on (1) large particles that occur randomly in the laboratory environment; (2) on analysis- or human-generated particles that occur in intimate proximity to the test being conducted; or (3) on particles adherent to samples received or on accessories used in testing.

Maintenance of conditions in the clean laboratory adequate for the analysis of a wide range of pharmaceutical products and related materials is critically dependent on laboratory design, particulate matter monitoring, housekeeping, and personnel training. Personnel access from less clean areas must be carefully controlled, and adequate protective garb must be worn. In addition to ongoing considerations of these factors, each and every step in the conduct and execution of an analysis must be executed with a level of cleanliness that is appropriate for the required data.

TRADEMARKS

The following trademarks used in this chapter are the property of the companies indicated. A brief product description is included; additional product information may be obtained from the vendor.

Flanders®. Filters, Flanders Filters Inc., Washington, NC 27889.
Climet®. Climet Instruments, 1320 W. Cotton Ave., Redlands, CA 92374.
Particle Measuring Systems®. Particle Measuring Systems Inc., 1855 S. 57th Court, Boulder, CO 80301.
Petrislide®. Membrane Filter Viewing Container, Millopore Products, Bedford, MA 01730.
Millipore®. Millipore Corp., Bedford, MA 01730.
Gelman®. Gelman Sciences Inc., Ann Arbor, MI 48103.
NYLAflo®. Nylon Media Capsule Filter, Gelman Sciences Inc., Ann Arbor, MI 48103.
Milli-Q®. Water Purification System, Millipore Products, Bedford, MA 01730.
Pall®. Pall Corporation, Glen Cove, NY 11542.
Ultipor®. Nylon Media Capsule Filter, Pall Corp., Glen Cove, NY 11542.
Ultipor® N_{66}. Nylon Microporous filter medium, Pall Corp., Glen Cove, NY 11542.
Coulter®. Coulter Electronics, Inc., Hialeah, FL 33010.
Teflon®. Fluorocarbon resin, DuPont Chemicals, Wilmington, DE 19898.
Tyvek®. Speenbonded polyolefin, DuPont Tyvek, Wilmington, DE 19880.
Gore-Tex®. Expanded PTFE, W. L. Gore and Associates Inc., 555 Paper Mill Road, Newark, DE 19714.
HIAC/Royco®. Pacific Scientific Co., Silver Spring, MD 20904.
Dimple Pleat™. Pleated high efficiency air filtration medium, Flanders Filters Inc., Washington, NC 27889.

REFERENCES

1. P. P. DeLuca, S. Bodapatti, and S. Im, Guideline for the identification of particles in parenterals, *FDA Bylines, 3* (July 1980).
2. *Federal Standard 209E,* Federal standards for clean room and work station requirements, controlled environment, (November 1992).
3. *British Standard BS 5295,* Environmental cleanliness in enclosed spaces. Part 1. Specification for clean rooms and clean air devices, British Standards Institution, London, 1989.
4. *Flanders Technical Bulletin No. 581D,* "HEPA filters and filter testing, (3rd Ed.), Flanders Filters, Inc., Washington, DC, 1989.
5. W. W. Szymanski, and P. E. Wagner, *J. Aerosol Sci., 21(3)*: 441–451 (1990).
6. H. Umhauer, and M. Bottlinger, The effect of particle shape and structure on the results of single particle light scattering analysis, Proceedings of the 2nd International Congress on Optical Particle Sizing, Tempe, AZ, 1990, pp. 425–434.
7. A. W. Lieberman, *Pharm. Technol.,* 71–77 (1979a).

8. A. W. Lieberman, *Pharm. Technol.*, 61–66 (1979b).
9. A. W. Lieberman, A new 0.05m 0.1 CFM optical particle counter, Proceedings of the 20th DOE/NRC Nuclear Air Cleaning Conference, Boston, MA, *22,* 1988, pp.1137–1143.
10. H. T. Sommer, and C. E. Montague, Design and performance of optical aerosol particle counters with different light sources, Proceedings of the 3rd Chinese Aerosol Conference, Beijing, China, (per HIAC, Silver Springs, MA) 1990.
11. B. Y. H. Liu, W. W. Szymanski, and K. O. Ahn, *J. Environ. Sci.,* 19–24 (May/June 1985).
12. W. Montague, and H. T. Sommer, *Filt. News,* 26–30 (Nov./Dec., 1989).
13. R. J. Knollenberg, *J. Appl. Meteorology, 9*: 86–103 (1970).
14. D. Bemer, J. F. Favries, and A. Renoux, J. Aerosol. Sci. *21(5)*: 689–700 (1990).
15. H. Buettner, *Aerosol. Sci. Tech., 12(2)*: 413–421 (1990).
16. R. Caldow, and J. Blesner, *J. Parenteral Sci. Technol., 43*: 174–179 (July/Aug. 1989).
17. H. T. Sommer, Optical sizing of single particles (HIAC laser), Proceedings of the 2nd International Congress of Optical Particle Sizing, Tempe, AZ (per HIAC, Silver Springs, MA), 1990, pp.612–618.
18. *National Fluid Power Association Document B93-20,* Procedure for qualifying and controlling cleaning methods for hydraulic power fluid sample containers, NFPA, Milwaukee, WI 1972.
19. *Millipore Corporation Technical Bulletin AD030,* Detection and analysis of particulate contamination, Millipore Corp., Bedford, MA 1991.
20. *American Society for Testing and Material Document No. 311,* Standard practice for processing aerospace liquid samples for particulate contamination analyzes using membrane filters, ASTM, Philadelphia, PA, 1989.
21. A. Teetsov, *Microscope, 25*: 103–113 (1977).
22. A. M. Dixon, Human contamination issues in cleanroom manufacturing, Proceedings of the PDA International Conference on Particle Detection, Metrology and Control, Washington, DC, 1990, pp.65–81.
23. Quintin T. Phillips, *J. Environ. Sci.*, 27–31 (Sep./Oct. 1983).
24. T. E. Munson, and R. L. Sorenson, *Sterile Pharmaceutical Manufacturing Applications for the 1990s* (M. J. Groves, W. P. Olson, and M. H. Anisfelt, eds.), Interpharm Press, Buffalo Grove, IL., 1990, pp.163–184.
25. Å. L. Möller, *Med. Dev. Technol., 3(7)*: 24 (1992).

21

Particles in Fluids and on Surfaces: Specialized Bibliography

Alvin Lieberman

Particle Measuring Systems, Inc., Fremont, California

I. INTRODUCTION

A. Basis for Bibliography Presentation

Bibliographies are typically combined with discussion of a specific subject area and provide supporting technical information on the subject of concern. The overall subject area of this book is a prime example of technology which is of concern in many technical disciplines. The chemical, physical and engineering problems of contaminant generation, transport, deposition, control, and measurement are discussed in many areas. This is particularly true for the situations where the contaminant materials are particulate in nature. Both the technology and the application of that technology are of concern.

As one observes technology development in many areas, it often becomes apparent that developments in one technical discipline are often unfamiliar to practitioners in other areas, even if the technology is nearly identical. Some examples may aid in understanding the situation:

1. Heterogeneous nucleation is of concern in both cloud physics and crystallization. The same phenomena control rain and snow formation as well as product crystal formation and growth phenomena, but many chemical engineers and atmospheric scientists are not familiar with the similarity of their efforts.

2. Hydraulic fluid power engineers have solved many of the problems concerned with movement of viscous liquids through small apertures, but pharmaceutical production system designers are usually not familiar with the latest work in the fluid power industry.

3. Semiconductor manufacturing process tools use deionized water that has been cleaned to the point where the concentration of particles larger than 0.1 μm is no more than 10 per liter, but the particulate control procedures used in the semiconductor industry are seldom seen when WFI is produced. At the same time, semiconductor industry procedures for prevention and control of microbiological contamination would be inadequate in most pharmaceutical processing areas.

These few examples illustrate how information in one source may be directly applicable to problems in another area. If one is concerned with the technology and not the technical discipline in which it is described, then many solutions to problems can be found in the literature in a different technical area.

Literature in many technical disciplines was scanned, and a small sample of the pertinent documents from those sources is presented here. The technical areas covered include atmospheric physics, chemical engineering, chemistry, civil engineering, contamination control, electrical engineering, food processing, fluid mechanics, mechanical engineering, nuclear energy, optics, pharmacology, powder technology, and statistics. Some documents and papers in the area of particle technology are presented in the journals which covers each of those areas. It is estimated that the literature shown in this chapter cover no more than a few percent of the pertinent information. However, it is believed that the information shown here describes a major part of the technology of concern. Subject areas not included in this chapter are referred to within the abstracted documents.

This chapter presents some examples of literature from a database directed to examination of phenomena associated with particle physics and engineering. Those aspects of particle physics which are of particular interest to personnel involved with the definition, control, and characterization of particulate contamination are reported. The literature was found in a wide variety of journals, reports, proceedings, etc., as will be seen by examining the sources for the several documents reported in any particular technical area.

B. Bibliography Format

Each item in this bibliography has a similar format:

1. The author(s) or the agency which produced the item is identified.
2. The title of the paper, report, standard procedure, etc., is stated.
3. The source of the item is identified.
4. The content of the item is abstracted and presented.

II. TECHNICAL SUBJECT AREAS

A. Adhesion on Substrates

Ashley, K. C., Precision cleaning using high pressure freon, Proceedings of 30th Annual Technical Meeting, Institute of Environmental Sciences, Orlando, FL, 1984.

Requirements to efficiently remove radioactive particles from various surfaces and objects led to development of a high-pressure solvent spray technique. The physics of small-diameter particles suggest the necessity of application of shear forces to the particle to overcome adhesive inter-molecular force. Equipment developed generates high-pressure, high-velocity Freon TF spray for surface cleaning.

Ayala, R. E., P. A. Hartley, and G. D. Parfitt, The relevance of powder liquid wettability to the cohesiveness of carbon black agglomerates, *Particle Characterization 3*(1):26–31 (1986).

Particle size enlargement operations often use a liquid to improve powder cohesiveness. Capillary suction and surface tension forces acting through liquid menisci between particles are the primary source of cohesive bonding. Strength of these forces, and the wet agglomerate strength, depends upon the total liquid content, its distribution within the agglomerate and the powder wettability. Data show a compact strength increase with increasing surface tension for liquids with zero contact angle.

Bansal, I. K., Control of surface contamination of silicon wafers in the semiconductor industry, Proceedings of the 29th Annual Technical Meeting, Institute of Environmental Sciences, Los Angeles, CA, 1983.

As IC geometries become smaller, the need for ultrapure processing liquids for wafer cleaning is ever increasing. This paper compares extent of particulate contamination on wafers by using several grades of chemicals and defining the contaminant levels by several different analytical methods.

Blitshteyn, M., The use of blow-off guns in semiconductor manufacturing, *Microcontamination 4*(1):20–28 (1986).

Blow-off or nitrogen guns are used in semiconductor manufacturing for the removal of particles from surfaces. Some of these guns also include ionizers to neutralize static charges on surfaces. This paper examines whether the addition of ionizing devices improves the performance of blow-off guns, as well as how these guns may become particle generators. It is concluded that the ionizers do assist in cleanliness and that the problem areas arise from incorrect installation and operation.

Braaten, K. A., and K. T. Paw U., A net deposition model, *Aerosol Science and Technology 17*(4):289–302 (1992).

A net deposition model (dependent on friction velocity) that accounts for ideal deposition of aerosols to a surface, rebound of particles impacting the surface, and reentrainment of particles from the surface has been developed. The model time step is based on a scaled time interval between coherent boundary layer structures (bursts) with deposition, rebound, and reentrainment determined for each time step. Deposition and rebound are functions of friction velocity and reentrainment is based on a Monte Carlo simulation using a surface renewal concept.

Busnaina, A. A. et al., An experimental and numerical study of microcontaminant particle removal process, Proceedings of 34th Annual Technical Meeting, Institute of Environmental Science, King of Prussia, PA, 1988, PP. 663–678.

For removal of particles from silicon wafers using drag and centrifugal forces, the entire wafer surface should be covered with a water film. Flow visualization techniques are used. A nondimensional coverage parameter accounts for effects of shielding and allows extension of data to additional values of the governing parameters. A 3D turbulent time-dependent prediction procedure for swirling flow is used to predict flow between two rotating wafers.

Busnaina, A. A., and F. W. Kern, A fluid dynamics study of a wafer rinsing process, *Solid State Technology, 30*(11):111–114 (1987).

Wafer rinsing is an effective method to remove microcontamination particles from silicon wafers. Flow visualization is used here to observe the effect of rotational speed, jet flow rate and nozzle type on the flow pattern. Some novel chemical flow visualization techniques are used.

Busnaina, A., J. Taylor, and I. Kashkoush, Measurement of the removal force and the removal efficiency of submicron particles on silicon substrate, Proceedings Microcontamination Conference, San Jose, CA, 1991, pp. 462–480.

Removal force of submicron particles is estimated using hydrodynamic and centrifugal forces. The removal force is correlated with the percentage of removed particles to estimate the required adhesion force for particle 0.1 to 1.0 μm in size. Removal efficiency and effectiveness of hydrodynamic force removal technique in removing these particles is studied. Effect of particle size, concentration, rotational speeds and deposition time on removal effectiveness is considered.

Canovas, G. V. B., and M. Peleg, Detachment of dusts by exploiting the selective affinity of fine powder to particulates of different species, *Powder Technology 42*(2):185–186 (1985).

The mechanisms and kinetics of fine particles and dust adhesion to surfaces and the absorption of fine particles on particulate surfaces is considered. It is pointed out that fine particles can have preferential affinity to different materials. The fact can result in migration of already absorbed fines to a new surface. Complete detachment of particles from original surface can be used to produce dry powder cleaning processes.

Carosso, P. A., and N. J. P. Carosso, Surface contamination monitoring by the measurement of scattering distribution functions, *Journal of Environmental Science 30*(3):34–37 (1987).

This paper describes the use of bidirectional scattering distribution functions (BSDF) to quantify surface degradation, providing thereby an effective method of contamination monitoring. The approach is based on the use of witness surfaces located next to the critical surfaces. Contamination accretion can be monitored during all phases of spacecraft fabrication, assembly and testing. This paper discusses the apparatus adopted to perform BSDF measurements and discusses some practical aspects of data collection. The theory and methods are described briefly.

Chawla, M. K., Thin film surface molecular contamination monitoring and control, Proceedings 11th Symposium of International Confederation of Contamination Control Societies, London, 1992, pp. 321–323.

Surface contamination can be detrimental to proper bond adhesion, coating, painting, etc., to surfaces. This paper describes a new technique to quantify surface cleanliness. Optically stimulated electron emission is used to define surface film presence quantitatively.

Derjaguin, B. V. et al., The role of the pressing-on in the adhesion of elastic particles, *Powder Technology* 37(1):87–93 (1984).

An analysis was carried out of the influence of the preliminary pressing-on components of the adhesion force of different natures for the case of the contact of elastic material particles with a rigid substrate. Two pressing-on components are considered: those realized under the effect of a dynamic load and of a static load. It is shown that an increase in the adhesion of elastic particles to the surface resulting from the preliminary pressing-on is attributable to an increase in the adhesion force component due to a double electric layer in contact.

Dzubay, T. G., and R. K. Barbour, A method to improve the adhesion of aerosol particles on Teflon filters, *Air Pollution Control Association Journal 33*(7):692–695 (1983).

This experiment was designed to test stability of coarse particle deposits on Teflon filters and to test effectiveness of application of a thin layer of mineral oil to the deposit sides of the filters before use. Sample stability was found to be enhanced.

El-Shobokshy, M. S., Experimental Measurement of Aerosol Deposition to Smooth and Rough Surfaces, *Atmospheric Environment 17*(3):639–644 (1983).

The deposition velocity of aerosol particles in turbulent flow to smooth and rough pipe wall surfaces was determined. Roughness sizes were represented by the centering average values. Data are quite closely in agreement with theory.

Gale, G. W. et al., A flow visualization and numerical study of a particle removal process from silicon wafers, Proceedings 34th Annual Technical Meeting, Institute of Environmental Sciences, King of Prussia, PA, 1988, pp.75–381.

Flow visualization techniques are used to study contaminant particle removal from wafer surfaces using drag and centrifugal force. Understanding the flow field around a rotating wafer is essential in assessing the process effectiveness. Experimental work supplements earlier study. A 3D turbulent time-dependent prediction procedure for swirling flows predicts the flow field between two rotating wafers. Water droplet trajectories from a peripheral jet are shown.

Hamberg, O., and E. M. Shon, Particle size distribution on surfaces in clean rooms, Proceedings 30th Annual Technical Meeting, Institute of Environmental Sciences, Orlando, FL, 1984.

Experimental particle size distributions on clean room surfaces, resulting from the gravity settling of airborne particles are presented and found to be significantly different from the data in MIL-STD 1246. Theoretical surface distributions, based on fallout from Fed. Std. 209 airborne distribution are derived and show good agreement with experimental data. It is shown that surface cleaning tends to make a PSD resulting from fallout approach a MIL-STD 1246 distribution. It is recommended that 1246 use be limited to cleaned surfaces.

Haynie, F. H., Size distribution of particles that may contribute to soiling of material surfaces, *Air Pollution Control Association Journal 35*(5):552–554 (1985).

Some questions had been raised at an EPA workshop as to what size of particle contributes most to soiling of surfaces and what ambient air measurement technique provides the best proxy for soiling potential.

An experiment was conducted which showed coarse particles dominate soiling of smooth surfaces; a sampler must collect particles over 10 μm; a hi-vol sampler may provide fair indication of soiling on vertical surfaces but would be poor for horizontal surface soiling indication.

Hecht, L., An introductory review of particle adhesion to solid surfaces, *Journal of the IES 33*(2):33–37 (1990).

This paper presents an up-to-date overview of particle adhesion to solid surfaces. The discussion consists of fundamental concepts of adhesion, the nature of solid surfaces, the physical properties of particles and comments on the various theories of adhesion. Some practical examples are also cited. A numeric example of adhesive forces as a function of particle size is given in detail.

Hovatter, W., and E. A. Hendrickson, Cleanroom wipers: a comparative study. *Microcontamination 2*(6):47–55 (1984).

This paper examines several different forms of contamination that are associated with wiping materials and it outlines some testing methods that can be used to determine the level of cleanliness of currently available wiping materials.

Khilnani, A., and D. Matsuhiro, Adhesion forces in particle removal from wafer surfaces, *Microcontamination 4*(4):28–30 (1986).

Adhesion forces that occur during removal of submicron particles from semiconductor wafer surfaces are discussed. Capillary adhesion, the strongest force binding the particles, can only be overcome by mechanical scrubbing. Further research is needed to establish the effect of processing on particle adhesion and to define cleaning practice better.

Kurz, M. R., A. Busnaina, and F. E. W. Kern, Measurement of detachment forces of particles on silicon substrate, Proceedings of 35th Annual Technical Meeting, Institute of Environmental Science, Anaheim, CA, 1989, pp. 340–344.

A study was made to quantify the required removal force for spherical particles on a flat surface. Removal is accomplished by subjecting the particle to a shearing force strong enough to overcome the adhesion force. Shear force can be generated using a wafer rinsing method, rotating the wafer about its axis while covered with a water film. The flow field results in a drag on the particles in addition to the centrifugal force. Force values can be calculated for several operating conditions.

Kwan, S. C. et al., Precision cleaning of very large structures, *Journal of Environmental Science 27*(4):27–30 (1984).

Performance and life time requirements for future space vehicles are very stringent. Contamination levels for both particles and molecular items must be controlled to appropriate MIL-STD-1246 levels. Methods are being developed for precision cleaning and verification. This paper discusses results of solvent wipe precision cleaning methods with inexperienced personnel to achieve the required 1246 levels.

Loo, C. E., and J. Bridgwater, Theory of thermal stresses and deposit removal, *Powder Technology 42*(1):55–66 (1985).

A theory is developed for the thermal stresses that are generated in deposits of foulant on a heat-exchange surface. Calculations of the stored elastic strain energy permits development of a criterion for deposit removal by spalling. The amount of material removed from a surface is about proportional to the deposit thickness, a theoretical corroboration of a result often used in describing the removal of solid from a heat-exchange surface.

Martin, R. E., L. M. Hanna, and E. J. Bouwer, Determination of bacterial collision efficiencies in a rotating disk system, *Environmental Science and Technology 25*(12): 75–82 (1991).

The probability of attachment of bacteria to glass surfaces for a range of concentrations of electrolyte on the glass. The collision efficiency was determined by deposition of cells on the rotating glass disk. Deposition could be limted by interaction of similarly charged electric double layers. Transport of cells to the disk surface was comparable to that of similarly sized PSL particles; however deposition for both decreased with increasing distance from the disk center. An explanation is suggested.

Mehrotra, V. P., and K. V. S. Sastry, A novel method for the calculation of pendular bond characteristics between unequal sized spheres, *Powder Technology 41*(3): 259–263 (1985).

This paper presents a novel and more accurate method to calculate the mean radius of curvature of the liquid meniscus and the forces of attraction between two unequal sized spherical particles. The method, using the fact that the forces exerted by two particles are the same, does not need an explicit mathematical description of the liquid meniscus.

Menon, V. B., A. C. Clayton, and R. P. Donovan, Removing particulate contaminants from silicon wafers: a critical evaluation, *Microcontamination* 7(6):31–36 (1989).

A study is described to evaluate the effects of cleaning solution chemistry and particle composition on a megasonic cleaner's efficiency at various power settings. The use of megasonics for rinsing hydrophobic wafers following a wet oxide etch bath was also studied. Megasonic cleaning is extremely effective in these cleaning operations.

Menon, V. B. et al., Megasonic cleaning: a comparative evaluation of water cleaning solutions, Proceedings of 35th Annual Technical Meeting, Institute of Environmental Science, Anaheim, CA, April, 1989, pp. 320–324.

Particle removal efficiencies of a megasonic wafer cleaner were evaluated for three cleaning solutions: DI water, Ammonia/hydrogen peroxide and a choline peroxide solution. Effects of particle size, type and concentration on cleaning efficiency were studied. Solution chemistry was important, especially for wafers with low initial particle counts.

Mittal, K. L., A multidisciplinary approach to surface contamination control, *Microcontamination* 6(5):16–24 (1988).

A general discussion is presented on surface adhesion, surface cleaning and particles on surfaces.

Montz, K. W., J. K. Beddow, and P. B. Butler, Adhesion and removal of particulate contaminants in a high-decibel acoustic field, *Powder Technology 55*(2):133–140 (1988).

Interaction of a high decibel acoustic field with micron sized particles adhered to a glass slide was studied. Experimental design used an acoustic field to overcome adhesion forces acting on contaminant materials. Image analysis was used to determine particle counts before and after cleaning. Removal rates for sound pressure levels below 160 dB were poor, but exceeded 98% at 165 dB.

Musselman, R. P., and T. W. Yarbrough, Shear stress cleaning for surface departiculation, *Journal of Environmental Science 30*(1):51–56 (1987).

A cleaning technique for removal of radioactive surface contamination has been proven effective for removing non-hazardous contaminant particles as small as 0.1 μm. The process uses a controlled high-velocity liquid spray inside a vapor containment to remove particles from a surface. The viscous drag force generated by the fluid applies a shear stress greater than the adhesion force that holds small particles to a substrate. Fluid mechanics and field tests indicate general cleaning parameters.

Nagarajan, R., Surface properties and cleanability: a rational correlation, *Journal of Environmental Science 34*(1):26–33 (1990).

Suitability of ultrasonic surface cleaning is studied in terms of the maximum cleaning potential after multistage cleaning and the first-stage cleanability. These parameters are linked to the nature of adhesive interaction between the particle, the substrate surface and any intervening liquids.

Nagarajan, R., and R. W. Welker, Precision cleaning in a production environment with high pressure water, *Journal of the IES 35*(4):34–44 (1992).

Two cleaners using high-pressure (to 3000 psig), high-flow water were procured for use at suppliers of critical disk drive components. Acceptance runoff, certification and production run cleanliness data are presented here in terms of liquidborne particle counts on contaminants extracted from substrates. Results indicate that residual particles on high-pressure-sprayed parts may be lower by more than a factor of 100 compared to those obtained by other cleaning methods. Cleaning efficiency increases with particle size.

Nelligan, J. D. et al., Particle adhesion and removal in model systems. V. Interpretation of the kinetics of particle detachment, *Journal of Colloid and Interface Science* *89*(1):9–15 (1982).

Theoretical interpretation of the kinetics of detachment of spherical colloidal hematite particles from steel is given. The experimentally established probability of escape is compared with the values calculated from Dahneke's models taking into consideration the energy barrier due to the double layer and dispersion forces and the distance of closest approach.

O'Donoghue, M., The ultrasonic cleaning process, *Microcontamination* *2*(5):63–67 (1984).

Efforts are made to describe the unique cleaning capabilities of the ultrasonic process and to outline how this practice can best be designed and implemented. When properly used, ultrasonic cleaning methods can provide a highly effective means of providing removal of insoluble particles from hard substrate surfaces.

Ohmi, T., H. Inaba, and T. Takenami, Research on adhesion of particles to charged wafers critical in contamination control, *Microcontamination* *7*(10):29–39 (1989).

Adhesion of floating particles to a charged surface is stimulated by electrostatic force. The rate of adhesion from electrostatic force increases as particle size decreases. The quantity of particle adhering depends on the airborne particle concentration, particle density and ion concentration and type of atmosphere. Primary means of reducing particle adhesion is to minimize wafer charge.

Paw, U. K. T., The rebound of particles from natural surfaces, *Journal of Colloid and Interface Science* *93*(2):442–452 (1983).

The basic theory behind particle rebound is reviewed and the results of various experiments reported. Biological and nonbiological particles rebound from surfaces at characteristic critical impact speeds, regardless of the surface type. These particles of 20–40 μm diameter are larger than those used in previous rebound work, but still support the general predictions of the theory. Estimated adhesion energies, based on Bradley-Hamaker theory, are too high by factors of 4–200.

Phillips, Q. T. et al., Parts cleaning: an evaluation of ultrasonic systems, Proceedings of the 7th International Committee of Contamination Control Societies Conference, Paris, 1984.

This paper summarizes several evaluations of ultrasonic cleaning systems for electronic parts. The evaluations were made using 300 series stainless steel and aluminum coupons. Efficiency and cost of the cleaning systems are compared. The appendix includes a discussion of the ASTM F24 test.

Ranade, M. B., Adhesion and removal of fine particles on surfaces. *Aerosol Science and Technology* *7*(2):161–176 (1987).

Adhesive forces result from molecular and electrostatic interaction and are influenced by the surrounding medium and the composition. The magnitude of adhesive forces relative to particle mass increases greatly for small particles; removal is difficult. Techniques for adhesion measurements are marginal and new developments are urgently needed.

Riley, D. J., and R. G. Carbonell, Investigating liquid-based particle deposition and the effects of double-layer interactions using hydrophilic silicon wafers, *Microcontamination* *8*(12):19–25, 60–61 (1990).

A series of tests were undertaken to add knowledge as to the mechanisms of liquid-based deposition in a cleanroom environment. Particles with both a + and a – zeta potential were used. Deposition onto wafer surfaces was analyzed as a function of particle concentration and duration of wafer submersion in the liquid. The importance of surface interactions is emphasized. Charge interactions affected both deposition and the mechanisms of particle adherence.

Schmidt, W. R., and D. C. Grant, Method for measuring the particle cleanliness of high purity liquid chemical containers, Proceedings of the 38th Annual Technical Meeting, Institute of Environmental Sciences, Nashville, TN, 1992, pp. 420–430.

A new container for storing and delivering chemicals and methodology for determining particle cleanliness are described. The packaging technology uses a bag-in-bottle concept in which a precleaned film liner, in a rugged overpack is used to store, transport and dispense chemical. Cleanliness testing involves filling the containers with very clean water and measuring the resulting increase in particle concentration. Containers were found to contain an average of 2.1 particles/ml (≥ 0.5 μm).

Sirch, E., Procedures for cleaning and decontamination of solid surfaces in pharmacy, Proceedings 10th Symposium of International Committee of Contamination Control Societies, Zurich, 1990, pp. 328–331.

According to VDI 2083, part 4, "Cleanliness of Surfaces," surface cleanliness is classified in five surface cleanliness classes. Each class corresponds to its number of particles < 1 μm, < 10 μm, etc., per square centimeter. Only particles > 10 μm are of interest with parenteral drugs. However, viable ones are of great importance, thus requiring safe and reproducible automatic cleaning of surfaces for this industry. Procedures are reviewed.

Stowers, I. F., Advances in cleaning metal and glass surfaces to micron-level cleanliness. *J. Vacuum Science and Technology 15*(2):751–754 (1978).

A high-pressure 6.9-MPs liquid-spraying technique was developed to remove small (> 5 μm) particles from large 100-m^2 surfaces. Cleanliness levels available by this method are described for various substrate materials and are found to be much better than obtained from ultrasonic methods or vapor degreasing.

Taylor J. et el., Measurement of removal force of submicron particles on a silicon wafer, *Proceedings of 36th Annual Technical Meeting, Institute of Environmental Science, New Orleans, LA, 1990,* pp. 422–426.

The removal force of submicron particles is estimated using hydrodynamic and centrifugal forces. The removal force is correlated with the percentage of particles removed to give an estimate of the required detachment force. Particles from 0.5 to 1 μm were considered.

Tsai, C.-J., D. Y. H. Pui, and B. Y. H. Liu, Elastic flattening and particle adhesion, *Aerosol Science and Technology 15*(4):239–255 (1991).

The effect of elastic flattening on particle adhesion to solid surfaces was studied. Attracting and repulsing molecular potential energies and forces are first determined for a flattened sphere-surface system. The energy conservation principle and force balance are then used to obtain the analytical equations that describe the relationship between the elastic flattening and the van der Waals force.

Umesaki, M., T. Hyohno, and T. S. Morita, CFC-113-free substrate cleaning in thin film magnetic disc manufacturing, *Proceedings of 11th Symposium of International Confederation of Contamination Control Societies, London, 1992,* pp. 145–149.

A cleaning system was developed with a drying process using hot ultrapure water instead of CFC-113 vapor. The substrates are dipped into the hot water (ca. 80°C) and are raised out of the water slowly. The substrates are completely dried owing to the surface tension of water and the temperature of the substrates. Effect of brush shapes during submersion is discussed.

Vinke, H., P. J. Hamersma, and J. M. H. Fortuin, Particle-to-bubble adhesion in gas liquid solid slurries, *AIChE Journal 37*(12):1801–1809 (1991).

Adhesion of small particles to gas bubbles is often encountered in slurry work, as during flotation of finely divided minerals and in slurry reactors, where this adhesion may result in enhanced gas absorption rates. The adhesion of small particles to gas bubbles is studied with a modified bubble pick-up method. Adhesion extent is revealed by the angle by which the gas bubble surface is covered by adhering particles under static conditions. It depends on the modified Eotvos number, the ratio of particle and bubble radii and the effective contact angle.

Whyte, W., and N. Donaldson, Cleaning a cleanroom, *Microcontamination* 5(11):49–54 (1987).

Cleaning of cleanroom areas is discussed. Wet wiping methods with various cleaning solutions are described as they remove particles and chemicals, disinfect surfaces or dissipate electrostatic charges. Cleaning procedures are discussed in some detail. Microbiological effects are also considered.

B. Entrainment and Saltation

Akiyama, T., and Y. Miyamoto, Solid particle transport from a trough in a wind tunnel, *Powder Technology* *57*(4):235–240 (1989).

A study was made of particle transport from a trough in a wind tunnel. The Einstein method for calculating the rate of sediment movement by water was successfully used, with introduction of a correction factor for describing solid particle transport from the trough under wind action. The correction factor was correlated in terms of two Richardson numbers based on velocities within the trough.

Akiyama, T., and Y. Tanijiri, Criterion for reentrainment of particles, *Powder Technology* *57*(1):21–26 (1989).

Criteria for reentrainment of fine particles, subjected to an air stream were studied. Ambient air relative humidity alone was found an insufficient criterion of the reentrainment threshold. Measurements of particle properties and experiments led to the conclusion that flowability, determined by angle of repose, compressibility, angle of spatula and cohesion force, is a good criterion to evaluate the reentrainment threshold of fine particles.

Braaten, D. A., K. T. Paw U., and R. H. Shaw, Particle resuspension in a turbulent boundary layer—observed and modeled, *Journal of Aerosol Science* *21*(5):613–628 (1990).

Particle resuspension from a surface solely by turbulent fluid forces was studied experimentally by observing the intermittent particle resuspension process. Experimental details are given. Resuspension was observed in discrete and intermittent events, with the fraction of particles suspended as a function of time generally following an exponential decay. A Monte Carlo model, which matches actual processes, was developed on the hypothesis that resuspension is associated with coherent burst structures.

Fan, B., A. R. McFarland, and N. K, Anand, Characterization of the aerosol particle lift force, *Journal of Aerosol Science* *23*(4):379–388 (1992).

In the analysis of particle despositional effects, we show that there is a characteristic parameter which can be used in evaluating the lift effects on particles in shear flow. Comparison of lift and drag forces are made to illustrate the impact of the lift force on particle deposition. An analytical solution is obtained for particle trajectories in sheared laminar flows. A stability condition is obtained for the equations which describe the particle trajectories.

Hamm, B. A., W. L. West, and G. B. Tatterson, Sludge suspension in waste storage tanks, *AIChE Journal* *35*(8):1391–1394 (1989).

The objective of this work is to determine how the cleaning radius and suspension rate change with time for a typical sludge material and to provide a model that explains the results in a general fashion. The use of a slurry pump to suspend solids is discussed and the mechanisms of particle removal from the sludge bank are pointed out.

Itano, M., M. Miyashita, and T. Ohmi, Particle deposition and removal on wafer surface in RCA cleaning process, Proceedings of Microcontamination Conference, San Jose, CA, 1991, pp. 521–543.

Tests were run using acid and alkaline solutions used in wet cleaning processes for semiconductor manufacturing, to investigae their ability to prevent particles from being deposited on wafer surfaces and to remove particles from contaminated wafers. It is seen that maximum particle deposition is observed in low pH solutions and the deposition increases with pH. This is ascribed to the fact that the zeta potential of particles decreases with increase of solution pH. Alkaline solutions are then adequate for removing particles. Further studies are described.

Kashkoush, I., and A. Busnaina, Numerical simulation of ultrasonic and megasonic cleaning prcesses, Proceedings of the 38th Annual Technical Meeting, Institute of Environmental Sciences, Nashville, TN, 1992, pp. 19–25.

The acoustic streaming potential associated with a sound field in liquids has proven to be capable of removing submicron particles from silicon substrates. The acoustic streaming generated from a 40–800 kHz sound source is numerically simulated. The Navier-Stokes equations governing the flow field are solved using a finite difference scheme. Preliminary results show that the acoustic streaming and associated shearing forces play a major role in removing submicron particles from substrates.

Mark, D. et al., Particle blow-off: a source of error in blunt dust samplers, *Aerosol Science and Technology* *1*(4):463–469 (1982).

From blunt sampler theory, it is possible to predict oversampling for simple single-orifice axisymmetric total dust samplers arising from particle blow-off from external surfaces. Wind tunnel experiments confirm the predicted broad trends, revealing the errors in question to be as large as 50% for sampling dry dusts under some conditions of practical relevance. Some suggestions for test aerosol definitions are made.

Noll, C. G., Ensemble theory for the electrostatic deposition and entrainment of dust, *Filtration and Separation* *18*(4):330–333 (1981).

The electrostatic deposition of particles depends on the competition between electrostatic deposition and entrainment of dust. This paper discusses the gas-stream-deposit interface in terms of classical ensemble theory. Particular emphasis is directed towards the canonical ensemble, which applies to Markov correlated events.

Seyfert, N., Collection, bouncing, and adhesion on crossed fibers in mono- and multilayer design, *Aerosol Science and Technology* *9*():201–212 (1988).

The author studies the theoretical aspects of filtration by fabrics with special attention of bouncing of particles on fibers. Differences in efficiency between single- and crossed-fiber flow fields are detailed. From these, the computation of fiber junction helps to predict fiber efficiency with respect to particle bouncing and leads to an advance over calculations based on a single-fiber model.

Taheri, M., and G. M. Bragg, A study of particle resuspension in a turbulent flow using a Preston tube, *Aerosol Science and Technology* *16*(1):15–20 (1992).

A technique was developed for the study of resuspension of particles from a flat surface using a Preston tube to determine the shear and flow forces required to suspend particles from a surface. The procedure was used to study resuspension of glass beads at various free-stream velocities. Results indicate that particles are resuspended according to a log normal distribution with respect to any free-stream velocity, surface shear stress and shear velocity.

Wen, H. Y., and G. Kasper, On the kinetics of particle re-entrainment from surfaces, *Journal of Aerosol Science* *20*(3):483–498 (1989).

A kinetic model of particle reentrainment is stated and compared with data from industrial high-purity gas systems and with controlled experiments using PSL, 0.4-1 μm. The model described reentrainment as a first-order reaction analogous to molecular desorption from an inhomogeneous surface with a rate constant based on adhesion force/removal force ratio. The model estimates widths of adhesion force distributions and the initial surface particle populations.

Wu, Y.-L., C. I. Davidson, and A. G. Russell, A stochastic model for particle deposition and Bounceoff, *Aerosol Science and Technology* *17*(4):231–244 (1992).

A new model for particle deposition and bounce-off that combines current knowledge of turbulent bursts with the stochastic properties of turbulent fluctuations is presented. The model predictions for deposition velocities agree with reported experimental data for low particle relaxation times. The model also provided the distribution of particle velocities on reaching the surface, used to compute the fraction of particle bounceoff.

C. Filter Evaluation, Gases

Anon, ASTM., F1215-1989; Standard test method for determining the initial efficiency of a flatsheet filter medium in an airflow using latex spheres. American Society for Testing and Materials, 1916 Race Street, Philadelphia, PA, 1989.

This method covers techniques for measuring the initial particle size efficiency of flatsheet filter media using monodisperse aerosols. It uses light scattering particle counting in the size range of 0.5 to 5.0 μm and airflow test velocities of 1 to 25 cm/s. The method is applicable for filter media with efficiencies less than 99.9% or penetrations greater than 0.001 at 1.0 μm sphere diameter.

Cadwell, G. H., Understanding HEPA filtration, *Medical Device and Diagnostic Industry* 5(1):39–42 (1983).

This article discusses HEPA filter terminology and construction and testing of HEPA filters for clean operations. The areas covered include rated flows, filtration mechanisms, flow straightening and testing procedures.

First, M. W., and J. M. Price, Performance of 1000 CFM and 1800 CFM HEPA filters on long exposure to low atmospheric dust loadings, III, Proceedings 17th DOE Air Cleaning Conference, Boston, 1982.

Comparative tests were made on standard and on high-volume HEPA filters when exposed to ambient dust loadings. Initial results reported previously indicated that an increase in service life of 1.6 times could be anticipated for the 1800-CFM filters when operated at 1000 CFM. This is only half of the theoretically expected increase. Additional tests have been conducted to determine the effect of prefilters on the service life of the 1800 CFM filters at both low rates.

Gogins, M. A. et al., Design and operation of optimized high efficiency filter element test systems, Proceedings of 33rd Annual Technical Meeting, Institute of Environmental Science, San Jose, CA, 1987.

This paper discusses improvements in the design and operation of high-efficiency filter test systems. The areas discussed are optimization of system hardware and application of statistical analysis to expedite testing. HEPA filters are tested in 15 s; ULPAs in 60 s.

Hanson, W. D., DOP Testing HEPA filter banks in series, Proceedings 17th DOE Air Cleaning Conference, Boston, MA, 1982.

A new method was developed to test large HEPA filter banks installed in series. A system is described for uniform concentration of DOP smoke within 5-ft space existing between filter banks with up to 42 HEPAs. A large area sampler inlet was used to scan more filter areas rapidly.

Herman, R. L., In-place HEPA filter aerosol test system, Proceedings 17th DOE Air Cleaning Conference, Boston, MA, 1982.

An in-place HEPA filter aerosol test system showing characteristics of a low-cost reliable, safe test method was designed. The system consists of a diffuser/ejector assemblies and single-point sample ports downstream of the filter stage under test.

Irie, T., et al., Study on the inplace test of particulate collection efficiency for air cleaning apparatus, Proceedings of 8th Symposium of International Committee of Contamination Control Societies, Milan, 1986.

This a final report of the working group of Japan Air Cleaning Association on Air Cleaning System Design. It shows some useful and practical methods to determine collection efficiency for filters. One is to use one relative dust concentration meter as a light scattering type and to measure concentrations up and downstream of the filter alternately for 20–30 min. Other recommended methods and results of field measurements are also reported.

Jacox, J. (1992): Review of HEPA filtration test standards and their application to nuclear applications. Proceedings 21st Air Cleaning Conference, San Francisco, CA, 1992.

This paper reviews HEPA filtration test standards as they relate to nuclear applications. It includes both nuclear and "nonnuclear" standards and references for completeness since they are closely related and overlap. Some 30 references are given on the subject.

Kasper, G., and H. C. Wang, Establishing a particle test sequence for evaluating performance of 10-inch process gas filters, *Microcontamination 8*(9):33–37 (1990).

Data are provided on experience with relevant test methods for 25-cm filter cartridges used in process gas lines. The discussion concentrates on particle impurities generated by the actual filter; the effects of moisture are not considered and the issue of particle generation from the housing is touched upon briefly. Test procedures include particle removal efficiency determination, particle shedding, effect of flow changes.

Kirchner, K. W. et al., In-situ continuous scanning high efficiency particulate air filter monitoring system, Proceedings 18th DOE Air Cleaning Conference, Baltimore, MD, 1984.

Using current technology, a continuous in-situ monitoring system was designed. XYZ sampling system was designed and built to transport a sample probe to any location on a given filter bank. Laser single-particle spectrometers were used and chosen as the most sensitive test system and a microprocessor was used to control the entire system. Results are discussed. Either ambient air or generated challenge aerosol is used.

Kirk, G. G., C. E. Childress, and D. D. Schmoyer, Methods of testing HEPA filters with short upstream approaches, Proc. 20th DOE/NRC Nuclear Air Cleaning Conference, Boston, MA, 1988, pp. 1210–1215.

A series of tests is described where aerosol injection points for HEPA filter testing could not allow 10 duct diameters for mixing. It was concluded that turning vanes and perforated plate baffles, alone, do not help to make the challenge uniform. A baffle near the injection point is an effective way of meeting the N-510 criterion.

Levy, R. V., K. S. Souza, and C. B. Neville, The matrix approach: microbial retention testing of sterilizing grade filters with final parenteral products. Part I, *Pharmaceutical Technology 14*(9):160–173 (1990).

A program is outlined to validate performance of sterilizing-grade filters under the FDA "Guidelines for Sterile Drug Products Produced by Aseptic Processing." The methods give consideration of the chemical and biochemical attributes of the drug product and to the parameters of the sterile filtration process of manufacturing. A classification system was used to categorize actual drug products and to select representative ones for use in a retention testing program.

Loughborough, D., Development of filter systems—injection systems and mixing device evaluation, Proceedings of 20th DOE/NRC Nuclear Air Cleaning Conference, Boston, MA, 1988, pp. 1216–1230.

One problem in testing HEPA filters is to ensure a uniform challenge across the filter face. Injecting with a single-point injector requires 30 hydraulic diameters from injection point to sampling for uniformity. Use of a Stairmand disk mixer reduces this to 10 diameters. It is shown that the mixing length requirement may increase for low aerosol exit velocities. The use of multiple-point injectors with a Stairmand disk reduces the mixing length to 5.5 diameters.

Matigot, C. et al., Application of laser spectrometer to the measurement of filter efficiency, Proceedings of 7th International Committee of Contamination Control Societies Conference, Paris, 1984.

This paper describes the method used at Sofiltra to measure number of efficiency of filter panels with a laser particle spectrometer. Work was done with a challenge aerosol that is polydisperse, enabling analysis of the distributions up and downstream in small size classes. Efficiency is given for each class small enough so that it can be considered as a monodisperse aerosol.

Molter, W., C. Helsper, and S. Kaminski, Fast, automated testing of different air filter media, *Filtration and Separation 29*(1):45–50 (1992).

A common test procedure has been developed for testing both HEPA and gas cleaning filters. Two different test rigs were built for this purpose. By simultaneously measuring fractional penetration and pressure drop, loading studies can be done within a time period of minutes. Measurements were made to show the influence of face velocity, upstream concentration, test dust material, filter material and electrostatic charge of particles and filters.

Notaro, J., and M. F. Patterson, High-purity gas applications of pleated-membrane filter cartridges, *Microelectronics Manufacturing Technology 10*(1):17–18 (1987).

Testing was carried out with OPCs on the particles from pleated-membrane gas filters used for gas cleaning. It was found that some particles are present in the effluent gas and some conclusions as to possible sources are made. However, adequate gas cleaning is shown possible. It is recommended that cartridge changeout not be done more frequently than every several years and that the cartridges should last much longer. Less than 4 PCR > 0.1 μm are seen.

Oikawa, S. et al., Filter performance certification test in clean room, Proceedings of 9th International Symposium on Contamination Control, Los Angeles, CA, 1988, pp. 175–178.

This study was conducted to verify the performance of the super ULPA filter and the ULPA filter when environmental conditions change. Cleanliness 30 cm downstream from the filters was measured in a vertical uniform flow clean room. Cleanliness does not change with flow velocity, heating downstream filter surface or vibration on the filters. As temperature, RH, upstream filter surface temperature and concentration increases, the ULPA downstream values change.

Sadjadi, R. S. M., and B. Y. H. Liu, Characteristics of pin-hole leaks in HEPA and ULPA filters, Proceedings of 36th Annual Technical Meeting, Institute of Environmental Science, New Orleans, LA, 1990, pp. 242–249.

Characteristics of pinhole leaks was studied theoretically and experimentally. Important parameters affecting pinhole leak detection were determined and their effects quantified. A sensitive leak detection method was developed. The method is capable of detecting pinholes as small as 2 μm diameter. Ultrafine aerosols are used and detected with a CNC.

Scripsick, R. C. et al., Evaluation of methods, instrumentation and materials pertinent to quality assurance filter penetration testing, Proceedings 18th DOE Air Cleaning Conference, Baltimore, MD, 1984.

Present DOE filter testing uses a hot DOP aerosol generator, an OWL size analyzer and a scattered-light photometer concentration monitor. Alternative test methods using a cold DOP polydisperse aerosol generator and a PMS aerosol spectrometer are described. Test results indicate that this alternative method is quite satisfactory. Advantages and limitations are given.

Sinclair, C. S., and A. Tallentire, Microbiological evaluations of papers: influence of dispersion flow rate on penetration, *Particulate and Microbiological Control* pp. 52–57 September (1982).

This article describes how penetration is affected by the rate at which airborne spores are presented to medical-grade filtration papers. The data will aid in defining the experimental variables that govern penetration of paper by microorganisms. It is concluded that medical-grade papers are not absolute barriers to airborne microorganisms. Apparatus and results are described in detail.

Tillery, M. I. et al., The effect of particle size variation on filtration efficiency measured by the HEPA filter quality assurance test, Proceedings 17th DOE Air Cleaning Conference, Boston, MA, 1982.

A theoretical model is used to calculate the filtration efficiency indicated by a standard photometer for challenge aerosols of different PSDs and HEPA filters with different efficiencies vs. particle size. The model compares calculated efficiency indicated by the photometer with efficiencies based on particle number and mass. The differences in efficiency measured by the QA test procedure and the efficiencies with respect to aerosol mass and number have been calculated for a range of different size particles.

Wright, M. W., Criteria and methods to evaluate performance with comparing filters to filters, *CleanRooms* *3*(5):26–31 (1989).

This article looks at the composition of filters and important criteria and newer means of evaluating the performance of HEPA & ULPA filters. A figure of merit considering penetration and pressure drop is considered.

Zeimer, W., In-situ test of ULPA-filters, Proceedings of 9th International Symposium on Contamination Control, Los Angeles, CA, 1988, pp. 115–122.

For technical and economic reasons, it is not possible to carry out in situ tests of HEPA filters in large areas with manual scanning methods. Pinholes can only be detected quickly and without manipulation by means of an automated scanning device. Traverse rate and aerosol concentration determines the minimum detectable pinhole size. Pinholes larger than 70 μm can be detected at the building site, using equipment described here. In filter manufacture, this can be reduced to 30 μm.

D. Filter Evaluation, Liquids

Anon, ASTM, *Standard Practice for Determining the Performance of a Filter Medium Employing a Single-Pass, Constant-Rate, Liquid Test,* ASTM F795-88, American Society for Testing and Materials, Philadelphia, PA, 1988.

This practice guides the investigator into designing and subsequently using a test where a feed stream, flowing at a constant rate, makes one pass through the filter medium. The filter medium, held in an appropriate housing, can be any sort of medium, except for those with preformed beds of filter aid. The practice addresses measurement of permeability of the medium, efficiency and capacity.

Anon, ASTM, *Standard Practice for Determining the Performance of a Filter Medium Employing a Single-Pass, Constant-Pressure, Liquid Test,* ASTM F796-88, American Society for Testing and Materials, Philadelphia, PA, 1988.

This practice guides design and use of a test where a feed stream, at constant pressure, makes one pass through the filter medium. Any filter medium can be used, except for those with performed filter and beds. The practice addresses permeability, efficiency and life of the filter medium. Efficiency can be defined by measuring turbidity, mass concentrations and particle size distributions.

Anon, ASTM, *Standard Practice for Determining the Performance of a Filter Medium Employing a Multipass, Constant-Rate, Liquid Test,* ASTM F797-88, American Society for Testing and Materials, Philadelphia, PA, 1988.

This practice guides the investigator into designing, and subsequently using, a test where a feed stream, flowing at constant rate and after emerging from the filter medium as the filtrate is recirculated back to the feed tank while fresh, or makeup, test particles are continuously added to the feed tank. The filter medium may be any sort of porous material, but not one with a preformed filter aid bed. The practice addresses methods of measuring permeability, efficiency and life of the medium.

Chrai, S. S., Integrity test for filter systems, *Pharmaceutical Technology 12*(10):62–71 (1988).

This article reviews nondestructive tests for verifying integrity filter systems, including bubble-point, diffusion, forward-flow and pressure-hold tests. Advantages and problems of each test are discussed. A combination of the bubble point and diffusion tests appears to provide adequate information for integrity testing of filter systems.

Goldsmith, S. H. et al., A method for measuring particle shedding from microporous membrane filter cartridges in liquid streams, *Microcontamination* 2(3):47–52 (1984).

The method provides for uniform and realistic evaluation of filter cartridge cleanliness. Results are repeatable because of the apparatus and specification of operating parameters. These simulate typical process conditions. The method is described in detail, along with apparatus and the methodology to be used. Typical data are shown.

Goldsmith, S. H., and G. P. Grundelman, In situ method for evaluating filter cleanliness, *Pharmaceutical Manufacturing* 1(9):37–43 (1984).

A test method was developed for measuring initial filter cleanliness and the ease of flushing filters clean. The apparatus circulates high-purity water through the test filter, whose effluent is monitored for both number and size of particles. Details of the method are given; the method yields meaningful and repeatable information.

Goldsmith, S. H., and G. P. Grundelman, Validation of pharmaceutical filtration products, *Pharmaceutical Manufacturing* 2(11):31–37 (1985).

Even when a liquid-stream filter is not used for sterilization, the importance of filter validation cannot be overlooked. This paper discusses several techniques for validating the retentive efficiency, compatibility and life expectancy of filters with absolute pore size ratings above 0.2 μm. Filter efficiency test methods are described and several test method standards are mentioned.

Grant, D. C. et al., Shedding characteristics of filters in liquids, *Microcontamination* 5(7):30–36 (1987).

New test methods are described for accurate measurement of filter cleanliness and results obtained on studies carried out. Very low submicron particle concentrations are shown achievable for shedding data. Constant and pulsating flow were both used. Some filter variability was seen.

Hargesheimer, E. E. et al., Pilot scale evaluation of filtration processes using particle counting, Proceedings American Water Works Association Conference, San Diego, CA, 1990.

An experiment is described where three different OPCs and a turbidimeter were installed on-line to monitor filtered water quality in a 2.9-GPM plant. Several objectives were carried out. Turbidity and cumulative particle counts were closely correlated. Particle count changes were much more sensitive as indicators of potential filter media breakthrough. OPCs generate much information and can be used as stand-alone, reliable indicators of filter performance.

Johnston, P. R., The most probable pore-size distribution of fluid filter media. I. Evidence of such a distribution from results of extended bubble-point measurements, *Journal of Testing and Evaluation* 11(2):117–121 (1983).

This theoretical distribution addresses filter media with a random arrangement of pores. It is based on the concept of a smallest pore, the size of which is a function of the size of the building blocks making up the filter medium. The distribution is strictly a function of the porosity (ratio of void volume to bulk volume). The greater the porosity, the wider the pore size distribution.

Lee, J. Y., Auditing an aseptic filtration process, *Pharmaceutical Technology* 13(2):55–72 (1989).

This article discusses techniques for performing an audit of an aseptic filtration process using the approach of assured coverage of technical issues and compliance issues as well as areas that the FDA auditors may review during inspection. Subjects covered include filtration parameters, materials compatibility, filter sterility assurance, validation, etc.

Levy, R. V. et al., The matrix approach: Microbial retention testing of sterilizing grade filters with final parenteral products. II, *Pharmaceutical Technology* 15(5):58–68 (1991).

This article describes filter retention test procedures used for microbial testing performed with actual drug products under scaled-down simulated processing conditions. Summaries of these data are planned for use to support validation of this membrane filter type for use with other chemically similar products that are processed under comparable test conditions with the material used as challenge.

Lukaszewicz, R. C., and T. H. Meltzer, Concerning filter validation, *Journal of the Parenteral Drug Association 33*(4):187–194 (1979).

Particulates, including organisms, can be retained by filters by at least one of two mechanisms: sieve retention or adsorptive sequestration. Where particle capture is the result of sieving, the largest pores are of consequence to the smallest particles. The bubble point measurement serves as an index of retention in that situation and offers a basis for filter validation. Adsorptive particle arrest may be much more complex.

Meltzer, T. H., An investigation of membrane cartridge shedding. I. A quantitative comparison of four competitive filters, Proceedings 6th Pure Water Conference, Santa Clara, CA, 1987.

A study was carried out on groups of competitive cartridge filters to determine the degree of shedding in tests with clean water passage and stress by varied pressure. It was found that variability between cartridges is too great to permit differentiation solely on the basis of manufacturer with regard to particle shedding. Differentiation on the basis of retention can be determined. The test system and the statistical basis for conclusions are given.

Meltzer, T. H., A critical review of filter integrity testing. I. The bubble point method: assessing filter compatibility: initial and final testing, *Ultrapure Water 6*(4):40–51 (1989).

Bubble-point testing is discussed and the need for testing with the correct product liquid under correct process conditions is emphasized. Problem areas are pointed out. Utility of the method for assessing structure of leaks and carrying out initial and final testing are pointed out.

Meltzer, T. H., A critical review of filter integrity testing. II. The diffusive airflow and pressure-hold methods: initial and final testing, *Ultrapure Water 6*(5):45–55 (1989).

This article discusses diffusive airflow and pressure-hold methods of filter integrity testing, their use in assessing structural compatibilities of filters and appropriateness of initial and final integrity testing of filters.

Olson, W. P. et al., Problems of industrial filtration and membrane filter testing, Proceedings III World Filtration Congress, Downington, PA, 1982.

Effective filtration requires validating filter effectiveness and Bacterial retention must relate to a routine and simple nondestructive test of filter integrity. Process systems must be sited so that grow-through does not occur. Many of the problems might be resolved if the bubble-point test values closely approximated calculated values of the filter pore size. Since hydrophobicity of a membrane filter matrix is inversely related to the bubble point of water-wet filters, the inconsistency between theoretical and measured bubble points may be resolved if the contact angle of water with the matrix is not assumed to be zero and an accepted published value is substituted.

Olson, W. P. et al., Diffusion and bubble point testing of microporous cartridge filters: electromechanical methods, *Journal of Parenteral Science and Technology 37*(4):117–124 (1983).

The Automation Diffusion Tester, Electronic Forward Flow Test Kit and Satocheck Filter Tester were evaluated for diffusion/bubble-point testing of pleated membrane cartridge filters. Each machine performed according to manufacturer's claims and was validated by means of simple tests. Each machine was unique in some particular and each can be considered useful for the evaluation and control of filters.

Roberts, K. L., and D. J. Velasquez, Characterizing the rating and performance of membrane filters for liquid applications using latex spheres, *Swiss Contamination Control 3*(4b):71–74 (1990).

Studies using latex spheres as the challenge for characterizing membrane filters are discussed. The method provides a reproducible test which can be discriminate between different membrane pore sizes based on sieve retention. Challenging membranes with a wide-range of spheres sizes provides a detailed profile of the membrane retention capabilities and of retention under worst-case conditions.

Trompler, T., Qualification of high purity filter cartridges for ultraclean liquid filtration, Proceedings 10th Symposium of International Committee of Contamination Control Societies, Zurich, 1990, pp.156–158.

A new test rig is described for testing 10-in. membrane filter elements in ultraclean water. Comparative statements can be made concerning rinse-up behavior, particle retention and the effects of pressure shocks.

E. Filtration of Liquids

Darby, J. L., and D. F. Lawler, Ripening in depth filtration: effect of particle size on removal and head loss, *Environmental Science and Technology 24*(7):1069–1079 (1990).

Suspensions were filtered in laboratory scale deep-bed tests. Detailed measurement of PSDs at different depths and times and continuous monitoring of pressure at different depths during ripening were performed. Ripening is the improvement in removal efficiency occurring as previously retained particles begin to serve additional collectors for suspended particles. Effects of particle size of the suspension on the collection process are described.

Grant, D. C. et al., Particle capture mechanisms in gases and liquids: an analysis of operative mechanisms in membrane fibrous filters, Proceedings of 34th Annual Technical Meeting, Institute of Environmental Science, King of Prussia, PA, 1988, pp. 465–472.

This paper reviews theories of particle capture mechanisms for membrane filters and discusses effectiveness in various fluids. Theory indicates that membrane filters remove essentially all particles from gas streams by interception and diffusion. The operative mechanisms in liquids depend on the physical chemistry of the particle-fluid-filter media system. Retention efficiencies are also considered.

Grant, D. C., and J. G. Zahka, Sieving capture of particles by microporous membrane filters from clean liquids, Proceedings of 10th Symposium of International Committee of Contamination Control Sciences, Zurich, 1990, pp. 160–164.

This paper describes the particle retention efficiency of several types of microporous membrane filters in liquids challenged with low particle levels in a model developed to predict retention. This model predicts retention by sieving alone, as a function of particle diameter, filter pore size, filter thickness and filter loading.

Gyure, D. C., Set realistic goals for cross-flow filtration, *Chemical Engineering Progress 88*(11):60–66 (1992).

Cross-flow filtration offers key advantages in development of safer, milder, and more selective methods of producing many materials. It offers high specificity, unique function and resistance to fouling in the presence of biological mixtures. Challenges include selecting the best filter system for each application, developing novel processes for biological mixtures, designing large-scale systems, and operating systems cost effectivity in a production environment.

Jaisinghani, R., and B. M. Verdegan, Electrokinetics in hydraulic oil filtration—The use of antistatic additive, Proceedings World Filtration Congress III, 1983, pp.618–624.

Experimental evidence showing the importance of particle-filter media double layer interactions in hydraulic oil filtration is presented. These interactions are shown to be important even at very low ionic concentrations in oil containing 0 and 50 ppm of ionic antistatic. The filtration ratio correlates with the media zeta potential for constant particle charge. The effects of antistatic agent on particle contaminant and filter media charge are discussed.

Johnston, P. R., Liquid filtration, *Chemical Engineering Progress 84*(11):18–26 (1988).

Definitive testing of a filter medium involves simply trying it in specific process streams and seeing what happens at different velocities. The uselessness of the "micron rating" is pointed out, as is the need for data on filter permeability and effects of materials and pore uniformity.

Kaye, B. H., Describing filtration dynamics from the perspective of fractal geometry, *KONA 9*():218–236 (1991).

This paper reviews the basic concepts of fractal geometry used in the study of the filtration process, describes recent application of fractal geometry to the study of filtration systems, and indicates where fractal geometry may be of interest to the practicing technologist.

Liu, Y. et al., Removal of fine particles from nonaqueous media by cross-flow electrofiltration—Hydrodynamics and mass transfer, *Particulate Science and Technology 1*(1):27–53 (1983).

An improved method of removal of fine particles suspended in nonaqueous media by the application of a high-voltage field was previously described. However, to better understand the cross-flow electrofilter for future design improvements, a mathematical analysis of the filter was performed and more experiments were conducted with filter tubes of various porosities. Outlet concentrations were measured with gamma ray spectrometer. A description of the flow field in the devices was obtained by solving the Navier-Stokes equations for annular flow with suction through an outer porous wall. A perturbation solution for the flow field was then used in a model which describes the concentration distribution in the electrofilter. Cake thickness and outlet slurry concentrations were also computed.

Raistrick, J. H., The relevance of zeta potential to the filtration of small particles from potable liquids, *Filtration and Separation 20*(2):124–126 (1983).

Asbestos/cellulose and "Zeta Plus" filter sheets were challenged with aqueous suspensions containing small negatively charged particles. When the particles were smaller than the filter pore sizes, removal efficiencies were found to be high when the particles were suspended in water and low when suspended in aqueous liquids such as beer. It is concluded that in potable liquid filtration, the zeta potential of the filter medium is irrelevant because the organic molecules found in potable liquids adsorp upon positively charged surfaces and prevent subsequent particle adhesion.

Rege, S. D., and H. S. Fogler, A network for deep bed filtration of solid particles and emulsion drops, *AIChe Journal 34*(11):1761–1772 (1988).

A model was developed to simulate flow of emulsions and solids through porous media. Particle deposition due to direct interception and pore plugging by straining are accounted for. The effects of the particle to pore size ratio and of fluid velocity on particle deposition are shown. The value of the model is its ability to predict effluent concentration profiles, permeability changes during deep-bed filtration and changes in the filter coefficient.

Tiller, F. M., and C. S. Yeh, The role of porosity in filtration. 10. Deposition of compressible cake on external radial surfaces, *AIChE Journal 31*(8):1241–1248 (1985).

The theory of deposition of compressible particulate beds on cylindrical surfaces is developed for both constant-pressure and constant-rate filtration. Radial flow is seen in tubular filters and in both filtering and sedimenting centrifuges. It differs mathematically from filtration on planar surfaces by the manner in which the frictional drag balance on the particle structure is coupled with Darcy's equation.

Wrasidlo, W., and K. J. Mysels, The structure and some properties of graded highly asymmetrical porous membranes, *Journal of Parenteral Science and Technology 38*(1):24–31 (1984).

Some properties of a new type of membrane, a graded highly asymmetrical skinned microporous membrane, are described and some of their rationale reviewed. The asymmetry of the new membranes as measured by the ratio of pore sizes of the two sides exceeds 100 and the pore size varies exponentially across the thickness. As a result, retention is mainly due to mechanical sieving and relatively independent of the surface properties of the particles to be retained, the nature of the carrier liquid and the rate of filtration.

Zimmels, Y., Probabilistic design approach for filtration systems, *Powder Technology 67*(2):175–185 (1991).

Probabilistic aspects of filtration systems that capture particles along a filtration path are considered. The use of the geometric, exponential and Poisson distribution functions in probabilistic representation of filters is reviewed, and an extended multiple-capture, scope of the filtration process is introduced. The effects of random variables that characterize the filter as well as those pertaining to the particle population are presented. An approach in which filter performance at a prescribed level of confidence can be achieved is given, followed by design equations of the filter length and the filter parameter.

F. Light and Electron Microscopy

Burdett, G. J., and A. P. Rood, Membrane-filter, direct-transfer technique for the analysis of asbestos fibers or other inorganic particles by transmission electron microscopy, *Environmental Science and Technology 17*(11):643–648 (1983).

A new method to transfer inorganic particles collected on a membrane filter onto an evaporated carbon film is described. This allows the advantages of membrane filter collection to be combined with analytical capabilities of the TEM. This is part of a suite of new techniques that allow optical SEM and TEM work to be carried out on the same sample. Details of the technique are given along with some comparative data.

Carpenter, S., K. Hall, and K. Saul, Playing the numbers game: achieving accurate bacteria counts using epifluorescence, *Microcontamination 8*(1):27–30, 58 (1990).

Special problems exist in ensuring that sampling of DI water for epifluorescence measurements results in accurate counts when bacterial levels are under 200/liter. This article describes the steps taken to meet the problem. Statistical procedures and calculations are shown. The importance of the number of fields counted is pointed out.

Cheah, P. K. P., A microscope graticule for sizing near-homogeneous particles, *Journal of Aerosol Science 14*(1):47–48 (1983).

For counting and sizing near-homogeneous particles, the standard May or Porton graticules have progressions of size descriptors which are too large. These increase by a factor of $\sqrt{2}$. The graticule described here is useful for near-homogeneous particles with standard deviation of 1.5 or less. The factor used is approximately 1.08.

Coplen, R., R. Weaver, and R. Welker, Correlation of ASTM F312 microscope counting with liquidborne optical particle counting, Proceedings of 34th Annual Technical Meeting, Institute of Environmental Science, King of Prussia, PA, 1988, pp. 390–394.

Particle counting errors caused by air bubbles in liquidborne OPCs were examined. Duplicate analyses were made by measurement of membrane filters with light microscopy using ASTM F312. Means of establishing correlation and counting methods are described. Caution must be used when comparing results since operator experience with F312 can affect the absolute concentrations reported. PSDs of particles from solid surfaces compare well with PSDs in MIL-STD-1246.

Ender, U. J., Counting small particles by optical microscope, Proceedings 10th Symposium of International Committee of Contamination Control Societies, Zurich, 1990, pp. 437–439.

This paper describes a simple microscopic method that is very rapid and can be used to count and size particle of 0.2 μm. Using Nuclepore-type filters, a few droplets of water are placed below the filter to displace air within pores. The contrast in refractive indexes between the filter and air and between the filter and water is reduced, resulting in a more transparent filter, which reduces the scattering by the filter medium. Comparisons are made to other measurement methods.

Hartman, A. W., Investigations of array sizing. Accuracy of the sizing process, *Powder Technology* *39*(1):49–59 (1984).

The techniques of two-dimensional array sizing by optical microscopy is studied for use in determination of the average diameter of 0.9-μm PSL spheres. Error sources for the diameter measurement are discussed, including the effects of microcracks, anisotropy and dissolved impurities. Data from a particle calibration program at NBS are discussed. Comparisons with other methods are given and found excellent.

Hartman, A. W., Investigations in array sizing. The center distance finding technique, *Powder Technology* *46*(2-3):109–120 (1986).

A new technique, called center distance finding is described to find the mean size and PSD of monosize microspheres, by optical microscopy. Two-dimensional contacting microspheres structures on a microscope slide are illuminated with parallel light. The spheres act as positive lenses, refracting the light into small focal spots containing the diameter information of interest. In support of these distance measurements, a precision technique for finding the microscope image magnification and distortion is given. It is shown that the emulsion of the Polaroid film that was used has the needed dimensional stability.

Hostomsky, J. et al., Size analysis of non-spherical particles. A correlation between the results of photosedimentation and microscopy, *Powder Technology 49*(1):45–51 (1986).

Theoretical considerations and published experimental data are summarized for the sedimentation behavior of nonspherical particles. PSDs of the projected surface area of the particles were determined by the sedimentation method with photoelectric measurements of concentration and by evaluating the photomicrographs of the samples. Mathematical transformations of the surface area distribution with respect to different characteristic particle dimensions were derived to check result consistency.

King, R. P., Measurement of particle size distribution by image analyzer, *Powder Technology 39*(2):279–289 (1984).

Application of image analysis of particle size distribution determination is best for very small grains (1–25 μm) or when the grains are imbedded in a solid matrix. Techniques of measurement are described and the analytical methods for the transformation of random chord measurements made on section to three-dimensional size are provided in convenient form. An example is given using data observed with an image analyzer. It is shown how the technique can be used for particles of irregular shape.

McCrone, W. C., Dispersion staining, *American Laboratory 15*(12):52–55 (1983).

Dispersion staining is based on characteristic interference colors caused by differences in refractive index between the sample and the surrounding fluid medium. Problem-solving applications have been made in many areas. Some typical effects are described for many materials.

McCrone, W. C., Microscopy and ultramicroanalysis, *American Laboratory 18*(4):22–30 (1986).

The application of microscopy in carrying out microanalytical procedures is discussed. Limitations of automated analytical methods are discussed. Examples of data only obtainable through the use of morphological and skilled observations are given.

McCrone, W. C., Future light microscopy, *American Laboratory 20*(4):21–28 (1988).

Applications for light microscopy are summarized. Means of improving contrast, of better measurement of particle properties and of general miscellaneous improvements in microscopy are given.

McCrone, W. C., Why use the polarized light microscope? *American Laboratory 24*(6):17–21 (1992).

The application of polarized light microscopy to morphological identification of microscopic particulate matter is defined for a number of applications.

Mittelman, M. W., Rapid enumeration of bacteria in purified waters by direct counts, Proceedings of 9th International Symposium on Contamination Control, Los Angeles, CA, 1988, pp. 531–536,

Direct epiflourescence examination of purified water produces total microorganism counts in < 30 min. The technique is described for polycarbonate filters. Recoveries of purified water bacteria are frequently greater than currently used culturing methods. A positive correlation between direct counts and actual endotoxin levels is shown.

Mittelman, M. W. et al., Rapid enumeration of bacteria in purified water systems, *Medical Device and Diagnostic Industry* 7(6):144–149 (1985).

Use of a direct count epifluorescence microscopic measuring method for enumeration of bacteria in water is described. Materials and methods are described in detail and it is shown that the currently used viable-count method seriously underestimates the total number of bacteria. The use of epifluorescence allows determination of pyrogens as well as of viable bacteria. Real-time data are also obtained.

Olan-Figueroa, E. et al., Flattening coefficients for DOP and oleic acid droplets deposited on treated glass slides, *American Industrial Hygiene Association Journal, 43*(6):395–399 (1982).

The flattening coefficient, *F,* which is the ratio of apparent ratio droplet diameter as measured on a microscope slide to the true diameter of a spherical drop, was determined for particles of oleic acid and DOP aerosols collected on treated slides. The slides were made oleophobic by immersion in a fluorocarbon surfactant (NYE BAR Type CT or 3M FC-721). The mean value of F for oleic acid was 1.34 and for DOP, 1.35. *F* is constant in from 2.7 to 29.1 μm. Techniques for measurement are discussed.

Rooker, S. J. et al., On the visibility of fibers by phase contrast microscopy, *American Industrial Hygiene Association Journal 43*(7):505–515 (1982).

The factors affecting the visibility of fibers under the phase contrast microscope are discussed with reference to the visual acuity of the eye, the optical performance of the microscope and the optical properties of the prepared samples. It is possible to measure the limits of detection of small fibers collected on membrane filters. Full experimental methods are given and the results expressed in terms of phase shifts and minimum detectable brightness differences. Fibers down to about 0.15 μm diameter are visible with good optics provided the phase shift exceeds about 4°.

Russ, J. C., and W. D. Stewart, Quantitative image measurement using a microcomputer system, *American Laboratory 15*(12):70–75 (1983).

Evaluation of microscopic and photographic images has long relied on the acquired skill of human judgment. Introduction of the microcomputer into the laboratory enables image analysis to be performed easily and provides an economic alternative to manual systems. The system described here uses a combination of video and graphics table input and is able to solve most applications problems when neither measurement by itself is adequate.

Warnecke, H. J., and B. Klumpp, Particle measurements on technical surfaces, *Swiss Contamination Control 3*(4b):34–36 (1990).

Surfaces measured are those such as process chamber walls, transport container interiors, work table surfaces, etc. A measuring system consists of an image processing system on a microscope combined with statistical analysis and an *x-y* stage. On smooth surfaces detection sensitivity is 0.3 μm; on rough surface, it is 2–3 μm.

Yang, M., and D. L. Tolliver, Ultrapure water particle monitoring for advanced semiconductor manufacturing, *Journal of Environmental Science 32*(4):35–42 (1989).

Intensive use of the Nomura microscience method disclosed significant differences between two water distribution systems. A data collection and management system is necessary as on-line monitoring instruments are installed. Performance of modern filters can only be shown accurately by using particle counting methods with better than 0.1 μm sensitivity.

G. Liquid-Borne Particle Measurement

Anon, BSI, Determination of particle size distribution, BS3486, Part 7, British Standards Institution, London, 1988.

This document gives recommendations on methods of determining PSD and concentration by single particle counting instruments using light interaction effects. It also provides information to assist in making comparisons between different laboratories and/or by different instruments.

Atkinson, C. M. L., and R. Wilson, Artifact peaks in particle size distribution measured by the electrical sensing zone (Coulter counter) method, *Powder Technology 34*(2):275–284 (1983).

Artifact peaks on the coarse side of PSDs measured on the electrical sensing zone device for a series of copolymer latices were examined. These artifacts result from particles entering the orifice mouth at the periphery. It is shown that the "volume" shift of the artifact peak relative to the main or normal peak may be calculated from an equation relating the axial potential of the voltage gradient across the orifice mouth and the orifice size. These artifacts can cause difficulty in the use of the "half-count" calibration method.

Barman, B. N., M. N. Myers, and J. C. Giddings, Rapid particle size analysis of ground minerals by flow hyperlayer field-flow fractionation, *Powder Technology 59*(1):53–63 (1989).

Flow/hyperlayer field flow fractionation is capable of rapid separation of particles over 1 μm in size and was applied to characterization of ground coal and limestones. Results of analysis were found to be consistent and reproducible, agreeing with those obtained from microscopic examination and from elution profiles of latex particle standards. Validity of the size-dependent, density-dependent separation by the method was demonstrated by comparison of elution profiles for a range of materials of different properties.

Barth, H. C., and S.-T. Sun, Particle size analysis, *Analytical Chemistry 57*(5):151R–175R (1985).

Papers published from late 1981 to the end of 1984 were searched for relevant reports on particle size analysis. In addition to discussing the methods, some details on data handling are included. A few papers on comminution and on particle generation are also discussed.

Borchert, S. J. et al., Particulate matter in parenteral products: a review, *Journal of Parenteral Science and Technology 40*(5):212–241 (1986).

This paper discusses several aspects of the topic of particulate matter in parenteral products. It includes the use and limitations of the inspection and counting methods for visible and subvisible particles, the identification of particles and the elucidation of sources, mechanisms of formation and particle reduction techniques.

Daniels, C. A., and A. A. Etter, Latex particle size analysis by the Coulter nanosizer: response to synthetic mixture of sizes, outsize particles and comparison with other methods, *Powder Technology 34*(1):113–119 (1983).

Use of the Coulter nanosizer is described. Results of the study show that the device gives excellent size data correlations with other accepted methods on latices of narrow or broad distributions. Further, the method displays a sensitivity to outsize particle content, although this sensitivity varies with chemical type of latex. Response to various types of latex, to latex containing outsize particles and to two or more families of particles is documented.

Fitch, E. C., Control of hydraulic fluid contamination. II: Analysis of contaminant in fluid systems, *Hydraulics and Pneumatics 36*(5):68–72 (1983).

Analytical and sampling methods for contaminants in hydraulic fluids are discussed. The use of present data is described. A variety of analytical instruments is described for specific contaminant parameters.

Jager, J. et al., On-line particle size measurement in dense slurries, *Powder Technology 62*(2):155–162 (1990).

An automatic unit for fast and representative dilution of a process stream, down to the solids concentration for Fraunhofer diffraction size measurements was developed. The unit was built and tested. The combined system for dilution and sizing is integrated into a supervisory control system. Operating experience is summarized.

Knapp, J. Z., P. P. Deluca, and B. Conti, Review of commercially available particulate measurement systems, *Journal of Parenteral Science and Technology 42*(1):Suppl. (1987).

An objective review of the parenteral particle measurement literature and the available instruments in use is presented. The goal of the review was to resolve the present confusion of measurements in effect in this industry. It is concluded that the present lack of agreement is the result of several factors, most traced to instrument limitations and a combination of shortcomings in both sample handling technique and method. A bibliography is presented with an overview of particles in parenteral solutions emphasizing instrument evaluations and comparisons.

Lieberman, A., Fine particle characterization methods in liquid suspensions, *Particle Characterization in Technology* (J. K. Beddow, ed.), CRC Press, (1984), pp. 187–232.

This discussions covers some of the reasons why fine particles are characterized in terms of size, composition or concentration. Problems in handling particles before and after characterization are described, including some of the procedures in common use for sampling and the physical phenomena which affect materials in the sample. Methods and instruments in use are described in terms of their operating principles and procedures. Ancillary elements for operation are described and some application areas discussed.

Martyak, J. E., J. C. Carmody, and T. P. Clancy, Particle metrology; identification of particles in deionized water, Proceedings Microcontamination Conference, San Jose, CA, 1991, pp. 52–80.

Particle contaminant measurement was carried out in a deionized water system. Enumeration and identification were both carried out. On-line OPC measurements to 0.05 μm. SEM/EDS, TEM/EDS and Augur electron spectroscopy data are reported. In all cases, the contaminants were analyzed for elemental particle composition. Microorganism identification was also carried out. As a result of the work, the DI water distribution and polishing systems were upgraded.

Nicoli, D. F., J. S. Wu, and Y. J. Chang, On-line submicron particle sizing: carrying the analytical laboratory to the process facility, *American Laboratory 22*(15):70–79 (1990).

The dynamic light scattering technique is reviewed and shown to be well suited to on line applications due to its intrinsic speed and simplicity. A system for automatic sample acquisition and dilution is described for use with a modified DLS instrument in a pilot plant area.

Novak, J. W., and J. R. Thompson, Extending the use of particle sizing instrumentation to calculate particle shape factors, *Powder Technology 45*(2):159–167 (1986).

Data sets were cross-correlated from several particle sizing methods to determine particle shape factors for powders. Calculating shape factors from sizing data has a significant advantage over the more conventional microscopic methods because larger, more representative samples are used. Most powders showed significant difference in shape between the coarse and the fine ends of the size distribution.

Pohl, M. C. Analysis of reference samples by photosedimentation, Proceedings 14th Powder/Bulk Solids Handling Conference, Rosemont, IL, 1989, pp. 420–429.

Operation of photosedimentation devices is reviewed with emphasis on the Horiba system. Data from reference materials are reported.

Schindler, U., Particle measurement methods, *Chem. Ing. Techn. 54*(12):156–1159, (1982).

A survey is made of new particle measurement methods shown at the 1982 ACHEMA show. Both gas-borne and liquid-borne systems are discussed.

Somasundaran, P., Y. B. Huang, and C. C. Gryte, CAT scan characterization of sedimentation and flocs, *Powder Technology 53*(1):73–77 (1987).

The application of a CAT scanner to characterization of sedimentation and flocculating systems is described. Pixel size of 0.75 mm × 0.75 mm is used. Even though this system is quite coarse compared to potential microdevices, applications for real time data production are pointed out.

Sugure, S., T. Oja, and S. Bott, Advances in simultaneous multiangle Doppler electrophoretic light scattering analysis, *American Laboratory 21*(2):98–109 (1989).

Electrophoretic light scattering makes direct velocity measurements of particles moving in an electric field. It is analogous to determining speed of a moving object by radar. Instruments are available for particle sizing together with electrophoretic mobility determination. Operating principles, system design and data are presented.

Yang, F.-S. et al., Colloid characterization by sedimentation field-flow fractionation. II. Particle-size distribution, *Journal of Colloid and Interface Science 92*(1):81–91 (1983).

It is shown that the accurate theoretical relationship between size and retention volume in sedimentation FFF provides a raw size or mass distribution spectrum directly from each experimental run. Advantages of this method in obtaining PSD curves include the theoretical traceability which obviates the need for calibration standards, the controlled centrifugal field strength which provides flexibility and redundancy, the high resolution and the capability of collecting narrow fractions for addition analyses. Experimental data are shown to verify the applicability of the method.

H. Membrane Filters

Anon ASTM, D 3863-80; Retention characteristics of 0.40 to 0.45 μm membrane filters used in routine filtration procedures for the evaluation of microbiological water quality, American Society for Testing and Materials, Philadelphia, PA, 1980.

This method covers a procedure to test membrane filters for their ability to retain bacteria whose diameter is equal to or slightly larger than membrane filter pore size of 0.40 to 0.45 μm. The method is based on cultivation of organisms of diameters equal to or slightly larger than those of the pores of the filter to be tested and filtering aliquot samples containing organisms through the membrane. The filtrate is examined for sterility after incubation.

Chan, K. et al., Interfacial forces, average pore size and size distribution of ultrafiltration membranes, *Industrial & Engineering Chemistry 21* Prod R&D(4):605–612 (1982).

A variety of ultrafiltration membranes working with very dilute aqueous solutions were used here. The strengths of the solute-solvent-membrane interfacial forces were quantitatively determined from retention volume data from liquid chromatography. From those data from analysis of solute separation in ultrafiltration, the average pore size and pore size distribution in the membrane were determined.

Dzubay, T. G., and R. K. Barbour, A method to improve the adhesion of aerosol particles on Teflon filters, *Air Pollution Control Association Journal 33*(7):692–695 (1983).

This experiment was designed to test the stability of coarse particle deposits on Teflon filters and to test the effectiveness of application of a thin layer of mineral oil to the deposit sides of the filters before use. Sample stability was greatly enhanced.

Emory, S. F., Principles of integrity-testing hydrophilic microporous membrane filters. Part I. *Pharmaceutical Technology 13*(9):68–77 (1989).

This article addresses the physics (capillary phenomena and molecular diffusion) that are at work in the two most common filter integrity tests: the bubble point and diffusion tests. Some general practical guidance is provided for conducting the tests and for interpreting the results.

Emory, S. F., Principles of integrity-testing hydrophilic microporous membrane filters. Part II. *Pharmaceutical Technology 13*(10):36–46 (1989).

This article, prepared for nonengineers involved with integrity testing, addresses in simple terms, the physics operating in diffusion testing of filters. Practical guidance is offered for conducting the tests and interpreting results.

Hango, R. A., Optimization of membranes in high purity water systems for semiconductor manufacturing, *Journal of Environmental Science 31*(4):49–53 (1988).

Large volumes of ultrapure water are needed for semiconductor manufacturing. Numerous stages of treatment are needed to produce this water.Cost savings have been realized by improving the membrane units at several stages. This presentation shows operating experiences and water quality for the optimum system and describes the test methods to verify performance.

Johnston, P. R., and T. H. Meltzer, Microbial control: the rating of membrane filter pore sizes, *Ultrapure Water 4*(8):14–17 (1987).

Since pores of solution-type membrane filters are convoluted structures of complex and uncertain architecture, pore-size ratings have been inferred from flow rates of fluid flow or porosimetry methods. Use of PSL spheres to determine penetration is pointed out to aid in reducing uncertainty.

Krygier, V., Rating of fine membrane filters used in the semiconductor industry, Proceedings of the 5th Pure Water Conference, San Francisco, CA, 1986.

Methods for rating performance of membrane filters for use in the semiconductor industry are reviewed. The methods include microbial challenge, liquid displacement (as a more precisely bubble point test), zeta potential rating, latex sphere penetration and colloidal silica removal capability.

Lee, J. K. et al., Particulate retention by microporous membrane filters in liquid filtration: a predictive model and experimental study, Proceedings of the 38th Annual Technical Meeting, Institute of Environmental Sciences, Nashville, TN, 1992, pp. 297–308.

Particulate retention by microporous membranes during liquid filtration was studied theoretically and experimentally. A model to predict particle retention using a multilayer concept for filter retention by sieving and using flow resistance for cake filtration was developed to predict particle retention as a function of particle size and loading. The filtration sequence during particle loading was a sieving dominant stage within the membrane followed by transition to a cake filtration stage on the filter surface. Filter efficiencies were determined vs. particle size.

Meltzer, T. H., Membranes: The utilitarian significance of membrane pore-size distributions, *Ultrapure Water 4*(2):16–24 (1987).

This paper explores the implications of filter pore size distributions and considers advantages. The relationship between particle PSD and pore size PSD is pointed out.

Meltzer, T. H., Microfiltration: the dependability of microporous membranes and the utility of repetitive filters, *Ultrapure Water 4*(4):21–23 (1987).

Operation of microporous membrane filters is discussed; the concept of filter absoluteness is defined as the freedom of filter efficiency from such operational influences as applied differential pressure, particle challenge density and the makeup of the suspending vehicle. It derives from the relationship of the PSD to the pore size distribution. The value of repetitive filtration is pointed out for applications where integrity testing of the ensemble will always prove positive, even if any single member fails.

Meltzer, T. H., The pore structures of microporous filters, *Ultrapure Water 5*(5):49–55 (1988).

Particle retention by filters depends on relative sizes and shapes of both filter pores and particles. Sizes and shapes of various commercially available filter types differ, depending on filter manufacturing methods. Pore structures of various filters are reviewed, including track-etch, porous PTFE, matted fiber, and gels. Charge-modified filters and anisomorphic types are also discussed.

Nagasaki, I., Charge-enhanced membranes for water filtration, *Solid State Technology 31*(8):97–100 (1988).

A recent development in ultrapure water filtration involves bonding of molecular groups to the membrane surfaces. These enhanced surfaces have sufficient charge to efficiently adsorb submicron particles smaller than the pore size without excessive flow obstruction. Efficiency, stability and service life are discussed.

Roberts, K. L., D. J. Velasquez, and D. M. Stofer, Challenging the membrane filter ratings, *Semiconductor International* (Nov. 1990).

Membrane filter ratings for pore size have long been established by observing penetration of microorganisms. It is shown that penetration of hard latex spheres larger than the bacterial pore size is significant. It is suggested that membranes should be rated by latex particle testing.

Ropion-Granger, J. et al., Filtration of suspensions of monosize latex particles, Proceedings of 7th International Committee of Contamination Control Societies Conference, Paris, 1984.

The ability of dilute suspensions to clog filters can be estimated by filterability tests on membranes. An experimental study was made using calibrated latex suspensions from 0.1 and 0.2 μm and membranes of pore size 0.45 to 8.2 μm. Variation of the flow rate in the initial period filtration can be described by any of four models, each leading to a coefficient of filterability. The time constant which characterizes the clogging ability of the suspension is shown to be a function of concentration. It is also shown that the membranes allow a significant fraction of particles larger than the nominal pore size to penetrate.

Shoemaker, W., Membrane filtration systems for industrial uses, *Filtration News 5*:4–8 (1987).

A brief summary is given on the use of membrane filters for industrial use. Microfiltration, ultrafiltration and reverse osmosis uses are pointed out. The flow paths used in cartridges, especially cross-flow methods, are described.

Simonetti, J. A. and H. G. Schroeder, Evaluation of bacterial grow through, Proc. 30th Annual Technical Meeting, Institute of Environmental Sciences, Orlando, FL, 1984.

A variety of different commercial submicron rated membrane filter cartridges were inoculated with a known number of viable cells and placed in service at a simulated DI water point of use. Pseudomonas diminuta was selected as the model organism. Both continuous and intermittent use were simulated. Since the downstream bacterial counts varied considerably for the cartridges tested, lab experiments were carried out to obtain insight into the possible mechanisms which may be associated with proliferation of bacteria through various types of membrane materials.

West, R., and J. Ruddock, Performance of a new dual asymmetric PTFE membrane, Proceedings of 37th Annual Technical Meeting, Institute of Environmental Sciences, San Diego, CA, 1991, pp. 841–848.

This report describes features of a new dual asymmetric PTFE membrane filter and a test method used to assess its performance in a bulk chemical delivery environment. Using a model system, test filters were challenged with a steady-state particle concentration. The filter was able to reduce the particle concentration by a factor of 6.

Williams, R. E., and T. H. Meltzer, Membrane structure, the bubble point and particle retention: a new theory, *Pharmaceutical Technology* (May, 1983).

It is shown that the bubble-point measurement for membranes may define pore structures that, despite their presence, do not generally govern a filter retentivity, even when such retentivity is restricted to the sieve retention mechanism. In effect, the particle capture ability of membranes is seen to exceed the predictions based upon the "largest pore" size as defined by the bubble-point measurement.

Wrasidlo, W, and K. J. Mysels, The structure and some properties of graded highly asymmetrical porous membranes, *Journal of Parenteral Science and Technology 38*(1):24–31 (1984).

Some properties of a new type of membrane, a graded highly asymmetrical skinned microporous membrane, are described and some of their rationales reviewed. The asymmetry of the membranes as measured by the ratio of pore sizes of the two sides exceeds 100 and the pore size varies exponentially across the thickness. As a result, retention is mainly due to mechanical sieving and relatively independent of the surface properties of the particles to be retained, the nature of the carrier liquid and the rate of filtration.

I. Optical Particle Counters

Barber, T. A., Limitations of light-obscuration particle counting as a compendial test for parenteral solutions, *Pharmaceutical Technology 31*(10):34–52 (1988).

Light obscuration particle counting is suitable for use under carefully controlled conditions as a survey or screening assay for parenteral solutions. In some cases, however, significant count errors can result when this assay is used to count and size pharmaceutical contaminant particles. Defects in this method that result in count errors may prevent the method from being satisfactory as a compendial assay method. This article discusses principal instrument-related and application-related limitations.

Batchelder, J. S., and M. A. Taubenblatt, Real-time single particle composition detection in liquids, *Solid State Technology 35*(10):S1–S9 (1992).

An optical technique is described that measures the amplitude and phase of forward scattered light from a particle. Using this information in real time, the instrument used can assign the particle to one of several composition classes and accurately compute its size. The instrument is based on Nomarski interferometer operation.

Hargesheimer, E. E., C. M. Lewis, and C. M. Yentsch, Selecting particle counters for water quality sampling, *Waterworld News July/Aug.* pp. 16–17 (1991).

The usefulness of OPCs in water treatment evaluation is discussed. They provide direct measurement of particles as well as improved operational control. Practical use requires clear formulation of objectives and goals, establishment of anticipated particle sizes and concentrations of concern, examination and evaluation of instruments that can meet the objectives of the study.

Hiraoka, M. et al., A new instrument for measuring particles in high purity water, *Ultrapure Water 7*(4):37–45 (1990).

This paper discusses counting particles in high-purity water, using an air-cooled argon ion laser. A scanning-type laser-beam scattering system is used to minimize effects of beam intensity range across the laser beam. The minimum size measurement is 0.7 µm. Design and performance data are given for the commercial instrument.

Knollenberg, R. G., The importance of media refractive index in evaluating liquid and surface microcontamination measurements, *Journal of the IES 30*(2):50–58 (1987).

Results of theoretical and laboratory studies of the effects of media refractive index on the amount of light scattered by particles. The media include liquids and surfaces. Exact solutions are given for a number of contaminant/liquid binaries. Appropriate solutions are given for scattering by surface particles. Results of the study suggest methods of limiting errors associated with both substrate conditions.

Lieberman, A., Calibration methods for liquid-borne particle counters, *Pharmaceutical Manufacturing 2*(11):39–44 (1985).

Procedures for calibrating optical particle counters with monosize particles are reviewed and evaluated. Problems and advantages of the several methods are discussed. It is concluded that use of a pulse-height analyzer is best for facilities which can obtain one and that a manual or microprocessor controlled moving-window differential method is best for all others.

Lieberman, A., Microprocessor controlled particle counter system, *Journal of Parenteral Science and Technology 44*(3):133–138 (1990).

Operation of an optical particle counter with microprocessor control of particle size calibration and data recording is described. Sensors are available to measure particles 0.1 μm and larger. A system with a range from 2 to 125 μm is described. Operating procedures and data capabilities and data processing abilities are summarized.

Lieberman, A., Calibration and correlation problems for liquid-measurement OPCs. Proceedings 11th Symposium of International Confederation of Contamination Control Societies, London, 1992, pp. 517–521.

Calibration and correlation procedures for operation of liquid-borne OPCs are summarized. The bases for calibration are traceability for the standard method, verification of OPC resolution, sizing and counting accuracy. Standard methods are discussed. Problems and approaches to inter-OPC correlation are also discussed. Standard methods and materials are described.

Montgomery, C. N., In-situ particle sizing in liquids, Proceedings of 33rd Annual Technical Meeting, Institute of Environmental Science, San Jose, CA, 1987, pp. 353–365.

Techniques for monitoring particles in liquid media tend to fall into two general categories: sizing and sampling. This paper discusses two techniques, available in each category. An in situ system is described which allows measurement of particles > 0.3 μm in liquids without sample extraction from a variety of liquids. Effects of refractive index on sizing sensitivity is discussed. Details of system design and operation are described.

Raasch, J. and H. Umhauer, Errors in the determination of particle size distributions caused by coincidence in optical particle counters, *Particle Characterization 1*(1):53–58 (1984).

When performing PSD analyses by scattered light, coincidence in the measuring volume of the OPCs not only cause too small a particle concentration to be recorded, but also a PSD to be obtained which is systematically shifted toward coarser particle sizes. It is pointed out that this shift can be precalculated when the true distribution is given. The results show under which conditions a correction of the measured distribution is indispensable.

Sommer, H. T., Physics and performance of light extinction particle counters, Proceedings 10th Symposium of International Committee of Contamination Control Societies, Zurich, 1990, pp. 23–27.

The physics of obscuration OPCs are illustrated and explained. The difference between white and monochromatic light sensors are shown; sensors are compared and correlations established. Advantages of laser diode light sources are discussed. Interpretation of size measurements is more complex and requires detailed knowledge of the pertinent physics. Mie theory for these systems is discussed.

Sommer, H. T., Reliability and count accuracy of optical particle counters, Proceedings of 36th Annual Technical Meeting, Institute of Environmental Science, New Orleans, LA, 1990.

This paper addresses the concerns in count correlating OPCs. The physics of OPCs is briefly described to show limitations due to the nature of the measuring process. Calibration and alignment procedures show that instrument configuration can affect data quality. Data are given to show reliability and consistency of present-day OPCs. Data show that OPCs can also be count correlated with high reproducibility.

Strauss, C. M., To spec or not to spec (particles in process chemicals), Proceedings of 31st Annual Technical Meeting, Institute of Environmental Sciences, Las Vegas, 1985.

Problems in setting up specifications based on size distributions in process chemicals are described. Rather than a classification system at several sizes, a maximum concentration at and greater than a specified particle size is suggested. Experimental data on instrument variability are presented to show the effects on establishing specifications. Five key factors in improving precision, accuracy, and correlation between laboratories in particle counting are suggested. These are calibration procedures, matrix effects for the chemicals, aliquot validation, standardized specifications and a lot acceptance basis.

Tricome, J. H. et al., A comparison of the HIAC particle size analyzer, an automated image analyzer and the PMA microscope method of particle size analysis, *Journal of Parenteral Science and Technology* *40*(2):58–60 (1986).

The USP has proposed particle limits for SVPs using a HIAC particle size analyzer. This paper compares results from the HIAC, an automated image analyzer and the PMA membrane microscope method. Ten lots of a dry powder antibiotic were used for analysis. No correlation is evident among the different methods. The HIAC results are usually lower than the PMA method and those of the image analyzer.

J. Particle Agglomeration and Coalescence

Brockmann, J. E. et al., On simultaneous coagulation and diffusional loss of free molecule aerosols in turbulent pipe flow, *Aerosol Science and Technology* *1*(2):163–178 (1982).

The population balance equations for simultaneous coagulation and deposition of free molecule aerosols in turbulent pipe flow were cast in dimensionless form and integrated to obtain a simplified two-equation model. The relative importance of coagulation and deposition is shown to depend on a dimensionless parameter which is a ratio of a characteristic deposition time to a characteristic coagulation time. Solutions to these equations are presented.

Chen, J. D., P. S. Hahn, and J. C. Slattery, Effects of electrostatic double-layer forces on coalescence, *AIChE Journal* *34*(1):140–143 (1988).

As buoyancy forces a drop or bubble to approach a fluid-fluid interface, a film of the continuous phase is formed. Effects of electrostatic forces upon the drainage and the stability of the film formed as a drop or bubble reaches the interface are analyzed.

Cohen, R. D., Equilibrium distributions of size and weight in linear flocculation, *Powder Technology* *59*(2):145–148 (1989).

The use of statistical mechanics and the Boltzmann distribution of energy states was considered as a tool for studying flocculation. The primary outcome is the number of flocs at equilibrium, useful for information on the expected size and weight distribution of aggregates. This was applied here to the case of linear flocculation. Assumptions made are given and results indicate excellent potential for study of fine particle flocculation.

Glasgow, L. A., and J.-P. Hsu, Floc characteristics in water and wastewater treatment, *Particulate Science and Technology* *3*(3):285–303 (1984).

Behavior of an aggregate of colloidal particles or floc in a water-suspended treatment process is dependent on its physical characteristics, including size, strength, density, and permeability. The success of a water treatment operation can be gauged by the properties of the aggregates which are produced. Sizes and density are important in sedimentation processes where the rate of solids subsidence constrains plant operation. An experimental study of how the floc characteristics are interrelated is described.

Gregory, J., Flocculation in laminar tube flow, *Chemical Engineering Science* *36*:1789–1794 (1981).

The extent of orthokinetic flocculation of suspensions as a result of laminar tube flow was calculated using Smoluchowski theory and shown to depend mainly on the tube dimensions and not on the flow rate. When a correction factor allowing for hydrodynamic resistance between approaching particles is introduced, the calculated flocculation rate can be much lower than the Smoluchowski value and depends inversely on flow rate and particle size. A further effect which must be included is the nonuniformity of shear rate in the tube, which leads to less flocculation than expected on the basis of uniform shear rate.

Hsu, J.-P., and L. A. Glasgow, Floc size reduction in the turbulent environment, *Particulate Science and Technology 1*:205–222 (1983).

The dynamic behavior of the PSD in wastewater treatment processes results from the interaction of a random growth process with nonlinear, stochastic hydrodynamics. Successful modeling of the PSD response to high dissipation conditions depends on accurate quantification of the breakage mode, the total dissipation rate, transition probability and the daughter particle size distribution. The inferential determination of these quantities from PSD data is difficult. Direct photographic observations of flow breakup in a turbulent jet apparatus and in the impeller stream of a baffled, agitated tank allows measurement of both the breakage rate and the daughter PSD for individual parent flocs.

Jiang, Q., and B. E. Logan, Fractal dimensions of aggregates determined from steady-state size distributions, *Environmental Science and Technology 25*(12):2031–2038 (1991).

Aggregates formed by Brownian motion, shear coagulation and differential sedimentation have fractal geometries. In order to model coagulation of fractal aggregates, a set of collision functions containing a fractal dimension for use in a general coagulation equation was derived. These functions predict greater collision frequencies than models based on aggregates with Euclidean properties. Fractal dimensions of inorganic aggregates are in the range expected for coagulation by Brownian motion and shear coagulation.

Neumann, A. W. et al., The concept of negative Hamaker coefficients. III: Determination of the surface tension of small particles, *Powder Technology 37*(2):229–234 (1984).

The main forces responsible for the interaction of small particles suspended in a liquid will normally be van der Waals forces of attraction and electrostatic double-layer forces of repulsion. This paper describes results of studies where we have used the concept of double zero or repulsive van der Waals forces to determine the surface tension and subsequently the Hamaker coefficient of small particles by a number of different techniques. This is based on the fact that there is a close relationship between van der Waals interaction and the surface tensions.

Overbeek, J. T. G., Interparticle forces in colloid science, *Powder Technology 37*(2):195–208 (1984).

The nature of the forces between the polymolecular particles is the same as that of the forces between atoms or ions, but their dependence on distance may be different. Colloid systems offer many opportunities for study. The paper discusses: van der Waals forces, electrical double-layer forces and steric forces, connected with the presence of layers of large molecules on the particles. Some comments are given regarding magnetic and structural forces.

Pailthorpe, B. A., and W. B. Russel, The retarded van der Waals interaction between spheres, *Journal of Colloid and Interface Science 89*(2):563–566 (1982).

We present and assess a relatively simple approximation suggested elsewhere for retarded interactions between spheres in the presence of electrolyte. A theoretical analysis is given of requirements for use.

Parker, M. R. et al., Particle aggregation in colloids in high magnetic fields, *J. Magnetism and Magnetic Materials 27*:250–256.

The influence of high magnetic fields on the stability of colloidal dispersions of para- and diamagnetic particles has been examined. The theory of Svoboda was extended to allow examination of the dynamics of particle aggregation in colloids. Particle trajectories are calculated leading to binary pair formation. Experimental results are presented on particle aggregation in high magnetic fields (flocculation). Here, binary pairs seem to form and to be stable at enhanced settling rates.

Sakka, S., Formation of particles in sol-gel process, *KONA 7*:106–118 (1989).

Formation of particles in starting solutions and its significance for the sol-gel processing oxide of materials are reviewed on the basis of previous works. First, hydrolysis and polycondensation reactions of metal alkoxides in alcohol-water solutions are briefly discussed. Then, it is shown that powders prepared by the sol-gel method using metal alkoxides are suitable as starting powders for high-performance glasses and ceramics, because the particles are fine, monodisperse and uniform in chemical composition.

Tobin, T., and D. Ramakrishna, Coalescence of charged droplets in agitated liquid-liquid dispersions, *AIChE Journal 38*(8):1199–1205 (1992).

Experimental studies were performed to assess the effect of drop charge on turbulent drop coalescence. It is well known that surface charge can affect coalescence rate of Brownian particles in a quiescent medium; no studies were found in the literature which explicitly examine charge effects on larger drops in turbulent media. It was found that drop charge can substantially inhibit coalescence between drops in agitated dispersions. The effect can arise solely from latent drop charges, attributed to preferential adsorption of hydroxide ions onto a water-organic phase interface.

Williams, H. A., S. J. Peng, and A. Naylor, In situ measurement of particle aggregation and breakage kinetics in concentrated suspension, *Powder Technology 73*(1):75–83 (1992).

A study is presented on formation of silica flocs and their subsequent breakage in a stirred tank, as a function of agitation rate, solids concentration, and polymer flocculant dosage. Evolution and disintegration of aggregates (1–500 μm) in concentrated aqueous slurries were measured in situ using a scanning laser microscope. The value of the method and the types of information procurable are discussed.

K. Particle Physical and Chemical Property Effects

Anderson, J. L. et al., Motion of a particle generated by chemical gradients. I:non-electrolytes, *Journal of Fluid Mechanics 53*(4):107–121 (1982).

When a particle is placed in a fluid in which there is a nonuniform concentration of solute, it will move to a higher or lower concentration depending on whether the solute is attracted or repelled from the particle surface. Understanding this phenomenon requires that the equations representing conservation of mass and momentum within the fluid near the particle are solved.

Batchelder, J. S., and M. A. Taubenblatt, Real-time single particle composition detection in liquids, *Solid State Technology 35*(10):S1–S9 (1992).

An optical technique is described that measures that amplitude and phase of forward scattered light from a particle. Using this information in real time, the instrument used can assign the particle to one of several composition classes and accurately compute its size. The instrument is based on Nomarski interferometer operation.

Beddow, J. K., and D. W. Luerkens, Effects of powder production and of material processing on powder particle morphology of adipic acid, *Powder Technology 68*(2):163–173 (1991).

It is well known that the particle formation mechanism can affect the product particle morphology. A study is reported which shows how morphological analysis can be used to obtain detailed information of the morphology of the adipic acid particles made by three different methods: atomization, comminution and crystallization. Data are reported on single particles and on powder samples.

Carlon, H. R., M. S. Guelta, and B. V. Gerber, Some candidate replacement materials for dioctyl phthalate in "hot smoke" aerosol penetrometer machines, *Aerosol Science and Technology 14*(2):233–246 (1991).

Research was carried out using ATI Q-127 aerosol penetrometer test machines to identify relatively innocuous and inexpensive replacement materials for DOP. A synthetic hydrocarbon (poly-alpha olefin) can be used to replace DOP directly with minimum impact on existing hardware and procedures. Isostearic acid and oleic acid also performed well as DOP replacements. All three materials also show promise for cold smoke testing applications.

Chang C.-R. et al., Particle morphology and bulk properties, *Particulate Science and Technology 1*(4):433–449 (1983).

Experimental results relating bulk properties to newly developed shape terms are presented here. The properties studied are tap density, apparent density, processing parameter, flow time and angle of internal friction of several materials. The correlations obtained show the utility and the potential of morphological analysis.

Geoffroy, J. M., and J. T. Carstensen, Effects of measurement methods on the properties of materials, *Powder Technology 68*(1):91–96 (1991).

Different methods have been suggested for the determination of the compressional properties of materials. The Heckel equation has been suggested as a method which may be useful for this purpose. However, the manner in which the measurements are made and the method of deformation of the material may affect the shape of the Heckel plot.

Kousaka, Y., Behavior of submicron particles suspended in a fluid, *KONA 3*:69–75 (1985).

Some topics on submicron particles suspended in a gaseous or liquid medium are introduced. First, some typical size-dependent properties of a particle are considered and then topics on coagulation and deposition which are important in industrial processes are introduced.

LeBlanc, S. E., and H. S. Fogler, Dissolution of powdered minerals: the effect of polydispersity, *AIChE Journal 35*(5):865–868 (1989).

The effect of polydispersity on dissolution processes is discussed. Polydisperse mineral systems with some particles dissolving simultaneously where the mass transfer and surface reaction rate limit dissolution are essentially equal. Particle size distribution effects on the specific rate limiting phenomena are discussed.

Leschonski, K., Particle characterization, present state and possible future trends, *Particle Characterization 3*(3):99–103 (1986).

The investigations of single and combined properties of particles still belongs to the most important areas of present research and practical applications of particle characterization. Methods to measure single-particle properties, such as size, shape, composition and velocity, are of much higher sophistication than before and they still aim at better accuracy and higher resolution. There is much lack of understanding and of theoretical interpretation and of experimental evidence of the differences seen in the correlation of different equivalent diameters, obtained when analyzing the same irregularly shaped particles using different physical principles. The need for further work in particle characterization is emphasized.

Marlow, W. H., Aerosols chemical physics, *Aerosols Microphysics 11*(1):1–6 (1982).

The physics of clusters and very small particles on heterogeneous or surface processes, homogeneous (nucleation) and aerosol thermodynamics is considered. A brief summary of the problem is made.

Matijevic, E., Monodispersed colloids: art and science, *Langmuir 2*(1):12–20 (1986).

Procedures for preparation of inorganic, organic and mixed colloids, consisting of particles uniform in size and shape are described. The use of such well-defined systems in the characterization of their optical, magnetic and electrokinetic properties is shown. Interactions of monodisperse systems with various solutes, with other particles and with plane surfaces are discussed. A refined treatment of the electric double layer of unlike particles is described and theory compared with experimental data.

Nikolakakis, I., and N. Pilpel, Effect of particle shape on the tensile strength of powders, *Powder Technology 42*(3):279–283 (1985).

Previous work has shown that the tensile strength of an assembly of fine particles depends on their nature, packing fraction, moisture content, particle size and shape, temperature and length of the consolidation period. Studies were carried out on the effect of particle shape. It was found that shape effects become increasingly important as the particle size is reduced.

Overbeek, J. T. G., Interparticle forces in colloid science, *Powder Technology 37*(2):195–208 (1984).

The nature of the forces between polymolecular particles is the same as that of the forces between atoms or ions, but their dependence on the distances may be different. Colloid systems offer many opportunities for study. The paper discusses van der Waals forces, electrical double-layer forces and steric forces, connected with the presence of layers of large molecules on the particles. Some comments are given regarding magnetic and structural forces.

Rader, D. J. et al., Evaporation rates of monodisperse organic aerosols in the 0.02–0.2 μm diameter range, *Aerosol Science and Technology* 6(3):247–260 (1987).

The tandem differential mobility analyzer (TDMA) technique was used to measure evaporation rates of monodisperse organic particle droplets in the 0.02–0.2 μm size range. Transition regime mass transfer expressions were used to determine vapor pressures from measured evaporation rates. These were found to be in good agreement with reported values. The data provide direct evidence for the validity of the Kelvin equation for the materials tested.

Schiffer, R., and K. O. Thielheim, Light reflection for the randomly oriented convex particles with rough surface, *Journal of Applied Physics 53*(4):2825–2830, (1982).

Light reflection from rough spheres large compared to the incident wavelength is considered; the resulting reflection pattern is identical to that from randomly oriented convex particles. Only the limiting case of short wavelength is treated; the laws of geometrical optics are assumed valid. It turns out that the decisive quantity is the probability distribution of the directions of the surface normal vectors. The effect of increasing roughness is discussed.

Stanley-Wood, N. G. et al., The variation in, and correlation of, the energetic potential and surface areas of powders with degree of uniaxial compaction stress, *Powder Technology 60*(1):15–26 (1990).

Particles of different sizes, shapes, surface and hardness were characterized in terms of surface area, surface energy and interparticle porosity before and after uniaxial compaction. Compacts of several materials were diametrally fractured to show the physico-mechanical behavior of the compacts and to assess the particle-particle bond.

Weng, H.-L., and E. L. Parrott, Effect of some processing variables on the kinetics of the solid-state addition of sulfacetamide and succinic anhydride, *Powder Technology 39*(1):1–5 (1984).

Thermal and chemical analyses were used to study the solid-state addition reactions of sulfacetamide and succinic anhydride. A pelletization or compaction method was used to determine effects of compressional pressure, molar ratios of the reactants, particle size and temperature on the kinetics of the reaction.

Wilkinson, M. C. et al., The stability of latex particles in aqueous suspensions, *Particulate Science and Technology 5*(1):65–82 (1987).

Monodisperse PSL of 2-, 4- and 9-μm diameter as supplied for intended use as particle standards with respect to both particle size and number density were studied by a variety of techniques and under a variety of storage conditions. The sizing and counting techniques used light and EM microscopy, and Coulter counting. Polymeric stabilizer and bactericide additives were used at different levels and their effect on three-year stability monitored at different number densities. Accelerated aging was used for tests. The latices were shown to be remarkably stable in the presence of the selected stabilizer/bactericide combination.

L. Particle Shape Considerations and Measurement

Ah Chin, A. D., P. B. Butler, and D. W. Luerkens, Influence of particle shape on size measured by the light-blockage technique, *Powder Technology 54*(2):99–105 (1988).

Five alumina particles, each of different shape, were independently and repeatedly dropped through the light beam in a HIAC/ROYCO light extinction particle counter. The particles were also measured by a Quantimet image analyzer which determines size from the area projected by a particle lying "flat." As expected, when the diameter of a flake-like particle is measured by the H-R counter, some variability is seen. For other particles, close correlations are shown between the two methods of measurement.

Arakawa, M., et al. Comparative data of particle size distribution of flake like particles by various methods, *KONA 3*:10–16 (1985).

Results of particle size measurements by several methods based on different principles are almost equal to each other as far as these methods apply for spherical powder. However, there are questions for irregular shape particles. Using mica powders as samples of the thin plate, we compared the results of measurements based on optical microscope, sieve, sedimentation balance, photosedimentometer, Coulter counter and microtrac. Comparisons were made against a spherical equivalent diameter.

Beddow, J. K., Morphological analysis, ten years of progress, *Particulate Science and Technology 1*(1):101–123 (1983).

The need for and the principles of morphological analysis of particulate matter are briefly discussed. Applications of analysis to a number of problems are cited. Topics include relating morphology to the mechanisms of powder production; wear debris analysis and kinetic modeling of crystallization processes. Relationships between the behavior of particles and their morphology are discussed in connection with abrasive wear of metals, sedimentation in low-flow regimes, measurement of bulk properties. The use of the particle image analyzer to analyze color as a morphic feature is presented. Future work is also suggested.

Chang, C.-R. et al., Particle morphology and bulk properties, *Particulate Science and Technology 1*(4):433–449 (1983).

Experimental results relating bulk properties to newly developed shape terms are presented here. The properties studied are tap density, apparent density, processing parameter, flow time and angle of internal friction of several materials. The correlations obtained show the utility and the potential of morphological analysis.

De Silva, R. L., and D. E. Creasy, Shape factors for barium chromate crystals in the sieve size range, *Powder Technology 35*(2):131–134 (1983).

The use of a simple microscopic counting technique, which was applied to sieve fractions of a standard fourth root of two sieve progressions to determine particle shape factors for polydisperse samples is described. Results obtained in the sieve size range up to 295 μm, for the ratio of mean equivalent spherical volume diameter to mean sieve diameter, are presented and discussed.

Jones, S., The closure problem in the Zahn-Roskies method of shape interpretation, *Powder Technology 34*(1):93–94 (1983).

A mathematical treatment of the Zahn-Roskies method for pattern recognition is described. Details of the process are given and its application for use with reentrant shaped particles is discussed.

Kaye, P. H. et al., Airborne particle shape and size classification from spatial light scattering profiles, *Journal of Aerosol Science 23*(6):597–611 (1992).

A new instrument for study of airborne particle is described. A laminar flow system is used to constrain aerosol particles to traverse a focused laser beam singly. Scattered light is collected as two-dimensional photon distributions, representing light scattered over 84% of the total sphere of scattering. The potential for classifying airborne particles on the basis of shape and size is discussed.

Leith, P., Drag on nonspherical objects, *Aerosol Science and Technology 6*(2):153–161 (1987).

Stokes' law describes drag forces on a sphere in creeping flow. This law can be extended to a nonspherical object by allocating interactions of the fluid with the object into its interaction with two analogous spheres, one with the same projected area and one with the same surface area as the object. Agreement between data and the equations for shape factor based on this approach was excellent for prisms. This equation was also found to predict shape factors better than empirical equations, as well.

Leong, K. H., Thermophoresis and diffusiophoresis of large aerosol particles of different shapes, *Journal of Aerosol Science 15*(4):511–517 (1984).

Thermophoresis and diffusiophoresis of nonspherical aerosol particles is examined. The particle shapes studied are cylinders and oblate and prolate spheroids. Phoretic velocities are derived for large particles of different shapes in the slip region of flow. Diffusiophoresis is apparently independent of particle shape in contrast to thermophoresis, which varies greatly with shape. For nonspherical particles, it depends on orientation to the thermal gradient.

Parrott, E. L., Micromeritics in dosage form design and development, *Pharm. Manufacturing* 2(4):31–37 (1985).

In many pharmaceutical tests, the discussion of fine particles is limited to definitions of diameters, size distributions and measurement methods in a chapter on powders or size reduction. A thorough consideration should list other topics, up to 12 or so, most of which are addressed here.

Saito, F. et al., Free settling of a non-spherical particle in a stagnant liquid, *Particulate Science and Technology* 2(3):247–258 (1984).

The free settling behavior of the cylindrical particles made of four kinds of materials with horizontal motion in stagnant water observed with an image sensor anemometer in the Reynolds number range from about 150 to 1000 was studied. The settling behavior of the particles was found to be simply expressed by an equation for a simple harmonic motion. By correcting the drag coefficient for cylindrical particles with a factor obtained from simple harmonic motion, the relationship between the drag coefficient and the Reynolds number for cylindrical particles was found to be parallel to that for spherical particles.

Tohno, S., and K. Takahashi, Shape analysis of particles by an image scanner and a microcomputer: application of agglomerated aerosol particles, *KONA* 6:2–14 (1988).

Algorithms are presented for feature extraction of geometrical particle shape. An image scanner scans the particle micrograph and presents compressed image data to a computer. Shape analysis methods are (1) calculation of basic particle shape parameters (Feret diameter, area, perimeter, first and second moments); (2) fractal analysis, (3) opening method, (4) separation of circular primary particles from agglomerates. The methods are applied to shape analysis of agglomerates.

Yin, M.-J. et al., Effects of particle shape on two-phase flow in pipes, *Powder Technology* 46(1):53–60 (1986).

It is shown that particles with the same sphericity but of different shapes behave differently in slurry flow. A quantitative relationship between pressure drop and particle morphological descriptors is determined. By changing solids concentration for each of several sets of particles, the pressure drop characteristics of a variety of particle shape systems were obtained.

M. Particle Size Change

Berglund, K. A., and M. A. Larson, Modeling of growth rate dispersion of citric acid monohydrate in continuous crystallizers, *AIChE Journal* 30(2):280–287 (1984).

A mathematical model for prediction of the crystal PSD from a continuous crystallizer is presented. The kinetic data used for the model were obtained from batch contact nucleation experiments with citric acid monohydrate. The distribution of growth rates, as well as the initial PSD were estimated here. Results from the model indicate that the excess number of crystals usually present at small sizes in continuous crystallizers is due to growth rate dispersion and not size-dependent growth.

Brown, R. A., Theory of transport processes in single crystal growth from the melt, *AIChE Journal* 34(6):881–911 (1988).

The quality of large semiconductor crystals grown from the melt for electronic and optical devices is strongly affected by the intricate coupling of heat and mass transfer and melt flow in growth systems. This paper reviews the present understanding of the processes starting from the simplest descriptions of solidification processes to numerical calculations needed for quantitative modeling of processing with solidification. Some model descriptions are included.

Duggirala, S. K., and L. T. Fan, Stochastic analysis of attrition—a general cell model, *Powder Technology* 57(1):1–20 (1989).

A stochastic treatment of attrition behavior of particles in chemical and mineral processing units is developed. A cell model was formulated which can be used for batch or continuous systems, to yield the mean particle size distributions and the fluctuations arising from the random nature of attrition in these processes.

The model considers multiple flow streams and involves the dependency of the attrition rate on particle size and number. Process dynamics are handled as the transient evolution of the particle size distribution.

Hiller, F. C. et al., Physical properties of therapeutic aerosols, *CHEST 80*(6):901S–903S, (1981).

This work describes results of measurements made to determine the PSD of commonly used medical aerosols and the effect of high humidity on PSD of some of these aerosols. If therapeutic aerosol particles change size considerably at high humidity, deposition quantity and site in the respiratory tract could be altered. If there is little change in size, deposition could be estimated using theoretical deposition methods devised for of stable aerosol particles. Results of this work suggest that aerosol produced by most bronchodilator devices does not change size with humidity.

Hsu, J.-P., and L. A. Glasgow, Floc size reduction in the turbulent environment, *Particulate Science and Technology 1*:205–222 (1983).

The dynamic behavior of the PSD in wastewater treatment processes results from the interaction of a random growth process with nonlinear, stochastic hydrodynamics. Successful modeling of the PSD response to high dissipation conditions depends on accurate quantification of the breakage mode, the total dissipation rate, transition probability and the daughter particle size distribution. The inferential determination of these quantities from PSD data is difficult. Direct photographic observations of floc breakup in a turbulent jet apparatus and in the impeller stream of a baffled, agitated tank allows measurement of both the breakage rate and the daughter PSD for individual parent flocs.

Kawashima, Y., et al., An experimental study of the kinetics of the spherical crystallization of sodium theophylline monohydrate, *Powder Technology 34*(2):255–260 (1983).

Direct agglomeration of crystals produced by salting out in a liquid was carried out in a stirred vessel. Components to produce spherically agglomerated crystals are shown. The crystal diameters decreased with an increase in agitation speed of the chloroform fraction in the mixture. The spherical crystallization process was described by first-order kinetics. The rate constant was found to be a function of the agitation speed of the system and the differential between residual concentration of material in initial and equilibrium states.

Tom, J. W., and P. G. Debenedetti, Particle formation with superficial fluids: a review, *Journal of Aerosol Science 22*(5):555–584 (1991).

The rapid expansion of supercritical solutions (RESS) is a new and promising method of particle formation. The process allows fast attainment of uniform conditions and of high supersaturation in the carrier fluid, which favor formation of small, monodisperse particles. A wide variety of materials can be used for formation. RESS can be used to comminute shock-sensitive solids, to produce intimate mixtures of amorphous materials, to form polymeric microspheres and to deposit thin films. The fundamentals, experimental methods, applications and results from studies are reported here.

N. Particulate Contamination in Liquids

Anon, ASTM, ASTM F 1095-88; *Rapid Enumeration of Bacteria in Electronics-Grade Purified Water Systems by Direct-Count Epifluorescence Microscopy,* American Society for Testing and Materials, Philadelphia, PA 1988.

This method describes a procedure for counting total numbers of bacteria in electronics-grade purified water systems. An on-line, large-volume sampling technique is used with acridine orange epifluorescence microscopy. Sensitivity to <20 cells/100 ml can be obtained routinely. A sample of 5 l or more is required. Some detritus, as algae and fungi, may also fluoresce.

Blackford, D. B., G. J. Sem, and T. A. Kerrick, Residue analysis in high purity water using a condensation nucleus counter, Proceedings PDA/IES International Conference on Particle Detection, Metrology and Control, Arlington, VA, 1990.

A continuous monitoring technique was developed for measuring ppb concentrations of nonvolatile residue in ultrapure water. It is based on determining residue after evaporation of atomized droplets. A CNC monitors the concentration of ultrafine NVR particles. Information on organic material content is possible by changing drying temperature. The higher the temperature, more organic material is evaporated.

Brown, J., N. Jayawardena, and Y. Zelmanovich, Water systems for pharmaceutical facilities, *Pharmaceutical Engineering 11*(4):15–23 (1991).

This paper reviews the various processes for production of pharmaceutical-grade water and their design. Deep bed filtration, activated carbon filtration, water softening, decarbonization, UV sterilization, RO, deionization, distillation, electrodeionization and final filtration techniques are discussed briefly. Some typical systems and their performances are described.

Carpenter, S. E., Improving epifluorescence monitoring methods in estimating numbers of bacteria in ultrapure water, Proceedings of 36th Annual Technical Meeting, Institute of Environmental Science, New Orleans, LA, 1990, pp. 212–216.

Monitoring of bacteria in ultrapure water using epifluorescent preparations can be time-consuming and present difficulties in obtaining accurate information. This paper gives an alternative and easier method of preparing samples for counts and to supply important information on obtaining accurate statistical information on the counts themselves.

Clancy, T. P., and R. A. Hango, Response of laser particle monitor to bacterial challenges in high-purity water, *Ultrapure Water 9*(4):51–54 (1992).

Response of a PMS HSLIS 0.05 μm OPC to challenges of known bacterial concentrations was determined. Test procedures are given and some discrepancies between the OPC data and epifluorescence microscopy data are found.

Fitch, E. C., Control of hydraulic fluid contamination: I. Contaminant identification and description, *Hydraulics and Pneumatics 36*(3):72–78 (1983).

The complete scope of contamination in fluid systems has been given here and the balance is shown diagrammatically. The scope is the same regardless of the fluid. In order to control contamination, it is essential that we learn to describe, analyze, exclude and remove the contaminant, as well as to protect contaminant-sensitive components and manage the system. Identification and description methods are available to reflect the state of a system. Only the analysis and degree of contamination must be defined.

Governal, R. A., and F. Shadman, Combining ozone and the UV-185 to oxidize and remove organic particles from the ultrapure water environment, *Microcontamination 10*(2):25–29 (1992).

The organic particle removal mechanisms from an ultrapure water system using UV-185, dissolved ozone and a combination of the two are the focus of this article. The particle-removing mechanisms are complex and competing; complete oxidation is preferred since incomplete oxidation can produce many small particles. The UV-185/ozone combination is effective into the low ppb levels. The synergistic effects are thought to result from enhanced generation of highly reactive radicals during decomposition of dissolved ozone in water by UV radiation.

Grant, D. C., Measurement of particle concentrations in central chemical delivery systems, *Journal of the IES 33*(4):32–37 (1990).

Recently developed high-sensitivity light-scattering OPCs for measuring particles in high-purity semiconductor prices liquids are evaluated. Results indicate that these low-flow-rate counters are appropriate for obtaining accurate measurements. Results indicate that particles may come from passage through filters and from sources downstream of the filter.

Hango, R. A., T. P. Clancy, and B. J. Eldred, Examination of particles less than 0.1 μm in semiconductor pure water by SEM and on-line particle counting methods, *Ultrapure Water 8*(4):44–48 (1991).

This paper covers recent accomplishments in detection, measurement and elemental classification of particles < 0.1 μm in diameter. Procedures were SEM, energy dispersive spectroscopy and laser particle counting. Experimental data are shown for the systems and found extremely useful.

Hunt, J., Using particle counters in drinking water treatment plants, *Waterworld News, July/Aug,* pp. 13–14 (1991).

A test program to evaluate an OPC for drinking water treatment plants is described. Filter bed performance both at steady state and variable conditions is shown. Advantages over turbidimeters are pointed out.

Hurd, T. Q. (1992): Atmospheric effects on particle levels in liquid chemicals. Proceedings Semiconductor Pure Water and Chemistry Conference, Santa Clara, CA, 1992.

Experiments were carried out measuring particle contribution of an uncontrolled atmosphere to bottled chemical particle counts. A novel method for batch sampling, using a peristaltic pump, was developed and experimentally verified. All experiments showed an increase in contamination levels caused by both bottle transport and atmospheric additions.

Long, M. L., Photoresist particle control for VLSI microlithography, *Solid State Technology 27*(3):159–161 (1984).

Particle control as required by VLSI microlithography is difficult both conceptually and practically. As linewidths decrease toward the 1 μm and less dimensions, particles that are a tenth to a fourth the size of minimum dimensions can affect the integrity of the resists and the function of the circuit. Particle control here includes not only the cleanliness of the resist in the container, but must also include all the handling procedures during process steps. It is important to understand all the reactions of all materials used in the process sequence.

O'Sullivan, T. J. et al., Applications of ultrafiltration in biotechnology, *Chemical Engineering Progress 80*(1):68–75 (1984).

Ultrafiltration is a membrane separation technique which is quite effective in many areas. The process relies on a permeable membrane to separate macromolecular and suspended solids from smaller molecular compounds in a solution on the basis of size. The method is most effective with particles from 1 to 100 nm. Operating parameters and application areas are discussed.

Pate, K. T., Examining the Design, Capabilities and benefits of bulk chemical delivery systems, *Microcontamination 9*(10):25–30 (1991).

This paper summarizes various bulk chemical delivery systems and present results of evaluation of a system recently installed at an Intel fab. Performance data compiled under different operating conditions are compiled and results compared with particle levels of low-particulate bottled chemicals. The bulk delivery system showed particle levels <0.1% of those of bottled chemicals.

Pawlaczyk, J. et al., Investigation of insoluble contaminations of infusion solutions, *Acta. Polon. Pharm. 39*(1–3):151–157, (1982).

Levels of contamination of various infusion solutions with insoluble particles produced at hospital pharmacies, a Polish blood donor station and in the pharmaceutical industry were estimated using Groves' index. No significant differences were found between contamination of solutions in glass containers from industry or at hospital pharmacies. Low levels of insoluble contamination, complying with the limits of British Pharmacopeia-1973, were only found in infusion solutions in plastic containers.

Sinha, D., and J. Van Winkle, Nonvolatile residue and high-purity water quality, *Ultrapure Water 8*(9):40–44 (1991).

Using a CNC to examine spray droplet residue, NVR measurements are made on-line, continuously in high-purity water. The importance of this measurement in defining water quality is pointed out in developing successful water quality management programs. For example, NVR variation as a function of temperature can indicate the level of organics in the water.

Taylor, S. A., Particulate contamination of sterile syringes and needles, *Journal of Pharmacy and Pharmacology 34*:493–495 (1982).

Commercially available sterile needles and syringes were examined for particulate contamination using the HIAC counter. The number of particles delivered was small compared with the total number permitted for LVPs. Where syringes are used in particle counting techniques, their contribution of particles should be considered.

Tolliver, D. L., Detection of microcontaminants in semiconductor process fluids using an acoustic technology, *Solid State Technology 25*(9):116–123 (1982).

Various methods of detecting solid particles in process fluids are compared. An acoustic technique that detects particles in clear, translucent or opaque fluids is described. Calibration of the acoustic system is described and data collected on DI water, HF, sulfuric acid and photoresists before and after filtration are presented.

Tsuchizaki, M., Y. Saitoh, and N. Saitch, Investigation of bacteria in ultra-pure water, Proceedings 11th Symposium of International Confederation of Contamination Control Sciences, London, 1992, pp. 123–125.

Bacteria in ultrapure water were surveyed and characterized for their growth in the water as well as taxonomy. Major species were identified. Hot water treatment was suggested as an effective means of control for semiconductor ultrapure water supplies.

Verdegan, B. M. et al., Field dependence of particle electrophoretic mobility in nonpolar liquids, *Particulate Science and Technology 3*:115–126 (1985).

Recent studies show that particle electrophoretic mobility in nonpolar liquids depends on field strength at high fields. The study provides independent confirmation using cross-flow electrofiltration and explores possible explanations for the phenomenon. Mobility was found independent of field strength at low and high fields, but not at intermediate fields. This is explained on the basis of the relaxation effects and partial detachment of the diffuse double layer from the particle. Further details are given.

Wetegrove, R. L., and F. C. Pocius, Controlling microorganisms in cooling-water systems, *Chemical Engineering 90*(10):75–78 (1983).

Microbes can clog systems and hinder performance. Often, chlorination is the preferred treatment, with continuous application preferred to stock doses. Reaction with other chemicals is also discussed.

Wickert, K., Endotoxin control using charge modified filter media, *Pharmaceutical Engineering 11*(4):25–27 (1991).

This paper reviews some historical data on endotoxin control and removal by charge modified filter media, answers some questions regarding practical use, discusses recent work on quantification of removal capabilities in typical systems and proposed practical operating systems for FDA specifications. Positively charged media is shown to be most effective and it is pointed out that bacterial grow through does not occur in high-efficiency membranes. The effects of pH, ionic species and some proteins are still to be defined.

O. Sampling Considerations

Anon, ASTM, F303-78; *Sampling Practices for Sampling Aerospace Fluids from Components,* American Society for Testing and Materials, Philadelphia, PA, 1989.

Three practices, differing on the basis of component geometry, are described. They cover sampling procedures for use in determining the cleanliness of liquids from components. A static fluid sampling method is used to sample fluids from cavities; a flowing fluid sampling method is used to sample fluid which is flowing directly through components; a rinse fluid sampling method is used to sample contaminants rinsed from component surfaces.

Harris, C. C. et al., Concentration profiles in a laboratory flotation machine: development of a simple sampling technique, *Powder Technology 35*(2):211–221 (1983).

The sampling method uses a pipet with pitot-tube-shaped sampling orifice. Error analysis, statistical methods, and consideration of hydrodynamic factors have led to a simple and accurate sampling technique. The incremental volume method used to verify the pipet method confirms agreement between procedures, but the pipet method provides more detail with the same effort. Some experimental data for the method are given.

Harris, G. E., M. R. Fuchs, and L. J. Holcombe, A guide to environmental testing, *Chemical Engineering 99*(11):98–108 (1992).

Careful sampling and analysis of air, water, and wastes will keep your company ahead of regulations and public scrutiny. Necessary testing objectives are laid out and EPA reference methods are summarized. The bases for meaningful analytical procedures are summarized.

Heidenreich, E., and M. Stintz, Dispersion of samples and sample-feeding for counting methods, *Particle Characterization 1*(1):28–31 (1984).

This paper reviews investigation of two different measuring system principles (light extinction and electrical resistance) by use of the same signal handling equipment. The two techniques and units are compared with regard to the resulting destiny distributions of particle size below 0.25 mm. The final conclusion including recommendations for the use of ultrasonic treatment and for improvements to sampling and sample feeders are valid for all counting methods.

Hertzson, L., Liquid particle sampling in semiconductor manufacturing, *Solid State Technology 31*(10):7–9 (1988).

Procedures for counting and sizing liquid-borne particles are discussed very generally.

Hurd, T. Q., Atmospheric effects on particle levels in liquid chemicals, Proceedings Semiconductor Pure Water and Chemistry Conference, Santa Clara, CA, 1992.

Experiments were carried out measuring particle contribution of an uncontrolled atmosphere to bottled chemical particle counts. A novel method for batch sampling, using a peristaltic pump, was developed and experimentally verified. All experiments showed an increase in contamination levels caused by both bottle transport and atmospheric additions.

Kitami, K. et al., Analytical technology for trace impurities in ultrapure water. Proceedings 11th Symposium of International Confederation of Contamination Control Societies, London, 1992, pp. 119–122.

To detect trace amounts of impurities, precision sampling and analytical procedure for counting particles > 0.05 μm, contamination from preconcentration technology for metallic impurities and silica, and contamination-free sampling bottles for TOC were developed. Requirements for 16 M DRAM systems are estimated. Electron microscopy is used widely.

Nivens, D. E. et al., Detection of microbial biofilms using on-line monitoring techniques, Proceedings Microcontamination Conference, San Jose, CA, 1991, pp. 81–92.

Rapid, in situ detection of microbial biofilm formation is possible with two analytical techniques: attenuated total reflection-Fourier transform IR spectroscopy and quartz crystal microgravimetry. Flow cells were used to monitor on-line and nondestructively the initial biofilm formation. Both techniques provide linear dynamic range of two orders of magnitude. The first technique detection limit is 5×10^5 and the second limit is 3×10^4 cells/cc, respectively.

P. Sedimentation in Liquids

Bailey, C., M. Cross, and D. Edwards, Numerical modeling of multiphase particulate flow and its application to sedimentation, *Particulate Science and Technology 5*(4):357–370 (1987).

A numerical model was developed to simulate batch settling of a suspension of particles initially with a uniform concentration in a host field. Various particle size fractions are represented as separate phases with distinct concentrations and velocities. This permits larger particles to move to the vessel bottom and pack. Packing behavior and numerical implementation of the model is based on the control volume techniques of Patankar.

Batchelor, G. K., and C.-S. Wen, Sedimentation in a dilute polydisperse system of interacting spheres. 2. Numerical results, *Journal of Fluid Mechanics 124*:495–528, (1982).

Analytical formulae for effects of interaction between pairs of rigid spherical particles on the mean velocity of each species in a statistically homogeneous dilute polydisperse system were evaluated numerically. We have calculated the pair-distribution function and the associated value of the sedimentation coefficient for a wide variety of conditions of the two interacting species, including different values of the radii of the spheres, different values of the ratio of their (reduced) densities, values of the Peclet number of the interaction and different forms of the potential of the mutual force exerted directly between the two spheres.

Bustos, M. C., and F. Concha, Boundary conditions for the continuous sedimentation of ideal suspensions, *AIChE Journal 38*(7):1135–1138 (1992).

This work applies results of studies on the existence and uniqueness of the solution for a conversation of material law with initial and boundary conditions to the continuous sedimentation of ideal suspensions in a thickener and to show that the values assumed by boundary concentrations are quite restrictive.

Clark, N. N. et al., Drag coefficient of irregular particles in Newton's settling regime, *Powder Technology 59*(1):69–72 (1989).

This paper presents data on terminal velocity of particles settling in Newton's regime and correlates the drag coefficient with particle shape characteristics other than sphericity. Terminal velocity was measured and related to a shape characteristic, developed to describe large-scale macroscopic shape.

Karamanov, D. G., and L. N. Nikolov, Free rising spheres do not obey Newton's law for free settling, *AIChE Journal 38*(11):1843–1846 (1992).

The specific behavior of free-rinsing particles can be explained by the effect of turbulence on the particle movement. The drag coefficient of such light particles is a constant and can be described by the laws of free settling, only when Re >230. The coefficient depends on both Re and density for certain ranges of properties.

Mariow, B. J., and R. L. Rowell, Sedimentation potential in aqueous electrolytes, *Langmuir 1*:85–90 (1985).

The validity of the Smoluchowski treatment of the sedimentation potential (Dorn effect) is explored in relation to the recent cell model treatment of Levine et al. which is applicable for higher-volume fractions where both electrical and hydrodynamic interaction must be considered. An apparatus was built for the rapid successive measurement of the sedimentation potential gradient, the particle volume concentration and the specific volume conductivity of the medium. Measurements on a model system showed that the Smoluchowski theory was followed below a volume fraction of 0.018.

Selim, M. S. et al., Sedimentation of multisized particles in concentrated suspensions, *AIChE Journal 29*(6):1029–1038 (1983).

A model is developed for predicting the sedimentation velocity in suspension of multisized nonflocculating solids, in which the retarding effect of the smaller particles on the settling velocity of the larger ones is taken into account. Tests of the model, with comparison to other models, show that it provides improved prediction of data on suspensions comprising both discrete particle size mixtures and continuous ones, and that it is applicable to continuous countercurrent solid-liquid operations.

Tory, E. M., M. T. Kamel, and C. F. C. M. Fong, Sedimentation is container-size dependent, *Powder Technology 73*(7):239–250 (1992).

In very dilute, bounded suspensions of identical spheres, the initial distributions of particle and fluid velocities in small containers differ enormously from those in very large ones. In the latter, local interactions are dwarfed by the massive collective influence of distant spheres. Large regions move downward or upward virtually en masse. Since particles and fluid in these clusters move with almost the same velocity, the mean of the initial distribution is scarcely affected by the much larger container size, but individual velocities can exceed the mean greatly.

Q. Slurry and Colloid Physics

Reiner, E. S., and C. J. Radke, Electrostatic interactions in colloidal suspensions: tests of pairwise additivity, *AIChE Journal 37*(6):805–824 (1991).

The electrical double-layer free energy is computed numerically for a model colloidal suspension of charged, hexagonally packed monodisperse cylinders in symmetric electrolye solutions using the Poisson-Boltzmann equations. Comparison is made to the energies estimated from sums of pair interactions and from two versions of a radially symmetric model; one in which the outer boundary of the cylindrical cell is chosen to make the particle volume fraction in the approximate model agree with that of the full lattice and the other in which the outer boundary position is chosen as the half the nearest-neighbor separation in the lattice.

Roco, M. C., and C. A. Shook, Critical deposit velocity in slurry flow, *AIChE Journal 31*(8):1401–1404 (1985).

The critical deposit velocity in pipes and channels is the lower limit at which slurry systems operate safely. For a given average slurry concentration, the critical velocity depends on the distribution of the velocity and concentration in the whole cross section. Using a model which computes the profile of the solids velocity at the pipe bottom, the critical velocity can be determined. In this paper, we introduce a criterion for the threshold velocity using results for fine uniform sand slurry flow in a pipe.

Russel, W. B., Theoretical approaches to the rheology of concentrated suspensions, *Powder Technology 51*(1):15–25 (1987).

This review highlights work on stable dispersions, specifically hard spheres, and flocculated systems, emphasizing the effects of structure created by many-body interactions. Predictions of weak flow behavior, e.g., low shear viscosity, linear viscoelastic response and plastic yield stresses, conform closely with experimental results and provide insight into the mechanisms responsible for shear thinning at higher rates.

Sugrue, S., Predicting and controlling colloid suspensions stability using electrophoretic mobility and particle size measurements, *American Laboratory 24*(6):64–71 (1992).

Electrostatic stabilization as monitored by zeta potential distributions can prevent aggregation and can be applied to nonaqueous or aqueous dispersions. Multiangle Doppler electrophoretic light scattering represents a rapid high-resolution technique for determination of zeta potential strategies to enhance stability and to assess the success of these strategies. Photon correlation spectroscopy offers an ideal means of detecting aggregation.

Van de Ven, T. G. M., Interactions between colloidal particles in simple shear flow, *Advances in Colloid Interface Science 17*:105–127, (1982).

Recent progress in experimental and theoretical developments dealing with colloidal interactions between two spheres in shear flow is reviewed. A systematic comparison is made between spheres in simple electrolyte and in cationic polyelectrolyte solutions. Microrheological observation performed with the traveling microtube make in-depth study possible of the colloidal forces and the nature of polymer bridge formation. Reasons are discussed for aging of colloidal aggregates. Coagulation rates are presented for systems with weak and strong Brownian motion.

Yamazaki, H. et al., Concentration profiles of solids suspended in a stirred tank, *Powder Technology 48*(3):205–216 (1986).

The solids concentration profiles in a slurry mixing tank with mechanical agitation were measured with a photoelectric method. The concentration profile was explained by the sedimentation-dispersion model in which the slurry flow in a stirred tank with a flat-disk turbine or a marine propeller is assumed to be dominated mainly by the upflow. The model parameters, modified Peclet numbers in terms of the falling velocity of the solid particles; one, in terms of the liquid flow velocity, was markedly influenced by suspended solids.

R. Suspension and Multiphase Flow

Baker, J. R., Motionless mixers stir up new uses, *Chemical Engineering Progress 87*(6):32–38 (1991).

Use of motionless mixing systems in shown, including applications as diverse as absorption, extraction and reaction enhancement. Mixer operation and some designs are shown and application areas for different designs are pointed out.

Graham, D. P. et al., Measurement and prediction of flow patterns of granular solids in cylindrical vessels, *Powder Technology 50*(1):65–76 (1987).

A kinematic model of gravity-induced flow was applied to the prediction of flow patterns of granular solids in cylindrical vessels. Experimental evidence is given which validates the use of such a model to calculate steady-state velocity profiles and residence time distributions in axisymmetric geometries. A reliable technique for measuring particle residence times is described and a method for determination of the kinematic constant is demonstrated.

Lieberman, A., Fluid contamination analysis using particle counters. I. Preliminary preparation, *BFPR Journal 16*(1):215–221, (1982).

When analyses are conducted on fluid samples to determine levels of contamination, samples must be obtained and transported to a laboratory. The basics of sample container selection and cleaning and of fluid handling and storage prior to particle counting are reviewed. Particle counter calibration is discussed along with recent revisions to the currently accepted ACFTD particle size distribution used for the ISO 4402 calibration method. The parameters that must be controlled to maximize counting accuracy are considered.

Mifflin, R. T., and W. R. Schowalter, Flocculation of colloids suspended in a viscoelastic fluid undergoing shear, *AIChE Journal 34*(11):1839–1849 (1988).

The effect of viscoelasticity on the kinetics of the shear-induced flocculation of colloids was studied. A numerical technique for such three-dimensional problems called global residual minimization was developed as an alternate to finit difference methods. In the method minimization of the residual of the momentum balance is used as the selection criterion for the velocity-field solution. Trajectory simulations were carried out for spheres.

Oldshue, J. Y., Fluid mixing technology and practice, *Chemical Engineering 90*(12):82–108 (1983).

Fluid mixing can be either a simple process or one that is extremely complex, with many independent steps involving a variety of fluid phenomena. This report provides guidance for designing readily definable processes, such as blending, solids suspension and heat transfer; it present an up-to-date treatment of the more complicated processes entailing mass transfer, chemical reaction and shear-rate sensitivity.

Soo, S. L., Comparisons of formulations of multiphase flow, *Powder Technology 66*(1):1–7 (1991).

Methods of approach toward the formulation of the dynamics of multiphase flow include those of single free particles, continuum mechanics including interfaces with volume and time averaging and statistical averaging. Unresolved aspects are transfer integrals and stress systems. Basic premises and experimental bases are compared. Physical meaning of mathematical operations are discussed.

Stamatakis, K., and C. Tien, Dynamics of batch sedimentation of polydispersed suspensions, *Powder Technology 56*(2):105–117 (1988).

Dynamics of batch sedimentation of polydispersed suspensions was studied, assuming that polydispersed particles in a suspension segregate and group themselves into zones of constant composition. These zones move downward with constant velocity. Equations were derived to describe the particle concentration profile throughout the suspension phase and the sediment composition profile.

Storms, R. F., B. V. Ramaroa, and R. H. Weiland, Low shear rate viscosity of bimodally dispersed suspensions, *Powder Technology 63*(3):247–259 (1990).

The low shear rate viscosity was determined for suspensions of bimodally distributed polymer spheres dispersed in silicone oil over a large range of particle volume fractions. Analysis of the data showed that the concentration dependence of the reduced viscosity would be predicted by a modification of the empirical equation proposed by Eilers for monodispersed suspensions.

Tory, E. M., and D. K. Packard, On the divergence problem in calculating particle velocities in dilute dispersions of identical spheres, *Powder Technology 47*(1):39–42 (1986).

It is well known that the divergence problem in calculating the mean particle velocity in a dilute dispersion of identical spheres can be overcome. The difficulty arising from the principal interaction between two spheres is overcome by using the fact that the mean velocity at a point in the dispersion is zero. A divergence problem in calculating the variability of particle velocity arises from the same term and cannot be overcome by this method. Consequently a more restrictive conditions on return flow is needed, but the exact nature of this condition is not yet known.

Turian, R. M. et al., Settling and rheology of suspensions of narrow-sized coal particles, *AIChe Journal 38*(7):969–987 (1992).

Settling rates, yield stresses and shear-stress/shear rate dependencies were determined as functions of solids concentrations for suspensions in water of coal particles of narrow PSD. Analytical techniques are defined for all parts of the work. Settling rate data observations show that hindered settling due to concentration effects is compounded by particle shape effects, also. Several useful relations are developed and discussed.

Zimmels, Y., Hindered motion of particulate system in Newtonian fluids. Acceleration and steady regimes, *Powder Technology 34*(1):191–202 (1983).

The hindered motion particles in Newtonian fluids is considered as in a finite system of constant total volume. Equations of motion are derived using updated hindrance factors as a function of the particle concentration. Extended solutions in the Stokes, intermediate and turbulent flow regimes are derived for gravitational settling systems. It is shown that analytic solutions obtainable by use of the Dallavalle approximations are in good agreement with those calculated from the general correlation. In this context, particle concentrations that yield maximum flux are evaluated for different size fractions.

S. Viable Particulate Materials

Carmody, J. C., and J. E. Martyak, Controlling bacterial growth in an ultrapure deionized water system, *Microcontamination 7*(1):28–35 (1989).

Profiling the central DI water production system and the primary and secondary DI water distribution loops to manufacturing, one can identify the areas of bacterial proliferation. Removing, cleaning and reducing process distribution locations identified as contributors lowers both viable and nonviable bacterial contamination.

Janes, J. B. et al., Bacterial and endotoxin retention by charge-modified filters, *Semiconductor International 11*(4):80–82 (1988).

This study provides information on the absolute bacterial and endotoxin retention capabilities of positively charged membrane filters and strongly supports long-term point-of-use service in high-purity water systems. The importance of the charge modification in combination with use of rated pore sizes of 0.1 μm is pointed out for bacterial retention.

Khan, S., Automatic flexible aseptic filling and freeze-drying of parenteral drugs, *Pharmaceutical Technology 13*(10):24–34 (1989).

This article describes processes and technology required for a fully automated and flexible "people-less" system for filling, freeze-drying and packaging small volume parenterals. Product quality improvements, process and economic considerations are addressed, as well as capability for changeovers and sterility assurance.

Ljungqvist, B., and B. Reinmuller, Supplementary microbiological assessment with the aid of particle challenge test and visualization technique, Proceedings 11th Symposium of International Confederation of Contamination Control Societies, London, 1992, pp. 541–544.

Supplementary methods are described for microbiological assessment in clean zones. The hazards of entrainment of ambient air into unidirectional air flow units supplied with HEPA filters are discussed. A particle challenge test is described with resultant risk factor definition. Visualization of air movements is also considered.

Martyak, J. E., J. C. Carmody, and G. R. Husted, Characterizing biofilm growth in deionized ultrapure water piping systems, *Microcontamination 11*(1):39–44 (1992).

Biofilm, the microbiologically induced fouling of an aquatic system, occurs in DI water systems at the interface between the liquid and the solid component wall. Rather than the past procedure of maintaining turbulent flow to prevent buildup, a study was carried out to find if different piping materials and varying flow rates better prevented biofilm formation. Ozonation, point-of-use filtration and dynamic hydraulic operations were also studied.

Norquist, R. W., Filtration of serum protein products, *Medical Device and Diagnostic Industry 4*(11):37–41 (1982).

A discussion is given of problems of final product filtration where the fluid is already relatively free of microcontaminants. Reasons for filtration are summarized and filter clogging problems are mentioned. The need for good manufacturing practices so as to minimize contamination that will affect filters is emphasized.

Peterson, A. J., and D. O. Hankner, Assessment of long-term bacterial retention, *Medical Device and Diagnostic Industry 7*(6):150–155 (1985).

A test was carried out to determine the long-term bacterial retention of a filter module as well as its ability to meet product requirements. A test system is described and data are shown. Bacterial grow-through was not observed over a 90-day test period.

Quebbeman, J. F. Microbiological surveillance in aseptic production facilities, Proceedings of 37th Annual Technical Meeting, Institute of Environmental Sciences, San Diego, CA, 1991, pp. 674–678.

Microbial surveillance techniques for aseptic production facilities are described for airborne and deposited bacteria. The need for a good monitoring plan is emphasized. Advantages and problems with common sampling and measurement systems are discussed.

Shadman, F., R. Governal, and A. Bonner, Interactions between UV and membrane filters during removal of bacteria and TOC from DI water, Proceedings of 36th Annual Technical Meeting, Institute of Environmental Science, New Orleans, LA, April 1990, pp. 221–223.

Sequencing of various units in the polishing loop of a DI water system is discussed for optimum TOC removal. It is indicated that TOC can be removed more effectively if the filter is placed before the UV unit. This occurs since the UV unit may degrade bacteria and organics to sizes too small for effective filtration.

Simonetti, J. A., and H. G. Schroeder, Evaluation of bacterial grow through, Proceedings of 30th Annual Technical Meeting, Institute of Environmental Sciences, Orlando, FL 1984.

A variety of different commercial submicron rated membrane filter cartridges were inoculated with a known number of viable cells and placed in service at a simulated DI water point of use. Pseudomonas diminuta was selected as the model organism. Both continuous and intermittent use were simulated. Since the downstream bacterial counts varied considerably for the cartridges tested, lab experiments were carried out to obtain insight into the possible mechanisms which may be associated with proliferation of bacteria through various types of membrane materials.

Whyte, W., and L. Niven, Airborne bacteria sampling: the effect of dehydration and sampling time, *Journal of Parenteral Science and Technology 40*(6):182–187 (1986).

Lengthy airborne sampling is thought to cause desiccation of bacterial media and loss of viability of collected bacteria. Experiments reported here show that microorganisms commonly found in room air are little affected (ca 8% viability reduction) by a 13% reduction in the water content of the agar in Petri dishes. Acceptance of this reduction would allow bacterial plates to be exposed for about 24 h in still air, over 6 h in a laminar flow bench and over 1 h in a 30-LPM bacterial slit sampler.

Wolber, P., and M. Dosmar, Depyrogenation of pharmaceutical solutions by ultrafiltration: aspects of validation, *Pharmaceutical Technology 11*(9):38–46 (1987).

This article discusses ultrafiltration as a means of depyrogenate pharmaceutical solutions and presents considerations that must be addressed when validating ultrafiltration systems used for this purpose. The areas of concern are biological safety of the elements, extractables, compability of product and filter, integrity of the membrane, selection of the system and membrane components, pyrogen challenge and sanitization and sterilization. These points are all discussed here.

Index

Abbe:
number or dispersion, 48
resolution criteria, 68
-type condenser lens, 33
Aberration:
chromatic, 36
severest, 42
spherical, 36
"Absolutely clean", 732
Absorption, 748, 754, 774
classical, 264
differential, 247
infrared spectra, 260
light, 247
X-ray, 259
Accelerating voltage, 251
Accelerators, 640
Accurate contaminating particle count requirements, 466
AC fine test dust (ACFTD), 462, 587
agglutination, immunoassays, 692
batch sample measurement, 619
Acetone, 242
Acid(s):
free, 643
penicillinoic thiolactones in penicillin powders, 645
Acoustic detection systems:
acoustic response, 16
application areas, 17
beam convergence effects, 17
particle size analysis capability, 16, 174
Adhesion, 826-833
Administration sets, 654
Admixtures:
pharmacy-produced, 642
precipitation linked to particulate matter in, 643
Adsorption phenomena, 644
Adverse Inspection Observation, (a 483), 815
Aerosol OT, 595, 616, 632
Aerosol particles (monitoring of), 793
Aerospace, 593-594
Agglutination tests (microsphere), 687, 689, 691 (*see also* Particle[s], clinical applications)
detection requirements, 690
test sensitivity, 690
Agglomeration, coalescence, 853 (*see also* Particle[s], clinical applications)
Airborne:
counts, 795
manual particle count chronic problems, 803
particles, 239
viable and nonviable particle contaminants, 819
Air bubbles in liquids and removal with ultrasound, 806 (*see also* False counts)
Air conditioning system, 793
Air filtration, 792
Air pollution sample, 42
Air quality (maintained), 793
Airy discs (pattern), 12, 66
Airy function, 167
Alcohol lamp, 44, 242
Algae, 42
Algorithm (s), 84
calipering, 75, 97
computer, 74
floating point, mathematic, 75
grid, 81
intersecting chords, 81, 83
in modern digital imaging systems, 81
scaling, 95
sizing, 81, 84
Alignment, 768

Aluminum, 288, 646
  from phosphate-containing pH-adjusting reagents, 642
American National Standards Institute, 594-600, 620
  ANSI/(NFPA) T2.9.6R1, 594-600, 620
American Society of Testing and Materials (ASTM):
  index of standard X-ray diffraction data, 283
  powder diffraction data, 268
American standards (*see* Standards, American)
Amici-Bertrand lens, 37 (*see also* Lens, supplementary)
Amorphous particles, 643
Amplifier(s), 33
Ampules, 31, 645, 651
  cleaning of, 816
  opening, 815
  opening with an oxypropane torch, 816
  tubular glass, 646
Amyl acetate, 242, 243, 256
Analog image (*see also* Image and image analysis, Image processing):
  correction, 72
  to digital conversion, 96
  and shading, 72
Analysis:
  advantage of X-ray with simultaneous SEM or TEM data, 256
  of bulk samples (X-ray spectrometry), 259
  chemical, 252
  compositional, 253
  digital, 61, 63
  EDXS (electron dispersive X-ray spectroscopy), 257
  morphologic (shape), 61, 253
  mounting stubs, 259
  particle isolation, 243
  particle, predominance of volume evaporation, 265
  population, 61
  qualitative, 257, 258, 259
  qualitative, with X-ray spectrometry, 258
  quantitative, with X-ray spectrometry, 259
  X-ray crystallography and electron diffraction, 269, 276
Analyst, particle:
  analysis area cleanliness, 791
  collection of particles of interest, 237
  crystal spectrometer operation, 257
  electron diffraction, special skills, 280
  electron microscopy as a tool, 244
  time requirements for WDXS analysis, 257
  tools or other equipment, 792
Analytical methods, 745
  ensemble, 745
  localization of organic, inorganic, biological samples, 263
  non-ensemble, 745
Analyzer, polarized light:
  module, 37
  rotatable, 37
  units, 36
  use of second polarizer, 37
Ancillary equipment for polarized light microscopy, 30, 33, 41
Angel hair, 655
Angle(s):
  condenser controls illuminating cone, 42
  crystal, 34
  extinction, 34
Angular:
  dispersion, 81
  orientation, 248
Anions, 29, 645
Anisotropic crystal, 37
ANSI (*see* American National Standards Institute)
Antibiotic(s):
  cephalosporin, 643
  manufacturer of generic, 645
Antioxidant, 644
Aperture, 247 (*see also* Objective[s])
  angle, 246
  angular, 35
  of condenser, controls angle of illuminating cone, 42
  diaphragm, 33, 39, 40-42 (*see also* Diaphragm)
  cautionary note: low numerical aperture, 40X polarizing microscope objective, 36
  numerical angle of illuminating cone, 33–35
  numerical, controlled by diaphragm, 42
Apochromats (APO), 36, 42
Apparatus (crystal rotation – spindle stages), 34
Aragonite, 268
Archimedes, 94
Area equivalent diameter, 803
Aroclor 5442, 44, 242
Artifactual counts, 806
Asbestos, 42, 285
Ascorbic acid, 642
Aseptic fill, 642, 793
ASME, 809
Aspect ratio, 82
Astigmatism, a factor in resolution, 247 (*see also* Light microscope, resolution)
ASTM (American Society for Testing and Materials):
  index, 283
  powder diffraction data, 268
Atmospheric contamination, 792
Atom/atomic:
  dimensions, 43
  electrons, 246
  elemental data, 236
  evaporated metal, 248

[Atom/atomic]
excited state, 252
innermost (K) shell, 252
ionization, 252
ions and mass spectrometry, 263
lithium drift diffusion, 253
number, 252, 259
orbital vacancies filled by electronic relaxation, 252
orbital vacancy, 252
outer shells, 252, 253
parent, 252
shells, 252
silicon, 253
Attachments, polarizing microscope (*see* Polarizing microscope)
Auditors (regulatory), 815
Auger:
electron, 252
electron spectroscopy, 255
electrons produced by internal conversion of characteristic X-ray, 255
well suited to studying surface characteristics, 256
Autocorrelation function, 163
Automated IA systems, 62, 89 (*see also* Image and image analysis, Image processing)
Automotive, 593-594 (*see also* Engine)

Back focal plane, 41
Background counts, sources:
clean glassware, 805
opening containers, 813
smooth nonshedding surfaces, 812
Background substrate, 259
Backscattered electrons, 250
Bacteria, 793
Bacterial:
contamination, 658
growth control, 868
grow through, 869
Ball bearing stages (*see* Polarizing microscope)
Barium, 646
sulfate, 645
Base (free), 643
Baseline correction, 590, 614
Batch sampler, 603, 619
Bausch and Lomb, 96 (*see also*, Omnicon)
Beaker (ultraclean), 811
Beam:
collimated, 269
cone, 285
currents (high), 259
diffracted beam of maximum intensity, 254, 283
direction, 285
dump, 766
[Beam]
electron, 246, 250, 251, 259
energy, 257
focused collimated of high-energy ions, 265 (*see also* SIMS)
impact, 248
incident, 250, 253, 285
infrared (IR), 260
intensity, 246
laser, 263, 265
penetration, 259
primaries, 248
raster, 248
raster scan on specimen surface (SEM), 248
scanned in synchrony with CRT, 253
secondary electron signal, 248
spot, 249
static spot, 248
of a visual CRT, 248
Beam spot swells (SEM), 248
Becke line, 48, 50, 53, 56
Bertrand lens, 37, 39, 49, 55-56 (*see also* Lens, supplementary and Nikon Optiphot)
Beryllium:
block, 32
planchette, 242
stubs, 243
support, 257
toxicity cautions, 259
Bessel function, 166
Beta ratio, 621
Binocular observing tubes, 37
Biochemical, clinical type microscopes (except for Nomarski), 31
Biofilm growth, 869
Biological:
cells, 114, 136
microscope condensers, 34
solutions, 807
Biomedical:
microscopists, 44
-type objectives, 35
Biophysical vision data (*see* Visual inspection, basics)
Birefringence, 31, 35, 50
Bismuth, 644
Bisulfite, 644
Blank counts:
criteria, 813, 488
low, 813
reduced background counts, 805
in USP XXII, 812
in USP XXIII, 813
Bloom, 640, 654, 657
Blow molded/molding:
char, 641
Japan, (blow-fill seal), 652

[Blow molded/molding]
  plastic, 641
Bonding energy, 252
Bottles (glass), 31
Boundaries (*see* Edges)
Bragg, Sir Lawrence, 253
Bragg('s):
  angle, 280
  criteria, 273
  equation, 253, 273
  law, 254, 281, 285
  reflection, 255
  relation(ship), 280, 281
Bremsstrahlung, 252
British Pharmacopoeia (BP), 643, 807
British Standard (BS), 794-795
Brown amorphous material (BAM), 643
Brownian motion, 13, 162
Bubble-point testing, 840
Buffers, 646
Bulb filament:
  burned out, 42
  effect of positioning, 41
  position adjustment, 40
Bulb replacement, 42
Bulk materials:
  particle infrared spectrum, 260
  pulverized to make particle standards, 259
  raw material, 644
Bunsen:
  burner, 242
  flame, 44
Bureau Community of Reference (BCR) (*see* Standard reference materials)

Calcite, 268
Calcium, 646
  carbonate, 268
  fluoride, 36
  oxalate, 642
  from phosphate-containing pH-adjusting reagents, 642
Calculations (computer-based), 259
Calibrated eyepiece micrometer, 51
Calibration, 42-43, 118-119, 120-122, 147-148, 585, 587, 589, 592-602, 607, 610, 616, 618, 620, 624, 626, 633, 710, 718, 768
  ACFTD fine test dust method, 594-600, 620 (*see also* International Standards Organization 4402 and ANSI/[NFPA] T2.9.6R1)
  flow cytometry, 704
  half-count method, 595
  hybrid method, 599-600, 638
  ISO 4402, 592-594, 597, 599, 600-611, 620, 625
  latex in oil method (*see* ANSI/[NFPA] T2.9.6R1)
  microscope eyepiece micrometer, 42
[Calibration]
  on-line, 601, 620-625, 629, 633
  optical particle counter, 18, 26 (*see also* Particle counters and Particle counter types)
  [For data analysis, *see* Jaenicke Geometric and Particle-Triggered Poisson model entries]
  verification, 594, 597, 624-625
Camera:
  cathode, 64 (*see also* electron gun)
  electron beam, 64
  electron gun (heated cathode), 64
  electron-hole pairs, 63
  LCD, 98
  negative terminal, 64
  photoconductive layer, 63-64
  photographic, 32
  raster, 64
  relay lens, 98
  resistor, 64
  target electrode, 63, 64
  target voltage supply, 64
  transducer, 63
  video,
    cathode ray tube (CRT or monitor), 64
    display, 63, 77, 97
    output signal, 64
Canada balsam, 44
Carbide scriber, 44
Carbon, 256, 257
  adhesive, 259
  coating (sputtering or vacuum evaporation), 258
  evaporated film, 259
  grids, 259
  paint, filter mounting, 259
  paste versus paint, 258
  planchette, 242
  stubs, 243
  support, 257
Carbonization (LAMMA), 265
Cargille Labs, 56
Cassegranian optical microscope, 260
Cast film filters (*see* Filters)
Cathode ray tube (CRT), 248, 253
  electron beam, 249
  image (intensity-modulate), 249
  modulate brightness of spot, 253
  modulation using element of interest, 265
  monitor, video, 249, 263
Cations, 29
CEA (*see* Controlled environment area)
Cedarwood oil, 36
Cells, 42 (*see also* Particles)
  Blood, 114, 119, 121, 125, 133-136, 145-147
  Other, 114, 136-138, 146
Centrifugal sedimentometer, 10
  resolution, 160

[Centrifugal sedimentometer]
size sensitivity, 160
Cephalosporins, 640
Certified reference materials (*see* Standard reference materials)
Chalnicon camera, 97
Chamot and Mason chemical reagents, 56
Channel coincidence, multiple strike zones, 726
Characterized (particles), 30
Charcoal treatment, 643
Chemical:
bond, 267
delivery module, 576
delivery systems, 570, 576, 583
elements, distribution of particular, 253
reagents, 30
Chord(al), 75
array, 83
fixed, 82
maximum or minimum, 75
measurement, 84, 97
precision, 82
resolution, 82
rotating, 83
static, 83
Chromatic aberration, 36 (*see also* Light microscope, resolution)
Chromatograph, 709
Class (clean area), 652
Class 10:
hood and <788> methodology, 815
sensitive systems and, 799, 800
Class 100:
air, 237
bench top, 821
environments, 792
hood, 795
laminar airflow hoods, 792, 811
laminar flow areas, 795
new counter data and previous baseline, 803
sensitive systems, 799
systems for verifying and monitoring, 799
Class 1000, 10,000 and higher:
areas, 795
and HEPA filters, 793
main laboratory areas, 821
sensitivity requirements at lower flow rates, 799
Clean:
area, 652
garments, 652
molding process, 654
Cleaning, 651
of glass and plastic ware, 805
methods, 602-603, 607-608
Clean laboratory:
design, 820
[Clean laboratory]
ultrasonic bath applications, 806
Cleanliness:
classes, 795
level of, 791, 792 (*see also* Contamination levels)
Clean room:
clothing, 818
zones, 820-821
Cleanup system, 607-608
Clearance of particulates (physiological):
clearance kinetics, 669, 673
macrophage response, 677
phagocytosis, 669, 673, 677
reticuloendothelial response, 670, 677
Climet image analyzer system models, 138-139, 496-498, 501, 502, 518, 519
Clinical diagnostic testing, 687
active tests, 691
after agglutination, 693
agglutination tests, 687, 689, 691
detection requirements, 690
test sensitivity, 690
antigen/antibody testing, 687
capture ELISA, 687
immunoassays, 687
instrumental evaluation, 692,693
qualitative agglutination, 689, 695
quantitative assays, 691, 695
single microsphere assays, 705
solid-phase immunoassays, 687
strip (chromatographic), 696
Closures, 639
silicone oils, 640
Clutch, microscope stage, 35
Coaxial illumination (lighting), 32
Coefficient of variation (CV) (*see* Errors)
Coil, 244
Coils (filament), 32, 39-40
Coincidence error, 587, 591-592, 594-595, 597, 599-600, 601, 607-614, 619-620, 622, 625-627
concentration limit estimate for oil samples, 591-592, 594, 595, 597, 599-600
Coulter Counter, empirical correction, 126-128
definition, 591
demonstration, 468-471
determination,
calculation of parameters, 523-525
calculation of single particle count accuracy, 534
effective detector volume, 536
geometric detector volume, 468-473
model-based dilution experiment, 554-559
parameter accuracy limits, 525
effect on particle size distribution, 463, 599, 611-614
effects, 465-466

[Coincidence error]
evaluation, step-by-step procedure, 527-534
horizontal coincidence, 502-504
Jaenicke Geometric Poisson model, 540
Particle-Triggered Poisson model,
raw count, 542, 543
single particle counts, 547, 548
universal curve,
raw counts, 492
single particle counts, 509
vertical coincidence, 502-504
Coldstage, 48
Collecting optics:
double lens, 763
Fourier, 762
reverse Fourier, 762
Collection of particles, 237
Collection transmission, 268
Collector (secondary), 250
Collision, 251
Collodion, 242, 243, 256, 259
film, 32
Colloids, 866
Color systems, 62, 101
true image color advantages, 101
Column:
electron-optical, 248
as a multiple condenser lens system in SEM, 248
Combustible materials:
size measurements, 4
Combustion, powder, 2
Commercial inspection systems, 443-448
Common main object type (single lens) for stereomicroscope, 33
Comparator, 588, 595
Compendial:
assays, 644
groups, 638
particle analyses, 815 (*see also* USP XXIII <788> SVI Assay)
requirements, 791
Compensator(s), 37
Components and solubility limits, 642
Composition (elemental), 258
Compound microscope, 242
incident or reflected illumination, 242
used to locate particles on prepared stubs, 259
Compounds:
organic, 267
uniquely descriptive chemical, 270
Computer(s), 62
algorithm, 74
-based data collection and reduction, 799
control, 819
data bases, 269, 804
-driven spectrometers, 258
[Computer(s)]
external, 793
people, 793
hardware, 84
memory, 95
microcomputers, 97, 98
off-the-shelf software, 97
software, 84, 95, 96, 101
surface, 804
Concave mirror, 41
Concentration measure, 534-538 (*see also* Coincidence error)
effect of analysis threshold selection, 537-538
particle size dependent effective detector volume, 535-537
particle size independent geometric detector volume, 534-535
Condensation of positive and negative ions, 267
Condenser:
Abbe resolution criteria, 68
Abbe-type lens, 33
adjustable lamp, 41
angular control of illuminating cone, 42
electrostatic ion lens, 263
fixed focus, 39
functions as a multiple lens system (SEM), 248
lamp, 39
lens, 245, 246
mount (substage), 39
numerical aperture, 42
optical axis (polarizing microscope), 33
substage, 33, 34, 39, 245
Conductive:
coating, 257
material, 256, 257
path, 257
Conductor, poor and negative charge, 257
Cone (illuminating), 42
Conjugate focus, 39, 41
Conoscopy (quantitative), 35
Container(s), 639, 641
Container cleaning, 832
Container defect inspection, 442, 443
Contaminants:
AC coarse test dust, 622
AC fine test dust, 587, 592-594, 597-600, 606, 611-612, 615, 618-619, 622, 624-625, 630
immiscible droplets, 595, 601, 615-616, 618, 630, 632
ISO fine test dust, 593
ISO medium test dust, 599-600
ISO ultrafine test dust, 600
latex, 588, 591-592, 594-600, 611-614, 616
nature of, 601, 607, 614-615
particle properties, 587
SAE 5-80 micrometer test dust, 622

[Contaminants]
silica, 587, 593, 614, 630
size distribution of, 587, 591-592, 593, 594, 597-598, 599, 600, 601, 603, 606-607, 611-614, 616, 618, 620, 622-623, 624, 625, 626, 630-632, 633
soot, 587, 614, 630
water droplets, 595, 614-616, 632
wear metals, 587, 588, 604, 606, 614, 630
Contaminating particles, 259
improved microscope assay, 723 (*see also* Compendial assays and USP XXIII <788> SVI Assay)
Contamination:
composition, 687
of containers, 576
of particle samples,
particle-free gloves, 237
risk, 241
touch, 237
particulate, 860-863
Contamination level, 585, 604, 610, 627, 629, 630, 633
Contrast:
effects,
in light microscope, 247
in TEM, 247
information, 243
limiting detection, 42
pattern (TEM) from differential electron scattering, 247
resolution (*see also* Gray scale),
minimum for human eye detection, 71, 321-322
Controlled environment area (CEA):
and clean laboratory design, 820
clean room in larger area, 821
laboratory-type, 795
standard, 794
verifying and modifying, 799
Cooling (particle), 31
Coordinate profile:
Cartesian, 92
polar, 92
Copper, 645
Cork, 44
Correction factors for particle size effects, 259
Correlation problems between particle counters, 26
Cost reduction with improved quality, 642
cleanliness and cost balance, 791
Coulter, 807
Coulter Counter, 709, 711, 712, 719, 745
accuracy, 129-131, 133-140, 144-147
method, 113-133
models, 138-140, 145-146
principle, 114-115
Count (systematic), 34
Count accuracy, 465 (*see also* Count efficiency, Particle count accuracy)
coincidence counting errors, 463, 464, 467, 468
determination limits, 557-559
effect of model selected, 480
Jaenicke Geometric Poisson model, 540
linearity under increasing load, 467
particle size dependence, 464, 470, 471, 472
Particle-Triggered Poisson model,
raw count, 542, 543
single particle counts, 547, 548
prerequisites, 466
rate of change, 479
raw counts, 477-480, 490-501
single particle counts, 508-510
versus reproducibility, 459, 465, 722
Count balance equation, 506-507
Count efficiency:
applied to false counts and undercounts, 544-546
normalized relationship, 538-543
Counter (*see also* Particle counters and Particle counter types):
[For data analysis, *see* Jaenicke Geometric and Particle-Triggered Poisson model entries]
Coulter, 445
optical, 445
Count error, 477
horizontal coincidence, 502-504
vertical coincidence, 502-504
Coupled plasma (CP) spectroscopy, 644
Cover glass fragments, 44
Cover slip, 242
Critical micelle concentration (CMC), 642, 643
Cross-contamination, 793
Crossed Nicols, 31
Crossed polars, 31, 50, 53–55
anisotropic crystal viewed between, 37
background not black when using, 35
and certain micro-chemical tests, 31
fully-, 37
Cross-flow electrofiltration, 842
Crosshair, 38
Crosshair graticule, 34, 37
CRT (*see* Cathode ray tube)
Crystal(s), 239, 243, 277
angles, 34, 57
angles of axes, 270
anisotropic, 37
array, three-dimensional, 270
atomic arrangement in solid materials, 270
atoms, uniform separation of, 269
axes, three coordinate, 270
band structures, 269
cell or basic unit structure, 270

[Crystal(s)]
complex unit cells, 270
computer data base interpretation of crystal structure from diffraction analysis, 269
cooling, 277
cubic system, 273
of defined interplanar spacing, 255
density, 277
dielectric constant, 277
diffracting planes, 255
evaporation, 277
face-centered cubic, 283
faces (Miller indices), 270
fingerprint, 270
glass fibers (fine), 277
grid lines, 269
growing, 277
identical composition, distinctive morphology, 268
interatomic spacing of component atoms, 270
intercept reciprocals, 273
lattice(s), 270
14 possible for six crystal structures, 270
simplest, 270
spacing, 281
macrostructure (shape) different, 270
of the material of interest, 268
mechanical,
shock, 277
vibration (free of), 277
microscopic, 277
mixed solvent system, 277
morphology, 49, 269
origination, 277
parallelepiped sides, basic unit cell, 270
planes, 273, 281, 283
planes through lattice points, 270
poorly defined, 277
radiation interaction result, 273
reciprocal lattice, 283. 284, 285
reciprocals of the intercepts, 273
recrystallization, 277
reference, 37, 55-56
regular arrays give rise to symmetry, 270
seeding the solution, 277
single pattern, 283, 284
single pure or essentially pure, 269. 274
six systems, 270
solubility, 277
solvation, 277
spacings of lattice points, 270
supersaturation, 277
surface "crusts," 277
symmetrical, 645
symmetry of (internal, external), 270
temperature-controlled oven, 277

[Crystal(s)]
two-dimensional,
lattice or grid, 269
point array, 269
unit-cell,
parameters, 284
spacings, 273
vapor diffusion, 277
velocities, 57
viscosity, 277
well-formed, 278
X-ray diffraction patterns, 273
Crystalline:
characteristics, 270
compounds with metastable forms, 268
morphology, 278
particles, 37
planes (Bragg's law), 254
plates (compensator), 37
pure form, required for crystallography, 268
state, distinguishing feature, 270
Crystallization, 642
Crystallography(phic), 29, 49, 268 (*see also* X-ray crystallography)
analyses, 278
axes, 270
critical analysis application, 268
single line isolation, 253
useful for identification,
in electron mode, 235
in X-ray mode, 235
Crystal rotation apparatus (spindle stages), 34
Crystal spectrometers, 255, 257
complexity of operation, 257
detection limits, 255
versus EDXS, 255
Crystal systems, 270
Crystobalite, 268
Cubic unit cell, 273
Cumulants method, 164
Current (dark), 253
Current Good Manufacturing Practice (cGMP), 638, 653
Cylindrical electromagnets, 246

Darkfield, 57 (*see also* Epi-darkfield and Reflected light)
on a polarizing microscope, 41
stages, 33
Data:
bases (computer), 269, 804
collection (computer-based), 799
compositional, 250
error, 791
fitting, 769
measured, 268

[Data]
morphologic, 250
outlier, 803
reduction (computer-based), 799
retrieving, 769
X-ray, 259
Deagglomeration, 601, 605-607, 619, 624, 630, 632
Delamination, 646
Deliquescent, 44
Demagnify, 248
Density differences, 259
Depolarizer (*see* Polarizing microscope)
Design (process), 639
Desorption:
by heating, 263
laser, 265
Detector array, 762
ring pattern, 765
spacing, 767
X pattern, 766
Detectors, suspension:
turbidity, 157
X-ray, 157
Detector specification parameters (light extinction):
effective scan beam height, 489
geometric detector volume, 489
Detergent molecules, 642
Device(s), 653, 792
elastomeric components, 654
manufacture, 655
Dewar vessel, 277
Dextrans, 642
Dextrose, 642
Diaphragm (*see* Polarizing microscope)
DIC (differential interference contrast), 34
Dichroic studies, 260
Differential absorption of light, 247
Differential geometry, arc length as natural parameter, 81
Diffracted light, 247
similarity to X-ray diffraction (Bragg's equation), 253
Diffraction (*see also* Scattering intensity):
analysis, 269
angle, 254
anomalous, 756
diffraction analysis (crystal structures), 269
effects, 247 (*see also* Light microscope, resolution)
electron patterns, 269
halo, 48
patterns, 247, 268, 269, 284, 285
planes, 255
ring, 283
to wavelength relationship (Bragg's law), 254
Diffuser (*see* Polarizing microscope)
Diffusion, 797
Diffusion constant, 13
Digitized image processing (*see also* Image and image analysis, Image processing):
data (discrete), 71
digital subtraction, 72
image, 70, 83, 98
image processor, 71, 81
picture elements, 70
process, 70, 71
resolution, 70
vertical and horizontal lines, 70
Diluent, 601, 603, 607-610, 616, 619, 624, 626-627, 632
handling, 608
Dilution:
detector parameter evaluation, 527-534
fluid, 611-614, 617, 620, 622-625, 627-632 (*see also* Diluent)
method of, 601, 608-609, 619, 632
model experiment, 523-525
on-line system for, 624-625, 627-629
ratio determination, 631
system calibration, 625
of used oil samples, 603, 607-610, 630-632
DIN standard, 33, 41
Diopter adjustment ring, 37
Disc:
scintillator, 248
storage, 63
Dispersion staining, 30, 31, 49, 844
optical particle counters, 18
Dissolution (selective), 269
Distilled water lines, 807, 808
Documentation of particle contamination problem solving, 650
DOP (*see* Hot dioctyl phthalate)
Drive screw, 264
Drug(s), 646
amino acid, 643
ampiphilic in nature, 642
micelles, 642
nonpolar (hydrophobic), 642, 643
peptide-containing, 644
polar (hydrophilic), 642
protein-containing, 643
raw material control, 642
Dry powder dosage forms, 644, 646
Duco cement, for particle mounting, 277
Dynamic light scattering (DLS), 13, 156 (*see also* Photon correlation spectroscopy)
Brownian motion, 162
particle size limitation, 162
quasi-elastic light scattering (QELS), 163
size range, 162, 165

Edge(s) (*see also* Reproducibility):
  agglomerated particles, 74
  boundaries, disjointed or incomplete, 74
  boundary growing, 74
  complex particles, 74
  convex perimeter, 75
  expansion/contraction routine, 74
  fiber, 74
  segments, 74
  sharp edge contrast, 69
  uncertainty, 67, 69
EDXS (*see* Electron dispersive X-ray spectroscopy)
EELS (*see* Electron energy loss spectrometry)
Effect of particulates (physiological):
  factors, 669
  granulomatous reactions, 675, 676
  inflammatory response, 675, 676, 677
  thrombus formation, 675, 679, 684
Elastomer, 640
Electrical sensing zone:
  aperture blockage, 123
  applications,
    blood components, general, 136
    cells and other organisms, 136
    parenteral liquids, 138
    particle contamination evaluation, 140
    powders and other solids, 142
    red blood cells, 135
  calibration, mass balance, 120
  calibration, monosize particles, 121
  coincidence, empirical correction, 126-128
  comparison with optical particle counters, 124
  electrolyte choice, 123
  hydrodynamic focusing, 129
  method, 113-154
  operation, 15
  particle resistivity effects, 117
  particle size range, 15, 122
  pulse particle measurement, 15
  standardization, 144
  viscosity adjustment for large/dense particle samples, 124
  volumetric response, 117
Electrodeposited thin foil particle mounting, 285
Electromagnetic lens(es), 244
  cylindrical, 246
  to demagnify, 248
  to focus, 248
  a powerful solenoid, 244
  weak, 245
Electron(s), 243, 244, 245, 246, 252
  acceleration across potential difference, 247
  atomic, 246
  Auger, 252
  backscattered, 250, 255
  beam, 246, 252, 257
[Electron(s)]
  bound, 253
  core, 252
  current, 252
  decelerate, 251
  dense, 247
  dry emissive specimen surface, 248
  emitting specimen points, 249
  energy, 251, 252
    levels, 255
    loss, 255
    signal proportional to energy of X-ray, 253
  ground state, 246
  impact the surface of the scintillator, 249
  inelastic collisions, 255
  inner shell excitation level, 287
  multiplier, 265
    secondary, 264
  penetration and speed, 250
  potential energy levels, 252
  primary beam (SEM), 248, 250
  radiation, 269
  removal (TEM), 247
  scattering, 247, 248
  secondary, 249, 250, 257, 265
  secondary signal, 248
  shells, 252
  SEM signal with simultaneous X-ray signals, 253
  split screen images, 253
  transition appears as an X-ray, 252
  trapped, 253
Electron beam, 243, 245, 252
  CRT (cathode ray tube), 249
  incident electron, 255
  intensity, 246
  microscope, 249
Electron diffraction, 270, 273, 279 (*see also* X-ray diffraction)
  analysis, special training required, 280
  Bragg's law application to lattice spacing of crystal, 281
  camera, standardization of length in, 281
  current applications in particle analysis, 285
  homogeneous populations, 279
  interplanar spacing of crystal lattice, 281
  magnification standardized with excitation voltage, 281
  microscope, focusing, 281
  patterns, 279, 280
  phenomena, 279
  scattering interaction, 279
  TEM under diffraction conditions, 280
Electron dispersive X-ray spectroscopy (EDXS), 257, 640, 644
Electron energy loss spectrometry (EELS), 236, 286, 287, 288

[Electron energy loss spectrometry (EELS)]
basic components, 288
collective excitations (plasmons), 286
electron energy spectrum, 255
electron gun, 288
energy regions, 286
excited states produced by energy loss, 286
illumination system, 288
inner shell loss edge, 287
interfaced to an electron microscope, 288
intra- and interband transitions, 286
magnetic deflection detector, 255
magnetic-sector energy analyzer, 288
minicomputer-based multichannel analyzer (MCA), 288
primary excitation, 286
selectively collects electrons, 255
serial detectors, 288
specimen chamber, 288
spectrometer, 288
vibrational excitations (phonons), 286
Electronic(s):
components (rinsing of), 808
manufacturing, 804
Electronic particle counting (EPC) method, 779 (*see also* Particle counters and Particle counter types) [For data analysis, *see* Jaenicke Geometric and Particle-Triggered Poisson model entries]
HIAC/Royco particle counter, 781
measurement deficiencies, 781,782
Electronic relaxation, 252
Electron illumination (TEM), 247
Electron microscope/microscopy, 61, 235, 246, 843
applications, 243
focusing with lens current, 246
image magnification, 245
and imaging electrons, 246
lenses, 246
operating components, 244
and optical microscopy, 244
resolution determining factors, 247
astigmatism, 246
chromatic aberration, 247
diffraction effects, 247
spherical aberration, a critical limit, 247
scanning type, 248
simple solenoid as weak electronic lens, 244
single particle isolation and mounting, 243
specimen image formation, 247
transmission type, 246
Electron-optical column, source, 248
Electron signal, 248
Electron source:
field emission gun, 248
gun, 246
[Electron source]
heated filament, 248
Electrophoresis, 709, 720
Electrostatic forces, 649, 796
double-layer forces, 853, 866
Elemental constituency, 260
Element light detection capability, 257
Elements, 29
identifying, 253, 256 (*see also* Line[s], element)
Elongation, 31
Emission:
emission intensity versus excitation, 252
intensity of X-ray emission, 251
produced in cold cathode field, 265
secondary pattern, 248
thermionic, 265
Emissivity, 248, 257
Emulsions, 140, 143 (*see also* Contaminants)
Energy:
dispersive spectra, 252
energy transfer process, 264
excitation, 257
melting point determination, 31
Energy dispersive X-ray spectrometry (EDXS), 243, 253, 257, 259:
advantages, 253
cost comparison with WDXS, 258
detection limit compared to crystal spectrometry, 255
electron diffraction analysis for small particles, 285
with FTIR to identify large majority of particles, 260
inorganic elemental analysis, 260
micro-FTIR, use of, 259
operated at liquid nitrogen temperatures, 253
solid state detectors, 253
versus crystal spectrometer, 255
versus X-ray crystallography, 278
wideband detection capability, 253
Engine, 585-587, 603-605, 611, 613, 614, 629-631
oil (*see* Lubricating oil)
Ensemble averaging (EA), 155
defined, 155
Fraunhofer diffraction (FD), 156
Bessel function, 166
Fraunhofer limit, 158
Mie scattering theory, 157
peak height measurement, 163
intensity detection limitation, 169
particle size limitation, 170
repeatability vs. accuracy, 170
scattering wave vector, 163
peak resolution, 155
"linewidth", 163
testing, 160

[Ensemble averaging (EA)]
sedimentation (SD), 156
techniques, 155
transmittance (T), 157
resolution, 155
Entrainment, 833-835
criteria, 833
Environment, 641, 791
Environmental:
particle control mechanisms, 792
particles, 792, 804
witness plates, 804
Epi-darkfield lighting, 31 (*see also* Top light)
Epifluorescence, 843, 845, 860
Epi-illumination (*see* Top light)
Epoxy, 44
Equipment, 30
contact with water and other solvents, 807
design and maintenance, 638
Equivalent circular diameter, 66, 81, 84
Erlenmeyer flask, 277
Erosion rate, 267
Errors, coefficient of variation (CV), 595, 599, 600, 616-618, 619-620, 627
sources of,
agglomeration, (*see* Deagglomeration)
bubbles, 615-616, 618
calibration, (*see* Calibration)
coincidence, (*see* Coincidence error)
contamination, 595, 601-603, 607-610, 618-619, 624, 632
dilution (*see* Dilution)
flow rate (*see* Flow rate)
human, 585, 587, 594, 598, 601, 604, 611, 614, 633
immiscible droplets, 595, 601, 615-616, 618, 630, 632
instrumentation, 585-588, 589-591, 594, 596-599, 600, 601, 607, 614, 625, 630, 633
interpretation, 585, 601, 611, 619-620, 625
mixing, (*see* Resuspension)
particle statistics, (*see* Statistics, particle)
resuspension, (*see* Resuspension)
sampling, (*see* Sampling)
settling, (*see* Resuspension)
volume measurement, 600, 601, 608-611, 617-618, 624, 626, 627-629, 632
Error sensitivity analysis, USP <788> SVI Assay:
calculations and derivations, 559-566
laser detector, 517-518
summary of results, 519-520
white-light detector, 518-519
Ethylene dichloride, 256
Evaporated metal atoms, 248
Evaporation, 265
Ewald sphere, 285
Excipients, 646
Excitation potential, 252
Exit pupil (*see* Eyepoint)
Extinction:
angle(s), 31, 34
coefficient (optical), 11, 19, 23, 157
efficiency, 598
Extraneous particles, 791-792
Eyepiece(s), microscope, 41–42
adjustable eye lenses, 37
Bertrand lens, 39
common, 38
crosshair, 37
magnification range, 38
magnified real image, 37
measuring, 33
micrometer, 42, 43, 51
most useful, 38
pinhole, 39
scales, 43
Eyepoint (exit pupil, Ramsden disc), 41
Eyes (*see* Vision), 37
Eye tube(s), 36–37, 41

Factors, shape, 61
False count(s):
cavitation bubbles and immiscible droplets, 460, 529
and undercount analysis, 510-511, 515-523
False reject rate (*see* RAG)
Far-field in-line holographic recording systems:
single-lens, 216
dual-lens, 217
Fast atom bombardment, 263
Fast Fourier transform, 74 (*see also* Fourier)
FDA vs. Bristol Laboratories, 1949 challenge, 305
Federal Standard (FS) 209D, 794-795, 800
Feret's diameters, 77, 90
orientation dependence, 77
parallel tangents, 77
sieving, 77
Fermi level energy, 287
Fiber(s), 74, 81, 85, 88, 90 (*see also* Edges and Particles)
and guard region, 85
straight, 82
Fiber (glass) spindle-stage needle, 31
Fiber-optic:
bundles, 32
illuminator, 31
Fiber optics, 760
Fiber optic sieve:
assemblies, 738-740
durability, 737
experimental results,
FDA trial, 743-744
test results, replicability, 743-744

Field:
camera, 70
lens, 70
Field-flow fractionation, 846, 848
Filament, 39
aperture diaphragm, 41
centering, 39
coils, 40
different location in new bulb, 42
focus, 41
illuminating ray path, 41
image, 248
lamp condenser controls focus of, 41
position adjustment, 39
three conjugate foci of, 41
Fillers (fill equipment), 647, 649
Films:
laminated, 641
production, 256, 259
transparent, 243
Filterability index, 8
Filtered:
distilled water, 805
water, 804, 813
Filters, physical, 68, 239, 643-644, 647, 648
average efficiencies, 797
cartridge, 807, 808
cast film, 237, 239
cellulose acetate, 87
cost, 87
Dacron,
blood, 658
isolation, 649
defect rate of, 88
depth filters,
polymeric, 725
silver, 725
evaluation,
gases, 834, 836
liquids, 838
green membrane filters, 87-88
HEPA, 649, 651, 835
high-surface-area capsule, 807
media shedding, 581, 839, 840
membrane, 84, 89, 96
membrane-type, 31, 43-44, 649
fragile nature, 237
material with adherent particles, 258
Nuclepore, 243
sample, 241
surface, 242
thinness, 237
paper, 31
particle, 84
and phthalates in a cellulose ester, 649
poly(diethyleneglycol) isophthalate (PDEGI), 649
[Filters, physical]
sintered plastic media, 654, 655
surface defects, 87, 96
testing, 585, 587, 593-594, 598-599, 601, 604, 607, 618-629, 633
track-etched or nucleation (random), 237, 243
blinding or blocking, 239
cylindrical pores, 239
difficulty in handling, 239
durability, 239
flexibility, 239
focusing and scattering effects, 260
lower fluid flow rates than cast film, 239
made by chemical etching, 239
primary use, 239
scanning electron microscope reveals structure, 237
with SEM, X-ray microprobes, 239
sieve like function, 239
static electric potential, 239
thinness, 239
versus cast film, 237
ULPA, 838
variable quality of, 87
vendors, 87
Filters, visual
background filter, 71
image sharpening, 72, 74
Laplacian, 72, 74
polarizing 31
prescreening, 88
quality of, 87
round, 85
Filtration, 43, 638
equipment cleaning, 811
flasks, 808
funnels, 811, 812
importance of, 647
for isolation of particles in liquid, 237
liquids, 841
for "particle-free" water and solvents, 807
pressure cans, vessels, 808
products, 807
solution, 812
terminal in-line, 646
Final image considerations, electron microscope, 247
Fine dust test, 719
Fingerprint region, 260
FL (*see* Fluorite)
Flash, 655
Flexible-collodion film thickness, 32
Floors, 805
Flow cytometry, (*see* Clinical diagnostic testing)
Flow rate, 587, 589-591, 594, 595, 600, 607, 610-611, 618, 619, 624, 626, 627

[Flow rate]
errors, 589-591, 607, 610-611, 618, 619, 624, 626, 627
limit, 587, 589-591, 594, 600, 627
Flow passage:
boundary wall interference effects, 175
clogging, 175
Fluids (clean), 807
Fluorescence, 259
Fluorescent screen, 246, 247
Fluorite(s) (FL), 36, 42
Fluotar, 36 (*see also* Fluorite)
Flush solutions, 791
Flux lines, 244
Fly ash, 38
Focal:
circle, 255
length, 43
length determined by lens coil current, 246
plane, 37
point, 265
Focus:
electron microscope, 246
and power density (irradiance), 264
secondary electrons, 249
filament, 41
second conjugate, 39
and slide changes, 42
on a specimen, 39
Focusing capability (microscope), 41
Focusing eye lens, 33
Food and Drug Administration (FDA), 779, 814, 820
Center for Antibiotics Analysis, 781
Forceps, 239, 256, 812
particle sample collection, 237
Form factor, 753
Form-fill-seal, 641, 652
Formvar, 242, 256, 259
Forward scattering (*see* Diffraction)
Fourier, 29
digital image analysis, 74
equation, 92
frequencies, 74
image formation, 74
lens, 167
from spatial to frequency domain, 74
transform, 74, 99
Fourier transform infrared spectrometer, 30
Fourier transform infrared spectrophotometer (FTIR), 243, 259 (*see also* Micro-Fourier)
complements the use of SEM with EDXS, 259
with EDXS to identify large majority of particles, 260
infrared absorption spectra, 260
interference signals, 259
[Fourier transform infrared spectrophotometer (FTIR)]
interferometric detection avoids overheating, 260
spectra with sharply resolved absorbances, 243
versus X-ray crystallography, 278
Fourier transform infrared spectroscopy (FTIR), 640
Fractal analysis, 94
Fractionation, 709
Frame averaging, 72
Fraunhofer, 47–48
diffraction, 12, 166, (*see also* Scattering intensity)
analysis, 12
condition, 200
pattern, 200, 201, 203
resolution, 168
size distributions, 168
size range, 166
hologram, 230
limitations, 754
Fresnel diffraction, 199, 200, 203, 230, 754
FTIR (*see* Fourier transform infrared spectroscopy)
Functional groups, 29
Fusion methods, 57

Gamma radiation, 654
Garments as particle sources, 652, 657, 658
Garvin, J. M., 661
Gas bubble(s):
false particle counts, 529
removal, 615
and ultrasound, 806
Gas chromatography (GC), 807
GC (*see* Gas chromatography)
Gelatinous particles, 643
Gelman, 807, 811
Geologist (petroleum), 37
Geometry, 800
Geometry X-ray collection, 259
Glare, 39, 42
Glass, 646
fiber, 31
flakes, 32
ground, 39
lens, 245, 246
microspheres, 265
particles, 816
slides, 242, 243
types (I and II), 641, 646
vials, 644
Glassware, 805
cleaning of, 815
dedicated, 806
rinse laboratory, 808
Gloves, 647, 811, 817, 819
particle-free, 237

Glycerin, 36
GLP (*see* Good laboratory practice)
GMP (*see* Good manufacturing practice)
Gold, 248, 257
Good laboratory practice (GLP), 793, 820
Good manufacturing practice (GMP), 295, 638, 658
  validation prerequisites, 295
Gowning anteroom, 822
Grain size, 42
Granule formation, 826
Graticule(s), 33-34, 37, 42 (*see also* Measuring scale)
  microscope eyepiece, 37
Gray-scale (gray level or threshold) 62-63, 71 (*see also* Image processing)
  brightness (*see* Contrast)
  contrast enhancement of, 71
  histogram, 72
  impacts measurement, 71
  impacts particle detection, 71
  level value, 73
  manipulation, 71
  nontransparent medium, 71
  resolution, 71
  threshold, 71, 96
  tonal range, 71
Greenough type of objective, 33
Grid (prefilmed), 256
Ground glass, 39
Ground state, 246
Gun:
  chamber assemblies, 248
  electron source, 246
  primary ions for SIMS analysis, 265
Gunner, B. W., 661
Gun-type solvent dispensers, 808, 810

Handle, polarized light module, 37
Handling of particles for analysis, contamination and loss risk, 241
Haze, in liquid products, due to stopper extractables, 645
Head (monocular, binocular, or trinocular), 37
Helical path (spiral), 244
Helix, 244
HEPA filters (*see* High efficiency particulate air)
HIAC/Royco models, 138-141, 500-502
High efficiency particulate air (HEPA) filters, 237, 239, 649, 651, 656, 657
  ceiling airflow, 821
  control of the environmental air particle burden in the laboratory, 822
  critical functions in the clean laboratory, 793
  description of construction, 796
  fibrous, 797
  filtered hood, 805, 818
[High efficiency particulate air (HEPA) filters]
  filtered laminar flow hood, 804
  filter protection, 805
  filters/filtration, 792, 797-798, 804
  hood, 810
  integrity of, 815
  laminar flow enclosure, 810
  minimum efficiency, 797
  particle control mechanisms, 795-796
  pinhole leaks, 797
  protected hood, 804
  provides reduced particle burden areas, 819
High performance liquid chromatography (HPLC), 807
High-voltage scanning microscopes, 257, 258
Hologram:
  ampule, particle contamination detection, 223, 224
  analysis, 226-230
  data on particle size versus detectability, 409-417
  examples, 205,220
  formation, 203
  field of view, 213
  fringe effect, 209
  recording, 204
  re-illumination, 204
  resolution,
    particle cross section, 213
    particle position, 214
Hologram imaging system, 218
  image analysis, 226
  velocimetry, 226
Holographic system design, 209
  droplet size distributions, 222
  field of view, 213
  fringe visibility, 209
  particle measurement, 216-218
  reference beams, 212
  relay optics, 213
  resolution, 213
Holography, general, 197
  exposure time limits, 198
  three-dimensional particle imaging, 198
Holography, in-line far-field, 203 (*see also* Fraunhofer hologram)
Home pregnancy tests, 693, 696
Homogeneous particle isolation prior to analysis, 237
Hood, laminar flow, 239, 651, 652, 731, 792, 805, 810
Hot dioctyl phthalate (DOP): 797
Hotstage, 48, 57
HPLC (*see* High performance liquid chromatography)
Human:
  activity, 792
  analyst, 66
    and microscope, 66
    results from, 97

[Human]
contribution to particle burden, 793
eye, 246 (*see also* Particle detection accuracy)
detection versus image analysis systems, 71
minimum detectable contrast limit, 71, 321
inspection performance benchmark (nondestructive testing), 295
FDA vs. Bristol Laboratories 1949 challenge, 305
and RCA automated inspection system, 305
Huygen's Principle, 754
Hydraulic, 585, 587, 595, 604, 607, 614, 618, 621-622, 624, 630, 633
fluid, MIL-H-5606, 607
fluids, 140
shock effect, 581
system wear, 585
5-Hydroxymethyl furfural, 458, 643

IA systems (*see* Image and image analysis, Image processing)
ICP (*see* Inductively coupled plasma)
I.E.S. lighting recommendations, 305, 325
Illumination:
aperture, 42
brightfield episcopic, 76
condenser, illuminating cone, 42
correct, 39
darkfield episcopic, 76
episcopic, 75, 76
even illumination on the retina, 41 (*see also* Köhler illumination)
fiber optic, 76
filament, 41
foot switches, 32
incident, 263
Köhler method, 39
lateral reflected, 76, 88
microscope, 39
ray path, 41
reflected, 75, 88
systems, 41
in TEM (248)
transmitted, 75
Image(s) and image analysis:
accuracy, 71
analog (visual), 62
analysis in morphology, 76
background (filter or microslide), 62
center, 39
classic, 92
color, 62, 63
defined in spatial resolution term, 71
digital subtraction, 72
digitized, 62-63, 69-71
discrete, 62
dot map, 253
[Image(s) and image analysis]
electronic, 63
element, smallest, 70
enhancement (*see* Gray scale, histogram)
features of interest in, 62
filament, 248
final, 247
fineness (*see* accuracy)
formation (SEM), 248
gray level (*see* Gray scale)
illuminating radiation for specimen, 247
image plane, 245
instrument,
programming, 95
purchase, 95, 101
vendor support, 95
light optical, 63
memory, 71, 95
modern, 77, 81
plane, 41
produced on the screen of TEM, 248
quality, 42
reduced contrast in, 39
secondary electron, 265
secondary ion mass spectrometer (SIMS), 265
segmentation, 63, 72
shading, 71
shape analysis applications, 95, 99
calculus, 92, 98
harmonic functions, 92
infinite series, 92
moment functions, 92
statistics 92
symmetry operations, 92
small particle location, 266
split screen for composition and morphology analysis, 253
spot array, 249
three-dimensional, 31
using high speed computers, 76
video,
camera, 66
image, 63, 248, 249
visual (*see* analog)
Image processing and Image analysis systems:
considerations for purchase of image analysis systems, 95-96
contrast enhancement, 63
convolution,
mask or kernel, 72
processing of image, 72-74
digital, 74
digital filtering, 63
expansion-contraction (erosion-dilation), 74
filtering process,
independent of input pixel, 72

[Image processing and Image analysis systems:
considerations for purchase of image analysis systems]
output pixel proportional to input pixel, 72
spatial, 74
Fourier technique, 74
frame averaging (noise suppression), 71-72
histogram,
sliding, 72
stretching, 72
image processing for contrast enhancement, 72
instrument,
point resolution, 66
precision, 66
measurement errors using, 67 (*see also* Measurement routines)
miniature head, color camera, 101
oblique extinction and the engagement lever, 35
operating parameters, 89
optical characterization by microscopy, 29
optimization and aperture diaphragm adjustment, 42
particle,
calipering, 83
counting, 76
populations, 76
shape analysis, 98
size measurement with chordal arrays, 82, 83
point processing, 72
routines, 74
shading, shadow correction, 71, 72
from spatial to frequency domain, 74
speed, 67
storage, 97
tied to gray scale resolution, 71
versus manual method, 62, 90
Imaging:
beam, 249
electrons, 246
lensless, scanning microscope, 248
mode (secondary electron), 248
picture points, 249
Immersion processors (*see* Wet sinks)
Immiscible liquid droplet counts, 615
Impaction, 796
Improved membrane/microscopic assay, 786
Impurities, 643
Index/indices:
contrast, 572, 575
liquid, 44
of unknown crystals, 37
refractive, 37, 42, 57
Inductively coupled plasma, 807
Inflammatory response to particulates:
bradykinin response, 675, 676
macrophage, 675, 676, 677
Infrared (IR), 260, 261
Infrared microspectrophotometry, 259
Infrared microspectroscopy, 242
molecule with strong polar bond vibrations, 261
Raman complementary use, 261
salt plates, 242
spectral data for particle identification, 259
spectrophotometer, 259
spectroscopy, 235
use in pharmaceuticals, aerospace, and microelectronics, 259
Infrared spectrometers, 792
Infrared spectroscopy, 649
Infusion phlebitis:
characteristics, 682
incidence, 661, 683
in-line filtration, 683
Insecticide properties and particle size, 3
In situ:
counters, 20
measurement application areas, 177
particle measurements, 132
sensor, 175
calibration, 177
gas/liquid flow effects, 176
sensor volume, 176
masking volume, 181
particle size effect, 176
polarized light system, 182
Insolubles, 639
Inspection booth (*see* Inspection station):
Inspection flow graphs:
automated,
"accept in three" (A3-A/3) inspection, 337
"reject in two" (A2-R/2), 372
manual,
standard and cull inspection (M2-A/2), 304
standard single inspection (M1-A/1), 303
Inspection methodology, 296
"eye-breaks", 305
histogrammic analysis, 311-317
performance validation, 295
probabilistic methodology, 296
procedures, 651
standard and cull inspection sequence, 304
standard single inspection, 303
statistical replicability, 296
training period, 305
Venn diagram analysis, 306
Working Standard Test Method, 296
Inspection Observation, Adverse (a 483), 815
Inspection replication for validation demonstrations (*see also* Multiple sequential inspections, Serial accept and Serial reject inspection systems):
manual inspection minimization, 349-352, 368-370
mortality table comparison, 309

[Inspection replication for validation demonstrations]
multiple inspections, requirements and results, 346, 347
requirements, 347, 348
RZNI Table description 349-352, 359
RZNI Tables, 355-367
statistical basis, 308, 348
validation demonstration design, 387-398
Inspection security:
Reject Zone Efficiency, 316, 317
validation results with 0.05 significance level, 295, 348, 389-398
container clarity effect, 309-310
different probabilistic inspection techniques, 337-345
extension to serial reject systems, 345, 346
GMP requirements, 349, 350
practical considerations, 438-442
practical procedure, 422-438
Inspection station:
design considerations, 297
I.E.S. lighting recommendations, 305, 325
lighting,
advances, 304
flicker effect, 303, 304
nondestructive inspection requirements, 303
light source, 303-304
pacer, 304
training period, 305
Instrumental:
analysis, 241, 242
counting, 795
Instrumentation and particle identification, 649
analysis area cleanliness requirements, 795
Insulators, 250
Intensity:
cut-off, 251
modulation, 249
peak, 251
Intensity transform function (ITF):
compress gray-scale values, 72
to enhance histogram, 72
two level ITF, 72
Interatomic spacing, 283
Interception, 797
Interference, 754
Interference figures, 31
observing, 39
Interferometer, 572, 573
International Standards Organization (ISO), 592-594, 597, 599-600, 603, 611, 620, 624, 630, 633
ISO 4402, 592-594, 597, 599, 600-611, 620, 625
Interparticle forces, 854
Interpretation of data, 236
Interpupillary distance, 43
Intravenous therapy, 646
Ion(s):
analysis, 265
effective for small particles, 268
bombardment, 265, 268
combinational effects with primary beam, 267
common mounting methods for surface analysis barred, 268
detector, 264
high-energy, 265
microprobe, 29
microscope (*see* Secondary ion mass spectrometry [SIMS])
most sensitive surface analytical method available, 268
mounting technique, 268
neutralized by collision with electrons, 267
plasma cloud reactions, 267
primary, 265
reflector, 264
surface chemical composition of particles, 268
Ionization, 264
chemical, 263
uncontrolled, 265
Ionizing grids, 656
Ion microprobe, 29
IPA (*see* Isopropanol)
Irradiance/irradiation, 259, 264
Irradiated particle sample, 264
ISO (*see* International Standards Organization)
Isoaxial sampling, 185
ISO fine test dust, 593
Isokinetic sampling, 184
flow rate effects, 179
particle size limit, 185
Isomers, optically active, 239
Isopropanol (IPA), 798, 805, 807, 817

Jackson:
candle turbidimeter, 12
turbidity units (JTU), 4, 12
Jaenicke Geometric Poisson model, 476-477
count error calculation, 464, 468, 469, 476, 538-540
count linearity versus detector volume, 476-477
derivation, 538-540
detector volume versus linear counting span, 476, 477
maximum raw counts, 476-477, 538-540
monotonic counting response, 476-477
with USP XXIII <788> coincidence count specification, 513, 549-551
Japan (ultrasonic treatment), 652
JEOL, 290
Jet, filtered solvent dispenser, 805
Jet, pressurized cleaning with ultra-clean water, 731, 732

Juliet III image analyzer system:
  boundary function, 98
  calculus, 98

Kamack equation, 160
  radial concentration, 160
  radial dilution, 160
Kevex, 257
Kevlar fibers, 47
Killer defects, 569
Kinetics, 645
Knurled ring, 38–39
Köhler, August, 39
Köhler illumination:
  contrast reduction problem, 39
  difficulties, 39
  filament and specimen image separation, 41
  retinal illumination, 41
  set up, 39, 42, 58
Kratel models, 138-141

Laboratory:
  analyses, 792
  constant temperature, 48
  control of environmental particles in, 793
  OPC data and new technology, 803
  particle identification and measurement, 29
  particles from surrounding areas, 793
  pure water, 807
  work stations, 793
Laminar flow, 237, 651, 652, 731, 792, 805, 810
LAMMA (*see* Laser microprobe mass analyzer)
Laplace transform deconvolution, 164
Large volume injection (LVI), 638, 642, 811
  blow-molded units, 641
  fill process, 647
  Japan, 652
Laser, 746, 760
  argon ion, 761
  beam, 263
  desorption, 265
  diode, 761
  hard-seal, 800
  helium-neon, 761
  Nd-YAG, 263
  particle counters, 799, 800
  pilot, 263
  pulsed YAG, 264
  red target (He-Ne), 264
Laser microprobe mass analyzer (LAMMA):
  for chemical nature of particles, 265
  delimitation, 263
  dependence on sample preparation/geometry of sample laser beam, 265
  detection of trace contaminants in single particles, 265
[Laser microprobe mass analyzer (LAMMA)]
  disadvantages of, 265
  fast data acquisition, 265
  objectives/diffracting elements from UV transparent material, 263
  optical window, 263
  peak deformations from time and energy spreads of ions, 265
  quartz slide, 263
  redisposition of material, 265
  saturation effects, 265
  screw drive focusing, 264
  selective or partial evaporation, 265
  size range amenable to, 263
  time-of-flight spectrometer, poor mass resolution, 263-265
  uncontrolled ionization or carbonization, 265
Latex particles, and particle counter size calibration, 18
Leica Quantimet image analyzer system, 99, 101
  color image processing, 99, 101
  gray-tone image processor, 99
  high speed image processing, 99
  logic array, 99
  morphologic image,
    analysis, 99
    mathematical transformation, 99
  motorized stages, 99
  software, 99
  Windows user-interface, 99
Leitz, 36
Lens(es) (*see also* Eye lens):
  Abbe-type, 33
  accelerating, 264
  biconvex glass, 245
  condenser lens, 245, 246
  current, 245, 246
  cylindrical electromagnets in electron microscope, 246
  electromagnetic, 248
  electrostatic, 264
  eye, 41
  field, 245
  focusing, 264
  focusing achieved with the objective lens, 246
  focusing eye, 33
  front, 33
  objective (second or third) lens, 248
  relay, 66
  resolution, 70
  supplementary, 33, 37
  system, 247
  system (TEM), 247
Lens coil, 246, 248
  axial extension, 248
Light:
  absorption, 247

[Light]
diffracted, 247
exit port, 39
incident, 263
ordinary, 44
path, 37
rays, 37
reflected, 68, 248
transmitted, 68, 263
undiffracted, 247
yellow-green, 246
Light blockage, 19, 23 (*see also* Particle counters and Particle counter types)
[For data analysis, *see* Jaenicke Geometric and Particle-Triggered Poisson model entries]
method, 138-141
Light extinction (*see also* Particle counters and Particle counter types):
[For data analysis, *see* Jaenicke Geometric and Particle-Triggered Poisson model entries]
counter historical development, 21
counting, 644
clean area for, 822
of solution products, 792
and sonication, 806
of an SVI, 813
and USP XXII, 814
per USP <788>, 815
versus visual inspection, 650
particle counters, 18, 451-568, 643
collection angle effect on response, 189
count accuracy, single particle, 520, 546-548
illumination sources, 187
particle size calibration, 191
refractive index effects, 187
Light flux, 762
Lighting (stereomicroscope), 31
Light microscope/microscopy, 61, 71, 843
brightness resolution, 71
contrast effects and differential absorption of light, 247
factors determining resolution, 247
astigmatism, 246
chromatic aberration, 247
diffraction effects, 247
spherical aberration, 247
maximum chord, mean measurement of, 74-75
particle location for SEM analysis, 259
point-to-point resolution, 247
polarization in particle identification, 31
preliminary step in particle identification, 236, 243
resolution testing, 75
specimen illumination, 245
Light obscuration particle counter (*see* Particle counters and Particle counter types)
Light optical microscopy, 649
evaluation of witness plates, 804
sizing and counting, 804
Light scattering, 709, 749
angular pattern, 752
Brillouin, 746
collaborative study, 773
devices, 797, 799
diffusive wave, 746
elastic, 746
electric field, 746
electrophoretic, 746
Fraunhofer, 746, 754
geometry, 749
laser, 804
method, 138-141
Mie, 746
multiple, 747, 748
optical particle counters, 19 (*see also* Particle counters and Particle counter types)
[For data analysis, *see* Jaenicke Geometric and Particle-Triggered Poisson model entries]
high sensitivity, 189
particle size response oscillation, 189
phase Doppler, 746
photon correlation spectroscopy, 746
photon migration measurement, 746
quasi-elastic, 746
Raman, 746
Rayleigh, 746, 752
surface Raman, 746
transient, 746
variables, 748
Light sources:
concave mirror, 41
inhomogeneous (coiled filament lamp), 39
modifications, 41
separate, 41
Line(s) (element):
characteristic, 257
confirming, 257
identifications, 257
principal, 257, 263
second, 257
spectrum, 256
variation greater than 10% from published value, 257
Lintless tissue, 44
Liquid-borne particle measurement, 846-848
Liquid filter:
evaluation, 838-841
testing, 838
Liquid filtration theory, 841-843
Liquids, 44, 235 (*see also* Particulate matter, suspended in liquid)
Low magnification stereomicroscopy (LMS as first step for SEM or TEM), 244

Low power objective, 38
Low power stereomicroscope, 29-31 (*see also* Stereomicroscope)
  lighting, 32
Low-voltage scanning electron microscopy and accelerating voltages, 257
Lubricants, 646
Lubricating oil, 585-587, 592, 604-607, 609, 611, 614, 615, 618, 621, 630-632, 633
Luerkens' theory of morphology:
  geometric capabilities for powders, 98
  grand radial distribution, 98
  microroughness, 98
  moments of radial distribution, 98
  partial symmetries, 98
LVI (*see* Large volume injection)
Lyophilized (lyo):
  formulations, 643
  materials, 644
  parenterals, 640
  powders, 644
  stoppers, 641

McCrone, 57
Machineability, 652
Macromolecule, 754
Magnetic field (external, internal), 244, 248
Magnetic flux lines, 244
Magnification (*see also* Objective[s]):
  common main, 33
  eyepieces range in, 38
  Greenough, 33
  highest, 38
  objective, 35
  stepped (higher image quality), 33
  true magnification, 35
  zoom (most versatile), 33
Makeup air, 793
Maltese cross figure, 55-56
Manual inspection for visible particles:
  Accept/Reject decision, 296
    discrimination, 318
  ambient light, 305
  contamination quality regions,
    rule-of-thumb, 314
    statistically defined, 311-314
  discrimination, 318
  effect of container clarity, 309, 310
  "eye-breaks", 305
  GMP requirements, 295
  histogrammic analysis, 311-319
  human inspection standard, 295, 302
  inspection security basis for validation, 295
  inspection strategy effect, 376-387
  inspector selection criteria, 326-333
    durability of selection criteria, 333-336
[Manual inspection for visible particles]
    dynamic selection criteria, 332
    training period, 305
  probabilistic methodology, 296
  RAG, 318
  reject rate analysis, 310-314
  rejects compared to census mortality estimates, 308, 309
  RZE, 316, 317
  statistical replicability, 296
  training period, 305
  USP XXIII requirements, 295
  Venn diagram inspection results analysis, 306
  Working Standard Test Method, 296
Manual microscopic method, 88, 90
Manufacturing: 39
  of generic antibiotics, 645
  improved process, 642
  of powder dosage forms, 644
Martin's diameter, 78, 90
Mass:
  to charge ratios, 267
  particle, 247, 253
  peak, 268
  resolution (LAMMA), 265
Mass spectra from particles, 263
Mass spectrometer(try) (MS), 29, 263, 643, 807
  laser microprobe mass analyzer (LAMMA), 263-265
  optical window, 263
  quadrupole, 265
  secondary ion obtained by adjusting, 265
Matrices (transparent), 31
Matrix, 31, 70
  conversion, 769, 771
  convolution image filtering, 72-74
  maximum field width, 70
  scan grid, 85
  square field, 85
Matter (suspended), 42
Maximum linear dimension, 66, 81
Mean size, 746, 774
Measuring:
  large optic axial angles in interference, 36
  scale, 37, 42 (*see also* Graticule)
Mechanical alignment of microscope, 42
  objective, 39
  parts, 38
  stage, 39
Mechanical stage, 33-34
Mechanical tube length, 36, 43
Medical device:
  assembly, process design, 657
  design, 658
  execution, process design, 642
Medical grade tubing, 654

Medium, 42
Meeker burner, 44
Megasonic cleaning, 830, 834
Membrane(s):
  analytical, 812
  filter examination, 14
Membrane filters, 848-851 (*see also* Filters, membrane)
  asymmetric, 850
  cast film types, 237
  particle collection procedure, 14
  performance rating, 849
  pore size, 850
  track-etch types, 239
Membrane/microscopic (M/M) procedure, 779
Mercaptobenzothiazole (MBT), 640, 645
Mercury-cadmium-tellurium (MCT) detector and noise, 260
Metal atoms:
  evaporated, 248
  gold, 248, 257
  palladium, 248, 257
  platinum, 248
Metallurgical (*see* Reflected-light microscope)
Metal-organic complex, 645
Metals, 250
Met One models, 138-141
Mica flakes, 90
Micelle(s), 642, 643
  reverse, 616
Michel-Lèvy chart, 50-54
Microanalytical instruments, sensitivity limits of, 243
Microchemical testing, 31, 649
Microchemistry, 29, 56
Microcrystallites (NaCl), 265
Microfiltration unit, 808
Micro-Fourier transform infrared spectrometer, 29
Micro-Fourier transform infrared spectrophotometer (FTIR), 243:
  application, 260
  micro-Raman, 260
  revolutionized chemical analysis of particulate matter, 259
  signal/noise capability, 260
  spectra with sharply resolved absorbances, 243
  usable spectra for subnanogram particles, 260
Microinfrared techniques, 235
Micromanipulation, 237
Micrometer, 43
Micrometry, 42
Micropipettes:
  collodion, 242, 243
  Formvar, 242
Microporous membrane filters:
  and hydraulic shocks, 581, 582
  Log Reduction Value, 580
  particle shedding, 581, 582
[Microporous membrane filters]
  pore size rating, 580, 581
  and semiconductor process chemicals, 576, 577, 579-582
Microprobe(s), 29, 237, 792:
  analysts and single particles, 243
  current-generation multiple spectrometer, 257
  ion advantages for surface and particle analysis, 267
  system resolution capability, 266
Micro-Raman, 29
Micro-Raman spectroscopy, 260 (*see also* Raman)
  complementary to IR, 261
  elastic scattering by sample molecules, 260
  laser beam, source for, 261
  nonpolar bonds and molecular vibrations, 261
  qualitative localization of organic, inorganic, biological samples, 263
  Raman line analysis in monochannel mode, 263
Microsampling accessory, 260
Microscope(s), 29, 793
  alignment, 38
  base, 39
  biological condensers, 34
  compound, 242
  condenser, controls angle of illuminating cone, 42
  critical setting, 42
  design, 38
  diffuser hot spot, 40
  illumination, 29, 39
  illumination, general, 32
    Kohler, 39
    Rheinberg, 57
  illuminator(s):
    above objective, 36
    adjustable, 32
    separate, 32
  infrared, in transmission mode, 260
  lamp, 41
  lens, supplementary, 33, 37
  machine shop modifications, for full centering and focusing capability, 41
  mechanically aligned, 42
  mechanical tube length, 36, 43
  neither nosepiece nor stage centerable, 38
  numerical aperture, 35
  objectives, 33, 36
    centerable in the nosepiece, 38
  polarizing, 31, 34, 57
  reflecting, 260 (*see also* Cassegranian optical)
  selection for particle identification, 31
  set-up, alignment, illumination, and use, 30
  slide, 31, 34, 242, 263
  stage refinements, 35
  true polarizing, 31
  turret, 260

Microscopic:
  evaluation of witness plates, 804
  image analysis, 86, 88
    automated focusing, 87, 97
    automated stage drive, 87, 97
    color systems, 62
    and computer data processing, 88
    manual, 88
    mechanical scanning, 86, 87
    operational factors in, 62
    parameters, 88
    for particle identification, 62
    speed, 88, 90
  measurement method, 14
  particle analysis:
    resolution, 247
    used as an initial step, 235
    versus instrumental (nonmicroscopic) method, 235
  particle counting, 792, 795, 805, 812 (*see also* Coincidence error)
  scale, 263
Microscopist:
  basic tools, 31
  experienced, 29
  and flexible-collodion film, 32
  and new technology, 29
Microscopy, 709, 711, 719, 843-845 (*see also* Microscopic image analysis)
  light, 247
  polarized light, 29-59
  scanning electron, 248
  ultraviolet (UV), 263
Microspatula, 237
Microsphere(s), 687, 711, 713, 715
  calibrators, 687
  characterization, 688
  cleaning, 687, 688
    colloidal stability, 688
    methods, 688
    surface surfactant, 687,688
    water requirements, 688
  clinical diagnostics, 687
    agglutination tests, 687, 689, 691
      detection requirements, 690
      test sensitivity, 690
    antigen/antibody testing, 687
    capture ELISA (enzyme-linked immunosorbent assay), 694, 687
    immunoassays, 687
    single microsphere assays, 705
    solid-phase immunoassays, 687
  coating,
    covalent coupling, 689
    simple adsorption, 689
  critical coagulation concentration, 688
[Microsphere(s)]
  diagnostic applications, 687
  dyed microsphere
    Clearblue Easy (pregnancy test), 696
    markers, 703
    sandwich test, 696, 700, 701
    stains, 703
  electrokinetics, 688
  flow cytometry, (*see* clinical diagnostics)
  glass, 265
  home pregnancy tests, 693, 696
  latex particles for size calibration, 687
  particle concentration fluorescence immunoassay, 699
  sensitivity, 690
  silica, 687, 688
  size range (six orders of magnitude), 687, 688
  solid phase assays (bound/unbound separation), 698
  superparamagnetic, 687, 688, 699
    assays, 700-702
  surface chemistry, 687, 688
  tests,
    active tests, 691
    after agglutination, 693
    clinical agglutination tests, 689
    instrumental evaluation, 692,693
    qualitative agglutination, 689, 695
    quantitative assays, 691, 695
    strip (chromatographic), 696
Mie scattering, generalized, 756 (*see* Scattering intensity)
Miller indices, 270, 273
Millipore:
  filter jet solvent dispenser, 809
  (Milli-Q) system, 807
  stainless support screen, 811
Mirror (concave for filament image reflection), 41
Mixing technology, 867 (*see also* Resuspension)
Mold, 42
Molecule/molecular:
  polarizability, 260
  short lived fragments, 267
  structural data, 236
  symmetrical, 260
Monitoring:
  airborne counts, 795
  of large particles, 804
  remote, 804
Monsanto, 44
Morphology/morphologic, 239, 253, 792
  analyses, 61-62, 70, 92, 98, 858
  boundary, 95
  crystalline, 278
  data, 236
  high speed computers, 76

[Morphology/morphologic]
irregular boundary function, 92
mathematical descriptors, 76 ,99
microroughness, 62
rotational symmetries, 62
roughness, 62, 95
shape, 62, 90 (*see also* Analysis, shape)
size, 62
unique, 259
Motionless mixers, 867
Mountant:
adhesive material, 256
amyl acetate, 242, 243
collodion, 242, 243
film (transparent), 243
Formvar plastic, 242, 243
plastic tape (two-sided), 256
Moving particle (wavelength), 247
MS (*see* Mass spectrometry)
Multiphase flow, 867-868
Multiple inspection strategy for containers (*see also* Inspection replication, Serial accept and Serial reject inspection systems):
link to inspection security and false reject rate, 346
relationship of automated, manual and total validation inspections required, 350-351

National Fluid Power Association (NFPA), 589, 592, 594, 599-600, 620
ANSI/(NFPA) T2.9.6R1, 594-600, 620
National Institute of Standards and Technology (NIST), 595, 598, 599, 600
(*see also* Standard reference materials)
Needle and needle probes:
coated tip, 242
dissecting holder, 43
fabrication, 44
finely pointed tungsten, 241, 242, 243
fine needle (tungsten wire), 43
glass fiber spindle-stage, 31
micro, 260
noncoring, 792, 814
particle nucleation site, 642
pencil point, 816
probe sizes, 242
resharpening by chemical etching, 242
sampling, 641
tungsten, 241, 242, 243, 257, 259
tungsten (red hot), 44
Neofluar (*see* Fluorite)
Nephelometric turbidity units (NTU), 11, 12
Newtonian (color series), 50
NFPA (*see* National Fluid Power Association)
Nicol(s):
crossed Nicols, 31
prisms, 29
Nikon microscope, 35, 43, 48, 88
Optiphot-Pol and II Pol, 37 (*see also* Bertrand)
photodocumentation caution, 37
NIST (*see* National Institute of Standards and Technology)
Noise:
electronic (*see* Noise level)
level, 588-589, 594, 595, 596, 599, 614, 626
reduced by frame averaging, 72
"snow" appearance, 72
summing or averaging, 72
thermal, 253
Nomarski differential interference contrast (DIC) microscopes, 34, 58
Nomograph, 56
Noncentered condition, 40
Nonconducting materials, 257
Noncrystalline (amorphous), 643
Nonenvironmental background particulate matter sources, 804, 805
Nonintrusive measurement, 186
Nonmorphologic analysis, 71
Nonpolarized transmitted light microscopy magnification limits, 235
Nonvolatile residue, 22
measurement, 20, 860
Noran 8502 image analyzer system (Tracor), 75, 83, 97
Noran Voyager image analyzer system (Tracor),
hardware, 100, 101
SEM, TEM, 101
SPARC workstation, 101
Nosepiece(s), 35–36, 38
Nuclear magnetic resonance (NMR), 643
Nucleation:
centers, 642
reactions, 645
sites, 642
Nuclepore, 237, 243
membranes, 278
Number crunching, 74
Numerical aperture, 33-35 (*see also* Aperture)
controlled by aperture diaphragm, 42
Nylaflo, 807
Nyquist criterion, 69

Objective(s), 32
aberration, 36
achromatic, 36
angular aperture, 35
annular stops, 36
aperture, 35
apochromats (APO), 36, 42
back focal field, 40
back focal plane, 37, 39, 41
biomedical-type not suitable, 35
centering, 41

[Objective(s)]
relative to stage rotation, 38
central stops, 36
correction, 36
cover-glass thickness, 36
diameter of the front lens, 35
filament image, 41
fixed, 38
fluorite, 36, 42
focal length, 35, 43
front lens determines focal length, magnification, and working distance, 35
German manufacture, 36
for highest accuracy (true magnification), 35
highest magnification, 39
irises, 36
knurled rings surrounding the, 38
magnification,
engraved on are nominal, 35
lowest magnification, 35
low power, 38
square root of two series, 35
traditional, 35
manufacturers of, 35
mechanically aligned, 39
mechanical tube length, 36
metallurgical, 36
most important features of, 35
no cover glass (NCG), 36
numerical aperture, 35, 36
o.D. (ohne Decke – objective magnification without cover glass), 36
particle identification work, 36
phase plates, 36
for polarizing microscopes, 36
prisms, 36
radius of curvature, 35
recenter, 38
single carriers, 35
stage rotation relative to, 38
strain free, 36
and substage condenser, 41
true magnification with stage micrometer, 35
two series of magnification, 35
variable magnification, 35
working distance, 35
Objective design (in stereomicroscope), 33
Objective immersion, 263
Objective lens, 246
in electron microscope, 247
spherical aberrations, 247
focusing in light microscope, 246
initial magnified image in TEM, 247
magnetic field, 248
resolution,
achromatic, 67-68
[Objective lens]
apochromatic, 67-68
power, 68
Ocular micrometer, 67
Oil(s), 645
additives, 587, 589, 607, 614, 615, 632
analysis, 585-587, 590, 592, 594, 601-633
dilution (*see* Dilution)
interpretation (*see* Error, sources of)
methodology, 618-633
filter, 587, 594, 598-599, 604, 607, 618, 621-622, 629-630, 633 (*see also* Oil filter test methodology)
fuel, 585, 607, 614-615
hydraulic (*see* Hydraulic)
lubricating (*see* Lubricating oil)
particle counting of, 585-633
sampling 601-602, 603-604, 629-630 (*see also* Sampling and Oil sample handling)
Oil filter test methodology, 618
dual sensor calibration, 624, 625
flow rate,
control, 626
errors, 611
multipass filter test methods, 621
sample,
collection, 629
content, 630
feeder operation, 610
Oil-immersion work, 34
Oil sample handling, 618
contaminant,
deagglomeration, 607
resuspension, 605
magnetic stirring, 606
oil sample content, 614
on-line sample measurement, 623-624
procedures, 603
Olympus polarizing microscope, 35, 41, 43, 48
Omnicon image analyzer system, 87-88, 96-97
green membrane filters, 87
use of reflected light, 88
On-line particle measurements, 132
Opaque, 42
OPC (*see* Optical particle counter)
Open vial access systems, 646
Optical:
axis, 245
condenser, removed for mass spectrometer use, 263
microscopy and low magnification stereomicroscopy, 244
properties, 242, 264
train, 260
Optical model, 771
Optical particle counter (OPC), 792, 851-853 (*see also* Particle counters and Particle counter types)

[Optical particle counter (OPC)]
[For data analysis, *see* Jaenicke Geometric and Particle-Triggered Poisson model entries]
accuracy of size distribution, instrument-dependent, 799
advantages, 18
calibration, 18, 26, 851, 852
coincidence (*see also* Coincidence error)
collection optics, 800
correlation, 26, 585
disadvantages, 18
dynamic size range, 193
electronic noise level, 588
electronic saturation, 613
electronic system capabilities, 194
evolution, 2
generalized function of, 799
geometry, 800
illumination,
source effects, 187
wavelength effects, 19
interferometric optics, 22, 572
Jaenicke Geometric Poisson model, 540
light extinction system,
calibration, 19, 592, 594, 599
coincidence error (*see* Coincidence error)
error sources during measurement, 601
flow rate limitations, 588
historical background, 22
minimum detection limit, 588
operating principle, 18
resolution, 591, 596
sensitivity, 20
light scattering system,
historical background, 23
in situ systems, 20
sensitivity, 20
operating principles, 19
particle size sensitivity, 25
volumetric systems, 20
light source, 799
new technology and baseline changes, 803
operation basis, 17
operating principles, 17, 587
particle shape, orientation impact on size measurement, 803
particle size distribution effects, 611
Particle-Triggered Poisson model,
raw count, 542, 543
single particle counts, 547, 548
portable, 799
pulse height response, 800
purchaser concerns and validation, 803
refractive index ,
differences, 803
ratio effects, 18, 190

[Optical particle counter (OPC)]
response curve, 800
sensing volume ranges, 18
sensitivity dependent on, 800
scattering angle effects, 189
signal-to-noise ratio, 193
source illumination, 800
Optics (wide-angle collecting), 803
ion microprobe advantages, 267
reflectance, 260
visualization of area of analysis, 267
Orbital shells, 253
Orientation (angular), 248
Organic species and mass spectrophotometry, 263
Organisms, 136-138
Original image, 71, 73
O-ring, 808
Outlier data, 803
Out-of-limits conditions, 803
Ovens, 641
Oversaturation, 642
Oxacillin and NMR, mass spectrometry, X-ray crystallography, 278
Oxygen, 257

Pacific Scientific, 792
Paint film and particles, 43
Palladium, 248, 257
Pall filter, 807, 808
Parallel-detection electron energy loss spectrometer (PEELS):
channeling conditions, 290
combined EELS-SAED, 290
crystals, 290
electron beam, 288
minimum detectable limit versus EDXS, 288
minimum detectable mass fraction (MDF), 289
photodiode array, 288
sample mounting on carbon-filmed grids, 289
signal-to-noise ratio (S/N), 289
valence of an atomic species, 289
Parenteral(s), 637
fluids, 138-142
freeze-dried, 645
products and glass flakes, 646
surfactant nature of many, 642
Particle(s) (*see also* Particulate matter):
analysis, 256
compendial, 815
detection probability histogram 310-316, 337-345
laboratories, 791
laser micro-Raman molecular analysis of single particle, 263
modes, 236
and pure laboratory water, 807

[Particle(s) (*see also* Particulate matter)]
resolution improvement with PIDS, 767
shape, 117-120
size analysis (*see* size distribution)
size distribution, 113, 114, 126, 129-131
size effects, 259
sizing accuracy, 198
submicron particle size analysis with PIDS, 757
analysts, 44
burden,
estimate as function of container surface/volume ratio, 641
washing to minimize, 651
clinical applications,
agglomeration and coalescence, 853-855
agglutination tests, 689
dyed microsphere chromatographic strip tests, 696
enzyme-linked immunosorbent tests and assays, 694
filter separation agglutination tests, 693
fluorescence immunoassay, 699
protein adsorption, 689
supermagnetic microsphere based assays, 699
surface titration, 688
coincidence (*see* Coincidence error)
concentration,
and accuracy of counting data, 466
constant error contours,
raw count concentration contours, 496-501
single particle concentration contours, 520
effective concentration measure, 472, 474
particle size dependent, 501, 502, 534, 535
effect of analysis threshold, 537, 538
geometric concentration measure, 472
particle size independent, 534, 535
geometric versus effective concentration measures, 474-476
conducting, 125
contamination, 131, 140-142
amorphous materials vs. particles, 458
assay interference, 460
in chemical delivery systems, 576
control of, 576, 577, 579-583
data, random spiking, 803
and filtration, 576, 577, 579-582
forecasts, 569, 571-572
gross problem indication, 649
hydraulic shocks and, 581-582
measurement prerequisites, 460
membrane/microscope assay, 457
monitoring, 132
polymers vs. particles, 458
problem-solving and documentation, 650
produced and inspected quality, 317
quality regions, 296

[Particle(s) (*see also* Particulate matter)]
Reject Zone Efficiency parameter, 317
statistically defined, 314-316
summarized, 318
"Safety of Large Volume Parenteral Solutions" conference, 457
in semiconductor process chemicals, 569-583
single particle count limits, 453
size and count limits, 453, 457
stability studies, 299-302
in water, 570, 571
count(s), counting, count accuracy, 88, 96, 134-140, 145-147, 453, 585-633
analytical models, 463, 504-505, 538-543
artifact counts, 806 (*see also* False counts)
automated counts, 88
balance equation, single particle counts versus particles counted, 509-510, 546-549
blank (background), 88
coincidence (*see* Coincidence error)
concentration effects, 463, 466, 474-475
(*see also* Jaenicke Geometric and Particle-Triggered Poisson models)
contaminant, 96
count efficiency,
applied to false counts and undercounts, 544-546
normalized relationship, 538-543
of raw counts, 491-493
of single particle counts, 509, 510, 547, 548
count error, 85, 477
horizontal coincidence, 502-504
raw counts, 491-493
single particle counts, 509, 510, 547, 548
vertical coincidence, 502-504
count loss,
raw counts, 492
single particle counts, 509, 544, 546, 547
counts, 66
drug powders, 96
effects of,
agitation, 85
dilution, 85
pouring, 85
error sensitivity (*see* USP<788>SVI Particle Contamination Assay)
for filter testing, 585, 601, 604, 620-629, 633
filtration, 85 (*see also* Filter)
graticules, microscope, 33
interference,
air bubbles, 781
silicone or other immiscible droplets, 781
Jaenicke Geometric Poisson model, 464, 468, 469, 476, 538-540
counting capability, 540
manual method, 90

[Particle(s) (*see also* Particulate matter)]
maximum count, 477
medical devices, 96
oil samples, 585-633
on-line, 601, 604, 620-627, 633
Particle-Triggered Poisson model,
raw count, 542, 543
single particle counts, 547, 548
Poisson models (*see* Jaenicke Geometric and Particle-Triggered Poisson models)
precision, electrical sensing zone counter data, 126, 133-137, 144-145
purpose of, 791
recombination routine, 88
shear force, 85
single particle count error analysis, 522
and spurious counts, 806
statistics, 616
variability sources, 618
system, on-line, 623
counters,
advantages, 454
agreement between, 585, 587, 591, 594, 597, 598-600, 622, 624-626, 629, 633
Coulter, 77
count efficiency, a key concept, 489-491
disadvantages, 455-457
electrical sensing zone, 77 (*see also* Coulter Counter)
flow cell bubble formation, 572
flow cell window, material considerations, 572
HIAC/Royco, 77, 458
initial validation, 452
light extinction, 18, 77, 187, 451
comparability, 457
count accuracy contour, 451
raw counts versus concentration, 496-501
single particle counts versus concentration, 520
replicability vs. accuracy, 459, 465
USP XXIII accuracy specification, 451
light obscuration (*see* light extinction)
limitations, 585-587, 589-592, 594, 597-598, 600, 614, 615, 616, 618, 619, 624, 628-629, 633
operating limits,
coincidence error limit (*see* Coincidence error and Particle, coincidence)
concentration limit (*see* Coincidence error and Particle, coincidence)
electronic noise level (*see* Noise level)
electronic saturation, 589-590
flow rate (*see* Flow rate)
resolution (*see* Resolution)
operating principles, 587-588
particle size calibration (*see* Calibration)

[Particle(s) (*see also* Particulate matter)]
universal count efficiency and count error curve,
raw counts, 492
single particle counts, 509
counter types: instruments and nomenclature,
acoustic pulse, 16, 174
Coulter Counter (conductimetric, electrical sensing zone), 15, 113-154
basic principles and operation, 114, 117, 118, 125
electro-sensing zone counter performance versus optical particle counters, 138-141
resistivity, 114, 117, 118, 125
electronic particle counter (EPC), 779, 781, 782
ensemble averaging (EA) (*see* Ensemble averaging)
filter observation, 14
holographic systems, 216-218
particle resolution limits, 213
light extinction (*see* Particle counters)
light obscuration particle counter (*see* light extinction)
light scattering, 19, 188
optical particle counter (OPC), (*see* Optical particle counter)
polarization intensity differential scattering (PIDS),
analysis showing peak separation for three particle size (sub-1μm) mixture, 760
effect of discrete detector array on scattering pattern detection in PIDS, 767
single particle counter (SPC), (*see* Single particle counter)
deposition model, 827
detection accuracy,
as background varies, 323, 324
as contrast varies, 321, 322
for fixed inspection times, 321-323
minimum contrast for detection, 321, 322
size versus visual acuity, 323, 324
detectors (*see also* Particle sensors)
basic characterization,
B_H, scan beam height, 493, 535-537
V_{DG}, Geometric detector volume, 493, 534, 535
count efficiency,
of raw counts, 491-493
of single particle counts, 509, 510, 547, 548
count error,
raw counts, 491-493
single particle counts, 509, 510, 547, 548
effective beam height, 472
effective detector volume, 493, 494
functional parameter description, 493, 494
geometric beam height, 472

[Particle(s) (*see also* Particulate matter)]
geometric detector volume, 493
maximum count capability,
raw counts, 477, 478
single particles, 509, 510
particle concentration at maximum count,
raw count error, 477
single particle count error, 509, 510,
elimination, as production cost reduction, 642
entrainment model for particle deposition studies (abstract), 827
formation, explained with simple chemistry, 649
identification (*see also* contamination)
characteristics for (*see* identifying characteristics)
first instruments used, low-power stereomicroscope and polarized-light microscope, 29
frequently applied methods, 235
isolation and scrutiny of individual process steps, 650
lacking unique morphology, 259
matrix check list assembly for particle identification, 44
microscope requirements, 30
most used objectives, 36
Particle Atlas, 43
and a polarizing microscope, 33
preliminary step, 236, 243
as primary objective, 257
and sophisticated instrumentation, 649
with stepwise magnification, 235
thermal properties, 31
identifying characteristics,
absorption of color, 45
agglomerated particles, 74, 85, 88, 96
analytical techniques, 236
background (filter) separated from, 71
birefringence, 30, 45, 50–51, 53
boundaries, 74
brittleness/elastomericity, 46
chart of types, 44
chemical composition, 30
color, 30, 43, 50
contrast enhancement, 72
contrast methods, 57–58
detection enhanced, 72
dispersion (staining), 30, 45, 48–50
edges (see Edges)
elongate, 84
elongation (see sign of elongation below)
extinction angle, 30, 45, 55
fibers (see also measurement), 74, 81-82, 84, 88
fluorescence, 30, 31
fusion methods, 57
hardness, 30, 45

[Particle(s) (*see also* Particulate matter)]
hydrodynamic diameter, 163
linear friction coefficient (Stokes' law), 156
of interest, needs to be recovered, 43
interference figure, 45, 49, 55
magnetism, 30, 43, 45
margins, 74
melting point, 30
microchemistry, 56
morphology, 30, 45
opaque, 29, 42
other, 30
pleochroism (many colors), 30, 45-46
porosity, 119, 125
property effects, 855-857
reflectivity, 30
refractive index/indices, 11, 30, 45, 47-49, 57
remarks, 45
roughness, 95
rugosity, 91, 94-95
shadows, 96
shape factor, 118, 857
considerations, 857-859
effects on motion, 859
effects on size, 455-457
sign of elongation, 30, 45, 53–55
surface texture, 30
table of characteristics, 30
translucency, 42
transparency, 30
image (*see* Image[s] and image analysis, Image processing)
isolation, 237, 239, 241, 242, 243
mounted for light microscopy examination, 43
mounting film preparation, 256
mounting for spectrographic examination, 237-243
techniques, 243, 259
manipulation,
and low-power stereomicroscopy, 31
picking procedures, 44, 239
measurement, 66, 77-84, 97, 99, 846 (*see also* Feret's and Martin's diameters)
absorbance, 260
area, 77, 84
boundary, 80
centrifugal frictional coefficient, 156
circularity, 91
concentration, 574, 581-583
dispersion, 124
dry-volume size, 163
effective diameter, 164
errors in, 572-575, 581, 583
extinction coefficient, 157
gas bubbles, 572-575, 581, 583
graticules, counting, 33

[Particle(s) (*see also* Particulate matter)]
internal, 132-133
linear dimension, 91
orientation, 81
overlapping, 85, 96
particle array measurement methods, 5
perimeter, 80-81
population analysis, 88
profile, 77
projected area diameter, 91
refractive index, effect of, 572-575
sampling, 574, 581-583
segmentation, 99
shear force, 85
of single particles, 131
size, 167, 572, 573
size/detector interaction, 472, 535-538
sphericity, 91-92
submicron particle size analysis, polarization and wavelength effects, 757
suspension for unbiased analysis, 122, 124-125
viscosity, effect on, 572, 573
monitoring, 692
background counts, 812-815
continuous cell growth monitoring, 132
floors and work surfaces, 805
instrumentation considerations, 803,804
laboratory, 799-804
"particle free" water and solvents, 807-810
size/detector interaction, 472, 535-538
technology, 799
morphology (*see* Particle shape and Morphology)
quality regions,
Reject Zone Efficiency parameter, 317
rule-of-thumb, 314
statistically defined, 314-316
summarized, 318
sampling (*see* Sampling)
sensor (*see also* Particle detector),
extinction, 587-588, 596, 598, 613-614, 630
resolution (*see* Resolution)
scattering, 587-588, 596, 598, 613-614, 630
shape, 76, 90
analysis, 90, 94
characteristic linear dimension, 90
computer-based analysis, 94
elongated, 82
factors, 91
geometrical particle property, 90
irregularly shaped, 82, 84
radial array, 98
size, "size", sizing, 45, 66, 76-77, 79, 88, 96
accuracy, 198, 454, 457
Airy function, 167
analysis (*see* size distribution)
automated, 76, 81, 83
[Particle(s) (*see also* Particulate matter)]
breadth, 79
calibration, 456, 457
calipering measurement, 83 (*see also* Algorithm, calipering)
change, 859-860
definition,
arrays, 4, 155
discrete particles, 14
size distribution (PSD) , 113, 114, 126, 129-131, 155, 463
applications, 593, 620-623, 630
and count data, 597. 501. 503. 509-600, 611-614
droplets, cationic-stabilized emulsions, 165
effect of concentration, 463
effects, 259
intensity-weighted distribution, 164
length, 79
limits exceed laser desorption capability, 265
liposomes, 165
maximum dimension, 721
method of cumulants, 164
numerical deconvolution, 164
orientation during measurement, 455
pharmaceutical powder example, 169
pulverized bulk material standards, 259
real resolution, 168
reference materials, 18, 147
refractive index effect, 454, 459
relative concentration, 156
reporting, 630
resuspension, 741
separation, 722
technique, 735
shape effect, 454, 455, 456, 721
size/detector interaction, 472, 535-538
"size measurement", 721
standard(s), 66, 259
step, 80
thickness, 79
trapping, 739
weighing, 77
suspension,
uniformity throughout sample volume, 122, 124-125
Particle counting reference standard (PCRS), 786 (*see also* USP XXIII <788> SVI Assay)
Particle Measurement Systems, Inc. (PMS), 499, 501, 502, 517-518, 520
Particle perimeter, 80-81, 95 (*see also* Image and image analysis, Image processing)
equispaced method, 95
random walk procedure, 94
Particle profile, 95 (*see also* Image and image analysis, Image processing)

[Particle profile]
Euclidean shapes, 95
intersection frequency, 95
uniform line matrix 95
Particle settling velocity, 10, 461, 462
Particle-Triggered Poisson model, 477-480, 541-544
application, 488- 494
concentration at maximum count occurrence,
raw count peak, 543
single particle count peak, 548
constant error,
raw count concentration contours, 496-501
single particle count concentration contours, 520
count efficiency, 489, 510
analysis applied to false counts and undercounts, 544-546
count linearity versus detector volume, 477, 478
count loss (for single particles), 509, 544, 546, 547
counts and single particle counts compared, 510
derivation, 541-544
detector volume versus linear counting span, 476, 477
false count and undercount analysis, 510, 511
initial particle concentration by linear regression analysis, 484, 485
maximum count capability,
computation, 477, 478
raw counts, 543-548
single particle counts, 546-549
multisized particle coincidence calculations, 501
particle concentration for peak count, computation, 477,478
particle counts vs. single particle counts, 505
ratio of multi-particle group counts to single particles, 543-544
raw count efficiency, 492, 494,495
raw count error, 492, 494,495
single particle count efficiency, 509, 510, 546-548
single particle count error, 509, 510, 546-548
single particle peak count capability, 548
single particle peak count capability, concentration for, 548
universal response curve,
raw counts, 494
single particle counts, 508-510
with USP XXIII <788> count error limit determination, 512-515
validation, 481-488
vertical and horizontal coincidence, 502-504
Particulate matter, 235, 637 (*see also* Sources of particulate matter)
accurate particle assay requirements, 804
additive contribution, 637, 638, 639
aerosol, 793
airborne, 637
[Particulate matter]
air contamination monitoring in pharmaceutical laboratories, 800
allowable or unavoidable, 638
allow detection of trace contaminants in (LAMMA), 265
background control measures, 804
biological material, 792
capture point, 796
characterizing, 709, 791
closure-related, 640
common sources, 637
stoppers, 299
of visible, 298
composition, 258
conducting material, 256
contaminant, 259, 792, 793
control, 795
corrective action, 649
cotton, 792
counting, 791
crystals, 239
effluent control, 3
eliminating, 637, 639
environmental, 792
excessive, 638
fall out, 804
fine metallic dust, 793
first step, identification; then isolation, 243
fluffy, 646
forceps, 239, 241
formation in finished product, 639
fragments: glass, plastic, metal, or biological material, 792
generation and sonication, 806
glass, 815
glass slides, 242
handling contamination and loss risks, 241
heterogeneous composition, 638
heterogeneously sourced contaminant, 792
hood, 239
human activity, 792
identification, 639, 644, 791
identity and source relationship, 650
in injectable product, 637, 638
inorganic, 793
for instrumental analysis, 241
instrument operations, 792
isolated for analysis, 237, 239, 241, 242, 243
isolation, 795, 817-818
lost on the stub, 259
manipulation, 239. 241
mass, 247, 253
measuring and size, 799
in medical devices, 638
membrane filter sample, 241

[Particulate matter]
microinstruments, 239
microneedle, 239
micropipettes, 242
microslide, 239, 242
monitoring, 795, 799, 804
mountant, 242, 243
moving and wavelength, 247
needle (finely pointed tungsten), 241
needle probes, 241
nonenvironmental sources of, 805
nonspherical, 803
nonviable, 638, 793
orientation, 803
paper, 792
particle-free test materials, 804
personnel, 818
pharmaceutical contaminants, X-ray spectra (table), 258
picking, 239
polyethylene pipettes, 242
probe, 239
process- or point-generated, 792, 804, 805
versus airborne, 651
randomly sourced, 638 (*see also* Source(s) of particulate matter)
shape, 799, 803
source identification, 258
storage, 639
submicron, 791, 792
subvisible size, 792
suspended in liquid, 237
total burden, 637
uncoated (visualization of), 257
velocity/wavelength relationship, 247, 253
wavelength of illumination, 803
work surfaces, 805
Patch testing, 21
Pattern:
raster, 249
secondary emission, 248
PCS (*see* Photon correlation spectroscopy)
PDA meeting, 1983 annual, 779
Peak overlap, shape, 257
PEELS (*see* Parallel-detection electron energy loss spectrometer)
Penicillin:
drug powder, 643
powders, 278, 645
People, particle sources and control, 258
controlling particles and, 792, 818-819
garb, 823
greatest source of particles, 793
training, 819
Perimeter map, 83 (*see also* Image and image analysis, Image processing)
Petri dishes, 31
Petrislide(s), 804, 812
Petroleum geologists, 37
pH, 642
control to reduce particle contamination, 643
of glass solutions, 646
phosphate-containing pH-adjusting reagents, 642
sieving, 648
and stopper/iron interaction, 645
and type of glass used, 646
Pharmaceutical:
companies and pure water, 807
device(s),792, 806
injectable solutions, 257
materials, 791, 806
particulate matter,
analytical standards, 297-298
reject rate analysis, 310-314
and use of X-ray crystallography, 278
visual inspection for, 297-298
practical contamination monitoring, 804
product assay, 804
research, 57
research and EDXS along with FTIR, 260
solution, 792
ultrasonic baths, 806
X-ray spectra of contaminant particles in products, 258
Pharmaceutical Manufacturers Association, 780
Phase contrast, 41, 58
microscope (versus true polarizing), 31
telescope in Köhler alignment procedure, 39
Phase transition process, 264
Phlebitis:
definition, 675
incidence, 683
Phosphor screen, 246
Photoelectron shower proportional to X-ray energy, 253
Photometer, 11, 12
Photomicrography, 35, 42
Photomultiplier tube (PMT), 248, 249
signal intensity modulates beam of CRT, 248
Photon correlation spectroscopy (PCS), 163 (*see also* Dynamic light scattering)
second-order autocorrelation function (ACF), 163
Photorecording, 37
Photosedimentation, 847
Photosedimentometer, 157
Physiology and particulates:
clearance of,
clearance kinetics, 669, 673
macrophage response, 677
phagocytosis, 669, 673, 677
reticuloendothelial response, 670, 677

[Physiology and particulates]
effect of,
factors, 669
granulomatous reactions, 675, 676
inflammatory response, 675, 676, 677
thrombus formation, 675, 679, 684
inflammatory response to,
bradykinin response, 675, 676
macrophage, 675, 676, 677
physiological repositories for,
lungs, 671
spleen, 671
liver, 671
Picture points, 249
PIDS (*see* Polarization intensity differential scattering)
Pinhole leak(s), 797, 798
Pixel(s) and pixel array 61, 62, 64, 66, 69-70, 72-74, 77, 81, 83, 92
(*see also* Image processing and Resolution)
center, 73
coordinates, 71
counting, 97
digital, 66
edge pixel boundary, 69
error, 84
fractional, 84
gray-scale resolution, 66
individual, 71
input pixel, 72
linear dimension, 66, 90
magnification of the relay (projection) lens, 66
number, 71
numerical aperture (NA), 66
numeric value, 71
output pixel, 72
peripheral, 84
resolution of digitizing process, 66
sawtooth effect, 75
smallest image element, 70
target pixel, 73
Planchette(s) (*see also* Stubs):
beryllium,
polishing, 243
toxicity, 258
carbon, 243
carbon or beryllium, 242, 258
Planck's constant, 247, 253
Plane polarized-light, 31
Planimeter, 77
Plasma, 267
Plastic:
containers, 644
granules, 652
ware, 805
Plasticizers, 640
Plenums, 792
Pleochroism, 31
PMS (*see* Particle Measurement Systems, Inc.)
PMT (*see* Photomultiplier tube)
Point-generated particles, 237
Point of generation, 250
Polarization intensity differential scattering (PIDS)
Polarized light, 243
crosshairs, 37
work, 36
Polarizer:
crossed Nicols, 31
crossed polars, 31, 53
crosshairs, 37
interference with diaphragm observation, 39
Olympus, 41
removable and darkfield, Rheinberg illumination, and phase contrast, 41
rotatable, 33
second, for vector analysis of crystalline materials, 37
Polarizing:
attachments, 33
filters commonly provided, 31
Polarizing microscope(s), microscopy, 29-58, 235, 818, 822
Abbe-type condenser lens, 33
adjustment(s),
ring (diopter), 37
some continuous, 42
analysis technique at high magnifications, 35
analyzer, module, 37
and anisotropic crystal, 37
aperture (low numerical), 36
built-in illumination for, 33
camera,
electronic, 33
Polaroid, 33
crosshair graticule, 34
crystal rotation apparatus (spindle stages), 34
darkfield use limitations, 41
depolarizer, 37
diaphragm,
adjustable iris, 33
closed field, 39
focusing adjustment, 39
function of, 39
image relayed to eyepoint, 41
leaves, 39, 41
second conjugate focus of aperture, 39
substage, 33
tube, 37
diaphragm aperture, 33, 39, 48
closed about 15%, 40
control, 42
effect, 42

[Polarizing microscope(s), microscopy]
diffuser, 39, 40
built in, 33
DIN standard, 33
field depth, 42
field diaphragm, 33, 39
changes with objective, 42
focused in plane of specimen, 42
leaves, 41
ray path trace of, 41
refocusing due to slide thickness, 42
setting, 42
substage condenser adjustment, 41, 42
three conjugate foci of, 41
filament (bulb), 33
filters (inserting), 33
foot switch control of illumination, 32
graduations, 34
instrument cost, 30
interference figures, 37
laboratory setup, 29
lamp condenser, 33
light exit port, 33
mechanical alignment technique, 38
nosepiece, 35
numerical aperture requirements, 36
objectives,
magnification series, 35
numerical aperture, 36
optical axis, 35, 37, 38, 41
strain-free lenses, 35
for particle identification, 31, 33
particle mounting for, 43
phase contrast with polarizer removed, 41
refinements, 33
Rheinberg illumination with polarizer removed, 41
ring(s),
clamp, 32
knurled, 38, 39
light, 32
rotating a, 37
rotatable polarizer, 33
rotational axis, 38
separate light sources in, 33
slide, refocusing required for each, 34
stage(s),
ball bearing, 34
click stop at "zero" setting, 33
grease stage, 34
multigimbal stage(s), 34
rotatable and graduated, 34
screw holes, openings, 34, 38
screws, 35, 39
spindle stages (crystal rotation apparatus), 34
wrenches, 38
substage condenser requirements, 33

[Polarizing microscope(s), microscopy]
supplementary lens, 37
vibration direction, 33
Polars (slightly uncrossed), 37, 44
Polycrystalline, 283, 285
Polydispersity, 712, 713, 718, 719
Polyethylene, 242, 654
Polygon:
exscribed, 94
inscribed, 94
Polymer technology, 640
Polymorphism, 57
Pooled contents, 815
Port (exit), 39
Powders, 142-144 (*see also* Particles)
appearance, 2
combustion, 2
crystallized, 644
dry, 646
lyophilized, 644
size measurement methods, 2
spray-dried, 644
transport, 2
Precipitate(s)/precipitation, 269, 642, 643, 645
Preservative, 644
Pressure:
rinser, 808
vessel, 805
Probabilistic evaluation of contaminating particle inspection data, 307-317
summary, 318
Probe, 250
needle probe fabrication, 44
plate, 31
wire, 263
Processor, image, 66
digital image, 71
Product inspection and product quality, 317
Productivity improvement with inspection automation, 295
Product properties, 2
PTFE (TEFLON, inert coating), 640
Pulse height analyzer, 588, 595
Pumps, 647
Pupil (exit), 41
Pure water guidelines, 570
Purification systems, 807
Pyrogens (absence of), 637

Quality:
standards, 637
with regard to particulate matter, 642
Quality Control Section of PMA, 780
Quality determination, sampling accuracy limits, 418, 419
Quantimet image analyzer system, 101

Quartz, 50-51, 268
Quartz-halogen bulbs: 800
Quasi-elastic light scattering, 163

Radial vectors, 92
Radiation:
  Bremsstrahlung, 252
  continuum of, 251
  energizing, 236
  illuminating, 244, 246, 247
  infrared, 260
  low energy (infrared), 236
  other types of, 250
  possible damage, 257
  second electron, 250
  specimen image formed by, 247
  white, 251
  X-rays, 236
Radius, 35, 283
RAG (Reject and Gray Zone false reject rate), 318
Raman, 29, 261 (*see also* Micro-Raman)
Ramsden disc, 41
Raster(ed):
  controlled scan frame (SEM), 248, 259
  pattern, 243, 249
  point, 249
  primary ion beam is, 265
  scan instead of point excitation, 259
  screen raster size, 250
Ratio comparison, 253
Raw material, 643:
  chemical constituency of, 645
  grades, 644
  vendors, 645
Rayleigh criteria, 68
Rayleigh scattering (*see* Scattering intensity)
Ray path, 39, 41
RCA clean, modified liquid sequence, 576
Recrystallization, 269
Redispersion, (*see* Resuspension)
Referee particle counting methodology, membrane/microscopic proposal, 785
Reference:
  beams, 212
  crystals, 37
  materials, 709, 714 (*see also* Standard reference materials)
Reflection, 747
Reflective light (metallurgical) microscope, 31
Refocus and slide changes, 42
Refraction, 747
Refractive index (indices), 31, 42, 57, 587, 588, 595, 598, 611-614, 748, 751, 771
  of air, 35
  anisotropic (one having more than one), 37
  differential, 19
[Refractive index (indices)]
  effect of, 595, 598, 613-614
  high, 37
  imaginary (absorption) component, 803
  liquids, 44
  low, 37
  of particles and liquid, 588, 595, 598, 611-614
  ratio, 574
  real (scattering) component, 803
  response of counter affects, 799
  standard, 44
Regulatory agencies, 638, 653
Relay lens magnification, 69
Relay optics, 213
Remote:
  monitoring, 804
  sampling, 799
  sensor(s), 799
Reproducibility, 87:
  and differences in illumination and contrast, 67
  and edge uncertainty, 67
Requirements:
  cost, time, 236
  meet but not exceed, 791
Resolution, 587, 591, 594, 595-597, 598, 600, 625 (*see also* Pixel array)
  adequate, suitable. 66, 69-70
  brightness (*see* Microscopes, light)
  calculation basis, 68
  decrease, 42
  edge contrast, 68
  effect on particle boundary definition, 68
  and electron microscope, 247
  empirical test of, 74-75
  frequency response, 69
  gray scale (contrast) (*see* Gray scale)
  improves with reduced image change, 71
  increase, 36, 42
  light microscope limits, 247
  magnification and resolution, 66-70
  mass, 265
  point-to-point, 247
  poor, 257
  requirement, 89
  response, 69
  smallest image element, 70
  spatial, 71
  spatial resolution of, 71
  TEM versus light microscope, 247
  testing of SEM, 71, 75
  theoretical resolution parameter, 66
  two specimens appearing as one, 247
  versus illumination wavelength, 247
  video resolution, 66
  white light, 88
Resolution of particle velocity, 213

Resolve, vectorially, 37
Resolving power, 42, 247
Resuspension, 601, 603, 605-607, 615, 618, 619, 624, 629, 632
  error, 601, 603, 605-607, 615, 618, 619, 624, 629
  methods of, 605-607
Reticle, 769
Reticuloendothelial system:
  and cellular immunity, 677
  and humoral immunity, 677
  macrophage response, 677
  phagocytosis, 677, 678
Retina in Köhler light, 41
Reynolds,
  criterion, 158
    terminal velocity, 158
    turbulent flow, 158
  number,
    liquid, 179
    particles, 158
Rheinberg illumination:
  modification of darkfield, 57
  on a polarizing microscope, 41
Rheostats, 57
Robotics, 652
Rotating stages, 34, 38
RZE (Reject Zone Efficiency), 316, 317 (*see also* Inspection security)
RZNI tables, 355-367
  design notes, 287
  overview and examples, 373-376

SAE (*see* Society of Automotive Engineers)
Saltation, 833
Salt plate, 242
  circular, 260
Sample, 252
  agitation effect, 462
  air pollution, 42
  containers, 601-603, 605, 616
    cleanliness, 173, 576, 601
    closures, 602
  feeder, batch, 610
  field, 594, 601-602, 604, 614-616, 629-633
  flattening, 260
  flow velocity, 461
  handling considerations, 2
  inhomogeneous particle, 263
  multipass filter test, 592, 599-600, 607, 618-629, 630, 632, 633
  neutral buoyancy, 461
  observation, 20, 174
  point location bases, 179
  points, 248
  ports, 604
    location/accessibility, 179, 604
[Sample]
  preparation for microscopy, 43, 44
  quantity,
    particle size distribution relationship, 180
    statistical requirements, 180
  scheduling, 184
  size, statistical requirements, 180
  support surface, 248
  surface, 257
  thickness, microtome sectioning, 260
  used oil, 585-586, 588, 590, 592-593, 594, 600, 605-607, 609, 611, 614, 616, 618-619, 620, 629-633
  withdrawal problems, 173, 863-864
Sample cell:
  dry powder, 761
  liquid dispersion, 761
Sampling, 461, 601-604, 621, 623-624, 629-630
  engine data, 603-604, 629-630
  error, 601-604, 621, 624, 630
  filter test stands data, 603, 621, 623-624, 629-630
  in situ, 863
    application areas, 177, 582, 603
    particle size sensitivity, 175
    problem areas, 176
    sample scheduling, 175
    sample volume determination, 174, 182
    sidestream sampling, 583
    velocity profile effects, 181
  on-line, 603-604, 621, 623-624, 629
  operating system line, 603-604
  reservoir data, 603-604
Saturation, electronic, 589-590
Scale(s), 43
Scanning electron microscopy (SEM), 30, 74, 237, 242, 243, 252, 256, 259
  analysis, particle location, 259
  analysts and single particles, 243
  beam spot pattern, 248
  bulk particle mounting, disadvantages, 256
    agglomeration, 256
    particle size/shape effect, 256
  chemical analysis, 252
  column versus TEM, 248
  electromagnetic lenses, 244
  electron diffraction analysis alternative, particles too small for, 285
  electron-optical column, 248
  electron source, 248
  high magnification capability, 235
  high voltage, 248
  illuminating radiation, 244
  lensless imaging, 248
  low magnification stereomicroscopy, a first step, 244
  mass spec particle mounting, 263

[Scanning electron microscopy (SEM)]
  maximum chord, 74-75
  micro-FTIR powerfully complements the use of, 259
  particle location verification, 259
  photon generation, 248
  probe of specimen, 250
  raster pattern, 248
  sample preparation versus TEM, 256
  scintillator generation of photons, 249
  and X-rays, 253
  X-ray/atomic number, 252
Scanning laser light scattering devices, 804 (*see also* Monitoring, airborne counts)
Scattering counter historical development, 23, 24 (*see also* Particle counters and Particle counter types)
Scattering intensity, 749
  angular pattern, 749
  fine structure, 757
  Fraunhofer, 754
  nonspherical particle, 750, 753
  polarization effect, 752, 757, 759
  polystyrene latex, 751
  Rayleigh-Debye-Gans, 752
  refractive index effect, 751
  wavelength effect, 748, 752
Schlieren effects, 640
Scientific evaluation, 779 (*see also* Validation)
Scintillator:
  electron to photon conversion, 248
  preparation, 248
  structure, 249
Screen, video
  anode, 248, 249
  dark areas on, 247
  forms an image on the, 263
  phosphor, 246
Screening (straining), 797
Screen mesh numbers, 5
Secondary electrons, 249, 250, 257
  and artifactual emission of, 257
  Bragg's equation, 253, 254
  emission, 248
Secondary ion mass spectrometry (SIMS), 243, 265
  analysis limited by particle size/number of atoms, 268
  design differences in column and ion detection, 265
  microprobe system (higher resolution), versus microscope, 266
  particle preparation, 268
  primary ion beam for bulk or surface mass analyses, 265
  principal ion microscope, ion microprobe designs, 265
[Secondary ion mass spectrometry (SIMS)]
  quadrupole, 265
  secondary ion images, 265
  spatial and mass resolution, 265
Second conjugate focus, 39
Sedimentation (SD), 156, 864-866
  centrifugal, 10, 156
    homogeneous start, 157
    line start, 157
  gravimetric, 9, 10, 156
  homogeneous starting (HOST), 157
  line start (LIST), 157
  measurement time, 159
  multiple detectors, 159
  particle size limit estimate, 158
  peak detection, 157
  scanning, 159
    turbidity, 157
    X-ray, 157
  size range, 158
Segmentation:
  consistent, 68
  effect of microscopic resolution on, 68
  key to precise and accurate particle sizing, 68
Selected area electron diffraction (SAED), 243, 279, 380 (*see also* Electron diffraction)
SEM (*see* Scanning electron microscopy)
Semiconductor process chemicals:
  chemical properties, 572
  contamination in, 564-583
  disposal of, 583
  physical properties, 572-575, 580
  refractive index, 572-575
  in wafer cleaning, 571, 576-579
Semiconductor process liquids, 569
Semiconductor wafer cleaning process, 577
Sensitivity limits of instruments, 243
  microanalytical, 243
Sensor comparability and resolution:
  HIAC/Royco, 784
  Russell Labs, 784
  validation of counting accuracy, 451-568
Serial accept inspection systems, 371
  "Accept in Three" (A3-A/3) inspection, 337
  standard and cull inspection (M2-A/2), 304
Serial reject inspection systems, 371,372
  "Reject in Two" (A2-R/2), 372
Shakespeare Corporation, Juliet III image analyzer system 98
Shell vial, 44
Sidestream sampling, 184, 185
  sampling probe configurations, 185
  stream flow, 187
Sieves, screens, 5
  aperture configurations, 6
  cascade assemblies, 5

[Sieves, screens]
mesh size, 5
micromesh, 7
particle size measurement range, 6
surface sieves,
fiber optic bundle, 723-726
photo lithography, 726, 727
radioactive source exposure (track etched or nucleation filters), 731
woven wire, 6
Sieving, 709, 725, 757
effect of depth filtration, 725
experimental requirements for accuracy, 731
procedures, 13
retention, 648
surface filter, 726, 727
Silica gel, 44
Silicon:
lattice, 253
wafers or chips, 804
Silicone, 657
alternatives, 640
baked-on, 640
high levels, 640
oil, 655
supplier, 641
Silt content determination, 9
SIMS (*see* Secondary ion mass spectrometry)
Single particle counter (SPC) (*see also* Particle counters and Particle counter types),
[For data analysis, *see* Jaenicke Geometric and Particle-Triggered Poisson model entries]
count accuracy contour for laser sourced detector, 520
Single particle counting, 155
Single particle counts, 546-549
maximum capability, 548
Size distribution, 709, 713, 718, 746, 760, 770
Size of elongation, 31
Skill for particle isolation, 243
Skin cells, 658, 793
Slide, microscope 31 (*see also* Tools)
adjustments required, 42
handling, 34
permanent preparations, 44
refocusing of field diaphragm, 42
Slip agents, 646, 657
Slot (compensator for polarized light microscope), 37
Slurry and colloid physics, 866-867
Small volume injectables (SVI), 638, 779 (*see also* USP XXIII <788> SVI Assay)
dissolved material, 643
LEC testing of, 813
lyophilized product, 641
penicillinoic acid, 642
[Small volume injectables (SVI)]
pooling vials, 811
in production of aseptically-filled, 652
silicone use in SVI products
sulfur in particles, 645
Society of Automotive Engineers (SAE), 593, 599, 605, 622
Sodium nafcillin and oxacillin, 643
Sodium nitrite, 43-44, 242
Software, 257
Solenoid fields, 245
Solenoids, 244
Solid phase assays (bound/unbound separation), 698
Solid-state material, 264
Solubles/solubility, 639
limits of components, 642
Solute concentration, 642
Solution(s), 791,792
contact path, 655
container interactions, 646
opaque, 29
particles may be in, 43
suitable, 256
Solvent(s), 242
flask (squeeze-bottle action), 808
gun-type dispensers, 808
higher rate of flow, 807
providing, 791
wiping down working surfaces with, 798, 805
Source(s) (of particulate matter), 637
agglomeration, 638
aggregation, 638
closures (stoppers), 298, 299, 300, 639, 640
extractables, 645
fragments from, 640
containers, 639
crystallization, 638
degradents, 638
depyrogenation ovens, 641
determining the, 639
device components, 637, 638
drying ovens, 641
elastomeric closures (stoppers) of glass containers, 645
elimination of, 637
environment, 637
external to the product, 638, 641
extractables, 639
extraneous, 638
fillers, 640
filters (through fragmentation), 648
finished product, 639
formulation, 637, 638
glass, 641
gloves (powdered), 647

[Source(s) (of particulate matter)]
human,
handling, 640, 647
skin cells, 647
hypodermic syringe needle, 641
interaction of formulation with its container, 638
needle, 642
nonenvironmental, 637, 638, 647
packaging materials, 637, 640
personnel activities, 638
point-generated, 638
potential, 640
precipitates/precipitation, 638, 643
process,
-derived, 639
-related, 637, 638, 639, 645
product-package interactions, 637, 638
rubber fragments, 640
sedimentation, 638
solution, 637
stopper-related, 299, 639, 640
of visible, 298
washing effect, pharmaceutical containers, 641
Spacing:
interatomic, 283
interplanar, 253, 254, 255
known, 283
Spatial:
convolutions, 72
convolution mask (or kernel), 72
point-for-point multiplication, 72
domain, 74
filter, 760
frequencies, 74 (*see also* Fourier, frequencies)
resolution, 71
Specifications, 795
Specimen(s):
field diaphragm focused in plane of, 42
focus on a, 39
image formation, 247
light absorption, 247
opaque, 33
plane of the, 39, 42
points, 247, 249
ray path, image forming, 41
separated from filament, 41
size dispersive of NaCl microcrystallites, 265
support, 256
surface, 248, 249, 250, 251
Spectra/spectrum, 257, 260
Spectral libraries, subtraction, 260
Spectrometer(s):
contemporary design, 258
multiple computer-driven, 258
time-of-flight, 264, 265
Spheres, 710, 714
Spherical aberration, 36 (*see also* Electron microscope, resolution and Light microscope, resolution, 247)
Spikes, 803
Spindle-stage(s):
crystal rotation apparatus, 34
needle (glass fiber), 31
Spongiform media, 647, 648
elimination of, 637
Spray nozzle droplet measurement methods, 3
Spray processors, 577, 579
Sputtering, 267
Stability formulation and problems, 642
Stabilized filtrate distribution, 581-583
Stage(s), microscope, 34:
attachable mechanical, 34
ball bearing, 24
center, 38
clips, 34
clutch, 35
crystal rotation apparatus (spindle stages), 34
darkfield, 33
elaborate, expensive, 35
engagement lever, 35
grease (cost-saving measure), 34
high magnification, 35
holes (many of which are threaded), 34
insert, 31
lock, 34
mechanical, 33-34
micrometer, 43
minimum required, 35
multigimbal, 34
rotate, 38
rotating, 34
rotation,
and centering of objectives, 38
lock, 34
and objective (accidental bumping of), 41
screw openings, 38
slow-motion rotation, 35
spindle, 34
universal multigimbal, 34
verniers, 34
Staining (dispersion), 31
Standard, 709, 713, 714, 718
Standardization:
of count accuracy, 144-147 (*see also* Single particle counts)
of size accuracy, 147-148
Standard procedures (*see* Standards)
Standard reference materials, 121, 147
Standards (comparison), 258
American, 147
British, 138, 147
I.S.O., 147

Static, 651
  charges, 655
  electricity, 243
  electric potential, 656
Statistics, particle, 601, 616-618, 627-629, 630-631
Stereo dissecting microscope, 248
Stereomicroscope, 31, 239, 241, 242, 257
  accessories, 33
  audiovisual format application, 33
  boom stand, 32
  camera, 33
  definition, 31
  electronic image documentation, 33
  flexible use mounting, 32
  image capture system, 33
  lighting, 31
  low power, 29, 31
  objective design, 33
  particle recovery, 43
  Polaroid camera recording, 33
  set-up, alignment, illumination, and use, 30
  unconventional techniques, 33
Sterility, 637
Sterilization, 640:
  may alter the particle burden, 641
  steam, 639
Sterilizing tunnel, 652
Stirring (*see* Resuspension)
Stokes-Einstein equation, 163
Stokes' Law, 156, 793
Stopper(s), 811, 815
  butyl, 645
  and cephalosporin drug powder antibiotics, 645
  composition, 645
  halobutyl rubber, 639, 640, 645
  lyophilized (lyo), 641
  major particulate source, 299-300
  manufacture, 640
  particle shedding stability, 299-300
  rubber and high chemical purity, 640
  rubber with reduced leachables, 652
  silicone resin coating, 639
  sulfur-cured rubber, 639
  transfer to sterilizing tunnel, 652
Stoppering machinery, 640
Storage, 640, 642, 643
Stub(s), 31 (*see also* Planchettes)
  beryllium,
    polishing, 243
    toxicity, 258
  carbon, 243, 258
  identify location before X-ray, 259
    except on beryllium, 259
Subpicogram quantity detection limits, 256 (*see also* X-ray combined with SEM or TEM)
Substage condenser, 33-34, 41, 245
  condenser mount, 39
  field diaphragm focused in plane of specimen, 42
  focus knob, 39
  and objective together relay image of filament, 41
  refocusing after slide change, 42
  strain-free lenses, 35
Substrate, 31, 43, 259, 260 (*see also* Background)
  adhesion, 826
Superparamagnetic microspheres, 687, 688, 699
  assays, 700-702
Support(s):
  carbon specimen, 256
  specimen, 243
Surface:
  analysis, the quality of light optics essential, 266
  chemistry, 687, 688
  cleaning processes, 830
  concentration, 644
  contamination, 804
  contour, 248
  irregular, 259
  particle buildup, worst case, 793
  roughness, 259, 734-8
  soiling, 828
  specimen, 249, 250, 251
Surfactant(s):
  above CMC, 643
  monomeric, 642
  nature, 642
Suspension:
  detectors,
    turbidity, 157
    X-ray, 157
  flow through tubes, 2
  rheology, 866, 868
SVI (*see* Small volume injection)
Synchronization, 249
Syringe, 655

Tablet, 31
Tape sampling, 804
Technology (particle monitoring), 799
Teflon, 808
Television:
  closed-circuit system, 248
  picture generated on the monitor, 263
TEM (*see* Transmission electron microscopy)
Temperature, 642, 643, 657
Terminal velocity for particle sedimentation, 156, 865
Test dust, (*see* Contaminants)
Test tube, 44
Thermal degradation, 264
Thermocouples, 57
Thermometers, 57
Thietan-2-one, 643

Thin window detectors for carbon and oxygen detection, 257
Three-dimensional:
  image, 31
  information (TEM), 248
  pseudo image, 243
Three-stage assay proposal, USP<788>, 786
Threshold (*see* Gray scale)
Thrombus:
  development, 679
  fragmentation to emboli, 679
  role of particulate matter, 679
Time-of-flight spectrometer, components, 264
Tissue culture, 807
Titanium dioxide (rutile, anatase, brookite), 268
Tonal range, 71
Tools for isolating particles (*see also* Particle, isolation):
  acrylic cement, 256
  adhesive, 256
  brushes (glass fiber), 239
  drop of fluid, 239
  electron microscopy as analyst's, 244
  film (transparent), 243
  forceps, 239, 241, 256
  glass slides, 242, 243
  masking tape, 256
  micropipettes, 239
  needle probes, 239, 241
  newspaper, 256
  planchette, carbon or beryllium, 242
  polyethylene pipettes, 242
  primary tool for particle isolation, 242
  probe, 250
  razor blade, 256
  salt plate, 242
  shadowing particles with evaporated metal atoms, 248
  slides (quartz), 263
  specimen support, 243
  two-sided plastic tape, 256
Top light, 31-32 (*see also* Epi-darkfield)
Touch contamination, 657
Traceability, 594, 598, 599, 600, 709, 711, 719
Track etched filters (*see* Filters)
Translucent, 42
Transmission electron microscopy (TEM), 235, 243, 245, 246, 248, 256
  analysts and single particles, 243
  bulk particle mounting and undesirable features, 256
    agglomeration, 256
    particle size/shape effect, 256
  column longer than in SEM, 248
  comparison to transmission light microscope, 246
  contrast pattern from differential electron scattering, 247
[Transmission electron microscopy (TEM)]
  contrast provided by electron scattering, 247
  depth penetration limit of electrons (in TEM), 248
  electromagnetic lenses, 244
  electrons, weak penetration of, 248
  excited phosphor imaging, 246
  high magnification capability, 235
  illuminating radiation, 244
  image degradation with buildup of negative charge, 257
    avoid by increasing mounting conductivity, 259
    increases artifactual emission of secondary electrons, 257
  image formation, 247
  imaging lens system, 247, 248
  low magnification stereomicroscopy as a first step, 244
  sample preparation, 256
  resolution versus light microscope, 247
  under diffraction conditions, 280
  versus transmission light microscope, 246
Transmitted light applications, 31, 33
Trapezoid rule, 92
Trinocular:
  head, 37
  phototube, 37
Troubleshooting, 637, 638, 655, 656
True polarizing microscope (*see* Polarizing microscope)
Tryptophan, 643
Tube diaphragm, 37
Tubing (cutting of), 656
Tungsten:
  lamp, 760
  needle, 44
  wire, 43
Turbidimeter, 4, 11, 12
Turbidity:
  of potable water, 4
  spectroscopy, 746
Tyvek, 819

Uhlir, particle contamination quality analysis, 305, 419-421
Ultrapor:
  filter, 807
  N_{66} capsule figure, 808
Ultrasonic bath(s), 733-734 (*see also* Deagglomeration and Cleaning)
  cleaning of glassware and, 805-806
  hot spots, 806
  power ratings, 806
Ultrasonication, 43
Ultrasound removal of air or gas bubbles, 806
Ultraviolet:
  light, 247

[Ultraviolet]
  microscopy, 263
Uncrossing polars, 37
United States Pharmacopeia (*see* USP)
United States Pharmacopeial Forum (PF), 781
Universal count accuracy curves:
  raw counts, 491, 492
  single particle counts, 492-495
Universal multigimbal stage(s), 34
Used oil testing, 629
  sample handling, 630
  sample line purging, 604
  sample port location, 604
USP XXIII, 88, 96, 643, 779
  filtration and, 807
  human performance yardstick defined, 295
  inspection system validation requirement, 295
  modifying blank procedure, 814
  polysulfone filter funnels, 812
  sensitivity to error injection, 515-520
  <788> SVI Contaminating Particle Assay, 514
    assay blank limit revision, 813
    calculation of concentration for 10%
      coincidence count error for 10μm particle, 548-549
        Jaenicke Geometric Poisson model error, 548-550
        validated Particle Triggered Poisson model calculation, 549-551
    cleansing of particle assay containers, 805
    count accuracy testing, 521
    count error analysis, 522, 523
    count standard (PCRS, 15 μm particles, nominal 4000 particles/ mL.), 526
    effect of model choice on coincidence calculation,
      Jaenicke Geometric Poisson model, 549-552, 559-566
      Particle-Triggered Poisson model, 549-551
    following compendial particle analysis instructions, 815, 816
    particle assay station cleansing, 805
    single particle counts required, 505
    10μm 10% coincidence limit count accuracy specification model-based calculation
      conversion of USP count accuracy specification to count efficiency terms, 551-552
      effect of model selection, 551
    XXIII SVI analysis error sensitivity analysis, 559-566
    water required for single compendial analysis, 808
  single particle functional tests, 526
  size-related effect of coincidence, 515
  translation of SVI specification into count efficiency terms, 552-554
[USP XXIII]
  "visibility" criterion, 295

Vacuum flange, seal, 263
Validation:
  concerns, 784
  of particle inspection security, 295 (*see also* Inspection security)
    container clarity effect, 309-310
    mortality table comparison, 309
    statistical basis, 308
    training, 305
  sample preparation,
    Duke Scientific, 783
    Ben Venue, 783
  samples
    DVB spheres, 783
    pollen suspension, 781
  statistics,
    R-charts, 782
    X-charts, 782
Valves, 807
van der Waals forces, 649, 651
Vapor, 285
Variability (*see* Errors)
Vector, 244
Velocity:
  particle, 247, 253
  profiles, 181
Vendor(s), 470, 471, 638, 640, 641, 644, 656, 657, 807
  of nondestructive particle inspection systems, 443-448
  of raw materials, 645
  of stoppers, 652
Venn diagram analysis, 306
  failure of FDA vs. Bristol Labs enforcement action described, 305 (*see also* Histogram analysis, Inspection replications for validation demonstrations)
Vernier(s), 34, 37
Vessel, pressure, 805
Viable particulate materials, 868-870
Vials, 31
  cleaning of, 817
  dry powder, 817
  may be emptied using a cleaned blanked syringe, 814
  products, 792
  small volume dry powder, 815
  transfer to sterilizing tunnel, 652
  tubular glass, 646
Vial stoppers (*see* Stoppers)
Vibration directions (polarizer and analyzer), 37
Vibrations (molecular), 261
Video assisted semi-automatic inspection, 335-344
  inspection security comparison, 344-345

[Video assisted semi-automatic inspection]
validation development, 341-343
Video camera, 804
Video rate, 71
Visible particle contamination:
correlogram of manual versus automated inspection results, 302
particle size visibility determined holographically, 417
produced and inspected quality, 317
quality categories, 296
statistically defined, 318
stability studies, 299-302
with video inspection system, 335-345
Vision (*see also* Eye[s]):
focus on cross or scale, 37
problems, 37
unaided human, 29
Visual analysis:
contrast ratio requirements, 322, 722
effect of background clarity or contrast, 321-324, 724
field coverage limits for accuracy, 722
interferences, 722
linear ocular scales, 67
particle sizing, 67
variability in results, 67 (*see also* Venn diagram)
Visual inspection, 236, 644, 646, 651, 795
basics,
Accept/Reject decision discrimination evaluation, 318-319
accuracy and scene illumination, 319
biophysical vision data, 319-324
visual angle, 320
nondestructive inspection for particle contamination in sealed container, 816
unsatisfactory quality, gross production problem manifestation, 649
Volatiles, 645
Voltage:
instrument accelerating, 251
potential difference, 247
Volume evaporation, 265
Volumetric counters, 20

Washing:
clean manufacture vs. cleaning, 652
a cost-effective procedure, 656
cycle, 640
and particle burden, 651
and particle size, 652
as a source of contamination, 641
ultrasonic, 652
variable effectiveness of, 654
Water, 36
clean, importance to particle analyst, 807
[Water]
droplets (*see* Contaminants)
filtered, 804
filtered distilled, 805
high purity, 807
reagent-grade, 807
tap, 807
treatment, 142
ultraclean, 805
Wavelength, 36, 803
Bragg reflection of X-rays, 255
moving particle, 247 (*see also* Planck's constant)
range, 251
relation between X-rays and atomic number, 252
short (TEM), 247
X-ray penetration versus accelerating voltage, 252
Wavelength dispersive X-ray spectrometry (WDXS), 243, 257
advantages and disadvantages, 253, 257
compared to EDXS, 258
crystal X-ray spectrometer, 253, 255
design, 255
detector, 255
diffraction relationship (Bragg's law), 254
examines single x-ray line, 253
expertise required, 257
focal circle, 255
geometrical requirements, 255
identifies/counts X-rays, 253
in phase response, 254
limitation (principal), 255
and particle identification, 257
peak-to-background ratio, 257
resolution compared to EDXS, 257
sensitivity compared to EDXS, 257
source, 255
time requirements, 257
WDXS (*see* Wavelength dispersive X-ray spectrometry)
Wear metals (*see* Contaminants)
Wet benches (*see* Wet sinks)
Wet sinks, 577, 578
contamination in, 577, 581-582
White light particle counters, 800
Wipes, 805
Witness plates, 804
Working Standard Test Method, 296
Work surfaces, 805

X-ray(s), 236
absorption effects, 259
analytical configurations, 29
background that limits sensitivity, 252
beam, 253
capability added, 30
characteristic (elemental), 252

[X-ray(s)]
charge, 253
collected by bias voltage, 253
collection in combination with SEM, 253
collection geometry, 259
data, 259
decreasing energy, 252
detector entered, 253
diffraction, 273
data, 268
phenomena, 253
with discrete energy levels, 252
dot map, 253, 263
electronic relaxations resulting in X-ray photon emission, 252
electrons freed, 253
elemental, 253
elemental analysis, 253
emission, 257
emitted, 252, 253
energy, 253
excitatation, 251
fluoresce lower energy X-rays, 252
fluorescence effects, 259
fluorescent yield, 252
focusing, 255
generate lower energy, 251
geometric requirements of focusing, 255
geometry (collection) and irregular surfaces, 259
higher energy, 251, 252
lines (major), 257
monochromatic, 255
photoelectrons, 253
photon, 252, 253
principal quantum numbers, 252
quanta, 251
radiation, 252
reflection phenomena, 253
signals and simultaneous collection of electron signals, 253
signal used to modulate brightness of CRT spot, 253
silicon lattice, 253
solid angle of, 255
spectral lines, 252
spectroscopy, 251
split screen images, 253
striking the detector, 253
wavelengths (of different), 255
X-ray analysis, 32, 235, 649
absorption effects, 259
X-ray crystallography, 643 (*see also* Crystallography)
analytical mode of last resort, 278
diffraction pattern analysis, 268
essential criterion for analysis, 269
novel degradants, 278
for particle analysis, 269
[X-ray crystallography]
particle location prior to, 259
penicillin powders, 278
and pharmaceutical particulate matter, 278
raw materials, 278
special information required, 269
versus EDXS and FTIR, 278
X-ray diffraction, 29, 283
arc length, 275
basic physical phenomena, 273
Bragg's equation, 253, 254, 255, 273, 276, 281
camera, 275, 276
camera diameter fixed for particle sizing, 281
camera system X-ray background reduction, 276
collimated beam, 274
using grating, 274
comparison standards, 276
crystalline,
lattice, 273
planes, 273
differentiates identical chemical composition and distinctive crystal morphology, 268
diffracted beam source, 281
diffraction lines, 276
glass fiber support for crystalline particle sample, 277
in-phase waves, 274
interference pattern generation, 274
interplanar d-spacings, 276
nickel filter, 276
patterns, 243, 273, 276
photographic film, 275, 277
powder,
camera, 277
and single-crystal studies, 276
projecting/recording patterns, 275
radiation at atomic or molecular positions, 273
reflection of radiation at successive crystal planes, 273
ring and spot patterns, 273
single crystal analysis, 276
single particles, 243
small-angle, 276
studies relate to basic crystal structure, 270
three-dimensional,
array of atoms, 274
crystal lattice, 273
two-dimensional interferences of reflected rays, 273
vacuum, 276
wavelength, 276
wave motion, 273
X-ray microprobe, 242
X-ray spectra:
contaminant particles in pharmaceutical products, 258
elements (specific), 258

[X-ray spectra]
  proportional concentrations, 258
X-ray spectrometry/spectroscopy, 235, 243, 252, 256, 818
  absorption effects, 259
  applied in combination with other techniques, 235, 256
  background sensitivity limits, 252
  general particle analysis, 256
  qualitative analysis mode for particle identification, 258
[X-ray spectrometry/spectroscopy]
  solid state energy dispersive, 253
  used to generate a dot map, 253
  viewing with, 256

ZAF, 259
Zeiss polarizing microscope (Standard GFL or Universal), 35, 37, 43
Zernike (Dutch physicist), 58 (*see also* Phase contrast)
Zoom systems, 33, 94
Z range, 258